T0344786

Karst Hydrogeology, Geomorphology and Caves

Karst Hydrogeology, Geomorphology and Caves

Jo De Waele
Department of Biological, Geological and Environmental Sciences
University of Bologna
Italy

Francisco Gutiérrez
Department of Earth Sciences
University of Zaragoza
Spain

The right of Jo De Waele and Francisco Gutiérrez to be identified as the authors of this work has been asserted in accordance with law.

Registered Offices
John Wiley & Sons, Inc., 111 River Street, Hoboken, NJ 07030, USA
John Wiley & Sons Ltd, The Atrium, Southern Gate, Chichester, West Sussex, PO19 8SQ, UK

Editorial Office
The Atrium, Southern Gate, Chichester, West Sussex, PO19 8SQ, UK

For details of our global editorial offices, customer services, and more information about Wiley products visit us at www.wiley.com.

Wiley also publishes its books in a variety of electronic formats and by print-on-demand. Some content that appears in standard print versions of this book may not be available in other formats.

Limit of Liability/Disclaimer of Warranty

While the publisher and authors have used their best efforts in preparing this work, they make no representations or warranties with respect to the accuracy or completeness of the contents of this work and specifically disclaim all warranties, including without limitation any implied warranties of merchantability or fitness for a particular purpose. No warranty may be created or extended by sales representatives, written sales materials or promotional statements for this work. The fact that an organization, website, or product is referred to in this work as a citation and/or potential source of further information does not mean that the publisher and authors endorse the information or services the organization, website, or product may provide or recommendations it may make. This work is sold with the understanding that the publisher is not engaged in rendering professional services. The advice and strategies contained herein may not be suitable for your situation. You should consult with a specialist where appropriate. Further, readers should be aware that websites listed in this work may have changed or disappeared between when this work was written and when it is read. Neither the publisher nor authors shall be liable for any loss of profit or any other commercial damages, including but not limited to special, incidental, consequential, or other damages.

Library of Congress Cataloging-in-Publication Data applied for

Hardback ISBN: 9781119605348

Cover Design: Wiley
Cover Image: © Radoslav Husák

Set in 9.5/12.5pt STIXTwoText by Straive, Pondicherry, India

SKYA5211104-2250-43F9-B048-19531BD42F61_062322

To our wives, Delia and Marina, and our children, Francho, Isabel, Nicholas, and Thomas.

Contents

Preface

When five years ago Wiley asked us to write a book inspired by the "bible" of karst science – a new "Ford and Williams" – we certainly felt very honored, but we also realized what an enormous job that would be. Karst science has grown rapidly over the last decades as illustrated by the publication of excellent books, including not only the Ford and Williams (2007) "Karst Hydrogeology and Geomorphology", but also other recent works such as "Cave Geology" by Arthur Palmer (2007); the "Karst Geomorphology" volume in the "Treatise on Geomorphology" edited by Amos Frumkin (2013); the "Hypogene Karst Regions and Caves of the World" edited by Klimchouk et al. (2017); the three editions of the "Encyclopedia of Caves" edited by William White, David Culver, and Tanja Pipan (2005, 2019), or the "Encyclopedia of Caves and Karst Science" edited by John Gunn (2004). We have seen an exponential increase in the number of karst-related papers and a growing cave and karst science community, so it is very difficult to keep up-to-date on the many aspects of this complex and multidisciplinary science. We decided to take some time to consider the offer, and in 2019 we finally were ready for this huge endeavor. It may seem strange, but the pandemic, with its lockdowns over several long periods during the last two years, has contributed to make the accomplishment of this task a little bit easier and feasible.

Our book "Karst Hydrogeology, Geomorphology and Caves" is not a new version of the Ford and Williams "bible," and we slightly changed the original title to include "caves". Following an introductory chapter, we deal with the karst rocks in Chapter 2 and devote Chapters 3 and 4 to dissolution processes and denudation in karst areas, respectively. Chapter 5 addresses the very special hydrogeology of the karst systems. Surface landforms are treated in detail in Chapters 6, 7, and 8. Chapters 9, 10, and 11 shed light onto the underground world, describing the typical morphology of caves, their physical and chemical deposits (detrital sediments and speleothems), and the many ways through which these natural underground voids can form (speleogenesis).

In our book, in contrast to the Ford and Williams volume, we do not address hazards in karst, nor do we deal with resources, tourism, management, conservation, and restoration of caves and karst areas. Unfortunately, there was no space, which is a pity, but we might include these important applied topics in a second edition of our book.

Although it is practically impossible to read all the rapidly increasing articles and books dealing with karst, we have tried to illustrate the main features and concepts with examples and case studies from all over the world, in an attempt to select the key literature on the different topics. Regretfully, because of space constraints, many important publications are lacking, because we had to make choices, or simply because we missed them! We apologize for this and would be happy to receive your valuable inputs.

We are complementary karst scientists: Francisco develops most of his studies above the ground, whereas Jo prefers to conduct research underground. We are indebted to many friends and colleagues, experts in their fields, both at the surface and under our feet. Our knowledge is largely rooted in the many discussions and collaborations we have had over the years with scientists, students, and cavers. From the surface side, we are greatly indebted to superb geologists and geomorphologists such as Mateo Gutiérrez (Francisco's father), Paul Williams, Derek Ford, Ugo Sauro, Ángel Ginés, Tony Cooper, Ken Johnson, Mehdi Zarei, Ahmed Youssef, Rogelio Linares, Alfonso Benito, just to mention some of them.

Regarding the fascinating world of caves, the illuminating company of great cavers has been essential in our academic careers: Arthur Palmer in different settings, from the Guads to Slovenia; John Mylroie in the Bahamian flank margin caves; Philippe Audra and Lukas Plan in the European hypogene (sulfuric acid) caves; Alexander Klimchouk in different caves of hypogene nature; Andrej Mihevc for the many interesting talks on Dinaric karst; Giovanni Badino (who left us too soon) concerning the role of physical and atmospheric processes in caves; Paolo Forti for secondary cave minerals and speleothems, and gypsum caves of course; Bartolomeo Vigna for what concerns underground water flow and monitoring; José Maria Calaforra for gypsum caves; Francesco Sauro for quartzite caves; Augusto Auler for tropical caves; and all the other colleagues and friends we simply cannot mention because of space constraints.

We are greatly indebted to colleagues that have reviewed several portions of the book: Concha Arenas, Luis Auqué, Andrea Columbu, Anthony Cooper, Ilenia Maria D'Angeli, Ugur Doğan, Ángel García-Arnay, María José Gimeno, Ángel Ginés, Ergin Gökkaya, Mateo Gutiérrez, Vince Matthews, James McCalpin, Derek Mottershead, Alessia Nannoni, Ana Navas, Clifford Ollier, Luca Pisani, Jorge Sevil, Mehdi Zarei.

This book is nicely illustrated thanks to the involvement of many cavers, some of them among the best cave photographers in the world: Rosangela Addesso, Luciana Alt, Alireza Amrikazemi, José Antonio Arz, Marek Audy, Darko Bakšić, Daniela Barbieri, Pavel Bella, Jean-Yves Bigot, Peter Bosted, Richard Bouda, Andrea Brogi, Laurent Bruxelles, Dave Bunnell, Mark Burkey, Didier Cailhol, José Maria Calaforra, Victor Carvajal, Iztok Cencič, Weihai Chen, Ataliba Coelho, Anthony Cooper, Carla Corongiu, Vittorio Crobu, Philippe Crochet, Guy Decreuse, Riccardo De Luca, Danilo Demaria, Gloria Desir, Marcel Dikstra, Ugur Doğan, Kevin Downey, Csaba Egri, Javier Elorza, Victor Ferrer Rico, Michal Filippi, Paolo Forti, Amos Frumkin, Stefano Furlani, Peter Gedei, Ángel Ginés, Ergin Gökkaya, Lluís Gómez-Pujol, Andrey Gorbunov, Roberto F. Gracía, Francesco Grazioli, Josipa Grbin, Michael Gruber, Adrian Harzyna, Monica Hölzel, Andrew Hounslea, Chris Howes, Radoslav Husák, Mauro Inglese, Stéphane Jaillet, Stephan Kempe, Alexander Klimchouk, Marcelo Krause, Matej Krzic, Jurij Kunaver, Sergio Laburu, Orlando Lacarbonara, Michael Lace, Lewis Land, Stein-Erik Lauritzen, Chien C. Lee, Matej Lipar, Robert Loucks, Borut Lozej, Piero Lucci, Joyce Lundberg, Yu Manman, Francesco Maurano, Donald McFarlane, Tony Merino, Piotr Migoń, Andrej Mihevc, Vito Moura, Mohammad Nazari, Gildas Noury, Federico Ortí, Muhammed Öztürk, Arthur Palmer, Paolo Petrignani, Leonardo Piccini, Lukas Plan, Natasa Ravbar, Michel Renda, Joaquín Rodríguez-Vidal, Alessio Romeo, Antonio Rossi, Antonio Santo, Francesco Sauro, Ugo Sauro, Andreas Schober, Sandro Sedran, Robbie Shone, Neil Silverwood, Michael Simms, Roberta Tedeschi, Nicola Tisato, Paola Tognini, Scott Trescott, Pablo Valenzuela, Marco Vattano, Adrián Vázquez, Etienne Venot, Márton Veress, Jeff Wade, Tony Waltham, Chen Weihai, Mirjam Widmer, Peter Wilson, Max Wisshak, Robert Wray, Serdar Yeşilyurt, and Nadja Zupan Hajna.

Some of the figures were prepared with the help of Centro Italiano di Documentazione Speleologica (CIDS), Franco Anelli Bologna, Veronica Chiarini, Andrea Columbu, Paolo Forti, Silvia Frisia, Surányi Gergely, Ergin Gökkaya, Gruppo Speleologica Bolognese, Karst Research Institute at Postojna, NASA/JPL/University of Arizona, National Library of France, Paul Reich, Gibran Romero-Mujalli, Laura Sanna, Jorge Sevil, Kevin Stafford, and University of Chicago Library.

Without all these contributions, this book would not have been so attractive.

Francisco and Jo

References

Culver D.C., White W.B. eds. (2005). *Encyclopedia of Caves*. Amsterdam: Academic Press-Elsevier.

Ford, D.C. and Williams P.W. (2007). *Karst Hydrogeology and Geomorphology*. Chichester, UK: Wiley.

Frumkin A. ed. (2013). *Treatise on Geomorphology*. Volume 6, Karst Geomorphology. Amsterdam: Elsevier.

Gunn J. ed. (2004). *Encyclopedia of Cave and Karst Science*. New York: Fitzroy Dearborn.

Klimchouk A.B., Palmer A.N., De Waele J. et al. eds. (2017). Hypogene karst regions and caves of the world. Cham: Springer.

Palmer, A.N. (2007). *Cave Geology*. Dayton, Ohio: Cave Books.

White W.B., Culver D.C. eds. (2011). *Encyclopedia of Caves*. Waltham: Academic Press-Elsevier.

White W.B., Culver D.C., Pipan T. eds. (2019). *Encyclopedia of Caves*. London: Academic Press- Elsevier.

1

Introduction to Karst

1.1 The Term Karst. Definition and Origin

Karst is a widely accepted term in the scientific community referring to a special type of landscape with distinctive landforms and hydrology, mainly arising from dissolution of various types of rocks. Karst terrains commonly display enclosed depressions, sinking streams, caves and major springs, and are characterized by prevalent subsurface drainage. Aquifers developed in karst rocks typically have a well-developed secondary porosity, mainly resulting from the solutional enlargement of discontinuities. Groundwater circulates through these interconnected subsurface channels much more rapidly than in aquifers with intergranular and fracture porosity, and may reach turbulent flow regime, which contributes to enhance dissolution rates and mechanical erosion processes. Karst geomorphology and hydrogeology are closely interrelated. Surface water tends to infiltrate preferentially through enclosed depressions and swallow holes, which are connected to integrated networks of solutional conduits, caves, and ultimately springs. The long-term evolution of these complex karst systems is controlled by changes in the regional or local base level of erosion. Moreover, underground channels may be responsible for the formation of the typical surface landforms (e.g., sinkholes generated by cave roof collapse).

Descriptions of karst phenomena date back to the antiquity, but the term karst has been used since the nineteenth century. In the seventh century BC, the Greek philosophers, who lived in a region largely underlain by karstified carbonate rocks, proposed two novel explanations for the hydrological cycle. One group, including Thales, Plato, and Pliny, sustained that sea water penetrates into the rocks and is driven upward losing its salts, until it reaches the ground surface through springs. The other group, among them Aristotle, proposed that springs are related to water condensation in caves. The Greek geographer Strabo (60 BC to 28 AD), in the second book of his 17 volumes *Geographica*, described various karst features such as caves, underground streams and springs. The Roman philosopher Seneca (4 BC to 65 AD), in his book *Naturales Questiones,* explained solution processes, cave development, as well as disappearance and resurgence of streams. The Peutinger's Tabula, a thirteenth century parchment copy of a previous Roman map, ascribed to Marcus Vipsanius Agrippa (died 12 BC), reports "Fonte Timaus", the famous karst spring of the Reka-Timavo River (Figure 1.1). This is probably the first known map locating a karst feature. A review on historical references to karst phenomena can be found in LaMoreaux (1995).

The capitalized word "Karst" is the German form of the regional place names "*Kras*" (Slovenian) or "*Carso*" (Italian), derived from the pre-Indo-European words *karra* (*gara*), *krs* or *kar*, meaning "stone" or "barren stony ground" (Roglić 1972; Gams 1973b; Kranjc 2018). This is the region in the northwestern sector of the Dinaric area around Trieste (Slovenia and Italy) traditionally regarded

Karst Hydrogeology, Geomorphology and Caves, First Edition. Jo De Waele and Francisco Gutiérrez.
© 2022 John Wiley & Sons Ltd. Published 2022 by John Wiley & Sons Ltd.

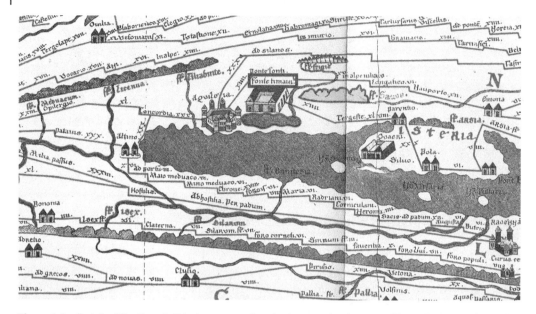

Figure 1.1 Detail of Peutinger's Tabula representing the karst region between Trieste (Italy) and Slovenia and locating the spring "Fonte Timaus," on the northern coast of the Adriatic Sea. *Source:* Own work/Wikimedia Commons/CC0 1.0.

as the Classical Karst, where much of the pioneering karst investigations were developed. In the second century BC, during the Roman occupation of the region, the maps reported the name "*Carsus*", subsequently transformed into "*Karstia*" in Valvasor's famous book (1687) (Figure 1.2). The area became part of the Austro-Hungarian Empire in the second half of the nineteenth century. In this period, geographers of the University of Vienna, including Albrecht Penck, focused their attention to this limestone area and the word was germanized into "karst," gaining popularity among the scientific community. Jovan Cvijić, a pupil of Albrecht Penck, published a comprehensive review of karst landforms entitled "*Das Karstphänomen*" (Cvijić 1893). This pioneering monograph is believed to mark the beginning of proper karst studies in the Western world (Sweeting 1972). von Sawicki (1909) expanded Cvijić's work to a global scale, describing tropical karst landscapes.

Ford and Williams (2007), in their second edition of the book *Karst Hydrogeology and Geomorphology*, defined karst as "*a terrain with distinctive hydrology and landforms that arise from a combination of high rock solubility and well developed secondary (fracture) porosity.*" A more general definition of karst is proposed in the book *Speleogenesis* (Klimchouk et al. 2000, p. 46), adapted from a previous definition by Huntoon (1995): "*The karst system is an integrated mass-transfer system in soluble rocks with a permeability structure dominated by conduits dissolved from the rock and organized to facilitate the circulation of fluid.*" This definition is sufficiently broad to include any kind of rock, fluid, dissolution process, and geological context, and is not restricted to the classic karst landforms.

1.2 Classification of Karst

The evolution and characteristics of karst systems are controlled by a number of factors, notably: lithology, stratigraphy, geological structure, topography, precipitation, temperature, chemical composition of the water, and base level changes, which may be governed by various processes such as sea-level variations, tectonism, or diapirism (e.g., White 1988) (Figure 1.3). Karst has been classified in multiple ways, mainly using criteria related to the factors indicated above.

Figure 1.2 Historical map of the Kras region between Trieste (Italy) and Slovenia indicating the name "Karstia". *Source:* From Valvasor (1687).

Figure 1.3 The factors that control the development of karst landscapes.

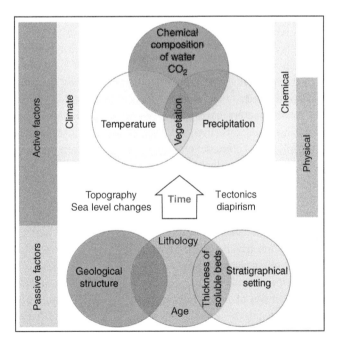

Probably, the first attempt to classify karst was made in the early works of Jovan Cvijić, inspired by his studies in the Dinaric karst (Sweeting 1972). He used the term holokarst for a perfectly developed solutional karst landscape lacking river valleys. This type of situation mainly occurs in thick successions of soluble rocks (normally pure limestones) extending well below base level.

Merokarst is an imperfect or partial karst landscape in which fluvial activity and mechanical erosion still play an important role. It mainly develops in areas underlain by successions of soluble rocks with interbedded insoluble or less soluble lithologies. In later works, Cvijić realized that this classification worked well in the Dinaric and Moravian karsts, but he was forced to add a third "transitional" class for the many examples he encountered in France, Greece, and Cuba (Ford 2007). Nowadays, the concepts of merokarst and transitional karst have been substituted with fluviokarst, referring to landscapes and hydrological systems in which fluvial activity, mechanical erosion, and surface drainage play a significant role. The terms autogenic and allogenic fluviokarst are used to specify the source of the water, either the in situ karst aquifers or external areas with insoluble rocks.

Karst features can be grouped, according to the position in which they form, into exokarst and endokarst, developing at and beneath the surface, respectively. One of the most commonly used classifications is based on the main lithology in which the karst system is formed (e.g., limestone karst, evaporite karst, and quartzite karst) (Figure 1.4). Most karsts are developed in carbonate rocks, including limestone, dolostone, chalk, and marble (Figure 1.5a–c). Evaporite rocks, such as gypsum and rock salt (halite), much more soluble than carbonate rocks, also display well-developed karst systems, although generally with a smaller preservation potential and a lower degree of exposure due to their lower mechanical strength and higher erodibility (Figure 1.5d–f, h). The importance of dissolution in the formation of karst porosity in quartz sandstones (quartzites) has been confirmed in recent years (Wray and Sauro 2017) (Figure 1.5g).

Soluble sediments may be affected by karst processes during or soon after their deposition leading to syngenetic or penecontemporaneous karst. These early dissolution processes are generally related to subaerial exposure phases of soluble sediments in marine and lacustrine basins. Dissolution of young, poorly lithified carbonate, and evaporite sediments is often called eogenetic karst, which displays features similar to the ones created by syngenetic karst (Grimes 2006). Intergranular porosity in these rocks with limited compaction, cementation, and fracturing, offers the principal paths for water flow. These types of karst can be considered as the opposite to the so-called telogenetic karst, developed in fully lithified hard rocks, in which flow mainly occurs along discontinuities.

Climatic factors have a profound impact on the intensity of dissolution processes and the resulting landforms (Figure 1.3). This is why karst is often classified from a climate perspective. The following climatic types of karst are often identified (e.g., Jennings 1985): temperate, Mediterranean, tropical, humid, arid, semiarid, glacial (or alpine), and periglacial (or nival). This classification has significant limitations, since similar landforms can form in different climates, and the role played by litho-structural factors and time can be much more important than climate. Other classifications and definitions are based on topographic or environmental factors. The distinction between mountain karst and lowland karst, although readily understandable, is probably too trivial. The term coastal karst is often used for dissolution occurring in carbonate rocks in the seawater–freshwater mixing zone (Lace and Mylroie 2013). Biokarst refers to karst landforms created or influenced to a significant degree by biological processes (Viles 2004a, 2004b).

Another way of classifying karst is based on the stratigraphic position of the soluble rocks and their degree of exposure (Figure 1.4). The terms exposed or bare karst are used where the soluble rocks crop out at the surface (Figure 1.5a, b, f). Open karst typically refers to areas in which soluble rocks are exposed extensively. Mantled or covered karst refers to dissolution of soluble rocks overlain by unconsolidated deposits, either transported allochthonous sediments (e.g., alluvium, loess, volcanic ash), residual soils related to bedrock dissolution (e.g., terra rossa), or plant litter (Figure 1.5c–e). Some European geomorphologists use the terms subsoil karst (Gvozdeckij 1965) and cryptokarst

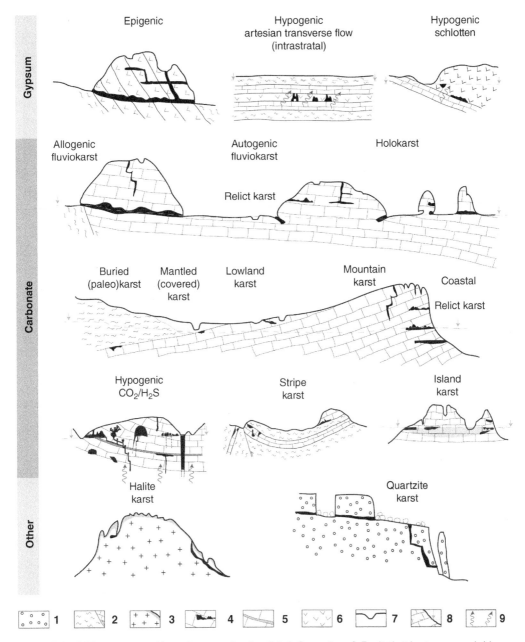

Figure 1.4 Different types of karst (see text for details): 1, Quartzites; 2, Fault that juxtaposes soluble rocks against less soluble rocks; 3, Halite with surficial insoluble residue; 4, Carbonate rocks with caves; 5, Less permeable bed within a carbonate sequence; 6, Gypsum; 7, Permeable unconsolidated sediments; 8, Carbonate rocks with local groundwater level or the mean sea level; 9, Ascending flow.

(Lacroix 2004) for describing the subcutaneous dissolution that occurs in the epikarst zone under a thin cover of surficial formations. The term interstratal karst is used when the karst rock is overlain by insoluble or less soluble rock formations, which are designated by some authors as caprock. This type of subjacent karst is particularly important in evaporitic formations, which may be affected by

Figure 1.5 Images of various types of karst features developed in different lithologies and settings. (a) Bare karst landscape in UNESCO World Heritage Torcal de Antequera (Málaga, Spain) formed on subhorizontal stratified limestone showing well developed structural karren and lack of surface drainage network. *Source:* Francisco Gutiérrez (Author). (b) Partial view of an alpine carbonate massif, formed by the stacking of thrust sheets, that functions as the recharge area of a major spring. Note recumbent fold and lake in overdeepened glacial basin (Ordesa National Park, Spain). (c) Pinnacles and solutionally enlarged fractures in Early Cretaceous dolomitic limestones overlain and filled by iron-rich red clays.

Figure 1.6 Panoramic view of the Ojos Negros iron mine (Iberian Chain, Spain), which used to supply around 20% of the iron production in Spain. The iron exploited in this mine occurs in a deep paleokarstic weathering mantle related to the dissolution under oxidizing conditions of the underlying Mg-Fe-rich Ordovician dolomites and the concomitant concentration of iron oxides (gossan-type deposit). The irregular rockhead exhumed by mining shows pinnacles, tower-like features, and solutionally enlarged joints. *Source:* Francisco Gutiérrez (Author).

widespread deep-seated dissolution and generate large-scale dissolution-collapse breccias and subsidence structures (Warren 2016). Some authors propose the use of the term intrastratal karst where there is preferential dissolution of a particular bed or rock unit within a sequence of soluble rocks (Ford and Williams 2007). Buried karst indicates inactive karst features unconformably overlain by sediments deposited during or after the development of the karstification. Note that in an interstratal karst the karstification is younger than the capping rocks, whereas in a buried karst the karstification is older than the overlying sediments. Exhumed karst refers to a previously buried karst that has been exposed by erosion (Figure 1.6).

The term paleokarst, frequently used by stratigraphers and sedimentologists, is typically an inert karst of considerable age decoupled from the contemporary conditions (Figure 1.5h). A paleokarst may be buried or exhumed. Some paleokarsts hold valuable mineral deposits (e.g., James and Choquette 1988) (Figure 1.6). This is not to be confused with relict karst, referring to karst landforms that are now abandoned by dissolving waters, such as perched passages in multilevel cave systems or drowned coastal karsts. Some authors also use relict or inherited karst to describe essentially inactive features developed under past environmental conditions, for instance caves found in hyperarid regions formed during past pluvial periods.

Klimchouk and Ford (2000b) proposed several terms to describe the morphologic and hydrologic evolution of confined interstratal karst systems affected by fluvial downcutting and base level lowering. Their deep-seated karst occurs when the formation affected by dissolution is completely

---

Figure 1.5 (Continued) Dissolution beneath the karstic residue is enhanced by the oxidation of iron sulfides in the weathering mantle and the consequent acidification of the percolating waters. This subsoil karst has been exhumed by iron mining (Cabárceno Park, Santander, Spain). Bear in the central and lower part of the image for scale. (d) Drone view of active collapse sinkholes in the partially mantled gypsum karst of the Ebro River valley, NE Spain. Large doline with trees in the bottom is 75 m across. (e) Collapse sinkholes related to eogenetic dissolution of Quaternary salt deposits beneath an unconsolidated cover (mantled karst) in the Dead Sea, Israel. Salt dissolution has been induced by the anthropogenic decline of the water level in this terminal lake. The sinkhole alignments are controlled by concealed faults. (f) Saline spring in Jahani salt extrusion, Zagros Mountains, Iran. The salt exposure, locally covered by residual capsoils, corresponds to the flank of a namakier (salt glacier). (g) Golondrinas Cave developed in well stratified and vertically jointed quartzites in Auyan Tepui, Venezuela. *Source:* Photo by Vittorio Crobu, La Venta Esplorazioni Geografiche. (h) Example of paleokarst recorded by a clay-filled cave of probable pre-Quaternary age developed in Triassic gypsum (Chodes, Iberian Chain, Spain). *Source:* Photos except g by Francisco Gutiérrez (Author).

covered by the confining non-karst rocks. The subjacent karst stage starts when incision locally breaches the confining unit and the karst aquifer is brought into direct hydraulic connection with the surface. The entrenched karst phase starts when the fluvial systems incise below the karst formations, although it is still partially overlain by caprock. In the denuded karst, the insoluble overburden is completely removed and the karst formation is fully exposed.

Some terms are used to describe the cartographic distribution of the karst phenomena and their geometrical relationship with adjacent non-karst formations. The karst unit entirely surrounded by insoluble rocks is often called impounded karst or karst barré. A stripe karst is a narrow belt of soluble rocks bounded by non-karst formations. This type of situation commonly occurs in moderately to steeply dipping successions. Contact karst indicates areas of intense karstification associated with the contact between karst and non-karst rocks (Gams 2001). In these settings, unsaturated aggressive water coming from the insoluble formations may produce significant dissolution in the karst rocks along the contact zone (Figure 3.49). The contact between the different formations, either a stratigraphic discontinuity, a fault or an intrusive contact, may also act as a hydrogeological barrier forcing the convergence and upwelling of groundwater flow enhancing dissolution.

A notion that has received great attention in recent times is the distinction and characterization of karst systems according to the source of the dissolving water and/or aggressiveness. Epigene (or epigenetic) karsts are dominated by downward groundwater flow from an overlying or adjacent recharge area. Hypogene (or hypogenic) dissolution is related to groundwater coming from below (Klimchouk 2007) or renewed aggressiveness generated at or below the water table (Palmer 2007). Hypogenic caves can be generated by fluids enriched in CO_2 or H_2S, generally thermal, but also by undersaturated waters entering the soluble rocks (often evaporites) from below, giving rise to extensive maze caves or schlotten-type isolated cavities.

Pseudokarst refers to karst-like features produced by other processes than dissolution or dissolution-induced subsidence (Figure 1.7). These areas often host shallow caves and enclosed depressions similar to sinkholes, considered by some authors as the diagnostic features of karst. The term pseudokarst links to the geomorphological concept of equifinality or morphologic convergence, whereby different processes may generate identical or similar landforms. It would be advisable to abandon the term pseudokarst, since it does not provide any information about the genetic process, and use already existing specific terms instead (Eberhard and Sharples 2013). Examples are glacier caves generated by the melting and movement of ice or lava tubes produced by low viscosity lava flows. The use of "karst" in these cases is misleading and should be avoided. Similarly, thermokarst, referring to enclosed depressions resulting from thawing of ground ice or buried glacier ice and the subsidence of the overlying deposits, is an established term, but has nothing to do with karst (Figure 1.7a, b). The use of vulcanokarst, although less common, is highly discouraged (Figure 1.7c, d). Significant cave systems and dense fields of holes also develop in fine-grained detrital deposits such as loess or valley fills, but their genesis is related to subsurface water erosion of fine particles (piping) and the collapse of the resulting conduits (Bernatek-Jakiel and Poesen 2018) (Figure 1.7e, f).

1.3 Global Distribution of Karst

Karst, as described above, develops in soluble rocks under suitable geological, hydrological, and climatic conditions. This includes most carbonate rocks, evaporites and, in some cases, quartzites. The global distribution of carbonate and evaporite rocks at or near the surface amounts to approximately 20% of the Earth's ice-free landmasses. Not all of these soluble formations are located in regions where karst can be active today. In some areas, the availability of liquid water is too scarce

Figure 1.7 Images of different types of karst-like features not related to dissolution processes. (a) Water-filled enclosed depressions (kettle holes) beyond the frontal moraine of a glacier generated by the thawing of buried glacial ice and the subsidence of the overlying sediments. (b) Subsidence depression generated by the melting of a large buried ice block (thermokarst) accumulated by a flood (jökulhlaup) induced by the 1727 Oraefajökull eruption in Iceland. Note person for scale on the right edge. (c) Cueva del Viento lava tube generated by a basaltic eruption of the Teide Volcano (Tenerife, Canaries, Spain) at around 27 ka. (d) Collapsed lava tube in the eighteenth-century lavas of the Timanfaya National Park, Lanzarote island, Canaries, Spain. (e) Drone image of collapse depressions related to conduits and caves produced by subsurface water erosion in fine-grained valley-fill deposits (Aguarales de Valpalmas, Spain). The elongated depression on the left is 25 m long. (f) Valley fill dominated by fine-grained deposits with a dense network of conduits generated by piping, producing a peculiar terrain riddled by pinnacles and collapse holes (Aguarales de Valpalmas, Spain). Persons for scale in the bottom right. *Source:* Francisco Gutiérrez (Author).

for dissolution processes to occur, either because most of the water is permanently frozen (permafrost regions), or because precipitation is too low (hyperarid regions). In the current global warming scenario, the extent of karst is expected to increase at high latitudes in the near future due to the degradation of the permafrost. Wang et al. (2019) estimate that a global surface temperature rise of

2 °C may lead to the disappearance of one-third of the permafrost in the northern hemisphere. Recent estimates report 12–13% of the continental areas to hold currently active karst (Hollingsworth 2009). This estimate is very close to that reported by the most recent world karst map of Williams and Fong (2016) and differs little from the earlier assessments by Palmer (2007) and Ford and Williams (2007). Most of these maps are based on the available geological data, which are very heterogeneous and the documented presence or absence of karst caves. Obviously, these are rough approximations biased by the lack of data in many areas worldwide.

Dissolution of soluble rocks may occur in the subsurface by unsaturated groundwater flows. Deep-seated interstratal dissolution of salt and Ca-sulfate formations affects extensive and difficult-to-assess areas in large evaporitic basins, commonly by the migration of dissolution fronts (e.g., Devonian Prairie evaporites in Canada, Permian salts in the Delaware Basin in New Mexico, Hith Formation in the Arabian Peninsula) (Gutiérrez and Cooper 2013; Warren 2016). It is estimated that c. 25% of the continental surface is underlain by evaporitic formations (Kozary et al. 1968), and it is reasonable to believe that in a significant proportion of those areas karst processes are active today. If we take into account those karst areas with hidden deep-seated dissolution, the actual extent of karst in Earth's ice-free continental areas could be around or higher than 15% (Figure 1.8). Recent works document the production of digital nation-wide karst maps that illustrate the importance of karst. For instance, karst is found in more than 30% of mainland China (Lei et al. 2015) and 18% of the United States is underlain by soluble rocks (Weary and Doctor 2014). Several countries are dominated by outcrops of karst bedrock (e.g., countries of the Dinaric Alps).

To better understand the spatial and temporal distribution of soluble rocks in the stratigraphic record as well as the variable intensity and extent of karstification processes throughout geological time, it is essential to consider the climatic history of the Earth. The Quaternary and late Neogene are characterized by a relatively cold climate controlled by waxing and waning polar ice sheets. The cyclic transfer of water from the sea to the continental ice masses and vice versa (i.e., glacial and interglacials), with frequencies of c. 100 ka over the past 780 ka, were accompanied by sea level oscillations with amplitudes of around 100 m. The sea level drops during the lowstands involved the emergence of extensive platforms underlain by carbonate rocks and a substantial base level decline, creating adequate conditions for new or rejuvenated dissolution. The resulting karst features were submerged and/or covered by younger sediments during the subsequent sea level rise, becoming an inactive relict karst or a buried paleokarst, respectively. Today, some 10% of the world's surface is covered by glacial ice, representing about 80% of the total volume of surface freshwater, and an additional 20% is affected by permafrost. In the last glacial maximum (c. 20 ka), around 25% of the Earth's surface was covered by ice and the sea level was approximately 125 m lower than today. In contrast, during most of the Phanerozoic, especially between the Eocene and the Triassic, the Earth was a "greenhouse" planet with a CO_2-rich atmosphere and largely devoid of glacial ice. This warm climate and the relatively higher and more stable sea level favored the accumulation of extensive carbonate formations. These factors also explain why the extensive marine evaporites deposited during the Phanerozoic have no similar-scale counterparts in the Quaternary saline systems, developed under the current cooler "icehouse" climate. The greatest volumes of evaporites were accumulated from Neo-Proterozoic to Cretaceous times in marine basin-wide and platform settings. The earlier evaporites typically consist of halite-rich successions, hundreds of meters thick, whereas the more recent ones are dominated by Ca-sulfate units, 10–50 m thick associated with carbonates (Warren 2010, 2016). A world map showing the distribution of the main saline basins can be found in Warren (2016, figure 5.30).

Summarizing, globally karst occurs over approximately 15% of the continental ice-free land surface (Goldscheider et al. 2020). These karst areas are especially abundant in the northern hemisphere (North America, Europe, and Asia), where outcrops of soluble rocks represent around 20%

Karst areas of the world

Continuous carbonate rocks
Discontinuous carbonate rocks
Continuous evaporite rocks
Mixed carbonate and evaporite rocks
Discontinuous evaporite rocks

Figure 1.8 Global karst map, based on the WOKAM map. *Source:* Adapted from Chen et al. (2017) and Goldscheider et al. (2020).

of the land surface. In contrast, the landmasses derived from the Gondwana supercontinent (South America, Africa, and Australia) have sparser outcrops, except along their continental margins. For instance, the carbonate outcrop of the Nullarbor Plain in southern Australia covers an area almost as large as Great Britain ($200\,000\,\text{km}^2$).

1.4 Karst Terminology

Karst features have been known since prehistoric times, and since humans have developed languages in different cultures, several names have been used to describe similar karst landforms. This is why many different words are used to designate features such as caves, shafts, dolines, etc., even in the same language and region. The International Union of Speleology has a rather updated online multi-lingual dictionary of caving and speleological terms (UIS 2019). This glossary was based on earlier works in different languages, such as Trimmel (1965) (in German), Kósa (1967) (Hungarian-English), Monroe (1970) (for Northern America), Gèze (1973) (in French), Gams (1973b) (Slovene), Roglić (1974a) (Croatian), Jennings (1979) (for Australia), and many others. These were condensed in Kósa (1995/96) and Panoš (2001), then by Lowe and Waltham (2002) and ultimately by Field (2002b) for the English terms only. Nowadays, although certain words are widely recognized and used at a global scale (e.g., doline, polje), certain karst features have two different names (e.g., karren or lapiés), or even more (e.g., anemolite, anthodite, eccentric, and helictite). The international karst literature can sometimes be rather confusing indeed. Hopefully, this book will help readers to identify the most proper terms.

References

Bernatek-Jakiel, A. and Poesen, J. (2018). Subsurface erosion by soil piping: significance and research need. *Earth-Science Reviews* 185: 1107–1128.

Chen, Z., Auler, A.S., Bakalowicz, M. et al. (2017). The World Karst Aquifer Mapping project: concept, mapping procedure and map of Europe. *Hydrogeology Journal* 25 (3): 771–785.

Cvijić, J. (1893). *Das Karstphänomen*, Versuch einer morphologischen Monographie, Geographische Abhandlungen herausgegeben von A. Penck, Wien, Band V, Heft 3, 1–114.

Eberhard, R.S. and Sharples, C. (2013). Appropriate terminology for karst-like phenomena: the problem with 'pseudokarst'. *International Journal of Speleology* 42 (2): 109–113.

Field, M.S. (2002b). *A Lexicon of Cave and Karst Terminology with Special Reference to Environmental Karst Hydrology*, U.S. Environmental Protection Agency Report EPA/600/R-02/003 (Supersedes 1999 edition). Washington DC: U.S. Environmental Protection Agency.

Ford, D.C. (2007). Jovan Cvijić and the founding of karst geomorphology. *Environmental Geology* 51: 675–684.

Ford, D.C. and Williams, P.W. (2007). *Karst Hydrogeology and Geomorphology*. Chichester, UK: Wiley.

Gams, I. (1973b). *Slovenska Kraska Terminologija (Slovene Karst Terminology)*. Ljubljana: Zveva Geografskih Institucij Jugoslavije, Univerza v Ljubljani.

Gams, I. (2001). Notion and forms of contact karst. *Acta Carsologica* 30 (2): 33–46.

Gèze, B. (1973). Lexique des termes français de Spéléologie physique et de karstologie. *Annales de Spéléologie* 28 (1): 1–20.

Goldscheider, N., Chen, Z., Auler, A.S. et al. (2020). Global distribution of carbonate rocks and karst water resources. *Hydrogeology Journal* 28 (5): 1661–1677.

Grimes, K.G. (2006). Syngenetic karst in Australia: a review. *Helictite* 39 (2): 27–38.

Gutiérrez, F. and Cooper, A.H. (2013). Surface morphology of gypsum karst. In: *Treatise of Geomorphology. Karst Geomorphology*, vol. 6 (ed. A. Frumkin), 425–437. Amsterdam: Elsevier.

Gvozdeckij, N.A. (1965). Types of karst in the U.S.S.R. problems of the speleological research. In: *Proceedings of the International Congress of Speleology*, 47–54. Brno, Czech Republic.

Hollingsworth, E. (2009). Karst regions of the World (KROW) - populating global karst datasets and generating maps to advance the understanding of karst occurrence and protection of karst species and habitats worldwide. Master thesis. University of Arkansas.

Huntoon, P.W. (1995). Is it appropriate to apply porous media groundwater circulation models to karstic aquifers? In: *Groundwater Models for Resources Analysis and Management* (ed. A.I. El-Kadi), 339–358. Boca Raton, Florida: Lewis Publishers.

James, N.P. and Choquette, P.W. (ed.) (1988). *Paleokarst*. New York: Springer-Verlag.

Jennings, J.N. (1979). Cave and karst terminology. *Australian Speleological Federation Newsletter* 83: 1–13.

Jennings, J.N. (1985). *Karst Geomorphology*. Oxford and New York: Blackwell.

Klimchouk, A.B. (2007). *Hypogene Speleogenesis: Hydrogeological and Morphogenetic Perspective*. Special Paper no. 1, 106pp. Carlsbad, NM: National Cave and Karst Research Institute.

Klimchouk, A.B. and Ford, D.C. (2000b). Types of karst and evolution of hydrogeologic setting. In: *Speleogenesis Evolution of Karst Aquifers* (ed. A.B. Klimchouk, D.C. Ford, A.N. Palmer, et al.), 45–53. Huntsville: National Speleological Society.

Klimchouk, A.B., Ford, D.C., Palmer, A.N. et al. (2000). *Speleogenesis: Evolution of Karst Aquifers*. Huntsville: National Speleological Society.

Kósa, A. (1967). *Szpeleológiai szakkifejezések, angol-magyar (kézirat gyanánt)*. Budapest.

Kósa, A. (1995/96). *The Caver's Living Dictionary*. Budapest, Hungary: ER-PETRO.

Kozary, M.T., Dunlap, J.C., and Humphrey, W.E. (1968). Incidence of saline deposits in geologic time. *Geological Society of America Special Paper* 88: 43–57.

Kranjc, A. (2018). A short history of "Kras". *Natura Sloveniae* 20 (2): 65–67.

Lace, M.J. and Mylroie, J.E. (ed.) (2013). *Coastal Karst Landforms*. Dordrecht: Springer Science + Business Media.

Lacroix, M. (2004). Cryptokarst. In: *Encyclopedia of Geomorphology*, vol. 1 (ed. A.S. Goudie), 205. London: Routledge.

LaMoreaux, P.E. (1995). Historical references to karst studies. In: *Proceedings of the Fifth Multidisciplinary Conference on Sinkholes and the Engineering and Environmental Impact of Karst* (ed. B.F. Beck), 11–16. Rotterdam: Balkema.

Lei, M., Gao, Y., and Jiang, X. (2015). Current status and strategic planning of sinkhole collapses in China. In: *Engineering Geology for Society and Territory* (ed. G. Lollino, A. Manconi, F. Guzzetti, et al.), 145–151. Dordrecht: Springer.

Lowe, D.J. and Waltham, T. (2002). *A Dictionary of Karst and Caves*. Caves Studies 10. London: British Cave Research Association.

Monroe, W.H. (1970). *A Glossary of Karst Terminology*. US Geological Survey Water Supply Paper 1899-K. Washington D.C.: US Geological Survey.

Palmer, A.N. (2007). *Cave Geology*. Dayton, Ohio: Cave Books.

Panoš, V. (2001). *Karstological and Speleological Terminology*. Žilina, Slovakia: Slovak Caves Administration and Geological Institute of the Czech Academy of Sciences.

Roglić, J. (1972). Historical review of morphologic concepts. In: *Important Karst Regions of the Northern Hemisphere* (ed. M. Herak and V.T. Stringfield), 1–18. Amsterdam: Elsevier.

Roglić, J. (1974a). Contribution to croatian karst terminology. *Carsus Iugoslaviae* 9 (1): 1–72.

von Sawicki, L.R. (1909). Ein Beitrag zum geographischen Zyklus im Karst. *Zeitschrift für Geographie* 15 (185–204): 259–281.

Sweeting, M.M. (1972). *Karst Landforms*. London: McMillan Press.

Trimmel, H. (1965). Speläologisches Fachwörtenbuch. In: *Third International Speleological Congress, Wien-Obertraun-Salzburg, Austria, 1965*. Vienna: Landesverein für Höhlenkunde in Wien und Niederösterreich.

UIS (2019). The Caver's Multi-Lingual Dictionary. http://www.uisic.uis-speleo.org/lexintro.html (accessed 27 March 2019).

Valvasor, J.W. (1687). *Die Ehre des Hertzogthums Crain*, vol. 4. Lubljana: Endter.

Viles, H.A. (2004a). Biokarst. In: *Encyclopedia of Geomorphology*, vol. 1 (ed. A.S. Goudie), 86–87. London: Routledge.

Viles, H.A. (2004b). Biokarstification. In: *Encyclopedia of Caves and Karst Science* (ed. J. Gunn), 304–306. New York: Fitzroy Dearborn.

Wang, C., Wang, Z., Kong, Y. et al. (2019). Most of the northern hemisphere permafrost remains under climate change. *Scientific Reports* 9 (1): 3295.

Warren, J.K. (2010). Evaporites through time: Tectonic, climatic and eustatic controls in marine and nonmarine deposits. *Earth-Science Reviews* 98: 217–268.

Warren, J.K. (2016). *Evaporites: A Geological Compendium*. Dordrecht: Springer.

Weary, D.J. and Doctor, D.H. (2014). Karst in the United States: A digital map compilation and database. *USGS Open-File Report* 2014-1156: 1–32.

White, W.B. (1988). *Geomorphology and Hydrology of Karst Terrains*. New York: Oxford University Press.

Williams, P.W. and Fong, Y.T. (2016). World map of carbonate rock outcrops v3.0. https://crc806db.uni-koeln.de/layer/show/296/ (accessed 23 November 2021).

Wray, R.A.L. and Sauro, F. (2017). An updated global review of solutional weathering processes and forms in quartz sandstones and quartzites. *Earth-Science Reviews* 171: 520–557.

2

Karst Rocks

2.1 Karst Rocks Within the Rock Classifications

There are three main types of rocks: igneous, metamorphic, and sedimentary (Figure 2.1). Most karst rocks belong to the sedimentary group, but karst development may be influenced by other associated non-karst rocks. Moreover, some metamorphic and igneous rocks may also display karst features. Igneous rocks are generated from molten or partially molten material (i.e., magma), and are divided into extrusive (or volcanic) and intrusive (or plutonic), depending on whether they form at the Earth's surface or at some depth. Metamorphic rocks result from the transformation of preexisting rocks by compositional and textural changes, essentially in the solid state, in response to marked variations in temperature, pressure, and/or the chemical environment, generally at significant depths in the Earth's crust. Sedimentary rocks are those that form in a surface environment by: (i) the mechanical accumulation of fragments derived from older rocks, (ii) inorganic or biogenic precipitation from a solution, (iii) the accumulation of organic material, and (iv) the chemical alteration of other rocks. Two main groups of sedimentary rocks are differentiated (Selly 2005): allochthonous or detrital, and autochthonous or chemical (Figure 2.1). The formation of detrital rocks involves the erosion of particles from preexisting rocks (detritus) and their transport and deposition by various agents (e.g., water, wind, and ice). Autochthonous or chemical rocks mainly result from precipitation of minerals from a solution. This is a rather artificial division, since many sedimentary rocks form by the coeval accumulation of chemical and detrital components. Sedimentary rocks commonly exhibit two properties that differentiate them from igneous and metamorphic rocks and that have an important influence on karst development: (i) they show stratification (bedding, layering) related to successive episodes of deposition (i.e., law of superposition); and (ii) they generally consist of particles, including fossils, with intervening pores that, when interconnected, contribute to the permeability of the rock.

Detrital rocks may have a highly variable mineralogy, depending on the source of the material and their maturity. They are classified according to the predominant size of the particles (gravel, sand, silt, and clay) into conglomerate, sandstone, siltstone, and claystone, respectively (Figure 2.1). Mudstone or mudrock collectively includes siltstones and claystones. Shale is typically used for indurated mudstones with fissility. The terms rudite (rudaceous rock), arenite (arenaceous rock), and argillite (argillaceous rock) are synonyms of conglomerate, sandstone, and claystone, respectively, although scarcely used at the present time. Conglomerates are mainly composed of gravel-sized particles larger than 2 mm in diameter. This size limit includes a broad range of dimensions. The Udden-Wentworth grain-size scale divides gravel into granule (2–4 mm), pebble (4–64 mm), cobble (64–256 mm), and boulder (25.6 cm to 4.1 m). There are extended grain-size classifications

Karst Hydrogeology, Geomorphology and Caves, First Edition. Jo De Waele and Francisco Gutiérrez.
© 2022 John Wiley & Sons Ltd. Published 2022 by John Wiley & Sons Ltd.

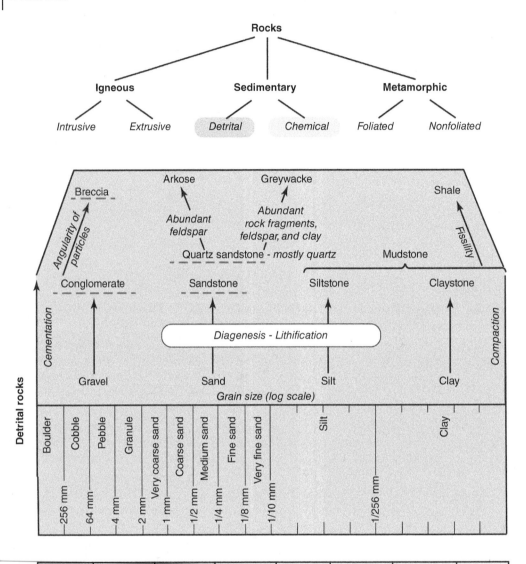

Figure 2.1 General classification of rocks and classification of the two types of sedimentary rocks. The description of detrital or clastic rocks and chemical rocks is largely based on their grain size and composition, respectively. The main karst rocks are underlined. *Source:* Adapted from Siever (1982).

for particles larger than boulders (megaclasts; e.g., Blair and McPherson 1999). The word breccia, with no genetic meaning, is used for deposits and rocks consisting of angular gravel-sized particles. Sandstones are composed of grains with an average size between 0.0625 mm and 2 mm. They may include particles composed of more than one mineral or crystal, termed rock fragments or lithic grains. Conglomerates and sandstones typically have four components: grains (the skeleton or framework), matrix, cement, and porosity. The matrix is finer grained material that may infill the space between the framework grains, and both were accumulated simultaneously (syndepositional). The cement refers to minerals that have precipitated, preferentially in pores, after sediment deposition (postdepositional). With increasing calcium carbonate content, sandstones grade into calcareous sandstones, sandy limestone, and finally pure limestone. Similarly, with increasing lime content, mudstones and shales grade into marls, argillaceous limestones, and limestones. This gradation illustrates that in practice the differentiation between detrital and chemical rocks is somewhat artificial.

Chemical rocks are those that principally result from the precipitation of minerals out of a solution. However, they may also be related to the accumulation of biogenic material and its diagenetic transformation, or to chemical weathering in a near-surface environment. They are mainly classified according to their chemical or mineralogical composition (Figure 2.1). Chemical sediments form in the environment of deposition (autochthonous), whereas detrital sediments are mainly eroded from elsewhere and brought to the depositional environment (allochthonous). By far, the most widespread groups of chemical rocks are carbonates (e.g., limestone, dolomite) and evaporites (gypsum/anhydrite, halite). There are other groups of less abundant chemical rocks, some of which may have high economic importance, including kerogenous rocks, ironstones, phosphate rocks, siliceous rocks, and residual sediments. The kerogenous or organic rocks are mainly related to the accumulation of organic matter under anoxic conditions (e.g., peat) and their diagenetic transformation or maturation (e.g., lignite, coal, and petroleum). Ironstones and phosphate rocks are dominantly of shallow marine origin. However, they may form in various ways, including the accretion of phosphatic excreta from birds and bats (guano). Siliceous rocks, mainly composed of silica (SiO_2), may result from the accumulation of siliceous organic components (e.g., radiolarite, diatomite) or precipitation of cryptocrystalline silica (e.g., chert nodules and horizons).

Potential karst rocks are those with soluble components susceptible to significant dissolution. These principally include the two main groups of chemical rocks: carbonates and evaporites. Significant karst features may also develop in detrital rocks such as conglomerates and sandstones when they are composed of a high percentage of soluble minerals, typically calcite. Under exceptional conditions, well-developed karst features may also occur in quartz sandstones (quartzites). Metamorphic rocks such as marbles and other metacarbonates, derived from the recrystallization of limestones and dolomites by the effect of pressure and heat, may also host karst. Carbonatite is a very rare igneous rock with a high proportion of carbonate minerals that may also display spatially restricted karst features.

2.2 Carbonate Rocks and Minerals

Since carbonate rocks are extremely important from the scientific and applied perspective, they receive major attention from professionals of different fields (Tucker and Wright 1990; Schlager 2005; Ford and Williams 2007; Flügel and Munnecke 2010; James and Jones 2015). They are the product of biochemical systems that have been active since the Archean (ca. 3.5 Ga). According to James and Jones (2015), carbonate sedimentary rocks comprise roughly 20% of the exposed rocks, whereas the more detailed analysis of Goldscheider et al. (2020) reports 15.2%. Stratigraphic successions dominated by carbonate formations may reach several kilometers in thickness and extend across thousands of kilometers. Carbonate rocks, with a largely biogenic origin, are crucial archives for reconstructing the evolution of life on Earth. They can also be used as proxies for the geochemistry of the water from which they precipitated in the past, and as paleoclimate archives.

Carbonate rocks have a multifaceted economic importance. Permeable, typically karstified formations constitute valuable aquifers and it has been estimated that around 10% of the global population is supplied by karst waters (Stevanović 2019), although some authors give greater estimates of up to one quarter of the world's water deriving at least partly from karst aquifers (Smart and Worthington 2004a). About 30–40% of the known hydrocarbon reserves (petroleum and natural gas) in the world occur in carbonate reservoirs (Xia et al. 2019), especially in dolomites, since dolomitization is accompanied by a significant increase in porosity. Some of the most highly productive wells exploit reservoirs related to collapsed paleocave systems underlying major composite unconformities. These reservoirs are characterized by high interconnected porosity related to both dissolution and collapse processes (Loucks 2004). Carbonate rocks host a large proportion of the Earth's metallic ores (e.g., Mississippi-Valley-Type Pb-Zn ores, siderite) (Sverjensky 1986) and magnesite ($MgCO_3$), the name referring to both the mineral and the rock, which are important economic resources. Lime from low magnesium limestone is the main component for cement manufacturing. Hard, dense, and low-porosity carbonates are one of the preferred rocks for aggregate production, mainly used as roadstone. In agriculture, pulverized limestone (agricultural lime) and the products resulting from calcining carbonate rocks (quicklime: calcium oxide; slaked lime: calcium hydroxide), are extensively used to correct the acidity of the soils and provide them with calcium. Carbonates with high mechanical strength and durability are prized for building stones. Travertine and calcareous tufa have been important building and ornamental stones since ancient times (Herrero and Escavy 2010). Carbonate sediments are also important in many chemical-processing industries (glass, paper, and paint). They are used in the processing of a variety of metals (e.g., iron, copper, lead, and zinc) where limestones and limes serve as fluxes for segregating the impurities of metal ores. Carbonate formations are increasingly used in environmental protection for the removal of pollutants. Limestone is utilized in power stations for the desulfurization of fossil fuel emissions to prevent acid rain (Gunn 2004). Carbon dioxide, the main gas responsible for the greenhouse effect and global warming, can be captured via the precipitation of human-induced carbonates, a process known as "carbon sequestration by mineral carbonation" (Lal 2007).

Carbonate rocks are characterized by a simple mineralogy dominated by calcite (high- and low-Mg), aragonite, and dolomite. Aragonite is a thermodynamically unstable form of calcium carbonate that tends to transform into calcite. Aragonite is an important mineral in carbonate deposits that are accumulating today (e.g., Bahamas), but it is very scarce or absent in old lithified carbonates. Dolomite is principally a diagenetic mineral resulting from the dolomitization of calcium carbonates (Machel 2004). Penecontemporaneous or syndepositional dolomites are proportionally rare (Vasconcelos et al. 1995). There are two principal pure carbonate end members, limestone composed of calcite and/or aragonite, and dolomite or dolostone, consisting of dolomite. The calcite:dolomite ratio shows a general decreasing trend with the age of the rocks, although dolomitization is not only a cumulative process related to burial, but can be related to a multitude of processes (Given and Wilkinson 1987); the ratio is about 80 : 1 in Cretaceous strata and 3 : 1 in Lower Paleozoic formations. Figure 2.2 shows a simple mineralogical classification of carbonate rocks considering the proportion of the two main carbonate minerals (calcite and dolomite) and

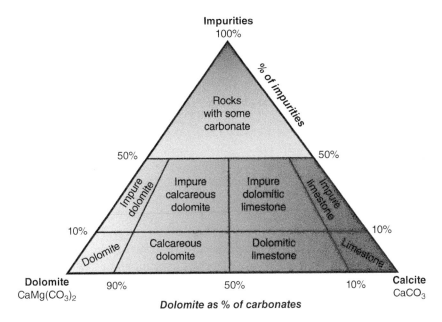

Figure 2.2 Compositional classification of carbonate rocks based on the weight percentage of calcite, dolomite, and impurities. The overall solubility of the rock decreases toward the calcite vertex and away from the vertex of the non-soluble impurities.

impurities. This is a useful classification for karst studies, since it reflects the dissolution susceptibility of carbonate rocks, which increases toward the vertex of calcite, which dissolves faster than dolomite in typical natural settings.

A distinctive feature of carbonate rocks is that they are essentially autochthonous sediments largely related to biological activity. They are not made, but born in a "carbonate factory" (James and Jones 2015). Moreover, because of the soluble nature of their components, they typically experience significant diagenetic changes such as mineralogical transformations (aragonite–calcite–dolomite), rapid loss of primary intergranular porosity by cementation, and development of secondary porosity by dissolution or dolomitization. These alterations may completely obliterate the original depositional texture and fabric of the sediment.

Volumetrically, the vast majority of the carbonate rocks accumulate in shallow marine environments within the neritic zone; i.e., from the seashore to the edge of the continental shelf. These are mainly carbonate shelves or shallow banks in open oceans with warm water and low detrital input that may cover extensive areas. For instance, the modern Great Barrier Reef complex in Australia covers almost 350 000 km². Carbonates also form in the upper part (ca. 200 m) of the water column in pelagic (open ocean) environments. Here, most sediment forms as suspended calcareous plankton that subsequently sinks to the ocean floor. At some depth, due to the high hydrostatic pressure and low temperature, carbonate minerals start to dissolve (below the lysocline), and at higher depths they are completely dissolved (carbonate compensation depth). The aragonite and calcite compensation depths are at around 2 km and 4.5 km depth, respectively, although these values vary depending on the ocean (deeper in the Atlantic than in the Pacific) and latitude (deeper in warmer tropical waters than in polar ones). A wide variety of carbonate rocks also forms in terrestrial or continental settings including lakes, palustrine environments, fluvial systems, springs, caves, or carbonate soils (Alonso-Zarza and Tanner 2010).

2.2.1 Carbonate Minerals

Some of the main characteristics and diagnostic features of the principal soluble minerals that form carbonate and evaporite rocks are indicated in Table 2.1. The most abundant minerals in carbonate rocks are calcite ($CaCO_3$), aragonite ($CaCO_3$), and dolomite ($CaMg(CO_3)_2$). Other carbonate minerals include magnesite ($MgCO_3$), ankerite ($CaFe(CO_3)_2$), and siderite ($FeCO_3$). The triangular diagram of Ca-Mg-Fe-carbonates shows that carbonate minerals with intermediate compositions are rare due to immiscibility, with the exception of the complete range between magnesite and siderite (solid solution) (Figure 2.3).

There are two main anhydrous polymorphs of $CaCO_3$: calcite (trigonal, rhombohedric) and aragonite (orthorhombic, pseudohexagonal). Amorphous calcium carbonate (ACC), unstable at normal conditions, can also occur ($CaCO_3.nH_2O$). The abundance of the polymorphs calcite and aragonite in modern sediments is determined by the radius of the Ca^{2+} cation (0.98 Å), which is close to the ionic radius that dictates the formation of trigonal (<0.99 Å) and orthorhombic crystals (>0.99 Å). The Ca^{2+} radius is so close to the threshold value that changes in the precipitation conditions can easily shift the crystalline structure. Magnesium in seawater plays a significant role in this change (Sun et al. 2015). In pure calcite, layers of carbonate anions alternate with layers of calcium cations (Figure 2.4). Each Ca^{2+} cation is in octahedral coordination with six CO_3^{2-} anions forming trigonal crystals. Cations smaller than Ca^{2+}, such as Fe^{2+} (0.74 Å) or Mg^{2+} (0.66 Å), may substitute Ca^{2+} ions (diadochy) or form Ca-free trigonal carbonate minerals (siderite, magnesite). Sr^{2+}, with a larger radius (1.12 Å), forms orthorhombic (pseudohexagonal) crystals (strontianite). Magnesium cations can incorporate into Ca positions of the crystalline structure forming Mg-calcite. Some authors differentiate between low-Mg calcite (LMC), with <4 mol% of $MgCO_3$, and high-Mg calcite (HMC), with >4 mol% of $MgCO_3$. This is an important variable, since the solubility of calcite shows a general increase with rising Mg content, which contributes to reduce the stability of the crystal lattice. Mg-calcite with approximately 12 mol% of $MgCO_3$ has the same solubility as aragonite, which is more soluble than pure calcite (Morse and Mackenzie 1990). Calcite crystals are usually white or colorless and display variable habits such as rhombohedra, scalenohedra, prisms, and bipyramids (Figure 2.5).

In pure aragonite, Ca^{2+} is coordinated to nine oxygens in six surrounding $(CO_3)^{2-}$ groups, forming orthorhombic arrangements (Figure 2.4). Cation positions tend to be occupied by ions larger than Ca^{2+}, such as Sr^{2+}. Although aragonite is a metastable polymorph of $CaCO_3$, it is an abundant component in modern carbonate sediments. This apparent contradiction is attributed to metabolically mediated precipitation; many plants and animals incorporate aragonite rather than calcite into their calcareous skeletons. For instance, pearls are mainly composed of aragonite. Aragonite has higher density and around 8% lower volume than calcite. Therefore, the aragonite-to-calcite inversion implies a volume increase and a reduction in porosity. Aragonite crystals typically display acicular, tabular, prismatic, and fibrous habits (Figure 2.5), and similar to calcite, easily effervesces in dilute HCl acid.

In an ideal stoichiometric dolomite, with equal proportions of Ca^{2+} and Mg^{2+} cations, layers of Ca^{2+}, CO_3^{2-}, and Mg^{2+} alternate regularly forming hexagonal crystals (Figure 2.4). Given the high Mg:Ca ratio of seawater (normally around 5 : 1), it would be expected that dolomite should be the main primary carbonate precipitate instead of calcite. This is partially due to the fact that Ca-OH radicals dehydrate more easily than the Mg-OH radicals, with a stronger bond with the hydroxyl anion. Fe^{2+}, with an intermediate ionic radius between Ca^{2+} and Mg^{2+}, can easily fit into the cationic positions of the dolomite structure. This explains why dolomite typically contains more Fe^{2+} than calcite and shows, upon weathering, a beige or pinkish color, related to the oxidation of Fe^{2+}

Table 2.1 Properties and diagnostic features of the main minerals of karst rocks. Halides are rarely observed at the surface due to their high solubility, and anhydrite hydrates into gypsum under meteoric conditions.

Group	Mineral composition	Hardness/Specific gravity	Crystallography	Identification
Carbonates	Calcite $CaCO_3$	3/2.71	Hexagonal (rhombohedral)	Usually white or colorless, hexagonal prisms, rhombohedra, scalenohedra, easily effervesces with dilute HCl acid, vitreous to dull luster, hard, but does not scratch iron steel tools, double refraction
	Aragonite $CaCO_3$	3.5–4/2.94	Orthorhombic	Usually white or colorless, acicular, tabular, prismatic, fibrous, easily effervesces with dilute HCl acid, vitreous luster
	Dolomite $CaMg(CO_3)_2$	3.5–4/2.85	Hexagonal (rhombohedral)	Commonly white, pink, colorless, rhombohedral crystals, effervesces slightly in dilute HCl when powdered, vitreous luster
	Magnesite $MgCO_3$	3.5–5/3.0	Hexagonal (rhombohedral)	Usually white, gray or yellow, rare crystals, usually massive, does not react to cold HCl acid, vitreous luster
Sulfates	Gypsum $CaSO_4 \cdot 2H_2O$	2/2.32	Monoclinic	Usually white, gray, colorless, typical crystals are tabular, can be scratched with a fingernail, clear variety is called selenite
	Anhydrite $CaSO_4$	3–3.5/2.98	Orthorhombic	Usually colorless, massive, granular or fibrous, harder and denser than gypsum, vitreous luster
Halides	Halite $NaCl$	2.5/2.16	Cubic	Typically white, cubic crystals, rapidly dissolves in water
	Sylvite KCl	2/1.99	Cubic	White to colorless, cubic crystals, light, bitter taste than halite, rapidly dissolves in water
	Carnallite $KMgCl_3 \cdot 6H_2O$	2.5/1.6	Orthorhombic	White, red, colorless, fibrous, light, bitter taste, rapidly dissolves in water, absorbs moisture form the air (deliquescent)
Silicates	Quartz SiO_2	7/2.65	Hexagonal (rhombohedral)	Variable color, often white, pseudohexagonal prismatic crystals with rhombohedral terminations are common, transparent, lack of cleavage, conchoidal fracture, hard (scratches iron steel tools). Only soluble in hydrofluoric acid.
	Opal $SiO_2 \cdot nH_2O$	5.5–6/2–2.25	Amorphous	Amorphous variety of SiO_2 (mineraloid), highly variable color, irregular veins, masses, nodules, vitreous to resinous luster, conchoidal to irregular fracture, insoluble in most acids

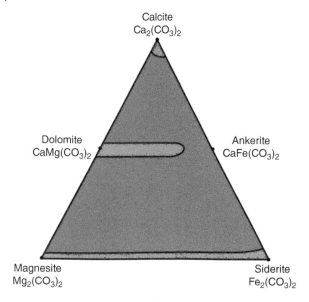

Figure 2.3 Triangular diagram with the principal end members of Ca-Mg-Fe-carbonates. Natural compositions falling in the purple region are rare due to immiscibility. Formulas of calcite, magnesite, and siderite have been doubled for consistency with the formulas of dolomite and ankerite. *Source:* Adapted from Perkins (2002).

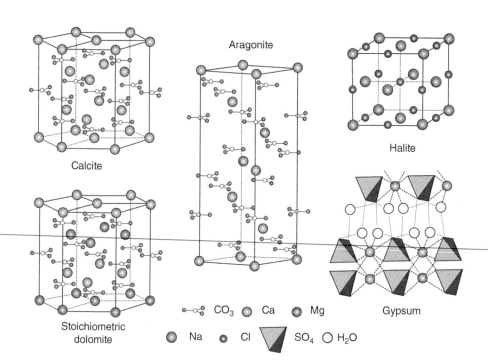

Figure 2.4 Configurations of cations and anions in calcite, aragonite, dolomite, halite, and gypsum.

into Fe^{3+}. Dolomite is characterized by rhombohedral habit and cleavage (Figure 2.5). It can be differentiated from calcite and aragonite by the lack or limited effervescence in cold dilute HCl acid (acid test) and by staining with Alizarin Red S solution, which stains calcite red while leaving dolomite unstained.

Figure 2.5 Specimens of some of the main constituent minerals of karst rocks. Top-left: White translucent halite crystal with cubic cleavage. Top-right: Prismatic crystals of aragonite, twinned in large specimen. Bottom-left: Rhombohedral transparent crystal of calcite (Iceland spar). Bottom-center: Rhombohedral crystals of Fe-rich dolomite (teruelite). Bottom-right: Gypsum crystals (arrow head and Roman sword forms). *Source:* Francisco Gutiérrez (Author).

2.2.2 Depositional Environments and Components of Marine Limestones

There are two main "carbonate factories" in marine environments (James and Jones 2015): (i) The benthic factory (i.e., sea floor), where carbonate sediment accumulates in place, although it can be reworked by waves and currents. This is the main carbonate-producing system in shallow neritic environments. (ii) The pelagic factory, where fine-grained carbonate particles from the near-surface parts of the water column, mainly skeletons of calcareous phytoplankton and zooplankton, sink to the ocean floor. This is the principal source of carbonate in open-ocean, deep-water environments, where benthic carbonate production at the sea floor is very limited (i.e., low temperature, aphotic conditions, and high hydrostatic pressure). Pelagic sediments are important source- and reservoir rocks for hydrocarbons.

Great part of the carbonate formations of the rock record were deposited in marine platforms, especially under clear warm waters where the carbonate production rate reaches the highest values. There are three end members of platforms, which can be attached to the continent or form isolated banks in the open ocean (Figure 2.6): (i) Rimmed platforms with elevated reefs, islands, or sand shoals in their outer edge that absorb most waves and swells protecting a relatively quiet-water lagoon. (ii) Open platforms characterized by the lack of rims and lagoons and the presence of higher-energy grainy facies. (iii) Inclined platforms, usually designated as ramps, where nearshore wave-affected facies grade outboard into deeper-water and lower-energy facies. Extensive shallow-water platforms developed on a flooded craton that may reach as much as thousands of kilometers across are called epeiric platforms. Banks are platforms surrounded by deep-ocean water and disconnected from terrestrial clastic input. Atolls are a special type of bank typically developed on subsiding volcanoes.

There are a number of marine carbonate depositional systems with characteristic facies associations (James and Jones 2015): peritidal systems, neritic systems, reefs, carbonate slopes, and deep-water pelagic systems. A large variety of sedimentary models, showing the distribution and relationships of carbonate facies in marine environments controlled by different factors, can be found in books on carbonate sedimentology (Tucker and Wright 1990; Schlager 2005; Flügel and Munnecke 2010; James and Jones 2015). The variable characteristics of the different facies (e.g., mineralogy, non-soluble components, texture, and bed thickness) and their spatial relationships may play an important role in the karstification processes. The peritidal systems comprise three main depositional environments: the intertidal zone influenced by daily tides, the supratidal zone

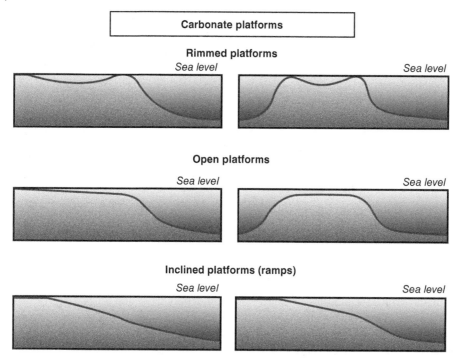

Figure 2.6 Sketch of the three main types of carbonate platforms. Rimmed and open platforms can be attached to the continent or form isolated banks in open ocean. *Source:* Adapted from James and Jones (2015).

that is inundated by seawater during severe storms, and the subtidal zone below the low tide. Sediments accumulated in these mud-rich nearshore tidal systems typically show rapid alternations of facies and generally include features related to subaerial exposure. Regarding the neritic systems, there are some differences between those with warm and cool waters. In the warm-water tropical systems, characterized by clear water with temperatures >20 °C year-round, carbonate is mainly produced by photozoan biota; i.e., organisms that are photosynthetic (e.g., calcareous algae) or contain photosymbionts (e.g., scleractinian corals). These systems have generated the world's thickest and most widespread carbonate successions. In the cool-water systems, carbonate sediments are mainly produced by heterozoan biota (e.g., benthic foraminifers, mollusks, and echinoderms) and are restricted to areas of low siliciclastic input due to lower carbonate production rates. The carbonate slopes are the inclined regions that stretch from the outer edge of the platform to the basin floor, which may reach depths of thousands of meters. These systems are characterized by deposition of talus derived from the platform edge or rim (peri-platform talus) and a variety of re-sedimented and pelagic facies. In the pelagic systems, the sediments are usually uniform and are fed by the planktonic biota developed in the shallow part of the water column and by sediment-gravity flows sourced in the neritic and slope zones. The most important pelagic sediment in modern oceans is biogenic ooze. This is a fine-grained and soft deposit dominated by planktonic foraminifers and cocoliths. Chalk is a white pelagic carbonate sediment predominantly of Late Cretaceous age formed mainly of remains of planktonic foraminifers and algae.

Carbonate sediments are formed of calcareous allochem particles or clasts held in a matrix and/ or cement (Figures 2.7 and 2.8). The allochems include precipitates and calcareous marine organisms formed in the basin, that may have experienced some transport and retexturing. The allochems range in size from silt to boulder and are either sediment grains or large in-place skeletons or structures such as reef corals or stromatolites (Figure 2.7c, d, f). The matrix is mainly mechanically deposited calcareous mud (Figures 2.7e and 2.8a–c), whereas the cement is largely composed of postdepositional calcite crystals that have precipitated between the allochems. The different components of the carbonate sediments can be rapidly modified by biogenic (bioerosion, burrowing), chemical (cementation, dissolution), and hydrodynamic processes (storms, tides). The main types of allochems include (Jones 2005; James and Jones 2015):

- Coated grains, ooids, and pisoids. These are grains with a nucleus of variable nature (e.g., sand grain, bioclast) surrounded by a laminated cortex. Ooids and pisoids are coated grains with diameters smaller and larger than 2 mm, respectively (Figure 2.8c). Marine ooids are abundant in shallow agitated seawater environments. Oncolites refer to coated grains whose growth is mediated by algae.
- Pellets and peloids. Pellets are typically sand-sized micrite (microcrystalline calcite or aragonite) mainly excreted by benthic animals (e.g., sea cucumbers, worms, and gastropods) that ingest mud. Peloids are similar ovate micrite grains produced by inorganic diagenetic alteration of other grains (Figure 2.8a).
- Bioclasts are grains of any dimension derived from the calcareous skeletons of animals and plants (Figures 2.7e and 2.8b, d). They may be entire skeletons or fragments related to: (i) postmortem disintegration of skeletal component by the decay of the bonding organic tissue, (ii) fragmentation by biological activity (biobreakage), and (iii) physical breakdown. Their composition depends on the mineral that precipitates in the skeletons of the different organism: aragonite (e.g., corals, bivalves, and calcareous algae); LMC (e.g., brachiopods, bryozoans); HMC (e.g., crinoids, echinoids, and red algae). Tropical sediments are dominated by metastable aragonite highly susceptible to invert into calcite. In contrast, temperate carbonates, dominated by LMC and HMC, are more stable.
- Microbialites, stromatolites, and mats. These are structures generated by carbonate precipitation mediated by microbes (Figure 2.8). Stromatolites are typically laminated domal or columnar structures associated with shallow, quiet-water conditions (Figure 2.7f). The formation of mats, typical in peritidal and lake-margin environments, is related to photosynthetic microbes.
- Lithoclasts are particles of a wide range of sizes derived from lithified or semi-lithified carbonate substrates. They are typically tabular clasts that are ripped up and transported from the sea floor by storm waves. Their formation may also be related to tidal channel avulsion and seismic events. The terms extraclast and intraclast refer to lithoclasts derived from outside or inside the depositional basin.

The matrix is typically formed of grains <4 μm long of microcrystalline calcite or aragonite, known under the shortened name of micrite (Figure 2.8a). Micrite is generally a mechanically deposited component (allomicrite), but can also form as cement on the sea floor or within the sediment (automicrite). Allomicrite may generate in multiple ways, including fallout of calcareous plankton and precipitated carbonate crystals from the water column, disintegration of allochems, bioerosion, and secretion by fish.

Carbonate rocks may also include large in-place structures such as reefs (Figures 2.7c, d). These are biologically constructed reliefs that rise above the surrounding sea floor. Reefs are composed of complex associations of organisms with carbonate precipitates and sediments, and typically have a high primary porosity, making reef limestones important hydrocarbon reservoirs. At the present time, the main reef builders are scleractinian corals that house symbiotic photosynthetic cyanobacteria or microalgae and may grow at rates of ~0.2–1 cm yr^{-1}. Optimum conditions for the development of these coral reefs include seawater temperatures between 25 and 29 °C, transparent and

Figure 2.7 Images of carbonate sediments. (a) Thick carbonate succession affected by contractional structures (folds and thrusts). Gray- and light-colored units are Paleogene limestones and the tan unit in the background is Cretaceous calcarenite. Ordesa National Park, Pyrenees, Spain. (b) Dolomitic massif at Corvara, Italian Alps. In this region, the French geologist Déodat de Dolomieu was the first to identify the mineral and the rock subsequently called dolomite after him, also giving the name to the Dolomite mountains (UNESCO Natural World Heritage Site since 2009). (c) A modern reef with branching corals in the Red Sea, Hurgada, Egypt. (d) A recent and emerged dome-shaped coral in the Red Sea, Hurgada, Egypt. (e) Poorly lithified sediment of the Pliocene Jackson Bluff Formation, consisting of shells within a dark carbonate-rich mud matrix. Apalachicola River, Florida. (f) Exhumed dome-shaped stromatolite with concentric structure in the Miocene Chipola Formation. Apalachicola River, Florida. (g) Well-bedded beachrock with syndepositional seaward dip, mainly consisting of cemented bioclasts. Volcanic island of La Graciosa, Canary Islands, Spain. (h) Stratification plane at the top of a reddish hardground developed on a strongly cemented and bioturbated lumachella (or coquina) of mollusks with iron oxides. Cretaceous Dariyan Formation, Zagros Mountains, Iran. *Source:* Francisco Gutiérrez (Author).

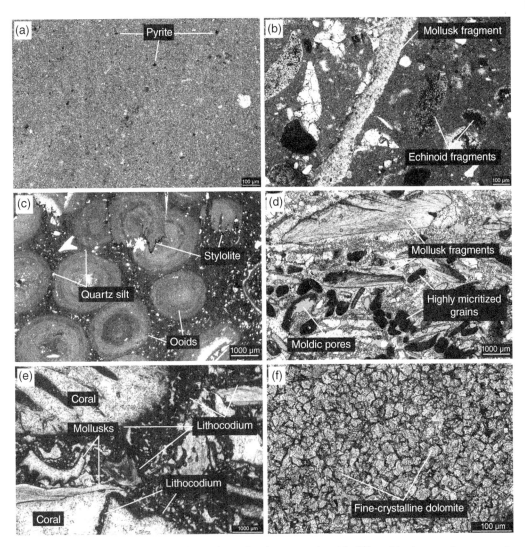

Figure 2.8 Microphotographs of thin sections of carbonate rocks with different textures and components. (a) Mudstone (Folk's classification: micrite) with traces of pyrite and skeletal fragments. Mudstone has a very faint texture of merged soft peloids. Lower Cretaceous, south Texas. (b) Wackestone (Folk: biomicrite) with mollusk and echinoid fragments. Lower Cretaceous, south Texas. (c) Ooid packstone (Folk: oomicrite) with some quartz silt in matrix. Stylolite at top of image. Upper Jurassic, East Texas. (d) Skeletal grainstone (Folk: biosparite) with abundant moldic pores shown by blue mounting medium. Skeletal grains are predominantly mollusk fragments. Lower Cretaceous, south Texas. (e) Boundstone (Folk: biolithite) of coral and mollusk fragments bound together by the microbialite *Lithocodium* (grain coater with pores). The binder *Lithocodium* actually fills the pores making this a bindstone. From the Stuart City reef. Lower Cretaceous, south Texas. (f) Crystalline dolostone where dolomitization has destroyed the original depositional texture. Permian, West Texas. *Source:*Images and descriptions courtesy of Robert Loucks.

low-nutrient (oligotrophic) waters, salinities between 25 and 35‰, and low detrital input, which would reduce light penetration and cause abrasion. Reefs typically show a zonation with five distinctive facies: (i) The reef crest that may emerge at low tide and is dominated by organisms adapted to high-energy hydrodynamic conditions (e.g., sheet-like encrusting organisms). (ii) The reef flat on the leeward side, with abundant bioclasts derived from the erosion of the crest. (iii) The protected backreef dominated by low-energy mud facies. (iv) The seaward reef front, between the base of the wave action and 100 m depth, with a large diversity of reef builders. (v) The deeper forereef below the coral and algal growth zone, where reef-derived coarse fragments accumulate.

Two main types of bioconstructions are differentiated in the rock record, lens-shaped bioherms and tabular biostromes. Bioconstructions made up of skeletal components larger than 1 cm in size are termed reefs, whereas in reef mounds the skeletal components are less than 1 cm. Several types of reefs are differentiated according to their position and overall geometry. Barrier reefs are linear features developed along the shelf margin, they can exceed 100 km in length and have an associated inner lagoon. Fringing reefs are linear bioconstructions that grow near the shore and lack an intermediate lagoon. Atolls are ring-shaped reefs that surround an inner lagoon. Patch reefs are isolated reefs generally <100 m in diameter with subcircular geometry, developed in lagoon environments. Pinnacle reefs are characterized by a high aspect ratio, with a vertical dimension that can be one order of magnitude higher than the horizontal.

2.2.3 Limestone Classification Schemes

There are several sedimentological classifications of limestones. Limestones with clastic texture may be named according to the dominant grain size. The terms calcirudites, calciarenites, calcisiltites, and calcilutites (micrite mud) designate rocks mainly consisting of gravel-, sand-, silt-, and clay-sized particles, respectively. See grain size scale in Figure 2.1. Dunham (1962) proposed a simple classification that first differentiates between rocks in which the original texture has been destroyed (e.g., crystalline limestone) and those consisting of non-bound components (allochthonous limestones) that preserve the original depositional texture. The latter are described according to the nature of the supporting framework (grain-, matrix-supported) and the proportion of grains. This scheme was later expanded by Embry and Klovan (1971) adding terms for allochthonous limestones and autochthonous reef rocks (boundstones) (Figure 2.9). This widely used classification recognizes six types of allochthonous limestones that essentially indicate a gradation from lower to higher energy conditions (mudstone, wackestone, packstone, grainstone, floatstone, and rudstone) and three autochthonous limestones composed of organically bound components: framestone, bindstone, and bafflestone. Framestones consist of large skeletons, such as framework corals, that are stacked on top of each other. Bindstones are bioconstructions of laminated or sheet-like anastomosing growth forms with cavities filled by sediment, cement or both. Bafflestones comprise upright branching growth forms that baffle water movement favoring the deposition of suspended sediment. The rock names derived from this

Allochthonous limestone Original components not bound during deposition						Autochthonous limestone Original components bound during deposition		
Less than 10% >2 mm components				Greater than 10% >2 mm components		Organisms that build a rigid framework	Organisms that encrust and bind	Organisms that act as baffles
Contains lime mud (<0.03 mm)		No lime mud		Matrix-supported	>2 mm component-supported			
Mud-supported								
Less than 10% grains (>0.03 mm, <2 mm)	Greater than 10% grains	Grain-supported					Boundstone	
Mudstone	Wackestone	Packstone	Grainstone	Floatstone	Rudstone	Framestone	Bindstone	Bafflestone

Figure 2.9 Embry and Klovan's (1971) classification of limestones mainly based on the depositional texture. This classification expands the previous scheme proposed by Dunham (1962).

classification are commonly complemented by prefixes that indicate the dominant type of component (e.g., oolitic grainstone).

The other widely used classification is that proposed by Folk (1962), which is focused on the types of allochems (particles), the surrounding material (matrix or calcite cement), and some textural features (Figure 2.10). This classification includes two schemes. The first one differentiates between autochthonous reef rocks, designated as biolithite, orthochemical rocks, including micrite (pure carbonate mud) and dismicrite (micrite with spar-filled burrows), and allochemical rocks. The latter are described indicating the type of allochems (intraclasts, ooids-pisolites, bioclasts, and pellets) and whether they are surrounded by micrite matrix or calcite cement (e.g., biomicrite,

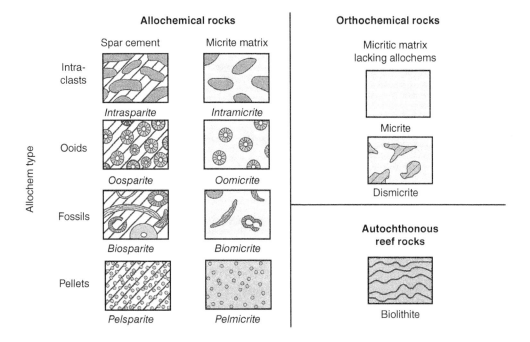

Figure 2.10 Classification of limestones proposed by Folk (1962), mainly focused on the type of allochems and surrounding material and some textural features of the particles that reflect energy conditions. *Source:* James and Jones (2015). Reproduced with permission of Wiley.

biosparite). Combinations of the different terms can be used to classify the rock (e.g., biopelmic-rite). The second scheme is used to add terms referring to the textural features of the rock, including percentage of allochems and matrix, sorting, and rounding. This scheme reflects a progressive increase in the energy conditions of the depositional environment.

Insalaco (1998) designed a classification mainly focused on scleractinian coral reefs. It differentiates between autochthonous rocks made up of bioconstructions with different geometries (platestone, sheetstone, domestone, pillarstone, and mixtone) and allochthonous rocks consisting of particles derived from the destruction of the reefs. Wright (1992) developed an entirely different classification considering depositional, biological, and diagenetic features. According to James and Jones (2015), all these classifications have limitations in describing some carbonate rocks that may be important for karst studies, including dolostones, in which the original texture has been destroyed by dolomitization, rocks consisting of carbonate and siliciclastic components typical of transitional environments, and carbonate breccias. An additional caveat of some classifications is that their application requires the use of a microscope. Images of thin sections of a wide variety of limestone rocks accompanied by detailed petrographic and sedimentological descriptions can be found in Adams et al. (1984), Adams and MacKenzie (1998), and Melgarejo (2003). Some representative examples are illustrated in Figure 2.8.

2.2.4 Carbonate Sequence Stratigraphy

Sequence stratigraphy is an important integrating approach in sedimentary geology conceived for analyzing the relationships between the marine stratigraphic record (stratigraphy, sedimentology, and seismic data) and the principal dynamic controlling factors, including sea-level changes and the interplay between sedimentation and accommodation, which is the space available for sediment to fill (Tucker 1991; Schlager 2005; James and Jones 2015). Tectonic activity can cause local- and basin-scale relative sea-level variations, whereas eustatic changes that may be related to glacial and interglacial periods (glacioeustatic) have a global impact. The sedimentation pattern in a carbonate platform will be mainly determined by the ratio between the sedimentation rate and the rate at which the accommodation space is created (or reduced). Transgression occurs when the available accommodation space does not limit sedimentation, a situation frequently associated with sea-level rise and/or low carbonate production. It may be expressed in the stratigraphic record as: (i) retrogradation, when the rate of sea-level rise exceeds sedimentation, and facies shift shoreward (i.e., progressively deeper facies accumulate); and (ii) aggradation, when sedimentation keeps pace with sea-level rise, and the lateral distribution of facies does not experience significant changes. Regression occurs when the accommodation space limits sedimentation, and the depositional environments experience progradation, with accumulation of progressively shallower facies at each section.

The depositional sequence is the basic unit in sequence stratigraphy (Figure 2.11). This unit is limited by top and bottom sequence boundaries, which are unconformities with evidence of subaerial exposure in the shallow part of the sequence and that grade into a conformable discontinuity in the deeper zones. Each depositional sequence comprises a series of systems tracts, which are contemporaneous sediments accumulated in spatially associated depositional environments (e.g., peritidal, lagoon, platform margin, and slope). The different systems tracts can be ascribed to sections of a curve that represents the variations of the relative sea level, which is the space between the sea surface and the top of the sediment pile (Figure 2.11). The transgressive systems tract (TST) develops when sea-level rise is more rapid than sedimentation and the depositional systems move shoreward or retrograde. The highstand systems tract (HST) develops when the rising

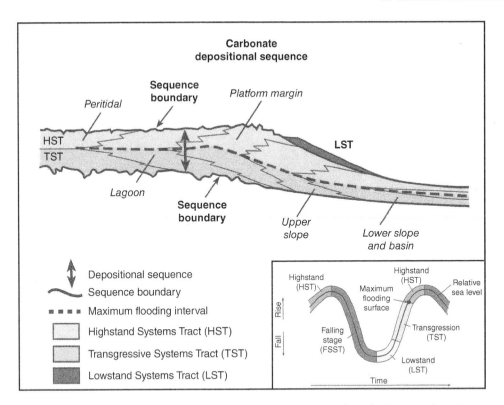

Figure 2.11 Elements of a typical carbonate depositional sequence bounded by unconformable contacts in the shallow zone. In the TST depositional facies shift shoreward (retrogradation) as the relative sea level rises. In the HST facies move basinward (progradation). The change from the TST to the HST is marked by the maximum flooding interval that records the crest of the sea level cycle. The LST is formed during the lowest point of sea-level fall. The inset figure sketches the different system tracts that develop during a cycle of sea-level change. *Source:* James and Jones (2015). Reproduced with permission of Wiley.

sea-level is reaching the crest of the cycle. During this period the sea-level rise slows down, sedimentation exceeds the creation of accommodation space, and consequently the depositional systems move seaward or prograde. The boundary between the TST and the HST, which is recognized as the change from transgressive to regressive systems, is termed the maximum flooding surface. The falling stage systems tract (FSST) and the lowstand systems tract (LST) develop when sea level falls (progradation) and when the relative sea level is at the lowest point of the curve, respectively. Systems tracts can be divided into conformable successions known as parasequences, which are typically meter-scale shallowing upward sequences. The development of these sequences may be related to external factors (allogenic) such as tectonics and eustatic changes, or to internal factors (autogenic) such as carbonate production. Of special importance for karst studies are the Quaternary stratigraphic cycles controlled by glacioeustatic changes. These cycles provide information on the relative position of the sea level, which is the hydrological base level for karst systems in coastal environments. The sea level dropped and rose abruptly during the glacial and interglacial cycles respectively, at intervals of about 100 kyr since the middle Quaternary (Figure 2.12).

As a result of the common cyclic pattern of sea level changes, carbonate successions frequently display stratigraphic cyclicity that may play an important role in karst development. For instance, carbonate formations may consist of a succession of shallowing-upward sequences bounded at the

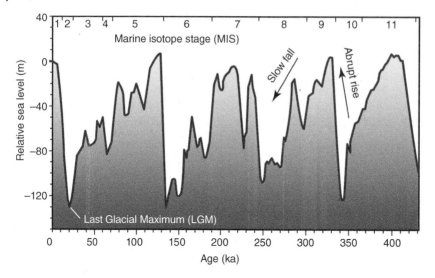

Figure 2.12 Late Quaternary sea level curve. *Source:* Adapted from Johnson and Watt (2012).

top by discontinuities with evidence of subaerial exposure, including erosion and solution features, more pronounced cementation, desiccation structures, paleosols. Tinker et al. (1995) recognized multiple paleokarst events in the San Andres Formation (Yates oil field, Texas) related to recurrent subaerial exposure episodes associated with the cyclic stratigraphy of the formation. This information on the distribution of solutional porosity is important for hydrocarbon exploration and production. Carbonate successions may also include hardgrounds, which are synsedimentary indurated surfaces developed on the sea floor during periods of very low sedimentation. These stratigraphic discontinuities that record a depositional hiatus, typically display strong bioturbation, high concentration of fossils including specific organisms adapted to the hard surface, mineralizations of iron oxides and calcium phosphates (Figure 2.7h). They are used as marker horizons for correlation and as indicators of flooding events. The lithological, permeability and erodibility changes associated with sequence boundaries may serve as preferential flow paths for karst development, and may also control the exhumation of stratigraphic surfaces by the stripping of the overlying sediments. Filipponi et al. (2009) analyzed the 3D spatial relationship between more than 1500 km of conduits from 18 large caves in the world and the associated stratigraphy. The results of their statistical analysis revealed the existence of "inception horizons," defined as parts of a rock succession (e.g., discontinuities, layers) that are particularly susceptible to the development of karst conduits, because of physical, lithological and/or chemical deviations from the predominant carbonate facies within the limestone successions (Lowe 1992b).

2.2.5 Limestone Diagenesis

Diagenesis refers to the suite of physical and chemical changes that experience sediments after deposition. These changes may occur immediately after accumulation, during burial by younger sediments before they reach the high temperature and pressure conditions of the metamorphic realm, and when the sediments return to the surface or near-surface meteoric environment. The diagenetic transformations are related to changes in temperature, pressure, water chemistry, groundwater flow, and biological activity. Typically, soft, highly porous, water-saturated, unconsolidated carbonate deposits are buried by a progressively thicker pile of sediments and transformed into denser and less porous lithified rocks. These brittle rocks may be affected by fracturing and faulting that guide groundwater flow and eventually karstification. Diagenesis, and especially the chemical changes associated with it, are particularly important in carbonate sediments because of their soluble nature (James and Choquette 1990; Moore 2001; Moore and Wade 2013; James and Jones 2015; AAPG 2020). Different diagenetic stages are recognized (Figure 2.13). The eogenetic stage (early diagenesis) covers the temporal span from deposition to the moment in which the sediment is buried below the zone of influence from surface processes. The mesogenetic stage (middle diagenesis) encompasses the time during which the sediment is out of the influence of surface diagenetic processes. The telogenetic stage (late diagenesis) occurs when carbonate rocks are brought back to surface or near-surface meteoric conditions. Other authors use more intuitive terms for the different diagenetic environments such as synsedimentary, burial, and meteoric diagenesis. The principal diagenetic processes include compaction, dissolution, precipitation, neomorphism, and dolomitization.

Compaction. Buried sediments subject to progressively higher overburden load experience compaction, which produces the rearrangement and deformation of particles and matrix with the consequent changes in texture (shape and arrangement of particles) and fabric (orientation of components). Compaction involves multiple physical changes, including tightening of the framework packing, porosity reduction, dewatering, density increase, and layer thickness decrease. Shallow-water carbonate sediments typically compact to approximately half their original thickness with just 100 m of overburden, with a reduction of around 50–60% of their initial pore volume (James and Jones 2015).

Carbonate dissolution can take place at any diagenetic stage. In the meteoric environment pores are partially (vadose zone) or completely filled (phreatic zone) with freshwater. Carbonate components dissolve when the pore water is undersaturated with respect to the carbonate minerals.

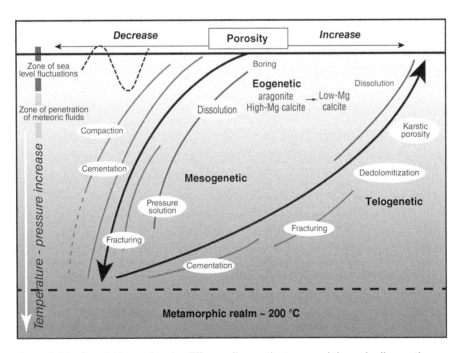

Figure 2.13 Sketch illustrating the different diagenetic stages and the main diagenetic processes. *Source:* Adapted from Wright (2006).

Undersaturation can be related to various mechanisms: (i) water acidification (pH lowering) by the incorporation of dissolved CO_2 by microbial activity, oxidation of organic matter, diffusion from the atmospheric air, increasing pressure or decreasing water temperature, (ii) mixing with waters with higher CO_2 contents or different chemical composition, and (iii) localized stress (pressure solution) related to lithostatic pressure (overburden load) or deviatoric tectonic stress. Dissolution can be fabric specific, when each component is affected selectively in a different way, and non-fabric specific, when dissolution creates pores cutting across particles and cements and there is fabric destruction.

Carbonate precipitation in open spaces leads to the cementation and lithification of the carbonate sediment. Carbonate minerals, including dolomite, can precipitate in a wide variety of voids, such as interparticle pores, intraparticle spaces (voids inside skeletons), cavities within reefs (framework voids), intercrystal spaces, dissolution voids (molds and vugs produced by fabric and non-fabric specific dissolution, respectively), and fenestral pores typical of muds deposited in tidal flats. The characteristics of the cements largely depend on water composition, crystallization rate, and hydrogeological conditions. Cements precipitated from saline water tend to be dirty and inclusion-rich, whereas those from freshwater are typically clear and inclusion-free. Rapid precipitation produces microcrystalline cements, whereas large crystals generally record slow precipitation. Cements formed under phreatic conditions drape voids and particles with a uniform thickness (isopachous), whilst those precipitated within the vadose zone show meniscus or stalactitic (geopetal) morphology influenced by the surface tension of water and gravity, respectively.

Neomorphism refers to transformations in the carbonate minerals of any component of the sediment (allochems, matrix, and cement) that result in the neoformation of crystals of the same mineral or a polymorph. The two main types of neomorphism include: (i) polymorph transformation, also known as inversion, such as aragonite to calcite inversion; and (ii) recrystallization, which involves alteration without any mineralogical change, such as crystal enlargement. Differentiation between neomorphic sparry calcite, derived from the transformation of preexisting minerals, and cement, related to precipitation of minerals from a solution, may be a challenging task for carbonate petrologists and sedimentologists. The alteration of aragonite and high-Mg calcite (HMC) to diagenetic low-magnesium calcite (LMC) is a very common process, involving dissolution and precipitation from an oversaturated solution. Aragonite is an unstable polymorph and both aragonite and HMC are more soluble than LMC. The aragonite-LMC inversion (calcitization) entails crystallographic change, from orthorhombic to trigonal (Figure 2.4), and macroscopic dissolution, and consequently tends to be fabric destructive. In contrast, the HMC-LMC alteration, mediated by microscopic dissolution without crystallographic change, tends to preserve the original fabrics.

The importance of the processes described above varies depending on the diagenetic environment. In the **synsedimentary marine diagenetic environment**, which includes the sea floor and the sediments situated a few meters below, diagenetic processes are mainly driven by changes in water chemistry and temperature, as well as microbial activity. In the neritic zone and especially in warm-water environments, cementation is the dominant process together with bioturbation. The cements, precipitated from seawater with a high Mg:Ca ratio are mostly composed of aragonite or high-Mg calcite. This early cementation process leads to the formation of hardgrounds and beachrock. The latter are well cemented and layered sandy to gravelly beach deposits with a syndepositional seaward dip that form in shallow wave-influenced environments (Figure 2.7g).

Carbonate sediments deposited on the sea floor may change their relative position to terrestrial environments due to sea level fluctuations and tectonic activity, moving into the **meteoric diagenetic environment**. Here, the interstitial seawater is replaced by freshwater, leading to substantial mineralogical and physical transformations. The development of solutional karst features at the ground surface and in the subsurface is the dominant process and is one of the main topics of this book. Moreover, aragonite and high-Mg calcite typically transform into the less soluble low-Mg calcite. Commonly, aragonite allochems (e.g., bioclasts) are dissolved generating moldic porosity (Figure 2.8d), while the released ions may precipitate as low-Mg calcite cement reducing interparticle porosity. Precipitation of speleothems, carbonate soils (i.e., calcrete or caliche) and crusts are also common. Diagenetic processes (dissolution, cementation) in the meteoric environments are strongly influenced by the position of the water table that bounds the vadose and phreatic zones, as well as by the presence of a freshwater–saltwater mixing zone associated with the shoreline. The vadose zone is dominated by downward gravity-controlled flow (infiltration and percolation), whereas in the phreatic zone, where all the available pore space is occupied by water, groundwater tends to flow subhorizontally toward the base level, following the local hydraulic gradient. The fluctuating zone where phreatic and vadose waters mix is a zone of intense corrosion. In coastal areas, there is a seaward-tapering freshwater lens lying above denser seawater, with a landward-dipping transition or mixing zone with brackish water in between. This is also a zone of enhanced dissolution (see also Figure 5.26). The impact of meteoric diagenetic processes is largely determined by the amount of the unstable aragonite mineral and climate, which determine the amount of water that flows through the vadose zone and how much biogenic CO_2 incorporates into the infiltrating waters while traversing the soil cover.

The **burial diagenetic environment** embraces the area from the base of the meteoric zone to the top of the low-grade metamorphic realm, with temperature and pressure ranges of around 50–200 °C and 700–200 000 kPa (ca. 7–2000 times atmospheric pressure), respectively. Burial diagenesis is generally a prolonged period with multiphase alterations during which carbonate sediments interact with fluids of various compositions under high temperature and pressure conditions. The main overall changes include volume and porosity reduction by compaction, pressure-solution and cementation. Dissolution, dehydration of hydrated minerals (e.g., gypsum) and alteration and maturation of organic matter also occur during burial diagenesis. Pressure-solution involves dissolution of carbonate minerals at specific points or zones where high stress values (lithostatic, tectonic) contribute to reduce the solubility of carbonate minerals (e.g., intergranular contacts). The dissolved ions may migrate to areas of lower stress (e.g., intergranular pores) and reprecipitate as calcite cement. The most common products are stylolites or sutured seams, which are interdigitated jagged surfaces, frequently layer-parallel and containing residual insoluble clays (Figure 2.8c). The amplitude of the stylolites (i.e., the height of the waves) provides a measure of the minimum amount of rock mass removed by dissolution. This is a chemical compaction process that generally occurs at depths greater than 300 m once the rock is lithified, and may result in an additional volume reduction of 20–30%, following previous physical compaction. Burial cements are generally composed of clear and coarse sparry calcite that covers previous synsedimentary and meteoric diagenetic features. These cements that precipitate under reducing conditions typically have high Fe^{2+} and Mn^{2+} contents. Dissolution in the burial environment is generally induced by CO_2 and organic acids released by the thermal alteration of organic matter (kerogen).

2.2.6 Dolomite and Dolomitization

Dolomites (dolostones) are as abundant as limestones in the geological record and have special economic interest because numerous hydrocarbon reservoirs and mineral deposits, mainly lead and zinc, are hosted in porous and permeable dolomites. Dolomite was first discovered in 1791, when Count Déodat de Dolomieu identified rocks in the Italian Alps with the appearance of limestone but that did not effervesce in weak acid. The word dolomite designates both the mineral and the rock, which is also named dolostone. Despite being known for over 200 years, the formation of dolomite is still poorly understood. Dolomite may form as a synsedimentary authigenic mineral under special, often microbially mediated, physicochemical conditions (Vasconcelos et al. 1995; Bontognali et al. 2010; Petrash et al. 2017), but in most cases, it is a diagenetic product derived from the replacement of precursor calcite (Warren 2000; James and Jones 2015). This can happen when fluids from overlying evaporites such as gypsum move down into underlying carbonates as sea level falls and lowstand sequence stratigraphical units are deposited (Tucker 1991).

The mineral dolomite has a trigonal crystalline structure equivalent to that of calcite, but with approximately 50% of the cation sites occupied by Mg^{2+} instead of Ca^{2+} (Figure 2.4). Not all dolomites are stoichiometric ($Ca_{50}Mg_{50}(CO_3)_2$). Excess in Ca contributes to increase the solubility of the mineral. These calcian dolomites predominate in synsedimentary dolomites. There is a continuous spectrum between pure limestone (100% calcite/aragonite) and pure dolostone (100% dolomite), with dolomitic limestones and calcareous dolomites in between (Figure 2.2). The diagenetic dolomitization of limestones tends to obliterate the original fabrics (fabric destructive) (Figure 2.8f). The classification schemes conceived for limestones (Figures 2.7 and 2.8) are not applicable to these crystallized rocks, which are commonly described according to the crystal size, from very finely crystalline ($>100\,\mu$m) to very coarsely crystalline ($>1000\,\mu$m).

Dolomite precipitation, depending on the form in which carbonate is present in the water, which is mainly a function of its pH (Figure 3.10), can be expressed as

$$Mg^{2+} + Ca^{2+} + 4HCO_3^- \leftrightarrow MgCa(CO_3)_2 + 2CO_2 + 2H_2O \tag{2.1}$$

$$Mg^{2+} + Ca^{2+} + 2CO_3^- \leftrightarrow MgCa(CO_3)_2 \tag{2.2}$$

and dolomite replacement of calcite as

$$2CaCO_3 + Mg^{2+} \leftrightarrow MgCa(CO_3)_2 + Ca^{2+} \tag{2.3}$$

Dolomite precipitation rarely occurs in modern seawater despite it typically having a Mg:Ca ratio of ~5 : 1 and is largely supersaturated with respect to this mineral phase in the shallow-water zone. This is because there are a number of factors that inhibit dolomite precipitation, which can be counteracted by specific processes and conditions, eventually allowing synsedimentary dolomite formation:

- Mineral kinetics: dolomite precipitation requires more thermodynamic energy than calcite or aragonite, which readily precipitate in seawater. This limiting factor can be counteracted if the Mg:Ca ratio raises to around 10 : 1. This can occur by Ca depletion through evaporation-driven precipitation of calcium carbonates and sulfates.
- Hydration: Mg^{2+} has a smaller cationic radius (0.66 Å) than Ca^{2+} (0.98 Å), and consequently much higher electrostatic attraction to the hydroxyl radical (OH^-). The hydroxyl anions that surround the Mg^{2+} cations (hydration) make their incorporation into the dolomite crystal lattice difficult. Increasing temperature and microbial activity contribute to weaken the Mg-OH bonds and reduce hydration.
- Crystallization rates: dolomite is characterized by slow crystallization rates. Rapid crystallization under supersaturated conditions favors the incorporation of higher proportions of Ca^{2+} in the crystal structure, leading to precipitation of high-Mg calcite and Ca-rich dolomite. Crystallization rate can be decelerated by dilution of the fluid through the incorporation of fresh water, favoring the formation of more stoichiometric dolomite.

- Carbonate-bicarbonate activity: most carbonate ions under normal pH conditions (7–8) are in the form of HCO_3^-. However, Mg^{2+} and Ca^{2+} cations combine more easily with the form CO_3^{2-}. Some highly basic environments with pH ≥ 9 in which the form CO_3^{2-} dominates are more favorable for dolomite precipitation.
- Sulfate activity: sulfate (SO_4^{2-}), which is present at high concentrations in seawater, inhibits dolomite precipitation. This constraint can be overcome by the sulfate depletion through various mechanisms, including precipitation of sulfate minerals (e.g., gypsum), sulfate reduction by bacterial activity, water dilution, and microbial activity.

Synsedimentary (authigenic) dolomite forms by direct precipitation or replacement of carbonate below the sediment surface soon after deposition. It is commonly associated with high-salinity waters and the most common settings include humid and arid peritidal environments, evaporitic lakes and environments with organic-rich sediments. In humid peritidal settings (e.g., Bahamas), dolomitization is promoted by the increase of the Mg:Ca ratio associated with aragonite precipitation from interstitial fluids. In arid peritidal settings affected by strong evaporation such as the sabkhas of the Persian Gulf and the Red Sea, dolomite formation is related to a sequence of processes: (i) storms incorporate marine water into the sabkhas, (ii) the salinity and density of the water in the sabkha increases by evaporation, (iii) the dense fluid sinks into the underlying sediment, (iv) precipitation of gypsum and anhydrite increases the Mg:Ca ratio and reduces sulfate concentration, and (v) calcite and aragonite are replaced by dolomite generally when the Mg:Ca ratio exceeds 10 : 1. In evaporitic lakes located in coastal and terrestrial environments, dolomite precipitation may be induced by pH increase caused by the metabolic activity of sulfate-reducing bacteria that lower the concentration of sulfate. This microbially mediated dolomite typically has the form of microcrystals that precipitate along the external biofilm of the bacteria. Authigenic dolomite also forms associated with anoxic and organic-rich (kerogenous) sediments in which both the pH and CO_3^{2-} activity increase by microbial activity.

Most dolomites of the rock record are secondary diagenetic products. Extensive dolomitization requires large amounts of seawater, which is the main source of Mg. Several hydrological and geochemical models have been proposed to explain the formation of dolostones in the shallow-burial early-diagenetic environment and in the deep-burial late-diagenetic environment (Warren 2000; James and Jones 2015). In the **early-diagenetic dolomitization**, surface or meteoric waters may play an instrumental role. Moreover, the mineralogy of the sediment may be important, since aragonite and high-Mg calcite are easier to dolomitize than the less soluble low-Mg calcite. Three models are proposed for shallow early-diagenetic dolomites (Figure 2.14): (i) brine reflux, (ii) seawater–freshwater mixing, and (iii) thermal convection. **Brine reflux** occurs when shallow marine water on a rimmed platform is concentrated by evaporation, generating a progressively more saline and denser brine, and eventually leading to precipitation of sulfates. The Mg-rich and sulfate-depleted water descends into the underlying carbonate sediment and toward the outer zone of the platform, mixing with marine connate fluids resulting in the formation of dolomite by replacement and cementation. A similar situation may occur when the backrim environments are disconnected from the open ocean by a sea-level drop and/or rim aggradation. The **seawater–freshwater mixing model** postulates that dolomite should form in the mixing zone. Here, seawater supplies Mg and the dilution effect of the freshwater reduces the activity of the sulfate anion favoring dolomite precipitation. It has been argued that dolomite precipitation is negligible in most modern mixing zones. However, the sea level has been in its current position for only the last 6000 years, probably a time span insufficient for significant dolomite formation. **Thermal convection** occurs at the steep seaward margin of carbonate platforms, where a significant lateral thermal

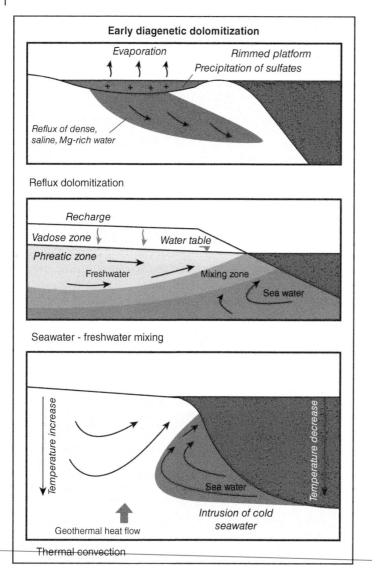

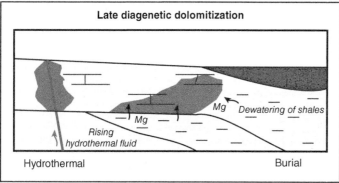

Figure 2.14 Sketch illustrating hydrological and geochemical models of early and late diagenetic dolomitization. *Source:* Adapted from Ford and Williams (2007) and James and Jones (2015).

gradient exists. The temperature in the continent increases with depth due to the geothermal gradient, whereas in the adjacent deep marine water the temperature decreases downward with depth. This leads to a convective half-cell flow, whereby cold seawater intrudes into the platform at depth and is expelled back into the ocean at lower depths. The interaction of seawater and freshwater promotes dolomitization in a zone that wedges out inland.

The formation of **deep-burial late-diagenetic dolostones** is typically associated with hot and saline fluids. These dolostones are porous rocks that have lost their original fabrics and that may host hydrocarbons and mineral deposits. Two genetic models are proposed with markedly different spatial extents: (i) Burial; and (ii) Hydrothermal. **Burial** is a regional process driven at depth by high-temperature conditions. It is suggested that Mg is largely derived from the compaction and dewatering of shales. **Hydrothermal** dolomitization is induced by hydrothermal fluids that rise along faults or fractures, generating highly porous (vuggy) and coarsely crystalline dolomitic bodies of limited lateral extent (Davies and Smith 2006). The Mississippi-Valley-type Pb-Zn mineral deposits may form in this hydrothermal setting. These are sulfide deposits, mainly sphalerite and galena, that occur associated with voids, collapse breccias, and/or as replacement of the carbonate rock.

2.2.7 Terrestrial Carbonates

Carbonate sediments can form in a wide variety of terrestrial (or continental) settings (Figure 2.15). The modern environments in which these deposits precipitate may have outstanding scientific and aesthetic value. For instance, the hypersaline Dead Sea, lying at the Earth's lowest elevation on land, the travertines associated with perched hot springs of Pamukkale (Turkey) (Figure 2.15e) and Mammoth Hot Springs (USA), the tufa mounds constructed at springs fed by rising groundwater of Takht-e Soleyman (Iran) (Figure 2.15b) and Mono Lake (USA), the fluviatile freshwater tufas of Plitvička Jezera National Park (Croatia), or the countless caves with striking speleothems (Figure 2.15g). Terrestrial carbonate sediments are far less important than the marine counterparts for karst development due to several reasons: (i) these deposits did not become important until the Devonian, when plants widely colonized the terrestrial environments favoring soil development, dissolutional weathering and the incorporation of ions in the meteoric waters (Alonso-Zarza and Tanner 2010), (ii) terrestrial carbonate deposits tend to be spatially restricted and of limited thickness, although some carbonate lacustrine formations may reach large volumes and areal extent, and (iii) terrestrial carbonate successions typically include layers or packages of non-soluble rocks (e.g., claystones, marls) that inhibit karst development. Nonetheless, some continental carbonate deposits are the sedimentary outcome of "upflow" dissolution processes in karst systems (e.g., carbonate spring deposits, speleothems precipitated under vadose and phreatic conditions, tufa terraces), recording valuable climatic, hydrologic, geochemical, and paleogeographic information. The volume edited by Alonso-Zarza and Tanner (2010) provides a comprehensive overview of terrestrial carbonates, differentiating four types of depositional environments (lacustrine, palustrine, fluvial, and springs), plus carbonates that precipitate in caves (speleothems), and secondary deposits related to edaphic or meteoric diagenetic processes (calcretes).

The terms (calcareous) tufa, travertine and sinter are used to designate a wide variety of terrestrial carbonate sediments, although frequently with unclear and overlapping meanings. According to Jones and Renault (2010), the word sinter is usually restricted to silica spring deposits (e.g., Ford 1989) precipitated from high-temperature hot springs and geysers, and consequently should not be used for carbonate deposits to avoid confusion. Viles (2004) indicates that tufa and travertine, which can be used synonymously, refer to terrestrial freshwater accumulations of calcium

Figure 2.15 Examples of terrestrial carbonate deposits. (a) Oblique aerial view of the Gallocanta Lake (Aragón, Spain), located in the bottom of a karst polje underlain by Triassic evaporites. Primary (authigenic) dolomite precipitates in this lake with hypersaline and Mg-rich waters. (b) Tufa mound related to an artesian paleospring ca. 70 m high and 1 km across at Takht-e Soleyman, Iran. Center of inner paleolake at the top at 38S 700815E, 4052437N. *Source:* Photo by Alireza Amrikazemi. (c) Cluster of pinnacle-like stromatolites on the shore of the Dead Sea, exposed by the rapid decline of the lake level. (d) Varved deposit of light aragonite and dark siliciclastic laminae of the Lisan Formation, formed in the Lisan Lake, pluvial precursor of the current Dead Sea. The central package is affected by a lake-verging seismically induced slump, bounded at the top and bottom by nondeformed deposits. Coin for scale (23 mm) in the center of the image. (e) Rimstone pools of travertine associated with a perched hot spring at Pamukkale, Turkey. (f) Miocene rhythmic palustrine succession of cycles comprising limestone–dolostone layers and marl-clay beds, including Mg-rich sepiolite. These sequences record changes from humid to arid climate controlled by the orbital Milankovich cycles. Calatayud Neogene Basin, NE Spain. (g) Flowstones and draperies strongly degraded by condensation corrosion in the Marmol Cave, Ricla, NE Spain. (h) Calcrete developed on an alluvial deposit made up of basaltic clasts. Calcium carbonate is mainly supplied by aeolian input. Sotavento Beach, Fuerteventura Island, Spain. *Source:* Photos except b by Francisco Gutiérrez (Author).

carbonate whose formation often involves a degree of organic involvement. The word tufa derived from *tophus* or *tufo*, used in Roman times to describe crumbly white deposits, normally referring to volcanic ashes but also used for soft limestones (Pentecost 2004). To avoid confusion with the volcanic tufa, we suggest to add calcareous to the word tufa. The word travertine comes from the Latin *lapis tiburtinus*, or Tibur stone, that was originally used to designate the massive and typically laminated hot spring deposits around Rome (Tivoli) and used as building stone (e.g., the Colosseum, the largest Roman monument entirely made of travertine blocks) (Pentecost 2005). Pedley (1990) and Ford and Pedley (1996) proposed a commonly adopted differentiation between calcareous tufa and travertine. According to these authors, the term travertine should be applied to hard crystalline deposits that lack macrophytes or invertebrates and formed mainly from hydrothermal waters (Figure 2.15e). In contrast, calcareous tufas form in cool water environments (near ambient temperature), tend to be more porous and friable, and typically contain remains of microphytes, macrophytes, invertebrates, and bacteria. Nonetheless, this terminological division has some caveats, including the lack of a clear temperature boundary and the difficulty of its application to relict deposits that formed associated with hydrological systems that are no longer active (Jones and Renault 2010).

Gierlowski-Kordesch (2010) provides an extensive review of lacustrine carbonate sediments, largely based on the analysis of more than 250 modern and ancient lakes. Lakes with carbonate deposition are highly diverse due to the numerous variables that control sedimentation. The lakes may be perennial (permanently filled with water) or ephemeral (with periods of desiccation and exposure of the sediments) (Figure 2.15a), endorheic (closed drainage) or exorheic (open drainage), deep or shallow, with fresh or saline water, with highly variable oxygen and nutrient contents, and with diverse biota that mediate in the precipitation of carbonate deposits. This array of factors explains the environmental diversity of carbonate lakes and the numerous types of facies, with more than 30 different types of carbonate minerals being reported in lacustrine sediments. According to James and Jones (2015), the main differences between lacustrine and marine depositional systems include: (i) the lack of tidal activity in lakes, (ii) lacustrine systems are not necessarily affected by sea level changes, and the position and evolution of the lake outlet in open systems plays a fundamental role, (iii) higher sensitivity of the lakes to rapid environmental changes due to their lower water volume, and (iv) significant contribution of the detrital input to sedimentation in many lakes. These two latter factors explain why thick lacustrine limestone sequences devoid of insoluble units and adequate for extensive karst development are rare.

The lake environments are divided into littoral, sublittoral, profundal, and pelagic zones. The littoral zone is the shallow sector adjacent to the shore, which may correspond to a well-defined bench or a ramp. The profundal zone is the deep aphotic part of the lake bottom. The sublittoral zone is the transition region between the previous ones and the pelagic zone is the water column above the profundal zone. The lake waters can be stratified with layers having different temperatures and/or salinity, and consequently different chemistries. This layering generally develops due to differential heating of the surface water by solar radiation and because water is densest at a temperature of 3.98 °C. In temperature-stratified lakes, the upper warm epilimnion layer is separated from the lower colder hypolimnion by the thermocline or metalimnion. In salinity-stratified lakes, the chemocline separates the saline and dense monimolimnion at the bottom from the upper less saline and less dense mixolimnion. The seasonal or permanent stratification of the lakes may create anoxic conditions that precludes habitation by most organisms and bioturbation, facilitating the preservation of laminated facies. Mixing of the layers may occur when water in the epilimnion attains a temperature of ca. 4 °C and sinks to the lake bottom, displacing the warmer water of the hypolimnion to rise to the surface (turnover). Meromictic lakes are permanently stratified, whereas monomictic, dimictic, and polymictic lakes experience one, two, or several mixing episodes each year, respectively. Carbonate deposition can be related to three main types of processes: (i) abiotic precipitation of carbonate minerals from oversaturated waters, (ii) biologically mediated formation of carbonates, and (iii) input of detrital carbonate grains mainly by surface waters and aeolian activity. Sedimentation in lakes is controlled by a number of interrelated extrinsic (climate, tectonics) and intrinsic factors (hydrology, climate, and sediment input). The hydrological conditions, including the endorheic or exorheic regime of the lake, determine the amount of water input and its chemical composition, largely dependent on the lithologies of the catchment. The incorporation of water through subaqueous springs may produce peculiar carbonate deposits, such as the tufa mounds of Mono Lake in California. Climate controls the net water balance, and temperature has a strong influence on numerous processes, including water stratification, biogenic activity, and the solubility of carbonate minerals.

Gierlowski-Kordesch (2010) differentiates five general facies types in the carbonate sediments that form in saline and freshwater lakes: laminated carbonates, massive carbonates, microbial carbonates, marginal carbonates, and open-water carbonates. Carbonate facies can also be separated according to their distribution within the lake into marginal facies and lake-center facies. Deposition of marginal facies in the strandline and littoral zone is largely governed by hydrodynamic conditions and the geometry of the lake bottom. Characteristic facies of these shallow environments include coarse-grained shoals and beaches, palustrine muddy deposits with strong bioturbation and evidence of subaerial exposure, and fine-grained micritic deposits accumulated in benches. The latter typically contain abundant bioclasts and may also include oncoids, ooids, and microbial buildups such as stromatolites (laminated) (Figure 2.15c) and thrombolites (non-laminated). In the deep central parts of perennial lakes, beyond the influence of shoreline processes, sedimentation is dominated by fine-grained facies. Massive or structureless limestones are attributed to oxygenated lake floors inhabited by burrowing organism. Laminated limestones, largely related to the settling out of particles from suspension, record temporal variations in the amount or type of sediment and their preservation generally indicates anoxic conditions that preclude the development of burrowing biota. Some laminated facies comprise annual couplets related to seasonal stratification cycles (varves) (Figure 2.15d), while other are related to rhythmic changes with other frequencies (rhythmites) (Figure 2.15f).

Palustrine carbonates can be considered as part of a continuous spectrum between groundwater calcretes and lacustrine carbonates (Alonso-Zarza 2003) (Figure 2.16). The depositional environments in which they form are characterized by very low gradients, low energy, and shallow water, which determine frequent subaerial exposure of the carbonate deposits initially accumulated under subaqueous conditions. These palustrine sediments may form not only associated with lake basins, but also in fluvial and alluvial systems and in coastal marine environments. The development of carbonate palustrine facies involves: (i) an initial depositional stage, in which carbonate muds, typically containing charophytes, mollusks, and ostracods, accumulate in a shallow freshwater environment; and (ii) a transformation stage related to subaerial exposure and pedogenic processes that give rise to a number of facies and microfabrics (Freytet and Verrecchia 2002). Alonso-Zarza and Wright (2010a) recognize a number of palustrine facies that reflect the amount of time spent by the primary carbonate deposit under subaerial conditions and the degree of secondary transformation (Figure 2.16): (i) mottled limestones, with yellow-orange-red mottling related to remobilization of iron due to changes in the Eh (oxidation potential) induced by water table oscillations, (ii) nodular and brecciated limestones, consisting of centimeter-scale rounded to angular micrite nodules generated by fragmentation and fissuring of the primary deposit by desiccation, (iii) limestones with vertical root cavities up to several

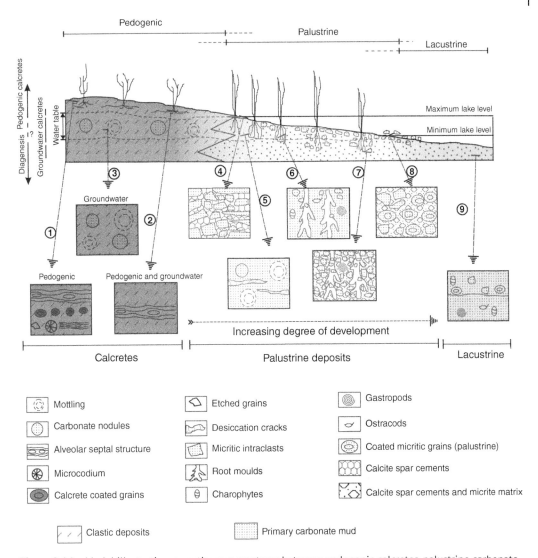

Figure 2.16 Model illustrating a continuous spectrum between pedogenic calcretes, palustrine carbonate sediments developed within the lake level oscillation zone, and lacustrine carbonates. The boxes show common petrographic features of the different facies. Palustrine facies 4–8 indicate a gradation toward progressively higher postdepositional transformation. *Source:* Alonso-Zarza (2003). Reproduced with permission of Springer.

decimeters long, which tend to be relatively more indurated than other facies, (iv) pseudomicro-karst facies, corresponding to limestones with centimeter-sized subvertical cavities, mainly related to mechanical disturbance (desiccation, root activity) and secondarily to dissolution, and (v) granular limestones, including peloids, coated grains, and intraclasts generated by desiccation and bioturbation, plus subsequent reworking and coating of the particles during flooding events. Other associated facies include tufa deposits, carbonate-filled channels, organic-rich marlstones and claystones, and green-brown marlstones.

Calcretes, also known as caliche, are near-surface accumulations of secondary calcium carbonate in soils, weathered material, unconsolidated deposits, or rocks (Figure 2.15h). Two main types of calcretes are differentiated from the genetic perspective (Carlisle 1983): pedogenic and non-pedogenic. Calcretes are estimated to underlie around 13% of the Earth's land surface and are most widespread in semiarid regions (Nash 2004). The pedogenic calcretes are typically white- and cream-colored mineral deposits that accumulate in soil profiles. The precipitated carbonate may be derived from the parent material, airborne dust, and rainfall. The calcium and bicarbonate ions are leached by vadose downward flow and accumulate preferentially in the B illuviation soil horizon. Bk horizons are weak calcic horizons, whereas K horizons refer to prominent layers of carbonate accumulation, in which secondary carbonate occurs as a continuous medium. Calcretes evolve through multiple morphological stages of calcium carbonate accumulation that can be identified in chronosequences (e.g., terrace sequences). These stages, together with some indexes of soil development, such as the amount of secondary carbonate, can be used as tools for relative dating and correlation (Gile et al. 1966; Machette 1985). Nonetheless, it should be taken into account that the accumulation of calcium carbonate forms part of a multivariable equation, including, in addition to time, the nature of the parent material, the input of carbonate from above (rainfall, dust), climate, and biological activity. In gravelly deposits, initially, the clasts are coated by discontinuous and then continuous carbonate. In a subsequent stage, the secondary carbonate forms massive accumulations between clasts producing a plugged hardpan. Then, calcium carbonate accumulates as laminae on the upper part of the hardpan. The more evolved stages are characterized by multiple generations of brecciation, pisolith formation, and accumulation of lamina around the blocks. Pedogenic calcretes developed on fine-grained deposits grade from a powdery to a nodular texture until the hardpan is developed. Non-pedogenic calcretes encompass a wide variety of carbonate accumulations, from crusts developed on rock surfaces (e.g., case hardening) to groundwater calcretes related to the abiotic cementation of sediments in the shallow phreatic zone and the capillary fringe. These non-pedogenic groundwater calcretes form in numerous environments, such as deltas, fluvial and alluvial deposits, or lake margins. They can be differentiated from pedogenic calcretes by their massive character and the lack of internal horizons, sharp basal and top contacts, and absence of vertical root traces and biogenic microfabrics (Alonso-Zarza and Wright 2010b). Deposits indurated by caliche may behave as resistant protective layers similarly to caprocks, favoring the preservation potential of deposits and geomorphic surfaces, as well as relief inversion; a surface once developed on lowlands forms the top of prominent reliefs. Moreover, pedogenic calcretes characteristic of dry conditions may be used as paleoclimate indicators.

2.2.8 Porosity of Carbonate Rocks

The porosity of a material refers to the percentage by volume of pore spaces within the rock. The voids within a rock may be isolated or interconnected. Effective porosity is a measure of the percentage of voids that are hydrologically interconnected. Permeability indicates the capacity of a rock to transmit a fluid. These parameters have decisive importance from the applied perspective, since they determine the ability of the rocks to host and transmit valuable resources (e.g., freshwater, hydrocarbons, and ore deposits), as well as the susceptibility of the rock to experience dissolution when it interacts with undersaturated water. The porosity and permeability of carbonate rocks (see also Figure 5.4) have some peculiarities compared to other lithologies: (i) The porosity in carbonate rocks, composed of soluble minerals, may be related to a very complex multiphase diagenetic history of porosity destruction and creation. (ii) Porosity and permeability are scale-dependent and can be highly heterogeneous (Kiraly 1975). The overall permeability of a rock massif or aquifer may be highly different to those estimated from a borehole or a rock sample, due to the variable sizes of the permeability features (e.g., large caves versus small pores) (see also Figure 5.5).

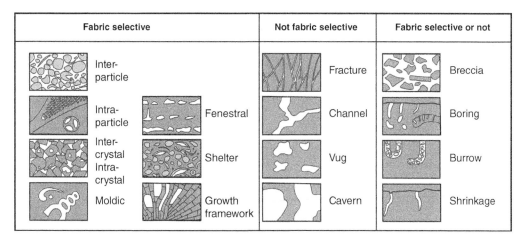

Figure 2.17 Sketches illustrating the main types of porosity in carbonate rocks. *Source:* Adapted from Choquette and Pray (1970) and James and Jones (2015). Reproduced with permission of Wiley.

Primary porosity refers to the pore spaces present in the sediment at the time of deposition, whereas secondary porosity indicates the voids generated during the postdepositional history of the rock. Choquette and Pray (1970) proposed a classification for the porosity of carbonate rocks based on the spatial relationship between the pores and the fabrics, differentiating between: (i) fabric-selective porosity, (ii) non-fabric selective porosity when the pores cross-cut the depositional components and fabrics of the rock, and (iii) types of porosity that may be either fabric-selective or non-fabric-selective (Figure 2.17).

Fabric-selective porosity can be interparticle, intraparticle, intercrystalline, intracrystalline, moldic, fenestral, shelter, and growth framework. The interparticle porosity is characterized by voids between particles, and depends on the textural characteristics of the grains (size, geometry, and packing). The permeability in the Jurassic carbonates of the Ghawar Oil Field in Saudi Arabia, which is considered the largest conventional oil field in the World, is mainly related to intergranular porosity. Intraparticle porosity refers to pores within the grains, which are mainly skeletal components of organisms. In the intercrystalline porosity, common in dolostones, the pores are between crystals. Intracrystalline porosity instead shows solution voids within crystals. In the moldic porosity the pores are generated by the selective dissolution of specific components (e.g., biofragments), whereas in the fenestral one, typical in muddy tidal-flat facies, pores are millimeter-sized, elongated, and lamination-parallel, attributed to gases derived from the decay of organic matter. In the shelter porosity, the voids are created under particles, typically shells, and in the growth framework porosity, the unfilled spaces are created by the growth of colonial organisms (e.g., corals).

Fractures, channels, vugs, and caverns are classified as non-fabric-selective porosity. Fractures are discontinuities generated by brittle deformation with (faults) or without shear displacement (joints). These are critical permeability features for karst development, which may show a wide range of dimensions, geometries, apertures, densities, and degree of interconnection. The high production rates of the giant oil fields associated with the Oligocene-Miocene Asmari Limestone in the Zagros Mountains, Iran, is largely related to its high fracture porosity. Channel, vug, and cavern are terms with poorly defined meanings used by sedimentologists and petrologists to designate solutional voids with different dimensions and patterns.

Breccias, borings, burrows, and shrinkage belong to the fabric-selective-or-not type of porosity. Breccias contain secondary porosity associated with angular clasts generated by deposition or brittle deformation of lithified sediments by various mechanisms, including interstratal dissolution of evaporites and the collapse of overlying carbonate rocks (solution-collapse breccias), collapse of cave roofs (cave-collapse breccias), faulting (fault breccias), or hydrofracturing by pressurized and corrosive hydrothermal fluids (hydrothermal breccias). Borings and burrows are perforations of tubular holes by organisms in hard (borings) and soft substrates (burrows). Surficial cracks and fissures, typically with polygonal patterns and generated by desiccation, create the so-called shrinkage-related porosity.

The evolution of the porosity in carbonate sediments through time can be very complex and largely depends on their geological history. The burial of the sediments to progressively greater depths is generally accompanied by an overall decrease in porosity by compaction and cementation, despite some processes such as fracturing and dissolution may create new pore spaces. Deep-burial dissolution may be related to CO_2 released from hydrocarbon maturation, dolomitization, or the circulation of fluids under high pressure and temperature conditions. Dissolution is the most important process responsible for the generation of secondary porosity, which is particularly significant when the rocks are at the surface or in near-surface conditions (meteoric environment), either in the eogenetic or telogenetic diagenetic environments. Solutional porosity developed in the telogenetic diagenetic stage, typically affecting lithified rocks, tends to be largely controlled by discontinuity planes such as bedding, joints, or faults that function as preferential pathways for groundwater flow.

2.2.9 Other Carbonate Rocks

2.2.9.1 Carbonate Conglomerates and Sandstones

Conglomerates and sandstones form approximately 25% of the stratigraphic record (Blatt et al. 2005). Conglomerates are coarse-grained detrital sedimentary rocks composed predominantly of rounded gravel-sized particles (>2 mm in diameter). The size of the skeletal grains of conglomerates may range from small granules (2–4 mm) to boulders (>25.6 cm) (Figure 2.1). The intergranular space in conglomerates is typically filled by fine-grained matrix and cement. Monomictic and oligomictic conglomerates are composed of a single or a few rock types, respectively, whereas polymictic conglomerates include a wide variety of lithologies. Monomictic/oligomictic conglomerates that are mainly composed of carbonate minerals may display well-developed karst geomorphology and hydrology. Surface and subsurface weathering in these rocks may occur through two main types of processes that may operate in combination (Göppert et al. 2011): (i) chemical weathering, whereby dissolution evenly wears all the components of the rocks; and (ii) mixed chemical and mechanical weathering, in which dissolution contributes to reduce the cohesion of the rock and its disintegration, creating adequate conditions for the mechanical weathering and erosion of loose particles. Conglomerates are typically derived from the erosion of high-relief source areas affected by uplift, such as the margins of fault-controlled sedimentary basins, and are generally proximal syntectonic sediments deposited during major phases of deformation. Carbonate conglomerates occur in regions where the source area is dominated by limestone and/or dolomite. Common sedimentary environments in which they form include alluvial fans, fluvial systems, and fan deltas.

Bergada et al. (1997) reported several areas associated with the marginal Paleogene conglomerates of the Ebro Cenozoic basin (NE Spain) with surface and subsurface karst features. Karst features in these carbonate-rich conglomerates, several hundred meters thick and with intercalations of fine-grained beds, are largely controlled by subvertical joints. Preferential dissolution and mechanical erosion along the fractures lead to the formation of spectacular mazes of corridors and intervening towers and pinnacles, as well as caves. Caves, up to 12 km long, mainly comprise shafts guided by fractures and connected subhorizontal passages that may be controlled by impervious beds or the base level (Figure 2.18a, b). Similar joint-controlled towers also occur in conglomerates dominated by siliciclastic rocks, but with carbonate cement (e.g., Meteora, Greece, Figure 2.18c). Göppert et al. (2011) described the karst geomorphology and hydrology developed in folded Oligo-Miocene carbonate conglomerates with marl intercalations in the Northern Molasse Basin of the Alps (Austria and Germany). These authors found that the inventoried karst features (karren, dolines, poljes up to 250 m long, swallow holes, and caves controlled by fractures and bedding planes) reach smaller dimensions than in limestone karst terrains. In the Italian Fore-Alps, the Montello Hill, underlain by around 2000 m of Miocene calcareous conglomerates, displays densely packed dolines resembling a polygonal karst developed on a stepped sequence of seven rock-cut terraces (Ferrarese and Sauro 2005). In Slovenia, Lipar and Ferk (2011) documented eogenetic karst features, including caves, dolines, pocket valleys, and springs, developed in a fractured and cemented Pleistocene fluvioglacial terrace with around 90% of soluble components and underlain by claystone bedrock. The authors differentiate four types of caves and passages: joint-controlled stream caves, shelter caves, breakdown caves, and vadose shafts. Lapaire et al. (2007) provided a detailed account of caves encountered in carbonate conglomerates in the Jura Mountains of Switzerland that created difficult ground conditions for the construction of a motorway. Large, still partially unexplored, cave systems also occur in the Late Cretaceous carbonate conglomerates of the Valle de Angeles Group in Honduras, hosting the deepest conglomerate cave in the world (Finch and Pistole 2009). According to Filippov (2004), the longest known conglomerate cave in the world is Bol'shaya Oreshnaya Cave (Siberia, Russia), with more than 58 km of surveyed phreatic passages formed in an Ordovician conglomerate.

As explained above, the differentiation between detrital and chemical rocks is somewhat artificial. Limestones can be composed primarily of sand-sized transported particles (sandy limestones) and can be described as sandstones (i.e., calcarenite, calcareous sandstone). These carbonate rocks may host well-developed karst systems. However, thick successions of allochthonous detrital sediments dominated by carbonate sandstones are rare. Sandstone typically contains a high proportion of silicate minerals, particularly quartz, due to its low solubility and high resistance to erosion (quartz has a hardness of 7 on the Mohs scale, calcite only has 3). Sand-sized carbonate detrital particles transported in undersaturated water degrade rapidly by dissolution and abrasion. Moreover, sandstone successions typically include a significant proportion of impervious beds (mudstones, marls) that inhibit karst development. Nonetheless, individual beds or packages of carbonate sandstones may function as susceptible rocks for the initiation of karst development (inception horizon), which may subsequently evolve and expand through positive feedback mechanisms. One of the longest caves in India, Krem Puri in Meghalaya, has formed in a carbonate-cemented quartz sandstone bed within a prevalently siliciclastic sedimentary sequence (Sauro et al. 2020).

2.2.9.2 Carbonate Breccias

Breccias are by definition composed of gravel-sized angular clasts, typically rock fragments. This is a non-genetic term that is generally accompanied by a prefix to define the origin of the deposit or rock. The breccias that occur in sedimentary successions are related to two main types of processes: (i) primary detrital breccias composed of clasts with poor textural maturity that were not transported over a sufficient distance to become rounded (e.g., colluvium, talus, some landslide deposits, and proximal glacial till); and (ii) secondary breccias generated by the in situ fragmentation of brittle sediments. The brecciation and comminution processes involved in the formation of the latter breccias may be caused by several mechanisms, including shearing along faults (fault breccia) and diapir rims (diapiric breccia), the impact of meteorites (suevite breccia, impact breccia, and impactite), and subsurface dissolution and collapse (solution-collapse breccia). The Chicxulub meteorite impact, occurred in the Yucatán Peninsula, Mexico at the Cretaceous-Paleogene boundary (K/Pg boundary), generated a complex carbonate-rich impact breccia as much as 800 m thick that extends across hundreds of kilometers (Figure 2.18d). This complex

Figure 2.18 Images of conglomerates, breccias and marbles. (a) Smooth surface on dipping carbonate conglomerates with interbedded fine-grained beds in the Salnitre (or Collbató) Cave, Monserrat Massif, NE Spain. (b) Pinnacles and large grikes generated by preferential dissolution and mechanical erosion along joints in the Montserrat Massif, NE Spain. Circle indicates location of hermitage. (c) Conglomerate towers and corridors in Meteora, Greece. In the largest tower, with a local relief of around 200 m, the entrance of the joint-controlled Monks Prison Cave. (d) Carbonate-rich polymictic chaotic floatbreccia of the Carajícara Formation, generated by the Chicxulub meteorite impact at the K/Pg boundary. Image taken at the type locality of the formation, Carajícara, Cuba. *Source:* Photo by José Antonio Arz. (e) Carbonate chaotic packbreccia in the Late Jurassic Arab Formation (Member B), generated by interstratal dissolution of anhydrite and the collapse of the overlying limestone strata. Hawtat Bani Tamim, Saudi Arabia. (f) Transtratal breccia in a Jurassic limestone related to the collapse of a concealed cave. The crackle packbreccia in the sagged beds grades into a chaotic packbreccia in the right sheared margin of the collapse structure. A-2 highway near Ricla village, Iberian Chain, Spain. (g) White coarsely crystalline Devonian marble and siliciclastic metasediments in the metamorphic aureole of the Panticosa granodioritic batholith, Pyrenees, Spain. Granodiorite rocks in the background and a residual cirque glacier in the foreground. (h) Marble and protruding layers of carbonate-silicate rocks showing differential corrosion and small structural karren. Contact aureole of the Posets granodioritic batholith, Pyrenees, Spain. *Source:* Photos except d by Francisco Gutiérrez (Author).

oil-producing breccia includes facies generated by various mechanisms, such as impact brecciation, accumulation of ejecta, or large mass movements triggered by the meteorite impact (e.g., Grajales-Nishimura et al. 2000; Stinnesbeck et al. 2004).

Solution-collapse breccias form by subsurface dissolution of karst rocks, typically evaporites, and the consequent collapse of the overlying and interbedded sediments. Warren (2016) provides a comprehensive review on these karstic diagenetic breccias that can be syndepositional or form at depths greater than one thousand meters. Two main types of solution-collapse breccias can be differentiated: transtratal and interstratal. Transtratal breccias are related to the upward propagation of localized cavities by progressive roof collapse (i.e., stoping), resulting in the development of chimney-like structures filled by brecciated material (breccia pipes) that cut across the strata (Figure 2.18f). These breccia pipes may reach hundreds of meters across and in depth, and typically display sharp lateral boundaries with shear fabrics. Some breccia pipes do not reach the surface (blind breccia pipes). The stoping process may cease if the cavity roof and the breakdown pile come into contact, because of the bulking effect (i.e., volume increase) associated with the collapse process. Other breccia pipes penetrate up to the surface, generating subsidence sinkholes. Collapse pipes filled by resistant breccias, generally cemented, may be expressed at the surface as positive-relief features due to differential erosion and relief inversion.

Interstratal breccias generate by regional dissolution of evaporite units and the fragmentation of the overlying and intercalated beds, frequently limestones and dolomites. This process entails the condensation of the stratigraphic succession and the subsidence of the overlying sediments. The resulting stratabound breccias tend to be monomictic or oligomictic, frequently contain a matrix of insoluble material, and may reach hundreds of meters in thickness and extend across extensive areas (Figure 2.18e). Because of the low internal permeability of most evaporitic rocks, especially salt, the dissolution responsible for the development of the interstratal breccias generally advances laterally along a dissolution front from the updip or feather edge of the soluble unit. It may also affect progressively deeper units of the stratigraphic succession, producing multiple breccia horizons. Solution-collapse breccias can be divided into packbreccias and floatbreccias, depending on whether they display clast- or matrix-supported texture, respectively. They can be also classified according to their fabrics; crackle indicates limited relative displacement of the clasts, mosaic refers to fragments with significant displacement, but their edges can be roughly matched, and chaotic (or rubble) denotes completely disordered fragments (Stanton 1966; Warren 2016 and references therein) (Figure 2.19).

Solution-collapse breccias have several important implications from the karst and applied perspective. (i) They provide evidence of the previous existence of evaporitic units removed by dissolution. Relatively thin horizons of breccias and insoluble residues may pass laterally beyond the

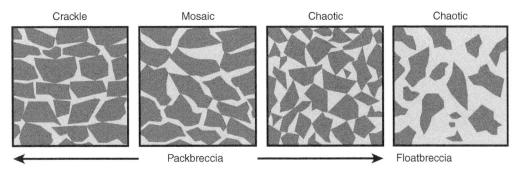

Figure 2.19 Classification of breccias according to their texture and fabrics. *Source:* Adapted from Warren (2016).

dissolution fronts into thick salt and Ca-sulfate units. (ii) The porous and permeable breccias may host and/or transmit valuable resources (e.g., freshwater, hydrocarbons, metallic ore deposits). (iii) High-permeability solution-collapse breccias may pose difficult conditions for some engineering projects (e.g., leakage in dams, dangerous water inrushes in mines and tunnels). (iv) The groundwater that flows through carbonate breccias may cause substantial dissolution, adding solution porosity to the previous collapse-related porosity and renewing collapse processes. The void space may be also filled by cement.

2.2.9.3 Marbles

Metacarbonate rocks include a range of metamorphic rocks derived from the transformation of carbonate and carbonate-bearing protoliths (Figure 2.18g, h). Limestones and dolomites are the precursors of marble. The metamorphism of impure carbonate rocks with silicate minerals is characterized by decarbonation reactions that entail the partial or total consumption of the original carbonate minerals. Thus, a precursor impure carbonate rock may be converted into a carbonate-silicate rock, and a carbonate-bearing mudstone into a calc-silicate rock mainly consisting of calcium-bearing silicates. Rosen et al. (2007) proposed a simple classification scheme for metacarbonate rocks based on the modal content of carbonate minerals (calcite, dolomite, and aragonite), calc-silicate minerals, and all other silicates (Figure 2.20). The content of Ca- and Mg-carbonate minerals defines the boundaries between pure marbles (>95%), impure marbles (95–50%), and carbonate-silicate rocks (50–5%) (Figure 2.18h). Metacarbonate rocks with less than 5% of carbonate minerals are divided into calc-silicate and carbonate-bearing silicate rocks, with a content in calc-silicate minerals higher or lower than 50%, respectively. Obviously, marbles are important rock types from the karst perspective. They are common lithologies in metamorphic terrains associated with orogenic belts and also occur in contact aureoles developed around magmatic intrusions (Figure 2.18g). They can be classified according to the main mineral into calcite marbles and dolomite marbles. Pure calcite marbles are rather unusual, in contrast with the less soluble dolomite marbles. The latter display a wide range of calcium–magnesium silicate minerals. The

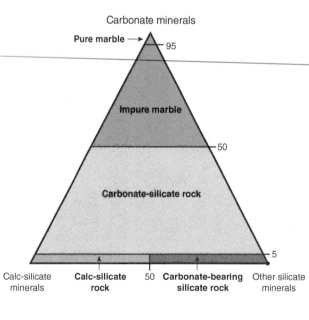

Figure 2.20 Classification of metacarbonate rocks based on the content of Ca- and Mg-carbonate minerals, calc-silicate minerals and other silicate minerals. *Source:* Adapted from Rosen et al. (2007).

mineral assemblages and the presence of specific minerals can be used as indicators of the metamorphic grade. For instance, the occurrence of talc, tremolite, and diopside + tremolite ± wollastonite in dolomite marbles are indication of low, middle, and high grades of regional metamorphism, respectively. The main modifications associated with the conversion of limestones and dolomites into marble (marmorization) include recrystallization, grain coarsening, and porosity reduction. This process results in the formation of a mosaic of progressively larger interlocking crystals with phaneritic texture (i.e., eye-visible crystals). Recrystallization also involves the obliteration of the primary sedimentary textures and structures of the protolith (e.g., fossils, sedimentary structures, and bedding).

There are a number of works that document karst features in marbles, but generally the mineralogical composition of the metacarbonate rocks, either dominated by dolomite or calcite, is not specified. In the South Island of New Zealand, Williams and Dowling (1979) investigated karst solution processes and rates in a catchment underlain by Ordovician marble and dominated by autogenic subsurface drainage. De Waele and Follesa (2003) reported karst features in a plateau partially underlain by Precambrian dolomitic marbles in Lusaka, Zambia, including small caves and rounded karren developed beneath a laterite residual cover. In Patagonia, Chile, Maire (2004) described a marble karst formed under extreme environmental conditions, characterized by an annual precipitation greater than 7000 mm and strong winds. Here, rapid differential hydro-aeolian corrosion produces peculiar parallel, tapering ridges in the downwind side of non-soluble boulders. Skoglund and Lauritzen (2011) documented the Nonshaugen maze caves in Norway, that occur in a stripe karst developed in marble horizons intercalated within mica-schists, probably formed under subglacial conditions. These marble stripe karsts are widespread in Norway (Lauritzen and Skoglund 2013). One of the longest and most complex cave systems in Italy, Corchia (over 60 km long and almost 1200 m deep), is entirely carved in Triassic-Lower Jurassic dolomitic and calcite marbles (Piccini 2011).

2.2.9.4 Carbonatites

Carbonatites are rare peralkaline igneous rocks (i.e., $Al_2O_3 < Na_2O+K_2O$) with at least 50% of carbonate minerals, mainly calcite, but also dolomite, magnesite, and sodium carbonate. Bell (1989) indicated that there are more than 450 sites with exposed carbonatites in the world, with individual areas below 20 km^2. They occur in small shallow intrusive complexes accompanied by alkali-rich rocks such as nepheline syenite (plutonic rock rich in alkali-feldspar, feldspathoids, and a quartz content <5%). The carbonatite rock is generally emplaced in the core of the intrusive complex at a late stage. Extrusive carbonatite lavas and pyroclasts also exist (e.g., the active Oldoinyo Lengai Volcano, Tanzania). The origin of carbonatite rocks, which contain magmatic carbonate minerals, is attributed to partial melting of peridotite rocks in the mantle with carbonate minerals or a CO_2-rich fluid phase. Carbonatites host multiple minerals such as those of the apatite group ($Ca_5(PO_4)_3(F,Cl,OH)$; repository of most rare earth elements) and magnetite. High-calcite carbonatites are suitable rocks for karst development. Their chemical weathering, mainly by dissolution of the carbonate minerals, produces economically valuable residual deposits with high concentrations of phosphates, base metals and rare earth elements. Erdosh (1979) examined the mineral potential of some carbonatite complexes of the so-called Ontario Carbonatite Province in northern Ontario and western Quebec, Canada. Apatite content ranges from 5 to 25% and is significantly enriched in the weathering profiles. At the Cargill complex, carbonate dissolution has produced a well-developed karst topography and high-grade phosphate residual deposits, mainly concentrated in sinkholes and troughs. Moreover, concentrations of rare earth minerals occur as discontinuous horizons at the top of the residuum. Lottermoser (1990) documented exceptionally high concentrations of rare earth elements in thick laterite residual deposits developed atop the Mt. Weld carbonatite, Western Australia. Rare earth elements experienced significant vertical and lateral mobility during weathering and eventually incorporated into secondary phosphates and aluminophosphates, especially in topographic lows.

2.3 Evaporite Rocks and Minerals

According to Warren (2016), an evaporite is a salt rock in its broadest sense (e.g., halite, gypsum; Table 2.1, Figure 2.5) originally precipitated from a supersaturated surface or near-surface brine in hydrochemical conditions driven by solar evaporation. These chemical sediments mainly form in hot and cold hyperarid to semiarid climatic environments with an overall negative water balance (evaporation exceeds water input). At the present time, the distribution of arid regions in which evaporites precipitate is controlled by regional factors, such as the subtropical belts dominated by dry descending air masses and local factors, including continentality (distance from the ocean), rain shadow effect, or the presence of nearby cold oceanic currents. The location and extent of the arid regions have experienced significant variations throughout geological time. For instance, arid subtropical belts repeatedly changed their latitudinal distribution during the Pleistocene glacial and interglacial periods in concert with the expansion and contraction of the ice sheets. Moreover, rain shadow deserts did not exist before the buildup of the associated mountain belts (e.g., deserts and salars in the Andean region). An arid climate is not a sufficient requisite for the accumulation and preservation of significant evaporite formations in the rock record. The evaporitic system also needs a long-term source of salts, such as the inflow of seawater or the dissolution of preexisting evaporites, and the creation of accommodation space by subsidence.

Evaporitic minerals may also precipitate from brines that are concentrated by other processes not related to solar heating, although these deposits have very limited importance for karst studies. In cold environments (e.g., lakes in Antarctica), the loss of liquid water by freezing may result in concentrated brines and the precipitation of salts (cryogenic salts). Evaporitic minerals may also precipitate in the subsurface by diagenetic or metamorphic processes (burial salts). The formation of these minerals typically occurs in the vicinity of an evaporitic mass that experiences dissolution and reprecipitation. Supersaturation conditions may be driven by temperature changes and the mixing of fluids with different compositions. Anhydrite is the most common burial salt. This mineral is characterized by retrograde solubility, whereby the solubility of $CaSO_4$ decreases with increasing temperature. Salts may also precipitate in the subsurface from hydrothermal waters under temperature conditions higher than those of the diagenetic realm (hydrothermal salts). These evaporitic minerals tend to fill fractures and voids. A spectacular example are the giant transparent gypsum crystals as much as 11 m long discovered in the dewatered Cueva de los Cristales in Naica Mine, Chihuahua, Mexico (García-Ruiz et al. 2007).

As evaporation proceeds, the residual brine increases its salinity and reaches a state of supersaturation with respect to a series of evaporitic minerals that precipitate sequentially, from low- to high-solubility mineral phases. The order of precipitation and the resulting suite of evaporitic minerals are controlled by the chemical composition of the parent brine as well as potential variations related to water inflow and mixing processes. The ionic composition of modern seawater is rather consistent worldwide, leading to a characteristic precipitation sequence. Nonetheless, the composition of seawater has changed through the geological history, as revealed by studies on fluid inclusions and the mineralogy and isotopic signature of marine evaporitic formations of multiple ages. For instance, the absence or limited occurrence of Mg-sulfates in ancient marine evaporites, which are the main bittern salts at the present time, is attributed to temporal variations in the ionic composition of seawater. This is one of the reasons why modern marine evaporitic systems are a limited key for reconstructing the past.

The chemistry of nonmarine brines is much more diverse, and consequently produces a wide variety of minerals and precipitation sequences (Eugster and Hardie 1978). The composition of nonmarine brines depends on the lithologies that interact with the surface and underground waters that feed the lakes.

Modern seawater is dominated by Na and Cl, with lesser amounts of SO_4, Mg, Ca, K, CO_3, and HCO_3, has a density of $1.03\,g\,cm^{-3}$ and a salinity of 35‰ (Table 2.2). Alkaline earth carbonates, principally $CaCO_3$, are the first minerals that precipitate from seawater subject to evaporation. This precipitation stage starts in mesohaline waters when the brine reaches a salinity of 40–60‰. Gypsum ($CaSO_4\cdot2H_2O$) saturation is reached once the seawater has been concentrated by approximately four times and reaches a salinity of around 140‰ (penesaline brine). Gypsum and halite may precipitate coevally when the brine has a salinity within the range of 250–350‰ and has been concentrated 7–10 times. At this stage, the initial solution has lost 85–90% of the water by evaporation. After the gypsum precipitation stage, the brine is impoverished in calcium but still contains around two-thirds of the original sulfate. Halite salt mainly forms when the salinity rises above 350‰ and the brine is concentrated more than 10 times (supersaline brine). At about 60 times seawater concentration, the bittern salts (K- and Mg- sulfates and chlorides) precipitate. Epsomite ($MgSO_4\cdot7H_2O$) and carnallite ($KMgCl_3\cdot6H_2O$) are the dominant bittern precipitates (Kendall 2005; Warren 2016). At this stage, the brine has a density higher than $1.3\,g\,cm^{-3}$ and a viscous oily feel.

Evaporation induced by solar heating can occur either at the surface of a standing water body or in interstitial water within a sediment at shallow depth. The progressive concentration of the brine involves the increase in its density and specific heat capacity (i.e., heat required to raise the temperature 1 °C). The salinity increase contributes to slow down the evaporation process, whereas factors such as the presence of dry wind and the increase in temperature accelerate evaporation. At very high salinity, brine evaporation is very slow or may even stop. This explains why the accumulation of beds of bittern salts is so rare, since they require an extreme degree of aridity to reach the necessary concentration (ca. 99% water loss).

The depositional environments in which evaporites accumulate may be associated with marine, nonmarine, or hybrid brines. Two main groups of evaporite hydrologic systems can be differentiated

Table 2.2 Precipitation sequence of evaporite minerals from modern seawater subject to progressive evaporative concentration. Degree of evaporation refers to the times the solution has increased its original concentration.

Brine stage	Mineral precipitate	Salinity (‰)	Degree of evaporation	Water loss (%)	Density (g cm^{-3})
Normal or euhaline	Skeletal carbonate	35–37	1	0	1.03
Mesohaline	Alkaline earth carbonates	35–140	1–4	0–75	1.04–1.1
Penesaline	Gypsum/anhydrite	140–250	4–7	75–85	1.1–1.214
	Gypsum/ anhydrite ± halite	250–350	7–10	85–90	1.214–1.126
Supersaline	Halite	>350	>10	>90	>1.126
	Bittern salts (K-Mg salts)	Extreme	>60	≈99	>1.290

Source: Adapted from Warren (2016).

that produce formations with contrasting textural and stratigraphic features: systems with ephemeral saline waters and systems with perennial saline waters. The ephemeral systems include coastal sabkhas (e.g., Arabian Peninsula) and ephemeral continental saline lakes. The latter are designated with a wide variety of terms in the literature, such as playa-lakes, continental sabkhas, pans (Gutiérrez 2013). There are also a large number of local names used to identify these continental lakes that remain dry during most of the time; e.g., salar (Bolivia, Chile), salada (United States, Spain), and chott (northern Africa). The ephemeral continental saline lakes are fed by groundwater and surface runoff. Sabkhas may also receive significant inflow of marine water during storm events via breached channels. These hydrologic systems are characterized by flat, vegetation-free surfaces underlain by muddy sediments or salt crusts. The water table of the saline groundwater is situated at shallow depth most of the year. Capillary evaporation leads to the precipitation of secondary intrasediment (early diagenetic) displacive or replacive salts in the capillary fringe or immediately below the water table. The water table eventually rises above the flat bottom of the lake for a short period, allowing the precipitation of subaqueous primary textures, including beds of bottom-nucleated and aligned crystals. Once the coastal or continental ephemeral lake is desiccated again, the exposed sediments are affected by secondary overprints, mainly related to dissolution and aeolian erosion. Common early diagenetic or eogenetic karst features include solution pits and depressions and planar erosional bevels. Aeolian deflation tends to excavate the low-cohesion deposits situated above the capillary fringe, producing extremely flat surfaces (Stokes surfaces) that represent erosional hiatuses. The sabkha successions have a high proportion of insoluble fine-grained facies and the evaporites, mainly precipitated within the sediment, typically represent less than 50% of the sediment pile. Evaporite sediments accumulated in the deeper parts of ephemeral continental saline lakes, where the brine ponds more frequently, may be much thicker (>100 m), purer and with larger lateral extent (e.g., Salar de Uyuni, Bolivia). There are typically stacks of crusts or beds of subaqueous primary textures with abundant microkarst overprints that record alternating flooding and drying periods in these deposits.

Two types of perennial saline systems are identified in the Quaternary and modern evaporites: perennial coastal saline lakes and perennial continental saline lakes. The coastal saline lakes with permanent subaqueous conditions, often termed salinas, are sub-sea-level depressions that do not have a permanent surface connection to the sea, but are largely fed by inflows of seawater during storms. The continental saline lakes are generally located in the topographically lowest parts of endorheic lacustrine basins. These perennial lakes with evaporite deposition may be shallow (e.g., Great Salt Lake, Utah; 4180 km^2; maximum depth 10 m) or deep (e.g., Dead Sea; 810 km^2; maximum depth ≈300 m). The evaporites deposited in these environments are dominated by stacked beds with primary subaqueous textures (e.g., bottom-nucleated crystals, laminites) and lack features related to subaerial exposure. These terrestrial evaporite formations may reach significant thickness and the proportion of insoluble matrix is generally low. Warren (2010, 2016) identifies two additional perennial evaporite depositional systems in the stratigraphic record that do not have Quaternary counterparts or modern analogs: platforms and basin-wide evaporite systems. These were sub-sea-level and hydrologically isolated systems fed by marine waters of huge dimensions (saline giants) compared with the modern saline systems. These environments account for the large-scale ancient evaporite formations in the rock record, with lateral extents of hundreds or thousands of kilometers (Figure 2.21). The Hormuz (Precambrian) and the Zechstein (Permian) are examples of basin-wide evaporites, whereas the Hith Anhydrite (Jurassic) represents a formation deposited in an ancient shallow platform. The main reason why these giant saline systems do not have Quaternary equivalents is because their development requires relatively stable sea-level conditions, typical of the periods in which the Earth was dominated by a greenhouse (warm) climate.

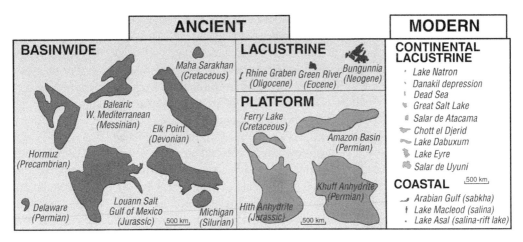

Figure 2.21 Extent of ancient and modern evaporites. The vast evaporite formations deposited in basin-wide and platform environments in the geological past do not have counterparts in the Quaternary and modern record. *Source:* Adapted from Warren (2016).

In contrast, the Quaternary has been an icehouse climatic period, characterized by climate-driven high-amplitude and high-frequency eustatic changes. The continuous sea level fluctuations preclude the development of large semi-isolated brine systems with long-sustained supply of marine waters.

A peculiarity of the evaporitic sediments related to their highly reactive nature (e.g., high solubility, instability under burial conditions), is that they rapidly lose their primary depositional texture, making their sedimentological interpretation difficult. Primary minerals may precipitate in various hydrological settings. Crystals may form at the brine-air interface as rafts and subsequently sink to form cumulate beds and pelagic laminites. Bottom-nucleated crystals may also precipitate at the base of the water column. Bottom crystals can be reworked by waves and currents to form dunes, ripples, and intraclast beds. Primary evaporitic deposits may be also reworked by various types of subaqueous mass movements and accumulate in deeper environments as olistoliths and turbidites. Warren (2016) differentiates between secondary diagenetic textures, developed during the burial eogenetic and mesogenetic stages, and tertiary textures that form in the telogenetic stage, when the evaporites return to a near-surface environment by uplift and/or exhumation. The evaporites that precipitate within the sediment in the shallow subsurface, including the gypsum and anhydrite nodules of sabkha environments, are considered secondary diagenetic sediments. During the burial diagenesis, evaporitic sediments are affected by extensive recrystallization, involving a substantial loss in porosity and permeability. This explains why subsequent dissolution in the telogenetic stage tends to be concentrated at the edges of the evaporitic mass, like a melting block of ice. This is especially evident in the case of salt formations that typically have very limited fracture porosity due to their plastic rheology.

Evaporites represent approximately 2% of the world's sedimentary rocks. However, these highly soluble formations may reach thousands of kilometers in lateral extent (Figure 2.21). It is estimated that around 25% of the continental surface is underlain by evaporites (Kozary et al. 1968). Evaporites typically attain lower primary thicknesses than carbonate successions, but they may be locally thickened by halokinetic (salt flow) and tectonic processes, reaching as much as 10 km in vertical extent in salt diapirs. Halite-bearing formations, characterized by a very low yield strength and plastic rheology, can flow laterally and vertically under differential loading conditions, deform the overlying sediments, and produce a wide variety of salt structures (Jackson and Hudec 2017), including giant active landslides (e.g., the Grabens of Canyonlands, Utah). Evaporite formations, due to their low mechanical strength, may function as detachment levels controlling the development of thrust sheets and contrasting deformation styles in the supra- and infra-evaporitic rocks. Ancient and modern evaporites are extremely important formations from the scientific and economic perspective. They constitute valuable archives of past arid-semiarid conditions, although they generally lack fossils. Their mineralogical and chemical features, including fluid inclusions, can be used to infer the composition of the brines from which they precipitated.

Halite and gypsum are resources of prime importance. Other relevant industrial salts include sodium carbonates (e.g., trona), potash, and borates. Hypersaline brines in some contemporaneous evaporitic environments contain high concentrations of some elements subject to exploitation, such as iodine, bromine, or lithium. The latter, due to its high electrochemical potential and light weight, has become a strategic resource for the manufacture of batteries for electric vehicles. The largest reserves are located in the so-called Lithium Triangle (the Salars of Atacama, Uyuni, and Hombre Muerto, respectively in Chile, Bolivia, and Argentina) (An et al. 2012; Grosjean et al. 2012; Steinmetz and Salvi 2021).

Evaporites are also important reservoir seals and fluid aquitards that receive special attention from the petroleum industry. It is estimated that around 50% of the world's known hydrocarbon reserves occur in reservoirs sealed by evaporites. The largest oil field is Ghawar in the Middle East, sealed by Jurassic anhydrites of the Arab and Hith formations. North Field in offshore Qatar is the largest known gas field, sealed by evaporitic dolomites of the Permian Khuff Formation. Interestingly, the majority of the recently discovered giant hydrocarbon fields are associated with carbonate reservoirs sealed by evaporites (Warren 2016). Salt in stratiform and halokinetic formations, due to its low overall permeability and despite its high solubility and high mobility, is considered by some geoscientists as a suitable lithology for the storage of wastes (e.g., radioactive and industrial waste, and CO_2 sequestration) and resources (e.g., hydrocarbons). The repositories or stores used are mainly caverns created by solution mining and underground spaces generated by conventional underground excavations. An example of the latter case is the WIPP site in New Mexico, a deep geological radioactive waste repository excavated at 650 m depth in the Permian Salado Formation (Holt and Powers 2010).

2.3.1 Gypsum and Anhydrite

Gypsum ($CaSO_4 \cdot 2H_2O$) is the principal Ca-sulfate mineral that precipitates in near-surface conditions (Figure 2.22). The reason why gypsum almost always precipitates in depositional environments instead of anhydrite ($CaSO_4$), despite their similar solubilities, is because of the higher activation energy required for anhydrite nucleation. Primary gypsum is converted into secondary anhydrite upon burial during diagenesis (Figure 2.23). Subsequently, anhydrite rocks that return to near-surface conditions by uplift or exhumation (telogenesis) transform into "tertiary" gypsum. Most of the mature gypsum formations found in the geological record that contain karst features have experienced the double gypsum-anhydrite-gypsum conversion history.

The word gypsum derives from the Greek word "γψπσοσ" (gypsos) and the Arabic "جص" (jibs), meaning plaster. Anhydrite comes from the Greek "anhydrous," because of the lack of water. Gypsum crystallizes in the monoclinic crystalline system. Its structure comprises layers of H_2O that alternate with layers containing Ca and SO_4 (Figure 2.4). Anhydrite crystallizes in the orthorhombic system. Both Ca-sulfate minerals tend to have white and gray colors, vitreous luster and are transparent to translucent (Figure 2.5). They can be differentiated by the lower hardness of

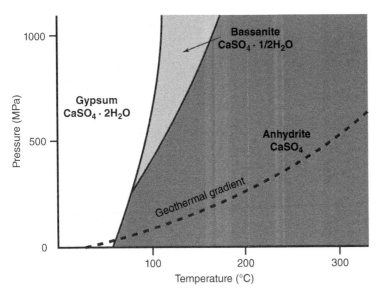

Figure 2.22 Thermobaric stability fields of gypsum, bassanite, and anhydrite. *Source:* Adapted from Zanbak and Arthur (1986).

gypsum (2 on the Mohs scale), which can be scratched with the fingernail, versus 3–3.5 for anhydrite. In addition, anhydrite has a higher specific gravity of 3.9 and is heavier compared to gypsum, which has a specific gravity of 2.32 (Table 2.1). Moreover, gypsum commonly occurs as tabular and prismatic crystals (Figure 2.5), whereas anhydrite is usually massive, granular, or fibrous. Twinning and crystal intergrowths in gypsum are frequent and common varieties include selenite (large transparent crystals), alabaster (finely crystalline) and satin spar (fibrous, generally filling veins, and fractures) (Figure 2.24).

Gypsum typically precipitates from an evaporating brine after the alkali earth carbonates. The gypsum crystals that grow on the bottom of a subaqueous saline lake or seaway typically display two main habits: prismatic growth-aligned crystals that coalesce to form beds, and lenticular crystals, typically sand-sized (Figure 2.23). The bottom-nucleated growth-aligned gypsum crystals that precipitate in perennial water bodies may reach large dimensions (Figures 2.24d, e and 2.25a). During periods of dilution, the brine may become undersaturated with respect to gypsum, and the gypsum surface at the bottom experiences dissolution to form a planar surface. The gypsum crystals deposited at the bottom may be reworked by waves and currents to produce ripples and ooid-like grains designated as gypsolites. Gypsum may also be deposited in deeper environments as pelagic laminites (e.g., saline-stratified meromictic lakes) (Figure 2.25c) or redeposited by sediment-gravity flows (e.g., turbidites) and mass movements (e.g., olistostromes, slumps) (Figure 2.24g). In shallow environments, during periods of desiccation, the gypsum crystals may be exported by the wind to form lunettes in the downwind marginal zones or even dune ergs (e.g., White Sands National Park, New Mexico) (Figure 2.26). Gypsum and anhydrite may also precipitate in the form of nodules within the capillary zone from saturated interstitial brines (Figures 2.24f and 2.25b). These nodules may coalesce, displace the surrounding sediment and form enterolithic and chickenwire textures characteristic of supratidal sabkha deposits and evaporitic mudflats. The enterolithic texture is characterized by complex folding and the chickenwire texture consists of polygonal nodules separated by thin darker stringers of other minerals. In these environments

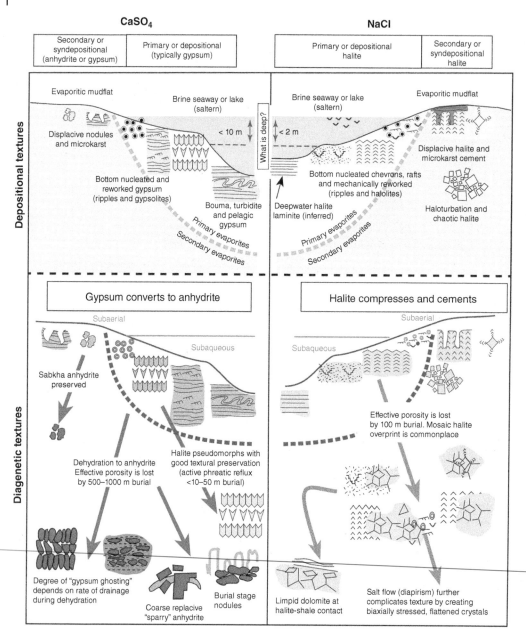

Figure 2.23 Main gypsum/anhydrite and halite depositional (upper) and diagenetic (lower) textures. *Source:* Warren (2016). Reproduced with permission of Springer.

anhydrite may precipitate instead of gypsum when the interstitial brine reaches high temperatures and is highly enriched in Na and Cl (Warren 2016).

Gypsum becomes unstable upon burial and heating and converts into anhydrite (Figures 2.22 and 2.23). This transformation involves a 39% volume reduction, a four-fold increase in thermal conductivity, and consumption of heat (Sonnenfeld and Perthuisot 1984; Jowett et al. 1993). Azimi and Papangelakis (2011) investigated the gypsum-anhydrite conversion in an autoclave. At

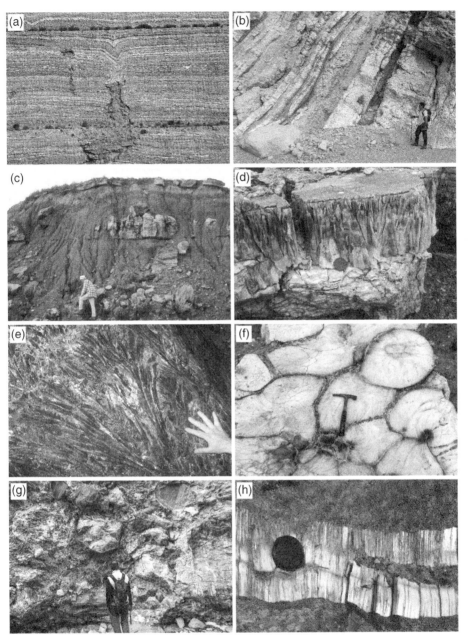

Figure 2.24 Images of sedimentary and secondary structures in gypsum. (a) Horizontally bedded gypsum with marl partings in the continental Miocene Zaragoza Formation, Ebro Cenozoic Basin, Spain. This is "tertiary" gypsum which has been converted into anhydrite during burial and then back into gypsum. Note local synsedimentary dissolution-induced subsidence structure. (b) Outcrop of the Miocene Gachsaran Formation, consisting of gypsum beds with chickenwire texture and interbedded marls and mudstones. Gezeh Anticline, Zagros Mountains, Iran. (c) Gypsum bed in the Permian Blaine Formation (Elm Fork Member) intercalated within a mudstone succession. Note differential dissolution in the gypsum layer and sagging of the overlying strata. Person for scale. (d) Gypsum bed consisting of large bottom-nucleated and growth-aligned gypsum crystals in the Permian Blaine Formation, NW Oklahoma. (e) Dark brown decimeter-scale gypsum megacrystals with radial fabrics that form dome structures in the Middle Badenian Tyrassky Formation, Kristalnaya Cave, Western Ukraine. (f) Large coalescing nodules of alabastrine secondary gypsum with greenish fibrous gypsum in the internodular space. Miocene continental mudflat sediments, Fuentes de Jiloca, Calatayud Basin, NE Spain. (g) Intraclastic megabreccia (olistostrome) in the Eocene Vallfogona Fm. consisting of unsorted, unstratified, chaotic, and subrounded to angular clasts of re-sedimented gypsum and dark carbonate marls, ranging from granule size to more than 10 m long. Eastern Pyrenees, Spain. (h) Vein filled by secondary fibrous gypsum (satin spar) in Miocene red mudstones deposited at the margin of a playa-lake. Crystal orientation perpendicular to fracture. Ebro Cenozoic Basin, Spain. *Source:* Photos by Francisco Gutiérrez (Author).

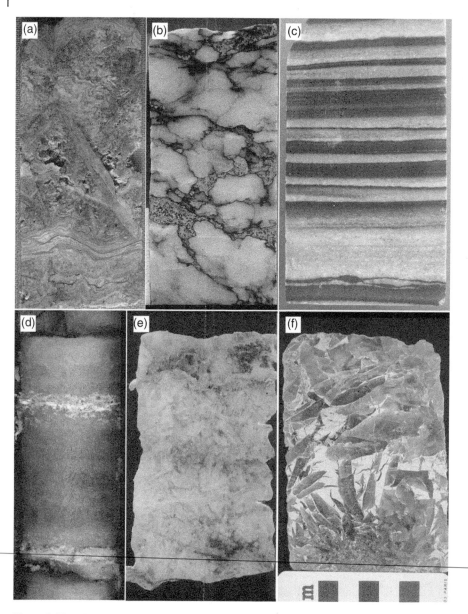

Figure 2.25 Photographs of various types of evaporite sediments in borehole cores. Width of cores is about 8–10 cm. (a) Subvertical selenitic gypsum crystals and laminated microcrystalline gypsum. Middle Miocene, Wiazownica, Poland. (b) Nodular anhydrite and dolomicritic light brown matrix. Late Triassic, Iberian Chain, Spain. (c) Laminated anhydrite. Alternation of light gray secondary anhydrite and dark organic-rich claystone. Zechstein, Late Permian, Poland. (d) Recrystallized transparent halite with slight banding and thin irregular partings of micritic carbonate. Zechstein, Late Permian, Poland. (e) Primary halite with hopper texture and milky white color related to high concentration of fluid inclusions. Neogene Lorca Basin, Betic Cordillera, Spain. (f) Large glauberite crystals embedded in micritic carbonate matrix. Early Miocene Zaragoza Formation, Ebro Basin, Spain. *Source:* Photos by Federico Ortí.

temperatures below 100 °C and under atmospheric pressure, the transformation occurs through direct dehydration of gypsum, resulting in the formation of anhydrite nuclei followed by the growth of the nuclei through dissolution of gypsum and precipitation of stable anhydrite.

Figure 2.26 White Sands National Park in New Mexico, an erg of dunes composed of gypsum crystals transported downwind of Lucero playa-lake in the Tularosa Basin, New Mexico. *Source:* Photo by Philippe Crochet.

Above 100 °C the conversion consists of step-wise dehydration of gypsum to hemihydrate (bassanite) and then to anhydrite (Figure 2.22). The gypsum-anhydrite conversion (anhydritization) involves substantial textural changes, commonly the formation of anhydrite nodules that may include some relics (ghosts) or pseudomorphs of the original gypsum (Figure 2.23). The release of water may be accompanied by high pore-fluid conditions and the weakening of the converting bed (Heard and Rubey 1966), favoring the development of contorted enterolithic textures and deformation features (Sonnenfeld and Perthuisot 1984). The migration of $CaSO_4$-saturated water derived from gypsum dehydration may lead to the precipitation of sparry anhydrite cements in adjacent non-evaporitic units (e.g., veins and fractures in mudstones) (Figure 2.24h). The gypsum-anhydrite conversion generally occurs when the temperature in the diagenetic environment rises above 50–60 °C (Figure 2.22). The depth of transformation may vary significantly depending on factors such as the local geothermal gradient, the salinity of the interstitial brine, the fluid and lithostatic pressures, and the thermal conductivity of the overlying formations (e.g., Jowett et al. 1993). When the pore fluid is close to halite saturation, gypsum may convert into anhydrite at lower temperatures (~35–45 °C). These conditions may occur a few meters below the depositional surface in hot environments. However, the gypsum-anhydrite conversion typically occurs at burial depths of hundreds of meters, where the gypsum beds have lost most of their intercrystalline porosity by compaction. Porosity loss in anhydrite beds at depths of around 500 m is almost complete due to compaction and cementation, and consequently they function as aquitards and seals for cross-formational flows of groundwater or hydrocarbons (Warren 2016).

Anhydrite converts again into gypsum when it is brought back to near-surface conditions (Figure 2.22). This transformation can be considered as a weathering process, whereby meteoric waters interact with an anhydritic sediment that progressively converts into gypsum by hydration. The anhydrite-gypsum conversion typically occurs through the downward migration of a "weathering front," which may have a very irregular geometry controlled by the topography and spatial variations in permeability (e.g., fractures). This is an essential preparatory process for karst development, since dissolution features mainly form in the gypsum rock, rather than in the less soluble and less permeable anhydrite strata. Most of the authors believe that the conversion occurs via anhydrite dissolution and gypsum precipitation from a gypsum-saturated brine (James 1982). When anhydrite fully converts into gypsum, the molar volume of the solid phase increases by a factor of 1.6, suggesting that it may cause expansion. However, this potential volume increase does not necessarily occur in natural systems at significant depths due to the following factors (Klimchouk, 1996a and references therein): (i) Part of the anhydrite may be removed from the system in solution. Open system conditions are expected since anhydrite-gypsum conversion generally requires the ingress of external water. This is supported by the fact that the porosity of gypsum is significantly higher than that of the precursor anhydrite. (ii) Hydration pressure may be counterbalanced by overburden load. Nonetheless, significant expansion may occur where anhydrite is suddenly (in geological sense) exposed at the surface and released from the confining pressure by underground and opencast excavations (James 1982, 1992). The associated human-induced heaving may create difficult ground conditions for the construction of linear infrastructure in tunnels (e.g., Butscher et al. 2016; Ramon et al. 2017). Moreover, rapid anhydrite hydration caused by ground source heat pump installation with boreholes may result in a rather widespread uplift in urban areas (e.g., Staufen town, Germany; Goldscheider and Bechtel 2011; Fleuchaus and Blum 2017).

Gypsum and anhydrite formations occur in marine, transitional, and continental successions and may reach hundreds of meters thick (Figure 2.24a, b). Klimchouk and Andrejchuk (1996) indicate that the area covered by gypsum/anhydrite may represent approximately 5% of the continents. These rocks are more abundant in the northern hemisphere, underlying around 35–40% of the United States (Johnson 1996) and around 20% of the former USSR (Gorbunova 1977). Gypsum outcrops are generally more restricted than limestone, but extensive dissolution frequently occurs in mantled and interstratal conditions. Karst development in anhydrite units is believed to be preceded by its conversion into gypsum. Ca-sulfates are frequently associated with carbonate rocks and their interstratal dissolution may result in the development of carbonate solution-collapse breccias and large-scale subsidence structures (e.g., the Jurassic Arab and Hith formations in Saudi Arabia; Memesh et al. 2008). They also occur in association with halite, whose dissolution may contribute to enhance gypsum karst. The solubility of gypsum increases with the presence of other dissolved salts, particularly halite (i.e., saline effect; Ponsjack 1940).

The main uses of gypsum are as retarding agent in the cement manufacture (60%) and the production of plaster products, notably wallboard (30%) (Kogel et al. 2006, Warren 2016). When gypsum is calcined at 160 °C, it loses 1.5 mol of its water to form calcium sulfate hemihydrate ($CaSO_4 \cdot 1/2H_2O$), popularly known as Plaster of Paris. This name comes from old gypsum quarries located in the Montmartre district of Paris. The hemihydrate, when mixed with water, forms a viscous and cohesive mass that can be spread, molded and cast before it sets to a solid and resistant material. Gypsum is also used for soil conditioning, as filler in a variety of products (e.g., paint, paper), or as sculpting rock in the case of the alabastrine variety called alabaster (Figure 2.24f). Gypsum is mainly extracted from open-pit mines (quarries). Global gypsum mine production in 2019 was 140 million metric tons (Mt), and the main producers were USA (20 Mt), China (16 Mt), Iran (16 Mt) and Turkey (10 Mt).

2.3.2 Halite

Halite (NaCl), together with gypsum/anhydrite, is the most common evaporitic mineral and rock. It typically precipitates after gypsum from brines subject to progressive evaporative concentration. The word halite comes from the Greek "ηαλοσ" (halos), meaning salt. Halite crystallizes in the cubic system (Table 2.1). Its structure comprises Na^+ cations in cubic arrangement and Cl^- anions at the center of the cell and at the center of the edges. Each Cl is surrounded by six Na in octahedral coordination (Figure 2.4). Halite is generally colorless or white, transparent to translucent, and displays perfect cubic cleavage (Figure 2.5). It has low specific gravity (2.16) and is slightly harder than gypsum (2.5 in the Mohs scale), so that it cannot be scratched with the fingernail. Cubic crystals are common, including hopper crystals related to more rapid growth at the edges than at the faces of the cubes. It may be confused with sylvite (KCl), but this less common halide mineral has a bitter taste.

Halite that precipitates on the floor of shallow hypersaline brine environments typically forms upward-growing crystals with chevron, cubic and cornet-like geometries (Figure 2.23). Aligned chevrons are the most common primary texture in ancient halite formations. Their development is related to the quicker growth of the crystal edges pointing upward, which overwhelm the growth of other adjacent crystals with less favorable orientations (i.e. competitive growth). Halite chevrons typically display a cloudy or milky appearance due to the presence of abundant fluid inclusions that define chevron-shaped bands reflecting successive growth stages. Other less common bottom-nucleated crystals include upward-widening cornets and aggregates of hopper crystals generated by overgrowths of sunken rafts, initially precipitated at the brine-air interface (Figure 2.25e). Hoppers are cubic crystals with stepped hopper-shaped hollows on the sides of the cubes related to faster crystal growth at the edges. Halite may also precipitate within sediments of mudflats and sabkhas from interstitial brines subject to capillary evaporation. These are generally hopper crystals that displace the surrounding soft muddy deposits.

Microscopic to small crystals (crystallites) may precipitate in the uppermost part of a standing body of brine, where salinity may rapidly increase by evaporation. Halite rafts are floating crystals that grow at the brine-air interface and that are sustained by surface tension. These crystals eventually settle to the bottom when their weight exceeds the holding capacity of the surface tension, or when the brine surface is disturbed by waves or droplets. This "rain of crystals" produces laminae, bands or beds of cumulate crystals. Periodic changes in the salinity of the brine (i.e., freshwater influxes, evaporative concentration), as well as variations in detrital input, may produce alternations of laminae or layers with different compositions (e.g., halite, gypsum, and clay) (Figure 2.27c, d). In shallow environments, crystals accumulated at the bottom of the brine may be reworked by currents to form crossbeds and ripples. Small crystals may also act as the nuclei to concentric ooids composed of halite (i.e., halolite). When ephemeral shallow lakes and coastal salinas desiccate, the exposed halite deposits may experience a broad range of secondary processes, including: (i) Erosion of particles by deflation, which may result in the formation of eolian deposits in the downwind margin of the depression (e.g., lunettes, nebkhas). The eolian excavation of the bottom sediments may advance down to the water table and form very planar truncation surfaces. (ii) Polygonal cracking of salt crusts, which may be accompanied by preferential precipitation of salts at the edges of the polygons by capillary pumping. These laterally expanding crusts may produce buckle structures, called tepees (Figure 2.27a), and overthrust edges, commonly related to more rapid precipitation along the wind-facing edges. (iii) Dissolution and reprecipitation of salts, which may involve the expansion and bulging of crusts to form irregular wrinkles, blisters, and tumuli. (iv) Dissolution by rainfall and runoff waters, which may result in the formation of a wide variety of dissolution features (Figure 2.27b), small subsidence pits, and the complete removal of beds and crusts. The broad range of disturbances experienced by the salt crusts related to dissolution and precipitation processes, generally induced by wetting and drying cycles, is collectively called haloturbation.

Figure 2.27 Images of salt sediments. (a) Linear buckle structure (tepee) developed on a halite crust exposed by the rapid decline of the water level in the Dead Sea (Israel). (b) Dissolution features (eogenetic karst) developed on exposed halite deposits in the Dead Sea (Israel). The surface shows a "karst pavement" with grikes and rillenkarren (solution flutes) along the polygon margins and solution pipes within the polygons. (c) Alternation of thin beds and laminae of halite and mud in Holocene sediments in the southern sector of the Dead Sea (Zeelim alluvial fan, Israel). (d) Halite beds with marl intercalations in Miocene lacustrine sediments (Remolinos Mine, Ebro Cenozoic Basin, NE Spain). (e) Dipping halite beds in the Precambrian-early Cambrian Hormuz Formation (Cave N2, Namakdan Diapir, Qeshm Island, Iran). *Source:* Photos a to e by Francisco Gutiérrez (Author). (f) Sylvinite including beds of clayey material (gray), halite (light pink), and sylvite (red). Late Eocene potash sediments in the Subiza Mine, Pyrenees, Spain. *Source:* Photo by Federico Ortí.

Two main types of halite facies may accumulate at the deep bottom of perennial brine systems. In salinity-stratified meromictic environments, halite crystals precipitate at the air-brine interface and in the upper part of the brine column, where supersaturation conditions are reached by evaporation. These crystals settle to the deep bottom to form finely layered and laminated bottom cumulates. In deep non-stratified holomictic environments, where the brine is saturated all the way from the surface to the bottom, massive meshworks of clear and poorly oriented coarse halite crystals form, like those in the present-day deep bottom of the Dead Sea (Warren 2016). A peculiarity of halite is that it may aggrade at extremely high rates in subaqueous basins, rapidly counterbalancing the creation of accommodation space. For example, a succession of interbedded halite and clay more than 1000 m thick has accumulated in less than 10000 years (>10 cm yr^{-1}) in the Danakil Depression of the Ethiopian Rift (Warren 2016).

The diagenetic transformation of halite deposits is characterized by rapid loss of effective porosity, primarily by cementation and recrystallization (Figure 2.23). Once the halite beds are fully cemented and cannot be subjected to fluid flow, they barely experience any alteration. Widespread occlusion of pores by cementation typically occurs within a few tens of meters of burial depth. At shallow depths, in the synsedimentary diagenetic environment, secondary halite cement may precipitate by various mechanisms, including dissolution and reprecipitation or supersaturation of dense influxing brines that incorporate into the lower-temperature subsurface environment. Cements are typically coarse-grained mosaics of halite spar. Halite may also be affected by deep burial recrystallization (Figure 2.25d), which produces sutured mosaic textures in which crystals meet at triple junctions.

Since ancient times, halite has been an essential commodity for human life and frequently the main motive for commercial relationships and conflicts. It was such a precious resource as a food preservative and spice, that it was used in the past as currency. In fact, the Latin word "salarium" (meaning salary) has its roots in the word salt. Nowadays, it is the most commonly exploited industrial evaporite, with an annual production that has continuously increased in modern times up to more than 280 billion tonnes. This rise in salt production has been especially significant since the 1950s, coinciding with the start of the so-called "great acceleration" and the Anthropocene, a proposed geological epoch during which humans have produced profound alterations on the Earth's surface processes, climate, and ecosystems. Today, the main producers are China and the United States, with about 26 and 14% of the total world output, respectively. Around one third is derived from hard mining of rock salt, another third is produced by solar evaporation of seawater and inland brines, and solution mining accounts for most of the remaining share. The advantages of salt manufacture in the marine-fed solar evaporation ponds located in hot and arid coastal settings include the low energy cost and the ability to produce very pure salt. Nowadays, the main usages are feedstock for the chemical industry (60%), human consumption as edible product (30%), and as a de-icing agent on roads, mainly in the northern hemisphere during winter (10–15%) (Warren 2016).

2.3.3 Other Salts

Evaporite formations may include beds or packages, meters to tens of meters thick, of other salts such as sodium sulfates and potash salts. The dissolution of these evaporite minerals may play an important role in the karstification of evaporite successions and other associated phenomena (e.g., subsidence, chemical modification of underground and surface waters). However, their presence is commonly unnoticed due to multiple reasons: (i) these minerals rarely crop out at the surface due to their very high to extremely high solubility, (ii) borehole data are generally scarce in evaporitic terrains, which are characterized by poor-quality groundwaters, (iii) conventional boreholes are not suitable for detecting high-solubility salts that require drilling with saturated brines to prevent their dissolution, and (iv) these salts are commonly associated with easier-to-identify and thicker halite and Ca-sulfate beds, and consequently dissolution tends to be entirely ascribed to the latter.

Sodium sulfate salts occur in continental evaporite formations deposited in ephemeral and perennial lakes developed in arid endorheic basins. They may form packages several tens of meters thick and frequently occur as assemblages of several minerals. The most frequent ones are thenardite (Na_2SO_4), mirabilite ($Na_2SO_4 \cdot 10H_2O$), glauberite ($Na_2Ca(SO_4)_2$), and blödite ($Na_2Mg(SO_4)_2 \cdot 4H_2O$), also known as astrakanite. Thenardite and mirabilite, respectively, are the anhydrous and decahydrate single-sodium sulfate salts and the commercially most important Na-sulfate salts. They are hypersoluble minerals (e.g., thenardite 660 g L^{-1}; mirabilite 764 g L^{-1} at 25 °C; mass of solute in 1 L of solution), and consequently are rapidly leached by undersaturated water. Thenardite is a colorless to white mineral that it is extremely hygroscopic. It may convert via hydration into mirabilite, which contains 55.9% of water. This transformation involves high crystallization pressures and substantial volume increases that cause historical building decay and swelling problems in artificial excavations that abruptly expose thenardite to meteoric–atmospheric conditions. Glauberite, which crystallizes in the monoclinic system, commonly occurs as prismatic crystals, generally showing rhomboidal geometry in cross-sections (Figure 2.25f), and has a bitter salty taste. In near surface conditions, where it interacts with meteoric water, it typically transforms into gypsum by hydration and incongruent dissolution, as revealed by secondary gypsum pseudomorphs after glauberite (Salvany et al. 2007). In the middle reach of the Ebro River valley, NE Spain, interstratal dissolution of glauberite within a Miocene continental evaporite formation has generated thickenings in Quaternary terrace deposits (>50 m) and kilometer-sized flat bottom subsidence depressions. Here, glauberite forms packages up to 30 m thick and reaches an aggregate thickness of around 100 m (Guerrero et al. 2013).

Potash salts collectively designate evaporite minerals containing potassium. These minerals precipitate at the bittern end of the crystallization sequence (>60 times seawater concentration) and typically require very restricted and arid hydrological conditions. The most common potash salts are carnallite ($KMgCl_3 \cdot 6H_2O$) and sylvite (KCl), with solubilities at 25 °C of 1118 g L^{-1} and 411 g L^{-1}, respectively (mass of solute in 1 L of solution). Sylvinite is a mixture of sylvite and halite and constitutes the most exploited potash ore. Potassium is an essential fertilizer used for improving agricultural production in a world with a rapidly growing population and food demand. The fertilizer industry uses more than 90% of the produced potash. Carnallite and sylvite are invariably associated with halite and may occur as beds up to a few meters thick. Due to their extremely high solubility, these salts may be dissolved by NaCl-rich brines and replaced by halite. In the Cretaceous Maha Sarakham Formation, Thailand, Warren (2016) inferred early diagenetic dissolution of carnallite atop a rising salt pillow and coeval deposition of sylvinite in the topographically lower and subsiding rim of the salt structure. The experience gained by underground potash mining indicates that sylvinite beds locally show interruptions or thinnings up to hundreds of meters across related to dissolution processes. These salt-dissolution anomalies are known as barren zones or "salt horses" by miners. Three main types are differentiated (Boys 1993; Warren 2016): (i) syndepositional dissolution depressions filled by halite, related to water table drops and karstification in the vadose zone, (ii) leach anomalies up to several hundred meters across generated by crossflows of NaCl-saturated brines that produce the incongruent dissolution of sylvite and its transformation into halite (Hovorka et al. 2007), and (iii) solution-collapse breccias resulting from dissolution by transformational flows from below (hypogene) or above (epigene), and the collapse of the overlying and associated sediments. The margins of these collapse anomalies may include sylvinite-enriched recrystallization zones that grade toward the center of the collapse structure into breccias with progressively higher proportions of insoluble components (Boys 1993). These collapse structures are very important in terms of ore quality and mining safety. The intersection of a collapse structure by the mine galleries may establish a connection with an overlying or underlying pressurized aquifer and the incontrollable inflow of undersaturated water.

2.4 Quartz Sandstones and Quartzites

Quartz sandstones or quartz arenites are detrital sedimentary rocks in which the mineralogy is dominated by quartz; according to Pettijohn et al. (1987) they should contain >95% quartz and <5% feldspars and lithics (Figure 2.28a, b). These clastic sediments may have the intergranular spaces filled by secondary siliceous cements formed during diagenesis. Quartz, due to its resistance to physical and chemical weathering, tends to concentrate during sedimentary processes, while other minerals such as feldspars or calcite are rapidly weathered away. Quartz-rich rocks generally correspond to mature sediments formed by particles that have experienced significant transport. Pettijohn et al. (1987) estimate that approximately 35% of the sandstones are quartz arenites. Quartzites are the metamorphic equivalent of quartz sandstones. They are typically made up of a low-porosity interlocking mosaic of quartz crystals related to the recrystallization of the original grains. Quartz is one of the less soluble minerals in the Earth's crust. Consequently, these essentially monomineralic quartzose rocks were traditionally considered immune to solutional weathering and the associated karst-like features were initially interpreted as pseudokarst (i.e.,

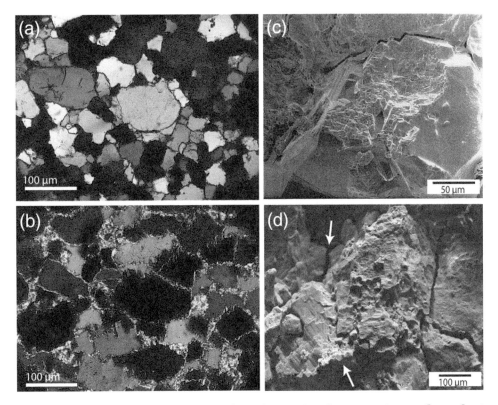

Figure 2.28 (a, b) Thin sections at crossed nicols of unweathered quartz sandstones. *Source*: Courtesy of Francesco Sauro. (a) Almost pure quartz sandstone composed of quartz grains cemented by syntaxial overgrowths. (b) Quartz grains with a thin coating of phyllosilicates. Interdigitated contacts correspond to microstylolites generated by pressure solution during burial metamorphism. (c, d) SEM images showing different dissolution features on quartz grains. *Source*: Courtesy of Francesco Sauro and Robert Wray. (c) Etched quartz grains with highly corroded overgrowths. Permian Snapper Point Formation at Jervis Bay, south of Sydney, Australia. (d) Quartz grains with V-shaped dissolution pits and notches (arrows). Mataui Formation, Roraima Group, Venezuela.

morphologic convergence or equifinality). However, the investigations carried out in some regions with outstanding geomorphological and hydrological karst features contributed to recognize the important morphogenetic role played by chemical weathering in quartzose rocks. This change in paradigm was mainly stimulated by the explorations carried out during the mid-1960s to late-1970s in the tablelands (tepuis) of the Guyana Shield in tropical Venezuela (Wray 2013; Wray and Sauro 2017). Here, vertically-walled mesas composed of massive Precambrian quartzites of the Roraima Group display shafts over 300 m deep, joint-controlled corridors and towers, caves more than 10 km long, karren and internal drainage (Wray 2010). Similar landforms developed in quartzose rocks were subsequently investigated and largely ascribed to chemical weathering in other regions, notably in Brazil (Auler and Sauro 2019), and in Africa and Australia (Young et al. 2009).

Silica (SiO_2) occurs naturally in several crystalline (minerals) and noncrystalline (mineraloids) forms or allotropes. Quartz, which is one of the most abundant minerals in the Earth's crust (plagioclase and alkali feldspar being the other two), is the main crystalline form of silica. It is an essential mineral in many acid igneous rocks such as granites and granodiorites and their volcanic equivalents. Quartz can be identified by its high hardness (7 on Mohs scale), lack of cleavage, conchoidal fracture, and vitreous luster. Quartz sand particles tend to be white or colorless, but are frequently coated by iron oxides and/or hydroxides attaining a yellowish color. Quartz has low-temperature (α-quartz) and high-temperature (β-quartz) polymorphs, with a transition at around 573 °C. Other crystalline forms include the rare high-pressure phases coesite and stishovite, mostly associated with meteorite impact craters, and the high-temperature forms tridymite and cristobalite, typical of volcanic rocks. Chalcedony, with multiple varieties (e.g., chert, flint, agate, and jasper), is composed of minute quartz crystals. The noncrystalline forms of silica include amorphous silica (SiO_2; volcanic glasses, silica cements) and several hydrated mineraloids including opal ($SiO_2 \cdot nH_2O$). Its water content may range from 3 to 21% by weight, but is usually between 6 and 10%. The most important silica species in the study of quartz sandstone and metaquartzite weathering are α-quartz, amorphous silica, and opal (Wray and Sauro 2017).

All the silica forms have a very low solubility under near surface conditions, especially when compared with carbonate and evaporite minerals. However, there are large differences among the different polymorphs, with much larger solubilities for the amorphous forms (Krauskopf 1956; Siever 1962). For instance, the equilibrium solubilities of quartz and amorphous silica in pure water at 25 °C are 6–14 mg L^{-1} and 100–140 mg L^{-1}, respectively (mass of solute in 1 L of solution). The solubility of the different forms increases considerably with temperature (prograde solubility), remains essentially stable between pH 3 and 8, and rises exponentially in alkaline solutions as the pH rises above 8 (Young et al. 2009).

Martini (1979), in his seminal work on the quartzite karst developed in Eastern Transvaal, South Africa, introduced the concept of "arenization," which reconciles the apparent contradiction related to the occurrence of well-developed karst in rocks characterized by very low solubility and slow dissolution kinetics. Slow dissolution along crystal and grain boundaries reduces the coherence of the rock and increases its porosity (chemical weathering) (Figure 2.28c, d). Subsequently, the loose particles and/or crystals are eroded and transported by surface and subsurface flowing water (mechanical erosion). According to this concept, dissolution is not responsible for the removal of a significant rock mass, but plays a critical precursory or preparatory role (Ford 1980; Jennings 1983; Mecchia et al. 2014, 2019). In evaporite and carbonate karst systems, most of the geomorphic work (i.e., rock mass removal) is achieved by dissolution and solute transport, whereas in quartzose rocks it is produced by the mechanical erosion of rock previously weakened by chemical weathering. The idea of arenization is very similar to the "phantomization" concept proposed to explain cave development in highly impure limestones and crystalline rocks such as granite or gneiss. According to Dubois et al. (2014), in an initial stage under low hydrodynamic energy conditions, dissolution leads to the removal of the more soluble components, resulting in a weakened ghost rock including unweathered bedrock and residual material (alterite). In a subsequent stage with a higher hydrodynamic energy, flowing water removes the easily erodible ghost-rock particles to create galleries or corridors. The main differences between arenization and phantomization is that the latter, which affects polymineralic rocks, involves the generation of secondary residual material by dissolution (Wray and Sauro 2017).

The fact that quartz sandstones and quartzites display karst-like features in very specific areas indicates that their development requires the concurrence of a number of intrinsic and extrinsic favorable factors, such as (Doerr 1999; Ford and Williams 2007): (i) Thick to massive bedding and widely spaced subvertical joints. Formations affected by a dense network of discontinuity planes have low rock mass strength and may be affected by rapid mechanical erosion. (ii) High mineral purity, so that dissolution does not produce a significant volume of weathering residue that may block fissures and conduits. (iii) Limited soil and surficial formations that could clog the slowly enlarging passages. (iv) High rainfall that counteracts the low solubility of the rock. (v) Tectonic and geomorphic stability (e.g., planation surfaces in cratons) lacking significant competing morphogenetic processes that could overwhelm the combined effects of chemical weathering and mechanical erosion. (vi) Very long periods of exposure of the rocks to chemical weathering. (vii) Sufficient local topographic and hydraulic gradients for the water flows to evacuate the rock material altered by dissolution.

References

AAPG (2020). *AAPGWiki. An Encyclopedia of Subsurface Science*. Tulsa, OK: America Association of Petroleum Geologists Open access: https://wiki.aapg.org/Main_Page.

Adams, A.E. and MacKenzie, W.S. (1998). *A Color Atlas of Carbonate Sediments and Rocks under the Microscope*. London: Manson.

Adams, A.E., MacKenzie, W.S., and Guilford, C. (1984). *Atlas of Sedimentary Rocks under the Microscope*. New York: Wiley.

Alonso-Zarza, A.M. (2003). Paleoenvironmental significance of palustrine carbonates and calcretes in the geological record. *Earth-Science Reviews* 60: 261–298.

Alonso-Zarza, A.M. and Tanner, L.H. (ed.) (2010). *Carbonates in Continental Settings. Facies, Environments and Processes*, Developments in Sedimentology, vol. 61. Amsterdam: Elsevier.

Alonso-Zarza, A.M. and Wright, V.P. (2010a). Palustrine carbonates. In: *Carbonate Sediments in Continental Setting. Facies, Environments and Processes* (ed. A.M. Alonso-Zarza and L.H. Tanner), 103–131. Amsterdam: Elsevier.

Alonso-Zarza, A.M. and Wright, V.P. (2010b). Calcretes. In: *Carbonate Sediments in Continental Setting. Facies, Environments and Processes* (ed. A.M. Alonso-Zarza and L.H. Tanner), 225–267. Amsterdam: Elsevier.

An, J.W., Kang, D.J., Tran, K.T. et al. (2012). Recovery of lithium from Uyuni salar brine. *Hydrometallurgy* 117: 64–70.

Auler, A.S. and Sauro, F. (2019). Quartzite and quartz sandstone caves of South America. In: *Encyclopedia of Caves* (ed. W.B. White, D.C. Culver and T. Pipan), 850–860. New York: Academic Press.

Azimi, G. and Papangelakis, V.G. (2011). Mechanism and kinetics of gypsum-anhydrite transformation in aqueous electrolyte solutions. *Hydrometallurgy* 108: 122–129.

Bell, K. (ed.) (1989). *Carbonatites: Genesis and Evolution*. London: Unwin Hyman.

Bergada, M., Cervello, J., and Serrat, D. (1997). Karst in conglomerates in Catalonia (Spain): morphological forms and sedimentary sequence types recorded on archaeological sites. *Quaternaire* 8: 267–277.

Blair, T.C. and McPherson, J.G. (1999). Grain-size and textural classification of coarse sedimentary particles. *Journal of Sedimentary Research* 69: 6–19.

Blatt, H., Tracy, R.J., and Owens, B.E. (2005). *Petrology. Igneous, Sedimentary, and Metamorphic*. New York: W.H. Freeman and Company.

Bontognali, T.R., Vasconcelos, C., Warthmann, R.J. et al. (2010). Dolomite formation within microbial mats in the coastal sabkha of Abu Dhabi (United Arab Emirates). *Sedimentology* 57 (3): 824–844.

Boys, C. (1993). A geological approach to potash mining problems in Saskatchewan, Canada. *Exploration and Mining Geology* 2: 129–138.

Butscher, C., Mutschler, T., and Blum, P. (2016). Swelling of clay-sulfate rocks: a review of processes and controls. *Rock Mechanics and Rock Engineering* 49: 1533–1549.

Carlisle, D. (1983). Concentrations of uranium and vanadium in calcretes and gypcretes. In: *Residual Deposits: Surface Related Weathering Processes and Materials* (ed. R.C.L. Wilson), 185–195. London: Geological Society of London.

Choquette, P.W. and Pray, L.C. (1970). Geologic nomenclature and classification of porosity in sedimentary carbonates. *American Association of Petroleum Geologists Bulletin* 54 (2): 207–250.

Davies, G.R. and Smith, L.B. (2006). Structurally controlled hydrothermal dolomite reservoir facies: an overview. *American Association of Petroleum Geologists Bulletin* 90: 1641–1690.

De Waele, J. and Follesa, R. (2003). Human impact on karst: the example of Lusaka (Zambia). *International Journal of Speleology* 32: 71–83.

Doerr, S.H. (1999). Karst-like landforms and hydrology in quartzites of the Venezuelan Guyana shield: Pseudokarst or "real" karst? *Zeitschrift für Geomorphologie* 43 (1): 1–17.

Dubois, C., Quinif, Y., Baele, J.M. et al. (2014). The process of ghost-rock karstification and its role in the formation of cave systems. *Earth-Science Reviews* 131: 116–148.

Dunham, R.J. (1962). Classification of carbonate rocks according to their depositional texture. In: *Classification of Carbonate Rocks* (ed. W.E. Ham) Memoir 1, 108–121. Tulsa, OK: America Association of Petroleum Geologists.

Embry, A.F. and Klovan, J.E. (1971). A late devonian reef tract on northeastern Banks Island, N.W.T. *Bulletin of Canadian Petroleum Geology* 19: 730–781.

Erdosh, G. (1979). The Ontario carbonatite province and its phosphate potential. *Economic Geology* 74: 331–338.

Eugster, H.P. and Hardie, L.A. (1978). Saline lakes. In: *Lakes* (ed. A. Lerman), 237–293. New York: Springer.

Ferrarese, F. and Sauro, U. (2005). The Montello hill: the "classical karst" of the conglomerate rocks. *Acta Carsologica* 34 (2): 439–448.

Filipponi, M., Jeannin, P.Y., and Tacher, L. (2009). Evidence of inception horizons in karst conduit networks. *Geomorphology* 106 (1-2): 86–99.

Filippov, A.G. (2004). Siberia, Russia. In: *Encyclopedia of Caves and Karst Science* (ed. J. Gunn), 645–647. New York: Fitzroy Dearborn.

Finch, R. and Pistole, N. (2011). Honduras: caving in conglomerate. *NSS News* 5: 4–9.

Fleuchaus, P. and Blum, P. (2017). Damage event analysis of vertical ground source heat pump systems in Germany. *Geothermal Energy* 5: 1–15.

Flügel, E. and Munnecke, A. (2010). *Microfacies of Carbonate Rocks: Analysis, Interpretation and Application*. Berlin. Heidelberg: Springer-Verlag.

Folk, R.L. (1962). Spectral subdivision of limestone types. In: *Classification of Carbonate Rocks* (ed. W.E. Ham) Memoir 1, 62–84. America Association of Petroleum Geologists: Tulsa, OK.

Ford, D.C. (1980). Threshold and limit effects in karst geomorphology. In: *Thresholds in Geomorphology* (ed. D.R. Coates and J.D. Vitek), 345–362. London: Allen and Unwin.

Ford, T.D. (1989). Tufa - the whole dam story. *Cave Science* 16: 39–49.

Ford, T.D. and Pedley, H.M. (1996). A review of tufa and travertine deposits of the world. *Earth-Science Reviews* 41: 117–175.

Ford, D.C. and Williams, P.W. (2007). *Karst Hydrogeology and Geomorphology*. Chichester, UK: Wiley.

Freytet, P. and Verrecchia, E.P. (2002). Lacustrine and palustrine carbonate petrography: an overview. *Journal of Paleolimnology* 27: 221–237.

García-Ruiz, J.M., Villasuso, R., Ayora, C. et al. (2007). Formation of natural gypsum megacrystals in Naica, Mexico. *Geology* 35 (4): 327–330.

Gierlowski-Kordesch, E.H. (2010). Lacustrine carbonates. In: *Carbonate Sediments in Continental Setting. Facies, Environments and Processes* (ed. A.M. Alonso-Zarza and L.H. Tanner), 1–102. Amsterdam: Elsevier.

Gile, L.H., Peterson, F.F., and Grossman, R.B. (1966). Morphological and genetic sequences of carbonate accumulation in desert soils. *Soil Science* 101: 347360.

Given, R.K. and Wilkinson, B.H. (1987). Dolomite abundance and stratigraphic age; constraints on rates and mechanisms of Phanerozoic dolostone formation. *Journal of Sedimentary Research* 57 (6): 1068–1078.

Goldscheider, N. and Bechtel, T.D. (2011). Editors' message: the housing crisis from underground - damage to the historic town by geothermal drillings through anhydrite, Staufen, Germany. *Hydrogeology Journal* 17: 491–493.

Goldscheider, N., Chen, Z., Auler, A.S. et al. (2020). Global distribution of carbonate rocks and karst water resources. *Hydrogeology Journal 28* (5): 1661–1677.

Göppert, N., Goldscheider, N., and Scholz, H. (2011). Karst geomorphology of carbonatic conglomerates in the Folded Molasse zone of the Northern Alps (Austria/Germany). *Geomorphology* 130 (3-4): 289–298.

Gorbunova, K.A. (1977). *Karst in Gypsum of the USSR*. Perm: Perm University (in Russian).

Grajales-Nishimura, J.M., Cedillo-Pardo, E., Rosales-Domínguez, C. et al. (2000). Chicxulub impact: the origin of reservoir and seal facies in the southeastern Mexico oil fields. *Geology* 28: 307–310.

Grosjean, C., Miranda, P.H., Perrin, M. et al. (2012). Assessment of world lithium resources and consequences of their geographic distribution on the expected development of the electric vehicle industry. *Renewable and Sustainable Energy Reviews* 16 (3): 1735–1744.

Guerrero, J., Gutiérrez, F., and Galve, J.P. (2013). Large depressions, thickened terraces, and gravitational deformation in the Ebro River valley (Zaragoza area, NE Spain): Evidence of glauberite and halite interstratal karstification. *Geomorphology* 196: 162–176.

Gunn, J. (2004). Limestone as a mineral resource. In: *Encyclopedia of Caves and Karst Science* (ed. J. Gunn), 481–483. New York: Fitzroy Dearborn.

Gutiérrez, M. (2013). *Geomorphology*. Boca Raton: CRC Press.

Heard, H.C. and Rubey, W.W. (1966). Tectonic implications of gypsum dehydration. *Geological Society of America Bulletin* 77: 741–760.

Herrero, M.J. and Escavy, J.I. (2010). Economic aspects of continental carbonates and carbonates transformed under continental conditions. In: *Carbonates in Continental Settings. Geochemistry, Diagenesis and Applications* (ed. A.M. Alonso-Zarza and L.H. Tanner), 275–296. Amsterdam: Elsevier.

Holt, R.M. and Powers, D.W. (2010). Evaluation of halite dissolution at a radioactive waste disposal site, Andrews County, Texas. *Geological Society of America Bulletin* 122: 1989–2004.

Hovorka, S.D., Holt, R.M., and Powers, D.W. (2007). Depth indicators in Permian Basin evaporites. *Geological Society of London*, Special Publication 25: 335–364.

Insalaco, E. (1998). The descriptive nomenclature and classification of growth fabrics in fossil scleractinian reefs. *Sedimentary Geology* 118: 159–186.

Jackson, M.P.A. and Hudec, M.R. (2017). *Salt Tectonics. Principles and Practice*. Cambridge: Cambridge University Press.

James, A.N. (1982). Engineering properties of evaporitic rocks. *Bulletin of the International Association of Engineering Geology* 25: 125–126.

James, A.N. (1992). *Soluble Material in Civil Engineering*. New York: Ellis Horwood.

James, N.P. and Choquette, P.W. (1990). Limestone: the meteoric diagenetic environment. In: *Sediment Diagenesis* (ed. I. McIlreath and D. Morrow), 35–74. St John's, ND, Canada: Geological Association of Canada Reprint Series 4.

James, N.P. and Jones, B. (2015). *Origin of Carbonate Sedimentary Rocks*. Chichester: Wiley.

Jennings, J.N. (1983). Sandstone pseudokarst or karst? In: *Aspects of Australian Sandstone Landscapes*, vol. 1 (ed. R.W. Young and G.C. Nanson), 21–30. Australian and New Zealand Geomorphology Group Special Publication.

Johnson, K.S. (1996). Gypsum karst in the United States. *International Journal of Speleology* 25: 183–193.

Johnson, S.Y. and Watt, J. (2012). Influence of fault trend, bends, and convergence on shallow structure and geomorphology of the Hosgri strike-slip fault, offshore central California. *Geosphere* 8: 1632–1656.

Jones, B. (2005). Carbonates. In: *Encyclopedia of Geology*, vol. 5 (ed. R.C. Selley, L.R.M. Cocks and I.R. Plimer), 522–532. Amsterdam: Elsevier.

Jones, B. and Renault, R. (2010). Calcareous spring deposits in continental settings. In: *Carbonate Sediments in Continental Setting. Facies, Environments and Processes* (ed. A.M. Alonso-Zarza and L.H. Tanner), 177–224. Amsterdam: Elsevier.

Jowett, E.C., Cathles, L.M. III, and Davis, B.W. (1993). Predicting depths of gypsum dehydration in evaporitic sedimentary basins. *American Association of Petroleum Geologists Bulletin* 77: 402–413.

Kendall, A.C. (2005). Evaporites. In: *Encyclopedia of Geology*, vol. 5 (ed. R.C. Selley, L.R.M. Cocks and I.R. Plimer), 94–97. Amsterdam: Elsevier.

Kiraly, L. (1975). Rapport sur l'état actuel des connaissances dans le domaine des caractères physiques des roches karstiques. In: *Hydrogeology of Karstic Terrains (Hydrogéologie des Terrains Karstiques)*, vol. 3 (ed. A. Burger and L. Dubertret), 53–67. International Union of Geological sciences.

Klimchouk, A.B. (1996a). The dissolution and conversion of gypsum and anhydrite. *International Journal of Speleology* 25: 21–36.

Klimchouk, A.B. and Andrejchuk, V. (1996). Sulfate rocks as an arena for karst development. *International Journal of Speleology* 25: 9–20.

Kogel, J.E., Trivedi, N.C., Barker, J.M. et al. (ed.) (2006). *Industrial Minerals and Rocks. Commodities, Markets, and Uses*. Littleton (CO): Society of Mining Metallurgy and Exploration.

Kozary, M.T., Dunlap, J.C., and Humphrey, W.E. (1968). Incidence of saline deposits in geologic time. *Geological Society of America Special Paper* 88: 43–57.

Krauskopf, K.B. (1956). Dissolution and precipitation of silica at low temperatures. *Geochimica et Cosmochimica Acta* 10: 1–26.

Lal, R. (2007). Carbon sequestration. *Philosophical Transactions of the Royal Society B: Biological Sciences* 363: 815–830.

Lapaire, F., Becker, D., Christe, R. et al. (2007). Karst phenomena with gas emanations in Early Oligocene conglomerates: risks within a highway context (Jura, Switzerland). *Bulletin of Engineering Geology and the Environment* 66 (2): 237–250.

Lauritzen, S.E. and Skoglund, R.Ø. (2013). Glacier Ice-Contact Speleogenesis in Marble Stripe Karst. In: *Treatise on Geomorphology. Karst Geomorphology*, vol. 6 (ed. A. Frumkin), 363–396. Amsterdam: Elsevier.

Lipar, M. and Ferk, M. (2011). Eogenetic caves in conglomerate: an example from Udin Boršt, Slovenia. *International Journal of Speleology* 40 (1): 53–64.

Lottermoser, B.G. (1990). Rare-earth element mineralisation within the Mt. Weld carbonatite laterite, Western Australia. *Lithos* 24: 151–167.

Loucks, R.G. (2004). Hydrocarbons in karst. In: *Encyclopedia of Caves and Karst Science* (ed. J. Gunn), 431–433. New York: Fitzroy Dearborn.

Lowe, D.J. (1992b). The origin of limestone caverns: An inception horizon hypothesis. PhD thesis. Manchester Metropolitan University.

Machel, H.G. (2004). Concepts and models of dolomitization: a critical reappraisal. *Geological Society of London, Special Publications* 235 (1): 7–63.

Machette, M.N. (1985). Calcic soils of the southwestern United States. *Geological Society of America, Special Paper* 203: 1–21.

Maire, R. (2004). Patagonia marble karst, Chile. In: *Encyclopedia of Caves and Karst Science* (ed. J. Gunn), 572–573. New York: Fitzroy Dearborn.

Martini, J.E.J. (1979). Karst in Black Reef quartzite near Kaapsehoop, Eastern Transvaal. *Annals of the South African Geological Survey* 13: 115–128.

Mecchia, M., Sauro, F., Piccini, L. et al. (2014). Geochemistry of surface and subsurface waters in quartz-sandstones: significance for the geomorphic evolution of tepui table mountains (Gran Sabana, Venezuela). *Journal of Hydrology* 511: 117–138.

Mecchia, M., Sauro, F., Piccini, L. et al. (2019). A hybrid model to evaluate subsurface chemical weathering and fracture karstification in quartz sandstone. *Journal of Hydrology* 572: 745–760.

Melgarejo, J.C. (ed.) (2003). *Atlas de Asociaciones Minerales en Lámina Delgada*. Barcelona: Publicacions de la Universidad de Barcelona.

Memesh, A., Dini, S., Gutiérrez, F. et al. (2008). *Evidence of large-scale subsidence caused by interstratal karstification of evaporites in the Interior Homocline of Central Saudi Arabia*. European Geosciences Union General Assembly. Geophysical Research Abstracts 10, A-02276.

Moore, C.H. (2001). *Carbonate Reservoirs: Porosity, Evolution and Diagenesis in a Sequence Stratigraphic Framework*, Developments in Sedimentology, vol. 55. Amsterdam: Elsevier.

Morse, J.W. and MacKenzie, F.T. (1990). *Geochemistry of Sedimentary Carbonates*, Developments in Sedimentology, vol. 48. Amsterdam: Elsevier.

Nash, D.J. (2004). Calcrete. In: *Encyclopedia of Geomorphology*, vol. 1 (ed. A.S. Goudie), 108–111. London: Routledge.

Pedley, H.M. (1990). Classification and environmental models of cool freshwater tufas. *Sedimentary Geology* 68: 143–154.

Pentecost, A. (2004). Travertine. In: *Encyclopedia of Caves and Karst Science* (ed. J. Gunn), 1574–1578. New York: Fitzroy Dearborn.

Pentecost, A. (2005). *Travertines*. Berlin: Springer.

Perkins, D. (2002). *Mineralogy*. New Jersey: Prentice Hall.

Petrash, D.A., Bialik, O.M., Bontognali, T.R. et al. (2017). Microbially catalyzed dolomite formation: from near-surface to burial. *Earth-Science Reviews* 171: 558–582.

Pettijohn, F.J., Potter, P.E., and Siever, R. (1987). *Sand and Sandstone*. New York: Springer-Verlag.

Piccini, L. (2011). Speleogenesis in highly geodynamic contexts: the Quaternary evolution of Monte Corchia multi-level karst system (Alpi Apuane, Italy). *Geomorphology* 134 (1-2): 49–61.

Ponsjack, E. (1940). Deposition of calcium sulphate from sea water. *American Journal of Science* 239: 559–568.

Ramon, A., Alonso, E.E., and Olivella, S. (2017). Hydro-chemo-mechanical modelling of tunnels in sulfated rocks. *Géotechnique* 67: 968–982.

Rosen, O., Desmons, J., and Fettes, D. (2007). Metacarbonate and related rocks. In: *Metamorphic Rocks. A Classification and Glossary of Terms* (ed. D. Fettes and J. Desmons), 46–50. Cambridge: Cambridge University Press.

Salvany, J.M., García-Veigas, J., and Ortl, F. (2007). Glauberite-halite association of the Zaragoza Gypsum Formation (Lower Miocene, Ebro Basin, NE Spain). *Sedimentology* 54: 443–467.

Sauro, F., Mecchia, M., Tringham, M. et al. (2020). Speleogenesis of the world's longest cave in hybrid arenites (Krem Puri, Meghalaya, India). *Geomorphology* 359: 107160.

Schlager, W. (2005). *Carbonate Sedimentology and Sequence Stratigraphy*. Denver: SEPM Society for Sedimentary Geology.

Selly, R.C. (2005). Sedimentary rocks. Mineralogy and classification. In: *Encyclopedia of Geology*, vol. 5 (ed. R.C. Selley, L.R.M. Cocks and I.R. Plimer), 25–37. Amsterdam: Elsevier.

Siever, R. (1962). Silica solubility, 0–200 °C and the diagenesis of siliceous sediments. *Journal of Geology* 70: 127–150.

Siever, R. (1982). *Earth*. San Francisco: W.H. Freeman and Company.

Skoglund, R.Ø. and Lauritzen, S.E. (2011). Subglacial maze origin in low-dip marble stripe karst: examples from Norway. *Journal of Cave and Karst Studies* 73: 31–43.

Smart, C.C. and Worthington, S.R.H. (2004a). Karst water resources. In: *Encyclopedia of Caves and Karst Science* (ed. J. Gunn), 481–483. New York: Fitzroy Dearborn.

Sonnenfeld, P. and Perthuisot, J.-P. (1984). *Brines and Evaporites*. New York: Academic Press.

Stanton, R.J. (1966). The solution brecciation process. *Geological Society of America Bulletin* 7: 843–848.

Steinmetz, R.L.L. and Salvi, S. (2021). Brine grades in Andean salars: When basin size matters A review of the Lithium Triangle. *Earth-Science Reviews* 2017: 103615.

Stevanović, Z. (2019). Karst waters in potable water supply: a global scale overview. *Environmental Earth Sciences* 78 (23): 662.

Stinnesbeck, W., Keller, G., Adatte, T. et al. (2004). Yaxcopoil-1 and the Chicxulub impact. *International Journal of Earth Sciences* 93: 1042–1065.

Sun, W., Jayaraman, S., Chen, W. et al. (2015). Nucleation of metastable aragonite $CaCO_3$ in seawater. *Proceedings of the National Academy of Sciences* 112: 3199–3204.

~~Sverjensky, D.A. (1986). Genesis of Mississippi Valley-type lead-zinc desposits.~~ *Annual Review of Earth and Planetary Sciences* 14 (1): 177–199.

Tinker, S.W., Ehrets, J.R., and Brondos, M.D. (1995). Multiple karst events related to stratigraphic cyclicity: San Andres Formation, Yates field, west Texas. *American Association of Petroleum Geologists Memoir* 63: 213–237.

Tucker, M.E. (1991). Sequence stratigraphy of carbonate-evaporite basins: models and application to the Upper Permian (Zechstein) of northeast England and adjoining North Sea. *Journal of the Geological Society* 148: 1019–1036.

Tucker, M.E. and Wright, V.P. (1990). *Carbonate Sedimentology*. Oxford: Blackwell Scientific Publications.

Vasconcelos, C., McKenzie, J.A., Bernasconi, S. et al. (1995). Microbial mediation as a possible mechanism for natural dolomite formation at low temperatures. *Nature* 377 (6546): 220–222.

Viles, H.A. (2004). Tufa and travertine. In: *Encyclopedia of Geomorphology*, vol. 2 (ed. A.S. Goudie), 595–596. Routledge/Taylor & Francis: London-New York.

Warren, J.K. (2000). Dolomite: occurrence, evolution and economically important associations. *Earth-Science Reviews* 52: 1–81.

Warren, J.K. (2010). Evaporites through time: Tectonic, climatic and eustatic controls in marine and nonmarine deposits. *Earth-Science Reviews* 98: 217–268.

Warren, J.K. (2016). *Evaporites: A Geological Compendium*. Dordrecht: Springer.

Williams, P.W. and Dowling, R.K. (1979). Solution of marble in the karst of the Pikikiruna Range, northwest Nelson, New Zealand. *Earth Surface Processes* 4: 15–36.

Wray, R.A.L. (2010). The Gran Sabana: the world's finest quartzite karst? In: *Geomorphological Landscapes of the World* (ed. P. Migon), 79–88. Amsterdam: Springer.

Wray, R.A.L. (2013). Solutional weathering and karstic landscapes on quartz sandstones and quartzite. In: *Treatise on Geomorphology*, vol. 6 (ed. A. Frumkin), 463–483. San Diego: Academic Press.

Wray, R.A.L. and Sauro, F. (2017). An updated global review of solutional weathering processes and forms in quartz sandstones and quartzites. *Earth-Science Reviews* 171: 520–557.

Wright, V.P. (1992). A revised classification of limestones. *Sedimentary Geology* 76: 177–185.

Wright, V.P. (2006). *Carbonate Depositional Systems: Reservoir Sedimentation and Diagenesis*. Hermitage, UK: Geoscience Training Alliance (unpublished course notes).

Xia, L.W., Cao, J., Wang, M. et al. (2019). A review of carbonates as hydrocarbon source rocks: basic geochemistry and oil–gas generation. *Petroleum Science* 16 (4): 713–728.

Young, R.W., Wray, R.A.L., and Young, A.R.M. (2009). *Sandstone Landforms*. Cambridge: Cambridge University Press.

Zanbak, C. and Arthur, R.C. (1986). Geochemical and engineering aspects of anhydrite/gypsum phase transitions. *Bulletin of the Association of Engineering Geologists* 23: 419–433.

3

Dissolution of Karst Rocks

3.1 Introduction

Dissolution is the main process responsible for the development of the specific geomorphic and hydrologic features found in karst terrains. Most karst rocks are essentially monomineralic, hence the problem of their karstification mainly concerns the dissolution of the dominant constituent mineral. Soluble minerals that interact with undersaturated water (solvent) progressively decompose into individual chemical components (solutes), mainly ions that contribute to increase the content of dissolved solids in the water. The detached chemical components transfer from the surface of the mineral to the aqueous solution through a boundary layer by diffusion. Water that has the ability to dissolve a mineral is said to be **undersaturated** with respect to that mineral, or to be **solutionally aggressive** toward it. As the solid mineral dissolves, the concentration of its dissolved components progressively increases. Eventually, the water may not be able to hold more dissolved material and becomes at equilibrium (**saturated**) with that mineral. The amount of mineral that can be dissolved in a certain amount of water depends on the mineral type, temperature, pressure, and the chemical composition of the solution, including its pH. This quantity is the **saturation concentration** of the mineral, which expresses its **solubility** in a specific solvent. These chemical reactions may be reversed. If the amount of dissolved material exceeds the saturation concentration, the solution becomes **supersaturated** and the mineral may precipitate. The supersaturation state may be reached in multiple ways, such as by water evaporation, addition of dissolved ions, mixing of solutions, and changes in temperature, pressure or pH. If the water in contact with the mineral is continuously replaced by newly arriving undersaturated fluids, dissolution does not stop until the mineral is fully leached. If the water in contact with the mineral is static and not renewed, the solution may reach saturation conditions and dissolution halts. The rate at which the mineral is dissolved and transferred to the solution depends on chemical (solubility), kinetic (dissolution rate), and hydrodynamic factors. The latter, which play a critical role, especially in high-solubility minerals, mainly refer to flow velocity and regime (e.g., laminar versus turbulent flow).

Dissolution is said to be **congruent** when the entire mineral dissolves and the composition of the dissolved solute and the mineral stoichiometrically match. Table 3.1 provides the dissolution reactions and equilibrium solubilities of the main constituent minerals of karst rocks, plus gibbsite ($Al(OH)_3$) for comparison. Gibbsite is a virtually insoluble mineral found in residual chemical rocks (e.g., bauxite) resulting from intense weathering of Al-bearing silicates. **Incongruent** dissolution occurs when alteration of the primary mineral phase involves the release of ions to the solution and the simultaneous formation of a secondary less soluble mineral. Consequently, there is no

Karst Hydrogeology, Geomorphology and Caves, First Edition. Jo De Waele and Francisco Gutiérrez.
© 2022 John Wiley & Sons Ltd. Published 2022 by John Wiley & Sons Ltd.

Table 3.1 Constituent minerals of karst rocks grouped by equilibrium solubilities, dissolution reactions, and solubilities in pure water at 25 °C, pH 7 and 1 atm pressure (except for carbonate minerals), plus density, water activity, and ionic strength of the solutions in equilibrium with the corresponding mineral. The solubilities of calcite and dolomite refer to solutions in pure water at 25 °C with carbon dioxide partial pressures (P_{CO_2}) of $10^{-3.42}$ atm (a) and $10^{-1.3}$ atm (b). These are the P_{CO_2} in the atmosphere ($CO_2 = 0.038\%$ by volume) and soils with a high CO_2 content ($CO_2 = 5\%$ by volume), respectively. Values obtained with the geochemical code PHREEQC (Parkhurst and Appelo 2013) and the databases of thermodynamic data PITZER for halite, glauberite, sylvite, thenardite, mirabilite, and carnallite, and WATEQ4F for gibbsite, quartz, amorphous silica, calcite, and dolomite. Solubilities expressed as mmol kg⁻¹ and g kg⁻¹ refer to amount of solute dissolved in 1 kg of solvent, whereas g L⁻¹ refers to mass of dissolved solute in 1 L of solution. Note that solubilities for the hyper-soluble minerals expressed as g kg⁻¹ and g L⁻¹ are markedly different due to the fact that the solutions have densities significantly higher than $1\,g\,cm^{-3}$.

	Mineral	Dissolution reaction	Mineral solubility (25 °C; pH 7; 1 atm)				Water (solution)		
			(mmol kg⁻¹)	(g kg⁻¹)	g L⁻¹	log K	Density (g cm⁻³)	Water activity	Ionic strength (molal)
Low	Gibbsite	$Al(OH)_3 \rightarrow Al^{3+} + 3OH^-$	$3.04 \cdot 10^{-5}$	$2.37 \cdot 10^{-6}$	$2.37 \cdot 10^{-6}$	-33.89	1.0	1.0	$1.28 \cdot 10^{-7}$
	Quartz	$SiO_2 + H_2O \rightarrow H_4SiO_4$	0.1047	0.0063	0.0063	-3.98	1.0	1.0	$1.60 \cdot 10^{-7}$
	Amorphous silica	$SiO_2 + H_2O \rightarrow H_4SiO_4$	1.942	0.1167	0.1167	-2.71	1.0	1.0	$5.45 \cdot 10^{-7}$
Moderate	Aragonite	$CaCO_3 \leftrightarrow Ca^{2+} + CO_3^{2-}$	(a)0.5919 (b)3.359	(a)$5.92 \cdot 10^{-2}$ (b)0.3359	(a)$5.92 \cdot 10^{-2}$ (b)0.3359	-8.34	(a)1.0 (b)1.0	(a)1.0× (b)1.0×	(a)$1.75 \cdot 10^{-3}$ (b)$9.7 \cdot 10^{-3}$
	Calcite	$CaCO_3 \leftrightarrow Ca^{2+} + CO_3^{2-}$	(a)0.521 (b)2.970	(a)$5.21 \cdot 10^{-2}$ (b)0.297	(a)$5.21 \cdot 10^{-2}$ (b)0.297	-8.48	(a)1.0 (b)1.0	(a)1.0 (b)1.0	(a)$1.56 \cdot 10^{-3}$ (b)$8.6 \cdot 10^{-3}$
	Dolomite	$CaMg(CO_3)_2 \leftrightarrow Ca^{2+} + Mg^{2+} + 2CO_3^{2-}$	(a)0.319 (b)1.813	(a)$5.87 \cdot 10^{-2}$ (b)0.334	(a)$5.87 \cdot 10^{-2}$ (b)0.334	-17.09	(a)1.0 (b)1.0	(a)1.0 (b)1.0	(a)$1.89 \cdot 10^{-3}$ (b)$1.047 \cdot 10^{-2}$
High	Gypsum	$CaSO_4 \cdot 2H_2O \rightarrow Ca^{2+} + SO_4^{2-} + 2H_2O$	15.23	2.62	2.59	-4.58	0.99	1.0	0.061
Hyper-soluble	Halite	$NaCl \rightarrow Na^+ + Cl^-$	6094.0	356.15	423.81	1.57	1.19	0.755	6.094
	Glauberite	$CaNa_2(SO_4)_2 \rightarrow Ca^{2+} + 2Na^+ + 2SO_4^{2-}$	423.5	117.81	129.59	-5.31	1.10	0.978	2.964
	Sylvite	$KCl \rightarrow K^+ + Cl^-$	4791.0	357.17	410.74	0.9	1.15	0.843	4.791
	Thenardite	$Na_2SO_4 \rightarrow 2Na^+ + SO_4^{2-}$	3495.0	496.42	660.23	-0.3	1.33	0.877	10.49
	Mirabilite	$Na_2SO_4 \cdot 10H_2O \rightarrow 2Na^+ + SO_4^{2-} + 10H_2O$	1977.0	636.66	763.99	-1.21	1.20	0.935	5.931
	Carnallite	$KMgCl_3 \cdot 6H_2O \rightarrow K^+ + Mg^{2+} + 3Cl^- + 6H_2O$	3145	873.55	1118.1	4.33	1.28	0.582	12.58

Source: Courtesy of Luis Auqué.

stoichiometric match between the composition of the mineral and the dissolved solutes. The classical example is the chemical alteration of aluminosilicate minerals. These minerals release ions to the solution, and the atoms remaining in the solid phase are reorganized into less soluble minerals. For instance, the alteration of the plagioclase albite ($NaAlSi_3O_8$) to the aluminum hydroxide gibbsite ($Al(OH)_3$) involves the release of Na^+ cations and silicic acid (H_4SiO_4).

$$NaAlSi_3O_8 + H^+ + 7H_2O \rightarrow Al(OH)_3 + Na^+ + 3H_4SiO_4 \tag{3.1}$$

Other common cases are related to some carbonate phases and salts. For instance, the dissolution of dolomite ($CaMg(CO_3)_2$) in calcite-saturated solutions may occur incongruently, with the release of magnesium cations and the simultaneous precipitation of calcite. This "dedolomitization" process, usually driven by gypsum dissolution and the consequent increase in calcium concentration (Bischoff et al. 1994), is thought to occur across a migrating dissolution–precipitation front akin to that invoked to explain the transformation of aragonite into the more stable calcite. High-Mg calcite is a solid solution ($Ca_{1-x}Mg_xCO_3$) that exhibits a complex dissolution pattern in distilled water. It shows an initial stage of congruent dissolution and a subsequent stage in which the concentration of Ca^{2+} in the solution decreases and a new calcite with lower-Mg content precipitates. Appelo and Postma (2005) indicate that this process of high-Mg calcite incongruent dissolution, together with the transformation of aragonite into calcite, occurs extensively in recent carbonate sediments when they change from a marine-water to a fresh-water environment due to eustatic sea-level changes. The sulfate salt glauberite ($Na_2Ca(SO_4)_2$), which forms packages as much as tens of meters thick in terrestrial evaporite formations, may transform by incongruent dissolution and hydration into gypsum ($CaSO_4 \cdot 2H_2O$), releasing Na^+ cations. This is a common near-surface alteration process that explains the gradation from glauberite in the non-weathered rock situated in the subsurface to secondary gypsum in the surficial alteration zone (Salvany et al. 2007).

Table 3.1 illustrates the enormous solubility range of the main constituent minerals of rocks that can hold well-developed karst features. The solubility of halite in g L^{-1} is 67 300 times higher or 4.8 orders of magnitude greater than that of quartz (i.e., Log_{10} halite solubility/quartz solubility). The solubilities of halite and gypsum are ca. 1400 and 9 times higher than that of pure calcite in a slightly acidic solution with a P_{CO_2} of $10^{-1.3}$ atm (corresponding to P_{CO_2} in the air of soils with high CO_2 content, around 5% by volume). Karst rocks and minerals, according to their solubility and the geological effects resulting from their dissolution, can be grouped into four categories: hyper-soluble evaporites, high-solubility evaporites, moderately soluble carbonates, and low-solubility quartzose rocks. The **hyper-soluble** evaporites are those composed of salts with solubilities above $100 g L^{-1}$ (e.g., halite, carnallite, sylvite, glauberite, and thenardite). They rarely crop out at the surface because they are readily dissolved. Moreover, these salts, with the exception of halite, hardly occur as packages meters or tens of meters thick. The limited halite exposures that exist in the world essentially correspond to salt extrusions associated with active salt diapirs located in arid regions (e.g., Zagros Mountains in Iran; Mount Sedom in the Dead Sea, Israel). In these outcrops, the evolution of surface and subsurface karst landforms can be extremely rapid at a human time scale; a single rainfall event may produce substantial changes (Frumkin and Ford 1995; Bruthans et al. 2008; Frumkin 2013). The gross dissolution of these largely buried formations mainly occurs by underground water flows beneath other insoluble or less-soluble formations (i.e., interstratal karstification). This subjacent dissolution, which typically advances through the long-term migration of dissolution fronts, may affect areas in the order of hundreds of kilometers across, generating large non-tectonic subsidence structures, and may even result in the formation of supra-salt dissolution-induced sedimentary basins (Warren 2016). The development of solution features and caves controlled by discontinuities in these rocks at depth is very rare since permeability features tend to be annealed by salt flow. The content of dissolved ions (total dissolved solids; TDS) in waters associated with salt karst is typically in the range 250–350 g L^{-1}. Concentrations in runoff water with limited contact with salt may be as low as 15 g L^{-1} (Bruthans et al. 2017).

The **high-solubility** category is represented by the calcium sulfate minerals. Anhydrite essentially transforms into gypsum before calcium sulfate dissolution occurs. Gypsum solubility is significantly higher than the solubilities of carbonate minerals, but is much lower than that of halite (164 times). Gypsum, in contrast to the hyper-soluble salts, forms extensive outcrops in some regions, especially in arid and semiarid environments (Fig. 4.10). The exposed gypsum formations may occur in elevated reliefs (e.g., mesas, cuestas, hogbacks, and plateaus), but their solubility and erodibility generally inhibit the formation of prominent massifs like those found in carbonate formations. Karst developed in gypsum shares some characteristics with those of the hyper-soluble salts and of the moderately soluble carbonates (Gutiérrez and Cooper 2013). Interstratal karstification, although much slower, locally plays a significant role even in hyperarid areas, and may generate large-scale dissolution-induced subsidence structures (e.g., dissolution of the Arab and Hith formations in the Interior Homocline of Saudi Arabia; Memesh et al. 2008). Dissolution features in gypsum generally evolve much more rapidly than those found in carbonate rocks, but changes at a human-time scale and under natural conditions are rather moderate. Discontinuity planes such as stratification and jointing play a decisive role in the development of permeability features, including caves. Gypsum caves tend to evolve and degrade more rapidly than those formed in carbonate rocks due to the higher dissolution rate and lower mechanical resistance of gypsum. The TDS in evolved waters associated with gypsum karst generally falls within the 2–5 g L^{-1} range (D'Angeli et al. 2017a).

The **moderately soluble** carbonates (limestones and dolostones) are extensively exposed and frequently form prominent massifs and mountain ranges thanks to their lower solubility and high mechanical resistance. These topographic conditions plus the large thickness that carbonate successions may attain allow the formation of deep cave systems, with no equivalent vertical dimensions in other rocks. The deepest known cave in carbonate rocks, Veryovkina Cave, Georgia, reaches 2212 m in vertical dimension (Minton and Droms 2019), whereas the 265-m deep Monte Caldina Cave, northern Apennines, Italy, is the deepest in gypsum (Franchi and Casadei 1999; De Waele et al. 2017c). Dissolution-induced subsidence related to interstratal karstification in carbonate rocks has a negligible impact compared to evaporites. In most cases, the role played by dissolution at a human- or engineering-time scale is insignificant. Discontinuities, with higher densities than in the more ductile evaporites, play a prime importance in the development of dissolution features. Caves evolve less rapidly than in gypsum formations, but have a much higher preservation potential. TDS in carbonate karst waters is generally within the range 40–1000 mg L^{-1} (Han and Liu 2004).

Finally, the **low-solubility** quartzose rocks (quartz sandstone and quartzites) are highly resistant formations that tend to form prominent outcrops. Despite their very low solubility, they locally display well-developed karst features (Wray and Sauro 2017). In these rocks, slow and persistent dissolution plays a preparatory role by loosening the constituent particles (arenization), which are subsequently removed by mechanical erosion (Martini 1979). The content of silica (SiO_2) in meteoric waters sampled in quartzose rock terrains is typically below 10 mg L^{-1} and TDS is commonly <10 mg L^{-1} (Mecchia et al. 2014; Sauro et al. 2020).

3.2 Basic Concepts and Parameters

3.2.1 Water, an Exceptional Dipolar Molecule, and Solvent

Water behaves differently from most other chemical compounds because it has a polar character. No other substance on Earth is commonly found as solid, liquid, or gas. Water's anomalous properties play a critical role in the Earth's climate and contributed in creating suitable conditions for life. Because the water dipoles tend to hold together by electrostatic forces, water has extremely high heat capacity (i.e., amount of heat to be supplied to a mass of material to produce a unit change in its temperature), contributing to the climate inertia (i.e., resistance or slowness to changes). Water has anomalously high freezing and boiling points and absorbs a large amount of heat before it boils. Water vapor in the atmosphere plays a crucial role in the heat budget of the Earth, since it is responsible for about two-thirds of the natural greenhouse effect.

The main implication of the polar character of water in the field of karst is its ability to dissolve ionic compounds (Bland and Rolls 1998; Hiscock 2005). The water molecule (H_2O) is made up of two hydrogen atoms and one oxygen atom. Different hydrogen and oxygen isotopes with a variable number of neutrons may combine in the water molecules (^{1}H: protium, ^{2}H: deuterium, ^{3}H: tritium, ^{16}O, ^{17}O, ^{18}O). Eighteen combinations are possible, but the most common is $^{1}H_2^{16}O$. The atoms in the molecule have an angular and asymmetric arrangement, with a bond angle of 104.5° (Figure 3.1). The oxygen atom has eight electrons, six of which are located in an external orbit, and the hydrogen

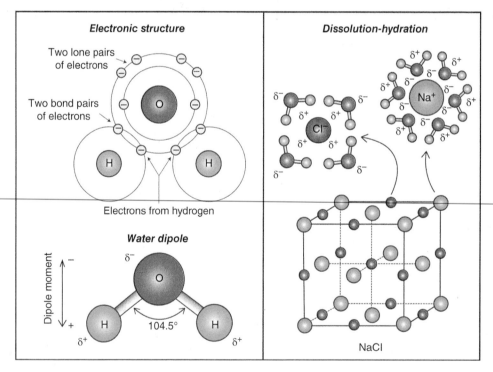

Figure 3.1 Electronic structure of the water molecule, in which the oxygen atom, with higher electronegativity than the hydrogen atoms, attracts the shared pairs of electrons (covalent bond) generating polar bonds and a polar asymmetric molecule with dipolar moment. Dissolution in water of an ionic compound such as halite releases ions that are hydrated by the polar water molecules, with the edges of the opposite charge facing toward the cation or anion.

atoms have one electron each. There are two bond pairs of electrons, each consisting of one electron provided by the oxygen and another one by a hydrogen atom. The four non-bonding outer electrons of the oxygen are arranged in pairs, known as lone pairs. Oxygen and hydrogen have different electronegativity, which indicates the power to attract the shared pair of electrons. The electronegativity of oxygen is 3.44 (Pauling scale), which is the second highest of all the elements. Hydrogen has an electronegativity of 2.20, which is around the middle of the scale. The oxygen pulls the bonding electrons of the covalent bonds toward itself and away from the corresponding hydrogen atom. Therefore, the O—H bonds in the water and the water molecule itself are polar. There is a small negative charge on the oxygen (δ^-) and a corresponding positive charge on the hydrogen atoms (δ^+). In three dimensions, the covalent pairs and lone pairs of electrons around the oxygen are oriented toward the vertex of a tetrahedron. The unequal sharing of the bonding electron pairs and the asymmetric structure of the water molecule result in a dipole moment. Other molecules have polar bonds (e.g., CO_2), but do not result in polar molecules (null dipolar moment) because they have a linear and symmetric arrangement (i.e., polar bonds are balanced).

When a soluble mineral such as halite (NaCl) dissolves, the Na^+ cations in the solution attract the negative edge (oxygen) of the polar water molecules (Figure 3.1). This results in the hydration of the cation, which is surrounded by water molecules with the oxygen poles pointing inwards. Similarly, the Cl^- anion is also hydrated, but in this case the positive hydrogen edges of the polar water molecules are attracted. Hydration energy or enthalpy of hydration is defined as the amount of energy released when 1 mol of ions undergo hydration. The strength of the interaction between hydrated ions and the water molecules significantly influences the ease with which ions from a mineral can be transferred to a solution by water molecules (i.e., dissolution). The values of hydration energy depend on the size and charge of the ions. The smaller the size of the ions, the more strongly they attract water molecules, due to the higher charge density. For example, the Na^+ ion ($-406\,kJ\,mol^{-1}$) has higher hydration energy than the larger K^+ ion ($-322\,kJ\,mol^{-1}$). The doubly charged cation Ca^{2+} ($-1577\,kJ\,mol^{-1}$) has a much higher hydration energy than the singly charged Na^+ ($-406\,kJ\,mol^{-1}$). Finally, cations are generally smaller than anions and consequently are more strongly hydrated.

The opposite charges of the water molecules attract each other through a type of interaction known as hydrogen bonding. The negatively charged oxygen atom of a water molecule attracts the hydrogen atom of a neighboring water molecule. This type of cohesive forces restricts the relative movement between the molecules and causes the anomalous boiling and freezing points of water, as well as the peculiar density maximum of liquid water. In most liquid substances, the atoms and molecules move closer together as the temperature decreases, ultimately solidifying. However, pure water has its highest density in its liquid phase at 3.98 °C ($1.0\,g\,cm^{-3}$) because the molecules are packed closest together at this temperature. Surprisingly, to reach the solid ice phase, with a density of $0.92\,g\,cm^{-3}$, the water molecules move further apart. This is an important temperature and phase change for karst development, since liquid water may dissolve rocks, but ice obviously cannot. Moreover, freezing of liquid water leaves the dissolved ions in the remaining solution, resulting in an increase in their concentration, eventually leading to the precipitation of cryogenic salts.

A small proportion of the H_2O molecules in pure water dissociates into hydrogen (H^+) and hydroxide ions (OH^-), through a reaction known as self-ionization:

$$H_2O(l) \leftrightarrow H^+(aq) + OH^-(aq) \tag{3.2}$$

where (l) and (aq) indicate the state of the species; liquid and aqueous solution. One of the O—H covalent bonds of the water molecule breaks and the bond pair of electrons remains with the oxygen of the OH^- anion. The hydrogen has no electrons and has a positive charge (H^+), while the hydroxide group has one extra electron and has a negative charge (OH^-). Most of the hydrogen ions combine with a water molecule to form the H_3O^+ ion, called either hydronium or oxonium ion. However, for simplicity, it is generally assumed that they are all H^+. The self-ionization of water only occurs to a very limited extent and the equilibrium of Reaction 3.2 lies well to the left. At room temperature (25 °C or 298 K), in pure water the concentrations of H^+ and OH^- are equal (only 10^{-7} mol L^{-1}, corresponding to a neutral pH of 7). Since in 1 mol of H_2O, hydrogen ion, and hydroxide ion are, respectively, 18, 1, and 17 g (atomic and molecular weights), the small concentration of H^+ and OH^- means that 1×10^7 (10 million) liters of liquid water contain 1 g of dissociated H^+ (or 19 g of H_3O^+ hydronium ions) and 17 g of dissociated OH^- (see definition of mole and molar mass in Section 3.2.2). A cubic tank with an edge of 21.54 m is necessary to hold that amount of water.

3.2.2 Concentration Units and Related Parameters

The concentration of a substance is the quantity of solute, expressed in various units, present in a given quantity of solution or solvent (Table 3.2). Geoscientists dealing with weathering processes and karst waters generally use mass concentrations of dissolved ions, commonly stated as milligrams per liter ($\mathbf{mg\,L^{-1}}$), which indicates the milligrams of solute per liter of solution. Another frequent mass concentration unit is parts per million (**ppm**), which indicates parts by weight of solute in 1 million parts by weight of solution, which is equivalent to milligrams of dissolved material per kilogram of water ($\mathbf{mg\,kg^{-1}}$). These units indicate concentrations in mass by volume (m/V) and mass by mass (m/m), and are appropriate to report the concentrations of chemical species in dilute solutions. In solutions with a low content of dissolved solids, typical in carbonate karst systems, the density of the solution is approximately the same as the density of the solvent, or, in other words, 1 kg of water closely approximates 1 L of water. Under these circumstances, ppm can be considered as equivalent to mg L^{-1} (i.e. ppm = mg kg$^{-1} \approx$ mg L^{-1}). However, in concentrated solutions, common in evaporite karst settings, and evaporitic sedimentary environments, the two scales become increasingly different as the proportion of dissolved solids increases. See the solubilities of gypsum and halite in Table 3.1 expressed as g kg^{-1} and g L^{-1}. For instance, evaporated marine water that has reached the gypsum-precipitation stage has a density of $\geq$1.1 g cm^{-3} (1 L weighs $\geq$1.1 kg) (Table 2.2).

Table 3.2 Common concentration units and some conversion relationships.

Name (symbol)	Unit	Definition	Conversion
Milligrams per liter	mg L^{-1}	milligrams of solute per liter of solution	ppm $\approx$ mg L^{-1a}
Parts per million	ppm	milligrams of solute per kilogram of solvent	mg L$^{-1} \approx$ ppm^a
Molarity (M)	mol L^{-1}	moles of solute per liter of solution (moles: g/M mass)	mg L^{-1} = mmol L$^{-1} \times$ molar mass
Molality (m)	mol kg^{-1}	number of moles of solute per kilogram of solvent	$M \approx m^a$
Normality (N)	eq L^{-1}	(or equivalent concentration) equivalents of solute per liter of solution	eq = moles $\times$ valence meq L^{-1} = mg L$^{-1} \times$ (valence/molar mass)
Volume per liter	cm^3 L^{-1}	volume of dissolved mineral per liter	cm^3 L^{-1} = g L$^{-1} \div$ density

^{a}Equivalence valid for diluted solutions in which the density of the solvent closely approximates that of the solution ($\approx$1 kg dm^{-3}).

Mass is not always the most convenient unit to express concentration. Seawater contains more Mg^{2+} ions than SO_4^{2-}, but sulfate reaches a greater mass. For the interpretation of chemical processes, it is easier to consider the number of ions or molecules per liter. In the SI system, the standard unit of molar concentration is moles m^{-3}, although more practical units are used, such as **moles L^{-1} (molarity, M)**, or mmoles L^{-1}. By convention, a **mole** is the mass of substance containing the same number of fundamental units (e.g., ions, molecules) as there are atoms in exactly 12 g of ^{12}C isotope. Such number of atoms is 6.022×10^{23}, known as the Avogadro's Number (N_A). A mole of substance contains N_A fundamental units.

An alternative more intuitive definition of mole is the atomic, molecular, or formula weight of a substance expressed in grams (molar mass). Thus, the mole allows us to determine the weight of a substance according to the number of atoms or molecules (fundamental units), and vice versa. Moles provide a bridge between the atoms involved in the reactions and the macroscopic amounts of material measured in the laboratory. The mass in grams of 1 mol of substance is given by the molar mass of that substance, which can be obtained from any chemistry book. For instance, a solution with a concentration of $1 \, mol \, L^{-1}$ of Na^+ (or 1 M), contains 22.9 g of dissolved sodium per liter (atomic mass of Na = 22.9). If pure halite (NaCl) is dissolved in distilled water, the solution will have the same molar concentrations of Na^+ and Cl^-, or the same number of dissolved Na^+ cations and Cl^- anions. However, their masses will not be the same, given that their atomic weights are different. Note than molarity, which is molar concentration by volume, changes slightly with temperature variations, as does the volume and density of solutions. To convert from moles L^{-1} to mg L^{-1} (from molar to mass concentration):

$$\frac{mg}{L} = \frac{mol}{L} \times 1000 \times molar\,mass \tag{3.3}$$

Palmer (2007) presents a didactic exercise to better understand these concentration units and their relationship. Suppose that the analysis of a sample of water that has only interacted with calcite ($CaCO_3$) gives a Ca^{2+} concentration of $85 \, mg \, L^{-1}$. Given that the molar mass of Ca = 40.08 g mol^{-1} = 40080 mg mol^{-1}, the molar concentration of $Ca^{2+} = 85 \, mg \, L^{-1} \div 40080 \, mg \, mol^{-1} = 0.0021 \, mol \, L^{-1}$. Conveniently, this is also the number of moles of dissolved calcite, because one molecule of calcite needs to be dissolved to release one Ca^{2+} ion. The molar mass of CO_3 is 60.01 g mol^{-1} (12.01 for carbon plus 16.00 for each of the three oxygens). The total molar mass of calcite is 100.09 g mol^{-1} (adding the molar mass of Ca), and Ca represents 40% of it. Therefore, the total concentration of dissolved calcite in the solution is $85 \, mg \, L^{-1} \div 0.40 = 213 \, mg \, L^{-1}$ of $CaCO_3$.

Molality, denoted with the symbol m, is the number of moles of solute dissolved in 1 kg of solvent (mol kg^{-1}). Compared with molarity, molality uses mass of solvent, rather than volume of solution. Unlike molarity, molality is independent of temperature, because the mass of solvent does not change with temperature. In diluted solutions, in which most of the mass corresponds to the solvent, molarity and molality closely approximate ($M \approx m$). However, molality becomes increasingly larger than molarity as the amount of dissolved solids in the solution increases.

The concept of equivalent concentration takes into account ionic charge and is useful when investigating the proportions in which substances react (stoichiometry). The **equivalent concentration** or **normality** (N), is the molar concentration of a solute multiplied by the valence of the solute (ionic charge). Common units are eq. L^{-1} and meq L^{-1}. To convert meq L^{-1} into mg L^{-1} (from equivalent to mass concentration):

$$\frac{mg}{L} = \frac{meq}{L} \times \left(\frac{molar\,mass}{valence} \right) \quad or \quad N\left(\frac{eq}{L} \right) = \frac{g}{L} \times \left(\frac{valence}{molar\,mass} \right) \tag{3.4}$$

the ratio between the molar mass (or molar weight) and the valence is known as the equivalent weight. Molar and equivalent weights of the single charged Na^+ ion are equal. The equivalent weight of the double charged Ca^{2+} ion is half its molar mass. Suppose a stoichiometric cation exchange reaction between Ca^{2+} and Na^+. Each divalent Ca^{2+} will be replaced by two monovalent Na^+, involving a number of Na^+ equivalents twice the number of Ca^{2+} equivalents.

A frequent problem addressed in karst studies is the assessment of the **volume of mineral** (or rock) removed by dissolution over a period of time. This can be used to measure the rate at which porosity and/or void space is created by subsurface dissolution, which has important scientific and practical applications (e.g., aquifer evolution, cave enlargement, and creation of solutional cavities at a dam site). For these calculations, cubic centimeters per liter ($cm^3\ L^{-1}$) is the most appropriate unit. To convert from $g\ L^{-1}$ to $cm^3\ L^{-1}$, divide by the density of the mineral. Suppose a dam built on evaporitic bedrock. The impoundment of the reservoir has been accompanied by leakage and the occurrence of a corresponding spring with a flow rate of $864\,000\ L\,day^{-1}$ ($10\,L\,s^{-1}$). The dissolved NaCl in the spring water related to dissolution of halite by leakage water is $50\,g\,L^{-1}$, which is equivalent to $23.15\,cm^3\ L^{-1}$ of halite (halite density $= 2.16\,g\,cm^{-3}$). This means that halite dissolution is creating an alarming volume of void space at the dam site of around $20\ m^3$ per day ($23.15\,cm^3\ L^{-1} \times 864\,000\,L\,day^{-1} \times 10^{-6}\ m^3\ cm^{-3}$), and approximately $7300\ m^3$ per year, which is equivalent to a cube with an edge of about $19.4\,m$.

The **ionic strength**, I, of a solution is a function of the concentration of all ions present in that solution, generally expressed in $mol\ kg^{-1}$ or $mol\ L^{-1}$ (molar ionic strength), and their charge:

$$I = \frac{1}{2}\left(\sum_{i=1}^{n} c_i z_i^2\right) \tag{3.5}$$

where 1/2 is because both cations and ions are introduced, c_i is the concentration of ion i, and z_i is the charge number of that ion. Ionic strength emphasizes the increased contribution of species with charges greater than one to solution nonideality (effective concentration reduced due to interactions among ions). For example, the molar ionic strength of a solution containing $0.01\ mol\ L^{-1}$ of dissolved $CaSO_4$ ($Ca^{2+} + SO_4^{2-}$) and $0.02\ mol\ L^{-1}$ of dissolved $MgCl_2$ ($Mg^{2+} + 2Cl^-$) is:

$$I = \frac{1}{2}\left[\left(0.01 \times 2^2\right) + \left(0.01 \times 2^2\right) + \left(0.02 \times 2^2\right) + \left(2 \times 0.02 \times 1^2\right)\right] = 0.1\left(molar\right) \tag{3.6}$$

In most karst waters, there are only seven or eight ionic constituents in significant concentration:

$$I = \frac{1}{2}\left\{\left[Na^+\right] + \left[K^+\right] + 4\left[Ca^{2+}\right] + 4\left[Mg^{2+}\right] + \left[HCO_3^-\right] + \left[Cl^-\right] + 4\left[SO_4^{2-}\right] + \left[NO_3^-\right]\right\} \tag{3.7}$$

The ionic strength of the water may have an important impact on karstification processes. As the ionic strength of a solution rises, the "interference" between ions increases due to electrostatic interactions. This lowers the effective concentration (or activity) of the ions, so that the solubility of ionic compounds in that solution will tend to increase. Other factors being equal, the ability of such a solution to dissolve a rock will be enhanced. As explained in Section 3.2.4, the activity coefficient of an ion in a solution, which relates the concentration of the ion and its effective concentration (or activity), is a function of the ionic strength of the solution. Fresh waters typically have ionic strength between 10^{-3} and $10^{-4}\ mol\ L^{-1}$. The ionic strength of seawater is $0.7\,mol\ L^{-1}$. Spring waters in salt karsts reach ionic strengths as high as $8\text{--}10\,mol\ L^{-1}$.

TDS concentration is the measure of the mass of solid material dissolved in a given volume of water, and is commonly measured in $mg\ L^{-1}$. The dissolved solids may include small amounts of organic

Table 3.3 Common chemical categories of water according to the TDS concentration. Drinking water generally has a TDS below 500 mg L^{-1}. Higher TDS fresh water is drinkable but taste may be objectionable.

Category	TDS (mg L^{-1})
Soft water	<60
Hard water	>120
Fresh water	<1000
Brackish water	1000–1010 000
Saline water (seawater)	10 000–35 000 (35 000)
Hypersaline water	>35 000

matter, but generally TDS concentration is essentially the sum of the inorganic ions dissolved in water (principally Ca^{2+}, Mg^{2+}, K^+, Na^+, HCO_3^-, Cl^-, SO_4^{2-}). It provides a measure of the degree of salinization of the water and is widely used for categorizing waters (Table 3.3). The measurement of the **electrical conductivity** (EC) of a solution also provides a relative indication of the amount of dissolved salts, given that water is an electrolytic solution. EC is commonly expressed in microsiemens cm^{-1} (μS cm^{-1}) and is referred to a temperature value, commonly 25 °C. TDS concentration can be related to the EC of the water, but the relationship is not constant:

$$\text{TDS}\left(\frac{\text{mg}}{\text{L}}\right) = k_e \cdot \text{EC} \cdot \left(\frac{\mu\text{S}}{\text{cm}}\right) \tag{3.8}$$

where the correlation factor k_e is typically between 0.5 and 0.8, and is largely a function of the type and concentration of dissolved ions.

Water hardness is generally defined as the aggregate concentration of Ca^{2+} and Mg^{2+} in water expressed as equivalent $CaCO_3$. Water hardness can be calculated using the formula:

$$\text{TH} = 2.5 \cdot \left[Ca^{2+} \right] + 4.1 \cdot \left[Mg^{2+} \right] \tag{3.9}$$

where TH is total hardness in mg L^{-1} equivalent $CaCO_3$, $[Ca^{2+}]$ and $[Mg^{2+}]$ are concentrations in mg L^{-1}, and the preceding figures correspond to the ratio of the molar masses of $CaCO_3$ (100.09) and the corresponding cations (Ca^{2+} = 40.08; Mg^{2+} = 24.32). Hardness may be a useful measure of the amount of dissolved limestone and dolostone in karst waters without significant quantities of other ions (e.g., Na^+, Cl^-, and SO_4^{2-}). In these "clean" carbonate karst waters, the total hardness shows a good correlation with electrical conductivity. Krawczyk and Ford (2006) illustrated the relationship between electrical conductivity (their specific conductivity; SpC) and total hardness using the analysis of 2309 samples from karst waters worldwide (springs, wells, and streams) (Figure 3.2). The regression derived from the 1949 samples with SpC < 600 μS cm^{-1} yields an acceptable correlation coefficient (R^2 = 0.93). However, the correlation decreases drastically for the 360 samples with SpC > 600 μS cm^{-1}, which contain a significant proportion of ions (sulfate, nitrate, and chloride) related to the dissolution of other lithologies. The authors propose that for predominantly bicarbonate waters with SpC < 600 μS cm^{-1} ("clean" carbonate karst waters), simple field measurements of electrical conductivity may be used to roughly estimate total hardness as a proxy for the dissolved load of carbonate rocks. The obtained general regression is shown in Figure 3.2. The validity of the SpC versus TH correlation in a given geographical area should be checked with

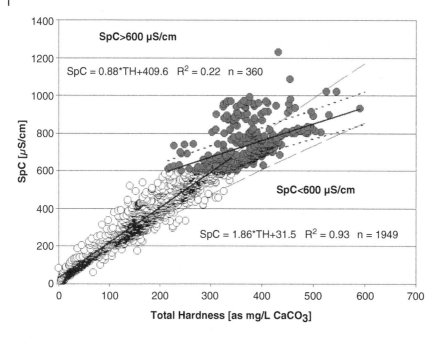

Figure 3.2 Correlation between electrical conductivity (SpC) and total hardness (TH expressed as mg L^{-1} CaCO$_3$) generated with 2309 published analyses of limestone and dolomite karst worldwide. Open circles are waters with SpC < 600 μS cm^{-1} and filled circles waters with SpC > 600 μS cm^{-1}. PHREEQC curves for calcite or aragonite dissolution are marked by dashed lines for ion activities (lower) and ion concentrations (upper). The one standard error bands of best fits above SpC = 600 μS cm^{-1} are depicted with short dashes. *Source:* From Krawczyk and Ford (2006). Published with permission of Wiley.

comprehensive analyses before the field worker relies on SpC alone. This can be a useful approach for karst denudation research.

3.2.3 Equilibrium Constant and Gibbs Free Energy of Reaction

In chemical reactions, reactants or reagents interact to give products. A model reaction can be expressed as:

$$A + B \leftrightarrow C + D \tag{3.10}$$

The double arrow indicates a bidirectional reaction, whereby A + B can combine to give C + D and vice versa. A common example, which is significant for carbonate dissolution, is the hydrolysis of the carbonate ion:

$$CO_3^{2-}(aq) + H_2O(l) \leftrightarrow HCO_3^-(aq) + OH^-(aq) \tag{3.11}$$

Note that in hydrolysis, an O—H bond of H$_2$O is broken, whereas in hydration, ions are associated with polar water molecules (see Section 3.2.1). At the beginning of the reaction there is no C and D present, and the only reaction is of A and B to give C and D. However, as soon as C and D are formed, they can start to react and re-form A and B. There are simultaneous forward and reverse reactions. Eventually, a dynamic equilibrium is reached, whereby forward and reverse reactions proceed at the same rate and the concentrations of A, B, C, and D remain unaltered. At this point, there is balance between the concentrations of products and reactants (denoted with square brackets), defined by the equilibrium constant of the reaction (K) for specific temperature and pressure conditions:

$$K = \frac{[C][D]}{[A][B]} \tag{3.12}$$

Alterations in the system that involve changes in the concentration of one of the chemical species would disturb the equilibrium. If more A is added (e.g., CO_3^{2-} in Reaction 3.11), the rate of the forward reaction increases and more C and D will be formed. The reaction is displaced to the right, but K remains constant. The disturbance of the equilibrium and its reestablishment is an example of Le Chatelier's Principle: "when a stress is applied to a system in dynamic equilibrium, the equilibrium adjusts to minimize the effect of the stress."

The relationship of concentrations of reactants and products in equilibrium can be expressed according to the Law of Mass Action for the reaction:

$$a\text{A} + b\text{B} \leftrightarrow c\text{C} + d\text{D} \tag{3.13}$$

as

$$K = \frac{[C]^c[D]^d}{[A]^a[B]^b} \tag{3.14}$$

where A, B, C, and D are the concentrations of dissolved elements or the partial pressure of gases. For example, oxygen in air has a partial pressure of 0.21 atm, because oxygen has a proportion of 21% by volume in the atmosphere. If the material is in solution, the concentration is usually expressed as molarity (moles of solute per liter of solution) or molality (moles of solute per kg of solvent).

Chemical systems tend to equilibrium, that is to a state of minimum energy. The energy involved in chemical processes is measured using the Gibbs free energy (G), which depends on the contribution of the heat content (enthalpy, H), the degree of randomness (entropy, S), and temperature T in kelvin:

$$G = H - TS \tag{3.15}$$

The change in Gibbs free energy for a chemical reaction is given by:

$$\Delta G = \Delta H - T\Delta S \tag{3.16}$$

where Δ denotes "change in." Reactions tend to be spontaneous when there is a decrease in the G value or, in other words, ΔG is negative. These situations are favored by negative values of ΔH (exothermic reactions) and positive values of ΔS (randomness increases). The variation of the Gibbs free energy for a reaction (ΔG_r°) can be calculated subtracting the values of ΔG_f° for the formation of products and reactants:

$$\Delta G_r^\circ = \Delta G_f^\circ(\text{products}) - \Delta G_f^\circ(\text{reactants}) \tag{3.17}$$

where ΔG_f° for a compound is the energy change for its formation from the elements under standard conditions (25 °C, 1 atm). Note that ΔG_f° for elements is zero. Reactions with negative ΔG_f° values are energetically favored and tend to be spontaneous. However, some changes or reactions that result in an overall decrease in energy do not always occur spontaneously, because reactants need to overcome an activation energy to give products. A significant example in karst is the metastable mineral aragonite, which can persist in the stratigraphic record for long periods because its conversion to the more stable (lower ΔG_f°) polymorph calcite requires an activation energy.

The values of the variation of the Gibbs free energy of a reaction (ΔG_r°) and the equilibrium constant of the reaction (K) are related by the equation:

$$\Delta G_r^\circ = -2.303RT \log K \tag{3.18}$$

where R is a constant with the value of $8.314\,\mathrm{J\,K^{-1}\,mol^{-1}}$, and T is the temperature in kelvin. The energy variation is related to the change in the concentration of reactants and products during the reaction to reach equilibrium. This formula allows the calculation of the K constant from the Gibbs free energy of the reaction (standard values if calculated for 25 °C, or 298.15 K).

$$\log K = -\frac{\Delta G_r^\circ}{2.303RT} \tag{3.19}$$

3.2.4 Activity

Ions in aqueous solutions are affected by the presence of other ions and the associated electrostatic interactions. Cations tend to be surrounded by anions and vice versa. Therefore, the proportion of potentially reactive ions of a given species is always less than the molar sum of ions of that species in the solution. In other words, the effective concentration is lower than the actual concentration. To account for this nonideal behavior, we use the concept of activity instead of concentration. Activity is generally symbolized by "a". Standard brackets ( ) are commonly used for activity, whereas square brackets [ ] signify concentration. The activity (a_i) is calculated multiplying the concentration [c_i] by an activity coefficient (γ_i):

$$\gamma_i = \frac{(a_i)}{[c_i]} \tag{3.20}$$

In diluted solutions, the activity closely approximates the concentration ($\gamma_i \approx 1$). Figure 3.3 illustrates the relationship between the activity coefficient of common ions in karst systems and ionic strength, which depends on the concentration of ions and their charge. The activity of the ions

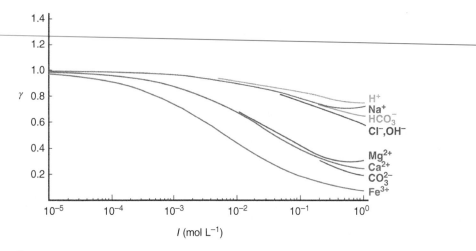

Figure 3.3 Variation of the activity coefficient of common ions in karst waters as a function of the ionic strength of the solution. *Source:* Adapted from Hiscock (2005).

rapidly decreases as the salinity of the solution increases. However, in solutions with high ionic strength the activity coefficient of some ions may reach values higher than 1. Determination of activity is essential for the computation of equilibrium conditions for soluble minerals in solutions and to predict their behavior (e.g., saturation index). A number of equations are used for estimating activity coefficients, depending on the ionic strength of the solution. For instance, the extended Debye–Hückel equation provides estimates of γ_i to a maximum ionic strength of about 0.1 M or a TDS content of approximately 5000 mg L^{-1}:

$$\log \gamma_i = \frac{-A z_i^2 \left(I\right)^{\frac{1}{2}}}{1 + B \, r_i \left(I\right)^{\frac{1}{2}}} \tag{3.21}$$

where A and B are temperature-dependent constants, z is the valence of the ion, I ionic strength, and r_i is the radius of the hydrated ion in centimeters. Other activity models are used to deal with higher salinity waters.

Water has an activity of 1 in diluted solutions, but as the salinity of the solution increases, its value decreases due to electrostatic interactions with the dissolved ions. For instance, water has an activity of 0.75 in a solution saturated with respect to halite. Consequently, this parameter must be taken into account when dealing with saline waters or brines (e.g., salt karst). The activity of water can be calculated using different approaches. An option is to use the osmotic coefficient (φ) of a solution of sodium chloride, assuming that the activity of the water in a complex solution is equal to that of a sodium chloride solution with the same ionic strength. The complex formulation can be found in Lietzke and Stoughton (1961), and Wolery (1992). Simpler methods have been proposed by Garrels and Christ (1965), Truesdell and Jones (1974), and Fritz (1981).

3.2.5 Saturation Index

According to the Law of Mass Action and using activities as a measure of effective concentration, the equilibrium constant for the dissolution reaction of a mineral such as halite can be expressed as:

$$NaCl_{(s)} \leftrightarrow Na^+_{(aq)} + Cl^-_{(aq)} \tag{3.22}$$

$$K_{eq} = \frac{aNa^+ \; aCl^-}{aNaCl} \tag{3.23}$$

Since the activity of solid halite is 1, the equilibrium constant can be expressed as $K_{eq} = aNa^+ \cdot aCl^-$. The product of the activity of the ions of this equation in a specific solution is designated as the ion activity product (IAP = $aNa^+ \cdot aCl^-$). When the system is at equilibrium, IAP = K_{eq}, the ion activity product is called the solubility product (K_{sp}). These concepts allow us to assess the deviation of the solution from equilibrium with respect to one or more minerals:

1) IAP = K_{sp}: There is dynamic equilibrium. The solution is **saturated** with the mineral.
2) IAP < K_{sp}: The solution is **undersaturated** or **aggressive** toward the mineral, and tends to dissolve it. Net precipitation is not feasible.
3) IAP > K_{sp}: The solution is **supersaturated** with the mineral, and the mineral is likely to precipitate. Net dissolution is not possible.

There are different parameters for calculating the disequilibrium degree or saturation degree of a solution with respect to a mineral:

$$\Omega = \frac{IAP}{K_{sp}} \tag{3.24}$$

$$SI = \log\left(\frac{IAP}{K_{sp}}\right) \tag{3.25}$$

In the case of the saturation ratio (Ω), values equal to 1 indicate equilibrium. The more intuitive and more commonly used saturation index (SI), is 0 at equilibrium, positive values indicate supersaturation and negative values undersaturation.

3.2.6 pH and the Acidity of Karst Waters

In chemistry, pH is a numerical scale used to assess how acidic or basic a water-based solution is. It indicates the activity of the hydrogen ion (H^+), or its effective concentration, which is slightly less than the true concentration. It is defined as the decimal logarithm of the reciprocal of the hydrogen activity in a solution, or the negative decimal logarithm of the hydrogen activity:

$$pH = \log\left(\frac{1}{aH^+}\right) = \log 1 - \log\left(aH^+\right) = 0 - \log\left(aH^+\right) = -\log\left(aH^+\right) \tag{3.26}$$

The p stands for "decimal cologarithm of," which is the negative logarithm of a number (colog $x = -\log x$). It allows the use of more easily manageable numbers, rather than the inconveniently small figures associated with dilute solutions. Because the pH is logarithmic and inversely indicates the activity of H^+ ions: (i) a lower pH indicates higher activity of the H^+ ion in the solution; and (ii) a difference of 1 pH unit is equivalent to a tenfold difference in the H^+ ion activity. A solution is said to have a neutral pH when it is neither acidic nor basic, but this neutral value depends on the temperature. At 25 °C the neutral pH value is 7, so that solutions with a pH less than 7 are acidic and those with a pH greater than 7 are basic. Pure water at 25 °C (298 K) has a neutral pH value of 7, with equal activities for the hydrogen and hydroxide ions; $(H^+) = (OH^-) = 10^{-7}$. The neutral pH value is lower than 7 if the temperature increases and vice versa. The pH scale is traceable to a set of standard solutions whose pH are established by international agreement. The pH value can be less than 0 for very strong acids, or greater than 14 for very strong bases.

Any substance that causes an increase in the activity of H^+ in the solution will cause the pH to fall and is called an acid. An acid is a proton donor. Similarly, a substance that causes a decrease in the activity of H^+ is called a base. A base is a proton acceptor. For instance, rain water is mildly acidic because it absorbs carbon dioxide (CO_2) from the atmosphere, which reacts with the water molecules and is slowly converted into bicarbonate and hydrogen ions, essentially creating strongly dissociated carbonic acid:

$$CO_2 + H_2O \leftrightarrow HCO_3^- + H^+ \tag{3.27}$$

Typical rain water has a pH of about 5.6. This means that the activity of the hydrogen ion in the solution is $10^{-5.6}$. When water infiltrates and percolates through the soil, it absorbs additional CO_2 decreasing the pH down to about 4.5–4.8. Such acidic water dissolves carbonate minerals vigorously, which function as bases that contribute to decrease the activity of H^+ in the solution and tend to neutralize the pH of the water:

$$CaCO_3 + H^+ \leftrightarrow Ca^{2+} + HCO_3^- \tag{3.28}$$

$$CaMg(CO_3)_2 + 2H^+ \leftrightarrow Ca^{2+} + Mg^{2+} + 2HCO_3^- \tag{3.29}$$

One might expect the dissolution process to stop when the pH reaches the neutral value of 7. However, the attraction between H^+ and CO_3^{2-} is so strong, that H^+ continues to be used until the pH rises well above 7. Oceanic water has an average pH of 8.0–8.3, mainly due to dissolved calcium carbonate (Bland and Rolls 1998). Waters in carbonate settings, even if they are not in equilibrium with the dissolved carbonate minerals, typically have pH values between 7 and 8 (Palmer 2007). The water in contact with peat, which is acidic, has a pH of 3–4. This explains the strong karstification of carbonate bedrocks overlain by peat soils. For instance, in some regions of Scotland (e.g., Skye Island), the Cambro-Ordovician Durness Limestone shows a soft and strongly corroded alteration zone beneath peat soils.

Carbonic acid derived from the uptake of soil CO_2 by percolating waters is designated in karst terminology as an epigenic acid. This is a widespread source of acidity located at or near the land surface that is constantly replenished by biological activity in the soils (Palmer 2007). This epigenic acid is responsible for the development of the vast majority of the surface karst landforms and a large proportion of the explored caves in carbonate terrains (epigenic karst). The acidity of karst waters may also be related to hypogene acids that originate deep beneath the surface by various processes (Palmer 1991, 2007; Dublyansky 2000a, 2000b; Auler 2013), including gases such as CO_2 and H_2S related to volcanic activity, CO_2 released when carbonate rocks are metamorphosed, CO_2 and organic acids produced by the transformation of organic matter into oil and natural gas, and H_2S originated by redox reactions that involve the reduction of sulfur compounds (e.g., dissolved sulfate) and the oxidation of organic carbon compounds (e.g., organic matter). Hydrogen sulfide that reacts with oxygen produces sulfuric acid, substantially increasing the aggressiveness of the water. Sulfuric acid can also be produced by the oxidation of sulfides (e.g., pyrite), which might be considered as a hypogene source of acidity, since sulfides typically form in oxygen-poor zones beneath the surface (Palmer 2007). Dissolution by hypogene fluids has a limited impact on the development of surface karst landforms, but can produce significant caves, which tend to be more localized than those developed in epigene settings. Dissolution by these fluids can occur in arid regions, since they do not depend directly on rainfall and surface infiltration waters.

3.3 The Dissolution of Carbonate Rocks in Normal Meteoric Waters

Limestones and dolostones display well-developed karst features in most climatic environments, despite the very low solubilities of calcite and dolomite in rainwater with values of around $0.05–0.06 \, g \, L^{-1}$ (Table 3.1). This is largely because the aggressiveness of surface water increases substantially when it percolates through the soil, uptakes CO_2, and becomes more acidic. This section is focused on the dissolution of carbonate rocks in shallow meteoric environments, where CO_2 from the atmosphere and especially from the soil is the main source of water acidity (Langmuir 1997; Appelo and Postma 2005; Ford and Williams 2007). Carbon dioxide concentrations are reported in volume percent, as concentration in mg L^{-1}, and more frequently as the partial pressure of CO_2 in the total gas phase. The partial pressure of a gas is the part of the total pressure exerted by a mixture of gases related to the gas of interest. The relationships between these units are:

$$P_{CO_2} = \left[\frac{V(CO_2)}{V(gas)} \right] P_{gas} \qquad (3.30)$$

$$P_{CO_2} = \frac{\left[CO_2 \left(mg\, L^{-1} \right) \times 22.4 \right]}{1000 \times 44.01} = 5.092 \times 10^{-4}\, CO_2 \left(mg\ L^{-1} \right) \tag{3.31}$$

where 22.4 is the molar volume (L) of CO_2 and 44.01 the molar mass (g) of CO_2.

3.3.1 Carbon Dioxide in the Atmosphere

Today, CO_2 constitutes about 0.040% by volume of the atmospheric air. Since the total air pressure at sea level is one atmosphere (1 atm), the CO_2 partial pressure (P_{CO_2}) is 0.00040 atm ($10^{-3.4}$ atm). The main constituents of the atmosphere are nitrogen (N_2: 78%) and oxygen (O_2: 21%). If the water in contact with the atmosphere absorbs sufficient CO_2 gas to reach equilibrium, the CO_2 concentration in the water would also be 0.00040 atm. The volume percentage of CO_2 in the atmosphere remains almost constant with increasing elevation, apart from local modifications near the ground where the main CO_2 sinks and sources are located. Consequently, its partial pressure or concentration in mass per volume decreases as the barometric pressure falls with increasing elevation. Atmospheric pressures at 3000 m and 6000 m in elevation are approximately 10% and 20% lower than that at sea level.

The concentration of CO_2 in the world's atmosphere has increased systematically from the beginning of the Industrial Revolution and especially after World War II, largely due to burning of fossil fuels (Figure 3.4). This rise is related to the perturbation of the global carbon cycle caused by anthropogenic activities, with an increasing positive balance between the CO_2 sources and sinks. The principal sources include emissions related to combustion of fossil fuels (carbon, natural gas, and oil) and a number of alterations collectively designated as land-use changes, including deforestation and forest fires. The main net sinks are the oceans that absorb CO_2 from the atmosphere, the soils with the associated vegetation and its photosynthetic activity, and weathering

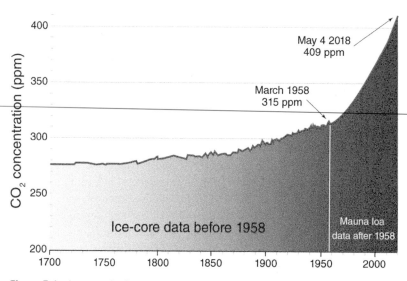

Figure 3.4 Increase in the atmospheric CO_2 concentration since 1700. Data from air trapped in ice cores (pre-1958) and from the Mauna Loa observatory (post-1958). The latter shows the general trend after removing seasonal oscillations. Note the great increase after World War II and the proposed starting time for the Anthropocene period (nuclear Trinity Test of 1945). *Source:* Adapted from Scripps Institution of Oceanography.

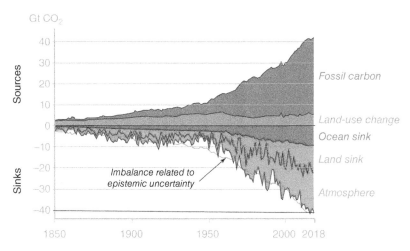

Figure 3.5 Evolution of the global CO_2 budget since 1850. The diagram shows CO_2 emissions derived from fossil fuels and land-use changes, as well as independent assessments for the different sinks (atmosphere, ocean, and land). Note the progressive increase in the atmosphere sink, especially since the mid-twentieth century, principally related to growing CO_2 emissions caused by the combustion of fossil fuels. The difference between the estimated sources and sinks is the budget imbalance, related to limitations in the data and understanding of the contemporary carbon cycle (epistemic uncertainty). *Source:* Adapted from Le Quéré et al. (2018).

processes, notably the dissolution of carbonates and silicates. Le Quéré et al. (2018) estimate a positive averaged global budget for the atmospheric CO_2 in the 2009–2018 decade of 18 Gt CO_2 yr^{-1} (Gt = billion tons). Their budget calculation has an imbalance of 2 Gt CO_2 yr^{-1} due to gaps in understanding of the global carbon system (Figure 3.5).

Some authors postulate that the rapid human-induced rise in the atmospheric CO_2, together with the accompanying global temperature increase, may enhance the karstification of carbonate rocks, and that this dissolution process might contribute to buffer the CO_2-related greenhouse effect. Jeannin et al. (2016), based on the statistical analysis of three independent karst groundwater datasets from the Swiss Jura covering over 20 years (1990–2011), identified several trends consistent with climate warming and enhanced dissolution of carbonate rocks: (i) increase in the temperature of spring water (ca. 0.5 °C in 25 years); (ii) decrease in pH (0.01 pH units); (iii) increase in bicarbonate concentration (~5%); and (iv) P_{CO_2} increase in the atmosphere of caves. The authors ascribe the inferred increase in carbonate dissolution to higher CO_2 content in the atmosphere and soils associated with global warming. They postulate that the intensification of carbonate dissolution may contribute: (i) to slow down the atmospheric rise in CO_2, acting as a carbon sink; and (ii) to attenuate the acidification of oceanic water by buffering the increase of CO_2 absorbed from the atmosphere. A number of authors have assessed the role played by karstic dissolution as carbon sink worldwide, which essentially consumes atmospheric and edaphic CO_2, ranging between 0.1 and 0.4 Gt C yr^{-1} (Liu and Zhao 2000; Gombert 2002; Amiotte Suchet et al. 2003; Binet et al. 2020). It is generally assumed that dissolution of 1 mol of $CaCO_3$ mineral consumes 1 mol of CO_2 gas. In a recent paper, Romero-Mujalli et al. (2019) state that although carbonate weathering and transfer of carbon toward the coastal zone is one of the relevant sinks for atmospheric CO_2, there is insufficient field data on a number of factors to reliably assess the impact of climate change on global calcite dissolution rates. These authors stress that a significant knowledge gap is the flux of weathering products between the soil-rock system and river systems.

It is important to note that over geological time scales the cycle of carbonate dissolution and precipitation does not involve a net loss of atmospheric CO_2. Carbonate dissolution by carbonic acid temporarily removes CO_2 and generates dissolved Ca^{2+} and carbonate species (CO_3^{2-}, HCO_3^-), that ultimately precipitate as calcite or aragonite, mainly in the oceanic "carbonate factories" and mediated by biological activity. The stoichiometry of the net reaction indicates that the amount of CO_2 consumed is equal to the amount of CO_2 released (Eq. 3.32). In contrast, carbonate dissolution induced by sulfuric acid even yields a net release of CO_2 to the atmosphere (Berner et al. 1983). Silicate weathering, which also consumes CO_2, involves a net loss of this gas from the atmosphere. The dissolved Ca^{2+} and Mg^{2+} released from silicate weathering may ultimately participate in the precipitation of carbonates, but in this case, as indicates the overall reaction (Eq. 3.33), there is net loss of atmospheric CO_2 (Berner et al. 1983). In those reactions, H_2CO_3 is carbonic acid generated by CO_2 gas dissolved in water. See more information on the different reactions in Sections 3.3.3 and 3.3.4.

$$CaCO_3(s) + H_2CO_3(aq) \rightarrow CaCO_3(s) + H_2O + CO_2 \tag{3.32}$$

$$CaSiO_3(s) + H_2CO_3(aq) \rightarrow CaCO_3(s) + SiO_2 + H_2O \tag{3.33}$$

3.3.2 Carbon Dioxide in Soils

P_{CO_2} is several orders of magnitude higher in the soil than in the atmosphere and may reach as much as 0.1 atm (Brook et al. 1983). This large extra CO_2 is derived from microbially mediated decomposition of organic matter and respiration in the plant root system (Figure 3.6). Around 90% of the CO_2 generated in the soil is released upward to the atmosphere. The rest is retained or transported downward, dissolved in percolating water that increases its acidity and aggressiveness with respect to carbonate minerals. The CO_2 concentration in the soil air may show large spatial and temporal variations depending on multiple factors, including temperature, moisture conditions, biological activity, and the physical characteristics of the soils (Miotke 1974). The production of CO_2 by biological activity increases with temperature and water availability, and reaches the highest values in warm-wet environments (Figures 3.6 and 3.7). In regions with seasonality, it peaks during the growing period in which both biomass production and degradation are higher. For instance, in soils of the subtropical karst of southern China, Yuan (2001) measured mean monthly CO_2 concentrations at 0.5 m depth ranging from around 500 ppm in the dry and cooler winter months, to 26 000–40 000 ppm during the hot and rainy monsoon season. Clay-rich and peat soils tend to have higher CO_2 contents because the low porosity inhibits the escape of gases. Moreover, when soils dry out, CO_2 spells out of the soil more rapidly, decreasing its partial pressure (Miotke 1974). Brook et al. (1983) found a reasonable correlation between soil CO_2 field data from arctic to tropical environments, and mean annual actual evapotranspiration. The latter variable is considered to account for the combined effect of temperature, precipitation, and water availability in soils:

$$\log P_{CO_2} = -3.47 + 2.09 \left(1 - e^{-0.00172 \, AET}\right) \tag{3.34}$$

where P_{CO_2} is the mean soil CO_2 partial pressure in the growing season and AET the calculated mean annual actual evapotranspiration. Based on this regression, Brook et al. (1983) produced a global map of soil CO_2 partial pressure. This map, combined with maps of groundwater recharge and soil pH, was used by Kessler and Harvey (2001) to estimate the global flux of dissolved

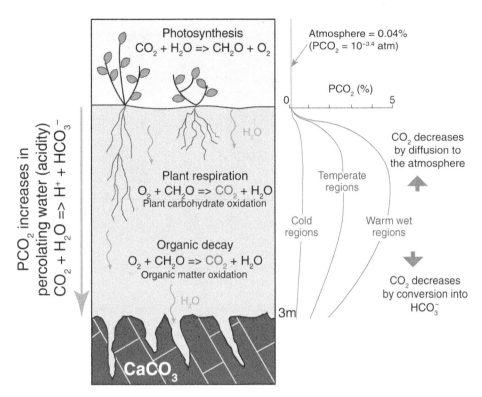

Figure 3.6 Concentration of CO_2 in the soil air varies with depth and depends on climatic conditions. High temperature and humidity enhance both biomass production and degradation. The concentration of CO_2 in soils is typically orders of magnitude higher than in the atmosphere due to organic matter decay and root respiration. Surface water that infiltrates and percolates through the soils increases its P_{CO_2} and acidity. *Source:* Adapted from Railsback (2009).

carbonate moving downward into groundwater. Recently, Romero-Mujalli et al. (2019) published a more detailed global map of average P_{CO_2} in soils using a more complex equation that incorporates mean annual surface temperature and the water content of soils calculated from the degree of saturation and soil porosity, applying remote sensing methods (Figure 3.7).

It is important to note that soils with high CO_2 production may have low contents of organic matter and vice versa. This largely depends on the balance between organic matter accumulation and oxidative degradation with the associated release of CO_2, which is largely influenced by climatic conditions. Figure 3.8 shows the soil organic carbon map of the world, with estimates of the organic C content in soils at 1 m depth. In warm-humid tropical and subtropical environments both biomass production and degradation are high, catalyzed by favorable temperature and moisture conditions, resulting in soils with high CO_2 contents but low content in organic matter (Eswaran et al. 1993). In cold regions (e.g., boreal forests, taiga) organic matter degrades slowly and tends to build up. A special case are the peat soils, which, although occupying only 2–3% of the total continental areas, contain approximately one-quarter of the world's carbon stored in soils, exceeding the amount stored in vegetation (Turetsky et al. 2015). Although organic matter undergoes slow anaerobic decomposition, peat typically grows by the accumulation of partially decayed vegetation and organic matter in these soils.

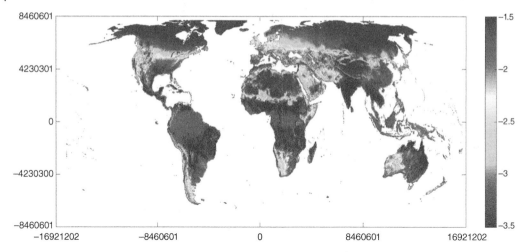

Figure 3.7 Global map of average P_{CO_2} in soils, expressed as the logarithm of the partial pressure of CO_2 in the soils ($\log_{10} P_{CO_2}$). Note higher values in warm and wet tropical and subtropical regions. Values estimated using an equation that considers the variables: (i) atmospheric P_{CO_2}; (ii) mean annual surface temperature ($1\,km^2$ spatial resolution); and (iii) daily surficial (depth > 5 cm) water content of soils (0.25° spatial resolution). Map projection Eckert IV. Additional details on the data and procedure used to produce the map can be found in Romero-Mujalli et al. (2018, 2019). *Source:* Courtesy of Gibran Romero-Mujalli.

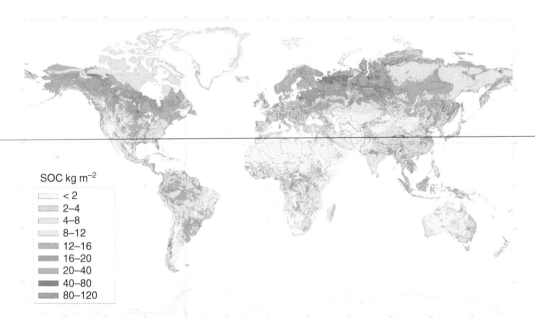

Figure 3.8 Soil Organic Carbon (SOC) world map which indicates the organic carbon content in soils at 1 m depth ($kg\,m^{-2}$). Note the abundance of soils with high organic carbon content (i.e., peatlands) in cold regions of the northern hemisphere. *Source:* Soil Survey Division of the US Department of Agriculture. Courtesy of Paul Reich (Soil and Plant Science Division, USDA-NRCS).

Groundwater may also increase its CO_2 content beneath the soil by the oxidation of organic matter in sediments (Keller and Bacon 1998), and the decay of dissolved organic solids carried by recharge waters (Albéric and Lepiller 1998). In a worldwide analysis of stalagmites covering the whole climatic spectrum, van Beynen et al. (2001) found that all samples contained considerable amounts of organic particles ($>0.07\,\mu m$), often even amenable to radiocarbon dating (Blyth et al. 2017), and very useful for paleoenvironmental reconstructions (e.g., vegetation changes, Blyth et al. 2008).

3.3.3 Dissolved Carbon Dioxide in Water and the Carbonic Acid System

Carbon dioxide is the most soluble of the standard atmospheric gases (e.g., 64 times more soluble than N_2). The solubility of gases in water is described by the Henry's law, which relates the solubility of a gas in a liquid to the partial pressure of the gas above the liquid. Henry's Law constant (K_H) for CO_2 is equal to the activity at equilibrium of CO_2 in the water divided by the partial pressure of CO_2 in the gas phase at a given temperature:

$$K_H = \frac{\left(CO_2^{aq}\right)}{P_{CO_2}} \quad \text{or} \quad P_{CO_2} = K_H \cdot \left(CO_2^{aq}\right) \tag{3.35}$$

where $\left(CO_2^{aq}\right)$ is the activity of carbon dioxide in water ($mol\,L^{-1}$), P_{CO_2} is the partial pressure of CO_2 in the air (atm), and K_H represents the Henry's constant ($mol\,L^{-1}\,atm^{-1}$). It may be more convenient to express P_{CO_2} by the activity of CO_2 in the atmosphere. In this case, Henry's law is expressed as

$$\left(CO_2^{aq}\right) = K_H \cdot R \cdot T \cdot \left(CO_2^{g}\right) \tag{3.36}$$

where R is the ideal gas constant ($R = 0.08206\,atm\,L\,K^{-1}\,mol^{-1}$) and T is temperature in Kelvin. The solubility of CO_2 in water and the corresponding K_H constant decrease significantly with temperature. Consequently, being the rest of the factors constant, the solubility of carbonate minerals in the H_2O-CO_2 system is higher in cold waters. Figure 3.9 shows equilibrium solubilities of CO_2 in water at different temperatures and in contact with a gas phase with variable P_{CO_2} values. Note that the solubility of gases in water increases with pressure.

Figure 3.9 Equilibrium solubilities of CO_2 (ppm) in water at different temperatures and P_{CO_2} in the gas phase. Note that a decrease in temperature from 30 to 0 °C implies an increase in solubility by a factor of 2.6. To convert into mmol L^{-1}, mg L^{-1} = mmol L^{-1} × 44.01. *Source:* Adapted from Bögli (1980).

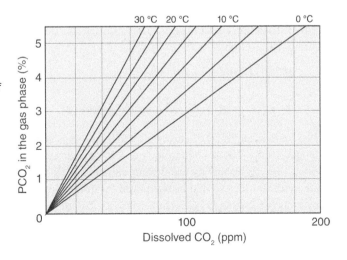

When CO_2 from the gas phase $\left(CO_2^g\right)$ dissolves in water, it becomes aqueous CO_2 $\left(CO_2^{aq}\right)$, and some of the dissolved CO_2 associates with water molecules to form carbonic acid (H_2CO_3)

$$CO_2(g) \rightarrow CO_2(aq) \tag{3.37}$$

$$CO_2(aq) + H_2O \rightarrow H_2CO_3 \tag{3.38}$$

Carbonic acid dissociates rapidly, and $CO_2(aq)$ is much more abundant than H_2CO_3 (at $25\,°C$, 600 times more abundant). To facilitate calculations, the two reactions are combined adopting the convention whereby the two species are summed up as $H_2CO_3^*$ $(H_2CO_3^* = CO_2(aq) + H_2CO_3)$. The overall reaction becomes (Appelo and Postma 2005):

$$CO_2(g) + H_2O \leftrightarrow H_2CO_3^* \tag{3.39}$$

with the equilibrium constant

$$K_{CO_2} = \frac{\left(H_2CO_3^*\right)}{P_{CO_2}} \quad \left(K_{CO_2} = 10^{-1.5} \text{ at } 25°C\right) \tag{3.40}$$

The carbonic acid experiences a stepwise dissociation, releasing two protons or hydrogen ions (H^+) that contribute to increase the acidity of the water. The first-order dissociation and the equilibrium constant are

$$H_2CO_3^* \leftrightarrow H^+ + HCO_3^- \tag{3.41}$$

$$K_1 = \frac{\left(H^+\right)\left(HCO_3^-\right)}{\left(H_2CO_3^*\right)} \quad \left(K_1 = 10^{-6.3} \text{ at } 25°C\right) \tag{3.42}$$

The second-order dissociation of the bicarbonate ion and its equilibrium constant

$$HCO_3^- \leftrightarrow H^+ + CO_3^{2-} \tag{3.43}$$

$$K_2 = \frac{\left(H^+\right)\left(CO_3^{2-}\right)}{\left(HCO_3^-\right)} \quad \left(K_2 = 10^{-10.3} \text{ at } 25°C\right) \tag{3.44}$$

The dissociation (self-ionization) of water is also involved in the equilibrium of the CO_2-H_2O solution (see Section 3.2.1)

$$H_2O(l) \leftrightarrow H^+(aq) + OH^-(aq) \tag{3.45}$$

$$K_w = \left(H^+\right)\left(OH^-\right) \quad \left(K_w = 10^{-14.0} \text{ at } 25°C\right) \tag{3.46}$$

The ionic strength (I) of carbonate karst waters is generally below 0.05, and consequently the extended Debye–Hückel equation (Eq. 3.21) can be used for calculating activity coefficients (γ_i). Even at $I \leq 0.05$, the activity coefficients can change by a factor of 2. Consequently, activities rather than concentrations should be used even for dilute karst waters (Dreybrodt 2000). For uncharged species such as CO_2 and H_2CO_3 the activity coefficients are given by (Plummer and Mackenzie 1974):

$$\gamma_i = 10^{0.1 \cdot I} \approx 1, \text{ if } I < 0.1 \tag{3.47}$$

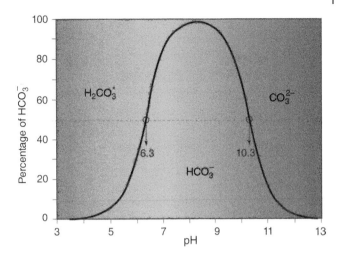

Figure 3.10 Percentage of HCO_3^- of total dissolved carbonate as a function of pH in a CO_2-H_2O system at 25 °C. At pH = 6.3, $[H_2CO_3^*] = [HCO_3^-]$ and at pH = 10.3, $[HCO_3^-] = [CO_3^{2-}]$. Color ramp reflects pH; green for neutral, red for strongly acidic and dark blue for strongly alkaline. *Source:* Adapted from Appelo and Postma (2005).

The dissociation of carbonic acid (H_2CO_3) into bicarbonate (HCO_3^-) and carbonate (CO_3^{2-}) and the proportion of the different species in the solution is governed by the pH, as illustrates Figure 3.10. Eq. 3.44 can be rearranged considering the equilibrium constant at 25 °C as

$$\frac{\left(CO_3^{2-}\right)}{\left(HCO_3^-\right)} = \frac{10^{-10.3}}{10^{-pH}} \tag{3.48}$$

Hence, at a pH of 10.3 the activities of CO_3^{2-} and HCO_3^- are equal. Similarly, using Eq. 3.42 the activities of HCO_3^- and $H_2CO_3^*$ are equal at a pH of 6.3. At pH < 4 virtually no HCO_3^- and CO_3^{2-} are present in the solution, and only $H_2CO_3^*$ exists. As the pH increases, more carbonic acid dissociates, forming bicarbonate, and bicarbonate dissociates producing carbonate. At pH > 12, CO_3^{2-} is essentially the only existing ion. However, since karst waters rarely have pH values higher than 8.3, the activity of the species CO_3^{2-} can be neglected (Dreybrodt 2000).

3.3.4   The Dissolution of Calcite and Dolomite

Three different reactions are involved in the dissolution of calcite by water containing CO_2 and considering that the amount of CO_3^{2-} at the pH of meteoric carbonate waters is negligible (Plummer et al. 1978):

$$CaCO_3\left(s\right) + H^+ \leftrightarrow Ca^{2+} + HCO_3^- \tag{3.49}$$

$$CaCO_3\left(s\right) + H_2CO_3 \leftrightarrow Ca^{2+} + 2HCO_3^- \tag{3.50}$$

$$CaCO_3\left(s\right) + H_2O \leftrightarrow Ca^{2+} + CO_3^{2-} + H_2O \tag{3.51}$$

and the overall reaction between carbon dioxide and $CaCO_3$ is (Appelo and Postma 2005):

$$CaCO_3\left(s\right) + CO_2 + H_2O \leftrightarrow Ca^{2+} + 2HCO_3^- \tag{3.52}$$

The stoichiometry of this reaction indicates that for each Ca^{2+} ion dissolved, one molecule of CO_2 is needed, which is converted into HCO_3^-. These reactions show that dissolution of $CaCO_3$ in the H_2O-CO_2 system is controlled by the activities of H^+, Ca^{2+}, HCO_3^-, and H_2CO_3 in the solution in contact with the mineral surface. The temperature-dependent equilibrium constant of calcite is given by the product of the activities of Ca^{2+} and CO_3^{2-}

$$K_c = \left(Ca^{2+}\right)\left(CO_3^{2-}\right) \tag{3.53}$$

The dissolution reaction of dolomite is

$$CaMg(CO_3)_2 \leftrightarrow Ca^{2+} + Mg^{2+} + 2CO_3^{2-} \tag{3.54}$$

the global reaction combining the reactions related to the H_2O-CO_2 system is

$$CaMg(CO_3)_2 + 2CO_2 + 2H_2O \leftrightarrow Ca^{2+} + Mg^{2+} + 4HCO_3^- \tag{3.55}$$

and the equilibrium constant of dolomite

$$K_d = \left(Ca^{2+}\right)\left(Mg^{2+}\right)\left(CO_3^{2-}\right)^2 \tag{3.56}$$

Note that two molecules of CO_2 are needed for releasing each Mg^{2+} and Ca^{2+} cation.

In the carbonate system, Ca^{2+} may react with HCO_3^- and CO_3^{2-} to form ion pairs ($CaHCO_3^+$ and $CaCO_3^0$), but in natural karst waters the activity of these species is negligible (Dreybrodt 2000). In a pure H_2O-CO_2-$CaCO_3$ system, and neglecting ion pairs, the water contains five dissolved ions: Ca^{2+}, H^+, HCO_3^-, CO_3^{2-}, and OH^-. To calculate equilibria observing the neutrality of electrical charges in the solution, the following charge-balance equation can be formulated (sum of positive equivalents = sum of negative equivalents)

$$z_i m_i \left(cations\right) = z_i m_i \left(anions\right) \tag{3.57}$$

where z_i is ionic charge and m_i is the molar concentration

$$2\left[Ca^{2+}\right] + \left[H^+\right] = \left[HCO_3^-\right] + 2\left[CO_3^{2-}\right] + \left[OH^-\right] \tag{3.58}$$

Conveniently, for the pH range of 4–8.4 and with an error margin of $\leq 2\%$, this equivalence can be simplified to

$$2\left[Ca^{2+}\right] = \left[HCO_3^-\right] \tag{3.59}$$

This simplification can be justified by Figure 3.10, which shows that for pH < 8, $[CO_3^{2-}] << [HCO_3^-]$. Therefore, the ionic strength of meteoric karst waters can be approximated by

$$I = 3\left[Ca^{2+}\right] \tag{3.60}$$

Suppose that $[Ca^{2+}] = x$ moles, then $[HCO_3^-] = 2x$. In the equation of the ionic strength (Eq. 3.5), we would have $I = 1/2 \cdot (2^2 x + 1^2 2x) = 3x$. In the presence of Mg^{2+} derived from dolomite or magnesian calcite

$$I = 3\left(\left[Ca^{2+}\right] + \left[Mg^{2+}\right]\right) \tag{3.61}$$

These are practical relations that allow estimating ionic strength directly from the concentration of metal ions and facilitate the calculation of activity coefficients (Dreybrodt 2000).

3.3.5 The Solubility of Carbonate Minerals

There are two common polymorphs of $CaCO_3$ with different crystalline structure: aragonite (orthorhombic) and calcite (hexagonal). Although aragonite is a metastable form that tends to transform into calcite (aragonite–calcite inversion), it is an abundant component in modern carbonate sediments and is more soluble than pure calcite (Figure 3.11). The crystalline structure of calcite consists of alternating layers of CO_3^{2-} anions and Ca^{2+} cations (Figure 2.4). Mg^{2+} cations, which have a smaller radius than Ca^{2+} cations (0.66 Å versus 0.98 Å), can randomly substitute for Ca^{2+}, forming magnesian calcites. The solubility of magnesian calcites varies considerably with Mg^{2+} content. Figure 3.11 displays the "best-fit" curve based on solubility data from biogenic magnesian calcites, which show significant variability depending on the Mg content. This curve provides a rationale for the boundaries proposed by some authors for classifying magnesian calcites: (i) Low-Mg calcites (LMC; <4 mol% of $MgCO_3$) with solubilities lower than that of pure calcite; (ii) Intermediate-Mg calcites (IMC; 4–12 mol% of $MgCO_3$), mostly with solubilities between those of pure calcite and aragonite; and (iii) High-Mg calcites (HMC; >12 mol% of $MgCO_3$), with solubilities greater than that of aragonite. Other authors simply differentiate into LMC and HMC, placing the boundary at 4 mol% of $MgCO_3$. The important implication for karst studies is that the $CaCO_3$ minerals that form limestones, and especially their Mg^{2+} content, may influence considerably the chemical karstification susceptibility of the rocks.

In an ideal stoichiometric dolomite mineral, with equal proportions of Ca^{2+} and Mg^{2+} ($Ca_{50}Mg_{50}(CO_3)_2$), layers of Ca^{2+}, CO_3^{2-}, and Mg^{2+} alternate (Figure 2.4). This mineral is slightly less soluble than calcite and aragonite. However, not all dolomites are stoichiometric, and excess Ca^{2+} contributes to increase the solubility of the mineral. These calcian dolomites predominate in synsedimentary dolomites. Additionally, disordered dolomites, found in Holocene deposits and characterized by

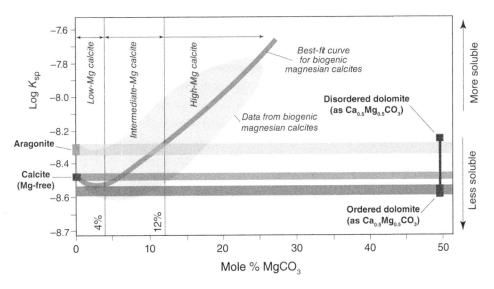

Figure 3.11 Logarithm of the solubility product of the main Ca-Mg-carbonate minerals (aragonite, calcite, and dolomite) with different Mg contents. The best-fit curve for magnesian calcites is derived from data of biogenic magnesian calcites, which show considerable dispersion. *Source:* Adapted from Railsback (2009).

high densities of crystallographic dislocations and substitutions, have higher solubility than ordered dolomites, even exceeding that of aragonite (Figure 3.11).

3.3.6 Open and Closed CO$_2$-Dissolution Systems

The carbon dioxide dissolved in karst waters (CO_2 (aq)) forms carbonic acid (H_2CO_3), which largely dissociates into H^+ and HCO_3^- Eqs. (3.38) and (3.41). The dissolution of carbonate minerals consumes H^+ and H_2CO_3 and depletes the amount of dissolved CO_2 Eqs. (3.49) and (3.50). As shown in Eq. (3.52), the dissolution of 1 mol of calcite ($CaCO_3$) consumes 1 mol of CO_2 and produces 1 and 2 mol of Ca^{2+} and HCO_3^-, respectively. The condition of open or closed system depends on whether the CO_2 consumed by carbonate dissolution can be replenished or not (Langmuir 1997; Appelo and Postma 2005; Ford and Williams 2007):

- In an ideal open system, the CO_2 partial pressure remains constant, because the CO_2 (and H^+, H_2CO_3) consumed by carbonate dissolution is resupplied. The dissolution of the carbonate mineral proceeds until the water reaches saturation with respect to the carbonate mineral and under the fixed P_{CO_2} value conditions. In the meteoric zone of a karst system, CO_2 replenishment essentially occurs through diffusion of CO_2 from the atmosphere and the soil air. Root respiration and degradation of organic matter contributes significantly to maintain high CO_2 levels in soils. An open-air pool on limestone, a stream flowing along a bedrock channel carved in limestone, or limestone bedrock underlying a thin porous soil within the vadose zone would be examples of open conditions in which there is a large permanently available gaseous CO_2 reservoir (Figure 3.12).
- In an ideal closed system, the dissolved CO_2 initially present in the water (and H^+, H_2CO_3) is not replenished as it is consumed by carbonate mineral dissolution. Consequently, the system reaches saturation conditions with respect to the carbonate mineral at a P_{CO_2} level lower than the initial one, dissolving less carbonate than in an open system. Dissolution in a water-filled narrow conduit (within the phreatic zone) proceeds under closed-system conditions (Figure 3.12).

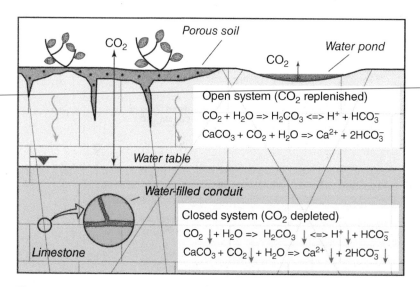

Figure 3.12 Sketch illustrating open-system and closed-system conditions in a limestone karst. In an air-free pond on bare limestone and in the limestone bedrock underlying a thin porous soil, the CO_2 consumed by dissolution is replenished by the incorporation of CO_2 from the atmosphere and soil air. In a deep phreatic conduit, there is no CO_2 replenishment, reducing the ability of the water to dissolve calcite or aragonite.

Given the same initial P_{CO2} and rest of the variables being equal, a larger amount of carbonate mineral is dissolved in an open system than in a closed system at saturation (Langmuir 1997). Figure 3.13 illustrates the pathways of calcite dissolution in ideal open systems with constant P_{CO_2} values (CO_2 replenishment), and in ideal closed systems in which no CO_2 is added. The pathways that start with equal values of P_{CO_2} and HCO_3^- show that when saturation is reached, the amount of calcite dissolved, expressed as concentration of bicarbonate in the water, is significantly higher in the open system than in the closed system. The closed systems attain higher pH due to CO_2 depletion.

Carbonate dissolution by groundwater may occur under open conditions, mainly in the recharge zone, and under closed conditions along most of its flow path within the phreatic zone. A given spring or well water may be a mixture of waters that have flowed under both conditions. The ability of the system in the near-surface vadose zone to maintain a constant P_{CO_2} depends on multiple factors, such as the soil porosity, the rate of water recharge, the CO_2 production in the soil, which may experience significant seasonal variability, or the depth at which percolating water first starts to interact with carbonate minerals. Descending percolating water dissolves carbonate minerals, resulting in an increase in Ca^{2+} concentration with depth up to a saturation level. This process can be envisioned as a downward-migrating dissolution front (Appelo and Postma 2005). In a closed system, a vertical profile of P_{CO_2} would show a decrease associated with the increase in the Ca^{2+} concentration (Jakobsen and Postma 1999), whereas in an open system the P_{CO_2} level would remain constant across the section of the profile in which Ca^{2+} increases (Reardon et al. 1980).

Dreybrodt (2000) provides a formulation for the evolution of the chemical composition of an idealized H_2O-CO_2-$CaCO_3$ system during the process of dissolution in a general intermediate-system

Figure 3.13 Pathways of calcite dissolution at 12 °C and 1 bar total pressure in ideal open systems with fixed P_{CO_2} values ($10^{-2.0}$, $10^{-2.5}$, $10^{-3.0}$, $10^{-3.5}$), and in ideal closed systems (no CO_2 added) with initial P_{CO_2} values of $10^{-2.0}$, $10^{-2.5}$. The open-system pathways (green lines) indicate a higher amount of calcite dissolution at saturation, expressed as concentration of bicarbonate, than in the closed-system pathways (red lines) starting from the same pH neutral value and HCO_3^- concentration. In the open system, the CO_2 consumed by calcite dissolution is replenished, allowing the formation of additional carbonic acid. Note that dissolution of 1 mol of $CaCO_3$ in the H_2O-CO_2 system produces 2 mol of bicarbonate (Eq. 3.59). *Source:* Adapted from Langmuir (1997).

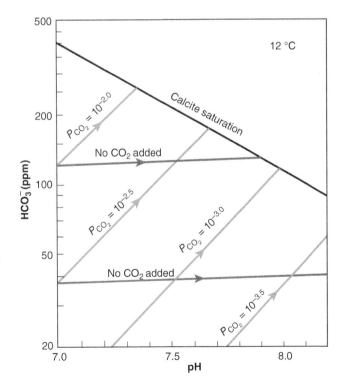

condition. This is a situation in which there is a gaseous phase in contact with the solution, but the transfer of CO_2 across the air–water interface is not sufficient to replenish all the CO_2 consumed by carbonate dissolution. The gradation from open- to closed-system conditions is incorporated through the ratio between the volume of gas acting as CO_2 source and the volume of solution (Vg/Vs). This ratio is 0 in closed systems and tends to ∞ in open systems. When water percolates through unsaturated soils, Vg/Vs may be close to 1. Dreybrodt (2000) also formulates the evolution of the chemical composition of systems that evolve from open or intermediate conditions to closed conditions. The resulting graph illustrates the expected chemical evolution of waters that infiltrate in limestone terrain, dissolve calcite initially under open or intermediate conditions within the three-phase unsaturated zone, and subsequently under closed conditions along the two-phase saturated zone (Figure 3.14). These graphs illustrate that the frequent high calcium concentrations of about 2 mmol L^{-1} measured in spring waters in temperate regions, cannot be explained alone by dissolution under closed conditions. This would require an excessively high initial P_{CO_2} value of around 0.05 atm (characteristic of tropical soils). Instead, percolation water dissolves a significant amount of $CaCO_3$ in the unsaturated zone under open or intermediate conditions, before it enters into the closed-system conditions of the saturated zone (Dreybrodt 2000). Figure 3.15 shows saturation concentration of dissolved calcite in ideal open and closed systems with variable initial P_{CO_2}, and the resulting CO_2 concentration and pH in the calcite-saturated solution.

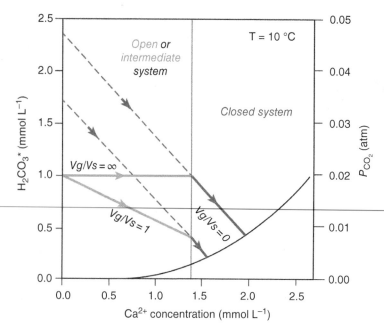

Figure 3.14   Evolution of $H_2CO_3^*$ ($H_2CO_3^* = CO_2(aq) + H_2CO_3$) and P_{CO_2} for changing conditions as a function of the Ca^{2+} concentration produced by $CaCO_3$ dissolution. In an initial stage, the solution, starting with a P_{CO_2} of 0.02 atm (typical of soils in temperate regions), evolves under open (Vg/Vs = ∞) and intermediate conditions (Vg/Vs = 1). When both waters have reached an arbitrary Ca^{2+} concentration of ca. 1.5 mmol L^{-1}, they enter into narrow joints within the saturated zone and dissolution proceeds under closed conditions (Vg/Vs = 0) until equilibrium, given by the lower limiting curve. The extrapolation of the closed-system trajectories to $[Ca^{2+}]$ = 0 (dashed lines) indicates the initial $H_2CO_3^*$ concentration and the chemical evolution if dissolution would have occurred under closed conditions from the beginning. Source: Adapted from Dreybrodt (2000).

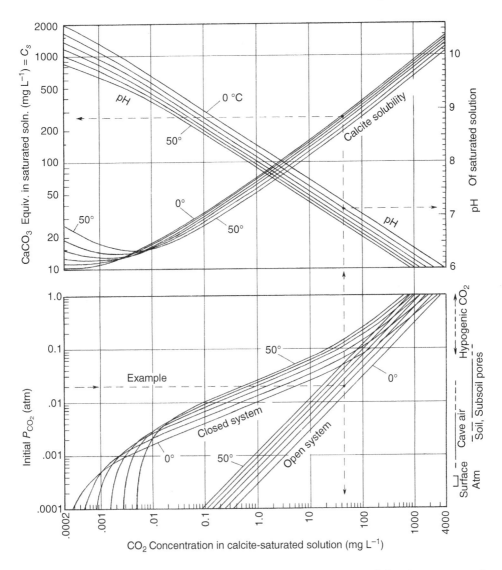

Figure 3.15 Saturation concentration of dissolved calcite in ideal open and closed systems at various temperatures and 1 atm versus the initial P_{CO_2}, and the CO_2 concentration and pH in the calcite-saturated solution. For example (dashed lines), in an open system at 10 °C with initial P_{CO_2} = 0.02 atm, dissolved calcite at saturation reaches a concentration of 273 mg L^{-1}, the concentration of CO_2 = 47 mg L^{-1} in the calcite-saturated solution, with a pH = 7.12. The graph shows that in closed systems with very low CO_2, a situation that almost never exists, the temperature effect reverses as CO_2 decreases below 0.01 mg L^{-1}. *Source:* Palmer (1991). Reproduced with permission from the Geological Society of America.

3.4 The Dissolution of Carbonate Rocks by Sulfuric Acid

Dissolution of carbonate rocks and cave development by sulfuric acid (H_2SO_4) are known since the first half of the twentieth century. However, the first comprehensive model was published much later by Egemeier (1981) in his seminal work on the Lower and Upper Kane Caves in Wyoming. Soon after, Kirkland (1982) and especially Hill (1981b, 1987, 1990, 1995, 2000, and references therein) developed the fundamental basis that has inspired recent investigation through their studies on Carlsbad Cavern and subsequently Lechuguilla Cave (discovered in 1986), both located in the Delaware Basin, New Mexico. Sulfuric acid dissolution is responsible for the development of the most spectacular hypogenic caves, with outstanding dimensions, complexity, and mineral decoration. In contrast, the role played by sulfuric acid on the development of surface karst landforms is rather limited, other than the formation of spectacular sinkholes through the collapse of caves. Bottrell et al. (2000) and Palmer (2007) explain the main physicochemical aspects of carbonate dissolution by sulfuric acid and Palmer (2013), and Palmer and Hill (2019) provide reviews of sulfuric acid caves.

Sulfuric acid responsible for carbonate karst development may be produced in several ways. The principal process is the oxidation of dissolved hydrogen sulfide (H_2S) in rising groundwater when it reaches the oxygenated part of the karst aquifer, to form sulfuric acid (H_2SO_4). Hydrogen sulfide of magmatic origin may be present in juvenile and meteoric waters (e.g., Cueva de Villa Luz, Mexico; Hose and Pisarowicz 1999), but the main source is the reduction of sulfate in oxygen-poor diagenetic environments in the presence of organic carbon compounds, a redox reaction commonly associated with hydrocarbon reservoirs. Sulfuric acid can also be produced through the oxidation of metal sulfide minerals (e.g., pyrite, Tisato et al. 2012). This process tends to be localized and mainly produces scattered porosity, but in areas with abundant sulfide minerals this process has created or initiated large cave systems (Auler and Smart 2003; Palmer 2013; Klimchouk et al. 2016). An additional mechanism that only contributes to boost local carbonate dissolution occurs in caves in the coastal mixing zone, by oxidation of hydrogen sulfide derived from sulfate reduction below the halocline (Stoessell et al. 1993; Ritter et al. 2019).

The gas hydrogen sulfide (H_2S) is produced by sulfate reduction during the diagenetic transformation of organic matter in sedimentary basins. The abundant organic matter contained in some formations such as mudstones and shales (source rocks) initially transforms into kerogen, a term that collectively designates a wide variety of solid and insoluble organic compounds. At higher depths and temperatures, kerogen partially converts into crude oil and natural gas through a process called maturation (Dembicki 2017). The association of dissolved sulfate and hydrocarbon is thermodynamically unstable in diagenetic environments. Hydrogen sulfide is created in oxygen-poor conditions by redox reactions that involve the simultaneous exchange of electrons between chemical species. Oxidation is electron loss whereas reduction is electron gain. Dissolved sulfate, mainly derives from the dissolution of anhydrite, is reduced to hydrogen sulfide, while hydrocarbons are oxidized degrading into simpler organic compounds (e.g., methane; CH_4) and producing bicarbonate or carbon dioxide. The production of hydrogen sulfide may be either bacterially mediated (bacterial sulfate reduction; BSR) or induced by heating (thermochemical sulfate reduction: TSR). Empirical evidence reveals that BSR and TSR occur in two well-differentiated thermal environments whose depth range is determined by the geothermal gradient, generally between 20 and $40\,°C\,km^{-1}$ (Machel 2001) (Figure 3.16). BSR occurs in diagenetic settings at temperatures from $0\,°C$ up to about 60–80 °C, corresponding to depths less than 2–2.7 km at normal geothermal gradients. The metabolic activity of sulfate-reducing bacteria ceases at higher temperatures. TSR occurs at temperatures of about 100–140 °C and mainly within the depth range of 3.3–4.7 km. The following reaction exemplifies the reduction of sulfate derived from gypsum/anhydrite dissolution in the presence of hydrocarbons (HC) (Palmer 2007):

$$Ca^{2+} + SO_4^{2-} + HC + H_2O \rightarrow H_2S(aq) + Ca^{2+} + HCO_3^-$$
(3.62)

In the presence of alkaline earth metals (mainly Ca and Mg), the redox reaction can be accompanied by the precipitation of calcite or dolomite. In the Delaware Basin, Hill (1995) inferred the

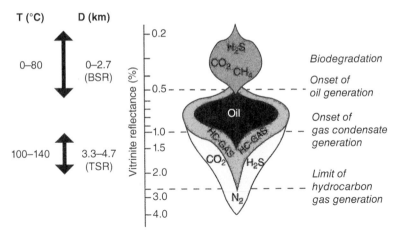

Figure 3.16 Sketch illustrating the generation of oil and gas by the diagenetic maturation of organic matter and the release of hypogene acidic gasses. Hydrogen sulfide gas is produced by bacterial sulfate reduction (BSR, $T = 0$–$80\,°C$) and thermochemical sulfate reduction (TSR, $T = 100$–$140\,°C$) in two independent thermal environments. Depth ranges indicate approximate values considering a constant geothermal gradient of $25\,°C\,km^{-1}$. HC stands for hydrocarbons. Vitrinite reflectance is a measure of the degree of diagenetic alteration and maturation of the organic matter. *Source:* Adapted from Machel (2001).

replacement of anhydrite in the Castile Formation from the presence of large masses of calcite, in combination with the bacterial reduction of sulfate by methane rising from an underlying hydrocarbon reservoir:

$$Ca^{2+} + 2SO_4^{2-} + 2CH_4 + 2H^+ \rightarrow 2H_2S(aq) + CaCO_3 + 3H_2O + CO_2 \tag{3.63}$$

The H_2S produced can migrate as a dissolved gas elsewhere. Hydrogen sulfide is a mild acid that can dissociate releasing hydrogen ions, and dissolve some calcite or dolomite:

$$H_2S \leftrightarrow H^+ + HS^- \tag{3.64}$$

However, this gas typically forms in a deep environment where water is generally saturated with respect to those carbonate minerals. As illustrated by Hill (1995) in the Delaware Basin, migrating H_2S-rich fluids may produce economic mineral deposits. Native sulfur (S^0) deposits are attributed by Hill (1995) to the bacterially mediated oxidation of H_2S in a redox interface between deep flows and oxygenated meteoric waters in confined aquifers:

$$H_2S + \frac{1}{2}O_2 \rightarrow S^0 + H_2O \quad \left(\text{at } pH < 6-7\right) \tag{3.65}$$

$$2HS^- + O_2 \rightarrow 2S^0 + H_2O \quad \left(\text{at } 7 < pH < 9\right) \tag{3.66}$$

The dissolved H_2S can also interact with metal-rich basinal fluids in carbonate formations forming the typical Mississippi Valley Type (MVT) deposits. These are epigenetic stratabound sulfide bodies predominantly composed of lead and zinc sulfides (e.g., galena, sphalerite) and hosted in limestones and dolostones. This complex overall redox reaction can be expressed as (Hill 1995):

$$H_2S + CO_2 + MeCl + Mg^{2+} + 2CaCO_3 + H_2O \rightarrow MeS + Ca^{2+} + CaMg(CO_3)_2 + HCO_3^- + Cl^- + 3H^+ \tag{3.67}$$

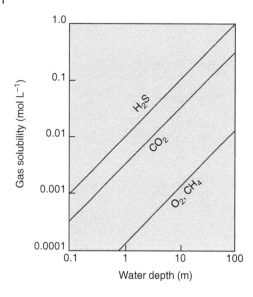

Figure 3.17 Solubility of various gases in water at 25 °C and within the water depth range of 0–100 m. At a water depth of 10 m, the solubilities are $H_2S = 0.1\,mol\,L^{-1}$, $CO_2 = 0.034\,mol\,L^{-1}$, O_2, and CH_4 (methane) $= 0.0013\,mol\,L^{-1}$. At those concentrations, some gases would escape from the solution as the water ascends above a depth of 10 m. *Source:* Adapted from Palmer (2007).

where Me indicates metals, mainly Pb, Zn, and Fe. Note that the stoichiometry of the reaction indicates net dissolution of carbonate minerals that creates the porosity in which the sulfide minerals precipitate.

The solubility of gases in water increases with pressure and depth (Figure 3.17). Consequently, when the water carrying dissolved H_2S ascends toward the surface, the gas may become supersaturated and bubble out of the solution, such as carbon dioxide that escapes from a carbonated drink.

Hydrogen sulfide has a characteristic smell of rotten eggs. Nonetheless, H_2S is more soluble in water than most other gases and tends to remain in the aqueous solution all the way to the upper part of the saturated (phreatic) zone (Figure 3.17). When hydrogen sulfide interacts with oxygen in the aerated part of the aquifer, it oxidizes producing sulfuric acid (H_2SO_4). The conversion takes place in one or more steps (Engel et al. 2004; Palmer 2007):

$$2H_2S + O_2 \rightarrow 2S + 2H_2O \tag{3.68}$$

$$2S + 3O_2 + 2H_2O \rightarrow 4H^+ + 2SO_4^{2-} \tag{3.69}$$

$$S_2O_3^{2-} + H_2O + 2O_2 \rightarrow 2H^+ + 2SO_4^{2-} \tag{3.70}$$

or by the following direct reaction when the process is facilitated by sulfur-oxidizing bacteria

$$H_2S + 2O_2 \rightarrow 2H^+ + SO_4^{2-} \tag{3.71}$$

The oxidation of the ascending hydrogen sulfide can occur in two different settings of speleogenetic importance: (i) the more oxygenated upper part of the saturated zone of the aquifer, close to the water table and (ii) above the water table in relation with vapor condensation (De Waele et al. 2016). Sulfuric acid is a strong acid that increases substantially the aggressiveness of the water with respect to carbonate minerals. Figure 3.18 shows the increase in calcite solubility when H_2S in solution completely oxidizes to sulfuric acid in an aqueous solution in which there is no loss of CO_2.

Figure 3.18 Increase in calcite solubility when H_2S in solution experiences complete oxidation to sulfuric acid and assuming that there is no CO_2 loss. The initial carbon dioxide level in the water is $0.001\,mol\,L^{-1}$ ($P_{CO_2} \sim 0.6\,atm$). For example, a solution with $50\,mg\,L^{-1}$ of H_2S can contain at equilibrium $0.065\,cm^3\,L^{-1}$ of dissolved calcite. This value rises to $0.12\,cm^3\,L^{-1}$ if all the H_2S is oxidized to sulfuric acid with no loss of CO_2. Calcite dissolution by sulfuric acid involves the production of CO_2 (see Eq. 3.72). *Source:* Adapted from Palmer (2007).

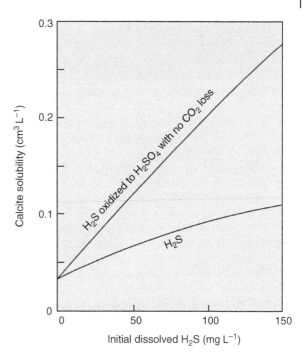

The formation of most sulfuric acid caves in the world is related to the oxidation of hypogene hydrogen sulfide at or close to the water table (Palmer 2013; De Waele et al. 2016). Sulfuric acid rapidly dissolves carbonate minerals:

$$CaCO_3 + SO_4^{2-} + 2H^+ + 2H_2O \rightarrow CaSO_4 \quad 2H_2O + CO_2 + H_2O \tag{3.72}$$

The dissolution of calcite by sulfuric acid produces CO_2 which contributes to promote carbonate dissolution in case it is not released from the dissolution system. It also causes the precipitation of gypsum (speleogenetic gypsum), the most common by-product of carbonate dissolution by sulfuric acid. Isotopic signatures help to elucidate the origin of these sulfates. For instance, oxidation of deep-sourced H_2S versus dissolution of gypsum in the meteoric zone (Onac et al. 2011 and references therein). Dolomite reacts in a similar way and may be accompanied by the precipitation of gypsum and epsomite. Thick massive gypsum may precipitate in water-table ponds and on cave floors (e.g., Carlsbad Cavern and Lechuguilla Cave, New Mexico) (Figure 3.19a). It may also form crusts on cave walls, both below and above the water table (Figure 3.19b). These precipitates may replace the host limestone rock preserving its original texture through a process designated by Egemeier (1981) as replacement solution. Gypsum, due to its high solubility can be dissolved by vadose cave waters, percolate toward deeper cave levels, and reprecipitate forming extraordinary speleothems such as the so-called chandeliers of Lechuguilla Cave (Palmer 2007, 2013) (Figure 3.19c).

In the shallow and oxygenated meteoric zone, the oxidation of metal sulfides, chiefly pyrite (FeS_2), produces sulfuric acid contributing to intensify the dissolution of carbonate minerals. Pyrite is a relatively common constituent in marine sedimentary rocks, mainly forming during the diagenesis by bacterial sulfate reduction processes (Berner 1984). This mineral is commonly found in shale and mudstone interbedded within carbonate successions. Most metal sulfides are mono-sulfides with sulfur in the 2-minus oxidation state, and do not produce acidity (H^+) when oxidized:

$$MeS + 2O_2 \rightarrow Me^{2+} + SO_4^{2-} \tag{3.73}$$

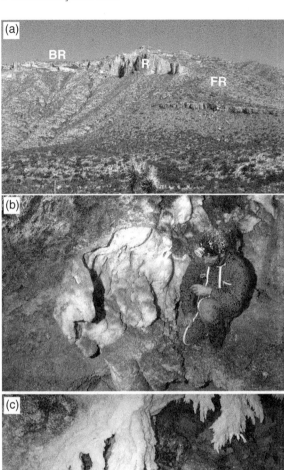

Figure 3.19 Images related to sulfuric acid speleogenesis and their geological context. (a) View of the Capitan Reef in the Guadalupe Mountains N.P., the rim of a Permian carbonate platform in the Delaware Basin exhumed by differential erosion. Image taken in McKittrick Canyon, west Texas. These are the carbonate formations that host the paradigmatic Carlsbad and Lechuguilla sulfuric acid caves. BR: backreef, R: reef, FR: forereef. *Source:* Francisco Gutiérrez (Author). (b) Example of gypsum crust on a cave wall as by-product of dissolution of carbonate rock by sulfuric acid (replacement solution) in Provalata Cave, Republic of Macedonia. *Source:* Photo by Jean-Yves Bigot. (c) Gypsum speleothems in Lechuguilla Cave, New Mexico, known as chandeliers. These speleothems are generated by evaporation of seeping waters that have dissolved gypsum related to dissolution of limestone by sulfuric acid in an above lying passage. *Source:* Photo by Jean-Yves Bigot.

However, iron sulfide minerals such as the polymorphs pyrite and marcasite (FeS_2) are disulfides with sulfur dianions in the 1-minus oxidation state, associated with iron in the 2-plus oxidation state. The oxidation of these minerals and the production of sulfuric acid can be described by the following reaction (Bottrell et al. 2000):

$$FeS_2 + 3.5O_2 + H_2O \rightarrow Fe^{2+} + 2SO_4^{2-} + 2H^+ \tag{3.74}$$

This reaction contributes to increase the acidity of the water, but in an oxygen-rich environment the reduced Fe^{2+} forms ferric hydroxide through a reaction that involves the oxidation of the metal and the hydrolysis of water molecules

$$Fe^{2+} + 0.25O_2 + 2.5H_2O \rightarrow Fe(OH)_3 + 2H^+ \tag{3.75}$$

This reaction also produces acidity. Where calcite is present, the oxidation of the metal disulfide is accompanied by calcite dissolution that contributes to neutralize the acidity of the water (Figure 3.17d). This set of geochemical processes can be described by the following overall reaction:

$$FeS_2 + 4CaCO_3 + 3.75O_2 + 3.5H_2O \rightarrow Fe(OH)_3 + 4Ca^{2+} + 4HCO_3^- + 2SO_4^{2-} \tag{3.76}$$

The main factors that control the impact of sulfide oxidation on carbonate dissolution are the availability of O_2, which has a low solubility in water (Figure 3.17), and the amount of pyrite that can contribute to the acidity of the water (Bottrell et al. 2000). The stoichiometry of the reaction indicates that the oxidation of one mole of pyrite has the potential to cause the dissolution of 4 mol of calcite. It also shows that the coupled oxidation and dissolution reactions involve the production of calcium and sulfate that may result in the precipitation of gypsum.

Dissolution of carbonate rocks related to the oxidation of pyrite is mostly a localized process that produces scattered secondary porosity within the oxygenated meteoric zone of karst systems. However, in areas with formations containing abundant pyrite, it may be the main mechanism responsible for the creation or initiation (inception horizons) of cave systems (Auler and Smart 2003; Auler 2013): examples are Lower Crevice Cave, Iowa (Morehouse 1968), and the giant maze cave systems in Campo Formoso, NE Brasil (Auler and Smart 2003; Klimchouk et al. 2016; Auler et al. 2017). Precipitation of gypsum is relatively common in these caves and contributes to enhance breakdown processes by crystallization pressures (White and White 2003; Tisato et al. 2012; Garrecht Metzger et al. 2015). Important-related topics from the environmental and engineering perspective include acid mine drainage, that locally enhances karstification with a buffering effect (Webb and Sasowsky 1994), and distress in structures including heaving and concrete degradation related to gypsum precipitation (Hawkins 2014).

Sulfuric acid dissolution may also occur in drowned caves in the coastal mixing zone (Bottrell et al. 1991; Stoessell et al. 1993). Here, the seawater that inundates the cave is overlain by a layer of less dense freshwater, both separated by a landward tapering interface called the halocline. The sulfate (SO_4^{2-}) dissolved in the salt water below the halocline may be reduced to hydrogen sulfide (H_2S) by sulfate-reducing bacteria in the presence of surface-derived organic debris trapped at the interface between the saltwater and freshwater. This hydrogen sulfide may diffuse upward and come in contact with oxygenated waters above the halocline, producing sulfuric acid. This strong acid contributes to enhance the development of secondary porosity in cave walls, together with the freshwater–saltwater mixing effect. Note that these processes occur below the water table, in the phreatic zone, and may partly explain the abundance of solution holes observed by divers around the halocline.

3.5 The Dissolution of Gypsum and Halite

Gypsum dissolution involves the release of water molecules from the crystalline structure:

$$CaSO_4 \cdot 2H_2O \leftrightarrow Ca^{2+} + SO_4^{2-} + 2H_2O \tag{3.77}$$

In normal meteoric conditions, anhydrite ($CaSO_4$) is rarely affected by karstification processes. Instead, it converts into secondary gypsum by hydration within a weathering zone that may reach hundreds of meters in thickness. The existing literature supports the concept that true anhydrite karsts barely exist in the world despite the widespread occurrence of this lithology in the subsurface. The dissolution of gypsum at microscopic scale is influenced by its crystalline structure, comprising layers of Ca and SO_4 units, linked by layers of H_2O through weak hydrogen bonding (Figure 2.4). This layered structure allows the development of perfect cleavage planes. Fan and Teng (2007) investigated dissolution on cleavage planes using fluid cell atomic force microscopy (AFM). They found that dissolution occurs through a layer-by-layer process, and that etch pits are much shallower than those developed in other minerals such as calcite, due to the rapid solutional migration of unstable steps (Figure 3.20).

The equilibrium solubility of gypsum in pure water at 25 °C, 1 atm pressure and pH 7 is 2.6 g L^{-1} (mass of dissolved solute in 1 L of solution) (Table 3.1). This value is 160 times lower than the solubility of halite in the same conditions, and around 50 and 9 times higher than the solubility of calcite in water with P_{CO_2} values of $10^{-3.42}$ (P_{CO_2} in the atmosphere) and $10^{-1.3}$ atm (P_{CO_2} in soils

Figure 3.20 Atomic force microscopy images (15 × 15 μm) showing the morphology generated by dissolution on the surface of a gypsum crystal (cleavage plane). Etch pits expand laterally by the solutional migration of steps controlled by the layered structure of the gypsum. *Source:* Fan and Teng (2007). Reproduced with permission of Elsevier.

with high CO_2 content), respectively. The solubility of gypsum is temperature-dependent, with a maximum value at 43 °C (Blount and Dickson 1973; Sonnenfeld and Perthuisot 1984) (Figure 3.21). From 0 to 30 °C, the temperature range encompassing most karst waters, the solubility of gypsum increases by about 20%, and at temperatures above 43 °C it decreases gradually (Blount and Dickson 1973; James 1992). Cigna (1985) discusses the possible effects on gypsum solubility of water mixing processes involving high temperature waters (40–100 °C). The pressure variations that occur within depths of a few hundred meters have a negligible impact on gypsum solubility. Moreover, at greater depths (>450 m), where pressure could have some influence on gypsum solubility, Ca-sulfate mainly occurs as anhydrite (Zanbak and Arthur 1986; Newton and Manning 2005).

The solubility of gypsum may increase substantially in waters with dissolved ions other than Ca^{2+} and SO_4^{2-} (foreign ion or saline effect). Those foreign ions contribute to reduce the activities of Ca^{2+} and SO_4^{2-} by ion pairing and the increase in the ionic strength of the solution (see Section 3.2.5). Of special significance is the presence of dissolved NaCl, which is a relatively frequent situation since gypsum and halite rocks are frequently associated in evaporitic successions. According to experimental data obtained by Ponsjack (1940) at 35 °C, gypsum solubility reaches a maximum of $7.3 \, g \, L^{-1}$ at $139 \, g \, L^{-1}$ of dissolved NaCl, then it decreases but remains above the solubility in pure water (Figure 3.22). This ionic-strength effect explains the relatively frequent presence of waters in evaporite karst systems with dissolved contents of gypsum above $2.6 \, g \, L^{-1}$. In the Poechos Dam, Perú, James (1992) documents waters undersaturated with gypsum, but with dissolved concentrations of this mineral of $6.2 \, g \, L^{-1}$, due to high contents of foreign ions. A similar situation has been reported by Frumkin (2000a) in Mount Sedom salt diapir, Israel. This factor may play an important role in the renewal of water aggressiveness by mixing solution, whereby a saturated water may become undersaturated when it mixes with another water with a different composition and/or temperature.

Other chemical reactions such as sulfate reduction or dedolomitization may contribute to enhance or maintain gypsum dissolution by reducing the concentration of SO_4^{2-} or Ca^{2+} in the solution (Reaction 3.77 tends to displace to the right). According to Klimchouk (2000a), sulfate reduction by anaerobic bacteria is an important mechanism in maintaining the gypsum-dissolution

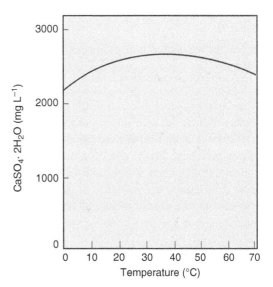

Figure 3.21 Solubility of gypsum in pure water as a function of temperature, with a maximum at 43 °C. Note that solubility in this graph is expressed as dissolved $CaSO_4 \cdot 2H_2O$ and in Figure 3.22 as dissolved $CaSO_4$. *Source:* Experimental data of Blount and Dickson (1973) and adapted from Klimchouk (2000a).

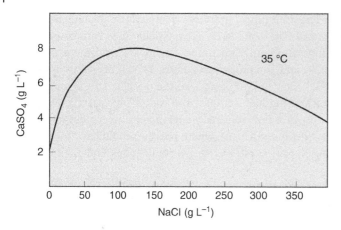

Figure 3.22 The solubility of gypsum in water with different concentrations of dissolved halite. *Source:* Adapted from Ponsjack (1940).

potential of groundwater in confined conditions, especially in basal injection (artesian) cave systems with cross-formational flows. The reduction of sulfate in aquifers with dispersed organic matter can be described with the following simplified reaction:

$$SO_4^{2-} + 2CH_2O \rightarrow H_2S + 2HCO_3^- \tag{3.78}$$

this process, which entails an increase in bicarbonate concentration, may be accompanied by the precipitation of calcite. Klimchouk (1997), in the Pre-Carpathian region, Ukraine, interpreted sulfur deposits associated with a cavernous Miocene gypsum formation to be controlled by a basal injection (artesian) gypsum karst system, that provides the necessary reactants and hydraulic conditions for the development of the deposit.

The dedolomitization process associated with gypsum dissolution can be described by three reactions:

$$MgCa\left(CO_3\right)_2\left(dolomite\right) + 2H^+ \rightarrow Ca^{2+} + Mg^{2+} + 2HCO_3^- \tag{3.79}$$

$$CaSO_4 \cdot 2H_2O\left(gypsum\right) \rightarrow Ca^{2+} + SO_4^{2-} + 2H_2O \tag{3.80}$$

$$Ca^{2+} + HCO_3^- \rightarrow CaCO_3\left(calcite\right) + H^+ \tag{3.81}$$

hence, the net dedolomitization reaction is

$$dolomite + gypsum = 2calcite + Mg^{2+}\left(aq\right) + SO_4^{2-}\left(aq\right) + 2H_2O\left(l\right) \tag{3.82}$$

these coupled reactions result in the retention of undersaturated conditions of groundwater with respect to gypsum, contributing to maintain gypsum dissolution. The dedolomitization process consumes calcium ions derived from the dissolution of gypsum, converting dolomite into calcite (Schoenherr et al. 2018). Moreover, they involve a net loss of solid mass that may lead to secondary porosity development. Raines and Dewers (1997a) indicate that this slow mechanism may have some contribution to gypsum karst development in aquifers with slow-flow conditions and with groundwaters nearly saturated with respect to gypsum. Bischoff et al. (1994) proposed that dedolomitization reactions may have favored the development of the hypogene gypsum karst of Bañolas Lake, Spanish Pyrenees, although its relative contribution is most probably rather limited considering the high-flow rate of groundwater derived from an underlying carbonate aquifer and largely undersaturated with respect to gypsum (Gutiérrez et al. 2019b).

Halite dissolution occurs through a simple reaction resulting in a solution with equal molar concentrations of Na^+ and Cl^-:

$$NaCl \rightarrow Na^+ + Cl^- \tag{3.83}$$

Waters in salt karst systems with Na/Cl molar ratios different from 1 suggest other sources for those ions. For instance, dissolution of saline formations including potash salts (e.g., carnallite: $KMgCl_3 \cdot 6H_2O$; sylvite: KCl) would produce molar ratios lower than 1. Halite solubility in pure water at $25\,°C$, pH 7, and 1 atm pressure is $424\,g\,L^{-1}$ (referred to as the volume of solution) or $356\,g\,kg^{-1}$ (referred to as the mass of solvent; Langer and Offermann 1982) (Table 3.1). The high density of the brines resulting from salt dissolution may result in salinity-stratified water bodies with influence on dissolution processes and cave development, as documented in Mount Sedom diapir, Israel (Frumkin 2000a, 2013). The waters of the Upper Gotvand Reservoir in the Karun River, Iran, submerged an exposed salt pillow, resulting in the incorporation of more than 41 million metric tons of dissolved salt in the reservoir. The salinity of the 150-m deep reservoir water is stratified, with concentrations as high as $200\,g\,L^{-1}$ at the bottom (Gutiérrez and Lizaga 2016, Jalali et al. 2019).

The solubility of halite shows a linear increase with temperature, but the increments within the normal temperature conditions in meteoric environments are negligible (McMurry 2004). The same consideration applies to pressure variations (Frumkin 2000b). According to Krumgalz et al. (1999), the solubility of halite at 100 bars pressure, equivalent approximately to the overburden load at 400 m depth, is just 2% higher than that at the surface. Halite solubility may increase or decrease due to the foreign-ion and common-ion effects, respectively. Dissolution of other chlorides (sylvite, carnallite), which often occur in association with halite, may contribute to reduce the solubility of NaCl (Lerman 1970).

3.6 The Dissolution of Silica

Quartz is very resistant to chemical weathering, except under some extreme environmental conditions, such as those found in the humid tropics, or at specific sites with high temperatures and/or waters with very high-pH (alkaline) (Figure 3.23). Silica, either in the form of quartz (density $2.65\,g\,cm^{-3}$) or amorphous silica, such as opal (density $\sim 2.1\,g\,cm^{-3}$), can dissolve in water by hydrolysis forming the monomer $Si(OH)_4$, mainly existing as monomolecular silicic acid (H_4SiO_4). This acid comprises four OH groups attached to a central silicon atom.

$$SiO_2\left(quartz, amorphous\ silica\right) + 2H_2O \rightarrow H_4SiO_4\left(aq\right) \tag{3.84}$$

The dissolution of silica has several features relevant to karst studies: (i) Solubilities in water under normal environmental conditions are very low, but the solubility of amorphous silica is significantly higher than that of quartz. Equilibrium solubilities of quartz and amorphous silica in pure water at $25\,°C$, 1 atm pressure and pH 7 are $6.3\,mg\,L^{-1}$ and $116.7\,mg\,L^{-1}$, respectively, according to values obtained with the geochemical code PHREEQC and the WATEQ4F thermodynamic database (Table 3.1). Similar values have been proposed by a number of authors, reviewed in Wray and Sauro (2017). (ii) The solubility of silica in its different forms increases substantially with temperature. (iii) The solubilities of quartz and amorphous silica increase abruptly at pH values above 8–9. (iv) The rate of quartz dissolution is extremely slow, complicating experimental investigations of silica dissolution at low temperature conditions.

(a)

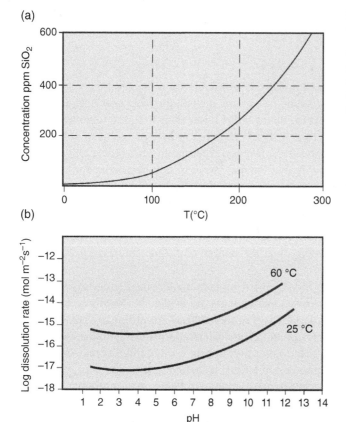

(b)

Figure 3.23 (a) Dependence of quartz solubility on temperature in pure water. *Source:* Adapted from Rimstidt (1997) and Wray and Sauro (2017). (b) Dependence of quartz dissolution rate on the pH at 25 °C and 60 °C. *Source:* Adapted from Wray and Sauro (2017).

Rimstidt (1997), by means of experimental laboratory studies, analyzed the equilibrium constant for quartz dissolution (K_q) within the 0–300 °C temperature range, 1 atm and neutral pH, obtaining the expression:

$$\log K_q = -0.0254 - \left(\frac{1107.2}{T_{kelvin}}\right) \tag{3.85}$$

Figure 3.23a depicts the graphic representation of quartz solubility as a function of temperature based on this equation. It shows that within the normal temperature range of earth surface conditions, quartz solubility slightly increases with temperature. A substantial increase occurs at temperatures above 100 °C, characteristic of the diagenetic and metamorphic realms.

The solubility of silica is pH dependent due to the dissociation of silicic acid. At neutral pH conditions, silicic acid H_4SiO_4 is mostly undissociated. However, at pH above 8–9, it experiences four consecutive dissociations that contribute to increase in the solubility of silica:

$$H_4SiO_4\left(aq\right) \rightarrow H_3SiO_4^-\left(aq\right) + H^+ \tag{3.86}$$

$$H_3SiO_4^-\left(aq\right) \rightarrow H_2SiO_4^{2-}\left(aq\right) + H^+ \tag{3.87}$$

$$H_2SiO_4^{2-}\left(aq\right) \rightarrow HSiO_4^{3-}\left(aq\right) + H^+ \tag{3.88}$$

$$HSiO_4^{3-}\left(aq\right) \rightarrow SiO_4^{4-}\left(aq\right) + H^+ \tag{3.89}$$

These reactions involve the reduction of the activity of H_4SiO_4 in Reaction 3.84, which is displaced to the right, and hence contribute to increase the solubility of silica. Wray and Sauro (2017) quote a number of works carried out in natural settings that substantiate the correlation between alkaline pH and higher amounts of dissolved silica in the water (Piccini and Mecchia 2009; Mecchia et al. 2014; Sauro et al. 2020). Figure 3.23b illustrates the dependence of quartz dissolution rate on the pH at different temperatures in pure water. It shows an abrupt increase of up to four orders of magnitude for values of pH higher than 9. However, such highly alkaline conditions are rarely found in quartzite and siliceous sandstone terrains, where soils and waters are generally acidic (Ford and Williams 2007). A decrease in the pH of alkaline waters, for example by the addition of CO_2, may induce the precipitation of amorphous silica ($SiO_2 \cdot nH_2O$). This amorphous silica can slowly expel water through a process known as ageing, transforming into a hydrous cryptocrystalline form of cristobalite and then into chalcedony (microcrystalline quartz) (Bland and Rolls 1998). Wray and Sauro (2017) provide a comprehensive review on works that explore the influence of dissolved inorganic cations, metal ions, and organic acids on the solubility and dissolution rate of quartz. They conclude that in most cases, the relationships are unclear and results from laboratory experiments are not always consistent with field measurements.

3.7 Factors that Influence the Solubility and Saturation State

This section describes chemical and physical processes that affect the solubility and saturation state of soluble minerals in water (see Section 3.2.5), and therefore influence their dissolution and precipitation. A change in a solution from undersaturated to supersaturated state, or vice versa, may induce net precipitation or dissolution, respectively. An increase in the degree of undersaturation enhances the aggressiveness of the water and may boost dissolution. The general effects of these processes are mainly illustrated with calcite, the main constituent of limestones, but aragonite and dolomite are similarly affected. Most of those mechanisms also affect the dissolution of the main minerals that constitute evaporite karst rocks (e.g., gypsum, halite). Some of the processes and their effects on the saturation state of natural waters with respect to calcite are graphically illustrated in Figure 3.24, which shows the calcite saturation line expressed as Ca^{2+} activity at equilibrium for variable P_{CO_2} values. Waters above and below that line are supersaturated or undersaturated with respect to calcite, respectively. The change in CO_2 content is one of the main factors that controls the dissolution and precipitation of carbonate minerals. A decrease in P_{CO_2} related to exsolution (CO_2 degassing) or photosynthetic activity (which consumes CO_2) may cause the solution to become supersaturated in calcite and induce its precipitation. This is represented schematically by horizontal arrows pointing to the left that reach the supersaturation area, and arrows pointing downward up to the calcite saturation line (Figure 3.24). Groundwater typically experiences CO_2 exsolution when it discharges into an open system with lower P_{CO_2}, such as in a cave or at the surface. This is one of the main mechanisms involved in the precipitation of speleothems and travertine deposits, and CO_2 loss may be aided by turbulence. The CO_2 content may also be depleted by photosynthesis, which involves the consumption of CO_2 (see simplified reaction in Figure 3.6). This diurnal process that occurs in the shallow photic zone of aquatic environments participates in the precipitation of carbonate deposits, such as stromatolites or phytoherms. The aerobic decay of organic matter and plant respiration increase the CO_2 and may inhibit precipitation or boost dissolution (Figure 3.24). Evaporation increases the concentration of the dissolved ions in the solution and may cause carbonate precipitation. This is illustrated by a vertical arrow pointing upward and positioned at

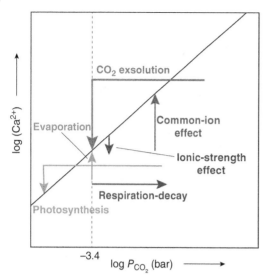

Figure 3.24 Schematic plot showing the effects of various processes on the saturation state of natural waters with respect to calcite. Waters are supersaturated and undersaturated with respect to calcite above and below the calcite saturation line, respectively. The value $10^{-3.4}$ corresponds to the atmospheric CO_2 pressure. *Source:* Adapted from Langmuir (1997).

the atmospheric CO_2 pressure. Calcite precipitation may occur if saturation conditions are reached according to a reaction that releases CO_2 gas (Figure 3.24):

$$Ca^{2+} + 2HCO_3^- \rightarrow CaCO_3 + CO_2(g) + H_2O \qquad (3.90)$$

3.7.1 Effects Related to Temperature and Pressure Changes

Carbonates, unlike most other minerals, have an exothermic heat of dissolution, and hence their solubility decreases with increasing temperature (retrograde solubility) (Langmuir 1997). Figure 3.25 illustrates the effect of temperature on the solubility products of calcite, aragonite, and

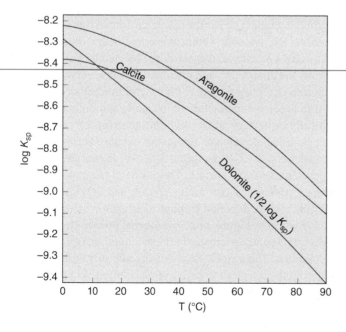

Figure 3.25 Solubility products of aragonite, calcite, and dolomite from 0 to 90 °C. *Source:* Adapted from Langmuir (1997).

ordered dolomite (less soluble than disordered dolomite). It shows that between 0 and 90 °C the solubilities decrease by about five to six times for calcite and aragonite, and about 13 times for dolomite. The retrograde solubility of calcite is also illustrated in Figure 3.15, which shows that even at constant CO_2 levels, the solubility increases with decreasing temperature; the calcite solubility lines move toward higher saturation concentrations as the temperature of the system decreases. The inverse relationship between solubility and temperature is magnified by the fact that the solubility of CO_2 gas, a major source of acidity, also increases with decreasing temperature, as illustrated in Figure 3.9. The combined effects of temperature and CO_2 pressure on the solubility of the main carbonate minerals are shown in Table 3.4. It indicates the solubilities of calcite, aragonite, and dolomite at various temperatures (10, 25, 40, and 50 °C) in solutions with CO_2 partial pressures of $10^{-3.42}$ atm and $10^{-1.3}$ atm, corresponding to the P_{CO_2} of the atmosphere and soils with high CO_2 content, respectively. The solubilities increase around 6 times with the 130-fold increase in CO_2. The temperature effect is important, but less pronounced than the CO_2 effect. For instance, the solubilities of the carbonate minerals increase around 1.9–2.3 times when the temperature drops from 50 °C to 10 °C, and 1.3 times when the temperature falls from 25 °C to 10 °C.

Because of the retrograde solubility of carbonate minerals, cooling may boost their dissolution. Saturated water may become undersaturated by a temperature decline, increasing its aggressiveness. Conversely, the temperature rise may cause a solution to become supersaturated inducing carbonate precipitation. Bögli (1980) indicated that enhanced dissolution due to cooling (cooling corrosion) is an active mechanism in the vadose zone of karst systems, especially in regions with strong daily thermal variations. Water warmed during the day may cool as it infiltrates into the ground increasing its aggressiveness. However, temperature has an antagonistic effect when soil CO_2 is the main source of acidity. The production of CO_2 in soils by degradation of organic matter increases with temperature and peaks during the growing season. Figures 3.6 and 3.7 illustrate the higher CO_2 content in soils associated with warm and wet environments.

As Figure 3.17 illustrates, the solubility of CO_2 in water increases with increasing pressure and depth. This pressure effect plays an important role in ascending thermal waters, which frequently have high concentrations of dissolved CO_2. In hydrogeology, water is qualified as thermal when its temperature at the emergence point is significantly warmer (>5 °C) than the average air temperature in the region. As the water ascends and the pressure falls, the thermal water may become supersaturated with respect to CO_2, and some of the dissolved gas is released as bubbles

Table 3.4 Solubility (mg L^{-1}) of calcite, aragonite, and dolomite starting with pure water at pH = 7 and CO_2 partial pressures of $10^{-3.42}$ and $10^{-1.3}$ atm, corresponding the P_{CO_2} of the atmosphere and soils with high CO_2 content, respectively. Values obtained with the code PHREEQC and the thermodynamic database WATEQ4F. For dolomite, the intermediate order phase has been considered (log K = −17.09 at 25 °C).

	Calcite		Aragonite		Dolomite	
P_{CO_2}	$10^{-3.42}$	$10^{-1.3}$	$10^{-3.42}$	$10^{-1.3}$	$10^{-3.42}$	$10^{-1.3}$
10 °C	66.9	383.1	76.1	438.1	82.4	479.6
25 °C	52.4	297.0	59.2	335.9	58.7	334.3
40 °C	41.5	230.3	46.5	258.3	43.9	244.7
50 °C	35.7	194.8	39.8	217.2	37.3	203.6

Source: Courtesy of Luis Auqué.

(exsolution). Consequently, the solution may become supersaturated with respect to calcite due to the pressure effect on CO_2 solubility (Dublyansky 2000a, 2000b; 2013). The rising thermal water may also be affected by the opposing cooling effect explained above, whereby both the solubilities of the carbonate minerals and CO_2 increase with decreasing temperature. The equilibrium related to these antagonistic cooling and degassing effects may explain the complex spatial and temporal patterns of dissolution and precipitation documented in some thermal caves (e.g., caves in the Rózsadomb, Budapest, Hungary; (Leél-Őssy 2017)). Thermal waters experience rapid CO_2 degassing when they emerge at the surface in hot springs. This CO_2 exsolution, frequently aided by stream turbulence, creates supersaturated conditions and the formation of extraordinary travertine deposits (Altunel and Hancock 1993; Yoshimura et al. 2004; Jones and Renault 2010; Figure 2.15b, e).

Palmer (1991) explored the impact of temperature decrease on the solutional aggressiveness of ascending thermal waters by a finite-difference analysis, in which CO_2-rich water rises through a 100-m long fissure in limestone, cooling down in the flow direction. The water was assumed to be saturated with calcite at the entrance of the fissure and with concentrations of dissolved CO_2 of $0.005\,M$ ($220\,mg\,L^{-1}$). Various temperature drops and discharge values were considered, selecting conditions comparable with those reported in the thermal caves of Budapest (Bolner-Takács and Kraus 1989). The analytical experiments indicated that the solutional widening rates in the fissure increase in the direction of flow, attaining a constant value at some distance from the entrance at low discharge values. The results of the experiments also indicated: (i) a 10-fold increase in the widening rate when doubling either the thermal gradient (cooling rate) or the CO_2 concentration; and (ii) an increase in the widening rate with discharge up to a critical value, above which enlargement of fissures by cooling water is independent of discharge.

Andre and Rajaram (2005) also analyzed numerically the solutional widening of a fissure along a cooling flow path using both equilibrium and kinetic models and coupling fluid flow, heat transfer, and reactive transport. The slightly different results obtained in these computational analyses with respect to those presented by Palmer (1991) are attributed by Andre and Rajaram (2005) to the fact that Palmer's analysis did not consider the feedbacks between aperture growth, flow rate, and the temperature field. The computational model of Andre and Rajaram (2005), which simulates the early development of hydrothermal caves and hot springs in hypogene settings, considers a 500-m long fracture with an initial uniform width of 0.05 mm and a constant hydraulic gradient of $0.1\,m\,m^{-1}$. The fluid and rock temperature at the entrance are 60 °C, and along the lateral boundaries of the rock mass (1 m thick) a geothermal gradient of 0.025 °C m^{-1} is specified. Water entering at the bottom of the fracture is saturated with calcite at a total dissolved CO_2 concentration of 1.646 mmolal. The hot groundwater that flows along the fracture exchanges heat with the surrounding rock. The results indicate that feedbacks between flow rate, temperature, and dissolution rate govern the behavior of the system.

The authors differentiate three main stages (Figure 3.26): (i) A long early stage in which fracture aperture growth is slow and relatively uniform along the fracture length. In this stage, the dissolution rate is largely controlled by water cooling along the flow path and the retrograde solubility of calcite. (ii) A relatively short stage, designated as maturation time, in which there is an abrupt and rapid increase in the flow rate and the fracture aperture grows much faster and nonuniformly along the fracture length. Fracture widening near the entrance slows down and almost stops, while further into the fracture it continues to increase. This maturation stage is related to a positive feedback between aperture growth and flow rate, when the fluid temperature field is largely conduction-controlled. (iii) After maturation, the flow rate increases at a slower rate and the aperture growth significantly decreases near the entrance, while fracture growth rate slows down, with values

Figure 3.26 Results of a computational model carried out by Andre and Rajaram (2005) that simulates the widening of a fracture by rising thermal water along a cooling flow path using equilibrium and kinetic models, and coupling fluid flow, heat transfer, and reactive transport.
(a) Evolution of the mass flow rate through time, showing an abrupt and rapid acceleration during the "maturation time." (b, c) Evolution of the fracture aperture through time showing three stages: (i) slow and relatively uniform widening prior to maturation; (ii) fast nonuniform growth during maturation (around 6600–6800 years); and (iii) a significant decrease in the aperture growth near the entrance after maturation, while rapid growth occurs farther into the fracture, where there is still a significant temperature gradient. *Source:* Adapted from Andre and Rajaram (2005).

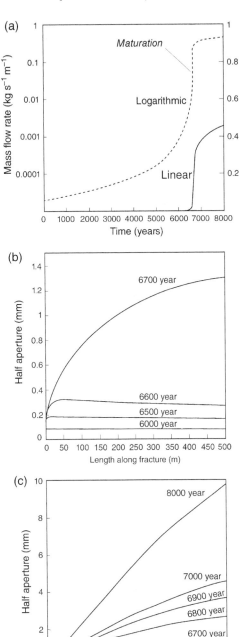

increasing away from the entrance, while rapid growth occurs farther into the fracture. This behavior is ascribed to negative feedbacks that arise because of higher flow rates; the water maintains a high temperature for a greater distance along the fracture, reducing the cooling corrosion effect in the section of the fracture located close to the entrance. At this stage, the heat transfer regime in the fluid switches from conduction- to convection-controlled. Andre and Rajaram (2005) indicate that their maturation time is analogous to the breakthrough time concept applied in meteoric karst systems, since both correspond to a stage in which there is an abrupt increase in the flow rate and the solutional widening of the fracture. However, in meteoric systems, fracture widening is a kinetically controlled process that advances as a dissolution front from the fracture entrance. In contrast, in the hydrothermal system, dissolution growth occurs almost uniformly along the fracture before maturation, and subsequently it increases away from the entrance, at higher rate during maturation and at slower rate after maturation. These authors also indicate that as the flow rates continue to increase, the fluid becomes increasingly isothermal along the length of the flow system. If the flow is saturated with respect to calcite at the entrance and the temperature gradient along the fracture is very small, the effect of retrograde solubility and cooling corrosion will be diminished and aperture growth rates may become very small. These conditions may be sufficient to justify the development of a hot spring, but not a large cave system, that may require additional mechanisms such as mixing with shallow groundwater as proposed by Bakalowicz et al. (1987) for the polygenetic caves of the Black Hills of South Dakota.

3.7.2 Common-Ion Effect

The principle of the common-ion effect is that the solubility of a mineral is reduced when any of the ions created by the dissolution of that mineral is added to the solution from other source (e.g., dissolution of a different mineral). The saturation state with respect to a mineral is given by the ratio between the ion activity product and the solubility product (see Eqs. 3.24 and 3.25). The incorporation of a common foreign ion increases the ion activity product and the degree of saturation. Therefore, the solution can attain equilibrium (saturation) with the dissolution of a lower amount of mineral, or may become supersaturated, inducing the precipitation of the mineral. Changes in the equilibria of calcite, aragonite, and dolomite by the common-ion effect may be related to the dissolution of other minerals that increase the concentration/activity of Ca^{2+}, Mg^{2+}, HCO_3^-, and/or CO_3^{2-}. Other carbonate and magnesium minerals are rare, thus the most frequent situation is the introduction of Ca^{2+} into the water from gypsum dissolution, which is significantly more soluble than calcite. The association of Ca-sulfates with dolostones and limestones is relatively common in carbonate successions. At $10\,°C$, the addition of $100\,mg\,L^{-1}$ of Ca^{2+} from gypsum causes a reduction in the solubility of calcite of around 34% (Ford and Williams 2007). Figure 3.24 illustrates the effect of the addition of Ca^{2+} on the equilibrium of calcite. The vertical red arrow represents the increase in the activity of Ca^{2+} up to the calcite saturation line. Further dissolution of gypsum would increase Ca^{2+} even more, causing the water to become supersaturated and eventually leading to calcite precipitation.

As water infiltrates from the surface, calcite is generally the first mineral to approach saturation. Dolomite is slightly less soluble and dissolves more slowly. If the groundwater encounters gypsum along its flow path, its rapid dissolution provides abundant Ca^{2+} to the solution and calcite precipitates from the supersaturated solution due to this common-ion effect. This process has been documented in a number of karst aquifers, such as the Floridan aquifer (Wicks and Herman 1996). The common-ion effect may also participate in the dedolomitization process that occurs in stratigraphic successions with associated dolostones and gypsum (see Eqs. 3.79–3.81). The dissolution of gypsum by water in equilibrium with calcite and dolomite fosters calcite precipitation due to increased Ca^{2+} concentration (common-ion effect). As calcite precipitates, the decrease in the concentration of the carbonate ions induces the dissolution of dolomite. The net result is that the dissolution of gypsum induces the transformation of dolomite into calcite (incongruent dissolution) and produces waters enriched in Mg^{2+}, Ca^{2+}, and SO_4^{2-} (see net reaction in Eq. 3.82) (Appelo and Postma 2005). Overall, these reactions involve the creation of secondary porosity. This is because the dissolved gypsum and dolomite occupied greater volumes than the calcite that precipitates mainly from the calcium and carbonate ions derived from the dissolution of the former minerals (Palmer 2007). This mechanism may contribute to facilitate gypsum karst development in aquifers with interbedded limestone, dolomite, and gypsum (Back et al. 1983; Bischoff et al. 1994; Raines and Dewers 1997a) and may allow gypsum caves to reach greater dimensions; calcite precipitation retards gypsum saturation (Palmer 2007). Other relatively common manifestations of the common-ion effect in karst environments include the rapid cementation of surficial deposits (e.g., alluvium) associated with gypsum bedrock, the formation of calcareous tufa deposits at locations where waters saturated or nearly saturated with calcite flow across gypsum formations or receive waters derived from them (Nicod 1993; Sancho et al. 1997; Ordóñez and Benavente 2014), and the precipitation of calcite speleothems in gypsum caves (De Waele et al. 2017c).

3.7.3 Ionic-Strength Effect

The addition of foreign ions to the water, different from those produced by the dissolution of a mineral, contributes to decrease the saturation state of the solution with respect that mineral. For instance, the incorporation of a significant amount of dissolved NaCl to water in equilibrium with calcite increases the ionic strength of the solution (Eq. 3.6) and therefore reduces the activity or effective concentration of the ions involved in the solubility product of calcite (Eqs. 3.20 and 3.21). The ion activity product decreases and the solution becomes undersaturated, renewing its aggressiveness with respect to calcite (Eqs. 3.24 and 3.25). This is illustrated in Figure 3.24, which shows a decrease in the activity of Ca^{2+} caused by the addition of foreign ions. This effect generally has a modest influence in the freshwaters of carbonate aquifers characterized by low concentrations. The addition of $250\,mg\,L^{-1}$ of dissolved NaCl causes an increase in the solubility of calcite of approximately $10\,mg\,L^{-1}$ (Ford and Williams 2007). However, mixing of freshwater or brackish carbonate groundwater with seawater in coastal zones may have a strong effect on dissolution, precipitation, and transformation of carbonate minerals (see Section 3.7.5). The impact of the ionic-strength effect may be substantial in karst systems developed in evaporite bedrocks including the common association of gypsum and halite. The rapid dissolution of the hyper-soluble halite may significantly boost gypsum dissolution by substantially increasing the ionic strength of the solution. As explained in Section 3.5, the addition of $140\,g\,L^{-1}$ of NaCl to a solution may increase the solubility of gypsum by around three times (Ponsjack 1940) (Figure 3.22).

3.7.4 Ion-Pair Effect

Dissolved ions of opposite electrical charge are attracted to each other by electrostatic forces. These forces, according to Coulomb's Law, are directly proportional to the charges of the ions and inversely proportional to the square of the distance between the ions and the dielectric constant of the medium. Some ions only associate weakly, forming cation–anion pairs within the solution that also contains many free ions. Ion pairing increases as ionic strength increases. This reduces the effective concentration or activity of the free ions. Therefore, the ion activity product of the mineral of interest decreases, increasing its solubility and affecting the saturation index. In the $CaCO_3$-CO_2-H_2O system, Ca^{2+} associates with HCO_3^- and CO_3^{2-} to form $CaHCO_3^+$ and $CaCO_3^0$ ion pairs (Plummer and Busenberg 1982). However, according to Dreybrodt (1988, 2000), in natural karst waters the activity of these species can be safely neglected. In the presence of SO_4^{2-}, a considerable amount of $CaSO_4^0$ ion pairs are generated, decreasing the activity of Ca^{2+} and increasing the solubility of calcite. For instance, at $20\,°C$ and with concentrations of $[Ca^{2+}] = 80\,mg\,L^{-1}$ and $[SO_4^{2-}] = 45\,mg\,L^{-1}$, the proportion of Ca^{2+} ions associated in ion pairs is around 10% of the total. Wigley (1971), in a study of brackish spring water with TDS of $1700\,mg\,L^{-1}$ from a gypsum and carbonate basin in British Columbia, estimated that 26.7% of the Ca^{2+} ions were paired with SO_4^{2-}, considerably affecting the saturation indices of calcite and gypsum. Buhmann and Dreybrodt (1987), in their study on the effect of foreign ions on the dissolution kinetics of calcite, show theoretically and experimentally that, while ion-pairing affects the solubility of calcite, its influence on the kinetics of the dissolution process is very limited.

The single-charged chloride ion does not pair to an appreciable extent at low temperatures, thus in bicarbonate and sulfate karst waters, the main ion associations involve the cations Ca^{2+}, Mg^{2+}, K^+, Na^+, H^+ and the anions CO_3^{2-}, HCO_3^-, SO_4^{2-}, and OH^-. The most important ion pairs in these hydrochemical environments include $CaSO_4^0$, $CaHCO_3^+$, $CaCO_3^0$, $MgHCO_3^+$, $MgCO_3^0$, and $MgSO_4^0$ (Ford and Williams 2007). Ion pairing increases the solubility of gypsum and carbonate minerals (generally <10%), and its effect is particularly important for the calculation of saturation indices, especially for waters in which the TDS exceeds $100\,mg\,L^{-1}$. If pairing is not considered, the calculated saturation indices are overestimated. The importance of the incorporation of ion pairs in the calculations can be illustrated by estimating the solubility of gypsum in distilled water at $25\,°C$ and 1 atm ($K_{gypsum} = 10^{-4.6}$) considering the formation of the ion pair $CaSO_4^0$ ($K_{ion\,pair} = 10^{2.23}$) and neglecting it. In both cases the activity of Ca^{2+} and SO_4^{2-} are taken into account using the extended Debye–Hückel equation (activities of water and gypsum are assumed to be 1). The calculation of the solubility of gypsum allowing for the ion pair $CaSO_4^0$ yields $2.54\,g\,L^{-1}$, and an unrealistic value of $1.76\,g\,L^{-1}$ when obviated. Thus, the ion pair has a contribution of around 30% to the total solubility value (Luis Auqué, pers. comm.).

3.7.5 Water Mixing

Mixing of two different freshwaters saturated with calcite may generate an undersaturated mixture capable of causing further calcite dissolution. This mechanism was designated as mixing corrosion (*mischungskorrosion*) by Bögli (1964, 1980, pp. 35–37). The potential role of water mixing on karst development was described before by Laptev (1939), but it became a widely used concept in the karst literature thanks to Bögli's work (Dreybrodt 2000; Palmer 2007). The underlying reason for this mechanism of renewed aggressiveness by water mixing is that in the dissolved $CaCO_3$ range of $0–350\,mg\,L^{-1}$, common in most carbonate karsts, the equilibrium relationships between calcite or dolomite with the P_{CO_2} and P_{H_2S} are nonlinear. Note the convex-upward shape and the progressive decrease in the slope of the saturation curves of calcite and dolomite as a function of P_{CO_2} and P_{H_2S} in Figure 3.27.

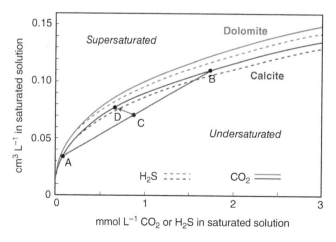

Figure 3.27 Variation of calcite and dolomite saturation concentration as a function of P_{CO_2} and P_{H_2S} and the effect of water mixing on the saturation state. Mixing of waters A and B, both saturated with calcite, results in an undersaturated mixture C lying below the saturation line. The CD arrow indicates the dissolution induced by the renewed aggressiveness. Its slope is related to the fact that CO_2 is consumed by calcite dissolution. See additional explanation in text. *Source:* Adapted from Palmer (2011).

The areas above and below those curves indicate supersaturation and undersaturation with respect to the corresponding minerals, respectively. If waters with compositions A and B mix in a closed system, the composition of the resulting mixture will fall somewhere on the straight line A-B, which lies in the undersaturation area (Figure 3.27). This water is capable of dissolving additional calcite. The position of the mixture on the line will depend on the relative proportion of the waters that mix. If the mixture falls on point C (equal proportions of the mixing waters), further dissolution in a closed system would occur, following the sloping line C—D up to saturation, because dissolution of each molecule of $CaCO_3$ consumes one molecule of CO_2. This mixing effect is greatest when the two initial solutions span the left part of the curve with higher slope change. The effect is diminished or nonexistent if either solution is supersaturated. This mechanism does not work in gypsum or halite because their solubilities do not depend on dissolved gases (Palmer 2007).

Bögli (1980) proposed that mixing corrosion plays an essential role in the development of phreatic caves in carbonate rocks because water, assuming linear dissolution kinetics, rapidly loses its aggressiveness. However, the importance of this process was reconsidered by subsequent work revealing that dissolution changes from linear (first-order) to nonlinear (higher-order) kinetics when the water is close to equilibrium. In a fracture fed by undersaturated water, during the initial stage of linear dissolution kinetics, both the dissolution rate and the widening of the fracture drop exponentially with the distance from the entrance. Subsequently, once a critical concentration is reached and breakthrough conditions are established, the flow rate is drastically accelerated and significant widening occurs along the entire length of the fracture (Palmer 1984; White 1988; Dreybrodt and Eisenlohr 2000). Gabrovšek and Dreybrodt (2000) modeled the evolution of the aperture widths of a confluence of two fractures into a third one that receives waters at equilibrium with different or equal P_{CO_2} values (with and without mixing corrosion), and considered linear and nonlinear dissolution kinetics. They found that: (i) the combination of mixing corrosion and linear dissolution kinetics does not create large conduits; (ii) dissolution by nonlinear kinetics can generate large conduits, and their development is accelerated by the interplay of mixing corrosion; (iii) mixing corrosion is only active in the early stages of karstification, before breakthrough conditions, subsequently its effect is overwhelmed by undersaturation conditions maintained by nonlinear kinetics; and (iv) mixing corrosion influences the initial pattern of conduits controlled by the fracture network. Romanov et al. (2003a) also illustrated how mixing corrosion may influence the pattern and evolution of dissolutional conduits. Kaufmann (2003) modeled mixing corrosion in a natural karst system recharged by sinking streams and in an artificial karst system with a fixed head controlled by a reservoir. In the natural system, mixing corrosion accelerates dissolution and favors the preferential enlargement of conduits, whereas in the artificial system, mixing corrosion is less important but may propitiate substantial leakage even if the impounded water is almost saturated with calcite. This author contends that mixing corrosion may play a significant role in the evolution of karst aquifers during both their early and mature stages when recharge waters are close to saturation.

The speleogenetic role played by mixing corrosion in porous carbonate rocks dominated by diffuse flow, rather than fracture permeability, may be significantly different. Dreybrodt et al. (2010) modeled dissolution in a porous limestone related to mixing corrosion that occurs where percolating surface water mixes by diffusion with phreatic water, both saturated with calcite but with different P_{CO_2} values. The modeled dissolution process, controlled by the position of the water table, results over a period of ca. 10 kyr in the development of horizontal blind-ended passages along the diffusion-mixing zone that resemble the isolated caves described in the eogenetic limestone karst of Florida (Florea 2006; Florea et al. 2007) or in the Caribbean islands (Mylroie and Carew 1990). Some authors propose that mixing corrosion may participate in the formation of solution pockets controlled by fractures in phreatic passages; the aggressiveness is renewed when the fracture and passage waters mix. However, Palmer (2007) contends that the most probable genetic mechanisms is localized dissolution by water injected into fractures during flooding.

The influence of freshwater and seawater mixing in coastal zones has a more complex influence on the dissolution of carbonate minerals (Plummer 1975; Wigley and Plummer 1976; Hanshaw and Back 1979; Palmer 2007). This typically involves a significant contrast in salinity and CO_2 content. Freshwater generally has higher CO_2 concentrations than seawater. Figure 3.28 illustrates the effect on calcite saturation of the mixture of different proportions of seawater and freshwater, both saturated with calcite, and the latter with variable P_{CO_2} values. The curves indicate that the mixture becomes undersaturated with small proportions of seawater and with a wide range of seawater proportions if the freshwater has a high CO_2 content. Therefore, renewed aggressiveness due to mixing occurs at the freshwater side of the mixing zone. The marked difference shown by the different curves suggests that the influence of the CO_2 content may be as important as the salinity contrast (Palmer 2007). Carbonate rocks in coastal areas may be affected by several mixing-corrosion cycles related to sea level oscillations. In the Yucatán Peninsula, Mexico, a karst plain underlain by fractured Cenozoic limestones, Back et al. (1984) showed that groundwater flows along 100 km toward the coast and becomes saturated at ca. 250 mg L^{-1} of dissolved calcite. However, in the final 1 km of their flow path close to the coast, an additional 120 mg L^{-1} of dissolved $CaCO_3$ is added to the groundwater attributable to the effects associated with freshwater and seawater mixing. They interpreted that the serrated morphology of the coast, with coves, lagoons, and headlands, is related to the collapse of fracture-controlled caves developed within the freshwater–seawater mixing zone. Dreybrodt and Romanov (2007) modeled porosity changes in the freshwater–saltwater mixing zone of carbonate islands and coastal aquifers with homogeneous and heterogeneous porosity assuming rapid dissolution (i.e., solution in equilibrium with calcite) and solving the advection-transport equation for salinity. Their results indicate that flank margin caves (Mylroie and Carew 1990, 2000) may develop in time scales of the order of tens of thousands to a hundred thousand years. They also point out that this time may be reduced by feedback mechanisms, such as enhanced flow and dissolution rates in zones of increased permeability. Baceta et al. (2001) document an ancient example of mixing corrosion in a Paleocene carbonate formation in the Spanish Pyrenees. The observed paleokarst features,

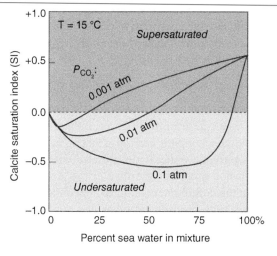

Figure 3.28 Variation in calcite saturation index where different proportions of freshwater and seawater mix at 15 °C. Both solutions are assumed to be saturated with calcite and P_{CO_2} values apply to the freshwater. At low CO_2 contents, the mixture becomes undersaturated with small proportions of seawater. At high CO_2 contents calcite undersaturation increases and occurs over a wider range of seawater proportion in the mixture. *Source:* Adapted from Palmer (2007).

interpreted as evidence of a freshwater–seawater paleomixing zone, including sponge-like porosity, transtratal collapse breccias (blind breccia pipes), solutionally enlarged joints with sandstone fillings, and dolomitized lenses.

3.7.6 Exotic Inorganic Acids

Other inorganic acids generated from sources different from the CO_2 and sulfur systems may enhance the dissolution of carbonate minerals, although their contribution is in most cases anecdotic (Ford and Williams 2007). The hydrochloric acid (HCl) significantly increases the solubility of calcite, although it rarely participates in the dissolution of carbonate rocks in natural conditions.

$$CaCO_3 + HCl \rightarrow HCO_3^- + Cl^- \tag{3.91}$$

The solubility of calcite is doubled by the addition of around $100\,mg\,L^{-1}$ of HCl. This strong acid occurs in volcanic gases, but at very low concentrations compared to acidic gases such as CO_2, SO_2, and H_2S. Interestingly, it is the most commonly used acid for enhancing the productivity of hydrocarbons in carbonate reservoirs due to its low cost and high dissolving power (Chang et al. 2008).

The oxidation of metals such as Fe^{2+} (see Eq. 3.75) or Mn^{2+} accompanied by the hydrolysis of water molecules is another potential source of acidity. Manganese in its reduced oxidation state (II) is a relatively common cation in oxygen-poor environments, but typically occurs at very low concentrations, and rapidly oxidizes forming precipitates of pyrolusite (MnO_2).

$$Mn^{2+} + \frac{1}{2}O_2 + H_2O \rightarrow MnO_2 + 2H^+ \tag{3.92}$$

Metal carbonates other than the common Ca-Mg-carbonates may enhance the acidity of the water. The most frequent ones are siderite ($FeCO_3$) and rhodocrosite ($MnCO_3$), but their solubilities are around 100 times lower than that of calcite (Langmuir 1997). Siderite is a relatively common diagenetic mineral in carbonate rocks and frequently forms speleothems. Two alternative reaction paths are:

$$2FeCO_3 + \frac{1}{2}O_2 + 5H_2O \rightarrow 2Fe(OH)_3 + 2HCO_3^- \tag{3.93}$$

or

$$FeCO_3 + H^+ \rightarrow Fe^{2+}HCO_3^- \tag{3.94}$$

$$Fe^{2+} \rightarrow Fe^{3+} + e^- \tag{3.95}$$

$$2Fe^{3+} + 6H_2O \rightarrow 2Fe(OH)_3 + 6H^+ \tag{3.96}$$

Kempe (2009) documents a maze-cave complex in siderite-rich Middle Devonian reef limestones in the Harz Mountains, Germany. The author interprets that weathering (dissolution and oxidation) of siderite in the oxic zone above the water table generates acidity that produced isolated caves around the siderite ores, which are partially filled by the by-product goethite ($Fe^{3+}O(OH)$).

3.7.7 Acid Rain

The corrosive effect of polluted acidic air and precipitation on limestone and marble stones was noted in the seventeenth Century by the English writer John Evelyn. The term acid rain was coined in 1872 by the Scottish chemist Robert Angus Smith. Acid rain is caused by emissions of sulfur dioxide (SO_2) and nitrogen oxides (NO and NO_2), generally referred to as NO_x, that experience oxidation reactions to form acids (sulfuric and nitric). Sulfur dioxide is mainly derived from the combustion of sulfur-bearing fossil fuels (e.g., coal with pyrite) and the smelting of sulfide ores. The main natural sources of this gas, much less important than the artificial ones, are emissions in volcanic areas (Figure 3.29) and wildfires (Camuffo 1992; Berresheim et al. 1995). Anthropogenic nitrogen oxides are mostly derived from high-temperature combustion in engines, largely from automobiles, and the manufacture of inorganic fertilizers. The emissions of these pollutants increased significantly after the Industrial Revolution. However, the global concentrations of SO_2 in the atmosphere (main source of acid rain) are decreasing, thanks to pollution regulation policies (Rodhe et al. 2002), whereas pollution by nitrogen oxides keeps on increasing (Charola and Ware 2002). The pH of rain water is typically between 5.6 and 6.4. In industrialized areas and in large regions situated hundreds of kilometers downwind, the pH often reaches values below 4.5, and measurements in rain and fog as low as 2.4 have been reported. The use of tall smokestacks reduces local pollution, but contributes to the spread of acid rain.

Sulfur dioxide (SO_2) transforms into sulfur trioxide (SO_3) through various oxidation reactions. In the presence of water, SO_3 converts rapidly into sulfuric acid (H_2SO_4), whose strong dissociation produces hydrogen ions and acidifies the water:

$$SO_3(g) + H_2O(l) \rightarrow H_2SO_4(aq) \rightarrow 2H^+ + SO_4^{2-} \tag{3.97}$$

Nitrogen dioxide reacts with water under oxidizing conditions to form nitric acid, which is also a strong acid:

$$NO_2(g) + \frac{1}{2}H_2O + \frac{1}{2}O_2 \rightarrow HNO_3(aq) \rightarrow H^+ + NO_3^- \tag{3.98}$$

Figure 3.29 Laguna Caliente crater lake in the summit of the active Poas Volcano, Costa Rica, which is considered to be one of the world's most acidic lakes, sometimes reaching pH values close to 0. The emission of gases, mainly SO_2, and its transformation into sulfuric acid, creates acid fog and rain, preventing the growth of vegetation, mainly in the downwind side (left). *Source:* Francisco Gutiérrez (Author).

Acid rain may be produced by wet and dry deposition. Dry deposition results from the transfer of pollutant gases and/or particles from the atmosphere to a surface in the absence of rain. Wet deposition involves the participation of any form of precipitation. Dry deposition is far more important than wet deposition in the acid deterioration of stone, and may have severe effects in areas located close to the source of the pollutants (Charola and Ware 2002). Acid rain produces multiple adverse impacts on forests, freshwater environments, soils, life-forms, and human health. It can cause the rapid degradation of building stone (stone decay), specially limestone, marble, and detrital rocks with carbonate cements. The largest contributor is SO_2, according to a reaction whereby sulfuric acid dissolves calcite and results in the precipitation of Ca-sulfates (sulfation reaction) and the release of CO_2:

$$CaCO_3 + SO_4^{2-} + 2H^+ + 2H_2O \rightarrow CaSO_4\ 2H_2O + CO_2 + H_2O \tag{3.99}$$

Note that this is the same reaction described for the replacement of carbonate by gypsum in sulfuric acid caves (Eq. 3.72). Gypsum, due to its high solubility, is generally washed away by rainwater. The etching of calcite crystals and the concomitant precipitation of gypsum reduces rock cohesion, favoring its progressive disintegration. The surface of carbonate rocks in sheltered zones protected from direct rainfall and affected by acid rain typically displays gypsum-rich black crusts underlain by a more friable layer (Camuffo et al. 1982). This layering facilitates the detachment of weathered planar fragments (flaking or spalling) (Charola and Ware 2002 and references therein). Bonazza et al. (2009) predict annual surface recession in carbonate building stones in Europe during the twenty first century using the Lipfert damage function and a global climate model. The Lipfert model allows the estimation of surface recession related to clean rain, wet acid rain, and dry deposition of sulfur and nitrogen oxides. Their results indicate that chemical dissolution of carbonate stones via the karst effect will increase with future CO_2 concentration, whereas recession related to wet and dry acid rain will experience a general decrease. This work reviews the different damage functions that can be used to estimate surface recession on carbonate building stones.

Sulfuric and nitric acids supplied by acid rain replace carbonic acid in some weathering reactions and reduce the bicarbonate flux (alkalinity) in surface waters. For instance, this is illustrated by the hydrolysis of albite to kaolinite, which involves the consumption of CO_2 and the production of bicarbonate:

$$2NaAlSi_3O_8 + 2CO_2 + 3H_2O \rightarrow Al_2Si_2O_5(OH)_4 + 2Na^+ + 2HCO_3^- + 4SiO_2 \tag{3.100}$$

According to Amiotte-Suchet et al. (1995), acid rain may contribute to reduce CO_2 consumption by weathering at a local scale, but it has a weak influence on the global CO_2 budget. Note that dissolution of carbonate minerals by sulfuric acid does not consume CO_2, but releases it (Eq. 3.99), and that acid rain is rapidly neutralized in areas with carbonate outcrops. Acid rain also affects the amount of dissolved organic matter (DOM) in surface waters. A number of studies document DOM increases in natural waters from across Europe and North America due to reductions in the emissions of sulfur oxides (SanClements et al. 2012). Liao et al. (2018) investigated temporal variations of DOM in the vadose zone of a karst aquifer affected by acid rain in China, applying fluorescence spectroscopy in two drip sites in Henshang Cave. They found that variations are controlled by monsoonal rains and that acid rain has no discernible effect on the quantity and quality of DOM due to the strong neutralizing power of carbonates.

While there are numerous investigations that explore the impact of acid rain on the degradation of carbonate stones, the effects on carbonate formations in natural environments has not received much attention. As Ford and Williams (2007) indicate, researchers dealing with karst water equilibria and erosion rates should be alerted to its potential effect. According to Charola and Ware (2002), it is extremely difficult to assess how much of the deterioration of a stone is due to acid deposition, since decay is due to the interaction of various natural and human-induced mechanisms. Most probably, the same consideration is applicable to natural systems.

3.8   Dissolution and Precipitation Kinetics of Karst Minerals

The dissolution of a mineral in water is achieved through the net transfer of mass from the solid phase to the aqueous phase. The chemical components detached from the mineral surface move as dissolved species toward the bulk of the solution through a diffusion boundary layer (DBL) (Figure 3.30). This is a heterogeneous reaction that involves two phases separated by an interface, in contrast to homogeneous reactions that take place in only one phase (e.g., $CO_2\,(aq) + H_2O \rightarrow H_2CO_3$). The dissolution rate will largely depend on the degree of undersaturation or disequilibrium of the solution. As dissolution proceeds, there is a quantifiable net reduction in the mass of solid and a concomitant net increase in the concentration of the chemical species derived from the dissolving mineral. At equilibrium, the rates of the forward (dissolution) and back reaction (precipitation) are equal, so that no net change is observable with time despite both processes are continually occurring. The solubility and equilibrium concepts in dissolution systems mainly refer to the quantity of mineral that can be dissolved in an amount of water under certain conditions, until it becomes saturated with respect to that mineral. The central problem of dissolution kinetics is the rate at which the dissolution reaction takes place, and how it is affected by multiple chemical and physical factors such as the degree of saturation, composition of the solution (e.g., pH, foreign ions), hydrodynamic conditions, dislocations and defects in the mineral surface, specific mineral surface area in contact with water, temperature, pressure, etc. Dissolution rate equations allow karst scientists to develop numerical simulations to better understand and predict the evolution of karst systems. For instance, the time required for the formation of a phreatic passage, or the expected increase in permeability in the foundation of a dam built on karst rocks.

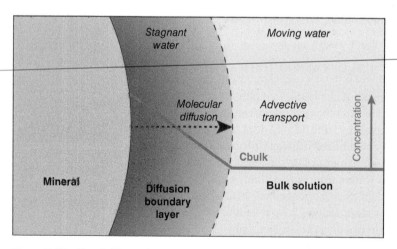

Figure 3.30   Sketch illustrating the diffusion boundary layer associated with the surface of a mineral in contact with flowing water. The dissolved components diffuse across the stagnant diffusion boundary layer toward the bulk solution, controlled by a transverse concentration gradient, from equilibrium at the mineral surface (C_{eq}), to the lower saturation levels of the bulk solution (C_{bulk}).

In a static aqueous solution, dissolved ions and molecules move slowly from regions of higher concentration to regions of lower concentration through the process of **molecular diffusion**. If the solution is moving, the dissolved components are dispersed by **advective transport**, which is much faster than molecular diffusion. In most karst situations, the water is in motion, and consequently advective transport dominates in the bulk solution. However, at the mineral–water interface there is a thin **diffusion boundary layer** (DBL) where the water is relatively static and the dissolved components move across it by molecular diffusion (Figure 3.30). In the boundary layer associated with a mineral undergoing dissolution there is a transverse concentration gradient, decreasing from saturation (or equilibrium; C_{eq}) at the mineral surface to the lower saturation levels of the bulk solution (C_{bulk}). The DBL can be envisioned as a more concentrated and semi-static water film attached to the surface of the minerals and rocks in contact with water that may contribute to slow down the overall dissolution process. The flux of chemical components through the DBL is explained by Fick's first law. In one dimension, it can be expressed in a molar basis as:

$$J = -D \left(\frac{C_{eq} - C_{bulk}}{X} \right) \tag{3.101}$$

where J is the diffusion flux given by the amount of substance per unit area per time, D is the diffusion coefficient or diffusivity, whose dimension is area per unit time, C_{eq} is the concentration of the species at the mineral–water interface, which is assumed to be the equilibrium concentration (moles per unit volume), C_{bulk} is the concentration in the bulk solution, and X is the thickness of the DBL. In karst dissolution systems, values of D are $1-2 \cdot 10^{-5}$ cm^2 s^{-1} at 25 °C, falling to about one half of these rates at 0 °C (Dreybrodt 1988). Fick's law indicates that diffusion flux is higher in situations in which the bulk solution is highly undersaturated (e.g., diluted solutions) and the DBL is thin (i.e., high concentration gradient). The thickness of the boundary layer depends on factors such as surface roughness, fluid viscosity, temperature, and especially the flow velocity and regime (laminar versus turbulent) in the bulk liquid. As explained below, high fluid velocities and turbulent flow conditions contribute to reduce the thickness of the DBL and accelerate the dissolution process in diffusion-controlled systems.

Traditionally, researchers have addressed the dissolution kinetics problem through laboratory experiments (Dreybrodt 1988). In the rotating disc experiment, a flat circular disc of mineral or rock rotates around an axis perpendicular to its plane and the surface of the disc interacts with a current of water directed perpendicularly to the mineral surface. As the flow approaches the disc, the flow lines bend over and finally proceed parallel to the surface of the disc. Transport of dissolved components occurs by molecular diffusion across the DBL attached to the disc surface, whose thickness can be reduced by increasing the spinning rate. This experiment is suitable to investigate diffusion-controlled (first order) kinetics under controlled hydrodynamic conditions, but has the disadvantage of the small ratio between the surface area of material exposed to dissolution and the volume of solvent (A/V). Therefore, a very long time may be required to reach near-equilibrium conditions. In the batch experiments, the area of the dissolving material can be greatly increased by adding tiny crystals or particles to a solution, which is stirred to keep the dissolving material in suspension. Dissolution rate can be increased by rising the stirring speed. The main drawback of this method is that the hydrodynamics of the system is poorly defined. In both experiments, the dissolution rates can be estimated by measuring variations in the chemical composition of the solution (e.g., Ca^{2+} in calcite and gypsum dissolution experiments).

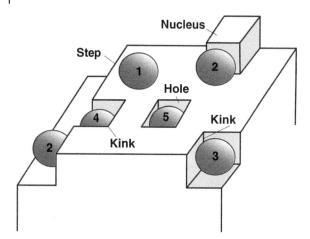

Figure 3.31 Sketch illustrating a mineral surface at the atomic scale showing different types of surface sites (steps, kinks, holes, and nuclei), as well as adsorbed ions (spheres). Numbers indicate the number of likely chemical bonds. *Source:* Adapted from Morse and Arvidson (2002).

During the past 30 years, Atomic Force Microscopy (AFM) has enabled direct observation of mineral surfaces at molecular resolution during dissolution and precipitation processes (Figures 3.20). This technique has contributed to better understand the mechanisms and rates of both processes, and the influence of multiple chemical and physical factors. Ruiz-Agudo and Putnis (2012) provide a comprehensive review for calcite. AFM shows that dissolution at the mineral surface preferentially occurs at irregularities such as steps, kinks, and holes controlled by the layered structure of the crystals (Figure 3.31). At these sites, atoms and molecules are attached to the solid by a lower number of bonds and consequently have higher free-energy available for dissociation (high-energy sites). Holes, typically controlled by defects and dislocations, are enlarged to form etch pits that may expand and get deeper, eventually coalescing with other nearby pits (Figure 3.32). Dissolution commonly occurs through the layer-by-layer removal of material via the retreat of steps of etch pits. High-energy sites are also favorable adsorption sites for reaction inhibitors, which can

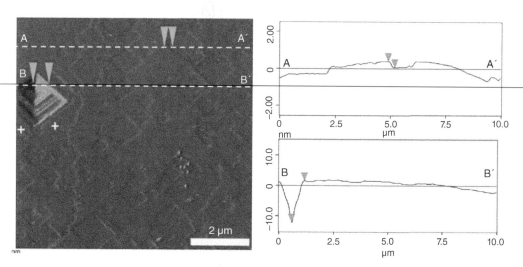

Figure 3.32 AFM deflection image showing deep (dislocation-nucleated) and shallow etch pits developed by dissolution on calcite cleavage surfaces in contact with a highly undersaturated solution. Transverse profiles show the local relief of shallow (AA′) and deep etch pits (BB′). The vertical distance between points marked with arrows is 0.3 and 16 nm, equivalent to ~1 and ~45 unit cells, respectively (1 nm = 10^{-9} m, 1 μm = 10^{-6} m). *Source:* Ruiz-Agudo and Putnis (2012). Published with permission of Cambridge.

substantially reduce dissolution rates even at low percentages of surface coverage (Morse and Arvidson 2002). Overall dissolution rates in AFM studies can be estimated from sequences of images of sufficient size, measuring the volume changes related to the retreat of steps and the expansion and deepening of etch pits. Under similar experimental conditions, there are significant differences between the dissolution rates estimated using AFM and those determined by conventional laboratory experiments. For instance, Arvidson et al. (2003) report AFM-derived dissolution rates for calcite one order of magnitude lower than those obtained in laboratory experiments (Figure 3.35a). The reasons for this large deviation can be attributed to the fact that AFM measurements are commonly obtained from smooth cleavage surfaces, which are less reactive than the irregular surface of natural mineral samples, with large steps and deep etch pits (Ruiz-Agudo and Putnis 2012). A more recent technique that allows the observation of changes on mineral surfaces at atomic scale is optical interferometry. This method provides sequential topographic maps of the mineral surface derived from time-lapse scans.

Dissolution reactions comprise a series of physical and chemical processes that can be separated into the following sequence of steps (Morse and Arvidson 2002): (i) Diffusion of reactants through the solution to the solid surface across the boundary layer; (ii) Adsorption of the reactants on the solid surface; (iii) Migration of the reactants along the surface to an "active" site (e.g., step, kink); (iv). Chemical reaction between the absorbed reactant and solid, which may involve the breakage of bonds and the hydration of ions; (v). Migration of products away from the reaction site; (vi). Desorption of the products to the solution; and (vii) Diffusion of products from the mineral surface to the bulk solution across the boundary layer. For instance, in the case of calcite dissolution, a H^+ ion that has diffused through the boundary layer adsorbs on the mineral surface, moves across until encountering a CO_3^{2-} molecule at an active site. The HCO_3^- ion created in the reaction will diffuse away, exposing a Ca^{2+} atom, which dissociates in its turn. The dissolution rate will be determined by whichever step of the sequence is the slowest. The reaction cannot proceed faster than that rate-limiting step. When the dissolution rate is limited by steps 1 and/or 7, which involve the diffusion of reactants and products across the DBL, the reaction is said to be diffusion-controlled. When any of the processes that occur at the surface of the mineral (steps 2–6) limits the rate, the reaction is said to be surface-controlled. Consequently, Berner (1978) differentiated two main types of dissolution systems: **diffusion-controlled** (or transport-controlled) systems, and **surface-controlled** (or surface-reaction-controlled) systems. Dissolution of high-solubility minerals (e.g., halite), which tend to release dissolved products very rapidly, are dominated by diffusion-controlled kinetics. The rate is limited by how rapidly the dissolved material is transferred through the DBL away from the solid surface. In contrast, relatively insoluble minerals (e.g., calcite at $pH > 6$) are dominated by surface-controlled dissolution kinetics. The dissolution rate is mainly limited by the slow pace at which the dissolution reaction occurs at the surface of the mineral. Dissolution rate has a linear relationship (first order) with the saturation state or ratio (Ω) in diffusion-controlled systems, and a nonlinear relationship (higher order) in surface-controlled systems (Figure 3.33). In the latter ones, dissolution rate may decrease drastically as the solution approaches saturation. This has relevant implications for karst development, since it allows water to keep on dissolving rocks along a considerable flow distance before reaching saturation.

3.8.1 Dissolution Kinetics of Calcite, Aragonite, and Dolomite

Morse and Arvidson (2002) published a comprehensive review on the dissolution kinetics of the main carbonate minerals (calcite, high-Mg calcite, aragonite, and dolomite) based on laboratory studies, mostly carried out with simple dilute solutions. They indicate that dissolution rates may be

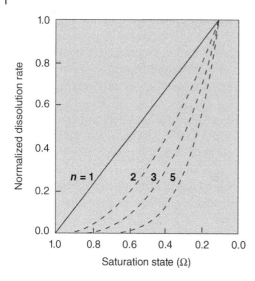

Figure 3.33 Plot of normalized dissolution rate versus saturation state showing the shape of the dissolution rate equation (Eq. 3.102) for first order (n = 1; diffusion-controlled) and higher order (n > 1; surface-controlled) kinetics. *Source:* Adapted from Morse and Arvidson (2002).

considerably different in the chemically more complex natural systems. The dissolution rate of carbonate minerals is commonly expressed using the empirical equation:

$$R = -\frac{dm_{calcite}}{dt} = \left(\frac{A}{V}k\right)(1-\Omega)^n = k^*(1-\Omega)^n \tag{3.102}$$

where R is the dissolution rate (e.g., μmol m^{-2} h^{-1}), m is moles of calcite, t is time, A is the total surface area of the solid in contact with the solution, V is the volume of the solution, k is the rate constant, Ω is the saturation ratio (ratio between the ion activity product and solubility product), and n is a positive constant known as the order of the reaction. Values of n and k vary with saturation ratio, temperature, and P_{CO_2} (Palmer 2007). Note that $(1-\Omega)$ indicates the degree of disequilibrium or undersaturation, since $\Omega = 1$ at saturation, and $\Omega < 1$ in undersaturated solutions. In plots of dissolution rate versus saturation ratio, the rate equation is expressed as a straight line for $n = 1$ (diffusion-controlled or transport-controlled dissolution systems) (Figure 3.33). In this first-order kinetics, the dissolution rate has a linear relationship with the saturation ratio. In contrast, as the order of the reaction increases above 1 (surface-controlled systems), the rate equation in the plot becomes increasingly less linear. There is a more rapid decrease in the dissolution rate as the saturation ratio of the solution increases and the deviation from the linear relationship increases with the order.

Plummer et al. (1978) proposed that the dissolution kinetics of calcite in CO_2-H_2O systems is controlled by three processes that occur at the surface of the mineral: (i) reaction with H^+; (ii) reaction with $H_2CO_3^*$ ($H_2CO_3^* = CO_2(aq) + H_2CO_3$); and (iii) dissociation of $CaCO_3$. These authors, using the temperature-dependent equilibrium constants for the different reactions, proposed a theoretical net dissolution rate equation; the "PWP equation" (PWP stands for Plummer–Wigley–Parkhurst):

$$R = k_1 \cdot aH^+ + k_2 \cdot aH_2CO_3^0 + k_3 \cdot aH_2O - k_4 \cdot aCa^{2+} \cdot aHCO_3^- \tag{3.103}$$

where a represents activity at the surface of the mineral and k_4 refers to the back reaction of calcite precipitation. The aim of this equation was to understand the processes of calcite dissolution rather than to apply it to field problems. In fact, it considerably deviates from experimental data (Fig. 3.35).

Palmer (1991) recalculated the original PWP data of Plummer et al. (1978) to produce a simple and practical equation applicable to karst conduits and valid for any flow condition, P_{CO_2}, or temperature:

$$S = 31.56 k \left(1 - \frac{C}{C_s} \right)^n / \rho \qquad (3.104)$$

where S is the rate of solutional wall retreat ($cm\,yr^{-1}$), k and n are respectively the reaction coefficient and the reaction order with variable values that can be obtained from Palmer (1991, table 1), C/C_s is the saturation ratio and ρ bedrock density. The constant 31.56 converts the results to $cm\,yr^{-1}$.

Numerous studies have demonstrated that at extreme undersaturation conditions and low pH values (pH <4–5), the dissolution of calcite is dominated by diffusion-controlled (or transport-controlled) kinetics. This is designated as the H^+-dependent regime, since dissolution rate is mainly governed by diffusion of H^+ across the diffusion boundary layer (Figure 3.34). As equilibrium is approached and the pH value rises, the importance of surface-controlled kinetics increases. Note that calcite dissolution has a buffering effect, producing a rise in the saturation ratio, and an accompanying increase in the pH. This is the so-called H^+-independent regime, in which dissolution rate mainly depends on the speed of the reaction that occurs at the surface of the mineral. Between the H^+-dependent and H^+-independent regimes there is a transition region with complex kinetics. The boundaries between H^+-dependent, transitional, and H^+-independent regimes move to lower pH values with increasing temperatures (Morse and Arvidson 2002). Figure 3.35 shows the variation of calcite dissolution rate in relation to pH according to the theoretical PWP rate model, some dissolutional experiments carried out with calcite powder, and AFM surface measurements. Despite the discrepancies, the functions show a similar form with two main segments: (i) a steep linear section at low pH values (transport-controlled kinetics), indicating a rapid decrease in dissolution rate as dissolution proceeds and the pH rises; and (ii) a gently sloping to flat segment at circumneutral and alkaline pH values, in which the dissolution rate barely changes. Note that within the range of pH values commonly observed in karst environments (ca. 5–8.9), there is a deviation of about one order of magnitude between the rates

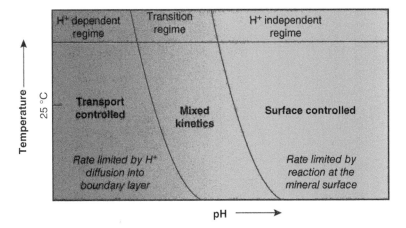

Figure 3.34 Sketch showing the mechanisms that control calcite dissolution rate as a function of pH and temperature. The boundaries between the different regimes move to lower pH values with increasing temperatures. *Source:* Adapted from Morse and Arvidson (2002).

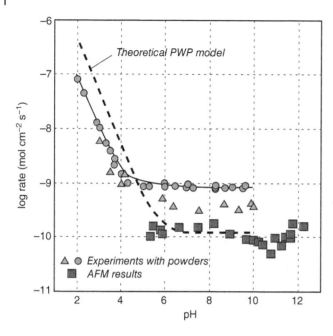

Figure 3.35 Comparison of theoretical and experimental calcite dissolution rates in relation to pH, obtained by various authors. The theoretical model (dashed line) corresponds to the PWP equation of Plummer et al. (1978). Experimental rates include results from dissolution experiments with powders and rates estimated from microscale AFM studies. Note that the latter, although in accordance with the theoretical curve, are up to one order of magnitude lower than the experimental results. *Source:* Adapted from Morse and Arvidson (2002).

measured in experiments using powders and the theoretical and AFM values. Figure 3.36 illustrates the decrease in calcite dissolution rates with saturation in an open system with turbulent flow. This graph produced by Palmer (1991) recalculating data from Plummer et al. (1978), indicates a significant decrease in the dissolution rate at a saturation ratio (C/C_s) of 0.7–0.8, changing from a reaction order n of 1.5 to 4.4. The initial high dissolution rates at $C/C_s < 0.1$ are obviated due to disagreement with other experimental data. A similar decrease in the dissolution rate has also been observed in experiments with laminar flow through artificial fractures in limestone (Palmer 2007). This deceleration of the dissolution rate at near-equilibrium conditions is very important for the development of karst aquifers and caves. High-order dissolution allows the enlargement of discontinuities rather uniformly over long flow distances before reaching saturation. Without this decrease in the dissolution rate, dissolution in most cases would be restricted to short distances from the input points of the water.

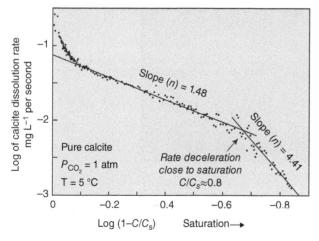

Figure 3.36 Plot of calcite dissolution rate versus saturation ratio produced by Palmer (1991) using data obtained by Plummer et al. (1978) in open-system experiments with calcite suspended in turbulent water. ($1-C/C_s$) indicates the degree of undersaturation. The logarithmic scale is used to produce straight lines. Dissolution rate significantly decelerates at a saturation ratio of 0.7–0.8, changing into higher order kinetics. The initial points at $C/C_s < 0.1$ indicating high-order rate are ignored due to disagreement with other empirical data, suggesting that they may be an artifact of the experimental method. *Source:* Adapted from Palmer (1991, 2007).

The H^+-dependent dissolution regime occurs under strong acidic conditions with pH values (<4–5) lower than those commonly found in most natural carbonate karst systems (Figure 3.34). This transport-controlled dissolution kinetics may apply in systems with sulfuric acid dissolution, acidic water derived from peat deposits, and areas with anthropogenic sources of acidity (e.g., acid mine drainage, highly acidic rains). Dissolution rate is mainly governed by diffusion of H^+ across the DBL (k_1 constant of Eq. 3.103). Therefore, the hydrodynamic conditions, which control the thickness of the boundary layer, largely determine the dissolution rates. The evolution of the system in this transport-controlled regime depends on whether the solid or the solution dominates the mass action. If the solid dominates (high A/V ratio; A/V: area/volume), the solution composition changes rapidly (pH and saturation ratio increase) and the system moves promptly toward surface-reaction conditions (e.g., limited flow of acidic water along a narrow joint). If the solution dominates (low A/V ratio), there is limited change in the solution composition and the solid may be fully dissolved without changing into surface-controlled conditions (e.g., a small calcite particle blown into an acidic lake). In the transition region to surface-controlled dissolution, characterized by mixed kinetics, the processes described by the constants k_1, k_2, and k_3 (diffusion and surface reaction components) in the PWP equation (Eq. 3.103) play a controlling role. There is intermediate dependence of dissolution rate on the hydrodynamic conditions and the type and density of surface defects may have a significant influence. In the H^+-independent regime, dissolution rate is dominated by surface-controlled kinetics. According to Palmer (2007), this is the regime in which most limestone dissolution occurs. The solution is close to saturation and has a low H^+ concentration, with the pH approaching the upper limit. Diffusion across the boundary layer does not retard dissolution rate, which is therefore independent on the hydrodynamic conditions. A number of experiments reveal that within this field, dissolution rate is significantly influenced by the type and density of surface defects that provide reactive sites highly susceptible to dissociation. Consequently, as Morse and Arvidson (2002) indicate, there is no general dissolution rate equation applicable to all calcites, since there may be significant variations depending on the density of defects. For instance, Schott et al. (1989) measured dissolution rates two to three times higher in strained calcites with a high density of dislocations than in unstrained calcites under near-equilibrium conditions.

According to Ruiz-Agudo and Putnis (2012), dissolution in calcite cleavage surfaces at very high undersaturation mainly occurs through the retreat of surface steps associated with rhombohedral etch pits, either preexisting or created upon contact with a solution. These sites provide energetically favored step edges in comparison with atomically flat surfaces. The location of the etch pits is primarily controlled by defects: point defects in the case of shallow etch pits and dislocations in the case of deep etch pits (Figure 3.32). In contrast, in near-equilibrium aqueous solutions, etch pits do not form. Teng (2004) suggested that calcite dissolution occurs by three different mechanisms depending on the degree of undersaturation: (i) etch-pit formation on defect-free surfaces at extreme undersaturation; (ii); defect-assisted etch-pit development at medium undersaturation; and (iii) dissolution at preexisting steps close to equilibrium. This author also identifies critical saturation ratios for the transition between the different dissolution mechanisms ($\Omega = 0.007$ for stage 1–2 and $\Omega = 0.4$–0.5 for stage 2–3).

Laboratory experiments and AFM studies demonstrate that the presence of some foreign ions may significantly inhibit calcite dissolution, especially under near-equilibrium conditions. This inhibiting effect, which increases with the concentration of the foreign species, is ascribed to the adsorption of ions and ion pairs to the crystal surface, especially at dislocations that function as dissolution sites (Figure 3.31). Adsorbed impurities pin steps at specific points, reducing the rates of step retreat. Phosphate is the main inhibitor with significant impact on calcite dissolution in seawater (Morse and Arvidson 2002). An increase in the reaction order up to about 16 at phosphate concentrations as low as 10 μM has been observed, ascribed to adsorption of the ions to reactive sites. Microscale studies using AFM show that Mg^{2+} at low concentrations (<1 mM) also inhibits calcite dissolution. At conditions far from equilibrium, the inhibitory effect is attributed to the reduction in the etch-pit spreading rate, but the effect is higher at near equilibrium conditions. Dreybrodt and Eisenlohr (2000) propose that inhibitors released from the dissolution of carbonate minerals (intrinsic impurities) progressively adsorb on the surface of the mineral, reducing dissolution rates. They indicate that this may explain the discrepancies observed between experimental studies obtained from freshly broken samples and the slower rates observed in surfaces already affected by significant dissolution, which are a better approximation to natural conditions. This introduces the concept whereby the rate laws may depend on the history of dissolution. Other geologically relevant divalent cations including Mn^{2+}, Sr^{2+}, and Zn^{2+} also act as dissolution inhibitors (Ruiz-Agudo and Putnis 2012). The inhibitory effect of heavy metals has also been explored in laboratory experiments and AFM studies (e.g., Ni^{2+}, Cd^{2+}, Hg^{2+}, and Co^{2+}). These investigations are generally aimed at optimizing industrial processes, exploring the role played by calcite in reducing the bioavailability of toxic elements, and improving our understanding of the long-term behavior of geological environments (e.g., salt formations) used for the storage of hazardous products (radiogenic and chemical wastes) that may create anomalous geochemical conditions (Terjesen et al. 1961). However, it is unlikely that trace elements have a significant effect on dissolution processes in most natural karst systems.

Dreybrodt and Eisenlohr (2000) indicate that the theoretical rate equation proposed by Plummer et al. (1978) (PWP Eq. 3.103) is valid for pure synthetic calcite and can be applied if the concentrations of the ions at the surface of the mineral are known. However, this is a major problem, since only concentrations in the bulk solution are confidently known. Moreover, to estimate dissolution rates in natural hydrogeological systems other processes should be considered: (i) Diffusional mass transport of all the species involved in the reactions, which is largely influenced by the hydrodynamic conditions. Under highly turbulent conditions, the thickness of the DBL decreases and mass transport by diffusion is enhanced. (ii) Reaction rates involving the conversion of CO_2 to H^+ and HCO_3^-, which is a relatively slow process that provides the H^+ ions for converting carbonate to bicarbonate ions. The release of each Ca^{2+} ion involves the consumption of one CO_2 molecule. Consequently, the condition of the system as open or closed with respect to CO_2 can be a rate-limiting factor. Taking these processes into account, a dissolution model for both open and closed conditions was developed (Dreybrodt 1988; Dreybrodt and Eisenlohr 2000). In the closed system, two parallel calcite planes separated by a distance of 2δ confine the water. The ratio of the volume V of the solution to the area A of the surfaces is $V/A = \delta$. The CO_2 concentration drops as the Ca^{2+} concentration increases, because CO_2 is not replenished. Laminar versus turbulent flow conditions influence diffusional mass transport by controlling the thickness of the DBL. The rate laws can be approximated by the simple linear relation:

$$F = \alpha \left(C_{eq} - C \right) \tag{3.104}$$

where α is a transport coefficient that depends on the temperature, the P_{CO_2} of the solution, δ, which is the thickness of the DBL, and the type of flow (i.e., from laminar to highly turbulent). Figure 3.37 depicts the values of α as a function of δ for closed systems with laminar and highly turbulent flow conditions. This graph shows how the V/A ratio and hydrodynamic conditions influence dissolution rates. At low δ, where the volume of the solution is limited compared to the surface area, there is a linear increase of α, equal for laminar and turbulent flow. This pattern

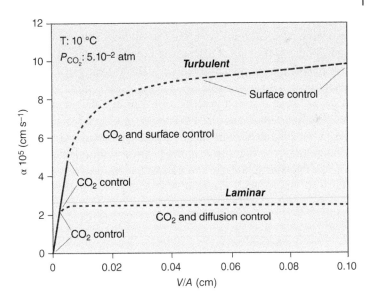

Figure 3.37 Transport coefficient α of calcite dissolution in a CO_2-closed system as a function of $\delta = V/A$ for laminar and highly turbulent flow. The annotations indicate the different dissolution rate controls for the different fields. *Source*: Adapted from Dreybrodt and Eisenlohr (2000).

illustrates a dissolution regime controlled by the CO_2 conversion. For $\delta > 2\cdot10^{-3}$ cm the laminar flow values of α become constant due to mixed rate control by CO_2 conversion and diffusional mass transport. This is a joint with a very small aperture of 0.04 mm. For turbulent flow an additional linear increase is observed, followed by a region of increasing α, where the rate is controlled by both CO_2 conversion and surface reactions. For $\delta > 0.05$ cm (0.1 cm wide joint) the rates become constant and are controlled entirely by the surface reactions. In this upper region, rates are given by the PWP equation. This figure illustrates that the PWP equation can lead to overestimated rates, since the theoretical equation does not incorporate the rate limiting processes of CO_2 conversion and diffusional mass transport (see also Palmer 1991, 2007).

Aragonite, which is slightly more soluble than calcite and dolomite (Table 3.1), shows reaction kinetics and rate constants similar to those of calcite (Morse and Arvidson 2002 and references therein). The dissolution of aragonite, which is the dominant carbonate mineral that precipitates in modern shallow-water tropical environments, plays a fundamental role during its diagenetic inversion to calcite (polymorph transformation). Magnesian calcites, common in modern biogenic carbonate sediments, are solid solutions in which a considerable proportion of the Ca^{2+} sites are occupied by smaller-radius Mg^{2+} cations. These substitutions contribute to reduce the stability of the crystal lattice and to increase solubility. Intermediate-Mg calcite (4–12 mol% $MgCO_3$) has higher solubility than calcite, and high-Mg calcite (>12 mol% $MgCO_3$) is even more soluble than aragonite (Figure 3.11). According to Morse and Arvidson (2002), despite the difficulties in analyzing the dissolution kinetics of magnesian calcite, available data indicate that reaction orders are quite similar to those reported for calcite.

Dolomite dissolution kinetics has received much less attention than calcite, partially due to the slowness of the process (Morse and Arvidson 2002). Commonly, there is a direct relationship between the solubility of minerals and their dissolution rate. That is, the higher the solubility, the faster the dissolution process. Dolomite is somehow anomalous. At typical conditions in karst systems, dolomite has a slightly lower solubility than calcite (see Table 3.1), but it dissolves much more slowly. Busenberg and Plummer (1982) obtained laboratory measurements of dolomite dissolution rates with hydrothermal and sedimentary dolomite. Their experiments covered a wide range of conditions: pH = 2–10; P_{CO_2} = 0–1 atm; T = 1.5–65 °C, but always far from equilibrium due to the slow dissolution rates. Forward rate was found to be dependent on pH at pH < 6, on P_{CO_2}

over the entire pH range, and to be constant at pH > 8. Dissolution was nonstoichiometric (incongruent dissolution), with the Ca/Mg ratio released to the solution higher than in the bulk solid. Similar to the case of calcite, the process was described with a rate equation including three forward reactions of dolomite with H^+, $H_2CO_3^*$, and H_2O, and a backward reaction related to the adsorption of HCO_3^- at positively charged surface sites (see Eq. 3.103). Figure 3.38a depicts dolomite dissolution rate as a function of pH from different sources, including data from Busenberg and Plummer (1982) and the PWP curve of calcite for comparison. The graph shows a similar shape to that of calcite, with an H^+-dependent behavior under acidic conditions. Dissolution rates are more than one order of magnitude lower than calcite, but unfortunately data for pH 5.0–8.9, which is the typical range in most karst systems, are very scarce. In an ideal stoichiometric dolomite $(Ca_{50}Mg_{50}(CO_3)_2)$, layers of Ca^{2+}, CO_3^{2-}, and Mg^{2+} alternate regularly. Busenberg and

(a)

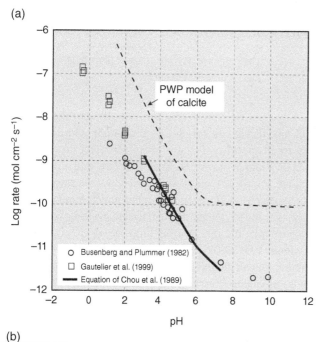

Figure 3.38 (a) Dolomite dissolution rates as a function of pH measured by various authors and the PWP curve for calcite of Plummer et al. (1978). Data for dolomite show significant deceleration at around pH = 6 and rates more than one order of magnitude lower than those of calcite. *Source:* Adapted from Morse and Arvidson (2002). (b) Variation of saturation index during a dissolution run with a rotating disc of microcrystalline sedimentary dolomite and turbulent flow. Note the rapid decrease in the dissolution rate at conditions far from equilibrium (SI ~ −3; Ω ~ 10^{-6}). *Source:* Adapted from Herman and White (1985).

(b)

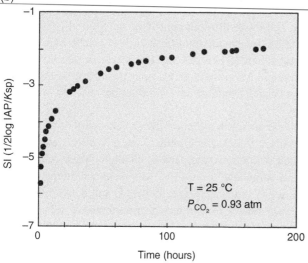

Plummer (1982) attributed the slowness of the dolomite dissolution process to two factors. One of them is that the $Mg\text{-}CO_3$ bonds are stronger than the $Ca\text{-}CO_3$ bonds, so the slow dissociation in the $MgCO_3$ crystal lattices is rate-limiting. Additionally, as the concentration of the solution increases ($SI < -3$), the back reaction related to the absorption of HCO_3^- on positively charged sites contributes to reduce the net dissolution rate.

In a subsequent work, Herman and White (1985) carried out dissolution kinetics experiments using three types of stoichiometric dolomite (hydrothermal single crystal, microcrystalline sedimentary rock, and coarse-grained dolomite marble) with the rotating disc approach, which allows controlling hydrodynamic conditions. The dissolution runs were performed with deionized water, $P_{CO_2} \sim 1\,atm$, $T = 0, 15, 25\,°C$. These authors obtained the following main results: (i) at conditions far from equilibrium (saturation ratio $\Omega \sim 10^{-6}$) dissolution rate slows down significantly, and subsequently the approach to saturation is extremely slow (Figure 3.38b); (ii) a considerable increase in the dissolution rate is observed at the transition from laminar to turbulent flow and the influence of flow conditions increases as the saturation ratio increases; (iii) in general, the observed rates are lower than those predicted by the equation of Busenberg and Plummer (1982); and (iv) dissolution rates for all the three types of samples were similar in form and rate, but rate values increased with decreasing grain size; by a factor of 1.5 between a single large crystal and microcrystalline rocks.

3.8.2 Dissolution Kinetics of Gypsum and Halite

As explained above, the dissolution process comprises two steps: (i) the release of dissolved chemical species at the mineral surface (i.e., chemical reaction); and (ii) their diffusion toward the bulk solution through the DBL (i.e., transport) (Figure 3.30). The kinetics of the dissolution process at each time is controlled by the slowest step, either the surface reaction (surface-controlled kinetics) or the transport of the dissolved material away from the mineral surface (transport-controlled kinetics). Dissolution rate in the latter regime is strongly influenced by hydrodynamic conditions. High flow velocities and turbulence contribute to reduce the thickness of the DBL, accelerating dissolution. Where the reaction rate at the surface is extremely high, as occurs with high-solubility minerals such as halite (Alkattan et al. 1997a), the kinetics of the dissolution process is transport-controlled. Where the surface reaction is slower than diffusional transport, the process is surface-controlled. There is an intermediate case of mixed kinetics, in which the rate-limiting step varies depending on the saturation state of the solution. Close to equilibrium, the surface-reaction becomes slower than diffusional transport and the regime changes from transport- to surface-controlled. According to experimental studies (Frenkel et al. 1989; Raines and Dewers 1997b; Jeschke et al. 2001), gypsum is an example of mixed kinetics, in contrast to previous researchers that proposed that the dissolution rate was exclusively transport-controlled (James and Lupton 1978; Berner 1978; White 1988).

Jeschke et al. (2001) investigated gypsum dissolution kinetics applying two experimental procedures: (i) free-drift batch experiments with powders of synthetic gypsum, selenite ("marienglas"), alabaster, and gypsum rock starting with ultrapure water; (ii) rotating disc experiments with selenite. The former allows measuring dissolution rates close to saturation, whereas the second one provides good control of the hydrodynamic conditions and the thickness of the boundary layer. The surface-reaction rate constant and the reaction order can be estimated by combining results of both experiments. The results of the batch experiments revealed two dissolution regimes, with an almost linear rate law ($n = 1\text{--}1.2$) up to $0.9\,C_{eq}$ (saturation ratio $\Omega \sim 0.9$) and highly nonlinear kinetics close to equilibrium with a reaction order $n = 4\text{--}5$. That is, when the solution was close to

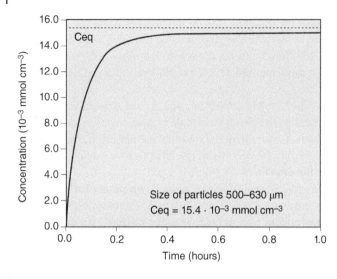

Figure 3.39 Free-drift batch experiment carried out with natural selenite powder. The variation of Ca^{2+} concentration with time shows a rapid deceleration of the dissolution rate at ca. 0.9 C_{eq}, indicating a change from transport- to surface-controlled kinetics. The latter is attributed to the inhibiting effect caused by the accumulation of impurities on the mineral surface. *Source:* Adapted from Jeschke et al. (2001).

saturation, the dissolution rate decelerated significantly and changed from transport- to surface-controlled kinetics (Figure 3.39). Jeschke et al. (2001) also carried out batch experiments with pure synthetic gypsum, but in this case dissolution showed a linear rate law up to saturation conditions. The authors proposed that the near equilibrium nonlinear kinetics in natural samples is related to the accumulation of impurities on the mineral surface that inhibit the dissolution reaction.

Raines and Dewers (1997b) investigated gypsum dissolution kinetics with a mixed flow reactor and the rotating disc technique using gypsum rock and different types of solutions with low ionic strengths and a wide range of saturation indexes. They observed a sharp increase in the dissolution rate at the transition from laminar to turbulent conditions, followed by a rise up to a constant value independent of the spinning velocity. The authors interpreted that this could indicate that the hydrodynamical boundary layer has reached a minimum thickness under turbulent conditions, or that surface reaction becomes rate limiting. Raines and Dewers (1997b) illustrate with a coupled flow and dissolution model the importance of mixed kinetics for gypsum karst development. Similar to limestone, the net effect is that due to the marked decrease in dissolution rate close to saturation, water can enlarge conduits and discontinuities along much longer distances than those predicted by the models that assume transport-controlled kinetics over the full range of concentrations (James and Lupton 1978).

Colombani (2008), using holographic interferometry, estimated the pure surface reaction rate constant of gypsum (k_s) in conditions unambiguously free from the influence of diffusional mass transport. This is the rate constant of the equation

$$R = k_s \left(1 - \frac{C}{C_{eq}}\right)^n \tag{3.106}$$

where R is dissolution rate, C is concentration of dissolved species at the mineral surface, C_{eq} their solubility, and n a constant. Holographic interferometry allows observing dissolution in the absence of any convective flow and measuring variations in the concentration of the solution attached to the dissolving mineral by changes in the refraction index. This author also deduced a pure surface reaction rate constant by extrapolating the inverse of the slope of the $R(c)$ curves (dissolution rate versus concentration) from multiple experimental results with a boundary layer thickness equal to 0, obtaining a striking agreement with the holographic interferometry measurements.

Halite has a solubility in pure water of $424\,g\,L^{-1}$ at 25 °C and 1 atm (mass of dissolved solute in 1 L of solution). This high-solubility mineral is characterized by a very rapid dissolution reaction at the mineral surface and therefore the dissolution rate is governed by a diffusion-controlled regime and the associated hydrodynamic conditions. Alkattan et al. (1997a) illustrate the inverse linear relationship between dissolution rates and the NaCl concentrations of the solutions. Stiller et al. (2016) investigate the rate of salt dissolution by the chemically complex Dead Sea brines diluted with groundwater, to gain insight into the development of sinkholes in the Dead Sea coasts. Experiments were carried out with halite slices from Mount Sedom and various dilutions of Dead Sea brines under stirring and no stirring conditions. Interestingly, they observed that dissolution rates in diluted Dead Sea brines under stirring conditions were as much as three times slower than those measured by Alkattan et al. (1997a) in a similar experimental set-up with pure NaCl solutions of the same salinity. The differing rates are ascribed, in addition to the different set-ups and calculation methods, to the high concentrations of Mg and Ca in the Dead Sea brines, and the activity of the water. Alkattan et al. (1997a, 1997b) investigated the influence of trace metals and anions on the dissolution rate of halite using compressed halite powders and the rotating disc method. Trace quantities of Cd, Pb, Cr, Co, and Zn, in order of importance, significantly lowered dissolution rates, whereas Fe and Zn had a negligible effect on dissolution kinetics. The metal Cd^{2+} has a dramatic effect, with a sevenfold decrease in the dissolution rate with the addition of just $0.01\,mol\,kg^{-1}$ of $CdCl_2$. It is interpreted that metals such as Cd^{2+} form ion pairs ($CdCl^+$) that attach to the crystal surface reducing the number of Na—Cl bonds exposed to dissociation (Alkattan et al. 1997a). In order of their degree of inhibition, trace concentrations of dissolved F^-, Br^-, and I^- also reduced the dissolution rate of halite, with a direct relationship between the concentration of each anion and its inhibiting effect (Alkattan et al. 1997b).

3.9 Geological Controls on Karst Development

There is frequently a large gap between dissolution processes investigated in laboratory experiments and numerical models, and those that operate in real karst systems over geological time scales, as the comparison of some results with field data reveals. In addition to the spatial and temporal scale problems (Viles 2001), a number of geological controls that have a critical impact on karst development may be difficult to explore in experiments and simulations. Some of those factors include the multiple characteristics of the karst and non-karst rocks of the stratigraphic successions (e.g., composition, texture, mechanical resistance, and erodibility), the presence of heterogeneous networks of discontinuity planes affected by solutional widening (e.g., fractures, bedding planes), geological structures with complex geometrical relationships, the participation of mechanical erosion processes, or hydrological and hydrochemical conditions that are experiencing continuous spatial and temporal changes. A comprehensive characterization of the geological setting and its evolution is essential to understand the hydrology and morphogenesis of karst systems.

3.9.1 Rock Composition and Purity

The proportion of soluble minerals and impurities in the rocks largely determines the feasibility of karst development, dissolution rates, and the type and dimensions of karst features that may form. The most common impurities in karst rocks are clay minerals and quartz. These non-soluble components inhibit dissolution and at certain proportions may prevent the development of karst features. In the case of carbonate rocks, it is commonly accepted that well-developed karst mostly occurs in bedrocks with bulk purities (percentage of $CaCO_3$ and $CaMg(CO_3)_2$) greater than 80–90% (Bögli 1980; Ford and Williams 2007). Klimchouk and Ford (2000a) indicate that most limestone and dolostone caves are associated with bulk purities greater than 90%. The distribution of impurities within the rock is also an important factor. Limestones or dolostones with a significant proportion of silica in the form of scattered chert nodules may be a good arena for karst, whereas carbonate rocks with the same proportion of silica, but present as dispersed fine particles (e.g., quartz grains), may not be suitable for karst. The ternary diagram shown in Figure 2.2 provides a good reference for classifying carbonate rocks according to the proportion of calcite (plus aragonite), dolomite, and impurities. Most carbonate karsts are expected to occur in lithologies at the basal part of the diagram, opposite to the impurities vertex, and especially close to the calcite vertex. Calcite has a more rapid dissolution kinetics than dolomite, despite their similar solubility (Figure 3.40a). Rauch and White (1970) investigated the lithological control on the development of caves in a heterogeneous Ordovician–Cambrian carbonate succession in the folded Appalachians of central Pennsylvania. They conducted chemical and petrographic analyses of the different stratigraphic units and estimated the volume of the accessible caves longer than 30 m as a proxy for conduit permeability. The total volume of all caves in each unit divided by the thickness of the unit provided a weighted cave frequency value for each unit. They found that: (i) the bulk of the cave volume is concentrated in a few units; (ii) most caves occur in limestone, and are extremely rare in dolomite; (iii) clay content and other impurities inhibit cave development; and (iv) caves are proportionally more abundant in micritic limestones than in coarser sparry limestones. Consistent findings were obtained in a subsequent paper by Rauch and White (1977) via dissolution rate experiments carried out with rock samples from the same stratigraphic succession, circulating aqueous solutions through cylindrical holes (23 °C, $P_{CO_2} = 1\,atm$). They observed decreasing dissolution rates with increasing contents of dolomite and impurities. Dissolution rates at 22% saturation for dolomites were at least two times lower than those for pure limestones, with the lowest values obtained for impure dolomites. Shtober-Zisu et al. (2015) document half-tube-shaped subhorizontal notches up to hundreds of meters long on carbonate slopes of Mount Carmel (Israel) generated by differential weathering along specific beds with higher purity, as revealed by the amount of insoluble residue left after 5% HCl attack. Plan (2005) measured dissolution rates at field sites of the Austrian Alps using tablets of different lithologies. He found an overall direct relationship between the dissolution rates and the Ca/Ca + Mg ratio of the rock, as a measure of the relative content of calcite and dolomite. Some variations were attributed to the content of insoluble minerals, textural factors, and the mechanical removal of grains which lead to overestimated dissolution rates.

The type of $CaCO_3$ mineral may also play a significant role in young carbonate rocks. As illustrated in Figure 3.11, aragonite is more soluble than the polymorphs low-Mg, pure and intermediate-Mg calcite, whilst high-Mg calcite (>12% mol% of $MgCO_3$) is more soluble than aragonite. The Mg^{2+} cations that occupy the Ca^{2+} positions contribute to reduce the stability of the crystal lattice and increase the solubility of the mineral (Morse and Mackenzie 1990). This essentially affects young carbonate sediments, since aragonite, a metastable mineral, and high-Mg calcite, typically transform into low-Mg calcite during the early diagenesis (see neomorphism in Chapter 2).

The classification of sedimentary rocks into detrital (allochtonous) and chemical (autochtonous) is somewhat artificial. Intermediate rocks composed of carbonate components and a considerable amount of non-soluble detrital particles are relatively common. The triangular diagram shown in Figure 3.41 can be used to classify carbonate rocks according to the proportion of sand- and mud-sized impurities, the latter including clay and silt particles. The lithologies more suitable for karst development are those that fall close to the $CaCO_3$ vertex, away from the non-soluble sand- and

Figure 3.40 Images showing some geological controls on karst development. (a) Mushroom-shaped rocks generated by differential dissolution and physical weathering on subhorizontal Cretaceous dolomite strata with different composition and texture (Ciudad Encantada, Cuenca, Spain). (b) Exposure of the Precambrian-Early Cambrian Hormuz Salt with solution flutes and sparsely covered by red residual clays (capsoil). Note salt pillar capped by clayey soil in the skyline. Circle indicates lens cap 8 cm in diameter for scale (Hormuz diapir, Iran). (c) Evidence of differential dissolution in a gypsum block at the margin of the Ebro River, NE Spain. Dissolution is more rapid in the coarsely crystalline and impure matrix than in the microcrystalline and pure alabaster nodules. (d) Outcrop of well-bedded and jointed Cretaceous limestone showing secondary porosity (fractures, bedding planes) and tertiary porosity (solutionally enlarged discontinuity planes, conduits). Sonabia Beach, Cantabria, Spain. (e) Road-cut showing steeply dipping Cretaceous limestone with a highly heterogeneous porosity, including localized dissolution cavities filled with residual clays (A-2 Highway, Alhama de Aragón, Spain). (f) Vertical image of a mesa capped by subhorizontal Cretaceous limestone crisscrossed by a complex joint system with variable geometries (straight, curved) and densities. The joints control the development of structural karren (corridors or giant grikes). Image from GoogleEarth centered at 30T 397230E4734240N (Las Tuerces, Palencia, Spain). (g) Sequence of thrusts affecting Paleogene limestones and controlling the development of an internally drained glacio-karstic depression and the location of a spring (blue circle). Ponors indicated with blue squares (Rivereta Stream, Ordesa National Park, Spain). Source: Francisco Gutiérrez (Author).

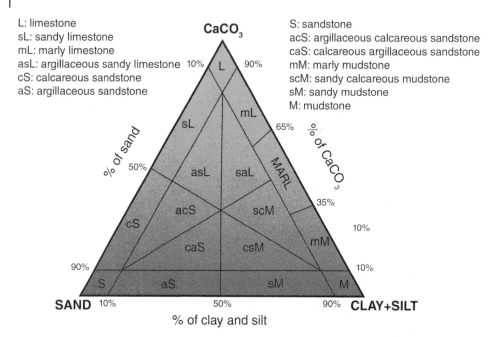

L: limestone
sL: sandy limestone
mL: marly limestone
asL: argillaceous sandy limestone
cS: calcareous sandstone
aS: argillaceous sandstone

S: sandstone
acS: argillaceous calcareous sandstone
caS: calcareous argillaceous sandstone
mM: marly mudstone
scM: sandy calcareous mudstone
sM: sandy mudstone
M: mudstone

Figure 3.41 Classification of intermediate sedimentary rocks consisting of CaCO₃ and insoluble sand and clay-silt particles. The descriptors preceding the main component are indicated in reverse order according to their proportion. *Source:* Adapted from Vatan (1967).

mud-sized impurities. The dissolution of these impure carbonate rocks entails the generation of an insoluble residue, the amount of which depends on the rock volume affected by dissolution and the proportion of impurities contained in the rock. This insoluble residue may coat the rock surface inhibiting dissolution. It may also be transported away from the dissolving surface, eventually clogging permeability features and protoconduits, inhibiting water circulation. Clay particles, characterized by high cohesion, tend to remain adhered to the rock surface forming a protecting coating, whereas cohesionless sand particles are more easily transported by water. Eisenlohr et al. (1999), in an enlightening article, measured the surface-controlled dissolution rates of limestone and marble rocks with high purity (>90%) in H_2O-CO_2 solutions by using free drift batch experiments under closed-system conditions for CO_2 (10 °C, initial $P_{CO_2} = 5 \cdot 10^{-2}$ atm). They inferred that impurities released by dissolution were directly absorbed at the reaction surface, acting as inhibitors that progressively decrease the dissolution rates. As dissolution proceeds and when all possible absorption sites are occupied, a condition is reached at which the inhibiting effect remains constant. This interpretation was corroborated by Auger spectroscopy that revealed the presence of non-soluble components on the surface of the limestone. The inhibiting effect played by impurities is most probably one of the main reasons for the large discrepancies shown between different dissolution rate experiments. For instance, large differences can be expected between data obtained from pure synthetic calcite or almost pure freshly broken Iceland spar, and limestone rock with impurities and subject to long-sustained dissolution.

Impurities play a similar inhibiting role on the dissolution of evaporite rocks, although it seems that higher proportions of non-soluble components are necessary to prevent karst development due to the higher solubility of salt and gypsum. The interface between gypsum bedrocks and Quaternary detrital covers typically displays a relatively thick clay-rich karstic residue related to the progressive solutional lowering of the rockhead (Figures 3.40b and 4.16). The degradation of salt extrusions by dissolution is hindered by the development of capsoils and caprocks consisting of the insoluble or less soluble fraction of the saline formation that restrict the salt exposures (Bruthans et al. 2009; Zarei et al. 2012) (Figure 3.40b). The solutional removal of salt in the subsurface through the migration of dissolution fronts results in the formation of progressively thicker, low-permeability clayey residues that limit the water–salt interaction.

Most karst studies include descriptions of the local or regional stratigraphy, with general information on the lithological succession. However, in most cases, quantitative data on the composition of the rock formations are lacking, despite their importance. It would be advisable to add information on the mineralogical composition of the key units and the proportion of impurities. These data can be obtained from complete mineralogical analyses; e.g., mineral counting on thin sections with a polarizing microscope. Alternatively, crushed carbonate rocks can be attacked with a relatively weak acid (e.g., HCl 0.6 N at ca. 50 °C) to remove the carbonate fraction. This allows the assessment of the percentage of impurities, whose grain size distribution and mineralogy can be determined in subsequent analyses (e.g., sieving and X-ray diffraction; XRD).

3.9.2 Grain Size and Texture

Texture refers to the geometric features of the constituent grains and crystals of rocks and deposits and their arrangement, including size, shape, sorting (uniformity of size), and packing. These features may have a large impact on dissolution rates since they influence the specific surface area of the rock-forming components (i.e., total surface area per unit mass) and the available reactive surface for dissolution. The finer the grain size, the higher is the potential reactive surface for dissolution. In batch experiments, dissolution rates are increased by reducing the size of the particles or crystals. This allows to increase the ratio between the surface area of soluble material exposed to dissolution and the volume of solvent (A/V), and to reduce the time required to reach near-saturation conditions. For instance, if a cubic particle with a volume of $1 cm^3$ (and surface area of $6 cm^2$, or $600 mm^2$) is divided into 1000 cubic particles of $1 mm^3$ in volume, the total surface area increases by a factor of 10 ($6000 mm^2$). A number of laboratory and field studies illustrate that dissolution rate is inversely proportional to grain/crystal size and sparite content (Ford and Williams 2007 and references therein). The boundary between micrite (microcrystalline calcite or aragonite) and coarser sparite is generally established at $4 \mu m$ (0.004 mm), although some authors use other values. For a region of the Appalachians with Paleozoic carbonate rocks, Rauch and White (1977) plotted the relative cave volume developed in each lithological unit, against percentage of sparite, showing a general decrease in cave development with increasing sparite content. These authors placed the micrite–sparite boundary at $15 \mu m$.

The potential surface area that may interact with a solution also depends on the geometry of the particles. The surface available for dissolution is larger in particles with irregular shapes and small-scale irregularities. Think about a spherical particle with a smooth surface in contrast with a highly angular particle of the same volume and with complex micro-roughness. Walter and Morse (1984) demonstrated that the surface roughness of the grains (i.e., microstructure) has a strong effect on dissolution rate. They conducted batch dissolution tests using rhombic calcites with smooth surfaces, and various types of biogenic skeletal carbonates with porous surfaces and complex microstructure. Dissolution rates for the skeletal grains, with higher reactive surface areas, were significantly higher (up to a factor of 7) than those measured for the rhombic calcite crystals of the same size and at the same saturation level. They introduced a surface roughness factor to account for the microstructure effect on dissolution rate. Plan (2005), using tablets of carbonate rocks to measure dissolution rates in the field, obtained significantly different dissolution rates in polished surfaces, ground surfaces and cut surfaces with higher roughness, in increasing order.

Other relevant textural features are sorting and packing, which are interrelated. Sorting is a measure of the range of grain sizes in the sediment. Well-sorted sediments have a restricted range of grain sizes. They are also said to be poorly graded because they have a limited size gradation. Packing indicates the closeness among sediment grains. This depends on factors such as the size and shape of the particles. Loosely packed sediments have large openings between the grains, which may correspond to interconnected pores through which water can circulate, favoring dissolution. In poorly sorted sediments, the space between the large particles tends to be filled by finer particles, contributing to reduce the porosity. Karst rocks composed of tightly packed fine-grained particles or crystals (e.g., interlocking texture) may offer limited surface for dissolution, showing lower dissolution rates that other rocks with the same chemical composition but with coarser grains. This is the case of very fine-grained and compact carbonate rocks with aphanitic texture and gypsum strata with alabastrine texture (Figure 3.40c). In Gaping Gill Cave, England, a fine-grained bed of micrite designated as the Porcellaneous Band restricts cave development and controls the position of perched passages (Glover 1974). Rauch and White (1977), in laboratory dissolution experiments using rock samples with different mineralogical and textural characteristics, obtained higher dissolution rates for pure limestones than those of very pure limestones. They attributed this apparent contradiction to the presence of silty streaks in the pure limestone, that incorporate textural changes (sorting, packing, and porosity) in the rock and contribute to increase the surface area exposed to dissolution.

Grain and crystal size can indirectly influence dissolution features by controlling the fracture network that guides water flow. A striking example has been thoroughly documented in the large gypsum maze caves of the Podil'sky region in western Ukraine, developed in a Miocene gypsum unit up to 25 m thick (Klimchouk et al. 2009). This gypsum lithosome with subhorizontal attitude includes three subunits with variable crystal sizes (Figure 2.24e) that control the fracture pattern and the development of cave storeys with different morphological features. The upper subunit includes giant elongated domes up to 8–10 m high made up of large crystals with a radial and concentric structure. The fractures, guided by the subvertical edges of the gypsum domes, show a wide dispersion in their orientation and passages have a dominant slot geometry. Moreover, giant selenite crystals commonly protrude from the cave ceilings as pendants, suggesting slower dissolution than the surrounding finer grained matrix. The 2–3 m thick middle part is characterized by smaller dome structures, 0.5–3 m in diameter, carved in micro- and macro-crystalline gypsum. The fracture network has a higher density and passages with quadrangular shape are the most common. The lower subunit, 6–8 m thick, mainly consists of micro-crystalline gypsum, has a widely spaced joint network with a dominant orthogonal pattern, and passages tend to have rhomb-like cross sections.

3.9.3 Porosity

The nature, scale and distribution of voids of karst rocks are among the most important physical parameters controlling dissolution processes, and hence the hydrology and morphogenesis of karst environments. There is a close relationship between the development of solutional porosity and the evolution of karst aquifers and cave systems. The porosity (n) of a rock is defined as the percentage of pore space within a given volume of rock

$$n = \frac{V_v}{V} \times 100 \tag{3.107}$$

where V_v is the volume of voids and V is the total volume of rock. A closely related parameter is the void ratio (e), which is the ratio of the volume of voids to the volume of the solids (V_v/V_s). The effective porosity is the total pore volume minus the occluded pore volume, and may be regarded as the interconnected pore space through which water can effectively flow. When a saturated rock volume is drained by gravity (e.g., water table decline), part of the water is released from the effective porosity, and the rest is retained in interconnected pores by capillary and molecular forces and as water filling occluded pores. The specific yield, also known as the drainable porosity, is the volume of water that can be drained by gravity from the total volume of rock or soil, expressed in percentage. The specific retention is the ratio of the water that will hold against the pull of gravity to the total volume of rock or soil, expressed as a percentage. This parameter is related to the total surface area of the pore spaces, which depends on the size of the particles (i.e., specific surface area) and the density of pores. The intergranular porosity in a rock or soil composed of particles lacking cement largely depends on the particle size distribution, the sorting, and the shape of the grains. The highest porosity is attained when all the particles are of the same size, while the presence of particles of different sizes reduces the porosity. For instance, a material made up of spherical particles of equal dimensions (i.e., perfect sorting) and with cubic or rhombohedral packing have porosities of 47.6% and 26%, respectively. The porosity is reduced with the addition of smaller spherical particles (i.e., lower sorting) that occupy part of the space between the larger spheres.

Sedimentologists dealing with chemical rocks classify porosity using various criteria. According to the time of formation, they differentiate between primary porosity, formed at the time of deposition, and secondary porosity, developed during the postdepositional history of the sediment. They also classify porosity according to the geometrical relationships between the voids and the fabrics (Choquette and Pray 1970; see Section 2.2.8 and Figure 2.17): fabric-selective (e.g., interparticle, intercrystalline); fabric-selective or not (e.g., borings, shrinkage); and nonfabric selective that cross-cuts the depositional components of the rock (e.g., fractures, conduits). The porosity of carbonate and evaporite rocks may experience substantial and complex changes throughout their diagenetic history due to the soluble nature of their components. The inversion of the unstable aragonite polymorph into the lower density calcite implies a reduction in porosity. Dolomitization of limestone is typically accompanied by an increase in porosity due to the smaller molar volume of dolomite compared to that of the precursor calcite. The recrystallization processes associated with the conversion of limestones and dolomites into marble by metamorphism causes substantial reductions in porosity. Evaporites experience rapid reductions in porosity during the burial diagenesis by recrystallization and cementation. Halite typically loses its effective porosity within a few tens of meters of burial, becoming an impervious rock in the absence of solutional porosity. Karst hydrology and speleogenesis are essentially controlled by secondary, large-scale, interconnected, nonfabric selective porosity, which is largely related to the solutional widening of discontinuity planes (e.g., bedding, fractures) (Ford and Williams 2007). Fabric-selective porosity can be important in the eogenetic karst associated with young and poorly lithified soluble rocks, in which interparticle porosity offers the principal paths for water flow (Lipar and Ferk 2011). Fabric-selective porosity may also have some influence on the characteristics and distribution of small-scale features such as some karren, stalactites and even small caves formed in diagenetically mature rocks. In the quartzites and quartz sandstones of the Akopán Tepui, Venezuela, Aubrecht et al. (2019) document weathering features such as small depressions, shelters and small caves restricted to a cross-bedded layer intercalated within a succession dominated by horizontal bedding. Based on a petrographic analysis, they attribute this selective weathering to the higher porosity of the cross-bedded layer, that facilitates the ingress of water and a number of alteration processes:

(i) dissolution of the quartz cement; (ii) etching of the quartz grains and their interparticle contacts; and (iii) leaching and removal of phyllosilicates.

For hydrogeologists, primary porosity includes all types of small-scale voids present in the portions of the rock between discontinuity planes, designated as matrix porosity by some authors. Secondary porosity includes the fractures that develop in the rock after its lithification. Tertiary porosity refers to solutional conduits, mainly related to the widening of discontinuity planes by dissolution (Figure 3.40d). The tertiary porosity and permeability components are typically highly heterogeneous (Figure 3.40e). A carbonate or evaporitic rock formation with limited matrix porosity and fractures, may have a high overall porosity and permeability due to the presence of cave systems that function as underground water channels (Hartmann et al. 2014). Kaufmann and Braun (2000) illustrate the evolution of porosity and groundwater flow in a porous and fractured limestone aquifer by numerical modeling using the finite-element method. During the early stage of karstification, the system is dominated by a rather homogeneous flow carried by the porous rock matrix and the narrow fractures. As some fractures are preferentially enlarged by dissolution, the flow becomes strongly heterogeneous and starts to be dominated by conduit flow. Somehow, the fractures compete for the water flow and those that increase their contributing area experience more rapid enlargement through a self-accelerating mechanism (i.e., greater discharge, higher dissolution rate). Some fractures or faults may have a more favorable orientation and larger initial width, concentrating flow from the start. The development of this tertiary porosity entails an increase in the overall hydraulic conductivity of several orders of magnitude.

3.9.4 Bedding Planes and Stratigraphic Contacts

Bedding planes, together with the various types of mechanical discontinuities (e.g., joints, faults), play a critical role in the development of the distinctive hydrological and geomorphic features that characterize karst systems. They are the planar breaks that can be penetrated by water and guide groundwater flow in the initial evolutionary phases of karst aquifers and caves. These narrow discontinuities and their intersections may experience solutional widening evolving into protoconduits, conduits, and accessible cave systems (tertiary porosity), leading to a substantial increase in the overall permeability of the rock formations and the prevalence of subsurface drainage. Most conduits and cave passages are controlled by bedding planes and fractures. These primary and secondary discontinuities also function as planes of weakness that contribute to reduce the rock mass strength and control instability processes with an important morphogenetic role, such as cave-roof collapse or slope movements. Fracture porosity in hydrogeology commonly refers to all the discontinuity planes present in the rock that serve as pathways for water circulation, including those of mechanical and depositional origin. This is a simplification, since bedding planes are not fractures, but primary depositional features. Geoscientists dealing with rock instability processes commonly use the term joint to designate all the discontinuity planes, regardless of their origin (e.g., bedding, fractures; Selby 1993).

Beds (layers, strata) are the basic lithostratigraphic units of sedimentary successions. They are laterally traceable three-dimensional sedimentary bodies bounded by bedding planes with relatively uniform physical and mineralogical characteristics. They display a wide range of geometries (tabular, wedge-shaped, lenticular, and sigmoidal) and may be subdivided into laminae, generally less than 1–3 cm thick. Higher rank lithostratigraphic units, frequently used in the description of karst rocks, include member, formation, and group. Formations are bodies of rock with some internal homogeneity that are mappable at the surface and traceable in the subsurface. Members are lithologically separable parts of a formation, and groups are a suite of contiguous formations with common features.

Bedding planes (layering, stratification) are the surfaces that mark the boundary between successive beds. They may display multiple geometries (i.e., planar, undulated, channel-shaped, and sutured), which may change within the same plane over short distances. Individual stratification planes in carbonate successions may reach huge lateral extents, of the order of hundreds of kilometers, and may be extremely planar, in contrast with other types of sedimentary formations (e.g., alluvial sediments). These characteristics have important implications for the development of extensive and laterally continuous caves and the generation of large rapid landslides. The term parting is commonly used to designate very thin beds or laminae intercalated between two thicker beds. For instance, a shale parting a few millimeters thick between two limestone beds. Bedding planes may be related to two main types of changes in the depositional environment: (i) variations in the sedimentary processes; and (ii) interruption of the sedimentation (i.e., hiatus). Changes in the depositional processes may be accompanied by variations in the characteristics of the deposits, including the lithology, mineralogy, grain size, texture, or sedimentary structures, which may significantly influence dissolution processes. A storm event in a carbonate platform may lead to the accumulation of a coarse-grained bed (tempestite) on an erosional surface carved on a younger fine-grained carbonate deposit. A halite bed on top of a gypsum bed in a lacustrine succession records an increase in the salinity of the lake brine. Interruptions in the sedimentation may be related to a decrease in sediment supply and to sharp hydrological changes, such as the emergence of the sea floor or the desiccation of a lake. Hiatuses may represent longer time periods than the units preserved in the stratigraphic record. Hardgrounds are indurated discontinuities with concentrated fossils and minerals that develop on the sea floor during long hiatuses or periods of very slow aggradation rate (Figure 2.7h). The emergence of depositional surfaces is commonly accompanied by the development of erosional and dissolution features (Figure 2.27b), which may be subsequently buried by renewed submergence. These prominent stratigraphic discontinuities with paleokarst features may play a significant role as preferential groundwater-flow pathways and karstification zones (Palmer and Palmer 1995). Some bedding planes may represent major stratigraphic contacts that are depicted in geological maps, such as the boundaries of formations and members, commonly defined by easily recognizable lithological changes. They may also represent major breaks in the sedimentation history (hiatuses), plus erosional, and deformation phases. Angular unconformities are erosional surfaces that separate two groups of rocks whose bedding planes are not parallel or concordant, recording deformation and erosion between the deposition of the two stratigraphic series. Paraconformities and disconformities separate sedimentary groups with a concordant relationship and represent long interruptions in the sedimentation. The disconformities have an irregular geometry related to a major phase of erosion during the hiatus, in contrast with the non-erosional nature of the paraconformities. Non-conformities are erosional surfaces that bound sedimentary rock resting upon unstratified igneous or metamorphic rocks.

The thickness of beds and hence the spacing of the bedding planes may significantly influence the development of surface and subsurface karst landforms. Table 3.5 includes a classification for bed thickness widely used by stratigraphers and sedimentologists. Groundwater flow and solutional attack is more dispersed in thinly bedded rocks with multiple penetrable planes, eventually inhibiting the formation of significant caves and their preservation due to low rock mass strength (Waltham 2002). Moreover, fractures in thinly bedded rocks tend to have less lateral extent than those in massive rocks. Closely spaced bedding planes also limit the development of some karren, especially the hydraulically controlled types. It is widely recognized that the finest karst landforms require medium to massive bedding (Ford and Williams 2007).

Table 3.5 Classification and terminology for bed thickness (Campbell 1967) and joint spacing (Selby 1993).

Bed thickness (cm)	Description	Joint spacing (cm)	Description
>100	Very thick or massive	>300	Very wide
100–50	Thick	300–100	Wide
50–10	Medium	100–30	Medium
10–3	Thin	30–5	Close
<1	Laminated	<5	Very close

Bedding planes penetrable by water can be envisioned as two roughly parallel rock surfaces with some small-scale morphological complexity, including widenings, channels and constrictions related to depositional and diagenetic features. The actual initial geometry and lateral extent of the bedding contacts may be significantly different to those usually considered in laboratory and numerical dissolution models. They may also hold slickensides, including linear striations and polished surfaces developed by bedding-parallel slippage related to flexural-slip folding. This is the type of relative displacement that occurs when a pile of paper sheets is folded. Knez (1998), in a very thorough investigation conducted in Škocjanske Jama, a UNESCO World Heritage cavern in Slovenia, found that cave passages are controlled by three specific bedding planes out of the 62 existing ones in the thick-to-massive limestone succession. The main common feature of these contacts is the presence of shearing features generated by relative slippage between beds, probably accompanied by some bedding-normal separation. The aperture and water-accessibility of bedding planes may increase by dilation induced by unloading related to the erosion of overlying and adjacent rocks or the retreat of glaciers. Cave passages controlled by bedding planes tend to be sinuous and concordant with the bedding, in contrast to caves guided by fractures, typically comprising multiple straight sections with angular intersections (Palmer 2007).

The important role played by bedding planes and lithological changes is emphasized by the inception horizon concept. It postulates that at the earliest stages of speleogenesis (inception), conduit development is focused at specific beds or stratigraphic contacts with physical and mineralogical attributes that favor groundwater flow and dissolutional activity (Lowe 1992b, 2004; Lowe and Gunn 1997). Inception horizons may correspond to contacts that bound rocks with contrasting solubility and/or permeability: limestone–dolomite, shale and limestone, halite bed within a gypsum succession, micrite limestone–sparry limestone, etc. Potential inception horizons also include beds and contacts that function as sources of acidity, such as layers with sulfides that release sulfuric acid by oxidation, or dolomite-shale contacts releasing CO_2 under metamorphic conditions (Pezdič et al. 1998). Filipponi et al. (2009) carried out a 3D statistical analysis of the spatial relationships between phreatic conduits of the Siebenhengste Cave system, Switzerland, and geological features. The cave, more than 177 km in length and with a depth range of 1340 m, is developed in a dipping carbonate formation 180 m thick. Their results revealed that accessible phreatic conduits are concentrated at specific horizons within the stratigraphic succession. Around 80% of the conduits are associated with seven horizons (Figure 3.42). The same approach was also applied by these authors to other 17 large epigenic cave systems worldwide, obtaining comparable results (table 1 in Filipponi et al. 2009). These include the Mammoth Cave system, Kentucky (676 km long), where Palmer (1989) identified that cave passages are preferentially developed along five horizons. These works illustrate the importance of the stratigraphic background for a good understanding of cave development.

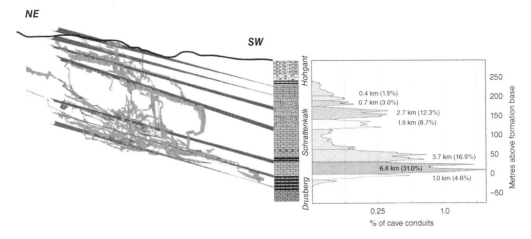

Figure 3.42 Projection of a portion of the Siebenhengste–Hohgant Höhlensystem and frequency distribution of phreatic conduits, mainly concentrated at specific stratigraphic positions, interpreted as inception horizons. *Source:* Filipponi et al. (2009). Published with permission of Elsevier.

3.9.5 Joints

The word fracture is commonly used by structural geologists as a general term that designates any break in a material generated by brittle deformation, including joints and faults. Faults are discrete surfaces, or zones with multiple surfaces and internal deformation (fault zone, shear zone), that separate two blocks of material that have slid past the other (shear displacement). Joints are planar breaks in a rock without any appreciable shear displacement. Lineaments refer to relatively long linear surface features, generally with geomorphic expression, commonly, but not necessarily, attributable to structural features. For instance, sinkhole alignments controlled by a fault, a fold axis, or a stratigraphic contact between soluble and non-soluble rocks (Figures 3.45, 3.47, and 3.49). Lineament mapping is conducted in some areas for groundwater (e.g., siting water-supply wells) and mineral exploration. The origin of these features may be controversial and in general terms lineaments should be treated as potential fracture-trace locations, unless they are confirmed. In a study conducted in a reservoir site near Tampa, Florida, Upchurch et al. (2019) illustrated the importance of ground truthing following photolineament mapping (e.g., geophysics, penetration tests, and trenching) and the significant proportion of false positives that may be identified.

Despite joints being less prominent than faults, they are commonly more important for karst development due to their higher spatial frequency. They are the most abundant structural elements in the crust of the Earth (Davies 1984). Joints tend to have a planar geometry with a straight trace in plan view, although they can also be curved and sinuous (Figure 3.40f). In well-bedded rocks they tend to be perpendicular to the stratification. Joints commonly display systematically preferred orientations. Pervasive parallel joints that define a clear trend constitute a joint set. Two or more kinematically related joint sets intersecting at relatively regular angles form joint systems. Conjugate systems are sets intersecting with an acute dihedral angle.

Joint surfaces may display irregularities at different scales, including plumose marking. These are feather-like structures comprised of a series of tiny ridges and troughs that splay symmetrically from a central axis, that is commonly parallel to bedding. These features are rarely observed in carbonate rocks because they best develop in very fine-grained or aphanitic rocks and are rapidly obliterated by dissolution. Joints may show some plane-normal opening related to dilational displacement, although in soluble rocks it may be difficult to discriminate the contribution of mechanical separation and dissolution. Joints are generally cohesionless planes of weakness with variable separation, including local widenings and constrictions that influence water flow at the early stages of karstification (speleo-inception). Joints may also function as preferential sites for mineral deposition by open-space filling or/and replacement, forming veins. These secondary minerals, frequently calcite, gypsum or quartz, contribute to reduce/occlude the penetrability of the joints. Stylolithic joints are a peculiar type of joints along which rock has been removed by pressure-induced dissolution (pressure-solution). From the kinematic perspective their formation has the opposite effect to that of dilational joints, since they accommodate contraction and involve volume loss. Stylolithic joints display sutured traces formed by the interpenetration of the adjacent rock faces. They tend to form along preexisting discontinuity planes (bedding, joints) through which dissolving fluids can circulate, and commonly display a submillimetric seam of insoluble material (clay, iron hydroxides) that was not flushed out of the rock by solution.

In addition to their width, two important attributes of joints from the hydrological and morphogenetic perspective are density and orientation. Joint density can be measured and described in a number of ways: (i) average spacing of joints; (ii) number of joints per area; (iii) total length of joints in a specified area; and (iv) total surface area of joints within a volume of rock. Table 3.5 includes a classification of joint spacing proposed by Selby (1993). Similar to bedding, high densities of joints inhibit the development and preservation potential of numerous karst features, due to more diffuse groundwater flow and the limited mechanical strength of the rock mass. Joint spacing is influenced by lithology and bed thickness. Incompetent rocks like mudstones are typically more densely jointed than competent rocks such as well-lithified limestone. Thick beds commonly have more widely spaced joints than thin beds. Joints tend to be less penetrative in stratigraphic successions comprising different lithologies with variable bed thickness, in contrast to the throughgoing planes commonly developed in more homogeneous stratigraphic series. Orientation data of joints, as well as other structural elements (bedding, faults, intersections of discontinuities, and fold axes), can be depicted and analyzed using several methods. When three-dimensional strike and dip/plunge data are available, stereographic pole diagrams and pole-density diagrams can be used. Two-dimensional strike values can be displayed in rose diagrams and strike histograms, which provide quantitative data on the frequency of structural features for orientation intervals. The comparison of the available structural data with spatial and geometrical attributes of surface and subsurface karst features (e.g., strike and inclination of cave passages) allows the assessment of the role played by the different structural controls on their development. Projecting cave surveys on geological maps and geological cross sections is highly useful for visually identifying lithological and structural controls. Ballesteros et al. (2014) reviews the approaches used to compare the geometrical attributes of structural elements and caves, and propose a method to quantitatively assess in a GIS environment the control exerted by structural factors on the direction and inclination of cave passages. Their "SpeleoDisc" method is based on the projection of the cave surveys of geological maps and cross-sections showing the main litho-structural features of the cave area, and the comparison between conduit groups and families of discontinuities, previously defined from structural data and cave surveys (strike and inclination of survey shots). Table 3.6 includes a number of studies that analyzed the spatial and genetic relationships between fractures and various types of karst features.

Rocks may be affected by different generations of joints formed during different periods and in response to various mechanisms, often difficult to determine. Most joints are considered to be of tectonic origin that accommodate imperceptible movement related to deviatoric stress fields. Joint systems may develop coevally to folding. A higher density of joints often develops along the hinge zone (zone of greater curvature of folds) due to layer-parallel stretching and shortening in the

Table 3.6 Examples of investigations that analyze the relationships between a wide variety of karst features and structural elements. The main parameters and findings are indicated.

Karst features	Parameters	Main findings	References
Fifty **caves** in a variety of structural settings (unfractured, fractured, folded)	Deviation of vadose passages from the local dip of the strata and the dip direction Deviation of phreatic passages from the strike of the strata	Greatest deviation of phreatic passages from the strike direction in massive fractured rocks	Palmer (1999)
Fractures and **dissolutional cavities** in limestone, Apennines, central Italy	Orientation of joints, faults and dissolutional cavities	Two modes of dissolution are identified: (i) diffuse flow and weathering in the epikarst zone, controlled by closely spaced joints and bedding planes; and (ii) focused deep dissolution guided by highly permeable fault damage zones. The latter mode, ascribed to ascending CO_2-rich hypogene fluids of volcanic origin, control the distribution of large and deep sinkholes in the area	Billi et al. (2007)
Sulfuric acid caves in the Majella anticline, Apennines, Central Italy	Position of strike-slip faults, fracture cluster zones and related damage zones	Vertically extended strike-slip faults allow cross-formational H_2S flow from depth, and fracture cluster zones along the anticline hinge control karst macro-scale porosity.	Pisani et al. (2021)
Caves carved in quartz-sandstones in tepui mesas of Venezuela and rock features found in the caves; vertical to inclined pillars, pendants and floor bumps (Figure 3.43)	Orientation of cave features and joints, both thoroughgoing and strata-bound	Horizontal caves developed at specific stratigraphic units controlled by silt and iron hydroxide layers (inception horizons) and strata-bound joint networks. Linear passages and erosional features within the caves controlled by jointing	Sauro (2014)
Pleistocene flank-margin limestone **caves** associated with the neotectonic NW-SE-oriented Kanawinka Fault and a paleoshore area in southern Australia	Orientation of cave passages	The anomalous linear orientation of the flank-margin caves is related to the structural control imposed by a NW-SE joint set genetically related to the normal fault.	White and Webb (2015)

(Continued)

Table 3.6 (Continued)

Karst features	Parameters	Main findings	References
Eight epigene gypsum **caves** developed in folded and faulted Messinian evaporites in the northern Apennines, Italy	Orientation of cave sections and strike and dip of structural elements	Good correlation between main directions of cave development and the regional structural trends. Macro-scale structures (major faults, fold axes) control underground drainage basins and mesoscale features (bedding, steeply dipping joints) guide the development of local collapse structures	Pisani et al. (2019)
10652 solution **sinkholes** in a limestone massif of the Taurus Mountains, Turkey, that correspond to a NW-SE oriented neotectonic horst	Orientation of major axes of depressions	Sinkholes are mostly elongated and show a prevalent NW-SE orientation controlled by joints and faults consistent with the structural grain. Secondary orientations vary spatially depending on the structural pattern of each sector	Öztürk et al. (2018a)
Sinkholes and **large depressions** related to limestone dissolution in the Salento Peninsula, Italy	Orientation of karst depressions and faults	Authors identify multiple fault sets, but the karst depressions are mainly guided by the most recent normal faults (Apenninic normal faults)	Pepe and Parise (2014)
513 closed depressions interpreted as evaporite-dissolution collapse **sinkholes** enlarged by eolian erosion in Kotido Crater, Arabia Terra, Mars	Orientation of major axes of depressions, fractures, and wind direction inferred from dune fields	Elongated depressions with a preferred orientation range consistent with the dominant fracture set and the prevailing wind direction	Parenti et al. (2020)
Structural **karren, caves, sinkholes, canyons, and rock falls** in the limestone karst of the Potiguar Basin, Brazil, controlled by fractures and bedding planes	Characterisation of landforms at different scales using LiDAR and UAV data	Network of primary and secondary discontinuity planes exert a major control on landform development at multiple spatial and temporal scales	Silva et al. (2017)

Karst corridors controlled by faults and joints on Jurassic limestones with a general homoclinal structure in the Carpathians, Romania	Orientation of faults, joints and karst corridors (fractures widened by dissolution and mass wasting processes)	Joints and fractures form an orthogonal system, with dominant strike-parallel and dip-parallel sets. Karst corridors mainly develop along strike-parallel extensional faults	Tîrlă and Vijulie (2013)
Residual limestone hills in northern Puerto Rico	Orientation of elongated hills and structural data	Major axes of elongated towers show three predominant trends controlled by the strike of the dipping limestones and fractures	Day (1978)
1903 Frank limestone **rock slide-avalanche** ($30 \cdot 10^6$ m^3) in the Canadian Rocky Mountains, which partially destroyed the town of Frank (70–90 residents killed)	Structural data of the failure surfaces and the discontinuities that controlled them	Sliding surface controlled by bedding in the steeply dipping limb of an anticline with reduced strength due to syntectonic flexural slippage. Headscarp and lateral margins controlled by joints perpendicular to bedding. Discontinuity planes most probably weakened by karstification	Cruden and Krahn (1978)
2009 Jiweishan limestone **rock slide-avalanche** in Wulong, China, that killed 74 people	Structural data from strongly karstified limestone with interbedded shales dipping 17–24°	Rock slide avalanche controlled by a stratigraphic contact between limestone and shale beds (sheared by bedding-parallel slippage) and cross joints, including a strongly karstified zone with dissolution conduits filled with residual clays	Zhang et al. (2018)

outer and inner parts of the fold, respectively (bending-moment fracturing). Joints may be also related to non-tectonic brittle deformation. During diagenesis, once the sediments have reached some degree of lithification, joints form due to burial loading and compaction. Subsequently, when the strata are brought close to the surface due to erosion and or tectonic uplift, the expansion related to the decrease in confining pressure (unloading) may result in the formation of a new generation of joints. These may be particularly prominent along escarpments, where there is also significant loss in lateral confining pressure (debuttressing).

Bedding planes and joints form relatively dense networks of discontinuities potentially penetrable by water; the fracture permeability of hydrogeologists. The great majority of the cave passages are guided by these ubiquitous partings and their intersections (Klimchouk and Ford 2000a; Palmer 2007) (Figure 3.43). The relative importance of these primary and secondary permeability features (bedding-controlled versus joint-controlled caves) varies in each cave or in different sectors of a cave system depending on multiple factors, such as aperture, orientation with respect to the hydraulic gradient, or spacing. Using finite-difference modeling, several authors independently demonstrated that the initial aperture of fissures (i.e., bedding planes, joints) has a prime influence in determining the most favorable paths for cave development in epigene systems (Palmer 1991, Dreybrodt 1996; Dreybrodt et al. 2005b). Their results indicate that the breakthrough time, which is the time needed for a fissure to reach its maximum widening rate along its entire length, has higher dependence on the initial fissure width than on hydraulic gradient and flow length. A different situation is commonly found in hypogene caves related to upward transtratal flow coming from an underlying aquifer under confined conditions. In these settings, there is no significant

Figure 3.43 Passage of the Imawarì Yeuta Cave developed in quartz sandstones in Auyán Tepui, Venezuela. The location of the passage is controlled by a layer of iron hydroxides marked by a recession right beneath the flat roof of the cave that functions as an inception horizon. The pillars at the right side are controlled by a strata-bound joint system. *Source:* Photo by Alessio Romeo, La Venta Esplorazioni Geografiche.

competition among the joints and all the available planes tend to be widened by an ascending and sluggish water flow, producing maze caves. Paradigmatic examples are the great gypsum caves of western Ukraine, which display multiple storeys of passages guided by joint systems, with variable geometries and spacing in distinct gypsum subunits controlled by the different texture of the crystals (Klimchouk et al. 2009; Rehrl et al. 2010).

There are important differences between bedding planes and joints: (i) bedding planes have much greater extent and lateral continuity than joints, which may be restricted to specific beds or sedimentary packages, and may die out laterally over short distances; (ii) bedding planes juxtapose different rock units that may have contrasting permeability and mineralogy, whereas joints cut across strata (transtratal structures); (iii) in stratigraphic successions with subhorizontal attitude, bedding planes are generally more favorable for the development of phreatic conduits, whereas steeply dipping joints are mainly exploited by downward vadose flow; and (iv) bedding planes can be much older than joints, and consequently water circulation and karstification can start earlier, during diagenesis. Caves controlled by joints and bedding-joint intersections tend to have angular patterns composed of straight segments that join at various angles, in contrast with the more sinuous bedding-controlled caves associated with specific rock units.

3.9.6 Faults

Faults are fractures along which there is visible offset by shearing. Faults can occur as discrete planes, or as relatively wide fault zones comprising a variety of internal structures, such as slip surfaces, fault rock assemblages (e.g., breccia, gauge, cataclastite, and mylonite) and subsidiary brittle and ductile deformation structures. Frequently, it is possible to differentiate within the fault-distorted volume of rock an intensely deformed shear zone or core zone that accommodates most of the slip, and damage zones at both sides consisting of kinematically related subsidiary structures (e.g., secondary faults and fractures, drag folds) (Choi et al. 2016). Faults are classified according to their relative displacement (Figure 3.46). Strike-slip faults are characterized by horizontal slip between the adjacent blocks; displacement is parallel to the strike of the fault plane. These faults tend to have a steep dip and their subvertical attitude results in rather straight cartographic traces. According to the relative sense of movement, they are classified into right-lateral (dextral) and left-lateral (sinistral) strike-slip faults. We can describe these faults standing on one block and facing toward the other block. If the opposite block has moved to the right or to the left, the fault is right- or left-lateral, respectively. In pure dip-slip faults, the displacement vector is parallel to the dip of the fault plane. The footwall and the hanging wall are the fault blocks situated below and above a dipping fault plane. In normal faults, the hanging-wall moves down with respect to the footwall. These faults accommodate extension and may be accompanied by horizontal dilational separation and the development of wide openings close to the surface, where the confining pressure is low. The main normal fault plane (master fault) is commonly accompanied by a number of smaller-throw normal faults (secondary faults) that may have opposite dip to that of the master fault (antithetic faults). In reverse faults, the hanging wall moves upward with respect to the footwall. This convergent displacement accommodates contraction. Low-dipping faults with reverse displacement are named thrust faults (Figure 3.40g). These faults may have very low dips, can be areally very extensive, and tend to be controlled by weak stratigraphic units. They may stack subconcordantly different stratigraphic units, functioning from the hydrogeological perspective similarly to stratigraphic contacts. Thrusts commonly develop along mechanically weak evaporitic formations that play the role of detachment zones (strain localizers). Dissolution of the evaporites along these tectonized units produces horizons composed of solution-collapse breccias and insoluble residues, known as rauhwackes (Warren 2016). Oblique-slip faults have strike-slip and dip-slip displacement components. The displacement vector lies at an angle with the strike of the fault surface between 0° and 90° (rake). The net displacement of a fault is given by the length of the virtual line that links points in the footwall and hanging wall fault surfaces that used to match before the onset of fault activity (piercing points). This displacement vector can be split into different components: vertical displacement or throw, strike-slip displacement, dip-slip displacement and horizontal displacement or heave. The apparent displacement observed along faults in outcrops and maps, which can be significantly lower than the actual net displacement, is called separation. Stratigraphic separation is the thickness of strata that originally separated two beds brought into contact at a fault. This is an important feature associated with most faults with important hydrogeological and speleogenetic implications; they truncate and offset stratigraphic units, usually juxtaposing rocks with different composition and permeability.

Rocks associated with fault zones can experience significant alterations by mechanical and chemical processes. Fault zones may include multiple subparallel fault surfaces separating masses of broken rock. Frictional sliding along fault planes, even at very shallow depths, produces polished surfaces and slickensides, including striations, grooves, ridges, and transverse steps (chatter marks). These may serve as kinematic indicators that allow inferring the direction and sense of displacement. Fault zones may include various secondary rock types with contrasting permeability. The competent wall rocks (e.g., well-lithified limestone) may become intensely fractured and sheared to form fault breccias consisting of angular clasts with preferred orientation (shear fabrics). These fault breccias can be highly porous and function as preferred pathways for groundwater flow. However, they may lose their porosity by mineral deposition and strong cementation, becoming low permeability zones or hydrogeological barriers. Incompetent wall rocks such as shales are transformed by shearing into a clayey crushed rock called fault gouge. Fault surfaces are frequently accompanied by secondary, fault-parallel, clay layers of low permeability known as clay smears, consisting of fine-grained particles accumulated along the shear zone from the wall rocks. Dike systems that cut across sedimentary rocks, commonly related to the injection of magma along joints and fractures, may also function as hydrogeological barriers that divide aquifers into compartments. The dolomitic limestone of the Far West Rand, South Africa, is the most important aquifer of the country. It reaches more than 1 km in thickness and is divided into a number of well-defined groundwater compartments by vertical dikes. The gold in these auriferous reef carbonates is mined by underground galleries after dewatering by pumping. The associated water table declines have induced a large number of collapse sinkholes, including the infamous 1962 West Driefontein Mine event, which caused the loss of 29 lives (Bezuidenhout and Enslin 1970).

Faults and their different zones may play a variable role from the hydrological perspective (Caine et al. 1996). They can behave as preferential, high-permeability groundwater flow paths and zones of intense karstification. They may also function as low permeability zones (e.g., clay smears, cemented shear zones) that deflect and guide the groundwater flow along the adjacent more permeable rocks. Springs are commonly spatially associated with faults (Figure 3.40g). Dublyansky and Kiknadze (1983), in a review on hydrological data from tunnel works in carbonate rocks, showed that most of the water inrushes occurred from the fractured damage zone of faults, rather than from the core zone. Faults with significant stratigraphic separation that juxtapose karst and non-karst rocks may function as hydrogeological barriers that define the boundaries of underground water catchments and control the location of the main discharge zones of aquifers. The discharge zone of the Garrotxa–Banyoles carbonate–evaporite aquifer in the eastern Pyrenees, Spain, is controlled by a normal fault that juxtaposes the soluble formations against impervious marls. This fault zone controls the rising of groundwater from the confined carbonate aquifer through the overlying evaporitic formation, generating a deep-seated hypogene gypsum karst, a high density of large sinkholes and the spring-fed Banyoles Lake (Gutiérrez et al. 2019b). Su Gologone in Supramonte (Sardinia, Italy), a 135-m deep vauclusian spring with flow rates ranging between $80\,Ls^{-1}$ during droughts to over $10\,m^3\ s^{-1}$ during severe floods, is controlled by a fault that puts in contact an overturned Mesozoic carbonate sequence with Paleozoic phyllites (De Waele 2008).

Faults may play an active or passive morphogenetic role, depending on whether they create landforms by active deformation, or control their development by guiding water flow, karstification and other processes such as rock collapse or erosion. The active versus passive role of faults on karst landforms is illustrated by the different types of structurally controlled poljes. Some poljes are associated with inactive faults that juxtapose karst formations against non-soluble rocks (border poljes). These poljes, commonly with a thin cover, form by differential dissolutional lowering and planation along the contact-karst zones that receive aggressive allogenic runoff from the outcrops of non-karst rocks. Other poljes are bounded by active faults and their flat floors are underlain by thick syntectonic sedimentary successions, including recent alluvial and lacustrine deposits. These enclosed depressions behave hydrologically as karst poljes with springs and ponors, but their topography is mainly related to the relative balance between tectonic subsidence and sediment aggradation (i.e., neotectonic poljes) (Mijatović 1984; Gracia et al. 2003; Sanz de Galdeano 2013; Doğan et al. 2017). The creation of relief by active faults with vertical displacement involves base level changes that may control the development of multilevel caves in the upthrown block and drowned caves in the downthrown block (Miller 1996). Sulfuric acid caves in the Guadalupe Mountains (New Mexico, Texas) decrease in age from high to low elevation (Polyak et al. 1998). DuChene and Cunningham (2006) attribute this age distribution to the water table decline caused by normal faulting related to the Basin and Range extension and the development of the Río Grande Rift in the late Cenozoic. Recent tectonics uplifted the Guadalupe Mountains and disconnected them from a large upland recharge area; i.e., normal faulting-induced drastic changes in the regional hydrogeology.

Faults, despite being able to reach great dimensions (>10 km depth, >1000 km long), are not as important as bedding planes and joints for cave development. Palmer (2007) estimates that less than 5% of all caves follow faults. They may control the overall trend of caves, specific passages, and may function as barriers or deflectors for cave development. Caves that are largely or entirely guided by faults are relatively scarce (e.g., Grotta di Castellana, southern Italy). A number of studies illustrate cave sections and speleothems offset by faults (Jeannin 1990; Bini et al. 1992; Vandycke and Quinif 2001; Plan et al. 2010; Szczygieł 2015). In Mount Sedom diapir (Israel), an active salt wall controlled by the Dead Sea transform fault system, Frumkin (1996, 2013) documents a syntectonic cave passage, whose cross section has been offset by dip-slip displacement on a vertical fault. Multiple solution notches and sloping facets on the upthrown side indicate a stepwise passage development, which could be related to episodic faulting and/or intermittent dissolution controlled by a creeping fault (Figure 3.44). The displacement of seismogenic faults and the associated earthquakes may have an impact on cave and sinkhole development. Speleoseismology is the investigation of earthquake records in caves (Forti 2001a). These can be classified into primary and secondary paleoseismic evidence. Primary evidence is the offset of a cave by coseismic displacement on the source fault. Secondary paleoseismic evidence, related to ground shaking and off-fault deformation, includes broken speleothems (Šebela 2008), growth anomalies in speleothems (e.g., tilting), deformed cave deposits (e.g., liquefaction), and collapse processes (Postpischl et al. 1991; Gilli 2005; Szczygieł et al. 2021). There is an increasing interest on the use of caves as archives of paleoearthquakes. However, a number of processes unrelated to seismic activity (e.g., gravitational

Figure 3.44 Cave passage in Mount Sedom diapir offset by dip-slip displacement along a subvertical fault. The stepwise development of this syntectonic passage may be related to episodic faulting and/or intermittent flooding in a passage controlled by a creeping fault. *Source:* Photo by Amos Frumkin.

deformation, underground ice, floods, and human disturbance) can generate the same and similar features (i.e., equifinality). These alternative mechanisms should be excluded, preferentially using multiple accurately dated geological archives, before the evidence is ascribed to large paleoearthquakes, in order to avoid seismic hazard overestimates (Becker et al. 2005, 2006).

Sinkholes frequently form alignments related to the active or passive control of underlying faults. Along the coasts of the Dead Sea, active halite dissolution of salt beds related to the rapid lowering of the lake and the collapse of the resulting cavities is leading to the development of numerous sinkholes that locally form rapidly expanding alignments (Figure 3.45). These are ascribed to concealed active faults of the Dead Sea pull-apart basin, as revealed by seismic reflection profiles and the analysis of structural and seismological data (Abelson et al. 2003; Closson and Karaki 2009). It is proposed that the sinkhole lineaments can be used to trace the location of subcropping active faults. Koša and Hunt (2006) conducted a thorough structural investigation in the Upper Permian carbonate platform of the Delaware Basin exposed in the Guadalupe Mountains (New Mexico and Texas), which hosts more than 300 caves, including Lechuguilla Cave and Carlsbad Cavern. They found that syndepositional faults have a profound role on speleogenesis. The authors differentiate two main generations of deformation structures and three karstification phases. The oldest structures, parallel to the ENE-WSW-oriented exhumed reef slope (Figure 3.19a), correspond to a belt of syndepositional gravitational normal faults, growth monoclines, and synclines developed along the SE sector of the platform and ascribed to differential compaction of the basinal sediments over which the platform prograded. The youngest structures are tectonic joints and small-throw normal

Figure 3.45 Sinkhole alignment in the Dead Sea coast, Israel, related to halite dissolution along a concealed active fault. Note the multiple shoreline features developed in the last decades by the decline of the lake level at ca. 1 m yr^{-1}. Red ellipse indicates a person for scale. *Source:* Francisco Gutiérrez (Author).

faults approximately perpendicular to the platform margin, and related to the late Cenozoic extension of the Basin and Range province. The Guadalupe Mountain Range and the Delaware Basin, located in the footwall of a major NNW–SSE normal fault system, have been uplifted by 1–2 km and tilted to the east. The first karstification phase is recorded by syndepositional fissure-like cavities controlled by the synsedimentary faults and fractures and filled by platform-derived sediments. The second karstification phase, which cross-cuts the previous cavity fills, are fissure-like cavities filled by post-Permian siliciclastic sediments and controlled by the youngest syndepositional faults and fractures located next to the reef escarpment. The third karstification phase includes the explorable sulfuric acid hypogene cave systems, which are controlled by the two generations of faults and fractures with roughly orthogonal trends. In northwest Yucatan, Mexico, there is a 180 km-diameter semicircular belt with a high density of cenotes (collapse sinkholes) that coincides with the buried structure generated by the Chicxulub impact crater at the Cretaceous-Paleocene boundary. This "ring of cenotes" is attributed to a higher permeability zone, probably related to faulting and surface deformation produced by the impact of the asteroid (Perry et al. 1995).

3.9.7 Folds

Folds are flexures in the strata with changes in the dip of the bedding, and also often variations in strike (Figures 3.46 and 3.47). An anticline is a fold in which the convexity points toward the younger beds of the folded sequence. A syncline is convex in the direction of the oldest beds. Older sediments occur in the inner and outer parts of the anticlinal and synclinal folds, respectively. The terms antiform and synform are used to describe folds with the form of an arc and a trough,

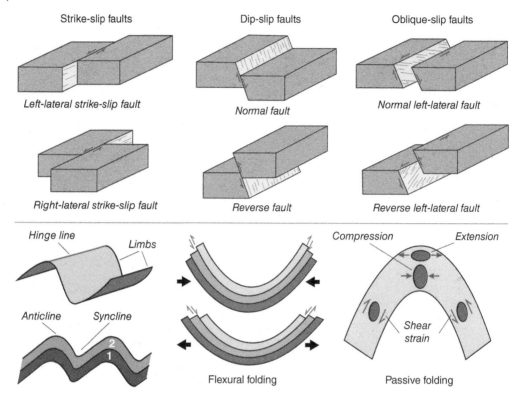

Figure 3.46 Main types of faults and folds and two extreme folding mechanisms. In flexural folding the flexing of layers is accompanied by bedding-parallel slip. Note reverse and normal slip along bedding planes in synforms affected by folding and "unfolding," respectively. In passive folding, layers experience internal deformation. The limbs undergo shear strain and the hinge zone is affected by compression (inner arc) and extension (outer arc).

respectively, when the relative chronology of the stratigraphic succession is not known. The limbs are the flanks of the fold, and the hinge zone is the portion of the fold where bedding planes show the greatest curvature. The hinge line on a folded bedding plane contains the points of maximum curvature (hinge points). The axial surface of a fold passes through successive hinge lines in a stacking of folded surfaces; this surface may be planar or curved. The plunge of the fold is the angle between the hinge line and a horizontal plane. The dip of the axial surface allows differentiating between upright folds, inclined, and recumbent, with vertical, dipping and subhorizontal axial surfaces, respectively. In overturned folds at least one of the limbs is upside down. The vergence of an inclined fold is the direction of inclination or of overturning of a fold. The interlimb angle is used to classify folds in terms of fold tightness: gentle (170–180°), open (90–170°), tight (10–90°), and isoclinal (0–10°; parallel limbs). Fold size is commonly characterized by wavelength (distance between adjacent hinges) and amplitude. The geometry of the folds can be extremely diverse: chevron, cuspate, circular, elliptical, box, etc.

Folding may be achieved by two main mechanisms, depending on the mechanical properties and rheology of the sediments (Davies 1984). Flexural folding operates when stratification planes play a strong mechanical influence (e.g., folding of a book). Folding is accommodated by the flexing of layers accompanied by bedding-parallel slippage (bedding-parallel faults with no stratigraphic separation). Passive folding is the dominant mechanism when the influence of bedding planes is

Figure 3.47 Castillo de Acher perched syncline in the Spanish Pyrenees, formed by Cretaceous carbonate formations underlain by red Permo-Triassic detrital rocks. The upper image is a shaded relief model of the WNW-plunging syncline, with numerous solution sinkholes and swallow holes concentrated along the fold axis. Red arrow indicates plunge direction and black arrow points to a moraine. Inset image shows the profile of the hanging syncline looking toward the ENE. *Source:* Francisco Gutiérrez (Author). The lower image shows the trough-like topography of the top of the mountain and the sinkhole alignment developed along the hinge zone of the fold. Note complex folding associated with a thrust sheet in the background. *Source:* Open Source Geospatial Foundation.

very weak and there is significant internal deformation within the beds (e.g., folding of soft ice cream layers). Carbonate successions with stiff beds are commonly affected by flexural folding, whereas the more ductile and plastic evaporite formations tend to undergo passive folding. In flexural-slip folds, the strata tend to retain their original thickness, but they may experience some internal deformation, mainly in the hinge zone where the curvature is greatest. Beds in the outer and inner arcs of the hinge zone undergo stretching and shortening, respectively. This may be accommodated by continuous layer-parallel deformation in soft layers, and the development of fractures in brittle beds: tension fractures and normal faults in the outer arc and minor thrusts in the inner arc (bending-moment fractures). These are usually zones of preferred groundwater flow and intense karstification (Figure 3.47).

A structural analysis carried out in the longest cave system of South America, comprising the hypogene Toca da Boa Vista and Toca da Barriguda (eastern Brazil) developed in folded carbonates, showed that the main passages occur along the hinge zone of anticlines and that conduits are mainly related to the enlargement of subvertical joints, which show greater density in these portions of the folds (Ennes-Silva et al. 2016). Similar results were obtained in a series of caves developed in the Salitre Formation in the Irecê and Una-Utinga Basins, Bahia (Brazil) (Pontes et al. 2021). Flexural-slip folding is commonly accompanied by the development of joints, typically showing three main trends with respect to the axis of folding: (i) cross joints perpendicular to the fold axis and the direction of lowest stress; (ii) longitudinal joints subparallel to the fold axis and perpendicular to the direction of greatest stress; and (iii) oblique joints comprising two conjugate sets interpreted as shear joints, in which the acute angle of intersection is bisected by the shortening direction. Bagni et al. (2020) and Pontes et al. (2021) identify a higher density of joints (syn-folding longitudinal and cross joints) along the hinge zone of gentle anticlines in Brazil, defined by belts with more abundant fracture-controlled karst features. They propose that this finding can be used for hydrocarbon exploration in karstified carbonate reservoirs. In karst environments, it is relatively common to find folds related to non-tectonic deformation and with complex geometries. For instance, gravitational basin structures with centripetal dips generated by downward passive bending of strata (sagging) due to the dissolution of underlying rocks (e.g., sagging sinkholes; Gutiérrez et al. 2008a). Monoclines are peculiar steplike folds in which subhorizontal strata show an abrupt bend to a steeper dip within a narrow zone. These are asymmetrical folds with two hinge zones, one anticlinal (upper) and one synclinal (lower) with the steep middle limb in between. Their formation can be related to dip-slip displacement along buried faults (fault-propagation fold) and the ductile flexure of the overlying cover (drape fold). They may also result from bending of strata above migrating dissolution fronts in dipping evaporitic formations (Gutiérrez and Cooper 2013; Warren 2016) (Figure 6.76d). These are gravitational folds restricted to the supra-evaporitic units (disharmonic structure). For instance, in the Interior Homocline of central Saudi Arabia, the down-dip migration of a dissolution front in the Late Jurassic anhydrite Hith formation and the subsidence of the overlying sediments have developed a monocline and a deformation belt with geomorphic expression that extends over more than 550 km (Powers et al. 1966; Memesh et al. 2008) (Figure 6.68e, f).

Folding structures may have a strong influence on the distribution of underground drainage basins, groundwater flow paths, and dissolution processes and landforms. In the vadose zone, where the flow is governed by gravity, water tends to circulate downward along bedding partings in the dip direction, toward the axis of the nearest syncline. The water flow may also change from one bedding plane to a deeper one following steeper transtratal joints that function as shortcuts. If the permeable bed is underlain by a low permeability unit within the vadose zone, the water may divert its path once it reaches the syncline axis, and flow along the plunge direction. Here, rocks may have a higher density of joints and higher permeability. Where the vadose water that circulates along dipping bedding planes reaches the water table, it tends to flow along the strike direction toward the nearest outlet. In folded terranes, vadose passages tend to be oriented in the dip direction (transverse flow), and show more sinuous and longer passages in gently dipping successions than in steeply dipping ones. In contrast, phreatic passages frequently follow the strike of the bedding and fold axes (longitudinal flow) (Palmer 2007). In the Austrian Alps, Goldscheider and Neukum (2010) studied the underground drainage pattern in a folded and faulted stratigraphic succession, including two karstified limestone formations 60–160 m thick separated by a marl unit 60 m thick. Hydraulic connections inferred from multi-tracer tests revealed that: (i) plunging synclines govern the main flow path in the upper aquifer, while anticlines act as water divides; (ii) near the base level and under phreatic conditions, the groundwater flows toward the anticlines, where springs are located; and (iii) the upper and lower aquifers are hydraulically connected via crossformational faults. This information is critical for delineating catchment boundaries and source protection zones for springs.

3.9.8 Interbedded Non-soluble Rocks

The best karst development typically occurs in thick successions of carbonate and/or evaporite sediments without significant intercalations of non-soluble rocks. An example is the fully developed solutional karst landscape lacking river valleys found in some sectors of the Dinaric region, where thick and pure limestone successions extend well below the base level (i.e., holokarst). However, it is more common to find successions of soluble rocks with interbedded non-soluble beds or units, generally with low permeability (e.g., clay, shale, and coal), in the stratigraphic record. These interbeds play an important role in the hydrogeology of karst systems and dissolution processes. They have an influence on the karstification susceptibility of the stratigraphic successions and may act as dissolution inhibitors. This is illustrated by the conditions found in the site initially proposed for the construction of the Upper Mangum Dam, Oklahoma, and those of the subsequently selected Lower Mangum Dam (Johnson 2003a, 2003b). At Upper Mangum, the intense karstification detected in the Blaine Formation, consisting of alternating gypsum and shale units with subordinate dolomite, led to the abandonment of the site. In contrast, evidence of karstification at Lower Mangum was very limited, where gypsum in the Flowerpot Shale is restricted to thin beds.

The presence of low-permeability and insoluble units controls the hydrostratigraphy and behavior of aquifer systems, as well as the groundwater conditions and types of karst. These units may function as aquicludes that separate different aquifer units within multilayer groundwater systems. A good example is the Floridan Aquifer System, considered to be one of the most productive carbonate aquifers in the world, which includes the upper and lower Floridan aquifers separated by an intermediate confining unit (Upchurch et al. 2019). The geometrical relationships between the aquifer and confining units determine different groundwater conditions with strong influence on circulation patterns and karst development; unconfined, concealed unconfined and perched groundwater with water table, and confined and artesian groundwater with potentiometric surfaces below and above the ground surface, respectively. Klimchouk and Ford (2000b) used these conditions to describe the different evolutionary stages of confined interstratal karst systems affected by fluvial downcutting and base level lowering.

The contacts between karst and non-karst rocks commonly function as preferential groundwater flow paths and zones of enhanced karstification (contact karst). A soluble bed underlain by a non-soluble bed (i.e., shale) may function as an inception horizon at which cave systems start their development (Filipponi et al. 2009). As the dissolutional conduits become larger and the velocity and turbulence of the water flow increases, the cavities may enlarge mainly by subsurface mechanical erosion of the basal non-soluble sediment. This process may be responsible for the creation of most of the void volume in some caves, as has been illustrated in subhorizontal, dipping, and folded stratigraphic series. The multilevel gypsum caves of Sorbas occur in an alternating succession of subhorizontal gypsum and marls. Initially, small phreatic conduits developed at the base of the gypsum units, which were subsequently enlarged by mechanical erosion of the underlying marls under vadose conditions and in relation to a lowering base level (Calaforra and Pulido-Bosch 2003; Figure 3.48). The Seso Cave in the Spanish Pyrenees is an inclined and linear vadose passage developed along a dipping marl unit (>30% carbonate) overlain by a limestone bed 1.5 m thick and controlled by a subvertical joint (Bartolomé et al. 2015). The Bossea Cave in the Ligurian Alps, northern Italy, is a remarkable example of an active cave system with a river developed along the contact between karst and non-karst rocks (Antonellini et al. 2019). Here, subparallel steeply dipping faults bound an elongated hydrogeological compartment surrounded by impermeable rocks and comprising a metacarbonate succession underlain by metavolcanic rocks. The contact between the marble and the volcanics is a mechanical detachment zone that shows the following relevant features: (i) a plunging antiformal transpressional structure subparallel to the bounding faults of the elongated compartment; (ii) a wide fault zone tens of meters wide, including highly fractured damage zones and a friable core zone; and (iii) contrasting structural styles on the stacked units

Figure 3.48 Vadose passage in Sorbas (SE Spain) carved in soft marls underlying a subhorizontal gypsum package. The irregular geometry of the cave top is related to differential erosion controlled by resistant cones of decimeter-scale selenite crystals with radiating fabrics at the base of the Messinian gypsum unit (protruding pendants). Note the inward sloping walls and the incised bottom channel excavated in the marls. *Source:* Francisco Gutiérrez (Author).

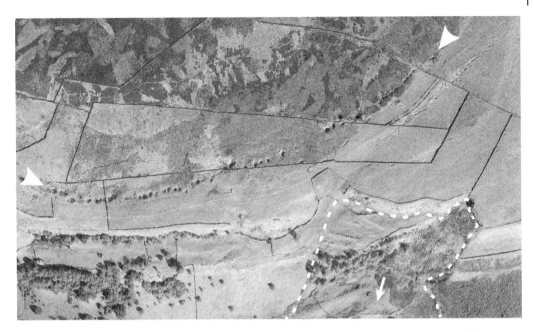

Figure 3.49 String of sinkholes developed along the contact (white triangles) between non-soluble sandstones and mudstones (upper part) and the Carboniferous Great Limestone Member (light green lower part), 2 km NW of Muker, North Yorkshire, United Kingdom. Dashed line and arrow indicate a landslide. *Source:* Image from GoogleEarth (centered at 30 U 554418 E 6 027 299 N).

(disharmonic structure), with tight symmetric flexural-slip folds in the marble and buckle folds with penetrative cleavage in the metavolcanics. The Bossea Cave developed along the plunging axial zone of the antiformal structure defined by the marble and volcanics contact, mainly exploiting bedding planes. Once it reached the detachment zone, speleogenesis mainly progressed by internal mechanical erosion of the strongly tectonized material and roof collapse. In fact, around 70% of the volume in the section of the cave associated with this contact is related to internal erosion of metavolcanic rocks.

The long-term morphogenetic effect of the enhanced karstification that affects karst rocks next to the contact with non-soluble and impervious rocks is illustrated by the development of large border poljes, that receive aggressive allogenic runoff from the adjacent formations (Ford and Williams 2007). The identification of these intensely karstified contact-karst zones has important practical implications. For instance, they help to define belts of higher sinkhole susceptibility (Figure 3.49). The grout curtains and cutoff walls built at dam sites to prevent water losses should be extended across these potentially more cavernous zones and indented into the impervious units.

3.10 Biokarst Processes

Biokarst landforms refer to geomorphic features, mostly of small scale, developed on karst rocks by weathering and erosion processes caused or mediated by biological activity (Viles 1984, 2004a). Some authors include within the biokarstification processes the formation of carbonate deposits aided by biological activity. However, according to this conception, most carbonate rocks should be considered as biokarstification products, since the deposition of a great proportion of these sediments is mediated by organic activity (e.g., coral reefs; James and Jones 2015). Here, biokarst is restricted to weathering and erosion processes and features.

3.10.1 Biokarst Processes Associated with the Surface Environment

Biokarst is mainly attributed to small organisms associated with rock surfaces, often forming biofilm communities, including cyanobacteria, lower plants (algae, mosses), fungi, and lichens (mutualistic associations between algae or cyanobacteria and fungi). These may live on the rock surface (epilithic) or within the substrate (endolithic). This latter group includes euendolithic organisms that actively bore in the rock and chasmoendoliths that dwell in preformed cavities associated with the surface (Golubić et al. 1981). Higher plants and the associated microorganisms can be also considered as biokarstification agents. The fragmentation of the bedrock by the expansion of the root systems increases the surface of rock exposed to dissolution in the epikarst. Faunal effects can be grouped into three categories: creation of holes and pits, abrasion, and the action of excrements. Biokarstification by animals is particularly important in the intertidal zone of limestone coasts, where the boring and grazing activity of a wide variety of organisms may significantly contribute to fret and lower the rock surface (Spencer 1988a, 1988b). A good and well-illustrated overview of the bioerosional and biocorrosional morphologies created by a wide variety of organisms in the subtropical and tropical coasts is given by Kázmér and Taborosi (2012).

According to Viles (2004b), the main erosional manifestations of biokarstification are related to three groups of processes: (i) biochemical contribution to carbonate dissolution; (ii) weathering activity related to lichens, cyanobacteria and algae; and (iii) biochemical and biophysical processes of boring and grazing. Some of these bioerosion processes, as well as other abiotic processes, may operate in combination and through complex synergistic interactions, and consequently it is frequently difficult to isolate the effects of specific organisms.

By far, the most important contribution of the biota to the dissolution of carbonate rocks is through the production of CO_2 in the soil by decomposition of organic matter and plant root respiration. These processes, as illustrated by the oxidation reaction of a simple carbohydrate (glucose), involve the release of H_2O, CO_2, and energy (E) (see Section 3.3.2 and Figure 3.6):

$$C_6H_{12}O_6 + 6O_2 \rightarrow 6CO_2 + 6H_2O + E \tag{3.108}$$

The biogenic CO_2 dissolved in water that percolates through the soil is the main driver of dissolution in epigene carbonate systems. Here, carbonic acid dissolution is an inorganic process, but the acidity has largely an organic source, although it may be physically distant (Ford and Williams 2007). Some organisms secrete substances such as organic acids that induce or enhance carbonate dissolution. The role played by organisms in the development of specific dissolution features is commonly established on the basis of spatial relationships, but in many cases the exact chemical and physical mechanisms and their relative contribution are not well understood. There seems to be a significant knowledge gap to be filled through multidisciplinary studies. Nonetheless, there are some exceptions. A recent study by using pH-sensitive foils demonstrates that the full-body bores created by the giant clam *Tridacna crocea* in coral reefs are generated by acids secreted by its pedal mantle, which lowers the pH of the seawater at the contact surface as much as 2 units (Hill et al. 2018). Faunal excrements deposited upon limestone surfaces may cause surface weathering or indirectly encourage it stimulating the growth of organisms (e.g., lichens). Bassi and Chiatante (1976), in a study on the marbles of the Milan Cathedral (Italy), found that pigeon excrements stimulate the development of fungal colonies that accelerate rock decay by secreting acidic substances. Gómez-Heras et al. (2004) show that acid attack and salt weathering (e.g., gypsum, halite crystals) by leachates derived from pigeon droppings contribute to the deterioration of limestone.

Rock surfaces exposed to daylight are frequently colonized by photosynthetic organisms such as cyanobacteria, lichens, and algae. Heterotrophic organisms such as fungi can also contribute to rock weathering (Sterflinger and Krumbein 1997). Microclimatic factors and orientation generally have a strong influence on their distribution (Danin and Garty 1983). The biochemical and biophysical processes associated with these organisms may produce local etching and boring in the rock, as well as general downwearing. Nonetheless, some authors contend that these biofilms may play a protecting role against some weathering and erosion processes (Jennings 1985; Fiol et al. 1996). Klappa (1979) documented via petrographic studies textural and fabric changes in calcretes (caliche) produced by the physical and chemical effects of endolithic lichens. Viles (1987), using SEM, identified micropores produced by lichens on limestone outcrops in the Mendip Hills, England, but their overall role in lowering or protecting the surface is not clear. Moses and Smith (1993) interpreted that lichens contribute to the growth of solution pans (kamenitzas) in limestone outcrops in Ireland through mechanical disturbance related to wetting and drying cycles and the resulting large volume changes of the thalli. Chen et al. (2000), in a review on weathering of rocks by lichen colonization, indicate that calcareous rocks are the most severely affected. The physical effects are related to the mechanical disruption caused by hyphal penetration, expansion and contraction of lichen thalli, and pressures induced by precipitation of organic and inorganic salts induced by lichen activity. They also contribute to chemical weathering by the excretion of organic acids, especially oxalic acid (Ascaso et al. 1982), that enhance the dissolution of carbonate minerals. Danin and Garty (1983) proposed a weathering model for limestone surfaces in the Negev Desert involving lichens, cyanobacteria, and snails to explain the development of microscale features, including holes and grooves ascribed to lichen activity.

Folk et al. (1973), in the Cayman Islands, suggested that endolithic cyanobacteria in black-coated limestone outcrops caused preferential dissolution of calcite over dolomite. They ascribed the intricate and jagged morphology of the rock surfaces, lacking any gravitational orientation, to the biochemical effects of these organisms, using the term phytokarst. However, most probably these intricate karren features are the result of multiple interacting physical and chemical processes. Viles et al. (2000) provided a record of cyanobacterial colonization on terrestrial limestone surfaces over 16 years in the Aldabra Atoll, Indian Ocean, but did not find a clear causal relationship between weathering attributable to the cyanobacteria and microscale landform development. Danin and Caneva (1990) showed that cyanobacteria darken and accelerate weathering and pit formation in limestone walls of Jerusalem and marble monuments in Rome. Fiol et al. (1996) proved that mechanical removal of limestone particles plays an important role on rillenkarren development in Mallorca, traditionally ascribed to inorganic dissolution by surface waters. Here, cyanobacteria cause the micritization of the rock, increasing the surface exposed to dissolution and reducing rock cohesion, favoring mechanical erosion processes.

Biokarst and bioerosion processes are particularly important in the photic intertidal and subtidal zones of limestone rocky shores (Spencer 1988a, 1988b; Naylor et al. 2012; De Waele and Furlani 2013). Here, the interaction of multiple trophically linked vagile and sessile organisms cause surface lowering and generate erosional features. The surface of the carbonate rocks is colonized by communities of microorganisms (cyanobacteria, algae, fungi, and lichens) that may produce micropits and irregularities in the substratum and reduce its cohesion by chemical and physical processes (Tudhope and Risk 1985) (Figure 3.50). For instance, euendolithic cyanobacteria bore micropores aided by secreted

(a)   (b)

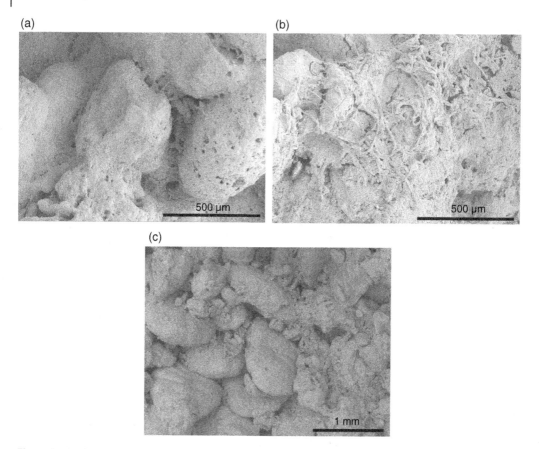

(c)

Figure 3.50 Scanning electron microscope images of limestone samples collected from a subhorizontal surface in the supralittoral zone of a limestone cliff with solution pans and densely colonized by microorganisms (s'Aladren, Mallorca, Spain). (a) Grains showing micro-bores with elliptical section ascribed to cyanobacteria. (b) Dense network of hyphae covering the grains. (c) Image of the fresh rock. Bioerosion features are restricted to a 2–4 mm thick cortex associated with the surface. *Source:* Photos by Luís Gómez-Pujol.

acid substances and those produced by the heterotrophic activity of associated bacteria (Schneider and Le Campion-Alsumard 1999). Pomar et al. (2017) analyze the spatial distribution and morpho-metric features of millimeter-sized biopits in a limestone coast in Mallorca. Here, biopits are best developed in shaded exposures and in areas sheltered from the prevailing winds and waves. A wide variety of grazing animals with different habitats (echinoids, mollusks, amphineurids, and fish) feed from these biofilms by mechanically rasping the rock surface and removing mineral components. Spencer (1988a, 1988b) provides compilations of erosion rates produced by sponges, worms (poly-chaetes), mollusks, echinoids, and grazing fish (e.g., parrotfish). Limpets and chitons graze the rock surface with their radulae, which have numerous teeth that may be hardened by specific minerals (Figure 3.51). In shore platforms in Sussex, England, Andrews, and Williams (2000), based on the $CaCO_3$ content of the feces of limpets, infer that these mollusks cause an average lowering rate of $0.15 \, mm \, yr^{-1}$ by ingesting chalk as they graze and excavate hollows. Depending on their population density, their activity may account for 12–35% of the overall lowering rate.

A number of animals excavate burrows that function as protecting extra-somatic armors. The basal teeth and spines of echinoids produce hemispherical burrows. Sponges produce burrows up

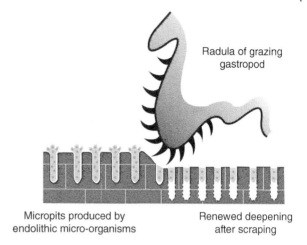

Figure 3.51 Bioerosion produced by the combined action of photosynthetic microorganism that bore micropits aided by secreted acidic substances, and grazing gastropods (e.g., limpets) that lower the limestone surface by their abrasive feeding activity. Subsequently, the endolithic microorganisms bore deeper up to the light compensation depth. *Source:* Adapted from Schneider and Le Campion-Alsumard (1999).

Radula of grazing gastropod

Micropits produced by endolithic micro-organisms

Renewed deepening after scraping

to tens of centimeters deep in coral reefs. Experimental studies carried out with a specific boring sponge by Zundelevich et al. (2007) suggest that dissolution prevails over mechanical erosion. Rock-boring bivalve mollusks and barnacles use a mixture of mechanical and chemical penetration processes in attacking rock, coral and shell. The date mussels *Lithophaga* (meaning "rock eater") bore holes in the rock with the aid of acidic secretions. The resulting ichnofossils serve as markers for the reconstruction of sea level changes, as well as local vertical deformation in coastal areas (Laborel and Laborel-Deguen 1994). The remaining marble columns of the Roman Serapeo market in the port of Pozzuoli, Naples area, show holes bored by *Lithodomus lithophagus* at 7 m above the current sea level (Figure 3.52). This is attributed to vertical movements caused by activity of the Campi Flegrei volcanic field, which according to radiocarbon dating on shells experienced significant subsidence between the third and fifteenth centuries, followed by progressive uplift (Morhange et al. 2006). Interestingly, an engraving of the Serapeo ruins was used by Charles Lyell in the frontispice of his seminal work "Principles of Geology", printed in 1830.

3.10.2 Biokarst Processes in Caves

A considerable number of works document solution features (speleogens) in caves ascribed to the corrosional effects of the metabolic products of bats and birds (guano, urine, CO_2, H_2O, and heat). These biocorrosion processes have higher impact in warm climates, where these animals are more abundant. The decomposition of guano through microbial activity produces acids (phosphoric, sulfuric, nitric, and carbonic), whereas the degradation of urea yields nitric and other weak organic acids. These leachates contribute to corrode the associated carbonate bedrock and speleothems. Gaseous nitrogen and sulfur compounds (NH_3, H_2S) spelled from the guano deposits can produce dissolution on cave ceilings and walls. These gases rise through warm air convections induced by the exothermic degradation of guano or the heating effect of bat colonies, and oxidize to form nitric and sulfuric acids. The acidic fluids and gases derived from guano can interact with rocks, sediments and minerals present in the caves, causing their dissolution and producing a wide variety of secondary minerals, mainly phosphates (over 60 different minerals), but also sulfates (e.g., gypsum) and nitrates (Hill and Forti 1997; Forti 2001b, Audra et al. 2019, and references therein). Two main groups of corrosion features related to the activity of bats are differentiated (Lundberg and McFarlane 2012): (i) subcutaneous or crypto-corrosion morphologies developed beneath guano accumulations; and (ii) features developed on exposed bedrock and speleothem surfaces by

Figure 3.52 Marble columns of the Serapeo Roman market at Pozzuoli port with holes bored by acids secreted by a *Lithofaga* bivalve mollusk. These ichnofossils record vertical displacements related to activity of the Flegrei volcanic field in Naples area (bradyseismic movements). A drawing of the Serapeo ruin was used by Charles Lyell in the frontispiece of his book "Principles of Geology". *Source:* Francisco Gutiérrez (Author).

thermal convection and condensation corrosion. The latter processes tend to be particularly intense at spots with large concentrations of bats (e.g., bat nurseries). Audra et al. (2019) report pH values associated with guano piles in Domica Cave, Slovakia, as low as 3.5, and an increase in acidity with age and depth. In Runaway Bay Caves, Jamaica, Lundberg and McFarlane (2009) measured the thermal effect of bats roosting in bell holes (blind vertical cylindrical cavities on cave roofs), and modeled the impact of their metabolism on dissolution. They measured higher temperatures in bell holes occupied by bats and estimated that 400 g of bats (around 10 individuals) yield in each 18-hour roost period 41 g of CO_2, 417.6 kJ of heat, and 35.6 g of H_2O. These metabolic products create a water film of around 0.44 mm with CO_2 at 5%, resulting in the dissolution of 0.005 cm^{-3} of $CaCO_3$ per day. This biokarstic mechanism alone could produce 1-m-deep bell holes in 50 000 years. The authors indicate that the rising heat ensures the vertical growth of the bell holes, overwhelming the effects of other geological controls with variable orientations.

In the Gomantong cave system (Malaysia), well known as a harvesting site for edible bird-nests, Lundberg and McFarlane (2012) estimate biogenic corrosion rates of 3–4.6 mm kyr^{-1} by modeling the metabolic outputs from bats and birds. Corrosion rates may rise to 34 mm kyr^{-1} in areas with high densities of bats. These authors also document subvertical dissolution flutes with semicircular cross section and hemispherical top (apse flutes) related to biogenically mediated condensation corrosion at sites with high concentrations of birds and bats (Figure 3.53). Calaforra et al. (2018) document rounded holes 4–5 cm across and up to 15 cm deep (guano holes) in flat limestone surfaces covered by a thin guano layer in Natuturingam Cave (Puerto Princesa Underground River

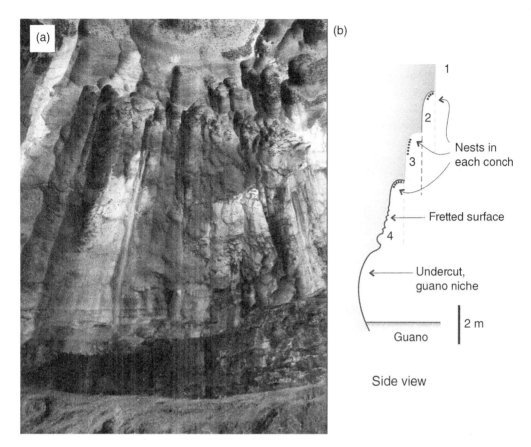

Figure 3.53 (a) Rock wall in Gomantong Cave, Malaysia, showing different generations of apse flutes generated by condensation corrosion induced by the metabolic activity of bats, and an undercut niche related to sub-guano corrosion. Guano pile at the bottom of the image. *Source:* Lundberg and McFarlane (2012). Published with permission of Elsevier. (b) Sketch illustrating different generations of flutes, from oldest to youngest (1–4), and a basal niche produced by acidic fluids derived from the guano accumulation.

Park, Philippines). Here, splash erosion by dripping water drills a hole in the soft guano, reaching the limestone bedrock, which is subsequently corroded by acids derived from the degradation of guano. In Drotsky's Cave, Bostwana, developed in dolostones and hosting bat colonies with a population between 30 000 and 90 000 individuals, Dandurand et al. (2019) describe a number of features on the cave walls and ceilings attributed to biogenic corrosion: ceiling spherical cupolas and bell holes, nested lateral notches and arches, smoothed walls, and intensely corroded stalagmitic pillars. Several mechanisms seem to be involved in the development of these speleogens, including: (i) condensation corrosion attributable to the metabolic products of bats (CO_2, H_2O, and heat), the decomposition of guano and probably convection currents; (ii) corrosive effect of the urea of the bat urine, whose bacterial decomposition generates nitrites and ammonia (CO_2 and ammonia produces carbonic acid); (iii) crypto-corrosion associated with guano accumulations through the production of phosphoric and sulfuric acids by the decomposition of organic products. Interestingly, the authors suggest that the lack of traces of human activity on the rock surface of this cave with significant Late Stone Age archeological remains, may be related to active weathering processes induced by bats. Barriquand et al. (2021) document the impact of biocorrosion caused by bat colonies on cave morphology in the Azé Cave, France. Here, the formation of a flowstone at ≥22 kyr isolated a portion of the cave preventing access to bats. This blocked cave section displays limited dimensions and shows the preservation of bear claw marks indicative of limited corrosion. In contrast, the cave portion associated with the entrance has been enlarged and overprinted by corrosion features induced by bat products: ceiling cupolas, cusp-like depressions and grooves on the walls, and guano potholes on the floor. Moreover, the presence of guano has resulted in the precipitation of phosphate minerals on archeological remains leading to their exceptional preservation conditions. The authors estimate cave wall retreat related to biocorrosion at 5–7 mm kyr^{-1}.

Growing attention is being paid to the interactions between cave microorganisms and their geochemical environment, which may result in the dissolution of the host rock and speleothems, and the precipitation of minerals (often called biothems; Palmer 2007). However, as Northup and Lavoie (2001) point out in their review on the geomicrobiology of caves, more quantitative investigations are needed to demonstrate and explain the role played by some microorganisms in the formation of dissolution features and precipitates. Jones (2001) indicates that the microbes have a low preservation potential and even if they are preserved, it may be difficult to determine the mechanisms through which they mediate in the development of the associated features. Barton (2013) uses the term biospeleogenesis, referring to the mechanisms by which biological activity can bring about the chemical dissolution of host rock, primarily through redox reactions mediated by microorganisms, especially bacteria.

Biospeleogenesis in caves, due to their isolation and nutrient-limited conditions, is mainly related to reactions mediated by microorganisms (i.e., life forms too small to see without the aid of a microscope). Moreover, the absence of light in most cave systems precludes the production of organic compounds by photosynthetic activity carried out by autotrophic life. Heterotrophic microorganisms, which use organic compounds produced by other organisms as a carbon and energy source, are scarce and mainly take advantage of the limited organic matter carried into the cave from the surface. Of special importance in the underground environment are the chemolithotrophic bacteria that use chemical compounds as a source of energy through redox reactions. Aerobic bacteria use oxygen as the terminal electron acceptor, whereas anaerobic bacteria use other compounds in the redox reactions. Metabolic processes of sulfur-, iron-, and manganese bacteria can generate considerable acidity and dissolve carbonate sediments. Microorganisms living at the interface between the host rock and cave passages can utilize reduced chemical species in the host rock such as S, Fe, or Mn, which are common in carbonate rocks.

Regarding bacterial activity, Ford and Williams (2007) differentiate three underground karst environments: (i) The anaerobic zone situated below the water table. (ii) Environments around the water table with both oxidizing and reducing species. These include underground streams and pools, where oxygen from the cave atmosphere diffuses into the groundwater. Heterotrophic organisms and oxidation reactions can consume oxygen creating reducing conditions. (iii) The relict portions of cave systems situated above the water table, dominated by oxidizing conditions and where the lack or scarcity of flowing water facilitates the development of stable bacterial biofilms. Andrejchuk and Klimchouk (2001) present an interesting case study on the geomicrobiological evolution of Zoloushka Cave, Ukraine. This gypsum cave changed progressively from phreatic-confined to partially inundated conditions during the Holocene due to fluvial downcutting, and it was rapidly dewatered by pumping since the 1940s in relation to quarrying operations. During the artesian phase, sulfate reduction mediated by anaerobic bacteria and in presence of hydrocarbons, produced H_2S and isotopically light bioepigenetic sulfur deposits and calcite by the replacement of gypsum. As the water table declined, the new oxidizing conditions brought about the oxidation of Mn^{2+} and Fe^{2+} and the massive deposition of the Fe- and Mn-hydroxides, probably induced by microbial activity.

In the deep environments where anoxic conditions prevail, organisms use electron acceptors to capture energy, including Fe^{3+} and SO_4^{2-}. The formation of most hypogene sulfuric acid caves (Carlsbad, Lechuguilla caves) is related to the oxidation of rising hydrogen sulfide (H_2S) when it reaches the oxygenated zone of the aquifer, close to and above the water table (see Section 3.4). A major source for H_2S is the reduction of sulfate by *Desulphovibrio* bacteria associated with the diagenetic transformation of organic matter in sedimentary successions (Machel 2001). This is most probably the main deep bacterial source of acidity for carbonate dissolution. In the Delaware Basin, bacteria obtain energy from the reduction of sulfate in deep anhydrite deposits, producing H_2S that is isotopically light due to biogenic fractionation (see Eq. 3.62) (Hill 1995, Barton 2013). Recent works sustain that the transformation of H_2S into H_2SO_4 in the caves of the Guadalupe Mountains, New Mexico, is a microbially mediated reaction (Barton 2013), rather than an abiotic auto-oxidation process, which occurs at slow rates (Palmer and Palmer 2000). This is supported by quantitative studies in Lower Cane Cave, Bighorn Basin, Wyoming, where most of the H_2S in groundwater is rapidly oxidized by filamentous sulfide-oxidizing bacteria attached to the carbonate substratum, contributing to boost limestone dissolution, and only 8% of the dissolved H_2S is released to the atmosphere of the cave (Engel et al. 2004). Similarly, in Cueva de Villa Luz, Mexico (Hose et al. 2000) and Frasassi Cave, Italy (Vlasceanu et al., 2000), bacteria gain energy from the oxidation of sulfur or sulfide to sulfuric acid and contribute to the dissolution of the carbonate bedrock. In Frasassi Cave, biofilms of acidophilic, sulfur-oxidizing bacteria, belonging to the genera *Thiobacillus* and *Sulfobacillus*, have pH values below 1. In Movile Cave, Romania, Sarbu et al. (1996) identified mats of chemoautotrophic bacteria that fix inorganic carbon using H_2S as an energy source, and function as the food base for a large number of invertebrates. In Cueva de Villa Luz, Mexico, Hose and Pisarowicz (1999) illustrate that microbial biofilms situated above the water/air interface indirectly promote the development of karren features. Here, sulfuric acid accumulates on rock surfaces coated by microbial biofilms that prevent buffering by limestone dissolution. The highly corrosive dripping acid etches the rocks below forming rillenkarren and spitzkarren. Gypsum is the main byproduct of carbonate dissolution by sulfuric acid (see Reaction 3.72). In the caves of the Guadalupe Mountains, gypsum locally occurs associated with elemental sulfur, forming alternating bands. Barton (2013) attributes these alternations to changes in oxygen availability within a biotic system. When there is sufficient oxygen in the environment sulfide-oxidizing bacteria transform H_2S into H_2SO_4, causing rapid carbonate dissolution and gypsum precipitation. When oxygen is depleted, H_2S is oxidized to form products with lower oxidation states, such as element sulfur. In Lechuguilla Cave, New Mexico, Cunningham et al. (1995) infer that chemolithoautotrophic bacteria associated with corrosion residues in the cave ceilings utilize iron, manganese or sulfur in the limestone and dolomite bedrock, eroding the substratum and producing residual floor deposits up to 1.5 m thick. These bacteria could serve as the base for a food chain, providing organic matter to the heterotrophic bacteria and fungi identified in the residues. Subsequently, Northup et al. (2000), in Lechuguilla and Spider caves, attribute corrosion residues rich in iron and manganese oxides to limestone corrosion along the cave walls influenced by iron- and manganese-oxidizing bacteria that use the reduced iron and manganese present in the limestone.

Bottrell et al. (1991), based on geochemical studies conducted in blue holes in Bahamas, proposed that in the anchialine caves where saline water is overlain by a layer of less dense freshwater, bacterial activity may generate acidity by two types of redox reactions. Microbially mediated reduction of the sulfate dissolved in the saline water produces H_2S and elemental S under oxygen-depleted conditions related to the presence of abundant organic matter, and pyrite is formed where pH is buffered by calcite dissolution. Subsequently, the reduced sulfur compounds are reoxidized at shallower depth, contributing to the enhanced karstification typically observed in the freshwater–saltwater mixing zone.

Jones (1995), in the poorly illuminated twilight zone of limestone caves in the Cayman Islands, documents etched calcite crystals associated with microbial biofilms, differentiating pit, spiky, and blocky morphologies controlled by the intercrystalline boundaries and crystallographic structure of the crystal (e.g., cleavage planes). The author indicates that most etching appears to take place beneath the copious mucus that is commonly associated with the epilithic microbes. Simms (1990) describes light-oriented phytokarst pinnacles and micro-terraces (or photokarst) in the twilight zone of unroofed passages in the Burren coast of Ireland, ascribed to the boring and/or solutional action of red and blue-green algae. These features are only found in areas with significant marine influence and their morphology varies depending on the distance above sea level, slope of the rock surface, and position relative to the light source. Bull and Laverty (1982) also document phytokarst features in the entrance of caves in Sarawak, Malaysia. Coombes et al. (2015) explored how light attenuation in a marine karst cave (Puerto Princesa Cave, Palawan, Philippines) affects biofilm development (extent, structure, and thickness) and the penetration depth of endoliths. They found that bioerosion is largely controlled by light availability, since endolithic boring is mainly related to the activity of phototrophic organisms.

Barton (2013), Parker et al. (2013), and Calapa et al. (2021) suggest that up to 100 m long caves developed in iron (III) ore deposits of Minas Gerais region, Brazil (Simmons 1963), might be related to biospeleogenetic processes involving the reduction of iron oxides by iron-reducing bacteria and the mobilization of Fe^{2+} under anaerobic conditions created by high organic input in this humid tropical area. A considerable number of the microbiological studies conducted in caves are motivated by the presence of valuable rock art (e.g., Altamira Cave in Spain; Lascaux Cave in France), which may be degraded by the colonization of microorganisms. This biodegradation is frequently exacerbated because of contamination produced by visitors and lighting, as well as inadequate restoration treatments. In Lascaux Cace, France, Bastian et al. (2010) documents the replacement of indigenous communities by microbial populations selected by biocide application, including the rapid development of black stains probably related to a fungus that synthesizes melanin.

References

Abelson, M., Baer, G., Shtivelman, V. et al. (2003). Collapse-sinkholes and radar interferometry reveal neotectonics concealed within the Dead Sea basin. *Geophysical Research Letters* 30: 1545.

Albéric, P. and Lepiller, M. (1998). Oxidation of organic matter in a karstic hydrologic unit supplied through stream sinks (Loiret, France). *Water Research* 32: 2051–2064.

Alkattan, M., Oelkers, E.H., Dandurand, J.L. et al. (1997a). Experimental studies of halite dissolution kinetics, I. The effect of saturation state and the presence of trace metals. *Chemical Geology* 137: 201–219.

Alkattan, M., Oelkers, E.H., Dandurand, J.L. et al. (1997b). Experimental studies of halite dissolution kinetics: II. The effect of the presence of aqueous trace anions and $K_3Fe(CN)_6$. *Chemical Geology* 143: 17–26.

Altunel, E. and Hancock, P. (1993). Morphology and structural setting of Quaternary travertines at Pamukkale. *Turkey. Geological Journal* 28 (3-4): 335–346.

Amiotte Suchet, P., Probst, J.L., and Ludwig, W. (2003). Worldwide distribution of continental rock lithology: Implications for the atmospheric/soil CO_2 uptake by continental weathering and alkalinity river transport to the oceans. *Global Biogeochemical Cycles* 17 (2): 1038.

Amiotte-Suchet, P., Probst, A., and Probst, J.L. (1995). Influence of acid rain on CO_2 consumption by rock weathering: local and global scales. *Water, Air, and Soil Pollution* 85: 1563–1568.

Andre, B.J. and Rajaram, H. (2005). Dissolution of limestone fractures by cooling waters: Early development of hypogene karst systems. *Water Resources Research* 41 (1): W01015.

Andrejchuk, V.N. and Klimchouk, A.B. (2001). Geomicrobiology and redox geochemistry of the karstified Miocene gypsum aquifer, western Ukraine: the study from Zoloushka Cave. *Geomicrobiology Journal* 18 (3): 275–295.

Andrews, C. and Williams, R.B. (2000). Limpet erosion of chalk shore platforms in southeast England. *Earth Surface Processes and Landforms* 25: 1371–1381.

Antonellini, M., Nannoni, A., Vigna, B. et al. (2019). Structural control on karst water circulation and speleogenesis in a lithological contact zone: The Bossea cave system (Western Alps, Italy). *Geomorphology* 345: 106832.

Appelo, C.A.J. and Postma, D. (2005). *Geochemistry, Groundwater and Pollution*. Rotterdam: Balkema.

Arvidson, R.S., Ertan, I.E., Amonette, J.E. et al. (2003). Variation in calcite dissolution rates: A fundamental problem? *Geochimica et Cosmochimica Acta* 67: 1623–1634.

Ascaso, C., Galván, J., and Rodríguez-Pascual, C. (1982). The weathering of calcareous rocks by lichens. *Pedobiologia* 24: 219–229.

Aubrecht, R., Lánczos, T., Schlögl, J. et al. (2019). Selective weathering of cross-bedded layers forming shelters and small caves on Akopán Tepui (Venezuela): Field, laboratory and experimental evidence about diagenesis and weathering of the Matauí Formation arenites (Roraima Supergroup, Middle Proterozoic). *Geomorphology* 325: 55–69.

Audra, P., De Waele, J., Bentaleb, I. et al. (2019). Guano-related phosphate-rich minerals in European caves. *International Journal of Speleology* 48 (1): 75–105.

Auler, A.S. (2013). Sources of water aggressiveness. The driving force of karstification. In: *Treatise of Geomorphology. Karst Geomorphology*, vol. 6 (ed. A. Frumkin), 23–28. Amsterdam: Elsevier.

Auler, A.S. and Smart, P.L. (2003). The influence of bedrock-derived acidity in the development of surface and underground karst: evidence from the Precambrian carbonates of semi-arid northeastern Brazil. *Earth Surface Processes and Landforms* 28: 157–168.

Auler, A.S., Klimchouk, A.B., Bezerra, F.H.R. et al. (2017). Origin and evolution of Toca da Boa Vista and Toca sa Barriguda Cave System in North-eastern Brazil. In: *Hypogene karst regions and caves of the world* (ed. A.B. Klimchouk, A.N. Palmer, J. De Waele, et al.), 827–840. Cham, Switzerland: Springer.

Baceta, J.I., Wright, V.P., and Pujalte, V. (2001). Palaeo-mixing zone karst features from Palaeocene carbonates of north Spain: criteria for recognizing a potentially widespread but rarely documented diagenetic system. *Sedimentary Geology* 139: 205–216.

Back, W., Hanshaw, B.B., Plummer, L.N. et al. (1983). Process and rate of dedolomitization: mass transfer and ^{14}C dating in a regional carbonate aquifer. *Geological Society of America Bulletin* 94: 1415–1429.

Back, W., Hanshaw, B.B., and Van Driel, J.N. (1984). Role of groundwater in shaping the eastern coastline of the Yucatan Peninsula, Mexico. In: *Groundwater as a Geomorphic Agent* (ed. R.G. LaFleur), 281–293. Boston, Massachussetts: Allen and Unwin.

Bagni, F.L., Bezerra, F.H., Balsamo, F. et al. (2020). Karst dissolution along fracture corridors in an anticline hinge, Jandaíra Formation, Brazil: implications for reservoir quality. *Marine and Petroleum Geology* 115: 104249.

Bakalowicz, M., Ford, D.C., Miller, T.E. et al. (1987). Thermal genesis of dissolution caves in the Black Hills, South Dakota. *Geological Society of America Bulletin* 99 (6): 729–738.

Ballesteros, D., Jiménez-Sánchez, M., García-Sansegundo, J. et al. (2014). SpeleoDisc: A 3-D quantitative approach to define the structural control of endokarst: an application to deep cave systems from the Picos de Europa, Spain. *Geomorphology* 216: 141–156.

Barriquand, L., Bigot, J.-Y., Barthèlemy, D. et al. (2021). Caves and bats: morphological impacts and archaeological implications. The Azé Prehistoric Cave (Saône-et-Loire, France). *Geomorphology* 388: 107785.

Bartolomé, M., Sancho, C., Moreno, A. et al. (2015). Upper Pleistocene interstratal piping-cave speleogenesis: the Seso Cave system (Central Pyrenees, Northern Spain). *Geomorphology* 228: 335–344.

Barton, H.A. (2013). Biospeleogenesis. In: *Treatise of Geomorphology. Karst Geomorphology*, vol. 6 (ed. A. Frumkin), 38–56. Amsterdam: Elsevier.

Bassi, M. and Chiatante, D. (1976). The role of pigeon excrement in stone biodeterioration. *International Biodeterioration Bulletin* 12: 73–79.

Bastian, F., Jurado, V., Nováková, A. et al. (2010). The microbiology of Lascaux cave. *Microbiology* 156 (3): 644–652.

Becker, A., Ferry, M., Monecke, K. et al. (2005). Multiarchive paleoseismic record of late Pleistocene and Holocene strong earthquakes in Switzerland. *Tectonophysics* 400: 153–177.

Becker, A., Davenport, C.A., Eichenberger, U. et al. (2006). Speleoseismology: a critical perspective. *Journal of Seismology* 10: 371–388.

Berner, R.A. (1978). Rate control of mineral dissolution under earth surface conditions. *American Journal of Science* 278: 1235–1252.

Berner, R.A. (1984). Sedimentary pyrite formation: an update. *Geochimica et Cosmochimica Acta* 48: 605–615.

Berner, R.A., Lasaga, A.C., and Garrels, R.M. (1983). The carbonate-silicate geochemical cycle and its effect on atmospheric carbon-dioxide over the past 100 million years. *American Journal of Science* 283: 641–683.

Berresheim, H., Wine, P.H., and Davis, D.D. (1995). Sulfur in the atmosphere. *Composition, Chemistry, and Climate of the Atmosphere* 8: 251–307.

Bezuidenhout, C.A. and Enslin, J.F. (1970). Surface subsidence and sinkholes in the dolomitic areas of the Far West Rand, Transvaal, Republic of South Africa. In: *Land Subsidence*, vol. 89, 482–495. International Association of Hydrological Sciences, Publication.

Billi, A., Valle, A., Brilli, M. et al. (2007). Fracture-controlled fluid circulation and dissolutional weathering in sinkhole-prone carbonate rocks from central Italy. *Journal of Structural Geology* 29: 385–395.

Binet, S., Probst, J.L., Batiot, C. et al. (2020). Global warming and acid atmospheric deposition impacts on carbonate dissolution and CO_2 fluxes in French karst hydrosystems: evidence from hydrochemical monitoring in recent decades. *Geochimica et Cosmochimica Acta* 270: 184–200.

Bini, A., Quinif, Y., Sules, O. et al. (1992). Les mouvements tectoniques récents dans les grottes du Monte Campo dei Fiori (Lombardie, Italie). *Karstologia* 19: 23–30.

Bischoff, J.L., Juliá, R., Shanks, W.C. III et al. (1994). Karstification without carbonic acid: Bedrock dissolution by gypsum-driven dedolomitization. *Geology* 22 (11): 995–998.

Bland, W. and Rolls, D. (1998). *Weathering. An Introduction to the Scientific Principles*. London: Arnold.

Blount, C.W. and Dickson, F.W. (1973). Gypsum-anhydrite equilibria in systems $CaSO_4 \cdot H_2O$ and $CaCO_3 \cdot NaCl \cdot H_2O$. *American Mineralogist* 58: 323–331.

Blyth, A.J., Baker, A., Collins, M.J. et al. (2008). Molecular organic matter in speleothems and its potential as an environmental proxy. *Quaternary Science Reviews* 27 (9-10): 905–921.

Blyth, A.J., Hua, Q., Smith, A. et al. (2017). Exploring the dating of "dirty" speleothems and cave sinters using radiocarbon dating of preserved organic matter. *Quaternary Geochronology* 39: 92–98.

Bögli, A. (1964). Mischungskorrosion, ein Beitrag zur Verkarstungs-problem. *Erkunde* 8: 83–92.

Bögli, A. (1980). *Karst Hydrogeology and Physical Speleology*. Berlin: Springer-Verlag.

Bolner-Takács, K. and Kraus, S. (1989). The results of research into caves of thermal water origin. In: *Karszt és Barlang*, 31–38. Special Issue.

Bonazza, A., Messina, P., Sabbioni, C. et al. (2009). Mapping the impact of climate change on surface recession of carbonate buildings in Europe. *Science of the Total Environment* 407 (6): 2039–2050.

Bottrell, S.H., Smart, P.L., Whitaker, F. et al. (1991). Geochemistry and isotope systematics of sulphur in the mixing zone of Bahamian blue holes. *Applied Geochemistry* 6 (1): 97–103.

Bottrell, S.H., Gunn, J., and Lowe, D.J. (2000). Calcite dissolution by sulfuric acid. In: *Speleogenesis. Evolution of Karst Aquifers* (ed. A.B. Klimchouk, D.C. Ford, A.N. Palmer and W. Dreybrodt), 304–308. Huntsville, Alabama: National Speleological Society.

Brook, G.A., Folkoff, M.E., and Box, E.O. (1983). A world model of soil carbon dioxide. *Earth Surface Processes and Landforms* 8 (1): 79–88.

Bruthans, J., Asadi, N., Filippi, M. et al. (2008). A study of erosion rates on salt diapir surfaces in the Zagros Mountains, SE Iran. *Environmental Geology* 53 (5): 1079–1089.

Bruthans, J., Filippi, M., Asadi, N. et al. (2009). Surficial deposits on salt diapirs (Zagros Mountains and Persian Gulf Platform, Iran): characterization, evolution, erosion and the influence on landscape morphology. *Geomorphology* 107: 195–209.

Bruthans, J., Kamas, J., Filippi, M. et al. (2017). Hydrogeology of salt karst under different cap soils and climates (Persian Gulf and Zagros Mts., Iran). *International Journal of Speleology* 46: 303–320.

Buhmann, D. and Dreybrodt, W. (1987). Calcite dissolution kinetics in the system H_2O-CO_2-$CaCO_3$ with participation of foreign ions. *Chemical Geology* 64: 89–102.

Bull, P.A. and Laverty, M. (1982). Observations on phytokarst. *Zeitschrift für Geomorphologie.* 26: 437–457.

Busenberg, E. and Plummer, N. (1982). The kinetics of dissolution of dolomite in CO_2-H_2O systems at 1.5 to 65 °C and 0 to 1 atm PCO_2. *American Journal of Science* 282: 45–78.

Caine, J.S., Evans, J.P., and Forster, C.B. (1996). Fault zone architecture and permeability structure. *Geology* 24 (11): 1025–1028.

Calaforra, J.M. and Pulido-Bosch, A. (2003). Evolution of the gypsum karst of Sorbas (SE Spain). *Geomorphology* 50: 173–180.

Calaforra, J.M., De Waele, J., Forti, P. et al. (2018). The guano holes: a new corrosion form from Natuturingam Cave (Palawan, Philippines). *Travaux de l'Institut de Spéologie "Émile Racovitza'* 57: 35–47.

Calapa, K.A., Mulford, M.K., Rieman, T.D. et al. (2021). Hydrologic alteration and enhanced microbial reductive dissolution of Fe (III)(hydr)oxides under flow conditions in Fe (III)-rich rocks: contribution to cave-forming processes. *Frontiers in Microbiology* 12: 1932.

Campbell, C.V. (1967). Lamina, lamina set, bed and bed set. *Sedimentology* 8: 7–26.

Camuffo, D. (1992). Acid rain and deterioration of monuments: how old is the phenomenon. *Atmospheric Environment* 26: 241–247.

Camuffo, D., Del Monte, M., Sabbioni, C. et al. (1982). Wetting, deterioration and visual features of stone surfaces in an urban area. *Atmospheric Environment* 16: 2253–2259.

Chang, F.F., Nasr-El-Din, H.A., Lindvig, T. et al. (2008). Matrix acidizing of carbonate reservoirs using organic acids and mixture of HCl and organic acids. In: *SPE Annual Technical Conference and Exhibition*. Denver, Colorado: Society of Petroleum Engineers.

Charola, A.E. and Ware, R. (2002). Acid deposition and the deterioration of stone: a brief review of a broad topic. *Geological Society, London, Special Publication* 205 (1): 393–406.

Chen, J., Blume, H.P., and Beyer, L. (2000). Weathering of rocks induced by lichen colonization - a review. *Catena* 39: 121–146.

Choi, J.H., Edwards, P., Ko, K. et al. (2016). Definition and classification of fault damage zones: a review and a new methodological approach. *Earth-Science Reviews* 152: 70–87.

Choquette, P.W. and Pray, L.C. (1970). Geologic nomenclature and classification of porosity in sedimentary carbonates. *American Association of Petroleum Geologists Bulletin* 54 (2): 207–250.

Cigna, A.A. (1985). Some remarks on phase equilibria of evaporites and other karstificable rocks. *Le Grotte d'Italia* 12: 201–208.

Closson, D. and Karaki, N.A. (2009). Salt karst and tectonics: sinkholes development along tension cracks between parallel strike-slip faults, Dead Sea, Jordan. *Earth Surface Processes and Landforms* 34: 1408–1421.

Colombani, J. (2008). Measurement of the pure dissolution rate constant of a mineral in water. *Geochimica et Cosmochimica Acta* 72: 5634–5640.

Coombes, M.A., La Marca, E.C., Naylor, L.A. et al. (2015). The influence of light attenuation on the biogeomorphology of a marine karst cave: a case study of Puerto Princesa Underground River, Palawan, the Philippines. *Geomorphology* 229: 125–133.

Cruden, D.M. and Krahn, J. (1978). Frank rockslide, Alberta, Canada. In: *Rockslides and Avalanches, 1. Natural Phenomena* (ed. B. Voight), 98–112. Amsterdam: Elsevier.

Cunningham, K.I., Northup, D.E., Pollastro, R.M. et al. (1995). Bacteria, fungi and biokarst in Lechuguilla Cave, Carlsbad Caverns National Park, New Mexico. *Environmental Geology* 25: 2–8.

Dandurand, G., Duranthon, F., Jarry, M. et al. (2019). Biogenic corrosion caused by bats in Drotsky's Cave (the Gcwihaba Hills, NW Botswana). *Geomorphology* 327: 284–296.

D'Angeli, I.M., Serrazanetti, D.I., Montanari, C. et al. (2017a). Geochemistry and microbial diversity of cave waters in the gypsum karst aquifers of Emilia Romagna region, Italy. *Science of the Total Environment* 598: 538–552.

Danin, A. and Caneva, G. (1990). Deterioration of limestone walls in Jerusalem and marble monuments in Rome caused by cyanobacteria and cyanophilous lichens. *International Biodeterioration* 26: 397–417.

Danin, A. and Garty, J. (1983). Distribution of cyanobacteria and lichens on hillsides of the Negev Highlands and their impact on biogenic weathering. *Zeitschrift für Geomorphologie* 27: 423–444.

Davies, G.H. (1984). *Structural Geology of Rocks and Regions*. New York: Wiley.

Day, M.J. (1978). Morphology and distribution of residual limestone hills (mogotes) in the karst of northern Puerto Rico. *Geological Society of America Bulletin* 89: 426–432.

De Waele, J. (2008). Interaction between a dam site and karst springs: the case of Supramonte (Central-East Sardinia, Italy). *Engineering Geology* 99 (3-4): 128–137.

De Waele, J. and Furlani, S. (2013). *Seawater and biokarst effects on coastal limestones*. In: *Treatise on Geomorphology. Karst Geomorphology*, vol. 6 (ed. A. Frumkin), 341–350. Amsterdam: Elsevier.

De Waele, J., Audra, P., Madonia, G. et al. (2016). Sulfuric acid speleogenesis (SAS) close to the water table: examples from southern France, Austria, and Sicily. *Geomorphology* 253: 452–467.

De Waele, J., Piccini, L., Columbu, A. et al. (2017c). Evaporite karst in Italy: a review. *International Journal of Speleology* 46 (2): 137–168.

Dembicki, H. (2017). *Practical Petroleum Geochemistry for Exploration and Production*. Amsterdam: Elsevier.

Doğan, U., Koçyiğit, A., and Gökkaya, E. (2017). Development of the Kembos and Eynif structural poljes: morphotectonic evolution of the upper Manavgat River basin, central Taurides, Turkey. *Geomorphology* 278: 105–120.

Dreybrodt, W. (1988). *Processes in Karst Systems: Physics, Chemistry, and Geology*. Berlin: Springer-Verlag.

Dreybrodt, W. (1996). Principles of early development of karst conduits under natural and man-made conditions revealed by mathematical analysis of numerical models. *Water Resources Research* 32 (9): 2923–2935.

Dreybrodt, W. (2000). Equilibrium chemistry of karst water in limestone terranes. In: *Speleogenesis. Evolution of Karst Aquifers* (ed. A.B. Klimchouk, D.C. Ford, A.N. Palmer, et al.), 130–135. Huntsville, Alabama: National Speleological Society.

Dreybrodt, W. and Eisenlohr, L. (2000). Limestone dissolution rates in karst environments. In: *Speleogenesis. Evolution of Karst Aquifers* (ed. A.B. Klimchouk, D.C. Ford, A.N. Palmer, et al.), 136–148. Huntsville, Alabama: National Speleological Society.

Dreybrodt, W. and Romanov, D. (2007). Time scales in the evolution of solution porosity in porous coastal carbonate aquifers by mixing corrosion in the saltwater-freshwater transition zone. *Acta Carsologica* 36: 25–34.

Dreybrodt, W., Gabrovšek, F., and Romanov, D. (2005b). *Processes of Speleogenessis: A Modeling Approach*. Ljubljana: Založba ZRC.

Dreybrodt, W., Romanov, D., and Kaufmann, G. (2010). Evolution of caves in porous limestone by mixing corrosion: a model approach. *Geologia Croatica* 63: 129–135.

Dublyansky, V.N. (2000a). A giant hydrothermal cavity in the Rhodope Mountains. In: *Speleogenesis: Evolution of Karst Aquifers* (ed. A.B. Klimchouk, D.C. Ford, A.N. Palmer, et al.), 317–318. Huntsville, Alabama: National Speleological Society.

Dublyansky, Y.V. (2000b). Dissolution of carbonates by geothermal waters. In: *Speleogenesis. Evolution of Karst Aquifers* (ed. A.B. Klimchouk, D.C. Ford, A.N. Palmer, et al.), 158–159. Huntsville, Alabama: National Speleological Society.

Dublyansky, Y.V. (2013). Karstification by geothermal waters. In: *Treatise of Geomorphology. Karst Geomorphology*, vol. 6 (ed. A. Frumkin), 57–71. Amsterdam: Elsevier.

Dublyansky, V.N. and Kiknadze, T.Z. (1983). *Hydrogeology of Karst of the Alpine Folded Region of the South of the USSR*. Moscow: Nauka.

DuChene, H.R. and Cunningham, K.I. (2006). Tectonic influences on speleogenesis in the Guadalupe Mountains, New Mexico and Texas. In: *Caves and Karst of Southeastern New Mexico* (ed. L. Land, V.W. Lueth, W. Raatz, et al.), 211–218. Socorro: New Mexico Geological Society.

Egemeier, S.J. (1981). Cavern development by thermal waters. *National Speleological Society Bulletin* 43: 31–51.

Eisenlohr, L., Meteva, K., Gabrovšek, F. et al. (1999). The inhibiting action of intrinsic impurities in natural calcium carbonate minerals to their dissolution kinetics in aqueous H_2O-CO_2 solutions. *Geochimica et Cosmochimica Acta* 63: 989–1001.

Engel, A.S., Stern, L.A., and Bennett, P.C. (2004). Microbial contributions to cave formation: new insight into sulfuric acid speleogenesis. *Geology* 32: 269–273.

Ennes-Silva, R.A., Bezerra, F.H., Nogueira, F.C. et al. (2016). Superposed folding and associated fracturing influence hypogene karst development in Neoproterozoic carbonates, São Francisco Craton, Brazil. *Tectonophysics* 666: 244–259.

Eswaran, H., Van Den Berg, E., and Reich, P. (1993). Organic carbon in soils of the world. *Soil Science Society of America Journal* 57: 192–194.

Fan, C. and Teng, H.H. (2007). Surface behavior of gypsum during dissolution. *Chemical Geology* 245: 242–253.

Filipponi, M., Jeannin, P.Y., and Tacher, L. (2009). Evidence of inception horizons in karst conduit networks. *Geomorphology* 106 (1-2): 86–99.

Fiol, L., Fornós, J.J., and Ginés, A. (1996). Effects of biokarstic processes on the development of solutional rillenkarren in limestone rocks. *Earth Surface Processes and Landforms* 21: 447–452.

Florea, L.J. (2006). Architecture of air-filled caves within the karst of the Brooksville Ridge, west-central Florida. *Journal of Caves and Karst Studies* 68: 64–75.

Florea, L.J., Vacher, H.L., Donahue, B. et al. (2007). Quaternary cave levels in peninsular Florida. *Quaternary Science Reviews* 26 (9-10): 1344–1361.

Folk, R.L., Roberts, H.H., and Moore, C.H. (1973). Black phytokarst from Hell, Cayman Islands, British West Indies. *Geological Society of America Bulletin* 84: 2351–2360.

Ford, D.C. and Williams, P.W. (2007). *Karst Hydrogeology and Geomorphology*. Chichester, UK: Wiley.

Forti, P. (2001a). Seismotectonic and paleoseismic studies from speleothems: the state of the art. *Netherlands Journal of Geosciences* 80: 175–185.

Forti, P. (2001b). Biogenic speleothems: an overview. *International Journal of Speleology* 30: 39–56.

Franchi, M. and Casadei, A. (1999). Il sistema carsico di Monte Caldina. Alta Valle del Fiume Secchia, Reggio Emilia. *Speleologia Emiliana* 10: 19–27.

Frenkel, H., Gerstl, Z., and Alperovitch, N. (1989). Exchange-induced dissolution of gypsum and the reclamation of sodic soils. *Journal of Soil Science* 40: 599–611.

Fritz, B. (1981). *Etude Thermodynamique et Modélisation des Réactions Hydrothermaux et Diagenétiques. Sciences Géologiques Mémoire* 65: 1–197.

Frumkin, A. (1996). Uplift rate relative to base-levels of a salt diapir (Dead Sea Basin, Israel) as indicated by cave levels. In: *Salt Tectonics*, vol. 100 (ed. G.I. Alsop, D.J. Blundell and I. Davison), 41–47. *Geological Society, London Special Publication*.

Frumkin, A. (2000a). Speleogenesis in salt. The Mount Sedom area, Israel. In: *Speleogenesis. Evolution of Karst Aquifers* (ed. A.B. Klimchouk, D.C. Ford, A.N. Palmer, et al.), 443–451. Huntsville, Alabama: National Speleological Society.

Frumkin, A. (2000b). Dissolution of salt. In: *Speleogenesis. Evolution of Karst Aquifers* (ed. A.B. Klimchouk, D.C. Ford, A.N. Palmer, et al.), 169–170. Huntsville, Alabama: National Speleological Society.

Frumkin, A. (2013). Salt karst. In: *Treatise on Geomorphology. Karst Geomorphology*, vol. 6 (ed. A. Frumkin and J. Shroder), 407–424. Amsterdam: Elsevier.

Frumkin, A. and Ford, D.C. (1995). Rapid entrenchment of stream profiles in the salt caves of Mount Sedom, Israel. *Earth Surface Processes and Landforms* 20 (2): 139–152.

Gabrovšek, F. and Dreybrodt, W. (2000). Role of mixing corrosion in calcite-aggressive H_2O-CO_2-$CaCO_3$ solutions in the early evolution of karst aquifers in limestone. *Water Resources Research* 36 (5): 1179–1188.

Garrecht Metzger, J., Fike, D.A., Osburn, G.R. et al. (2015). The source of gypsum in Mammoth Cave, Kentucky. *Geology* 43: 187–190.

Garrels, R.M. and Christ, C.L. (1965). *Solutions, Minerals and Equilibria*. San Francisco: Freeman Cooper and Co.

Gilli, É. (2005). Review on the use of natural cave speleothems as palaeoseismic or neotectonics indicators. *Comptes Rendus Geoscience* 337 (13): 1208–1215.

Glover, R.R. (1974). Cave development in the Gaping Ghyll System. In: *Limestones and Caves of Northwest England* (ed. T. Waltham), 343–384. Newton Abbot: David and Charles.

Goldscheider, N. and Neukum, C. (2010). Fold and fault control on the drainage pattern of a double-karst-aquifer system, Winterstaude, Austrian Alps. *Acta Carsologica* 39: 173–186.

Golubić, S., Friedmann, E.I., and Schneider, J. (1981). The lithobiontic ecological niche, with special reference to microorganisms. *Journal of Sedimentary Research* 51: 475–478.

Gombert, P. (2002). Role of karstic dissolution in global carbon cycle. *Global and Planetary Change* 33: 177–184.

Gómez-Heras, M., Benavente, D., de Buergo, M.Á. et al. (2004). Soluble salt minerals from pigeon droppings as potential contributors to the decay of stone based Cultural Heritage. *European Journal of Mineralogy* 16: 505–509.

Gracia, F.J., Gutiérrez, F., and Gutiérrez, M. (2003). The Jiloca karst polje-tectonic graben (Iberian Range, NE Spain). *Geomorphology* 52: 215–231.

Gutiérrez, F. and Cooper, A.H. (2013). Surface morphology of gypsum karst. In: *Treatise of Geomorphology. Karst Geomorphology*, vol. 6 (ed. A. Frumkin), 425–437. Amsterdam: Elsevier.

Gutiérrez, F. and Lizaga, I. (2016). Sinkholes, collapse structures and large landslides in an active salt dome submerged by a reservoir: the unique case of the Ambal ridge in the Karun River, Zagros Mountains, Iran. *Geomorphology* 254: 88–103.

Gutiérrez, F., Guerrero, J., and Lucha, P. (2008a). A genetic classification of sinkholes illustrated from evaporite paleokarst exposures in Spain. *Environmental Geology* 53: 993–1006.

Gutiérrez, F., Fabregat, I., Roqué, C. et al. (2019b). Sinkholes in hypogene versus epigene karst systems, illustrated with the hypogene gypsum karst of the Sant Miquel de Campmajor Valley, NE Spain. *Geomorphology* 328: 57–78.

Han, G. and Liu, C.Q. (2004). Water geochemistry controlled by carbonate dissolution: a study of the river waters draining karst-dominated terrain, Guizhou Province, China. *Chemical Geology* 204 (1-2): 1–21.

Hanshaw, B.B. and Back, W. (1979). The major geochemical processes in the evolution of carbonate-aquifer systems. *Journal of Hydrology* 43: 287–312.

Hartmann, A., Goldscheider, N., Wagener, T. et al. (2014). Karst water resources in a changing world: review of hydrological modeling approaches. *Reviews of Geophysics* 52 (3): 218–242.

Hawkins, A.B. (2014). Engineering implications of the oxidation of pyrite: An overview, with particular reference to Ireland. In: *Implications of Pyrite Oxidation for Engineering Works* (ed. A.B. Hawkins and M. Stevens), 1–98. Dordrecht: Springer.

Herman, J.S. and White, W.B. (1985). Dissolution kinetics of dolomite: effects of lithology and fluid flow velocity. *Geochimica et Cosmochimica Acta* 49: 2017–2026.

Hill, C.A. (1981b). Speleogenesis of Carlsbad Caverns and other caves in the Guadalupe Mountains. In: *Proceedings of 8th International Congress of Speleology* (ed. B.F. Beck), 143–144. Bowling Green: National Speleological Society.

Hill, C.A. (1987). Geology of Carlsbad Cavern and other caves in the Guadalupe Mountains, New Mexico and Texas. *Bulletin of the New Mexico Bureau of Mining and Mineral Resources* 117: 1–150.

Hill, C.A. (1990). Sulfuric acid speleogenesis of Carlsbad Cavern and its relationship to hydrocarbons, Delaware Basin, New Mexico and Texas. *American Association of Petroleum Geologists Bulletin* 74: 1685–1694.

Hill, C.A. (1995). Sulfur redox reactions: hydrocarbons, native sulfur, Mississippi Valley-type deposits, and sulfuric acid karst in the Delaware Basin, New Mexico and Texas. *Environmental Geology* 25: 16–23.

Hill, C.A. (2000). Sulfuric acid, hypogene karst in the Guadalupe Mountains of New Mexico and West Texas (U.S.A.). In: *Speleogenesis. Evolution of Karst Aquifers* (ed. A.B. Klimchouk, D.C. Ford, A.N. Palmer, et al.), 309–316. Huntsville, Alabama: National Speleological Society.

Hill, C.A. and Forti, P. (1997). *Cave Minerals of the World*, 2nde. Huntsville, Alabama: National Speleological Society.

Hill, R.W., Armstrong, E.J., Inaba, K. et al. (2018). Acid secretion by the boring organ of the burrowing giant clam. *Tridacna crocea. Biology Letters* 14: 20180047.

Hiscock, K. (2005). *Hydrogeology. Principles and Practice*. Malden: Blackwell.

Hose, L.D. and Pisarowicz, J.A. (1999). Cueva de Villa Luz, Tabasco, Mexico: reconnaissance study of an active sulfur spring cave and ecosystem. *Journal of Cave and Karst Studies* 61 (1): 13–21.

Hose, L.D., Palmer, A.N., Palmer, M.V. et al. (2000). Microbiology and geochemistry in a hydrogen-sulphide-rich karst environment. *Chemical Geology* 169: 399–423.

Jakobsen, R. and Postma, D. (1999). Redox zoning, rates of sulfate reduction and interactions with Fe-reduction and methanogenesis in a shallow sandy aquifer, Rømø, Denmark. *Geochimica et Cosmochimica Acta* 63: 137–151.

Jalali, L., Zarei, M., and Gutiérrez, F. (2019). Salinization of reservoirs in regions with exposed evaporites. The unique case of upper Gotvand Dam, Iran. *Water Research* 157: 587–599.

James, A.N. (1992). *Soluble Material in Civil Engineering*. New York: Ellis Horwood.

James, N.P. and Jones, B. (2015). *Origin of Carbonate Sedimentary Rocks*. Chichester: Wiley.

James, A.N. and Lupton, A.R.R. (1978). Gypsum and anhydrite in foundations of hydraulic structures. *Geotechnique* 28: 249–272.

Jeannin, P.Y. (1990). Néotectonique dans le karst du nord du Lac de Thoune (Suisse). *Karstologia* 15 (1): 41–54.

Jeannin, P.Y., Hessenauer, M., Malard, A. et al. (2016). Impact of global change on karst groundwater mineralization in the Jura mountains. *Science of the Total Environment* 541: 1208–1221.

Jennings, J.N. (1985). *Karst Geomorphology*. Oxford and New York: Blackwell.

Jeschke, A.A., Vosbeck, K., and Dreybrodt, W. (2001). Surface controlled dissolution rates of gypsum in aqueous solutions exhibit nonlinear dissolution kinetics. *Geochimica et Cosmochimica Acta* 65: 27–34.

Johnson, K.S. (2003a). Gypsum karst and abandonment of the upper Mangum damsite in southwestern Oklahoma. In: *Evaporite Karst and Engineering/Environmental Problems in the United States* (ed. K.S. Johnson and J.T. Neal), 85–94. Norman: Oklahoma Geological Survey.

Johnson, K.S. (2003b). Gypsum karst as a major factor in the design of the proposed lower Mangum dam in southwestern Oklahoma. In: *Evaporite Karst and Engineering/Environmental Problems in the United States* (ed. K.S. Johnson and J.T. Neal), 95–112. Norman: Oklahoma Geological Survey.

Jones, B. (1995). Processes associated with microbial biofilms in the twilight zone of caves; examples from the Cayman Islands. *Journal of Sedimentary Research* 65: 552–560.

Jones, B. (2001). Microbial activity in caves - A geological perspective. *Geomicrobiology Journal* 18: 345–357.

Jones, B. and Renault, R. (2010). Calcareous spring deposits in continental settings. In: *Carbonate Sediments in Continental Setting. Facies, Environments and Processes* (ed. A.M. Alonso-Zarza and L.H. Tanner), 177–224. Amsterdam: Elsevier.

Kaufmann, G. (2003). A model comparison of karst aquifer evolution for different matrix-flow formulations. *Journal of Hydrology* 283 (1–4): 281–289.

Kaufmann, G. and Braun, J. (2000). Karst aquifer evolution in fractured, porous rocks. *Water Resources Research* 36: 1381–1391.

Kázmér, M. and Taborosi, D. (2012). Bioerosion on the small scale - examples from the tropical and subtropical littoral. *Hantkeniana* 7: 37–94.

Keller, C.K. and Bacon, D.H. (1998). Soil respiration and georespiration distinguished by transport analyses of vadose CO_2, $^{13}CO_2$, and $^{14}CO_2$. *Global Biogeochemical Cycles* 12: 361–372.

Kempe, S. (2009). Siderite weathering as a reaction causing hypogene speleogenesis: the example of the Iberg/Harz/Germany. In: *Hypogene Speleogenesis and Karst Hydrology of Artesian Basins* (ed. A.B. Klimchouk and D.C. Ford), 59–60. Kiev: Ukranian Institute of Speleology and Karstology.

Kessler, T.J. and Harvey, C.F. (2001). The global flux of carbon dioxide into groundwater. *Geophysical Research Letters* 28: 279–282.

Kirkland, D.W. (1982). Origin of gypsum deposits in Carlsbad Caverns, New Mexico. *New Mexico Geology* 4: 20–21.

Klappa, C.F. (1979). Lichen stromatolites; criterion for subaerial exposure and a mechanism for the formation of laminar calcretes (caliche). *Journal of Sedimentary Research* 49: 387–400.

Klimchouk, A.B. (1997). The role of karst in the genesis of sulfur deposits, Pre-Carpathian region, Ukraine. *Environmental Geology* 31: 1–20.

Klimchouk, A.B. (2000a). Dissolution and conversions of gypsum and anhydrite. In: *Speleogenesis: Evolution of Karst Aquifers* (ed. A.B. Klimchouk, D.C. Ford, A.N. Palmer, et al.), 160–168. Huntsville, USA: National Speleological Society.

Klimchouk, A.B. and Ford, D.C. (2000a). Lithologic and structural controls of dissolutional cave development. In: *Speleogenesis: Evolution of Karst Aquifers* (ed. A.B. Klimchouk, D.C. Ford, A.N. Palmer, et al.), 54–64. Huntsville: National Speleological Society.

Klimchouk, A.B., Andreychouk, V., and Turchinov, I. (2009). *The Structural Prerequisites of Speleogenesis in Gypsum in the Western Ukraine*. Sosnowiec-Simferopol: University of Silesia - Ukranian Institute of Speleology and Karstology.

Klimchouk, A.B., Auler, A.S., Bezerra, F.H. et al. (2016). Hypogenic origin, geologic controls and functional organization of a giant cave system in Precambrian carbonates, Brazil. *Geomorphology* 253: 385–405.

Knez, M. (1998). The influence of bedding-planes on the development of karst caves (a study of Velika Dolina at Škocjanske jame Caves, Slovenia). *Carbonates and Evaporites* 13: 121–131.

Koša, E. and Hunt, D.W. (2006). The effect of syndepositional deformation within the Upper Permian Capitan platform on the speleogenesis and geomorphology of the Guadalupe Mountains, New Mexico, USA. *Geomorphology* 78 (3-4): 279–308.

Krawczyk, W.E. and Ford, D.C. (2006). Correlating specific conductivity with total hardness in limestone and dolomite karst water. *Earth Surface Processes and Landforms* 31: 221–234.

Krumgalz, B.S., Starinsky, A., and Pitzer, K.S. (1999). Ion-interaction approach: pressure effect on the solubility of some minerals in submarine brines and seawater. *Journal of Solution Chemistry* 28: 667–692.

Laborel, J. and Laborel-Deguen, F. (1994). Biological indicators of relative sea-level variations and of co-seismic displacements in the Mediterranean region. *Journal of Coastal Research* 10 (2): 395–415.

Langer, H. and Offermann, H. (1982). On the solubility of sodium chloride in water. *Journal of Crystal Growth* 60: 389–392.

Langmuir, D. (1997). *Aqueous Environmental Geochemistry*. Upper Saddle River, NJ: Pearson Education.

Laptev, F.F. (1939). *Aggressive Action of Water on Carbonate Rocks, Gypsum and Concrete: Trudy Spetsgeo*. Moscow-Leningrad, Russia: State Science-Technical Publishing.

Le Quéré, C., Andrew, R.M., Friedlingstein, P. et al. (2018). Global carbon budget 2018. *Earth System Science Data* 10: 2141–2194.

Leél-Őssy, S. (2017). Caves of the Buda thermal karst. In: *Hypogene Karst Regions and Caves of the World* (ed. A.B. Klimchouk, A.N. Palmer, J. De Waele, et al.), 279–297. Cham: Springer.

Lerman, A. (1970). Chemical equilibria and evolution of chloride brines. *Mineralogical Society of America Special Papers* 3: 291–306.

Liao, J., Hu, C., Wang, M. et al. (2018). Assessing acid rain and climate effects on the temporal variation of dissolved organic matter in the unsaturated zone of a karstic system from southern China. *Journal of Hydrology* 556: 475–487.

Lietzke, M.H. and Stoughton, R.W. (1961). The calculation of activity coefficients from osmotic coefficients data. *The Journal of Physical Chemistry* 65: 508–509.

Lipar, M. and Ferk, M. (2011). Eogenetic caves in conglomerate: an example from Udin Boršt, Slovenia. *International Journal of Speleology* 40 (1): 53–64.

Liu, Z. and Zhao, J. (2000). Contribution of carbonate rock weathering to the atmospheric CO_2 sink. *Environmental Geology* 39 (9): 1053–1058.

Lowe, D.J. (1992b). The origin of limestone caverns: An inception horizon hypothesis. PhD thesis. Manchester Metropolitan University.

Lowe, D.J. (2004). Inception of caves. In: *Encyclopedia of Caves and Karst Science* (ed. J. Gunn), 437–441. New York: Fitzroy Dearborn.

Lowe, D.J. and Gunn, J. (1997). Carbonate speleogenesis: an inception horizon hypothesis. *Acta Carsologica* 26 (2): 457–488.

Lundberg, J. and McFarlane, D.A. (2009). Bats and bell holes: The microclimatic impact of bat roosting, using a case study from Runaway Bay caves, Jamaica. *Geomorphology* 106 (1-2): 78–85.

Lundberg, J. and McFarlane, D.A. (2012). Post-speleogenetic biogenic modification of Gomantong caves, Sabah, Borneo. *Geomorphology* 157: 153–168.

Machel, H.G. (2001). Bacterial and thermochemical sulfate reduction in diagenetic settings - old and new insights. *Sedimentary Geology* 140: 143–175.

Martini, J.E.J. (1979). Karst in black reef quartzite near Kaapsehoop, eastern Transvaal. *Annals of the South African Geological Survey* 13: 115–128.

McMurry, J. (2004). *Chemistry*. New York: Prentice-Hall. Inc.

Mecchia, M., Sauro, F., Piccini, L. et al. (2014). Geochemistry of surface and subsurface waters in quartz-sandstones: significance for the geomorphic evolution of tepui table mountains (Gran Sabana, Venezuela). *Journal of Hydrology* 511: 117–138.

Memesh, A., Dini, S., Gutiérrez, F. et al. (2008). *Evidence of large-scale subsidence caused by interstratal karstification of evaporites in the Interior Homocline of Central Saudi Arabia. European Geosciences Union General Assembly. Geophysical Research Abstracts* 10: A-02276.

Mijatović, B.F. (1984). Karst poljes in the Dinarides. In: *Hydrogeology of the Dinaric Karst* (ed. B.F. Mijatović), 87–109. Hannover: International Association of Hydrogeologists.

Miller, T.E. (1996). Geologic and hydrologic controls on karst and cave development in Belize. *Journal of Cave and Karst Studies* 58: 100–120.

Minton, M. and Droms, Y. (2019). Exploration of caves. Vertical caving techniques. In: *Encyclopedia of Caves* (ed. W.B. White, D.C. Culver and T. Pipan), 420–425. New York: Academic Press.

Miotke, F.D. (1974). Carbon dioxide and the soil atmosphere. *Abhandlungen zur Karst-und Höhlenkunde, Reihe A, Speläologie* 9: 1–52.

Morehouse, D.F. (1968). Cave development via the sulfuric acid reaction. *National Speleological Society Bulletin* 30: 1–10.

Morhange, C., Marriner, N., Laborel, J. et al. (2006). Rapid sea-level movements and noneruptive crustal deformations in the Phlegrean Fields caldera, Italy. *Geology* 34: 93–96.

Morse, J.W. and Arvidson, R.S. (2002). The dissolution kinetics of major sedimentary carbonate minerals. *Earth-Science Reviews* 58: 51–84.

Morse, J.W. and MacKenzie, F.T. (1990). *Geochemistry of Sedimentary Carbonates*, Developments in Sedimentology, vol. 48. Amsterdam: Elsevier.

Moses, C.A. and Smith, B.J. (1993). A note on the role of the lichen *Collema auriforma* in solution basin development on a carboniferous limestone substrate. *Earth Surface Processes and Landforms* 18: 363–368.

Mylroie, J.E. and Carew, J.L. (1990). The flank margin model for dissolution cave development in carbonate platforms. *Earth Surface Processes and Landforms* 15 (5): 413–424.

Mylroie, J.E. and Carew, J.L. (2000). Speleogenesis in coastal and oceanic settings. In: *Speleogenesis. Evolution of Karst Aquifers* (ed. A.B. Klimchouk, D.C. Ford, A.N. Palmer, et al.), 226–233. Huntsville: National Speleological Society.

Naylor, L.A., Coombes, M.A., and Viles, H.A. (2012). Reconceptualising the role of organisms in the erosion of rock coasts: a new model. *Geomorphology* 157: 17–30.

Newton, R.C. and Manning, C.E. (2005). Solubility of anhydrite, $CaSO_4$, in $NaCl-H_2O$ solutions at high pressures and temperatures: applications to fluid-rock interaction. *Journal of Petrology* 46: 701–716.

Nicod, J. (1993). Recherches nouvelles sur les karsts des gypses et des évaporites associées (Seconde partie: géomorphologie, hydrologie et impact anthropique). *Karstologia* 21: 15–30.

Northup, D.E. and Lavoie, K.H. (2001). Geomicrobiology of caves: a review. *Geomicrobiology Journal* 18 (3): 199–222.

Northup, D.E., Dahm, C.N., Melim, L.A. et al. (2000). Evidence for geomicrobiological interactions in Guadalupe caves. *Journal of Cave and Karst Studies* 62 (2): 80–90.

Onac, B.P., Wynn, J.G., and Sumrall, J.B. (2011). Tracing the sources of cave sulfates: a unique case from Cerna Valley, Romania. *Chemical Geology* 288 (3): 105–114.

Ordóñez, S. and Benavente, D. (2014). Revisión de los modelos hidrogeoquímicos de génesis de tobas calcáreas. *Estudios Geológicos* 70: 1–18.

Öztürk, M.Z., Şener, M.F., Şener, M. et al. (2018a). Structural controls on distribution of dolines on Mount Anamas (Taurus Mountains, Turkey). *Geomorphology* 317: 107–116.

Palmer, A.N. (1984). Recent trends in karst geomorphology. *Journal of Geological Education* 32 (4): 247–253.

Palmer, A.N. (1989). Stratigraphic and structural control of cave development and groundwater flow in the Mammoth Cave region. In: *Karst Hydrology, Concepts from the Mammoth Cave Area* (ed. W.B. White and E.L. White), 293–316. Von Nostrand Reinhold: New York.

Palmer, A.N. (1991). Origin and morphology of limestone caves. *Geological Society of America Bulletin* 103 (1): 1–21.

Palmer, A.N. (1999). A statistical evaluation of the structural influence on solution-conduit patterns. In: *Karst Modeling* (ed. A.N. Palmer, M.V. Palmer and I.D. Sasowsky), 187–195. Charles Town: Karst Waters Institute Special Publication 5.

Palmer, A.N. (2007). *Cave Geology*. Dayton, Ohio: Cave Books.

Palmer, A.N. (2011). Distinction between epigenic and hypogenic maze caves. *Geomorphology* 134 (1-2): 9–22.

Palmer, A.N. (2013). Sulfuric acid caves. In: *Treatise of Geomorphology. Karst Geomorphology*, vol. 6 (ed. A. Frumkin), 241–257. Amsterdam: Elsevier.

Palmer, A.N. and Hill, C.A. (2019). Sulfuric acid caves. In: *Encyclopedia of Caves* (ed. W.B. White, D.C. Culver and T. Pipan), 1053–1062. New York: Academic Press.

Palmer, A.N. and Palmer, M.V. (1995). The Kaskaskia paleokarst of the northern Rocky Mountains and Black Hills, northwestern USA. *Carbonates and Evaporites* 10: 148–160.

Palmer, A.N. and Palmer, M.V. (2000). Hydrochemical interpretation of cave patterns in the Guadalupe Mountains, New Mexico. *Journal of Cave and Karst Studies* 62: 91–108.

Parenti, C., Gutiérrez, F., Baioni, D. et al. (2020). Closed depressions in Kotido crater, Arabia Terra, Mars. Possible evidence of evaporite dissolution-induced subsidence. *Icarus* 341: 113680.

Parker, C.W., Wolf, J.A., Auler, A.S. et al. (2013). Microbial reducibility of Fe (III) phases associated with the genesis of iron ore caves in the Iron Quadrangle, Minas Gerais, Brazil. *Minerals* 3 (4): 395–411.

Parkhurst, D.L. and Appelo, C.A.J. (2013). Description of Input and Examples for PHREEQC Version 3. A Computer Program for Speciation, Batch Reaction, One Dimensional Transport, and Inverse

Geochemical Calculations. In: *Techniques and Methods*, Book 6, Chap. A43. Denver, Colorado: U.S. Geological Survey.

Pepe, M. and Parise, M. (2014). Structural control on development of karst landscape in the Salento Peninsula (Apulia, SE Italy). *Acta Carsologica* 43: 101–114.

Perry, E., Marin, L., McClain, J. et al. (1995). Ring of cenotes (sinkholes), northwest Yucatan, Mexico: its hydrogeologic characteristics and possible association with the Chicxulub impact crater. *Geology* 23 (1): 17–20.

Pezdič, J., Šušteršič, F., and Mišič, M. (1998). On the role of clay-carbonate reactions inspeleo-inception: a contribution on the understanding of the earliest stage of karst channel formation. *Acta Carsologica* 27 (1): 187–200.

Piccini, L. and Mecchia, M. (2009). Solution weathering rate and origin of karst landforms and caves in the quartzite of Auyan-tepui (Gran Sabana, Venezuela). *Geomorphology* 106: 15–25.

Pisani, L., Antonellini, M., and De Waele, J. (2019). Structural control on epigenic gypsum caves: evidences from Messinian evaporites (Northern Apennines, Italy). *Geomorphology* 332: 170–186.

Pisani, L., Antonellini, M., D'Angeli, I.M. et al. (2021). Structurally controlled development of a sulfuric hypogene karst system in a fold-and-thrust belt (Majella Massif, Italy). *Journal of Structural Geology* 145: 104305.

Plan, L. (2005). Factors controlling carbonate dissolution rates quantified in a field test in the Austrian alps. *Geomorphology* 68 (3-4): 201–212.

Plan, L., Grasemann, B., Spötl, C. et al. (2010). Neotectonic extrusion of the Eastern Alps: constraints from U/Th dating of tectonically damaged speleothems. *Geology* 38 (6): 483–486.

Plummer, L.N. (1975). Mixing of sea water with calcium carbonate ground water. *Geological Society of America Memoir* 142: 219–236.

Plummer, L.N. and Busenberg, E. (1982). The solubilities of calcite, aragonite and vaterite in CO_2-H_2O solutions between 0 and 90 °C, and an evaluation of the aqueous model for the system $CaCO_3$-CO_2-H_2O. *Geochimica et Cosmochimica Acta* 46: 1011–1040.

Plummer, L.N. and Mackenzie, F.T. (1974). Predicting mineral solubility from rate data; application to the dissolution of magnesian calcites. *American Journal of Science* 274: 61–83.

Plummer, L.N., Parkhurst, D.L., and Wigley, T.M.L. (1978). The kinetics of calcite dissolution in CO_2 systems at 25 °C to 60 °C and 0.0 to 1.0 atm CO_2. *American Journal of Science* 278: 179–216.

Polyak, V.J., McIntosh, W.C., Güven, N. et al. (1998). Age and origin of Carlsbad Cavern and related caves from $^{40}Ar/^{39}Ar$ of alunite. *Science* 279 (5358): 1919–1922.

Pomar, F., Gómez-Pujol, L., Fornós, J.J. et al. (2017). Limestone biopitting in coastal settings: a spatial, morphometric, SEM and molecular microbiology sequencing study in the Mallorca rocky coast (Balearic Islands, Western Mediterranean). *Geomorphology* 276: 104–115.

Ponsjack, E. (1940). Deposition of calcium sulphate from sea water. *American Journal of Science* 239: 559–568.

Pontes, C.C., Bezerra, F.H., Bertotti, G. et al. (2021). Flow pathways in multiple-direction fold hinges: implications for fractured and karstified carbonate reservoirs. *Journal of Structural Geology* 146: 104324.

Postpischl, D., Agostini, S., Forti, P. et al. (1991). Palaeoseismicity from karst sediments: the "Grotta del Cervo" cave case study (Central Italy). *Tectonophysics* 193 (1-3): 33–44.

Powers, R.W., Ramirez, L.F., Redmond, C.D. et al. (1966). *Geology of the Arabian Peninsula; Sedimentary Geology of Saudi Arabia*, vol. 560-D, 1–147. US Geological Survey Professional Paper.

Railsback, L.B. (2009). *Some Fundamentals of Mineralogy and Geochemistry*. University of Georgia http://railsback.org/FundamentalsIndex.html.

Raines, M.A. and Dewers, T.A. (1997a). Dedolomitization as a driving mechanism for karst generation in Permian Blaine Formation, southwestern Oklahoma, USA. *Carbonates and Evaporites* 12: 24–31.

Raines, M.A. and Dewers, T.A. (1997b). Mixed transport/reaction control of gypsum dissolution kinetics in aqueous solutions and initiation of gypsum karst. *Chemical Geology* 140: 29–48.

Rauch, H.W. and White, W.B. (1970). Lithologic controls on the development of solution porosity in carbonate aquifers. *Water Resources Research* 6: 1175–1192.

Rauch, H.W. and White, W.B. (1977). Dissolution kinetics of carbonate rocks: 1. Effects of lithology on dissolution rate. *Water Resources Research* 13 (2): 381–394.

Reardon, E.J., Mozeto, A.A., and Fritz, P. (1980). Recharge in northern clime calcareous sandy soils: soil water chemical and carbon-14 evolution. *Geochimica et Cosmochimica Acta* 44: 1723–1735.

Rehrl, C., Birk, S., and Klimchouk, A.B. (2010). Influence of initial aperture variability on conduit development in hypogene settings. *Zeitschrift für Geomorphologie* 54: 237–258.

Rimstidt, J.D. (1997). Quartz solubility at low temperatures. *Geochimica et Cosmochimica Acta* 61: 2553–2558.

Ritter, S.M., Isenbeck-Schröter, M., Scholz, C. et al. (2019). Subaqueous speleothems (Hells Bells) formed by the interplay of pelagic redoxcline biogeochemistry and specific hydraulic conditions in the El Zapote sinkhole, Yucatán Peninsula, Mexico. *Biogeosciences* 16 (11): 2285–2305.

Rodhe, H., Dentener, F., and Schulz, M. (2002). The global distribution of acidifying wet deposition. *Environmental Science and Technology* 36: 4382–4388.

Romanov, D., Gabrovšek, F., and Dreybrodt, W. (2003a). The impact of hydrochemical boundary conditions on the evolution of limestone karst aquifers. *Journal of Hydrology* 276: 240–253.

Romero-Mujalli, G., Hartmann, J., Börker, J. et al. (2018). Ecosystem controlled soil-rock P_{CO_2} and carbonate weathering - constraints by temperature and soil water content. *Chemical Geology* 527: 118634.

Romero-Mujalli, G., Hartmann, J., and Börker, J. (2019). Temperature and CO_2 dependency of global carbonate weathering fluxes - Implications for future carbonate weathering research. *Chemical Geology* 527: 118874.

Ruiz-Agudo, E. and Putnis, C.V. (2012). Direct observations of mineral fluid reactions using atomic force microscopy: the specific example of calcite. *Mineralogical Magazine* 76: 227–253.

Salvany, J.M., García-Veigas, J., and Ortl, F. (2007). Glauberite-halite association of the Zaragoza Gypsum formation (lower Miocene, Ebro basin, NE Spain). *Sedimentology* 54: 443–467.

Sancho, C., Peña, J.L., and Meléndez, A. (1997). Controls on Holocene and present-day travertine formation in the Guadalaviar River (Iberian Chain, NE Spain). *Zeitschrift fur Geomorphologie* 41: 289–308.

SanClements, M.D., Oelsner, G.P., McKnight, D.M. et al. (2012). New insights into the source of decadal increases of dissolved organic matter in acid-sensitive lakes of the Northeastern United States. *Environmental Science and Technology* 46: 3212–3219.

Sanz de Galdeano, C.S. (2013). The Zafarraya Polje (Betic Cordillera, Granada, Spain), a basin open by lateral displacement and bending. *Journal of Geodynamics* 64: 62–70.

Sarbu, S.M., Kane, T.C., and Kinkle, B.K. (1996). A chemoautotrophically based cave ecosystem. *Science* 272: 1953–1955.

Sauro, F. (2014). Structural and lithological guidance on speleogenesis in quartz-sandstone: evidence of the arenisation process. *Geomorphology* 226: 106–123.

Sauro, F., Mecchia, M., Tringham, M. et al. (2020). Speleogenesis of the world's longest cave in hybrid arenites (Krem Puri, Meghalaya, India). *Geomorphology* 359: 107160.

Schneider, J. and Le Campion-Alsumard, T. (1999). Construction and destruction of carbonates by marine and freshwater cyanobacteria. *European Journal of Phycology* 34: 417–426.

Schoenherr, J., Reuning, L., Hallenberger, M. et al. (2018). Dedolomitization: review and case study of uncommon mesogenetic formation conditions. *Earth-Science Reviews* 185: 780–805.

Schott, J., Brantley, S., Crerar, D. et al. (1989). Dissolution kinetics of strained calcite. *Geochimica et Cosmochimica Acta* 53: 373–382.

Šebela, S. (2008). Broken speleothems as indicators of tectonic movements. *Acta Carsologica* 37 (1): 51–62.

Selby, M.J. (1993). *Hillslope Materials and Processes*. Oxford: Oxford University Press.

Shtober-Zisu, N., Amasha, H., and Frumkin, A. (2015). Inland notches: Implications for subaerial formation of karstic landforms - An example from the carbonate slopes of Mt. Carmel, Israel. *Geomorphology* 229: 85–99.

Silva, O.L., Bezerra, F.H., Maia, R.P. et al. (2017). Karst landforms revealed at various scales using LiDAR and UAV in semi-arid Brazil: Consideration on karstification processes and methodological constraints. *Geomorphology* 295: 611–630.

Simmons, G.C. (1963). Canga caves in the Quadrilátero Ferrífero, Minas Gerais, Brazil. *The National Speleological Society Bulletin* 25: 66–72.

Simms, M.J. (1990). Phytokarst and photokarren in Ireland. *Cave Science* 17: 131–133.

Sonnenfeld, P. and Perthuisot, J.-P. (1984). *Brines and Evaporites*. New York: Academic Press.

Spencer, T. (1988a). Coastal biogeomorphology. In: *Biogeomorphology* (ed. H. Viles), 255–318. Basil Blackwell: Oxford.

Spencer, T. (1988b). Limestone coastal morphology: the biological contribution. *Progress in Physical Geography* 12: 66–101.

Sterflinger, K. and Krumbein, W.E. (1997). Dematiaceous fungi as a major agent for biopitting on Mediterranean marbles and limestones. *Geomicrobiology Journal* 14: 219–230.

Stiller, M., Yechieli, Y., and Gavrieli, I. (2016). Rates of halite dissolution in natural brines: Dead Sea solutions as a case study. *Chemical Geology* 447: 161–172.

Stoessell, R.K., Moore, Y.H., and Coke, J.G. (1993). The occurrence and effect of sulfate reduction and sulfide oxidation on coastal limestone dissolution in Yucatan cenotes. *Groundwater* 31: 566–575.

Szczygieł, J. (2015). Quaternary faulting in the Tatra Mountains, evidence from cave morphology and fault-slip analysis. *Geologica Carpathica* 66: 245–254.

Szczygieł, J., Sobczyk, A., Hercman, H. et al. (2021). Damaged speleothems and collapsed karst chambers indicate paleoseismicity of the NE Bohemian Massif (Niedźwiedzia Cave, Poland). *Tectonics* 40: e2020TC006459.

Teng, H.H. (2004). Controls by saturation state on etch pit formation during calcite dissolution. *Geochimica et Cosmochimica Acta* 68: 253–262.

Terjesen, S.G., Erga, O., Thorsen, G. et al. (1961). Phase boundary processes as rate determining steps in reactions between solids and liquids: the inhibitory action of metal ions on the formation of calcium bicarbonate by the reaction of calcite with aqueous carbon dioxide. *Chemical Engineering Science* 14: 277–288.

Tîrlă, L. and Vijulie, I. (2013). Structural-tectonic controls and geomorphology of the karst corridors in alpine limestone ridges: southern Carpathians, Romania. *Geomorphology* 197: 123–136.

Tisato, N., Sauro, F., Bernasconi, S.M. et al. (2012). Hypogenic contribution to speleogenesis in a predominant epigenic karst system: a case study from the Venetian Alps, Italy. *Geomorphology* 151: 156–163.

Truesdell, A.H. and Jones, B.F. (1974). A computer program for calculating chemical equilibria of natural waters. *Journal Research U.S. Geological Survey* 2: 233–248.

Tudhope, A.W. and Risk, M.J. (1985). Rate of dissolution of carbonate sediments by microboring organisms, Davies Reef, Australia. *Journal of Sedimentary Research* 55: 440–447.

Turetsky, M.R., Benscoter, B., Page, S. et al. (2015). Global vulnerability of peatlands to fire and carbon loss. *Nature Geoscience* 8 (1): 11–14.

Upchurch, S., Scott, T.M., Alfieri, M.C. et al. (2019). *The Karst Systems of Florida*. Cham, Switzerland: Springer.

Van Beynen, P., Bourbonnière, R., Ford, D.C. et al. (2001). Causes of colour and fluorescence in speleothems. *Chemical Geology* 175 (3-4): 319–341.

Vandycke, S. and Quinif, Y. (2001). Recent active faults in Belgian Ardenne revealed in Rochefort Karstic network (Namur province, Belgium). *Netherlands Journal of Geosciences* 80: 297–304.

Vatan, J.M. (1967). *Manuel de Sedimentologie*. Paris: Technip.

Viles, H.A. (1984). Biokarst: review and prospect. *Progress in Physical Geography* 8 (4): 523–542.

Viles, H.A. (1987). A quantitative scanning electron microscope study of evidence for lichen weathering of limestone, Mendip Hills, Somerset. *Earth Surface Processes and Landforms* 12: 467–473.

Viles, H.A. (2001). Scale issues in weathering studies. *Geomorphology* 41: 63–72.

Viles, H.A. (2004a). Biokarst. In: *Encyclopedia of Geomorphology*, vol. 1 (ed. A.S. Goudie), 86–87. London: Routledge.

Viles, H.A. (2004b). Biokarstification. In: *Encyclopedia of Caves and Karst Science* (ed. J. Gunn), 304–306. New York: Fitzroy Dearborn.

Viles, H.A., Spencer, T., Teleki, K. et al. (2000). Observations on 16 years of microfloral recolonization data from limestone surfaces, Aldabra Atoll, Indian Ocean: implications for biological weathering. *Earth Surface Processes and Landforms* 25: 1355–1370.

Vlasceanu, L., Sarbu, S.M., Engel, A.S. et al. (2000). Acidic cave-wall biofilms located in the Frasassi Gorge, Italy. *Geomicrobiology Journal* 17: 125–139.

Walter, L.M. and Morse, J.W. (1984). Reactive surface area of skeletal carbonates during dissolution: effect of grain size. *Journal of Sedimentary Petrology* 54: 1081–1090.

Waltham, T. (2002). The engineering classification of karst with respect to the role and influence of caves. *International Journal of Speleology* 31: 19–35.

Warren, J.K. (2016). *Evaporites: A Geological Compendium*. Dordrecht: Springer.

Webb, J.A. and Sasowsky, I.D. (1994). The interaction of acid mine drainage with a carbonate terrane: evidence from the Obey River, north-central Tennessee. *Journal of Hydrology* 161: 327–346.

White, W.B. (1988). *Geomorphology and Hydrology of Karst Terrains*. New York: Oxford University Press.

White, S. and Webb, J.A. (2015). The influence of tectonics on flank margin cave formation on a passive continental margin: Naracoorte, southeastern Australia. *Geomorphology* 229: 58–72.

White, W.B. and White, E.L. (2003). Gypsum wedging and cavern breakdown: studies in the mammoth cave system, Kentucky. *Journal of Cave and Karst Studies* 65 (1): 43–52.

Wicks, C.M. and Herman, J.S. (1996). Regional hydrogeochemistry of a modern coastal mixing zone. *Water Resources Research* 32: 401–407.

Wigley, T.M.L. (1971). Ion pairing and water quality measurements. *Canadian Journal of Earth Sciences* 8: 468–476.

Wigley, T.M.L. and Plummer, L.N. (1976). Mixing of carbonate waters. *Geochimica et Cosmochimica Acta* 40 (9): 989–995.

Wolery, T.J. (1992). EQ3NR, a Computer Program for Geochemical Aqueous Speciation-Solubility Calculations: Theoretical Manual, User's Guide, and Rrelated Documentation (Version 7.0). Lawrence Livermore National Laboratory Report UCRL-MA-110662 PT III.

Wray, R.A.L. and Sauro, F. (2017). An updated global review of solutional weathering processes and forms in quartz sandstones and quartzites. *Earth-Science Reviews* 171: 520–557.

Yoshimura, K., Liu, Z., Cao, J. et al. (2004). Deep source CO_2 in natural waters and its role in extensive tufa deposition in the Huanglong Ravines, Sichuan, China. *Chemical Geology* 205: 141–153.

Yuan, D. (ed.) (2001). *Guidebook for Ecosystems of Semiarid Karst in North China and Subtropical Karst in Southwest China, IGCP 448*. Guilin, China: Karst Dynamics Laboratory.

Zanbak, C. and Arthur, R.C. (1986). Geochemical and engineering aspects of anhydrite/gypsum phase transitions. *Bulletin of the Association of Engineering Geologists* 23: 419–433.

Zarei, M., Raeisi, E., and Talbot, C. (2012). Karst development on a mobile substrate: Konarsiah salt extrusion, Iran. *Geological Magazine* 149: 412–422.

Zhang, M., McSaveney, M., Shao, H. et al. (2018). The 2009 Jiweishan rock avalanche, Wulong, China: precursor conditions and factors leading to failure. *Engineering Geology* 233: 225–230.

Zundelevich, A., Lazar, B., and Ilan, M. (2007). Chemical versus mechanical bioerosion of coral reefs by boring sponges-lessons from *Pione cf. vastifica*. *Journal of Experimental Biology* 210: 91–96.

4

Denudation in Karst. Rates and Spatial Distribution

4.1 Basic Concepts

Denudation or erosion, commonly used as synonyms, refer to the group of physical, chemical, and biotic processes that produce the removal of mass as solid particles and dissolved material, respectively (Gaillardet 2004; Lupia-Palmieri 2004). In non-karst terrains, these processes essentially occur at the surface and result in the lowering of the ground surface. However, in karst areas, characterized by significant subsurface drainage, a significant part of the mass removal occurs by underground solutional erosion, which does not have a direct effect on surface lowering (Gunn 2013). Let us consider a carbonate karst watershed dominated by underground drainage (i.e., negligible surface runoff) with its outlet at a major spring. Drainage, mostly underground flow, increases the overall rock mass permeability and contributes to the solute load of the spring, but produces very limited surface lowering. Over geological time scales, the downwearing of the land surface reduces the thickness of cavity roofs, which may eventually collapse causing the catastrophic lowering of the ground at specific locations. Consider also a salt formation affected by interstratal dissolution through the migration of a dissolution front. The removal of salt as solutes by groundwater flow is accompanied by the progressive subsidence of the overlying rock formations and the ground surface. In this case, the reduction in topographic elevation is related to subsidence, a gravitational deformation process, but the underlying cause is underground solutional erosion. In denudation rate studies conducted in karst areas, it is essential to specify whether the rates correspond to the total amount of material removed from both the surface and the subsurface (denudation rates) or solely from the land surface (surface denudation rates). Denudation rate is the sum of both chemical and mechanical erosion processes, whereas solutional denudation rate refers to the chemical component or the mass removed by dissolution.

Solutional erosion in karst terrains is in most cases more important than mechanical erosion. In fact, most denudation studies conducted in karst environments neglect mechanical erosion, despite it can considerably contribute to the overall mass flux and play an important preparatory role for chemical erosion (e.g., rock fragmentation). Kaufmann and Braun (2001) developed a numerical model of surface erosion and landscape evolution for a limestone plateau rimmed by a steep escarpment 1000 m in local relief, incorporating fluvial erosion, hillslope diffusion, and chemical dissolution on limestone surfaces. This model significantly simplifies the real world disregarding internal drainage and subsurface dissolution, resulting in a fluvial landscape, rather than in the typical karst landscape characterized by enclosed depressions and negligible surface drainage (holokarst versus fluviokarst). Nonetheless, it provides interesting qualitative results, indicating that landscape evolution can be much faster in areas underlain by soluble bedrock due

Karst Hydrogeology, Geomorphology and Caves, First Edition. Jo De Waele and Francisco Gutiérrez.
© 2022 John Wiley & Sons Ltd. Published 2022 by John Wiley & Sons Ltd.

to widespread surface chemical denudation. Three fundamental features of denudation in karst areas that should always be kept in mind are that: (i) a great proportion of the erosion occurs in the subsurface, (ii) chemical (dissolutional) erosion is in most cases more important than mechanical erosion, and (iii) most of the mass flux is produced in solution.

In non-karst terrains, denudation rate studies refer to surface areas whose boundaries can be easily defined by topographic criteria (e.g., watershed divides). In contrast, solutional denudation in karst is a three-dimensional problem in which erosion occurs at the surface and within a rock volume. Ideally, denudation rates would be best expressed as the volume of material removed per unit volume of rock over a period of time ($m^3 km^{-3} yr^{-1}$). However, the horizontal and vertical boundaries of the rock volumes affected by dissolution (karst groundwater basins) are often difficult to establish, even after conducting thorough water tracing studies. The boundaries can cross-topographic divides and their position may change depending on groundwater flow conditions (see Section 5.6.3). Consequently, denudation rates are commonly expressed as the volume of material removed per unit area and per unit time ($m^3 km^{-2} yr^{-1}$), or more frequently as the average thickness of rock removed across a horizontal surface over a period of 1000 years ($mm kyr^{-1}$). Note that $1 m^3 km^{-2} yr^{-1}$ is equivalent to $1 mm kyr^{-1}$, which is designated by some authors as the Bubnoff unit, following the Russian geologist Sergius Nikolajewitsch von Bubnoff. The use of the unit $mm kyr^{-1}$ can have some problems: (i) solutional erosion is frequently measured as dissolved mass, which is transformed into volume or thickness incorporating some error by estimating average rock density, (ii) most erosion rate studies are based on data covering time spans of years or decades, but the unit gives the false impression that the obtained values refer to long-term erosion, and (iii) the mass or volume of material removed is referred to a horizontal surface, and consequently does not take into account surface roughness (contact area in a highly irregular surface is much greater than in a perfectly planar surface).

An additional complication is that drainage basins may have runoff and solute loads derived from karst rocks (autogenic) and non-karst rocks (allogenic) (Figure 5.16). Consider three drainage basins with different lithological characteristics, being the rest of the attributes equal. Drainage basin 1 is entirely underlain by karst rocks. In drainage basin 2, karst rocks occur in the downstream half of the basin and in drainage basin 3 in the upstream half of the basin. The total solutional erosion rate will be highest in the purely autogenic basin 1, where there is higher and longer interaction between water and karst rocks. Total solutional denudation will be higher in basin 2 than in basin 3, because the runoff that interacts with the karst rocks in basin 2, located in the lower part of the basin, derives from a larger contributing area. In basin 3, karst rocks solely interact with the autogenic runoff. If we consider the amount of dissolved material versus the area covered by karst rocks (specific solutional denudation rate), the highest value would correspond to basin 2. This theoretical example illustrates the importance of differentiating between autogenic and allogenic runoff and solute loads and the important role played by the distribution of the karst rocks within the basin. The separation of autogenic and allogenic components in basins with mixed lithologies is essential to avoid the underestimation of solutional denudation rates and to obtain values that can be used to make consistent comparisons with other areas (Lauritzen 1990). For instance, Williams and Dowling (1979) in a $45.1 km^2$ drainage basin in New Zealand, with around half of the area underlain by calcite marbles and with an average annual precipitation of 2158 mm, calculated solutional denudation rates estimating inputs, throughputs, and outputs of water and dissolved calcium and magnesium in the autogenic and allogenic drainage systems. They obtained a mean rate of chemical denudation for the area underlain by marble of $265 \pm 64 t$ $km^{-2} yr^{-1}$ or $100 \pm 24 m^3 km^{-2} yr^{-1}$, considering a density of $2.65 g cm^{-3}$ for the marble. This is equivalent to $100 \pm 24 mm kyr^{-1}$. Of the total dissolved Ca and Mg recorded at the gauging and water quality station, 85.5% is attributed to dissolution of karst rocks by autogenic (68%) and allogenic waters (17.5%), 9.9% to chemical weathering of non-karst rocks, and 4.6% to allochthonous carbonate minerals incorporated in the basin by precipitation and dry fall out.

Ford and Williams (2007) indicate that studies on denudation rates based on solute loads should assess the contribution of the different sources to the dissolved material and take into account the potential effect of karst deposition by re-precipitation on the overall mass balance. Solute discharge at the outlet of the basin is given by the product of river discharge Q and the corresponding solute concentration C. For instance, if the discharge is $100\,L\,s^{-1}$ and the concentration of Ca^{2+} derived from $CaCO_3$ dissolution is $50\,mg\,L^{-1}$, the solute discharge is $5\,g\,s^{-1}$, equivalent to approximately $12.5\,g\,s^{-1}$ of $CaCO_3$ (molar mass of $CaCO_3$ around 2.5 greater than that of Ca). Both discharge and solute loads have different values for the autogenic and allogenic components of the system. The total discharge is given by

$$Q_T = \left(P - E\right)_{\text{autogenic}} + \left(P - E\right)_{\text{allogenic}} \pm \Delta S \tag{4.1}$$

where P is precipitation, E is evapotranspiration, P–E runoff from the karst (autogenic) and non-karst (allogenic) portions of the basin, and ΔS change in storage. The solute load recorded at the outlet of the basin may have various sources including airborne supply external to the basin by precipitation and dry fallout:

$$\text{Solute load} = \text{net karst solution} + \text{allogenic corrosion of non karst rocks} + \text{airborne supply} \tag{4.2}$$

$$\text{Net karst solution} = \text{autogenic dissolution} + \text{allogenic dissolution} - \text{karst deposition} \tag{4.3}$$

Gross karst solution is the total amount of karst rocks dissolved by autogenic and allogenic waters. However, part of the dissolved mass may re-precipitate when the water reaches oversaturation conditions, causing karst deposition (speleothems, tufa, travertines, and cements) and removing solutes from the water. Hence, net karst solution is obtained by subtracting karst deposition from the gross karst solution (Ford and Williams 2007). The difference between gross karst solution and net karst solution may be considerable in areas where there is rapid precipitation of chemical sediments. This rarely occurs in cold environments, whereas in tropical and warm temperate climates the rates of carbonate accumulation can be high, especially at sites where oversaturation conditions are maintained by continuous CO_2 depletion (e.g., physical degassing, photosynthetic activity) or water mixing (e.g., common-ion effect). A striking example is provided by the geomorphic and stratigraphic record of the Antalya karst system, in Turkey. Here, groundwater mainly recharged in a fault-controlled system of poljes has caused significant subsurface solutional denudation along its path toward the Mediterranean Sea (Doğan et al. 2019), but a considerable proportion of the dissolved carbonate rocks is re-precipitated in the discharge zone of the aquifer system, generating the Plio-Quaternary Antalya tufa platform, covering $600\,km^2$ and around 250 m in thickness (Glover and Robertson 2003). Solutional denudation is also overestimated if the contribution of airborne supply to the solutes is not subtracted. This external input to the solute load may be considerable in arid and semiarid areas, especially those located at relatively short distance from major sources of aeolian dust. In these regions, annual dust deposition rates are frequently in the order of tens to hundreds of $g\,m^{-2}$ and calcite contents of 20–50% are common in some regions (e.g., dust sourced from the Saharan Desert and the Arabian Peninsula) (Goudie and Middleton 2006). Geological evidence for the importance of airborne supply of carbonate minerals in some regions and during some geological periods is given by thick loess accumulations, which cover as much as 10% of the Earth's land surface (Muhs et al. 2004).

4.2 Controlling Factors and the Influence of Climate

Ford and Williams (2007), in their historical review, indicate that quantitative studies on solutional denudation using modern methods started by the end of the nineteenth century. Spring and Prost (1883) calculated the annual solute load of the River Meuse at Liège in Belgium (ca. 10^6 t), and Ewing (1885) estimated the annual average denudation of carbonate rocks in a river basin in Pennsylvania, United States (34 mm kyr^{-1}). The variations displayed by karst landscapes in the different climatic environments, and the increasing amount of data on solutional erosion rates from numerous regions, although rather heterogeneous, stimulated discussion on the role played by climatic factors on carbonate karst denudation. Corbel (1959), in a groundbreaking work based on data from thousands of samples, postulated that temperature is the main factor that controls solutional denudation. He contended that the highest rates of chemical erosion occur in cold regions due to the inverse relationship between CO_2 solubility and temperature (Figure 3.9). This idea was not in agreement with the dramatic geomorphic imprint of karst denudation in humid tropical regions and the general notion that chemical weathering processes are particularly intense in warm and humid environments (e.g., Peltier 1950; Strakhov 1967). A number of subsequent studies challenged Corbel's interpretation and proposed that runoff is the main climatic factor that controls solutional denudation in carbonate areas (Pulina 1971; Gams 1972). Smith and Atkinson (1976) explored the relationship between climate and chemical denudation rates using 134 estimates of solutional denudation rates in carbonate karst catchments from all over the world and 231 reports on the mean hardness of spring and river waters. Note that total hardness is the aggregate concentration of Ca^{2+} and Mg^{2+} in water expressed as equivalent $CaCO_3$ or $CaCO_3+MgCO_3$. It provides an approximate measure of the amount of limestone and dolostone dissolved in the absence of other significant sources for those ions (e.g., gypsum). Smith and Atkinson (1976) grouped the data into three broad climatic domains (tropical, temperate, and arctic/alpine) and produced regressions of solutional denudation rate versus mean annual runoff (Figure 4.1). They inferred that: (i) runoff, rather than temperature, is the main climate-related controlling factor; (ii) rates of solutional denudation are commonly higher in tropical regions than in arctic/alpine regions with the same average runoff; and (iii) the effect of the higher solubility of CO_2 in the cold waters of arctic/alpine environments is overwhelmed by the higher CO_2 content in the soils of temperate and tropical regions. The authors emphasized the high heterogeneity of their denudation rate data, derived from catchments with a wide range of dimensions and characteristics (e.g., soil coverage, topography) and a highly variable number of samples. Moreover, the grouping of the three climatic groups is in some cases rather ambiguous and the regressions shown in Figure 4.1 underestimate denudation since they do not take into account the proportion of limestone in the watershed. For instance, Crowther (1989), based on their data from Malaysia and those obtained by other authors, suggests that the regression of the tropical environments underestimates the actual solutional denudation.

It is unquestionable that climate plays a primary role on solutional denudation in karst areas. However, solutional erosion responds to a complex equation that includes multiple variables, some of them interrelated and others unrelated to climate. This is supported by the fact that denudation rates show significant variations in areas located within the same climatic region (Smith and Atkinson 1976). In the following pages we explain qualitatively how the main climatic and nonclimatic factors influence solutional denudation and some of their interrelationships. Water runoff, or the difference between precipitation and evapotranspiration, is the key agent for dissolution and solute transport, both at the surface and in the subsurface. This explains the limited or negligible geomorphic effectiveness of carbonate dissolution in arid and hyperarid areas. However,

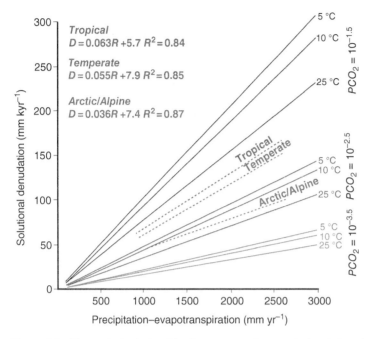

Figure 4.1 Theoretical relationships between maximum solutional denudation rates of limestone, water surplus, and P_{CO_2} considering open-system conditions and assuming that water reaches saturation (White 1984). Dashed lines are regressions obtained by Smith and Atkinson (1976) using solutional denudation rates from different morphoclimatic environments without considering the share of limestone in the catchments (*D*: solutional denudation, *R*: runoff or precipitation–evaporation). *Source:* Adapted from Ford and Williams (2007).

precipitation water has a limited capability to dissolve carbonate rocks due to its low CO_2 partial pressure. The acidity and aggressiveness of the water increase substantially when it infiltrates through a soil cover and absorbs CO_2 derived from the decay of organic matter and root respiration. As Table 3.1 shows, the solubility of calcite in water (25 °C, 1 atm) with the P_{CO_2} typical of soils with high CO_2 content (5% by volume) is five to six times higher than in water with the P_{CO_2} value of the atmosphere (0.04%), a CO_2 concentration 125 times lower than the former. However, the participation of the soil-derived CO_2 in the denudation equation depends on the presence of soils, regardless of the climatic conditions. Solutional denudation rates are much higher in a soil-covered karst area than in a bare karst with the same climatic conditions. The CO_2 concentration in the soil air, mainly related to biological activity, increases with temperature and water availability and reaches the highest values in warm-wet regions (Figures 3.6 and 3.7). In arctic and alpine environments, there is commonly high moisture in the soils, but the degradation of the organic matter is slow due to the low temperature, in contrast to humid tropical environments. The importance of temperature on the soil CO_2 partial pressure, the main source of acidity in epigenic carbonate karsts, is illustrated by some equations used to model this parameter. Brook et al. (1983) found an inverse relationship between soil P_{CO_2} and evapotranspiration, which accounts for the combined effect of temperature, precipitation, and water availability. Romero-Mujalli et al. (2019) used an equation including mean annual surface temperature and the water content of soils to estimate average P_{CO_2} in soils (Figure 3.7). The relationship between climate and solutional denudation in evaporites is much simpler, given the limited influence of temperature and biological activity on gypsum and salt dissolution; runoff is the critical parameter.

Other non-climatic factors can also play a significant role on karst denudation (Bakalowicz 1992). The lithology and mineralogy of the carbonate bedrock, which in most cases are not sufficiently characterized, may also influence the rates of solutional denudation. The percentage of non-soluble lithologies within the carbonate succession, internal impurities in the carbonate rocks, and high dolomite/calcite ratios can contribute to attenuate chemical erosion (Rauch and White 1977; Plan 2005). Dissolution at the surface or in aerated portions of the vadose zone occurs under open- or intermediate-system conditions, whereas it proceeds in a closed-system in the phreatic zone. In the latter, the CO_2 consumed by dissolution is not replenished and the capability of the water to dissolve limestone decreases as CO_2 is depleted. Consequently, the conditions under which dissolution occurs, open- versus closed-system, also influence the overall solutional denudation rates. This factor is linked to the proportion of runoff that drains underground, which may vary significantly. Another important factor is the relative contribution of diffuse and conduit flow in the groundwater movement. In the diffuse flow, the water tends to reach saturation conditions due to longer and greater interaction with the rock, whereas waters associated with rapid conduit flow with shorter residence time may not reach equilibrium and produce less denudational work. The geomorphology of the karst basin can have a very important influence by conditioning other relevant parameters. Consider two basins of equal size and climate underlain by the same carbonate formation. One of them has a nearly flat topography, whereas the other is characterized by a rugged terrain of bare limestone outcrops. The former can hold relatively stable soils with vegetation that supply CO_2 to the infiltrating waters, whereas the water recharged to the aquifer in the latter would have a much lower acidity and dissolution capability.

White (1984, 2000a) proposed an equation that relates limestone solutional denudation to three hydroclimatic variables: runoff, temperature, and CO_2 partial pressure:

$$D_{\max} = \frac{100}{\rho(4)^{1/3}} \left(\frac{K_c K_1 K_{CO_2}}{K_2} \right)^{1/3} P_{CO_2}^{1/3} (P - E) \tag{4.4}$$

where $D_{\max}$ is the maximum net autogenic solutional denudation rate (mm kyr^{-1} or $1\,\text{m}^3$ km^{-2} yr^{-1}). It is the maximum expected value because the equation assumes that the water reaches saturation with respect to calcite and that dissolution occurs under ideal open-system conditions. Rock density is ρ, K_c is the solubility product constant of calcite (see Eq. 3.53), K_{CO_2} is the equilibrium constant of the dissolution of CO_2 in water (see Eq. 3.40), K_1 and K_2 are the equilibrium constants of the first- and second-order dissociations of carbonic acid (see Eqs. 3.42 and 3.44), P_{CO_2} is the CO_2 partial pressure in water given in atm, P is precipitation and E is evapotranspiration, both in mm yr^{-1}. The relationship between maximum solutional denudation and runoff contained in the equation is represented in Figure 4.1 for different P_{CO_2} values. The equation, in agreement with the regressions of Smith and Atkinson (1976), predicts a linear relationship between denudation and runoff. The expression also illustrates the strong influence of the CO_2 partial pressure on solution, despite its inclusion as the cube root of its value (Gabrovšek 2009). Precipitation water that infiltrates through soils with high CO_2 levels can increase its P_{CO_2} by two orders of magnitude. An organic-rich soil has a CO_2 concentration around 100 times higher than that of the Earth's atmosphere, involving an increase in $D_{\max}$ by a factor of 5. The equation may give the impression that temperature is the least important variable and that, as suggested by Corbel (1959), it can have an inverse relationship with denudation. However, its influence is concealed in all the variables through different chemical, hydrological, and biological mechanisms. Temperature is contained in the equilibrium constants, and especially in K_{CO_2}, which accounts for the inverse relationship between CO_2 solubility and temperature. It is also subsumed in the evapotranspiration variable, which increases with temperature (e.g., Thornthwaite's potential evapotranspiration equation). Temperature is also included in the variable P_{CO_2}, which reaches its highest values in soils developed in humid and warm regions, with rapid production and degradation of organic matter (Bouma et al. 1997). This latter direct effect, which is related to the source of acidity, can largely overwhelm the two previous inverse effects of CO_2 solubility and evapotranspiration, as supported by the relative position of the curves for tropical, temperate, and arctic/alpine regions (Figure 4.1).

Solutional denudation rates have changed through geological times in concert with the variability of the complex suite of environmental factors that control both dissolution and karst deposition (runoff, temperature, soil CO_2, and vegetation). Thus, the exercise of extrapolating modern solutional denudation rates into the past, although attractive, can be highly speculative. The concentration of CO_2 in the world's atmosphere has experienced a very rapid rise in recent history due to human activity (burning of fossil fuels, deforestation, and forest fires), especially since the end of World War II. The CO_2 content of the atmosphere has increased by ca. 30% between 1958 (315 ppm) and 2020 (410 ppm) (Figure 3.4). It is estimated that the global annual temperature has increased at an average rate of 0.18 °C per decade since 1981, largely due to the anthropogenic greenhouse effect. These extremely rapid changes, which undoubtedly are having highly adverse consequences on our planet, may give the impression that we are living in a period characterized by particularly high CO_2 levels. However, temperatures and atmospheric CO_2 have been higher than today during most of Earth's history. Figure 4.2 shows that during the Cenozoic Era there were important changes in temperature and CO_2, but on average the globe was significantly warmer than during the Quaternary and today, and precipitation water was more acidic. There was a considerable increase in temperature and atmospheric CO_2 during the initial part of the Cenozoic, which peaked around 52–50 Myr ago at the Eocene thermal maximum. Atmospheric CO_2 was probably about five times higher than that of today. Subsequently, temperature and CO_2 experienced a significant decrease up to the onset of the Oligocene. In the late Eocene, temperatures were cold enough for the Antarctic ice sheet to begin its formation. That was the end of a greenhouse climate period and the beginning of the long icehouse climate period that extends

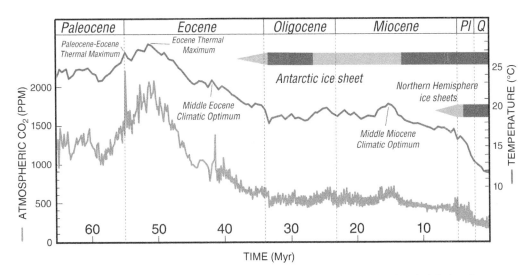

Figure 4.2 Variability throughout the Cenozoic of atmospheric CO_2 (orange line) and global surface temperature (red line) reconstructed from ^{18}O proxies in marine sediments. *Source:* Adapted from Hansen et al. (2008).

up to the present time. Climatic conditions during the Miocene and Pliocene were relatively stable, with warming periods like the Middle Miocene climatic optimum. Temperatures experienced a generally decreasing trend during the Pliocene, allowing the development of ice sheets at high latitudes in the Northern Hemisphere. The Quaternary period, covering the last 2.6 Myr, includes the Pleistocene and the Holocene epochs with their boundary at ca. 11 650 cal yr BP (years before 1 January 1950). The Pleistocene comprises a series of alternating glacial (stadials) and interglacial periods (interstadials) with average durations of around 100 kyr and tens of kyr, respectively (Figure 4.3). These cycles, that lasted in total around 2 Myr, were characterized by profound geomorphic and ecological instability with strong influence on the karst systems. The transfer of water between the oceans and the ice sheets and vice versa caused large sea level variations. Morphoclimatic zones and vegetation belts experienced latitudinal shifts in the order of tens of degrees. Large variations also affected the boundary of the permanently frozen ground or permafrost (Anderson et al. 2007). The Last Glacial Maximum (LGM) occurred around 22–19 kyr. At this time, the sea level was approximately 125 m lower than today. The interval between the LGM and the beginning of the Holocene, usually designated as Late Glacial, comprises several cold stadials and warm interstadials. The Holocene, which is the current interglacial, is characterized by fairly stable temperatures compared to the large variations of the Pleistocene, especially over the last 7 kyr. This succinct paleoclimatic chronicle makes it evident that solutional denudation rates have experienced continuous and substantial changes in the past. However, those variations are generally difficult to reconstruct due to the multiple problems associated with the erosional nature of the dissolution process, the scarcity of datable and continuous proxies, and the interplay of multiple overlapping factors. A topic of special interest, as discussed in Section 3.3.1, is the monitoring and prediction of how the rise in atmospheric CO_2 and global temperature may affect the rate at which carbonate rocks are dissolved, which in turn may function as buffering agents to the CO_2-related greenhouse effect (e.g., Jeannin et al. 2016). The rise in the concentration

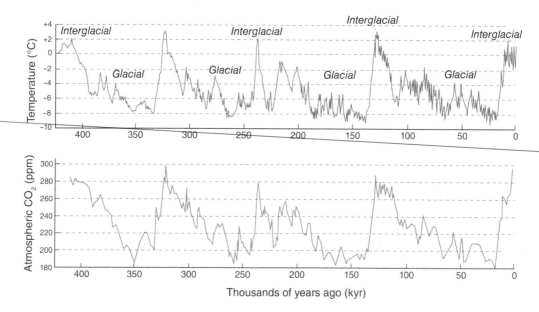

Figure 4.3 Graph of reconstructed temperature variation inferred from changes in isotope ratios and atmospheric CO_2 from the Vostok Station ice core, Antarctica, for the past 400 kyr. Note close correlation between temperatures and CO_2 values, higher during warmer periods (greenhouse effect). *Source:* Produced by William M. Connolley using data from the National Oceanic and Atmospheric Administration, U.S. Department of Commerce, Paleoclimatology branch.

of atmospheric CO_2 from 315 to 410 ppm ($10^{-3.5}$ to $10^{-3.38}$ atm) recorded between 1958 and 2020, which affects the acidity of precipitation water worldwide, has a positive effect on Eq. 4.4, where maximum dissolution varies with the cube root of P_{CO_2}, resulting in an increase of around 10% above the initial low value (Figure 4.1). Global warming may also enhance carbonate dissolution by increasing the production of CO_2 in the soils boosting organic matter decay and root respiration. On a regional scale, permafrost thawing also contributes to enhance dissolution by making the interaction between soluble rocks and liquid water possible (Ford 1987). In contrast, other impacts associated with anthropogenic global change, including soil erosion and deforestation, may contribute to a decrease in the important positive effect of soil CO_2 on carbonate dissolution.

4.3 Methods for Measuring Denudation Rates and the Carbonate Karst Experience

The quantitative assessment of solutional denudation is of great interest for understanding multiple aspects related to the functioning and evolution of karst systems: (i) the development of aquifer porosity and permeability, (ii) the formation and evolution of solution landforms at the surface and in the subsurface, (iii) the comparison of chemical erosion rates in different settings and climates, and (iv) the estimation of the amount of CO_2 consumed by carbonate dissolution and its contribution to the global carbon budget (e.g., Gao et al. 2009). Chemical erosion of karst rocks by dissolution occurs both at the surface and in the subsurface, is a relatively slow and persistent process, and has a widespread spatial distribution, in contrast to some mechanical erosion processes that affect the land surface at specific sites and in a catastrophic manner (e.g., large rapid landslides or sudden bedrock collapse sinkholes (Simms 2004)). Turowski and Cook (2017) differentiate two main groups of methods for measuring bedrock erosion: (i) Indirect methods, based on the amount of material discharged (sediment and/or solute load) from a defined area over a period of time, which include the approaches that use solute flux (Figure 4.4a). (ii) Direct methods that involve the measurement of changes occurred over a period of time (surface lowering, mass or volume loss). This latter group includes both short-term survey techniques and long-term dating techniques. The survey techniques are based on the comparison of accurate measurements carried out at specific points over times spanning from months to years (e.g., weight loss of standard tablets, micro-erosion meter [MEM], and terrestrial laser scanner [TLS]). The so-called dating techniques of bedrock erosion allow the estimation of long-term denudation rates by measuring the amount of topographic change occurred over long time scales, commonly using dated geological markers (limestone pedestals, cosmogenic ^{36}Cl). Table 4.1 includes solutional denudation rates in carbonate rocks obtained by different methods and in multiple environments.

4.3.1 Hydrochemical Measurements

Estimates of solutional denudation rates based on water and solute flux measurements are referred to three-dimensional hydrological systems with surface and subsurface runoff (karst drainage basin). As explained in Section 5.6.3, the karst drainage basin is an open system with quantifiable inputs, throughputs and outputs, and complex boundaries whose delimitation frequently requires the consideration of stratigraphic and structural relationships and the application of tracing tests. The key parameters that need to be recorded at a suitable output point (e.g., spring, river channel) with a well-established contributing area and internal lithology are water discharge and solute concentration (Figure 4.5). Solute flux is the product of water discharge and concentration. Some of the main

Figure 4.4 (a) Soil erosion plot in gypsum slopes installed for measuring chemical and mechanical denudation rates. The tanks at the outlet of the small drainage basin collect both sediment and solute load. Solute flux is measured with continuous records of electrical conductivity and analyzing water samples after each runoff event. Mediana de Aragón, NE Spain. *Source:* Photo by Gloria Desir. (b) Standard tablets of Lipica limestone (98–99% calcite) used to measure calcite dissolution and precipitation at Krizna Jama 2, Slovenia (see Prelovšek 2012). *Source:* Photo by Matej Krzic. (c) Tablets made of different limestone and dolostone rocks of the Triassic Hochschwab Unit (Northern Calcareous Alps, Austria) with different mineralogical and textural characteristics and pure marble in the lower left. *Source:* Photo Lukas Plan. (d) Traversing micro-erosion meter with digital dial gauge. *Source:* Photo by Lluís Gómez-Pujol.

methods used for measuring flow rate in karst hydrological systems are presented in Section 5.6.5. Solute concentration can be derived from continuous records of electrical conductivity, which often shows a good correlation with total hardness (Figure 3.2), especially in waters where most of the dissolved solids are derived from the dissolution of carbonate rocks. These correlations need to be checked and refined with a sufficient number of water samples and analyses of the solutes. Solute concentration and its variability can be measured from water samples that should be collected with a relatively high frequency and covering a wide range of flow conditions in every season. These data are used to generate mass-flux rating curves that model the relationship between water discharge and solute concentration or flux, despite its considerable variability (stage within a flow episode, seasonal, and interannual). The modern technology permits the installation of automatic stations that collect continuous measurements of discharge (hydrographs) and solute concentrations (chemographs) (see Section 5.6.7). Secondary stations are required in basins with mixed autogenic and allogenic runoff events to assess the contributions of the two components to the total solutional erosion of the karst bedrock. Secondary stations can also provide valuable information on the spatial variability of the solutional denudation. The representativeness of the available data can be assessed by comparing precipitation and discharge recorded over the monitoring period with longer meteorological and hydrological records obtained in the basin or in nearby areas with similar characteristics.

Table 4.1 Short- and long-term denudation rates in carbonate rocks obtained by various methods and in different environments. Rates derived from solute load correspond to overall values related to both surface and subsurface denudation in karst drainage basins, whereas the rest are local values of rock solutional downwearing.

Method	Precipitation	Location (climate)	Denudation rate (mm kyr^{-1})	References
Solute load	1500 (average)	Baget (Ariège, French Pyrenees)	48 (1973)	Bakalowicz (1979)
			83 (1974)	
			88 (1975)	
			56 (1976)	
			89 (1977)	
	1500	Ljubljanica catchment, Slovenia	60	Gams (2004)
	2100	Austrian Alps	95	Plan (2005)
	2160	Riwaka, New Zealand	85	Williams and Dowling (1979)
	2370	Waitomo, New Zealand	69	Gunn (1981a)
	3000	Alps, Slovenia	94	Kunaver (1979)
	2600	Svartisen, Norway	32	Lauritzen (1990)
Limestone tablets	1300	Tropical humid	4 (in air)	Gams (1986)
			2.4 (at the surface)	
			7.8 (in the ground)	
	1400	Temperate mountainous	5 (in air)	
			4.3 (at the surface)	
			14.6 (in the ground)	
	2100	Austrian Alps	11 (subaerial)	Plan (2005)
	2100	Austrian Alps	30–40 (subsoil)	
	3000	Julian Alps, Slovenia	6 (subaerial)	Kunaver (1979)
	797	Central Spain	1.75–2.48 (subsoil, dolostone)	Krklec et al. (2016)
		Japan	71–187 (allogenic stream)	Hattanji et al. (2014)
MEM/T-MEM		Dinaric karst	26–102	Kunaver (1979)
		Friuli-Venezia Giulia, Italy	10–190	Forti (1984)
		Malham, North Yorkshire, United Kingdom	10–50	Trudgill (1986)
	2600	Norway (subarctic)	25 (subaerial)	Lauritzen (1990)
	1800–3000	Julian Alps, Italy	10–40	Cucchi et al. (1994)
	600–1450	Southeastern Australia	8–26 (subaerial)	Smith et al. (1995)
	1350	Trieste, Italy	14–31	Cucchi et al. (2006)
	1700–2500	Alaska	25 (subaerial)	Allred (2004)
			71 (subsoil)	
	1350	Italy and Croatia	9–18 (inland)	Furlani et al. (2009)
Cosmogenic nuclides	600	Israel	20 (subaerial, flat)	Ryb et al. (2014b)
	50	Israel	1–3 (subaerial, flat)	

Note the temporal variability of rates obtained by Bakalowicz (1979) in different years.
MEM: micro-erosion meter, T-MEM: traversing micro-erosion meter.
Source: From Gabrovšek (2009), Stephenson and Finlayson (2009), and references therein.

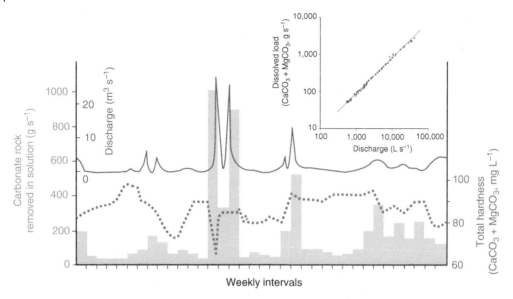

Figure 4.5 Discharge and solute load (total hardness) recorded between January and August 1974 in a mixed autogenic–allogenic basin in New Zealand and calculated flux of carbonate rock removed in solution. Note peak in water and solute discharge associated with low values of total hardness related to dilution. Inset corresponds to the rating curve obtained with measured values of discharge and dissolved load at the control station. *Source:* Adapted from Williams and Dowling (1979).

Corbel (1959) proposed a widely used formula for estimating solutional denudation rates using average values of solute concentration:

$$X = \frac{4\,ET}{100} \tag{4.5}$$

where X is the limestone dissolution rate in volume ($m^3\ km^{-2}\ yr^{-1}$), E is runoff (precipitation–evapotranspiration, $dm\ yr^{-1}$), and T is the average $CaCO_3$ content in water ($mg\ L^{-1}$). The number 4 in the equation is related to an assumed bulk density of $2.5\,g\,cm^{-3}$ for limestone. For basins with non-karst rocks with autogenic and allogenic runoff and dissolution, Corbel proposed the incorporation of the fraction of limestone in the drainage basin (n^{-1}) into the equation:

$$X = \frac{4\,ET\,n}{100} \tag{4.6}$$

These formulas were used in many of the earlier studies on solutional denudation, but they have a number of shortcomings (Ford and Williams 2007; Trudgill 2008): they (i) assume a density for carbonate rocks of 2.5, though it ranges between 1.5 and 2.9, (ii) ignore the potential contribution of gypsum dissolution to the concentration of dissolved Ca^{2+} and obviates Mg^{2+} derived from carbonate dissolution, unless total hardness is used for the estimation of T, (iii) do not consider off-basin solute introduced by precipitation and dry fallout, (iv) incorporate further significant errors in the difficult estimation of the average concentration of dissolved carbonate rock, (v) implicitly assume that in basins with non-karst rocks the second formula attributes all the solute content to autogenic dissolution, and (vi) be unable to separate autogenic and allogenic solution.

Significantly more accurate results can be obtained by averaging a series of measurements of water discharge and solute concentration, with the latter corresponding to direct measurements and/or estimates derived from discharge-solute concentration relationships (mass-flux rating curves):

$$D = \frac{\sum_i^m Q_i C_i}{m} \tag{4.7}$$

where D is the average solute discharge, i refers to equal-time intervals for which Q and C are assumed to be constant, and m represents the number of intervals. Continuous records of discharge and solute concentration can be integrated to calculate annual rates of solutional denudation (White 1988; Gabrovšek 2009):

$$D\left(\text{mmky}^{-1}\right) = 31.5 \frac{1}{ftA\rho} \int_o^t Q(t) C(t) \mathrm{d}t \tag{4.8}$$

where 31.5 is the unit conversion factor, f is the fraction of the basin underlain by karst rocks, t is the observation time (s), A is the surface area of the basin (km^2), ρ is the bulk rock density (kg m^{-3}), Q is the discharge (m^3 s^{-1}), and C is the solute content (mg L^{-1}).

In mixed basins with both karst and non-karst rocks, autogenic and allogenic denudation should be differentiated. This is an essential step in making valid comparisons between different basins and environments. A great part of the early studies that explored the role played by climate on solutional denudation were based on the comparison of data without removing the noise related to allogenic denudation, thus limiting the validity of their interpretations. An alternative is the separate determination of the contribution of allogenic and autogenic runoff events to the solute load using several control stations (Pulina 1971). Lauritzen (1990) proposed a procedure for calculating the autogenic and allogenic components of the solutional denudation with a linear model of uniform mixing, using data from two or more relatively homogeneous subcatchments with different proportions of carbonate rocks. This approach assumes that the autogenic and allogenic contributions to solute load are proportional to the area fraction of the karst rocks (f) and that autogenic and allogenic solutional denudation rates are equal in the different subcatchments:

$$D_{\text{auto}} = \frac{D_2 - D_1 + D_1 f_2 - D_2 f_1}{f_2 - f_1} \tag{4.9}$$

where D_1 and D_2 are the apparent denudation rates of each subbasin and f_1 and f_2 are the area fraction of karst rocks in each subbasin. The approach proposed by Lauritzen (1990) was tested in a subarctic basin in Norway with karst and non-karst rocks and an annual average precipitation of 2600 mm, obtaining an autogenic solutional denudation rate of 35.5 ± 10.2 mm kyr^{-1}. This estimation was consistent with other independent data, including denudation rates from comparable areas, MEM data (25 mm kyr^{-1}) and minimum postglacial surface lowering deduced from pedestals and protruding veins (13–23 mm kyr^{-1}). These results suggest that approximately 42–72% of the solutional denudation occurs at the surface in this particular area.

Studies on the magnitude and frequency relationships of solute flux in karst basins indicate that a significant proportion of the transport of the dissolved solids, but not necessarily the dissolution of the rock, occurs during high magnitude and low-frequency events. For instance, Groves and Meiman (2005) based on high-resolution (2-min) discharge and chemical data from the Cave City groundwater basin, Mammoth Cave system (Kentucky), estimated that high flow events occurring less than 5% of the year are responsible for 38% of the dissolved load leaving the system. However, in karst basins, generally with a less flashy behavior than non-karst drainage basins, high frequency and low magnitude flows play a much greater role on solute transport than they do on the transport of solid load (Williams and Dowling 1979; Gunn 1981a).

4.3.2 Weight Loss of Standard Tablets

Dissolution rates can be estimated by measuring the weight loss of rock samples with established geometry and dimensions (Figure 4.4b, c). Most researchers use circular tablets, but other geometries such as cubes (Urushibara-Yoshino et al. 1999) or spheres (Navas 1990) can be deployed as long as the surface exposed to dissolution is accurately known. The standard tablets can be prepared with multiple lithologies and are set up in three main types of locations: (i) in contact with the atmosphere at or above the ground surface, exposed to precipitation, runoff, and condensation water; (ii) buried within the soil; and (iii) submerged in standing or flowing water. After cutting the tablets, some researchers treat the surface by polishing and/or with a dilute acid to remove unnatural smoothness, a factor that can considerably affect solution rates (e.g., Plan 2005). The whole surface can be exposed to the interaction with water, which generally results in uneven dissolution with greater corrosion in the upper surface. Dissolution can be restricted to the upper side covering the rest of the surface (e.g., varnish, holder, and peripheral rubber tape; Plan 2005; Mulec and Prelovšek 2015). Tablets are accurately weighted prior to installation and are covered by a mesh bag to exclude processes such as bioerosion and abrasion in some studies. After a period of time, they are reweighted. Recovered samples are often rinsed with water and dried in an oven. Control samples can be used to assess the potential artificial mass loss caused by this treatment (e.g., Hattanji et al. 2014). Denudation rates are calculated from the mass loss divided by the density of the rock and the surface area. Results are often presented as a relative measure (percentage of weight loss).

The use of limestone tablets for assessing solution rates dates back to the mid-twentieth century (Chevallier 1953). An outstanding landmark regarding the application of this method was an international project launched at the seventh International Speleological Congress in 1977 (Shefield, United Kingdom), involving the installation of standardized limestone tablets around the world in at least three different positions: (i) 1.5 m above the ground, (ii) at the ground surface, either on rock or in grass, and (iii) in the soil at various depths below the A horizon. More than 1500 tablets were distributed at 59 stations in nine countries. The lithological factor was eliminated by preparing all the samples from the same rock formation exposed in a quarry close to Lipica village, Slovenia, a very pure limestone described as a recrystallized biopelmicrite. Some of the main results derived from this international collaboration include (Gams 1981a, 1986): (i) weight loss is generally greater in tablets placed within the soil than in the air or at the surface, (ii) solution rate seems to be directly dependent on water surplus (precipitation–evapotranspiration), rather than on temperature, (iii) solution rates show a distinct climatic pattern with the higher values recorded in the humid tropics, (iv) in arid climates weight loss in the air is often higher than on the ground surface, whereas the opposite seems to hold true for humid climates, and (v) mechanical weathering may significantly contribute to mass loss.

Denudation rates obtained from tablet measurements are not comparable with those derived from hydrochemical data, which are in most cases significantly higher (e.g., Gams 1981a; Crowther 1983; Plan 2005). For instance, Crowther (1983) in West Malaysia obtained, over a period of approximately one year, solutional denudation rates with the weight loss tablet method one or two orders of magnitude less than those calculated from solute load data (water hardness). The large difference is because the two methods measure essentially different phenomena. The hydrochemical data integrate solutional denudation that occurs in the whole hydrological karst system, including the underground karst aquifer, whereas measurements from tablets are point data related to more limited water–rock interaction. The weight loss method has other shortcomings that should be taken into consideration when interpreting the data: (i) uneven dissolution, with greater denudation on the upper side of the tablets (e.g., Urushibara-Yoshino et al. 1999), (ii) limited spatial and temporal representativeness, (iii) large error margins, although rarely computed. The errors estimated by Krklec et al. (2016) in a 1-year-long field experiment with dolostone tablets are around 40% of the estimated weathering rates, (iv) mechanical weathering can have a significant contribution (Plan 2005) and may even exceed the effect of dissolution (e.g., Krklec et al. 2016), and (v) results from rock tablets of the same lithology and located at the same site can show significant variability, attributable to sample differences (e.g., porosity, cracks). In spite of these limitations, the method is highly useful for estimating approximate denudation rates for specific sites and conditions, comparing rates from different settings and lithologies, and assessing the contribution of specific factors.

Urushibara-Yoshino et al. (1999) estimated solution rates at seven karst areas in Japan, covering a wide range of climatic conditions and using different types of limestone at each site. They reported marked interannual variations, solution rates in soils 1.5–5 times higher than those in subaerial conditions, and a high correlation between solution rates from tablets placed 1.5 m above the ground and water surplus and water deficit. Plan (2005), in the Austrian Alps, using tablets of different lithologies, found decreasing rates in dolomitic limestones with higher Mg/Ca ratios, and rates in dolostones around half of those obtained from limestone. Akiyama et al. (2015) explored the role played by soil moisture conditions on solution rates in soils. They installed limestone tablets at 15 and 50 cm depth in different positions along the side slope of a doline in the Akiyoshi-dai Plateau, Japan, and measured water saturation of the soil, soil P_{CO_2}, and solution rates, ranging between 3.9 and 21.9 mg cm^{-2} yr^{-1}. They found a tight direct correlation between solutional denudation rates and duration of water saturation in the soil, as a measure of the time during which the rock interacts with unsaturated water. Hattanji et al. (2014) measured solution rates up to 27 times higher in allogenic streams with highly undersaturated water with respect to calcite (saturation index between −2.8 and −2.5) than in autogenic streams (saturation index −0.5). Chemical denudation rates were equivalent to 150–187 and 71–72 mm kyr^{-1} in the allogenic streams. These results support the much faster karstification that occurs at the contact between non-karst and karst rocks (contact karst), and the potentially more rapid evolution of karst landscapes, controlled by a base level associated with an allogenic river. Mulec and Prelovšek (2015) demonstrate that illumination contributes to increased denudation rates in flowing freshwater in temperate environments by increasing biodissolution related to the formation of biofilms with epilithic phototrophs. They installed limestone tablets submerged in flowing water along a light gradient, from complete darkness (cave river) to direct exposure to sunlight (spring mouth). Measurements of weight loss, biomass dry weight, chlorophyll a, and photosynthetic photon flux densities reveal the significant contribution of biodissolution to the overall solution process. Higher light availability is accompanied by greater diversity of periphyton and higher dissolution rates. This work illustrates the potentially important role of biodissolution in the removal of rock fall deposits in pocket valleys headed by undermined and retreating limestone cliffs. The work by Krklec et al. (2016) demonstrates the potential substantial contribution of mechanical weathering. They buried tablets of dolostone and magnesite-rich dolostone in soil at 5–10 and 50–55 cm depth and found, using scanning electron microscopy (SEM) images, that most of the mass loss was related to crystal detachment. The higher weathering rates obtained at 5–10 cm depth, despite lower moisture and CO_2 content, is attributed to more intense physical weathering by thermal and moisture cycles. Luo et al. (2018) explored the influence of elevation on solution rate using limestone tablets set up at 1.5 m above the ground and at 20 and 50 cm below the surface in different sites covering an elevation difference of 1300 m. They computed higher solution rates on the surface tablets than in the subsurface tablets at high elevation, and the opposite pattern at low elevations. The authors suggest an elevation boundary at which surface and subsurface rates are equivalent. Obviously, elevation is a complex variable that contains multiple factors such as precipitation, temperature, CO_2 production in soils, or the effect of physical weathering, which can be significant at high altitudes with freeze and thaw cycles.

4.3.3 Micro-erosion Meter

MEM was originally produced by High and Hanna (1970) with the purpose of measuring erosion in limestone caves (Hanna 1966). Trudgill et al. (1981) then developed an improved version known as the traversing micro-erosion meter (T-MEM) (Figure 4.4c). The underlying concept is to obtain lowering or recession rates at specific points on rock surfaces by measuring and comparing their relative position at different times. It has also been used satisfactorily to measure deposition rates in tufa deposits in Australia (average rate 4.15 mm yr^{-1}; Drysdale and Gillieson 1997). The MEM consists of an equilateral triangular plate with legs at each corner that are locked precisely into stainless studs permanently fixed into the rock surface. The plate has a probe connected to a dial gauge or micrometer that accurately measures the position of the rock surface at a fixed point. The main advantage of the T-MEM is that it has a mobile dial gauge that allows repeated multipoint measurements at each site. The triangular base of the T-MEM that is locked into the permanent studs has a large inner hollow and a series of ball bearings fixed along the sides. The dial gauge, independent of the base, is mounted on three arms with radial distribution at angles of 120°. It can be displaced to a number of positions by placing each arm between different pairs of balls at each side of the plate. Up to 170 different measurements within an area of 45 cm^2 can be performed. Some T-MEMs are equipped with a digital dial gauge rather than an analogue one, which can be connected to a computer or tablet. The use of surface plotting software allows the volume of removed or deposited material to be estimated (Stephenson and Finlayson 2009).

Spate et al. (1985) conducted experiments under temperature-controlled conditions using two different T-MEMs to assess the impact of different sources of error on the reliability of the measurements and proposed an approach for introducing corrections. They identified the following error sources that may have a cumulative effect on the final measurements: (i) the impact of temperature changes varies depending on the type of instrument, which has differing temperature correction factors. (ii) Measurements are affected by differential expansion and contraction of the rock and the rock–stud interface. Rock swelling and contraction can reach high values, especially in coastal areas, and may be related to multiple processes such as temperature changes, wetting and drying, swelling clays, salt crystallization and hydration, or growth and hydration of epilithic and endolithic algae. Stephenson and Finlayson (2009) compiled measurements of rock swelling in coastal environments measured with MEMs, with values in limestone as high as 0.7 mm. (iii) Erosion of the rock by the tip of the probe, with potentially large impact on soft rocks (e.g., marly limestone). (iv) Operational irregularity of the instrument (e.g., dial gauge). (v) Wearing of the probe tip that leads to erosion overestimates. The error treatment approach was applied to measurements carried out over four years on inland limestone outcrops in New South Wales, Australia (average precipitation 950 mm). The calculated surface lowering rates ranged from 0.000 to 0.020 mm yr^{-1}, with an average value of 0.007 mm yr^{-1}. However, the errors reached similar magnitudes to those of the surface denudation rate, averaging 0.007 mm yr^{-1}. The main conclusions of this enlightening work were that (i) studies carried out with MEMs should include error assessments to gain insight into their reliability; (ii) the method has a limited usefulness for investigating microforms such as solution pans or solution flutes, which was one of the initial objectives of the technique. The impact of the errors on the reliability of the measurements depends on the speed of the denudation process and the length of the observation time. Surface lowering rates on bare limestone in inland environments are low and consequently their assessment requires long surveying periods, exceeding the time frame of most research projects (three to five years). Common rates in humid environments are typically lower than 0.040 mm yr^{-1} (Table 4.1), which explains the scarcity of studies in dry areas.

Cucchi et al. (1994), based on measurements carried out over 15 years in more than 50 sites in Pre-Alpine and Alpine regions of northeastern Italy with high precipitation (1800–3000 mm yr^{-1}), obtained surface denudation rates within the range of 0.01–0.04 mm yr^{-1}. Variations are related to climatic differences, lithological variability, and slope inclination. Reliable results can be obtained with relatively short surveying periods in areas with high denudation rates, such as coastal environments and particularly shore platforms, which have been the focus of a great part of the MEM investigations (Stephenson and Finlayson 2009). However, in these environments, as support the published lowering rates, dissolution commonly has a subordinate contribution to the overall denudation compared with other mechanical weathering and erosion processes such as wave quarrying, wetting and drying, salt weathering, wind abrasion, or bioerosion (grazing organisms). A compilation by Stephenson and Finlayson (2009) of published denudation rates measured with MEM and T-MEM on shore platforms carved in limestone shows values ranging between 0.1 and 1.2 mm yr^{-1}, mostly one order of magnitude higher than the common values reported in inland environments. Denudation rates in chalk are typically higher than 1 mm yr^{-1}. Cucchi et al. (2006) measured lowering rates using a T-MEM in and around the Gulf of Trieste (N Adriatic), including limestone outcrops located at inland sites and in the intertidal zone. Denudation rates in the coastal sites reach approximately 0.20 mm yr^{-1}, significantly higher than the values measured in similar limestones inland (ca. 0.02 mm yr^{-1}). Rates in marly limestones reach 2.42 mm yr^{-1}, at least ten times higher than those measured in neighboring limestones. Furlani et al. (2009) conducted a comparative study of lowering rates using MEM and T-MEM in coastal (cliffs, platforms) and inland environments (dolines, karren, and blocks) in the northeastern Adriatic region, including the classical karst and the drier Istrian karst. In the classical karst, coastal rates (0.14 mm yr^{-1}) were around eight times higher than those reported inland (0.018 mm yr^{-1}). In the Istrian karst, rates in the coastal areas (0.04 mm yr^{-1}) were approximately four times higher than those of the inland sites (0.009 mm yr^{-1}).

MEMs have obvious limitations for investigating the long-term evolution of solutional landforms (karren, dolines, and corrosion plains) due to restricted spatial and temporal representativeness of the data. Nonetheless, it is a useful method for assessing how rates vary over short time and spatial scales and allows exploring the influence of specific factors by isolating them in the investigation sites (e.g., lithology, slope, and precipitation; Smith et al. 1995). According to Stephenson and Finlayson (2009), the most innovative investigations will be those that combine T-MEM with other techniques that transcend the temporal and spatial constrains of the technique (e.g., laser scanner, photogrammetry, pedestals, and cosmogenic nuclides).

4.3.4 High-resolution 3D Surface Models

This group of techniques involves the comparison of high-resolution 3D models of rock surfaces from different dates. The surface models can be generated with LiDAR data acquired with terrestrial laser scanners (TLSs) or with 2D overlapping images processed with a photogrammetric range imaging technique such as Structure from Motion (SfM). These methods have been scarcely applied to the quantification of short-term denudation on carbonate rock exposures, mainly due to the low speed of the erosion processes compared with the accuracy of the measurements. Nonetheless, sound results could be obtained over relatively short monitoring periods in rapidly evolving salt and gypsum exposures. The ever-increasing improvements (resolution, accuracy, and precision) and cost-effectiveness of these methods will most probably open new possibilities in karst denudation studies. Some of the advantages of these techniques compared with MEMs include: (i) higher spatial density of data allowing the generation of high-resolution 3D surface models that can be combined with photographs, (ii) larger spatial coverage of the survey areas, (iii) no direct contact with the surface, and (iv) ability to analyze the evolution of rock surfaces with highly irregular and complex geometries including karrenfields and cliffs. TLSs use a laser source mounted on a tripod and record 3D coordinates of point clouds from the travel time of the laser light to the surface and back and its horizontal and vertical angles (Figure 4.6a) (e.g., Benito-Calvo et al. 2018). The scans can also include the acquisition of RGB color point data and photographs of the surface. Multiple overlapping scans are geometrically aligned and referred to a uniform coordinate system, often using artificial targets installed at fixed points (registration). The registered point cloud can be used to generate high-resolution 3D surface models that can be draped with color data and photographs to facilitate the analysis. The comparison of surface models from temporal surveys enable the identification and quantitative assessment of surface changes, including erosion and deposition (see review in Telling et al. 2017). Short range TLS with high ranging accuracy and high spatial resolution could be used to assess relatively slow denudation on rock outcrops. Gómez-Pujol et al. (2006) used a special TLS to assess millimeter-scale roughness on rock surfaces in carbonate coasts affected by splash and spray in Mallorca, Spain. They used a laser device mounted on a square frame with four legs in the corners, that can be locked precisely at the same site using stainless studs fixed to the rock (Figure 4.6b). This device with a laser source that moves in the *x* and *y* axes measures 3D coordinates of points at the rock surface with a horizontal resolution of 0.4 mm, covering an area of 40×40 cm, and with a vertical accuracy of 0.025 mm. The surface models captured at different times can be compared with the aid of permanent control points installed on the rock surface to measure the amount of denudation and the resulting micromorphologies. Laser devices combined with cameras have also been used satisfactorily to monitor

Figure 4.6 Application of the LiDAR technology to surface denudation. (a) Terrestrial laser scanner mounted on a tripod next to a gypsum exposure. *Source:* Francisco Gutiérrez (Author) (b) Laser scanner designed by Swantesson, which covers a maximum area of 40×40 cm with a spatial resolution of 0.4 mm and a vertical accuracy of 0.025 mm. *Source:* Photo by Lluís Gómez-Pujol.

the deterioration of samples of building stones subject to artificial aging tests in the laboratory, providing highly accurate data of the volumetric and morphological changes (e.g., Birginie and Rivas 2005). Inkpen et al. (2000) tested the practicality of close-range photogrammetry to assess relative changes in weathering features and surface lowering rates analyzing artificially weathered limestone blocks and a stone wall. The work illustrates the potential and problems of the technique and the important role of the observer when interpreting the surface models. Satisfactory results could be obtained by the low-cost and user-friendly photogrammetric technique called SfM, that can be used to generate 3D models from randomly acquired images (Tarolli 2014; James et al. 2017). According to Telling et al. (2017), with proper methodology and control, SfM can provide 3D models with a quality and resolution comparable to that of LiDAR, but more cost-effective and easier to use. An example is presented by Palmeri et al. (2020) that characterizes a limited number of solution flutes (Rillenkarren) on gypsum generating a 3D DTM with a spatial resolution of around 0.15 mm through an automatic close-range photogrammetry process designated as 3D-photo reconstruction, which couples SfM and MultiView-Stereo (MVS) techniques.

4.3.5 Long-term Surface Lowering Around Pedestals, Dikes, and Siliceous Nodules

Limestone pedestals are formed by differential surface lowering of carbonate bedrock around a caprock boulder that functions as an umbrella. They are designated as "karrentische" (table karren) in some works (Bögli 1961) and also develop on evaporite rocks (Figures 4.7 and 4.8). A great majority of the pedestals reported in the literature have formed in relatively flat limestone surfaces carved by glaciers (limestone pavement) and covered by scattered non-carbonate erratic boulders that exert a local shielding effect upon postglacial corrosional lowering. Limestone pedestals have also been documented in raised marine terraces underlain by carbonate rocks and locally capped by boulders transported by tsunamis, large waves, or rock falls (Matsukura et al. 2007; Mylroie and Mylroie 2017). These landforms have been traditionally used to estimate long-term solutional lowering rates by dividing the height of the pedestals, as a measure of the amount of differential surface lowering, by the time elapsed since the onset of dissolution. The chronological variable corresponds to the emplacement of the boulders and the exposure of the limestone surface to dissolution by glacier retreat, or the emergence of the bedrock surface in the case of raised marine terraces.

Differential surface lowering around low-solubility material embedded within the carbonate rock, such as siliceous dikes, veins, and nodules have also been used to estimate long-term lowering rates. These features provide minimum lowering rate estimates since veins and dikes are also affected by downwearing and nodules do not mark the position of the initial erosional surface (e.g., Lauritzen 1990) (Figure 4.7). André (1996, 2008) compiled published corrosional lowering rates calculated using limestone pedestals as well as protruding dikes and veins in areas affected by late Quaternary glacial erosion and distributed in a wide range of climatic environments. She found a direct relationship between precipitation and surface lowering rates (Figure 4.9): (i) arctic climate, 2–3 mm kyr^{-1} (Spitsbergen; 380–430 mm); (ii) dry subarctic, 5.3–6 mm kyr^{-1} (Lapland, Canada; 600–800 mm); (iii) temperate alpine, 8–17.5 mm kyr^{-1} (Alps, Pyrenees; 1500–3400 mm); (iv) tropical alpine, 32 mm kyr^{-1} (Irian Jaya, New Guinea; 4000 mm); and (v) hyperhumid subarctic, 60–100 mm kyr^{-1} (Chilean Patagonia, 7330 mm). The greatest rates have been reported in some islands of the Chilean Patagonia with an annual precipitation of around 7300 mm, where pedestals and emergent dikes reach meter-sized heights (Maire 1999; Hobléa et al. 2001) (Figure 4.8a).

The apparently simple method of using pedestals as geological gauges of the cumulative surface lowering, occurred over long time periods, has a number of problems and uncertainties:

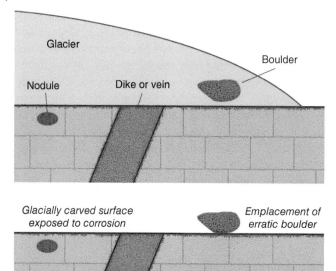

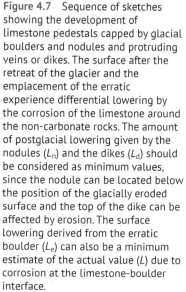

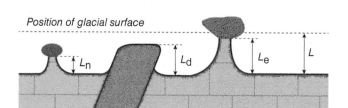

Figure 4.7 Sequence of sketches showing the development of limestone pedestals capped by glacial boulders and nodules and protruding veins or dikes. The surface after the retreat of the glacier and the emplacement of the erratic experience differential lowering by the corrosion of the limestone around the non-carbonate rocks. The amount of postglacial lowering given by the nodules (L_n) and the dikes (L_d) should be considered as minimum values, since the nodule can be located below the position of the glacially eroded surface and the top of the dike can be affected by erosion. The surface lowering derived from the erratic boulder (L_e) can also be a minimum estimate of the actual value (L) due to corrosion at the limestone-boulder interface.

(i) it is commonly assumed that the crown of the limestone pedestal marks the position of the reference surface. However, the limestone beneath the boulder can also be affected by dissolution, as reveals the presence of corrosion features in some examples of Spitsbergen and Norway (Lauritzen 2005). The boulder–pedestal interface is not a watertight contact and can be accessed by condensation water and water trickles that descend attached to the underside of the boulder by surface tension. Thus, the height of the pedestal should be considered as a minimum measure of the amount of the total denudation occurred around them (Figure 4.7). Lauritzen (2005) developed a simple mathematical model for pedestal growth considering lowering of the top of the limestone pedestal by condensation water. According to his model, the actual denudation rates can be 25–80% higher than those calculated from the pedestal height and their growth time. (ii) The height of the pedestals is difficult to measure in sloping surfaces, where there are large differences between the upslope and downslope sides. Moreover, it seems that the assessment of corrosional lowering in some earlier studies was based on the most prominent pedestals, rather than on measurements from large populations, resulting in overestimated values. This is illustrated by the differing denudation rates derived from pedestals at Norber, North Yorkshire, Northern England (Figure 4.7b): 41 mm kyr^{-1} (Sweeting 1966); 34 mm kyr^{-1} (Huddart 2002), 2–8.7 mm kyr^{-1} (Goldie 2005), and 46 mm kyr^{-1} (Parry 2007). (iii) The height of the pedestal can be influenced by the size and geometry of the capping boulder. Lauritzen (2005) observed in Spitsbergen and Norway a positive correlation between pedestal heights and the size of the perched boulder. It seems that the shielding effect increases with the size of the

Figure 4.8 Images of limestone and salt pedestals. (a) Limestone pedestal capped by an erratic boulder in Madre de Dios Island, Chilean Patagonia. *Source:* Photo by Stéphane Jaillet, Centre Terre, Edytem. (b) Pedestal formed beneath an erratic boulder of Silurian graywacke overlying Carboniferous limestone, Norber, Yorkshire, United Kingdom. *Source:* Photo by Peter Wilson. (c) Halite pedestal developed in the southern salt glacier (namakier) of Dashti (or Kuh-e-Namak) Diapir, Zagros Mountains, Iran. Note limestone boulder indented into the salt pillar due to halite dissolution beneath its underside. Prof. Mehdi Zarei, 1.8 m tall for scale. *Source:* Francisco Gutiérrez (Author).

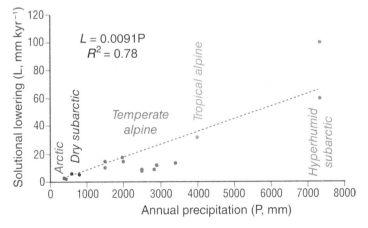

Figure 4.9 Relationship between approximate mean annual precipitation and approximate long-term surface lowering rates estimated from pedestals and protruding dikes and veins in previously glaciated environments. *Source:* Data from André (2008).

blocks. He also inferred a higher protective effect in the boulders with planar and concave geometry at the contact zone than those with convex shape that facilitates access to dissolving water. (iv) Corrosional lowering rates vary substantially depending on whether dissolution occurs on bare or on subsoil rock surfaces. Parry (2007) attributed high lowering rates estimated at the Norber pedestals (46 mm kyr^{-1}) to subsoil dissolution, (v) differential lowering can be the result of both dissolution and mechanical erosion. Goldie (2005) and Wilson et al. (2012a) contend that mechanical erosion of the densely fractured Carboniferous limestone at Norber, United Kingdom, played a significant role in the development of these profusely studied pedestals. (vi) Corrosional lowering has not necessarily operated continuously and at a uniform rate since the exposure of the limestone surface. Wilson et al. (2012a) propose that after the emplacement of the erratics at Norber, with a ^{36}Cl exposure age of ca. 18 kyr, surface lowering mainly occurred by nivation processes during cold intervals (e.g., 18–14.7 kyr, 12.9–11.7 kyr or Younger Dryas Stadial) and dissolution during temperate periods. (vii) The growth time of the pedestals is in most cases poorly constrained and based on rough estimates of the age of the deglaciation at the site of interest. The quality of this information is improving thanks to the application of site-specific geochronological methods such as cosmogenic nuclide (Wilson et al. 2012a) or radiocarbon dating.

Matsukura et al. (2007) calculated surface lowering rates in Kikai-jima, Ryukyu Islands, Southwestern Japan (annual rainfall 2280 mm) using pedestals developed on tectonically raised coral reef terraces. Here, the caprock boulders are large pieces of coral reef presumably deposited by tsunamis or typhoon-generated large waves on the reef flats before their emergence. The terraces consist of highly porous (4–36.4%) and highly fractured reef limestone mainly composed of aragonite (>93%). An average surface lowering rate of 205 mm kyr^{-1} was calculated using the height of the pedestals and the estimated age of emergence of the corresponding terraces (6.0, 3.0, and 1.5 kyr), constrained by radiocarbon dating on corals. The estimated denudation rate is two to three times greater than the rates estimated in the Chilean Patagonia at sites with an annual precipitation three times higher (André 2008), strongly suggesting that erosional lowering in Kikai-jima is related to the combined effect of mechanical weathering/erosion and eogenetic dissolution. This large difference is also consistent with the contrasting surface lowering rates obtained in coastal and inland sites with MEMs (Stephenson and Finlayson 2009 and references therein). Miklavič et al. (2012) and Mylroie and Mylroie (2017) estimated a minimum long-term surface lowering rate of around 50 mm kyr^{-1} using pedestals developed on a coral reef terrace at Guam Island, western Pacific Ocean. Here, pedestals up to 5 m in height form by differential surface lowering of an uplifted aragonite-rich reef terrace dated by U/Th at 126–132 kyr (maximum time for the pedestal growth). The capping boulders correspond to rock-fall blocks of calcitic limestone derived from an adjacent cliff carved in compact Plio-Pleistocene limestone. The height disparity of the pedestals is attributed to the variable timing at which the boulders fell from the cliff. These authors indicate that this cumulative denudation process should be taken into account when reconstructing glacio-eustatic sea-level changes and estimating tectonic uplift from the position of emerged carbonate benches. Important implications include that deposits related to old sea-level highstands may have been removed by karst denudation, as supported by the distribution of flank-margin caves (e.g., Bahamas), and that the existing surfaces are currently situated at positions below their original depositional elevation (indicators of minimum elevation). Mylroie and Mylroie (2017) indicate that solutional lowering is expected to be more rapid in highly porous eogenetic carbonates than in compact carbonate rocks, because the mass to be removed for an equivalent surface lowering is much less and the surface area exposed to weathering agents is much greater.

4.3.6 Long-term Erosion Rates Derived from Cosmogenic Chlorine-36 in Calcite

The application of terrestrial cosmogenic nuclides (TCNs) for dating geomorphic surfaces and estimating long-term erosion rates has resulted in important advances in the fields of Geomorphology and Quaternary sciences in recent decades (Cockburn and Summerfield 2004; Gosse 2007). Cosmic rays interact with target atoms in rocks and sediments underlying the Earth's surface producing in situ TCNs. The production of TCNs decreases exponentially with depth and is influenced by a number of factors (e.g., altitude, latitude, and slope). If the TCN production rate is known, then measured concentrations of cosmogenic nuclides can be translated into surface exposure ages, or how long a rock or a landform surface has been exposed. Corrections are introduced to account for processes that modify the surface (e.g., erosion) or interrupt/reduce the exposure (e.g., sediment accumulation, snow cover) (Schaefer and Lifton 2007). The assumption of zero erosion in a wearing rock surface leads to minimum exposure ages. The depth profiles of TCN concentrations are dependent on exposure time and erosion rate, and consequently can be used to estimate long-term erosion rates. Erosion removes the outermost portions of rock or sediment with the highest concentrations of cosmogenic nuclides. In a slowly eroding surface, minerals spend longer time near the surface than in a rapidly eroding landscape and build up larger concentrations of TCNs (Ivy-Ochs and Kober 2007). The ^{36}Cl-calcite system is used in carbonate areas to estimate surface exposure ages and long-term erosion rates. Cosmic radiation rapidly transforms ^{40}Ca into the radioactive nuclide ^{36}Cl, with a saturation time of about 10^5–10^6 years, when production and decay of nuclides become balanced. The patterns of ^{36}Cl production by different types of cosmic rays (hadrons, muons) at various depths are documented by Stone et al. (1994, 1998). A peculiarity of this isotope-mineral system is that at depths below a few meters ^{36}Cl production in calcite is almost entirely initiated by cosmic ray muons, with greater penetration than the hadrons that dominate production near the surface by spallation. Therefore, deep samples, which carry a memory of erosion on longer timescales, may be useful for testing the standard assumptions of constant erosion rates and for assessing temporal changes in denudation rates (erosion histories).

Stone et al. (1994) provided preliminary erosion rates from ^{36}Cl measurements of limestone surfaces at five sites in Australia and Papua New Guinea. The variability of the rates was roughly consistent with the present-day climatic conditions and water surplus of the different regions: 4.5 mm kyr^{-1} in the arid Nullabor Plain, 18–29 mm kyr^{-1} in the humid eastern highlands of Australia, and 184 mm kyr^{-1} in the extremely wet Strickland Ranges of Papua New Guinea. The surprisingly low value of 14 mm kyr^{-1} obtained in the Tindal karst of northern Australia, with monsoonal climate, was tentatively ascribed to past arid conditions during glacial periods. Stone et al. (1998), in a subsequent paper, lowered the rates indicated above by 20% considering revised production parameters based on a 20-m depth profile of ^{36}Cl concentrations obtained from an abandoned quarry located in the Wombeyan Caves Reserve, southeastern Australia. This work provides a long-term erosion rate of 26 mm kyr^{-1} for that site and suggests a change in the erosion rate from 23 mm kyr^{-1} to 100 mm kyr^{-1} at around 15 kyr, probably related to the transition from a cold and arid glacial stage climate to the more humid present-day conditions. Wilson et al. (2012b) determined the total amount and rate of surface lowering in a limestone pavement at Moughton, northwest England, by cosmogenic ^{36}Cl analysis and using the exposure age estimated at the Norber erratics-pedestals (17.9 kyr). They found that the glacially scoured limestone pavement has experienced an average postglacial surface lowering of 33 ± 10 cm, yielding a long-term rate of 18 mm kyr^{-1}. The authors stress that changes in climate and soil cover were most probably accompanied by substantial variations in the mechanical and chemical denudation processes and their rates. Matsushi et al. (2010) estimated long-term denudation rates along the convex slopes of a solution doline in the Akiyoshi karst, southwest Japan (~2000 mm annual precipitation), observing a direct relationship with the size of the topographic contributing area. The samples were collected from protruding pinnacles along a transect of a doline 150 m in diameter and 17 m deep, largely covered by fine-grained soil and vegetation, obtaining rates ranging from 63 to 256 g m^{-2} yr^{-1} that increase toward the center of the depression. These high values (23–94 mm kyr^{-1}) suggest that mechanical erosion significantly contributes to the surface lowering in the sloping surfaces. The rapid increase in the denudation rate toward the bottom of the depression can be attributed not only to greater water–rock interaction but also to progressively higher concentration of soil-derived CO_2. Based on an empirical relationship between denudation rate and contributing area, these authors produced a quantitative model that allows estimating the age of the dolines in their study area (in the order of 10^5 yr). They found that larger dolines are not necessarily older and that the depths appear to be determined by the catchment area, rather than by the age.

Ryb et al. (2014a) in a comprehensive study quantified denudation rates in the Judea Hills, Israel (500–630 mm mean annual precipitation) using cosmogenic ^{36}Cl (long-term surface chemical and mechanical erosion) and solute load (contemporaneous surface and subsurface solutional denudation). They obtained an average long-term denudation rate at bare-rock surfaces (21 ± 7 mm kyr^{-1}; 10^4 timescale) that was 1.4 times higher than the average of the annual rate estimated from solute load. This difference can be attributed to the contribution of mechanical erosion and greater surface dissolution during more humid periods in the past. In another study conducted on the eastern flank of the Judea Range, Ryb et al. (2014b), using ^{36}Cl measurements estimated erosion rates along a climatic gradient from the crest of the range to the Dead Sea Rift, covering a range in the mean annual precipitation between 600 and 50 mm. Rates from flat-lying bedrock outcrops vary from ~20 mm kyr^{-1} in the upper more humid zones to 1–3 mm kyr^{-1} in the hyperarid low elevation sites, showing a strong correlation with precipitation. In contrast, the samples collected from hillslopes in the hyperarid areas show higher rates than in the semi-arid zones and a strong dependency of slope gradient, indicating the dominance of mechanical erosion processes. Apparently contradicting results were obtained by Avni et al. (2019) from rock samples collected on hilltops (bare and soil-covered) at Mount Hermon, Israel, covering a climatic gradient from 1500 to 600 mm in average annual precipitation. Here, no correlation is observed between denudation rates and mean annual precipitation, and the highest erosion rates are obtained in the mid-altitude areas (42 ± 8 mm kyr^{-1}; 900–1200 mm). This pattern is ascribed to thicker soils and higher CO_2 concentrations below the tree line.

Yang et al. (2020), using in situ cosmogenic ^{36}Cl data, explored the influence of climate on denudation rates from 14 samples collected in diverse climatic zones in China (from 350 to 1105 mm of annual precipitation) and a 10-m depth profile at Guizhou. Plotting their denudation rates and those compiled from previous studies based on ^{36}Cl data (e.g., Xu et al. 2013; Ryb et al. 2014a, 2014b; Thomas et al. 2017, 2018; Avni et al. 2019), they proposed that hilltop erosion rates increase with average annual precipitation below 700 mm, whereas above 700 mm the denudation rates are largely restricted within 34.1 ± 11.7 mm kyr^{-1}. They suggest that solutional denudation in the humid areas is limited by the kinetics of the chemical weathering process. Obviously, the rate of solutional denudation on limestone exposures should increase with the amount of precipitation (Figures 4.1 and 4.9). The kinetics of the dissolution reaction should not play an important role, since it is mainly controlled by the saturation degree of the water with respect to the carbonate minerals, which is essentially constant in the highly undersaturated precipitation water that falls directly on the ground surface. Probably, the observed anomalous pattern is related to the fact that the dataset covers a relatively narrow precipitation range of 0–2500 mm (with most of the values below 1500 mm), the heterogeneities associated with the sampling points (e.g., lithology, vegetation), the variable contribution of mechanical erosion, and the assumption of constant erosion. As expected, hillslope samples commonly show higher denudation rates attributable to the contribution of mechanical erosion and regolith transport (Godard et al. 2016; Thomas et al. 2018). The data from the Guizhou profile admits contrasting interpretations, including a relatively constant high denudation rate, or a low denudation rate interrupted by a recent abrupt erosive event. Yang et al. (2020) highlight that caution should be exercised in interpreting apparent long-term denudation rates or exposure ages under the assumption of steady-state erosion.

The review presented above illustrates that denudation rates derived from cosmogenic ^{36}Cl, with values averaged for long time periods covering significant climatic variability, are highly useful for understanding the long-term evolution of landscapes and landforms. Nonetheless, the results often permit alternative interpretations due to a number of uncertainties including the temporal variability of erosion rates and the relative contribution of chemical and mechanical erosion. Probably, more robust conclusions could be obtained by improving the geological and geomorphological characterization of the sampling areas and points. This would provide insight into significant factors not related to climate (e.g., mineralogy, lithology, degree of fracturing, runoff contributing area, local morphology, presence of deposits generated by mechanical weathering and erosion).

Cosmogenic ^{36}Cl has also been applied to calculate retreat rates in limestone scarps affected by rock falls and to reconstruct coseismic surface ruptures in bedrock fault scarps. Domènech et al. (2018) calculated the long-term retreat rate of a limestone cliff in the Spanish Pyrenees (0.31–0.37 mm yr^{-1}), dating pre- and post-rock fall surfaces and assessing the volume of material released using point clouds obtained by TLS. Sharp variations in the concentration of cosmogenic ^{36}Cl along the height of limestone fault scarps can be used to obtain valuable data for assessing the seismogenic potential of the faults, including the timing and recurrence of rupture events (paleoearthquakes) and the amount of surface displacement per event (e.g., Benedetti et al. 2003).

4.4   Denudation Rates in Gypsum and Salt

Under similar conditions, solutional denudation rates in gypsum are much higher than in carbonate rocks due to the greater solubility of the former. In fact, rates are commonly reported using mm yr^{-1} rather than mm kyr^{-1}. The equilibrium solubility of gypsum in pure water at 25 °C and 1 atm pressure (2590 mg L^{-1}; mass of solute in 1 L of solution) is 50 times higher than that of calcite (52 mg L^{-1}) when the water has the P_{CO_2} of the atmosphere (ca. 0.04% by volume) (see Table 3.1). Gypsum and carbonate rock samples exposed to precipitation under the same conditions in a field laboratory in Trieste, Italy, yielded solutional denudation rates in gypsum (0.68–1.14 mm yr^{-1}) 30–70 times greater than those measured in the carbonate samples (0.010–0.035 mm yr^{-1}) (Klimchouk et al. 1996a). This difference is reduced considerably when the CO_2 content in the water increases (e.g., uptake from soils, deep-sourced CO_2), together with its capability to dissolve carbonate minerals. For instance, the solubility of calcite in water that is in equilibrium with a CO_2-rich soil (5% by volume) is 297 mg L^{-1}, that is, nine times lower than that of gypsum (Table 3.1).

Investigations dealing with denudation rates in gypsum are very scarce compared with those carried out in carbonate rocks, despite the important environmental and engineering implications of gypsum dissolution (James 1992; Cooper and Gutiérrez 2013) and the fewer difficulties encountered studying the dissolution of evaporite rocks: (i) gypsum and halite are much simpler dissolution systems than those related to carbonate minerals strongly influenced by dissolved gases or acidity, (ii) meaningful values with proportionally low errors can be obtained in relatively short monitoring periods thanks to the rapidity of the dissolution process, and (iii) waters in gypsum and salt karst rarely reach saturation conditions and therefore karst deposition, the last term of Eq. 4.3, can be neglected in most cases. A potential difficulty associated with the use of MEMs is that wearing caused by the probe tip during measurements is expected to have a higher adverse effect on gypsum and salt than in the harder carbonate rocks.

Klimchouk et al. (1996a) published a comprehensive study on gypsum dissolution rates compiling data obtained by various methods (tablets, MEM, volume loss, or wall retreat in river waters) in different regions: hypogene gypsum karst of western Ukraine, the epigene caves of Sorbas in SE Spain, multiple localities in Italy, and river waters in contact with gypsum in Russia and England. Standard tablets of Ukrainian Miocene gypsum were installed in different environments in Ukraine and Spain, including: (i) direct exposure to precipitation at the surface, (ii) focused percolation in the vadose zone (dissolution pipes beneath overburden), (iii) exposure to cave air in zones of condensation, (iv) cave lakes, (v) ephemeral streams in caves, and (vi) confined aquifer in gypsum and in the underlying non-karst rocks that feed the hypogene karst system from below. The data from Ukraine were subsequently expanded in Klimchouk and Aksem (2005) using measurements from 53 tablet stations covering eight years. Data from Italy comprise MEM measurements obtained on natural gypsum exposures with ten different lithologies and annual precipitation values ranging from 490 to 1350 mm, plus tablets of 12 different gypsum lithologies exposed in seven field laboratories.

Regarding the denudation rates of gypsum in permanent contact with surface water, Klimchouk et al. (1996a) quote a work by Pechorkin (1969) that reports values ranging from 79 to 190 mm yr^{-1} derived from the volume loss experienced by gypsum boulders during five years in the Kama River reservoir, Russia. James et al. (1981) estimated a rough retreat rate of 80–100 mm yr^{-1} from the evolution over 50–60 years of an undercut gypsum cliff in contact with the allogenic River Ure in England. Navas (1990) investigated denudation rates using spheres 16 cm in diameter of alabastrine gypsum suspended in turbulent flowing water under variable flow velocities (0.3–1.2 m s^{-1}), electrical conductivities (0.3–1.9 mS cm^{-1}), and gypsum saturation indices (−2.1 to −0.33). She obtained average loss mass values between 104 and 226 g m^{-2} h^{-1} (ca. 390–850 mm yr^{-1}). Multivariable regressions indicated that the main factor that controls dissolution rate in this transport-controlled system is flow velocity. Perhaps, one of the main problems associated with the interpretation of the rates obtained in running water is the uncertainty regarding the proportion of the denudation caused by mechanical processes, which may have a significant contribution under high-velocity and turbulent conditions, and in gypsum rocks that experience coherence loss, facilitating the physical detachment of crystals.

The largest dataset related to dissolution rates in gypsum directly exposed to precipitation is the one from Italy, with more than 3000 measurements from multiple stations covering an annual precipitation range between 490 and 1350 mm (Klimchouk et al. 1996a). Surface lowering rates (*D*, mm) show a good correlation (R^2 = 0.911) with the amount of liquid precipitation (*W*, mm), expressed by the regression:

$$D = 0.000750\,W + 0.1815 \tag{4.10}$$

According to this equation, a cumulative rainfall of 100, 500, and 1000 mm would cause a surface lowering of 0.26, 0.56, and 0.93 mm, respectively. Shaw et al. (2011) measured dissolution rates at 55 sites with triplicate standard gypsum tablets distributed across the Gypsum Plain of New Mexico and Texas (1500 km^2 in area), underlain by the Castile Gypsum Formation and with an average annual precipitation of 270 mm. The spatially distributed data spanning two years indicate an

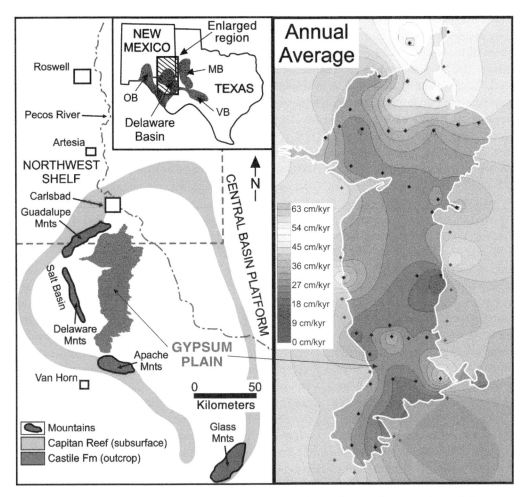

Figure 4.10 Location of the Gypsum Plain underlain by the Castile Gypsum Formation in the Delaware Basin (red polygon) and map of average annual denudation rates measured on standard gypsum tablets placed at 55 stations (dots) distributed across the Gypsum Plain. The general northward increasing trend is consistent with the precipitation gradient. *Source:* Figure courtesy of Kevin Stafford.

average surface denudation rate of 0.3 mm yr^{-1} and local maximum values above 0.5 mm yr^{-1}. The rates show a generally increasing trend toward the north, where rainfall is about twice that in the south (Figure 4.10). Interestingly, relatively higher denudation rates were recorded in the proximity of major streams, attributed to higher atmospheric moisture and greater contribution of condensation solution. Sanna et al. (2015), based on MEM measurements carried out at five stations in the gypsum outcrop of Sorbas (274 mm annual precipitation) and spanning 18 years, obtained an average surface lowering rate of 0.170 mm yr^{-1}. Note that this value is more or less half of the rate measured with tablets in the Gypsum Plain by Shaw et al. (2011) with a similar precipitation.

Dissolution caused by focused downward percolation of water through discontinuities in the gypsum and from an overlying non-karst formation is the process responsible for the development of solution fissures (e.g., grikes, cutters) and solution pipes in mantled and interstratal karst settings. Highly variable dissolution rates can be expected in this environment, mainly related to the greater amount of water that interacts with the rock. Tablets placed in vadose dissolution pipes in

Ukraine that interact with dripping water after flowing 1–2 m along the gypsum walls yielded average values of 0.66 mm yr^{-1} (Klimchouk et al. 1996a). Dissolution in cave lakes with sluggish flow at the bottom of the unconfined gypsum aquifer in Ukraine was assessed with standard tables at four stations. The lakes show salinity stratification, with an upper layer of more diluted water (TDS 1.4 g L^{-1}) underlain by denser water with higher saline content (TDS 2.13 g L^{-1}). This involves contrasting gypsum saturation indices, from −0.21 to −0.002. Tablets in the uppermost layer record intense dissolution, with an average value of 10.4 mm yr^{-1}, whereas the bulk water values are one order of magnitude lower (average 1.12 mm yr^{-1}) but with a wide variability (0.05–6.16 mm yr^{-1}), mainly depending on the intensity of the flow and thus the renewal of water. Dissolution rates in the confined gypsum aquifer of western Ukraine, which is traversed by hypogene cross-formational flows coming from an underlying carbonate aquifer, were explored using boreholes in two areas, one of them under natural conditions and another one with accelerated groundwater flow induced by dewatering for mining. Tablets installed within the gypsum aquifer (boreholes) that interact with slow groundwater flow under natural conditions (gypsum saturation index −0.21) provided an average dissolution rate of 0.22 mm yr^{-1}. Much higher rates (average 1.56 mm yr^{-1}) were obtained where the flow was accelerated by water pumping, despite the lower aggressiveness of the water (gypsum saturation index −0.06). Gypsum tablets placed in the underlying carbonate aquifer provided an estimate for the dissolution rate associated with the highly undersaturated groundwater (gypsum saturation index −1.7) when it first interacts with the gypsum unit at the base of the aquifer, with an average rate of 9.16 mm yr^{-1}. Klimchouk and Aksem (2005) provided similar values in a more recent study based on a larger dataset. The rates estimated for the three-dimensional confined aquifer, although comparatively low, indicate the large mass that can be removed in solution due to the large water–rock interaction area, and the high speed at which secondary porosity can be generated. These values have relevant implications for understanding the evolution of gypsum aquifers and the permeability variations that experience these high-solubility rocks at dam sites. In contrast, dissolution in the vadose zone, despite the locally high rates related to focused flow, makes a limited contribution to the overall karstification process.

Dissolution of gypsum exposed to cave air at sites where condensation occurs was measured with tablets at the entrance sector of caves in Spain as well as in Ukraine, obtaining similar average values of 0.004 and 0.003 mm yr^{-1}, respectively (Klimchouk et al. 1996a). A seasonal pattern was observed, with higher rates recorded during warm periods when there is greater condensation on the rock surface. The role of condensation solution of gypsum was further investigated by Sanna et al. (2015) with an 18-year-long record of MEM data from 14 stations in the Cueva del Agua, located in the semiarid gypsum karst of Sorbas (274 mm annual precipitation, average temperature 14.6 °C). This cave with multiple entrances is characterized by large thermal variations between the surface and the cave atmosphere, substantial air exchange, and the presence of abundant condensation-solution features such as bell-shaped cupolas and boxwork. Results indicate that this process is most active along the medium wall of the cave passages (0.020–0.040 mm yr^{-1}), slower at the roof (ca. 0.01 mm yr^{-1}), and absent or replaced by karst deposition close to the floor where evaporation prevails. This vertical zonation is mainly attributed to the thermal stratification of the air, with condensation in the upper part and evaporation in the lower part, and the downward increase in the amount of water that flows as a film attached to the cave walls and its progressive saturation with respect to gypsum. Tarhule-Lips and Ford (1998a) reported a much higher average rate of 0.36 mm yr^{-1} related to condensation solution on suspended gypsum tablets at the entrance zones of flank-margin caves in Cayman Brac and Isla Mona, Caribbean Islands. In this tropical marine environment, there is a large supply of condensation water from moist and warm air onto cooler walls related to diurnal variations in temperature and humidity. During daytime, when the outside temperature is higher than in the cave, warm air is drawn into the caves along the ceilings, cooled down in the interior, and expelled out along the floors. The wall temperature is less than the outside dew-point temperature and condensation occurs on the walls. Inside temperature is greater than the outside dew-point temperature at night, resulting in evaporation.

Works dealing with the calculation of the overall solutional denudation in gypsum karst systems using hydrochemical methods are scarce. Pulido-Bosch (1986) proposed a rough average rate of 0.26 mm yr^{-1} for the gypsum karst of Sorbas, Spain, considering the mean values of outflow rate and concentration of dissolved gypsum. Desir et al. (1995) assessed denudation rates on sparsely vegetated gypsum slopes with a very thin silty-loam gypsiferous regolith in the Ebro Basin, NE Spain (mean annual precipitation around 350 mm). They installed soil erosion plots (50–70 m^2, 15–24° slope angle) on north- and south-facing slopes, with more abundant vegetation cover in the former. The main parameters acquired include continuous records of rainfall, runoff volume, and sediment yield (solute and suspended load) for each runoff event using dynamic devices (collectors) at the outlet of the experimental plots (Figure 4.4a). The amount of gypsum removed by dissolution was estimated using the classical hydrochemical method, considering runoff volume and sulfate concentration. The data obtained from 56 rainfall events occurred over a period of four years showed a clear direct relationship between runoff volume and amount of solutes. They estimated mean annual erosion rates for the north- and south-facing slopes of 7.53 and 19.42 t ha^{-1} yr^{-1}, respectively (around 0.33 and 0.84 mm yr^{-1}), of which approximately 30–42% corresponds to mass removed in solution. Rates of solutional denudation expressed as dissolved mass per unit runoff and area ranged from 0.032 to 0.045 g L^{-1} m^{-2}.

Solutional denudation rates in salt rock can be extremely high due to the high solubility of halite. For comparison, the equilibrium solubility of halite in pure water at 25 °C and 1 atm pressure (424 g L^{-1}) is around 160 times higher than that of gypsum (2.6 g L^{-1}) (see Table 3.1). Note that 424 g L^{-1} is the solubility of halite in 1 L of solution, which corresponds to a brine with significantly higher density than pure water (1.19 g cm^{-3}). The solubility of halite expressed as mass of dissolved solute per unit mass of solvent is 356 g kg^{-1} (e.g., Langer and Offermann 1982) (Table 3.1). Dissolution in gypsum can produce considerable changes at human timescale, whereas the much rapid salt dissolution can result in substantial geomorphic alterations even during a single event, such as a storm, a flood, or the impoundment of a reservoir (e.g., Mottershead et al. 2007; Lucha et al. 2008b; Gutiérrez and Lizaga 2016). Considering a density for halite of 2.2 g cm^{-3}, an effective precipitation of 100 mm that dissolves halite until saturation removes a column of 16 mm of rock. Thus, the maximum chemical denudation reaches around 16% by volume of the water that flows in contact with salt. Moreover, dissolution of substantial amounts of salt increases the kinematic viscosity of the brines by as much as 10–20 times compared to carbonate karst waters. This implies a significant increase to a few centimeters in the threshold width of fissures and conduits at which laminar flow changes into turbulent flow, as opposed to 5–15 mm in carbonate karst waters (Frumkin 1994a).

Salt dissolution can occur in three main settings in which the solutional denudation process shows contrasting patterns and rates (Frumkin 2013) (Figure 4.11): (i) salt outcrops directly exposed to precipitation and surface runoff (bare karst); (ii) salt covered by non-consolidated deposits, either a residual soil (capsoil) or an allochthonous sediment (covered or mantled karst); and (iii) salt overlain by lithified material, either sedimentary formations (interstratal karst) or an indurated karstic residue generated by the dissolution of the salt, designated as caprock. Salt exposures where dissolution is caused by direct rainfall and surface runoff are very scarce due to the high solubility of the rock. They are mainly associated with rising diapiric structures located in arid and semiarid areas (Zagros Mountains, Iran; Red Sea Coast in the Arabian Peninsula; Atacama

Figure 4.11 Solutional denudation of salt. (a) Halite exposure of the Gachsaran Formation at Anbal ridge before the impoundment of the Upper Gotvand Reservoir, Iran. Note nested subhorizontal bevels generated during floods of the Karun River. *Source:* Photo by Mohammad Nazari. (b) Extensive exposures of salt with karrenfields in the Cardona Diapir, NE Spain. In the background, the salt is mantled by alluvium that inhibits dissolution (covered karst). *Source:* Francisco Gutiérrez (Author). (c) Northern salt glacier (namakier) of Dashti (or Kuh-e-Namak) Diapir, Iran, with extensive exposures of salt locally covered by a reddish dissolution residue. The capsoil protects the salt from direct dissolution and controls the development of humps. *Source:* Francisco Gutiérrez (Author). (d) Vertical salt exposure in the Hormuz Diapir, Iran, covered by a thin dissolution residue and colluvium on top. Note bedding-controlled fissures generated by the dilation of the extruded salt due to unloading and probably the expansion of clay partings. *Source:* Francisco Gutiérrez (Author). (e) Sharp discordant contact (salt table or salt mirror) between dipping salt and the overlying anhydrite-rich caprock at Mount Sedom, Israel. *Source:* Francisco Gutiérrez (Author). (f) Relationship between average slope-normal denudation per 100 mm of rainfall and the gradient of the slope, measured with 23 erosion pins on salt outcrops in Cardona Diapir. Note that data represent points of varying distance to the crest of the slope. *Source:* Adapted from Mottershead et al. (2007).

Desert, Chile; Cardona, NE Spain) and occasionally with escarpments at the margins of fluvial valleys that experience landsliding and rapid retreat (e.g., Povara et al. 1982). Karst in active diapiric extrusions with salt glaciers (i.e., salt sheets or namakiers) have the peculiarity that dissolution occurs in a mobile substratum, showing a gradation between the young salt extruded above the feeding vent, and the more mature karst developed in the distal sectors of the namakiers (Zarei and Raeisi 2010; Zarei et al. 2012). Salt beds contain impurities and can be associated with layers of other lithologies such as gypsum, claystone, marls, or limestone. The dissolution of halite in exposures releases the insoluble and less soluble components producing a weathering residue that can remain in place, especially in low-gradient areas, generating a residual soil. This autochthonous capsoil tends to protect the underlying salt and inhibit the dissolution process (Bruthans et al. 2009) (Figure 4.11c, d). On steep slopes, the residue is readily eroded and salt exposures affected by fast dissolution can persist. When the production of the capsoil is more rapid than its erosion, it progressively increases its thickness and can become indurated by cementation, transforming into a caprock. Caprocks can also form due to selective dissolution of salt beds at depth and the amalgamation and brecciation of the non- or less-soluble strata of the saline formation. The often sharp contact between the salt and the overlying caprock is termed salt table or salt mirror by some authors (Figure 4.11e). Interstratal salt dissolution caused by relatively deep groundwater flow commonly occurs through the migration of dissolution fronts at the edges of the salt formations. This subsurface solutional denudation, often called subrosion, is comparatively slow, but can operate in extensive areas and over long geological time scales. The solutional removal of the salt in the subsurface leads to the formation of a karstic residue composed of the non- or less-soluble components of the evaporitic formation, typically including residual clays and dissolution-collapse breccias. This process is accompanied by the subsidence of the overlying formations and the ground surface. Interstratal salt dissolution can be analyzed by multiple indirect evidence at the surface, including (e.g., Gutiérrez et al. 2014a): (i) hydrochemical evidence provided by the solute flux at springs and groundwater-fed drainages, (ii) geomorphic evidence such as large closed depressions, (iii) geodetic data indicating contemporaneous ground deformation rates (Kim et al. 2016), (iv) structural evidence given by non-tectonic ductile and brittle subsidence deformation structures (e.g., structural basins, monoclines, normal faults, and collapse structures), and (v) stratigraphic evidence offered by the sediments filling the accommodation space generated by dissolution-induced subsidence.

Few works reporting surface denudation rates in salt exposures have been published, and in most cases they are not directly comparable. This is because the rates depend not only on the amount of precipitation but also on other factors such as the gradient of the slope, the length of the slope above the control station and the runoff contributing area, the type of surface flow (e.g., sheet versus channeled), or the mineralogy and texture of the rock. In the Cardona salt diapir, NE Spain (average precipitation 500 mm yr^{-1}), Mottershead et al. (2007, 2008) measured denudation rates with a scanning reflectorless total station and erosion pins in salt slopes with variable gradients obtaining vertical erosion rates as high as 20 mm per 100 mm of rainfall (ca. 100 mm yr^{-1}). They found a good inverse relationship between denudation rates and slope angle (Figure 4.11f). Slower denudation on steep slopes is attributed to greater rock surface exposed to vertical precipitation and shorter water–rock contact time. The authors used the empirical relationship between denudation rates and slope angle to simulate the evolution of slopes with variable geometries. Bruthans et al. (2008) measured erosion rates over a period of five years in salt exposures and capsoils in various salt diapirs of the Zagros Mountains, Iran, with contrasting average annual precipitation: coastal diapirs (Hormuz, Namakdan; ca. 170 mm yr^{-1}) and mountain diapirs (Jahani; ca. 550 mm yr^{-1}). The NaCl content in the salt outcrops was higher than 96% and the weathering residue of the capsoils was mainly composed of gypsum, anhydrite, calcite, quartz, and dolomite. These capsoils have a significantly denser vegetation cover in the more humid mountain diapirs. Erosion was measured using plastic pegs fixed into the rocks and soils perpendicularly to the ground surface. The slope-normal erosion (E) was directly measured with the fixed plastic pegs and the vertical denudation (D) was calculated with the slope angle (A) using the formula $D = E/cosA$. The D value is higher than E on inclined surfaces and the difference increases with the slope angle. In the salt exposures, they obtained average solutional denudation rates standardized to long-term average precipitation of 30–40 mm yr^{-1} and 120 mm yr^{-1} in the coastal and mountain diapirs, respectively. The site-specific rates were directly proportional to the amount of precipitation and inversely proportional to the slope angle, with higher rates in subhorizontal surfaces where the ratio between the amount of rain and the exposed surface area is higher. The influence of the position of the pegs within the slopes, which determine the runoff contribution area, was not analyzed in this work. The rates in the salt exposures were much higher than the surface erosion rates measured in the capsoils, which are mainly related to mechanical processes (annual average around 3.5 mm yr^{-1}). This large contrast indicates that the relative coverage of salt exposures and capsoils plays an important role on the hydrochemical dynamics and geomorphic evolution of salt extrusions. Interestingly, the long-term uplift rates in the coastal diapirs (3–6 mm yr^{-1}; Bruthans et al. 2006, 2010) are one order of magnitude lower than the denudation rates in salt exposures and comparable with the surface erosion rates measured in the capsoils, which cover most of the salt extrusions. In the Cordillera de la Sal salt anticline (Atacama, Chile), with a highly irregular rainfall averaging ≤20 mm yr^{-1}, De Waele et al. (2009b, 2020) measured denudation rates with MEM of 0.40 and 1.73 mm yr^{-1} on vertical surfaces and a mean value of 0.80 mm yr^{-1} on subhorizontal exposures.

Regional solutional denudation rates have been estimated in salt diapirs in the Zagros Mountains, the Dead Sea, and in northern Spain using hydrochemical methods. Zarei et al. (2012) computed a regional solutional denudation rate of 4.9 mm yr^{-1} in Konarsiah diapir, Zagros Mountains (37.4 km^2, average annual precipitation 400 mm), using discharge and solute flux data. This salt extrusion is fed by two vents controlled by a strike-slip fault system and expressed as subdued salt fountains from which the salt flows laterally over the surrounding country rock forming salt sheets or namakiers. The salt in most of the extrusion is covered by a capsoil up to 30 m thick that increases its thickness away from the vent; salt is only exposed in 1.1% of the area. Monthly discharge and chloride concentration measured in the springs and streams associated with the diapir during a hydrological year indicated a mass of dissolved salt of 185 000 tonnes, equivalent to an overall solutional denudation rate of 2.3 mm yr^{-1}, which was corrected to 4.9 mm yr^{-1} considering that precipitation in the monitored year was 47% of the annual average. The dissolution rate was used to: (i) estimate the rate at which the capsoil is produced (0.25–0.30 mm yr^{-1}) considering the content of insoluble components in the salt formation (ca. 5%) and assuming no erosion; and (ii) the theoretical rate at which the salt should rise in the vents to keep the extrusion in a steady state equilibrium (ca. 7 cm yr^{-1}). According to the authors, in this diapir lacking known integrated cave systems, dissolution essentially occurs by diffuse flow beneath the capsoils within an outer zone of dilated and karstified salt a few tens of meters thick (Zarei and Raeisi 2010). They explain that the salt that extrudes at the surface experiences fracturing and dilation due to unloading or stress release, generating a permeable outer zone through which groundwater can flow diffusely and dissolve halite (Figure 4.11d). At greater depths (c. ≥40 m) with higher overburden load the plastic deformation of the salt tends to anneal fractures and other discontinuities. The proposed laterally continuous shallow zone with higher permeability and degree of karstification has similarities with the epikarst described in carbonate karsts. A different situation is found in diapirs covered by a thick caprock, such as the case of the elongated salt diapir (salt wall) of Mount Sedom, Israel, with a surface area of 13.7 km² and an average annual rainfall of 50 mm (Figure 4.11e). Here, Frumkin (1994a) estimated a minimum regional solutional denudation rate of 0.50–0.75 mm yr⁻¹. Dissolution is mainly related to rapid flow in integrated caves and slow diffuse groundwater flow, since the salt is protected by an anhydrite-rich caprock around 50 m thick. This overall karst erosion rate (minimum value) is one order of magnitude lower than the consistent long-term and short-term uplift rates estimated using multilevel caves and measured by DiNSAR in Mount Sedom (5–8 mm yr⁻¹; Frumkin 1996; Pe'eri et al. 2004; Weinberger et al. 2006). In the absence of continuous records of discharge and solute flux, the calculation of a minimum chemical denudation rate was based on a rough estimation of the effective precipitation (20–30% of the total rainfall) and the areal contribution of different types of runoff and their minimum solute load: surface runoff over caprock (43%, 4 g L⁻¹); flow through integrated cave systems, essentially during storm-derived events (31%, 85 g L⁻¹); and slow diffuse flow (26%, 320 g L⁻¹). The minimum solute discharge estimated is 15·10⁶ kg yr⁻¹. Since most of the denudation is related to halite dissolution (99%) and considering a density for rock salt of 2.2 g cm⁻³, the minimum solute flux is equivalent to the annual erosion of a volume of around 7000 m³ (a cube of salt with a side of approximately 19.1 m). Cardona and Viver (2002), using water and solute flux, estimated an overall denudation rate of 35 mm yr⁻¹ in the Cardona salt diapir, NE Spain (average precipitation 500 m yr⁻¹), which includes extensive exposures of salt, including in situ bedrock and slag heaps related to potash mining. Guerrero et al. (2019), based on solute flux data, estimated an average karstic erosional rate of 2.8 mm yr⁻¹ in the Salinas de Oro diapir, northern Spain, mainly related to salt dissolution beneath a thick caprock. This is most probably a maximum rate since the area of the evaporitic body that contributes to the dissolved loads is likely to be significantly larger than the cartographic extent of the diapiric rocks (18.4 km²), as supports the large extent of the subsidence structures related to interstratal karstification. This diapir of Triassic evaporites is expressed in the landscape as a deep depression rimmed by a crater-like limestone escarpment breached by a high-salinity surface drainage. This stream functions as the base level and outlet of the evaporitic karst aquifer and was used to assess the amount of dissolved salts. The diapir and the surrounding rocks are affected by subsidence related to interstratal karstification that generates or activates a complex system of gravitational structures, including an annular inward-facing monocline in the diapir rim, plus concentric and radial normal faults (Guerrero 2017).

An extreme case of human-induced salt dissolution has been documented at the Upper Gotvand Reservoir in the Karun River, Zagros Mountains, where a halite-rich and strongly karstified outcrop of the Gachsaran Formation has been submerged by the creation of a reservoir with a water depth of 130–140 m (Gutiérrez and Lizaga 2016; Jalali et al. 2019) (Figure 4.11a). The exposure of the Gachsaran Formation at Anbal (or Anbar) ridge, covering 4 km², corresponds to an active salt pillow with an estimated proportion of halite of around 35–40%. The impoundment of the reservoir, with a capacity of 4.5 billion cubic meters, started in July 2011 and by October 2014 around 66.5 million tonnes of dissolved halite had been accumulated in the reservoir, showing salinity stratification with concentrations in the bottom as high as 200 g L⁻¹. Jalali et al. (2019), based on the hydrological, hydrochemical, and isotope characterization of the different sources of salinity and a mass balance, estimated the mass of dissolved salt incorporated into the reservoir over a period of two years at 41 000 000 tonnes. Around 80% of this solute mass derives from direct dissolution by the reservoir water in contact with salt, mainly at Anbal ridge (ca. 16 400 000 tonnes yr⁻¹). This roughly indicates an average solutional denudation of ≤1.9 m yr⁻¹ considering that the majority of this contribution derives from the exposed salt pillow (4 km²) and a density of 2.2 g cm⁻³ for the salt. Dissolution mainly occurs by reservoir water that penetrates into the highly cavernous evaporitic formation and experiences circulation driven by the general downstream flow, density contrasts, and changes in the water level of the reservoir. This internal dissolution under phreatic conditions has resulted in the development of unprecedented collapse structures, hundreds of meters across that undergo rapid subsidence and expansion (Gutiérrez and Lizaga 2016).

4.5 Solutional Denudation of Quartz Sandstones and Quartzites

The dissolution of silica is addressed in Section 3.6. Quartz is a chemically and mechanically resistant mineral under normal near-surface conditions (hardness of 7 on the Mohs scale), whereas the noncrystalline amorphous silica (SiO_2) and the hydrous mineraloid opal ($SiO_2 \cdot nH_2O$) are more susceptible to solutional and mechanical erosion (5–6.5 on the Mohs scale). The equilibrium solubility of quartz in pure water at 25 °C, pH 7, and 1 atm pressure (6.3 mg L^{-1}) is around nine times lower than the solubility of calcite (59.2 mg L^{-1}) in water with atmospheric P_{CO_2}. In contrast, the equilibrium solubility at normal conditions of amorphous silica (116.7 mg L^{-1}) is around 19 and 1.9 times higher than those of quartz (in pure water) and calcite (in water with atmospheric P_{CO_2}), respectively (see Table 3.1). The solubility of quartz shows a general increase with rising temperature and pH, especially above 100 °C and pH>8 (e.g., Brady and Walther 1990) (Figure 3.23), but the increments are slight within the normal near-surface temperatures and the pH ranges commonly found in quartzose rock environments, characterized by acidic to circum-neutral waters (e.g., Mecchia et al. 2014; Sauro et al. 2019). Brady and Walther (1990) carried out batch experiments with ground quartz crystals to assess the effects of changes in pH, temperature, and ionic strength of the solution on the kinetics of quartz dissolution at temperatures between 25 and 60 °C. Dissolution rates at 60 °C were around one order of magnitude higher than at 25 °C. At 25 °C, they observed a significant increase in the dissolution rate at pH values above 8 and a slight decrease as the pH lowers from 7 to 3. They also found that the effect of increasing ionic strength on dissolution rate is very limited under near-neutral conditions. Bennett (1991) explored the influence of organic acids and inorganic electrolytes on the kinetics of quartz dissolution in dilute aqueous solutions between 25 and 70 °C with batch experiments. He observed that inorganic electrolytes increase the rate of quartz dissolution without increasing the solubility. This is attributed to the adsorption of cations to anionic surface sites at the mineral surface (e.g., alkali earths and $Si\text{-}O^-$ sites) that accelerates the dissolution process. Organic acids increase both the dissolution rate and solubility of quartz due to complexation reactions between organic acid anions and silicic acid (H_4SiO_4) in the solution. These reactions decrease the activity of the latter and cause the formation of organic-silica complexes at the mineral surface that destabilizes framework Si-O bonds (Si denotes a silica site in the solid surface). At 25 °C, citric and oxalic acids can increase quartz solubility by as much as 100%. Bennett (1991) also quotes a number of works that document biomediated quartz dissolution (e.g., bacteria, fungi). The role of microorganisms and bioalkalinization on quartz dissolution is addressed by Büdel et al. (2004) and Brehm et al. (2005).

It is widely accepted that dissolution plays an essential role in the development of karst landforms and hydrology in quartz sandstones and quartzites, although it quantitatively represents a limited proportion of the removed rock mass. Dissolution weakens the rock and predisposes it for mechanical erosion, which volumetrically is the main denudation process. In an initial preparatory phase dominated by chemical weathering, dissolution operates at the contacts and edges of quartz grains (Figure 2.28c, d), silica cement, or other silicate minerals (phyllosilicates). This slow dissolution increases the porosity and reduces the cohesion of the rock, which is progressively transformed into a friable arenaceous material through a process termed arenization

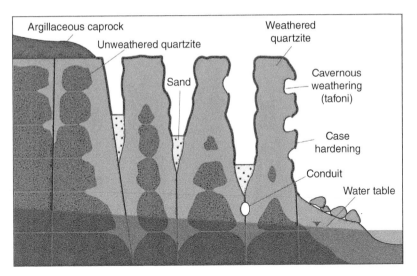

Figure 4.12 Sketch illustrating the arenization model. The quartz sandstone or quartzite is transformed into a low-cohesion weathered rock by dissolution of quartz along the grain edges and contacts. This chemical weathering process is guided by discontinuities (fractures, bedding planes) that function as permeability features and provide access to the water. Subsurface flows with sufficient gradient and velocity can mechanically erode the friable weathered rock and generate conduits. Evaporation and other physicochemical changes can induce the precipitation of minerals (silica, iron oxy-hydroxides, and carbonate) in an outer shell forming a hardened crust (case hardening), which can be interrupted by cavernous weathering features (tafoni). *Source:* Adapted from Martini (2004).

(Martini 1979, 2000, 2004). The chemical attack preferentially advances from fractures and bedding planes that provide suitable pathways for water ingress and circulation (Figure 4.12). In places with sufficient gradient and flow velocity (e.g., close to scarps), the softened weathered rock is mechanically eroded leading to the enlargement of fractures and bedding planes both at and below the ground surface (Figure 4.13a). As the discontinuities become wider, more rapid and turbulent flow can circulate through them, accelerating the creation of underground voids by subsurface mechanical erosion (piping or suffosion) through a positive feedback mechanism. Recently, an alternative model to arenization has been proposed to explain the origin of caves and a number of landforms developed in the South American tepuis, involving differential mechanical erosion controlled by selective lithification (Aubrecht et al. 2011, 2017). According to this model, descending silica-bearing fluids during the diagenesis split into finger-like flow paths at stratigraphic boundaries between fine-grained beds and underlying coarse-grained layers. These are capillary breaks below which the flow is dominated by gravitational rather than capillary forces. The localized descending flow produces selective cementation in the coarse-grained beds, forming resistant finger-like pillars. Selective mechanical erosion of the less indurated intervening zones is claimed to produce the cave systems and other features such as the peculiar pillars found in the tepuis (Figure 3.43). This research group (Aubrecht et al. 2011), in contrast to the advocates of the arenization model (e.g., Sauro et al. 2013a; Sauro 2014; Wray and Sauro 2017), contends that the caves in the tepuis are essentially related to the selective erosion of poorly cemented rock and the incongruent dissolution of aluminosilicates (laterization of feldspars and micas). In short, in the arenization model differential mechanical erosion occurs preferentially along chemically weathered zones, whereas in the selective lithification model erosion is focused on the less cemented portions of the rock mass.

Figure 4.13 Weathering and erosion features developed in quartzose rocks. (a) Vertical view of large grikes or corridors with trees developed along a set of widened joints in Triassic quartz sandstones (Buntsandstein Facies, Iberian Chain, Spain). (b) Case hardening and cavernous weathering (tafoni) developed in the quartz sandstone of the Late Cambrian Umm Ishrin Formation at Petra, Jordan. In the foreground "the Monastery" carved by the Nabateans during the first century BC. Circle indicates a person for scale. (c) Pedestal with a stem of pegmatitic quartz-rich granite capped by in situ schists (country rock of the intrusion) in the Rocky Mountain N.P., Colorado, United States. Probably, the large crystal size of the granite enhances disintegration by physical weathering processes (e.g., frost shattering) and more rapid erosion than in the capping schists. *Source:* Francisco Gutiérrez (Author).

Two important differences between karsts developed in detrital quartz-rich rocks with respect to the more soluble chemical rocks (carbonates and evaporites) include that a great part of the eroded mass is transported as solid load rather than as a solute and that karst features mainly occur close to high topographic gradients where subsurface flows have sufficient competence to evacuate sand grains. Some scientists contend that the term karst should be restricted to environments in which the development of landforms is essentially related to dissolution, and prefer to use the word pseudokarst. Others support the widely accepted opinion whereby karstic processes and landforms are found on any rock type where the process of solution is critical, although not necessarily dominant (Jennings 1983; Doerr and Wray 2004). Other chemical alteration processes in addition to quartz dissolution can participate in the softening of quartzose rocks. Tripathi and Rajamani (2003) attribute the weathering of the Delhi quartzites of India to the presence of small amounts of pyrite. The oxidation of this mineral produces an acidic solution that attacks minor amounts of detrital aluminosilicate minerals and sparry calcite cement, reducing the coherence of the rock and increasing its susceptibility to mechanical erosion (see Eq. 3.74). Salt weathering related to the precipitation of salts from the sulfate derived from the oxidation of pyrite could also be a significant process in these contexts. Dissolved components such as silica, iron, or carbonate can precipitate due to evaporation or other physicochemical changes (e.g., pH lowering, biological activity) in the outer shell of the rock, forming crusts of weathered rock hardened by secondary cements (e.g., opal, iron oxi-hydroxides, and calcite). This case hardening process creates a more resistant outer shell that inhibits erosion processes and favors the development of cavernous weathering or tafoni (Goudie et al. 2002; Dorn 2004; Turkington and Paradise 2005) (Figures 4.12 and 4.13b). Reviews on landforms developed on sandstones can be found in Wray (1997), Härtel et al. (2007), Young et al. (2009), and Wray and Sauro (2017).

Studies assessing solutional denudation rates in quartz sandstones and quartzites are very scarce due to the low solubility and slow dissolution kinetics of quartz. Paradise (1995) estimated surface recession rates in Paleozoic quartz sandstones at Petra, Jordan (average annual precipitation 130 mm) of 13–66 mm kyr^{-1} on horizontal surfaces and 7–18 mm kyr^{-1} on vertical surfaces. Here, denudation is mainly related to physical disintegration, which is inhibited by iron oxide crusts and shows higher rates in south-facing slopes with more intense diurnal heating–cooling and wetting–drying cycles (Figure 4.13b). Mecchia et al. (2014) collected extensive data on discharge, pH, electrical conductivity, temperature, and dissolved H_4SiO_4 (expressed in SiO_2 equivalent) in surface and cave waters at three tepuis in the Gran Sabana, Guyana Shield, Venezuela (average total and effective annual rainfall of 3000 and 2400 mm, respectively). The local relief of these table mountains exceeds 1000 m and they host fracture-controlled collapse sinkholes up to 350 m deep and caves as much as 18.8 km in length (Sauro et al. 2018, 2019). They found common features in each hydrodynamic environment with relatively well constrained average pH values and SiO_2 contents, showing an evolutionary trend along the flow paths: (i) rainwater 5.36 pH, 0.00 mg L^{-1}; (ii) stagnant surface water in peat, ponds, and swamps 4.7 pH, 0.12 mg L^{-1}; (iii) streams on the summits of the tepuis 4.2 pH, 0.20 mg L^{-1}; (iv) cave streams and resurgences 4.6 pH, 0.17–1.2 mg L^{-1}; (v) cave drips derived from rapid infiltration 3.7–5.5 pH, 0.15–3.4 mg L^{-1}; (vi) cave drips fed by slow infiltration 5.0–6.5 pH, 5.6–8.6 mg L^{-1}; (vii) stagnant cave pools fed by drippings 6.1–6.6 pH, 8.1–8.6 mg L^{-1}; (viii) streams in the lowlands surrounding the tepuis and underlain by arkosic sandstones 6.4 pH, 5–21 mg L^{-1}. They inferred limited dissolution of silica on the top surface of the tepuis, probably related to the presence of hardened iron hydroxide crusts. This is consistent with the slow long-term denudation rate of 1 mm kyr^{-1} calculated using the cosmogenic nuclides [10]Be and [26]Al by Brown et al. (1992). Most of the quartz dissolution occurs in the subsurface, especially by slow percolation along fractures, and the significant increase in the silica content observed in the lowlands is ascribed to the hydrolysis of silicates in the feldspar-rich arkosic sandstones. Mecchia et al. (2014, 2019) suggest that films of condensation water can play a significant speleogenetic role causing the slow and continuous diffusional transport of dissolved silica from the rock to the undersaturated solution, increasing the porosity of the rock associated with cave walls and reducing its cohesion (i.e., arenization). The authors made rough calculations of subsurface solutional denudation in three cave systems with data restricted to the dry season. For instance, at the Akopan-Dal Cin cave system, with an estimated catchment area of 0.8 km^2, they measured an average increase in dissolved silica between the inflow and outflow waters of 0.80 mg L^{-1}. This value together with an average discharge of 90 L s^{-1} yield a flux of dissolved silica of 72 mg s^{-1}, of which 4 mg s^{-1} (5%) are estimated to correspond to silica dissolved at the surface of the tepui (infiltration-derived discharge of 20 L s^{-1} and 0.2 mg L^{-1} of dissolved silica in the surface stream). The value of subsurface solutional denudation of the dry season of 68 mg s^{-1} is doubled to roughly assess the average annual value, yielding 4300 kg a^{-1} or 1.6 m^3 a^{-1}. Considering the catchment area (0.8 km^2) this is equivalent to an average denudation of 2 mm kyr^{-1}. This value is one and three orders of magnitude lower than those calculated in carbonate areas (Table 4.1) and in salt extrusions located in arid regions, respectively. In a work specific to the Aonda Cave system in the Auyán tepui,

Piccini and Mecchia (2009) estimate contributions to the dissolved silica load from surface and subsurface solution of 15 and 85%, respectively. These authors substantiate the important role played by several factors in the development of karst landforms in the tepuis, namely the flat topography of the top planation surface, the vast time span elapsed since the initiation of the formative processes, estimated at around 20–30 Ma, stress release in the peripheral zones of the table mountains and the margins of large collapse sinkholes (locally known as simas), and rates of retreat in the massive scarps significantly slower than cave development. Other critical elements for the generation and preservation of the karst features in the tepuis include the abundant rainfall, the mechanical resistance of the massive quartzites, and the mineralogical purity of the rock, which prevents the formation of significant clayey residual deposits that could clog subsurface flow paths.

In an interesting theoretical article, Simms (2004) analyzed the contrasting features of denudation processes in carbonate and silicate rocks and their implications on the long-term evolution of landscapes. Denudation mechanisms in these lithologies show fundamental differences regarding their rates, activation thresholds, and enhancing/inhibiting factors. Silicate rocks are mainly removed by mechanical erosion caused by surface runoff when a minimum threshold velocity for particle entrainment is exceeded. This is a relatively high energy and highly episodic process that is significantly inhibited by the vegetation cover and favored by seasonal runoff. In contrast, limestone denudation mainly occurs by dissolution through a low energy process that can occur at any flow velocity or even in static water, without the interplay of any threshold. This chemical denudation process tends to be more continuous and spatially homogeneous, with significant mass removal occurring beneath the surface, especially within a relatively shallow zone where the water reaches near-saturation conditions (epikarst). Moreover, denudation of carbonate rocks is enhanced by vegetation and favorable thermal regimes for the production of CO_2 in the soils by organic matter decay and root respiration. Numerous studies illustrate that in silicate terrains a significant proportion of the geomorphic work (mass transfer) and the geomorphic change occurs during short-lasting high magnitude and low frequency events (Wolman and Miller 1960). Walling and Webb (1986) found in a river catchment including limestone and siliciclastic rocks in Devon, England, that 90% of the solute load was transported during 56% of the time, while the equivalent proportion of suspended load was transported during high flow periods representing 6% of the time. Gunn (1982), in an analysis of the magnitude and frequency properties of dissolved-solid transport in several catchments underlain by carbonate rocks, found that flows greater than the mean discharge account for 45–74% of the solute load transport. As expected, larger volume flows carry larger amounts of dissolved solids. Simms (2004) uses the analogy of the "tortoise and hare race" to put forward the concept whereby the slow but continuous denudation of limestone may in some circumstances exceed the often more rapid but localized and episodic erosion of silicate rocks. Obviously, this is a very general conception, since limestones in some settings are affected by substantial mechanical erosion, and the resistance to erosion of silicate rocks is highly variable. To illustrate these contrasting erosional regimes, the author plotted the curves of (Figure 4.14): (i) mean annual sediment yield versus average effective precipitation (runoff) of Langbein and Schumm (1958) and Ohmori (1983) for silicate outcrops; and (ii) average solutional denudation of carbonate rocks in karst regions with soils and vegetation of Atkinson and Smith (1976). The mechanical denudation curves show a steep rise from a minimum sediment yield in hyperarid areas to a peak in semiarid zones with an average effective precipitation of around 250–350 mm, where there is typically a combination of sparse vegetation and flashy runoff. Beyond this peak, the classical curve of Langbein and Schumm (1958) suggests a rapid decline, whereas that of Ohmori (1983) indicates that rates may rise again in areas of very high runoff. In contrast, the solutional

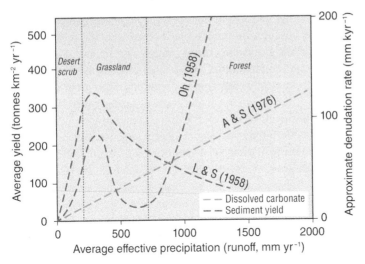

Figure 4.14 Relationship between annual effective precipitation (runoff), annual sediment yield in outcrops of silicate rocks according to Langbein and Schumm (1958) and Ohmori (1983), and annual solutional denudation of carbonate rocks in soil-covered karst regions after Atkinson and Smith (1976). The latter regression should be considered as a minimum estimate, since it was derived from average chemical denudation rates in a number of samples uncorrected for the proportion of carbonate rocks in the watershed. *Source:* Adapted from Simms (2004).

denudation curve of Atkinson and Smith (1976) shows a direct positive relationship between denudation rate and runoff. The comparison of this curve with that of Langbein and Schumm (1958) suggests that in humid tropical environments with an average effective precipitation higher than 1000 mm yr^{-1}, denudation rates in limestone may exceed those of silicate rocks. An important element associated with this comparison is that denudation in the siliceous terrains essentially refers to surface lowering, whereas in carbonate environments a significant proportion of the denudation occurs beneath the surface. The work by Simms (2004) also includes an interesting discussion on how more rapid corrosional denudation of limestone in Ireland may contribute to induce regional uplift by isostatic compensation at a rate higher than the denudation rate on inliers and outliers of silicate rocks, which tend to increase their relative relief. Note that isostatic compensation in response to erosional unloading may reach as much as 80% of the denudation (Gilchrist et al. 1994; Burbank and Anderson 2011).

4.6 Interpretation and Integration of Denudation Rates

The correct interpretation and integration of denudation rate measurements requires (i) understanding their actual meaning and the extent to which they are comparable, (ii) taking into account their limitations including the potentially high errors, and (iii) properly constraining their spatial and temporal representativeness to avoid inadequate extrapolations. Some methods provide short-term areal estimates that integrate surface and subsurface solutional denudation (mass-flux approach), while others offer point measurements of surface denudation spanning short-term periods (MEM, tablets) or long-term intervals of denudation (pedestals, cosmogenic nuclides). An important limitation of a considerable number of denudation studies is that the uncertainties associated with the assumptions, measurements, and calculations are not incorporated in the results, which in some cases can reach a significant percentage or even exceed the obtained rates, especially in the case of the low-solubility rocks. Another frequent limitation is the implicit assumption of no mechanical erosion, which can result in unsound interpretations. Some carbonate rocks, due to their textural characteristics (e.g., calcarenites, grainstones) and structure (e.g., dense jointing), are highly prone to granular disintegration and fragmentation, thus a large proportion of the surface lowering or recession may be ascribed to physical weathering and erosion, rather than solutional denudation. Denudation rates can show great variations across short distances controlled by the textural and mineralogical composition of the different strata within a karst formation. Moreover, in some environments, some physical weathering processes may have a larger impact on denudation than solution. For instance, frost shattering in periglacial environments with numerous freeze and thaw cycles, or salt weathering in areas associated with a significant source of salts (e.g., coastal environments). This disintegration and fragmentation of the rocks contributes to increase the water–rock interaction surface, thereby accelerating the dissolution process.

The mass-flux or hydrochemical method provides areal estimates of the dissolved mass derived from a three-dimensional karst hydrological basin over short observation periods. These denudation rates include the contribution of the solutional work achieved by both surface and subsurface water and are not comparable with those that assess surface denudation alone. Major potential sources of epistemic uncertainty include the definition of the catchment boundaries, the estimation of the proportion of soluble rocks, the contribution of allogenic dissolution, and in studies lacking discharge data, the estimated average effective precipitation. One of the main limitations of this approach is the representativeness of the data, which is dependent on the meteorological and hydrological characteristics of the monitoring period. Results can be significantly affected by large interannual variability in the solute flux related to multiple factors, such as the amount of precipitation, the characteristics of the precipitation events (snowfall versus rainfall, short high-intensity storms versus prolonged low-intensity rains), or the seasonal distribution of the precipitation and flood events (CO_2 content in soils may change significantly throughout the year). For instance, Bakalowicz (1992) in a drainage basin in the French Pyrenees computed an interannual variability of 25% in the flux of dissolved carbonate (Table 4.1). The obtained rates can be standardized by considering the average annual precipitation and/or discharge computed from long-term records (Bruthans et al. 2008; Zarei et al. 2012). This type of corrections generates significant errors, especially when the available water and solute flux data are restricted to a specific season, as is the case of the studies carried out in the South American tepuis, where data are restricted to the dry season (Mecchia et al. 2014).

MEM measurements and tablet weight loss provide short-term point-specific values of surface denudation related to both chemical and physical processes. These methods suffer from limited spatial and temporal representativeness. Hopefully, the development of high-resolution and high-accuracy techniques for generating and comparing 3D surface models (e.g., close-range LiDAR and photogrammetry) will help to partially overcome these limitations. Those techniques could provide the means for the detailed 3D monitoring of the evolution of specific karst features such as karren. Regarding the MEM, the important work by Spate et al. (1985) demonstrated the large impact of several error sources on the measured denudation rates, especially in low solubility rocks. These authors calculated errors with similar magnitudes to the surface lowering rates measured on inland limestone outcrops in Australia over four years. Nonetheless, subsequent works rarely incorporate comprehensive error assessments. Moreover, the significantly greater denudation rates observed in coastal limestone outcrops respect to those from inland exposures under similar precipitation illustrate the potential important contribution of mechanical weathering and erosion processes to the denudation observed on rock surfaces (Cucchi et al. 2006; Stephenson and Finlayson 2009). Assessing the relative role played by chemical and mechanical processes is probably one of the main challenges for denudation studies in karst regions. The tablet weight loss method also provides point denudation rates with considerable error margins when applied to carbonate rocks. An additional limitation is that denudation occurring on artificially cut surfaces could be considerably different to those occurring on natural surfaces affected by previous dissolution (e.g., Dreybrodt and Eisenlohr 2000) and the development of biofilms. Two important strengths of the weight loss method are the possibility of using tablets of a standard rock type to isolate specific variables and explore their influence on the denudation process, and the ability to retrieve the samples to examine under the microscope the micromorphologies generated by the denudation processes and the potential role of mechanical erosion (e.g., grain or mineral detachment; Krklec et al. 2016).

Pedestals and other protruding features controlled by non-soluble rock provide minimum surface denudation rates averaging erosion occurred locally over long geological periods. These estimates may incorporate large and difficult-to-assess uncertainties related to the time elapsed since the start of the denudation, the actual value of the surface lowering, or the amount of dissolution occurred at the crown of the stem of the pedestal (e.g., Goldie 2005; Lauritzen 2005). Moreover, these are black box geological gauges that record a geomorphic change, but not the morphogenetic processes and conditions, which can experience substantial changes through time (bare rock versus subsoil dissolution or solutional versus mechanical erosion; Wilson et al. 2012a).

TCNs have a great potential since they permit the assessment of average denudation rates at specific points covering longer geological periods. Some of the main problems include the high cost and complex processing of the samples, which result in a limited amount of data, the uncertainties associated with the computations, and the difficult-to-assess representativeness of the sampling points. Comprehensive geological and geomorphological characterization of the sampling points is essential to understand the meaning of the results and to elucidate to what extent they admit valid comparisons. It is important to bear in mind that denudation depends on multiple factors in addition to precipitation and that those factors can show substantial spatial and temporal variations.

Ideally, it is desirable to obtain denudation data using multiple methods that cover different spatial and temporal scales. This allows crosschecking measurements covering similar spatial and temporal scales, and inferring the relative contribution of different environments to the overall karst denudation (e.g., surface versus subsurface solution). The advantage of combining different approaches is well illustrated by Lauritzen (1990) in a basin in Norway partially underlain by marble. Here, autogenic chemical denudation was estimated at 32.5 ± 10.2 mm kyr^{-1} on the basis of solute flux. Measurements with MEM over 10 years yielded a surface denudation rate of 0.025 ± 0.0027 mm yr^{-1}. Pedestals and veins that stand proud on surfaces exposed by glacial retreat approximately 9000 years ago indicate minimum denudation rates of 13 mm kyr^{-1} and 23 mm kyr^{-1}, respectively. The short- and long-term point estimates of surface denudation rates are apparently consistent and the comparison of the solute flux and MEM rates suggests that around 50–86% of the solutional denudation occurs at the surface. In the Austrian Alps Plan (2005), using tablets, obtained an average surface denudation rate in rock exposures (11 mm kyr^{-1}), similar to that estimated from a few pedestals (10 mm kyr^{-1}). Here, the overall denudation rate estimated from hydrochemical data at a spring is around five times higher than that of the average rate derived from the tablet weight loss method.

4.6.1 Vertical and Spatial Distribution of Karst Denudation and Deposition

Gaining insight into the vertical and spatial distribution of dissolution is essential for understanding the development of karst landforms and aquifers, as well as the differences they display in different regions and karst types. Critical aspects include the relative contribution of solutional denudation at different depths (epikarst, bulk vadose zone, epiphreatic zone, and phreatic zone) and the spatial variability within each vertical zone. Denudation rate in a specific rock mainly depends on the degree of saturation of the water with respect to the corresponding mineral, the water–rock contact area, the amount of water that interacts with the rock, and the flow conditions (e.g., slow and laminar diffuse flow versus rapid and turbulent conduit flow). The spatial patterns of karst denudation can show significant variability depending on the type of rock (carbonates, evaporites, and quartz sandstones) and their different solubilities and dissolution kinetics. Quantitative information on the distribution and relative contribution of the dissolution process in each zone can be obtained from hydrological and chemical parameters gathered along the different flow paths.

In epigene carbonate karst systems, as supported by numerous hydrochemical studies, a great part of the solutional denudation is achieved at the surface and by percolation water in the upper portion of the vadose zone. In areas where the carbonate bedrock is mantled by a vegetated soil that supplies abundant CO_2 to the infiltration water, dissolution is particularly intense at the cover-bedrock interface (rockhead) and the uppermost zone of the bedrock. Ford and Williams (2007) compiled data on the approximate contribution of different vertical zones to the overall solutional denudation (Table 4.2). This dataset indicates that around 50–90% of the autogenic dissolution takes place within the upper portion of the percolation zone, typically less than 10 m thick. The rapid expenditure of the carbonate dissolution potential that affects the water in the upper part of the vadose zone is related to various factors: (i) the dissolution kinetics of carbonate minerals under normal pH conditions is characterized by an initial phase of relatively rapid

Table 4.2 Vertical distribution of solutional denudation in different carbonate karst areas.

Locality/author	Overall rate $(m^3 km^{-2} yr^{-1})$	Distribution of solutional denudation
Fergus River, Ireland; Williams (1963, 1968)	55	60% at the surface, up to 80% in the top 8 m
Derbyshire, United Kingdom; Pitty (1968)	83	Mostly at the surface
Northwest Yorkshire, United Kingdom; Sweeting (1966)	83	50% at the surface
Jura Mountains; Aubert (1967, 1969)	98	35% on bare rock, 58% under soil, 37% in percolation zone, 5% in conduits
Cooleman Plain, Australia; Jennings (1972a, 1972b)	24	75% from surface and percolation zone, 20% from conduit and river channels, 5% from covered karst
Somerset Island, Canada; Smith (1972b)	2	100% above permafrost layer
Riwaka South Branch, New Zealand; Williams and Dowling (1979)	100	80% in top 10–30 m, 18% in conduits mainly allogenic
Waitomo, New Zealand; Gunn (1981a)	69	37% in soil profile, most of remainder in upper 5–10 m of bedrock
Caves Branch, Belize; Miller (1982)	90	60% on surface and in percolation zone, 40% in conduits (in large allogenic river passages)
Svartisen, Norway; Lauritzen (1990)	32.5	42–72% at surface

Source: Adapted from Ford and Williams (2007).

dissolution. After a relatively short interaction with the rock, the solution reaches near-saturation conditions (saturation ratio ~0.7–0.8) and the dissolution rate significantly decelerates (high-order surface-controlled kinetics) (Figure 3.36). This means intense dissolution in a short initial portion of the flow path, and slow but spatially sustained dissolution in the following sections of the flow path until the water reaches saturation. (ii) The precipitation water that infiltrates through the soil increases by several times its solutional denudation capability due to CO_2 uptake. However, as the water percolates and dissolves carbonate minerals, the dissolved CO_2 is rapidly depleted in a semi-closed system with limited CO_2 replenishment (Figures 3.13 and 3.14). (iii) Intense dissolution together with stress release and other weathering processes in the uppermost zone of the carbonate bedrock results in a relatively thin zone with high degree of karstification and fracturing, designated as epikarst or subcutaneous zone. The high permeability contrast between the epikarst and the underlying bulk vadose zone causes the temporary storage of water in the lower part of the epikarst, increasing the water–rock interaction time in that zone. At the base of the epikarst, water tends to flow laterally converging toward major, widely spaced fissures that function as preferential drains of the main vadose zone (Williams 1983, 2008; Klimchouk 2000b, 2004b) (Figure 5.25). The large and widespread solutional removal of rock that occurs in the upper few meters in carbonate rock exposures has important implications from the geomorphological and applied perspective: (i) a considerable proportion of the solutional work results in surface lowering; (ii) dissolution by focused flow in cave conduits, in both the vadose and phreatic zones, quantitatively represents a limited proportion of the overall solutional denudation, but plays an essential role in the development of the karst hydrology and geomorphology; and (iii) the intense dissolution that occurs in the upper few meters of the rock mass creates a highly permeable and mechanically weak weathering zone (epikarst) prone to a number of engineering problems (Waltham and Fookes 2003). The nearly saturated percolation water that enters vadose cavities with lower P_{CO_2} experiences rapid CO_2 exsolution, reaching oversaturation conditions and resulting in the precipitation of speleothems. This mainly occurs in caves that function as open systems connected with the atmosphere. This karst deposition can be aided by evaporation in ventilated caves where the humidity is below 100%. This re-precipitation of the dissolved mass (karst deposition of Eq. 4.3) can occur just a few meters below the ground surface, due to the rapid autogenic dissolution that occurs in the epikarst, and is particularly important in warm and humid regions due to the greater contrast between the CO_2 concentration in the soil and in the atmosphere (Ford and Williams 2007). However, in most cases this type of localized karst deposition represents a quantitatively small component in the overall solute mass budget. A similar situation frequently occurs at karst springs and downstream of these output points, where groundwater abruptly flows into the surface atmosphere experiencing rapid CO_2 degassing, a process that may be accelerated by turbulence and photosynthetic activity (Figure 3.24). This results in oversaturation conditions and accumulation of calcareous tufa and travertines. This karst deposition process, which may occur in relatively large areas (e.g., long fluvial reaches) and at significant rates (e.g., Drysdale and Gillieson 1997; Vázquez-Urbez et al. 2010), can constitute a substantial subtraction to the gross karst dissolution. Travertines of hydrothermal origin are characterized by higher accretion rates (cm to m yr^{-1}) than tufa deposits precipitated from ambient-temperature water (mm to cm yr^{-1}) (Capezzuoli et al. 2014 and references therein).

In three tropical karst areas of Peninsular Malaysia, Crowther (1989) analyzed autogenic solutional denudation, as well as Ca and Mg fluxes in the vegetation–soil–rock system. This author made an innovative contribution assessing Ca and Mg budgets incorporating the main components of the system, including cycling by vegetation: (i) input by precipitation, (ii) runoff water on bare rock surfaces, (iii) throughflow on rock surfaces overlain by thin organic soils and litter,

(iv) throughflow at the rockhead overlain by thicker mineral soil, (v) canopy throughfall and fine litter fall (excluding large wood litterfall and root litter); and (vi) dripping water in caves. It is found that annual rates of calcium and magnesium uptake by the tropical forest are similar in magnitude to the net solute outputs from limestone outcrops. This implies that vegetation has an important role in solute production by releasing assimilated cations via canopy leaching and litter decomposition. The calcium concentrations and fluxes in the different components of the vegetation–soil–bedrock system in hillslopes are shown in Figure 4.15. These authors found that Ca concentrations in cave drips were remarkably similar despite the highly variable thickness of limestone above the cave, suggesting that dissolution activity is largely confined to the near-surface zone. They also observed an increase in the Mg/Ca+Mg ratio along the flow path attributable to speleothem formation, which involves a relative depletion of Ca from the solution. Cenki-Tok et al. (2009) assessed the role of vegetation on Ca isotope fractionation in a continental forested ecosystem in NE France. By measuring $\delta^{44/40}$Ca they found that soil solutions were significantly depleted in lighter isotopes, whereas vegetation was strongly enriched, indicating preferential ^{40}Ca uptake by plants. This finding proves the potential of Ca isotopes as tracers of biochemical processes in rock–water–soil–plant systems.

Data on the vertical distribution of solutional denudation in other karst rocks are rather limited, and show a different picture than that observed in carbonate karsts, mainly due to the markedly different solubility and dissolution kinetics of the corresponding minerals. In evaporite karst systems, the water rarely reaches near-saturation conditions in the upper meters of the percolation

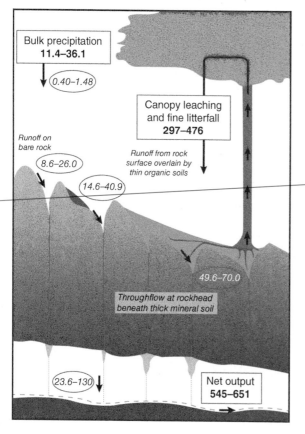

Figure 4.15 Calcium concentrations (ellipses; mg L^{-1}) and annual fluxes (squares; kg ha^{-1} yr^{-1}) within the vegetation–soil–bedrock system in hillslopes (30–45°) of an autogenic humid tropical karst in Malaysia. Note that the annual rate of calcium uptake by the tropical forest (assessed excluding large wood litterfall and root litter) is similar in magnitude to the net solute output from limestone outcrops. *Source:* Adapted from Crowther (1989).

zone due to the high solubility of gypsum and halite, and thus the groundwater can cause significant dissolution along long flow paths (see Section 3.8.2). Waters saturated with respect to halite are rarely observed, unless the solution has undergone prolonged interaction with halite or has been affected by evaporation. The development of the epikarst in evaporites is frequently inhibited by factors such as: (i) the lower density of fractures in these more ductile lithologies; (ii) the production of abundant fine-grained residual material that tends to clog permeability features (Klimchouk 2000b) (Figure 4.16); and (iii) the formation of crusts, especially in arid and semiarid environments, that reduce water infiltration (Ferrarese et al. 2003). In the hypogene gypsum karst of Ukraine, where the confined phreatic flow exploits all of the available fractures, deep-seated dissolution accounts for the great part of the solutional denudation responsible for the maze caves, which become explorable once they change into vadose conditions by fluvial entrenchment (Klimchouk and Aksem 2005). In the Cueva del Agua in the gypsum karst of Sorbas (Spain), Sanna et al. (2015) indicate that dissolution in the cave is mainly related to condensation waters (ca. 50%) and storm-derived floods, whereas slow infiltration water loses great part of its dissolution capability within the first few meters of its underground flow path. It should be noted that the caves in Sorbas are essentially related to mechanical erosion of argillaceous sediments underlying gypsum packages. Data from Konarsiah diapir, Iran, and Mount Sedom diapir, Israel, show a markedly different picture regarding the vertical distribution of salt dissolution. At Konarsiah, where the salt is covered by capsoils in around 99% of the area, most of the solutional denudation occurs by diffuse flow within a near-surface zone a few meters thick with higher permeability related to dilation and karstification (Zarei et al. 2012). At Mount Sedom (Israel), the main contribution is also related to slow diffuse flow beneath a thick caprock (ca. 74%), but localized dissolution by storm-derived flood events within integrated cave systems accounts for about 24% of the total budget (Frumkin 1994a).

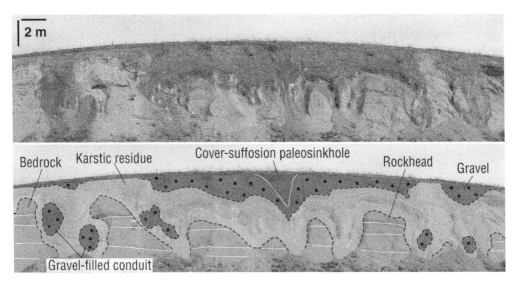

Figure 4.16 Epikarst developed in subhorizontal Miocene gypsum with marly partings overlain by a gravelly terrace of the Jarama River in the Madrid Cenozoic Basin, central Spain. Dissolution in the upper part of the bedrock has produced a highly irregular rockhead with solutional conduits and a relatively thick argillaceous residue. The karstic residue between the alluvial cover and the bedrock displays pseudobedding roughly concordant with the rockhead. Note disharmonic gravitational structure, with subhorizontal bedrock and deformed Quaternary alluvium. The solutional conduits are filled with karstic residue and gravels derived from the detrital cover by suffosion. The subcircular gravel pockets correspond to nearly transverse sections of inclined pipes filled by alluvium. *Source:* Francisco Gutiérrez (Author).

These differences in the denudation styles can be related to differing geomorphological, tectonic, and evolutionary features of the diapirs. Konarsiah diapir is a salt fountain with salt glaciers and a local relief of around 900 m. The 250 m high Mount Sedom is truncated on its eastern flank facing toward the Dead Sea by a steep escarpment, and its summit was planated by the pluvial Lisan Lake in the Late Pleistocene (ca. 14 kyr). The high local topographic gradient and the presence of active faults within the Mount Sedom diapir (Figure 3.44) most probably favored the development of the documented multilevel cave systems (Frumkin 1996).

Data on dissolved H_4SiO_4 from different environments of the tepuis in the Gran Sabana, Venezuela, indicate that most quartz dissolution, characterized by slow kinetics, occurs in the subsurface, especially along fractures with slow water percolation. Dissolution at the surface seems to be inhibited by the presence of crusts of iron hydroxides (Mecchia et al. 2014).

In the autogenic percolation system in carbonate karst, flow rates range over several orders of magnitude, from slow seepages to sinking streams at ponors, whereas solute concentrations rarely vary more than one order of magnitude. Consequently, solutional denudation in zones of high convergent flow can be many times greater than in zones where the flow diverges (Ford and Williams 2007). The large local differences of solution rates explain the distribution and morphology of karst landforms at different scales. The main factors that control the spatial variability of surface and near-surface autogenic solutional denudation include water availability, the aggressiveness of the water, topography, soil distribution and thickness (source of CO_2), lithological heterogeneities, and structural elements such as fractures. In an ideal horizontal outcrop of bare, massive, homogeneous, and unfractured limestone, surface dissolution is expected to cause an even lowering of the ground. Nonetheless, limestone formations contain discontinuities (fractures, bedding planes) and lithological variations that determine preferred water infiltration and dissolution at specific locations (e.g., joint intersections, more soluble beds). Differential solutional denudation results in an irregular topography with depressions (e.g., solution sinkholes) at the spots where surface erosion rates reach higher values. These topographic variations tend to be accentuated through positive feedback mechanisms, whereby dissolution tends to operate more intensely in the depressions: (i) hollows determine runoff convergence and greater water infiltration, (ii) soils tend to form and reach greater thickness in the depressions significantly enhancing the aggressiveness of the percolation water by increasing its CO_2 content, and (iii) the enlargement of the depressions involves the increase of their runoff contributing area and the amount of water available for solutional denudation. In contrast, elevated areas and topographic divides encourage runoff divergence, resulting in slower denudation. As explained above, the intense karstification in the upper few meters of the carbonate bedrock that interacts with highly undersaturated percolation water leads to the development of the epikarst, which tends to display greater development in the topographic lows. The permeability contrast between the epikarst and the underlying less permeable vadose zone also determines an additional convergent subcutaneous flow toward the more efficient percolation pathways at the base of the epikarst, commonly associated with the topographic lows (Figure 5.25). Progressive enlargement of these fissures by focused downward flow can result in the development of dissolution pipes and shafts (Klimchouk et al. 1996c), which eventually can lead to the formation of bedrock collapse sinkholes (Klimchouk 2000b).

The conceptual model that postulates greater dissolution rates at the bottom of the depressions and solution dolines has been substantiated in different works with short- and long-term quantitative data. Zambo and Ford (1997) monitored climatic and carbonate dissolution variables over two years in a solution doline in Aggtelek National Park, Hungary, and collected data from the side slopes of sinkholes covered by thin soils, and from a 7.5 m deep shaft excavated in the fine-grained fill at the sinkhole bottom. They estimated a significantly higher carbonate dissolution potential at the bottom of the sinkhole (17–30 g m^{-2} yr^{-1}) than beneath the soils on the side slopes (c. 3 g m^{-2} yr^{-1}). This difference was attributed to runoff and groundwater flow concentration at the sinkhole floor and more favorable conditions for the production of CO_2, including higher moisture and temperatures above freezing point year-round and with lower annual amplitudes. The available data were used to assess surface lowering rates of 40 mm kyr^{-1} and 70–100 mm kyr^{-1} on the steeper slopes of the doline and at the base of the doline fill, respectively. They proposed that the differential deepening rate (30–60 mm kyr^{-1}) can be used to roughly estimate the age of the sinkhole (3.5–7 Myr), assuming constant conditions. Matsushi et al. (2010) using in situ cosmogenic ^{36}Cl data from a solution doline in SW Japan estimated long-term denudation rates ranging from 23 to 94 mm kyr^{-1}, increasing toward the center of the depression. Similar results were obtained measuring subsurface CO_2 concentrations and soil water chemistry and using limestone tablets in different locations of a doline in the same area (Akiyama et al. 2015). The authors attribute this spatial trend to flow convergence and the progressive increase in the runoff contributing area toward the sinkhole center. Based on the relationship between denudation rate and contributing area, they proposed that sinkhole depth is largely influenced by the initial size of the embryonic depressions at the initial stage of development, and that sinkholes of the same age may have variable areas and depths. Ahnert and Williams (1997), through three-dimensional process-response modeling, show that the development of solution dolines and polygonal karst can be explained by higher solution rates at locations of flow convergence, related to topographic irregularities and/or permeability heterogeneities. The solution dolines enlarge reducing the interdoline areas, where sharp divides and peaks develop, controlled by reduced solution rates due to flow divergence. The deepening of dolines stops when their floor reaches the base level, determined by the water table, which introduces an additional control to the spatial distribution of solutional denudation. Then, the flat bottoms of the dolines expand laterally, the dolines coalesce, and the landscape is ultimately transformed into a corrosion plain with isolated conical hills as remnants of former interdoline zones (fenglin karst). When the rejuvenation effect is incorporated in the models by introducing a drop in the base level, new nested dolines develop in the former corrosion plain, which then expand to form a new corrosion plain with multi-tiered residual hills, including towers and cones with bimodal heights related to the different erosion cycles (Figure 6.55).

Spatial variations in solutional denudation can be more pronounced in basins with mixed autogenic and allogenic drainage. The highly undersaturated runoff water that flows on non-karst rocks tends to cause rapid solutional denudation when it first interacts with karst rocks (Figure 3.48). The outcrops of karst formations in contact with non-karst rocks and located on the downflow side are areas of enhanced dissolution, designated as contact karst. These belts typically show sharp morphological and hydrological variations, including a high density of enclosed depressions (sinkholes, poljes) and a change from surface to internal drainage via focused percolation at swallow holes and/or diffuse infiltration. Spatial variations of solutional denudation were assessed in a comprehensive study by Williams and Dowling (1979) in a mixed autogenic–allogenic basin in New Zealand, with around half of the catchment area underlain by calcite marble. Here, Ca concentrations in the different components of the autogenic and allogenic subsystems show that (Figure 4.17): (i) in both systems, great part of the solutional work is accomplished in the upper part of the outcropping carbonate rocks; (ii) allogenic dissolution is mainly achieved by focused flow in conduits with an important speleogenetic role; (iii) autogenic solution is a widespread process due to diffuse percolation and is responsible for most of the surface lowering; and (iv) concentrations of dissolved carbonate in the autogenic system are greater in percolation than in the cave stream waters, but the latter, with a higher discharge, transport higher amounts of solute.

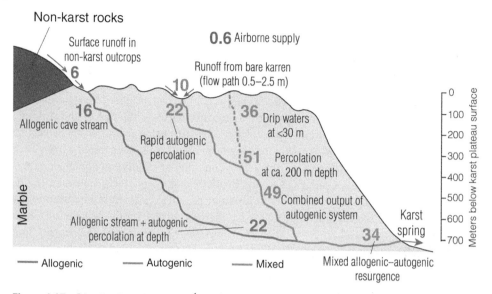

Figure 4.17 Dissolved calcium (mg L^{-1}) in the different components of the autogenic and allogenic subsystems in the Pikikiruna karst, New Zealand. *Source:* Adapted from Williams and Dowling (1979).

4.6.2 Long-term Rates of Base-Level Lowering and Downcutting

The base level of cave systems, which can be dictated by the river into which the cave streams discharge or the sea level in the case of coastal caves, controls the position of the water table. When the base level and the water table remain at a relatively stable position over a sufficiently long period of time, subhorizontal cave passages develop (Figure 4.18) (Palmer 1987; Columbu et al. 2015). These passages tend to form at or near the elevation of the valley floor, functioning as graded underground tributaries of the surface drainages. They commonly crosscut different strata

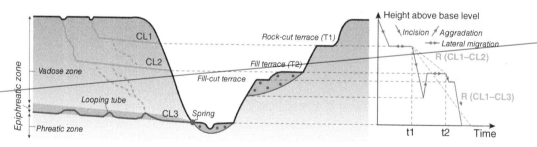

Figure 4.18 Hypothetical evolution of a multilevel cave associated with a fluvial system affected by a general downcutting trend, interrupted by episodes of base-level stability and aggradation. The sketch and the graph illustrate the following sequence of processes: (i) Interruption of fluvial incision and development of a rock-strath (or rock-cut) terrace (T1) and the first cave tier (CL1). (ii) Renewed incision followed by rapid valley aggradation with deposition of fill terrace T2. (iii) New prolonged phase of base level stability and development of the second cave storey (CL2). (iv) Fluvial downcutting up to the position of the active channel with a short interruption of base level stability recorded by a minor fill-strath terrace. (v) Development of the currently active phreatic passages within the epiphreatic zone related to present base level. The orange dashed line in the graph provides the long-term downcutting rate between the last phases of development of cave levels CL1 and CL2. The green dashed line depicts the long-term incision rate between CL1 and CL3 (or the present time). Note that the development of a cave storey can represent a long timespan and that not all fluvial terraces necessarily have correlative cave tiers and vice versa.

and karst formations. The largest passages are typically related to prolonged phases of base level stability. The subhorizontal pattern of these water-table-controlled passages reflects the very low hydraulic gradient in water-filled conduits, except during major floods, so that the phreatic-vadose transition zone lies at essentially the same elevation as the local spring (Palmer 1987, 2007). The water table can experience short-term vertical variations within the so-called epiphreatic zone, especially in aquifers characterized by irregular recharge and flashy response. Under these conditions phreatic passages tend to develop looping tubes (Gabrovšek et al. 2014) (Figure 4.18). Base level lowering causes a synchronic drop in the water table, and cave streams tend to adjust to the new base level by flowing at greater depths. This adjustment within the cave system can be achieved via vadose incision by chemical and mechanical erosion and through flow diversion toward deeper flow routes. New subhorizontal phreatic passages develop when the position of the base level experiences a new phase of relative stability. This interrelated succession of processes results in the development of multilevel cave systems (Harmand et al. 2017 and references therein) (Figure 4.18). Other frequently used terms for passages confined to a narrow elevation range include cave tiers or cave storeys, although these do not necessarily relate to base levels (Palmer 1987, 2007) (see Section 7.2.7).

Base level lowering associated with fluvial downcutting may be related to various driving mechanisms that can operate in combination: (i) regional tectonic uplift; (ii) local uplift controlled by specific tectonic structures (upthrown blocks associated with normal or reverse faults, growing anticlines, active diapirs), which can lead to complex variations in the development of cave systems controlled by the spatial variability of vertical movements (uplift and subsidence); (iii) regional uplift caused by isostatic adjustment of the crust to glacial or erosional unloading; (iv) eustatic sea-level changes mainly controlled by climate (glacial and interglacial periods); and (v) various types of adjustments in the fluvial systems related to variations in sediment supply and stream power, knick point migration, or rapid entrenchment induced by fluvial piracy leading to the abrupt disruption of the longitudinal profile of the captured river. The opposite vertical evolution occurs in the cave system when the base level rises together with the vadose-phreatic transition zone. Upward displacement of the base level is mainly related to sea-level rise and valley aggradation (alluvium, tufa deposits). The associated effects include: (i) flooding of cave passages deep in the phreatic zone that may discharge via Vauclusian springs into the valley alluvium, (ii) development of rising diversion conduits, and (iii) formation of new passages at higher levels. This is the *per ascensum* speleogenetic model, whereby the relative chronology of cave levels is reversed, with younger passages at higher elevation (Audra and Palmer 2011). A dramatic example is illustrated by cave systems developed in some regions of the Mediterranean coast (e.g., Audra et al. 2004; Mocochain et al. 2009). In the late Miocene, the connection between the Mediterranean Sea and the Atlantic was closed due to relative motions between the African plate and the Iberian microplate. This major paleogeographic change led to the Messinian Salinity Crisis between 6 and 5.3 Ma (late Miocene), in which the Mediterranean basin experienced partial dessication and the water level fell as much as 5 km. Some of the most important gypsum karst systems are developed in the Messinian evaporites deposited in this period (Sorbas in SE Spain, Northern Apennines, and Sicily in Italy). During this falling base-level phase, caves formed at progressively greater depths. When the Mediterranean re-established its connection with the Atlantic Ocean through the Strait of Gibraltar, the Zanclean flood at c. 5.3 Ma filled the basin. Many Miocene caves became flooded passages deep below the base level, locally discharging via Vauclusian springs, and new cave levels formed at higher elevations during the Pliocene and the Quaternary.

Fluvial downcutting tends to keep pace with tectonic uplift as well as with falls in sea level in areas close to the coast. Similarly, cave systems tend to adjust to base level lowering through the development of cave passages at progressively deeper levels. Dynamic equilibrium conditions are reached when uplift, fluvial entrenchment, and cave deepening operate at equivalent rates. In these situations, there is correspondence between the relative position of the valley floor and active phreatic passages, unless the presence of non-karst rocks prevents their development (Figure 4.18). During periods of base level stability, the valley floor tends to widen by lateral migration, forming strath surfaces on bedrock or alluvium (rock-cut or fill-cut terraces). Fluvial incision and aggradation largely depend on the balance between stream power and the power required to transport the sediment supplied to the stream (critical-power threshold). When the stream power exceeds that threshold, stream erosion prevails, whereas aggradation dominates when sediment supply overwhelms the stream power (Chorley et al. 1984; Bull 1991). Uplift and sea-level drop lead to an increase in stream gradient and stream power. Cave storeys and fluvial terraces are paleobase-level markers that can be used to reconstruct the vertical evolution of epigenic cave systems and the incision history of related rivers. When data on the numerical age of these markers is available, they can also be used to estimate long-term rates of fluvial incision as a proxy for uplift rate. Two key elements are necessary for these calculations: (i) relative elevation of cave levels and terraces above the active channel; (ii) the age of the geomorphic markers (terraces and cave storeys). An advantage of cave tiers is their high preservation potential, especially in deep narrow valleys (e.g., limestone canyons) where remnants of terraces are rapidly eroded. In contrast, confident dating of the formative phase of phreatic passages is a challenging task given their erosional nature, and commonly only minimum ages can be obtained. For instance, stalagmites start to accumulate in a cave once it becomes a fossil vadose passage and the floor is not submerged or affected by erosion. Consequently, the relative elevation of the cave storey above the valley floor divided by the basal age of the stalagmite (minimum age of the phreatic stage) provides a maximum long-term fluvial downcutting rate (Ford and Williams 2007). In some cases, detrital and chemical cave deposits can provide information on the timing of the last phase of phreatic activity in passages. Different dating methods are applied with variable age ranges: radiocarbon (ca. 50 kyr), U-series (ca. 450 kyr), Optically Stimulated Luminescence (OSL, ca. 300 kyr), Electron Spin Resonance (ESR; ca. 2 Myr), TCNs (ca. 5 Myr), and magnetostratigraphy (virtually unlimited but does not provide a direct numerical age). Cosmogenic-nuclide burial dating has considerably expanded the opportunities for dating cave deposits. This method is applied on quartz-bearing material previously exposed to cosmic rays at the surface and transported into caves, where it becomes shielded and the production of cosmogenic nuclides stops (Granger and Muzikar 2001).

The apparently simple approach of calculating long-term base-level lowering rates and uplift rates from the relative height of the geomorphic markers (cave storeys and terraces) and their numerical age has a number of complications that should be taken into consideration: (i) The development of cave storeys related to the position of the valley floors can be hindered or prevented by the presence of insoluble rocks and tectonic structures. (ii) Constraining the elevation of former base levels from fossil phreatic passages may have a considerable uncertainty (large elevation range related to a wide epiphreatic zone, intense modification during the vadose phase in old caves). (iii) Passages of the same cave tier lie at progressively lower elevation downvalley and sharp elevation variations may occur in relation to fluvial systems with knick points (non-graded longitudinal profiles), which are relatively frequent in streams carved in heterogeneous bedrock and with tufa deposition. (iv) The base level that controls the development of a cave system may correspond to a stream located in a different valley lying at the same or lower elevation. In these settings, phreatic conduits well below the valley floor can develop (Harmand et al. 2017). (v) The deepening of a valley recorded by a succession of cave storeys may not be related to tectonic uplift, but rather to local effects such as rapid fluvial incision induced by river piracy (e.g., Stokes et al. 2002; Harvey et al. 2014). (vi) Fluvial downcutting and the development of progressively lower cave storeys related to tectonic uplift or sea-level fall create a succession of geomorphic adjustments in interrelated systems that may have a considerable delay, especially in low sensitivity environments. For instance, downcutting rates in fluvial systems may change significantly depending on the variable resistance of the lithologies into which the river incises (e.g., Calvet et al. 2015). Entrenchment and diversion of passages in a cave system tend to slowly progress upstream away from the valley, like an underground headward retreating incision wave or knick point. The response time in gypsum and halite karst systems tends to be much shorter than carbonate karst due to their much higher solubility and erodibility (Frumkin 1996; Columbu et al. 2015, 2017; De Waele et al. 2020). (vii) Base-level lowering can show significant spatial variability within a fluvial system related to factors such as lithologically controlled knick points, glacial diversion of rivers (Westaway 2020), or active tectonic structures (e.g., McPhillips et al. 2016). (viii) Water table drop in an aquifer is not necessarily related to base level lowering (e.g., fluvial downcutting), and in some cases, it can be ascribed to a progressive increase in hydraulic conductivity over time, as Ford et al. (1993) demonstrated in the case of the Wind Cave aquifer in the Black Hills of South Dakota. (ix) Dated sediments frequently do not indicate the latest phase of activity of the passages, but younger episodes of detrital or chemical accumulation under vadose conditions, which may involve reworked material from higher passages (e.g., Häuselmann et al. 2020; Rixhon et al. 2020). Thorough sedimentological analyses, a good understanding of the morphological and stratigraphic relationships, and multi-dating approaches are essential in most cases. (x) Water depth during major floods in narrow canyons can reach several tens of meters (e.g., Granger et al. 2001; Benito and O'Connor 2013). These high-stage flood events can cause backflooding in old perched passages and accumulate younger fine-grained slackwater deposits.

Table 4.3 includes a compilation of long-term base-level lowering rates calculated in karst regions worldwide by dating, using various geochronological methods, cave storeys, and in some cases correlative fluvial terraces. The table distinguishes two broad types of geotectonic settings; stable continental regions (SCR) and non-stable continental regions (NSCR), defined using the world map by Wheeler (2016). The SCR category, broadly used in seismic hazard and neotectonic studies, refers to continental regions lacking major rifting in the last 23 million years (base of the Neogene) and without major orogeny, foreland deformation, or anorogenic intrusions in the last 100 million years (Wheeler 2016). The two categories provide a rough differentiation between regions with significant tectonic activity and long-sustained stability. Although several works indicate significant temporal variability, the data from carbonate karst areas show values above 100–150 mm kyr^{-1} in NSCR. Some of the highest rates are reported in active orogens associated with plate margins such as the Tibetan Plateau or the Southern Alps of New Zealand (Williams 1982; Wang et al. 2004; Liu et al. 2013). There are some exceptions, such as the Ariege Valley in the central French Pyrenees (13–50 mm kyr^{-1}; Sartégou et al. 2020) or the Têt Canyon in the Eastern Pyrenees (52–140 mm kyr^{-1}; Calvet et al., 2015). However, the Pyrenees mountain range is widely regarded as an inactive convergent boundary. The eastern sector of the Pyrenees is affected by post-orogenic extensional tectonics and GPS data suggest very slow N-S transverse extension in the central sector (Rigo et al. 2015; Nguyen et al. 2016) attributable to the southward displacement of the Iberian microplate with respect to stable Europe (Stich et al. 2006). Rates from carbonate karst regions located in SCR are mostly below 100–150 mm kyr^{-1}. The lowest value of 4 mm kyr^{-1} (4 m Myr^{-1}) has been reported in the Buchan karst located at the extremely stable southeastern sector of the Australian craton, with 2–3 m of fluvial incision over the last 760 ka (Webb et al. 1992). Regarding data from evaporite karst systems, Columbu et al. (2015) provide minimum ages from speleothems in multilevel gypsum caves of the Northern Apennines indicating a maximum

Table 4.3 Long-term rates of base-level lowering estimated from dated cave storeys and/or correlative terraces.

Location	Dated elements and geochronological method	Geological setting	Rate (mm kyr^{-1})	References
Mount Sedom diapir, Israel	Detrital cave deposits, C^{14}	AD	6000–7000	Frumkin (1996)
Gypsum caves, Northern Apennines	U-series	NSCR	920 (M)	Columbu et al. (2015)
Liuchong River, SE Tibetan Plateau	Detrital cave deposits, TCN	NSCR	480	Liu et al. (2013)
Qinling mountains, China	Speleothems, U-series	NSCR	190–510	Wang et al. (2004)
Mulu karst, Malaysia	U-series, ESR, magnetostratigraphy	NSCR	190	Farrant et al. (1995)
Southern Rocky Mountains, Canada	Speleothems, U-series, and magnetostratigraphy	NSCR	130–2070 (M) 40–70 (m)	Ford et al. (1981)
Domica-Baradla cave system, Western Carpathians	Detrital cave deposits, TCN	NSCR	311–561	Bella et al. (2019)
Southern Alps, New Zealand	Speleothems, U-series, C14	NSCR	280 (M)	Williams (1982)
Sierra Nevada, United States	Detrital cave deposits, TCN	NSCR	200, 30	Stock et al. (2004)
Yangtze River, Tibetan Plateau	Detrital cave deposits, TCN	NSCR	>110	McPhillips et al. (2016)
Apuan Alps, Italy	Speleothems, U-series	NSCR	80–1040	Piccini et al. (2003)
Toirano caves, Ligurian Alps, Italy	Speleothems and detrical cave deposits	NSCR	100	Columbu et al. (2021)
Central Styrian Karst, Eastern Alps	Detrital cave deposits, TCN	NSCR	100	Wagner et al. (2010)
Ardennes Massif, Belgium	Detrital cave deposits, TCN	SCR	29–150	Rixhon et al. (2020)
Seine Valley, NE France	Speleothems and detrital cave deposits, U-series, magnetostratigraphy	SCR	300 (M) 50–80	Nehme et al. (2020)
Atapuerca Range and Arlanzón River, N Spain	Terrace deposits correlative to cave levels	SCR	55–77	Moreno et al. (2012), Ortega et al. (2013)
Appalachian Plateaus, United States	Detrital cave deposits, magnetostratigraphy	SCR	60	Sasowsky et al. (1995)
Têt Canyon, Eastern French Pyrenees	Detrital cave deposits, TCN	NSCR	52–140	Calvet et al. (2015)
Peak District, Central England	Speleothems, U-series	SCR	55–175	Westaway (2020)
Mont Granier, French Alps	Detrital cave deposits, TCN	NSCR	7, 14, 90	Hobléa et al. (2001)

Table 4.3 (Continued)

Location	Dated elements and geochronological method	Geological setting	Rate (mm kyr^{-1})	References
Ardèche Valley, France	Magnetostratigraphy	SCR	50	Audra et al. (2001)
Craven District, N England	Speleothems, U-series	SCR	20–50 (M)	Gascoyne et al. (1983)
Ariège Valley, French Pyrenees	Detrital cave deposits, TCN	NSCR	13–50	Sartégou et al. (2020)
Gruta do Padre, E Brazil	Detrital cave deposits, magnetostratigraphy	SCR	25–34	Auler et al. (2002)
Mammoth Cave, Kentucky	Detrital cave deposits, TCN	SCR	30	Granger et al. (2001)
Buchan karst, SE Australia	Speleothems, U-series, magnetostratigraphy	SCR	4 (M)	Webb et al. (1992)

Data from evaporite karst are highlighted in gray, and the rest correspond to carbonate karst regions. Rates separated by commas correspond to different periods.
ESR: Electron Spin Resonance, C^{14}: radiocarbon dating, OSL: Optically Stimulated Luminescence, SCR: Stable continental region, AO: Active orogen, M: Maximum rate, m: Minimum rate.
Source: Downcutting rate from McPhillips et al. (2016) corresponds to an early Miocene phase of rapid incision. Rates from Stock et al. (2004) correspond to 2.7–1.5 Myr and 1.5 Myr-present, respectively.

downcutting rate of 920 mm kyr^{-1}. This area has experienced an average uplift rate of around 1 mm yr^{-1} since ~150 kyr B.P., with a three- to five-fold lower rate during the middle Pleistocene (Cyr and Granger 2008; Picotti and Pazzaglia 2008). In Mount Sedom salt diapir, Frumkin (1996) calculated a long-term base-level lowering rate of 6000–7000 mm kyr^{-1} (6–7 mm yr^{-1}), which compares well with the current uplift rate of the diapir measured by DiNSAR (5–8 mm yr^{-1}; Pe'eri et al. 2004; Weinberger et al. 2006).

References

Ahnert, F. and Williams, P.W. (1997). Karst landform development in a three-dimensional theoretical model. *Zeitschrift für Geomorphologie* 108: 63–80.

Akiyama, S., Hattanji, T., Matsushi, Y. et al. (2015). Dissolution rates of subsoil limestone in a doline on the Akiyoshi-dai Plateau, Japan. An approach from a weathering experiment, hydrological observations, and electrical resistivity tomography. *Geomorphology* 247: 2–9.

Allred, K. (2004). Some carbonate erosion rates of Southeast Alaska. *Journal of Cave and Karst Studies* 66: 89–97.

Anderson, D.E., Goudie, A.S., and Parker, A.G. (2007). *Global Environments Through the Quaternary.* Oxford: Oxford University Press.

André, M.F. (1996). Vitesses de dissolution aréolaire postglaciaire dans les karsts polaires et haut-alpins. De l'arctique scandinave aux Alps de Nouvelle Guinée. *Revue d'Analyse Spatiale Quantitative et Appliquée* 38–39: 99–107.

André, M.F. (2008). Quantifying Holocene surface lowering of limestone pavements in previously glaciated environments. *Geographica Polonica* 81: 9–17.

Atkinson, T.C. and Smith, D.I. (1976). The erosion of limestone. In: *The Science of Speleology* (ed. T.D. Ford and C.H.D. Cullingford), 151–177. London: Academic Press.

Aubert, D. (1967). Estimation de la dissolution superficielle dans le Jura. *Bulletin de la Société Vaudoise des Sciences Naturelles* 69: 365–376.

Aubert, D. (1969). Phenomènes et formes du karst jurassien. *Eclogae Geologicae Helvetiae* 62: 325–399.

Aubrecht, R., Lánczos, T., Gregor, M. et al. (2011). Sandstone caves on Venezuelan tepuis: return to pseudokarst? *Geomorphology* 132: 351–365.

Aubrecht, R., Lánczos, T., Schlögl, J. et al. (2017). Small-scale modelling of cementation by descending silica-bearing fluids: explanation of the origin of arenitic caves in South American tepuis. *Geomorphology* 298: 107–117.

Audra, P. and Palmer, A.N. (2011). The pattern of caves: controls of epigenic speleogenesis. *Géomorphologie: Relief, Processus, Environnement* 17 (4): 359–378.

Audra, P., Camus, H., and Rochette, P. (2001). Le karst des plateaux jurassiques de la moyenne vallée de l'Ardèche; datation par paléomagnetisme des phases d'évolution plio-quaternaires (aven de la Combe Rajeau). *Bulletin de la Société Géologique de France* 172: 121–129.

Audra, P., Mocochain, L., Camus, H. et al. (2004). The effect of the Messinian deep stage on karst development around the Mediterranean Sea. Examples from Southern France. *Geodinamica Acta* 17: 389–400.

Auler, A.S., Smart, P.L., Tarling, D.H. et al. (2002). Fluvial incision rates derived from magnetostratigraphy of cave sediments in the cratonic area of eastern Brazil. *Zeitschrift für Geomorphologie* 46: 391–403.

Avni, S., Joseph-Hai, N., Haviv, I. et al. (2019). Patterns and rates of 103–105 yr denudation in carbonate terrains under subhumid to subalpine climatic gradient, Mount Hermon, Israel. *Geological Society of America Bulletin* 131: 899–912.

Bakalowicz, M. (1979). *Contribution de la géochimie des eaux à la connaissance de l'aquifère karstique et de la karstification*. Paris: Université Pierre et Marie Curie.

Bakalowicz, M. (1992). Géochimie des eaux et flux de matières dissoutes: l'approche objective du rôle du climat dans le karstogénèse. In: *Karst et évolutions climatiques* (ed. J.N. Salomon and R. Maire), 61–74. Bordeaux: Presses Universitaires.

Bella, P., Bosák, P., Braucher, R. et al. (2019). Multi-level Domica-Baradla cave system (Slovakia, Hungary): Middle Pliocene-Pleistocene evolution and implications for the denudation chronology of the Western Carpathians. *Geomorphology* 327: 62–79.

Benedetti, L., Finkel, R., King, G. et al. (2003). Motion on the Kaparelli fault (Greece) prior to the 1981 earthquake sequence determined from ^{36}Cl cosmogenic dating. *Terra Nova* 15 (2): 118–124.

Benito, G. and O'Connor, J.E. (2013). Quantitative paleoflood hydrology. In: *Treatise on Geomorphology. Fluvial Geomorphology*, vol. 9 (ed. E. Wohl), 459–474. San Diego: Academic Press.

Benito-Calvo, A., Gutiérrez, F., Martínez-Fernández, A. et al. (2018). 4D monitoring of active sinkholes with a terrestrial laser scanner (TLS): a case study in the evaporite karst of the Ebro valley, NE Spain. *Remote Sensing* 10: 571.

Bennett, P.C. (1991). Quartz dissolution in organic-rich aqueous systems. *Geochimica et Cosmochimica Acta* 55: 1781–1797.

Birginie, J.M. and Rivas, T. (2005). Use of a laser camera scanner to highlight the surface degradation of stone samples subjected to artificial weathering. *Building and Environment* 40: 755–764.

Bögli, A. (1961). Karrentische, ein Beitrag sur Karstmorphologie. *Zeitschrift für Geomorphologie* 5: 185–193.

Bouma, T.J., Nielsen, K.L., Eissenstat, D.M. et al. (1997). Estimating respiration of roots in soil: interactions with soil CO_2, soil temperature and soil water content. *Plant and Soil* 195: 221–232.

Brady, P.V. and Walther, J.V. (1990). Kinetics of quartz dissolution at low temperatures. *Chemical Geology* 82: 253–264.

Brehm, U., Gorbushina, A., and Mottershead, D. (2005). The role of microorganisms and biofilms in the breakdown and dissolution of quartz and glass. *Palaeogeography, Palaeoclimatology, Palaeoecology* 219: 117–129.

Brook, G.A., Folkoff, M.E., and Box, E.O. (1983). A world model of soil carbon dioxide. *Earth Surface Processes and Landforms* 8 (1): 79–88.

Brown, E.T., Stallard, R.F., Raisbeck, G.M. et al. (1992). Determination of the denudation of Mount Roraima, Venezuela using cosmogenic [13]Be and [26]Al. *EOS* 73: 170.

Bruthans, J., Filippi, M., Geršl, M. et al. (2006). Holocene marine terraces on two salt diapirs in the Persian Gulf, Iran: age, depositional history and uplift rates. *Journal of Quaternary Science* 21: 843–857.

Bruthans, J., Asadi, N., Filippi, M. et al. (2008). A study of erosion rates on salt diapir surfaces in the Zagros Mountains, SE Iran. *Environmental Geology* 53 (5): 1079–1089.

Bruthans, J., Filippi, M., Asadi, N. et al. (2009). Surficial deposits on salt diapirs (Zagros Mountains and Persian Gulf Platform, Iran): characterization, evolution, erosion and the influence on landscape morphology. *Geomorphology* 107: 195–209.

Bruthans, J., Filippi, M., Zare, M. et al. (2010). Evolution of salt diapir and karst morphology during the last glacial cycle: effects of sea-level oscillation, diapir and regional uplift, and erosion (Persian Gulf, Iran). *Geomorphology* 121 (3-4): 291–304.

Büdel, B., Weber, B., Kühl, M. et al. (2004). Reshaping of sandstone surfaces by cryptoendolithic cyanobacteria: bioalkalization causes chemical weathering in arid landscapes. *Geobiology* 2: 261–268.

Bull, W.B. (1991). *Geomorphic Responses to Climatic Change*. New York: Oxford University Press.

Burbank, D.W. and Anderson, R.S. (2011). *Tectonic Geomorphology*. Chichester: Wiley.

Calvet, M., Gunnell, Y., Braucher, R. et al. (2015). Cave levels as proxies for measuring post-orogenic uplift: evidence from cosmogenic dating of alluvium-filled caves in the French Pyrenees. *Geomorphology* 246: 617–633.

Capezzuoli, E., Gandin, A., and Pedley, M. (2014). Decoding tufa and travertine (fresh water carbonates) in the sedimentary record: the state of the art. *Sedimentology* 61: 1–21.

Cardona, F. and Viver, J. (2002). *Sota la Sal de Cardona*. Barcelona: Espeleo Club de Gracia.

Cenki-Tok, B., Chabaux, F., Lemarchand, D. et al. (2009). The impact of water-rock interaction and vegetation on calcium isotope fractionation in soil- and stream waters of a small, forested catchment (the Strengbach case). *Geochimica et Cosmochimica Acta* 73: 2215–2228.

Chevallier, P. (1953). Érosion ou corrosion. First International Congress of Speleology, Paris, 1: 35–40.

Chorley, R.J., Schumm, S.A., and Sugden, D.A. (1984). *Geomorphology*. New York: Methuen Inc.

Cockburn, H.A.P. and Summerfield, M.A. (2004). Geomorphological applications of cosmogenic isotope analysis. *Progress in Physical Geography* 28: 1–42.

Columbu, A., De Waele, J., Forti, P. et al. (2015). Gypsum caves as indicators of climate-driven river incision and aggradation in a rapidly uplifting region. *Geology* 43 (6): 539–542.

Columbu, A., Chiarini, V., De Waele, J. et al. (2017). Late Quaternary speleogenesis and landscape evolution in the northern Apennine evaporite areas. *Earth Surface Processes and Landforms* 42 (10): 1447–1459.

Columbu, A., Audra, P., Gázquez, F. et al. (2021). Hypogenic speleogenesis, late stage epigenic overprinting and condensation-corrosion in a complex cave system in relation to landscape evolution (Toirano, Liguria, Italy). *Geomorphology* 376: 107561.

Cooper, A.H. and Gutiérrez, F. (2013). Dealing with gypsum karst problems: hazards, environmental issues, and planning. In: *Treatise on Geomorphology. Karst Geomorphology*, vol. 6 (ed. A. Frumkin), 451–462. Amsterdam: Elsevier.

Corbel, J. (1959). Érosion en terrain calcaire. *Annales de Geographie* 68: 97–120.

Crowther, J. (1983). A comparison of the rock tablet and water hardness methods for determining chemical erosion rates on karst surfaces. *Zeitschrift für Geomorphologie* 27: 55–64.

Crowther, J. (1989). Groundwater chemistry and cation budgets of tropical karst outcrops, Peninsular Malaysia, I. Calcium and magnesium. *Journal of Hydrology* 107: 169–192.

Cucchi, F., Forti, F., and Ulcigrai, F. (1994). Valori di abbassamento per dissoluzione carsiche. *Acta Carsologica* 23: 55–62.

Cucchi, F., Forti, F., and Furlani, S. (2006). Lowering rates of limestone along the western Istrian shoreline and the Gulf of Trieste. *Geografia Fisica e Dinamica Quaternaria* 29: 61–69.

Cyr, A.J. and Granger, D.E. (2008). Dynamic equilibrium among erosion, river incision, and coastal uplift in the northern and central Apennines, Italy. *Geology* 36: 103–106.

De Waele, J., Picotti, V., Cucchi, F. et al. (2009b). Karst phenomena in the Cordillera de la Sal (Atacama, Chile). In: *Geological Constraints on the Onset and Evolution of an Extreme Environment: The Atacama Area*, vol. 2 (ed. P.L. Rossi), 113–127. GeoActa Special Publication.

De Waele, J., Picotti, V., Martina, M.L. et al. (2020). Holocene evolution of halite caves in the Cordillera de la Sal (Central Atacama, Chile) in different climate conditions. *Geomorphology* 370: 107398.

Desir, G., Sirvent, J., Gutiérrez, M. et al. (1995). Sediment yield for gypsiferous degraded areas in the middle Ebro Basin (NE Spain). *Physics and Chemistry of the Earth* 20: 385–393.

Doerr, S.H. and Wray, R.A.L. (2004). Pseudokarst. In: *Encyclopedia of Geomorphology*, vol. 2 (ed. A.S. Goudie), 814–815. London: Routledge.

Doğan, U., Koçyiğit, A., and Yeşilyurt, S. (2019). The relationship between Kestel polje system and the Antalya tufa plateau: their morphotectonic evolution in Isparta Angle, Antalya-Turkey. *Geomorphology* 334: 112–125.

Domènech, G., Corominas, J., Mavrouli, O. et al. (2018). Calculation of the rockwall recession rate of a limestone cliff, affected by rockfalls, using cosmogenic chlorine-36. Case study of the Montsec Range (Eastern Pyrenees, Spain). *Geomorphology* 306: 325–335.

Dorn, R.I. (2004). Case hardening. In: *Encyclopedia of Geomorphology*, vol. 1 (ed. A.S. Goudie), 118–119. London: Routledge.

Dreybrodt, W. and Eisenlohr, L. (2000). Limestone dissolution rates in karst environments. In: *Speleogenesis. Evolution of Karst Aquifers* (ed. A.B. Klimchouk, D.C. Ford, A.N. Palmer, et al.), 136–148. Huntsville, Alabama: National Speleological Society.

Drysdale, R.N. and Gillieson, D. (1997). Micro-erosion meter measurements of travertine deposition rates: a case study from Louie Creek, Northwest Queensland, Australia. *Earth Surface Processes and Landforms* 22: 1037–1051.

Ewing, A. (1885). Attempt to determine the amount and rate of chemical erosion taking place in the limestone valley of Center County, Pennsylvania. *American Journal of Science* 3: 29–31.

Farrant, A.R., Smart, P.L., Whitaker, F.F. et al. (1995). Long-term Quaternary uplift rates inferred from limestone caves in Sarawak, Malaysia. *Geology* 23: 357–360.

Ferrarese, F., Macaluso, T., Madonia, G. et al. (2003). Solution and recrystallisation processes and associated landforms in gypsum outcrops of Sicily. *Geomorphology* 49: 25–43.

Ford, D.C. (1987). Effects of glaciations and permafrost upon the development of karst in Canada. *Earth Surface Processes and Landforms* 12: 507–521.

Ford, D.C. and Williams, P.W. (2007). *Karst Hydrogeology and Geomorphology*. Chichester, UK: Wiley.

Ford, D.C., Schwarcz, H.P., Drake, J.J. et al. (1981). Estimates of the age of the existing relief within the southern Rocky Mountains of Canada. *Arctic and Alpine Research* 13: 1–10.

Ford, D.C., Lundberg, J., Palmer, A.N. et al. (1993). Uranium-series dating of the draining of an aquifer: the example of Wind Cave, Black Hills, South Dakota. *Geological Society of America Bulletin* 105: 241–250.

Forti, F. (1984). Messungen des karsrabtrages in der Region Friul-Julisch-Venetien (Italian). *Die Höhle* 35: 125–139.

Frumkin, A. (1994a). Hydrology and denudation rates of halite karst. *Journal of Hydrology* 162: 171–189.

Frumkin, A. (1996). Uplift rate relative to base-levels of a salt diapir (Dead Sea Basin, Israel) as indicated by cave levels. In: *Salt Tectonics*, vol. 100 (ed. G.I. Alsop, D.J. Blundell and I. Davison), 41–47. Geological Society, London Special Publication.

Frumkin, A. (2013). Salt karst. In: *Treatise on Geomorphology. Karst Geomorphology*, vol. 6 (ed. A. Frumkin and J. Shroder), 407–424. Amsterdam: Elsevier.

Furlani, S., Cucchi, F., Forti, F. et al. (2009). Comparison between coastal and inland Karst limestone lowering rates in the northeastern Adriatic Region (Italy and Croatia). *Geomorphology* 104 (1-2): 73–81.

Gabrovšek, F. (2009). On concepts and methods for the estimation of dissolutional denudation rates in karst areas. *Geomorphology* 106: 9–14.

Gabrovšek, F., Häuselmann, P., and Audra, P. (2014). 'Looping caves' versus 'water table caves': the role of base-level changes and recharge variations in cave development. *Geomorphology* 204: 683–691.

Gaillardet, J. (2004). Denudation. In: *Encyclopedia of Geomorphology*, vol. 1 (ed. A.S. Goudie), 240–244. New York: Routledge.

Gams, I. (1972). Effect of runoff on corrosion intensity in the northwest Dinaric karst. *Transactions of the Cave Research Group of Great Britain* 14: 78–83.

Gams, I. (1981a). Comparative research of limestone solution by means of standard tablets. In: *Proceedings of the Eight International Congress of Speleology*, vol. I (ed. B.F. Beck), 273–275. Bowling Green: National Speleological Society.

Gams, I. (1986). International comparative measurements of surface solution by means of standard limestone tablets. *Razreda Sazu* 26: 361–386.

Gams, I. (2004). *Kras v Sloveniji v prostoru in Času*. Ljubljana: ZRC Publishing.

Gao, Q., Tao, Z., Huang, X. et al. (2009). Chemical weathering and CO_2 consumption in the Xijiang River basin, South China. *Geomorphology* 106: 324–332.

Gascoyne, M., Ford, D.C., and Schwarcz, H.P. (1983). Rates of cave and landform development in the Yorkshire Dales from speleothem age data. *Earth Surface Processes and Landforms* 8: 557–568.

Gilchrist, A.R., Summerfield, M.A., and Cockburn, H.A.P. (1994). Landscape dissection, isostatic uplift, and the morphologic development of orogens. *Geology* 22: 963–966.

Glover, C. and Robertson, A.H. (2003). Origin of tufa (cool-water carbonate) and related terraces in the Antalya area, SW Turkey. *Geological Journal* 38: 329–358.

Godard, V., Ollivier, V., Bellier, O. et al. (2016). Weathering-limited hillslope evolution in carbonate landscapes. *Earth and Planetary Science Letters* 446: 10–20.

Goldie, H.S. (2005). Erratic judgements: re-evaluating solutional erosion rates of limestones using erratic-pedestal sites, including Norber, Yorkshire. *Area* 37: 433–442.

Gómez-Pujol, L., Fornós, J.J., and Swantesson, J.O. (2006). Rock surface millimetre-scale roughness and weathering of supratidal Mallorcan carbonate coasts (Balearic Islands). *Earth Surface Processes and Landforms* 31: 1792–1801.

Gosse, J.C. (2007). Cosmogenic nuclide dating. In: *Encyclopedia of Quaternary Science*, vol. 1 (ed. S.A. Elias), 409–411. Amsterdam: Elsevier.

Goudie, A.S. and Middleton, N.J. (2006). *Desert Dust in the Global System*. Berlin: Springer.

Goudie, A.S., Migoń, P., Allison, R. et al. (2002). Sandstone geomorphology of the Al-Quwayra area of south Jordan. *Zeitschrift für Geomorphologie* 46: 365–390.

Granger, D.E. and Muzikar, P.F. (2001). Dating sediment burial with in situ-produced cosmogenic nuclides: theory, techniques, and limitations. *Earth and Planetary Science Letters* 188 (1-2): 269–281.

Granger, D.E., Fabel, D., and Palmer, A.N. (2001). Pliocene-Pleistocene incision of the Green River, Kentucky, determined from radioactive decay of cosmogenic ^{26}Al and ^{10}Be in Mammoth Cave sediments. *Geological Society of America Bulletin* 113 (7): 825–836.

Groves, C.G. and Meiman, J. (2005). Weathering, geomorphic work, and karst landscape evolution in the Cave City groundwater basin, Mammoth Cave, Kentucky. *Geomorphology* 67: 115–126.

Guerrero, J. (2017). Dissolution collapse of a growing diapir from radial, concentric, and salt-withdrawal faults overprinting in the Salinas de Oro salt diapir, northern Spain. *Quaternary Research* 87: 331–346.

Guerrero, J., Desir, G., Roqué, C. et al. (2019). The episodic rise, net growing rate and kinematics of radial faults of the Salinas de Oro diapir using paleoseismological techniques (NE Spain). Salt upwelling versus karstic subsidence. *Geomorphology* 342: 210–222.

Gunn, J. (1981a). Limestone solution rates and processes in the Waitomo district, New-Zealand. *Earth Surface Processes and Landforms* 6: 427–445.

Gunn, J. (1982). Magnitude and frequency properties of dissolved solids transport. *Zeitschrift für Geomorphologie* 26: 505–511.

Gunn, J. (2013). Denudation and erosion rates in karst. In: *Treatise on Geomorphology. Karst Geomorphology*, vol. 6 (ed. A. Frumkin), 72–81. Amsterdam: Elsevier.

Gutiérrez, F. and Lizaga, I. (2016). Sinkholes, collapse structures and large landslides in an active salt dome submerged by a reservoir: the unique case of the Ambal ridge in the Karun River, Zagros Mountains. *Iran. Geomorphology* 254: 88–103.

Gutiérrez, F., Carbonel, D., Kirkham, R.M. et al. (2014a). Can flexural-slip faults related to evaporite dissolution generate hazardous earthquakes? The case of the Grand Hogback monocline of west-central Colorado. *Geological Society of America Bulletin* 126: 1481–1494.

Hanna, F.K. (1966). A technique for measuring the rate of erosion of cave passages. *Proceedings of the University of Bristol Spelaeological Society* 11: 141–156.

Hansen, J., Sato, M., Kharecha, P. et al. (2008). Target atmospheric CO_2: where should humanity aim? *The Open Atmospheric Science Journal* 2: 217–231.

Harmand, D., Adamson, K., Rixhon, G. et al. (2017). Relationships between fluvial evolution and karstification related to climatic, tectonic and eustatic forcing in temperate regions. *Quaternary Science Reviews* 166: 38–56.

Härtel, H., Cílek, V., Herben, T. et al. (ed.) (2007). *Sandstone Landscapes*. Praha: Academia.

Harvey, A.M., Whitfield, E., Stokes, M. et al. (2014). The late Neogene to Quaternary drainage evolution of the uplifted sedimentary basins of Almeria. In: *Landscapes and Landforms of Spain* (ed. F. Gutiérrez and M. Gutiérrez), 37–61. Heidelberg: Springer.

Hattanji, T., Ueda, M., Song, W. et al. (2014). Field and laboratory experiments on high dissolution rates of limestone in stream flow. *Geomorphology* 204: 485–492.

Häuselmann, P., Plan, L., Pointner, P. et al. (2020). Cosmogenic nuclide dating of cave sediments in the Eastern Alps and implications for erosion rates. *International Journal of Speleology* 49 (2): 107–118.

High, C.J. and Hanna, F.K. (1970). A method for the direct measurement of erosion on rock surfaces. *British Geomorphological Research Group Technical Bulletin* 5: 1–25.

Hobléa, F., Jaillet, S., Maire, R. et al. (2001). Erosion et ruissellement sur karst nu: les îles subpolaires de la Patagonie chilienne (Magallanes, Chili). *Karstologia* 38 (1): 13–18.

Huddart, D. (2002). Norber erratics. In: *Quaternary of Northern England*, vol. 25 (ed. D. Huddart and N.F. Glasser), 200–203. Geological Convervation Review Series.

Inkpen, R., Collier, P., and Fontana, D. (2000). Close-range photogrammetric analysis of rock surfaces. *Zeitschrift für Geomorphologie Supplement Band* 120: 67–81.

Ivy-Ochs, S. and Kober, F. (2007). Exposure geochronology. In: *Encyclopedia of Quaternary Science*, vol. 1 (ed. S.A. Elias), 436–445. Amsterdam: Elsevier.

Jalali, L., Zarei, M., and Gutiérrez, F. (2019). Salinization of reservoirs in regions with exposed evaporites. The unique case of upper Gotvand Dam, Iran. *Water Research* 157: 587–599.

James, A.N. (1992). *Soluble Material in Civil Engineering*. New York: Ellis Horwood.

James, A.N., Cooper, A.H., and Holliday, D.W. (1981). Solution of the gypsum cliff (Permian, Middle Marl) by the River Ure at Ripon Parks, North Yorkshire. *Proceedings of the Yorkshire Geological Society* 43: 433–450.

James, M.R., Robson, S., and Smith, M.W. (2017). 3-D uncertainty-based topographic change detection with structure-from-motion photogrammetry: precision maps for ground control and directly georeferenced surveys. *Earth Surface Processes and Landforms* 42: 1769–1788.

Jeannin, P.Y., Hessenauer, M., Malard, A. et al. (2016). Impact of global change on karst groundwater mineralization in the Jura Mountains. *Science of the Total Environment* 541: 1208–1221.

Jennings, J.N. (1972a). The Blue waterholes, Cooleman Plain, NSW, the problem of karst denudation rate determination. *Transactions of the Cave Research Group of Great Britain* 14: 109–117.

Jennings, J.N. (1972b). Observations at the Blue waterholes, March 1965 to April 1969 limestone solution on Cooleman Plain, NSW. *Helictite* 10: 1–46.

Jennings, J.N. (1983). Sandstone pseudokarst or karst? In: *Aspects of Australian Sandstone Landscapes*, vol. 1 (ed. R.W. Young and G.C. Nanson), 21–30. Australian and New Zealand Geomorphology Group Special Publication.

Kaufmann, G. and Braun, J. (2001). Modelling karst denudation on a synthetic landscape. *Terra Nova* 13: 313–320.

Kim, J.W., Lu, Z., and Degrandpre, K. (2016). Ongoing deformation of sinkholes in Wink, Texas, observed by time-series Sentinel-1A SAR interferometry (preliminary results). *Remote Sensing* 8: 313.

Klimchouk, A.B. (1996a). The dissolution and conversion of gypsum and anhydrite. *International Journal of Speleology* 25: 21–36.

Klimchouk, A.B. (2000b). The formation of epikarst and its role in vadose speleogenesis. In: *Speleogenesis. Evolution of Karst Aquifers* (ed. A.B. Klimchouk, D.C. Ford, A.N. Palmer, et al.), 91–99. Huntsville: National Speleological Society.

Klimchouk, A.B. (2004b). Towards defining, delimiting and classifying epikarst: its origin, processes and variants of geomorphic evolution. *Speleogenesis and Evolution of Karst Aquifers* 2: 1–13.

Klimchouk, A.B. and Aksem, S.D. (2005). Hydrochemistry and solution rates in gypsum karst: case study from the Western Ukraine. *Environmental Geology* 48: 307–319.

Klimchouk, A.B., Cucchi, F., Calaforra, J.M. et al. (1996a). Dissolution of gypsum from field observations. *International Journal of Speleology* 25: 37–48.

Klimchouk, A.B., Sauro, U. and Lazzarotto, M. (1996c). "Hidden" shafts at the base of the epikarstic zone: a case study from the Sette Communi plateau, Venetian Pre-Alps, Italy. *Cave and Karst Science* 23(3): 101–107.

Krklec, K., Domínguez-Villar, D., Carrasco, R.M. et al. (2016). Current denudation rates in dolostone karst from central Spain: implications for the formation of unroofed caves. *Geomorphology* 264: 1–11.

Kunaver, J. (1979). Some experiences in measuring the surface karst denudation in high Alpine environment. In: Unione Internazionale de Spéléologie. In: *Actes du Symposium International sur l'Érosion Karstique*, 75–85. Aix-en-Provence-Marseille-Nimes: UIS Commision de érosion du karst.

Langbein, W.B. and Schumm, S.A. (1958). Yield of sediment in relation to mean annual precipitation. *Transactions American Geophysical Union* 39: 1076–1084.

Langer, H. and Offermann, H. (1982). On the solubility of sodium chloride in water. *Journal of Crystal Growth* 60: 389–392.

Lauritzen, S.E. (1990). Autogenic and allogenic denudation in carbonate karst by the multiple basin method: an example from Svartisen, North Norway. *Earth Surface Processes and Landforms* 15: 157–167.

Lauritzen, S.E. (2005). A simple growth model for allogenic pedestals in glaciated karst. In: *14th International Congress of Speleology*, paper O-52. Athen-Kalamos: Unione Internationale de Spéléologie.

Liu, Y., Wang, S., Xu, S. et al. (2013). New evidence for the incision history of the Liuchong River, Southwest China, from cosmogenic $^{26}Al/^{10}Be$ burial ages in cave sediments. *Journal of Asian Earth Sciences* 73: 274–283.

Lucha, P., Gutiérrez, F., and Guerrero, J. (2008b). Environmental problems derived from evaporite dissolution in the Barbastro-Balaguer Anticline (Ebro Basin, NE Spain). *Environmental Geology* 53: 1045–1055.

Luo, M., Zhou, H., Liang, Y. et al. (2018). Horizontal and vertical zoning of carbonate dissolution in China. *Geomorphology* 322: 66–75.

Lupia-Palmieri, E. (2004). Erosion. In: *Encyclopedia of Geomorphology*, vol. 1 (ed. A.S. Goudie), 331–336. New York: Routledge.

Maire, R. (1999). Les glaciers de marbre de Patagonie, Chili. Un karst subpolaire océanique de la zone australe. *Karstologia* 33: 25–40.

Martini, J.E.J. (1979). Karst in Black Reef quartzite near Kaapsehoop, Eastern Transvaal. *Annals of the South African Geological Survey* 13: 115–128.

Martini, J.E.J. (2000). Dissolution of quartz and silicate minerals. In: *Speleogenesis. Evolution of Karst Aquifers* (ed. A.B. Klimchouk, D.C. Ford, A.N. Palmer, et al.), 452–457. Huntsville: National Speleological Society.

Martini, J.E.J. (2004). Silicate karst. In: *Encyclopedia of Caves and Karst Science* (ed. J. Gunn), 1385–1393. London: Fitzroy Dearborn.

Matsukura, Y., Maekado, A., Aoki, H. et al. (2007). Surface lowering rates of uplifted limestone terraces estimated from the height of pedestals on a subtropical island of Japan. *Earth Surface Processes and Landforms* 32: 1110–1115.

Matsushi, Y., Hattanji, T., Akiyama, S. et al. (2010). Evolution of solution dolines inferred from cosmogenic ^{36}Cl in calcite. *Geology* 38 (11): 1039–1042.

McPhillips, D., Hoke, G.D., Liu-Zeng, J. et al. (2016). Dating the incision of the Yangtze River gorge at the first bend using three-nuclide burial ages. *Geophysical Research Letters* 43: 101–110.

Mecchia, M., Sauro, F., Piccini, L. et al. (2014). Geochemistry of surface and subsurface waters in quartz-sandstones: significance for the geomorphic evolution of tepui table mountains (Gran Sabana, Venezuela). *Journal of Hydrology* 511: 117–138.

Mecchia, M., Sauro, F., Piccini, L. et al. (2019). A hybrid model to evaluate subsurface chemical weathering and fracture karstification in quartz sandstone. *Journal of Hydrology* 572: 745–760.

Miklavič, B., Mylroie, J.E., Jenson, J.W. et al. (2012). Evidence of the sea level change since MIS 5e on Guam, tropical west Pacific. In: *Studia UBB Geologia Special Issue. National Science Foundation Workshop Sea-Level Changes into the MIS 5e: From Observation to Prediction*, vol. 30, 32. Cluj: *Studia Universitatas Babeş-Bolyai*.

Miller, T.E. (1982). Hydrochemistry, hydrology and morphology of the Caves branch karst, Belize. PhD thesis: McMaster University.

Mocochain, L., Audra, P., Clauzon, G. et al. (2009). The effect of river dynamics induced by the Messinian Salinity Crisis on karst landscape and caves: example of the Lower Ardèche river (mid Rhône valley). *Geomorphology* 106 (1-2): 46–61.

Moreno, D., Falguères, C., Pérez-González, A. et al. (2012). ESR chronology of alluvial deposits in the Arlanzón valley (Atapuerca, Spain): contemporaneity with Atapuerca Gran Dolina site. *Quaternary Geochronology* 10: 418–423.

Mottershead, D.N., Wright, J.S., Inkpen, R.J. et al. (2007). Bedrock slope evolution in saltrock terrain. *Zeitschrift für Geomorphologie, Supplementary Issues* 51 (1): 81–102.

Mottershead, D.N., Duane, W.J., Inkpen, R.J. et al. (2008). An investigation of the geometric controls on the morphological evolution of small-scale salt terrains, Cardona, Spain. *Environmental Geology* 53: 1091–1098.

Muhs, D.R., McGeehin, J.P., Beann, J. et al. (2004). Holocene loess deposition and soil formation as competing processes, Matanuska Valley, southern Alaska. *Quaternary Research* 61: 265–276.

Mulec, J. and Prelovšek, M. (2015). Freshwater biodissolution rates of limestone in the temperate climate of the Dinaric karst in Slovenia. *Geomorphology* 238: 787–795.

Mylroie, J.E. and Mylroie, J.R. (2017). Role of karst denudation on the accurate assessment of glacio-eustasy and tectonic uplift on carbonate coasts. In: *Advances in Karst Research: Theory, Fieldwork and Applications*, vol. 466 (ed. M. Parise, F. Gabrovšek, G. Kaufmann, et al.), 171–185. *The Geological Society, London,* Special Publication.

Navas, A. (1990). The effect of hydrochemical factors on the dissolution rate of gypsiferous rocks in flowing water. *Earth Surface Processes and Landforms* 15: 709–715.

Nehme, C., Farrant, A., Ballesteros, D. et al. (2020). Reconstructing fluvial incision rates based on palaeo-water tables in chalk karst networks along the Seine valley (Normandy, France). *Earth Surface Processes and Landforms* 45: 1860–1876.

Nguyen, H.N., Vernant, P., Mazzotti, S. et al. (2016). 3-D GPS velocity field and its implications on the present-day post-orogenic deformation of the Western Alps and Pyrenees. *Solid Earth* 7: 1349–1363.

Ohmori, N. (1983). Erosion rates and their relations to vegetation from the viewpoint of world-wide distribution. *Bulletin of the Department of Geography, University of Tokyo* 15: 77–91.

Ortega, A.I., Benito-Calvo, A., Pérez-González, A. et al. (2013). Evolution of multilevel caves in the Sierra de Atapuerca (Burgos, Spain) and its relation to human occupation. *Geomorphology* 196: 122–137.

Palmer, A.N. (1987). Cave levels and their interpretation. *National Speleological Society Bulletin* 49 (2): 50–66.

Palmer, A.N. (2007). *Cave Geology*. Dayton, Ohio: Cave Books.

Palmeri, V., Madonia, G., and Ferro, V. (2020). Capturing gypsum rillenkarren morphometry by a 3D-photo reconstruction (3D-PR) technique. *Geomorphology* 351: 106980.

Paradise, T.R. (1995). Sandstone weathering thresholds in Petra, Jordan. *Physical Geography* 16: 205–222.

Parry, B. (2007). Pedestal formation and surface lowering in the Carboniferous limestone of Norber and Scales Moor, Yorkshire, UK. *Cave and Karst Science* 34: 61–68.

Pechorkin, I.A. (1969). *Geodynamic of Coasts of the Kama Reservoirs. Part II*. Perm: Perm University Publications (in Russian).

Pe'eri, S., Zebker, H.A., Ben-Avraham, Z. et al. (2004). Spatially-resolved uplift rate of the Mount Sedom (Dead Sea) salt diapir from InSAR observations. *Israel Journal of Earth Sciences* 53: 99–106.

Peltier, L.C. (1950). The geographic cycle in periglacial regions as it is related to climatic geomorphology. *Annals of the Association of American Geographers* 40: 214–236.

Piccini, L. and Mecchia, M. (2009). Solution weathering rate and origin of karst landforms and caves in the quartzite of Auyan-tepui (Gran Sabana, Venezuela). *Geomorphology* 106: 15–25.

Piccini, L., Drysdale, R.N., and Heijnis, H. (2003). Karst morphology and cave sediments as indicators of the uplift history in the Alpi Apuane (Tuscany, Italy). *Quaternary International* 101: 219–227.

Picotti, V. and Pazzaglia, F.J. (2008). A new active tectonic model for the construction of the Northern Apennines mountain front near Bologna (Italy). *Journal of Geophysical Research: Solid Earth* 113: B08412.

Pitty, A.F. (1968). The scale and significance of solutional loss from the limestone tract of the southern Pennines. *Proceedings of the Geologist's Association* 79: 153–177.

Plan, L. (2005). Factors controlling carbonate dissolution rates quantified in a field test in the Austrian Alps. *Geomorphology* 68 (3-4): 201–212.

Povara, I., Cosma, R., Lascu, C. et al. (1982). Un cas particulier de karst dans les dépôts de sel (Slanic Prahova, Roumanie). *Travaux de l'Institut Spéologique Emile Racovitza* 21: 8793.

Prelovšek, M. (2012). *The Dynamics of Present-Day Speleogenetic Processes in the Stream Caves of Slovenia*. Ljubljana: Založba ZRC Publishing.

Pulido-Bosch, A. (1986). Le karst dans les gypses de Sorbas (Almeria). Aspects morphologiques et hydrogéologiques. *Karstologia Mémoires* 1: 27–35.

Pulina, M. (1971). Observations on the chemical denudation of some karst areas of Europe and Asia. *Studia Geomorphologica Carpatho-Balcanica* 5: 79–92.

Rauch, H.W. and White, W.B. (1977). Dissolution kinetics of carbonate rocks: 1. Effects of lithology on dissolution rate. *Water Resources Research* 13 (2): 381–394.

Rigo, A., Vernant, P., Feigl, K.L. et al. (2015). Present-day deformation of the Pyrenees revealed by GPS surveying and earthquake focal mechanisms until 2011. *Geophysical Journal International* 201 (2): 947–964.

Rixhon, G., Braucher, R., Bourlès, D.L. et al. (2020). Plio-Quaternary landscape evolution in the uplifted Ardennes: New insights from [26]Al/[10]Be data from cave-deposited alluvium (Meuse catchment, E. Belgium). *Geomorphology* 371: 107424.

Romero-Mujalli, G., Hartmann, J., and Börker, J. (2019). Temperature and CO_2 dependency of global carbonate weathering fluxes - Implications for future carbonate weathering research. *Chemical Geology* 527: 118874.

Ryb, U., Matmon, A., Erel, Y. et al. (2014a). Controls on denudation rates in tectonically stable Mediterranean carbonate terrain. *Geological Society of America Bulletin* 126: 553–568.

Ryb, U., Matmon, A., Erel, Y. et al. (2014b). Styles and rates of long-term denudation in carbonate terrains under a Mediterranean to hyper-arid climatic gradient. *Earth and Planetary Science Letters* 406: 142–152.

Sanna, L., De Waele, J., Calaforra, J.M. et al. (2015). Long-term erosion rate measurements in gypsum caves of Sorbas (SE Spain) by the Micro-Erosion Meter method. *Geomorphology* 228: 213–225.

Sartégou, A., Blard, P.H., Braucher, R. et al. (2020). Late Cenozoic evolution of the Ariège River valley (Pyrenees) constrained by cosmogenic [26]Al/[10]Be and [10]Be/[21]Ne dating of cave sediments. *Geomorphology* 371: 107441.

Sasowsky, I.D., White, W.B., and Schmidt, V.A. (1995). Determination of stream-incision rate in the Appalachian plateaus by using cave-sediment magnetostratigraphy. *Geology* 23: 415–418.

Sauro, F. (2014). Structural and lithological guidance on speleogenesis in quartz-sandstone: evidence of the arenisation process. *Geomorphology* 226: 106–123.

Sauro, F., Piccini, L., Mecchia, M. et al. (2013a). Comment on "Sandstone caves on Venezuelan tepuis: Return to pseudokarst?" by R. Aubrecht, T. Lánczos, M. Gregor, J. Schlögl, B. Smída, P. Liscák, Ch. Brewer-Carías, L. Vlcek. *Geomorphology* 132 (2011): 351–365. *Geomorphology* 197: 190-196.

Sauro, F., Cappelletti, M., Ghezzi, D. et al. (2018). Microbial diversity and biosignatures of amorphous silica deposits in orthoquartzite caves. *Scientific Reports* 8 (1): 17569.

Sauro, F., Mecchia, M., Piccini, L. et al. (2019). Genesis of giant sinkholes and caves in the quartz sandstone of Sarisariñama tepui, Venezuela. *Geomorphology* 342: 223–238.

Schaefer, J.M. and Lifton, N. (2007). Cosmogenic nuclide dating: methods. In: *Encyclopedia of Quaternary Science*, vol. 1 (ed. S.A. Elias), 412–419. Amsterdam: Elsevier.

Shaw, M.G., Stafford, K.W., and Tate, B.P. (2011). Surface denudation of the Gypsum Plain, west Texas and southeastern New Mexico. In: *U.S. Geological Survey Karst Interest Group Proceedings*, U.S. Geological Survey (ed. E.L. Kunjansky), 104–112. Scientific Investigations Report 2011-5031.

Simms, M.J. (2004). Tortoises and hares: Dissolution, erosion and isostasy in landscape evolution. *Earth Surface Processes and Landforms* 29: 477–494.

Smith, D.I. (1972b). The solution of limestone in an Arctic environment. In: *Polar Geomorphology* (ed. D.E. Sugden), 187–200. Institute of British Geographers Special Publication 4.

Smith, D.I. and Atkinson, T.C. (1976). Process, landforms and climate in limestone regions. In: *Geomorphology and Climate* (ed. E. Derbyshire), 369–409. Chichester: Wiley.

Smith, D.I., Greenaway, M.A., Moses, C. et al. (1995). Limestone weathering in eastern Australia. Part 1: erosion rates. *Earth Surface Processes and Landforms* 10: 427–440.

Spate, A.P., Jennings, J.N., Smith, D.I. et al. (1985). The micro-erosion meter: use and limitations. *Earth Surface Processes and Landforms* 10: 427–440.

Spring, W. and Prost, E. (1883). Étude sur les eaux de la Meuse. *Annales de la Société Géologique de Belgique* 11: 123–220.

Stephenson, W.J. and Finlayson, B.L. (2009). Measuring erosion with the micro-erosion meter - contributions to understanding landform evolution. *Earth-Science Reviews* 95 (1-2): 53–62.

Stich, D., Serpelloni, E., de Lis Mancilla, F. et al. (2006). Kinematics of the Iberia-Maghreb plate contact from seismic moment tensors and GPS observations. *Tectonophysics* 426: 295–317.

Stock, G.M., Anderson, R.S., and Finkel, R.C. (2004). Pace of landscape evolution in the Sierra Nevada, California, revealed by cosmogenic dating of cave sediments. *Geology* 32: 193–196.

Stokes, M., Mather, A.E., and Harvey, A.M. (2002). Quantification of river-capture-induced base-level changes and landscape development, Sorbas Basin, SE Spain. *Geological Society of London, Special Publications* 191: 23–35.

Stone, J.O.H., Allan, G.L., Fifield, L.K. et al. (1994). Limestone erosion measurements with cosmogenic chlorine-36 in calcite - preliminary results from Australia. *Nuclear Instruments and Methods in Physics Research Section B: Beam Interactions with Materials and Atoms* 92: 311–316.

Stone, J.O.H., Evans, J.M., Fifield, L.K. et al. (1998). Cosmogenic chlorine-36 production in calcite by muons. *Geochimica et Cosmochimica Acta* 62: 433–454.

Strakhov, N.M. (1967). *Principles of Lithogenesis*. Edinburgh: Oliver and Boyd.

Sweeting, M.M. (1966). The weathering of limestones, with particular reference to the Carboniferous Limestones of northern England. In: *Essays in Geomorphology* (ed. G.H. Dury), 177–210. London: Heinemann.

Tarhule-Lips, R.F. and Ford, D.C. (1998a). Condensation corrosion in caves on Cayman Brac and Isla de Mona. *Journal of Cave and Karst Studies* 60: 84–95.

Tarolli, P. (2014). High-resolution topography for understanding Earth surface processes: opportunities and challenges. *Geomorphology* 216: 295–312.

Telling, J., Lyda, A., Hartzell, P. et al. (2017). Review of Earth science research using terrestrial scanning. *Earth-Science Reviews* 169: 35–68.

Thomas, F., Godard, V., Bellier, O. et al. (2017). Morphological controls on the dynamics of carbonate landscapes under a mediterranean climate. *Terra Nova* 29: 173–182.

Thomas, F., Godard, V., Bellier, O. et al. (2018). Limited influence of climatic gradients on the denudation of a Mediterranean carbonate landscape. *Geomorphology* 316: 44–58.

Tripathi, J.K. and Rajamani, V. (2003). Weathering control over geomorphology of supermature Proterozoic Delhi quartzites of India. *Earth Surface Processes and Landforms* 28: 1379–1387.

Trudgill, S.T. (1986). Limestone weathering under a soil cover and the evolution of limestone pavements, Malham District, North Yorkshire, UK. In: *New Directions in Karst* (ed. K. Paterson and M.M. Sweeting), 461–471. Norwich: Geo Books.

Trudgill, S.T. (2008). Corbel, J. 1959: Érosion en terrain calcaire (vitesse d'érosion et morphologie). Annales de Géographie 68: 97-120. *Progress in Physical Geography* 32: 684–690.

Trudgill, S.T., High, C.J., and Hanna, F.K. (1981). Improvements to the micro-erosion meter. *British Geomorphological Research Group Technical Bulletin* 29: 3–17.

Turkington, A.V. and Paradise, T.R. (2005). Sandstone weathering: a century of research and innovation. *Geomorphology* 67: 229–253.

Turowski, J.M. and Cook, K.L. (2017). Field techniques for measuring bedrock erosion and denudation. *Earth Surface Processes and Landforms* 42: 109–127.

Urushibara-Yoshino, K., Miotke, F.-D., and Research Group of Solution Rates in Japan (1999). Solution rate of limestone in Japan. *Physics and Chemistry of the Earth (A)* 24: 899–903.

Vázquez-Urbez, M., Arenas, C., Sancho, C. et al. (2010). Factors controlling present-day tufa dynamics in the Monasterio de Piedra Natural Park (Iberian Range, Spain): depositional environmental settings, sedimentation rates and hydrochemistry. *International Journal of Earth Sciences* 99: 1027–1049.

Wagner, T., Fabel, D., Fiebig, M. et al. (2010). Young uplift in the non-glaciated parts of the Eastern Alps. *Earth and Planetary Science Letters* 295: 159–169.

Walling, D.E. and Webb, B.W. (1986). Solutes in river systems. In: *Solute Processes* (ed. S.T. Trudgill), 251–327. Chichester: Wiley.

Waltham, A.C. and Fookes, P.G. (2003). Engineering classification of karst ground conditions. *Quarterly Journal of Engineering Geology and Hydrogeology* 36: 101–118.

Wang, F., Li, H., Zhu, R. et al. (2004). Late Quaternary downcutting rates of the Qianyou River from U/Th speleothem dates, Qinling mountains, China. *Quaternary Research* 62: 194–200.

Webb, J.A., Fabel, D., Finlayson, B.L. et al. (1992). Denudation chronology from cave and river terrace levels: the case of the Buchan Karst, southeastern Australia. *Geological Magazine* 129: 307–317.

Weinberger, R., Lyakhovsky, V., Baer, G. et al. (2006). Mechanical modelling and InSAR measurements of Mount Sedom uplift, Dead Sea basin: implications for effective viscosity of rock salt. *Geochemistry, Geophysics, Geosystems* 7: Q05014.

Westaway, R. (2020). Late Cenozoic uplift history of the Peak District, central England, inferred from dated cave deposits and integrated with regional drainage development: a review and synthesis. *Quaternary International* 546: 20–41.

Wheeler, R.L. (2016). Maximum Magnitude (Mmax) in the Central and Eastern United States for the 2014 US Geological Survey Hazard Model. *Bulletin of the Seismological Society of America* 106: 2154–2167.

White, W.B. (1984). Rate processes: chemical kinetics and karst lanform development. In: *Groundwater as a Geomorphic Agent* (ed. R.G. LaFleur), 227–248. Boston: Allen and Unwin.

White, W.B. (1988). *Geomorphology and Hydrology of Karst Terrains*. New York: Oxford University Press.

White, W.B. (2000a). Dissolution of limestone from field observation. In: *Speleogenesis. Evolution of Karst Aquifers* (ed. A.B. Klimchouk, D.C. Ford, A.N. Palmer, et al.), 149–155. Huntsville: National Speleological Society.

Williams, P.W. (1963). An initial estimate of the speed of limestone solution in County Clare. *Irish Geography* 4: 432–441.

Williams, P.W. (1968). An evaluation of the rate and distribution of limestone solution and deposition in the River Fergus Basin, western Ireland. In: *Contributions to the Study of Karst*, Publication G5, Research School for Pacific Studies (ed. P.W. Williams and J.N. Jennings). Australian National University.

Williams, P.W. (1982). Speleothem dates, Quaternary terraces and uplift rates in New Zealand. *Nature* 298: 257–260.

Williams, P.W. (1983). The role of the subcutaneous zone in karst hydrology. *Journal of Hydrology* 61: 45–67.

Williams, P.W. (2008). The role of the epikarst in karst and cave hydrogeology: a review. *International Journal of Speleology* 37: 1–10.

Williams, P.W. and Dowling, R.K. (1979). Solution of marble in the karst of the Pikikiruna range, northwest Nelson, New Zealand. *Earth Surface Processes* 4: 15–36.

Wilson, P., Lord, T.C., and Vincent, P.J. (2012a). Origin of the limestone pedestals at Norber Brow, North Yorkshire, UK: a re-assessment and discussion. *Cave and Karst Science* 39: 5–11.

Wilson, P., Barrows, T.T., Lord, T.C. et al. (2012b). Surface lowering of limestone pavement as determined by cosmogenic (^{36}Cl) analysis. *Earth Surface Processes and Landforms* 37: 1518–1526.

Wolman, M.G. and Miller, J.P. (1960). Magnitude and frequency of forces in geomorphic processes. *Journal of Geology* 68: 54–74.

Wray, R.A.L. (1997). A global review on solutional weathering forms on quartz sandstones. *Earth-Science Reviews* 42: 137–160.

Wray, R.A.L. and Sauro, F. (2017). An updated global review of solutional weathering processes and forms in quartz sandstones and quartzites. *Earth-Science Reviews* 171: 520–557.

Xu, S., Liu, C., Freeman, S. et al. (2013). In-situ cosmogenic ^{36}Cl denudation rates of carbonates in Guizhou karst area. *Chinese Science Bulletin* 58: 2473–2479.

Yang, Y., Lang, Y.C., Xu, S. et al. (2020). Combined unsteady denudation and climatic gradient factors constrain carbonate landscape evolution: New insights from in situ cosmogenic ^{36}Cl. *Quaternary Geochronology* 58: 101075.

Young, R.W., Wray, R.A.L., and Young, A.R.M. (2009). *Sandstone Landforms*. Cambridge: Cambridge University Press.

Zambo, L. and Ford, D.C. (1997). Limestone dissolution processes in Beke doline Aggtelek National Park, Hungary. *Earth Surface Processes and Landforms* 22 (6): 531–543.

Zarei, M. and Raeisi, E. (2010). Karst development and hydrogeology of Konarsiah salt diapir, south of Iran. *Carbonates and Evaporites* 25: 217–229.

Zarei, M., Raeisi, E., and Talbot, C. (2012). Karst development on a mobile substrate: Konarsiah salt extrusion, Iran. *Geological Magazine* 149: 412–422.

5

Karst Hydrogeology

5.1 Introduction

Most of the water on Earth (96.5%) is stored in the oceans, 1% is saline groundwater or water in salty lakes, and only 2.5% is freshwater. Two-thirds of this freshwater (1.7%) are stored in ice caps and glaciers and the remaining 0.8% is the water we normally use in our daily life. Only 0.009% of the total world's water is the one we normally observe in lakes, rivers, and swamps. The "invisible" water includes water vapor in the atmosphere (0.001%), the water stored in plants and animals (0.0001%), soil water (0.001%), and, above all, fresh groundwater (0.76%). The freshwater resources related to the rain and snowfall, flowing in streams and rivers, into lakes, and infiltrating underground to form soil and groundwater, provide most of the world's population with drinkable water.

The water that evaporates from the surface of the oceans and continents rises into the atmosphere, cools down, and condenses or solidifies into droplets, snowflakes, or ice crystals, falling back to the surface of the Earth through various forms of precipitation. Some of this precipitated water flows on the surface (runoff), but part of it infiltrates into the subsurface through pores, fractures, and cavities. The study of the occurrence of groundwater, its movement and quality is designated as groundwater hydrology (or hydrogeology). The movement of water varies according to lithological and structural differences of the rock bodies involved and the laws governing groundwater flow are applicable in most situations. For a general overview on groundwater hydrology the reader is referred to textbooks such as Fetter (2014) and Singh (2016). Nonetheless, movement of water in karst can be very different from that in other rock types, and the general laws of groundwater hydrology are not always applicable. This chapter explains how groundwater moves in different karst settings, and which methods can be used to analyze karst drainage systems.

5.2 Brief Historical Overview on Karst Hydrogeology (<1900)

The first written record on the exploration of a cave with a karst spring is related to the Assyrian King Tiglatpilesar I (1113–1075 BC), who apparently visited the Tigris tunnel caves (SE Turkey) in 1110 BC, as the rock reliefs carved at the cave entrance indicate (Kusch 1993). Two centuries later, the Neo-Assyrian King Shalmaneser III revisited the same caves twice (in 852 and in 844 BC), leaving other rock reliefs in the cave entrances, as well as on an obelisk at Nimrud (cuneiform inscriptions) and on a statue of a bull. The same visit was reported on a bronze plate, which was part of the entrance door of a temple dedicated to Mamu, near the village of Tell Balawat, or ancient

Imgur-Enlil, 30 km southeast of Mosul, Iraq (King 1915). Two scenes are reported on this plate. The upper one shows the king in front of a cave entrance followed by several guards and a bull to be sacrificed. The cave is represented with four stalagmites, their feeding drips, and two people inside (a sculptor and a scribe). The lower scene shows a bull and a ram brought to the entrance of the cave for sacrifice, and some artists sculpting the rock reliefs at the entrance of the cave. Inside the cave a river is seen through three windows, with people wading waist-deep in the water, with a torch (?) in their hand and trees growing in the underground river (showing that the three windows are natural openings onto the subterranean river).

Several centuries later, underground rivers became part of the Greek mythology, being the north and south of the Gulf of Corinth intensely karstified areas. Karst water also drew the attention of Roman intellectuals, since karst springs were important sources of water supply in the Roman empire (Figure 5.1). The Greek and Roman philosophers were the first to formulate theories on the origin of springs and underground rivers (see Shaw 1992 for a detailed overview on these theories). Two main notions on the origin of spring water prevailed: a rainfall origin (Anaxagoras, fifth century BC; Plato and Aristotle, fourth century BC; and Vitruvius, first century AC) and a seawater source (Lucretius, first century BC; and Pliny the Elder, 77 AC). Plato, in reality, forwarded two contrasting ideas: a first that envisioned many subterranean channels, including a huge one named "Tartarus," and the water oscillating hence and forth, springing out forming rivers, and a second more advanced idea that envisaged rainwater feeding springs. In the same period, Aristotle believed that most water flowing from springs comes from rain, but some is also related to condensation occurring in the underground spaces (Shaw 1992).

The seawater origin of groundwater feeding springs was derived from the Bible, probably written in the third century BC. It was argued that despite the fact that rivers flow into the sea, the sea itself never overflows. This idea was adopted by Lucretius first, then by Pliny the Elder, and this belief

Figure 5.1 Pont du Gard. This famous aqueduct brought water from several karst springs (Fontaine d'Eure) to the Roman city of Nemausus (Nîmes) over a distance of 50 km for at least 5 centuries (first to sixth century AC). *Source:* Photo by Jo De Waele.

Figure 5.2 Drawing by Athanasius Kircher (1665) illustrating his theory on the seawater origin of springs and rivers (Kircher 1665). *Source:* Athanasius Kircher / Wikimedia Commons / Public domain.

persisted for many centuries. One of the most influential works on hydrology in the seventeenth century was "Mundus Subterraneus," written by Athanasius Kircher (1665). This book related underground rivers and karst springs with huge underground reservoirs (hydrophylacia), fed by seawater, distillation of vapor from inside the Earth, and minor amounts of rainwater (Shaw 1992) (Figure 5.2). Despite the increasing evidence that springs are fed by rainwater or sinking rivers, the notion of seawater origin persisted until the nineteenth century. Nonetheless, the general concept of the rainwater origin of springs became widely accepted during the eighteenth century.

For detailed information on the historical evolution of ideas regarding groundwater movement in karst areas the readers are referred to the excellent works of Pfeiffer (1963), Herak and Stringfield (1972), and Shaw (1992).

5.3 Definitions

A formation that is sufficiently porous and permeable to store, transmit, and yield a significant quantity of water to a borehole, well or spring is termed an **aquifer**. Typical examples are sands and gravels, but also fractured rocks can be good aquifers. Among fractured rocks, limestones represent important water reservoirs due to karstification phenomena. Aquifers can be **unconfined** (water table or phreatic aquifers), characterized by a saturated zone with an upper boundary named "water table" that is directly exposed to atmospheric pressure, or **confined** (also named artesian), when the permeable rock is contained between overlying and underlying rock units with significantly lower permeability, and **perched**, when local water saturation occurs in the unsaturated zone because of the existence of a spatially restricted low-permeability zone (Figure 5.3).

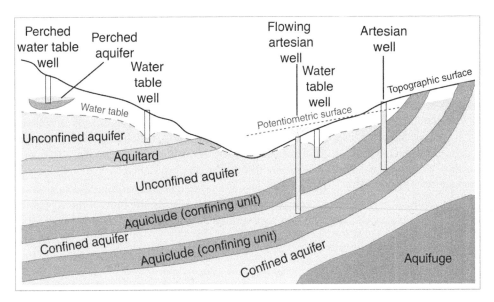

Figure 5.3 Schematic representation of a groundwater flow system showing units with different hydrogeological behavior.

An **aquifuge**, on the contrary, is a geological formation that cannot store nor transmit water (e.g., compact crystalline rocks in which there are no interconnected voids). Many rocks do have high porosity, thus can store important amounts of fluids, but permeability is too low to transmit the water. These rock units are designated as **aquicludes** (e.g., claystones, shales). When a rock formation has low permeability compared to the surrounding rock units, but stores and transmits water very slowly, it is classified as an **aquitard**, also called semi-confining bed or leaky confining bed. Good examples are the sandy clay beds within a sand sequence.

Water falling on the continental Earth surface under various forms (rain, snow, and ice) can penetrate underground through the pores and open spaces in the rock. Flow is driven downward by gravity until the water reaches the level where all the open spaces in the rock are filled with water. This level is the **water table** or groundwater level, defined as the top surface of the saturated zone in an unconfined aquifer, where the fluid pressure is equal to atmospheric pressure. This free-standing surface of water corresponds to the level of water in wells (and piezometers, being observation wells of groundwater levels), or in the fissures and pores in unconfined aquifers. The water table is the upper limit of the **saturated** (or **phreatic**) zone, and the lower limit of the **unsaturated** (or **vadose**) zone.

The water in confined aquifers is under pressure and boreholes intercepting these aquifers let water rise spontaneously to a certain level, sometimes even above the topographic surface and the well head. The imaginary surface defined by the elevation of the water levels in these boreholes is called the **potentiometric surface**, and the wells in which groundwater rises above the ground surface are called **artesian**. Note that the potentiometric surface is not to be confused with the water table (Figure 5.3).

The rate at which water infiltrates in the subsurface depends on a number of factors, including the type of material, slope, vegetation cover, presence of water prior to the infiltration event (antecedence), and the precipitation intensity (or flow intensity) itself. The physical properties of the material that control the quantity and the way in which water infiltrates underground are porosity and permeability.

The **porosity** φ of a rock or deposit is a measure of the voids (empty spaces) in a material, given by the ratio of the total volume of voids V_v over the bulk volume V_b, expressed in %:

$$\varphi = \frac{V_v}{V_b} \times 100 \tag{5.1}$$

Not all the pore spaces in a rock are interconnected, and thus able to let groundwater flow. For example, water bound to clay particles (bound water) or occluded in isolated "vuggy" porosity is contained in the rock but not released or drained. **Effective porosity**, which is the interconnected pore space through which water can effectively flow, is mainly influenced by grain size. Clays typically have high porosity (30–60%) but very low drainage ability and effective porosity, ranging between 0 and 5%. Highest values of effective porosity are attained in medium-to-coarse sands (20–30%), with slightly lower values for coarse gravels (10–20%). Thus, the ratio of the volume of interconnected pores in a rock V_w over the total rock volume V_b expressed in % is known as effective porosity φ_e and is given by:

$$\varphi_e = \frac{V_w}{V_b} \times 100 \tag{5.2}$$

Hydrogeologists make a distinction between primary, secondary, and tertiary porosity. **Primary** (or **matrix**) **porosity** is created during the deposition of the material, and is largely related to the textural characteristics of the particles or crystals (e.g., size, shape, and sorting), and their fabrics. This syndepositional (primary) porosity can be strongly modified by multiple processes (e.g., compaction, cementation, and dissolution) during the diagenesis of the rocks. In general, the porosity of karst rocks decreases during the burial diagenesis (mesogenetic stage) and increases in the telogenetic stage, when rocks come back to near-surface meteoric conditions (see Section 2.2.5).

The **secondary** (**fracture-** or **fissure-**) **porosity** refers to porosity created by rock fracturing, and should be distinguished from the **tertiary** (or **dissolution**) **porosity** mainly related to the solutional enlargement of discontinuity planes (fractures and bedding) that eventually may become conduits.

Some karst rocks such as rock salt and gypsum are formed by evaporation, and are thus composed of densely packed crystals. These rocks have a very low primary porosity (0–5%), but dissolution porosity can locally reach very high values. Quartz arenites (sandstones) can have high primary porosity (generally between 10 and 30%), depending on the grain size and on the degree of cementation. Limestones and dolostones also have a primary porosity ranging between 5 and 20%, but can reach over 50% of fracture and dissolution porosity (Figure 5.4). Metamorphic processes (recrystallization and compaction) normally reduce the porosity considerably, with marbles having less than 5%.

The average porosity of a rock volume is scale-dependent, and increases with the spatial scale of observation, typically being smaller at microscopic scale, higher when measured in boreholes, and greatest at the entire aquifer scale (Kiraly 1975) (Figure 5.5).

Porosity values in carbonate rocks decrease exponentially with depth due to the effects of burial diagenesis, mainly compaction and cementation (Schmoker and Halley 1982). This porosity reduction is more pronounced in limestones than in dolostones (Figure 5.6).

Permeability k is the ability of a material to transmit fluids (including gases), which depends not only on the physical properties of the material itself (grain size, sorting, cementation, etc.), but also on the interconnectedness between the different types of porosity. A highly porous material is not necessarily highly permeable (e.g., clays). Permeability is expressed in area (e.g., square centimeters), or as darcy units ($1\,\mathrm{d} = 10^{-8}\,\mathrm{cm}^2 \approx 1\,\mathrm{\mu m}^2$).

Porosity and permeability can vary depending on the location in which they are measured in an aquifer. In well-sorted sands and gravels, these parameters are relatively **homogeneous** across the aquifer.

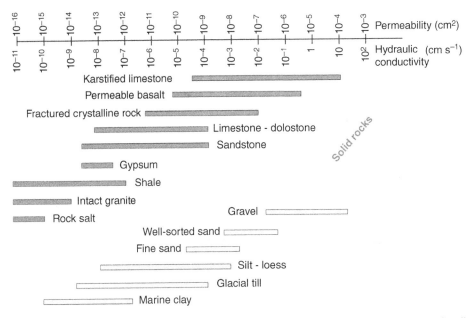

Figure 5.4 Permeability and hydraulic conductivity ranges of some common rock types and sediments.

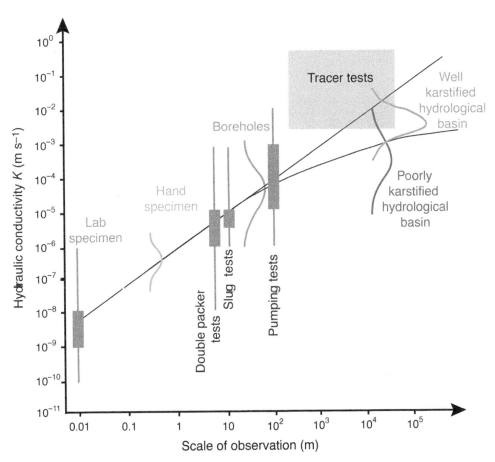

Figure 5.5 Scale effect of hydraulic conductivity in karst. *Source:* Modified from Kiraly (1975) and Hartmann et al. (2014). Boxplot data in gray are from Sauter (1992a) and represent data from a Jurassic karst aquifer in the Swabian Alps (Germany), whereas the gray box corresponds to a database of over 1800 dye-tracing experiments conducted in 25 countries. *Source:* Modified after Quinlan et al. (1992).

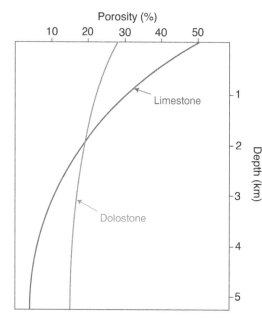

Figure 5.6 Porosity changes versus depth for limestones and dolostones. *Source:* Adapted from Schmoker and Halley (1982).

However, in most aquifers, porosity and permeability, and thus the ability of underground water to flow, are variable, and such aquifers are called **heterogeneous**. The permeability can be independent of the direction of water flow (**isotropic**) or variable according to the direction of flow (**anisotropic**) (Figure 5.7).

Different beds of coarse- and fine-grained well-sorted sands are a good example of heterogeneous and isotropic aquifers (Figure 5.7c). A sedimentary layer composed of well-sorted elongated or flattened grains will allow water to flow more easily along the elongation direction of the particles rather than perpendicular to it. This layer is a homogeneous and anisotropic aquifer (Figure 5.7b). Bedrock aquifers are generally anisotropic and heterogeneous (Figure 5.7d). A young porous limestone, such as a grainstone composed of equidimensional oolites, will essentially be a homogeneous and isotropic aquifer (Figure 5.7a). Once fractures are developed and dissolution starts to act

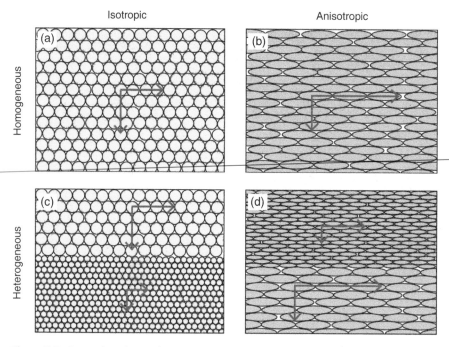

Figure 5.7 Isotropic, anisotropic, homogeneous, and heterogeneous aquifers. Blue arrows indicate water movement in horizontal and vertical directions. The length of the arrows reflects relative permeability values: (a) isotropic homogeneous; (b) anisotropic homogeneous; (c) isotropic heterogeneous; and (d) anisotropic heterogeneous.

upon these limestones, the aquifer will increasingly become heterogeneous and anisotropic, raising permeability as much as four orders of magnitude.

All three types of porosity (matrix, fissures, and conduits) are able to store water, but not all of this water is released by gravitational flow or artificial abstraction. Part of the water is retained by capillary and molecular forces and hardly contributes to groundwater flow. The ratio of the volume of water which can be drained by gravity from an initially saturated unconfined aquifer to the total volume of the porous medium is called **specific yield** S_y. In confined settings, the **specific storage** S_s is used instead, being the volume of water released from or taken into storage per unit volume of the aquifer per unit change in hydraulic head.

The storativity S (or storage coefficient) of a confined aquifer (or aquitard) is defined as the volume of water released from storage per unit surface area of the aquifer (or aquitard) per unit decline in hydraulic head. Storativity increases with thickness b of the aquifer (or aquitard):

$$S = S_s b$$

In unconfined aquifers, the storativity is mainly related to the specific yield S_y, with negligible amounts of water deriving from the specific storage, so $S \approx S_y$.

The transmissivity T is the ability of a water-bearing unit to transmit water, which obviously varies with direction in anisotropic aquifers. Transmissivity in confined aquifers depends on the properties of the aquifer transmitting the flow and its thickness b. The thickness in unconfined aquifers is that of the saturated portion of the aquifer, or the height of the water table above the lower boundary of the aquifer.

5.4 Groundwater Flow Fundamentals

Groundwater moves under the influence of gravity, similar to surface waters in rivers. But, with respect to river water, it moves at flow rates three or more orders of magnitude slower. In an isoline map depicting the water table (or groundwater level), the contour lines connecting points at which the water level is at the same elevation are called **equipotential lines** (Figure 5.8). They are very similar to contour lines on a topographic map. Equipotential lines connect points with equal hydraulic head and thus equal water pressure. Groundwater flows from areas of high hydraulic head to areas of lower hydraulic head and in certain conditions (when fluid pressure increases with depth), it flows in an upward direction. An example is the flow of groundwater into effluent surface rivers, where hydraulic pressure is lower (Figure 5.8).

Groundwater flows perpendicularly to the equipotential lines, in the same manner as surface water flows down a slope at right angles to the contour lines. The groundwater flow path is known as a **streamline**. The mesh formed by the intersection of equipotential lines and streamlines is called **flow net**.

5.4.1 Laminar Flow Through Granular Media

Water flow through a saturated porous medium was experimentally studied in 1856 by Henry Darcy, in his report on the water supply of the French city of Dijon (Darcy 1856). A cylinder with cross-section A was filled with sand, closed at both ends, and equipped with inflow and outflow tubes and two manometers (Figure 5.9).

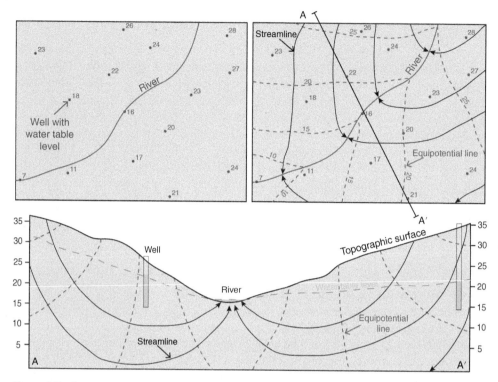

Figure 5.8 Equipotential lines, streamlines, and flow nets.

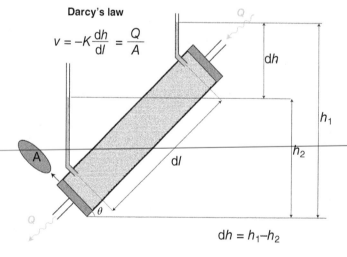

Figure 5.9 The experimental setup of Darcy's experiment.

Water was introduced saturating the sediment, until inflow and outflow were equal, corresponding to a constant flow rate (or discharge) Q. The specific discharge (volumetric flow rate per unit area) is defined as

$$v = \frac{Q}{A} \tag{5.3}$$

and is expressed as a velocity (e.g., centimeters per second), also called **Darcian velocity**.

Darcy's experiments showed that specific discharge v is directly proportional to the height difference between the inflow and outflow points ($dh = h_1 - h_2$), known as **hydraulic head**, and inversely proportional to the flow distance dl. Darcy's law can thus be written as

$$v = -K\frac{dh}{dl} = \frac{Q}{A} \tag{5.4}$$

where K is a coefficient of proportionality, known as **hydraulic conductivity** (or **coefficient of permeability**) and expressed as a velocity (e.g., meters per second), whereas dh/dl is the change of hydraulic head with distance, better known as **hydraulic gradient**. The conventional negative sign in this equation indicates that the flow is in the direction of decreasing hydraulic head. Note that specific discharge is independent of the angle of inclination of the cylinder, θ (if dh/dl and K are held constant), Darcy's law being valid for fluid flow in any direction in space.

It is worth noting that the Darcian velocity is not the true velocity of the water flow at microscopic scale, which has to be faster since water passes through interstitial pores, following a tortuous (and thus longer) pathway (Figure 5.10). The average measure of groundwater velocity can be obtained dividing the Darcian velocity by the effective porosity for a given cross-sectional area:

$$v_m = \frac{Q}{A\varphi_e} = \frac{v}{\varphi_e} \tag{5.5}$$

Hydraulic conductivity is a function of the physical properties of the material crossed by the fluid and the properties of the fluid itself. Hydraulic conductivity K is thus related to permeability k according to the equation

$$K = \frac{k\rho g}{\mu} \tag{5.6}$$

where ρ is the density of the fluid, μ is the dynamic viscosity of the fluid passing through the porous material, and g is the acceleration due to gravity. Darcy's law is applicable in **laminar flow** conditions with water moving along parallel flow lines, in the direction of flow, with no mixing or transverse motion.

5.4.2 Turbulent Flow Through Conduits

The Darcy equation holds for granular aquifers or narrow tubes and fissures, but is often not applicable in karst aquifers. Here, a significant proportion of the water moves through irregular channels and pipes, forming eddies and mixing rapidly, a movement known as **turbulent flow**.

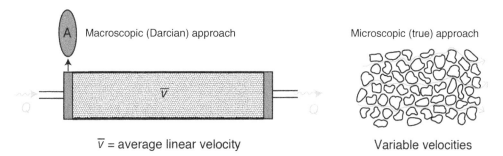

$\overline{v}$ = average linear velocity   Variable velocities

Figure 5.10 Macroscopic (Darcian) and microscopic approach to the analysis of groundwater flow. *Source:* Adapted from Hiscock and Bense (2014).

The transition between laminar and turbulent flow regimes occurs when the width of the opening exceeds 5–10 mm, largely depending on the flow velocity and the temperature. This threshold value can increase significantly in the case of brines with higher density and viscosity than pure water. This transition can be estimated using the Reynolds Number R, given by the formula

$$R = \frac{\rho v \left(\dfrac{A}{p} \right)}{\mu} \tag{5.7}$$

in which v is the mean velocity of the fluid (water), A is the cross-sectional area, and p is the wetted perimeter (the length of the water-conduit contact of the section). A/p is also known as the hydraulic radius. The viscosity μ of water decreases with rising temperature, whereas density is highest at 3.98 °C and decreases both below and above this temperature. Reynolds numbers lower than 500 are indicative of laminar flow, above 1000 flow usually becomes turbulent, but these values also depend on how irregular the walls of the fluid pathways are (flow can become turbulent even with $R < 500$ if pathways are very irregular).

If we idealize a cave conduit as a cylindrical pipe of diameter d, the discharge under laminar flow conditions can be expressed with the Hagen–Poiseuille equation

$$Q = \frac{\Pi d^4 \rho g}{128 \mu} \cdot \frac{dh}{dl} \tag{5.8}$$

This equation shows that discharge depends on the fourth power of the conduit diameter: large conduits are thus able to transmit water at extremely higher rates than smaller ones. This factor greatly controls the development of underground fluid flow networks, concentrating flow along a small number of large conduits and deactivating flow through the many smaller pathways.

Laminar flow in conduits is not the typical situation in karstified aquifers, and flow becomes turbulent with increasing water velocity, sinuosity, and wall roughness. In these conditions, the Darcy–Weissbach equation is the starting point

$$v^2 = \frac{Q^2}{A^2} = \frac{2dg}{f} \cdot \frac{dh}{dl} \tag{5.9}$$

with f being the friction factor, a dimensionless number that mainly depends on the conduit geometry, the size of the pathways, the roughness of the walls, and the presence of obstacles. Its value in caves is typically between 0.05 and 0.10, but can be as high as 300 in irregular conduits with extensive breakdown deposits.

More often, flow of water through caves occurs along open channels and in this case, the flow velocity can be estimated using the empirical Manning's equation

$$v = \frac{1}{n} \cdot \left(\frac{A}{p} \right)^{2/3} \cdot S^{1/2} \quad \left(\text{with } A / p, \text{or hydraulic radius, in meters} \right) \tag{5.10}$$

where n is the Manning's number (roughness coefficient) and S is the channel slope (a dimensionless number, derived from drop in altitude divided by flow distance). The roughness coefficient n can be estimated using reference river tracts where slope, flow velocity, and wetted perimeter (or hydraulic radius) can be measured. Values of n range between 0.014 for smooth steel conduits and 0.1 for highly irregular cobbled river channels.

5.4.3 Flow Through Fractured Media

Groundwater in karst does not only flow through the granular matrix and conduits, but also through narrow interconnected fissures such as solutionally enlarged bedding planes, joints, and faults. If a fracture is represented by two smooth parallel surfaces separated by a constant distance (aperture width b) and a fissure length W, laminar groundwater flow through this fracture is described by the so-called "cubic law" that replaces Darcy's law for granular aquifers (Snow 1969)

$$Q = \frac{W\rho g b^3}{12\mu} \cdot \frac{dh}{dl} \tag{5.11}$$

with the hydraulic conductivity of the fracture being represented by the first part of the equation

$$K = \frac{W\rho g b^3}{12\mu} \tag{5.12}$$

This formulation of flow through fractures assumes laminar, incompressible flow and impermeable fracture walls. This equation shows that flow rate increases with the cube of the fracture width. Hence, as a fracture is widened by solution, water flushed through it increases enormously, in a positive feedback mechanism between flow rate and dissolution.

Modeling real fracture systems is far more complex: a fractured aquifer contains different sets of fissures of different width and length, and with varying roughness and interconnectedness (National Research Council 1996).

5.5 Groundwater Flow in Karst Aquifers

Karst aquifers are groundwater flow systems in which water moves through different types of permeability features: the interconnected primary porosity of the rocks (often named matrix), the fissured system (joints, faults, and bedding planes), and the conduit system (Figure 5.11). Some authors therefore refer to the triple porosity (or permeability) nature of karst aquifers (Worthington 1999; White 2002, 2006).

More often researchers prefer to simplify karst aquifers as dual systems, in which porosity (permeability) related to pores and small fissures is referred to as the matrix porosity, whereas enlarged fissures and conduits are grouped as karst conduits (Kiraly 1998; Hartmann et al. 2014). Recharge, groundwater flow, and discharge are influenced by this dual behavior: slow and diffuse infiltration in the matrix and fast concentrated recharge through enlarged fractures and conduits; low flow (Darcian) velocity through small fissures and intergranular pores and fast (turbulent) flow in conduits (Figure 5.12); regular low discharge over extensive dry periods when the springs are mainly fed by the matrix flow, and flashy high discharge following rain events, when flow is dominated by water driven through conduits and enlarged fissures.

Matrix permeability is variable, from very low in old compacted and well-cemented limestones to rather high in young (eogenetic) limestones such as those in the Caribbean islands. Selenitic gypsum and well-crystallized rock salt are almost impermeable, but gypsarenites can have significant interconnected porosity. Fracture permeability is typical in most hard and brittle rocks including limestones, dolostones and of course sandstones, granites, basalts, and others. Fracturing is typically less dense in evaporites (gypsum and halite) since these rocks have a more ductile behavior. Conduit permeability can be important in low porosity and not too densely fractured soluble rocks, such as mature limestones and dolostones.

Matrix

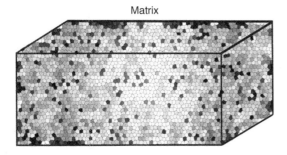

Matrix + fissures

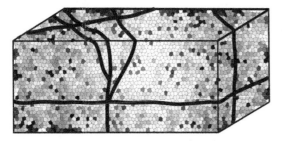

Matrix + fissures + conduits

Figure 5.11 The triple porosity model of karst aquifers.

Because of dissolution, these permeable pathways are constantly enlarged, progressively changing the karst aquifer properties. This positive feedback loop between dissolution and flow rate plays a central role in the hydrodynamic behavior of karst systems and their springs (Figure 5.13). Aquifers developing in soluble rocks ultimately tend to end up in low-porosity, high-permeability groundwater flow systems (Table 5.1) (Worthington 1999).

The large majority of water flow occurs along the conduit and enlarged fracture systems (well over 90%), whereas most of the water storage (over 90%) in the aquifer is in the matrix, no matter how concentrated or dispersed dissolution has acted on the rocks.

As indicated previously, groundwater flows from areas of high to low hydraulic pressure, in the downward direction of the hydraulic head gradient. In unconfined groundwater systems, the hydraulic head is often, but not always, similar to the topography of the ground surface. In confined flow systems, the direction of flow can be much more complicated. Hydraulic head is closely dependent on the boundary conditions of the groundwater flow system, such as the location where water enters and leaves the system (flow boundaries), or areas through which flow cannot occur (underlying, overlying, or lateral aquicludes or aquifuges). These boundary conditions can change over time, potentially resulting in drastic modifications in the groundwater flow pattern by modifying input and output areas or impervious boundaries. Entrenchment of rivers or glacial valleys lowers the local base level, increasing the hydraulic head gradient, whereas sea-level rise displaces output areas toward higher altitude, thus lowering the hydraulic head gradient. These regional topographic changes are controlled by an ensemble of geologic, climatic, and biological factors, and can be slow (e.g., tectonics) or fast (glacial-interglacial cycles) at a geological time scale.

We can describe a groundwater flow system by its three main components: where and how the water enters the system (inputs), where and how the water leaves the system (outputs), and the flow through the system (throughputs).

5.5.1 The Hydrological and Geological Water Cycle

Most groundwater flowing in karst aquifers has an atmospheric origin (rainfall and snow), but it is worth also mentioning other possible origins of groundwater. The hydrological cycle is the most

Figure 5.12 Schematic classification of karst aquifers and their dominant flow regimes. *Source:* Adapted from Atkinson (1985).

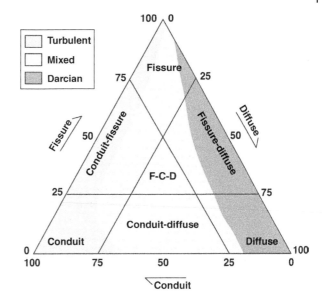

Figure 5.13 Schematic representation of the positive feedback of solutional enlargement of fractures on the hydrodynamics of the karst groundwater system. *Q* = flow rate; *P* = precipitation; *t* = time. *Source:* Adapted from Hartmann et al. (2014).

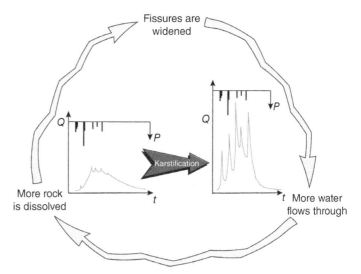

Table 5.1 Porosity, storage, hydraulic conductivity, and proportion of flow in Matrix (M), Fissure (F), and Conduit (C) permeability features in four contrasting karst aquifers: A. Silurian Niagara dolostone with poorly visible karstification. B. Lower Carboniferous limestone with extensive cave systems (Mammoth Cave area). C. Cretaceous chalk with high primary porosity. D. Eocene to Pleistocene young eogenetic limestone with large submerged cave systems close to the coast.

Area	Porosity (%)			Storage (%)			K (m s^{-1})			Flow (%)		
	M	F	C	M	F	C	M	F	C	M	F	C
A. Canada	6.6	0.02	0.003	99.7	0.3	0.05	1×10^{-10}	1×10^{-5}	3×10^{-4}	3×10^{-6}	3.0	97.0
B. Kentucky	2.4	0.03	0.06	96.4	1.2	2.4	2×10^{-11}	1×10^{-5}	3×10^{-3}	0.00	0.3	99.7
C. UK	30	0.01	0.02	99.9	0.03	0.07	1×10^{-8}	4×10^{-6}	6×10^{-5}	0.02	6.0	94.0
D. Yucatán	17	0.1	0.5	96.6	0.6	2.8	7×10^{-5}	1×10^{-3}	4×10^{-1}	0.02	0.2	99.7

Source: Adapted from Worthington et al. (2000).

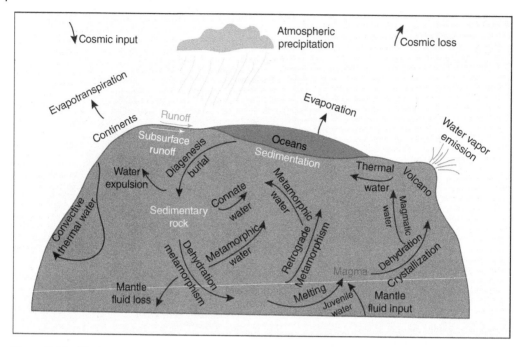

Figure 5.14 The hydrological and geological water cycles. *Source:* Adapted from Klimchouk (2017).

important one, but there is also a geological water cycle that may play an important role in some hypogene (deep-seated) conditions (Figure 5.14) (Klimchouk 2017).

The hydrological water cycle involves the atmospheric water, which can condense in clouds and create different types of precipitation, falling on land, or directly into the oceans. These precipitating waters are called **meteoric waters**. The water that falls on the continental surfaces creates runoff, or penetrates into the ground (infiltration waters) adding up to the groundwater flow. Some of these meteoric waters can have very deep flow paths, forming thermal groundwater circulation systems. The water in the ocean evaporates, whereas that on the land surface returns to the atmosphere directly (evaporation) or through the intermediation of plants (evapotranspiration). Some water in the atmosphere is lost or gained from the outer atmosphere (space).

Water is trapped in sediments (mainly in the oceanic depths) which get buried and undergo diagenesis. Part of this water held in sediments is expelled during burial due to lithostatic loading. The remaining water trapped in deeply buried sediments is called **connate water**, which, together with water released by metamorphic processes, make up most of the so-called basinal fluids. Also, crystallizing magmas release fluids rich in water, and these are termed **juvenile waters**, part of which are expelled through volcanic eruptions. Some of these juvenile fluids can also derive from deeper mantle sources.

The great majority of the waters involved in karst processes are of meteoric origin. Surface waters can act directly on the soluble rocks forming a characteristic set of surface landforms linked to an underground drainage system (e.g., sinking stream or dolines and related cave system). Meteoric waters can also follow long and deep flow paths rising to the surface at specific locations creating solutional porosity not directly related to the surface karst landforms. The former is categorized as **epigene karst**, whereas the latter is grouped in the **hypogene karst** systems (Klimchouk 2007). In the latter settings, especially deeper into the crust, connate, and juvenile waters can be involved, but most often the fluids interacting with the host rocks are of mixed origin.

5.5.2 The Energy and Forces Involved in Groundwater Flow in Karst

Groundwater flow in a karst aquifer and its evolution through time depends on the energy of the flow system and on the driving forces at play. Both kinetic and potential energies are important, under the form of chemical, thermal, and mechanical energies for the first, and gravitational energy for the second. Energy gradients (from high to low energy) provide the driving forces for the karst processes to take place (White 1988) (Figure 5.15).

Potential energy in karst depends on the vertical distance between recharge and discharge areas, which together with distance determine the hydraulic gradient. In other words, the potential energy is governed by the relief of the landscape. Water flows from high to low elevation areas, transforming potential energy into kinetic energy, and more specifically into mechanical and thermal energy. Flow of water through the unsaturated zone exerts mechanical work, most of the energy being used to overcome friction to flow, and only a minor amount of mechanical energy is spent by erosion and transportation of particles and solutes. The friction caused by descending water is partially transformed into heat, causing a small temperature rise, in ideal conditions around $2.34\,°C\,km^{-1}$ (Badino 2010).

Thermal energy in well-developed karst aquifer systems mainly depends on the temperature of the fluids (water and air) flowing through the system, being only marginally affected by geothermal heat flux. As a matter of fact, in aquifer systems where flow velocity is significant, such as in karst aquifers close to the water table, geothermal heat is transported by lateral flow of water, thus resulting in an almost negligible temperature increase with depth (i.e., the bottom of Veryovkina Cave in Abkhazia, with its 2212 m being the deepest in the world, is close to 1-2 °C, whereas the comparably deep gold and platinum mines in South Africa have wall temperatures in excess of 60 °C). In deep phreatic settings, far below the water table, flow diminishes thus allowing temperature to rise more significantly because of geothermal heat. The temperature increase of water flowing through a karst system due to geothermal energy, giving it enough time to equilibrate with the surrounding rock mass, mainly depends on the volume of water involved (and thus on recharge), but in alpine areas with 1000 mm of rainfall per year it can amount up to 0.5 °C, whereas in arid areas it can get much higher (Badino 2018a). These variations have a very small but noticeable effect on the solubility of several salts, such as gypsum and halite, as well as on quartz.

Water and air flowing through the vadose zone of a cave system will constantly exchange energy between themselves and the surrounding rock mass, ultimately reaching equilibrium. Although

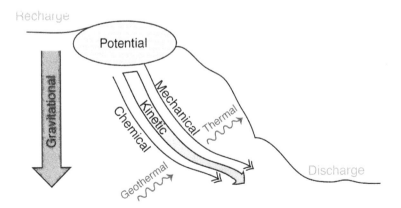

Figure 5.15 Energy fluxes in karst areas.

very difficult to evaluate and highly variable, infiltrating water is the main contributor to cave temperature, with air contributing only for a few percent (Badino 2010). Because of the different thermal capacities of water, air, and rock (that of water being five and four times higher than that of the rock and air, respectively), a rock massif may need hundreds to thousands of years to attain thermal equilibrium in response to climate change (Badino 2010). In a certain sense, caves record the temperature (and thus climate) of centuries ago. In vadose caves the thermal lapse rate in the cave atmosphere should be very similar to that of humid air masses in the lower troposphere ($4–6\,°C\,km^{-1}$), but is lower (between 2.8 and $4.0\,°C\,km^{-1}$), because of the energy release of descending waters and other factors (Luetscher and Jeannin 2004; Badino 2010, 2018a, 2018b).

Temperature also has a secondary effect on the fluids (especially water), with dynamic viscosity decreasing with increasing temperature. Water at $30\,°C$, typical of tropical areas, has a dynamic viscosity that is half of that at $5\,°C$, typical of alpine karst. This has a considerable effect on the transition from laminar to turbulent flow in narrow openings (see Eq. 5.7), which is extremely important during the initial stages of speleogenesis. This is one of the reasons why the breakthrough times for conduits in tropical areas are much lower than in cold temperature environments, other factors being kept identical (Dreybrodt et al. 1999). This positive effect on water flow with rising temperature (and lowering viscosity) is also important in deep phreatic settings, allowing karst development at great depths (Worthington 2001).

In general, the higher the energy in the karst system, the more efficiently the underground flow paths will be created and evolve. This energy depends on the combination of a wide variety of factors, but can mainly be summarized in a few points: (i) Topography: the altitudinal difference between recharge and discharge areas and their planimetric distance, both of which control the hydraulic gradient; (ii) The quantity (volume) of water that recharges the system; and (iii) The solutional aggressiveness of the water with respect to the minerals of the soluble rock, depending on factors such as the saturation degree of the solution, temperature, pH, and so forth.

Note that many of these factors are climate-dependent, and this brought several authors of the early years to classify karst accordingly (Lehmann 1954; Jennings 1985). It must however be noted that other factors, such as lithological and structural ones (Lowe 2000; Klimchouk and Ford 2000a; Skoglund et al. 2010), may have a decisive role in the development of karst aquifers and landscapes, limiting the applicability of the climate-based classifications.

5.5.3 Inputs

Inputs are of meteoric origin in epigene karst systems, directly on the soluble rocks, or by runoff derived from non-soluble terrains. Precipitation falls on soluble bedrock in the first case, such as a limestone mountain standing out in relief or an island underlain by coral reefs (Figure 5.16). This type of recharge is termed **autogenic**, and water tends to infiltrate through all the available outcropping fissures and pores. Recharge is normally diffuse, water chemistry is largely controlled by the composition of the karst rock, and flow rates at the multiple infiltration pathways are relatively low (Figure 5.17a). Autogenic infiltration can become more concentrated in places where the topography focuses runoff toward enclosed depressions such as solution dolines (Figure 5.17b). These large depressions commonly form at sites with initial greater permeability and infiltration rates, such as fracture zones. Here, greater percolation of surface waters enhances the dissolution rate through a positive feedback mechanism (Williams 1985). The wider the fractures become, the more water can infiltrate, enhancing dissolution, and removal of fines, further increasing flow rates. Commonly, the soluble bedrock is covered by more or less permeable soils that regulate the infiltration of water. Recharge through this surficial cover depends on the capacity of the soil of adsorbing

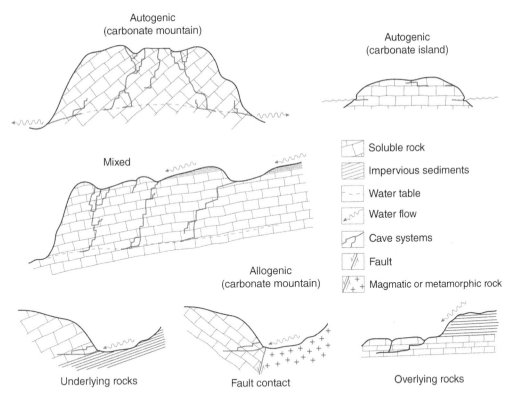

Figure 5.16 Different types of autogenic and allogenic recharge in karst areas.

Figure 5.17 Examples of karst inputs: (a) Karrenfield in Trentino, northern Italy. *Source:* Photo by Ugo Sauro. (b) Doline with some anthropogenic modifications close to Nikšić, Montenegro. See person in the center of the circular field for scale. *Source:* Photo by Jean-Yves Bigot. (c, d) The allogenic sinking stream of the Reka River below Škocjan village (Slovenia) in normal conditions (left) and during a flood. *Source:* Photos by Borut Lozej.

water, which depends on the intrinsic characteristics of the soil (e.g., grain size, sorting) and on the antecedent moisture conditions. Dry soils have a greater infiltration capacity than wet soils. Soils on soluble rock units mainly derive from the insoluble residues of the rock itself released by dissolution, and partly from airborne particles (aeolian dust, volcanic ashes). If these soils are permeable, infiltration rates are less attenuated, but if they are less permeable (more clayey) the soil mantle acts as an efficient infiltration regulator. Recharge through a soil cover is still classified as autogenic.

More often recharge to the soluble rock unit is from adjacent or overlying non-karstic rocks, generally occurring at discrete points in a rather concentrated way. This recharge is termed **allogenic**, and the typical examples are sinking streams sourced from non-karst terrain (e.g., the Reka River in the Classical Karst in Slovenia) (Figure 5.17c, d). Three main structural situations can be distinguished: (i) recharge from above through an erosional window at the edge of a stratigraphically higher retreating caprock; (ii) lateral recharge from a lower lying stratigraphic impervious rock unit; and (iii) recharge from an adjacent impervious rock unit juxtaposed to the soluble rocks by a fault (Figure 5.16). The volumes of water that enter the soluble rock depend on the contributing area of the sinking river, which is generally much larger than the area of a single, albeit very large, doline. Recharge from a more or less permeable non-karstic rock formation (caprock) overlying a soluble rock is also classified as allogenic. In this case, recharge will be regulated by the infiltration capacity of the upper non-karstic rocks, retarding the arrival of the meteoric waters. The dynamics and chemistry of allogenic recharge are very different from those of autogenic recharge, resulting in important differences in the development of the respective underground drainage networks (Palmer 2001).

Concentrated allogenic runoff disappears underground through input points known as **swallow holes** (ponors, swallets, and stream-sinks are synonyms). The caves associated with these focused flows can be vertical shafts, but more often they are nearly horizontal caves, especially if the allogenic flow comes from large, adjacent non-karstic areas. If the river is completely captured by the swallow hole the term lost river is sometimes used.

The capacity of a swallow hole to absorb the surface water flow mainly depends on the size of the underground passages, or more precisely on the narrowest constriction(s) occurring between the input point and the spring(s). This is also referred to as the **discharge carrying capacity** of a karst drainage system (White 2002). In mature systems, the size of the conduits tends to adjust to the largest flow rates that occur in the system. There is thus a **hydraulic control** on the amount of water that can flow through a karst groundwater system, which mainly depends on the size of the conduit, its length, tortuosity, and roughness (Palmer 1991). Ponding occurs if the water flow exceeds the inflow capacity of the swallow hole or the underground drainage system, and eventually part of the water will overflow the ponor or cause flooding upstream (backflooding).

The amount of water flowing through the system in mature karst systems mainly depends on the size of the recharge area (the bigger the area the larger the flow), and on the amount of atmospheric precipitation (mainly rain and snow) falling within it. This is called the **catchment control** by Palmer (1991). The amount of water flowing in most accessible cave systems is controlled by those factors, with water levels in the caves adjusting to the amount of water entering, and cave passage cross-sections slowly adjusting to the largest possible flows. During large floods the discharge carrying capacity of the conduits can be exceeded causing complete submersion of the cave passages, and water to rise from swallow holes. These swallets turning into resurgences, or vice versa, depending on the hydrological conditions, are named **estavelles**. Other examples of estavelles occur where karst groundwater systems discharge in a river, a lake, or the sea. The rise of the water level in these surface water bodies can lead to the local reversal of the hydraulic gradient and the groundwater flow, causing backflooding. The sea-, lake-, or river water flows into the conduits, penetrating over a certain distance into the cave passages. Once the flood pulse passes in the river (or lake), normal flow conditions will return and the backflooded water will flow downstream again.

5.5.4 Outputs

Karst aquifer systems end their subterranean pathways at karst springs, representing the main output points. Karst springs, together with those of some volcanic areas, are among the largest springs of the world, with mean flow rates that can be as high as $100\,\mathrm{m}^3\,\mathrm{s}^{-1}$ (a large river) (Table 5.2). They can be connected to an explorable underground river flowing out of a mountain, or to impenetrable fissures. The altitude of the spring mouth determines that of the water table at the output of the karst aquifer, whereas the hydraulic conductivity of the aquifer and the total discharge determine the slope of the water table: the higher the discharge and the lower the hydraulic conductivity, the steeper the water table gradient. Under increasing discharge conditions (floods) the water table will steepen faster with decreasing outflow rate and decreasing hydraulic conductivity.

Table 5.2 Large karst springs of the World.

| Spring | Flow rate ($\mathrm{m}^3\,\mathrm{s}^{-1}$) | | | Recharge area ($\mathrm{km}^2$) |
	Min.	Mean	Max.	
Tobio, Papua New Guinea	/	100		/
Matali, Papua New Guinea	20	90	240	350
Trebišnjica, Herzegovina	2	80	850	1140
Bussento, Italy	/	76	117	/
Dumanli, Turkey	25	50	/	2800
Galowe, Papua New Guinea	/	40	/	/
Ras el Ain, Syria	/	39	/	/
Ljubljanica, Slovenia	4	39	132	1100
Timavo, Italy	10	35	150	>1000
Ombla, Croatia	2	34	115	700
Chingshui, China	4	33	390	1000
Spring Creek, Florida, USA	/	33	/	>1100
Oluk Köprü, Turkey	/	30	/	>1000
Frìo, Mexico	6	28	515	>1000
Vaucluse, France	3	29	120	1160
Yedi Miyarlar, Turkey	/	25	/	>1000
Mchishta, Georgia	/	25	/	/
Coy, Mexico	13	24	200	>1000
Buna, Bosnia and Herzegovina	3	24	123	110
Silver, Florida, USA	/	23	/	1900

Source: After Smart and Worthington (2004b), Ford and Williams (2007), Kresic and Stevanović (2010), and White (2019).

The higher the altitude difference between the output (spring) and the water table upstream (recharge area), the greater the hydraulic head in the system. A raise in the hydraulic head involves an increase in the energy of the system, enabling the establishment of deeper groundwater circulation systems. In other words, the position of the springs exerts a fundamental control on karst groundwater circulation. The elevation of the outlets can change relatively rapidly by tectonic and/or climatic driven erosion/aggradation (e.g., fluvial incision, glacial advance, and retreat) and relative sea level changes.

Springs can be classified in many different ways, based on flow rate (large, small), flow duration and activity (perennial, intermittent, and paleospring), recharge area (emergence when the source is unknown, exsurgence for autogenic diffuse infiltration, and resurgence when the sinking river, often allogenic, is known), position (local base level spring, hanging, buried, submarine, intertidal, and sublacustrine), chemistry (fresh water, brackish, saline, mineralized, and sulfuric), temperature (thermal, lukewarm), ecological, and geological characteristics (Smart and Worthington 2004b; Springer and Stevens 2009). From a hydrological point of view three main types can be easily distinguished (Ford and Williams 2007): free-draining springs, dammed springs, and confined springs (Figure 5.18).

In free-draining springs the groundwater flows by gravity from the karst aquifer to the lower lying discharge point. This is a common type of descending (or gravity) spring, which can be of the conduit type or of the diffuse type, depending on whether the underground river can be explored upstream from the spring or not. Local geological conditions (faults, undulating basal contact with impermeable rocks) can cause local ponding of water, with the formation of localized sumps (drowned passages and isolated phreatic zones). Distinction can be made between underflow and overflow springs, the first being the lowest lying, generally perennial springs (often called local base level or graded springs), and the second being higher lying distributaries which are often intermittent, becoming active during high discharge events (sometimes called hanging springs).

The most common type of karst spring is of the dammed type, where water is forced to discharge because of the presence of a major hydrogeological barrier. This underground impoundment can be related to a normal lithological contact, a fault contact (Figures 5.18c and 5.19f), aggraded less permeable sediments (alluvium, glacial deposits) (Figure 5.18d), or the presence of a dense body of saline water (coastal areas) (Figure 5.18e). Further, in this case both underflow and overflow springs (Figure 5.19b) are generally present (Figure 5.18c) and when the karst system has been drowned or aggraded, springs can also occur below local base level (submarine springs or *vrulja*, sublacustrine springs). Springs at sea level can be influenced by tidal oscillations and are called intertidal springs (Figures 5.18e and 5.19c), whereas water outlets can also occur where karst springs are buried by more or less permeable alluvium (buried springs creating swamps, wetlands, and seeps).

Where the karst aquifer is confined by an overlying impermeable rock unit, the water flows under hydrostatic pressure, escaping from the water-bearing unit in areas where the hydraulic confinement is breached by erosion or along fault zones connecting the artesian aquifer to the surface. These artesian springs are also known as "vauclusian" springs (Figures 5.18f and 5.19a, d, e), from the *Fontaine de Vaucluse* in southern France (Figure 5.19e). Some of these springs can be very deep (over 300 m) and when the circulation pathways extend far beneath the surface, water can be thermal and mineralized.

5.5.5 Throughput

Underground water flow in karst is mainly controlled by the hydraulic gradient, with water following the most easily penetrable pathways such as bedding planes and interconnected fissures and pores. The density, size, and distribution of voids available for fluid flow determine the quantity of water that

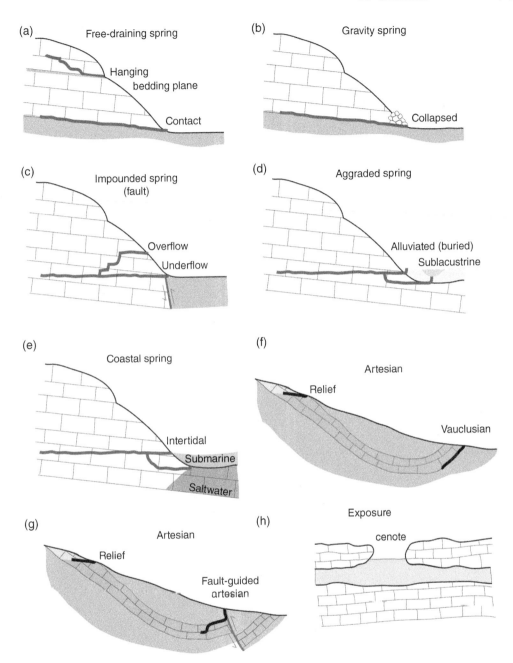

Figure 5.18 Types of karst springs: (a) free-draining spring; (b) gravity spring; (c) impounded spring; (d) aggraded spring; (e) coastal spring; (f) vauclusian spring; (g) fault-guided artesian spring; (h) cenote.

can be stored and transferred through the karst aquifer. The main drivers of the functioning of karst aquifer systems are gravity, and the position of the inputs and outputs, which are controlled by the topography. The altitudinal position of recharge and discharge areas and their horizontal distance dictate the hydraulic gradient: high elevation differences over short distances create steep hydraulic gradients, whereas long distances and small height differences cause the hydraulic gradient to be low.

Figure 5.19 Examples of karst springs: (a) The vauclusian spring of Su Gologone, Sardinia, Italy. *Source:* Photo by Vittorio Crobu. (b) Kläfferquelle, Hochschwab (Austria) after snowmelt. *Source:* Photo by Lukas Plan. (c) The Puerto Princesa Underground River flowing into the Philippine Sea. *Source:* Photo by Paolo Petrignani, La Venta Esplorazioni Geografiche. (d) The Cetina River spring in Croatia. *Source:* Photo by Arthur Palmer. (e) Fontaine de Vaucluse, France. *Source:* Photo by Philippe Crochet. (f) Spring of the Loue River, Doubs, southern France. *Source:* Photo by Philippe Crochet.

In isotropic aquifers, where the distribution of pores and fissures is homogeneous, water will move from input to output following the shortest flow route. In anisotropic aquifers, such as those developed in karst areas, this will not necessarily be true. Water will tend to follow the direction of the steepest hydraulic gradient, which is influenced by local changes in permeability and thus hydraulic conductivity.

Water will thus follow the shortest and steepest routes with the highest hydraulic conductivity. We can make a distinction between unsaturated (vadose) and saturated (phreatic) flow patterns.

Infiltrating water will tend to flow downward in the vadose zone, following the steepest permeable pathways. The flow is mainly driven by gravity and the openings that are enlarged by gravitational vadose waters will almost always be descending. Only where local geological conditions do not allow water to follow the downward direction (e.g., on less permeable beds), water can be obliged to form an underground pond limited downstream by a spillover, creating a perched phreatic section located above the local water table level. The most favorable flow paths are vertical fractures and steeply dipping bedding planes. Often the steepest available routes are too narrow to allow the entire amount of water to flow through, at least initially, so part of the water is obliged to follow other less steep openings. Eventually the steepest routes will be enlarged by dissolution and mechanical erosion, enabling all water to flow through, and the less steep segments will be progressively abandoned. This is why vadose cave segments are often characterized by a step-like longitudinal profile, with vertical shafts interspersed with less inclined canyon passages. Eventually, vadose waters reach the underlying phreatic zone. In some cases, however, the presence of less permeable rock units can force some vadose waters to create an output above the local water table (perched springs).

In contrast to the vadose zone, water does not necessarily follow the steepest descending openings in phreatic conditions. Water flows toward the outputs along a descending hydraulic gradient and following the shortest possible pathways with greatest hydraulic conductivity. In normal conditions, both large and small passages will be exploited by groundwater flow, with the larger passages carrying greater quantities of faster running waters, and smaller openings carrying minor amounts of slowly flowing waters. So, large conduits tend to concentrate most of the water, which is transmitted rapidly, whereas in the surrounding fractured bedrock the tiny fissures and smaller conduits transmit smaller amounts of water in a less efficient way. As a consequence, the local water table will be lowest in the larger conduits, and higher in the surrounding less permeable bedrock. In normal conditions water will flow from the fractured bedrock into the larger conduits, and from the vadose zone to the phreatic one. This causes underground water flow to concentrate along the most efficient pathways and this is why most cave systems are characterized by passages converging toward the main draining conduits, resulting in a branching pattern. The water level rises during floods, pressurizing both the conduits and the fractured bedrock. The larger conduits will be characterized by faster flowing and less mineralized waters, and these waters will be forced into the surrounding fractured bedrock. When the flood recedes, and the local water table level drops to its original position, the less mineralized waters previously pushed into the fractured bedrock will slowly flow back into the conduit (Figure 5.20).

Cave passages formed in the phreatic zone have an undulatory development, with rises and falls along their longitudinal profile. Unless the contrasts in hydraulic conductivity are very important, groundwater flow will mostly concentrate at or slightly below the water table level, since this is the shortest and less energy-consuming pathway. Only where highly conductive deep fractures or bedding planes are present, phreatic water can follow the longer flow routes at the expense of the tinier openings closer to the local water table. Some examples of underground water flow in caves in different lithologies are shown in Figure 5.21.

5.5.6 Lithological and Structural Control on Groundwater Flow

Groundwater in a karst aquifer flows under the influence of gravity connecting the inputs to the outputs and following the shortest and less resistant pathways. The geological and structural setting plays a fundamental role in determining these most efficient and feasible pathways (Figure 5.22).

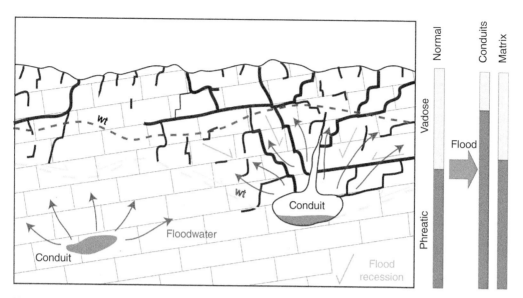

Figure 5.20 Vadose and phreatic flow before, during and after a flood.

Figure 5.21 Examples of underground water flows: (a) Lurgrotte, Austria, a river cave in dolostones. *Source:* Photo by Lukas Plan. (b) The underground river of Imawarì Yeuta, the longest quartzite cave in the world. *Source:* Photo by Vittorio Crobu, La Venta Esplorazioni Geografiche. (c) The river passage of Parks Ranch cave in the Permian gypsum of the Castile Formation, New Mexico. *Source:* Photo by Lukas Plan. (d) Underground river lakes in the Luigi Donini cave in Jurassic limestone, Sardinia, Italy. *Source:* Photo by Vittorio Crobu.

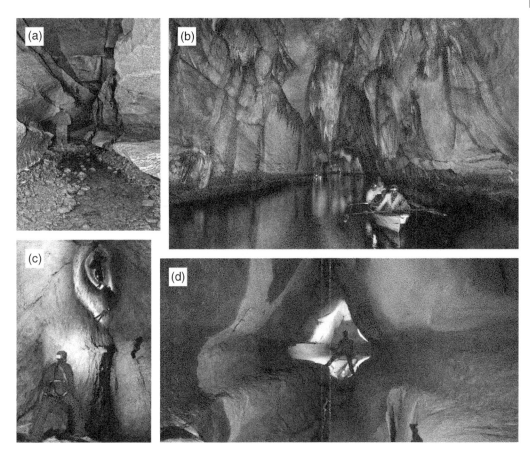

Figure 5.22 Structural control on underground drainage development: (a) Underground river controlled by the intersection of a vertical fracture and dipping bedding planes in Messinian gypsum, Rio Basino Cave, Italy. *Source:* Photo by Piero Lucci. (b) God's Highway in the Puerto Princesa Underground River, Palawan (Philippines), following a vertical 200-m long fracture zone. *Source:* Photo by Paolo Petrignani, La Venta Esplorazioni Geografiche. (c) A vertical vadose shaft in dolostones (the narrow upper part) and limestones (below), developed along a fracture, Su Eni 'e Istettai, Sardinia. *Source:* Photo by Vittorio Crobu. (d) Notches in a differentially eroded horizontal sandy limestone bed in the Mawpun sandstone cave in Meghalaya, India. *Source:* Photo by Daniela Barbieri, La Venta Esplorazioni Geografiche.

Water tends to follow the interconnected primary and secondary permeability structures, composed of the matrix porosity, the fissured system and the conduits. These permeability elements change through geological time, generally with fissures and conduits increasing their contribution to the total permeability. The main factors controlling the development of a karst groundwater flow network are lithology and structure.

Karst aquifers can develop in different soluble rocks, mainly carbonates, sulfates, salt rock, and, to a minor degree, quartz sandstones. The crystalline texture of evaporites causes their original porosity to be very low and permeability in gypsum is mainly related to discontinuities such as bedding planes and fractures. The latter display lower densities than in carbonate rocks because of the more ductile behavior of this lithology. Fractures in salt are very rare because of its extremely low yield strength and tendency to experience viscous deformation. Fracturing in salt is commonly restricted to a near-surface zone affected by stress-release and dilatation. Quartz sandstones have a medium to high primary porosity (mainly depending on grain size), which can be significantly reduced by cementation and recrystallization. Also in these rocks the most suitable permeability pathways are given by fractures and bedding planes. Carbonate rocks are much more complex, both from a chemical and a textural point of view. They can be composed of calcite-aragonite (limestones) or dolomite (dolostones), and can be crystalline, granular (fine to coarse), fossiliferous, and more or less pure. Being competent rocks, they can be densely fractured and are often well bedded (with highly variable thicknesses).

Differences in lithological composition in a sedimentary rock sequence can be significant for groundwater flow. Some rock types are more soluble than others (e.g., calcite is slightly more soluble than dolomite and dissolves at significantly faster rates) and some stratigraphic successions can contain insoluble and impervious beds. Clayey limestones, when dissolved, leave the insoluble clays, which often tend to clog and fill the embryonal solutional voids and protoconduits, eventually preventing water from flowing through. The lower the purity, the greater the difficulty with which permeability pathways are created by dissolution. In an alternation of limestones and shales, water flow often concentrates along limestone beds at the contact between these lithologies with contrasting hydrological and mechanical properties. Karst drainage is commonly subdued in thinly bedded alternations of limestone and shale.

The most efficient permeability pathways for groundwater flow in karst aquifers are of structural nature, mainly bedding planes, fractures, and faults. These can be grouped under the general term of fissures or fractures, which are planar breaks that can be penetrated and modified (dissolution, precipitation) by circulating groundwater. These planar openings are actually highly irregular, with widenings and constrictions along the groundwater flow path, sometimes completely closed in certain areas. Their aperture varies in size from 0 (completely sealed) to several meters, but generally ranges between 10 μm and 1 mm (Klimchouk and Ford 2000a).

Bedding (or parting) planes are created in sedimentary sequences by a change or a break in sedimentation. Changes can be minor (small variations in grain size) or more distinct (e.g., a thin clay layer between two limestone beds). Initially, bedding planes are often poorly permeable, and so not important for groundwater flow. Some processes such as layer-parallel slippage (i.e., flexural slip) causes bedding planes to become permeable, because of the creation of sheared zones and small voids along the moving surface. Also, rapid unloading can cause bedding planes to open because of pressure release (e.g., glacier retreat). The lateral (areal) extent of penetrable bedding planes depends on the thickness of the separated beds: in thinly bedded sequences, permeable bedding planes extend over a restricted area, whereas in thickly bedded carbonate series they can cover basin-scale extensions (hundreds of kilometers).

Joints are breaks of natural origin in a rock body with no appreciable shear displacement, whereas faults show a measurable offset by shearing. Fractures in sedimentary beds can be caused by a variety of processes, including shrinkage or expansion in the early phases of diagenesis, compression, loading and unloading, heating and cooling, and tectonics. More or less parallel joints make up a joint set, whereas two or more different sets cutting at regular angles are known as a joint system. In regularly bedded and poorly deformed limestone and dolostone sequences, an orthogonal joint system is often observed, but joint sets cutting at 60° angles are also common in such rocks. Joints can be confined to single beds, and are often perpendicular to the bedding planes. Some joints can cross multiple beds and reach considerable lateral extension. These are often called master joints, and are generally important features for groundwater flow (and thus cave formation). In general, the density of joints in a bed is largely dependent on the bed thickness: thin beds will have closely spaced joints, whereas thick beds will have more sparsely spaced joints and fractures. If the fractures are penetrable by fluids, thin soluble and densely fissured beds can be rapidly removed by dissolution.

When joints are filled with secondary minerals (often quartz or calcite) they are called veins, which often function as barriers for groundwater flow. Faults can also be characterized by low permeability because of the presence of mylonite, cemented and recrystallized fault breccias, or clay smears. However, the volume of the deformed rock surrounding the fault core (called the damage zone) (Kim et al. 2004) usually presents structural elements, including minor secondary faults and feathering features such as splay joints, that often create zones of increased permeability (Caine et al. 1996).

5.5.7 Karst Groundwater Basins

In general conditions, groundwater and surface waters are treated separately, because they flow at different velocities and obey to different laws. Groundwater flows through the interconnected pores and fractures of aquifers, at flow velocities in the order of some meters per year, whereas surface waters flow in stream channels at velocities that can reach a meter per second (eight orders of magnitude greater). This distinction becomes superfluous in karst, where surface and underground waters must be treated together, and flow velocities are extremely variable, from very fast in conduits, to very slow in the matrix porosity. Karst groundwater basins are composed of allogenic surface streams sinking underground, autogenic point recharge (through dolines or swallow holes) and diffuse infiltration, and an underground drainage system (interconnected fissures, conduits, and matrix porosity), all contributing to the flow of springs. On the other hand, drainage basins (or watersheds) at the surface are defined by topographic (or drainage) divides, which determine the quantity of precipitation contributing to the water flow in a river. The definition of the boundaries of a karst groundwater basin is a much more complicated task than defining watersheds on a topographic map. Rivers disappearing underground can pass underneath different topographic divides and feed a karst spring located in a distant topographic watershed. Underground drainage divides can change in different hydrological conditions, with springs being fed by groundwater basins with variable dimensions depending on the altitude of the changing water table level. The delineation of the boundaries of a karst drainage basin ideally relies on a combination of geological, hydrogeological, and structural investigations, the survey of the position of the water table, speleological explorations and mapping, and dye tracing.

Recharge in a karst groundwater basin can occur in a very concentrated way (autogenic and allogenic sinking streams), through a large number of small temporary streams feeding numerous dolines, or by diffuse infiltration in bare karrenfields or through soil-covered karst, giving rise to different responses at the karst springs. The epikarst plays an important hydrological role in the recharge of karst aquifers, forming a perched reservoir of water during long droughts and acting as a retarding filter for the sometimes impulsive nature of recharge (Williams 2008).

5.5.8 Hydrological Role of the Epikarst and the Transmission Zone

The epikarst is the uppermost part of the vadose (or unsaturated) zone in areas where soluble rocks occur (Williams 2008). It is characterized by highly weathered soluble bedrock, can be covered by a soil cover, and overlies the lower part of the vadose zone known as transmission zone. Although its thickness can vary widely, it is typically between 2 and 10 m. Its thickness depends on the balance between the surface lowering rate by denudation and the rate at which the development of the epikarst zone progresses downward by dissolution. Its permeability is given by the presence of fissures variously enlarged by dissolution, which aperture and density rapidly diminishes with depth from the surface. Overall permeability in the underlying transmission zone is concentrated along a few enlarged major fissures, and is much smaller than that in the overlying epikarst (Figure 5.23). This difference in permeability between the transmission zone (low permeability) and the epikarst (much higher permeability) results in the formation of a perched aquifer

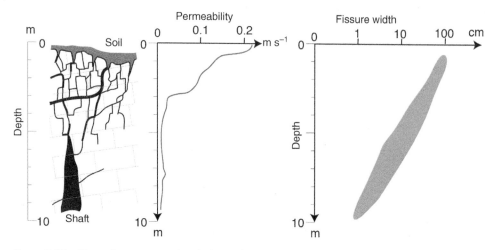

Figure 5.23 The epikarst zone and variation of permeability and fissure density with depth.

in the latter. From a hydrological point of view, the epikarst zone functions as a water storage that delays recharge from precipitation to the underlying transmission zone. The epikarst also constitutes a significant aquatic shallow subterranean habitat, with its own unique and specialized fauna (Culver and Pipan 2019). In limestone, it typically has a secondary porosity of 10–30%, which contrasts with the underlying less weathered rock in the transmission zone that generally has a porosity of less than a few percent (Figure 5.24a).

In gypsum rock, characterized by lower fracture densities, the development of the epikarst zone is limited, and so is the storage capacity (Ferrarese et al. 2002) (Figures 4.16 and 5.24b). Only some fractures funnel the percolating waters down to the phreatic zone. The infiltration capacity of the gypsum outcrops can be further diminished by secondary precipitation of gypsum and the development of crusts in arid climates, or by calcite precipitation in open fissures in more temperate zones.

In carbonate terrains, infiltrating water spends great part of its dissolution capability in the upper few meters. Only where concentrated infiltration occurs, inflowing waters can maintain

Figure 5.24 Epikarst in (a) limestone (quarry near Zavala, SW of Popovo Polje, Bosnia and Herzegovina). *Source:* Photo by Arthur Palmer, and in (b) gypsum (quarry near Bologna, Italy) (the wall is 7 m high). *Source:* Photo by Francesco Grazioli.

their aggressiveness over longer distances. It is generally believed that about 70% of the dissolution occurs within the first 10 m from the surface. As a consequence, the frequency and width of the solutionally enlarged joints diminish with depth, tapering out in the relatively unweathered bedrock below. Water is temporarily stored at the base of a well-developed epikarst zone because surface infiltration capacity is much higher than the percolation rate in the underlying transmission zone. This aquifer is perched above a leaky capillary barrier and its groundwater level is drawn down above the most permeable fractures (Williams 2008). Only few of the enlarged fissures penetrate into the underlying rock, forming the main flow paths for infiltration waters through the underlying transmission zone. The long-sustained focused percolation of water along the preferential leakage paths can give rise to the formation of vertical shafts at the base of the epikarst zone. These vertical voids, often called "hidden shafts," develop by downward flow coming from the perched aquifer of the epikarst. These enlarging shafts can lead to the formation of collapse sinkholes or can be intercepted by surface erosion (unroofing) (Klimchouk 1995; Klimchouk et al. 1996c).

The presence of only a few high conductivity flow paths in the transmission zone underlying the epikarst causes the water table to be undulated, with lows located above these preferential drainage routes. Groundwater flow and thus dissolution are focused in the cones of depression. This preferential dissolution leads to greater surface lowering above these points of efficient underground drainage, and a positive feedback is created between surface lowering, focused surface flow, and enhanced dissolution underground (Figure 5.25).

The geometry and the role of the epikarst are highly variable because many factors are involved in its presence and development. Karst aquifers do not always have an epikarst, either because it has not developed or because it has been removed (e.g. by glacial erosion). In such systems, storage and transmission through the vadose (unsaturated) zone are regulated by the permeability features of the rocks.

5.5.9 The Water Table in Karst

There has been much debate on whether a water table is present in karst aquifers or not. This is due to the fact that the water level observed in boreholes and in caves can be very irregular, at least in comparison to what hydrogeologists normally find in porous homogeneous aquifers. Also dye tests sometimes show different flow paths crossing each other at different altitudes without mixing (Zötl 1961). This is due to the existence of perched phreatic levels, representative of rather small portions of the entire karst aquifer. It is now widely accepted that the water table concept has its validity also in karst aquifers, as long as it is used at a large, groundwater basin scale. Its validity is more limited at a local, conduit-, or borehole-scale, where large variations in water level are mainly reflecting great spatial differences in permeability. The highly irregular water table level in karst rocks is mainly due to significant differences in hydraulic conductivity: flow is fast in conduits and enlarged fractures, whereas it is much slower in the tiny fissured system and the matrix porosity. This has great consequences on both the recharge and the discharge rate of the different portions of the karst aquifer system.

In well-karstified aquifers, the water table level is mainly controlled by the altitudinal position of the springs, which in turn depends on the local base level (river bed, lake, or sea level, a much less permeable lower lying rock unit). If the spring is fed by a large conduit, the water is drained in a very efficient way and the water level rises very slowly in the upstream direction. In these high-velocity flow paths the hydraulic gradient is very low. A main conduit will act as a low head in the karst aquifer, directing all water in the adjacent fissures, smaller conduits, and pores toward the conduit.

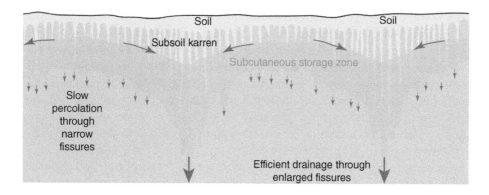

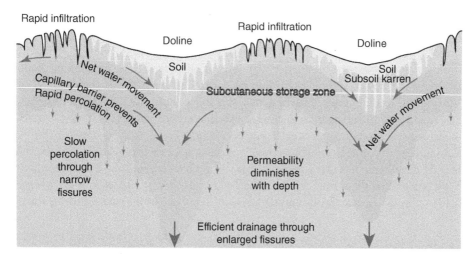

Figure 5.25 The epikarst zone and the development of solution dolines. *Source:* Adapted from Williams (2008).

In a map of equipotential lines, during low flow conditions, the water table would be lowest in correspondence with these main drains, and higher in the less permeable rock portions. During floods, water is rapidly transmitted through the most efficient conduits, and the water level rises accordingly. Often the water level in the conduits rises well above the altitude of their highest roof, and low mineralized floodwater can be injected into the cracks and small openings. This water will then return to the conduit when the flood recedes and the local water level drops to its normal position (Figure 5.20).

The highly irregular position of the water table in karst aquifers causes boreholes only meters apart to show sometimes extremely different groundwater levels. Whereas one borehole can be productive, because it has intercepted a widened fracture zone well connected to the conduit system, the borehole next to it can be completely dry, having intercepted only matrix porosity with very low hydraulic conductivity. Water table maps based on the groundwater level intercepted by many boreholes may highlight zones of lower water level, which correspond to areas of higher permeability, where master conduits (and sometimes caves) might be suspected. On the contrary, areas of high water table level revealed by borehole monitoring normally indicate the boundaries of the groundwater basin. The delineation of the boundaries of a karst groundwater basin should rely on a combination of tracer tests and knowledge on explorable cave systems (which provide insight into the minimum extent of the conduit system), and water table mapping (representing the matrix and fissured aquifer behavior).

5.5.10 Freshwater–Saltwater Interface

Above, we have described the behavior of karst aquifers in continental settings, where groundwater is generally fresh and has small differences in solute contents and density. This situation is not encountered in coastal areas, where both freshwater and saltwater are present, having very different chemical compositions and densities. These two different kinds of water tend to occur one on top of each other, with the higher density saline waters below and separated by an interface inclined landwards. The freshwater forms a sheet of water on top of the underlying salty water in coastal (anchialine) caves, and the saltwater–freshwater interface is often sharp, known as the halocline. The depth of this interface (Z_s) mainly depends on the differences in density between the two water bodies, and on the elevation of the water table above sea level, and is given by the Ghijben–Herzberg equation, named after the two European scientists who first described this phenomenon around 120 years ago (Badon-Ghijben 1888; Herzberg 1901), see Reilly and Goodman (1985):

$$Z_s = \left(\frac{\rho_f}{\rho_s - \rho_f} \right) \cdot h_f \tag{5.13}$$

where ρ_s and ρ_f are the densities of seawater and freshwater and h_f is the water table elevation (in meters above sea level).

The density of freshwater is 0.975 times that of seawater, corresponding to a freshwater/seawater ratio of 40/41. In practice, for every meter of freshwater standing above sea level, the halocline depth will be 40 m lower. Or, similarly, a 1-m-water table drop by pumping in a well will cause the halocline below it to rise 40 m. Whereas this rise occurs with a significant delay in porous aquifers (and the return of freshwater once pumping ceases in a salinized aquifer), it is considerably faster in the highly permeable coastal karst aquifers.

The Ghijben–Herzberg equation is generally valid if the boundary between saltwater and freshwater is well defined, the flow of the freshwater on top is horizontal and the underlying seawater body is static (Vacher 1988; Fratesi 2013). These conditions are hardly ever met in nature. The freshwater/saltwater contact is almost never sharp, and there is a gradual downward increase in salinity. This regular pattern can be disturbed at lithological and hydrogeological contacts because of the differences in permeability and hydraulic conductivity and the local flow of freshwater toward the saltwater zone and vice versa. The mixing zone can be very thick, and may even extend almost to the water table. Mixing between saltwater and freshwater occurs because of tidal fluctuations, seasonal and local variations in recharge, and anthropic influences (pumping). The tidal influence diminishes landward, and so does the mixing zone. The thickness of the mixing zone is proportional to the permeability of the limestones, and tidal influences can propagate for several kilometers inland in well-karstified limestones. For instance, in the Puerto Princesa underground river cave the tidal influence is visible up to 7 km from the coastline (Badino et al. 2018), whereas in Yucatán, Mexico, tidal oscillations can still be measured 9 km inland (Beddows et al. 2002).

In general, the thickness of the freshwater lens is proportional to the amount of recharge (which depends on climate, the size of the recharge area, etc.) and decreases with increasing permeability and hydraulic conductivity. Since hydraulic conductivity in karst areas can vary by orders of magnitude even over short distances, the geometry of the freshwater lens can be extremely variable. Permeability tends to increase with time due to dissolution, so that freshwater lenses tend to be

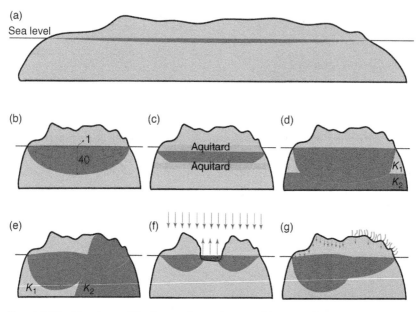

Figure 5.26 The shape of freshwater lenses in coastal karst aquifers in different recharge and permeability conditions. (a) The freshwater lens shown without vertical exaggeration. (b) Freshwater lens with vertical exaggeration. (c) Freshwater wedge in a confined aquifer. (d) Freshwater lens in a succession of rocks with different permeability ($K_2 > K_1$). (e) Freshwater lens in an island with a sharp lateral variation in permeability. (f) Freshwater lens influenced by an inland lake with strong evaporation. (g) Freshwater lens on an island with contrasting zones of infiltration (vegetation). *Source:* Adapted from Fratesi (2013).

thicker in younger limestones than in the more karstified and permeable older ones. Some general examples, including areas with changes in permeability and/or recharge, are shown in Figure 5.26.

Freshwater flows toward the sea according to the hydraulic gradient, whereas saltwater is introduced into the deeper parts of the coastal aquifer. This saltwater intrusion can be caused by a series of mechanisms, including the replacement of brackish and saltwater which is entrained by the seaward flow of freshwater on top, and convectional flow caused by differences in density (temperature and salinity).

5.6 Analysis of Karst Groundwater Basins

5.6.1 Defining Karst Aquifers

Karst aquifers have distinctive characteristics that make them very different from porous aquifers, where flow occurs through the interconnected intergranular porosity, and laminar flow equations can be used to approximate the behavior of the aquifer. In karst aquifers, the vast majority of the flow, often over 90%, occurs through enlarged fractures and conduits, whereas most of the water is stored in the matrix porosity and the epikarst (upper few meters of fractured and intensely weathered bedrock). Groundwater flow is extremely fast in the large flow pathways (conduits and enlarged fractures), relatively fast in the fissure network (fractures, joints, and bedding planes), and very slow in the matrix porosity (intergranular medium), resulting in a typical triple behavior (White 2002). These triple-porosity aquifers can sometimes be simplified in dual-porosity ones (Worthington and Ford 2009).

Because of their heterogeneous and anisotropic character, the investigation of karst aquifers cannot be comprehensively addressed with the classical hydrogeological investigation methods alone (Darcy's Law). Flow is Darcian in the matrix porosity and in the tiny fissures of the aquifer, but turbulent or non-Darcian in the conduits and enlarged fissures. Most boreholes drilled in karst areas do not intercept the enlarged flow paths (caves and conduits), and the data derived from them mostly represent the matrix part of the aquifer. In these conditions, and with some caution, Darcy's Law can be applied, but it will only represent the behavior of the aquifer at a local scale. At a basin-wide scale, groundwater flow cannot be based on Darcian principles, since most of the water flows through a few pathways at much greater velocities, often exceeding $100\,\text{m}\,\text{d}^{-1}$, and under turbulent regime. Hydrogeologists need to study the matrix and fracture components of the aquifer, applying methods similar to those used in porous aquifers, in combination with specific techniques suitable for the turbulent flow associated with enlarged fractures and conduits. But when can we really state that an aquifer is karstic or not?

Defining karst aquifers is a difficult task (Worthington et al. 2017). The geomorphological approach uses the association of the typical surface karst features (sinking streams, sinkholes, and karst springs) to identify the presence of a karst aquifer. This allows us to recognize rather easily if an aquifer must be regarded as karstic or not, but the lack of typical surface karst features does not imply that the underlying aquifer is not karstic. Many karst aquifers are characterized by an extensive network of interconnected solutionally enlarged fissures with an aperture of less than 1 cm, but with no evidence of caves or typical surface karst landforms. Most karst aquifers are characterized by turbulent flow, and Atkinson and Smart (1981) restrict the use of karstic groundwater flow to that occurring through conduits in which flow is turbulent. Freeze and Cherry (1979) distinguish a karstic carbonate formation from a non-karstic one on the basis of hydraulic conductivity values, being the karstified ones characterized by values greater than $10^{-6}\,\text{m}\,\text{s}^{-1}$ ($0.864\,\text{m}\,\text{d}^{-1}$). Huntoon (1995) uses a much broader and practical definition: *Karst is a geologic environment containing soluble rocks with a permeability structure dominated by interconnected conduits dissolved from the host rock, which are organized to facilitate the circulation of fluid in the downgradient direction, wherein the permeability structure evolved as a consequence of dissolution by the fluid.* This broad definition includes almost all limestone formations in the category of karst aquifers. In general, when studying karst aquifers (or presumed ones) it is important to keep in mind that flow through the rocks includes self-organized channel networks, in which groundwater can flow rapidly, with velocities often greater than several tens of meters per day (Worthington and Ford 2009).

5.6.2 Complementary Approaches in Karst Aquifer Studies

The anisotropy and heterogeneity of karst aquifers make them very difficult to study and to model compared to isotropic and homogeneous aquifers with intergranular or fracture permeability in non-soluble rocks. Storage systems and flow conditions in karst aquifers are almost impossible to generalize due to the extreme variability in geological contexts (hydrostratigraphy, structure, and topography). Conditions encountered in a borehole may be significantly different from those found in another borehole only a few meters apart. Classical modeling techniques in which the hydraulic properties such as hydraulic conductivity and storativity change predictably in space can seldom be used in karst.

A karst aquifer is sometimes treated as a black box, in which groundwater flow and transport behavior are predicted only on the basis of inflow and outflow parameters, ignoring the internal permeability structure of the aquifer (enlarged fractures and conduits). Often some characteristics of the karst aquifer are known, such as the position and size of conduits, sections of underground rivers, and the distribution and density of the fracture network. A gray box approach can be used in such cases. Water in karst is recharged through rainfall and snow melting that create runoff, water retention, and concentrated and/or diffuse infiltration. The infiltration water is temporarily stored in different compartments (soil, epikarst, vadose and phreatic caves, saturated fissures, and pores), flowing from one store to the other, and ultimately surfacing at various types of springs or recharging the deep regional groundwater system (Figure 5.27).

Different complementary approaches should be applied to study these complex groundwater systems, some used also in the study of conventional aquifers, others designed specifically for karst. First of all, a water budget assessment needs to be performed, and this requires the definition of the groundwater basin boundaries and the acquisition of spring hydrographs. Identifying the hydrogeological basin of a karst system often relies on tracer tests, with tracer injection in sinking streams, underground rivers, or boreholes, and tracer detection at springs or in boreholes. The local behavior of the karst aquifer can be studied in boreholes, but such data may have a limited representativeness for the characterization of the entire groundwater system. Additional data can be obtained by monitoring the hydrochemical behavior of springs, rivers, and minor underground water flows (trickles, small rivulets) in caves, and boreholes. The acquired data can feed groundwater models, which simulate flow and transport in the karst aquifer system.

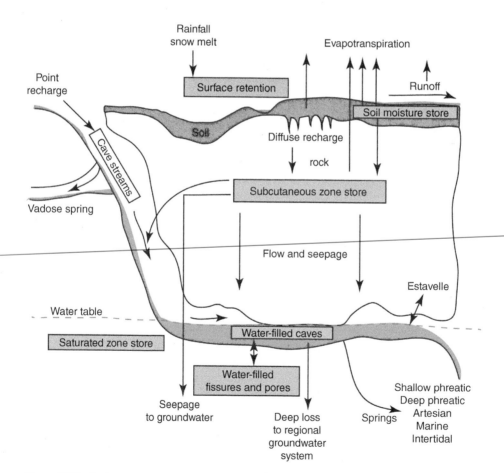

Figure 5.27 Recharge, storage compartments, and interconnections in a typical karst groundwater basin.

There is no single "best way" for studying a karst aquifer. The selection of the optimal approach for each situation depends on the characteristics of the aquifer, the available data and the objectives of the investigation. Some basic assumptions can be made to decide which methods should be used initially, but flexibility is needed to adjust the adopted methodologies in light of the results obtained. Often different complementary methods are combined in order to get better and more reliable results, and classical methods used in homogeneous and isotropic aquifers frequently lead to surprisingly biased results.

Some general methods are described in the following paragraphs, focusing on those applicable to karst aquifer systems. For details on general groundwater investigation techniques readers are directed toward excellent textbooks such as Freeze and Cherry (1979), Tóth (2009), Fetter (2014), Hiscock and Bense (2014), and Singh (2016), whereas some books are essentially devoted to karst hydrogeology, such as Milanović (2004b), Goldscheider and Drew (2007), Kresic and Stevanović (2010), and Stevanović (2015).

5.6.3 Groundwater Basin Definition and Water Balance

To be able to make an estimation of the water budget in a karst aquifer one needs to know the boundaries of the groundwater basin. These boundaries are both vertical and horizontal ones. Defining the vertical boundaries of a karst groundwater system relies on a combination of geological and topographical properties. These factors control where and how recharge and discharge occurs, and how deep groundwater flow and dissolution can occur. For instance, variability of precipitation with altitude, how much of the rainfall and snow melt will infiltrate, how concentrated this recharge will be, what controls the altitudinal position of discharge, where the water table will be, etc.

Karst aquifers can be laterally and vertically bounded by less permeable geological units (aquicludes). In karst aquifer systems with an allogenic input, the recharge boundaries can easily be defined by topographic and geological maps combined with ground truthing (e.g., contact between non-karst and karst lithologies). Some complications might be encountered in the case of losing or influent streams, where not all flow is conveyed underground, and part of the water crosses the entire karst area at the surface. In these cases, only accurate discharge measurements at different sections allow estimating the recharge to the karst groundwater basin by the losing stream.

Defining the karst groundwater basin limits, and especially the horizontal ones, is not an easy task. Whereas in non-karstic aquifers the horizontal boundaries often correspond to major topographic divides, which can easily be defined on contour maps and DTMs (digital terrain models), in karst groundwater basins this approach is not valid. Surface streams often disappear underground in one valley, passing beneath and across several topographic divides and neighboring valleys before reaching the spring, sometimes many kilometers away. In single karst massifs, different adjoining groundwater basins frequently drain more or less independent recharge areas feeding different springs. The position of the groundwater divides of these karst basins can vary in response to changing groundwater conditions; water draining toward one spring during low flow conditions can drain toward another spring during high water level conditions (floods).

In unconfined karst aquifers, the subterranean groundwater divides can be determined in different ways: (i) mapping of the groundwater level to locate areas in which groundwater diverges (ridges defined by the equipotential lines); (ii) tracing groundwater flow with artificial or natural tracers; (iii) constructing geological cross sections that show the spatial distribution of permeable lithological units. Generally, the combination of these three approaches gives the most reliable results.

5.6.4 Precipitation and Recharge

Input into a karst groundwater basin derives mainly from rainfall and snow melt that occurs directly onto the karst rocks (autogenic recharge), plus the input of indirect allogenic recharge. The estimations of the inputs can be confronted with those of the outputs, resulting in the assessment of the water budget of the karst groundwater basin. Knowing how much water enters and exits the system allows us to roughly estimate the changes in water storage over a certain period of time. The reference period generally used is the hydrological year, that runs from one dry season (period of lowest storage) to the next one. This time frame is not necessarily equal to 12 months, but changes from year to year depending on the hydrological conditions.

The main difficulties in assessing the input into the system are related to the estimation of the proportion of the precipitation contributing to the recharge, and the delineation of the catchment boundaries. Rainfall and snowfall are measured with rain gauges, which give punctual precipitation records. The distribution of rain gauges is often not sufficient to have a reliable estimate of the average precipitation over the entire catchment. Their altitudinal distribution is often biased toward lower elevations. These shortcomings can be overcome with altitudinal models, where rainfall is calculated using altitudinal gradients (i.e., rainfall and snowfall increase with elevation). The contribution of snowfall via snow melt is even harder to assess, because the measurements are often done with heated gauges that are not representative of natural snow melt; the snow cover may be significantly affected by sublimation. This is further complicated by substantial differences in snow accumulation (e.g., redistribution by deflation) and snow melting, depending on the orientation (exposure to sunlight and prevailing wind) of the slopes.

Not all snow and rain infiltrate and contribute to recharge, some being transferred to runoff and surface-water bodies, some returning to the atmosphere by evapotranspiration. The latter comprises evaporation from wet surfaces, transpiration from plants and interception losses (evaporation from soil and dead material), and depends on multiple factors such as temperature, wind, atmospheric humidity, the types of material involved, or the quantity and type of vegetation (Goldscheider and Drew 2007; Zhao et al. 2013). Evapotranspiration is often the most difficult component to estimate in water budget calculations. Infiltration depends on the type of bedrock, soil cover, vegetation, slope, and antecedent moisture conditions, along with the intensity of the meteorological event. A high moisture content in the soil and the epikarst reduces the infiltration capacity, in contrast to the conditions found at the end of the dry period, when voids are mainly occupied by air. Intense rainfall or massive snow melt often cannot infiltrate entirely and create substantial runoff.

5.6.5 Flow Measurements

In addition to the estimation of infiltration waters, measurements of flow rates in recharge points (sinking rivers), in the throughput (cave rivers and drips), and at discharge points (springs) can give useful insights into the water budget, the response of springs to recharge events and the behavior of the aquifer. There are many ways of measuring flow, depending on the discharge to be measured, logistical constraints (not all equipment can be carried into caves or to remote karst springs), the accuracy of data required, and the available resources; research centers can have expensive equipment, while cavers often use cheap methods.

The volumetric method is often the cheapest and most accurate approach. The time T needed to fill a known volume V of water is measured (e.g., a graduated bucket) (Figure 5.28a). The discharge is given by the simple calculation of V/T. This method can be used where streams form spatially restricted waterfalls, with potential errors related to the fact that it is often difficult to catch all the water in the container, especially on irregular river beds or at distributed drip sites (Figure 5.28b).

Another, rather rough way of estimating flow rate is the float method, in which a floating object is left in the stream and the time for the object to travel downstream along a distance is measured. It works best in steep and rather rectilinear reaches with fast flowing waters. The flow rate is obtained multiplying the cross-sectional area of the stream by the mean of a series of measured velocities. Floating objects often follow the fastest currents, so this measure provides an overestimate of the true discharge.

A more precise measure of flow can be obtained using the pitot tube, a device normally used in aircraft industry to measure air flow velocity, but easy to construct for in-cave measurements. A transparent L-shaped tube of at least 5 mm diameter with both ends open is placed with its short end into the water, its opening pointing upstream (against the current). The height H above the stream level of the water column created in the longer and vertical part of the tube is measured, and is used to calculate the stream velocity v in that section of the river with the formula

$$v = \sqrt{2gH} \qquad\qquad (5.14)$$

g being Earth's gravitational acceleration.

Flow velocities can also be measured with current meters (or flow meters) that resemble small-scaled windmills (Figure 5.28c). The faster the flow, the faster the rotation of these instruments, and velocities are obtained from calibration tables specific for each type of propeller and for fluids of known temperatures, densities, and dynamic viscosities. In both cases (pitot and flow meters), the discharge is obtained making a series of measurements along a cross section of the stream and at varying depths, summing the partial discharges of the subsections.

In cases where, for practical reasons, the above methods cannot be applied or are too imprecise, salt dilution tests can be used, using common table salt and a conductivity meter. Initially, conductivity is measured to know the background level of the stream, then dissolved salt is rapidly introduced in the stream pouring a liquid of known high concentration. Conductivity is measured at predefined time intervals from injection and at a certain distance, far enough downstream so that the salt is homogeneously mixed in the water. The underlying concept of this method is that the higher the flow, the more rapidly it dilutes the salt.

Since the height of the water level in a stream is correlated to the flow rate, water stage measurements can be used for estimating discharge rates at springs and in rivers. First, a rating curve has to be constructed correlating water level height with measured flow rates. A fixed graduated pole or bar can then be placed vertically in the cave stream, allowing the reading of the water heights, which can be converted into discharge values using the rating curve (Figure 5.28d).

In cases where continuous discharge measurements are required, it may be advisable to construct in-stream structures such as weirs and to install a continuous stage recorder upstream. Weirs are dams with sharp-edged notches or orifices through which the water is forced to flow. Their height must be regulated to pond enough water upstream, slowing the flow and preventing water to deviate around or below the dam. The height of the water level above the weir, measured at some distance upstream (elevation head on the weir), can be transformed into discharge using mathematical relationships. Weirs come in different shapes (rectangular, triangular at right angle), and sometimes it may be useful to use a combination of shapes and sizes to allow measuring discharge more accurately in different flow regimes. Small right-angled V-notched weirs are normally more suitable for small discharges (1–1000 L s^{-1}) (Figure 5.28e), whereas large rectangular ones, which are more difficult to construct, especially in caves, are required when flows exceed several cubic meters per second (Figure 5.28f).

Figure 5.28 Flow rate measurements in karst: (a) Direct measurement with a bucket at Tenda springs, northern Italy. (b) Rain gauge used to collect and measure water dripping from the ceiling in Bossea Cave, northern Italy. (c) Analogical current meter. (d) Flow measurements in a section with a linear weir and a pressure transducer upstream (right), Fuse Spring, northern Italy. (e) Triangular weir in Moncalvo gypsum mine, northern Italy. (f) Rectangular weir in Rio Martino Cave, northern Italy. *Source:* Photographs by Bartolomeo Vigna.

Another method for estimating flow rate in old (now abandoned) river channels and for obtaining indirect peak discharge estimates of previous floods makes use of the Manning's equation:

$$v = \frac{1}{n}\left(\frac{A}{p}\right)^{2/3} S^{1/2} \tag{5.15}$$

where n is the Manning's roughness coefficient, $\left(\dfrac{A}{p}\right)$ is the hydraulic radius (cross-sectional area A divided by the wetted perimeter p), and S is the channel slope (altitudinal drop over a certain distance). The roughness coefficient n of a certain channel section can be roughly estimated using tables. It is around 0.03 for smooth-walled channels with fine-grained sediments, 0.06 for a rough, cobble-floored channel, and reaches values as high as 0.1 in extremely irregular channels. This equation can be applied as long as sections are stable (not changed by erosion or aggradation) and open channel flow is guaranteed; this method is not applicable in entirely flooded cave sections where (phreatic) conduit flow occurs. For all the above mentioned methods, the reader is referred to the detailed explanations given in Groves (2007) and Palmer (2007).

5.6.6 Boreholes

In most types of aquifers, the hydraulic characteristics of the rock masses are obtained drilling a series of boreholes in which a number of physical tests can be carried out. In intergranular-porosity aquifers, three boreholes are usually sufficient to get a reliable idea of the main hydraulic parameters of the rock mass. Borehole tests are particularly reliable for understanding the hydrological behavior of isotropic and homogeneous granular aquifers, and are also very useful in fractured (non-karstic) aquifers. For details on conventional borehole analyses in hydrogeology the reader is referred to Misstear et al. (2006) and Fetter (2014). However, this type of analyses, if adopted alone, can give biased results in the study of karst aquifers. The heterogeneous and anisotropic nature of karst aquifers make hydrological tests conducted in a single borehole inappropriate to extrapolate the results to an extensive area, or even worse, to an entire karst aquifer. It is well known that in intensely karstified areas, one borehole can be completely dry, being located in homogeneous rock with insignificant solutional porosity and low matrix porosity, whereas another borehole only a few meters apart can intersect a major water-bearing karst conduit and display hydraulic head variations in response to heavy rain events of more than 10 m (Figure 5.29). Water levels measured in a 25-m spaced borehole transect perpendicular to the Trebišnjica River in Bosnia and Herzegovina show this random distribution of the water table level due to sharply changing permeability values even at short distances (Figure 5.30). This does not mean that borehole tests are completely worthless, or that groundwater cannot be obtained from them in well-developed karst aquifers. Sometimes, in karst areas where there is no direct access to caves and underground rivers, boreholes represent the only way of getting direct data on the underground water and the position of the water table level. The information obtained from boreholes should however be complemented with tracer tests, observations in caves, and a detailed knowledge on the geomorphological, stratigraphic, petrophysical, and structural characteristics of the entire karst hydrological system.

The inflow of water into a borehole drilled through a karstified rock mass occurs at discrete depth intervals, corresponding to the preferential flow paths associated with solutionally enlarged bedding planes, fractures, and conduits. Geophysical investigations in boreholes can give information on the location of these more permeable intervals in the rock sequence, as well as on the hydraulic conductivity and the specific storage of the rock immediately surrounding the borehole. Standard hydraulic borehole tests in karst aquifers give information on the hydraulic conductivity and the storage properties of the fissured matrix, but have a very limited usefulness for the hydraulic characterization of conduits (Sauter 1992a, 1992b).

Water in boreholes rises more or less rapidly following intense natural recharge events, depending on the degree of karstification, and similarly will drop when the recharge event ceases. In well-karstified aquifers the water level changes are large and fast, and as occurs in surface flashy streams, the hydrograph's recession limb decreases exponentially until pre-storm values are

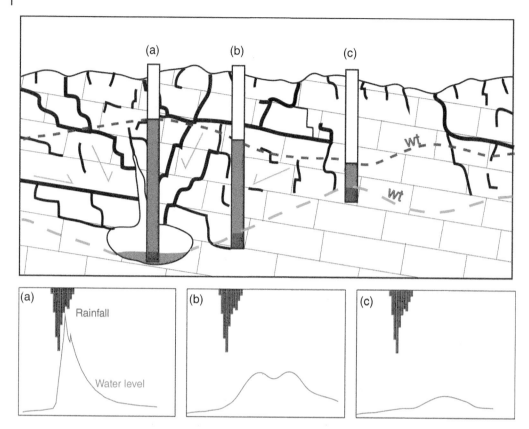

Figure 5.29 Hydraulic head measurements (blue curve) in response to an intense rainfall event (red hyetograph) in boreholes intercepting: (a) a conduit with rapid and short-response water flow; (b) fractured rock connected to the conduit network; and (c) inefficiently connected, low permeability rock with matrix porosity. Brown arrows indicate underground water flow. The green dashed line shows the water table (wt) during low flow conditions, whereas the blue dashed line shows water levels during floods.

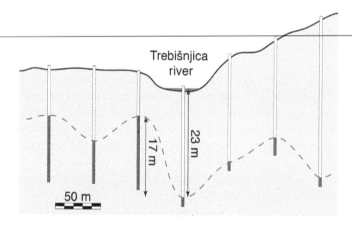

Figure 5.30 Water table measurements in a closely spaced set of boreholes along a transect perpendicular to the Trebišnjica River (losing or influent stream) in Bosnia and Herzegovina. *Source:* Adapted from Milanović (1981).

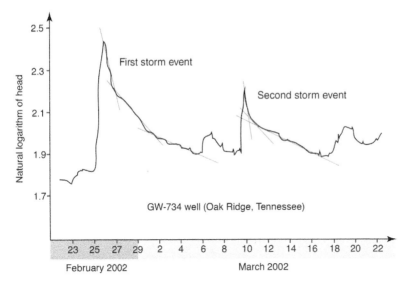

Figure 5.31 Natural logarithm of the water level measured in a borehole versus time recording the effects of two main rainfall events (Maynardville Limestone at Oakville, Tennessee). *Source:* Adapted from Shevenell (1996). Blue straight lines with decreasing slope in the recession limbs indicate segments with prevailing discharge from conduits, fractures, and the matrix porosity of the aquifer, respectively.

reestablished (see also Section 5.6.7). Detailed analysis of the recession limb often allows recognition of different segments with prevailing discharge from conduits, fractures, and the matrix porosity, in order of decreasing flow velocity (Shevenell 1996) (Figure 5.31).

The most common hydraulic tests in boreholes are pumping and recharge tests. In the first, a volume of water is withdrawn, in the second a volume of water is injected (or displaced), and in both cases changes in groundwater level are monitored for a certain period. Borehole tests can further be distinguished into slug, bail, packer, and tracer tests. In a slug test a fixed volume of water is instantly removed or displaced using a "slug" (e.g., a heavy and solid rod) causing an abrupt drop or rise in the water level, and the subsequent recovery of the hydraulic head in the borehole is measured (Figure 5.32a). Normally these are falling-head slug tests, in which a known volume of solid material (the slug) is introduced into the borehole. In the rising-head slug tests, also known as bail tests, a known volume of water is removed (Figure 5.32b). Both tests are carried out in open boreholes and give an approximate hydraulic conductivity measure of the rocks surrounding the borehole. The advantage of rising- and falling-head slug tests is that they are relatively cheap, quick and easy to perform (there are no associated problems related to contamination), and they do not require a fully equipped final borehole.

If isolated sections of a borehole need to be analyzed, packer tests can be employed. A borehole portion is isolated by using inflatable or mechanical devices at the bottom and top of the section of interest. The water pressure is increased in the area between the packers, and pressure decay is measured and analyzed using pressure gauges (Figure 5.32c). In hard and solid carbonate rocks, the tests are carried out from the bottom upward (ascending packer test). The length of the test section is variable, depending on the characteristics of the bedrock, but is often around 2 m.

A borehole tracer test involves injecting a tracer into the borehole and detecting the tracer in nearby wells or springs. The most frequently used tracers are fluorescent dyes, because of their detectability, low cost, and low toxicity. Although flow velocities in karst are orders of magnitude

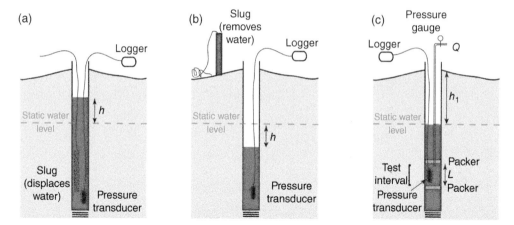

Figure 5.32 Boreholes tests: (a) Slug (falling head) test; (b) Bail (rising head) test; (c) Packer test. h = water level rise or fall; h_1 = depth from ground level; L = length of test interval; Q = flow rate.

larger than those in granular aquifers (in general in the range of 0.001–0.2 m s^{-1}), moderate pumping may be required to draw the tracer toward the monitoring well.

Pumping tests, although quite costly and complex, are widely used in hydrogeology to quantify well yields and the hydraulic properties of aquifers. Due to the anisotropic and heterogeneous nature of karst aquifers, their cost-benefit ratio is often too high to be suitable for hydrogeological studies in carbonate terrains. Depending on where the borehole is located (open karst conduit, fractured zone connected with the main conduits, or isolated rock with matrix porosity), different approaches need to be used to interpret the pumping test results. Pumping produces the drawdown of the water table creating a cone of depression, the shape and extent of which depend on the hydraulic properties of the surrounding rock mass and the duration of pumping. If the borehole solely intersects the rock matrix, conventional methods (Theis 1935) can be applied with some caution to obtain estimates of hydraulic parameters for this compartment of the karst aquifer. If a borehole intersects high-flow pathways (conduits or enlarged fractures), the approach described in Thrailkill (1988) is a good approximation for short-duration pumping tests, in which only the response of the conduit compartment is considered. When the pumping test is carried out over long periods, also the diffuse flow from the matrix-porosity compartment must be considered, and a double-continuum model representing the conduit and the matrix-porosity permeability features can be employed (Maréchal et al. 2008). A long-duration pumping test in a dual-porosity karst aquifer would give the typical response with three distinct segments in the drawdown curve (Figure 5.33). In such pumping tests the drawdown-time curve often shows deviations from the generalized pattern. These differences can be related to the increasing influence of impermeable boundaries or less permeable compartments of the aquifer, the flushing out of sediments from fractures and conduits, or the depression cone reaching more permeable parts of the aquifer (e.g., surface stream, karst conduit). In general, however, the assessment of the storage properties of different permeability compartments in a karst aquifer solely based on pumping tests is often a very risky approach. Pumping tests can sometimes be successfully applied in karst, but results will commonly be representative of limited areas around the boreholes, often less than 100 m (Sauter 1992a) (see also Figure 5.5).

Once a borehole is drilled, or the drilling operations are temporarily halted, variations in physical and chemical parameters along a depth profile in the hole can be recorded. Borehole logging records changes in different properties of the rock or the groundwater, in undisturbed and in situ

Figure 5.33 Schematic response of a dual porosity karst aquifer to a pumping test, showing three segments in the drawdown curve: (i) drainage derived mainly from the interconnected conduit and enlarged fracture network (left); (ii) drawdown during the transition period in which drainage comes from the conduits, fractures and the matrix porosity, which start to contribute significantly to the flow; (iii) water supplies come to an equilibrium state, and the curve stabilizes to a uniform slope (right).

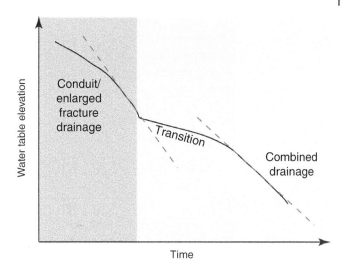

natural conditions. This provides much more reliable results for some parameters than visual and geological logging carried out on recovered core samples. They can also be the only data available in cases where core recovery during drilling was poor or simply not done.

A great range of variables has been successfully measured in borehole logging, including electrical resistivity, spontaneous potential, temperature, electrical conductivity, radioactivity, and sound. Miniaturized optical, acoustic, and video instruments can be introduced in the borehole, giving the opportunity to observe its walls and bottom, even underwater. Whereas visible images can only be acquired in clean waters, acoustic instruments can emit and record sonar pulses and create a virtual image of the borehole walls, showing the size of the hole, wall roughness, and the presence and direction of planar permeability features (fractures and bedding planes), even in extremely turbid water.

In uncased wells, electrical logging can be very useful for distinguishing rock units and constructing a stratigraphical section. Two electric properties can be measured: the spontaneous (or self-) potential, and the electrical resistivity (or the reciprocal; electric conductivity). The spontaneous potential log (SP) measures the natural or spontaneous electric potential difference that exists between an electrode in the borehole and a fixed one placed at the surface. A constant current is applied to a pair of electrodes for resistivity measurements, and the resulting electric potential is measured between another pair of electrodes at different depths in the borehole. The spontaneous or induced difference in potential is caused by differences in the way in which charged particles enter or leave the borehole, producing an electric current. It allows to distinguish between permeable and less permeable sections in the borehole. Electric measurements require the presence in the borehole of a conductive fluid (water-based borehole mud) that contrasts with the ionic concentration of the fluids in the surrounding rocks. Temperature logs are generally run together with electric logs, since electric conductivity (or resistivity) depends on temperature. Differences in temperature may indicate the inflow of colder or warmer waters from discrete horizons. Natural Gamma logging investigates changes in natural radioactivity along the borehole depth, enabling to distinguish argillaceous sediments, which are normally more radioactive, from sandstones and carbonate rocks. Gamma-Gamma (density) logging and Neutron logging are active systems, in which the borehole walls are bombarded with Gamma Rays and Neutrons, respectively, and the response of the bedrock at various depths is recorded. Low Gamma counts are characteristic of high density materials, so that Gamma-Gamma logs are a sort of rock density logs. In Neutron logs, a smaller counting at the detector indicates a higher concentration of hydrogen atoms (and thus water) in the surrounding rock, which in turn is an indication of water-bearing porosity. In general, well logging gives the best and most reliable results when multiple geophysical methods are used in combination.

In synthesis, borehole data can help in determining where and how much the aquifer is karstified, at least in the immediate surroundings of the boreholes. There are a number of observations that can help in identifying karstification and the presence of conduit or enlarged-fracture flow: (i) the borehole encounters large voids, losing drilling fluids and/or entire drill bits; (ii) borehole video-logging, or other geophysical techniques reveal karst porosity; (iii) tracer tests confirm fast water flow between boreholes or from borehole to spring; (iv) borehole packer tests show variations in conductivity by an order of magnitude when increasing the portions of investigated borehole (areas with karst permeability start to be influenced by the test); (v) segmented or non-linear response of water table drawdown during a pumping test (see Figure 5.33); (vi) irregular shape of the depression cone, indicating anisotropy; (vii) rapid changes in water quality and quantity in response to recharge events; (viii) very irregular water table contours, with troughs corresponding to areas of enhanced underground water flow; and (ix) water age inversion at depth, with younger conduit water underlying older matrix-porosity water. For more details on all kinds of borehole analyses the reader is referred to Keys (1997), or to Kresic (2007) when karst aquifers are the target.

5.6.7 Spring Hydrographs and Chemographs

Underground rivers respond to recharge events as do surface rivers, and the shape of the hydrograph at the spring is a measure of how the flow system behaves. Whereas borehole techniques can be used to get an idea of the aquifer properties at a local scale, the analysis of spring hydrographs allows getting a grasp on the behavior of the entire aquifer system. This is especially true in cases where groundwater flow is directed toward a single major spring. Major karst springs, in fact, are connected to an extensive and highly permeable conduit system that collects the water from the matrix, fissure, and conduit porosities typical of karst (Jeannin and Sauter 1998; Geyer et al. 2008).

A first analysis of spring hydrographs is based on the evaluation of the recession behavior related to an impulsive recharge event (a single short-duration storm). One such typical hydrograph is shown schematically in Figure 5.34, and the following features can be easily distinguished: (i) a lag time t_L between the rainfall peak and the highest spring discharge Q_{max}; (ii) a rapid rise in discharge, called the rising limb, up to the peak of the hydrograph; (iii) a less steep recession curve that brings the discharge back to baseflow Q_B. The time from peak flow Q_{max} to base flow Q_B is t_B. The rising limb of the hydrograph is composed of a concave and a convex segment separated by an inflexion point (ip). This inflexion point indicates the time at which infiltration is greatest. Similarly, the recession limb can be divided into a convex and a concave segment, the inflexion point here representing the end of the infiltration event, and Q_0 being the discharge at this moment (Bonacci 1993; Kovács et al. 2005). It is worth noting that the lag time is not the time needed for the flood water to physically reach the spring, but the time needed to transmit the flood pulse. Even in these ideal, single, short-lasting storm events, time variables t_L (lag time) and t_B (time needed to return to baseflow) are very difficult to determine with precision. What can be obtained with a certain degree of accuracy is the exponential function that best fits with the recession limb of the hydrograph. The most common (and simplest) fitting method is the Maillet exponential function

$$Q_t = Q_0 e^{-\alpha t}$$

$$(5.16)$$

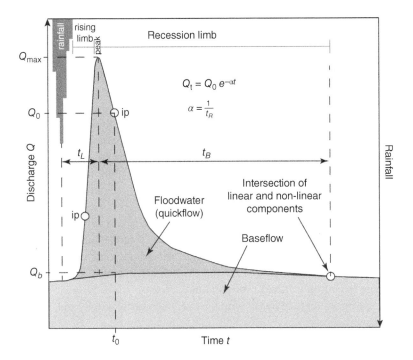

Figure 5.34 Typical hydrograph of a flow episode in a karst spring related to a single and short-duration storm recharge event (gray hyetograph). ip = inflexion point. See explanation in text. *Source:* Adapted from White (1988).

where Q_t is the discharge measured at time t, Q_0 is the discharge at the start of recession $t = 0$ (note that Q_0 is not peak flow Q_{max}), t is the time elapsed between Q_0 and Q_t, and α is the recession coefficient (Maillet 1905). From this equation, the response time t_R can be obtained, being $t_R = 1/\alpha$. In non-karstic springs, the recession coefficient allows getting an idea of storage and reserves of water that can be exploited from the aquifer.

Rainfall events commonly display different rain pulses. In large groundwater basins, recharge can come from different distant areas and can occur through different ways (intense rainfall, snow melt, allogenic recharge, and so forth). It has been emphasized earlier that flow paths and groundwater divides in karst can change between high flow and base flow situations, and not all water recharged upstream necessarily contributes to a single spring, but secondary outlets can be activated during floods. In the same manner, outputs can be disturbed by rising water levels downstream, such as in flooded poljes (Milanović 2004a). All these factors can complicate the interpretation of spring hydrographs considerably. Nonetheless, spring hydrograph analysis remains a valuable means for gaining insight into the behavior of the system in different hydrological conditions, and these methods are widely applied to both non-karstic and karstic aquifers.

In karst springs the recession curve can often be divided in two or more segments, representing in some way their dual (or triple) permeability components and behavior; an initial fast response related to the conduit flow, followed by much slower ones related to the diffuse flow in the fissure network and matrix porosity. Consequently, karst spring hydrographs are best analyzed using a multiple reservoir exponential model (Geyer et al. 2008), whereby discharge at the spring is the sum of the discharges from the multiple reservoirs, which can be treated separately using Eq. 5.16. This approach can lead to satisfactory results, but it is rather unrealistic to consider that the different reservoirs behave independently and are hydraulically isolated entities. For example, during floods, water from the conduits is pushed into the neighboring fissure network, and this water flows back into the conduits when hydraulic head decreases.

The classical Eq. 5.16 is often used in karst systems, but is valid only for homogeneous, isotropic, and unconfined aquifers (granular ones). In karst, this approach can be valid for analyzing short periods of time at low water stage, or poorly karstified aquifers (Mangin 1975). In a certain sense, the recession curve mathematically described by Eq. 5.16 corresponds to the drainage of the saturated, low permeability part of the karst aquifer, called the Matrix Restrained Flow Regime period in Bailly-Comte et al. (2010). Flow generated by this low permeability component is called baseflow.

Describing the drainage of the highly permeable part of karst aquifers (conduit system) is a much more complex matter. The upper part of the hydrograph (quickflow in Figure 5.34) represents the storm flow runoff, which is greatly influenced by recharge conditions, and the very rapid response of the different high-permeability compartments. An approximation to the drainage coming from the high-permeability compartments, named Conduit Flow Regime in Bailly-Comte et al. (2010), assumes that there is no diffuse recharge through the vadose zone, no exchange between the fissures and matrix porosity compartments and the conduit network, and that the spring only responds to the rise in hydraulic head in the conduits.

The total discharge at the spring would thus be the sum of baseflow and quickflow, and can be expressed using the following equation:

$$Q_t = \Psi_t + \phi_t \tag{5.17}$$

where ϕ_t corresponds to the baseflow, or in other words the delayed flow from the saturated matrix porosity and fissure aquifer compartments, which can be expressed by the Maillet's Eq. (5.16). The other term, Ψ_t, represents the quickflow, being the recharge through the unsaturated zone to the spring via the conduit network, which according to Mangin (1975) is best expressed as:

$$\Psi t = q_0 \frac{1 - \eta t}{1 + \varepsilon t} \tag{5.18}$$

where η characterizes the mean infiltration velocity (for very fast infiltration η is close to 1), and q_0 is the difference between the total spring discharge Q_0 at $t = 0$ (set at the recession inflexion point, corresponding to the end of the infiltration) and the baseflow component Q_b (Figure 5.34). The time interval between $t = 0$ and $t = \frac{1}{\eta}$ is assumed to be the duration of infiltration. In other words, infiltration stops at time $> \frac{1}{\eta}$. The parameter ε is a coefficient of heterogeneity, characterizing the importance of the concavity of the quickflow curve, and is positively correlated with infiltration. Mangin defined the function

$$Y_t = \frac{\Psi t}{q_0} \tag{5.19}$$

or

$$Y_t = \frac{1 - \eta t}{1 + \varepsilon t} \tag{5.20}$$

Note that Y_t is independent from discharge and thus allows the comparison between different karst systems.

Numerical simulations with finite-element models have shown hydrograph recession analysis in karst to give often equivocal results (Eisenlohr et al. 1997). Hydrograph analysis should therefore be coupled with detailed investigations that shed light into the provenance of the recharge waters and how these interact with the rocks.

Chemical compounds found in water not only allow us to assess its quality and identify the extent and origin of pollution, but can also be used as natural tracers that provide information on the structure of the karst aquifer and the dynamics of groundwater flow (Shuster and White 1971). The chemical composition of water in a karst area depends on multiple factors, such as the type of recharge (concentrated or diffuse, epigenic or hypogenic), climatic conditions, vegetation, and soil properties, land use, the mineralogy of the rocks, and the characteristics of the underground flow paths (pores, narrow fissures, enlarged fissures, conduits). The hydrochemical fingerprint at a karst spring results from the mixture of waters of different origins. The most useful information is obtained by recording the temporal variation of the concentration of the chemical compounds, especially during flow events related to significant recharge events. These graphs, showing the temporal variation of hydrochemical parameters, are termed chemographs. To understand the meaning of the hydrochemical variations measured at a spring, or along the flow path in karst, it is important to know the origin of the chemical compounds and the processes that control their concentrations. Four main categories of origins can be distinguished: precipitation, soil, host rock, and anthropogenic sources (Figure 5.35).

Water infiltrating in karst initially comes from the atmosphere as rainfall or snow, and has a characteristic chemical signature. Although precipitation waters mainly derive from water vapor produced by evapotranspiration at the Earth's surface, which later condenses in high parts of the troposphere because of the decreasing temperatures, it is never pure water. Water often condenses around microscopic particles suspended in the air (ashes, pollen, pollutants, salts). Its chemistry can also be influenced by marine spray in coastal areas, leading to increasing concentrations in Na^+, K^+, Cl^-, and SO_4^{2-}. Water, composed of H and O, also contains varying concentrations of heavy isotopes such as deuterium ^2H, tritium ^3H, and ^{18}O, which can be used as natural tracers.

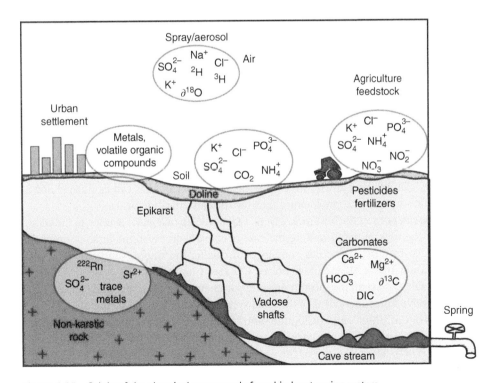

Figure 5.35 Origin of the chemical compounds found in karst spring waters.

Water deriving from rainfall or snow melt changes its chemical composition while passing through the soil first, and the epikarst next. The soil is rich in gases, mainly ^{222}Rn and CO_2. The former is a radioactive decay product of ^{226}Ra, which is naturally present in soils, especially in well-developed ones with abundant iron- and manganese-oxy-hydroxides. ^{222}Rn is highly soluble, and because of its short half-life (only 3.82 days), it is a very good natural tracer to study fast transport processes in karst systems. CO_2 is produced by the decay of organic matter and root respiration. Its concentration follows a seasonal cycle, with peak values during the growing period, in which both biomass production and degradation are higher. CO_2 concentrations in soils show lower values with decreasing temperatures, higher altitudes, and less vegetation (see Section 3.3.2).

Another chemical indicator derived from the soil is Dissolved Organic Carbon (DOC), a mixture of a series of complex organic molecules mainly produced by the decomposition of organic matter. DOC is typically defined as the amount of organic carbon that passes through a filter of 0.45 µm, in contrast to particulate organic carbon (POC). POC and DOC together make up the Total Organic Carbon (TOC). DOC plays an important role in the global carbon cycle, being the main source of organic carbon available for the growth of microorganisms. The quantity of DOC found in karst waters depends on its production, biodegradation, and adsorption in the soil; DOC in infiltration waters decreases with increasing soil thickness. High levels of DOC in spring waters indicate poor degradation and retention in the soils, and thus water derived from rapid infiltration. Furthermore, high levels of DOC often also indicate a higher probability of contaminants to be transferred through the aquifer, since organic complexes increase the mobility of metals and hydrophobic pollutants.

Karst aquifers are well-known for their limited depurating capacity, since pollutants can be rapidly transferred from their source to the springs. These substances can be introduced at specific locations and over long periods of time (landfills, sewage systems, leaking storage tanks), in which case they are considered as point sources, or across large areas at lower concentrations and over discrete time intervals (fertilizers and pesticides), which can be regarded as diffuse sources of pollution. If manure is applied, alongside the chemical substances, large amounts of microorganisms, including pathogenic ones, can also be introduced into the karst aquifer. Fertilizers are mainly composed of chemicals rich in macronutrients such as Nitrogen, Phosphorus, and Potassium, together with secondary nutrients (Calcium, Magnesium, Sulfur), and micronutrients (Copper, Iron, Manganese, Molybdenum, Zinc, Boron, and other metals). A part of these are taken up by plants, others are stored in the soil, whereas some are leached and transported in the infiltrating waters. Nitrates (NO_3^-) are very soluble compounds that can be incorporated into infiltrating waters the whole year round, showing higher concentrations when vegetation is lacking (after harvesting) and infiltration is high. Some elements are retained and concentrated in the soil moisture (K^+, SO_4^{2-}, PO_4^{3-}, Cl^-) during low infiltration periods (dry season), and can be flushed downward during the first storm events, producing characteristic concentration peaks.

Once infiltration waters enter the host rock, several reactions can take place leading to the chemical weathering of the rock minerals and, in the case of karst bedrock, to the dissolution of the constituent minerals. In pure carbonates composed of calcite, the water eventually will reach chemical equilibrium with $CaCO_3$ and the dissolved CO_2 (see Chapter 3).

In the $CaCO_3$-H_2O-CO_2 system the most interesting parameters used to shed light on the hydrochemical and hydrodynamic functioning of the karst system are the partial pressure of CO_2 (P_{CO_2}), the concentrations of Ca^{2+}, Mg^{2+}, and HCO_3^-, the saturation indices (SI) of the carbonate minerals (calcite, aragonite, dolomite) and the stable carbon isotope composition (^{13}C) of Dissolved Inorganic Carbon (DIC), being the sum of the concentrations in CO_2, H_2CO_3, HCO_3^-, and CO_3^{2-}. High-Mg calcite is more soluble and dissolves more rapidly than pure calcite (Figure 3.11), therefore higher Mg^{2+} concentrations are often considered as an indication of longer residence times of the waters (giving more time, proportionally more HMC will dissolve respect to pure calcite, thus increasing the Mg/Ca ratio).

In addition to dissolved ions, natural microparticles can give valuable insights into the hydrodynamic behavior of karst aquifers and can be treated as analogues of pollutants. Natural tracers that can be identified at karst springs are microorganisms and sediments. An extensive review on both natural and artificially introduced bacteria and bacteriophages is given by Harvey (1997), whilst sediment transport in karst aquifers is dealt with in detail in Mahler and Lynch (1999), Mahler et al. (2000), and Göppert and Goldscheider (2019).

One of the simplest ways of monitoring the physical and chemical behavior of karst springs in response to natural recharge events is the continuous monitoring of temperature (T), electrical conductivity (EC), and discharge (Q). Temperature and EC can be regarded as natural tracers during flood events, giving some valuable clues on the provenance of the water. To simplify, we can distinguish three types of water circulating in a karst system (Galleani et al. 2011; Banzato et al. 2017a): infiltration waters, circulation waters, and matrix porosity (slow-flow) waters. The first derive directly from rainfall or snowmelt, the second are waters flowing in the karst system, including the vadose zone and especially a thick saturated zone, whereas the last are the slow flowing waters in narrow low-permeability fracture systems and the matrix porosity of the rock.

Infiltration waters are normally characterized by low EC (low mineralization) and temperature values, depending on the season and the region, but often lower than the typical temperatures encountered in karst aquifers. In case of snowmelt, infiltration waters enter at around 0 °C. On the other hand, EC and T of the circulation waters that have resided long enough in the aquifer are in equilibrium with the host rock. The slow-moving matrix porosity waters, characterized by a very long water-rock interaction, have higher EC and T values than the infiltrating or rapidly flowing conduit waters. This is due to their slow motion and to the prolonged contact with the rock.

These different waters can interact with each other during significant infiltration events, giving rise to different responses at the springs. These interactions depend on the type of recharge (intensity, duration), the organization of the underground flow network (size and location of the main drainage paths, presence of a more or less thick saturated zone), and the connectivity between different compartments in the aquifer. Four end-member types of phenomena can be recognized: piston flow, substitution, mixing, and homogenization (Vigna and Banzato 2015; Filippini et al. 2018) (Figure 5.36). The piston flow is recognizable at the onset of a flood event, with a sudden increase in EC and T caused by the mobilization of resident (warmer and more mineralized) water. During substitution, the newly infiltrated water, low in T and EC, rapidly arrives at the spring without being mixed with resident water. This situation contrasts with the mixing phenomenon, whereby temperature gradually decreases due to the increasing influence of the newly infiltrated waters, whereas EC slowly increases because of the mixing with resident waters. If an aquifer responds with homogenization, both EC and T show very slight and delayed responses to the infiltration events. It must be taken into account that chemical and physical responses at the springs can be rarely categorized with just one of these end-member types, and that many intermediate responses are normally encountered, making the interpretation often more difficult than in the described ideal conditions.

Three main end-member types of karst aquifers can be distinguished according to their permeability characteristics and the drainage type (Vigna and Banzato 2015): systems with dominant drainage (Figure 5.37a), with interconnected drainage (Figure 5.37b), and with dispersive circulation (Figure 5.37c). Dominant drainage systems have a high permeability and a few main drainage conduits and secondary drains that are able to convey the infiltration waters very rapidly from

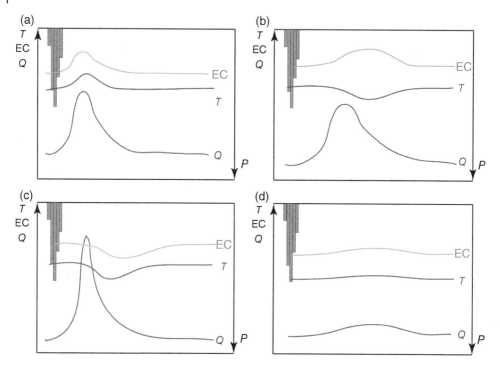

Figure 5.36 Main responses in electrical conductivity (EC), temperature (T), and discharge (Q) in spring water during a flood related to an infiltration event (hyetograph in gray): (a) piston flow; (b) mixing; (c) substitution; (d) homogenization.

the recharge areas to the springs. The saturated zone is limited, and water mainly flows under vadose conditions, similar to surface streams. The discharge during floods rises abruptly and has a rapid recession (flashy response). The contribution of resident and matrix water to the overall flow is negligible, and the chemical and physical imprint of the infiltration waters dominates. These systems are characterized by a substitution phenomenon, with a decrease at the springs of both EC and T (Figure 5.36c).

Systems with interconnected drainage are characterized by well-developed conduit and fracture permeability and by an extensive and deep saturated (phreatic) zone. The phreatic interconnected permeability systems form an important storage of underground water. During floods the discharge rises rapidly, but the recession is slower than in dominant drainage systems, because the water stored in the extensive phreatic zone regulates the spring flow. The peak of the flood occurs very shortly after the main recharge peak, and is caused by the transmission of the hydraulic pressure in the phreatic network, rather than by the arrival of the newly infiltrated waters. These resident waters are pushed out of the system, and cause a temporary increase in both T and EC, a phenomenon known as piston flow (Figure 5.36a). These systems are often characterized by substantial changes in water level (over 100 m) in the cave systems upstream of the springs, whereas the latter are often of the vauclusian type.

In systems with dispersive circulation, karst permeability is much less developed, and there are no enlarged fracture and conduit networks significant enough to concentrate the underground water flow. The phreatic zone is extensive, but characterized by slow water flow. Springs show a very regular behavior, with very slight and delayed seasonal changes in flow rate and temperature, and high values of EC throughout the year due to longer water–rock interaction.

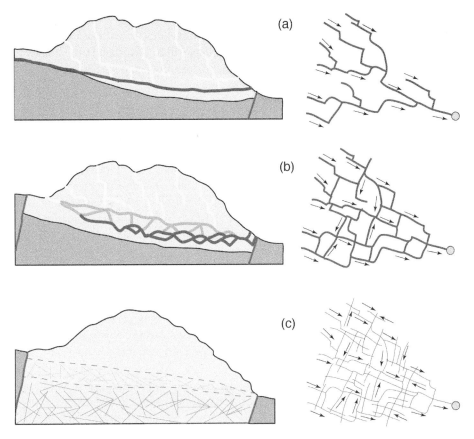

Figure 5.37 Main end-member types of karst aquifers and their drainage: (a) dominant drainage system; (b) interconnected drainage system; and (c) dispersive circulation system. Carbonate rocks are in blue and non karstified formations are shown in pink. *Source:* Adapted from Vigna and Banzato (2015).

5.6.8   Water tracing

Water tracing is a commonly used technique in hydrology, especially in karst areas, and many books and articles deal with the topic in great detail (Kranjc 1997; Käss 1998; Dassargues 2000; Field 2002a, 2003; Benischke et al. 2014; Goldscheider et al. 2008; Leibundgut et al. 2009; Leibundgut and Siebert 2011; Jones 2019). Tracing of water in karst is used to determine: (i) hydrological connections; (ii) flow paths and directions; (iii) groundwater catchment boundaries, preferentially in different hydrological conditions; (iv) recharge areas; (v) flow velocities; and (vi) sources of pollutants. Water tracing experiments can be qualitative and quantitative. The first allow us to reveal connections between injection points and downstream areas (caves, boreholes, or springs), and if analyses are carried out with sufficient frequency they allow the estimation of travel times. In quantitative tracing, the amount (mass) of dye injected is known and sampling and analyses are conducted with a high frequency or continuously, so that tracer concentrations can be monitored over regular time intervals from the first arrival, over the peak concentration, to the end of detection. The resulting chemographs allow obtaining data on the type of water circulation and properties of the aquifer. The percentage of dye detected can be calculated based on the dye recovery curve, also known as breakthrough curve (BTC).

Water tracing agents can be natural and artificial. The chemical compounds in the water, including the stable isotope composition, temperature, sediments, and the bacterial diversity can be regarded as natural tracers. Some have been explained earlier and others will be described in the following section. More frequently artificial tracers are used, including dyes, salts, and particulates (spores, microspheres, bacteriophages).

An ideal artificial tracer should have the following characteristics: (i) naturally absent in water; (ii) cheap and easy to obtain; (iii) readily soluble in water, resulting in a solution with a density very close to that of natural water (for particulates, of sufficiently fine particle size to prevent losses by filtration during transport through the karst aquifer); (iv) easily detectable and quantifiable at very low concentrations; (v) stable in natural conditions of illumination, temperature, and pH, and non-reactive with substances present in underground waters, and (vi) nontoxic for potential consumers and for the ecosystem. This last characteristic is probably the most important one, also from a legal point of view. Table 5.3 indicates the properties of twelve of the most frequently used artificial fluorescent tracers in karst, including their possible problems related to their use.

Uranine is the most commonly used tracer in karst studies, being inexpensive, ecologically safe, very soluble, with very low detection limits, and not adsorbed on clays (Figure 5.38c, e). It is, however, very sensitive to light and to strong oxidizing agents (e.g., chlorine bleach), so it cannot be used in daylight conditions or in chlorinated water. It is also present at low concentrations in urbanized areas (Uranine is used in medicine, in antifreeze liquids for cars, etc.), so its background concentration needs to be measured. Eosin is very similar to Uranine, giving a characteristic orange color. Because of its higher detection limit larger quantities are needed, but it is a good substitute to Uranine in organic-rich and acid waters.

Rhodamines are red fluorescent dyes, several of which are toxic and should not be used (e.g., Rhodamine B, Rhodamine WT, and, to a lesser extent, Sulforhodamine B). Sulforhodamine G is a widely-recommended dye (Figure 5.38d), especially for contemporaneous use with Uranine. It has a distinctive fluorescence peak and is less sensitive to light and low pH conditions than Uranine.

Pyranine has similar fluorescence characteristics to Uranine, but is highly pH-dependent (see Table 5.3), making its detection a little more difficult. Although it is safe and can be used in acidic waters, it is a rather unreliable dye that should be avoided. Naphthionate is also safe, has a blue-ultraviolet fluorescence, but interferes with DOC and is unstable over long time periods. Its higher detection limits also require high injection masses, and samples need to be analyzed shortly after collection, making it less friendly to use. The other blue-ultraviolet fluorescent dye Tinopal CBS-X is increasingly employed (Figure 5.38b, f), having clearly distinguishable fluorescence peaks. Its drawbacks are mainly the rather high detection limit (requiring large masses to be injected), its tendency to be absorbed on clays (so it should be used in large conduit systems), and its poor performance in acid waters.

Tracing of waters using fluorescent dyes normally aims at injecting a sufficient amount of tracer to be detected at the possible outlets, without causing the waters to be visibly colored, which could create alarm among the people. Concentrations even below $1 \mu g \, L^{-1}$ ($10^{-6} \, g \, L^{-1}$) are far above the detection limits (see Table 5.3), and are surely non-toxic for the environment and for the ecosystem. Dilution of most tracers used in hydrogeological studies is usually very fast, and concentration levels generally encountered during the tracer detection are well below toxic levels.

Salts are often used as tracers, since they are readily soluble in water and their presence can be monitored using simple EC measuring devices. Anions are less reactive than cations, the latter being easily adsorbed onto clays (e.g., cation exchange). In decreasing order of importance and performance, LiCl, NaCl, and KCl can be used, but their higher detection limits ($0.1 \mu g \, L^{-1}$ in ideal conditions) require large amounts to be injected. NaCl can have high background levels, whereas

Table 5.3 Fluorescent dyes frequently used in karst groundwater tracing experiments and their main properties.

Tracer	Color	CAS n°	Detection limit	Toxic	pH	Absorption wavelength (nm)	Fluor. (nm)	Fluor. intens. (%)	Analytical interfer. with	Problems
Uranine (Fluorescein)	Acid Yellow 73	518-47-8	0.001 µg L^{-1}	No	B	491	516	100	2, 7	Degrades at low pH and in organic-rich solutions
Eosin	Acid Red 87	17372-87-1	0.01 µg L^{-1}	No	A	515	540	11	1, 3	Very sensitive to light
Sulforhodamine G (Amidorhodamine G)	Acid Red 50	5873-16-5	0.005 µg L^{-1}	No	N	530	555	32	2, 4, 5, 6	
Sulforhodamine B	Acid Red 52	3520-42-1	0.007 µg L^{-1}	Mod	N	564	583	7	3, 5, 6	Ecotoxically unsafe
Rhodamine B	Basic Violet 10	81-88-9	0.006 µg L^{-1}	Yes	N	555	582	10	4, 6	Ecotoxically unsafe
Rhodamine WT	Acid Red 388	37299-86-8	0.006 µg L^{-1}	Yes	N	561	586	10	4, 5	Genotoxic
Pyranine	Solvent Green 7	6358-69-6	0.06 µg L^{-1}	No	A / B	407 / 455	512 / 510	6 / 18	1, 2 / 1, 2	Biodegradable / Biodegradable
Naphthionate	Acid Red 88	130-13-2	0.07 µg L^{-1}	No	Neutral	325	420	18	9, 10, DOC	Interferes with DOC
Tinopal CBS-X	Fluorescent Brightener 351	54351-85-8	0.01 µg L^{-1}	No	Neutral	355	435	60	9, 10, 11, DOC	Interferes with DOC
Tinopal 5BM GX	Fluorescent Brightener 22	12224-01-0	?	No	Neutral	355	435	50	8, 11, DOC	Interferes with DOC
Phorwite BBH Pure	Fluorescent Brightener 28	4404-43-7	?	No	Neutral	349	439	2	9, 10	Low detectability
Diphenyl Brilliant Flavine 7GFF	Direct Yellow 96	61725-08-4	?	No	Neutral	415	489	2	/	Low detectability

Fluorescence intensity is relative to that of Uranine (=100%).

DOC = Dissolved Organic Carbon. CAS n° is the Chemical Abstracts Service number of the chemical substances. Absorption wavelength refers to that of maximum excitation; Fluor refers to the maximum emission wavelength.

Source: After Field et al. (1995), Käss (1998), Field (2002a), Benischke et al. (2014), Ford and Williams (2007), and Goldscheider et al. (2008).

Figure 5.38 Some of the most popular fluorescent dyes used in tracer tests: (a) Orange fluorescein powder. *Source:* Photo by Bartolomeo Vigna. (b) Yellowish Tinopal CBS-X powder. *Source:* Photo by Bartolomeo Vigna. (c) Fluorescein in a cave stream in New Zealand. *Source:* Photo by Neil Silverwood. (d) Sulforhodamine G injection in a cave stream in France. *Source:* Photo by Philippe Crochet. (e) A sinking surface stream with fluorescein. *Source:* Photo by Bartolomeo Vigna. (f) Injection of Tinopal CBS-X in a sinking stream. *Source:* Photo by Bartolomeo Vigna.

K^+ and Li^+ are present at much lower concentrations, so that smaller tracer masses can be used (Käss 1998; Benischke et al. 2014).

Particulate tracers, used as analogues of natural particles such as clays, larger fine particles (up to 100 μm) and colloids (<1 μm), are often useful to study the transport of pollutants in karst aquifers (Auckenthaler et al. 2002). In earlier years, the spores of the club moss *Lycopodium clavatum*, having an average diameter of 33 μm, were used (Käss 1998). These spores are often dyed with different fluorescent colors, so that multiple-tracer experiments become possible. The spores are collected using conical plankton nets with a mesh of 25 μm and counted under a microscope. These spores have been replaced by fluorescent microspheres of variable diameter (0.5–90 μm), and especially by 1 μm uncharged polystyrene spheres to reproduce the transport behavior of most bacteria (having similar sizes), and microparticles <1 μm that simulate colloids. The simultaneous injection of particulates of different sizes and fluorescent dyes in solution allows getting an idea on the different transport mechanisms and flow velocities of large particles, colloids and dissolved components (Göppert and Goldscheider 2008). Lanthanide- and DNA-labeled clays have also been used to simulate the transport of small particles in karst groundwater systems (Mahler et al. 1998a, 1998b).

Bacteriophages (viruses that attack specific bacteria) have been used to study the transport mechanisms of viruses, which have sizes ranging between 0.02 and 0.35 μm. Mostly marine bacteriophages are used, being easy to produce, harmless, and neutral (Harvey 1997). Bacteria, both artificially introduced and natural ones, can also be used as tracers.

The first fluorescent dye tracing test was carried out soon after the discovery in 1871 of Fluorescein ($C_{20}H_{12}O_5$, CAS number 2321-07-5), when 10 kg of this substance, together with salt and shale oil, were injected into the sink of the Danube River in the Swabian Alps, Germany, on 9 October 1877, emerging at Aachtopf Spring 12 km away two days later (Knop 1878). This test confirmed a tracing experiment conducted two weeks earlier, in which 10 tons of NaCl were poured into the Danube sink and the Aachtopf Spring was monitored every hour. One year later, another tracing experiment, again using salt (NaCl), was carried out in the Swiss village of Lausen to verify the provenance of pollution that was causing a typhoid fever outbreak among villagers: hundreds of kilograms of NaCl were injected in a sinking stream close to a farm 700 m south of the village, and a strong chloride content was detected the day after in the water-supply spring of the village (Hägler 1873). Norbert Casteret carried out the first transnational tracer experiment in 1931, pouring 60 kg of Uranine in the Forau de Aigualluts ponor in the Spanish Pyrenees, giving a positive result 10 hours later at the Goueil de Joueou spring in France, 3.7 km away from the injection point. The large amount of dye colored the Garonne River along 50 km downstream (Freixes et al. 1997). An interesting tracing experiment was carried out between Skočjan caves (Slovenia) and the Timavo Springs (Italy) in 1928–1929, which made use of eels that were marked by a notch cut on their dorsal fins. Twenty-nine out of 494 marked eels were recovered at the springs during a one-year observation period (Sella 1929).

Some tracing experiments were accidental. A remarkable example involved the Pernod distillery at Pontarlier (Franche-Comté, France) on 11 August 1901. Following an accidental fire, the vats burst and poured a very large quantity of absinthe into the adjacent Doubs River. Two days after the spill, the smell of absinthe was perceived at the source spring of the Loue River (Figure 5.19f), 10 km away, and the day after reached the village of Mouthiers, 5 km downstream (Worthington and Gunn 2009). The connection between the Doubs and the Loue spring was proved again by Martel in 1910 using 100 kg of Fluorescein, coloring the river two days later along a distance of 100 km (Martel 1921). Radioactive substances were also used in the early years of tracing. For instance, 38 kg of uranium oxide were used in 1909 to verify the connection between the Reka sink in the Škocjan caves and the Timavo Springs west of Trieste (Vortmann and Timeus 1910). Of course, such kinds of experiments are no longer possible today because of environmental legislation, and although many fluorescent dyes are considered to be non-toxic at low concentrations, special care should be taken in determining the mass to be injected into the groundwater system. In general, and especially if springs are part of a water supply system, dyed water should remain colorless, and concentration at the spring should be kept below 30 μg L^{-1} for Uranine. A large number (over 30) of empirical equations exist to determine the most suitable mass of tracer to be injected (Field 2003), and Worthington and Smart (2003) tested several of these proposing two equations that appear to be the most suitable in most cases. These equations consider the mass of the tracer M (in grams), that is correlated with the bird fly distance L (m) between the injection point and the output, the spring discharge Q (m^3 s^{-1}), the concentration to be obtained at the spring C (mg L^{-1}), and the expected arrival time T of the dye (in seconds):

$$M = 19(LQC)^{0.95} \tag{5.21}$$

and

$$M = 0.73(TQC)^{0.97} \tag{5.22}$$

Before carrying out a tracer test, careful planning has to be done to ensure the success of the experiment and to avoid legal problems. A good geological, hydrogeological, hydrochemical, and speleological knowledge of the study area is a desirable starting point. These preliminary investigations allow us to make the best decisions regarding injection points, mass(es) of tracer(s) to be introduced, monitoring points and frequency of sampling, in case continuous monitoring is not foreseen. In many cases, it is wise to inform local people and stakeholders, and depending on the applicable legislation, previous authorizations must be obtained, especially when working in protected areas, or when spring waters are used for drinking. When asking for such permissions, the environmental and human safety of the selected tracers must be documented (Behrens et al. 2001).

Since tracer injection often requires large masses of dye to be dissolved, and their determination at the outputs can be achieved at very low concentrations, it is paramount to avoid contamination. As a general rule, people involved in tracer injection should not participate in collection of samples at the outputs. Fluorescent dye tracers often come in powders, which are easily dispersed into air, and can stick to skin, shoes, and clothes. Generally, it is advisable to use protection clothes, gloves, and glasses. Commonly, it is easier to dissolve the tracers in water before injection. When large tracer masses need to be dissolved, the powders should not be introduced in mass to avoid formation of insoluble lumps, but it should be moistened beforehand, creating a dense paste-like material.

The selection of injection points is another crucial decision. In the case of sinking rivers, the tracer is injected directly into the streams. Rivers forming large pools or with muddy bottoms should be avoided if possible, since tracers can be retained over long periods of time. The injection time should be reduced as much as possible, in order to obtain a clear breakthrough curve with a well-defined maximum, allowing the hydraulic characterization of the flow. Tracers can also be injected into "dry" sinkhole bottoms, enlarged fissures and shafts, but it is advised to introduce water, starting before and until well after the tracer is injected, in order to facilitate the tracer travelling to the underlying flowing groundwater. This is usually done by professional hydrogeologists, since tank trucks are often used in these cases, pouring thousands of liters into the injection point before and after tracer dilution. The same should be done when introducing tracers into boreholes to drive the tracer through the fissure network to the enlarged fissures or conduits below. This artificial flushing should however be balanced, avoiding to distort significantly the natural groundwater flow. Tracing a poorly permeable well often means introducing a dye that will stay, at least potentially, for a rather long time in the aquifer, so that the same dye cannot be used for many years in the same area. Dye can also be injected using the so-called "dry-set", placing it in a dry channel (or sinkhole bottom), ready to be flushed into the aquifer at the first important storm event that creates runoff and produces infiltration. This is usually done in remote areas, where the possibility of people or animals disturbing the dye are very low, and logistical problems do not allow a direct introduction during periods with natural runoff.

Sampling and tracer detection/measurement should be scheduled at the sites where the tracers are believed to arrive, based on previous investigations, but should also include as many as possible outputs, and even the ones that are considered less probable. This increases the chances of revealing surprising results, which are not that rare in karst studies, and gives more confidence in positive outcomes. There are three ways of sampling tracers at outputs: (i) integrative methods; (ii) discrete sampling, and (iii) continuous monitoring.

Integrative sampling makes use of charcoal or cotton detection bags (Figure 5.39a, b). Charcoal is used for most fluorescent dyes and cotton bags are suitable for Tinopal CBS-X, Phorwite BBH Pure, and Diphenyl Brilliant Flavine 7GFF. The property of charcoal (Figure 5.39c) to adsorb fluorescein is known since 1904 (Käss 1998), but it was only in the late 1950s that charcoal bags were

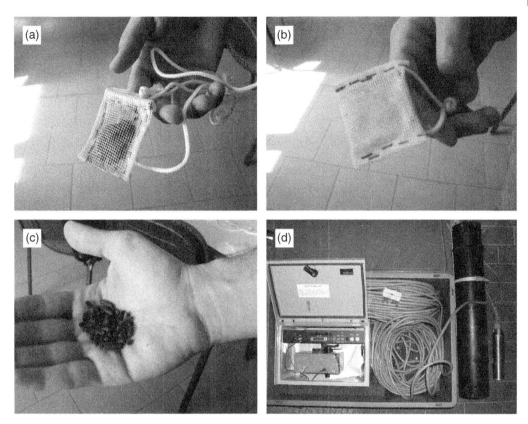

Figure 5.39 Types of tracer detection systems: (a) Charcoal detection bag. (b) Cotton detection bag for Tinopal CBS-X. (c) The grains of charcoal used for the preparation of charcoal bags. (d) Fluorimeter capable of measuring up to three dyes simultaneously with turbidity. *Source:* Photos by Bartolomeo Vigna.

first used by the caver's community in America (Dunn 1957). One sample bag (0-zero) needs to be placed before the tracing experiment starts (to have a blank). Then two numbered detector bags (1 and 2) are placed, one of them being replaced by a new one at regular time intervals (bag n°1 is replaced by bag n°3 the first time, bag n°2 is replaced by bag n°4 the second time, and so forth). The more often they are replaced, the better the arrival time will be resolved. The fluorescent dyes adsorbed onto the charcoal or cotton are successively eluted (extracted with a solvent) in the laboratory. See Wilson (1968), Smart and Simpson (2002), and Jones (2019) for details. To elute Uranine a solution of 5% potassium hydroxide (KOH) in 70% isopropyl alcohol (isopropanol) is often used, or a solution of 50% 1-propanol, 20% NH_4OH, and 30% of distilled water (called the "Smart" solution). If some time passes from detector collection to laboratory analysis, the samples should be dried and kept in a dark place. This method is simple, but only allows to have a very rough idea of the tracer quantities that pass through each monitoring site.

Discrete sampling requires manual or automatic sampling of water at specific time intervals and analyses of the water in the laboratory. Because many dyes are sensitive to light, sampling should take place close to the water emergence point, and samples should be stored in dark glass bottles and in dark and cool places. This method, if the sampling interval is tight enough, allows the reconstruction of a breakthrough curve (BTC). In the last decades, quantitative field spectrometers have been developed (Figure 5.39d), allowing continuous measurements of multiple tracers at a high temporal resolution (even seconds) (Schnegg and Costa 2003). In tracer experiments where multiple outlets need to be checked, the three methods are often combined, with the main possible outlets equipped with field fluorimeters, other important springs being sampled, and placing detector bags in all possible outlets as a backup, in case instruments fail.

Details on the analysis of tracers in samples and detectors can be found in Käss (1998). The standard method used is spectrofluorimetry. In case multiple dyes are used during a tracer experiment, their fluorescence peaks need to be clearly distinguishable from each other, with limited overlap between the spectra. A typical combination of suitable dyes is Uranine (green), Rhodamine (red) and Tinopal CBS-X (blue).

Discrete or continuous monitoring of a tracer allows us to reconstruct the arrival of the dye over time. The graphical representation of the tracer concentration vs. time is the breakthrough curve (BTC), in which the zero time corresponds to the moment of tracer injection. The combination of this BTC with a hydrograph allows estimating the quantity of dye that reached the monitored output point, often represented as the percentage of the injected tracer mass recorded at the output point through time. This value, named tracer recovery M_R is given by the numerical integration of tracer fluxes ($Q \cdot c$) in discrete time intervals dt:

$$M_R = \int_{t=0}^{\infty} \left(Q \cdot c \right) \mathrm{d}t \tag{5.23}$$

where Q is the time-dependent discharge, c is tracer concentration, and dt are the time intervals. BTCs allow us to determine the initial time of tracer arrival t_1, peak concentration Cp, peak time t_p, time when half of the recovered tracer has passed the sampling site $t_{R/2}$, and other significant values (Figure 5.40). Several computer programs can be used to analyze BTCs, QTRACER2 probably being the most widely used, enabling the estimation of hydraulic parameters, tracer recovery, mean residence time, mean flow velocity, longitudinal dispersion, and other characteristics (Field 2002a). Often BTCs are greatly disturbed, because flow in cave conduits and fractures is not homogeneous and highly influenced by the geometry of the flow paths. The main disturbing factors are retardation, dispersion and the presence of eddies. Eddies create "dead" and "slow" flow zones in sections of the underground rivers in which large pools are separated by waterfalls, for

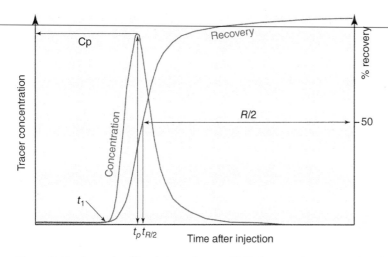

Figure 5.40 Example of breakthrough curve (BTC). See text for explanation of symbols.

example. These disturbing effects are described in detail by Hauns et al. (2001) based on modeling, laboratory, and field experiments conducted in Switzerland.

5.6.9 Isotopes

Not all atoms of an element are identical, but they can have different masses due to the fact that they contain different numbers of neutrons. Atoms of the same element having different atomic masses are called isotopes. Most of these isotopes occur in nature in significant amounts, and are called environmental isotopes, including those of hydrogen (H), carbon (C), nitrogen (N), oxygen (O), and sulfur (S). Some may be produced by anthropogenic activities, increasing their concentrations with respect to their natural background values. Some isotopes are radioactive (radioisotopes) and unstable, and undergo nuclear transformation emitting radiation in a natural way, with specific decay constants for each isotope.

The most useful isotopes in karst hydrological studies are those of hydrogen, oxygen, and carbon (Criss et al. 2014) (Table 5.4). Tritium (^{3}H) and radiocarbon (^{14}C) are both radioactive and are naturally produced in low quantities in the atmosphere by cosmic ray bombardment. The ^{14}C isotope derives from absorption of thermal neutrons by nitrogen. Their concentrations increased during nuclear bomb testing in the 1950s–1960s, peaking in 1963, when the moratorium on these nuclear tests was signed and their concentrations started to decline. All the other isotopes are called stable isotopes, since they do not undergo radioactive decay.

H, C, and O are natural constituents of waters in karst aquifers, and their stable isotope compositions can be used as natural tracers. The variation in the proportions of these isotopes are caused by natural processes, mainly fractionation occurring in the hydrological cycle during evaporation and condensation of water. For example, "heavier" water molecules (containing ^{18}O) will evaporate less easily than the "lighter" ones, and will condense more readily than the lighter molecules. Variations in stable isotope proportions are measured with modern mass spectrometers and compared with the stable isotope proportions in standard materials. Stable isotope ratios (R) are always expressed as percentage of the less abundant nuclide over the dominant one: Deuterium/Hydrogen (D/H), ^{13}C/^{12}C, or ^{18}O/^{16}O. Since these ratios are very small numbers, it is common practice to use the delta notation which is given by

Table 5.4 Isotopes often used in karst hydrogeology and their abundance on Earth.

Atomic number	Name	Symbol	Atomic mass	% abundance
1	Hydrogen	^{1}H	1.007825	99.985
	Deuterium	^{2}H	2.014102	0.0115
	Tritium	^{3}H	3.016049	Variable
6	Carbon	^{12}C	12.00000	98.93
		^{13}C	13.003355	1.07
	Radiocarbon	^{14}C	14.003242	Variable
8	Oxygen	^{16}O	15.994915	99.757
		^{17}O	16.999132	0.038
		^{18}O	17.999160	0.205

$$\delta = 1000 \times \left(\frac{R_{\text{sample}} - R_{\text{standard}}}{R_{\text{standard}}} \right) \tag{5.24}$$

in which R is the isotopic ratio of the examined element. This value is thus the deviation from the standard isotopic ratio expressed per thousand (per mil or ‰). The standards normally used in karst studies are the Vienna Standard Mean Ocean Water (VSMOW) for hydrogen and oxygen isotopes and Pee Dee Belemnite (PDB) for carbon isotope studies (Coplen 1988). Positive δ-values indicate a higher concentration of ^{18}O, ^{2}H or ^{13}C compared to those of the standards being used, whereas negative δ-values denote lower concentration of heavier isotopes in the sample.

Ocean water has a $\delta^{18}O_{\text{SMOV}}$ close to 0 ‰, and in most parts of the hydrological cycle, ^{18}O is depleted with respect to ocean water, giving negative δ-values, the most negative values being found in ice samples from cold, arctic regions (-50 to -25‰) (Figure 5.41). Temperature is the main factor controlling fractionation processes. As air masses rise or move toward the poles, temperature decreases, and water vapor turns into liquid droplets fractionating toward the heavier isotopes. The remaining air will thus be enriched in water vapor with lighter isotopes. As the residual air continues rising (or moving toward the poles), the water vapor that condenses becomes increasingly depleted in heavier isotopes. As a result of this fractionation process the amount of heavier isotopes in precipitation decreases with increasing latitude and altitude. This is related to previous cooling of the wet air masses that causes condensation of heavier waters and precipitation at lower latitudes and altitudes, leaving increasingly light vapor to condense at higher latitudes and altitudes. Other factors that influence the isotopic content of water are the continental and seasonal effects. Air masses generated above the oceans are depleted in heavier isotopes the farther they travel into continental areas, away from the original source of the vapor. There are also seasonal variations in the isotopic composition of rainwater related to the different provenances and trajectories of the humid air masses, and thus the variable condensation/evaporation fractionation processes. Another factor is the amount effect, related to phase transitions during rain events, with water condensing and evaporating along its way toward the ground. These processes mainly occur in light rain events or rain showers during warmer seasons. At a global scale, there is a linear relationship between the stable isotopes deuterium and ^{18}O, known as the Global Meteoric Water Line (GMWL) (Craig 1961), which globally can be expressed as:

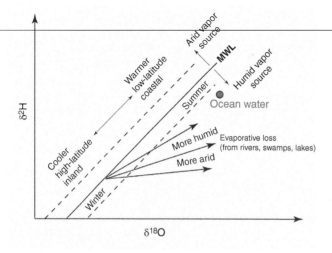

Figure 5.41 Graph showing $\delta^{18}O$ and $\delta^{2}H$ (deuterium) relationships in natural waters. MWL: Meteoric Water Line.

$$\delta D = 8\delta^{18}O + 10 \tag{5.25}$$

or in its revised version (Rozanski et al. 1992)

$$\delta D = 8.17(\pm 0.07)\delta^{18}O + 11.27(\pm 0.65) \tag{5.26}$$

Although this Global MWL works well on the Earth's scale, on regional scales different Local MWLs with differing slopes and intersects need to be constructed. Deviations from the GMWL are especially important for coastal areas, islands, and tropical mountainous regions.

The hydrogen and oxygen stable isotope ratios of shallow and cool groundwater reflects those of local precipitation, so they plot close to the LMWL. Differences in isotopic signature between various springs reflect the different sources of their recharge water (i.e., altitudinal, latitudinal, and geographical position): a coastal karst spring will have a higher $\delta^{18}O$ than a spring located in the interior of the continent, and a spring at lower altitude will have higher values than springs up in the mountains. Consequently, the stable isotope signature of spring waters can give us some clues about their origin, but caution must be taken, and it must be kept in mind that although a general provenance of the waters can often be recognized, the exact source can only be identified by tracer tests. When springs respond very rapidly to rainfall events, their stable isotope composition commonly shows a seasonal fingerprint: negative values during winter–spring (snow melt water) and more positive ones during summer–autumn. When measurements over several years reveal that stable isotope ratios do not match with the corresponding seasons, but show a delay time (related to the relatively low permeability of the aquifer and/or long travel distance), water transit times can be estimated. When isotopic values are essentially constant over time and lack seasonal pattern, a general mixing of waters recharged in cold and hot seasons can be suspected. Long and deep (often thermal) groundwater paths with prolonged water-rock contact cause isotopic values to shift toward heavier $\delta^{18}O$ values. Moreover, groundwater systems recharged by intensely evaporated water (lakes and swamps in arid areas) are characterized by heavier isotopes. Cave drips derived from soil water are also marked by slightly more positive values related to evapotranspiration processes (Lachniet 2009; Baker et al. 2019). Two springs apparently fed by the same karst aquifer can show differences in stable isotopes due to differences in the mean altitude of their respective recharge areas (more negative values indicating recharge at higher altitude).

Isotopes can also be used to date groundwaters, mainly through radioactive isotopes and their specific radioactive decay properties. Radioactive decay is given by the equation:

$$\frac{N}{N_0} = e^{-\lambda t} \tag{5.27}$$

where N is the number of radioactive atoms measured at time t, N_0 is the number of radioactive atoms present before decay started, and λ the characteristic decay constant. Each radioisotope is characterized by a specific half-life, which corresponds to the time required for half of the radioactive atoms to decay. Short-lived radionuclides (often thermonuclear fallout) can be used to date young groundwaters, whereas naturally occurring long-lived radionucleids (often cosmogenic) can be used to date old waters (Figure 5.42). Numerical dating of groundwater using these radioisotopes requires a deep understanding of the mixing phenomena in the groundwater system, and the chemical reactivity of the radioisotope considered. During storms, for example, newly infiltrated water reaches the spring more or less rapidly depending on the connectivity of the groundwater flow paths and their length. In some cases, a hydrostatic pressure wave will first reach the spring, pushing out older resident waters (piston flow), followed by neoinfiltration waters.

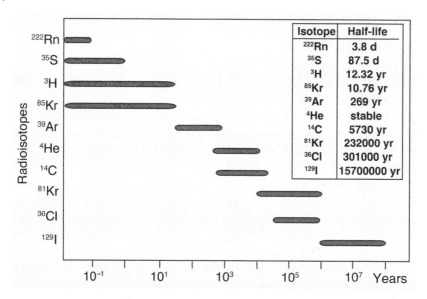

Figure 5.42 Radioisotopic dating systems in groundwater studies. In the half-life column "d" and "yr" indicates days and years, respectively.

During low flow conditions, mainly long-residence, matrix porosity water will reach the spring, so radiogenic ages will generally be older. Tritium ages of baseflow waters taken over many years can give an idea of mean residence times of waters that flow through matrix porosity. In general, it appears difficult to give a mean age to groundwater using these techniques in karst areas because of the intrinsic complex behavior of groundwater flow. For more details on groundwater dating the reader is referred to Kazemi et al. (2006).

5.6.10 Remote Sensing and Geophysical Techniques

Remote sensing and geophysical techniques are widely used methodologies by hydrogeologists, especially in the preliminary stages of a groundwater study. This is especially true for karst groundwater studies in which inputs and outputs can be studied, but information on water-bearing conduits and caves is scarce or completely lacking. Often there is no idea about the groundwater flow paths, and the anisotropy and heterogeneous nature of the permeability features of the aquifer make it difficult for hydrogeologists to select the right places for their monitoring or exploratory boreholes.

Remote sensing was traditionally based on the analysis of aerial photographs using stereoscopes that offer a three-dimensional view of the study areas. Photointerpretation allows mapping geological and geomorphological features such as geological contacts, lineaments, sinking streams, springs. Since groundwater flow in karst occurs preferentially along enhanced-permeability zones, geological mapping can help to identify these zones (e.g., fault zones, contact between soluble and insoluble rocks) in which boreholes can be planned with a greater probability of hitting major flow paths (Kresic 1995). Aerial photographs have progressively been replaced by satellite imagery starting from the first public (non-military) acquisitions in the 1970s, and more recently by airborne and Unmanned Aerial Vehicle (UAV) imagery (Silva et al. 2017). Nowadays, high-resolution multispectral images can be obtained at reasonable cost. A special mention must be made for thermal infrared imagery that allows identifying differences in water temperature or thermal contrasts related to warm or cold air coming out of cave entrances. The second method has been used to identify a shallow cave system in southern Spain during periods in which the thermal difference between the cave air and the outside were at a maximum (in this case during winter dawns) (Pérez-García et al. 2018). The instability of the air (i.e., winds) and the much lower specific heat of rocks (they heat and cool down rather rapidly) with respect to that of water make it very difficult to go beyond a mere qualitative use of this method in terrestrial applications. In contrast, thermal infrared images have successfully been used to locate submarine or sublacustrine karst springs in periods of great thermal contrast between the issuing groundwater and the surrounding waters (Mejías et al. 2012). Furthermore, these hidden discharges can be quantified with in situ measurements of EC, Radon, and Radium (Bejannin et al. 2017). Further details on remote sensing techniques applied in Earth Science applications can be found in Jensen (2015) and Lillesand et al. (2015).

Geophysical techniques can be applied to obtain information on the spatial variations of a number of physical parameters in the subsurface. These changes may be related to multiple factors, such as the composition of the rocks and soils, their texture and porosity, or the presence of water and its chemical composition. Geophysical investigations in karst can help in: (i) defining the overburden thickness; (ii) determining the boundaries of groundwater basins; (iii) identifying high porosity zones (fractured areas, voids, and conduits); (iv) detecting preferential infiltration zones (e.g., buried dolines) and groundwater flow paths (conduits).

When deeper features are targeted, such as karstified limestones in hydrocarbon reservoirs located kilometers below the ground surface or sea floor, three-dimensional seismic data combined with borehole data can allow the geometrical characterization of karst aquifers and associated features (e.g., paleokarstic surface). Good examples of such deep geophysical investigations are those carried out in the Adriatic (Dubois et al. 1993), in the Campos Basin in Brazil (Basso et al. 2018), and in the Tarim Basin oilfields in Northwest China (Tian et al. 2019).

Generally, geophysical surveys in hydrogeology have investigation depths ranging between a few meters and around 200 m. The physical properties that can be indirectly measured using geophysical techniques mainly include density, electrical conductivity (or its reciprocal, resistivity), and magnetic susceptibility.

Electrical surveys measure the electrical properties of the subsurface, which are greatly influenced by the nature of the soil and rock, porosity, the presence of air (poor electrical conductor) or water (with higher EC), and the chemistry of the fluids (the higher the ionic content the higher the EC). Electrical resistivity methods introduce electrical current into the ground with electrodes, and can be used successfully to image the topography of the rockhead (often highly irregular) overlain by an unconsolidated cover, paleosinkholes, or caves of significant dimensions (air-, water-, sediment-filled). These methods include Vertical Electrical Sounding (VES), Resistivity Profiling and Electrical Resistivity Tomography (ERT), also known as Electrical Resistivity Imaging (ERI). In karst areas one of the drawbacks of these methods is the difficulty of ensuring a good transmission of electricity into the ground, since the rock is often bare or mantled by a thin and often discontinuous layer of soil.

Another electrical survey technique is the Spontaneous Potential (SP) in which natural electrical currents are measured. These currents are created by water flow through capillary openings (electrofiltration), but are disturbed by other currents (telluric, meteoric, bioelectric). The method is cheap and easy to carry out, and can sometimes give interesting results on water flow in shallow karst (Chen et al. 2018).

Electromagnetic methods have also proved to be useful in the study of karst features. These include the passive ones that only need a receiver and use EM fields emitted from radio stations all around the world, and the active systems, in which both a transmitter and a receiver are needed. Very Low Frequency-Electromagnetics (VLF-EM), Radiofrequency-Electromagnetics (RF-EM), and the Radiomagnetotelluric method (RMT) are passive systems, whereas Time Domain Electromagnetics (TDEM) and Controlled Source Audio Magnetotellurics (CSAMT) are active ones. All these methods are useful for defining the upper boundary of the karst rock (rockhead) overlain by a relatively thin unconsolidated cover, the depth of the water table, the freshwater-saltwater boundary in coastal areas, electrically conductive structures such as steep faults, or moist sediment-filled cavities. These electromagnetic methods work well up to depths of some tens of meters.

Ground Penetrating Radar (GPR) is a high frequency electromagnetic method that transmits EM waves to the subsurface, and records the reflected or backscattered waves back to the surface due to changes in the dielectric constants of the materials (e.g., Rodríguez et al. 2014; Zarroca et al. 2017) (Figure 5.43). The high frequency of the EM waves (usually 50–500 MHz in applications related to karst) restricts their penetration depth, but in contrast to seismic methods, GPR offers a much greater spatial resolution. The penetration of GPR greatly depends on the moisture and clay content of the shallow layers, which attenuates the EM signal. For instance, this method can offer a good performance in an area where the karst bedrock is mantled by a dry sand-gravel soil.

Magnetic methods, which measure natural variations in the Earth's magnetic field, are rarely useful in karst, although they might be able to detect large air- or water-filled cavities. Microgravity is based on the measurement of very small variations in the Earth's gravitational field caused by changes in the subsurface density distribution (Styles et al. 2005). Since these variations are very small, their detection requires the use of high-precision instruments, and filtering out the much larger variations related to other factors such as topography, above-surface objects (e.g., buildings), latitude, position of the moon and sun (Earth tides), and regional geological variations. The gravity maps representing the corrected data of gravity anomalies are known as *Bouguer* anomaly maps and are able to locate subsurface voids, sharp and significant changes in the overburden thickness

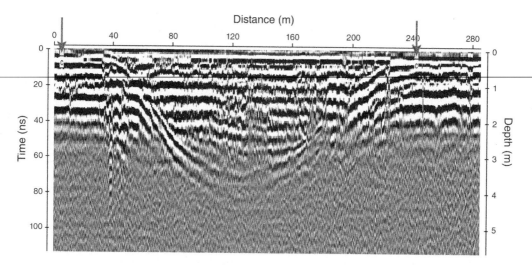

Figure 5.43 GPR profile acquired with a 100 MHz shielded antenna across a sagging sinkhole related to salt dissolution and buried by a highway (central sector of the Ebro Cenozoic Basin, NE Spain). Note collapse faults expressed as laterally truncated and offset reflections at the margins of the synform (basin structure in 3D). Arrows indicate location of marginal cracks in the road pavement.

(such as that covering pinnacles and cutters), and lateral variations in lithology that may correspond to hydrogeological barriers (e.g., faults). Microgravimetric maps, combined with resistivity surveys, may allow to better estimate the size of underground voids (McGrath et al. 2002). Placing two microgravimeters, one at the surface and the other in a cave below, and measuring microgravity changes over long time scales (over one hydrological year), allows quantifying the amount of water stored in the rock volume comprised between the surface gravimeter and the one placed in the cave (Champollion et al. 2018).

Geophysical investigations preferably should combine different complementary techniques. Geophysical surveys are often useful in preliminary investigations in karst areas to better plan drilling surveys. Their implementation is often cost-justified, since they allow considerable lowering of the cost of expensive direct (intrusive) exploration techniques. It must be borne in mind that interpretation of geophysical results is often ambiguous, and that the geological models based on these results must always be confronted with direct observations (profiles, drillholes, nearby outcrops) and geological knowledge. An overview of geophysical methods used in karst hydrogeology can be found in Bechtel et al. (2007), Chalikakis et al. (2011), and Benson and Yuhr (2015).

5.6.11 Karst Aquifer Modeling

A hydrogeological model has the aim to simplify a natural groundwater system in order to understand, predict, and manage water resources. Karst aquifer modeling can be space-dependent and time-dependent. The first treats the karst aquifer as a physical groundwater flow system that does not change through time, whilst the second considers a continuously changing system with a positive feedback that involves the progressive concentration of groundwater flow and the correlated development of karst conduits (larger pathways enlarge faster by dissolution, thus focusing increasingly higher flow). These latter models are used in speleogenesis, and will be dealt with later.

Space-dependent hydrogeological models can be used to predict the flow and the mass transport in karst groundwater systems. Flow models are meant to predict groundwater levels, underground water fluxes and discharges at outlets, whereas transport models have the aim of understanding and predicting the movement and distribution of substances (e.g., pollutants) in the groundwater system. Both models are based on a number of assumptions, especially regarding the subsurface structure of the karst system. The outcomes of each model are always an approximation to the real system, with uncertainties that must be kept in mind.

There are two main and complementary modeling approaches: the global (lumped parameter) model, and the distributed parameter model. The first is based on the mathematical analysis of input variables (rainfall and allogenic input) and the resulting output (spring hydrographs and chemographs) (dealt with in Section 5.6.7), whereas the second uses theoretical concepts and physical laws to describe and predict the hydraulic behavior of the karst groundwater system. Lumped parameter models require less data, and are thus easy to use, whereas distributed models need much more data, are complex to use and parameterize, and often require higher computing capability.

The starting point of any type of modeling effort is the description of the conceptual model of the karst aquifer, which represents the real system in a simplified way. This conceptual model needs to consider the duality of the karst aquifer (fast flow and low storage in conduits, and slow diffuse flow but large storage in the fissure network and matrix porosity), and the clear distinction of the three different zones of groundwater flow and storage (soil and epikarst, vadose or unsaturated zone, and phreatic or saturated zone) (White 2002). It may often be useful to conceptualize the karst aquifer as composed of different compartments interacting with each other: the epikarst-vadose compartment, where vertical flow dominates, and three phreatic compartments representing the matrix porosity, the fracture network, and the conduits (Mazzilli et al. 2019) (Figure 5.44).

Distributed parameter groundwater flow models can incorporate discrete channels (or fractures), matrix porosity, or a combination of both. Laminar flow in a continuum medium (matrix porosity) is described by the Darcy equation (Eq. 5.5), whereas flow in conduits and enlarged fractures can be described with the Darcy-Weissbach equation (Eq. 5.9). Five main alternative modeling approaches can thus be distinguished (Figure 5.45): the Equivalent Porous Medium Approach (EPM), the Double Continuum Approach (DC), the Discrete Fracture Network Approach (DFN),

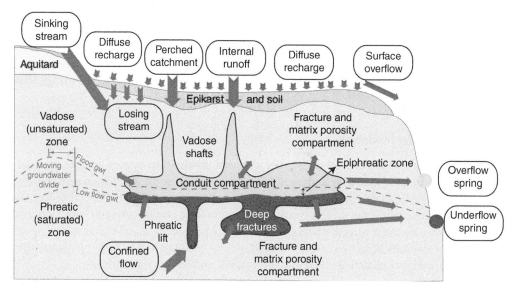

Figure 5.44 Conceptual scheme of a typical karst aquifer with its vadose and phreatic compartments gwt: groundwater table. *Source:* Adapted from White (1999).

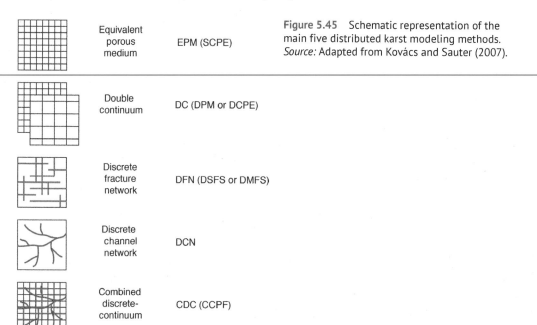

	Equivalent porous medium	EPM (SCPE)
	Double continuum	DC (DPM or DCPE)
	Discrete fracture network	DFN (DSFS or DMFS)
	Discrete channel network	DCN
	Combined discrete-continuum	CDC (CCPF)

Figure 5.45 Schematic representation of the main five distributed karst modeling methods. *Source:* Adapted from Kovács and Sauter (2007).

the Discrete Channel Network Approach (DCN), and the Combined Discrete-Continuum Approach (CDC) (Kovács and Sauter 2007; Ghasemizadeh et al. 2012). The choice of the most suitable model depends on the modeling objective (groundwater budget estimation, contaminant transport and groundwater flow reconstruction, drinking water management problems), the available data on the inputs, outputs and throughflow, and the dominant (supposed) flow mechanism (matrix, fracture or conduit flow).

The Equivalent Porous Medium Approach (EPM), sometimes also called Single Continuum Porous Equivalent (SCPE), or Distributed Parameter Approach, is the simplest approach, based on the assumption that flow occurs in an isotropic, homogeneous medium, so that laminar flow conditions are met and the Darcy Law can be applied. We know that this is almost never the case in karst aquifers, but this simple approach may give acceptable results at large (regional) scales in generally poorly karstified aquifers. These models are not adequate for predicting the local direction and rate of groundwater flow and the flux in point-source contaminant transport. In fact, these models neither integrate concentrated flow along conduits or fractures, nor take into account turbulent flow. Examples of studies where EPM approaches have given satisfactory results, at least at the regional scale, are the Barton Springs in the Edwards Aquifer in Texas (Scanlon et al. 2003), and the transboundary aquifer of the Western Mountain Aquifer Basin shared between Israel and the Palestinian Authority (Abusaada and Sauter 2013).

The Double Continuum Approach (DC), also called Double Porosity Model (DPM) or Double Continuum Porous Equivalent Approach (DCPE), takes into account the dual flow characteristics of karst. It considers the karst aquifer to be composed of one compartment characterized by low hydraulic conductivity and high storativity (matrix porosity component), and another compartment with higher hydraulic conductivity and lower storativity (fractures and conduits). Each compartment can be described by distinct Darcian or Darcian-Weissbach flow equations, and the two compartments are interconnected, with water interchange governed by differences in hydraulic potential. This model can also be applied when no information on the conduit network geometry is available. The flow dynamics in both the matrix porosity and the fractures or conduits is reproduced, and these models are thus capable of predicting contaminant flow and transport in the system, but not the spatial distribution of the contaminants in the aquifer. The DC approach is suitable in moderately-to-highly karstified systems and has been successfully applied in the Swabian Alb (Teutsch and Sauter 1998; Kordilla et al. 2012), in a karst area north of Montpellier, France (Maréchal et al. 2008), on eogenetic karst islands (Vacher and Mylroie 2002), and in a fractured and porous limestone aquifer in Burgundy, France (Robineau et al. 2018).

The Discrete Fracture Network Approach (DFN), with its two variants, Discrete Singular Fracture Set (DSFS) and Discrete Multiple Fracture Set (DMFS), can be applied to simulate flow and transport in fractured karst aquifers, where rock-matrix permeability is assumed to be negligible. Detailed geometrical information on the fracture network is required for its implementation. Depending on the characteristics of the fracture sets, flow conditions can be assumed laminar or turbulent. Good recent examples in which this method has been used are the Gomba Oilfield in Hungary (Bauer and Tóth 2017) and the Tahe Oilfield, western China (Mendez et al. 2019).

The Discrete Channel Network Approach (DCN) simulates turbulent flow in interconnected conduits, rather than in fractures embedded within an impermeable matrix. This modeling approach requires information on the geometrical characteristics of significant portions of the conduit system. A good example is given for the Hölloch Cave system in Switzerland (Jeannin 2001).

Finally, there are hybrid models that use a Combined Discrete-Continuum Approach (CDC) or the Coupled Continuum Pipe Flow (CCPF), integrating discrete models such as DFN and DCN, with EPM models. Enlarged fractures or conduits with turbulent flow are embedded in a low permeability matrix-porosity component with laminar flow. This approach was first introduced by

Kiraly and Morel (1976a, 1976b) for the Areuse Spring catchment in Switzerland. This type of models requires a good knowledge on the conduit network and its density, and the hydraulic parameters of both the matrix-porosity and the high-permeability compartments. In order to obtain satisfactory results, it is often necessary to add the epikarst compartment, which acts as a retarding storage and a concentrated-recharge element of the system (Kiraly 1998). A good example of such a continuum-pipe flow model is explained in Liedl et al. (2003). This type of model has also been used to predict groundwater flow and salt water intrusion in a coastal karst aquifer with conduits (Xu and Hu 2017). Presently, CCPF models are very theoretical, capable of resolving hypothetical problems related to flow, transport, and dissolution dynamics in karst aquifers. The USGS's MODFLOW, which is one of the most popular groundwater modeling packages, has included such a karst model in its offers (MODFLOW-CFP, Shoemaker et al. 2008). Nonetheless, its application to real-world situations is limited, since in most cases there is not enough detailed knowledge on the different compartments of the karst groundwater system and when sufficient information is available, the amount of data fed into the model is often beyond the computing capabilities of most computers.

References

Abusaada, M. and Sauter, M. (2013). Studying the flow dynamics of a karst aquifer system with an equivalent porous medium model. *Groundwater* 51 (4): 641–650.

Atkinson, T.C. (1985). Present and future directions in karst hydrogeology. *Annales de la Société géologique de Belgique* 108: 293–296.

Atkinson, T.C. and Smart, P.L. (1981). Artificial tracers in hydrogeology. In: *A Survey of British Hydrogeology*, vol. *1980*, 173–190. London: Royal Society.

Auckenthaler, A., Raso, G., and Huggenberger, P. (2002). Particle transport in a karst aquifer: natural and artificial tracer experiments with bacteria, bacteriophages and microspheres. *Water Science and Technology* 46 (3): 131–138.

Badino, G. (2010). Underground meteorology- "What's the weather underground?". *Acta Carsologica* 39 (3): 427–448.

Badino, G. (2018a). Geothermal flux and phreatic speleogenesis in gypsum, halite, and quartzite rocks. *International Journal of Speleology* 47 (1): 1–11.

Badino, G. (2018b). Models of temperature, entropy production and convective airflow in caves. In: *Advances in Karst Research: Theory, Fieldwork and Applications* (ed. M. Parise, F. Gabrovšek, G. Kaufmann, et al.), 359–379. London: Geological Society of London.

Badino, G., De Vivo, A., Forti, P. et al. (2018). The Puerto Princesa Underground River (Palawan, Philippines): some peculiar features of a tropical, high-energy coastal karst system. *Geological Society of London, Special Publications* 466 (1): 155–170.

Badon-Ghijben, W. (1888). Nota in verband met de voorgenomen putboring nabij Amsterdam. *Tijdschrift van het Koninklijk Instituut van Ingenieurs* 8–22.

Bailly-Comte, V., Martin, J.B., Jourde, H. et al. (2010). Water exchange and pressure transfer between conduits and matrix and their influence on hydrodynamics of two karst aquifers with sinking streams. *Journal of Hydrology* 386 (1–4): 55–66.

Baker, A., Hartmann, A., Duan, W. et al. (2019). Global analysis reveals climatic controls on the oxygen isotope composition of cave drip water. *Nature Communications* 10 (1): 1–7.

Banzato, C., Butera, I., Revelli, R. et al. (2017a). Reliability of the VESPA index in identifying spring vulnerability level. *Journal of Hydrologic Engineering* 22 (6): 04017008.

Basso, M., Kuroda, M.C., Afonso, L.C.S. et al. (2018). Three-dimensional seismic geomorphology of paleokarst in the Cretaceous Macaé Group carbonates, Campos Basin, Brazil. *Journal of Petroleum Geology* 41 (4): 513–526.

Bauer, M. and Tóth, T.M. (2017). Characterization and DFN modelling of the fracture network in a Mesozoic karst reservoir: Gomba oilfield, Paleogene Basin, Central Hungary. *Journal of Petroleum Geology* 40 (3): 319–334.

Bechtel, T.D., Bosch, F.P., and Gurk, M. (2007). Geophysical methods. In: *Methods in Karst Hydrogeology* (ed. N. Goldscheider and D. Drew), 171–199. London: Taylor & Francis.

Beddows, P.A., Smart, P.L., Whitaker, F.F. et al. (2002). Density stratified groundwater circulation on the Caribbean Coast of Yucatan peninsula, Mexico. In: *Karst Waters Institute Special Publication 7: Hydrogeology and Biology of Post-Paleozoic Carbonate Aquifers* (ed. J.B. Martin, C.M. Wicks and I.D. Sasowsky), 129–134. Charles Town: Karst Waters Institute.

Behrens, H., Beims, U., Dieter, H. et al. (2001). Toxicological and ecotoxicological assessment of water tracers. *Hydrogeology Journal* 9: 321–325.

Bejannin, S., van Beek, P., Stieglitz, T. et al. (2017). Combining airborne thermal infrared images and radium isotopes to study submarine groundwater discharge along the French Mediterranean coastline. *Journal of Hydrology: Regional Studies* 13: 72–90.

Benischke, R., Goldscheider, N. and Smart, C. (2014). Tracer techniques. In: Methods in Karst Hydrogeology (ed. N. Goldscheider, D. Drew), 164–184. London: Taylor & Francis.

Benson, R.C. and Yuhr, L.B. (2015). *Site Characterization in Karst and Pseudokarst Terraines: Practical Strategies and Technology for Practicing Engineers, Hydrologists and Geologists.* Dordrecht: Springer.

Bonacci, O. (1993). Karst springs hydrographs as indicators of karst aquifers. Hydrological Sciences Journal 38(1): 51–62.

Caine, J.S., Evans, J.P., and Forster, C.B. (1996). Fault zone architecture and permeability structure. *Geology* 24 (11): 1025–1028.

Chalikakis, K., Plagnes, V., Guerin, R. et al. (2011). Contribution of geophysical methods to karst-system exploration: an overview. *Hydrogeology Journal* 19 (6): 1169–1180.

Champollion, C., Deville, S., Chéry, J. et al. (2018). Estimating epikarst water storage by time-lapse surface-to-depth gravity measurements. *Hydrology and Earth System Sciences* 22: 3825–3839.

Chen, Y., Qin, X., Huang, Q. et al. (2018). Anomalous spontaneous electrical potential characteristics of epi-karst in the Longrui Depression, Southern Guangxi Province, China. *Environmental Earth Sciences* 77 (19): 659.

Coplen, T.B. (1988). Normalization of oxygen and hydrogen isotope data. *Chemical Geology: Isotope Geoscience Section* 72 (4): 293–297.

Craig, H. (1961). Isotopic variations in meteoric waters. *Science* 133 (3465): 1702–1703.

Criss, R., Davisson, L., Surbeck, H. et al. (2014). Isotopic methods. In: Methods in Karst Hydrogeology (ed. N. Goldscheider, D. Drew), 137–160. London: Taylor & Francis.

Culver, D.C. and Pipan, T. (2019). *The Biology of Caves and other Subterranean Habitats.* New York: Oxford University Press.

Dassargues, A. (ed.) (2000). *Tracers and modelling in hydrogeology*, IAHS Publ. 262. Wallingford: IAHS Press.

Dreybrodt, W., Gabrovšek, F., and Siemers, J. (1999). Dynamics of the early evolution of karst. In: *Karst Modeling* (ed. A.N. Palmer, M.V. Palmer and I.D. Sasowsky), 106–119. Charles Town, WV: Karst Waters Institute.

Dubois, P., Sorriaux, P., and Soudet, H.J. (1993). Rospo Mare (Adriatique): un paléokarst pétrolier du domaine méditerranéen. *Karstologia* 21 (1): 31–42.

Dunn, J.R. (1957). Stream tracing: Mid-Appalachian Region. *National Speleological Society Bulletin* 2: 7.

Eisenlohr, L., Király, L., Bouzelboudjen, M. et al. (1997). Numerical simulation as a tool for checking the interpretation of karst spring hydrographs. *Journal of Hydrology* 193 (1-4): 306–315.

Ferrarese, F., Macaluso, T., Madonia, G. et al. (2002). Solution and recrystallisation processes and associated landforms in gypsum outcrops of Sicily. *Geomorphology* 49: 25–43.

Fetter, C.W. (2014). *Applied Hydrogeology*. Harlow, UK: Pearson Education Ltd.

Field, M.S. (2002a). *The QTRACER2 Program for Tracer-Breakthrough Curve Analysis for Tracer Tests in Karstic Aquifers and other Hydrologic Systems*. National Center for Environmental Assessment-Washington Office, Office of Research and Development, US Environmental Protection Agency.

Field, M.S. (2003). A review of some tracer-test design equations for tracer-mass estimation and sample-collection frequency. *Environmental Geology* 43 (8): 867–881.

Field, M.S., Wilhelm, R.G., Quinlan, J.F. et al. (1995). An assessment of the potential adverse properties of fluorescent tracer dyes used for groundwater tracing. *Environmental Monitoring and Assessment* 38 (1): 75–96.

Filippini, M., Squarzoni, G., De Waele, J. et al. (2018). Differentiated spring behavior under changing hydrological conditions in an alpine karst aquifer. *Journal of Hydrology* 556: 572–584.

Ford, D.C. and Williams, P.W. (2007). *Karst Hydrogeology and Geomorphology*. Chichester, UK: Wiley.

Fratesi, B. (2013). Hydrology and geochemistry of the freshwater lens in coastal karst. In: *Coastal Karst Landforms* (ed. M.J. Lace and J.E. Mylroie), 59–75. Dordrecht: Springer.

Freeze, R.A. and Cherry, J.A. (1979). *Groundwater*. Englewood Cliffs, NJ: Prentice-Hall.

Freixes, A., Monterde, M., and Ramoneda, J. (1997). *Tracer tests in the Joeu karstic system (Aran Valley, Central Pyrenees, NE Spain)*. In: *Tracer Hydrology 97* (ed. A. Kranjc), 219–225. Rotterdam: A.A. Balkema.

Galleani, L., Vigna, B., Banzato, C. et al. (2011). Validation of a vulnerability estimator for spring protection areas: the VESPA index. *Journal of Hydrology* 396 (3-4): 233–245.

Geyer, T., Birk, S., Liedl, R. et al. (2008). Quantification of temporal distribution of recharge in karst systems from spring hydrographs. *Journal of Hydrology* 348 (3-4): 452–463.

Ghasemizadeh, R., Hellweger, F., Butscher, C. et al. (2012). Groundwater flow and transport modeling of karst aquifers, with particular reference to the North Coast Limestone aquifer system of Puerto Rico. *Hydrogeology Journal* 20 (8): 1441–1461.

Goldscheider, N. and Drew, D. (ed.) (2007). *Methods in Karst hydrogeology*, IAH: International Contributions to Hydrogeology, 26. London: Taylor & Francis.

Goldscheider, N., Meiman, J., Pronk, M. et al. (2008). Tracer tests in karst hydrogeology and speleology. *International Journal of Speleology* 37 (1): 27–40.

Göppert, N. and Goldscheider, N. (2008). Solute and colloid transport in karst conduits under low- and high-flow conditions. *Groundwater* 46 (1): 61–68.

Göppert, N. and Goldscheider, N. (2019). Improved understanding of particle transport in karst groundwater using natural sediments as tracers. *Water Research* 166: 115045.

Groves, C.G. (2007). Hydrological methods. In: *Methods in Karst Hydrogeology* (ed. N. Goldscheider and D. Drew), 59–78. London: Taylor & Francis.

Hägler, A. (1873). Beitrag zur Aetiologie des Typhus und zur Trinkwasserlehre. *Deutsches Archiv für klinische Medizin* 11: 237–267.

Hartmann, A., Goldscheider, N., Wagener, T. et al. (2014). Karst water resources in a changing world: review of hydrological modeling approaches. *Reviews of Geophysics* 52 (3): 218–242.

Harvey, R.W. (1997). Microorganisms as tracers in groundwater injection and recovery experiments: a review. *FEMS Microbiology Reviews* 20 (3-4): 461–472.

Hauns, M., Jeannin, P.Y., and Atteia, O. (2001). Dispersion, retardation and scale effect in tracer breakthrough curves in karst conduits. *Journal of Hydrology* 241 (3-4): 177–193.

Herak, M. and Stringfield, V.T. (1972). Historical review of hydrogeologic concepts. In: *Karst. Important Karst Regions of the Northern Hemisphere* (ed. M. Herak and V.T. Stringfield), 19–24. New York: Elsevier.

Herzberg, A. (1901). Die Wasserversorgung einiger Nordseebäder. *Journal für Gasbeleuchtung und verwandte Beleuchtungsarten sowie für Wasserversorgung* 44 (45): 815–819. and 842–844.

Hiscock, K.M. and Bense, V.F. (2014). *Hydrogeology: Principles and Practice*. Chichester, UK: Wiley.

Huntoon, P.W. (1995). Is it appropriate to apply porous media groundwater circulation models to karstic aquifers? In: *Groundwater Models for Resources Analysis and Management* (ed. A.I. El-Kadi), 339–358. Boca Raton, Florida: Lewis Publishers.

Jeannin, P.Y. (2001). Modeling flow in phreatic and epiphreatic karst conduits in the Hölloch cave (Muotatal, Switzerland). *Water Resources Research* 37 (2): 191–200.

Jeannin, P.Y. and Sauter, M. (1998). Analysis of karst hydrodynamic behaviour using global approaches: a review. *Bulletin Centre d'Hydrogéologie Neuchatel* 16: 31–48.

Jennings, J.N. (1985). *Karst Geomorphology*. Oxford and New York: Blackwell.

Jensen, J.R. (2015). *Introductory Digital Image Processing: A Remote Sensing Perspective*, 4e. New York: Pearson Education.

Jones, W.K. (2019). Water tracing in karst aquifers. In: *Encyclopedia of Caves* (ed. W.B. White, D.C. Culver and T. Pipan), 1144–1155. New York: Academic Press.

Käss, W. (1998). *Tracing Technique in Geohydrology*. Rotterdam: Balkema.

Kazemi, G.A., Lehr, J.H., and Perrochet, P. (2006). *Groundwater age*. Hoboken: Wiley.

Keys, W.S. (1997). *A Practical Guide to Borehole Geophysics in Environmental Investigations*. Boca Raton: Taylor & Francis.

Kim, Y.S., Peacock, D.C., and Sanderson, D.J. (2004). Fault damage zones. *Journal of Structural Geology* 26 (3): 503–517.

King, L.W. (1915). *Bronze reliefs from the Gates of Shalmaneser King of Assyria B.C. 860-825*. Order of Trustees. London: Oxford University Press.

Kiraly, L. (1975). Rapport sur l'état actuel des connaissances dans le domaine des caractères physiques des roches karstiques. In: *Hydrogeology of Karstic Terrains (Hydrogéologie des Terrains Karstiques)*, vol. 3 (ed. A. Burger and L. Dubertret), 53–67. International Union of Geological sciences.

Kiraly, L. (1998). Modelling karst aquifers by the combined discrete channel and continuum approach. *Bulletin d'Hydrogéologie* 16: 77–98.

Kiraly, L. and Morel, G. (1976a). Etude de régularisation de l'Areuse par modèle mathématique. *Bulletin du Centre d'Hydrogéologie, Neuchâtel* 1: 19–36.

Kiraly, L. and Morel, G. (1976b). Remarques sur l'hydrogramme des sources karstiques simulé par modèles mathématiques. *Bulletin du Centre d'Hydrogéologie, Neuchâtel* 1: 37–60.

Kircher, A. (1665). *Mundus Subterraneus*. Amsterdam: Jansson.

Klimchouk, A.B. (1995). Karst morphogenesis in the epikarst zone. *Cave and Karst Science* 21: 45–50.

Klimchouk, A.B. (2007). Hypogene speleogenesis: hydrogeological and morphogenetic perspective. *National Cave and Karst Research Institute*. Special paper 1: 106.

Klimchouk, A.B. (2017). Types and settings of hypogene karst. In: *Hypogene Karst Regions and Caves of the World* (ed. A.B. Klimchouk, A.N. Palmer, J. De Waele, et al.), 1–39. Cham: Springer.

Klimchouk, A.B. and Ford, D.C. (2000a). Lithologic and structural controls of dissolutional cave development. In: *Speleogenesis: Evolution of Karst Aquifers* (ed. A.B. Klimchouk, D.C. Ford, A.N. Palmer, et al.), 54–64. Huntsville: National Speleological Society.

Klimchouk, A.B., Sauro, U., and Lazzarotto, M. (1996c). "Hidden" shafts at the base of the epikarstic zone: a case study from the Sette Communi plateau, Venetian Pre-Alps, Italy. *Cave and Karst Science* 23 (3): 101–107.

Knop, A. (1878). Über die hydrographischen Beziehungen zwischen der Donau und der Aachquelle im Badischen Oberlande. *Neues Jahrbuch für Mineralogie, Geologie und Palaeontologie* 1878: 350–363.

Kordilla, J., Sauter, M., Reimann, T. et al. (2012). Simulation of saturated and unsaturated flow in karst systems at catchment scale using a double continuum approach. *Hydrology and Earth System Sciences* 16 (10): 3909–3923.

Kovács, A. and Sauter, M. (2007). Modelling karst hydrodynamics. In: *Methods in Karst Hydrogeology* (ed. N. Goldscheider and D. Drew), 215–236. London: Taylor & Francis.

Kovács, A., Perrochet, P., Király, L. et al. (2005). A quantitative method for the characterisation of karst aquifers based on spring hydrograph analysis. *Journal of Hydrology* 303 (1-4): 152–164.

Kranjc, A. (ed.) (1997). *Tracer hydrology 97*. Rotterdam: Balkema.

Kresic, N. (1995). Remote sensing of tectonic fabric controlling groundwater flow in Dinaric Karst. *Remote Sensing of Environment* 53 (2): 85–90.

Kresic, N. (2007). Hydraulic methods. In: *Methods in Karst Hydrogeology* (ed. N. Goldscheider and D. Drew), 65–92. London: Taylor & Francis.

Kresic, N. and Stevanović, Z. (2010). *Groundwater Hydrology of Springs*. Burlington: Elsevier.

Kusch, H. (1993). Die Tigrishöhlen in Ostanatolien (Türkei). Teil 2: Erforschungsgeschichte und Archäologie. *Die Höhle* 44 (4): 101–110.

Lachniet, M.S. (2009). Climatic and environmental controls on speleothem oxygen-isotope values. *Quaternary Science Reviews* 28: 412–432.

Lehmann, H. (1954). Das Karstphanomen in der verschiedenen Klimazonen. *Erdkunde* 8: 112–139.

Leibundgut, C. and Seibert, J. (2011). Tracer Hydrology. In: *The Science of Hydrology: Treatise on Water Science* (ed. S. Uhlenbrook), 215–236. Amsterdam: Elsevier.

Leibundgut, C., Maloszewski, P., and Külls, C. (2009). *Tracers in Hydrology*. Chichester: Wiley.

Liedl, R., Sauter, M., Hückinghaus, D. et al. (2003). Simulation of the development of karst aquifers using a coupled continuum pipe flow model. *Water Resources Research* 39 (3): 1057.

Lillesand, T., Kiefer, R.W., and Chipman, J. (2015). *Remote Sensing and Image Interpretation*. Hoboken: John Wiley & Sons.

Lowe, D.J. (2000). Role of stratigraphic elements in speleogenesis: the speleoinception concept. In: *Speleogenesis, Evolution of Karst Aquifers* (ed. A.B. Klimchouk, D.C. Ford, A.N. Palmer, et al.), 65–76. Huntsville: National Speleological Society.

Luetscher, M. and Jeannin, P.Y. (2004). Temperature distribution in karst systems: the role of air and water fluxes. *Terra Nova* 16 (6): 344–350.

Mahler, B.J. and Lynch, F.L. (1999). Muddy waters: temporal variation in sediment discharging from a karst spring. *Journal of Hydrology* 214 (1-4): 165–178.

Mahler, B.J., Bennett, P.C., and Zimmerman, M. (1998a). Lanthanide-labeled clay: a new method for tracing sediment transport in karst. *Ground Water* 36 (5): 835–843.

Mahler, B.J., Winkler, M., Bennett, P. et al. (1998b). DNA-labeled clay: a sensitive new method for tracing particle transport. *Geology* 26 (9): 831–834.

Maillet, E. (1905). *Essais d'Hydraulique Souterraine et Fluviale*. Paris: Herman et Companie.

Mangin, A. (1975). *Contribution à l'Étude Hydrodynamique des Aquifères Karstiques*. PhD thesis. University of Dijon.

Maréchal, J.C., Ladouche, B., Dörfliger, N. et al. (2008). Interpretation of pumping tests in a mixed flow karst system. *Water Resources Research* 44 (5): W05401.

Martel, E.A. (1921). *Nouveau Traité des Eaux Souterraines*. Paris: Librairie Octave Dion.

Mazzilli, N., Guinot, V., Jourde, H. et al. (2019). KarstMod: a modelling platform for rainfall-discharge analysis and modelling dedicated to karst systems. *Environmental Modelling & Software* 122: 103927.

McGrath, R., Styles, P., Thomas, E. et al. (2002). Integrated high-resolution geophysical investigations as potential tools for water resource investigations in karst terrain. *Environmental Geology* 42 (5): 552–557.

Mejías, M., Ballesteros, B.J., Antón-Pacheco, C. et al. (2012). Methodological study of submarine groundwater discharge from a karstic aquifer in the Western Mediterranean Sea. *Journal of Hydrology* 464-465: 27–40.

Mendez, J.N., Jin, Q., González, M. et al. (2019). Fracture characterization and modeling in karsted carbonate reservoirs: a case study in Tahe oilfield, Tarim Basin (western China). *Marine and Petroleum Geology* 112: 104104.

Milanović, P.T. (1981). *Karst Hydrogeology*. Littleton, CO: Water Resources Publications.

Milanović, P.T. (2004a). Dinaride poljes. In: *Encyclopedia of Caves and Karst Science* (ed. J. Gunn), 291–293. New York: Fitzroy Dearborn.

Milanović, P.T. (2004b). *Water Resources Engineering in Karst*. Boca Raton: CRC press.

Misstear, B.D., Banks, D., and Clark, L. (2006). *Water Wells and Boreholes*. Chichester: Wiley.

National Research Council (1996). *Rock Fractures and Fluid Flow: Contemporary Understanding and Applications*. Washington D.C.: National Academy Press.

Palmer, A.N. (1991). Origin and morphology of limestone caves. *Geological Society of America Bulletin* 103 (1): 1–21.

Palmer, A.N. (2001). Dynamics of cave development by allogenic water. *Acta Carsologica* 30 (2): 13–32.

Palmer, A.N. (2007). *Cave Geology*. Dayton, Ohio: Cave Books.

Pérez-García, J.L., Sánchez-Gómez, M., Gómez-López, J.M. et al. (2018). Georeferenced thermal infrared images from UAV surveys as a potential tool to detect and characterize shallow cave ducts. *Engineering Geology* 246: 277–287.

Pfeiffer, D. (1963). Die geschichtliche Entwicklung der Anschauugen über das Karstgrundwasser. *Beihefte zum Geologichen Jahrbuch* 57: 1–111.

Quinlan, J.F., Davies, G.J., and Worthington, S.R.H. (1992). *Rationale for the design of cost-effective groundwater monitoring systems in limestone and dolomite terranes: cost effective as conceived is not cost-effective as built if the system design and sampling frequency inadequately consider site hydrogeology*. In: *8th Waste Testing and Quality Assurance Symposium Proceedings*, 552–570. Washington D.C.: United States Environmental Protection Agency.

Reilly, T.E. and Goodman, A.S. (1985). Quantitative analysis of saltwater-freshwater relationships in groundwater systems - A historical perspective. *Journal of Hydrology* 80 (1-2): 125–160.

Robineau, T., Tognelli, A., Goblet, P. et al. (2018). A double medium approach to simulate groundwater level variations in a fissured karst aquifer. *Journal of Hydrology* 565: 861–875.

Rodríguez, V., Gutiérrez, F., Green, A.G. et al. (2014). Characterising sagging and collapse sinkholes in a mantled karst by means of Ground Penetrating Radar (GPR). *Environmental and Engineering Geoscience* 20: 109–132.

Rozanski, K., Araguás-Araguás, L., and Gonfiantini, R. (1992). Relation between long-term trends of Oxygen-18 isotope composition of precipitation and climate. *Science* 258 (5084): 981–985.

Sauter, M. (1992a). Assessment of hydraulic conductivity in a karst aquifer at local and regional scale. In: *Proceedings of the 3rd Conference on Hydrogeology, Ecology, Monitoring, and Management of Ground Water in Karst Terranes* (ed. J.F. Quinlan and A. Stanley), 39–56. Dublin (Ohio): National Groundwater Association.

Sauter, M. (1992b). Quantification and forecasting of regional groundwater flow and transport in a karst aquifer (Gallusquelle, Malm, SW. Germany). *Tübinger Geowissenschaftlichen Abhandlungen* C13: 1–150.

Scanlon, B.R., Mace, R.E., Barrett, M.E. et al. (2003). Can we simulate regional groundwater flow in a karst system using equivalent porous media models? Case study, Barton Springs Edwards aquifer, USA. *Journal of Hydrology* 276 (1-4): 137–158.

Schmoker, J.W. and Halley, R.B. (1982). Carbonate porosity versus depth: a predictable relation for south Florida. *American Association of Petroleum Geologists Bulletin* 66: 2561–2570.

Schnegg, P.A. and Costa, R. (2003). Tracer tests made easier with field fluorometers. *Bulletin d'Hydrogeologie* 20: 89–91.

Sella, M. (1929). Estese migrazioni dell'anguilla in acque sotterranee. *Le Grotte d'Italia* 3 (3): 97–109.

Shaw, T.R. (1992). *History of Cave Science. The Exploration and Study of Limestone Caves, to 1900.* Broadway: Sydney Speleological Society.

Shevenell, L. (1996). Analysis of well hydrographs in a karst aquifer: estimates of specific yields and continuum transmissivities. *Journal of Hydrology* 174 (3-4): 331–355.

Shoemaker, W.B., Kuniansky, E.L., Birk, S. et al. (2008). *Documentation of a Conduit Flow Process (CFP) for MODFLOW-2005.* Reston, Va: US Department of the Interior, US Geological Survey.

Shuster, E.T. and White, W.B. (1971). Seasonal fluctuations in the chemistry of limestone springs: a possible means for characterizing carbonate aquifers. *Journal of Hydrology* 14 (2): 93–128.

Silva, O.L., Bezerra, F.H., Maia, R.P. et al. (2017). Karst landforms revealed at various scales using LiDAR and UAV in semi-arid Brazil: Consideration on karstification processes and methodological constraints. *Geomorphology* 295: 611–630.

Singh, V.P. (2016). *Handbook of Applied Hydrology.* New York: McGraw Hill Professional.

Skoglund, R.Ø., Lauritzen, S.E., and Gabrovšek, F. (2010). The impact of glacier ice-contact and subglacial hydrochemistry on evolution of maze caves: a modelling approach. *Journal of Hydrology* 388 (1-2): 157–172.

Smart, C.C. and Simpson, B. (2002). Detection of fluorescent compounds in the environment using granular activated charcoal detectors. *Environmental Geology* 42 (5): 538–545.

Smart, C.C. and Worthington, S.R.H. (2004b). Springs. In: *Encyclopedia of Caves and Karst Science* (ed. J. Gunn), 699–703. New York: Fitzroy Dearborn.

Snow, D.T. (1969). Anisotropic permeability of fractured media. *Water Resources Research* 5: 1273–1289.

Springer, A.E. and Stevens, L.E. (2009). Spheres of discharge of springs. *Hydrogeology Journal* 17: 83–93.

Stevanović, Z. (ed.) (2015). *Karst Aquifers-Characterization and Engineering.* Cham: Springer.

Styles, P., McGrath, R., Thomas, E. et al. (2005). The use of microgravity for cavity characterization in karstic terrains. *Quarterly Journal of Engineering Geology and Hydrogeology* 38 (2): 155–169.

Teutsch, G. and Sauter, M. (1998). Distributed parameter modelling approaches in karst-hydrological investigations. *Bulletin d'Hydrogéologie* 16: 99–109.

Theis, C.V. (1935). The relation between the lowering of the piezometric surface and the rate and duration of discharge of a well using groundwater storage. *Eos Transactions of the AGU* 16 (2): 519–524.

Thrailkill, J. (1988). Drawdown interval analysis: a method of determining the parameters of shallow conduit flow carbonate aquifers from pumping tests. *Water Resources Research* 24 (8): 1423–1428.

Tian, F., Di, Q., Jin, Q. et al. (2019). Multiscale geological-geophysical characterization of the epigenic origin and deeply buried paleokarst system in Tahe Oilfield, Tarim Basin. *Marine and Petroleum Geology* 102: 16–32.

Tóth, J. (2009). *Gravitational Systems of Groundwater Flow: Theory, Evaluation, Utilization.* New York: Cambridge University Press.

Vacher, H.L. (1988). Dupuit-Ghyben-Herzberg analysis of strip-island lenses. *Geological Society of America Bulletin* 100 (4): 580–591.

Vacher, H.L. and Mylroie, J.E. (2002). Eogenetic karst from the perspective of an equivalent porous medium. *Carbonates and Evaporites* 17 (2): 182–196.

Vigna, B. and Banzato, C. (2015). The hydrogeology of high-mountain carbonate areas: an example of some Alpine systems in southern Piedmont (Italy). *Environmental Earth Sciences* 74 (1): 267–280.

Vortmann, G. and Timeus, G. (1910). L'applicazione di sostanze radioattive nelle richerche d'idrologia sotterranea. Le origini del Timavo. *Bollettino della Società Adriatica di Scienze Naturali in Trieste* 25: 247–260.

White, W.B. (1988). *Geomorphology and Hydrology of Karst Terrains*. New York: Oxford University Press.

White, W.B. (1999). Conceptual models for karstic aquifers. In: *Karst Modeling* (ed. A.N. Palmer, M.V. Palmer and I.D. Sasowsky), 11–16. Charles Town, WV: Karst Waters Institute.

White, W.B. (2002). Karst hydrology: recent developments and open questions. *Engineering Geology* 65 (2-3): 85–105.

White, W.B. (2006). Fifty years of karst hydrology and hydrogeology: 1953-2003. In: *Perspectives on Karst Geomorphology, Hydrology, and Geochemistry. A tribute Volume to Derek C. Ford and William B. White* (ed. R.S. Harmon and C.M. Wicks), 139–152. Boulder, CO: s.

White, W.B. (2019). *Springs*. In: *Encyclopedia of Caves* (ed. W.B. White, D.C. Culver and T. Pipan), 1031–1040. New York: Academic Press.

Williams, P.W. (1985). Subcutaneous hydrology and the development of doline and cockpit karst. *Zeitschrift für Geomorphologie* 29: 462–482.

Williams, P.W. (2008). The role of the epikarst in karst and cave hydrogeology: a review. *International Journal of Speleology* 37: 1–10.

Wilson, J.F. (1968). Fluorometric Procedures for Dye Tracing. In: *Techniques of Water-Resources Investigations of the U. S. Geological Survey*. Book 3. Chapter A12, 1–34. U.S. Geological Survey.

Worthington, S.R.H. (1999). A comprehensive strategy for understanding flow in carbonate aquifers. In: *Karst Modeling* (ed. A.N. Palmer, M.V. Palmer and I.D. Sasowsky), 30–37. Charles Town, WV: Karst Waters Institute.

Worthington, S.R.H. (2001). Depth of conduit flow in unconfined carbonate aquifers. *Geology* 29 (4): 335–338.

Worthington, S.R.H. and Ford, D.C. (2009). Self-organized permeability in carbonate aquifers. *Ground Water* 47: 326–336.

Worthington, S.R.H. and Gunn, J. (2009). Hydrogeology of carbonate aquifers: a short history. *Groundwater* 47 (3): 462–467.

Worthington, S.R.H. and Smart, C.C. (2003). Empirical determination of tracer mass for sink to spring tests in karst. In: *Sinkholes and the Engineering and Environmental Impacts of Karst* (ed. B.F. Beck), 287–295. Alabama: American Society of Civil Engineers.

Worthington, S.R.H., Ford, D.C., and Beddows, P.A. (2000). Porosity and permeability enhancement in unconfined aquifers as a result of solution. In: *Speleogenesis. Evolution of Karst Aquifers* (ed. A.B. Klimchouk, D.C. Ford, A.N. Palmer, et al.), 463–472. Huntsville: National Speleological Society.

Worthington, S.R.H., Jeannin, P.Y., Alexander, E.C. et al. (2017). Contrasting definitions for the term 'karst aquifer'. *Hydrogeology Journal* 25 (5): 1237–1240.

Xu, Z. and Hu, B.X. (2017). Development of a discrete-continuum VDFST-CFP numerical model for simulating seawater intrusion to a coastal karst aquifer with a conduit system. *Water Resources Research* 53 (1): 688–711.

Zarroca, M., Comas, X., Gutiérrez, F. et al. (2017). The application of GPR and ERI in combination with exposure logging and retrodeformation analysis to characterize sinkholes and reconstruct their impact on fluvial sedimentation. *Earth Surface Processes and Landforms* 42: 1049–1064.

Zhao, L., Xia, J., Xu, C.Y. et al. (2013). Evapotranspiration estimation methods in hydrological models. *Journal of Geographical Sciences* 23 (2): 359–369.

Zötl, J.G. (1961). Die Hydrographie des nordostalpinen karstes. *Steirische Beitrage zur Hydrogeologie* 1960-61: 53–183.

6

Karren and Sinkholes

6.1   Karst. A Special Geomorphic System

Geomorphology is generally defined as the study of landforms of the Earth's surface and the processes that create them. However, karst geomorphology encompasses a dual realm including the geomorphic features developed at the Earth's surface and in caves. Surface and underground landforms are addressed in Chapters 6, 7 and 8, and in Chapters 9 and 10, respectively.

Landforms can be created by the erosion of preexisting material (erosional/degradational landforms), accumulation of material (depositional/constructional landforms), or deformation (deformational landforms). Karst landforms can be also classified according to their position within the karst hydrological system as input, throughput, and output features. They show a wide range of dimensions, from millimeter-scale features such as microkarren, to poljes tens of kilometers long, and even subsidence morphostructures related to interstratal dissolution of salt that can reach hundreds of kilometers in length. The formation of landforms involves the movement of mass and the expenditure of energy. In geomorphic systems other than karst, the morphogenetic processes largely entail the displacement of mass as solid matter (e.g., solid load in rivers, sand particles in dune systems). However, the central process in karst landform development is the solutional denudation of rocks at the surface and in the subsurface, followed by the transport of the dissolved mass as solute load. Flowing liquid water undersaturated with respect to the constituent minerals of the rocks is the essential agent in karst, which can be considered as a coupled hydrological and geochemical system (Ford and Williams 2007). This explains why karst processes are not active or play a marginal morphogenetic role in hyperarid and permafrost regions, because of the scarcity of rainfall and the presence of frozen ground that prevents infiltration, respectively. Karst landforms found in those regions are typically relict features formed in the past under different climatic conditions, and can be used a paleoenvironmental indicators. Exceptions are found related to evaporite karst due to the high solubility of the constituent minerals.

The leading role of dissolution and the dominant subsurface drainage determine the special idiosyncrasy of karst geomorphology, with some notable variations depending on the type of soluble rock:

- Mechanical erosion plays a secondary role and sediment yield is often very low. Consequently, the surface geomorphology is dominated by erosional landforms produced by dissolution. An exception is found in salt because of its extremely high solubility and rare exposure. In these terrains, dissolution mostly occurs underground and the karst-related surface landforms are mostly deformational, related to subsidence.

- Precipitation of minerals from supersaturated waters in surface karst environments can generate constructional surface landforms, but these are rather scarce and of limited extent and diversity (e.g., calcareous tufa and travertine deposits). In contrast, chemical deposits in caves and the associated landforms have an enormous diversity.
- Drainage is dominated by subsurface flow and consequently a significant proportion of the denudational work is achieved underground.
- In terrains with carbonate rocks, most of the dissolution capability of the infiltration water is spent in the upper few meters. This results in a vertical zonation of the dissolution process and the resulting permeability features, with a shallow high-permeability epikarst zone underlain by less permeable bedrock. Curiously, due to the marked decrease in dissolution rate close to saturation, in more soluble evaporite bedrocks such as gypsum, groundwater can flow along longer distances before it reaches near-saturation conditions.
- The presence of cover deposits and soil plays an important and contrasting role in the solutional denudation of carbonate and evaporite rocks. In carbonate rocks the soil CO_2 contributes to substantially increase the aggressiveness of infiltrating water and its dissolution capability. In evaporite rocks, especially in salt, dissolution of which is not affected by the acidity of the water, residual soils, and other surficial formations can play a protecting role with important geomorphic influence.
- Solutional denudation of karst rocks, in contrast with the mechanical erosion that dominates on non-karst rocks, tends to be a gradual and continuous process that occurs without the interplay of any threshold. Dissolution occurs as long as there is flow of undersaturated water. Mechanical erosion in non-karst terrains tends to be more localized and episodic, is often constrained by thresholds, and rare catastrophic events may have a major morphogenetic contribution.
- Discontinuity planes, including fractures and bedding planes, are key elements for karst development. They are important permeability features that guide groundwater flow and dissolution. These discontinuities also determine rock mass strength and ground stability. A special situation is found in salt, that generally lacks fracture permeability due to its negligible yield strength and tendency to flow and anneal open discontinuities. In the scarce salt exposures, fractures related to unloading and thermal expansion and contraction, are restricted to the upper few meters. Here, subsurface flow and karst development tends to be confined to these near-surface weathering zones.
- Subsurface mass removal by dissolution can lead to the development of subsidence phenomena endemic to karst terrains, and the development of deformational landforms, notably subsidence sinkholes. The ground subsidence process can display a wide range of spatial and temporal patterns, from relatively small catastrophic sinkholes to large-scale slow-moving subsidence structures related to interstratal dissolution of evaporites.
- The solutional removal of mass in carbonate and gypsum rocks tends to occur differentially along flow paths controlled by discontinuities, which eventually transform into integrated conduit systems and caves. In contrast, evaporites, especially salt lacking fractures, can experience complete dissolution by the migration of dissolution fronts. Long-sustained subsurface salt dissolution can be accompanied by laterally migrating ground subsidence. Therefore, the contrasting solubility and permeability features of the various types of karst rocks can determine completely different solutional denudation styles and geomorphic processes.

6.2 Karren

The terms *karren* (German) and *lapiés* (French) collectively designate the large variety of small-scale sculpturing features developed on the surface of karst rocks, either exposed (bare or free karren) or beneath a cover (covered karren). The cover can correspond to residual soil, unconsolidated allochthonous deposits, organic litter and plants (e.g., moss), snow, or ice. The formation of karren is essentially related to the uneven or differential dissolution of the bedrock surface controlled by a number of factors (e.g., structural, textural, hydrodynamic, discontinuous cover, topographic, and biological), resulting in the development of depressions, clefts, channels, tubes, protruding features, and irregular patterns. Mechanical weathering and erosion processes can also contribute to the development of some sculpturing features. The size of most karren falls within the range from millimeters to a few meters, although solutionally enlarged fractures (Kluftkarren) and some runnels can reach tens or hundreds of meters in length. Ford and Williams (2007) use the term microkarren for features with a maximum or modal dimension below 1 cm. The formation of some karren is favored by or restricted to fine-grained, homogeneous, massive to thick-bedded rocks (e.g., solution flutes), whereas other karren types (e.g., grikes, solution pits, and pans) are controlled or can be aided by heterogeneities in the rock (e.g., discontinuity planes, more soluble components, textural variations). Karren occur in carbonate (limestone, dolostone) and evaporite rocks (gypsum, salt), as well as in carbonate detrital rocks and quartzose rocks. Karren in evaporites show some significant differences compared with those found in carbonate rocks (Macaluso and Sauro 1996a, 1996b; Gutiérrez and Cooper 2013): (i) karren in evaporite formations show less diversity, (ii) they display fewer variety in salt than in gypsum, but dominate the salt outcrops, notably solution flutes, (iii) the enhancing effect of some biological processes on carbonate dissolution due to water acidification does not influence the dissolution of evaporite minerals, which are characterized by simpler and more rapid dissolution, (iv) residual soils and weathering crusts generated by dissolution and reprecipitation in evaporites inhibit karren formation, especially in arid and semiarid areas, (v) the size and orientation of gypsum crystals may also play a relevant role on karren development, and (vi) some of the most common karren forms in carbonate rocks such as solution pans and grikes are quite unusual in gypsum and salt. According to White (1988), dolostones display a diversity of sculpturing features equivalent to those observed in limestones, but they tend to be more subdued because of the slower dissolution kinetics.

Karren, despite their limited size, are important landforms from multiple perspectives: (i) They are the most widespread landforms in karst areas, excluding interstratal karst settings, and tend to be more ubiquitous than sinkholes. (ii) They can play a significant role as input features for autogenic recharge, in some cases even more important than sinkholes and other enclosed depressions (Ginés 2009; Lundberg 2019). (iii) Karren can also function as pathways for the incorporation of detrital material into caves. (iv) They can be used as reduced-scale analogues for understanding the development of similar larger landforms (e.g., meandering bedrock channels, pediments), and in the case of evaporite rocks, they can be created over short time spans under controlled conditions. (v) Karren can provide evidence of environmental change. For instance, free karren overprinting covered karren reveal the presence of a soil mantle in the past and its subsequent erosion (Ginés 1990). (vi) Some karren offer suitable habitats for specific biota (Li et al. 2020a) and can be considered as geoecological indicators (Ginés 2002). (vii) Landscapes dominated by karren (karrenfields) can have a remarkable aesthetic and touristic value. A number of UNESCO World Heritage Sites have karrenfields as the main or an important natural value (e.g., stone forests of Yunnan in the South China Karst, Bemaraha tsingy of Madagascar, pinnacles of Gunung Mulu National Park in Malaysia, karrenfield of Torcal de Antequera in southern Spain, the Burren in Ireland, the Lena Pillars in Russia; Figure 6.1) (Williams 2011). (viii) The formation of some mineral deposits is associated with the production of weathering mantles by dissolution of carbonate rocks and the concomitant development of covered karren (Figures 1.5c and 1.6). (ix) Internal erosion of cover deposits through fissures and conduits in the epikarst, largely related to the solutional enlargement of fractures (Kluftkarren), leads to the development of cover-collapse and cover-suffosion sinkholes. These sinkhole types are responsible for the majority of the subsidence damage in karst areas because they tend to have a high probability of occurrence (Gutiérrez 2016).

Figure 6.1 Bemaraha tsingy of Madagascar, declared UNESCO World Heritage Site because of its karrenfield of universal value. The landscape is dominated by fracture-controlled giant grikes (Kluftkarren) and intervening pinnacles (Spitzkarren). *Source:* Photo by Chien C. Lee.

The first descriptions of karren were carried out in high-mountain environments of the German, Swiss, and French Alps during the last part of the nineteenth century and the beginning of the twentieth century, which explains the use of the general terms karren and lapiés (Ginés 2009; Veress 2010). Bögli (1960) published the first comprehensive classification of karren using a German nomenclature that has been widely used in the international literature (note that German nomenclature is capitalized; i.e., Rinnenkarren). Subsequently, this German terminology has been complemented by equivalent English terms (Pluhar and Ford 1970; Jennings 1985; Ford and Williams 1989). Karren exhibit a remarkable variety of morphological types. This diversity, combined with the large spectrum of depositional and denudational features encountered in caves, explains why karst is considered the geomorphic system with the greatest diversity of landforms (Gutiérrez and Gutiérrez 2016). A considerable number of karren classifications has been proposed, essentially based on genetic and morphologic criteria (McIlroy de la Rosa 2012). Bögli (1960, 1980) contended that a key factor for the development of karren is the way water interacts with the bedrock, which is largely influenced by the presence or absence of cover. Thus, he differentiated three main groups of single karren forms in his genetic classification: (i) *free karren* on bare rock surfaces on which the water interacts freely with the rock via direct rainfall, sheetwash, and channeled flow, (ii) *covered karren* in mantled bedrocks where percolation water enriched in biogenic CO_2 flows slowly because it is hindered by the soil mantle, and (iii) *half-exposed karren* in areas with discontinuous cover. Bögli's classification also includes grikes and karren tables (pedestals) as single forms that develop in both covered and bare karst settings, plus the categories of *complex karren* created by combinations of single forms, and *groups of complex forms* referring to large areas dominated by exposed karren (karrenfields). White (1988) proposed a genetic classification incorporating the role played by discontinuity planes (e.g., fractures, bedding planes) in the development of some karren. He also considered the condition of covered and bare bedrock surfaces and differentiated two main groups: (i) hydraulic forms generated by sheet flow and channeled flow and (ii) etched forms developed along structural weaknesses or on massive bedrock. Veress's (2010) classification is very similar to White's scheme. Ford and Williams (2007) adopted a morphological classification with subdivisions that incorporate genetic factors. This scheme distinguishes circular and linear forms, and the latter are subdivided into fracture-controlled and hydrodynamically controlled. It also includes karren assemblages and karrenfields under a general category of polygenetic forms. The morphological criteria proposed by Ford and Williams (2007) are very practical and intuitive but do not cover some karren that neither have circular nor linear geometry. In this work, we follow the genetic classification of Ginés (2004, 2009) with some

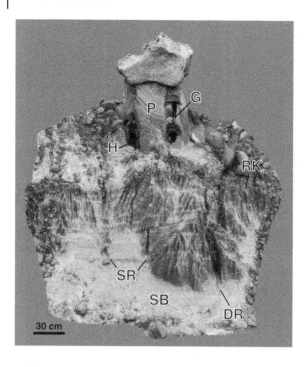

30 cm

Figure 6.2 Textured 3D model generated by Structure from Motion (SfM) photogrammetry of a karren assemblage developed on folded salt with thin clay partings. Cardona salt diapir, Spain. A pedestal (P) has developed where a sandstone boulder with cavernous weathering protects the underlying bedrock from dissolution. The stem of the pedestal shows a fracture-controlled grike (G) and holes and conduits (H). The salt exposure is sculpted by Rillenkarren (RK) that grade abruptly into a planar solution bevel (SB) locally carved by a solution runnel (SR) and a decantation runnel (DR). The latter is fed by a conduit that is affected by collapse and will transform into a solution runnel. *Source:* Francisco Gutiérrez (Author).

additions from the classification of covered karren by Zseni (2009). This scheme differentiates the following categories for elementary karren according to the solutional agent: (i) wetting by condensation water and spray, (ii) thin water films strongly influenced by water–rock adhesive forces, (iii) direct rainfall, (iv) channeled flow, (v) sheetwash, (vi) standing water, (vii) direct water infiltration, and (viii) soil water percolation. The last one refers to covered karren, whereas the others essentially correspond to free karren. Ginés (2009) distinguishes three levels of integration for the description and analysis of karren, including the *elementary karren* features related to specific processes and factors, spatial associations of different types of sculpturing features forming *karren assemblages* (Figure 6.2), and landscapes dominated by karren that can cover extensive tracts of land, known as *karrenfields* (Karrenfeld) (Figure 6.1). For instance, solutional widening along a joint by percolation water in exposed bedrock produces a grike (Kluftkarren), as an elementary karren feature. Dissolution acting in a system of subvertical and intersecting joints results in a network of grikes with intervening clints or pinnacles, which can display a number of other karren types on their surface, forming a karren assemblage. This complex suite of sculpturing features can cover large expanses of land forming a karrenfield such as limestone pavements or stone forests.

Some karren are polygenetic features generated by a combination of processes such as biokarst karren that result from processes mediated or enhanced by biological activity. It should be stressed that classifications are conceived for describing simple end-member karren, but complex features related to combinations of processes or the superposition of different forms are common. Moreover, the origin of some karren is still unclear and certain forms can be the result of different processes (e.g., pinnacles), representing cases of morphologic convergence or equifinality. The most common karren features are described below, together with the current understanding of their formative processes and controlling factors. Important publications that focus on karren include the conference volumes edited by Paterson and Sweeting (1986) and Fornós and Ginés (1996), the comprehensive book on karren edited by Ginés et al. (2009), which includes chapters on the main types of karren and on outstanding case studies, and the book that describes high-mountain karren by Veress (2010), largely based on his studies conducted in the Alps.

6.2.1 Bare Karren

The formation of bare or free karren essentially results from the interaction of water with exposed rock surfaces. In most cases, the water has not percolated through a soil cover, so that it has a limited capability to dissolve carbonate rocks due to its low P_{CO_2} content. Consequently, in carbonate rocks, bare karren tend to develop much more slowly than covered karren. This does not apply to evaporites. An additional difference is that mechanical erosion can play a significant role in the development of some bare karren features because of multiple water-erosion and weathering processes. The type of karren that develops in bare karst settings is strongly influenced by the way the water reaches the rock surface (condensation, spray, direct rainfall, overland flow, direct infiltration) and the hydrodynamic conditions of the water (water films governed by adhesive forces, gravity-controlled channeled or unconfined flow, standing water) (Table 6.1).

Irregular etching, ascribed to wetting by condensation, is a general term used to designate differential etching forms including hollows and protrusions with a broad variety of geometries, which are commonly less than 1 cm across. The rock surface displays an intricate microtopography characterized by the absence of any clear directional trend (Ford and Williams 2007; Ginés 2009) (Figure 6.3a). This feature reveals that its development is not related to gravity-controlled water flow. Condensation water from humid air that comes into contact with a cooler rock surface and reaches the dew point is considered to be the main solutional agent. The differential etching that produces the rough surface can be controlled by the variable susceptibility to dissolution and mechanical detachment of the different components of the rock. This is probably one of the most common types of free karren encountered in carbonate rocks, but they have received limited attention due to their restricted size and the lack of morphological regularity and diagnostic elementary forms. Crowther (1996) states that "despite being a fundamental property of limestone rock surfaces, roughness at the mm scale has been largely neglected in the investigation of karren features". It seems that irregular etching is most commonly found in arid and semiarid environments where the limited rainfall and overland flow favor its development and preservation.

Microrills (Rillensteine) are attributed to dissolution by thin water films not governed exclusively by gravity. These are tightly packed channels trending downslope, typically less than 1 mm in width, with a rounded cross-profile, and up to a few decimeters long. The ridges between the rills can be sharp or rounded. The plan-view pattern varies from straight, to sinuous, to tighly meandering, and tends to become straighter and more parallel with increasing slope (Ford and Lundberg 1987; Grimes 2007; Gómez-Pujol and Fornós 2009a) (Figure 6.3b). The microchannels can display some branching, both contributory and distributory depending on whether the slope is spreading or focusing the flow (Grimes 2007). The ridges can be partitioned into chains of elongated teeths due to channel branching or the local merging of adjacent grooves (Figure 6.3c). Grimes (2007), based on his research conducted in Australia, identified two main subtypes of microrills: (i) parallel and sharp-ridged microrills with regular widths that can be from straight to tortuous and (ii) microrills mainly found on the gently domed surfaces of cobbles that fan out and increase their width downslope. Gómez-Pujol and Fornós (2009a) in a thorough study on microrills from coastal and mountain sites in Mallorca and Menorca Islands, Spain, observed that the microchannels have a nongraded longitudinal profile characterized by a succession of enclosed depressions and intervening thresholds (Figure 6.3c). These concavities and convexities support the concept that dissolution by gravity-controlled runoff is not the main genetic agent. These authors also observed some textural control, with coarser grains on the walls of some channels and finer texture underlying their floor. Microrills mainly occur in homogeneous fine-grained carbonate rocks (e.g., mudstone) devoid of vegetation and biofilm colonization, and can be also found in microcrystalline gypsum. They appear to be more common in arid and semiarid environments and

Table 6.1 Classification of the main karren forms used in this work.

	Solutional agent	Karren forms	Common size and brief description
Bare bedrock	**Wetting by condensation**	Irregular etching	(<1 cm) Forms with no distinctive trend, including hollows and protrusions with a broad variety of geometries and commonly less than 1 cm across
	Thin water films governed by water-rock adhesive forces	Microrills (*Rillensteine*)	(<1 mm, width) Tightly packed channels typically less than 1 mm in width with rounded cross-profile and straight, sinuous, tightly meandering and branching patterns
	Direct rainfall	Rainpits (solution pits)	(<2 cm, width) Round-bottomed subcircular hollows mainly developed on low-gradient surfaces and commonly forming tightly packed clusters in association with Rillenkarren
		Solution flutes (*Rillenkarren*)	(1-2.5 cm, width) Narrow, tightly packed, straight to slightly sinuous solutional furrows that head at the crest of bare rock outcrops and extinguish downslope, giving way to a planar solution bevel known as Ausgleichsfläche. Troughs have a parabolic cross-profile with a rather constant width and are separated by sharp ridges
	Channeled water flow	Solution runnels (*Rinnenkarren*)	(5-50 cm, width and depth) Straight to low-sinuosity sharp-rimmed Hortonian channels developed in rock slopes that commonly widen and deepen downslope
		Wall karren (*Wandkarren*)	(2.5-35 cm, width) Straight channels carved on subvertical rock faces that typically head at the crest of the slope
		Meandering runnels (*Mäanderkarren*)	(2.5-35 cm, width) Highly sinuous asymmetric channels carved into sloping rock faces or along the floor of a larger runnel
		Decantation runnels	(2.5-35 cm, width) Channels generated by water released from an upslope point or linear store. Their cross-section reaches the largest size at or near the source, where the water has the highest degree of undersaturation
	Standing water	Solution pans (*kamenitzas*)	(5 cm-5 m, length) Depressions commonly subcircular to elliptical, but also with irregular plan geometry that can hold stagnant water. The bottom is commonly flat and the side walls are usually steep and can display a basal notch. Drainage may occur via overspill outlets connected to decantation runnels

			Description
Covered bedrock	**Sheetwash water flow**	Solution bevels (*Ausgleichsflächen*)	(0.2-1 m, length) Planar and relatively smooth surfaces generated by solutional sheetwash erosion and commonly found below the Rillenkarren extinction level. Corrosion terraces are meter-sized solutional bevels with a backwall more than 10 cm high
		Heelsteps (*Trittkarren*)	(5-30 cm, width) Subhorizontal and planar corrosion bevels enclosed by an arcuate backscarp with the concavity facing in the downslope direction
		Funmelkarren (*Trichterkarren*)	(5-30 cm, width) Concave features open downslope with funnel-shaped backwall and a comparatively small non-planar floor. Can be considered as a subtype of *Trittkarren*
		Cockling patterns	(1-10 cm, transverse size) Linear horizontal concavities that give a crinkled appearance to the rock surface
		Solution ripples	(1-10 cm, wavelength) Undulations perpendicular to the flow direction with significant lateral continuity and regularity
		Sharpened edges (*lame dentate*)	(10 cm – 1 m, length) Emergent parallel ridges on steep slopes and oriented in the flow direction comprising an upper flat bench and an underlying sharpened edge that extinguishes downslope
	Direct water infiltration	Grikes (Kluftkarren)	(1-100 m, length) Clefts resulting from solutional widening of discontinuity planes (joints, faults, bedding planes) in the epikarst
		Cutters	(1-100 m, length) Covered karst equivalent to grikes filled by residual soil or unconsolidated allochthonous deposits.
		Rounded runnels (*Rundkarren*)	(5-50 cm, width and depth) Linear furrows with rounded cross-section. When exposed, they become sharper by subaerial dissolution transforming into Rinnenkarren
		Cavernous karren	(10 cm-1 m) Irregular rounded voids with variable geometries and random orientations that are commonly interconnected
	Soil water percolation	Subsoil scallops and ripples	(15-50 cm, width, wavelength) Concavities and transverse-to-flow undulations developed on steep rock faces
		Subsoil notches	(10 cm -1 m, height, width) Subhorizontal indentations with semicircular section developed just below the ground surface
		Subsoil pits or cups	(5 cm-1 m) Hollows with rounded floor and a depth to width ratio of ≤1
		Karren wells and shafts	(10 cm-5 m) Vertical to steeply inclined cavities with depth to width ratios of 1-2 and ≥2, respectively

(Continued)

Table 6.1 (Continued)

Karren forms	Common size and brief description
Polygenetic karren	
Karren assemblages	
Karrenfields	
Clints (*Flachkarren*)	(1–2 m, length, width) Flat-topped rock blocks surrounded by grikes or cutters
Pinnacles	(1–10 m, heigth) Rock residuals with pointed tops isolated by grikes or cutters
Limestone pavement (*karrenfield*)	(>100 m) Subhorizontal or gently inclined rock exposures commonly concordant with bedding showing a network of intersecting grikes and intervening clints
Rock cities	(>1 km) Fracture-controlled landscape with urban pattern comprising rock blocks surrounded by corridors (streets)
Ruiniform landscape	(>1 km) Fields of scattered upstanding rock residuals with variable morphology. Can result from the retreat of side walls in rock cities. Has some overlap with stone forests
Biokarst	
Multiple forms	(0–1 mm) Large diversity of erosional features generated or promoted by organisms

Underlined terms indicate features controlled by discontinuity planes (e.g., fractures, stratification). The upper part of the classification indicates the elementary karren features, essentially related to specific genetic mechanisms. The lower part includes polygenetic karren, karren assemblages, and karrenfields.
Source: Adapted from Ginés (2004, 2009) and Zseni (2009), with some slight additions inspired by the classifications of Bögli (1960, 1980), White (1988), and Ford and Williams (2007).

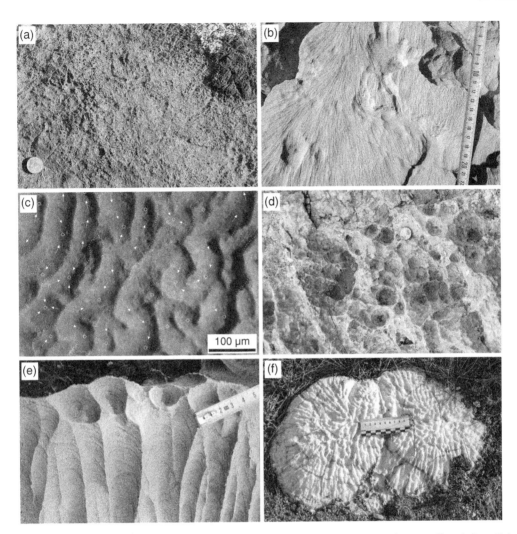

Figure 6.3 Free karren generated by condensation water (irregular etching), thin water films (microrills), and direct rainfall (rainpits and Rillenkarren). (a) Irregular etching developed on Jurassic limestone with an intricate pattern of millimeter-scale pits and protrusions. Calatorao, Iberian Chain, Spain. *Source:* Francisco Gutiérrez (Author). (b) Microrills with parallel and radial distribution on micritic limestone along the Murge coast, Italy. *Source:* Photo by Ugo Sauro. (c) Vertical image of microrills from the Balearic Islands, Spain showing the nongraded longitudinal profile of the grooves with a succession of depressions (circles) and thresholds. Arrows indicate surface gradient and dotted lines the crest of the ridges, locally forming isolated teeths associated with microrill branching. *Source:* Photo by Lluís Gómez-Pujol. (d) Tightly packed and coalescing rainpits in a limestone outcrop at El Colomer site, Tramuntana Range, Mallorca. Coin 16 mm in diameter for scale. *Source:* Francisco Gutiérrez (Author). (e) Large rainpits approximately 3 cm across developed along the crest of a ridge of finely laminated gypsum and associated with Rillenkarren (Verzino area, Calabria, Italy). *Source:* Photo by Ugo Sauro. (f) Karren assemblage developed on a slightly protruding nodule of white alabastrine gypsum, with rainpits in the crestal zone and radiating Rillenkarren on the sloping flanks (Rodén, NE Spain). *Source:* Francisco Gutiérrez (Author).

are rare in cold regions. It seems that their formation is largely related to thin water films supplied by dew or sea spray, which movement is strongly influenced by surface tension and water–rock adhesive forces. This explains the undulating long-profile of the grooves and their sinuous pattern, which decreases in sectors with higher gradients, where gravitational forces increase their relative influence on the flow of water films. Gómez-Pujol and Fornós (2009a) propose that dew solution reduces intergranular cohesion and grains are subsequently detached mechanically by other agents such as rain splash, runoff, or wind action.

Elementary karren forms related to direct rainfall include rainpits and solution flutes. *Rainpits* (solution pits) are subcircular hollows with semi-spherical, parabolic, or tapering cross-section 1–5 cm across, but in most cases with an average width below 2 cm. They can occur singly, forming alignments and more commonly as tightly packed clusters with adjoining pits sharing sharp edges (Figure 6.3d–f). Neighboring rainpits can coalesce to form larger and morphologically more complex depressions (Ginés and Lundberg 2009). They mainly develop on flat surfaces or low-gradient slopes and often occur spatially associated with solution flutes (Rillenkarren) in rock protuberances forming karren assemblages with a characteristic pattern; rainpits occur as densely packed clusters in the upper subhorizontal part of the protruding rock grading into solution flutes toward the sloping flanks (Dunkerley 1979, Macaluso and Sauro 1996b) (Figure 6.3e, f). This spatial association together with their strikingly similar morphometric features (width, cross-sectional profile) strongly suggest that rainpits and Rillenkarren result from similar genetic mechanisms. A compilation by Ginés and Lundberg (2009) on the limited available morphometric data indicate that rainpits have a characteristic width to depth ratio of ~2 and tend to be slightly deeper than solution flutes. Rainpits have been reported in bare outcrops of rather homogeneous carbonate rocks and gypsum (Macaluso and Sauro 1996a, 1996b) and appear to be particularly common in arid and semiarid environments. Ginés and Lundberg (2009) indicate that the distribution of rainpits in Mallorca Island, Spain shows an upper limit at ca. 500 m a.s.l., which corresponds to an average annual precipitation of 900 mm, whereas Rillenkarren display a much broader elevation range. Similar features can form beneath a soil cover (subsoil pitting), but tend to be smoother. Solution pans (kamenitzas) have a characteristic flat bottom, and pits produced by solutional etching and biokarstic boring typically display more irregular profiles. Rainpits form by dissolution and probably also mechanical erosion (detachment of grains loosened by solution) caused by rain drops that impact on rock exposures. They mainly occur on the summit of rock protuberances where the limited runoff contributing area prevents the development of sufficiently thick water films or sheetwash. It is believed that impinging raindrops, thanks to their kinetic energy, create rapid and turbulent water movement reducing the thickness of the boundary layer associated with the water–rock interface. The solvent is continuously renewed during rainfall events mainly by the ejection of water induced by rainsplash, which explains the sharp edges of the pits. The pits may also drain by overspill, basal infiltration and evaporation. Some dissolution may also occur by stagnant water.

Rillenkarren (solution flutes) are one of the karren types that has received greater attention. Lundberg and Ginés (2009) recommend the use of the term Rillenkarren because solution flute has been frequently used in a loose manner to describe other karren forms. They are narrow, closely packed solutional furrows oriented downslope that head at the crest of bare rock outcrops and extinguish downslope, frequently giving way to planar solution bevels known as Ausgleichsflächen (Figure 6.4). Their cross-section is characterized by a remarkably uniform succession of flutes with parabolic profile and sharp intervening cusp lines (Crowther 1998). In plan view, the troughs have a relatively constant width and tend to display a straight and parallel arrangement, but they can also branch, fanning out in convex slopes or converging in concave slopes, where they can merge into larger runnels (Lundberg and Ginés 2009). The furrows can also attain a slightly sinuous trace, a feature which seems to be particularly common in salt (Figure 6.4a). Rills can produce a herringbone pattern when developed on either side of a crest and, as explained above, can also grade in their upper termination into clusters of rainpits developed on the flat summit of rock bosses (Figures 6.3e, f and 6.4d). Mottershead (1996) documented concave longitudinal profiles in rills developed on limestone, with the deepest section lying in the upper half of the

Figure 6.4 Rillenkarren developed in different lithologies. (a) Sinuous rills on salt beds separated by partings of swelling clays. Cardona salt diapir, NE Spain. (b) Rillenkarren with some branching in a convex slope of limestone with lichen cover. Peracalç Range, Spanish Pyrenees. (c) Rillenkarren in a small salt protuberance with spikes in the crest. Dashti (or Kuh-e-Namak) Diapir, Zagros Mountains, Iran. (d) Rainpits and Rillenkarren developed on a gypsum ledge at the façade of San Juan de los Panetes Church, Zaragoza, Spain, which was built between the mid-sixteenth century to 1725 (average precipitation ca. $300\,mm\,yr^{-1}$). Contour gauge shows the cross-profile of some rills. *Source:* Francisco Gutiérrez (Author).

profile. Palmeri et al. (2020) observed concave–convex, linear and convex longitudinal profiles in a limited number of flutes on gypsum characterized by high-resolution close-range photogrammetry (3D-photo reconstruction). Small spikes and combs can develop at the nodal points of contiguous rills and rainpits (Macaluso and Sauro 1996b) (Figures 6.3f and 6.4d), in the upper termination of inter-rill ridges (Figure 6.4c), and as a result of the partitioning of the ridges into segments. Rillenkarren occur on bare rock slopes mostly with inclinations between 20° and 80°. They have been extensively documented in carbonate rocks and gypsum, and in salt outcrops they are the predominant karren type (Veress 2019). Solution flutes rarely occur on limestone in arid areas, suggesting a precipitation threshold for this lithology, and their development is favored by fine-grained textures lacking significant inhomogeneities.

Lundberg and Ginés (2009), in their comprehensive review on Rillenkarren, compiled morphometric data on flutes developed on carbonate outcrops from around 20 locations ranging from arctic to tropical climates. According to this dataset, the length of the rills, which tends to increase with slope gradient (Glew and Ford 1980), has a mean value of 19.23 cm, with dimensions ranging between ~10 and ~35 cm. The computed average depth and width of the flutes are 0.44 and 1.69 cm, respectively. Lundberg and Ginés (2009) also analyzed the characteristic dimensions of solution flutes developed on gypsum and salt using a limited amount of data from a previous compilation by Mottershead et al. (2000) and values from other works. The calculated mean length, depth, and width values for gypsum are 11.9, 0.31, and 0.92 cm, respectively. Mottershead et al. (2000) provided median (i.e., middle value) length, depth, and width values for rills in salt (excluding data from artificial and porous salt waste) of 23, 1.0, and 2.05 cm, respectively. Obviously, the restricted data from the evaporite rocks, especially salt, could be considerably biased by important factors unrelated to lithology, such as texture (Dunkerley 1983; Goudie et al. 1989), the slope of the rock exposures, or climate. Nonetheless, despite the limitations of these data, it seems that flutes in gypsum tend to be narrower (Stenson and Ford 1993), and flutes in salt are characterized by significantly higher depths, with a width to depth ratio of 2.07 (from median values), much lower than that calculated for carbonate rocks (4.37 from mean values). A practical finding derived from the worldwide morphometric analysis carried out by Lundberg and Ginés (2009) considering the three types of rocks is that average width of Rillenkarren shows a remarkably narrow range. Thus, these authors suggest that flute width, typically within the range of 1–2.5 cm, could be used as a diagnostic parameter for the identification of Rillenkarren.

Rillenkarren are the antithesis of erosional rills generated by the typical Hortonian flow. In the upper part of a slope, Horton overland flow is dominated by unconfined sheetwash with limited depth and erosion capability. With increasing distance from the divide, as the contributing area becomes larger, the flow depth and the basal shear stress increase to reach threshold values for focused rill erosion and the partitioning of the flow into confined threads. A key feature of Rillenkarren from the genetic perspective is that they head at the crest of the slopes, where there is no runoff contributing area and direct rainfall is the only possible solutional agent. The formation of Rillenkarren was explained by Glew and Ford (1980) with their widely accepted "raindrop impact and boundary layer model." These authors explored the development of Rillenkarren by exposing texturally uniform slabs of plaster of Paris (artificial gypsum) and salt with variable inclinations (22.5–60°) to constant rainfall and temperature. They observed the following morphological evolution: (i) rills with highly variable widths initiated in the crestal area and increased their width uniformity by a process of lateral coalescence, in which wider rills consumed adjoining narrower ones, (ii) rills deepened and lengthened downslope maintaining a stable width, (iii) rills reached maximum length grading downslope into a solution bevel, and (iv) both the fluted and planar surfaces evolved by parallel retreat. Maximum recession rates occurred with inclinations of around 45°, for which the amount of water intercepted and its residence time reach optimum conditions for dissolution. The amount of artificial rainfall required for the Rillenkarren on artificial gypsum to reach the equilibrium form was equivalent to the cumulative rainfall of 50 years with an average precipitation of 400 mm yr^{-1}. A similar morphological evolution of the cross-section of the flutes, with early width stabilization and longer deepening, was inferred by Mottershead and Lucas (2001) studying rills developed on gypsum surfaces of different known ages. Slabe (2009) reproduced the development of karren on plaster blocks with longer planar faces exposed to rainfall. In these experiments dissolution generated three types of forms with a clear vertical zonation: (i) Rillenkarren in the upper section, (ii) a relatively smooth surface in the intermediate zone, and (iii) Hortonian channels in the lower section. Glew and Ford (1980) explained that rill formation is essentially related to dissolution caused by impinging raindrops. The raindrops penetrate into the slow-moving laminar sublayer of flowing water associated with the rock surface, causing direct dissolution rather that dissolution by diffusion through the boundary layer. They indicate that the parabolic cross profile of the flutes is an effective shape for focusing raindrop denudation along the trough axis. Downslope, the flow depth increases to reach a critical value for which raindrops do not have sufficient kinetic energy to penetrate the laminar sublayer and the rills are replaced by the planar solution bevel (Ausgleichsfläche) dominated by uniform sheetwash dissolution. Given the short length of the rills, the dissolution caused by the water occurs in a very short period of time (a few seconds) and water remains far from saturation. This model has two morphometric implications (Lundberg and Ginés 2009) (Figure 6.5): (i) the larger the raindrops and their kinetic energy, the thicker the water film that can be fully

Figure 6.5 (a) Block diagram showing Rillenkarren heading at the slope crest and replaced by a planar solution bevel (Ausgleichsfläche), indicating a change in the denudation style, from differential dissolution caused by direct rainfall to uniform sheetwash solution. (b) Longitudinal profiles of slopes illustrating the influence of slope angle and raindrop size on the length of the rills. Rills tend to be longer on steeper slopes because of the slower downslope increase in the thickness of the water film. Larger raindrops produce longer rills because they can penetrate the basal laminar boundary layer in thicker water films. *Source:* Adapted from Lundberg and Ginés (2009).

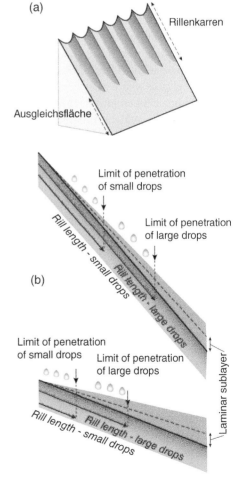

penetrated and the longer the rills, and (ii) the higher the slope angle, the thinner the water film for a given distance from the crest and the longer the rills (the rainfall for a given rock surface area decreases with the slope gradient, whereas flow velocity increases). Complementary mechanisms have been proposed by some authors to the direct raindrop dissolution model, which are thoroughly explained and discussed by Lundberg and Ginés (2009). Fiol et al. (1996), through scanning electron microscope observations, proved that biological activity can contribute to Rillenkarren growth in limestone. They found that bioerosion caused by cyanobacteria micritizes the rock increasing the contact surface area and reducing the intergranular cohesion, facilitating both rock dissolution and the mechanical detachment of grains by raindrop impact. Based on laboratory simulations, these authors estimated that the rate of particulate loss from a bare rock surface (lichen-free) amounts to 48% of the total mass loss, the remainder being in solution. Mottershead et al. (2000) indicated that the relative importance of solutional and mechanical erosion varies across the fluted surface, with a higher relative contribution of mechanical raindrop erosion in the fragile crest of the divides. This is supported by deeper flutes developed in high-solubility salt. Based on scanning electron microscope observations, Mottershead and Lucas (2004a) indicate that biological activity does not participate in the development of Rillenkarren on evaporite rocks. They also proposed that raindrops have the greatest mechanical effect on the steep sides of the rill divides, where the shear stress component is greatest and the likelihood of submergence of the surface by overland flow is minimum. The shear stress associated with the raindrop impact can cause the detachment of grains or crystals loosened by dissolution. Dreybrodt and Kaufmann (2007) presented a theoretical analysis of the physics and chemistry of dissolution by flowing water films on exposed rocks and applied their formulation to interpret field data of Rillenkarren. Nonetheless, the widely accepted raindrop impact and boundary layer explanation is still an incomplete model. It satisfactorily justifies the downslope change in the denudation style (rim effect), from regular round-bottomed flutes heading at the slope crest to planar solution surfaces where water films are thick enough to prevent the penetration of the laminar basal layer by impacting raindrops. However, it does not explain why randomly falling raindrops initiate the development of highly ordered flutes by differential erosion, rather than producing uniform surface recession.

Lundberg and Ginés (2009) reviewed data on the time of formation required for the development of Rillenkarren in different lithologies derived from observations made on surfaces of known age (Figure 6.4d), simulation studies, and solute load measurements. It would be more meaningful to indicate the cumulative rainfall rather than time of formation to account for the amount of solvent that interacts with the rock surface. The time required for the formation of specific flutes can be roughly estimated from the concentration of dissolved mineral in the water using the formula (Mottershead and Lucas 2001):

$$T = \frac{F}{PC} = \frac{A\gamma 10^6}{dw\cos\alpha C} \tag{6.1}$$

where T is time in years, F is flute mass removed in grams given by $F = lA\gamma$ (l is flute length in mm, A is the flute cross section in the slope-normal plane in mm^2, and γ is the density in g cm^{-3}), P is the volume of annual rainfall received in the horizontal plane (mm^3) given by $P = d\,w\,l\cos\alpha$ (d is the annual precipitation in mm, w is the flute width in mm, and α is the slope angle), *and* C is concentration of mineral removed (mg L^{-1}). The limited available data grossly indicate that rates of formation in limestone and gypsum are of the order of 10^2–10^3 and 10^1–10^2 years, respectively. Rillenkarren development can be extremely fast in high-solubility salt, where a few rainfall events could produce solution flutes. In salt slopes at Cardona diapir, NE Spain, Mottershead et al. (2007) measured erosion rates as high as 100 mm yr^{-1} (annual precipitation 500 m yr^{-1}). A similar value of around 120 mm yr^{-1} was estimated by Bruthans et al. (2008) in high-elevation salt extrusions of the Zagros Mountains, Iran (average precipitation 550 m yr^{-1}) (see Section 4.4).

The main sculpturing features generated by channeled water flow include the various types of free-karren runnels: solution runnels, wall karren, meandering runnels, and decantation runnels, which may have some overlap. These are hydrodynamically controlled linear karren according to the classification of Ford and Williams (2007).

Solution runnels (Rinnenkarren) are sharp-rimmed furrows generated by the concentration of unconfined overland flow into confined channeled flow. The overland flow may come from upslope areas and from the margins of the runnels. These Hortonian channels, in contrast to Rillenkarren, do not head always at the slope crest and commonly widen and deepen downslope as a result of the increase in the runoff contributing area and discharge (Figure. 6.6a, b). Their width and depth are typically between 5 and 50 cm and these features display a wide range of lengths from one to tens of meters (Ford and Lundberg 1987; Ford and Williams 2007; Veress 2009a, 2010). Exceptionally, large solution runnels up to 4 m in width and hundreds of meters in length have been documented on marble in the Chilean Patagonia, where the average precipitation reaches around 7000 mm yr^{-1} (Maire 1999; Hobléa et al. 2001; Veress et al. 2003; Maire et al. 2009). Rinnenkarren develop on slopes with gradients greater than 3° and the channels formed on subvertical slopes are called wall karren (Wandkarren). Veress (2009a) proposes a slope angle of 60° to differentiate between Rinnenkarren and Wandkarren, which often show a continuous gradation in slope. According to Ford and Lundberg (1987), many exposed rounded runnels (Rundkarren) formed under a cover and were subsequently exposed by the stripping of the soil. Once exposed, they were sharpened by subaerial dissolution resulting in a polygenetic form that acquires the morphology of Rinnenkarren.

The plan outline of solution runnels follows the slope gradient and is largely controlled by the topography and its local irregularities. They commonly display a linear pattern of straight to low-sinuosity subparallel runnels separated by broad interfluves (Figure 6.6a, b). There is a general inverse relationship between the slope gradient and sinuosity (Hutchinson 1996). Channels can also join forming a dendritic network or can show a centripetal arrangement converging toward a pit or a grike that functions as a swallet. Channels that propagate upslope can intersect a nearby channel, increasing the discharge in the capturing channel and beheading the captured channel. Solution runnels, unlike Rillenkarren, display a remarkable variety of cross-sectional geometries. Veress (2009a) differentiates simple and complex forms. The former include channels with V-shaped, U-shaped, rectangular cross-sections, and runnels with overhanging walls. A common complex form develops when the bottom of the channel is incised by a smaller nested channel, forming micro-strath terraces. The longitudinal profile can be graded or can be interrupted by a wide variety of irregularities, including steps and enclosed depressions with diverse geometries (kamenitzas, pits), or grikes. Crowther (1998), studying the longitudinal profile of Rinnenkarren at Lluc, Mallorca (Spain) differentiated three types of linear channel segments: (i) nonlinear, (ii) low-gradient steps (<15°) that form planar surfaces, and (iii) bevels that are less steep than the overall gradient of the profile. Veress et al. (2013), analysing networks of Rinnenkarren in the Austrian Alps, documented local enlargements in the cross-sectional area of trunk channels spatially associated with tributary junctions. They attribute these local widenings to the increase in turbulence and flow deflection that occurs just downstream of the confluence zones and the consequent enhanced dissolution. Veress et al. (2015) also analyzed the relationship between the slope angle and the density of two types of runnels; large channels with tributaries and greater catchments areas, and small channels with no tributaries and more restricted catchments. They observed that the density of large channels decreases with the slope angle, whereas small channels show the

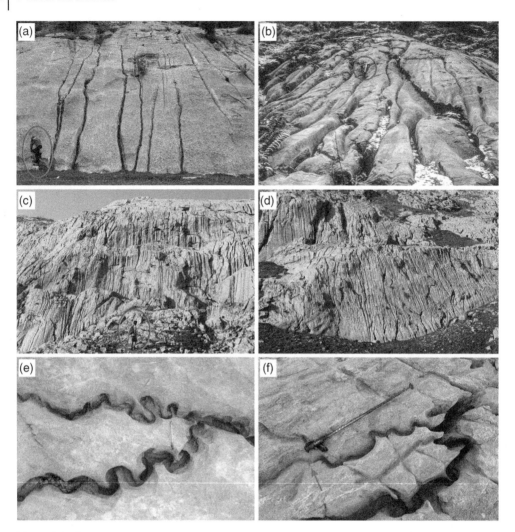

Figure 6.6 Examples of solution runnels, wall karren, and meandering runnels. (a) Straight to low-sinuosity solution runnels with contributory branching (left channels) developed on a ca 40° limestone slope. Cross-sectional area increases downslope. Linza, Spanish Pyrenees. Person for scale. (b) Solution runnels on marble that enlarge downslope. Benasque Valley, Spanish Pyrenees. Hammer for scale. (c) Limestone cliff furrowed by straight and parallel wall karren. Larra, Spanish Pyrenees. Person for scale. (d) Wall karren on steep slopes and lame dentate topped by triangular facets in the lower slope. Armeña-Cotiella Massif, Spanish Pyrenees. Person for scale. (e) Solution runnels that grade into converging highly tortuous and asymmetric meandering runnels in the Sennes Plateau, Dolomites, Italy. Pencil for scale. *Source:* Photo by Ugo Sauro. (f) Meandering runnels with variable degrees of entrenchment and some joint control. Arrow points to a knick point. Ordesa and Monte Perdido National Park. Walking stick for scale. *Source:* All photos by Francisco Gutiérrez (Author) except e.

opposite relationship. A number of factors can favor the development of solution runnels (Veress 2009a): (i) the dominantly turbulent regime of the channel flow, which increases in the steeper linear segments accelerating their headward retreat (Crowther 1998), (ii) the downslope increase in the contribution area and flow rate, (iii) rapid renewal of solvent, and (iv) open system conditions with respect to CO_2, which is replenished as it is consumed by the dissolution of carbonate minerals.

Wall karren (Wandkarren) are channels developed on subvertical slopes by water flowing down the rock face. These furrows are typically straight and oriented downslope, but may display some sinuosity (Figure 6.6c, d). Jennings (1985) describes them as a subtype of solution runnels that occurs on nearly vertical slopes. Ford and Lundberg (1987) indicate that they resemble Rinnenkarren, but they can be distinguished because wall karren become shallower and extinguish downslope. The channels can generate by water derived from a point or linear source upstream, and consequently there is some overlap between wall karren and decantation runnels, described below (Ford and Williams 2007). The channels head at the crest of the rock wall and can be abruptly interrupted by other karren such as solutionally enlarged discontinuity planes or holes. Veress (2009b, 2010) differentiates three main types according to the cross-sectional geometry of the channels: (i) grike-like, commonly V-shaped but often trapezoidal, with planar side walls and sharp edges, (ii) half-pipe, with curved side walls and semicircular to ellipsoidal geometry, and (iii) complex, with internal irregularities such as inset secondary channels. Wall karren are characterized by centimeter- to decimeter-scale widths and depths. Veress (2009b, 2010), in a morphometric study of wall karren in the Austrian Alps, measured widths between 2.5 and 34 cm, with dominant values between 4 and 12 cm. In some locations with favorable topographic and climatic conditions, they can reach lengths in excess of 100 m and meter-scale widths, like in the marble and limestone outcrops of some islands of the Chilean Patagonia (ca. 7000 mm yr^{-1} in average precipitation; Maire et al. 2009).

Meandering runnels (Mäanderkarren) are winding asymmetric channels carved into sloping rock faces or along the floor of a larger runnel (Figure 6.6e, f). They are considered by some authors as a variety of Rinnenkarren (Bögli 1960, 1980; Jennings 1985). Fluvial channels are considered of the meandering type when their sinuosity (ratio between channel length and straight distance) is higher than 1.5, but no threshold value has been defined for solution runnels. Mäanderkarren have been described in carbonate rocks and in gypsum (Macaluso and Sauro 1996a; Calaforra 1998). Veress (2009c, 2010) and Hutchinson (1996) suggest that true meandering runnels are characterized by both high sinuosity and markedly asymmetric cross sections. The channels typically display a subvertical to overhanging outer side and a gentle slip-off slope in the inner side. The outer concave wall commonly displays undercuts and may be carved by nested notches (Figure 6.6e). The asymmetry of the channels is attributed to higher flow velocity on the outer side of the meanders and the resulting more rapid solutional and mechanical denudation. This leads to the lateral and downstream migration of the channel line. The faster and more turbulent flow contributes to reduce the laminar boundary layer, increases the relative amount of solvent that interacts with the rock, and enhances mechanical erosion processes. The neck of meanders in tortuous incised channels can be traversed through the development of a cutoff tunnel and the meander loop may become abandoned and perched. Hutchinson (1996) documented meandering runnels in Mallorca (Spain) on slopes with gradients between 7 and 14° and observed an inverse relationship between channel sinuosity and slope. In a limestone karst in Switzerland, Zeller (1967) identified meandering runnels on slopes between 3.5 and 30°. Veress and Tóth (2004) in the Austrian Alps proposed a classification of karren meanders at different evolutionary stages based on their morphological and morphometric attributes.

Ford and Lundberg (1987) introduced the term *decantation runnels* to designate the channels generated by water released from an upslope store and to differentiate them from runnels created by runoff derived from direct rainfall (Figure 6.7a, b). The upslope store can be a point source (e.g., overspilling kamenitza, the stem of a tree, or a patch of moss) or a linear source (e.g., solutionally enlarged joint or bedding plane, snow banks, discontinuous soil cover). Because the channels may not receive additional water along their path, their cross section is largest at or

Figure 6.7 Images of kamenitzas and decantation runnels. (a) Two solution pans and associated overspill decantation runnels developed on roches moutonnées. Note Rillenkarren on the steep side wall of the kamenitza to the left. Lower Adige Valley, Italy. Boots for scale. *Source:* Photo by Ugo Sauro. (b) Large kamenitza with decantation runnel nested into a corrosion terrace with undercut headwall in Velebit Mountains, Croatia. *Source:* Photo by Nadja Zupan Hajna. (c) Kamenitzas with some detrital deposits at Kistanje, Croatia. Lens cap 8 cm in diameter for scale. *Source:* Francisco Gutiérrez (Author). (d) Succession of kamenitzas developed in the Lower Globigerina Limestone, Gozo Island, Malta. Lens cap 8 cm in diameter for scale. Azure Window natural arch in the background. *Source:* Francisco Gutiérrez (Author).

near the source, where the solvent has the highest degree of undersaturation. The water may experience some aggressiveness enhancement at the source by the uptake of CO_2 (e.g., organic matter degradation, root respiration, and higher solubility of CO_2 in cold melt water). Decantation runnels mainly form on bare rock surfaces, but are also found in areas with discontinuous cover. This type of runnels, classified on the basis of the source of the solvent, rather than on their morphology or the topography of the slope, shows some overlap with wall karren, solution runnels, and meandering runnels. For instance, runnels on vertical slopes fed by water issuing from a bedding plane can be classified as wall karren and decantation runnels at the same time. Ford and Lundberg (1987) indicate characteristic channel widths between 5 and 25 cm and lengths that may reach 25 m.

Kamenitzas (solution pans) are depressions that mostly develop on flat to gently sloping, bare rock surfaces where runoff does not drain readily and organic and inorganic material can easily be trapped (Figure 6.7). They are commonly subcircular to elliptical in plan view, but can also display highly irregular geometries, especially when they result from the amalgamation of adjacent pans. They have centimeter- to decimeter-scale depths and axial lengths that can reach several meters. Belloni and Orombelli (1970) measured a total of 83 kamenitzas in the limestone karst of Trieste, Italy, reporting average major and minor lengths of 43 and 21 cm, respectively, and a mean depth of 7.2 cm. Zwolinski (1996), analyzing the morphometry of 132 kamenitzas in the Dinaric Karst, calculated an average depth to length ratio of 0.4. Most pans have diameters ranging between 5 cm and 2 m (Cucchi 2009), but submerged kamenitzas as much as 15 m long developed on Pleistocene limestone have been documented in the Campeche Bank (Yucatán Peninsula, Mexico) by means of high-resolution multibeam bathymetry and side-scan backscatter (Goff et al. 2016). This type of bare karren has been also recognized in the stratigraphic record associated with a number of paleo-karst surfaces (Desrochers and James 1987; Di Stefano and Mindszenty 2000).

The side walls of kamenitzas are frequently undercut showing a basal corrosion notch and can be breached by overspill outlets connected to decantation runnels (Figure 6.7). The bottom tends to be flat and nearly horizontal, and is frequently covered by a thin veneer of detrital and organic material. They can also display nested depressions and stepped floors (Veress 2010) (Figure 6.7b). The initiation of kamenitzas can be related to the formation of subtle enclosed depressions by differential lowering, but they can also be inherited features from covered conditions (e.g., subsoil hollows). These depressions function as traps for water as well as for detrital and organic material focusing dissolution. The CO_2 released by the decay of the organic matter contributes to increase the dissolution capability of the stagnant water. Kamenitzas are believed to evolve by progressive dissolution acting on the floor and the walls, reaching the highest rates at the base of the margins. Mechanical processes such as frost shattering (Rose and Vincent 1986), salt weathering, and disintegration by biological activity (Cucchi 2009) can also contribute to the development of solution pans. The bottom is interpreted as a solution bevel that experiences slow solution lowering and that expands by back-migration of the margins, being somehow a micro-scale version of pediments that develop by lateral planation (Lundberg 2013). Several factors can be invoked to explain the more rapid dissolution that occurs at the base of the walls of the kamenitzas and a higher widening rate than the deepening rate: (i) the soil cover has an armouring effect reducing the rock–water contact surface on the floor, (ii) the water–rock interaction time is greatest at the base of the side wall, (iii) the runoff water that enters into the pans has its greatest aggressiveness at the margins, and (iv) the evaporation of the water that accumulates in the pans leads to the reprecipitation of secondary carbonate on the floor, thus water in the floor looses part of its aggressiveness dissolving the carbonate crust (Rose and Vincent 1986). Solution pans become relict features when their margin is cut by erosion or their floor intersects a penetrable feature such as a widened bedding plane or a fracture (i.e., grike) that allows the rapid drainage of the basin preventing water stagnation. Rose and Vincent (1986), based on the average content of dissolved $CaCO_3$ in the pans (74 10^{-5} M) at Gait Barrows, England, and considering the annual average precipitation (1100 mm yr^{-1}), estimated that the development of kamenitzas, 40 cm in diameter and 10 cm deep, requires a minimum of 3260 years. The actual time span should be significantly longer since the calculation assumes that all the water that enters into the pan via direct rainfall leaves the system carrying its share of dissolved $CaCO_3$ (i.e., there is no evaporation and the consequent precipitation of secondary carbonate). Cucchi et al. (1990) measured lowering rates in kamenitzas of 0.002–0.003 mm yr^{-1} near Trieste, northern Italy (average annual precipitation of 1100 mm), and estimated that the formation of a kamenitza 4–5 cm deep would need at least 2500 years.

Unconfined sheetwash water flow produces solution bevels, concavities such as Trittkarren and Trichterkarren with planar and rounded floors, respectively, flow-normal subhorizontal irregularities such as cockling patterns and solution ripples, and flow-parallel ridges with flat tops known as sharpened edges (lame dentate). *Solution bevels* (Ausgleichsflächen) are planar and relatively smooth surfaces usually less than 1 m long that are generated by solutional sheetwash erosion. They typically occur below the Rillenkarren extinction level and are attributed to unconfined water films that are too thick to allow raindrop impact to reach the rock surface and not thick enough for the separation of the water into channeled flow (Lundberg 2013) (Figures 6.2 and 6.8a). These surfaces, with no lithological control, can be from nearly horizontal to steeply inclined, when they develop below solution flutes. Presumably they expand by the backward retreat of their upslope edge, which can be the extinction line of Rillenkarren or a scarp (e.g., walls of kamenitzas, headwall of corrosion terraces). Consequently, their morphological evolution is envisioned as equivalent to that of pediments which evolve by backwearing of the slope situated at its upslope edge, and can resemble micropediments (Lundberg 2013). *Corrosion terraces* (mega-Ausgleichsflächen), typically found in high-elevation glacio-karst environments (Kunaver 2009), are considered as a variety of solution bevels. They are large gently sloping solution bevels several meters wide with a backwall more than 10 cm high cut into bedrock (Figures 6.7b and 6.8b). Corrosion terraces can display a stepped arrangement, forming a succession of shelves separated by scarps. Probably, the development of these large bevels is related to the presence of a snow cover that allows the sheet flow in the bedrock–snow interface to remain unbroken along longer distances (Lundberg 2013).

Trittkarren (heelsteps, steps) are an uncommon type of free karren that comprise a planar corrosion bevel and an arcuate backscarp with the concavity facing in the downslope direction (Figure 6.8c–e). They have been documented in carbonate rocks, gypsum (Macaluso and Sauro 1996a), and salt (Mottershead and Lucas 2004b), and appear to be more common in the case of carbonate rocks in regions with temporary snow cover (Lundberg 2013). The tread-like planar surface enclosed by the headwall typically has a horizontal or gently sloping attitude (Mottershead and Lucas 2004b) and can reach a few decimeters in width. The backwall has a sharp basal contact with the solution bevel and is commonly a few centimeters high, but can reach a decimeter scale (Vincent 1983). The riser that extinguishes downslope can be indented by Rillenkarren and its plan shape can be from semicircular to slightly curved or angular (Veress 2009d, 2010). The heelsteps can occur singly but are often associated with other adjoining Trittkarren or may form a stepped sequence along the slope. They can develop on rock exposures with a wide range of slopes. There seems to be some relationship between the slope gradient and the morphometric features of Trittkarren. In general, the steeper the slope, the smaller the bevel, and the higher the backwall (Veress 2010). There is broad consensus regarding their origin by turbulent sheet water flows, but the mechanisms responsible for the differential dissolution and the production of the backscarp-bevel sequence is not fully understood. Bögli (1960) suggested that the water sheet accelerates at the upper rim and that the consequent thinning of the water sheet increases the effectiveness of chemically aggressive rainfall and favors the diffusion of atmospheric CO_2 into the water film. Other authors propose that their formation may be related to detachment and reattachment of the boundary layer over a small preexisting step, through a mechanism analogous to that described for solution scallops (Ford and Williams 2007; Lundberg 2013). Veress (2010) proposes that in high

Figure 6.8 Various types of karren related to sheet wash water flow. (a) Solution bevel below Rillenkarren and cut across tightly folded salt with interlayered clay partings in the Cardona salt diapir. Scale 10 cm long. *Source:* Francisco Gutiérrez (Author). (b) Corrosion terraces with arcuate backwalls at 1100 m a.s.l. in Bojinac, southern Velebit, Croatia. Backpacks for scale. *Source:* Photo by Jurij Kunaver. (c) Assemblage of Trittkarren on gentle slope and associated with a scarp. Note Rillenkarren carved in risers, Muotatal, Swiss Alps. *Source:* Photo by Ugo Sauro. (d) Stepped Trittkarren developed on steep limestone slope showing relatively small treads, Zalmska River, close to the Biogradski ponor, Bosnia and Herzegovina. Pocket knife for scale. *Source:* Francisco Gutiérrez (Author). (e) Karren assemblage including stepped Trittkarren with cockling pattern and Rillenkarren feeding runoff to solution runnels, Tramuntana Range, Mallorca, Spain. *Source:* Photo by Ángel Ginés. (f) Trichterkarren resembling Trittkarren but with a nonplanar tread. Limestone outcrop at 2050 m a.s.l. in the Vallon des Morteys, Swiss Alps. Carabiner 10 cm long for scale. *Source:* Photo by Ángel Ginés. (g) Lame dentate or sharpened edges on a steep slope on pure and massive limestone at Mt. Mongioie, Ligurian Alps, Italy. *Source:* Photo by Bartolomeo Vigna.

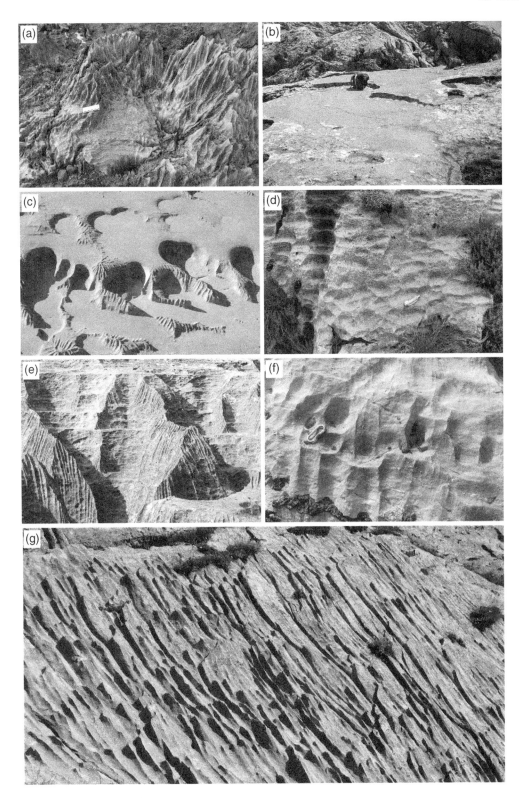

mountain areas they mainly develop by surface dissolution under sheet water flows fed by snow-melt. Mottershead and Lucas (2004b) examined thin sections under a petrographic microscope derived from the treads of Trittkarren (or stepkarren) developed on limestone, gypsum, and salt in Spain. They observed that the smooth surface of the steps carved into limestone and gypsum are underlain by a microcrystalline layer less than 0.25 mm thick showing an irregular basal contact with the host rock. In the limestone sites, this layer includes, in addition to fine calcite crystals, organic material and small voids of presumed biological origin (e.g., lichens). They observed rounded solutional voids associated with crystal boundaries and secondary epitaxial crystals in salt. The authors interpret that the microcrystalline layers provide evidence of dissolution and crystallization processes related to wetting and drying cycles and that this layer inhibits surface lowering in the steps. After each rainfall event, some water remains on the low-gradient step surface retained by its irregularities and surface tension. This solution leads to mineral precipitation by evaporation during the following drying phase. The resulting microcrystalline layer may be removed completely by dissolution during the subsequent precipitation event, depending on its magnitude. Mottershead and Lucas (2004b) proposed a genetic model involving continuous parallel retreat of the backscarp and episodic corrosional lowering or precipitation accretion on the tread, depending on the magnitude and temporal spacing of the wetting events. *Trichterkarren* (funnel karren) are considered by some authors a subtype of Trittkarren that lack a planar tread (Veress 2009d, 2010). They consist of a funnel-shaped backwall and a comparatively small concave floor (Figure 6.8f). Trichterkarren typically occur on steep slopes associated with Trittkarren. They likely result from the development of a channel in the bottom of preexisting Trittkarren (Lundberg 2013) and they seem to be associated with snow melting. Ginés (1998) documents funnel karren in the calcareous Swiss Alps at elevations between 1800 and 2200 m a.s.l. and attributes the lack of these features in the Tramuntana Range in Mallorca (Spain) to the scarcity of snow precipitation.

Cockling patterns and solution ripples are two types of transverse-to-flow linear karren that develop on slopes and are attributed to solution by sheet wash (Ford and Lundberg 1987; Ginés 2009). Both have characteristic centimeter-scale transverse dimensions (Sweeting 1972). *Cockling patterns* are horizontal concavities that give a crinkled appearance to the rock surface (Figure 6.8e). Ford and Lundberg (1987) indicate that on steep limestone slopes Rillenkarren tend to grade into cockling patterns. *Solution ripples* are flow-normal undulations with larger lateral continuity and regularity than cockling patterns (Wall and Wilford 1966; Jennings 1985). Their rhythmic form suggests systematic changes in the hydrodynamic conditions along the flow direction, resulting in evenly spaced changes in the dissolution process. *Lame dentate* (sharpened edges) are parallel ridges oriented in the flow direction comprising an upper flat bench and an underlying sharpened edge that extinguishes downslope (Figure 6.8g). They emerge from relatively planar steep rock slopes (>50°) and are associated with sheet water flow derived from snow melting (Perna and Sauro 1978; Ginés 2009; Lundberg 2013). They are also designated as tetrahedron karren because their upper flat surface commonly has the shape of a flat triangular facet (Choppy 1996). It is interpreted that the upper bench, where more persistent snow accumulates, deflects the water flow retarding dissolution in the underlying ridge. It is probable that dissolution is largely related to sheetflow along the rock–snow interface (Lundberg 2013).

6.2.2 Covered Karren

Solutional features developed beneath a soil cover or mantling deposits fall within the category of covered karren, also designated as subcutaneous karren, subsoil karren, and cryptokarren (Slabe 1999; Zseni 2009; Lundberg 2013). Dissolution of the rock surface in contact with a cover occurs under markedly different hydraulic and hydrochemical conditions than that acting on bare rock exposures, leading to distinctive karren features. In general, covered karren are characterized by smooth and rounded morphologies due to the more spatially continuous wet conditions, in contrast with the more angular and sharply etched subaerial surfaces. The flow along the rock–cover interface is slower and tends to be more widespread due to the influence of obstructing elements, and the influence of adhesion and capillary forces. Hence, there is longer water–rock interaction and the water tends to lose its aggressiveness over a shorter distance. Moreover, the water that infiltrates through a soil can increase its aggressiveness substantially by the incorporation of CO_2 derived from the degradation of organic matter and root respiration, as well as the incorporation of some organic acids. Consequently, dissolution of carbonate rocks is commonly more intense in subsoil conditions, especially in hot and humid tropical environments (see Section 3.3.2). This is supported by dissolution rates measured in different climatic environments by the weight-loss method using tablets placed at the surface and in soils (Table 4.1). Nonetheless, the dissolution capacity of the percolating water can be largely buffered in case the cover deposit has a significant carbonate content (Trudgill 1985). Factors such as the flow rate, the percentage of carbonate particles in the mantling deposit, their grain size (i.e., specific surface), the thickness of the cover, and the condition of the system as open or closed with respect to CO_2, determine the aggressiveness loss in the infiltration water before it reaches the rockhead. The classification of covered karren features is not an easy task because of the diversity of morphologies, the complications associated with the observation of these subsurface features, and the numerous terms used in the literature for the same type of morphology. Gams (1973a), Slabe (1999), and Zseni (2009) provided comprehensive reviews on subsoil karren. This section describes the most common elementary subsoil karren features (Rundkarren, cavernous karren, subsoil scallops and ripples, subsoil notches, pits or cups, and vertical wells and shafts). Karren assemblages related to the solutional widening of discontinuity planes in covered karst settings (cutters, pinnacles, and clints) are addressed in the following section on bare and covered structural karren.

Rundkarren are solution runnels developed by focused flow beneath a cover (Figure 6.11e). They have dimensions similar to those of Rinnenkarren and may deepen downstream, but display rounded lateral edges due to the combined action of channeled and widespread subsoil dissolution (Ford and Lundberg 1987). The furrows can have a nongraded longitudinal profile, with local reverse gradients and pits. Rundkarren often show a dentritic pattern and channels tend to become parallel on steep slopes. The rounded runnels are observed when cover deposits are removed by erosion, progressively acquiring the sharper geometry with angular shoulders of Rillenkarren by subaerial dissolution. These two types of karren can form coevally on slopes with a free-face section indented by Rillenkarren that connect with Rundkarren in a lower section covered by deposits (Ford and Lundberg 1987). Zseni (2009) proposes the term *cavernous karren* to describe centimeter- to decimeter-scale solutional voids with variable geometries (e.g., borings, enlarged porosity, irregular cavities) and random distribution and orientation. These openings with a spongework pattern that perforate the rock are generally filled with soil and may display variable degree of interconnection (Figure 6.9a). *Subsoil scallops and ripples* are regularly spaced concavities and flow-transverse undulations, respectively, with a steeper side in the upper part and with lengths or wavelengths within the 15–50 cm range. They develop by water sheets on steep to overhanging rock faces juxtaposed to cover deposits. According to Slabe (1999) their formation seems to be favored by a steep and permeable rock–cover contact. *Subsoil notches* are subhorizontal indentations with semicircular section that develop just below the ground surface in rock faces abutting cover deposits. They generally display a sharp upper edge and a rounded lower edge and can reach 1 m in width. Their formation is attributed to temporary retention of infiltration water just

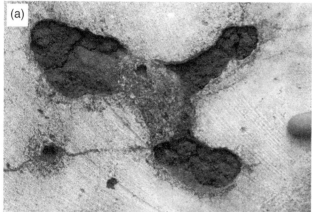

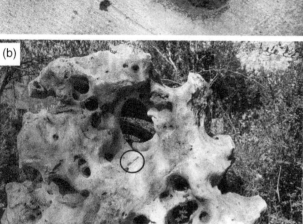

Figure 6.9 Examples of covered karren. (a) Cavernous karren exposed in a quarry in Menorca, Spain filled by red deposits washed down from the soil cover. Finger on the right for scale. *Source:* Francisco Gutiérrez (Author). (b) Ornamental limestone block in the Spanish Pyrenees showing sections of subsoil wells and shafts. The block has been rotated to a vertical position. Circle indicates pencil for scale. *Source:* Francisco Gutiérrez (Author). (c) Up to two meter high subsoil pipes with cemented rims developed in Late Pleistocene aeolian calcarenites and exhumed by differential erosion. Cape Bridgewater, southwestern coast of Victoria, Australia. *Source:* Photo by Matej Lipar.

beneath a stable ground surface during recharge events (i.e., water input exceeds percolation rate) and the higher aggressiveness of the water at shallow depth. The formation of these features has similarities with the flare slopes typically found in granites and sandstones, ascribed to more intense subsoil weathering at the foot of steep slopes and subsequent erosional exhumation of the notches by erosion (Twidale and Vidal Romaní 2005). *Solution pits* or cups are shallow holes with a depth to width ratio of ≤1 commonly found on subhorizontal surfaces. These subsoil depressions display rounded or tapering bottoms, in contrast with subaerial solution pans or kamenitzas characterized by flat floors. Their distribution can be controlled by weak spots on the rock surface (e.g., discontinuities, textural, or mineralogical changes). *Karren wells* and *karren shafts* are vertical to steeply inclined cavities with depth to width ratios of 1–2 and > 2, respectively (Zseni 2009) (Figure 6.9b). The morphometric criterion proposed to differentiate wells and shafts, although objective, may be difficult to apply because of the hidden nature of the conduits. These karren features are often controlled by discontinuities and have some overlap with the structural karren described below. The term subsoil pipe is frequently used as equivalent to wells and shafts, but Lundberg (2013) indicates that this term should be restricted to eogenetic and syngenetic subvertical conduits developed on poorly lithified and porous rocks (e.g., calcarenites). Their formation involves dissolution in the central part and carbonate precipitation along the edge to form an indurated rim (De Waele et al. 2009a, 2011; Lipar et al. 2015, 2021) (Figure 6.9c).

6.2.3 Bare and Covered Structural Karren

Structural karren collectively designate linear karren mainly resulting from the solutional enlargement of discontinuity planes (e.g., joints, faults, and bedding planes). Mechanical erosion processes can also contribute to their development. Downward water flow in the vadose zone along these penetrable planar features (fissure permeability) causes their progressive widening. This process is particularly effective along vertical to steeply dipping fractures with favorable orientation with respect to the downward flow direction. The fractures are progressively widened and deepened to form clefts that typically taper downward due to the reduction in the aggressiveness of the percolating waters (Palmer 2009b). The fissure-like karren controlled by fractures in bare karst settings are known as *grikes* (Kluftkarren). The equivalent clefts found in covered karst environments and commonly filled by clayey residual soil and/or unconsolidated deposits are designated as *cutters* (Figures 6.10–6.12). Grikes and cutters are typically a few centimeters or decimeters across and are normally longer and deeper than 1 m (Ginés 2009). The steep side walls of grikes can be sculpted by various types of free karren, such as Rillenkarren and Wallkarren, whereas those of cutters tend to display covered-karren features (e.g., Rundkarren) and more rounded geometries. In carbonate rocks, the development of subsoil cutters can be significantly more rapid than that of grikes related to direct water infiltration, due to aggressiveness enhancement in the infiltration water by CO_2 uptake from the soil.

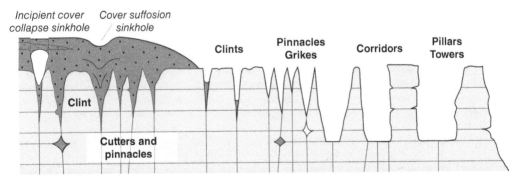

Figure 6.10 Diagram illustrating the development of grikes and cutters (subsoil grikes) by the solutional widening of joints, plus intervening pinnacles and clints. Further widening of grikes can result in the development of streets or corridors flanked by rock residuals with variable morphology (e.g., pillars, towers). The diagram shows how subsidence and internal erosion of cover deposits along cutters can lead to the development of cover-suffosion sinkholes and cover-collapse sinkholes.

Figure 6.11 Examples of grikes and limestone pavements. (a) Grikes in gently dipping Eocene limestone underlying the Great Pyramid at Giza, Egypt. Person for scale on the right. *Source:* Francisco Gutiérrez (Author). (b) Grikes controlled by two main sets of joints on a sloping and concordant limestone surface in the Larra karst, Pyrenees. Person in the right for scale. *Source:* Francisco Gutiérrez (Author). (c) Vertical view of solutionally enlarged joints and faults expressed as grikes and hollows. The most prominent fracture corresponds to a fault-line scarp. The outcrop is partially covered by scree (talus) generated by frost shattering. Larra karst, Pyrenees. *Source:* Francisco Gutiérrez (Author). (d) Limestone pavement comprising joint-controlled grikes and intervening flat-topped clints. Mignovillard, French Alps. *Source:* Photo by Guy Decreuse. (e) Limestone pavement at Malham Cove, North Yorkshire, England. Clints are indented by Rundkarren developed beneath a soil cover and subsequently exposed by the stripping of the mantling deposits. *Source:* Photo by Anthony Cooper.

Grikes and cutters constitute an essential component of the epikarst. White (1988) considers that Kluftkarren are among the principal landforms in karst, along with sinkholes and caves. These fissures, together with other shallow porosity features create a relatively thin high-permeability zone associated with the rockhead (epikarst) that can function as a perched aquifer (Williams 1983, 2008) (Figures 5.23–5.25). In mantled karst environments cutters can also play an important role in the development of cover-collapse and cover-suffosion sinkholes, acting as pathways for internal erosion (Figure 6.10). Moreover, the solutional enlargement of these subsoil fissures (cutters) in

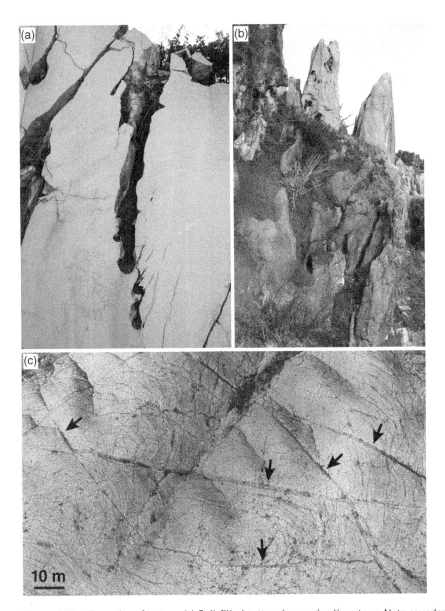

Figure 6.12   Examples of cutters. (a) Soil-filled cutters in massive limestone. Note rounded geometries related to subsoil dissolution and solutionally widened subsidiary joints with en-echelon arrangement in the lower part. Epikarst exposed in a limestone quarry near Zavala, SW of Popovo Polje, Bosnia and Herzegovina. (b) Solutionally enlarged fractures filled by iron-rich residual clays in one of the abandoned mines at Cabárceno Natural Park, northern Spain. The initially subsoil pinnacles in the background have been exhumed by iron mining and differential water erosion. (c) Vertical view of a gypsum exposure criss-crossed by joints enlarged by dissolution (pointed by arrows) and filled by residual clayey soil that favors vegetation growth. Calatayud Neogene Graben, Iberian Chain, Spain. *Source:* Francisco Gutiérrez (Author).

carbonate or evaporite rocks involves a gradual mass depletion and the creation of void space, which tends to be replaced by subsidence (ductile or brittle) and internal erosion of the overlying soil cover, resulting in ground settlement.

Joints, due to their high spatial frequency, are the prime structural elements that control the development of grikes and cutters. They generally have a planar geometry and tend to be perpendicular to the stratification, commonly with a density inversely proportional to bed thickness. Joints can be restricted to specific layers or penetrate through thick stratigraphic successions, and they can be dilated by various mechanical processes (e.g., tectonic stress, unloading). Rock formations are frequently criss-crossed by sets of subvertical joints intersecting at regular angles, forming a joint system. The solutional enlargement of these fractures can result in the development of a polygonal network of free or subsoil clefts with intervening blocks (Figures 6.11 and 6.12). Subvertical well-like hollows can form at joint intersections. The steep-sided rock compartments surrounded by grikes or cutters with pointed to sharp tops are designated as *pinnacles*, whereas those with flat tops are known as *clints* (Flachkarren) (Figures 6.10 and 6.13). Both features can coexist and clints can evolve into pinnacles by the recession of their walls.

The enlightening work by Palmer (2009b) on the cutters developed in a pure and jointed limestone underlying the Mitchell Plain, Indiana, coupled with a finite-difference dissolution model of a fissure, illustrates that the formation of the cutters and their geometry in limestone is closely tied to the dissolution kinetics of calcite. As explained in Section 3.8.1, this dissolution system is characterized by an initial phase of rapid dissolution until the water reaches a saturation ratio of around 70% (low-order kinetics), followed by a sharp reduction in the dissolution rate (high-order kinetics) that allows the water to flow along significant distances before it loses all its aggressiveness (Figure 3.36). This change in the reaction order and the dissolution rate can be expressed by a sharp narrowing in the cutters, showing a bulbous upper part leading to a narrower and more evenly tapered fissure. The transition depth between rapid and slow dissolution kinetics is strongly dependent on the discharge, which in turn varies greatly with time. Water associated with a limited flow rate rapidly loses its aggressiveness. Palmer (2009b) infers a direct relationship between discharge and cutter spacing, which determines the catchment area; i.e., larger joint spacing favors the development of wider cutters. The rock surrounding the cutters examined by Palmer (2009b) displays zones of solutionally enlarged primary pores and networks of subsidiary fissures, formed by periodic flooding of the epikarst. It seems that downward percolation and dissolution can be facilitated by gaps between the rock and the clay plug when the latter shrinks by drying.

Clints and grikes are the main components of *karst pavements*, a type of karrenfield (Figure 6.11). These occur in subhorizontal or gently inclined rock exposures commonly concordant with the bedding, and comprise a regular grid of grikes and intervening clints. The upper surface of the clints can be indented by various types of free karren including solution pans and runnels (Figure 6.11e). Karst pavements commonly occur on erosional surfaces scoured by glaciers and reshaped by solutional and mechanical denudation during post-glacial times. The erosional surfaces overprinted by these karrenfields can also be generated by other erosional agents such as wave action, differential water erosion leading to structural surfaces, or sheet-flooding in piedmont areas. Karst pavements are best developed in areas with subhorizontal, thick to massively bedded strata, traversed by regularly spaced subvertical joint sets intersecting at high angles (Vincent 2004). They are rarely observed in gypsum due to their high solubility and low mechanical resistance, and are also uncommon in low competence rocks lacking systematic jointing, such as chalk or eogenetic limestone (Goldie 2009). Goldie and Cox (2000) conducted a morphometric study of clints and grikes in limestone pavements of nine areas in Britain, Ireland, and Switzerland. They calculated median length and width for the clints of 235 and 99 cm, respectively, and observed that the most frequent clint width to length ratio is around 0.3–0.4. Data from a total population of more than 1400 grikes showed that 95% of grike widths is below 30 cm with a median value of 13 cm. Grike depth ranged from 4 to 274 cm, with a median of 74 cm. Vincent (1995, 2004, 2009) proposes that most of the classical pavements observed on the cyclic Lower Carboniferous limestones in northern England have gone through a complex evolution, including: (i) dissolution under

Figure 6.13 Examples of pinnacle karst. (a) The stone forest (shilin) of Yunnan, South China Karst. Note pinnacles and blades sculpted by solution flutes and runnels. *Source:* Photo by Yu Manman. (b) The pinnacles (tsingy) of Bemaraha in Madagascar. *Source:* Photo by Etienne Venot. (c) The Lena Pillars at an erosional escarpment excavated by the Lena River, Siberia. This landscape in a permafrost area is attributed to differential mechanical erosion acting on a fracture-controlled and soil-filled paleokarst. *Source:* Photo by Márton Veress. (d) The pinnacles more than 50 m high of Mount Gunung Api in Mulu National Park, Sarawak, Malaysia. *Source:* Photo by Chris Howes. (e) Pinnacles developed on Pleistocene aeolian calcarenite with cross bedding and a calcrete in the topmost part. Nambung National Park, Western Australia. *Source:* Photo by Francisco Gutiérrez. (f) Salt pinnacle generated by differential dissolution around a protective insoluble capsoil. Salt extrusion NE of Bastak, Zagros Mountains, Iran. Note person for scale at the bottom. *Source:* Francisco Gutiérrez (Author).

subaerial conditions, (ii) deposition of Late Glacial loess and development of peaty soils in wetter conditions, with widespread Rundkarren development under the loess mantle, and (iii) removal of the mantling deposits by surface and internal erosion, plus renewed subaerial dissolution. This author also indicates that some parts of the limestone pavements are paleokarst surfaces exhumed by glacial or coastal erosion and reshaped by subsequent dissolution. El Aref and Refai (1987) document relict grikes and limestone pavements developed during the past wet periods on Eocene limestones in the Giza Pyramids Plateau, Egypt, where the current annual rainfall is around 25 mm (Figure 6.11a). Silva et al. (2017) characterize limestone pavements on structural surfaces in the Potiguar Basin, Brazil using high-resolution airborne LiDAR data and orthoimages generated with photographs taken with an unmanned aerial vehicle (UAV).

The shilin terrain (shi: stone, lin: forest) of Yunnan, China (UNESCO World Heritage Site) includes one of the most popular pinnacle and pillar karst landscapes, with stone forests of variable dimensions distributed over an area of 350 km² (precipitation ca. 950 mm yr⁻¹) (Figures 6.13a and 6.15d). These fracture-controlled rock residuals occur on Permian limestones that have experienced several cycles of karstification, burial, and exhumation throughout the Mesozoic and Cenozoic. The stone forests started their development by interstratal (beneath a Permian basalt caprock) and subsoil dissolution. Within a context of tectonic uplift, differential erosion of the non-soluble material during the Quaternary, mainly red Tertiary clayey soils, has progressively exposed the pinnacles (rock teeth) and pillars, sharpening their morphology by subaerial dissolution with the consequent overprinting of the subcutaneous features by free karren (Sweeting 1995; Song and Liang 2009; Knez and Slabe 2013 and references therein). The tsingy of Bemaraha (UNESCO World Heritage Site) (Figures 6.1 and 6.13b) and Ankarana in Madagascar (1000–2000 mm yr⁻¹), developed on gently dipping Jurassic limestone, are among the most spectacular examples of giant-grike topography developed on bare limestone (Middleton 2004; Salomon 2009; Gilli 2019). These complex karrenfields comprise an intricate network of giant grikes bounded by rock remnants with variable geometries (pinnacles, blades, and towers), mainly controlled by the spacing and orientation of the fractures. The top and flanking surfaces of the rock blocks are fretted by various types of bare karren and their degradation by dissolution and mechanical processes (rock falls) can lead to the merging of adjacent grikes to form wide streets (corridors) (Veress et al. 2009). According to Veress et al. (2008), the development of the grikes of Bemaraha, and their exceptional dimensions up to 100 m deep, are largely related to joint-controlled phreatic caves. The cave passages were connected to the surface by dissolution along grikes by infiltration water and cave-roof collapse. Experiments with solidified sugar immersed in a static surrounding liquid show that gravity-driven flows of dense solute-laden fluids create shapes similar to those of the sharp pinnacles (Huang et al. 2020). Although the final morphologies are similar, the extremely more complex solutional kinetics of carbonates surely needs a more complicated modeling approach, but these experiments suggest that gravity-driven flows might be an essential element in pinnacle formation.

The Lena Pillars in Siberia form a spectacular landscape of vertically walled pillars and ridges separated by structurally controlled grikes and streets within a permafrost environment, where carbonate dissolution is negligible in present conditions (Figure 6.13c). Here, the discontinuous permafrost has an estimated thickness of 50–300 m and the rock surface in the pillars barely shows any conspicuous karren. These features occur on outcrops of densely fractured and well-bedded limestone along escarpments up to 150 m high excavated by the Lena River and some of its tributaries. Veress et al. (2014) propose that the grikes and pillars correspond to fracture-controlled paleokarst features formed under warmer conditions that were filled and buried by sediments. At the present time, they are in the process of being exhumed by differential erosion, favored by the adequate topographic conditions created by the incision of the Lena River. Active frost-shattering significantly contributes to reduce the size of the rock pillars. Other remarkable pinnacle terrains include Mount Kaijende in Papua New Guinea (Williams 2009) and Mount Gunung Api in Sarawak, Malaysia with pinnacles emerging over the rainforest as much as 50 m high (Ley 1980; Day and Waltham 2009) (Figure 6.13d).

A peculiar type of pinnacle landscape is found along an extensive coastal strip in Western Australia, characterized by dense fields of pinnacles up to 5 m high developed in Pleistocene aeolian calcarenite corresponding to a relict dune system. This is a cyclic formation comprising cross-bedded aeolian sands (average quartz content 42%), calcrete/microbialites, and paleosoils. The pinnacles are best exposed in the Nambung National Park (Figure 6.13e). Two modes of formation have been proposed by Lipar and Webb (2015) involving the development of an eogenetic epikarst in a poorly lithified and porous carbonate-rich deposit: (i) focused dissolution by downward vadose flow produces fissures and pipes that enlarge to form residual pinnacles dominated by conical geometries and (ii) the cemented sand infill of subvertical solution pipes, including microbialites with vertical lamination, are exhumed by differential solutional and mechanical erosion resulting in cylindrical pinnacles (Lipar et al. 2015). Lipar and Webb (2015), based on geochronological data and geometrical relationships situate the main dissolution phase at MIS 5 (ca. 130–180 kyr ago). Grimes (2009) attributes the pinnacles to focused cementation by downward finger flow of saturated water and subsequent differential erosion. Pinnacles, alike pedestals, can also form in salt rock by differential dissolution around a protective material (e.g., capsoil, boulder), which can be subsequently removed by mechanical erosion. In this case, dissolution is not controlled by discontinuity planes and these salt pinnacles constitute an example of equifinality (Figures 6.2 and 6.13f).

Grikes can be enlarged by rock face recession via dissolution and various mechanical erosion processes (e.g., rock falls, frost shattering) to form corridors of meter-scale width (Figures 6.14 and 6.15a, b). These large and frequently accessible passageways have been designated with numerous terms, including local names such as bogaz (Dinaric karst, Cvijić 1893), corridors (Western Australia, Jennings and Sweeting 1963), zanjones (meaning large trenches, Puerto Rico; Monroe 1968), and streets (Nahanni karst, Canada; Brook and Ford 1978; Judbarra karst, Northern Australia; Grimes 2012). Intersecting corridors with a street pattern and the intervening rock blocks form fracture-controlled landscapes known as *rock cities*. The retreat of the rock walls results in the expansion of the passageways at the expense of the rock compartments. At some stage, the low ground dominates over the upstanding rock blocks forming *ruiniform landscapes* of scattered rock residuals with variable morphologies (Migoń et al. 2017) (Figure 6.14). There is some overlap between the ruiniform landscape and the stone forests. Brook and Ford (1978), in their paper on the labyrinth karst of Nahanni National Park (Mackenzie Mountains, Canada; average temperature −4.5 °C; mean precipitation 566 mm yr^{-1}), propose the following evolutionary sequence: (i) water draining underground via vertical fissures creates strings of solution dolines, (ii) by enlargement and coalescence along the fractures they are converted to intersecting networks of karst streets, (iii) as streets deepen and widen, the intervening rock ridges are dissected and ultimately destroyed, (iv) large closed depressions of angular planform (karst platea) are formed, and (v) floors are alluviated and the platea become small poljes. Grimes (2012), in the Judbarra karst in Northern Australia, described a spatial and temporal zonation associated with a retreating non-karstic cover: grikes and karst pavements grading to rock cities and to rock residuals (towers, pinnacles). Migoń et al. (2017) propose a terminology for describing the various landforms associated with rock cities and ruiniform landscapes in different lithologies, essentially related to structurally controlled differential erosion (Figure 6.14). Passageways can be subdivided into avenues, streets, and lanes (<2 m), according to their scale. The open space at the intersections are designated as squares and plazas, which are wider than squares, and courtyards, which are semi-enclosed widenings inside blocks connected with avenues, streets or lanes by a single passageway. The positive features of the ruiniform landscapes commonly found in carbonate terrains can be subdivided into: towers (angular features with square base), walls (angular features with elongated base), spires (high towers not angular in their top parts), and pinnacles (towers with pointed tops).

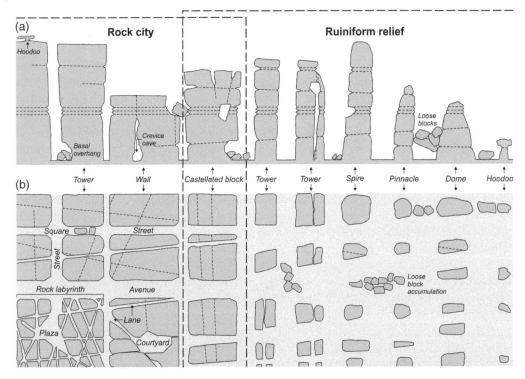

Figure 6.14 Diagrams in cross section (a) and plan view (b) illustrating the terminology proposed by Migoń et al. (2017) to describe rock cities and ruiniform relief developed in karst and non-karst bedrock. *Source:* Migoń et al. (2017) Reproduced with permission of Elsevier.

Remarkable examples of rock cities and ruiniform landscapes developed on carbonate rocks have been described in a wide range of morphoclimatic regions: Nahanni limestone karst, Mackenzie Mountains of Canada (Brook and Ford 1978); Sierra Maestra, Cuba (Panoš and Štelcl 1968); Bom Jesus da Lapa, Brazil (Tricart and Da Silva 1960); West Kimberley district in Western Australia (Jennings and Sweeting 1963; Goudie et al. 1990; Playford 2002); Judbarra-Gregory National Park, Northern Australia with both rock cities and ruiniform landscapes developed above a shale bed acting as the base level for solutional denudation (Grimes 2012); Tidirine area of the Middle Atlas in Morocco (De Waele and Melis 2009); outcrops of the Rosso Ammonitico Formation in the Venetian Pre-Alps (Sauro 2009); Torcal de Antequera, Betics, Southern Spain, with towers showing numerous bedding-controlled recesses (Durán et al. 2005; Martos 2010) (Figure 6.15c); outcrops of dolomitic limestones in Sicily (Di Maggio et al. 2012); Ciudad Encantada, Iberian Chain, Spain showing long streets, intervening walls with overhanging faces and mushroom-shaped residuals (Twidale and Centeno Carrillo 1993) (Figures 3.40a and 6.14a, b); Las Tuerces in northern Spain (Martín-Duque et al. 2012) (Figure 3.40f). Similar landscapes also develop by solutional and mechanical erosion along fractures in conglomerates and quartz sandstones (Figure 6.15e, f).

6.2.4 Coastal Karren

Rock coasts underlain by soluble rocks are complex geomorphic environments affected by a large variety of mechanical, biological, and chemical processes, including wave action, mass movements, abrasion, splash, spray, wetting and drying, salt weathering, frost shattering, bioerosion,

Figure 6.15 Examples of rock cities and rock residuals of ruiniform relief. (a, b) Vertical and ground view of corridors or streets developed on jointed Cretaceous dolomites at Ciudad Encantada, Iberian Chain, Spain. *Source:* Francisco Gutiérrez (Author). (c) Towers developed on well-bedded limestone with prominent recesses at bedding planes in El Torcal de Antequera, southern Spain. Circle indicates a climber for scale. *Source:* Francisco Gutiérrez (Author). (d) Rock residual in the stone forest of Yunnan, South China Karst, showing several undercuts at different elevations generated by subsoil dissolution. Note relatively smooth surface in the upper part indented by solution runnels (free karren) and more irregular covered karren morphologies in the recently exhumed lower part. *Source:* Photo by Yu Manman. (e) Joint-controlled towers in quartzose conglomerates at Meteora, Greece. *Source:* Francisco Gutiérrez (Author). (f) Fins or walls on carbonate conglomerates with a single penetrative joint set in Montserrat, NE Spain. *Source:* Francisco Gutiérrez (Author).

bioconstruction, and dissolution (Trudgill 1985) (Figure 6.16). Dissolution is frequently considered to play a secondary morphogenetic role in the development of erosional features in carbonate rocky coasts due to the fact that sea water is saturated or supersaturated with respect to calcite, especially in warm environments (the solubility of CO_2 in water decreases with rising temperature) (Schneider 1976) (Figure 3.28). This is supported by the occurrence of similar erosional features in non-karst rocks, the greater degree of development of the sculpturing features in the zones where

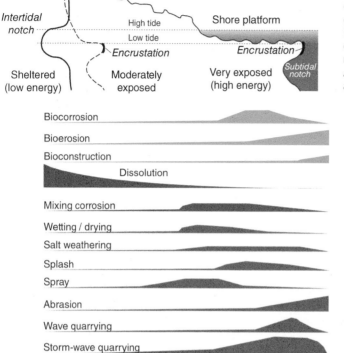

Figure 6.16 Typical profiles of limestone coasts under different exposure/energy conditions and general spatial distribution and intensity of the main biological, chemical, and physical morphogenetic processes. *Source:* Adapted from Lundberg (2009) and De Waele and Furlani (2013).

dissolution is less active (intertidal zone) (Gómez-Pujol and Fornós 2009b; Moses 2003), the fact that rock lowering rates in coastal sites are significantly higher than in nearby inland sites (Furlani et al. 2009; Stephenson and Finlayson 2009), and the limited karren development in cold environments where bioerosion has a very limited impact (Lundberg and Lauritzen 2002). However, the sea water can locally become aggressive by mixing with fresh water (see Section 3.7.5) and by biogenic acidification related to CO_2 input derived from respiration or organic matter degradation. The term karren in coastal environments is commonly used with a broader meaning than in inland settings, embracing not only small-scale sculpturing features related to rock dissolution, but also other karren-like morphologies generated by various processes, often difficult to identify and assess. Polygenetic forms and morphologic convergence are key concepts to understand sculpturing features in coastal karst environments.

There is broad consensus regarding the major role played by bioerosion in carbonate rocky coasts, referring to the mechanical and chemical removal of lithic substrate by direct organic activity (Spencer and Viles 2002; Viles 2004a, 2004b; De Waele and Furlani 2013). Bioerosion is particularly important in warm tropical areas and in the wetter zones within the coastal transect, especially if energy levels and erosional processes such as wave quarrying and abrasion are not high enough to inhibit colonization by biota. Small organisms such as algae and cyanobacteria that colonize the outer rock surface (epiliths) and the pores within a thin layer associated with the rock surface (endoliths) form biofilms. These can either weaken the rock by mechanical action and biomediated dissolution, or protect it from other weathering and erosional agents. Boring micro- and macro-organisms perforate the rock by mechanical action and/or dissolution induced by the secretion of acidic substances. Some invertebrates such as mollusks or echinoderms produce borings or depressions to shelter in (home scars) (Figure 3.52). Grazers mechanically rasp the weakened rock surface colonized by epiliths and endoliths, chiefly in the intertidal zone (Spencer 1988a, 1988b; Fox 2005) (Figure 3.51). Some bioconstructing organisms such as coralline algae, serpulid worms or barnacles can coat the surface forming resistant encrustations with a protective role (bioprotection), eventually leading to the development of positive and differential erosion forms (Naylor and Viles 2002). See additional information on biokarst in Section 3.10.1. A comprehensive compilation of biokarst examples from tropical and subtropical areas is given in Kázmér and Taborosi (2012).

The zone of operation and intensity of the different morphogenetic processes indicated above varies along a transect perpendicular to the coastline, which can be divided into three main sectors: the subtidal zone, the intertidal zone, and the supratidal zone that includes the sectors affected by splash and spray (Figure 6.16). Local variations in the relative contribution of the different mechanical, biological, and chemical processes depend on the following main controls (Lundberg 2009): (i) geometry of the rocky coast (e.g., notched plunging cliff, shore ramp, and/or shore platform), (ii) wetting regime and tidal range, (iii) energy level largely controlled by the degree of exposure to erosional agents such as waves and wind, (iv) climate, that has a direct influence on biological activity and processes such as wetting and drying, salt weathering, or gelifraction (frost weathering), and (v) lithological and structural characteristics of the rock. Significant differences are observed between porous Quaternary eogenetic carbonate rocks commonly found in tropical regions (e.g., emerged reef limestones), and diagenetically mature and well lithified limestones, relatively more frequent in temperate and cold regions. The geometry of the coast largely depends on the energy level, the degree of exposure to wave action and the potential protective role played by encrusting organisms (Figure 6.16). In low energy and sheltered environments, erosion by wave attack tends to produce linear horizontal notches typically incised into the rock, 1 dm to 5 m deep, which are particularly well developed in limestone (Trenhaile 1987) (Figure 6.17a). Notches form within the elevation range where backwearing processes are most effective, mainly in the intertidal zone (e.g., wave quarrying, abrasion, and bioerosion) (Kelletat 2005). Notches show better development in coast sections associated with submarine springs, suggesting a link between freshwater–seawater mixing and notch carving (Higgins 1980). They can also form in the subtidal zone below the portions protected by biological encrustations (subtidal notches). The geometry and size of tidal notches is influenced by factors such as the tidal range or the exposure level. A restricted tidal range (e.g., Mediterranean coast) and sheltered conditions favor the formation of deeply indented notches with limited height. In sloping coasts the top of the notch acquires the shape of a visor. Notching acting on all the sides of stacks separated from the coast produce mushroom rocks (Paskoff 2005) (Figure 6.17b). Erosional ramps and shore platforms tend to form in more exposed coast sections with higher energy levels. Erosional ramps reflect the landward decrease in the erosional effectiveness of wave action and are the most common form worldwide (Lundberg 2009). Karren typically display a clear zonation in sloping erosional ramps controlled by the relative contribution of different processes, with more pronounced pinnacles and more jagged morphologies toward the water surface. Shore platforms are nearly horizontal rock–cut surfaces mainly resulting from cliff retreat above an elevation determined by the vertical range where erosional processes are most active (Trenhaile 2002). These are time-transgressive straths progressively older toward the seaward edge of the platform. In warm waters, encrusting organisms tend to coat and protect the limestone surface in zones affected by frequent wetting (below mean tide level or in the surf zone in very exposed coasts). Progressive erosional backwearing above these armouring encrustations results in the development of shore platforms, that can be rimmed by biogenic ramparts or armoured rims (trottoirs). Greater erosion below these rock portions protected by encrustations

Figure 6.17 Examples of coastal karren. (a) Notch carved in Quaternary reef limestone in a sheltered area of Bahía Cochinos, Cuba. Note jagged and scoriaceous appearance of the rock surface. *Source:* Francisco Gutiérrez (Author). (b) Limestone stacks with basal notch forming mushroom rocks in Gozo Island, Malta. Divers for scale. *Source:* Photo by Stefano Furlani. (c) Biogenic borings in limestone with elongated bedding-parallel chert concretions. Coast Siant Jean de Luz, France. Lens cap 8 cm in diameter for scale. *Source:* Francisco Gutiérrez (Author). (d) Coalescing basins and pinnacles developed in Quaternary aeoleanites at Cala Blava, Mallorca. Hammer for scale. *Source:* Photo by Lluís Gómez-Pujol. (e) Tube karren (Röhrenkarren) and intervening downward projecting spires developed in the underside of a ledge at Lake Mask, Ireland. *Source:* Photo by Michael Simms. (f) Tube karren developed in Cretaceous limestone at Noja, northern Spain. *Source:* Photo by Javier Elorza.

can also produce differential backwearing and the development of subtidal notches. The platform surface may host wide, shallow, flat-bottomed pans with curved encrusted ridges, sometimes known as tidal pools, although this term should be used only for those pools located in the intertidal zone. They resemble kamenitzas of mountain karsts, especially in the coastal areas farther from wave and spray action, but become more irregular and jagged close to the coastline, where wave action, abrasion, salt weathering, and bioerosion become dominant. The different geomorphic elements (platform, ramp, notch) can coexist in a transect and their erosional evolutionary trend involves a general landward migration with the progressive and gradual transformation of each zone into the lower one.

Karren (or pseudokarren) features can form and be preserved where their rate of formation is not overwhelmed by the rates of other destructional or inhibiting processes such as mechanical erosion or sediment accumulation. According to Lundberg (2009), the most common coastal karren features include small *pinnacles* (Spitzkarren) and a wide spectrum of enclosed depressions that can be classified according to their width to depth ratio (W/D) and morphology. *Pans* and *basins* are depressions with a $W/D > 1$ and with flat or rounded bottoms, respectively. *Pits* are characterized by a $W/D < 1$. These depressions can coalesce to form more complex morphologies and their origin is frequently ascribed to the combined effect of multiple processes with a difficult-to-assess relative contribution. Pinnacles typically display a very jagged morphology and tend to be coated by a dark biofilm of algae and cyanobacteria, which led some authors to use the term (black) phytokarst (Folk et al. 1973; Viles 2004c). Gómez-Pujol and Fornós (2009b) interpret that pinnacles are erosional remnants resulting from the lateral growth and coalescence of flat-bottomed pans.

A peculiar karren form that can be found in coastal areas associated with marine and lake environments are *tube karren* (Röhrenkarren). These are vertical tubular forms tapering upward to a rounded apex that propagate upward from the underside of overhangs by condensation corrosion (Simms 2002; Drew 2009). Water vapor condenses on the colder rock causing its "antigravitative" dissolution (Dreybrodt et al. 2005a) through a mechanism similar to that invoked for the formation of bell holes in caves associated with tropical coasts (Tarhule-Lips and Ford 1998a). This type of karren also forms in limestones along freshwater lake shores (Plan et al. 2018). Tighly packed tubes can coalesce to form intervening downward projecting spires. Elorza and Higuera-Ruiz (2016) and Elorza et al. (2019) document two generations of tube karren associated with two marine straths (supratidal zone) in the northern coast of Spain. These display an average diameter of 2.5 cm and maximum lengths greater than 25 cm. In some special settings (twilight zone of caves or alcoves), light-oriented karren or *photokarren* have been documented (elongated holes and pinnacles), apparently related to selective colonization of photosynthetic biota controlled by the exposure to sunlight (Burren coast of Ireland, Simms 1990; coast of Sardinia, De Waele et al. 2009a).

The various types of karren developed in littoral settings and their morphological and morphometric attributes typically display a coast-parallel zonation that corresponds to the different hydrodynamic and biological zones (Trudgill 1985). This zonation is nicely illustrated by the flat-bottomed pans (tidal pools) up to several meters long and 1 m deep developed in the temperate microtidal environments of the Balearic Islands, western Mediterranean. Here, Gómez-Pujol and Fornós (2009b) identified four morphological zones: (i) zone A in the portion of the splash zone closer to the sea displays isolated and completely fretted pinnacles coated by cyanobacteria and lichens, corresponding to residual forms related to the expansion of preexisting pans, (ii) zone B shows shallow pans and intervening pinnacles joined at their bases by ridges, (iii) zone C in the spray zone is characterized by pans with a greater degree of coalescence and a rough interbasin surface and (iv) zone D displays isolated pans and smooth rock surfaces in between. Zonations with some similarities have been reported in other temperate and tropical regions (Moses 2003; De Waele et al. 2009a; Drew 2009). In general, the relief and jaggedness increases toward the sea, whereas the surface tends to be smoother and with a more subsued morphology landward. A very different picture is observed in the high-latitude coasts of Norway and Svalbard within a context of glacioisostatic uplift, where karren are absent in the intertidal zone and suites of pans, pits and basins are restricted to sloping marble outcrops in the supra-littoral zone (Lundberg and Lauritzen 2002).

6.3 Sinkholes (Dolines)

6.3.1 General Aspects and Classification

Sinkholes or dolines are enclosed depressions with internal drainage widely regarded as one of the most characteristic landforms of karst landscapes (Cvijić 1893). They are typically circular to sub-circular in plan and show wide morphological diversity (cylindrical-, conical-, bowl-, and pan-shaped) (Figure 6.18). In some cases, the variable geometry of the sinkholes can indicate different evolutionary stages and the relative age of the depressions. For instance, cylindrical collapse sinkholes may progressively evolve into broader conical- and pan-shaped basins by the degradation of their margins (Gökkaya et al. 2021). Sinkholes can reach hundreds of meters in length and are commonly a few meters to tens of meters deep, although they can attain depths of hundreds of meters. The term doline derives from the Slavic word *dolina*, which is used to designate small hollows in some regions of Carniola (Ljubljana area, Slovenia), while in the south Slavic languages it is used for valleys (Cvijić 1893). According to Sweeting (1972), doline was first used as a geomorphological term by Austrian geologists in the middle of the nineteenth century, when they were conducting studies in the Dinaric karst. Cvijić's (1893) seminal work *Das Karstphänomen* played a decisive role in the introduction of the term doline in the international literature (Kranjc 2013). In the past century, the word doline was widely used by European geomorphologists, whereas sinkhole was the preferred term among the American researchers and those dealing with engineering and environmental issues (Sowers 1996; Sauro 2003; Williams 2004a; Waltham et al. 2005; Gutiérrez et al. 2014b; Gutiérrez 2016). In the present century, sinkhole has become the dominant term in the scientific literature, probably because of the intuitive and self-explanatory character of the word and the significant applied component of a great proportion of the investigations. In this book we use the words sinkhole and doline as synonyms, although it should be noted that sinkhole is also used for depressions generated by subsidence processes not related to dissolution (e.g., piping, landsliding, and mining subsidence) (Cooper 2017). There is also a myriad of local names used to designate sinkholes in different regions (e.g., *cockpit* in Jamaica, *cenote* in Mexico, *obruk* in Turkey, *dahl* in the Arabian Peninsula, *daya* in northern Africa).

Figure 6.18 Examples of sinkholes. (a) The Forau de Aigualluts, a bedrock-collapse sinkhole 140 m long in Devonian marble that functions as a swallow hole at the headwaters of the Ésera River watershed in the Spanish Pyrenees. The water captured by the ponor crosses the Mediterranean-Atlantic topographic divide via underground conduits and issues at the Uelhs deth Joeu located 4 km apart. This connection was demonstrated by Norbert Casteret in 1931 using fluoresceine. *Source:* Francisco Gutiérrez (Author). (b) The natural entrance of Carlsbad Caverns, a bedrock-collapse sinkhole developed in Permian limestone set up with an amphitheatre to observe the emergence of the bats. *Source:* Francisco Gutiérrez (Author). (c) Zacil-Ha Cenote in Yucatán Mexico, a bedrock-collapse sinkhole that provides direct access to the freshwater aquifer. *Source:* Francisco Gutiérrez (Author). (d) A bedrock-collapse sinkhole 120 m long and 60 m deep, which represents an oasis in the semiarid mountains of Central Anatolia, Turkey. *Source:* Photo by Muhammed Öztürk. (e) Isolated and clustered sinkholes with permanent ponds related to gypsum dissolution. Nature Reserve of the Arcas Lakes, Cuenca, Spain. *Source:* Francisco Gutiérrez. (f) The Pito playa-lake around 850 m long in the floor of a solution sinkhole developed in Miocene gypsum, and enlarged and reshaped by wind erosion. Note yardangs in the downwind side of the basin (lower right). Saladas of Sástago-Bujaraloz, NE Spain, included in the Ramsar List. *Source:* Francisco Gutiérrez (Author). (g) Tighly packed solution sinkholes in the Sivas gypsum karst, forming a polygonal landscape. The fine-grained deposits of the sinkhole floors are used for cultivation. Road for scale in lower right. *Source:* Photo by Muhammed Öztürk. (h) Scarp-edged single and compound depressions in Kotido crater, Mars interpreted as collapse sinkholes related to evaporite dissolution. Arrows point to fissures at the foot of the cliffed margins. RGB HiRISE image ESP_016776_1810. *Source:* NASA/JPL/University of Arizona.

Sinkholes are best developed and preserved in flat areas, where the low topographic gradient favors water infiltration and inhibits surface runoff and mechanical erosion (Gutiérrez and Lizaga 2016; Öztürk et al. 2018a). They can occur as isolated depressions, forming clusters and alignments, or can completely riddle the ground surface (Figures 6.18 and 6.19). Complex geometries may result from the development of nested sinkholes or the coalescence of nearby dolines. Depressions resulting from the amalgamation of two or more adjacent dolines are commonly des-

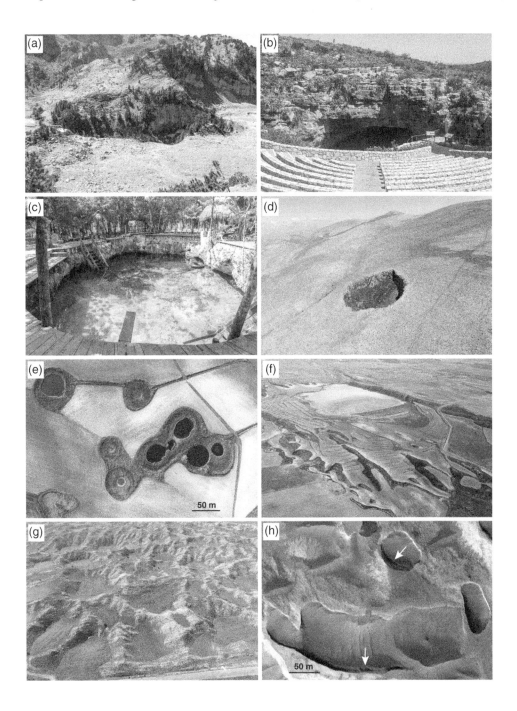

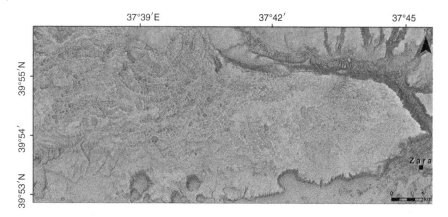

37°39′E   37°42′   37°45

39°55′N
39°54′
39°53′N

Figure 6.19 Red Relief Image Map (RRIM) derived from a digital elevation model showing the polygonal karst landscape formed by tightly packed solution sinkholes aroud Zara in the gypsum karst of Sivas. Large scarp-edged depressions in the lower part of the image are bedrock-collapse sinkholes. The RRIM, which combines slope, positive openness, and negative openness layers, facilitates the visualization of the enclosed depressions by shading their side slopes and lightening their edges and divides. *Source:* Model produced by Ergin Gökkaya.

ignated as compound sinkholes or uvalas. Some authors consider that the term uvala, introduced by Cvijić (1893), lacks a widely accepted definition and incorporates some ambiguity (Palmer 2007). According to Ćalić (2011), Jovan Cvijić erroneously considered uvalas as a transitional landform between doline and polje, founded on an evolutionary model whereby the expansion and coalescence of karst depressions leads to the sequential development of dolines, uvalas, and poljes. Ćalić (2011), based on the analysis of 43 uvalas more than 1 km in length and with irregular floors in the Dinaric karst and in the Carpatho-Balkanides of eastern Serbia, inferred that the studied basins are mainly related to areal corrosional lowering in zones with high fracture density, rather than to the coalescence of sinkholes. To avoid confusion, we propose the use of compound sinkholes instead of uvala to denote depressions related to sinkhole coalescence.

There are multiple reasons why sinkholes receive significant attention by karstologists, as well as scientists and practitioners from multiple disciplines. Sinkholes, together with the lack of an integrated fluvial drainage network, are one of the most distinctive geomorphic features of karst terrains (holokarst). This led some authors to go so far as to consider sinkholes as the main diagnostic feature of karst. Nonetheless, karst groundwater systems including well-developed caves can occur in areas devoid of sinkholes (Ford and Williams 2007). Dolines function as sinks that contribute to reduce the runoff connectivity between high and low areas (i.e., surface transfer of water and sediment). Sinkholes located in recharge areas play the role of water input features, often with a direct connection to the groundwater system. They can also function as focused recharge points for major streams (swallow holes or ponors) (Figures 5.17c, d and 6.18a). The rapid transfer of water from the surface to the phreatic zone via sinkholes and associated conduits also affects pollutants. Moreover, sinkholes are often used for uncontrolled waste disposal. Consequently, they significantly increase the vulnerability of aquifers to contamination (Lindsey et al. 2010). Sinkholes related to the foundering of cavity roofs (collapse sinkholes) or the interception of caves by surface erosional lowering (cave unroofing) are frequently the easiest or even only access for humans to cave systems (Figure 6.18b). Some sinkholes host permanent water-table lakes and these karst windows can be the sole direct access to drinkable water in vast regions (Figure 6.18c). This factor can dictate the distribution of settlements and transport routes, for instance the Maya civilization in the Yucatán Peninsula (Munro and Melo-Zurita 2011) or caravanserais of the Bedouins in the Arabian Peninsula (Youssef et al. 2016). Dolines can have a special microclimate (Iijima and Shinoda 2000; Whiteman et al. 2004) and constitute sites of remarkable ecological value (Figure 6.18d), including a large diversity of isolated wetlands (Tiner 2003) (Figure 6.18e, f). In rocky terrain, the deposits accumulated on the floor of solution sinkholes can be the only suitable land for cultivation (Kranjc 2013) (Figures 5.17b and 6.18g).

Ground subsidence related to the occurrence and activity of sinkholes is the main hazard associated with karst areas. Dolines are also prone to other hazards and geotecnical problems such as flooding by runoff concentration or water table rise, water leakage in hydraulic structures, and settlement by differential compaction of their sediment fill, frequently underlain by an irregular rockhead (Gutiérrez 2010, 2016; Gutiérrez et al. 2014b). Sinkholes function as traps for sediment and consequently can host valuable archives of geological processes and paleoenvironmental change in the dominantly erosional karst terrains. Moreover, subsidence can contribute to increase the preservation potential of sinkhole deposits and to increase the temporal length and completeness of the stratigraphic record. The analysis of sinkhole deposits in combination with geochronological data has been used to obtain information about various hazardous processes, including subsidence (Sevil et al. 2017; Gutiérrez et al. 2018), floods (Gutiérrez et al. 2017; Fabregat et al. 2019), hurricanes (Gischler et al. 2008; Lane et al. 2011; Brown et al. 2014), tephra fall-out (Siart et al. 2010), or soil erosion (Turnage et al. 1997; Hart 2014). Paleoenvironmental studies based on sinkhole deposits, including sinkhole lakes and paleolakes, have addressed a wide range of aspects such as climate variability (Barreiro-Lostres et al. 2014; Peros et al. 2017; Perrotti 2018; van Hengstum et al. 2018), sea-level changes (Kovacs et al. 2013) or the impact of past human activity on the landscape (Kulkarni et al. 2016). A number of remarkable paleontological and archeological sites are associated with sinkhole fills (Carbonell et al. 2008; Calvo et al. 2013; Daura et al. 2014; Zaidner et al. 2014; Gutiérrez et al. 2016). The high concentrations of paleontological and human remains often found in these localized environments can be related to various factors, such as accidental fall in natural traps (Klopmeier et al. 2018), camps, or the deliberate placing of elements (e.g., rituals, waste disposal). Sinkholes can also have remarkable aesthetic and recreational value (Figure 6.18c). For instance, the superlative polygonal karst landscapes of Jamaica (cockpit), the Philippines (Chocolate Hills in Bohol) (Figure 7.11e) or Sivas Basin in Turkey, consisting of extensive tracts riddled by densely packed solution sinkholes and intervening hills, foming a peculiar egg-box-like topography (Figure 6.19). Sinkholes also receive substantial attention in planetary geomorphology studies, especially those dealing with Mars. Since the first images of the Martian surface captured by the Viking Orbiter became available, a growing number of works have attributed enclosed depressions to sinkholes related to dissolution of both evaporite and carbonate sediments (Figure 6.18h). These depressions constitute lithological markers and provide evidence of the presence of liquid water in the past, recording valuable paleoclimatic and paleohydrological information (Baioni et al. 2009; Parenti et al. 2020 and references therein).

In the early geomorphological studies, sinkhole classifications were grounded on morphologic and morphometric criteria. Cvijić (1893), based on morphometric studies conducted in the Adriatic karst region and East Serbia differentiated three principal forms with characteristic width to depth ratios (W/D) and slope gradients (S): (i) bowl-shaped ($W/D \approx 10$; $S \approx 10$–12°), (ii) funnel-shaped ($W/D \approx 2$–3; $S \approx 30$–45°), and well-shaped ($W/D < 2$; $S \approx 90°$). Cvijić's (1893) work reveals the controversy that used to exist regarding the origin of dolines at that time and a tendency to ascribe all these landforms to a single process. Cramer (1941) made an outstanding contribution differentiating sinkholes related to solution and collapse. This work introduced the awareness that there are different types of sinkholes developed in different materials and by various mechanisms, but with some morphological convergence (equifinality) (Ford and Williams 2007). There are a number of genetic classifications of sinkholes, mostly inspired by the processes observed in carbonate karst areas (Williams 2004a; Beck 2005; Waltham et al. 2005). Gutiérrez et al. (2008a), based on the previous works and observations carried out in numerous exposures of paleosinkholes related to evaporite dissolution, proposed a more comprehensive genetic classification that covers the subsidence mechanisms that occur in both carbonate and evaporite karst areas (Figure 6.20). This scheme, adopted in other sinkhole reviews (Gutiérrez et al. 2014b, Gutiérrez 2016; Parise 2019), differentiates two groups of sinkholes: solution sinkholes and subsidence sinkholes. Solution sinkholes are related to the progressive differential solutional lowering of the ground where karst rocks are exposed at the surface or merely soil-mantled. These depressions are essentially generated by chemical denudation and therefore rarely pose ground instability problems. Subsidence sinkholes collectively designate a wide spectrum of dolines generated by subsurface dissolution and the deformation and/or internal erosion of the undermined overlying material. Here, the term subsidence is used in a broad sense to indicate the downward displacement of the ground, regardless of the rheology of the deformed material and the velocity of the process (Jackson 1997). Obviously, these are the most important sinkholes from the hazard perspective since their development involves the settlement of the ground (*terra infirma*). Subsidence sinkholes are classified using

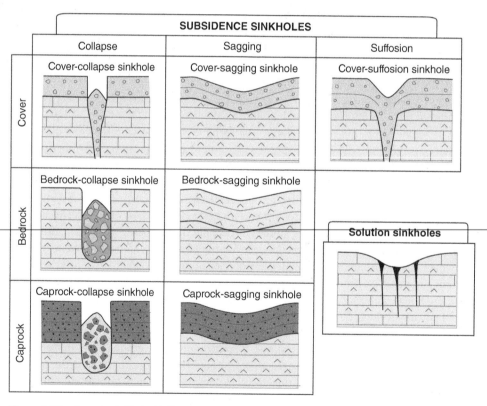

Figure 6.20 Genetic sinkhole (doline) classification applicable to both carbonate and evaporite karst terrains. Subsidence sinkholes are described indicating the type of material affected by gravitational deformation and/or internal erosion, and the subsidence mechanism. Complex sinkholes can be described using combinations of the proposed terms with the main material and/or subsidence process followed by the secondary one. *Source:* Adapted from Gutiérrez (2016).

two terms. The first descriptor refers to the type of material affected by downward displacement (cover, bedrock, or caprock), and the second term indicates the subsidence mechanisms (collapse, suffosion, or sagging). Cover denotes unconsolidated allogenic deposits (e.g., alluvium) and/or residual soil (e.g., insoluble residue), bedrock refers to karst rocks, either carbonates or evaporites, and caprock to non-karst rocks (e.g., sandstone, basalt). Collapse is the brittle deformation of rock or soil material through the development of well-defined failure planes or brecciation. Suffosion refers to the downward migration of unconsolidated cover deposits through voids accompanied by the progressive settling of the surface. Sagging, a subsidence mechanism commonly observed in evaporite karst areas and not included in the previous classifications, is the ductile flexing or passive bending of material related to the lack of basal support. Frequently, the development of subsidence sinkholes involves several subsidence mechanisms that may operate concomitantly or sequentially and may affect different types of materials (Figure 6.21). These complex sinkholes can

Figure 6.21 Examples of complex sinkholes. (a) Large elongated sagging sinkholes with marginal extensional fissures and nested collapses related to salt dissolution (cover-sagging and -collapse sinkholes). Dead Sea, Israel. Circle indicates person for scale. (b) Bedrock-sagging and -collapse paleosinkhole developed in Miocene evaporites. High-speed Madrid-Barcelona railway in Zaragoza city area, Ebro Cenozoic Basin, Spain. Thick and thin arrows indicate the edges of the sagging and collapse structures, respectively (c) Caprock-sagging and -collapse collapse paleosinkhole in an artificial exposure associated with a high-speed railway in Madrid area (Cañada Real), Spain. This subsidence structure is related to interstratal dissolution of a Miocene salt-bearing evaporite formation overlain by marls and limestone in the Madrid Cenozoic Basin. Note that the inactive paleosinkhole is truncated by an undisturbed erosional surface. *Source:* Francisco Gutiérrez (Author).

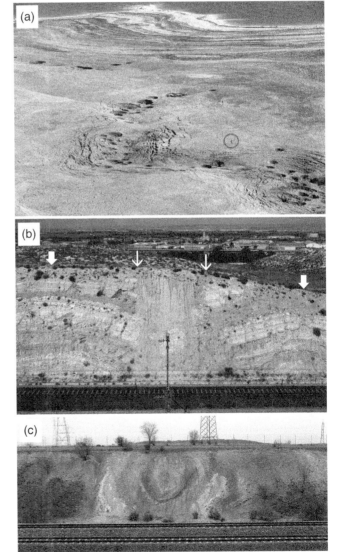

be described using combinations of the proposed terms with the main material and/or subsidence process followed by the secondary one (e.g., bedrock-sagging and -collapse sinkhole). The mechanisms that operate in the development of complex subsidence sinkholes could be plotted in a triangular diagram with the vertex representing the main processes (collapse, suffosion, and sagging).

6.3.2 Solution Sinkholes and Polygonal Karst

Solution sinkholes are the result of the differential lowering of the ground surface mainly by dissolution acting on bare-rock surface, at the bedrock–cover interface (rockhead), and within the more karstified and permeable upper portion of the bedrock (epikarst). The dissolved rock mass is removed as solute load by downward water percolation. The topography of the solution dolines indicates more rapid solutional downwearing in the deeper part than in the side slopes and margins (Figure 6.22). The differential and focused chemical erosion can be attributed to several factors acting individually or in combination: (i) greater water flow, (ii) higher water aggressiveness, and (iii) higher susceptibility of the rock to dissolution (e.g., fractures).

In a bare rock terrain, dissolution of the exposed bedrock surface is mainly produced by direct rainfall and overland flow, before it infiltrates into the ground. This surface water has limited capability to dissolve carbonate rocks due to its low CO_2 content, but it can achieve substantial erosional work in evaporite rocks such as gypsum, significantly more soluble and erodible than limestone and dolostone. Dissolution acting on a horizontal rock exposure will tend to lower the surface differentially, with faster downwearing at more susceptible spots controlled by heterogeneities in the rock mass (e.g., fractures and their intersections) (Figure 6.23). Gams (2000) relates the locally enhanced dissolution associated with solution sinkholes to the higher water–rock contact surface at sites with denser fracturing. Differential lowering of the surface produces concavities that tend to concentrate overland flow. The focused centripetal surface flow results in increasing runoff and dissolution toward the center of the depressions. This preferential solutional lowering of the bare rock surface is enhanced through a positive feedback as the radius of the depressions increases, leading to progressively larger contributing area and higher flow rate. Moreover, the increasing solutional erosion rate toward the center of the sinkhole tends to enhance sinkhole depth, steepen the side slopes, and reduce water infiltration along the slopes. During severe rainfall events, overland flow can rapidly concentrate in the lower part of the sinkhole (Day 1979) and infiltrate forming a recharge cone (Williams 1985).

In cold regions, pockets of melting snow can enhance dissolution by prolonged rock wetting and more aggressive water related to the higher solubility of CO_2 in cold water (Gams 2000). In mountain environments dominated by rock exposures and situated above the tree line, the accumulation of windblown snow in solution sinkholes by wind action tends to produce relatively small, scarp-edged depressions by focused dissolution at the bottom by meltwater (Figure 6.22c). The preferential direction of winds and the directional accumulation of snow in these depressions can cause dolines to have an asymmetric profile, as documented in the High Atlas in Morocco (Perritaz 1996). Their length to depth ratio typically falls between 1.2 and 1.5. The preferential corrosion in the floor of the dolines operates through a positive feedback, whereby sinkhole deepening allows the accumulation of greater amounts of snow, and thus increasing the amount of meltwater and the time it can interact with the bedrock (Veress 2017).

The growing diameter of the sinkholes and the steepening of their slopes can be also accompanied by surface flows with higher mechanical erosion capability (Figure 6.23). Physical weathering processes such as frost-shattering can also contribute to modify the geometry of the sinkholes. Mechanical erosion acting on the side slopes and accumulation of detrital material in the lower

Figure 6.22 Images of solution sinkholes and polygonal karst. (a) Shallow solution sinkhole with clayey fill used for cultivation. Villar del Cobo, Iberian Chain, Spain. *Source:* Francisco Gutiérrez (Author). (b) Air view of diffuse-edged solution sinkholes. Crop field in the sinkhole on the right is 280 m across. Rodenas, Iberian Chain, Spain. *Source:* Francisco Gutiérrez (Author). (c) Scarp-edged sinkholes (arrows) with low length to depth ratio in a periglacial environment. Windblown snow tends to accumulate in the depressions enhancing dissolution. Circle indicates person for scale. Armeña glacial cirque, Spanish Pyrenees. *Source:* Francisco Gutiérrez (Author). (d) Polygonal karst landscape consisting of adjoining conical depressions and rounded divides with a mesh pattern. Taşeli Plateau, Taurus Mountains, Turkey. *Source:* Photo by Muhammed Öztürk. (e) Polygonal karst in rugged terrain underlain by well bedded and densely fractured limestone. Giden Gelmez Mountains, Taurus Mountains, Turkey. *Source:* Photo by Muhammed Öztürk. (f) Peculiar polygonal karst with linear mesh pattern controlled by a penetrative joint set. Note elongation and alignment of sinkholes and the red clayey soil on the floor of the depressions. Evdilek Plateau, Taurus Mountains, Turkey. *Source:* Photo by Muhammed Öztürk.

part of the sinkholes contribute to increase the length to depth ratio of the depressions and to flatten their slopes (Figures. 6.18g and 6.22f). Moreover, the clastic material shed from the rock slopes has a higher specific surface than bedrock, and thus dissolves more rapidly (Gams 2000). Inorganic and organic particles can accumulate in the lower part of the sinkhole, mainly by sheetwash,

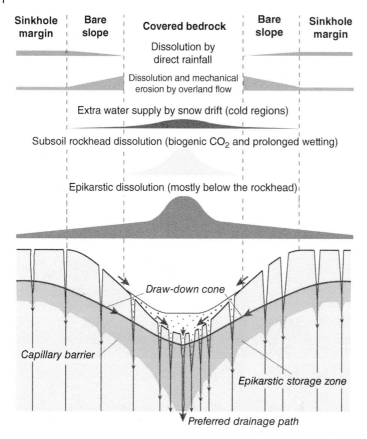

Sinkhole margin	Bare slope	Covered bedrock	Bare slope	Sinkhole margin

Dissolution by direct rainfall

Dissolution and mechanical erosion by overland flow

Extra water supply by snow drift (cold regions)

Subsoil rockhead dissolution (biogenic CO_2 and prolonged wetting)

Epikarstic dissolution (mostly below the rockhead)

Draw-down cone

Capillary barrier

Epikarstic storage zone

Preferred drainage path

Figure 6.23 Main processes involved in the development of solution sinkholes and their general spatial distribution. Dissolution can operate on bare rock surfaces, at the rockhead beneath a soil cover, and within the epikarst zone, mostly below the rockhead. The sketch illustrates the case of a cone-shaped sinkhole carved in a flat bare rock surface and with cover deposits in the lower part of the depression.

covering the bedrock. These deposits commonly show a concentric facies distribution, with colluvium and/or talus in the slopes that grade into fine-grained sheetwash deposits in the flat to gently sloping central part of the sinkhole. The contribution of dissolution on bare rock exposures to the development of solution sinkholes largely depends on the infiltration capacity of the ground. It is limited in rock exposures where the bedrock is affected by a high density of discontinuities widened by dissolution, which readily absorb rainfall water and inhibit overland flow (Gunn 1981b). However, bare-rock surface dissolution can play a significant role in evaporite outcrops and in carbonate exposures with limited infiltration capacity (e.g., massive and scarcely fractured rocks). Matsushi et al. (2010), in a solution doline in Japan (~2000 mm annual precipitation), estimated long-term denudation rates along the side slopes applying terrestrial cosmogenic nuclides on samples collected from rock exposures. They observed an increasing trend toward the center of the depression with values ranging from 63 to 256 g m^{-2} yr^{-1}. The high values (23–94 mm kyr^{-1}) suggest that mechanical weathering and erosion contribute to the surface lowering in the sloping surfaces. Using empirical relationships between denudation rate and contributing area, these authors modeled the evolution of doline topography and found that doline depth mainly depends on the catchment area, rather than on the estimated age of the sinkholes.

Dissolution at the rockhead beneath a cover of detrital deposit or residual soil can be also enhanced at the deeper part of solution sinkholes by various mechanisms. The cover tends to be restricted to the bottom of the sinkholes, where it reaches the greatest thickness and the most favorable topography and texture (i.e., fine-grained) for vegetation growth and soil development. The aggressiveness of the water with respect to carbonate rocks is greatly enhanced when it percolates through soils by uptake of biogenic CO_2. Subsoil rockhead dissolution in carbonate rocks can be significantly higher than in bare rock exposures (Table 4.1). Day (1976), in a limestone karst in Jamaica, found the highest soil carbon dioxide contents in the bottom of solution sinkholes. This mechanism does not influence sinkhole development in evaporites, where dissolution is not affected by water acidity. Sinkhole fill deposits, depending on their textural characteristics, can have lower hydraulic conductivity than the underlying bedrock and function as infiltration regulators, retarding water percolation and prolonging the water–rock interaction at the rockhead. These two mechanisms of dissolution enhancement by CO_2 uptake from soils and longer wetting of the rock surface in contact with damped soils tend to have their greatest impact in the central sector of sinkholes in carbonate rocks, and consequently contribute to their deepening (Figure 6.23). Low-permeability clayey cover deposits can also inhibit water infiltration inducing temporary ponding that reduces and retards water recharge and dissolution (Day 1979; Sauro 1991). Zambo and Ford (1997), in a soil-covered solution sinkhole in Aggtelek National Park (Hungary) estimated the carbonate dissolution potential of the percolation water at the soil-rock interface, observing a significant increase between the side slopes (ca. $3 \, \text{g} \, \text{m}^{-2} \, \text{yr}^{-1}$) and the bottom ($17–30 \, \text{g} \, \text{m}^{-2} \, \text{yr}^{-1}$), where the soil fill reaches 7.5 m in thickness. This increasing trend and the associated differential deepening of the sinkhole is attributed to concentration of runoff and through flow toward the sinkhole floor, plus more favorable moisture and thermal conditions in the soil for the production of CO_2. Veress (2017) proposes that the texture of the deposits accumulated in the sinkhole floors has influence on the relative contribution of the downward and lateral corrosion that operates at the rockhead and the morphological evolution of solution dolines. A low-permeability clay layer at the bedrock–cover interface can hamper downward water percolation and induce outward flow, favoring lateral growth. Water retained in fine-grained deposits by capillary forces also favors lateral corrosion. In contrast, downward percolation dominates in coarse-grained and permeable deposits favoring vertical growth. The overall dissolutional lowering of the rockhead must be accompanied by the gradual settling of the overlying cover and the land surface, although these processes seem to be too slow on a human timescale to constitute a detrimental hazard.

Dissolution in the epikarst significantly contributes to the development of solution sinkholes, especially in carbonate terrains (Williams 1985; Ford and Williams 2007). The solutional removal of mass that occurs in the epikarst beneath the rockhead does not lead to direct surface lowering, but considerably increases rock porosity, creating favorable conditions for faster dissolutional lowering at the rockhead. Water that percolates through carbonate bedrock loses great part of its dissolution capability within the first few meters of its underground flow path beneath the surface. Consequently, most fissures widened by dissolution (e.g., grikes, cutters) taper rapidly downward, creating a sharp vertical hydraulic conductivity contrast and a bottleneck effect for percolating water. Greater dissolution in this near-surface zone is also indirectly enhanced by physical weathering processes, including the development and dilation of discontinuity planes by stress release and the resulting increase in the water–rock contact surface (Klimchouk 2000b, 2004b). The water can easily penetrate into the highly corroded upper zone of the bedrock but drains slowly out of it, forming a temporary water store after severe rainfall events (Figure 6.23). This highly heterogeneous suspended aquifer associated with the upper part of the vadose zone was designated as the epikarst aquifer by Mangin (1974), and its highly irregular base can be considered as a leaky capillary barrier

(Williams 1983). According to Worthington et al. (2000), epikarst porosity is typically one to three orders of magnitude higher than that of the bulk rock of the vadose zone below. The hydraulic conductivity of the shallow epikarstic aquifer is also characterized by sharp horizontal variations. According to the conceptual model proposed by Williams (1985), preferred vertical leakage paths can develop by greater and deeper solutional widening at specific permeability features such as joints or joint intersections. These effective vertical drainage paths can create a depression in the water table of the suspended aquifer and govern the local flow paths. The water in the epikarst aquifer flows centripetally toward these drainage points, creating a zone of focused subsurface flow and enhanced dissolution (Figures 5.25 and 6.23). The progressive increase in permeability associated with the focused dissolution can contribute to enlarge the radius of the draw-down cone, and consequently the rates of water flow and dissolution, establishing a positive feedback loop. Williams (1985, 2008) proposes that this convergent flow pattern in the epikarst aquifer is the main process responsible for the development of solution sinkholes in carbonate rocks. It should be noted that a large part of the dissolution is expected to occur within the bedrock at some depth below the rockhead, with no direct impact on the topography. However, intense corrosion associated with the preferred vertical leakage paths facilitates focused water transfer from the soil to the bedrock and more rapid corrosional lowering at the top of a strongly karstified bedrock. Williams (1985, 2008) also suggests that the radius of the draw-down cones controls the size of the associated solution sinkholes. The dimensions of the depression cone depend on the hydraulic conductivity of the epikarst in each sector and the rate at which water drains down the leakage path at its base. The widening of fissures in the leakage path permits more rapid downward drainage, leading to the steepening of the hydraulic gradient and the expansion of the depression cone, stimulating the lateral growth of solution sinkholes. Moreover, the enlargement erosional lowering, subsurface dissolution, and breakdown of fissures and conduits results in more rapid and turbulent downward flows with greater capacity to cause internal mechanical erosion, potentially leading to the development of nested cover-suffosion sinkholes and cover-collapse sinkholes. Klimchouk (1995, 2000b, 2004b) indicates that the concentration of centripetal and downward flow at the base of the epikarst can result in the development of hidden shafts. The thinning and weakening of the roof of these cavities by surface erosional lowering, subsurface dissolution, and breakdown processes can eventually lead to the occurrence of collapse sinkholes. Evaporite rocks such as gypsum and halite are less prone to the development of an epikarstic aquifer than carbonate formations. These more ductile lithologies commonly display less pervasive fracturing and their dissolution tends to produce a greater amount of clayey insoluble residue that can clog the fissures widened by dissolution.

In carbonate rocks, solution sinkholes mainly occur in regions where precipitation exceeds evapotranspiration, and are more profusely developed under humid tropical and temperate conditions (Gams 2000). In contrast, gypsum outcrops can display striking polygonal karst landscapes even in semiarid areas due to the higher solubility and erodibility of this rock type (Figures 6.18g and 6.19). In arid and permafrost environments, the limited availability of liquid water hinders the formation of solution sinkholes. Nonetheless, Goudie (2010), using remote sensed imagery, reported numerous fields of shallow depressions in dryland areas worldwide underlain by limestone or caliche. He mainly attributes these depressions to dissolution (solution sinkholes) and designates them as dayas, which is the local name used for these dolines in northern Africa. The computed densities range between 0.88 and 11.44 depressions per km^2 and their distribution in some regions seems to be controlled by interdune swales (e.g., Weissrand Plateau, Namibia), where the bedrock is exposed or situated closer to the surface. Probably, some of these depressions located in areas with mean annual rainfall below 100–200 mm are relict features developed under more humid conditions in the past. Williams (1985) pointed out that solution sinkholes in temperate carbonate regions tend to be shallower than in tropical environments (cockpits), where they display greater relief and tend to reach much higher densities. These differences can be attributed to the greater intensity of the dissolution process in the humid tropical environments, because of greater effective rainfall and higher concentration of biogenic CO_2 in soils. Williams (1985) also identified the following unfavorable conditions for the development of solution sinkholes: (i) rocks with high vertical conductivity throughout the vadose zone that prevents the storage of water in a shallow suspended aquifer (e.g., aeolian calcarenites with high primary porosity), (ii) rocks with limited horizontal variations in permeability unsuitable for the development of preferred vertical leakage paths (e.g., nonfractured eogenetic reef limestones), and (iii) steep hillslopes characterized by lower infiltration, less effective vertical percolation, and higher mechanical erosion (Öztürk et al. 2018a). On slopes, solution sinkholes tend to display an elongated and asymmetric shape, oriented in the direction of the slope and with the deepest point displaced toward the downslope edge. This can be attributed to factors such as greater dissolution and mechanical erosion on the upstream side, which receives and interrupts the water flow from the uphill side, and the development of asymmetric draw-down cones in the water table of the epikarstic aquifer.

Solution sinkholes can completely pockmark the ground surface occupying all the available space and forming a landscape with an egg-box-like topography (Figures 6.19 and 6.22d–f). The topographic divides between the adjoining depressions display a cellular mesh pattern. This feature inspired the term polygonal karst, a type of karst landscape that was documented for the first time by Williams (1971, 1972a) in Papua New Guinea. Some authors use polygonal karst to describe areas with a high density of sinkholes, although the term should be restricted to landscapes in which the topographic catchment of the depressions covers the totality or most of the space (i.e., enclosed depressions share their edges). The divides typically display significant altitudinal variation, commonly defined by rounded ridges locally surmounted by protruding hills of various shapes (Sauro 1991). The depressions can be dissected by a centripetal system of gullies that focuses runoff toward the bottom of the basins, often with a sink (Williams 1972a). This drainage network, where present, imparts a characteristic star-shaped geometry to the contour lines. Polygonal karst landscapes essentially result from the development, growth, and coalescence of solution sinkholes, but it can also include polygenetic depressions such as solution sinkholes with nested subsidence sinkholes (e.g., cover and/or bedrock collapse). In carbonate rocks, polygonal karst has been documented in numerous humid tropical and temperate regions worldwide (Ford and Williams 2007). In contrast, this landscape also occurs in gypsum karsts located in semiarid regions, such as the remarkable case of the Sivas gypsum karst in Turkey (Doğan and Özel 2005; Doğan et al. 2019; Poyraz et al. 2021) (Figures 6.18g and 6.19).

From the hydrological perspective, the polygonal karst landscape constitutes an extremely efficient drainage system, in which the land surface is compartmentalized into a large number of microcatchments with centripetal drainage and a rather uniform distance between the divides and the sinks. Thus, rainfall is rapidly concentrated in the lower part of the depressions by overland flow, throughflow, and epikarstic flow and transmitted to the karst aquifer. Some authors compare these karst depressions characterized by intermittent, centripetal, and internal drainage with regular fluvial drainage basins (Day 1979). Williams (1972a), in his morphometric analysis of the polygonal karst of New Guinea, found a direct relationship between the area of the depressions and the rank of their stream network, given by the highest order following Strahler's (1957) scheme. This pattern is also observed in fluvial systems (Law of Drainage Areas; Schumm 1956).

Williams (1972b) proposed an evolutionary model for the polygonal karst of Papua New Guinea (Figure 6.24) that envisages an initial low-relief and subhorizontal surface underlain by limestone criss-crossed by a regular system of fractures. The formation of solution sinkholes is

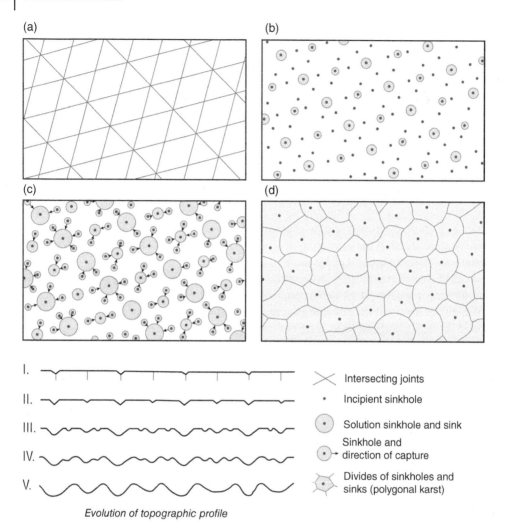

Figure 6.24 Evolutionary model proposed by Williams (1972a) for the polygonal karst of Papua New Guinea. Solution sinkholes and sinks form and grow at sites of greater permeability (intersecting joints). The larger and deeper sinkholes capture smaller sinkholes developed at less favorable sites. Sinkhole expansion and coalescence ultimately result in a karst landscape totally occupied by depressions that share divides with a polygonal pattern. *Source:* Figure adapted from Williams (1972a).

initiated at sites of greater fracturing and permeability (intersecting master joints). These depressions function as sinks focusing dissolution and encouraging the development of efficient vertical drainage paths. The growth of these sinkholes proceeds through a self-accelerating mechanism, since the enlargement of the catchments involves greater water flow, more rapid dissolution, and an increasing growth rate. Other solution sinkholes form at less favorable sites and grow at a slower rate. These smaller sinkholes are eventually captured by the deeper and more rapidly evolving depressions. These processes of sinkhole initiation, growth and coalescence ultimately result in a landscape totally occupied by depressions delimited by shared divides with a polygonal pattern. Subsequently, the topographic change is dominated by differential corrosional lowering, with greater dissolution in the central sector of the sinkholes (flow concentration, aggressiveness enhancement by biogenic CO_2) than in the divides (flow divergence), which experience limited variations in their position.

Ahnert and Williams (1997) developed a simplified three-dimensional process-response model to explain the formation of solution sinkholes and polygonal karst. The model assumes that the susceptibility of the bedrock to solutional lowering is enhanced at locations with greater permeability (e.g., fractures), and that runoff increases downslope, infiltrating at the bottom of the depressions. The different runs of the model that consider variable initial conditions and assumptions show that flow convergence is sufficient to explain the development of solution dolines and polygonal karst. The centripetal runoff simulated in the model somehow encapsulates the different mechanisms that contribute to increase water flow and dissolution in the central sectors of the depressions: (i) centripetal runoff, throughflow within the cover and epikarstic flow, (ii) boosted water aggressiveness by uptake of biogenic CO_2, and (iii) extra water supply by accumulation of windblown snow in cold regions. An additional relevant factor considered in the model is the lower dissolution rate that operates in the divides due to flow divergence. Morphometric studies carried out in multiple regions reveal that the mesh size of the polygonal network defined by the divides tends to be rather uniform, but show significant variations between areas. The regularity of the mesh size in each area can be attributed to two non-excluding factors: (i) competition for space by the growing sinkholes and (ii) the relatively uniform distribution of the high-permeability sites that govern sinkhole initiation (e.g., a nearly orthogonal and evenly spaced joint system). Differences in the mesh size and sinkhole density among the different regions can be explained by variations in the density of fractures and their openness and connectivity (Williams 1972a, 1993). In areas where dissolution is strongly controlled by a fracture set with a specific direction, the solution sinkholes can display markedly elongated geometry and tend to form well-defined alignments. Öztürk et al. (2017), in a polygonal karst landscape developed on a limestone plateau in the Taurus Mountains (Turkey), and underlain by a broad anticlinal structure, document markedly linear solution sinkholes with an average elongation ratio (length/width) of 14. These sinkholes appear to be controlled by a penetrative longitudinal joint set mainly associated with the axial zone of the fold (Figure 6.22f).

One of the main problems associated with the analysis of the evolution of polygonal karst landscapes, which are dominated by erosional landforms, is the difficulty of obtaining chronological information, either relative or numerical ages. Nonetheless, some areas offer the possibility of establishing age constraints. Ferrarese et al. (1998) documented the interesting case study of the Montello Hill, Southern Alps, Italy, underlain by conglomerates with a dominant carbonate composition and densely pitted by solution sinkholes. This gentle hill is a pop-up affected by active up-arching and displays several morphogenetic surfaces, including a stepped sequence of seven rock-cut terraces. The authors observe morphometric differences in the sinkholes depending on their geomorphic position (strath terraces, slopes) and the age of the different geomorphic surfaces. Sinkholes in older surfaces, due to their higher degree of development and coalescence, tend to reach larger areas and volumes and show greater areal density (i.e., percentage of area occupied by sinkholes). Williams (2004b) inferred that the polygonal karst of King Country, North Island, New Zealand (average rainfall 1618–2300 mm yr^{-1}) evolved during a relatively short time span of 1–3 Myr. The continuity in the development of solution sinkholes was repeatedly interrupted by the accumulation of pyroclastic deposits related to several volcanic eruptions, until the cover was stripped by erosion. The high density of depressions (ca. 55 depressions km^{-2}) and the small mesh size (ca. 160 m) of the polygonal landscape is attributed to the high fracture frequency in the carbonate bedrock.

6.3.3 Collapse Sinkholes

Collapse sinkholes are the morphological expression of the foundering of material above cavities. This mechanism can occur in bedrock (karst rocks), in non-karst rock or caprock underlain by cavernous bedrock (interstratal karst), and in nonconsolidated cover deposits (covered or mantled karst).

Dissolution is the main process responsible for the generation of subsurface cavities, but it can also foster collapse by reducing the mechanical strength and thickness of their roofs. For instance, the blowholes of the Australian Nullabor Plain, typically less than 2 m across, are ascribed to the intersection of solution cupolas in the roof of shallow caves by the gradual erosional lowering of the land surface (Doerr et al. 2012; Burnett et al. 2013). Collapse processes can contribute to the opening of these holes in the central part of the dome-shaped pockets when the roof reaches a critical thickness. Collapse sinkholes in their initial stages are characterized by a subcircular geometry in plan view and steep to overhanging edges (Figures 6.18 and 6.25). The length to depth ratio can be highly variable and often has values close to or below 1. The initial cylindrical morphology of collapse dolines can change progressively as the slopes degrade by various types of erosion processes and their bottoms are filled by detrital material, eventually attaining a bowl-shaped morphology similar to that of solution sinkholes. This morphological convergence can potentially lead to erroneous interpretations. Subsurface intrusive and/or non-intrusive methods such as trenching, boreholes, or geophysics may be necessary to unambiguously determine the origin of these dolines.

There are some distinctive features associated with collapse sinkholes that should be taken into account, some of them with important practical implications (Gutiérrez 2016): (i) The formation of cavities by subsurface dissolution is a slow process, whereas the collapse of cavity roofs up to the surface can occur in a very short time span. (ii) There can be a large time lag between the generation of the solutional cavities and the collapse processes responsible for the occurrence of sinkholes. The latter can be related to the collapse of relict cavities generated in the past under climatic, hydrological, and base-level conditions significantly different from the present ones. (iii) The volume of collapse sinkholes can be substantially lower than the volume of the underlying cavities due to the presence of unfilled voids and the bulking effect that affects the foundered material. In some situations, the opposite can also hold true when there is significant removal of breakdown material by mechanical and solutional erosion. (iv) A significant proportion of the diameter and area of mature sinkholes that have experienced considerable morphological modification can be related to mass wasting and erosion processes rather than to collapse. (v) Collapse sinkholes can form catastrophically and cause fatalities.

The collapse of cavity roofs consisting of rock or/and cover can occur via two principal modes. One of them is the progressive breakdown of the cavity roof through the detachment and fall of multiple blocks controlled by preexisting discontinuities and/or newly formed failure planes (Figure 6.26). This stoping process involves the upward propagation of the cavity roof and the

Figure 6.25 Images of collapse sinkholes. (a) Bedrock-collapse sinkholes in limestone that intersect the water table. Cañada del Hoyo sinkhole lakes, Iberian Chain, Spain. Largest depression is 220 m in length and contains a lake around 30 m deep. (b) Bedrock-collapse sinkhole 100 m long at Bsita, Al Jouf region, Saudi Arabia. Note concentric extensional fissures far from the sinkhole scarp (arrows) and disrupted drainage. (c) San Pedro caprock-collapse sinkhole 85 m in diameter and 108 m deep, including its 22 m deep lake, Oliete, Iberian Chain, NE Spain. This collapse has penetrated through a marl and limestone succession eventually disrupting a gully. (d) Caprock-collapse sinkhole developed in Cenozoic basalts underlain by cavernous limestone in Al Issawiah area, Al Jouf region, Saudi Arabia. (e) Recently formed cover-collapse sinkhole 16 m long in the Ebro Valley, close to La Puebla de Alfindén village, NE Spain. Geology students for scale. (f) Danvisky cover-collapse sinkhole, western Ukraine, formed on January 1998, above a hypogene gypsum cave. Arrow points to fault that offsets a buried soil exposed in the sinkhole wall. (g) Cover collapse sinkhole 44 m across and 47 m deep in the Hotamis Plain, central Anatolia, Turkey. This sinkhole cuts through deposits of the pluvial Konya Lake, underlain by cavernous limestone, and was induced by groundwater over-exploitation and the associated water table decline. Arrow points to geomorphologists for scale. *Source:* Francisco Gutiérrez (Author).

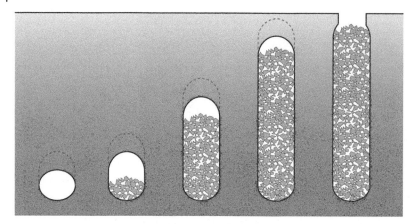

Figure 6.26 Sketch showing the upward propagation of a cavity by progressive roof collapse (stoping) and the generation of a collapse sinkhole at the surface. This stoping process generates a transtratal chimney-like pipe filled by breakdown material known as a breccia pipe. The initial solutional cavity occurs in soluble bedrock, but the upward propagation of the void by roof collapse can occur in bedrock, caprock, and cover. The color gradation reflects the increase in gravitational stresses with depth.

accumulation of a chaotic breakdown breccia below. Whether a stoping void can reach the surface or not depends on: (i) the volume of the cavity, (ii) the overburden thickness, (iii) the volumetric expansion associated with the breakdown process (bulking effect; Ege 1984; Andrejchuk and Klimchouk 2002), and (iv) the potential solutional or mechanical removal of the breakdown material (Lipar et al. 2019). The ascending collapse process may stop if the cavity roof and the breakdown pile meet. The bulking factor has a direct relationship with the degree of fragmentation and the reduction in size of collapse particles (comminution), with typical values within the range 0.2–0.4 (Benson and Yuhr 2015). The geometry of the failure planes may change as the cavity penetrates through different lithologies with variable mechanical characteristics and rheology.

Gravitational stresses deflect around cavities creating an arched compression zone (voussoir arch) underlain by a tension zone over the roof (Figure 6.27). This tension zone can control the development of fractures that typically attain the shape of a hemisphere or a blunted cone. Failures with more complicated geometries can occur in anisotropic material guided by preexisting planes of weakness such as bedding planes and fractures (Figure 10.9) (White 2019). The undermined material detached by these fractures eventually falls, generating cupola-shaped breakdown roofs (Figure 10.9). The roof of the cavity can propagate upward stepwise together with the associated compression and tension zones, and eventually intercepts the land surface creating a collapse sinkhole (Figure 6.26). The stability of the cavities largely depends on the ratio between the roof thickness and the cavity width (Al-Halbouni et al. 2018; Luu et al. 2019). Bed thickness is also an important parameter in stratified material. Moreover, the stability tends to increase as cavities propagate upward due to the decrease in gravitational stresses (Drumm et al. 2009; Shalev and Lyakhovsky 2012; Poppe et al. 2015).

The other collapse mode is the foundering of the cavity roof as a large coherent block with limited internal deformation (Gutiérrez et al. 2008a; Galve et al. 2015). This subsidence process is accommodated by the vertical displacement of blocks bounded by single or concentric ring faults with various geometries. Figure 6.28 shows some alternatives in which the annular faults have different geometries and the foundered blocks variable styles of internal deformation. In case A, arcuate failure planes controlled by compression arches determine the collapse of integral blocks with the geometry of a hemisphere or a blunted cone (Figure 6.29a, b). This type of failure leads to the

Figure 6.27 Sketch illustrating the development of a compression arch (light orange) and the underlying tension zone (grey) above a cavity roof by the deflection of gravitational stresses. S, span of the cavity; Tc, thickness of cavity roof; Tb, thickness of bed.

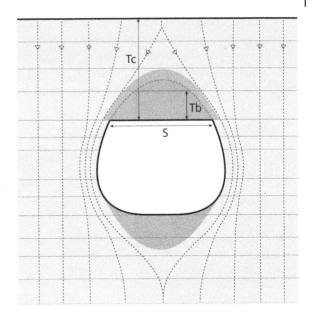

Figure 6.28 Collapse structures controlled by annular failure planes with different geometries that accommodate the foundering of large coherent blocks with limited internal deformation. See explanation in the text. df: drag folding; f: extensional fissure; b: downward bending; B: upward bulging; pf: pseudo-reverse faults.

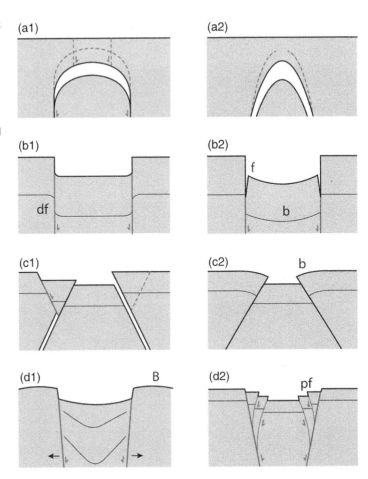

creation of concave-downward cavity roofs that thin toward the central part. New failures with variable geometries (e.g., vertical, outward dipping) can eventually occur restricted to the inner zone generating sinkholes with overhanging edges. In case B, the collapse is controlled by vertical and cylindrical faults (Figures 6.21b and 6.29c, g). The inner piston-like block can sink as a relatively rigid body with limited internal deformation causing shearing and drag folding in the strata

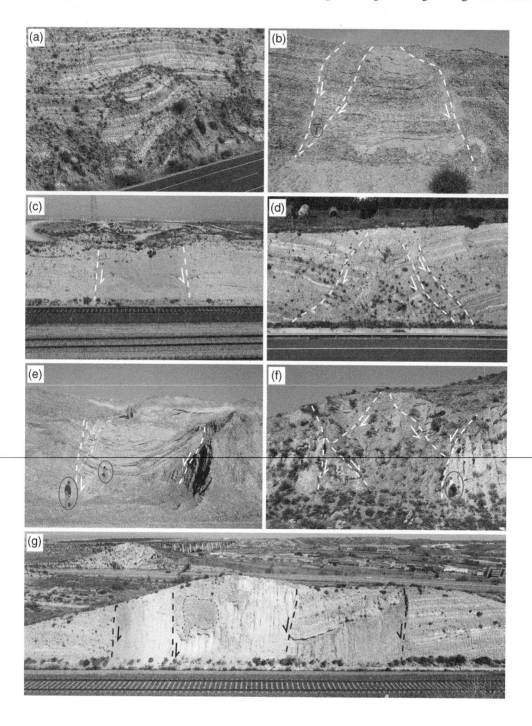

associated with the collapse fault (B1). The inner block can also experience some internal deformation by sagging (Al-Halbouni et al. 2018). This results in radial contraction, generating annular extensional fissures in the margin and reducing the mechanical coupling with the surrounding material (B2). In case C, block collapse occurs through the development of outward dipping ring faults with the geometry of a truncated cone. The collapse of the inner block tends to generate unstable overhanging scarps and creates void space beneath the wedge-shaped hanging wall. The undermined material at the margins can collapse through the development of secondary inward dipping faults (C1) and/or bend downward creating a marginal rollover and causing the inward rotation of the failure plane (C2) (Figures 6.29d). In C1, subsidence is accommodated by a conjugate system of outward and inward dipping faults, with the latter confined to the hanging wall of the former. Option D shows a less frequent geometry of inward dipping tronco-conical faults, which create a space problem for the downdropping block (Figure 6.29e, f). The inner block has to go through an annular space with progressively lower diameter. This space problem can be resolved by the disharmonic bending and contraction of the inner block creating a sheath fold, if it consists of soft and ductile material (D1) (Figure 6.21c). The foundered block experiences greater contraction as it penetrates through a progressively smaller section and causes a lateral push in the surrounding material potentially creating a marginal bulge (Sevil et al. 2020). Alternatively, the foundering of the block can be achieved through the development of secondary pseudo-reverse faults (oversteepened normal faults) at the margins that splay from the master normal fault (D2) (Figure 6.29e). These structures, when observed in outcrop, are very similar to those associated with normal tectonic faults (McCalpin 2009). However, collapse faults related to subsurface dissolution are shallow and spatially restricted gravitational structures with annular geometry and with no significant vertical offset across the collapse structures. Erroneous interpretations of these non-tectonic deformations have resulted in seismic hazard overestimates in some regions.

The type of subsidence mechanism in bedrocks and caprocks mainly depends on the geometry and size (span) of the cavity or dissolution zone, the thickness of the overlying rocks, and their mechanical resistance, largely determined by discontinuity planes. Cavities and karstification zones with large spans overlain by ductile rocks (low yield strength) tend to cause sagging subsidence, whereas competent rocks and cavities with limited span favor collapse by progressive stoping or the foundering of large coherent blocks. Deep cavities can propagate upward through the overlying rocks by progressive roof collapse (stoping) generating large breccia pipes, also called collapse chimneys or breakdown columns (Figure 6.26). These transtratal structures may reach

---

Figure 6.29 Images of exposures of collapse structures related to evaporite dissolution. (a) Hemispherical failure plane in Miocene gypsum, Calatayud, NE Spain. (b) Outward dipping failure planes with an overall blunted-cone geometry (upper part excavated). Terrace deposit of the Ebro River in Zaragoza city area, NE Spain. (c) Collapse controlled by vertical failure planes affecting Miocene gypsum and Quaternary pediment deposits. Note drag folding and/or sagging at the margins of the structure. Zaragoza city area, NE Spain. (d) Collapse block with some internal brecciation bounded by outward dipping failure planes. Note downward bending of undermined strata at the margins. Miocene gypsum in Calatayud, NE Spain. (e) Collapse controlled by vertical to inward dipping failure planes and associated pseudo-reverse faults. This is a synsedimentary paleosinkhole in a Quaternary gravel terrace filled by well-sorted fluvial sands with granules and fine gravel. Ebro Valley, NE Spain. (f) Paleosinkhole in a terrace deposit controlled by inward dipping master failure planes and secondary pseudo-reverse faults affecting sagged deposits with synformal structure. Villalba Baja, Alfambra River valley, NE Spain. (g) Bedrock collapse paleosinkhole controlled by nested steep to vertical ring faults. Dashed lines in the inner block indicate conduits in the strongly karstified gypsum bedrock filled by gravel derived from a terrace deposit that once stood at the surface. Zaragoza city, NE Spain. Source: Francisco Gutiérrez (Author).

hundreds of meters in vertical dimension, especially when they are rooted in thick evaporitic sequences (Warren 2006, 2016; Gutiérrez and Cooper 2013). Collapse breccias can be classified on the basis of the relative displacement of the clasts (crackle, mosaic, and chaotic), and whether they have a clast-supported (packbreccia) or matrix-supported texture (floatbreccia) (see Section 2.2.9.2; Kerans 1988; Loucks 1999; Warren 2016). Collapse breccias can be highly permeable and tend to act as preferential zones for groundwater flow and karstification when they are not cemented. They also have a high specific surface, favoring rapid mass depletion by dissolution and volume reduction (Gabrovšek and Stepišnik 2011; Hiller et al. 2014). Dissolution may gradually transform chaotic packbreccias into floatbreccias, consisting of corroded clasts embedded in a karstic residue (Gutiérrez et al. 2008a; Guerrero et al. 2013). This involves a volumetric reduction that may induce sustained or renewed subsidence. Obviously, the dissolution of breccias does not take place if they are derived from the breakdown of caprocks.

Bedrock-collapse sinkholes and caprock collapse sinkholes commonly occur catastrophically and are typically tens of meters in length (major axis). They have a very high damaging potential, but their probability of occurrence is typically much lower than that of the cover-collapse sinkholes. The degradation of the scarped slopes of bedrock-collapse sinkholes by mass wasting and erosion processes is commonly very slow. Nonetheless, old bedrock-collapse sinkholes can attain a bowl or truncated cone geometry by the erosional flattening of the slopes and the infill of their bottoms. This post-collapse morphological enlargement can be very significant and relatively rapid in gypsum bedrock, which is much more soluble and erodible than carbonate rocks (see Section 4.4). Gökkaya et al. (2021) analyzed the morphological evolution of bedrock collapse sinkholes in the Sivas gypsum karst (average precipitation $500\,\mathrm{mm\,yr^{-1}}$) applying the ergodic concept to an inventory of 295 depressions with a wide range of ages. Here, young bedrock collapse sinkholes display cylindrical geometry and major lengths of the order of tens of meters, constrained by the limited maximum size of the cavities that can be spanned by gypsum. In contrast, mature sinkholes show a tronco-conical geometry with greater average depth and major lengths typically above 200 m. The inferred morphological evolution indicates that great part of the volume of the mature sinkholes is related to the degradation of the side slopes by mechanical and chemical erosion and the solutional removal of large volumes of gypsum by downward vadose flows in these groundwater recharge features. Runoff and percolation water dissolve gypsum along the slopes and the deposits underlying the sinkhole floor, including collapse breccias. The solutional denudation process proceeds through a positive feedback mechanism that increases its geomorphic work as the area of the sinkhole grows. Caprock-collapse sinkholes have been reported in both evaporite (Cooper 1998; Gutiérrez et al. 2008a, 2016) and carbonate (Thomas 1974; Youssef et al. 2016) interstratal karsts overlain by various types of non-karst rocks such as sandstones, marlstones or basalts. Sinkholes in basalts may have a controversial interpretation since they could also be related to the collapse of lava tubes. Michelena et al. (2020) discuss the origin of a crater-like morphostructure 1.5 km long and 40 m deep that occurs in a 10 Myr basaltic plateau in Central Argentina and that was initially ascribed to an impact crater. The authors propose that it might correspond to a giant caprock-collapse sinkhole related to the subjacent dissolution of a volcanoclastic unit 40–85 m thick with a carbonate content above 60–90% and situated below the 100–150 m thick basaltic cap. An explosion crater related to a phreatomagmatic eruption is also considered as a likely option. The impact crater alternative is ruled out because of the absence of a continuous raised rim and the lack of high-pressure and -temperature minerals, impact melt/glass and textural features indicative of impact-related deformation.

The word cenote, derived from the Maya language, is used in the Yucatán Peninsula (Mexico) to designate steep-sided sinkholes related to the collapse of the roof of subhorizontal cave passages.

Figure 6.30 Images of sinkholes modified by slope degradation and infilling. (a) Mature bowl-shaped bedrock-collapse sinkhole 340 m long in Villar del Cobo, Iberian Chain, Spain. The slopes have been degraded mainly by frost-shattering which produces talus that accumulates on lower slopes and in the bottom of the sinkhole. *Source:* Francisco Gutiérrez (Author). (b) Cover-collapse sinkhole induced by groundwater pumping for irrigation, Al Jouf region, Saudi Arabia. Gullying in the friable sands at the sinkhole margin is causing rapid expansion of the depression. Note protection berm undermined by a retreating gully head. *Source:* Francisco Gutiérrez (Author). (c) Asagiekinli sinkhole in the Sivas gypsum karst. This 360 m long depression is an example of a mature tronco-conical bedrock-collapse sinkhole that has experienced significant expansion and probably deepening by mechanical and solutional denudation of gypsum. The doline has been partially truncated by the Kizilirmak River valley but remains as an internally drained depression. *Source:* Photo by Ergin Gökkaya. (d) Cover-collapse sinkhole in a mud flat of the Dead Sea that has undergone substantial enlargement by the development of retrogressive rotational slides in the margins, generating an irregular topography underlain by backtilted blocks. The sinkhole, originally ca. 17 m across, has expanded to reach 40 m in major length by the time the images were taken. See shaded relief model in Figure 6.33. *Source:* Francisco Gutiérrez (Author).

These caves are frequently flooded and hence the sinkholes constitute karst windows that provide direct access to the water table (Figure 6.18c). The development of the cave levels and the cenotes has been largely governed by repeated fluctuations in the water table as large as 130 m associated with glacio-eustatic sea-level changes. The caves developed before the present sea-level highstand as indicated by the abundance of drowned vadose speleothems (Beddows 2004; Smart et al. 2006). Sea level drops during glacial periods favored cave-roof collapse by the removal of buoyant support, while sea-level rise in the current interglacial led to the inundation of passages previously situated in the vadose zone. In northwest Yucatán, there is a 180 km-diameter semicircular belt with a high density of cenotes that coincides with the buried Chicxulub impact crater occurred at the K-Pg boundary. This is most probably a band with greater permeability and susceptibility for cave development and collapse (Perry et al. 1995). Cenote-like bedrock-collapse sinkholes are found in numerous low-lying coastal karst regions worldwide. The term blue hole is mostly used for collapse sinkholes situated in the sea floor (i.e., submerged cenotes) that may be as much as 125 m deep and 300 m across (Figure 6.31). These are common features in recent eogenetic reef limestones in the Caribbean and the Great Barrier Reef and their origin, similarly to cenotes, is tightly linked to sea-level fluctuations during the Pleistocene (Gascoyne et al. 1979; Mylroie et al. 1995; Mylroie 2004, 2019; Gischler et al. 2008). Locally, they seem to be controlled by gravitational faults developed along the steep margins of carbonate platforms and reefs, following the removal of buoyant and lateral support during sea-level lowstands. In the eastern Grand Bahama, these fractures guide the development of vertical caves during periods of eustatic fluctuations, and eventually lead to the occurrence of collapse sinkholes that provide access to the drowned caves for divers (Palmer and Heath 1985). Similar fractures with important speleogenetic role have been documented by Koša and Hunt (2006) in the Guadalupe Mountains.

Bedrock-collapse sinkholes are commonly less than 100 m in diameter, but they can reach hundreds of meters in diameter and depth. The largest known collapse sinkholes are the tiankengs of southeast China, meaning sky hole or heaven pit (Zhu and Waltham 2005; Zhu and Chen 2006; Zhu et al. 2019) (Figures 6.32a and 9.26c). These are giant dolines more than 100 m wide and deep that typically form in fengcong terrain above active cave streams with high erosional capability. They are characterized by subvertical to overhanging walls, floors largely covered by breakdown and rock-fall deposits, and commonly lack any relationship with the surface karst landscape; they can cut through hills and solution dolines. Zhu and Chen (2006) published a compilation of 50 tiankengs, three of them with lengths and depths greater than 500 m.

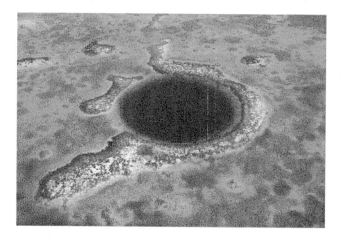

Figure 6.31 The Great Blue Hole of Belize, a UNESCO World Heritage Site. It is a submerged collapse sinkhole around 320 m wide and 125 m deep. *Source:* Photo by Andrew Hounslea.

The largest is Xiaozhai tiankeng (49R 353470E, 3 403 003 N) in Chongqing Province, which is 635 m long and 535 m wide, has a depth of 511 m from the lowest point of the rim and an estimated volume of 120 Mm^3 (equivalent to a cube with an edge of 230 m) (Figure 6.32a). This giant bedrock collapse sinkhole is traversed by a powerful and steep cave river with a gradient of 4.3% (Senior 2004; Zhu and Chen 2006). The deepest bedrock collapse sinkhole in the Dinaric karst is Crveno Jezero (Red Lake; 33 T 677921E, 4 813 649 N), located at the margin of Imotski Polje in Croatia (Figure 6.32b). This pit is more than 400 m across and reaches a depth of 518 m including the underwater portion of its lake, surrounded by vertical cliffs around 250 m high. The bottom of the lake is situated at 6 m below sea level and its depth fluctuates seasonally between approximately 315 and 250 m. Explorations carried out with autonomous underwater vehicles revealed the presence of lateral submerged passages at different depths and active water circulation across the pit close to the bottom (Bonacci and Roje-Bonacci 2000; Garašić 2012).

Waltham (2005) provided an inventory of collapse sinkholes larger than 100 m wide and deep from outside China. Regardless of the diverse local terminology, giant bedrock and caprock collapse sinkholes seem to be associated with specific conditions, some of which include: (i) Cave systems carved in massive and highly resistant limestones that can span large chambers (e.g., Mulu karst in Malaysia, Waltham 2005). (ii) Bedrock collapse in hydrologically active caves that are capable of removing substantial amounts of breakdown material by mechanical and chemical erosion (e.g., tiankengs in China, Zhu and Chen 2006; giant sinkholes in the Nakanai Mountains of New Britain and those associated with the Gruta do Janelão in Brazil, Waltham 2005). The erosion of collapsed material by cave streams explains why the volumes of the largest collapse dolines can be more than an order of magnitude greater than that of the largest underground chambers; Miao Room in China and Sarawak Chamber in Mulu, Malaysia have approximate volumes of 10.57 and 9.58 Mm^3, respectively. According to Palmer and Palmer (2005), in the Chinese tiankengs the breakdown material increases the hydraulic gradient in the cave stream enhancing the mechanical and solutional erosion capability of the flow. Moreover, the piles of rubble can block and divert the flow creating lateral passages that undermine the margins of the sinkhole inducing their expansion. Overall, a great part of the volume of tiankengs is related to the removal of collapse material by the underground stream, rather than to the initial void space created by subsurface dissolution in the bedrock. (iii) Collapse of deep isolated chambers generated by hypogene dissolution in limestone related to rising CO_2 and H_2S gases of volcanic origin (e.g., El Zacatón, Mexico, second deepest underwater shaft in the world, >329 m, Gary and Sharp 2006; obruks of central Anatolia, Bayari et al. 2009; Doğan and Yilmaz 2011; Figure 6.32c). (iv) Deep-seated hypogene dissolution of evaporite rocks overlain by thick caprocks (Sant Miquel de Campmajor Valley, NE Spain, Gutiérrez et al. 2019b). (v) Caprock-collapse sinkholes that puncture monoclinal scarps atop downdip migrating dissolution fronts in evaporites, in which large mass depletion can occur by rapid subjacent dissolution (Interior Homocline of Saudi Arabia, Memesh et al. 2008, Warren 2016; Bottomless State Park in New Mexico, Land 2003; Figure 6.32d). (vi) Mature depressions that have experienced substantial lateral and even vertical growth by mechanical and solutional erosion (e.g., Sivas gypsum karst, Gökkaya et al. 2021, Figure 6.30c; Doline Pozzatina, Gargano, Italy, Waltham 2005; sinkholes in the Iberian Chain, Spain; Gutiérrez et al. 2005, Figure 6.30a). Dissolution-induced paleocollapse structures (breccia pipes and foundered coherent blocks) hundreds of meters across and in vertical dimension have been extensively documented in the stratigraphic record (see review in a paper on large paleocollapse structures in Gozo, Malta by Galve et al. 2015). These features provide valuable information about the subsidence mechanisms that occurred beneath the surface and in some cases are useful pointers for ore and hydrocarbon exploration.

Figure 6.32 Images of bedrock-collapse sinkholes more than 100 m across. (a) Xiaozhai giant bedrock collapse sinkhole in the Xinlong fengcong karst, Chongqing Province, China. This is the largest known tiankeng, 635 m in length and 511 m deep from the lowest point of the rim. *Source:* Photo by Weihai Chen. (b) Crveno Jezero (Red Lake) in Croatia is a bedrock-collapse sinkhole around 400 m long in the upper part and 518 m deep, including its more than 250 m deep lake. *Source:* Francisco Gutiérrez (Author). (c) Kizoren obruk in central Anatolia, Turkey is a bedrock-collapse sinkhole 650 m long, 500 m wide and around 100 m deep. This doline has been attributed to the collapse of caverns created by hypogene dissolution of Pliocene limestone. The overhanging ledge in the middle part of the scarp is a shelfstone-type subaqueous bulbous tufa deposit accumulated before the recent lowering of the lake level (ca. 20 m). *Source:* Francisco Gutiérrez (Author). (d) Caprock-collapse sinkholes with lakes fed by underwater artesian springs in the Bottomless State Park, eastern margin of the Pecos River valley, New Mexico. Ascending water causes dissolution of evaporites, generating a monoclinal escarpment atop a dissolution front and alignments of caprock-collapse sinkholes. The largest sinkhole is around 170 m in length. *Source:* Image captured from GoogleEarth.

The development of cover-collapse sinkholes is commonly related to cavities rooted at the cover–bedrock interface and developed by downward internal erosion of surficial deposits through voids associated with the rockhead (Beck 1988) (Figure 6.10, left). The subsurface bedrock cavities are usually related to the solutional widening of discontinuity planes and their intersections (e.g., cutters), but can also correspond to other types of porosity such as the interparticle space of collapse breccias. When the undermined cover is cohesive and dominated by brittle rheology, arched failures tend to develop in the mantling deposits. The fallen material is accumulated on the floor of the cavity or is evacuated through the voids in the bedrock. Upward propagation of the soil cavities by successive roof collapse ultimately leads to the formation of cover-collapse sinkholes. Cover collapse can also occur by the downward movement of coherent blocks bounded by ring faults that extend up to the surface (Figure 6.29). The presence of a resistant material such as a duricrust or an artificial pavement favors the formation of larger sinkholes by the lateral expansion of the stoping cavities. Soil cavities can penetrate through thick cover, provided that there is sufficient volume of unfilled cavities in the bedrock to accommodate the collapsed material and its volumetric expansion (Doğan and Yilmaz 2011; Heidari et al. 2011; Taheri et al. 2015). Gutiérrez et al. (2008a) document breakdown chimneys tens of meters high in thick alluvium underlain by evaporites in NE Spain, filled by collapse deposits with shear fabrics at the margins. Cover-collapse sinkholes tend to form catastrophically and are usually less than 10 m across. The mechanics of the upward propagation of soil cavities has been analyzed in a number of works by analytical techniques (Tharp 1999; Augarde et al. 2003; He et al. 2004; Al-Halbouni et al. 2018, 2019) and physical laboratory models (Craig 1990; Abdulla and Goodings 1996; Lei et al. 2002; Poppe et al. 2015; Soliman et al. 2018; Luu et al. 2019; Xiao et al. 2020). Xiao et al. (2020) illustrate with physical models the potential impact of vacuum pressure induced by rapid water-table falls in mantled karsts with a low-permeability cover. They document how the magnitude, duration, and rate of the water-table decline can influence subsidence mechanisms and rates.

According to Shalev and Lyakhovsky (2012), in covers with a ductile component of deformation, surface collapse is preceded by slow ground settlement (creep deformation; Figure 6.33), whereas in highly cohesive and rigid cover, sinkholes typically occur without any precursory surface settlement. These factors have relevant implications for the potential anticipation of catastrophic collapse sink-

Figure 6.33 Incipient cover-collapse sinkhole 13 m across in soft lacustrine muds showing ductile sagging and marginal extensional fissures as precursory geomorphic indicators of collapse, Dead Sea, Israel. Insets are a shaded relief model showing the sag and an adjacent compound sinkhole that has experienced substantial expansion by slumping (shown in Figure 6.30d) and an orthoimage of the circular sag and the fissures. *Source:* Francisco Gutiérrez (Author).

holes and the implementation of early-warning systems. Cover-collapse sinkholes can also experience a long history of episodic activity, with prolonged periods of quiescence punctuated by events of sudden collapse. This kinematic style and information about the collapse events (displacement per event, timing, and recurrence) can be inferred via trenching. Discrete collapse events can be recorded by colluvial wedges shed from rejuvenated scarps, sedimentary packages with contrasting sedimentological features, and various geometrical relationships such as angular unconformities, upward fault truncations and nested collapse blocks or fissure fills (Carbonel et al. 2014, 2015; Youssef et al. 2016; Gutiérrez et al. 2017, 2018; Sevil et al. 2017). Collapse sinkholes developed in cover deposits can expand very quickly by mass wasting and water erosion, increasing their damaging potential (Figure 6.30). This can be prevented or decelerated by artificial filling. For instance, the collapse sinkholes that are currently developing in the emerged coastal areas of the Dead Sea underlain by mud deposits can experience very rapid expansion by retrogressive slumping. Great part of the area of some of those sinkholes is related to landsliding rather than to collapse (Figures 6.30d and 6.33). Cover-collapse sinkholes, despite their limited size, are responsible for the vast majority of the damage in most karst areas because of their high spatial–temporal frequency and their sensitivity to human influences (Waltham et al. 2005; Gutiérrez et al. 2007; Galve et al. 2011; Lei et al. 2015; Gutiérrez 2016). Other classifications designate cover-collapse sinkholes as dropout sinkholes (Williams 2004a; Waltham et al. 2005), although this term is not commonly used in the scientific literature.

6.3.4 Suffosion Sinkholes

Suffosion refers to the downward migration of unconsolidated cover material through subsurface voids in the bedrock and the simultaneous progressive subsidence of the soil mantle and the land surface (Figure 6.34). The subsurface voids through which the cover material is evacuated mainly correspond to covered or subsoil karren features, such as solutionaly enlarged discontinuity planes (cutters), cavernous karren, and karren wells and shafts (see Section 6.2.2). These voids associated with the rockhead can connect with larger voids and caves at some depth and hence may be able to absorb large volumes of surficial deposits by internal mechanical erosion. The drains developed in the epikarst that function as subsurface channels for sediment transport can have multiple orientations, geometries and connections. In outcrop, depending on the relative orientation of the dissolution features and the exposed face, they can be observed as vertical or inclined cutters, or in transverse sections as subcircular bodies of cover material surrounded by bedrock or residual material derived from dissolution of the bedrock (Figure 4.16). The internal erosion of the undermined cover may take place through a wide range of processes, collectively designated with terms such as suffosion, raveling, piping, or subsurface mechanical erosion. These processes include: (i) downwashing of particles by percolating waters (seepage erosion), which can be accompanied by the selective removal of finer particles (elutriation) conditioned by the competence of the flow and the size of the void space through which the water flows, (ii) cohesionless granular flow, like the downward movement of sand in a hourglass, (iii) non-Newtonian viscous sediment gravity flow of clayey deposits (Jancin and Clark 1993; Shalev and Lyakhovsky 2012), (iv) fall of particles, and (v) sediment-laden turbulent water flow. Gravitational forces and downward water percolation in the vadose zone are generally the main agents responsible for the internal erosion of the undermined cover. Water percolation in the karst bedrock beneath the cover tends to focus at preferred leakage paths, commonly related to fissures that have experienced greater solutional widening. These are the main leakage paths for the surficial aquifer and commonly the main channels for the internal erosion of the mantling deposits. During recharge events (e.g., heavy rainfall) these drains associated with the rockhead can create a draw-down cone in the water table of the temporary

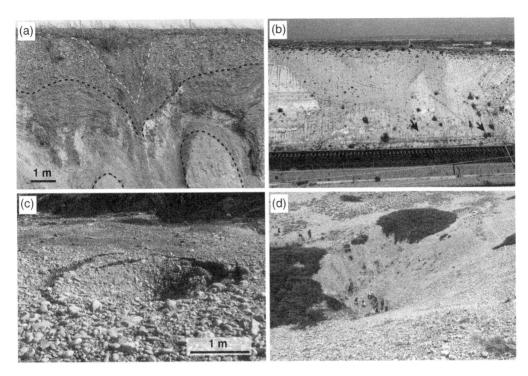

Figure 6.34 Cover-suffosion paleosinkholes and sinkholes. (a) Cover-suffosion paleosinkhole developed in a gravel terrace underlain by a clayey karstic residue and gypsum bedrock. Black dashed lines indicate the rockhead and the alluvium-residue contact. Jarama River valley, Madrid Cenozoic Basin, central Spain. See general view of exposure in Figure 4.16. (b) Cover-suffosion paleosinkhole in alluvium overlying Miocene gypsum with dissolutional conduits. Arrows indicate conduits that functioned as subsurface channels for the internal erosion of the cover deposits. High-speed Madrid-Barcelona railway in Zaragoza area, NE Spain. (c) Funnel-shaped suffosion sinkhole with extensional fissures along the margin on the floor of a river channel. Cardona salt diapir, NE Spain. (d) Suffosion sinkhole nested in the bottom of a mature bedrock collapse sinkhole. Villar del Cobo, Iberian Chain, Spain. The active suffosion sinkhole is related to the downward raveling of cohesionless scree produced by frost-shattering through subsurface voids. General view of sinkhole in Figure 6.30a. Students for scale. *Source:* Francisco Gutiérrez (Author).

suspended aquifer, focusing subsurface flow (Williams 1983, 1985). The resulting centripetal flow pattern tends to increase the flow rate along the subsurface drains enhancing both internal mechanical erosion and solutional widening through positive feedback mechanisms. The increase in the drainage capacity of the drains leads to the steepening of their gradient. Water and sediment flux in these conduits may eventually be inhibited by the accumulation of cohesive and low-permeability clay deposits.

Cover-suffosion sinkholes essentially develop when the undermined mantling deposits deform as a loose granular material or as a soft viscous layer without the development of persistent failure planes. The resulting gradual subsidence in the land surface produces funnel- or bowl-shaped sinkholes commonly less than 10 m across (Figure 6.34c, d). The subsurface erosion of a cohesionless sand cover would produce a funnel-shaped sinkhole with the angle of repose or internal friction angle of the material. A soft clay-rich deposit can flow downward by viscous deformation generating a bowl-shaped depression underlain by a sheath fold with centripetal dips. In pure cover-suffosion sinkholes, the deformation is spatially continuous and affects simultaneously the whole thickness of the mantling deposits. In contrast, in cover-collapse sinkholes the deformation is spatially discontinuous in as much as it is related to the development of discrete failure planes, and typically propagates toward the surface by episodic roof collapse (stoping). The development of suffosion sinkholes is inhibited by the presence of covers with large thickness and some degree of induration. Early diagenetic processes such as cementation or compaction contribute to increase the cohesion and yield strength of the deposits favoring brittle deformation and collapse. For instance, in a fluvial valley with a stepped sequence of terraces, the cohesionless or soft alluvium in the floodplain and low terraces may be prone to suffosion subsidence, whereas collapse may be the dominant process in older terraces with more competent deposits that have undergone some cementation. In nature, a complete spectrum between suffosion, collapse, and sagging sinkholes can be found in covered karst settings.

6.3.5 Sagging Sinkholes

Sagging refers to the progressive downward flexing of material (passive bending) induced by subsurface mass depletion and lack of basal support resulting from dissolution. This subsidence mechanism dominated by ductile deformation can occur in any kind of material (cover, bedrock, and caprock) and is generally related to evaporite dissolution (Gutiérrez et al. 2008a, 2016; Guerrero et al. 2013; Al-Halbouni et al. 2017) (Figures 6.35–6.37), although it has also been documented in carbonate karst terrains (Frumkin et al. 2015; Zhang and Chen 2019 and references therein). The common association of this type of subsidence process with evaporitic bedrocks can be attributed to several factors: (i) Dissolution tends to be more widespread (e.g., migration of dissolution fronts), so that the removal of mass and basal support commonly affects large areas favoring the ductile deformation of the overlying sediments. Dissolution is frequently focused along specific beds or units within the stratigraphic succession because of their higher solubility and/or permeability (e.g., halite or glauberite interbedded within gypsum or anhydrite). The solubility contrast between different units within evaporitic formations can be extremely high. (ii) The dissolution at the cover–bedrock interface (rockhead) in evaporites is significantly more rapid than in carbonate rocks. This process can contribute to the differential subsidence and ductile deformation of the overlying cover in relatively short time spans. (iii) Evaporite rocks have a more ductile rheology than carbonate formations. For instance, Young's modulus for gypsum is commonly three times lower than that of limestone (Waltham 1989). Two important features of sagging sinkholes with relevant practical implications are that subsidence occurs progressively, and that their development does not require the formation of subsurface cavities. The solutional removal of bedrock in the subsurface and the associated subsidence of the overlying material can occur simultaneously and at equivalent rates.

Sagging produces structural basins with centripetal dips in the sediments overlying the karstification zone. These structures are expressed as synforms in two dimensions, as is commonly observed in exposures (Figure 6.35). Sagging results in a peculiar disharmonic style of deformation, consisting of folded sediments overlying non- or less-deformed units. This is an apparently aberrant geometrical situation with greater strain in the younger sediments (Figures 6.36c and 6.37a). The simple explanation for this discordant structure is that deformation is related to a shallow gravitational process (i.e., non-tectonic dissolution-induced subsidence). Sagging of beds involves horizontal shortening; the length of a horizontal bed is greater than the horizontal length of the sagged bed. In three dimensions, the material affected by sagging experiences radial contraction, which can be counterbalanced in different ways depending on the rheology of the material: (i) layer-parallel lengthening and vertical contraction by continuous deformation (e.g., relative displacement between particles in detrital sediments), (ii) layer lengthening by dilation in widely distributed preexisting or newly formed joints (e.g., densely fractured rock) (Gutiérrez et al. 2018),

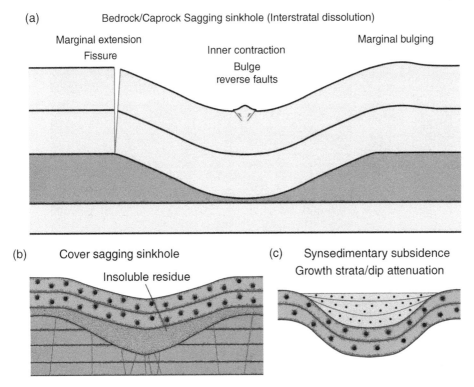

Figure 6.35 Sections of sagging sinkholes. (a) Bedrock- or caprock-sagging sinkhole related to differential interstratal dissolution of a soluble bed. Sagging involves radial contraction that can be counterbalanced by marginal extension. Marginal bulging may also occur due to moment stresses acting on elastic beds when there is no mechanical decoupling between the subsidence and peripheral areas. Contraction in the unconfined inner arc of the basin structure may cause local bulging and the development of moment-bending reverse faults. Note overall disharmonic structure, with greater deformation in the younger beds. (b) Cover-sagging sinkhole generated by differential solutional lowering of the rockhead, resulting in the development of an insoluble residue with thickness proportional to the amount of bedrock dissolved. Note the disharmonic deformation style. (c) Sedimentary fill in a sagging sinkhole deposited coevally to the subsidence phenomenon (synsedimentary subsidence). Note cumulative wedge-outs (growth strata) and upward dip attenuation in the synkinematic beds. Lower beds affected by postsedimentary subsidence do not show thickness changes across the sinkhole.

or (iii) development of extensional structures at the margins of the sagging basin, including concentric fissures, synthetic, and antithetic normal faults and keystone grabens (e.g., competent sediments) (Gutiérrez et al. 2011; Carbonel et al. 2013, 2014) (Figure 6.35a). The progressive steepening of the beds as sagging proceeds may lead to inward rotation of extensional faults acquiring the attitude of pseudoreverse faults. Zhang and Chen (2019) performed a numerical simulation of a sagging sinkhole developed above a shallow cavity in limestone during the construction of a railway embankment. The results of the simulation, considering the strain-softening behavior of the limestone observed in laboratory tests, revealed that the bedrock at the two ends of the cavity span yields from the top downward due to tensile load. As sagging proceeds, the central sector of the span yields by tension from the bottom to the top, leading to the formation of an inner collapse (i.e., bedrock-sagging sinkhole that evolves into a bedrock-collapse sinkhole). The authors attribute the observed and modeled sagging deformation to the strain softening behavior of the limestone that occurs between the elastic and residual states, as measured in laboratory tests.

Figure 6.36 Exposures of sagging sinkholes related to evaporite dissolution. (a) Caprock-sagging paleosinkhole generated by interstratal dissolution of salt-bearing evaporites and the consequent ductile bending of an overlying clay and marl succession with limestone intercalations. Miocene sediments of the Madrid Cenozoic Basin, Cañada Real, Spain. Circle indicates person for scale. (b) Sagging structures in Quaternary alluvium (mantled pediment deposit) underlain by Miocene evaporites. Ebro Valley in Zaragoza area, NE Spain. (c) Cover-sagging sinkhole related to the differential corrosional lowering of the rockhead in the gypsum bedrock. From base to top horizontal Miocene gypsum strata (G), insoluble residue (IR) with pseudostratification, and gravelly Quaternary alluvium (A) showing a synform offset by a small-throw collapse fault. Jalón River terrace in Calatayud Graben, Iberian Chain, Spain. (d) Cover-sagging and –collapse paleosinkhole in a Quaternary terrace. Note synform with inner collapse structure. Dashed lines indicate location of collapse faults. Fine-grained facies in the core of the synform most probably record synsedimentary subsidence and development of a low-energy depositional environment associated with the subsiding sinkhole in the alluvial plain. *Source:* Francisco Gutiérrez (Author).

Sagging of beds can be also accompanied by contraction in the upper and central part of the flexed layers, as well as extension in the underlying outer arc, both zones being separated by a neutral surface. The contraction in the upper part of the core of the basin, where the beds lack confining pressure (overburden load), can be accommodated by buckling folding and the development of small reverse

Figure 6.37 Internal structure and geomorphic expression of sagging sinkholes related to evaporite dissolution. (a) Cover- and bedrock-sagging and -collapse paleosinkhole exposed in the cuttings of the high-speed Madrid-Barcelona railway in the Ebro River valley, NE Spain. Note synform with inner collapse structure in the gypsum bedrock overlain by thick insoluble residue (IR) with pseudostratification and a small remnant of terrace alluvium (A). Arrow on the left points to outward dipping normal fault (pseudoreverse faults). Arrow on the right points to evidence of interstratal dissolution along specific beds (preexisting glauberite?). Note higher dip of layers above the karstification zone. (b) Cover sagging sinkholes in a terrace deposit filled by fine-grained palustrine facies (including fossils of aquatic fauna). The upward dip attenuation and the thickening of beds toward the central part of the synform records synsedimentary subsidence. Sagging subsidence eventually evolved to rapid collapse as revealed by marginal collapse faults and associated liquefaction structures. Terrace of the Jalón River in Calatayud Graben, Iberian Chain, NE Spain. (c) Bedrock-sagging sinkhole in "Ar" Ar, Saudi Arabia generated by interstratal dissolution of gypsum beds within the Cretaceous Badanah Formation. Note the dip of the strata toward the center of the depression (locally known as khabra). (d) Shippes Bowl sag is a ca. 2 km long caprock-sagging sinkhole related to interstratal dissolution of Carboniferous salt-bearing evaporites and the gradual passive bending of the overlying caprock of late Cenozoic basalts. Carbondale Collapse Center, Southern Rocky Mountain of Colorado, USA. *Source:* Francisco Gutiérrez (Author).

faults (moment-bending faults) (Figure 6.35a). The resulting local relative uplift has been documented in some active sinkholes by geodetic techniques (Desir et al. 2018). Extension in the outer arc can generate and dilate fractures. The greater amount of deformation and fracturing that commonly occurs in the central part of the basins (hinge zone in 2D) can create zones of preferred groundwater flow and dissolution. Flexing in stiff layers and paved surfaces can transfer moment stresses to the margins of the sagging basins and produce bulges by elastic flexure (Figure 6.35a). This marginal bulging has been documented by high-precision leveling in active sinkholes located in urban areas (Kersten et al. 2017; Desir et al. 2018; Gutiérrez et al. 2019a; Sevil et al. 2020). The subsidence zone at the surface can cover a larger area than the karstification zone or cavity at depth. This deviation is determined by the so-called angle of draw, which is the angle of inclination from the vertical of the line connecting the edge of the cavity or karstification zone and the edge of the subsidence area (NCB 1975). Progressive sagging subsidence in active depositional environments creates sediment traps and constantly renewed accommodation space. Deposition coeval to subsidence is recorded in the sinkhole fill deposits by stratigraphic arrangements known as growth strata or cumulative wedge outs, which display bed thickening toward the center of the basin and upward dip attenuation (Figures 6.35c and 6.37b).

Sagging in bedrock and caprock seems to be mainly associated with evaporitic formations including high-solubility units such as halite or glauberite. Interstratal dissolution of those beds in specific zones causes the sagging of the overlying soluble and/or non-soluble rocks, plus any existing cover. The sagging process is favored by the large span of the stratigraphically controlled karstification zones, the ductile rheology of the evaporites, and frequently also the presence of marl and clay partings interbedded within the evaporite formation. The alternation of competent and incompetent beds facilitates sagging by the flexural-slip mechanism (bedding-parallel slippage) (Gutiérrez et al. 2008a) (Figure 3.46). When sagging of bedrock is widespread and related to differential interstratal dissolution, it can produce a peculiar basin and dome morphostructure (egg-carton type structure), which is expressed as adjoining synforms and antiforms in cross-section (Gutiérrez and Cooper 2013) (Figures 6.36b and 8.1c). This type of morphostructure has been described in the Interior Homocline of Saudi Arabia (As Sulayyil area), related to the intrastratal dissolution of the anhydrite/gypsum units of the Late Jurassic Hith and Arab formations (Dini et al. 2009; Youssef et al. 2016) (Figure 8.3d). Excellent examples of bedrock sagging sinkholes and paleosinkholes have been documented in the central sector of the Ebro Basin associated with subhorizontally lying Miocene evaporites consisting of Ca-sulfates, glauberite, and halite (Gutiérrez et al. 2008a; Guerrero et al. 2008a, 2013) (Figure 6.37a). Here, interstratal dissolution of glauberite beds has produced large enclosed depressions. Perhaps surprisingly, the La Loteta Reservoir in the Ebro River valley is located in a 6-km long subsidence depression generated by deep-seated interstratal dissolution of halite (Gutiérrez et al. 2015). Artificial excavations in these basins reveal that the gypsum bedrock is affected by sagging structures related to interstratal dissolution. Locally, the sagging of the evaporite bedrock is accompanied by fracturing and the associated increase in permeability. The progressive deformation and dissolution of the fractured beds by preferred groundwater circulation initially leads to the formation of stratiform breccias with rough stratification and limited relative displacement among the blocks (crackle breccia). As dissolution progresses these packbreccias degrade to mosaic breccias, and then into chaotic packbreccias and floatbreccias made up of corroded clasts embedded in a fine-grained residue. In these large bedrock, sagging and collapse sinkholes both interstratal dissolution of salt beds plus the dissolution of the overlying deformed bedrock contribute to surface subsidence. A common situation found in this area is the superposition of bedrock collapses in the hinge zone of the sagging structures (Figure 6.21b). A likely explanation is that fracturing that occurs preferentially in the hinge zone of basins focuses groundwater flow leading to the formation of cavities and transtratal collapse structures (Guerrero et al. 2013). This sequence of subsidence mechanisms has been observed in gypsum caves of Russia, where sagging of cavity roofs preceded and accompanied collapse failure (Andrejchuk and Klimchouk 2002). Outstanding examples of caprock-sagging paleosinkholes have been exposed in the trenches excavated for the construction of transportation infrastructure in the Madrid Cenozoic Basin, central Spain. Here, highly ductile Miocene clays and marls with some limestone intercalations display sagging and collapse structures related to the interstratal dissolution of an underlying salt-bearing evaporitic unit (Gutiérrez et al. 2008b) (Figures 6.21c and 6.36a). In the Southern Rocky Mountains of Colorado, Kirkham et al. (2002) document hectometer- to kilometer-scale caprock-sagging depressions generated by interstratal dissolution of salt-bearing Carboniferous evaporites capped by Late Cenozoic basalts (Figure 6.37d). In the Judayyidat "Ar" Ar area of Saudi Arabia, extensive tracts of land are riddled by caprock- or bedrock-sagging sinkholes (locally known as khabras) generated by interstratal dissolution of gypsum within the Cretaceous Badanah Formation (Dini et al. 2007; Youssef et al. 2016). These basins showing centripetal dips in their margins were initially interpreted as solution sinkholes (Figure 6.37c).

In mantled karst settings, the differential solutional lowering of the top of the bedrock (rockhead) is accompanied by the gradual subsidence of the cover by slow vertical displacement and sagging. In areas with evaporitic bedrock, this subcutaneous dissolution process can lead to the development of a relatively thick residual material mainly consisting of clays and marls (Figures 4.16, 6.35b, and 6.36c). The thickness of this karstic residue sandwiched between the cover and the relatively unweathered bedrock is proportional to the amount of evaporite rock dissolved and its purity. Interestingly, karstic residues associated with the rockhead in gypsum commonly show a pseudostratification roughly concordant with the irregular geometry of the rockhead (Gutiérrez et al. 2008a). The resulting cover-sagging sinkholes are characterized by diffuse and difficult to map edges, with lengths that may reach hundreds of meters and very high length to depth ratios (Gutiérrez et al. 2007, 2018). Sagging in cover deposits is commonly accompanied by the development of collapse faults (cover-sagging and -collapse sinkholes). Trenching, together with geochronological data and retrodeformation analyses, allow determination of the relative contribution of the different mechanisms to the overall subsidence and assessment of long-term subsidence rates (Gutiérrez et al. 2009, 2018; Fabregat et al. 2017, 2019).

6.3.6 Factors that Control Subsidence Sinkholes and Human-Induced Sinkholes

Subsidence sinkholes, including those related to collapse, sagging, and/or suffosion, result from the simultaneous or sequential activity of two types of processes: subsurface dissolution (hydrogeological component), and downward movement of the overlying undermined material by gravitational deformation and/or internal erosion (mechanical component). In carbonate karst settings, the impact of active dissolution on sinkhole development over a human time-scale is negligible in most cases (Beck 2005; Gutiérrez 2016). In contrast, in evaporite formations, especially those including hyper-soluble salts, dissolution occurring over short time spans (e.g., lifetime of engineering projects) can significantly contribute to sinkhole development creating and enlarging cavities and reducing the mechanical strength of the evaporite formations (Cooper and Gutiérrez 2013; Frumkin 2013). Subsidence processes (i.e., the mechanical component), particularly collapse and suffosion, are not necessarily related to or accompanied by active dissolution. They can occur above preexisting relict cavities developed some time in the past, probably under different baselevel and paleohydrological conditions. These voids may have remained stable until a natural agent or anthropogenic alteration initiates the ground instability process, functioning as an activating or triggering factor. Collapse can be catastrophic, suffosion can generate sinkholes very rapidly, and sagging is typically a slow and gradual process. Moreover, the dominant subsidence mechanism in a sinkhole can change during its evolution. For instance, sagging sinkholes may eventually evolve into collapse sinkholes (Figures 6.21b, 6.33, and 6.37b).

The factors that control the dissolution of karst rocks and the resulting mass depletion in the subsurface are discussed in Chapter 3. The internal erosion and deformation processes acting on the undermined material are primarily controlled by geological, geotechnical, and hydrogeological factors. These are commonly difficult to investigate due to their subsurface nature and the heterogeneity of karst environments. The most important ones include (Gutiérrez et al. 2008c): (i) The thickness of sediments overlying the karstification zones and cavities. It is important to take into consideration that cavities can be related either to dissolution or to the upward migration of collapse processes (stoping) starting from deeper solutional voids. The latter cavities can be underlain by large breccia pipes rooted in deeply buried voids generated by dissolution. (ii) The mechanical properties of the material overlying the karstification zones and cavities, which may change through time by multiple processes such as dissolution, strain softening, variations in the water content, or diagenetic processes. (iii) Geometry and size of the subsurface voids, primarily the span of cavity roofs. (iv) Groundwater flow conditions and position and changes of the water table or piezometric level.

The processes involved in the generation of subsidence sinkholes (dissolution, subsidence) can be activated or accelerated by natural and anthropogenic changes in the karst environment. In many cases, the activity of sinkholes with detrimental effects is not strictly a natural process, but a human-induced hazard. Those sinkholes would not have occurred in that particular site and precise moment without the interplay of human activity. Nonetheless, the human contribution to the formation of damaging sinkholes, an issue with important legal implications, is frequently difficult to demonstrate. For instance, in China, where the impact of sinkholes has increased dramatically in recent times, around 75% of the damaging subsidence events are classified as human-induced (Lei et al. 2015). Table 6.2 presents the main natural and anthropogenic alterations that may promote the formation of subsidence sinkholes and their potential effects on the karst system (updated from Gutiérrez 2016). This table, which includes references of numerous case studies, can be used as a checklist to identify the potential factors that have contributed to the development of subsidence sinkholes.

The recent scientific literature on sinkholes indicates that water table decline is the main anthropogenic cause of sinkhole occurrence. These water table drops are mostly related to aquifer overexploitation, but can be also caused by dewatering from mining or tunneling. The adverse effects of groundwater pumping on the sinkhole hazard will keep on increasing in the near future. The impact of this activity seems to be particularly severe in dryland regions where aquifers are frequently the only available source of water and are subject to intense exploitation (Doğan and Yilmaz 2011; Taheri et al. 2015; Youssef et al. 2016, 2020) (Figures 6.25d, g and 6.30b). The main potential effects of lowering the groundwater level include: (i) the loss of buoyant support in the materials situated above cavities that change from phreatic to vadose conditions, (ii) increased hydraulic gradient and accelerated groundwater flow toward the center of cones of depression, which can enhance both internal erosion and dissolution, (iii) the replacement of sluggish lateral phreatic flow by downward vadose percolation with higher capability for internal erosion and with greater impact when the water table declines below the rockhead, and (iv) changes in water content and pore-fluid pressure associated with water table oscillations leading to variations in the mechanical properties of the materials. An additional effect that may operate in covered karst settings subject to rapid water table drops and with low-permeability surficial deposits is vacuum suction. When the water table declines, the effective porosity occupied by water is replaced by air that mainly permeates from the atmosphere. However, if the water level drop occurs with rapidity and the rate at which the air flows into the new vadose zone is not high enough to counterbalance the creation of water-free porosity, negative air pressure occurs. The associated vacuum effect can cause sediment compaction, reduce the stability condition of cavities, and control subsidence mechanisms and rates as illustrated by laboratory physical models (Xiao et al. 2020). Newton (1984), in his seminal work on human-induced sinkholes in Alabama, reported that more than 4000 sinkholes had resulted from groundwater withdrawal since 1900. In Florida, Aurit et al. (2013) demonstrated some temporal and spatial correlation between freeze events, strawberry farming regions, and sinkhole occurrence. Here, sinkholes are triggered by groundwater level drops related to groundwater pumping for sprinkler irrigation to protect the crops from freezing damage. In the Orlando area, central Florida, sinkholes are mainly related to cavities developed in the limestones of the Upper Floridan aquifer, which are overlain by a confining unit and a surficial aquifer. This is an interstratal karst and subsidence affects both caprock and cover. The water table in the surficial aquifer can be higher than the potentiometric level in the confined karst aquifer, leading to downward transtratal groundwater flow through the leaky aquitard. Xiao et al. (2016, 2018) found that one of the main factors that control the timing of sinkhole formation in this region is the large head difference between the surficial and confined aquifer, which are mainly caused by pumping and dewatering. These changes in the groundwater heads favor internal erosion and collapse processes by various mechanisms including buoyancy loss and enhanced downward flow. Similar results were obtained by Nam et al. (2020) in the same area by means of statistical analyses. Both rainfall events and water pumping can lead to significant increases in the head difference between the detrital surficial aquifer and the confined limestone aquifer. In a confined evaporite aquifer affected by hypogene dissolution in NE Spain, Linares et al. (2017) document higher frequency of sinkhole occurrence during drought periods in which the lower potentiometric level in the confined aquifer and the lower water table in the surficial aquifer favor cavity roof instability. These authors suggest that sinkhole hazard may increase in this Mediterranean region due to the intensification of the drought episodes predicted by climatic projections. Daoxian (1987) and Li and Zhou (1999) document some dramatic cases in China, where mining-related dewatering has induced thousands of sinkholes leading to severe consequences such as the abandonment of mines, relocation of a village, demolition of buildings, destruction of reservoirs and railways, sudden water inrushes in mines, and the sinking of a river into a mine via sinkholes acting as swallow holes. Miano and Keays (2018) document an interesting litigation case related to numerous sinkholes which occurred in a school campus in Philadelphia that posed a significant threat to human safety. The sinkholes were proved to be induced by dewatering for mining in a nearby quarry. A peculiar and dramatic example of human-induced sinkholes related to water table lowering is shown by more than 6000 sinkholes formed over the last 35 years in the Dead Sea shores of Israel and Jordan (Figures 6.21a, 6.30d, and 6.33). Here, the water table lowering is related to the rapid decline of the lake level (35 m since 1967) due to anthropogenic use of water in the catchment and the lake. This has led to the lakeward displacement of the brackish-saline water interface with the consequent circulation of unsaturated water through a salt unit around 20 m thick and overlain by ca. 20 m of unconsolidated deposits (inferfingered lacustrine muds and alluvial gravels). Salt dissolution acting preferentially at the salt edge and along concealed Quaternary faults is generating cover-collapse sinkholes and hectometer-to kilometer-scale cover-sagging sinkholes at very high rates (Yechieli et al. 2006; Closson et al. 2007; Frumkin et al. 2011; Abelson et al. 2017; Al-Halbouni et al. 2017). Sinkhole activity has resulted in significant damage, including the destruction of roads and tourist infrastructure. A 12-km long dike in a salt evaporation pond in the Lisan Peninsula, Jordan required remedial works at a cost of around 12 million dollars (Closson and Karaki 2015).

Sinkholes are also frequently induced by enhanced water input into the ground. Widespread or focused water infiltration promotes internal erosion processes and may adversely modify the mechanical strength and weight of sediments. The water input can be related to natural processes such as rainfall, floods, snow melting or ground ice thawing. Intentional or unintentional artificial water input can be very diverse including irrigation, leakage from hydraulic structures, runoff concentration (urbanization, soakaways, and drainage wells), runoff diversion, drilling operations, unsealed wells and boreholes, or injection of fluids (Table 6.2). Hyatt and Jacobs (1996) mapped and characterized 312 cover-collapse sinkholes triggered by flooding of the Flint River, Georgia in 1994. Around 90% of the sinkholes occurred within the limits of flooding. In the Loire River valley in France, where the unconsolidated alluvial cover is underlain by cavernous limestone, a severe rainfall episode and the associated flood of 2016 triggered numerous sinkholes damaging at least 20 houses, a flood-control dyke and two highways (Noury et al. 2018) (Figure 6.38a). Gutiérrez et al. (2007) calculated a minimum spatial–temporal frequency of

Table 6.2 Changes in the karst systems that may accelerate or trigger the development of sinkholes and their potential effects.

Type of change	Effects	1. Natural processes 2. Human activities
Water table decline	• Increases the effective weight of the sediments (loss of buoyant support). • Slow phreatic flow replaced by more rapid downward percolation favoring internal erosion, especially when the water table is lowered below the rockhead. • Accelerates groundwater flow in areas affected by cones of depression. • May reduce the mechanical strength by changes in water content. • Hydrofracturing of poorly drained deposits surrounding cavities (Tharp 2001) • Vacuum suction effect (Xiao et al. 2020). • Loose fine-grained deposits may be dragged with the pumped water causing internal erosion (Karimi and Taheri 2010; Khanlari et al. 2012).	1) Climate change (Linares et al. 2017; Meng and Jia 2018; Grube and Rickert 2019), sea level drop (Cooper and Keller 2001; Bastos et al. 2016), entrenchment of drainage network (Ortega et al. 2013), tectonic uplift, isostatic rebound, halokinetic uplift (Closson et al. 2007; Zarei et al. 2012). 2) Water pumping (Kemmerly 1980; Newton 1984; Beck 1986; LaMoreaux and Newton 1986; Daoxian 1987; Chen 1988; Waltham and Smart 1988; Currin and Barfus 1989; Chen and Xiang 1991; Destephen and Benson 1993; Shaqour 1994; Tihansky 1999; Lei et al. 2001; Kaufmann and Quinif 2002; He et al. 2003; Keqiang et al. 2004; Waltham 2008a; Karimi and Taheri 2010; Doğan and Yilmaz 2011; García-Moreno and Mateos 2011; Khanlari et al. 2012; Gao et al. 2013; Aurit et al. 2013; Taheri et al. 2015; Xiao et al. 2018), dewatering for mining and excavation operations (Foose 1953; Bezuidenhout and Enslin 1970; LaMoreaux and Newton 1986; Daoxian 1987; Chen 1988; Xu and Zhao 1988; Zhou 1997; Strum 1999; Li and Zhou 1999; De Bruyn and Bell 2001; Klimchouk and Andrejchuk 2005; Sprynskyy et al. 2009; Vigna et al. 2010; Pando et al. 2013; Jovanelly 2014), decline of water level in lakes (Yechieli et al. 2006; Frumkin et al. 2011; Abelson et al. 2017; Al-Halbouni et al. 2017), excavations acting as drainages (Fidelibus et al. 2011), ground-source heat pumps (Cooper et al. 2011).
Increased water input to the ground (cover and bedrock)	• Increases percolation accelerating suffosion. • Favors dissolution. • Increases the weight of sediments. • May reduce the mechanical strength and bearing capacity of sediments.	1) Rainfall (Keqiang et al. 2004; Gutiérrez-Santolalla et al. 2005; Zhao et al. 2010; Youssef et al. 2012; Lei et al. 2013; Jiang et al. 2017; Martinotti et al. 2017; Parise et al. 2018; Xiao et al. 2018; Luu et al. 2019; Shamet et al. 2020), floods (Hyatt and Jacobs 1996; Intrieri et al. 2018; Noury et al. 2018), snow melting, thawing of frozen ground (Satkunas et al. 2006). 2) Irrigation (Atapour and Aftabi 2002; Kirkham et al. 2003; Gutiérrez et al. 2007), leakage from pipes (Myers and Perlow 1984; Shaqour 1994; Scarborough 1995; McDowell and Poulsom 1996; Jassim et al. 1997; Abdulla and Mollah 1999; Gutiérrez and Cooper 2002; Dougherty 2005; McDowell 2005; Fleury 2009; Gökkaya and Tunçel 2019), canals (Swan 1978; Lucha et al. 2008b) or ditches (Moore 1988; Gutiérrez et al. 2007), impoundment of water (Milanović 2004b; Gutiérrez and Lizaga 2016; Jalali et al. 2019), runoff concentration (urbanization, soakaways, drainage wells) or diversion (Knight 1971; White et al. 1986; Crawford 2001), vegetation removal, drilling operations (Johnson 1989; Croxton 2003), unsealed wells and boreholes (Johnson et al. 2003; Johnson 2005; Lambrecht and Miller 2006; Liguori et al. 2008), injection of fluids, solution mining (Ege 1984).

Table 6.2 (Continued)

Type of change	Effects	1. Natural processes 2. Human activities
Impoundment of water	• May create extremely high hydraulic gradients leading to rapid turbulent flows favoring internal erosion and dissolution. • The base level rise may change groundwater flow paths and location of discharge zones. • Major and continuous changes in the water table causing repeated flooding and drainage of karst conduits. • Imposes a load.	1) Natural lakes (Day and Reynolds 2012). 2) Reservoirs (James 1992 and references therein; Milanović 2004b and references therein; Uromeihy 2000; Doğan and Cicek 2002; Jarvis 2003; Romanov et al. 2003b; Bonacci and Roje-Bonacci 2008; Johnson 2008; Bonacci and Rubinić 2009; Cooper and Gutiérrez 2013; Gutiérrez et al. 2015; Gutiérrez and Lizaga 2016; Frisbee et al. 2019; Jalali et al. 2019), underground dams and reservoirs (Roje-Bonacci and Bonacci 2013), ponds (Hunt et al. 2013), evaporation pans (Parise et al. 2015), sewage lagoons (Newton 1987; Alexander et al. 1993; Davis and Rahn 1997).
Erosion or excavation	• Reduces the thickness and mechanical strength of cavity roofs. • May concentrate runoff. • May create a new base level changing the path and rate of groundwater flows. • May create an outlet for internally eroded deposits.	1) Erosion processes (Cooper et al. 2011). 2) Excavations (Walker and Matzat 1999; Lolcama et al. 2002; Guerrero et al. 2008b; Fidelibus et al. 2011; Burke et al. 2020).
Underground excavations	• Disturb groundwater flows. • May intercept phreatic conduits and distort groundwater flow paths. • May cause sudden inrushes of water and flooding in underground openings involving accelerated internal erosion and karstification. • May weaken sediments over voids.	1) Biogenic pipes 2) Conventional and solution mining (Ege 1984; Dyni 1986; Daoxian 1987; Xu and Zhao 1988; Gongyu and Wanfang 1999; Kappel et al. 1999; Li and Zhou 1999; Andrejchuk 2002; Autin 2002; Gowan and Trader 2003; Johnson et al. 2003; Sharpe 2003; Yin and Zhang 2005; Lucha et al. 2008a; Warren 2006; Bonetto et al. 2008; Wang et al. 2008; He et al. 2009; Vigna et al. 2010; Mesescu 2011; Land 2013; Jones and Blom 2014; Zhang et al. 2019), tunneling (Milanović 2004b; Marinos 2001; Song et al. 2012).
Static loads	• Favors the failure of cavity roofs and compaction processes. Unloading favors the formation of fractures and dilation of preexisting ones.	1) Aggradation processes. Glacial loading and unloading (Anderson and Hinds 1997). 2) Engineered structures (Waltham 2008a; Zhang and Chen 2019), dumping (Fuleihan et al. 1997), heavy vehicles (Grosch et al. 1987; James 1993; Davis and Rahn 1997; Waltham et al. 2005).
Dynamic loads	• Favors the failure of cavity roofs and may cause liquefaction–fluidization processes involving a sharp reduction in the strength of soils. • Fracturing with the consequent increase in permeability and strength reduction.	1) Earthquakes (Closson and Karaki 2009; Del Prete et al. 2010a, b; Kawashima et al. 2010; Santo et al. 2011), impact of extraterrestrial bodies (Perry et al. 1995), explosive volcanic eruptions. 2) Artificial vibrations (blasting, explosions) (Daoxian 1987), punching pile construction (Meng et al. 2020).

Table 6.2 (Continued)

Type of change	Effects	1. Natural processes 2. Human activities
Drilling	• Weakens, punctures, and overloads cavity roofs. • Causes internal erosion favored by the holes, vibrations, drilling fluids, and pumping. • May induce localized and turbulent groundwater flows.	1) Borings (Meng et al. 2012; Benson and Yuhr 2015), water wells, horizontal directional drilling for the installation of pipelines (Smith and Sinn 2013).
Vegetation removal	• Reduces mechanical strength of cover deposits (root cohesion). • Increases infiltration.	1) Wild fires. 2) Vegetation clearance (Newton 1987; James 1993).
Thawing of frozen ground	• Favors dissolution. Significant reduction in the strength of the sediments.	1) Climate change (Satkunas et al. 2006). 2) Development, deforestation, water storage (Eraso et al. 1995; Trzhtsinsky 2002).

50 sinkholes km^{-2} yr^{-1} in a low terrace of the Ebro River, NE Spain underlain by gypsum where most of the subsidence events are induced by massive sheetflooding irrigation and leakage from unlined ditches and canals. In 2012, a heavy rainfall event triggered 41 sinkholes and 11 large subsidence areas with numerous fissures at Maohe village, Guangxi, China. Subsidence caused damage on 143 houses (69 collapsed) and 1830 people were relocated (Lei et al. 2013). In a dolomite karst area covering around $3.7 km^2$ in Gauteng Province, South Africa, 650 new sinkholes formed in 20 years between 1984 and 2004, of which 99% were associated with leaking water-bearing pipes (Buttrick et al. 2011).

The impoundment of water in revervoirs is a common cause of human-induced sinkholes (Figure 6.38e). The infill of a reservoir involves imposing a load by standing surface water and creating unnaturally high hydraulic gradients that may lead to rapid and turbulent subsurface flows with a high capability to flush out sediments from conduits and enlarge them by solutional and mechanical erosion (Romanov et al. 2003b; Milanović 2004b; Milanović et al. 2019). Moreover, continuous oscillations in the reservoir water level are accompanied by flooding and drainage cycles in the associated karst system, which favor both mechanical and chemical subsurface erosion. Preexisting and newly created sinkholes in reservoirs and in the foundation of dams may result in severe water leakages and stability problems, compromising the operation and safety of the hydraulic structure. Reviews on dam projects severely impacted or abandoned due to water losses through sinkholes functioning as ponors are presented by Milanović (2004b), Gutiérrez et al. (2015) and Milanović et al. (2019). The Anchor Reservoir, Wyoming, was built despite the fact that more than 50 cover-collapse sinkholes related to gypsum dissolution had been identified during the investigation phase. Immediately after filling the reservoir numerous new sinkholes formed, including a collapse 100 m in length and 18–30 m depth that required the construction of a large earth dyke to isolate the sinkhole from the reservoir water body (Jarvis 2003). Quail Creek Dike in Utah, an earth-fill embankment dam, failed catastrophically in 1989 due to leakage and internal erosion induced by the presence of karstified gypsum beds in the foundation. The new

Figure 6.38 Examples of sinkholes induced by anthropogenic and natural changes in the karst system.
(a) Cover-collapse sinkhole triggered by the 2016 Loire River flood in France. *Source:* Photo by Gildas Noury.
(b) Cover-collapse sinkhole induced by a sudden water table drop caused by the excavation of a tunnel for
a high-speed railway through a limestone aquifer in the Cantabrian Mountains, northern Spain. The
sinkhole captured a surface drainage functioning as a swallow hole. *Source:* Photo by Pablo Valenzuela.
(c) Megacollapse structure 450 m across caused by salt dissolution related to the partial submergence of a
salt ridge in the Upper Gotvand Dam, Zagros Mountains, Iran. The impoundment of the dam started in 2011
and the annotated lines in the image indicate new fissures and scarps formed between September 2013
and October 2016. *Source:* Image from Google Earth. (d) The JWS collapse sinkhole around 110 m in
diameter formed on July 2008 above a cavern created by solution mining in the Permian Salado Formation,
Eddy County, New Mexico. *Source:* Photo by Lewis Land. (e) Sinkholes in the Santa María de Belsué reservoir,
Spanish Pyrenees. The dam barely retains water due to leakage through sinkholes acting as ponors. An
additional dam with limited storage capacity (Cienfuens Dam) was built downstream of the spring largely
fed by the Belsué Reservoir. *Source:* Image from Google Earth.

concrete gravity dam with a cutoff wall has also suffered from leakage problems. The lowering of the water level to perform maintenance works revealed more than 200 sinkholes within 130 m upstream of the dam (Payton and Hansen 2003). The Mosul Dam in Iraq, built on Miocene evaporites, has induced numerous sinkholes in the reservoir (identified by bathymetric surveys) and downstream of the dam. This dam, considered as one of the most dangerous in the world, requires continuous grouting to counterbalance the voids created by water leakage in the foundation of the dam (Guzina et al. 1991; Sissakian et al. 2014; Al-Ansari et al. 2015). In the Upper Gotvand Reservoir, Zagros Mountains (Iran), the partial submergence of an exposed salt ridge of the Gachsaran Formation (salt pillow) has resulted in the development of unprecedented collapse structures hundreds of meters across related to subjacent halite dissolution. The diameter of one of them expanded from 270 to 450 m in just three years (Gutiérrez and Lizaga 2016; Jalali et al. 2019) (Figure 6.38c).

Surface and underground excavations can trigger sinkholes or create favorable conditions for their development. The main effects of lowering the ground surface by excavation (or natural erosion) include: (i) reducing the thickness and mechanical strength of cavity roofs, (ii) creating topographic depressions for runoff concentration and enhanced infiltration, and (iii) incorporating new local base levels and discharge zones that may modify the groundwater flow paths and rates. The long-term lowering of the ground surface by natural solutional and mechanical erosion results in a progressive thinning of cavity roofs. Eventually, the thickness of the roof may reach a critical value leading to the development of collapse sinkholes that contribute to unroof the cave passages (Knez and Slabe 2002). These sinkholes are designated by some authors as intersection sinkholes; i.e., the downwearing ground surface intercepts caves. Fidelibus et al. (2011) analyzed the complex interrelationships between the occurrence of collapse sinkholes impinging on a residential area in the Adriatic coast of Italy and the excavation of a canal dissecting unrecognized cavernous gypsum overlain by a loose sandy cover. The canal, aimed at improving the connection between a lagoon and the sea, and effectively acting as a drainage trench, caused various hydrological changes that triggered sinkhole development. The following alterations contributed to enhance internal erosion and gypsum dissolution: (i) lowering of the local water table, (ii) distortion of the groundwater flow paths, (iii) increase in flow velocity, (iv) amplification of the water table oscillations controlled by the tidal regime, and (v) creation of an outlet for the sediments filling the cavities and mobilized by internal erosion. In Tasmania, Australia, Burke et al. (2020) present a detailed analysis of the hydrological and geomorphic effects triggered by the interception of an active cave in a limestone quarry and the associated water table drop, including the occurrence of cover-collapse sinkholes.

The excavation of tunnels and mine galleries, apart from dewatering by pumping, may cause dramatic changes in the local hydrogeology, leading to the formation of sinkholes (Bonetto et al. 2008; Milanović 2004b; Vigna et al. 2010; Wang et al. 2008). The interception of conduits and cavernous rock by excavations performed below the water table may result in dangerous inrushes of water under pressure. The drainage of the aquifer toward underground artificial openings may lead to uncontrolled flooding of the excavation, rapid lowering of the water table, suspension of water supply from wells, enhanced internal erosion or collapse, and the development of sinkholes. In numerous mine districts in China, water inrushes and instability problems are particularly common when the excavation works intercept pervious and mechanically weak breccia pipes rooted in deep-seated solutional cavities developed in carbonate or evaporite formations (Daoxian 1987; Lu and Cooper 1997; Li and Zhou 1999; Yin and Zhang 2005; He et al. 2009; Dong et al. 2018). In northern Spain, the excavation of deep tunnels for a high-speed railway through a structurally complex karst aquifer caused the rapid lowering of the groundwater level and the occurrence of sinkholes in a river, functioning as ponors. The swallow holes beheaded the downstream section of the stream and captured the upstream portion which drained toward the tunnel (Valenzuela et al. 2015) (Figure 6.38b). Sinkhole hazards may be particularly severe when freshwater flows through salt units or even into salt mines coming from an overlying or adjacent aquifer (Kappel et al. 1999; Andrejchuk 2002; Gowan and Trader 2003), or from a surface water body (Autin 2002; Lucha et al. 2008a). The highly aggressive water can cause massive salt dissolution and uncontrollable sinkhole occurrence, leading to the abandonment of the mine. In the Cardona salt diapir in northeastern Spain, the interception of a phreatic conduit by a shallow gallery excavated for the disposal of hazardous wastes caused the inflow of freshwater from a nearby river, rapid dissolution, the formation of numerous sinkholes and ultimately the abandonment of the mine (Lucha et al. 2008a). Large collapse sinkholes can also form above cavities in salt formations created by solution mining, which involves the injection of fresh water and the recovery of a brine (Ege 1984; Dyni 1986; Johnson 1997; Andrejchuk 2002; Johnson et al. 2003; Mancini et al. 2009; Mesescu 2011; Warren 2016; Zhang et al. 2019). These cavities may propagate upward several hundred meters through overlying formations in a few decades, eventually leading to the sudden occurrence of sinkholes more than 100 m across (Ege 1984; Johnson 1997; Land 2013; Jones and Blom 2014). In 2008 and 2009, three large collapse sinkholes formed catastrophically above oversized caverns created by solution mining in the Upper Permian Salado Formation of the Delaware Basin; two in Eddy County, New Mexico (JWS sinkhole and Loco Hills sinkhole), and another one near Denver City, Texas (Denver city sink) (Land 2013) (Figure 6.38d). In 2012 Bayou Corne sinkhole, Louisiana, occurred above a cavity more than 1 km deep created by solution mining at the edge of a salt diapir. The sinkhole rapidly expanded to reach more than 350 m in diameter, and was accompanied by a methane gas emission that forced a long-term evacuation of residents (Jones and Blom 2014).

Natural and human-induced static and dynamic loading may trigger the collapse of preexisting cavities under marginal stability conditions. The load imposed by heavy vehicles, drilling rigs, dumped material, and engineering structures may cause sinkhole events (Meng et al. 2012; Zhang and Chen 2019). A similar effect may be expected from ground shaking related to explosions and earthquakes. Daoxian (1987) reports that exploration for groundwater in a limestone aquifer using explosives triggered 157 collapse sinkholes that resulted in the abandonment of Liangwu village, Guangxi, China. In the Apennines, Italy, Santo et al. (2011) identified a number of collapse sinkholes triggered or reactivated by destructive earthquakes (coseismic sinkholes) with intensities at the site higher than MCS VIII. Kawashima et al. (2010) documented two bedrock-collapse sinkholes triggered by the 2009 M_w 6.2 L'Aquila earthquake in the Italian Apennines. Meng et al. (2020) analysed the potential triggering role of punching pile construction on sinkhole development by creating continuous cycles of impact load and suction pressure and causing hydraulic fracturing by the water-hammer effect.

Other changes that may induce or favor sinkhole development are vegetation removal (loss of root cohesion) and thawing of frozen ground. Vegetation clearance increases infiltration and the susceptibility of cover deposits to internal erosion due to loss in root cohesion. Thawing of frozen ground (permafrost) involves a substantial change in the mechanical strength and permeability of sediments and the availability of liquid water to dissolve karst rocks. The Bratsk Reservoir, Siberia, has caused the partial thawing of the permafrost and the reactivation of a gypsum karst, resulting in the development of numerous collapse sinkholes along some coastal sectors. During reservoir impoundment (1963–1966), sinkhole occurrence reached a spatial–temporal frequency of 200 sinkholes km^{-2} yr^{-1} ranging from 2 to 30 m in diameter and causing severe damage to buildings and structures outside the reservoir area (Eraso et al. 1995).

6.3.7 Sinkhole Mapping

An essential step in most sinkhole studies is the construction of comprehensive cartographic sinkhole inventories. These maps are the basis for developing a number of analyses addressing issues of scientific and practical interest (Gutiérrez 2016), such as: (i) morphometry and size distribution, (ii) genesis of the depressions and controlling factors (Jennings 1975), (iii) spatial distribution and density, (iv) relationship with the hydrogeological component of the system, (v) development and evaluation of susceptibility, hazard, and risk models, and (vi) vulnerability assessments in karst aquifers (Lindsey et al. 2010; Jones et al. 2019). The quality of these analyses depends on the completeness, accuracy, and representativeness of the sinkhole maps and the associated databases (e.g., sinkhole type, status regarding their activity, age). Sinkhole maps are the most important data layer in prognostic studies aimed at: (i) identifying the most susceptible zones, (ii) predicting the spatial and temporal frequency of future sinkholes, and (iii) determining their frequency-size relationships. Preexisting sinkholes are commonly the best predictors for new sinkholes, especially when they show a clustered distribution and the controlling factors have remained essentially the same. Nonetheless, in certain circumstances the spatial distribution patterns of new sinkholes can be markedly different to those formed in the past (e.g., sinkholes induced by localized cones of depression).

Sinkhole mapping may not be a straightforward task and may pose a number of difficulties, in addition to those related to the accuracy and resolution of the data used (e.g., topographic maps, aerial photographs, and digital elevation models [DEMs]). Some frequent problems that affect the quality and usefulness of sinkhole maps include: (i) presence of dense tree vegetation that obscures the sinkholes, (ii) obliteration of depressions by natural and anthropogenic filling, (iii) presence of sinkhole-like natural and/or anthropogenic depressions not related to karst processes (i.e., morphologic convergence), (iv) ambiguity associated with the mapping of sinkholes with vague margins or different alternative edges (Jennings 1975; Šegina et al. 2018) (Figure 6.39a), and (v) a common mismatch between the extent of the topographic depressions and that of the areas affected by subsidence in the case of sinkholes related to subsidence processes (sagging, collapse, and/or suffosion) (Figure 6.39a). As illustrated below, the limitations associated with the presence of vegetation and anthropogenic infills can be partially overcome by filtering LiDAR data and using old imagery and maps, respectively. Unambiguously proving the non-karstic origin of some depressions may require complex investigations including the review of historical data and the application of on-site intrusive and non-intrusive methods (e.g., trenching, boreholes, and geophysics) (Gutiérrez et al. 2008d; Sorensen et al. 2017). As shows Figure 6.39a, the delineation of the edges of sinkholes, understood as topographic depressions with internal drainage, can be established alternatively following their watershed divide or along the rim of the steep inner slope. The meaning of these edges can be very different depending on whether the depressions correspond to solution or subsidence sinkholes, or in the case the area is completely occupied by sinkholes (e.g., polygonal karst) or not. Figure 6.39a illustrates two examples of subsidence sinkholes in which there is a large spatial deviation between the extent of the area affected by subsidence and that of the mapped sinkholes. Collapse sinkholes with degraded margins can have much larger extent than the collapse structures confined to the central sector of the depression (Gökkaya et al. 2021). In the case of sagging sinkholes with diffuse edges, mapping tends to restrict the sinkhole area to the inner part of the basin with obvious geomorphic expression, significantly underestimating the actual extent of the sinkholes. In the mantled evaporite karst of the Ebro Valley, Gutiérrez et al. (2018) demonstrated via trenching that the radii of sinkholes with a sagging component can be two to three times larger than that initially estimated by conventional geomorphic mapping and geophysical surveys. The latter methods tend to overlook a marginal sector affected by sagging with subtle topographic expression (Figure 6.21b).

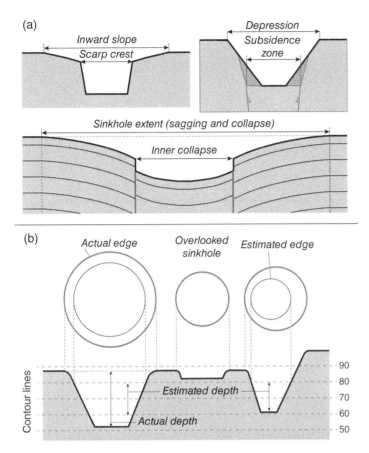

Figure 6.39 Potential problems associated with sinkhole mapping related to geometric features, the mismatch between the extent of the depression and the subsidence area, and the vertical resolution of topographic maps. (a) In the first example, the sinkhole edge could be delineated either at the scarp crest (rim or lip) or at the edge of a peripheral gentle slope with inward inclination, resulting in a much larger depression. In the second example, the extent of the collapse sinkhole with degraded margins is about two times larger than that of the area underlain by the subsidence structure. The third example shows a sagging and collapse sinkhole with diffuse edges. Restricting the sinkhole to the morphologically conspicuous inner collapse would lead to substantial underestimation of the actual size of the sinkhole, with a diameter around three times larger. (b) Sinkhole maps based on contour lines can overlook sinkholes with depths lower than the contour interval and underestimate the size of the depressions (length, area and depth).

This work discusses the importance of precisely identifying the edges of sinkholes for land-use planning and the establishment of set-back distances.

Cartographic sinkhole inventories should include, in addition to georeferenced data on the precise location of the depressions, information related to a number of attributes such as the genetic type, morphometry, numerical or relative chronology, state of activity, associated hydrological features, geological and geomorphic setting, conditioning and triggering factors, land use and interaction with human activity, and detrimental effects. Since some pioneering initiatives like that of the Florida Sinkhole Research Institute (Beck 1991), a number of institutions and associations has developed sinkhole and karst databases mostly integrated in a Geographical Information System (Gao et al. 2002; Florea 2005; Parise and Vennari 2013). The British Geological Survey is compiling a national karst database including features related to five different types of soluble rocks; limestone, dolomite, chalk, gypsum, and salt (Farrant and Cooper 2008; Cooper et al. 2011). This database is used in combination with other datasets to develop sinkhole susceptibility models by means of a heuristic scoring system. Gao and Weary (2008) present a conceptual model for constructing and integrating karst databases from national to local scales, to be used in the production of a national karst map of the United States. Kuniansky et al. (2016) present a review on nation- and state-wide sinkhole databases in the United States, some of them with associated sinkhole reporting mechanisms for the public. In this country, all the states with the exception of Delaware and Rhode Island have karst areas and the average annual cost of sinkholes in the nation exceeds 300 million dollars. Sinkhole and karst databases, together with the datasets and maps derived from them (e.g., sinkhole susceptibility and hazard models), are useful planning tools that may help decision-makers to manage karst areas minimizing environmental and engineering problems. It may be also a valuable source of information for private companies (insurance, geotechnical, and real state) and the general public, as well as a useful resource for researchers.

Topographic maps, especially those with large scales and small contour interval can be used to map sinkholes (Kasting and Kasting 2003; Angel et al. 2004; Brinkmann et al. 2008). Closed contour lines indicating lower ground inside are commonly depicted with hachures on the downslope side or as dashed lines. Old detailed topographic maps can be very useful for identifying and delineating buried sinkholes masked by human activity (Gutiérrez et al. 2011; Basso et al. 2013). Nonetheless, topographic maps have significant limitations that should be taken into consideration (Applegate 2003). Sinkhole mapping can be adversely affected by photogrammetric artifacts in the topographic maps associated with poor quality of the source images, scene boundary effects and tree canopy cover that leads to overestimated elevations. Moreover, as illustrated in figure 6.39b, sinkhole maps produced with contour lines can overlook a number of sinkholes with depths lower than the contour interval. Moreover, sinkhole maps derived from contour lines underestimate the actual extent of the sinkholes. The difference between the actual sinkhole area and that of the mapped depression has an inverse relationship with the gradient of the side slopes of the sinkhole. In addition to these planimetric constraints, contour lines always provide minimum depth values for the sinkholes (Figure 6.39b). In Barbados, Day (1983) mapped sinkholes in the field and found that only 54% of them were depicted in 1:10 000 scale topographic maps. In Mammoth Cave National Park (MCNP), Kentucky (295 km^2) and in a sector of the Apalachicola National Forest (ANF), Florida (300 km^2), Wall et al. (2017) compared previous maps of karst depressions produced manually using closed contour lines (6 m and 1.5 m contour intervals), with maps of karst depressions derived from filtered LiDAR data (DEMs with vertical accuracy of 0.37 m and pixel size of 1 m^2 and 2 m^2) by the automated fill-difference method. The number of putative karst depressions in MCNP increased from 820 to 1504 and in ANF from 822 to 4454.

Aerial photographs can be an extremely useful and cost-effective tool for mapping sinkholes, even in areas where there is availability of high-resolution orthoimages and DEMs (Zumpano et al. 2019). The interpretation of image pairs with sufficient resolution allows an experienced mapper to accurately delineate the edges of sinkholes (Williams 1972b), including shallow depressions. This work can be performed with a stereoscope or on the computer screen with a stereomapping GIS tool. Panno and Luman (2013) in an agriculturally-rich area in southern Illinois, using grey-scale aerial photographs from the 1940s identified approximately 30% more sinkholes than those delineated on topographic maps and recent airborne imagery. In the mantled evaporite karst around Zaragoza city, NE Spain, mapping active sinkholes requires the use of old aerial photographs since most of the depressions have been buried by man-made deposits (Gutiérrez et al. 2011; Sevil et al. 2021) (Figure 6.40). Aerial photographs from multiple dates, often covering several decades, can be used to (i) identify and map buried sinkholes (Brinkmann et al. 2007; Galve

Figure 6.40 Images of an area in the Ebro Valley evaporite karst, northern Spain with a large compound subsidence sinkhole. The upper image is an aerial photograph from 1957 showing the approximate edge of the compound sinkhole as mapped with a stereoscope. It hosts two wetlands with lakes in nested depressions (arrows). The lower image is an orthophoto from 2012 showing that the compound sinkhole has been buried and partially covered by human-structures, including the N-232a highway. This section of the highway has been affected by three sudden collapse sinkholes since 2006. *Source:* Francisco Gutiérrez (Author).

et al. 2009; Gutiérrez et al. 2011; Festa et al. 2012; Panno and Luman 2013; Kromhout and Alfieri 2018) (Figure 6.40), (ii) constrain the timing of recent sinkholes with bracketing ages, (iii) estimate the minimum probability of occurrence of sinkholes in an area; (iv) analyze the spatial–temporal evolution patterns of the recent sinkholes (Cooper 1998; Festa et al. 2012), and (v) identify potential relationships between land-use changes and sinkhole development. In a number of projects involving large budgets, old aerial photographs have allowed to demonstrate that the selected sites contain buried sinkholes that went unnoticed using expensive and site-specific high resolution data (e.g., LiDAR-derived DEMs). Currently, there is a growing number of international, national, and regional internet viewers that provide free online access to georeferenced satellite images and aerial photographs that can be extremely useful for sinkhole mapping. Some viewers such as Google Earth cover most of the globe and provide access to historical imagery, frequently with sufficient resolution to identify old and recently formed sinkholes.

The most significant methodological advance developed in recent years for sinkhole mapping is the acquisition of airborne LiDAR data (Light Detection and Ranging), often with national or regional coverage and freely accessible to the public. The LIDAR technique allows accurately and rapidly measuring the position of features on the earth's surface by emitting light pulses and recording their returns. The two-way travel time of the beam provides range. The backscattered light pulses can be derived from different reflection surfaces (e.g., tree canopy, building, ground surface). One of the great advantages of LiDAR is that the acquired point clouds can be classified and filtered to remove the returns from off-surface objects and produce DEMs of the bare ground (Gallay 2013) (Figures 6.41 and 6.42). The data from the light beams that penetrate through gaps in the tree cover allow capturing geomorphic features hidden by vegetation in

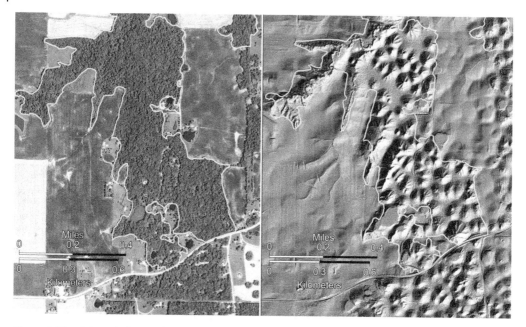

Figure 6.41 A densely forested area with a high density of cover-collapse sinkholes in Illinois shown in an orthoimage and in a shaded relief model derived from bare-ground LiDAR data. The removal of the returns from the vegetation in the LiDAR data allows imaging the geomorphic features masked by the forest. The yellow line encloses dense forest within the Antler-Annbriar groundwater basin. *Source:* From Panno and Luman (2018).

aerial photographs and satellite images (Filin et al. 2011; Miao et al. 2013; Kobal et al. 2014; Cigna et al. 2017). The DEMs derived from LiDAR data can be used to produce 3D representations of the terrain for the identification and manual mapping of sinkholes. Shaded relief models allow visualizing the topography by producing a 3D shading effect with varying greyscale tones determined by the light source and the aspect and slope of the ground surface. They facilitate the identification of the slope breaks associated with the sinkhole edges, commonly expressed as a relatively sharp change in the grey tone. The side slopes of the sinkholes are darkened or lightened in the unidirectional hillshade models, depending on their orientation (Figure 6.42). The multidirectional hillshades, which combine light from multiple sources, provide a more uniform representation of the sinkholes, with a darker tone all along the steep side slopes. Moreover, multidirectional hillshades improve the appearance of low-relief zones and display more detail in areas affected by over-lightening and deep shadows in the models generated by the unidirectional method. Improved sinkhole visualization and mapping can be achieved generating Red Relief Image Maps from DEMs. These models provide a 3D representation of the topography multiplying three geometric attributes: slope gradient, positive openness, and negative openness. The two latter attributes emphasize convex and concave features (Chiba et al. 2008). This visualization method represents the slope with chroma values of red, and an index (Ridge and Valley index) that combines positive and negative openness with a grey gradation. The resulting model eliminates the dependency of the incident light direction on the slope shading and emphasizes the concave inner part of the sinkholes and their convex rims (Gökkaya et al. 2021) (Figure 6.19).

Synthetic Aperture Radar Interferometry (InSAR) can be used to produce accurate DEMs, but the spatial resolution of the available data is in most cases insufficient for mapping small sinkholes (Jones 2020). An advantage of the SAR data is their large spatial and temporal coverage. In the

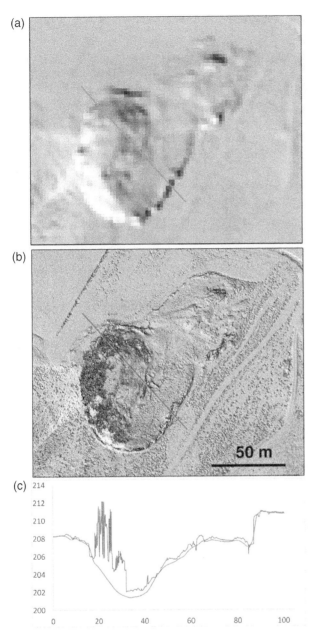

(a)

(b) 50 m

(c)

Figure 6.42 Shaded relief models and topographic profiles of a 150 m long cover- and bedrock-collapse sinkhole with trees in the left portion of its bottom, Ebro Valley evaporite karst, NE Spain. See aerial image of the compound sinkhole in Figure 1.5d. (a) Shaded relief model generated by the unidirectional hillshade method from a LiDAR-derived DEM (spatial resolution of 2 m) after filtering the vegetation. *Source:* Downloaded from the server of the Spanish Geographical Institute. Note that the model allows the identification of the sinkhole and the approximate mapping of its edges. (b) Shaded relief model generated from a DEM (spatial resolution of 2 cm) produced by SfM Photogrammetry with images captured by a drone with a 20 MP camera and flying at 80 m above the ground. This image captures with great detail the overall geometry of the depression and the numerous instability features such as marginal fissures, toppling blocks in the scarps, secondary inner scarps, and nested collapses. *Source:* Francisco Gutiérrez (Author). (c) Topographic profiles derived from the DEMs used for the production of the shaded relief models. The profile from the filtered LiDAR-derived DEM (2 m resolution) shows the overall topography of the bare ground surface. The profile constructed with the SfM-derived DEM (2 cm resolution) shows great detail capturing small features like the fissures, but depicts upward projections related to tree and shrub vegetation.

Hamedan Plain, Zagros Mountains, Iran, Vajedian and Motagh (2019) were able to map cover-collapse sinkholes larger than 20 m and deeper than 3 m using DEMs produced from high-resolution multi-temporal X-band SAR images acquired by the twin satellites TerraSAR-X and TanDEM-X, and processed by single-pass interferometry. An excellent alternative for mapping sinkholes in relatively small areas and with limited vegetation is Structure from Motion (SfM) Photogrammetry (Al-Halbouni et al. 2017; Cigna et al. 2017; Silva et al. 2017). It offers the possibility of generating 3D surface models with high accuracy and resolution from images taken by conventional digital cameras and through a fast and highly automated processing. The images are commonly acquired with UAVs that capture the same object from multiple viewpoints. SfM, in contrast with traditional photogrammetry, resolves the position, orientation, and internal parameters of the camera by identifying common points in the images and establishing their spatial relationships (Micheletti et al. 2015). Further processing allows generating dense point clouds, a polygonal mesh with texture, a DEM, and an orthoimage (Figures 6.2 and 6.42). The models can be accurately registered to a real-world coordinate system using the coordinates of ground control points or those recorded by an RTK antenna integrated in the UAV (Hackney and Clayton 2015). The main advantages of SfM with respect to airborne LiDAR data are the relatively low cost and simple processing of the former. In contrast, SfM is not a practical approach for covering large areas, does not allow filtering the vegetation, and can be affected by UAV operation restrictions in the target areas.

Generally, the best approach for producing sinkhole inventories as complete and accurate as possible is to map manually each depression using all the available resources: aerial photographs and orthoimages from multiple dates, old and recent topographic maps, 3D surface models derived from LiDAR data, InSAR data (Jones 2020) and/or Structure from Motion Photogrammetry, low-altitude oblique images (pictometry), multispectral images (Cooper 1998). These data layers, including aerial photographs for digital stereoview, can be used in a complementary fashion on a GIS platform (Alexander et al. 2013; Panno and Luman 2013; Zumpano et al. 2019; Parenti et al. 2020). Topographic profiles can be drawn automatically across the depressions for identifying and refining the position of their edges. The preliminarily mapped depressions, or at least those that pose doubts regarding their origin or extent, should be checked in the field. There is currently a tendency to map sinkholes using automated routines, but they commonly result in a significant number of false positives and overlooked positives. Mapping sinkholes manually is a rather rapid process that involves a thorough examination of the study area, providing the opportunity to grasp relevant features that can be critical for understanding the karst landscape. Automatic mapping can be a good alternative for scientific investigations covering large areas, but is not advisable for establishing the basis for decisions with potential economic and societal implications, unless it is complemented with a reliable validation and ground-truthing process. In the National Park of Slovak Karst, Slovakia, Hofierka et al. (2018) compared sinkhole maps produced by four methods: (i) outer closed contours from topographic maps (472 sinkholes), (ii) fill-difference method using DEMs derived from filtered LiDAR data (1618 sinkholes), (iii) accumulation of overland water flow (956 sinkholes), and (iv) manual mapping using models derived from the DEM such as hillshades and mean curvature maps (664 reference sinkholes). They found that the closed-contour method overlooked around 30% of the sinkholes and that around 60% of the depressions identified by the fill-difference method were false positives, largely located in areas with non-karst bedrock. Moreover, all the automatic methods (closed contour, fill-difference, and water flow simulation) significantly underestimated the size of most of the sinkholes (Figure 6.43).

LiDAR data and their derivative models can be used for the automatic delineation of sinkholes using several approaches: (i) outermost closed contour (Filin et al. 2011; Rahimi and Alexander 2013; Bauer 2015; Verbovšek and Gabor 2019; De Castro and Horta 2021), (ii) watershed-based delineation, and (iii) the fill-difference method (Doctor and Young 2013; Miao et al. 2013; Pardo-Igúzquiza

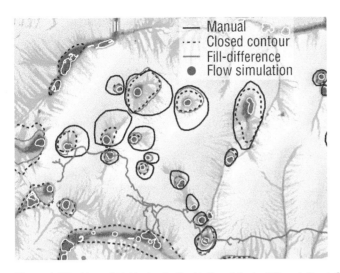

Figure 6.43 Map of sinkholes in the National Park of Slovak Karst, Slovakia delineated by various methods: (1) manual mapping (black line), (2) outermost closed contour (black dashed line), (3) fill-difference method (pink line), and (4) accumulation of overland water flow in the sinkhole floors (white line). Note that the different automated methods tend to underestimate the size of the sinkholes. The background map reflects water flow depth after a simulated rainfall. Red lines indicate the location of the Domica Cave system. *Source:* From Hofierka et al. (2018). Reproduced with permission of Elsevier.

et al. 2013; Zhu et al. 2014; Kobal et al. 2015; Wall et al. 2017; Zumpano et al. 2019). The latter technique is the most commonly used and is typically performed through several steps in a GIS environment: (i) generating a high-resolution DEM of the bare surface after removing the returns related to off-ground objects with a filtering algorithm (e.g., vegetation, buildings), (ii) extracting closed depressions by subtracting the original DEM from an additional DEM with filled depressions through a map algebra operation, (iii) separating true sinkholes from the extracted closed depressions by manual elimination (Panno and Luman 2018; Zumpano et al. 2019) or establishing thresholds for different morphometric parameters (e.g., depth, area/depth ratio, circularity, or elongation ratio) and considering spatial relationships with natural and anthropogenic features (e.g., streams, roads, and non-karst bedrock), and (iv) assessing the success rate by identifying true positives, false positives, and overlooked sinkholes. The evaluation of the automatic mapping can be carried out using detailed sinkhole maps produced by various methods, including visual examination of multiple remote sensed data and field inspection. Doctor and Young (2013) compared depressions mapped automatically using LiDAR data with manually delineated sinkholes in a covered karst area of Virginia. The map produced through the fill-difference method overlooked around 30% of the sinkholes identified by visual inspection of aerial photographs and elevation models, mainly shallow depressions. Zhu et al. (2014), using LiDAR data in Kentucky, increased by a factor of four the number of sinkholes previously mapped with low-resolution topographic maps, attaining a success rate of 80–93%. Similar results were obtained by Rahimi and Alexander (2013) in Minnesota, where the actual number of sinkholes may be around four times higher than those depicted in previous maps. Here, automatic mapping of sinkholes with a LiDAR-derived DEM yielded a success rate of 82%, and the proportion of missed sinkholes reached 9%. De Carvalho-Junior et al. (2014) applied a similar approach in Brazil, but using DEMs with much lower spatial resolution and accuracy generated from satellite data, obtaining a rather limited overall accuracy (40–50%). In several areas of the Bluegrass region of Kentucky, Zhu et al. (2020) explored several machine learning methods for separating sinkholes from automatically mapped depression using large datasets of morphometric data. Subsequently, the machine learning model with the best performance (neural network) was tested in a similar region combining the trained classifier and manual inspection. They found that the procedure saved more than 70% of the manual labor in visual inspection with a cost of missing a relatively small number of sinkholes. Nonetheless, the experiment was preceded by a complex statistical analysis and the machine learning classifiers may yield poor performance when applied to different areas. González-Díez et al. (2021) explored the use of filters based on Fast Fourier Transform applied to DEMs to emphasize changes in the general trend of the altimetry (e.g., inflection points) for the objective mapping of karst depressions. Attempts have been carried out to map sinkholes semi-automatically with high-resolution aerial images. Dou et al. (2015) followed three main steps: (i) Image segmentation grouping pixels into objects with homogeneous spectral and textural characteristics. (ii) Object classification considering several features (e.g., tone, texture, and shape). (iii) Extracting the objects attributable to sinkholes by comparing their features with those of a case library of known sinkholes. Although the authors obtained satisfactory results in their study areas in China, the method requires a highly specific technical qualification and may generate a large number of false positives.

6.3.8 Sinkhole Morphometry and Spatial Distribution

Geomorphometry, that refers to the measurement and analysis of morphometric properties of landscapes and landforms, is a highly useful approach in sinkhole investigation. The morphometric data extracted from sinkholes have multiple utilities: (i) describing and analyzing quantitatively specific sinkholes and sinkhole populations (Bondesan et al. 1992), (ii) classifying sinkholes on the basis of specific morphometric parameters or indexes (Cvijić 1893; Basso et al. 2013); (iii) interpreting their origin and controlling factors, (iv) inferring the role played by non-karstic processes on their development (e.g., expansion by erosion and mass wasting processes; Gökkaya et al. 2021), (v) determining the degree of maturity and relative age, (vi) inferring their morphological evolution (Day 1983; Ferrarese et al. 1998; Gökkaya et al. 2021); (vii) comparing the characteristics of sinkholes in different regions or zones within a study area (Williams 1993; Panno and Luman 2018; Gutiérrez et al. 2019a), (viii) estimating long-term rates of surface lowering caused by various processes (e.g., subsidence, chemical, and mechanical denudation) (Williams 2004a), (ix) deducing the relative age of geomorphic surfaces, (x) separation of sinkholes from automatically mapped closed depressions of non-karst origin (Zhu et al. 2020), (xi) inferring structural controls on their development (Öztürk et al. 2017; Verbovšek and Gabor 2019), and (xii) generating frequency-size relationships and hazard curves (Taheri et al. 2015; Gutiérrez et al. 2019a). Obviously, the soundness of the morphometric analyses depends not only on the quantity and quality of the data, but also on a good understanding of the corresponding karst system.

Traditionally, morphometric data have been obtained from topographic maps and aerial photographs, as well as through the collection of time-consuming measurements in the field (Bondesan et al. 1992; Ford and Williams 2007). Currently, most morphometric analyses are carried out with parameters extracted automatically from DEMs using a Geographical Information System. Some parameters are measured from topographic data and others are computed using various morphometric properties. Planimetric parameters describe the plan form of the sinkholes and can be obtained without elevation data. Relief attributes provide a three-dimensional description incorporating the vertical dimension. Bondesan et al. (1992) presented a comprehensive review on geomorphometry applied to sinkholes. Unfortunately, some morphometric parameters receive multiple names and a number of morphometric indexes are calculated using various formulas. This frequently hinders the possibility of conducting comparative studies. Here, we describe the main parameters and indexes and Table 6.3 presents a list of descriptors, codes, and definitions that might help to standardize

Table 6.3 Main morphometric parameters that can be used for the quantitative description and analysis of sinkholes.

Parameter	Code/Formula	Definition
Length	L	Length of the straight line between the most distant points of the perimeter (major axis)
Width	W	Maximum width normal to the major axis (minor axis)
Perimeter	P	Length of the mapped sinkhole edge
Area	A	Area enclosed within the mapped edge
Orientation	O	Azimuth of the straight line that defines sinkhole length
Elongation ratio	$E_r = L/W$	Ratio between the length and the width
Equivalent diameter	$DI_{eqa} = 2\sqrt{A/\pi}$	Diameter of a circle with an area equal to that of the sinkhole (DIeqa)
	$DI_{eqp} = P/\pi$	Diameter of a circle with a perimeter equal to that of the sinkhole (DIeqp)
Equivalent perimeter	$P_e = 2\sqrt{A\pi}$	Perimeter of a circle with an area equal to that of the sinkhole
Equivalent area	$A_e = P^2/4\pi$	Area of a circle with a perimeter equal to that of the sinkhole
Circularity ratio	$C_r = A/(P^2/4\pi)$	Ratio between the area of the sinkhole and the area of a circle having a circumference equal to the perimeter of the sinkhole
Sinuosity (complexity) index	$C_i = P/2\sqrt{\pi A}$	Ratio between the perimeter of the sinkhole and the perimeter of a circle with an area equal to that of the sinkhole
Depth	D (D_{max}) (D_{min}) (D_{av})	Elevation difference between the lowest point of the sinkhole and its edge. For sinkholes with altitudinal variation along the perimeter, maximum, minimum, and average depths are given by the vertical distance between the lowest point and maximum, minimum, and average elevation of the perimeter, respectively.
Length to depth ratio	$LD_r = L/D$	Ratio between sinkhole length and depth.
Equivalent depth	$D_{eq} = V/A$	Height of a cylinder with an area and volume equal to that of the sinkhole
Volume	V	Volume of the 3D space defined by the intersection between the topographic surface of the sinkhole and a surface interpolated from the sinkhole edge
Cylindricity index	$CY_i = D_{eq}/D$	Ratio between the equivalent depth and the actual depth

morphometric analyses. The main planimetric parameters are: (i) Length, which is the dimension of the straight line between the most distant points of the perimeter (major axis), (ii) Width, given by the longest axis perpendicular to the major axis (minor axis), (iii) Perimeter is the length of the mapped sinkhole edge, which in some cases, as discussed in the sinkhole mapping section, can have a considerable degree of uncertainty and ambiguity, and (iv) Area given by the surface area enclosed by the sinkhole perimeter. Some authors, when investigating sinkholes with a well-defined sink, measure length and width from the major axis and minor axis passing through the swallet, which is different to the conventional length and width (Williams 1972a). The orientation of the sinkhole refers to the azimuth of the major axis on noncircular depressions. The frequency distribution of the orientation values extracted from the sinkholes mapped in an area, commonly represented in rose diagrams or histograms, provides valuable information on the structural control on sinkhole development (e.g., dissolution guided by fracture sets). Some authors restrict these analyses to sinkholes with an elongation ratio above a specific value. The elongation ratio is a shape index given by the relation between the length and the width, indicating how many times the length is larger than the width. The elongation ratio can provide insight into the role played by structural factors and coalescence on the morphology and geometrical evolution of the depressions. In areas where dissolution is strongly guided by fractures, sinkholes tend to be elongated in the direction of the controlling structures and commonly form alignments. Aligned sinkholes eventually coalesce to form linear compound sinkholes with high elongation ratios (Öztürk et al. 2017) (Figure 6.22f). Orientation and elongation are complementary morphometric properties for investigating structural control. The equivalent diameter and the equivalent perimeter of a sinkhole are, respectively, the diameter and the perimeter of a circle with an area equal to that of the sinkhole. The equivalent diameter can be also referred to the diameter of a circle with a perimeter equal to that of the sinkhole. Both are different parameters and should be properly defined. The equivalent area is the area of a circle with a perimeter equal to that of the sinkhole. These equivalent parameters can be used to compute geometrical indexes that quantify the deviation of the plan form of the sinkhole from that of an equivalent circle, which is the polygon with the shortest length that encloses a given area. There are multiple ways to assess the circularity of sinkholes (Šegina et al. 2018). Here, we propose an easy-to-obtain circularity ratio given by the relation between the area of the sinkhole and the area of a circle having a circumference equal to the perimeter of the sinkhole (equivalent area), which has been widely used in drainage basin morphometry (Miller 1953; Chorley et al. 1984). A perfectly circular sinkhole has a circularity ratio of 1 and the value of the ratio decreases with the irregularity of the sinkhole perimeter. The sinuosity index, also designated as the complexity index by some authors, is the ratio between the perimeter of the sinkhole and the perimeter of a circle with an area equal to that of the sinkhole. The more sinuous the sinkhole perimeter, the higher the index. Note that this index is equivalent to the sinuosity index used for fluvial channels.

Sinkhole depth is the basic relief parameter and is an essential property for the three-dimensional characterization of sinkholes. This attribute was rarely considered in old morphometric analyses due to the difficulty of obtaining reliable values from topographic maps and aerial photographs. Provided there is high-quality topographic data available of the bare ground, it is a straightforward parameter when sinkholes have well-defined rims and occur on flat surfaces (Melis et al. 2021); elevation difference between the lowest point of the sinkhole and its edge. However, multiple depths can be measured in sinkholes with perimeters showing altitudinal variations; i.e., sinkhole perimeter cuts across contour lines. The main depth parameters that can be measured in these situations are maximum, minimum, and average depths, corresponding to the altitudinal difference between the lowest point of the sinkhole and the maximum, minimum, and average elevation of the perimeter, respectively. The length to depth ratio is a widely used index for characterizing the three-dimensional geometry of sinkholes and their relative horizontal versus vertical development. The adimensional length to depth ratio is high in pan-shaped sinkholes and low in shaft-like sinkholes. The equivalent depth is the height of a cylinder with an area and volume equal to that of the sinkhole. In a perfectly cylindrical sinkhole, the equivalent depth is equal to the actual depth. In a perfectly conical sinkhole the equivalent depth is one third of the actual depth. The ratio between the equivalent depth and the actual depth, which can be designated as the cylindricity index, indicates how much the sinkhole geometry approximates to a cylinder.

The volume of the sinkholes used to be estimated: (i) considering the area enclosing the different closed contour lines and the contour interval (Bondesan et al. 1992), (ii) using various formulas depending on the morphological characteristics of the sinkholes (e.g., cone, truncated cone), or (iii) multiplying area by average depth. Currently, volume can be measured from DEMs with sufficient accuracy and spatial resolution by computing the volume of the 3D space defined by the intersection between the topographic surface of the sinkhole and a surface interpolated from the sinkhole perimeter (Figure 6.44a). Šušteršič (2006a) contends that the semi-profile of sinkholes is the elementary 3D geometrical unit of the depressions, and illustrates with solution sinkholes from the Dinaric karst that it can be described with power functions. Péntek and Veress (2007) propose a mathematical approach to characterize the three-dimensional geometry of sinkholes using an "area function" that describes the variation of the area enclosed by contour lines at different depths. The function transforms the real depressions into bodies of rotational symmetry, substituting the enclosed contours by circles of the same area. The function includes three parameters that reflect the depth, diameter and position of the inflection point between the convex and concave section of the slope profile. These parameters can be used to identify geometrical evolutionary trends of dolines in a specific karst area. Péntek and Veress (2007) differentiate four morphological evolutionary paths for deepening sinkholes: (i) widening at base and rim, (ii) widening at base, (iii) widening at rim, and (iv) no widening.

The availability of relatively large and complete datasets of sinkhole-size parameters allows the analysis of frequency-size relationships. The most commonly explored parameters are length (major axis) and area. Volume is also a very meaningful dimension, but there are often limitations regarding the availability and accuracy of the data. The size range is given by the maximum and minimum values of the size parameter, and the orders of magnitude covered by the range can be calculated using the formula $\log_{10}$(maximum value/minimum value). The empirical size distribution of sinkholes can be affected by a number of factors that should be taken into consideration: (i) subsidence sinkholes in an area can be related to different mechanisms and each sinkhole type most probably has different scaling relationships, (ii) sinkholes can have multiple ages and the size of the sinkholes can be strongly influenced by their degree of maturity (e.g., growth), (iii) the dataset can be incomplete, especially for the small-sized sinkholes that are more difficult to identify, and (iv) sinkhole inventories commonly include single and compound depressions, that should be

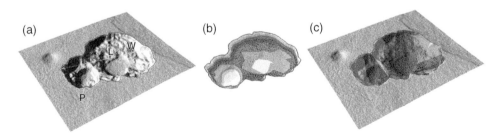

Figure 6.44 Example of sinkhole volume calculation with a DEM. See image in Figure 6.33. (a) Shaded relief model of a compound sinkhole in the Dead Sea generated from a DEM produced by Structure from Motion Photogrammetry. Red line indicates the mapped perimeter. *L*: Length; *W*: Width; *P*: Perimeter. Note incipient sinkhole to the left. (b) Extracted TIN surface of the sinkhole topography. (c) Surface interpolated from the sinkhole perimeter ("sinkhole cap") shown with transparency on the hillshade. The volume is given by the 3D space enclosed by the sinkhole TIN surface and the interpolated "sinkhole cap." *Source:* Models generated by Jorge Sevil.

separated depending on the aims of the analysis. Common graphic representations include size versus number of sinkholes equal or larger than a given size, or size in logarithmic scale versus cumulative frequency of sinkholes equal or larger than a given size. A number of studies shows that subsidence sinkholes follow a power-law frequency-size distribution at least for a significant range of magnitude values (Taheri et al. 2015; Gutiérrez and Lizaga 2016; Gutiérrez et al. 2016, 2019a; Yizhaq et al. 2017; Parenti et al. 2020; Gökkaya et al. 2021). This essentially means that the frequency of sinkholes drops exponentially as the size increases; i.e., highly frequent small sinkholes versus infrequent large sinkholes. This type of power-law distribution is observed for a number of hazardous geological processes including earthquakes, floods, slope movements, piping-related collapses (see reviews in Guzzetti et al. 2002; Corral and González 2018; Bernatek-Jakiel et al. 2019). Frequently, there are deviations in the upper and lower parts of the graphs between the empirical data and the functions. The points of the regression from which empirical data are no longer satisfactorily modeled by the power-law function are called cutoff or roll-over points. Figure 6.45 shows plots of sinkhole lengths (major axis) versus cumulative frequency of sinkholes equal or larger than a given length from multiple areas with different sinkhole types and karst settings: (i) cover-collapse sinkholes of the Val d'Orleans, France, developed on a weak and thin (6–10 m) sandy alluvial cover underlain by cavernous limestone (Gombert et al. 2015), (ii) cover collapse sinkholes in the Hamedan Plain, Zagros Mountains, Iran mainly induced by aquifer over-exploitation and the associated water table drop. Here, subsidence affects a thick and cohesive cover that reaches as much as 100 m in thickness (Taheri et al. 2015), (iii) collapse sinkholes developed at the Anbal salt pillow, Zagros Mountains, Iran related to cavities developed in salt and their upward stoping through a residual caprock, mainly consisting of distorted gypsum and clayey sediment (Gutiérrez and Lizaga 2016), (iv) multiple types of subsidence sinkholes developed on alluvium and various caprocks (marls, sandstones, and basalts) in the epigene evaporite karst of the Fluvia Valley, NE Spain (Gutiérrez et al. 2016), (v) various types of cover-subsidence sinkholes developed on

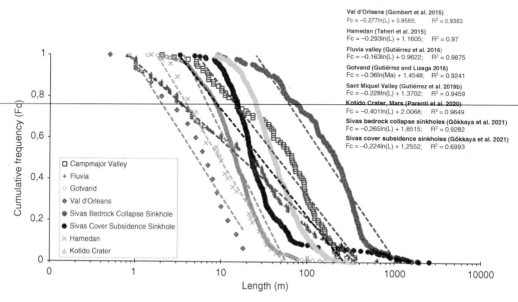

Figure 6.45 Cumulative frequency versus length of various types of subsidence sinkholes inventoried in different karst settings (carbonate and evaporite; covered, bare and interstratal; epigene and hypogene). See explanation in the text.

alluvium in the epigene Sivas gypsum karst, Turkey including both single and compound depressions (Gökkaya et al. 2021), (vi) closed depressions interpreted as collapse sinkholes related to evaporite dissolution in Kotido Crater, Mars, possibly enlarged by wind erosion and on a planet with a gravitational acceleration of $3.7\,m\,s^{-2}$ (Parenti et al. 2020), (vii) caprock collapse sinkholes in the deep-seated hypogene evaporite karst of Sant Miquel Valley, NE Spain (Gutiérrez et al. 2019b), and (viii) bedrock-collapse sinkholes in the Sivas gypsum karst that reach exceptionally large dimensions due to both coalescence and post-collapse enlargement by mechanical and chemical erosion (Gökkaya et al. 2021). Overall, the size distributions from the different areas show the following features: (i) power-law distributions that can be described satisfactorily with logarithmic functions over a wide size range, (ii) a general displacement of the different size distributions toward greater lengths as the resistance and thickness of the material affected by subsidence increases, (iii) larger size range in the inventory of the Fluvia Valley that includes multiple genetic types of both single and compound sinkholes, (iv) an upper roll-over showing lower empirical length than those predicted by the regressions attributable to both the incompleteness of the inventories for the small-size sinkholes and a mechanical threshold (i.e., minimum span of unstable cavities), and (v) a lower cutoff with larger empirical values than those modeled by the regressions ascribable to processes such as coalescence and expansion by erosion on the side slopes of the sinkholes. Inventories of new sinkholes that include information on their chronology and size at the time of formation can be used to produce hazard curves that indicate the annual probability of sinkholes reaching or exceeding a given dimension in a given area (probability of exceedance). An example is presented by Taheri et al. (2015) with sinkholes recorded over 22 years in the Hamedan area of the Zagros Mountains affected by aquifer over-exploitation. These frequency-size regressions allow estimating the recurrence (reciprocal of the annual probability of exceedance) of sinkholes equal or larger than a given size. These are commonly minimum or optimistic hazard values since sinkhole inventories are usually incomplete. For instance, in Hamedan (Iran), the estimated maximum recurrence for sinkholes with lengths equal or larger than 5 and 20 m were 0.8 and 2.1 years, respectively (Taheri et al. 2015). White and White (1995), analyzing published datasets of solution and subsidence sinkholes in different carbonate karst regions, observed an exponential distribution in the depth and length values. Pardo-Igúzquiza et al. (2020) analyzed the size distribution (area) of karst depressions mapped automatically from DEMs in four limestone massifs in Spain, two in the Betics, and two in the Pyrenees. In three of them, power-law regressions yielded a good fit, whereas in one of the massifs of the Betics (Sierra Gorda), neither log-normal nor power-law functions modeled satisfactorily the size distribution.

Surface roughness (or ruggedness), also designated as terrain roughness and topographic roughness, can be an interesting morphometric parameter to analyze and compare areas pockmarked by sinkholes. Surface roughness indicates how undulating or uneven the terrain is. It is essentially a function of the amplitude, wavelength and slope of karst landscape components. It can be computed using multiple approaches such as: (i) the ratio between the surface area of an analysis window and its planimetric area, (ii) variation of height in the cells of a given analysis window, (iii) the distribution and direction of the vectors normal to the surface at each cell of the analysis window, (iv) fractal analysis, or (v) surface curvature (Olaya 2009; Day and Chenoweth 2013). Both the size of the analysis window and the grid resolution play a key role. The grid should be sufficiently fine to represent the surface roughness associated with the landforms of interest (e.g., sinkholes). Surface roughness was used by Day (1979) in a pioneering work using topographic maps, aerial photographs and elevation data collected in the field as a discriminator of tropical karst styles in the Caribbean and Central America. Roughness values were generally lowest in doline landscapes, highest in cockpit terrain and intermediate in tower karst (fenglin). Brook and Hanson

(1991) analyzed wavelength variance within doline and cockpit landscapes in Jamaica, showing greater complexity of the latter and providing clues on the controlling role played by different fracture sets. Filho and Karmann (2007) computed surface roughness in the different geomorphic domains of the karst of the Serra da Bodoquena National Park in Brazil, obtaining the highest values in zones characterized by polygonal and labyrinth landscapes. Lyew-Ayee et al. (2007) compared the landscape and landforms in areas of Jamaica with cockpit and non-cockpit karst by morphometric analyses. They analyzed elevation range at multiple horizontal scales producing power-law functions of local relief versus window width. The components of the function corresponding to the intercept at the relief axis and the scaling exponent were used to assess topographic roughness. Not surprisingly, this rather sophisticated approach indicates that cockpit terrain has greater roughness.

The plan view location of a sinkhole can be represented by a point, commonly given by the centroid of the polygon that defines its edge or by the position of its deepest point or swallet. There may be a significant deviation between both points in sinkholes with internal asymmetry (Williams 1972a). Moreover, sinkholes may contain multiple sinks at similar elevation. Sinkhole density in a given area is calculated dividing the number of sinkholes by the surface area. However, this parameter does not take into consideration the areal extent of the sinkholes. In a region with small sinkholes, the depressions may occupy a limited proportion of the landscape, whereas in another region with the same sinkhole density but with much larger depressions, a great part or all of the space may be occupied by sinkholes (polygonal karst). For this reason, it is advisable to compute sinkhole density by number and by area (Pardo-Igúzquiza et al. 2016). The areal density is commonly expressed as the percentage of the area occupied by sinkholes. This parameter can be used to estimate the geomorphic imprint of sinkholes in an area, as a measure of the degree of pitting or "dolinization" (Bondesan et al. 1992). Table 6.4 includes sinkhole density data computed in different karst settings for a wide range of sinkhole types. The total volume of sinkholes divided by the surface of the study area provides an average measure of the differential surface lowering related to sinkhole development (chemical and mechanical erosion, subsidence). If the age of the geomorphic surface is known, the average lowering value can be used to estimate an average long-term lowering rate related to sinkhole development. The points assigned to each sinkhole (centroid, deepest sink) can be used to produce sinkhole density maps. These models are commonly generated calculating for each point of the territory the number of sinkhole-points that occur within a circular neighborhood of a given radius (Kernel density). These spatially distributed density models do not take into consideration the size of the sinkholes, giving the same weight to all the depressions regardless of their size. In areas where sinkholes have a wide range of dimensions, it may be interesting to produce areal-density maps displaying the spatial variation of the percentage area occupied by sinkholes within a search radius using a uniform Kernel function (Gutiérrez et al. 2019a; Parenti et al. 2020).

The points assigned to each sinkhole (centroid, deepest sink) can be used to analyze their spatial distribution (Williams 1972a, 1972b). An important feature with significant practical implications in the case of subsidence sinkholes is the degree of clustering versus dispersion. This property can be assessed by the Nearest Neighbor Index (NNI = L_a/L_e; Clark and Evans 1954), which compares the average distance between the nearest neighbors in the study area (L_a), and the mean theoretical distance between nearest sinkholes (L_e) in a field with random distribution and the same density (D). The theoretical distance L_e is given by $1/(2\sqrt{D})$, the derivation of which is explained by Clark and Evans (1954). The NNI ranges from 0 to 2.1491. A value of 0 indicates maximum aggregation or clustering, 1 random distribution, and 2.1491 a uniform pattern that is as evenly and widely spaced as possible (Figure 6.46). Note that this index quantifies the degree of dispersion of the points assigned to the sinkholes, rather than the polygons that define the sinkholes. Therefore, its actual meaning is affected by the size of the sinkholes. The distance between the edges of two

Table 6.4 Data of sinkhole density and nearest neighbor index calculated in a wide range of karst environments with solution sinkholes (gray pattern) and different types of subsidence sinkholes.

Location	Type of karst	Sinkhole type	Area (km²)	Number of depressions	Density of depressions (number km⁻²)	Density by area (%)	Nearest neighbor index	Pattern	Authors
Eleven limestone plateaus in the Taurus Mountains, Turkey	Bare limestone	Solution sinkholes	140 070	13 189	1.7–17.2				Öztürk et al. (2018b)
Northern Jamaica	Bare to covered limestone	Solution (and collapse?) sinkholes	13	37	2.85	41.6	1.294	Near random	Day (1976)
Hochschwab carbonate massif, Alps, Austria	Mostly bare carbonate bedrock	Mainly solution dolines	59	7151	3.91	6.7			Plan and Decker (2006)
Papua New Guinea (8 areas)	Bare limestone	Solution sinkholes		1228	10–22.1	~100	1.091–1.404	Near random	Williams (1972a)
Sierra Gorda, Betic Cordillera, southern Spain	Bare limestone	Karst depressions, mainly solution sinkholes	252	3100	12.3	3.24			Pardo-Igúzquiza et al. (2016)
Velebit Mountains, Croatia	Bare limestone	Solution sinkholes (mainly)	2274	40 000	17.6				Faivre and Reiffsteck (1999)
Krk Island, Croatia	Bare limestone	Solution sinkholes	68.5	–	20–57				Benac et al. (2013)
Sivas Basin, Turkey	Bare gypsum	Solution sinkholes	1609	42 127	26.2				Poyraz et al. (2021)
Montello Plateau, Pre-Alps, Italy (2 sectors)	Bare conglomerate	Solution sinkholes	27.47 and 33.3	1189 and 881	43.28 and 26.45	8.3–42.2 and 32.5–63.7			Ferrarese et al. (1998)

(Continued)

Table 6.4 (Continued)

Location	Type of karst	Sinkhole type	Area (km²)	Number of depressions	Density of depressions (number km⁻²)	Density by area (%)	Nearest neighbor index	Pattern	Authors
Matarsko podolje, Slovenia	Bare limestone and dolomite karst	Solution sinkholes	25	2350	94	11.7			Verbovšek and Gabor (2019)
Four karst massifs in Spain	Bare carbonate karst	Karst depressions, mainly solution sinkholes		324 to 3100			0.41–0.69	Clustered	Pardo-Igúzquiza et al. (2020)
Nullarbor Plain, Australia	Bare limestone	Collapse sinkholes	4200/11719	>615	>0.1–0.15				Doerr et al. (2012); Burnett et al. (2013)
Flint River, Albany, Georgia	Covered limestone	Cover-collapse sinkholes triggered by flooding	71	312	4.4		0.55	Clustered	Hyatt and Jacobs (1996)
Dougherty County, Georgia	Covered limestone	Cover-collapse and -suffosion sinkholes	183	3412 (all sinkholes) 275 (formed or enlarged between 1999 and 2011)	18.6			Clustered Random (new)	Cahalan and Milewski (2018)
Fluvia Valley, Pyrenees, Spain	Evaporite karst	Collapse and sagging sinkholes affecting cover, caprock and/or bedrock	45	135	3	1.5	1.1	Random	Gutiérrez et al. (2016)
Sant Miquel, Pyrenees, Spain	Hypogene evaporite karst	Mainly caprock-collapse sinkholes	6.7	94	14	10.3	1.3	Near random	Gutiérrez et al. (2019b)

Location	Karst type	Sinkhole type					Pattern	Reference
La Puebla, Ebro Valley, NE Spain	Covered evaporite karst	Cover- and bedrock-collapse sinkholes	0.25	158	633	22			Gutiérrez-Santolalla et al. (2005)
Kotido Crater, Mars	Evaporite karst	Bedrock-collapse sinkholes	21.7	513	23.6	3.8	0.82	Near random	Parenti et al. (2020)
Dead Sea, Jordan	Covered salt	Cover-collapse and -sagging	2.1	298	142		0.69	Clustered	Al-Halbouni et al. (2017)
Ambal salt pillow, Zagros Mountains, Iran	Salt covered by gypsum-rich caprock	Collapse sinkholes	4	693	170	6.7	0.3	Highly clustered	Gutiérrez and Lizaga (2016)
Konarsiah salt extrusion, Iran	Covered and bare salt karst	Various types of subsidence sinkholes	37.4	2631	70.2	11.2	1.007	Random	Zarei and Raeisi (2010)

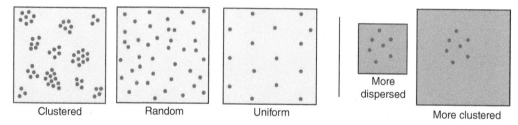

Figure 6.46 Sketches showing sinkhole populations represented as points with clustered, random, and uniform distribution. The diagrams on the right illustrate the impact of the boundaries and extent of the study area on the calculation of the NNI. The two study areas have the same sinkhole population but would yield significantly different nearest neighbor indexes.

large adjacent sinkholes may be small, whereas the distance between the centroids can be large. Moreover, NNI can yield markedly different values for the same sinkhole population depending on the area selected for its computation (e.g., minimum bounding polygon, quadrilateral polygon) (Figure 6.46). In the limestone karst of Minnesota, Gao et al. (2005) obtained significantly different NNI by changing the spatial scale of the analysis. Table 6.4 includes NNI calculated in a number of studies conducted in different karst environments with a wide range of sinkhole types. Solution sinkholes tend to display a more widespread and uniform distribution (e.g., polygonal karst; Williams 1972a), whereas sinkholes related to the different types of subsidence mechanisms often form clusters associated with the areas with more favorable conditions for their development. The NNI can be also used to explore whether new sinkholes tend to occur in the vicinity of the preexisting sinkholes or not, and therefore assess the prediction utility of the sinkholes (Drake and Ford 1972; Kemmerly 1982; Hyatt et al. 1999). This can be explored using the distance between each new sinkhole and the nearest old sinkhole and vice versa (Hyatt et al. 1999). Where new sinkholes preferentially form in the vicinity of previously existing ones, simple susceptibility models based on proximity could yield high prediction rates. There are other mathematical approaches that have been applied to describe the spatial distribution of sinkholes represented as points like the Ripley K-function (Pardo-Igúzquiza et al. 2020). Graphs showing the cumulative frequency of sinkholes (or new sinkholes) versus the nearest neighbor distance can be useful for assessing the width of the buffers around sinkholes that are expected to predict a certain percentage of new sinkholes (Gutiérrez et al. 2019a) (e.g., 70% of the new sinkholes occur at a distance less than 50 m from the older sinkholes). Hyatt et al. (1999), within an area of 71 km² of the covered limestone karst of Dougherty County, Georgia, examined the spatial relationships between preexisting sinkholes and new observed suffosion and cover-collapse sinkholes triggered by a major flood. The nearest neighbor analyses revealed that new sinkholes were not clustered near old sinkholes, most probably related to more diverse factors and triggers than the events associated with a single flooding event. In a mantled limestone karst area covering 183 km² in Georgia, Cahalan and Milewski (2018) mapped 3412 cover-collapse and cover-suffosion sinkholes and differentiated those that were formed or enlarged between 1999 and 2011. The complete set of inventoried sinkholes displays clustered distribution, in contrast with the random pattern of the recently active dolines. Good understanding of the general geological context and the geomorphological evolution of the region is essential to interpret the spatial distribution of sinkholes. For instance, in the Nullabor Plain of Australia, Burnett et al. (2013) identified a band 20–30 km wide and located 75 km inland with a higher concentration of blowholes. This belt with a high blowhole density was ascribed to relict flank margin caves developed along the freshwater-saltwater mixing zone during the Late Miocene (ca. 6 Myr).

In most cases, morphometric analyses are performed with a static picture of the sinkholes. However, it is important to bear in mind that morphometric attributes change through time due to multiple processes including the formation of new sinkholes, the growth of the existing ones, and the coalescence of adjoining depressions. Figure 6.47 shows a simple exercise that illustrates how some morphometric properties change through time by the lateral expansion of sinkholes that eventually coalesce and compete for space. It also illustrates the impact of the initial spatial distribution pattern on the evolution of the depressions and their morphometric features. The starting snapshots are two zones of equal area (0.41 ha) including 10 sinkholes each with initial diameters of 4.9 m. In one area, sinkholes display random distribution (NNI = 1) whereas in the other the spatial distribution is dispersed and close to uniform (NNI = 1.8). The following nine snapshots show the plan view and distribution of the depressions considering that the diameters of the depressions experience a 20% increase in each time-lapse. Sinkhole expansion results in the coalescence of adjoining depressions resulting in spatial and temporal patterns strongly influenced by the initial distribution of the sinkholes. The sketches and the computed parameters show the following remarkable aspects: (i) Sinkhole coalescence starts at an advanced stage in the dispersed population (stage 8), whereas it shows an early onset in the random population (stage 3). This has considerable impact on the temporal evolution of a number of parameters. (ii) The number of depressions and their density falls progressively since an early stage in the random population until it reaches a constant value. From stage 8 onward, there is only one compound depression and no competition for space. In the dispersed population, the number of sinkholes and their density remains constant until an advanced stage (stage 7). Then competition for space starts and the number of sinkholes and sinkhole density fall rapidly until there is just one compound depression. From stage 8 to 9 the number of depressions drops from 7 to 1. (iii) The percentage area covered by sinkholes increases more rapidly and reaches higher values in the dispersed population since there is less competition for space. This is illustrated by the contrasting evolution of the cumulative perimeter of the depressions, which determines the rate at which the area occupied by the depressions can increase by outward expansion. The cumulative perimeter of the random population shows a slow increase followed by a slow decrease. It rises rapidly in the dispersed population and from stage 8 onward, when coalescence starts, it drops sharply. Coalescence consumes perimetral length. Note that the decline of this parameter is also affected by the boundary effect of the study area. (iv) The average equivalent diameter, here corresponding to the diameter of a circle with a circumference equal to the perimeter of the sinkhole, shows a more rapid increase and higher values in the random population due to the early coalescence of sinkholes and the more sinuous geometry of their perimeter. This parameter is only computed up to stage 8, when sinkhole edges would progress beyond the boundaries of the study area. (v) The average elongation ratio in the random population rises to high values because of the early start of coalescence and the formation of more elongated compound depressions conditioned by the initial distribution. The average elongation ratio in the dispersed population is 1 until stage 8 when coalescence starts, followed by the development of a rather equidimensional compound depression with a ratio close to 1.

Morphometric techniques can be used to address the temporal evolution of sinkholes, including attributes such as their geometry, the degree of coalescence, or the spatial distribution patterns. However, these changes are in most cases very slow at the human-life time scale. Exceptions can be found in special situations such as sinkholes related to salt dissolution and developed in mechanically weak and easily erodible cover deposits (e.g., Dead Sea; Figure 6.30d). This limitation can be partially overcome by applying the ergodic hypothesis, whereby sampling across space is equivalent to sampling through time. Temporal changes can be analyzed studying landforms of differing ages at different locations substituting space for time. Its validity rests on the assumption

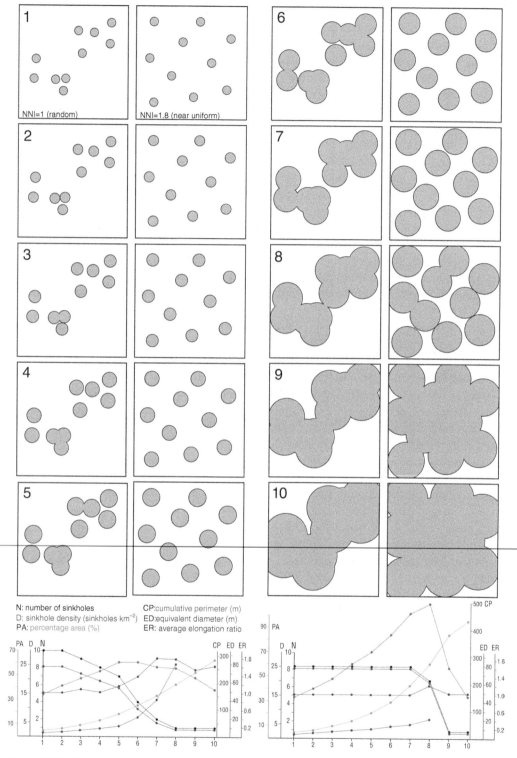

Figure 6.47 Sequence of sketches illustrating the impact of the initial spatial distribution (random versus dispersed) on the morphological and morphometric evolution of sinkholes affected by progressive expansion. The experiments simulate study areas of 0.41 ha, 10 sinkholes with initial diameters of 4.9 m, and an inter-stage diameter growth of 20%.

that geomorphic changes have followed a similar pattern in the different places considered in the analysis. Day (1983) compared solution sinkholes developed on a sequence of raised reef terraces in Barbados to infer long-term temporal changes. He found that doline density increases up to a maximum relative height of 150 m, decreasing above that elevation. This pattern can be attributed to sinkhole expansion and coalescence, involving an increase in the size of the depressions at the expense of the number of sinkholes and their density. In the Republic of Vanuatu, South Pacific Ocean, Strecker et al. (1986) analyzed the temporal evolution of karst topography on a chronosequence of uplifted coral terraces, with cone karst first appearing on surfaces dated at >149 to 500 kyr BP. Ferrarese et al. (1998) explored morphometric variations among different temporal populations of solution sinkholes developed on seven rock-strath terraces cut-across carbonate conglomerates in the Montello Hills, Italian Pre-Alps. They inferred a general deepening of the dolines with time. In a covered limestone karst in Georgia, Hyatt et al. (1999) compared the morphometric features of the cover-collapse and cover-suffosion sinkholes triggered by a flood with those of the preexisting sinkholes representing a population of degraded depressions that integrates a large chronological range. They found that old sinkholes are significantly larger and display higher length to depth ratios. Zarei and Raeisi (2010) and Zarei et al. (2012) investigated morphometric features of sinkholes in the active Konarsiah salt extrusion in Iran. This is a very special situation in which the karst bedrock is continuously flowing from the protrusion (summit dome) above the feeding vent toward the distal edge of the namakiers (salt glaciers) where capsoils reach higher thicknesses. This setting offers the opportunity to examine a complete temporal spectrum of sinkholes, which display increasing lengths and decreasing densities away from the summit of the salt extrusion. In the Sivas gypsum karst, Gökkaya et al. (2021) inventoried 295 bedrock collapse sinkholes showing a broad chronological spectrum, from fresh cylindrical depression tens of meters across, to mature tronco-conical basins with degraded slopes reaching more than 500 m in length. The morphometric data allowed the authors to infer a morphological evolution involving a relatively rapid post-collapse enlargement and deepening of the depressions by the solutional removal of large volumes of gypsum as solute load by downward vadose flows.

References

Abdulla, W.A. and Goodings, D.J. (1996). Modeling of sinkholes in weakly cemented sand. *Journal of Geotechnical Engineering* 122: 998–1005.

Abdulla, W.A. and Mollah, M.A. (1999). Detection and treatment of karst cavities in Kuwait. In: *Hydrogeology and Engineering Geology of Sinkholes and Karst. Proceedings of the Seventh Multidisciplinary Conference on Sinkholes and the Engineering and Environmental Impacts of Karst* (ed. B.F. Beck, A.J. Pettit and J.G. Herring), 123–127. Rotterdam: Balkema.

Abelson, M., Yechieli, Y., Baer, G. et al. (2017). Natural versus human control on subsurface salt dissolution and development of thousands of sinkholes along the Dead Sea coast. *Journal of Geophysical Research: Earth Surface* 122: 1262–1277.

Ahnert, F. and Williams, P.W. (1997). Karst landform development in a three-dimensional theoretical model. *Zeitschrift für Geomorphologie* 108: 63–80.

Al-Ansari, N., Adamo, N., Issa, I. et al. (2015). Mystery of Mosul Dam the most dangerous dam in the world: karstification and sinkholes. *Journal of Earth Sciences and Geotechnical Engineering* 5 (3): 33–45.

Alexander, E.C., Broberd, J.S., Kehren, A.R. et al. (1993). Bellechester Minnesota lagoon collapses. In: *Applied Karst Geology. Proceedings of the Fourth Multidisciplinary Conference on Sinkholes and the Engineering and Environmental Impacts of Karst* (ed. B.F. Beck), 63–72. Rotterdam: Balkema.

Alexander, S.C., Rahimi, M., Larson, E. et al. (2013). Combining LiDAR, aerial photography, and Pictometry® tools for karst features database management. In: *Proceedings of the Thirteenth Multidisciplinary Conference on Sinkholes and the Engineering and Environmental Impacts of karst* (ed. L. Land, D.H. Doctor and J.B. Stephenson), 441–448. Carlsbad: National Cave and Karst Research Institute.

Al-Halbouni, D., Holohan, E.P., Saberi, L. et al. (2017). Sinkholes, subsidence and subrosion on the eastern shore of the Dead Sea as revealed by a close-range photogrammetric survey. *Geomorphology* 285: 305–324.

Al-Halbouni, D., Holohan, E.P., Taheri, A. et al. (2018). Geomechanical modelling of sinkhole development using distinct elements: model verification for a single void space and application to the Dead Sea area. *Solid Earth* 9: 1341–1373.

Al-Halbouni, D., Holohan, E.P., Taheri, A. et al. (2019). Distinct element geomechanical modelling of the formation of sinkhole clusters within large-scale karstic depressions. *Solid Earth* 10: 1219–1241.

Anderson, N.L. and Hinds, R.C. (1997). Glacial loading and unloading: a possible cause of rock salt dissolution in the western Canada basin. *Carbonates and Evaporites* 12: 43–52.

Andrejchuk, V.N. (2002). Collapse above the world's largest potash mine (Ural, Russia). *International Journal of Speleology* 31: 137–158.

Andrejchuk, V.N. and Klimchouk, A.B. (2002). Mechanisms of karst breakdown formation in the gypsum karst of the fore-Ural region, Russia (from observations in the Kungurskaja Cave). *International Journal of Speleology* 31 (1): 89–114.

Angel, J.C., Nelson, D.O., and Panno, S.V. (2004). Comparison of a new GIS-based technique and a manual method for determining sinkhole density: an example from Illinois' sinkhole plain. *Journal of Cave and Karst Studies* 66: 9–17.

Applegate, P. (2003). Detection of sinkholes developed on shaly Ordovician limestones, Hamilton County, Ohio, using digital topographic data: dependence of topographic expression of sinkholes on scale, contour interval and slope. *Journal of Cave and Karst Studies* 65: 126–129.

Atapour, H. and Aftabi, A. (2002). Geomorphological, geochemical and geo-environmental aspects of karstification in the urban areas of Kerman city, southeastern Iran. *Environmental Geology* 42: 783–792.

Augarde, C.E., Lyamin, A.V., and Sloan, S.W. (2003). Prediction of undrained sinkhole collapse. *Journal of Geotechnical Geoenvironmental Engineering* 129: 197–205.

Aurit, M.D., Peterson, R.O., and Blanford, J.I. (2013). A GIS analysis of the relationship between sinkholes, dry-well complaints and groundwater pumping for frost-freeze protection in winter strawberry production in Florida. *PLoS One* 8: 1–9.

Autin, W.J. (2002). Landscape evolution of the Five Islands of south Louisiana: Scientific policy and salt dome utilization and management. *Geomorphology* 47: 227–244.

Baioni, D., Hajna, N.Z., and Wezel, F.C. (2009). Karst landforms in a Martian evaporitic dome. *Acta Carsologica* 38: 9–18.

Barreiro-Lostres, F., Moreno, A., Giralt, S. et al. (2014). Climate, palaeohydrology and land use change in the Central Iberian Range over the last 1.6 kyr: the La Parra Lake record. *The Holocene* 24: 1177–1192.

Basso, A., Bruno, E., Parise, M. et al. (2013). Morphometric analysis of sinkholes in a karst coastal area of southern Apulia (Italy). *Environmental Earth Sciences* 70: 2545–2559.

Bastos, A.C., Amado-Filho, G.M., Moura, R.L. et al. (2016). Origin and sedimentary evolution of sinkholes (buracas) in the Abrolhos continental shelf, Brazil. *Palaeogeography, Palaeoclimatology, Palaeoecology* 462: 101–111.

Bauer, C. (2015). Analysis of dolines using multiple methods applied to airborne laser scanning data. *Geomorphology* 250: 78–88.

Bayari, C.S., Pekkan, E., and Ozyurt, N.N. (2009). Obruks, as giant collapse dolines caused by hypogenic karstification in central Anatolia, Turkey: analysis of likely formation process. *Hydrogeology Journal* 17: 327–345.

Beck, B.F. (1986). A generalized genetic framework for the development of sinkholes and karst in Florida, USA. *Environmental Geology and Water Sciences* 8: 5–18.

Beck, B.F. (1988). Environmental and engineering effects of sinkholes. The processes behind the problems. *Environmental Geology Water Science* 12: 71–78.

Beck, B.F. (1991). On calculating the risk of sinkhole collapse. In: *Proceedings of the Appalachian Karst Symposium* (ed. E.H. Kastning), 231–236. Rotterdam: Balkema.

Beck, B.F. (2005). Soil piping and sinkhole failures. In: *Enyclopedia of Caves* (ed. W.B. White), 523–528. New York: Elsevier.

Beddows, P.A. (2004). Yucatán Phreas, Mexico. In: *Encyclopedia of Caves and Karst Science* (ed. J. Gunn), 786–788. New York: Fitzroy Dearborn.

Belloni, S. and Orombelli, G. (1970). Osservazioni e misure su alcuni tipi morfologici nei campi solcati del Carso Triestino. *Atti della Società Italiana di Storia Naturale e Museo Civico di Storia Naturale di Milano* 110: 317–372.

Benac, Č., Juračic, M., Matičec, D. et al. (2013). Fluviokarst and classical karst: examples from the Dinarics (Krk Island, Northern Adriatic, Croatia). *Geomorphology* 184: 64–73.

Benson, R.C. and Yuhr, L.B. (2015). *Site Characterization in Karst and Pseudokarst Terraines: Practical Strategies and Technology for Practicing Engineers, Hydrologists and Geologists.* Dordrecht: Springer.

Bernatek-Jakiel, A., Gutiérrez, F., Nadal-Romero, E. et al. (2019). Exploring the frequency-size relationships of pipe collapses in different morphoclimatic regions. *Geomorphology* 345: 106845.

Bezuidenhout, C.A. and Enslin, J.F. (1970). Surface subsidence and sinkholes in the dolomitic areas of the Far West Rand, Transvaal, Republic of South Africa. In: *Land Subsidence*, vol. 89, 482–495. Wallingford, England: International Association of Hydrological Sciences.

Bögli, A. (1960). Kalklösung und Karrenbildung. *Zeitschrift für Geomorphologie* 2: 4–21.

Bögli, A. (1980). *Karst Hydrogeology and Physical Speleology.* Berlin: Springer-Verlag.

Bonacci, O. and Roje-Bonacci, T. (2000). Interpretation of groundwater level monitoring results in karst aquifers: examples from the Dinaric karst. *Hydrological Processes* 14: 2423–2438.

Bonacci, O. and Roje-Bonacci, T. (2008). Water losses from the Ričice reservoir in the Dinaric karst. *Engineering Geology* 99: 121–127.

Bonacci, O. and Rubinić, J. (2009). Water losses from a reservoir built in karst: the example of the Boljuncica reservoir (Istria, Croatia). *Environmental Geology* 58: 339–345.

Bondesan, A., Meneghel, M., and Sauro, U. (1992). Morphometric analysis of dolines. *International Journal of Speleology* 21: 1–55.

Bonetto, S., Fiorucci, A., Formaro, M. et al. (2008). Subsidence hazards connected to quarrying activities in a karst area: the case of the Moncalvo sinkhole event (Piedmont, NW Italy). *Estonian Journal of Earth Sciences* 57: 125–134.

Brinkmann, R., Wilson, K., Elko, N. et al. (2007). Sinkhole distribution based on pre-development mapping in urbanized Pinellas County, Florida, USA. In: *Natural and Anthropogenic Hazards in Karst Areas: Recognition, Analysis and Mitigation*, vol. 279 (ed. M. Parise and J. Gunn), 5–11. London: Geological Society special publications.

Brinkmann, R., Parise, M., and Dye, D. (2008). Sinkhole distribution in a rapidly developing urban environment: Hillsborough County, Tampa Bay area, Florida. *Engineering Geology* 99: 169–184.

Brook, G.A. and Ford, D.C. (1978). The origin of labyrinth and tower karst and the climatic conditions necessary for their development. *Nature* 275: 493–496.

Brook, G.A. and Hanson, M. (1991). Double Fourier series analysis of cockpit and doline karst near Browns Town, Jamaica. *Physical Geography* 12: 37–54.

Brown, A.L., Reinhardt, E.G., van Hengstum, P.J. et al. (2014). A coastal Yucatan sinkhole records intense hurricane events. *Journal of Coastal Research* 30: 418–428.

Bruthans, J., Asadi, N., Filippi, M. et al. (2008). A study of erosion rates on salt diapir surfaces in the Zagros Mountains, SE Iran. *Environmental Geology* 53 (5): 1079–1089.

Burke, B., Slee, A., McIntosh, P.D. et al. (2020). A reactivated cave system induces rapidly developing cover-collapse sinkholes in Tasmania, Australia. *Journal of Cave and Karst Studies* 82: 31–50.

Burnett, S., Webb, J.A., and White, S. (2013). Shallow caves and blowholes on the Nullarbor Plain, Australia. Flank margin caves on a low gradient limestone platform. *Geomorphology* 201: 246–253.

Buttrick, D.B., Trollip, N.Y.G., Watermeyer, R.B. et al. (2011). A performance based approach to dolomite risk management. *Environmental Earth Sciences* 64: 1127–1138.

Cahalan, M.D. and Milewski, A.M. (2018). Sinkhole formation mechanisms and geostatistical-based prediction analysis in a mantled karst terrain. *Catena* 165: 333–344.

Calaforra, J.M. (1998). *Karstologìa de Yesos*. Monografìa Ciencias y Tecnologìa. Almeria: University of Almeria.

Ćalić, J. (2011). Karstic uvala revisited: Toward a redefinition of the term. *Geomorphology* 134: 32–42.

Calvo, J.P., Pozo, M., Silva, P.G. et al. (2013). Pattern of sedimentary infilling of fossil mammal traps formed in pseudokarst at Cerro de los Batallones, Madrid Basin, central Spain. *Sedimentology* 60: 1681–1708.

Carbonel, D., Gutiérrez, F., Linares, R. et al. (2013). Differentiating between gravitational and tectonic faults by means of geomorphological mapping, trenching and geophysical surveys. The case of the Zenzano Fault (Iberian Chain, N Spain). *Geomorphology* 189: 93–108.

Carbonel, D., Rodríguez, V., Gutiérrez, F. et al. (2014). Evaluation of trenching, ground penetrating radar (GPR), and electrical resistivity tomography (ERT) for sinkhole characterization. *Earth Surface Processes and Landforms* 39: 214–227.

Carbonel, D., Rodríguez-Tribaldos, V., Gutiérrez, F. et al. (2015). Investigating a damaging buried sinkhole cluster in an urban area (Zaragoza city, NE Spain) integrating multiple techniques: geomorphological surveys, DInSAR, DEMs, GPR, ERT, and trenching. *Geomorphology* 229: 3–16.

Carbonell, E., Burmúdez de Castro, J.M., Pares, J.M. et al. (2008). The first hominin species of Europe. *Nature* 452: 465–470.

Chen, J. (1988). Karst collapse in cities and mining areas, China. *Environmental Geology and Water Sciences* 12: 29–35.

Chen, J. and Xiang, S. (1991). Sinkhole collapse resulting from pumping of karst groundwater: a problem and its solutions. In: *Land Subsidence, International Association of Hydrological Sciences Publ*, vol. 200 (ed. A.I. Johnson), 313–322. International Association of Hydrological Sciences.

Chiba, T., Kaneta, S.I., and Suzuki, Y. (2008). Red relief image map: new visualization method for three dimensional data. *The International Archives of the Photogrammetry, Remote Sensing and Spatial Information Sciences* 37: 1071–1076.

Choppy, J. (1996). Les cannelures et rigoles sont des indicateurs climatiques. In: *Karren Landforms* (ed. J.J. Fornós and A. Ginés), 137–148. Palma de Mallorca: Universitat de les Illes Balears.

Chorley, R.J., Schumm, S.A., and Sugden, D.A. (1984). *Geomorphology*. New York: Methuen Inc.

Cigna, F., Banks, V.J., Donald, A.W. et al. (2017). Mapping ground instability in areas of geotechnical infrastructure using satellite InSAR and Small UAV surveying: a case study in Northern Ireland. *Geosciences* 7: 51.

Clark, P.J. and Evans, F.C. (1954). Distance to nearest neighbor as a measure of spatial relationships in populations. *Ecology* 35: 445–453.

Closson, D. and Karaki, N.A. (2009). Salt karst and tectonics: sinkholes development along tension cracks between parallel strike-slip faults, Dead Sea, Jordan. *Earth Surface Processes and Landforms* 34: 1408–1421.

Closson, D. and Karaki, N.A. (2015). Earthen dike leakage at the Dead Sea. In: *Engineering Geology for Society and Territory*, vol. 5 (ed. G. Lollino, A. Manconi, F. Guzzetti, et al.), 461–464. Dordrecht: Springer.

Closson, D., LaMoreaux, P.E., Abou-Karaki, N. et al. (2007). Karst system developed in salt layers of the Lisan Peninsula, Dead Sea, Jordan. *Environmental Geology* 52: 155–172.

Cooper, A.H. (1998). Subsidence hazards caused by the dissolution of Permian gypsum in England: geology, investigation and remediation. In: *Geohazards in Engineering Geology*, Engineering Geology Special Publications, vol. 15 (ed. J.G. Maund and M. Eddleston), 265–275. London: Geological Society.

Cooper, A.H. (2017). Voids. In: *Encyclopedia of Engineering Geology* (ed. P.T. Bobrowsky and B. Marker). Cham: Springer.

Cooper, J.D. and Keller, M. (2001). Paleokarst in the Ordovician of the southern Great Basin, USA: implications for sea-level history. *Sedimentology* 48: 855–873.

Cooper, A.H., Farrant, A.R., and Price, S.J. (2011). The use of karst geomorphology for planning, hazard avoidance and development in Great Britain. *Geomorphology* 134: 118–131.

Corral, A. And González, A., (2018). Power-law distributions in geoscience revisited. *Earth and Space Science* 6: 1–31.

Craig, W.H. (1990). Collapse of cohesive overburden following removal of support. *Canadian Geotechnical Journal* 27: 355–364.

Cramer, H. (1941). Die Systematik der Karstdolinen. *Neues Jahrbuch für Mineralogie, Geologie, und Palaontologie* 85: 293–382.

Crawford, N.C. (2001). Environmental problems associated with urban development upon karst, Bowling Green, Kentucky. In: *Geotechnical and Environmental Applications of Karst Geology and Hydrology. Proceedings of the Eight Multidisciplinary Conference on Sinkholes and the Engineering and Environmental Impacts of Karst* (ed. B.F. Beck and J.G. Herring), 397–424. Rotterdam: Balkema.

Crowther, J. (1996). Roughness (mm-scale) of limestone surfaces: examples from coastal and sub-aerial karren features in Mallorca. In: *Karren Landforms* (ed. J.J. Fornós and A. Ginés), 149–159. Palma de Mallorca: Universitat de les Illes Balears.

Crowther, J. (1998). New methodologies for investigating rillenkarren cross-sections: a case study at Lluc, Mallorca. *Earth Surface Processes and Landforms* 23: 333–344.

Croxton, N.M. (2003). Subsidence on Interstate 70 in Russell County, Kansas, related to salt dissolution. A history. In: *Evaporite Karst and Engineering/Environmental Problems in the United States*, vol. 109 (ed. K.S. Johnson and J.T. Neal), 397–424. Oklahoma Geological Survey Circular.

Cucchi, F. (2009). Kamenitzas. In: *Karst Rock Features. Karren Sculpturing* (ed. A. Ginés, M. Knez, T. Slabe, et al.), 139–150. Ljubljana: ZRC Publishing.

Cucchi, F., Radovich, N., and Sauro, U. (1990). I campi solcati di Borgo Grotta Gigante nel Carso Triestino. *International Journal of Speleology* 18: 117–144.

Currin, J.L. and Barfus, B.L. (1989). Sinkhole distribution and characteristics in Pasco County, Florida. In: *Engineering and Environmental Impacts of Sinkholes and Karst. Proceedings of the Third Multidisciplinary Conference on Sinkholes and the Engineering and Environmental Impacts of Karst* (ed. B.F. Beck), 97–106. Rotterdam: Balkema.

Cvijić, J. (1893). *Das Karstphänomen. Versuch einer morphologischen Monographie*, Geographische Abhandlungen herausgegeben von A. Penck, Wien, Band V, Heft 3, 1-114.

Daoxian, Y. (1987). Keynote address: Environmental and engineering problems of karst geology in China. In: *Karst Hydrogeology: Engineering and Environmental Applications. Proceedings of the*

Second Multidisciplinary Conference on Sinkholes and the Environmental Impacts of Karst (ed. B.F. Beck and W.L. Wilson), 1–11. Rotterdam: Balkema.

Daura, J., Sanz, M., Fornós, J. et al. (2014). Karst evolution of the Garraf Massif (Barcelona, Spain): doline formation, chronology and archaeopalaeontological archives. *Journal of Cave and Karst Studies* 76: 69–87.

Davis, A.D. and Rahn, P.H. (1997). Karstic gypsum problems at waste water stabilization sites in the Black Hills of South Dakota. *Carbonates and Evaporites* 12: 73–80.

Day, M. (1976). The morphology and hydrology of some Jamaican karst depressions. *Earth Surface Processes and Landforms* 1: 111–129.

Day, M.J. (1979). The hydrology of polygonal karst depressions in northern Jamaica. *Zeitschrift für Geomorphologie* Supplement Band 32: 25–34.

Day, M.J. (1983). Doline morphology and development in Barbados. *Annals of the Association of American Geographers* 73: 206–219.

Day, M.J. and Chenoweth, S. (2013). Surface roughness of karst landscapes. In: *Treatise of Geomorphology: Karst Geomorphology*, vol. 6 (ed. A. Frumkin), 157–163. Amsterdam: Elsevier.

Day, M.J. and Reynolds, B. (2012). Five Blues Lake National Park, Belize: a cautionary management tale. *Journal of Cave and Karst Studies* 74: 213–220.

Day, M.J. and Waltham, T. (2009). The pinnacle karrenfields of Mulu. In: *Karst Rock Features. Karren Sculpturing* (ed. A. Ginés, M. Knez, T. Slabe, et al.), 423–432. Ljubljana: ZRC Publishing.

De Bruyn, I.A. and Bell, F.G. (2001). The occurrence of sinkholes and subsidence depressions in the Far West Rand and Gauteg Province, South Africa, and their engineering implications. *Environmental and Engineering Geoscience* 7: 281–295.

De Carvalho-Junior, O.A., Fontes-Guimaraes, R., Montgomery, D.R. et al. (2014). Karst depression detection using ASTER, ALOS/PRISM and SRTM-derived digital elevation models in the Bambuí Group, Brazil. *Remote Sensing* 6: 330–351.

De Castro, T. and Horta, P.C. (2021). Assessment of a semi-automatic spatial analysis method to identify and map sinkholes in the Carste Lagoa Santa environmental protection unit, Brazil. *Environmental Earth Sciences* 80: 1–15.

De Waele, J. and Furlani, S. (2013). Seawater and biokarst effects on coastal limestones. In: *Treatise on Geomorphology. Karst Geomorphology*, vol. 6 (ed. A. Frumkin), 341–350. Amsterdam: Elsevier.

De Waele, J. and Melis, M.T. (2009). Geomorphology and geomorphological heritage of the Ifrane-Azrou region (Middle Atlas, Morocco). *Environmental Geology* 58 (3): 587–599.

De Waele, J., Mucedda, M., and Montanaro, L. (2009a). Morphology and origin of coastal karst landforms in Miocene and Quaternary carbonate rocks along the central-western coast of Sardinia (Italy). *Geomorphology* 106: 26–34.

De Waele, J., Lauritzen, S.-E., and Parise, M. (2011). On the formation of dissolution pipes in Quaternary coastal calcareous arenites in Mediterranean settings. *Earth Surface Processes and Landforms* 36: 143–157.

Del Prete, S., Di Crescenzo, G., Santangelo, N. et al. (2010a). Collapse sinkholes in Campania (southern Italy): predisposing factors, genetic hypothesis and susceptibility. *Zeitschrif für Geomorphologie* 54: 259–284.

Del Prete, S., Iovine, G., Parise, M. et al. (2010b). Origin and distribution of different types of sinkholes in the plain areas of Southern Italy. *Geodinamica Acta* 23: 113–127.

Desir, G., Gutiérrez, F., Merino, J. et al. (2018). Rapid subsidence in damaging sinkholes: measurement by high-precision leveling and the role of salt dissolution. *Geomorphology* 303: 393–409.

Desrochers, A. and James, N.P. (1987). Early Paleozoic surface and subsurface paleokarst: Middle Ordovician carbonates, Mingan Island, Québec. In: *Paleokarst* (ed. N.P. James and P.W. Choquette), 183–210. New York: Springer-Verlag.

Destephen, R.A. and Benson, C.P. (1993). Assessment of ground water withdrawal impacts in a karst area. In: *Applied Karst Geology. Proceedings of the Fourth Multidisciplinary Conference on Sinkholes and the Engineering and Environmental Impacts of Karst* (ed. B.F. Beck), 157–165. Rotterdam: Balkema.

Di Maggio, C., Madonia, G., Parise, M. et al. (2012). Karst of Sicily and its conservation. *Journal of Cave and Karst Studies* 74: 157–172.

Di Stefano, P. and Mindszenty, A. (2000). Fe-Mn-encrusted "Kamenitza" and associated features in the Jurassic of Monte Kumeta (Sicily): subaerial and/or submarine dissolution? *Sedimentary Geology* 132: 37–68.

Dini, S.M., Al-Khattabi, A.F., Wallace, C.A. et al. (2007). *Geologic Map of Parts of the Judayyidat 'Ar' Ar Quadrangle, Sheet 31F, Kingdom of Saudi Arabia*. Jeddah: Saudi Geological Survey.

Dini, S.M., Wallace, C.A., Halawani, M.A. et al. (2009). *Geologic Map of the As Sulayyil Quadrangle, Sheet 20H, Kingdom of Saudi Arabia*. Jeddah: Saudi Geological Survey.

Doctor, D.H. and Young, J.A. (2013). An evaluation of automated GIS tools for delineating karst sinkholes and closed depressions from 1-meter LIDAR-derived digital elevation data. In: *Sinkholes and the Engineering and Environmental Impacts of Karst* (ed. L. Land, D.H. Doctor and J.B. Stephenson), 449–458. Carlsbad: National Cave and Karst Research Institute.

Doerr, S.H., Davies, R.R., Lewis, A. et al. (2012). Origin and karst geomorphological significance of the enigmatic Australian Nullabor Plain "blowholes". *Earth Surface Processes and Landforms* 37: 253–261.

Doğan, U. and Cicek, I. (2002). Occurrence of cover-collapse sinkholes in the May Dam reservoir area (Konya, Turkey). *Cave and Karst Science* 29: 111–116.

Doğan, U. and Özel, S. (2005). Gypsum karst and its evolution east of Hafik (Sivas, Turkey). *Geomorphology* 71: 373–388.

Doğan, U. and Yilmaz, M. (2011). Natural and induced sinkholes of the Obruk Plateau and Karapinar-Hotamis Plain, Turkey. *Journal of Asian Earth Sciences* 40: 496–508.

Doğan, U., Koçyiğit, A., and Yeşilyurt, S. (2019). The relationship between Kestel Polje system and the Antalya Tufa Plateau: their morphotectonic evolution in Isparta Angle, Antalya-Turkey. *Geomorphology* 334: 112–125.

Dong, S., Wang, H., and Zhou, W. (2018). Comprehensive investigation and remediation of concealed karst collapse columns in Renlou coal mine, China. In: *Proceedings of the 15th Multidisciplinary Conference on Sinkholes and the Engineering and Environmental Impacts of Karst and the 3rd Appalachian Karst Symposium* (ed. I.D. Sasowsky, M.J. Byle and L. Land), 289–296. Carlsbad: National Cave and Karst Research Institute.

Dou, J., Li, X., Yunus, A.P. et al. (2015). Automatic detection of sinkhole collapses at finer resolutions using a multi-component remote sensing approach. *Natural Hazards* 78: 1021–1044.

Dougherty, P. (2005). Sinkhole destruction of Corporate Plaza, Pennsylvania. In: *Sinkholes and Subsidence* (ed. T. Waltham, F. Bell and M. Culshaw), 304–308. Chichester: Springer.

Drake, J.J. and Ford, D.C. (1972). The analysis of growth patterns of two-generation populations: the examples of karst sinkholes. *Canadian Geographer* 16: 381–384.

Drew, D. (2009). Coastal and lacustrine karren in western Ireland. In: *Karst Rock Features. Karren Sculpturing* (ed. A. Ginés, M. Knez, T. Slabe and W. Dreybrodt), 503–512. Ljubljana: ZRC Publishing.

Dreybrodt, W. and Kaufmann, G. (2007). Physics and chemistry of dissolution on subaerially exposed soluble rocks by flowing water films. *Acta Carsologica* 36: 357–367.

Dreybrodt, W., Gabrovšek, F., and Perne, M. (2005a). Condensation corrosion: a theoretical approach. *Acta Carsologica* 34 (2): 317–348.

Drumm, E.C., Akturk, O., Akgun, H. et al. (2009). Stability charts for the collapse of residual soil in karst. *Journal of Geotechnical and Geoenvironmental Engineering* 135: 925–931.

Dunkerley, D.L. (1979). The morphology and development of rillenkarren. *Zeitschrift für Geomorphologie* 23: 332–348.

Dunkerley, D.L. (1983). Lithology and micro-topography in the Chillagoe karst, Queensland, Australia. *Zeitschrift für Geomorphologie* 27: 191–204.

Durán, J.J., Andreo, B., Carrasco, F. et al. (2005). Central Andalusia: Karst, Paleoclimate and Neoseismotectonics. In: *Sixth International Conference on Geomorphology* (ed. G. Desir, F. Gutiérrez and M. Gutiérrez). Zaragoza: Field trip guide A-6.

Dyni, R.C. (1986). Subsidence investigations over salt-solution mines, Hutchinson, KS. *Bureau of Mines Information Circular* 9083: 1–23.

Ege, J.R. (1984). Mechanisms of surface subsidence resulting from solution extraction of salt. Geological Society of America. *Reviews in Engineering Geology* 6: 203–221.

El Aref, M.M. and Refai, E. (1987). Paleokarst processes in the Eocene limestones of the Pyramids Plateau, Giza. *Egypt. Journal of African Earth Sciences* 6: 367–377.

Elorza, J. and Higuera-Ruiz, R. (2016). Cilindros huecos relictos (tipo Röhrenkarren) generados en las calizas urgonianas de la costa oriental cántabra: morfometría, mecanismo de formación y consideraciones paleoambientales. *Revista de la Sociedad Geológica de España* 29: 59–77.

Elorza, J., Higuera-Ruiz, R., and Pascual, A. (2019). Karstificación litoral en calizas-dolomías urgonianas (Aptiense-Albiense) de la costa oriental de Cantabria: meteorización química, actividad biológica y abrasión mecánica. *Revista de la Sociedad Geológica de España* 32: 3–24.

Eraso, A., Trzhtsinsky, Y., and Castrillo, A. (1995). Dolinas de colapso y karst en yeso en la plataforma cámbrica del este de Siberia. *Boletín Geológico y Minero* 106: 373–378.

Fabregat, I., Gutiérrez, F., Roqué, C. et al. (2017). Reconstructing the internal structure and long-term evolution of hazardous sinkholes combining trenching, electrical resistivity imaging (ERI) and ground penetrating radar (GPR). *Geomorphology* 285: 287–304.

Fabregat, I., Gutiérrez, F., Roqué, C. et al. (2019). Subsidence mechanisms and sedimentation in alluvial sinkholes inferred from trenching, electrical resistivity imaging (ERI) and ground penetrating radar (GPR). Implications for subsidence and flooding hazard assessment. *Quaternary International* 525: 1–15.

Faivre, S. and Reiffsteck, P. (1999). Spatial distribution of dolines as an indicator of recent deformations on the Velebit mountain range (Croatia). *Géomorphologie: Relief, Processus, Environment* 5: 129–142.

Farrant, A.R. and Cooper, A.H. (2008). Karst geohazards in the UK: the use of digital data for hazard management. *Quarterly Journal of Engineering Geology and Hydrogeology* 41: 339–356.

Ferrarese, F., Sauro, U., and Tonello, C. (1998). The Montello Plateau. Karst evolution of an alpine neotectonic morphostructure. *Zeitschrift für Geomorphologie* Supplement Band 109: 41–62.

Festa, V., Fiore, A., Parise, M. et al. (2012). Sinkhole evolution in the Apulian karst of southern Italy: a case study, with some considerations on sinkhole hazards. *Journal of Cave and Karst Studies* 74: 137–147.

Fidelibus, M.D., Gutiérrez, F., and Spilotro, G. (2011). Human-induced hydrogeological changes and sinkholes in the coastal gypsum karst of Lesina Marina area (Foggia Province, Italy). *Engineering Geology* 118: 1–19.

Filho, W.S. and Karmann, I. (2007). Geomorphological map of the Serra da Bodoquena karst, west-central Brazil. *Journal of Maps* 3: 282–295.

Filin, S., Baruch, A., Avni, Y. et al. (2011). Sinkhole characterization in the Dead Sea area using airborne laser scanning. *Natural Hazards* 58 (3): 1135–1154.

Fiol, L., Fornós, J.J., and Ginés, A. (1996). Effects of biokarstic processes on the development of solutional rillenkarren in limestone rocks. *Earth Surface Processes and Landforms* 21: 447–452.

Fleury, S. (2009). *Land Use Policy and Practice on Karst Terrains. Living on Limestone*. Chichester: Springer.

Florea, L.J. (2005). Using state-wide GIS data to identify the coincidence between sinkholes and geological structure. *Journal of Cave and Karst Studies* 67: 120–124.

Folk, R.L., Roberts, H.H., and Moore, C.H. (1973). Black phytokarst from Hell, Cayman Islands, British West Indies. *Geological Society of America Bulletin* 84: 2351–2360.

Foose, R.M. (1953). Ground-water behaviour in the Hershey Valley, Pennsylvania. *Geological Society of America Bulletin* 64: 623–645.

Ford, D.C. and Lundberg, J. (1987). A review of dissolutional rills in limestone and other soluble rocks. *Catena Supplement* 8: 119–140.

Ford, D.C. and Williams, P.W. (1989). *Karst Geomorphology and Hydrology*. London: Unwin Hyman.

Ford, D.C. and Williams, P.W. (2007). *Karst Hydrogeology and Geomorphology*. Chichester, UK: Wiley.

Fornós, J.J. and Ginés, A. (ed.) (1996). *Karren Landforms*. Palma de Mallorca: Universitat de les Illes Balears.

Fox, W.T. (2005). Bioerosion. In: *Encyclopedia of Coastal Science* (ed. M.L. Schwartz), 191–192. Dordrecht: Springer.

Frisbee, M.D., Meyers, Z.P., Miller, J.B. et al. (2019). Processes leading to the re-activation of a sinkhole in buried karst and the subsequent drying of waterfalls in a small catchment located in northern Indiana, USA. *Journal of Cave and Karst Studies* 81: 69–83.

Frumkin, A. (2013). Salt karst. In: *Treatise on Geomorphology. Karst Geomorphology*, vol. 6 (ed. A. Frumkin and J. Shroder), 407–424. Amsterdam: Elsevier.

Frumkin, A., Ezersky, M., Al-Zoubi, A. et al. (2011). The Dead Sea sinkhole hazard: Geophysical assessment of salt dissolution and collapse. *Geomorphology* 134: 102–1117.

Frumkin, A., Zaidner, Y., Na'aman, I. et al. (2015). Sagging and collapse sinkholes over hypogenic hydrothermal karst in a carbonate terrain. *Geomorphology* 229: 45–57.

Fuleihan, N.F., Cameron, J.E., and Henry, J.F. (1997). The hole story: how a sinkhole in a phosphogypsum pile was explored and remediated. In: *Engineering Geology and Hydrology of Karst Terranes. Proceedings of the Sixth Multidisciplinary Conference on Sinkholes and the Engineering and Environmental Impacts of Karst* (ed. B.F. Beck and J.B. Stephenson), 363–369. Rotterdam: Balkema.

Furlani, S., Cucchi, F., Forti, F. et al. (2009). Comparison between coastal and inland Karst limestone lowering rates in the northeastern Adriatic Region (Italy and Croatia). *Geomorphology* 104 (1-2): 73–81.

Gabrovšek, F. and Stepišnik, U. (2011). On the formation of collapse dolines: a modelling perspective. *Geomorphology* 134 (1-2): 23–31.

Gallay, M. (2013). Direct acquisition of data: airborne laser scanner. In: *Geomorphological Techniques*. London: British Society for Geomorphology.

Galve, J.P., Gutiérrez, F., Lucha, P. et al. (2009). Sinkholes in the salt-bearing evaporite karst of the Ebro River valley upstream of Zaragoza city (NE Spain). Geomorphological mapping and analysis as a basis for risk management. *Geomorphology* 108: 145–158.

Galve, J.P., Remondo, J., and Gutiérrez, F. (2011). Improving sinkhole hazard models incorporating magnitude-frequency relationships and nearest neighbor analysis. *Geomorphology* 134: 157–170.

Galve, J.P., Tonelli, C., Gutiérrez, F. et al. (2015). New insights into the genesis of the Miocene collapse structures of the island of Gozo (Malta, central Mediterranean Sea). *Journal of the Geological Society* 172: 336–348.

Gams, I. (1973a). Forms of subsoil karst. In: *Proceedings of the 6th International Congress of Speleology* (ed. V. Panoš), 169–179. Prague: Academia.

Gams, I. (2000). Doline morphogenetic processes from global and local viewpoints. *Acta Carsologica* 29: 123–138.

Gao, Y. and Weary, D.J. (2008). A conceptual database model for spatial analysis and resource management in karst. In: *Proceedings of the Eleventh Multidisciplinary Conference on Sinkholes and the Engineering and Environmental Impacts of Karst* (ed. L. Yuhr, E.C. Alexander and B.F. Beck), 131–145. Reston: American Society of Civil Engineering.

Gao, Y., Alexander, E.C., and Tipping, R.G. (2002). The development of a karst feature database for Southeastern Minnesota. *Journal of Cave and Karst Studies* 64: 51–57.

Gao, Y., Alexander, E.C., and Barnes, R.J. (2005). Karst database implementation in Minnesota: analysis of sinkhole distribution. *Environmental Geology* 47: 1083–1098.

Gao, Y., Luo, W., Jiang, X. et al. (2013). Investigations of large scale sinkhole collapses, Labin, Guangxi, China. In: *Proceedings of the Thirteenth Multidisciplinary Conference on Sinkholes and the Engineering and Environmental Impacts of karst* (ed. L. Land, D.H. Doctor and J.B. Stephenson), 327–331. Carlsbad: National Cave and Karst Research Institute.

Garašić, M. (2012). Crveno jezero, the biggest sinkhole in Dinaric Karst (Croatia). *Geophysical Research Abstracts*. EGU General Assembly 2012, Vol. 14, EGU2012-7132-1.

García-Moreno, I. and Mateos, R.M. (2011). Sinkholes related to discontinuous pumping: susceptibility mapping base on geophysical studies. The case of Crestatx (Mallorca, Spain). *Environmental Earth Sciences* 64: 523–537.

Gary, M.O. and Sharp, J.M. Jr. (2006). Volcanogenetic karstification of Sistema Zacatón, Mexico. In: *Perspectives on Karst Geomorphology, Hydrology and Geochemistry. A Tribute Volume to Derek C. Ford and William B. White*, vol. 404, 79–89. Geological Society of America, Special Paper.

Gascoyne, M., Benjamin, G.J., Schwarcz, H.P. et al. (1979). Sea-level lowering during the Illinoian glaciation: evidence from a Bahama "Blue Hole". *Science* 205: 806–808.

Ginés, À. (1990). Utilización de las morfologías de lapiaz como geoindicadores ecológicos en la Serra de Tramuntana (Mallorca). *Endins: publicació d'espeleologia*: 27–40.

Gilli, É. (2019). *The Ankarana Plateau in Madagascar: Tsingy, Caves, Volcanoes and Sapphires*. Cham, Switzerland: Springer.

Ginés, A. (1998). Dades morfomètriques sobre les estries de lapiaz dels Alps calcaris suïssos i la seva comparació amb les estries de la Serra de Tramuntana. *Endins* 22: 109–118.

Ginés, A. (2002). Geoecología de las formas de lapiaz y correlación entre ecosistemas kársticos. In: *Karst and Environment* (ed. F. Carrasco, J.J. Durán and B. Andreo), 401–407. Nerja, Spain: Fundación Cueva de Nerja.

Ginés, A. (2004). Karren. In: *Encyclopedia of Caves and Karst Science* (ed. J. Gunn), 470–473. New York: Fitzroy Dearborn.

Ginés, A. (2009). Karren field landscapes and karren forms. In: *Karst Rock Features. Karren Sculpturing* (ed. A. Ginés, M. Knez, T. Slabe, et al.), 13–24. Ljubljana: ZRC Publishing.

Ginés, A. and Lundberg, J. (2009). Rainpits. In: *Karst Rock Features. Karren Sculpturing* (ed. A. Ginés, M. Knez, T. Slabe, et al.), 185–210. Ljubljana: ZRC Publishing.

Ginés, A., Knez, M., Slabe, T. et al. (ed.) (2009). *Karst Rock Features. Karren Sculpturing*. Ljubljana: ZRC Publishing.

Gischler, E., Shinn, E.A., Oschmann, W. et al. (2008). A 1500-yr Holocene Caribbean climate archive from the Blue Hole, Lighthouse Reef, Belize. *Journal of Coastal Research* 24: 1495–1505.

Glew, J.R. and Ford, D.C. (1980). A simulation of the development of rillenkarren. *Earth Surface Processes* 5: 25–36.

Goff, J.A., Gullick, S.P.S., Perez-Cruz, L. et al. (2016). Solution pans and linear sand bedforms on the bare-rock limestone shelf of the Campeche Bank, Yucatán Peninsula, Mexico. *Continental Shelf Research* 117: 57–66.

Gökkaya, E. and Tunçel, E. (2019). Natural and human-induced subsidence due to gypsum dissolution: a case study from Inandik, central Anatolia, Turkey. *Journal of Cave and Karst Studies* 81: 221–232.

Gökkaya, E., Gutiérrez, F., Ferk, M. et al. (2021). Sinkhole development in the Sivas gypsum karst. Turkey. *Geomorphology* 386: 107746.

Goldie, H.S. (2009). Case studies of grikes in the British Isles. In: *Karst Rock Features. Karren Sculpturing* (ed. A. Ginés, M. Knez, T. Slabe, et al.), 275–290. Ljubljana: ZRC Publishing.

Goldie, H.S. and Cox, N.J. (2000). Comparative morphometry of limestone pavements in Switzerland, Britain and Ireland. *Zeitschrift für Geomorphologie*, Supplement Band 122: 85–112.

Gombert, P., Orsat, J., Mathon, D. et al. (2015). Role des effondrements karstiques sur les desordres survenus sur les digues de Loire dans le Val D'Orleans (France). *Bulletin of Engineering Geology and the Environment* 74: 125–140.

Gómez-Pujol, L. and Fornós, J.J. (2009a). Microkarren. In: *Karst Rock Features. Karren Sculpturing* (ed. A. Ginés, M. Knez, T. Slabe, et al.), 73–84. Ljubljana: ZRC Publishing.

Gómez-Pujol, L. and Fornós, J.J. (2009b). Coastal karren in the Balearic islands. In: *Karst Rock Features. Karren Sculpturing* (ed. A. Ginés, M. Knez, T. Slabe, et al.), 487–502. Ljubljana: ZRC Publishing.

Gongyu, L. and Wanfang, Z. (1999). Sinkholes in karst mining areas in China and some methods of prevention. *Engineering Geology* 52: 45–50.

González-Díez, A., Barreda-Argüeso, J.A., Rodríguez-Rodríguez, L. et al. (2021). The use of filters based on the fast Fourier transform applied to DEMs for the objective mapping of karstic features. *Geomorphology* 385: 107724.

Goudie, A.S. (2010). Dayas: distribution and morphology of dryland solutional depressions developed in limestones. *Zeitschrift für Geomorphologie* 54: 145–159.

Goudie, A.S., Bull, P.A., and Magee, A.W. (1989). Lithological control on Rillenkarren development in the Napier Range, Western Australia. *Zeitschrift für Geomorphologie* 75: 95–114.

Goudie, A.S., Viles, H., Allison, R. et al. (1990). The geomorphology of the Napier Range, Western Australia. *Transactions of the Institute of British Geographers* 15: 308–322.

Gowan, S.W. and Trader, S.M. (2003). Mechanism of sinkhole formation in glacial sediments above Retsof Salt Mine, Western New York. In: *Evaporite Karst and Engineering/Environmental Problems in the United States*, vol. 109 (ed. K.S. Johnson and J.T. Neal), 321–336. Oklahoma Geological Survey Circular.

Grimes, K.G. (2007). Microkarren in Australia - a request for information. *Helictite* 40: 21–23.

Grimes, K.G. (2009). Solution pipes and pinnacles in syngenetic karst. In: *Karst Rock Features. Karren Sculpturing* (ed. A. Ginés, M. Knez, T. Slabe, et al.), 513–524. Ljubljana: ZRC Publishing.

Grimes, K.G. (2012). Surface karst features of the Judbarra/Gregory National Park, Northern Territory, Australia. *Helictite* 41: 15–36.

Grosch, J.J., Touma, F.T., and Richards, D.P. (1987). Solution cavities in the limestone of eastern Saudi Arabia. In: *Karst Hydrogeology: Engineering and Environmental Applications. Proceedings of the Second Multidisciplinary Conference on Sinkholes and the Environmental Impacts of Karst* (ed. B.F. Beck and W.L. Wilson), 73–78. Rotterdam: Balkema.

Grube, A. and Rickert, B.H. (2019). Karstification on the Elmshorn salt diapir (SW Schleswig-Holstein, Germany). *Zeitschrift der Deutschen Gesellschaft für Geowissenschaften* 169: 547–566.

Guerrero, J., Gutiérrez, F., and Lucha, P. (2008a). Impact of halite dissolution subsidence on Quaternary fluvial terrace development: Case study of the Huerva River, Ebro Basin, NE Spain. *Geomorphology* 100: 164–179.

Guerrero, J., Gutiérrez, F., Bonachea, J. et al. (2008b). A sinkhole susceptibility zonation based on paleokarst analysis along a stretch of the Madrid-Barcelona high-speed railway built over salt-bearing evaporites (NE Spain). *Engineering Geology* 102: 62–73.

Guerrero, J., Gutiérrez, F., and Galve, J.P. (2013). Large depressions, thickened terraces, and gravitational deformation in the Ebro River valley (Zaragoza area, NE Spain): Evidence of glauberite and halite interstratal karstification. *Geomorphology* 196: 162–176.

Gunn, J. (1981b). Hydrological processes in karst depressions. *Zeitschrift für Geomorphologie* 25: 313–331.

Gutiérrez, F. (2010). Hazards associated with karst. In: *Geomorphological Hazards and Disaster Prevention* (ed. I. Alcántara and A. Goudie), 161–175. Cambridge: Cambridge University Press.

Gutiérrez, M. (2013). *Geomorphology*. Boca Raton: CRC Press.

Gutiérrez, F. (2016). Sinkhole hazards. In: *Oxford Research Encyclopedia of Natural Hazard Science*. Oxford: Oxford University Press. https://oxfordre.com/naturalhazardscience/view/10.1093/acrefore/9780199389407.001.0001/acrefore-9780199389407-e-40.

Gutiérrez, F. and Cooper, A.H. (2002). Evaporite dissolution subsidence in the historical city of Calatayud, Spain: damage appraisal and prevention. *Natural Hazards* 25: 259–288.

Gutiérrez, F. and Cooper, A.H. (2013). Surface morphology of gypsum karst. In: *Treatise of Geomorphology. Karst Geomorphology*, vol. 6 (ed. A. Frumkin), 425–437. Amsterdam: Elsevier.

Gutiérrez, F. and Gutiérrez, M. (2016). *Landforms of the Earth. An Illustrated Guide*. Dordrecht: Springer.

Gutiérrez, F. and Lizaga, I. (2016). Sinkholes, collapse structures and large landslides in an active salt dome submerged by a reservoir: the unique case of the Ambal ridge in the Karun River, Zagros Mountains. *Iran. Geomorphology* 254: 88–103.

Gutiérrez, F., Gutiérrez, M. and Gracia, J. (2005). Karst, neotectonics and periglacial features in the Iberian Range. *Sixth International Conference of Geomorphology*, Field Trip Guide C-5, 55. https://geomorfologia.es/sites/default/files/C5%20Teruel_0.pdf

Gutiérrez, F., Galve, J.P., Guerrero, J. et al. (2007). The origin, typology, spatial distribution, and detrimental effects of the sinkholes developed in the alluvial evaporite karst of the Ebro River valley downstream Zaragoza city (NE Spain). *Earth Surface Processes and Landforms* 32: 912–928.

Gutiérrez, F., Guerrero, J., and Lucha, P. (2008a). A genetic classification of sinkholes illustrated from evaporite paleokarst exposures in Spain. *Environmental Geology* 53: 993–1006.

Gutiérrez, F., Calaforra, J.M., Cardona, F. et al. (2008b). Geological and environmental implications of the evaporite karst in Spain. *Environmental Geology* 53: 951–965.

Gutiérrez, F., Cooper, A.H., and Johnson, K.S. (2008c). Identification, prediction and mitigation of sinkhole hazards in evaporite karst areas. *Environmental Geology* 53: 1007–1022.

Gutiérrez, F., Guerrero, J., and Lucha, P. (2008d). Quantitative sinkhole hazard assessment. A case study from the Ebro Valley evaporite alluvial karst (NE Spain). *Natural Hazards* 45: 211–233.

Gutiérrez, F., Galve, J.P., Lucha, P. et al. (2009). Investigation of a large collapse sinkhole affecting a multi-storey building by means of geophysics and the trenching technique (Zaragoza city, NE Spain). *Environmental Geology* 58: 1107–1122.

Gutiérrez, F., Galve, J.P., Lucha, P. et al. (2011). Integrating geomorphological mapping, trenching, InSAR and GPR for the identification and characterization of sinkholes: a review and application in the mantled evaporite karst of the Ebro Valley (NE Spain). *Geomorphology* 134: 144–156.

Gutiérrez, F., Parise, M., De Waele, J. et al. (2014b). A review on natural and human-induced geohazards and impacts in karst. *Earth-Science Reviews* 138: 61–88.

Gutiérrez, F., Mozafari, M., Carbonel, D. et al. (2015). Leakage problems in dams built on evaporites. The case of La Loteta Dam (NE Spain), a reservoir in a large karstic depression generated by interstratal salt dissolution. *Engineering Geology* 185: 139–154.

Gutiérrez, F., Fabregat, I., Roqué, C. et al. (2016). Sinkholes and caves related to evaporite dissolution in a stratigraphically and structurally complex setting, Fluvia Valley, eastern Spanish Pyrenees. Geological, geomorphological and environmental implications. *Geomorphology* 267: 76–97.

Gutiérrez, F., Zarroca, M., Castañeda, C. et al. (2017). Paleoflood records from sinkholes. An example from the Ebro River floodplain, NE Spain. *Quaternary Research* 88: 71–88.

Gutiérrez, F., Zarroca, M., Linares, R. et al. (2018). Identifying the boundaries of sinkhole and subsidence areas and establishing setback distances. *Engineering Geology* 233: 255–268.

Gutiérrez, F., Benito-Calvo, A., Carbonel, D. et al. (2019a). Review on sinkhole monitoring and performance of remediation measures by high-precision leveling and terrestrial laser scanner in the salt karst of the Ebro Valley, Spain. *Engineering Geology* 248: 283–308.

Gutiérrez, F., Fabregat, I., Roqué, C. et al. (2019b). Sinkholes in hypogene versus epigene karst systems, illustrated with the hypogene gypsum karst of the Sant Miquel de Campmajor Valley, NE Spain. *Geomorphology* 328: 57–78.

Gutiérrez-Santolalla, F., Gutiérrez-Elorza, M., Marín, C. et al. (2005). Spatial distribution, morphometry and activity of La Puebla de Alfindén sinkhole field in the Ebro River valley (NE Spain), applied aspects for hazard zonation. *Environmental Geology* 48: 360–369.

Guzina, B.J., Sarić, M., and Petrović, N. (1991). Seepage and dissolution at foundations of a dam during the first impounding of the reservoir. In: *7th International Congress on Large Dams*, 1459–1475. Vienna: ICOLD.

Guzzetti, F., Malamud, B.D., Turcotte, D.L. et al. (2002). Power-law correlations of landslide areas in central Italy. *Earth and Planetary Science Letters* 195: 169–183.

Hackney, C. and Clayton, A.I. (2015). Unmanned aerial vehicles (UAVs) and their application in geomorphic mapping. In: *Geomorphological Techniques*. London: British Society for Geomorphology.

Hart, E.A. (2014). Legacy sediment stored in sinkholes: a case study of three urban watersheds in Tennessee, United States of America. *Physical Geography* 35: 514–531.

He, K., Liu, C., and Wang, S. (2003). Karst collapse related to over-pumping and a criterion for its stability. *Environmental Geology* 43: 720–724.

He, K., Wang, B., and Zhou, D. (2004). Mechanism and mechanical model of karst collapse in an over-pumping area. *Environmental Geology* 46: 1102–1107.

He, K., Yu, G., and Lu, Y. (2009). Palaeo-karst collapse pillars in northern China and their damage to the geological environments. *Environmental Geology* 58: 1029–1040.

Heidari, M., Khanlari, G.R., Beydokhti, A.T. et al. (2011). The formation of cover collapse sinkholes in North of Hamedan, Iran. *Geomorphology* 132: 76–86.

Higgins, C.G. (1980). Nips, notches, and the solution of coastal limestone: an overview of the problem with examples from Greece. *Estuarine and Coastal Marine Science* 10: 15–30.

Hiller, T., Romanov, D., Gabrovšek, F. et al. (2014). The creation of collapse dolines: a 3D modeling approach. *Acta Carsologica* 43: 241–255.

Hobléa, F., Jaillet, S., Maire, R. et al. (2001). Erosion et ruissellement sur karst nu: les îles subpolaires de la Patagonie chilienne (Magallanes, Chili). *Karstologia* 38 (1): 13–18.

Hofierka, J., Gallay, M., Bandura, P. et al. (2018). Identification of karst sinkholes in a forested karst landscape using airborne laser scanning data and water flow analysis. *Geomorphology* 308: 265–277.

Huang, J.M., Tong, J., Shelley, M. et al. (2020). Ultra-sharp pinnacles sculpted by natural convective dissolution. *Proceedings of the National Academy of Sciences* 117 (38): 23339–23344.

Hunt, B.B., Smith, B.A., Adams, M.T. et al. (2013). Cover-collapse sinkhole development in the Cretaceous Edwards limestone, central Texas. In: *Sinkholes and the Engineering and Environmental Impacts of Karst* (ed. L. Land, D.H. Doctor and J.B. Stephenson), 89–102. Carlsbad: National Cave and Karst Research Institute.

Hutchinson, D.W. (1996). Runnels, rinnenkarren and mäanderkarren: form, classification and relationships. In: *Karren Landforms* (ed. J.J. Fornós and A. Ginés), 209–223. Palma de Mallorca: Universitat de les Illes Balears.

Hyatt, J.A. and Jacobs, P.M. (1996). Distribution and morphology of sinkholes triggered by flooding following Tropical Storm Alberto at Albany, Georgia, USA. *Geomorphology* 17 (4): 305–316.

Hyatt, J., Wilkes, H., and Jacobs, P.M. (1999). Spatial relationship between new and old sinkholes in covered karst, Albany, Georgia, USA. In: *Hydrogeology and Engineering Geology of Sinkholes and Karst. Proceedings of the Seventh Multidisciplinary Conference on Sinkholes and the Engineering and Environmental Impacts of Karst* (ed. B.F. Beck, A.J. Pettit and J.G. Herring), 37–44. Rotterdam: Balkema.

Iijima, Y. and Shinoda, M. (2000). Seasonal changes in the cold-air pool formation in a subalpine hollow, central Japan. *International Journal of Climatology* 20: 1471–1483.

Intrieri, E., Fontanelli, K., Bardi, F. et al. (2018). Definition of sinkhole triggers and susceptibility based on hydrogeomorphological analyses. *Environmental Earth Sciences* 77: 4.

Jackson, J.A. (1997). *Glossary of Geology*. Alexandria, Virginia: American Geological Institute.

Jalali, L., Zarei, M., and Gutiérrez, F. (2019). Salinization of reservoirs in regions with exposed evaporites. The unique case of Upper Gotvand Dam, Iran. *Water Research* 157: 587–599.

James, A.N. (1992). *Soluble Material in Civil Engineering*. New York: Ellis Horwood.

James, J.M. (1993). Burial and infilling of a karst in Papua New Guinea by road erosion sediments. *Environmental Geology* 21: 144–151.

Jancin, M. and Clark, D.D. (1993). Subsidence-sinkhole development in light of mud infiltrate structures within interstratal karst of the coastal plain, Southeast United States. *Environmental Geology* 22 (4): 330–336.

Jarvis, T. (2003). The Money Pit: Karst failure of Anchor Dam, Wyoming. In: *Evaporite Karst and Engineering/Environmental Problems in the United States*, vol. 109 (ed. K.S. Johnson and J.T. Neal), 271–278. Oklahoma Geological Survey Circular.

Jassim, S.Z., Jibril, A.S., and Numan, N.M.S. (1997). Gypsum karstification in the middle Miocene Fatha Formation, Mosul area, northern Iraq. *Geomorphology* 18: 137–149.

Jennings, J.N. (1975). Doline morphometry as a morphogenetic tool: New Zealand examples. *New Zealand geographer* 31: 6–28.

Jennings, J.N. (1985). *Karst Geomorphology*. Oxford and New York: Blackwell.

Jennings, J.N. and Sweeting, M.M. (1963). The limestone ranges of the Fitzroy Basin, Western Australia. A tropical semiarid karst. *Bonner Geographische Abhandlungen* 32: 1–60.

Jiang, X., Lei, M., and Gao, Y. (2017). Formation mechanism of large sinkhole collapses in Laibin, Guangxi, China. *Environmental Earth Sciences* 76: 1–13.

Johnson, K.S. (1989). Development of the Wink Sink in Texas, USA, due to salt dissolution and collapse. *Environmental Geology and Water Sciences* 14 (2): 81–92.

Johnson, K.S. (1997). Evaporite karst in the United States. *Carbonates and Evaporites* 12: 2–14.

Johnson, K.S. (2005). Salt dissolution and subsidence or collapse caused by human activities. *Reviews in Engineering Geology* 16: 101–110.

Johnson, K.S. (2008). Gypsum-karst problems in constructing dams in the USA. *Environmental Geology* 53: 945–950.

Johnson, K.S., Collins, E.W., and Seni, S.J. (2003). Sinkholes and land subsidence owing to salt dissolution near Wink, West Texas, and other sites in western Texas and New Mexico. In: *Evaporite Karst and Engineering/Environmental Problems in the United States* (ed. K.S. Johnson and J.T. Neal), 183–195. Norman: Oklahoma Geological Survey.

Jones, E.J. (2020). Capabilities, limitations and opportunities for studying sinkholes using synthetic aperture radar interferometry. In: *Proceedings of the 16th Multidisciplinary Conference on Sinkholes and the Engineering and Environmental Impacts of Karst* (ed. L. Land, C. Kromhout and M.J. Byle), 68–74. Carlsbad: National Cave and Karst Research Institute.

Jones, C.E. and Blom, R.G. (2014). Bayou Corne, Louisiana, sinkhole: Precursory deformation measured by radar interferometry. *Geology* 42: 111–114.

Jones, N.A., Hansen, J., Springer, A.E. et al. (2019). Modeling intrinsic vulnerability of complex karst aquifers: Modifying the COP method to account for sinkhole density and fault location. *Hydrogeology Journal* 27: 2857–2868.

Jovanelly, T.J. (2014). Sinkholes and a disappearing lake: Victory Lake case study. *Journal of Cave and Karst Studies* 76: 217–229.

Kappel, W.M., Yager, R.M., and Todd, M.S. (1999). The Retsof salt mine collapse. In: *Land Subsidence in the United States* (ed. D. Galloway, D.R. Jones and S.E. Ingebritsen), 111–120. US Geological Survey Circular 1182.

Karimi, H. and Taheri, K. (2010). Hazards and mechanism of sinkholes on Kabudar Ahang and Famenin plains of Hamadan, Iran. *Natural Hazards* 55: 481–499.

Kasting, K.M. and Kasting, E.H. (2003). Site characterization of sinkholes based on resolution of mapping. In: *Proceeding of the 9th Multidisciplinary Conference on Sinkholes and the Engineering and Environmental Impacts of Karst*, Geotechnical Special Publication 122 (ed. B.F. Beck), 72–81. Huntsville, Alabama: ASCE.

Kaufmann, O. and Quinif, Y. (2002). Geohazard map of cover-collapse sinkholes in the 'Tournaisis' area, southern Belgium. *Engineering Geology* 65: 117–124.

Kawashima, K., Aydan, O., Aoki, T. et al. (2010). Reconnaissance investigation on the damage of the 2009 L'Aquila, Central Italy earthquake. *Journal of Earthquake Engineering* 14: 817–841.

Kázmér, M. and Taborosi, D. (2012). Bioerosion on the small scale - examples from the tropical and subtropical littoral. *Hantkeniana* 7: 37–94.

Kelletat, D.H. (2005). Notches. In: *Encyclopedia of Coastal Science* (ed. M.L. Schwartz), 729–730. Dordrecht: Springer.

Kemmerly, P.R. (1980). A time distribution study of doline collapse: Framework for prediction. *Environmental Geology* 3: 123–130.

Kemmerly, P.R. (1982). Spatial analysis of a karst depression population: Clues to genesis. *Geological Society of America Bulletin* 93: 1078–1086.

Keqiang, H., Bin, W., and Dunyun, Z. (2004). Mechanism and mechanical model of karst collapse in an overpumping area. *Environmental Geology* 46: 1102–1107.

Kerans, C. (1988). Karst-controlled reservoir heterogeneity in Ellenburger group carbonates of west Texas. *American Association of Petroleum Geologists Bulletin* 72: 1160–1183.

Kersten, T., Kobe, M., Gabriel, G. et al. (2017). Geodetic monitoring of subrosion-induced subsidence processes in urban areas: Concept and status report. *Journal of Applied Geodesy* 11: 21–29.

Khanlari, G., Heidari, M., Momeno, A.A. et al. (2012). The effect of groundwater overexploitation on land subsidence and sinkhole occurrences, western Iran. *Quarterly Journal of Engineering Geology and Hydrogeology* 45: 447–456.

Kirkham, R.M., Streufert, R.K., Kunk, M.J. et al. (2002). Evaporite tectonism in the Lower Roaring Fork river valley, westcentral Colorado. In: *Late Cenozoic Evaporite Tectonism and Volcanism in West-Central Colorado*, Geological Society of America Special Paper 366 (ed. R.M. Kirkham, R.B. Scott and T.W. Judkins), 73–99. Denver: Geological Society of America.

Kirkham, R.M., White, J.L., Sares, M.A. et al. (2003). Engineering and environmental aspects of evaporite karst in west-central Colorado. In: *Evaporite Karst and Engineering/Environmental Problems in the United States* (ed. K.S. Johnson and J.T. Neal), 279–292. Norman: Oklahoma Geological Survey.

Klimchouk, A.B. (1995). Karst morphogenesis in the epikarst zone. *Cave and Karst Science* 21: 45–50.

Klimchouk, A.B. (2000b). The formation of epikarst and its role in vadose speleogenesis. In: *Speleogenesis. Evolution of Karst Aquifers* (ed. A.B. Klimchouk, D.C. Ford, A.N. Palmer, et al.), 91–99. Huntsville: National Speleological Society.

Klimchouk, A.B. (2004b). Towards defining, delimiting and classifying epikarst: Its origin, processes and variants of geomorphic evolution. *Speleogenesis and Evolution of Karst Aquifers* 2: 1–13.

Klimchouk, A.B. and Andrejchuk, V. (2005). Karst breakdown mechanisms from observations in the gypsum caves of the western Ukraine: implications for subsidence hazard assessment. *Environmental Geology* 48: 336–359.

Klopmeier, N.W., Pesi, S.M., Morris, G. et al. (2018). Sinkholes as a Source of Wildlife Mortality. *Southeastern Naturalist* 17: 64–67.

Knez, M. and Slabe, T. (2002). Unroofed caves are an important feature of karst surfaces: examples from the classical karst. *Zeitschrift für Geomorphologie* 46: 181–191.

Knez, M. and Slabe, T. (2013). Stone forest and their rock relief. In: *Treatise of Geomorphology: Karst Geomorphology*, vol. 6 (ed. A. Frumkin), 139–146. Amsterdam: Elsevier.

Knight, F.J. (1971). Geologic problems of urban growth in limestone terrains in Pennsylvania. *Bulletin Association Engineering Geologists* 8: 91–101.

Kobal, M., Bertoncelj, I., Pirotti, F. et al. (2014). LIDAR processing for defining sinkhole characteristics under dense forest cover: a case study in the Dinaric Mountains. In: *The International Archives of the Photogrammetry, Remote Sensing and Spatial Information Sciences*, ISPRS Technical Commission VII Symposium, Istambul, Turkey, vol. XL-7, 113–118. International Society of Photogrammetry and Remote Sensing.

Kobal, M., Bertoncelj, I., Pirotti, F. et al. (2015). Using lidar data to analyse sinkhole characteristics relevant for understory vegetation under forest cover - Case study of a high karst area in the Dinaric Mountains. *PLoS One* 10 (3): e0122070.

Koša, E. and Hunt, D.W. (2006). The effect of syndepositional deformation within the Upper Permian Capitan Platform on the speleogenesis and geomorphology of the Guadalupe Mountains, New Mexico, USA. *Geomorphology* 78 (3-4): 279–308.

Kovacs, S.E., van Hengstum, P.J., Reinhardt, E.G. et al. (2013). Late Holocene sedimentation and hydrologic development in a shallow coastal sinkhole on Great Abaco Island, The Bahamas. *Quaternary International* 317: 118–132.

Kranjc, A. (2013). Classification of closed depressions in carbonate karst. In: *Treatise of Geomorphology: Karst Geomorphology*, vol. 6 (ed. A. Frumkin), 104–111. Amsterdam: Elsevier.

Kromhout, C. and Alfieri, M.C. (2018). Assesment of historical aerial photography as initial screening tool to identify areas of possible risk to sinkhole development. In: *Proceedings of the 15th Multidisciplinary Conference on Sinkholes and the Engineering and Environmental Impacts of Karst and the 3rd Appalachian Symposium* (ed. I.D. Sasowsky, M.J. Byle and L. Land), 89–96. Carlsbad: National Cave and Karst Research Institute.

Kulkarni, C., Peteet, D., Boger, R. et al. (2016). Exploring the role of humans and climate over the Balkan landscape: 500 years of vegetational history of Serbia. *Quaternary Science Reviews* 144: 83–94.

Kunaver, J. (2009). Corrosion terraces, a megaausgleichsfläche or a specific landform of bare glaciokarst. In: *Karst Rock Features. Karren Sculpturing* (ed. A. Ginés, M. Knez, T. Slabe, et al.), 161–168. Ljubljana: ZRC Publishing.

Kuniansky, E.L., Weary, D.J., and Kaufmann, J.E. (2016). The current status of mapping karst areas and availability of public sinkhole-risk resources in karst terrains of the United States. *Hydrogeology Journal* 24: 613–624.

Lambrecht, J.L. and Miller, R.D. (2006). Catastrophic sinkhole formation in Kansas: a case study. *The Leading Edge* 38: 342–347.

Lamoreaux, P.E. and Newton, J.G. (1986). Catastrophic subsidence: An environmental hazard, Shelby County, Alabama. *Environmental Geology and Water Sciences* 8: 25–40.

Land, L. (2003). Evaporite karst and regional groundwater circulation in the Lower Pecos Valley of Southeastern New Mexico. In: *Evaporite Karst and Engineering/Environmental Problems in the United States*, vol. 109 (ed. K.S. Johnson and J.T. Neal), 227–232. Oklahoma Geological Survey Circular.

Land, L. (2013). Evaporite karst in the Permian basin region of west Texas and southeastern New Mexico: the human impact. In: *Sinkholes and the Engineering and Environmental Impacts of Karst* (ed. L. Land, D.H. Doctor and J.B. Stephenson), 113–121. Carlsbad: National Cave and Karst Research Institute.

Lane, P., Donnelly, J.P., Woodruff, J.D. et al. (2011). A decadally-resolved paleohurricane record archived in the late Holocene sediments of a Florida sinkhole. *Marine Geology* 287: 14–30.

Lei, M., Jiang, X., and Yu, L. (2001). New advances of karst collapse research in China. In: *Geotechnical and Environmental Applications of Karst Geology and Hydrology. Proceedings of the Eight Multidisciplinary Conference on Sinkholes and the Engineering and Environmental Impacts of Karst* (ed. B.F. Beck and J.G. Herring), 145–151. Rotterdam: Balkema.

Lei, M., Jiang, X., and Yu, L. (2002). New advances in karst collapse research in China. *Environmental Geology* 42: 462–468.

Lei, M., Jiang, X., and Guan, Z. (2013). Emergency investigation of extremely large sinkholes, Maohe, Guangxi, China. In: *Sinkholes and the Engineering and Environmental Impacts of Karst* (ed. L. Land, D.H. Doctor and J.B. Stephenson), 145–151. Carlsbad: National Cave and Karst Research Institute.

Lei, M., Gao, Y., and Jiang, X. (2015). Current status and strategic planning of sinkhole collapses in China. In: *Engineering Geology for Society and Territory* (ed. G. Lollino, A. Manconi, F. Guzzetti, et al.), 145–151. Dordrecht: Springer.

Ley, R.G. (1980). The pinnacles of Gunung Api. *Geographical Journal* 146 (1): 14–21.

Li, J. and Zhou, W. (1999). Subsidence in karst mining areas in China and some methods of prevention. *Engineering Geology* 52: 45–50.

Li, K., Zhang, M., Li, Y. et al. (2020a). Karren Habitat as the Key in Influencing Plant Distribution and Species Diversity in Shilin Geopark, Southwest China. *Sustainability* 12: 5808.

Liguori, V., Manno, G., and Mortellaro, D. (2008). Evaporite karst in Sicily. *Environmental Geology* 53: 975–980.

Linares, R., Roqué, C., Gutiérrez, F. et al. (2017). The impact of droughts and climate change on sinkhole occurrence. A case study from the evaporite karst of the Fluvia Valley, NE Spain. *Science of the Total Environment* 579: 345–358.

Lindsey, B.D., Katz, B.G., Berndt, M.P. et al. (2010). Relations between sinkhole density and anthropogenic contaminants in selected carbonate aquifers in the eastern United States. *Environmental Earth Sciences* 60: 1073–1090.

Lipar, M. and Webb, J.A. (2015). The formation of the pinnacle karst in Pleistocene aeolian calcarenites (Tamala Limestone) in southwestern Australia. *Earth-Science Reviews* 140: 182–202.

Lipar, M., Webb, J.A., White, S.Q. et al. (2015). The genesis of solution pipes: evidence from the middle-late Pleistocene Bridgewater formation calcarenite, southeastern Australia. *Geomorphology* 246: 90–103.

Lipar, M., Stepišnik, U., and Ferk, M. (2019). Multiphase breakdown sequence of collapse doline morphogenesis: an example from Quaternary aeolianites in Western Australia. *Geomorphology* 327: 572–584.

Lipar, M., Szymczak, P., White, S.Q. et al. (2021). Solution pipes and focused vertical water flow: Geomorphology and modelling. *Earth-Science Reviews* 218: 103635.

Lolcama, J.L., Cohen, H.A., and Tonkin, M.J. (2002). Deep karst conduits, flooding and sinkholes: lessons for the aggregate industry. *Engineering Geology* 65: 151–157.

Loucks, R.G. (1999). Paleocave carbonate reservoirs: origins, burial depth modifications, spatial complexity and reservoir implications. *American Association of Petroleum Geologists Bulletin* 83: 1795–1834.

Lu, Y. and Cooper, A.H. (1997). Gypsum karst geohazards in China. In: *Engineering Geology and Hydrology of Karst Terranes. Proceedings of the Sixth Multidisciplinary Conference on Sinkholes and the Engineering and Environmental Impacts of Karst* (ed. B.F. Beck and J.B. Stephenson), 117–126. Rotterdam: Balkema.

Lucha, P., Cardona, F., Gutiérrez, F. et al. (2008a). Natural and human-induced dissolution and subsidence processes in the salt outcrop of the Cardona Diapir (NE Spain). *Environmental Geology* 53: 1023–1035.

Lucha, P., Gutiérrez, F., and Guerrero, J. (2008b). Environmental problems derived from evaporite dissolution in the Barbastro-Balaguer Anticline (Ebro Basin, NE Spain). *Environmental Geology* 53: 1045–1055.

Lundberg, J. (2009). Coastal karren. In: *Karst Rock Features. Karren Sculpturing* (ed. A. Ginés, M. Knez, T. Slabe, et al.), 249–266. Ljubljana: ZRC Publishing.

Lundberg, J. (2013). Microsculpturing of solutional rocky landforms. In: *Treatise of Geomorphology: Karst Geomorphology*, vol. 6 (ed. A. Frumkin), 121–138. Amsterdam: Elsevier.

Lundberg, J. (2019). Karren, surface. In: *Encyclopedia of Caves* (ed. W.B. White, D.C. Culver and T. Pipan), 600–608. London: Academic Press.

Lundberg, J. and Ginés, A. (2009). Rillenkarren. In: *Karst Rock Features. Karren Sculpturing* (ed. A. Ginés, M. Knez, T. Slabe, et al.), 169–184. Ljubljana: ZRC Publishing.

Lundberg, J. and Lauritzen, S.-E. (2002). The search for an arctic coastal karren model in Norway and Spitzbergen. In: *Landscapes of Transition* (ed. K. Hewitt), 185–203. Amsterdam: Kluwer Academic Publishers.

Luu, L.H., Noury, G., Benseghier, Z. et al. (2019). Hydro-mechanical modeling of sinkhole occurrence processes in covered karst terrains during a flood. *Engineering Geology* 260: 105249.

Lyew-Ayee, P., Viles, H.A., and Tucker, G.E. (2007). The use of GIS-based digital morphometric techniques in the study of cockpit karst. *Earth Surface Processes and Landforms* 32: 165–179.

Macaluso, T. and Sauro, U. (1996a). The karren in evaporitic rocks: a proposal of classification. In: *Karren Landforms* (ed. J.J. Fornós and A. Ginés), 277–293. Palma de Mallorca: Universitat de les Illes Balears.

Macaluso, T. and Sauro, U. (1996b). Weathering crust and karren on exposed gypsum surfaces. *International Journal of Speleology* 25: 115–126.

Maire, R. (1999). Les glaciers de marbre de Patagonie, Chili. Un karst subpolaire océanique de la zone australe. *Karstologia* 33: 25–40.

Maire, R., Jaillet, S., and Hobléa, F. (2009). Karren in Patagonia, a natural laboratory for hydroaeolian dissolution. In: *Karst Rock Features. Karren Sculpturing* (ed. A. Ginés, M. Knez, T. Slabe, et al.), 329–348. Ljubljana: ZRC Publishing.

Mancini, F., Stecchi, F., Zanni, M. et al. (2009). Monitoring ground subsidence induced by salt mining in the city of Tuzla (Bosnia & Herzegovina). *Environmental Geology* 58: 381–389.

Mangin, A. (1974). Contributions à l'étude hydrodynamique des aquifères karstiques. *Annales de Spéléologie* 29: 283–332.

Marinos, P.G. (2001). Tunnelling and mining in karstic terrain; an engineering challenge. In: *Geotechnical and Environmental Applications of Karst Geology and Hydrology. Proceedings of the Eight Multidisciplinary Conference on Sinkholes and the Engineering and Environmental Impacts of Karst* (ed. B.F. Beck and J.G. Herring), 3–16. Rotterdam: Balkema.

Martín-Duque, J.F., García, J.C., and Urquí, L.C. (2012). Geoheritage information for geoconservation and geotourism through the categorization of landforms in a karstic landscape. A case study from Covalagua and Las Tuerces (Palencia, Spain). *Geoheritage* 4: 93–108.

Martinotti, M.E., Pisano, L., Marchesini, I. et al. (2017). Landslides, floods and sinkholes in a karst environment: the 1–6 September 2014 Gargano event, southern Italy. *Natural Hazards and Earth System Sciences* 17: 467–480.

Martos, F.M. (2010). Condicionantes genéticos de las formas kársticas de El Torcal de Antequera (Cordillera Bética). *Geogaceta* 48: 23–26.

Matsushi, Y., Hattanji, T., Akiyama, S. et al. (2010). Evolution of solution dolines inferred from cosmogenic ^{36}Cl in calcite. *Geology* 38 (11): 1039–1042.

McCalpin, J.P. (2009). Paleoseismology in extensional tectonic environments. In: *Paleoseismology* (ed. J.P. McCalpin), 171–270. Amsterdam: Elsevier.

McDowell, P.W. (2005). Geophysical investigations of sinkholes in chalk, UK; case study 9. In: *Sinkholes and Subsidence* (ed. T. Waltham, F. Bell and M. Culshaw), 313–316. Chichester: Springer.

McDowell, P.W. and Poulsom, A.J. (1996). Ground subsidence related to dissolution of chalk in southern England. *Ground Engineering* 29: 29–33.

McIlroy de la Rosa, J. (2012). Karst Landform Classification Techniques. In: *Geomorphological Techniques* (ed. S.J. Cook, L.E. Clarke and J.M. Nield). London: British Society of Geomorphology.

Melis, M.T., Pisani, L., and De Waele, J. (2021). On the Use of Tri-Stereo Pleiades Images for the Morphometric Measurement of Dolines in the Basaltic Plateau of Azrou (Middle Atlas, Morocco). *Remote Sensing* 13 (20): 4087.

Memesh, A., Dini, S., Gutiérrez, F. et al. (2008). Evidence of large-scale subsidence caused by interstratal karstification of evaporites in the Interior Homocline of Central Saudi Arabia. European Geosciences Union General Assembly. *Geophysical Research Abstracts* 10: A-02276.

Meng, Y. and Jia, L. (2018). Global warming causes sinkhole collapse. Case study in Florida, USA. *Natural Hazards and Earth Systems Science Discussions* 1–8.

Meng, Y., Lei, M.T., Lin, Y.S. et al. (2012). Models and mechanisms of drilling-induced sinkhole in China. *Environmental Earth Sciences* 67: 1961–1969.

Meng, Y., Jia, L., and Huang, J.M. (2020). Hydraulic fracturing effect on punching-induced cover-collapse sinkholes: a case study in Guangzhou, China. *Arabian Journal of Geosciences* 13: 1–8.

Mesescu, A.A. (2011). The Ocnele Mari salt mine collapsing sinkhole. A NATECH breakdown in the Romanian Sub-Carpathians. *Carpathian Journal of Earth and Environmental Sciences* 6: 215–220.

Miano, S.T. and Keays, P.V. (2018). When sinkholes become legal problems. In: *Proceedings of the 15th Multidisciplinary Conference on Sinkholes and the Engineering and Environmental Impacts of Karst and the 3rd Appalachian Karst Symposium* (ed. I.D. Sasowsky, M.J. Byle and L. Land), 37–42. Carlsbad: National Cave and Karst Research Institute.

Miao, X., Qiu, X., Wu, S.-S. et al. (2013). Developing efficient procedures for automated sinkhole extraction from Lidar DEMs. *Photogrammetric Engineering & Remote Sensing* 79: 545–554.

Michelena, M.D., Kilian, R., Baeza, O. et al. (2020). The formation of a giant collapse caprock sinkhole on the Barda Negra plateau basalts (Argentina): Magnetic, mineralogical and morphostructural evidences. *Geomorphology* 367: 107297.

Micheletti, N., Chandler, J.H., and Lane, S.N. (2015). Structure from Motion (SfM) Photogrammetry. In: *Geomorphological Techniques* (ed. S. Cook, L. Clark and J. Nield) Chapter 2, Section 2.2. London: British Society for Geomorphology.

Middleton, G. (2004). Madagascar. In: *Encyclopedia of Caves and Karst Science* (ed. J. Gunn), 493–495. New York: Fitzroy Dearborn.

Migoń, P., Duszyński, F., and Goudie, A. (2017). Rock cities and ruiniform relief: forms, processes, terminology. *Earth-Science Reviews* 171: 78–104.

Milanović, P.T. (2004b). *Water Resources Engineering in Karst*. Boca Raton: CRC press.

Milanović, P.T., Maksimovich, N., and Meshcheriakova, O. (2019). *Dams and Reservoirs in Evaporites*. Dordrecht: Springer.

Miller, V.C. (1953). *A Quantitative Geomorphic Study of Drainage Basin Characteristics in the Clinch Mountain Area* Technical Report No. 3. New York: Department of Geology, Columbia University.

Monroe, W.H. (1968). The karst features of northern Puerto Rico. *Bulletin of the National Speleological Society* 30: 75–86.

Moore, H. (1988). Treatment of karst along Tennessee highways. *American Society Civil Engineers Geotechnical*, Special Publication 14: 133–148.

Moses, C.A. (2003). Observations on coastal biokarst, Hells Gate, Lord Howe Island, Australia. *Zeitschrift für Geomorphologie* 47: 83–100.

Mottershead, D. (1996). Some morphological properties of solution flutes (Rillenkarren) at Lluc, Mallorca. In: *Karren Landforms* (ed. J.J. Fornós and A. Ginés), 225–238. Palma de Mallorca: Universitat de les Illes Balears.

Mottershead, D. and Lucas, G. (2001). Field testing of Glew and Ford's model of solution flute evolution. *Earth Surface Processes and Landforms* 26: 839–846.

Mottershead, D.N. and Lucas, G.R. (2004a). Observations on stepkarren formed on limestone, gypsum and halite terrains. In: *Stone Decay: Its Causes and Control* (ed. B.J. Smith and A.V. Turkington), 247–272. London: Routledge.

Mottershead, D. and Lucas, G.R. (2004b). The role of mechanical and biotic processes in solution flute development. In: *Stone Decay: Its Causes and Control* (ed. B.J. Smith and A.V. Turkington), 273–291. London: Routledge.

Mottershead, D., Moses, C.A., and Lucas, G.R. (2000). Lithological control of solution flute form: a comparative study. *Zeitschrift für Geomorphologie* 44: 491–512.

Mottershead, D.N., Wright, J.S., Inkpen, R.J. et al. (2007). Bedrock slope evolution in saltrock terrain. *Zeitschrift für Geomorphologie, Supplementary Issues* 51 (1): 81–102.

Munro, P.G. and Melo-Zurita, M.L. (2011). The role of cenotes in the social history of Mexico's Yucatan Peninsula. *Environment and History* 17: 583–612.

Myers, P.B. and Perlow, M. (1984). Development, occurrence and triggering mechanisms of sinkholes in the carbonate karst of the Lehigh Valley, eastern Pennsylvania. In: *Sinkholes: Their Geology, Engineering and Environmental Impact. Proceedings of the First Multidisciplinary Conference on Sinkholes* (ed. B.F. Beck), 111–115. Rotterdam: Balkema.

Mylroie, J.E. (2004). Blue holes of the Bahamas. In: *Encyclopedia of Caves and Karst Science* (ed. J. Gunn), 155–156. New York: Fitzroy Dearborn.

Mylroie, J.E. (2019). Coastal caves. In: *Encyclopedia of Caves* (ed. W.B. White, D.C. Culver and T. Pipan), 301–307. New York: Academic Press.

Mylroie, J.E., Carew, J.L., and Moore, A.I. (1995). Blue holes: definition and genesis. *Carbonates and Evaporites* 10: 225–233.

Nam, B.H., Kim, Y.J., and Youn, H. (2020). Identification and quantitative analysis of sinkhole contributing factors in Florida's Karst. *Engineering Geology* 271: 105610.

Naylor, L.A. and Viles, H.A. (2002). A new technique for evaluating short-term rates of coastal bioerosion and bioprotection. *Geomorphology* 47: 31–44.

NCB (1975). *Subsidence Engineers' Handbook*. London: National Coal Board Mining Department.

Newton, J.G. (1984). Sinkholes resulting from groundwater withdrawals in carbonate terrains, an overview. *Reviews in Engineering Geology of the Geological Society of America* 6: 195–202.

Newton, J.G. (1987). *Development of Sinkholes Resulting from Man's Activities in the Eastern United States*. Denver (CO): US Geological Survey.

Noury, G., Perrin, J., Luu, L.-H. et al. (2018). Role of floods on sinkhole occurrence in covered karst terrains: Case study of the Orléans area (France) during the 2016 meteorological event and perspectives for other karst environments. In: *Proceedings of the 15th Multidisciplinary Conference on Sinkholes and the Engineering and Environmental Impacts of Karst and the 3rd Appalachian Karst Symposium* (ed. I.D. Sasowsky, M.J. Byle and L. Land), 251–258. Carlsbad: National Cave and Karst Research Institute.

Olaya, V. (2009). Basic land-surface parameters. In: *Geomorphometry: Concepts, Software, Applications* (ed. T. Hengl and H.I. Reuter), 141–170. Amsterdam: Elsevier.

Ortega, A.I., Benito-Calvo, A., Pérez-González, A. et al. (2013). Evolution of multilevel caves in the Sierra de Atapuerca (Burgos, Spain) and its relation to human occupation. *Geomorphology* 196: 122–137.

Öztürk, M.Z., Şimşek, M., Utlu, M. et al. (2017). Karstic depressions on Bolkar Mountain plateau, Central Taurus (Turkey): distribution characteristics and tectonic effect on orientation. *Turkish Journal of Earth Sciences* 26: 302–313.

Öztürk, M.Z., Şener, M.F., Şener, M. et al. (2018a). Structural controls on distribution of dolines on Mount Anamas (Taurus Mountains, Turkey). *Geomorphology* 317: 107–116.

Öztürk, M.Z., Şimşek, M., Şener, M.F. et al. (2018b). GIS based analysis of doline density on Taurus Mountains, Turkey. *Environmental Earth Sciences* 77: 1–13.

Palmer, A.N. (2007). *Cave Geology*. Dayton, Ohio: Cave Books.

Palmer, A.N. (2009b). Cutters and pinnacles in the Salem limestone of Indiana. In: *Karst Rock Features. Karren Sculpturing* (ed. A. Ginés, M. Knez, T. Slabe, et al.), 349–358. Ljubljana: ZRC Publishing.

Palmer, R.J. and Heath, L.M. (1985). The effect of anchialine factors and fracture control on cave development below eastern Grand Bahama. *Cave Science* 12: 93–97.

Palmer, A.N. and Palmer, M.V. (2005). Hydraulic processes in the origin of tiankengs. *Cave and Karst Science* 32: 101–106.

Palmeri, V., Madonia, G., and Ferro, V. (2020). Capturing gypsum rillenkarren morphometry by a 3D-photo reconstruction (3D-PR) technique. *Geomorphology* 351: 106980.

Pando, L., Pulgar, J.A., and Gutiérez-Claverol, M. (2013). A case of man-induced ground subsidence and building settlement related to karstified gypsum (Oviedo, NW Spain). *Environmental Earth Sciences* 68: 507–519.

Panno, S.V. and Luman, D.E. (2013). Mapping palimpsest karst features on the Illinois sinkhole plain using historical aerial photography. *Carbonates and Evaporites* 28: 201–214.

Panno, S.V. and Luman, D.E. (2018). Characterization of cover-collapse sinkhole morphology on a groundwater basin-wide scale using lidar elevation data: a new conceptual model for sinkhole evolution. *Geomorphology* 318: 1–17.

Panoš, V. and Štelcl, O. (1968). Physiographic and geologic control in development of Cuban mogotes. *Zetischrif für Geomorphologie* 12: 117–173.

Pardo-Igúzquiza, E., Valsero, J.J.D., and Dowd, P.A. (2013). Automatic detection and delineation of karst terrain depressions and its application in geomorphological mapping and morphometric analysis. *Acta Carsologica* 42: 17–24.

Pardo-Igúzquiza, E., Dowd, P.A., and Telbisz, T. (2020). On the size-distribution of solution dolines in carbonate karst: Lognormal or power model? *Geomorphology* 351: 106972.

Pardo-Igúzquiza, E., Pulido-Bosch, A., López-Chicano, M. et al. (2016). Morphometric analysis of karst depressions on a Mediterranean karst massif. *Geografiska Annaler: Series A, Physical Geography* 98: 247–263.

Parenti, C., Gutiérrez, F., Baioni, D. et al. (2020). Closed depressions in Kotido crater, Arabia Terra, Mars. Possible evidence of evaporite dissolution-induced subsidence. *Icarus* 341: 113680.

Parise, M. (2019). Sinkholes. In: *Encyclopedia of Caves* (ed. W. White, D. Culver and T. Pipan), 934–942. Amsterdam: Elsevier.

Parise, M. and Vennari, C. (2013). A chronological catalogue of sinkholes in Italy: the first step toward a real evaluation of the sinkhole hazard. In: *Sinkholes and the Engineering and Environmental Impacts of Karst* (ed. L. Land, D.H. Doctor and J.B. Stephenson), 383–392. Carlsbad: National Cave and Karst Research Institute.

Parise, M., Closson, D., Gutiérrez, F. et al. (2015). Anticipating and managing engineering problems in the complex karst environment. *Environmental Earth Sciences* 74: 7823–7835.

Parise, M., Pisano, L., and Vennari, C. (2018). Sinkhole clusters after heavy rainstorms. *Journal of Cave and Karst Studies* 80: 28–38.

Paskoff, R.P. (2005). Karst coasts. In: *Encyclopedia of Coastal Science* (ed. M.L. Schwartz), 581–586. Dordrecht: Springer.

Paterson, K. and Sweeting, M.M. (ed.) (1986). *New Directions in Karst*. Norwich: Geobooks.

Payton, C.C. and Hansen, M.N. (2003). Gypsum karst in southwestern Utah: failure and reconstruction of Quail Creek Dike. In: *Evaporite Karst and Engineering/Environmental Problems in the United States*, vol. 109 (ed. K.S. Johnson and J.T. Neal), 293–303. Oklahoma Geological Survey Circular.

Péntek, K. and Veress, M. (2007). A morphometric classification of solution dolines. *Zeitschrift für Geomorphologie* 51: 19–30.

Perna, G. and Sauro, U. (1978). *Atlante delle Microforme di Dissoluzione Carsica Superficiale del Trentino e del Veneto*. Trento: Memorie del Museo Tridentino di Scienze Naturali.

Peros, M., Collins, S., G'Meiner, A.A. et al. (2017). Multistage 8.2 kyr event revealed through high-resolution XRF core scanning of Cuban sinkhole sediments. *Geophysical Research Letters* 44: 7374–7381.

Perritaz, L. (1996). Le "karst en vagues" des Aït Abdi (Haut-Atlas central, Maroc). *Karstologia* 28 (1): 1–12.

Perrotti, A.G. (2018). Pollen and Sporormiella evidence for terminal Pleistocene vegetation change and megafaunal extinction at Page-Ladson, Florida. *Quaternary International* 466: 256–268.

Perry, E., Marin, L., McClain, J. et al. (1995). Ring of cenotes (sinkholes), northwest Yucatan, Mexico: its hydrogeologic characteristics and possible association with the Chicxulub impact crater. *Geology* 23 (1): 17–20.

Plan, L. and Decker, K. (2006). Quantitative karst morphology of the Hochschwab plateau, Eastern Alps, Austria. *Zeitschrift für Geomorphologie* Supplement Band 147: 29–54.

Plan, L., Stöger, T., Draganits, E. et al. (2018). A Pleistocene landslide-dammed lake indicated by karren features (Eastern Alps, Austria). *Geomorphology* 321: 60–71.

Playford, P. (2002). Palaeokarst, pseudokarst, and sequence stratigraphy in Devonian reef complexes of the Canning Basin, Western Australia. In: *The Sedimentary Basins of Western Australia* (ed. M. Keep and S.J. Moss), 763–793. Perth: Petroleum Exploration Society of Australia.

Pluhar, A. and Ford, D.C. (1970). Dolomite karren of the Niagara Escarpment, Ontario, Canada. *Zeitschrift für Geomorphologie* 14: 392–410.

Poppe, S., Holohan, E., Pauwels, E. et al. (2015). Sinkholes, pit craters, and small calderas: analog models of depletion-induced collapse analyzed by computed X-ray microtomography. *Geological Society of America Bulletin* 127: 281–296.

Poyraz, M., Öztürk, M.Z., and Soykan, A. (2021). GIS based analysis of doline density in Sivas gypsum karst. *Jeomorfojik Araştirmalar Dergisi* 6: 67–80. (in Turkish).

Rahimi, M. and Alexander, E.C. (2013). Locating sinkholes in LIDAR coverage of a glacio-fluvial karst, Winona County, Mn. In: *Sinkholes and the Engineering and Environmental Impacts of Karst* (ed. L. Land, D.H. Doctor and J.B. Stephenson), 469–480. Carlsbad: National Cave and Karst Research Institute.

Roje-Bonacci, T. and Bonacci, O. (2013). The possible negative consequences of underground dam and reservoir construction and operation in coastal karst areas: an example of the hydro-electric power plant (HEPP) Ombla near Dubrovnik (Croatia). *Natural Hazards and Earth System Science* 13: 2041–2052.

Romanov, D., Gabrovšek, F., and Dreybrodt, W. (2003b). Dam sites in soluble rocks: a model of increasing leakage by dissolutional widening of fractures beneath a dam. *Engineering Geology* 70: 17–35.

Rose, L. and Vincent, P.J. (1986). The kamenitzas of Gait Barrows National Nature Reserve, north Lancashire, England. In: *New Directions in Karst* (ed. K. Paterson and M.M. Sweeting), 473–496. Norwich (UK): Geobooks.

Salomon, J.N. (2009). The tsingy karrenfields of Madagascar. In: *Karst Rock Features. Karren Sculpturing* (ed. A. Ginés, M. Knez, T. Slabe, et al.), 411–423. Ljubljana: ZRC Publishing.

Santo, A., Ascione, A., Del Prete, S. et al. (2011). Collapse sinkholes distribution in the carbonate massifs of central and southern Apennines. *Acta Carsologica* 40: 95–112.

Satkunas, J., Taminskas, J., and Dilys, K. (2006). Geoindicators of changing landscape. An example of karst development in North Lithuania. *Geological Quarterly* 50: 457–464.

Sauro, U. (1991). A polygonal karst in Alte Murge (Puglia, Southern Italy). *Zeitschrift für Geomorphologie* 35: 207–223.

Sauro, U. (2003). Dolines and sinkholes: aspects of evolution and problems of classification. *Acta Carsologica* 32: 41–52.

Sauro, U. (2009). The rock cities of Rosso Ammonitico in the Venetian Prealps. In: *Karst Rock Features. Karren Sculpturing* (ed. A. Ginés, M. Knez, T. Slabe, et al.), 469–475. Ljubljana: ZRC Publishing.

Scarborough, J.A. (1995). Risk and reward: Pipes and sinkholes in East Tennessee. In: *Karst Geohazards. Engineering and Environmental Problems in Karst Terrane. Fifth Multidisciplinary Conference on Sinkholes and the Engineering and Environmental Impacts of Karst* (ed. B.F. Beck), 349–354. Rotterdam: Balkema.

Schneider, J. (1976). Biological and inorganic factors in the destruction of limestone coasts. *Contributions to Sedimentology* 6: 1–112.

Schumm, S.A. (1956). Evolution of drainage systems and slopes in badlands at Perth Amboy, N.J. *Geological Society of America Bulletin* 67: 597–646.

Šegina, E., Benac, Č., Rubinić, J. et al. (2018). Morphometric analyses of dolines—the problem of delineation and calculation of basic parameters. *Acta Carsologica* 47: 22–33.

Senior, K. (2004). Di Feng Dong, China. In: *Encyclopedia of Caves and Karst Science* (ed. J. Gunn), 285–287. New York: Fitzroy Dearborn.

Sevil, J., Gutiérrez, F., Zarroca, M. et al. (2017). Sinkhole investigation in an urban area by trenching in combination with GPR, ERT and high-precision leveling. Mantled evaporite karst of Zaragoza city, NE Spain. *Engineering Geology* 231: 9–20.

Sevil, J., Gutiérrez, F., Carnicer, C. et al. (2020). Characterizing and monitoring a high-rik sinkhole in an urban area underlain by salt through non-invasive methods: detailed mapping, high-precision leveling and GPR. *Engineering Geology* 272: 105641.

Sevil, J., Benito-Calvo, A., and Gutiérrez, F. (2021). Sinkhole subsidence monitoring combining terrestrial laser scanner and high-precision levelling. *Earth Surface Processes and Landforms* 46 (8): 1431–1444.

Shalev, E. and Lyakhovsky, V. (2012). Viscoelastic damage modeling of sinkhole formation. *Journal of Structural Geology* 42: 163–170.

Shamet, R., Soliman, M., Kim, Y.J. et al. (2020). Sinkhole investigation after Hurricane Irma. In: *Proceedings of the 16th Multidisciplinary Conference on Sinkholes and the Engineering and Environmental Impacts of Karst* (ed. L. Land, C. Kromhout and M.J. Byle), 155–164. Carlsbad: National Cave and Karst Research Institute.

Shaqour, F. (1994). Hydrogeologic role in sinkhole development in the desert of Kuwait. *Environmental Geology* 23: 201–208.

Sharpe, R.D. (2003). Effects of karst processes on gypsum mining. In: *Evaporite Karst and Engineering/Environmental Problems in the United States*, vol. 109 (ed. K.S. Johnson and J.T. Neal), 31–40. Oklahoma Geological Survey Circular.

Siart, C., Hecht, S., Holzhauer, I. et al. (2010). Karst depressions as geoarchaeological archives: the palaeoenvironmental reconstruction of Zominthos (Central Crete), based on geophysical prospection, sedimentological investigations and GIS. *Quaternary International* 216: 75–92.

Silva, O.L., Bezerra, F.H., Maia, R.P. et al. (2017). Karst landforms revealed at various scales using LiDAR and UAV in semi-arid Brazil: Consideration on karstification processes and methodological constraints. *Geomorphology* 295: 611–630.

Simms, M.J. (1990). Phytokarst and photokarren in Ireland. *Cave Science* 17: 131–133.

Simms, M.J. (2002). The origin of enigmatic, tubular, lake-shore karren: a mechanisms for rapid dissolution of limestone in carbonate-saturated waters. *Physical Geography* 23: 1–20.

Sissakian, V.J., Al-Ansari, N., and Knutsson, S. (2014). Karstification effect on the stability of Mosul Dam and its assessment, North Iraq. *Engineering* 6: 84–92.

Slabe, T. (1999). Subcutaneous rock forms. *Acta Carsologica* 28: 255–271.

Slabe, T. (2009). Karren simulation with plaster of Paris models. In: *Karst Rock Features. Karren Sculpturing* (ed. A. Ginés, M. Knez, T. Slabe, et al.), 47–54. Ljubljana: ZRC Publishing.

Smart, P.L., Beddows, P.A., Coke, J. et al. (2006). Cave Development on the Caribbean Coast of the Yucatan Peninsula, Quintana Roo, Mexico. In: *Perspectives in Karst Geomorphology, Hydrology, and Geochemistry* (ed. R. Harmon and C. Wicks), 105-128. Geological Society of America, Special Paper 404.

Smith, T.J. and Sinn, G.C. (2013). Induced sinkhole formation associated with installation of a high-pressure natural gas pipeline, west-central Florida. In: *Sinkholes and the Engineering and Environmental Impacts of Karst* (ed. L. Land, D.H. Doctor and J.B. Stephenson), 79–88. Carlsbad: National Cave and Karst Research Institute.

Soliman, M.H., Perez, A.L., Nam, B.H. et al. (2018). Physical and numerical analysis on the mechanical behavior of cover-collapse sinkholes in central Florida. In: *Proceedings of the 15th Multidisciplinary Conference on Sinkholes and the Engineering and Environmental Impacts of Karst and the 3rd Appalachian Karst Symposium* (ed. I.D. Sasowsky, M.J. Byke and L. Land), 405–416. Carlsbad: National Cave and Karst Research Institute.

Song, L. and Liang, F. (2009). Two important evolution models of Lunan shilin karst. In: *Karst Rock Features. Karren Sculpturing* (ed. A. Ginés, M. Knez, T. Slabe, et al.), 453–460. Ljubljana: ZRC Publishing.

Song, K.I., Cho, G.C., and Chang, S.B. (2012). Identification, remediation and analysis of karst sinkholes in the longest railroad tunnel in South Korea. *Engineering Geology* 135-136: 92–105.

Sorensen, P.B., Lykke-Andersen, H., Gravesen, P. et al. (2017). Karst sinkhole mapping using GIS and digital terrain models. *Geological Survey of Denmark and Greenland Bulletin* 38: 25–28.

Sowers, G.F. (1996). *Building on Sinkholes*. New York: ASCE Press.

Spencer, T. (1988a). Coastal biogeomorphology. In: *Biogeomorphology* (ed. H. Viles), 255–318. Basil Blackwell: Oxford.

Spencer, T. (1988b). Limestone coastal morphology: the biological contribution. *Progress in Physical Geography* 12: 66–101.

Spencer, T. and Viles, H. (2002). Bioconstruction, bioerosion and disturbance on tropical coasts: coral reefs and rocky limestone shores. *Geomorphology* 48 (1-3): 23–50.

Sprynskyy, M., Lebedynets, M., and Sadurski, A. (2009). Gypsum karst intensification as a consequence of sulphur mining activity (Jaziv field, Western Ukraine). *Environmental Geology* 57: 173–181.

Stenson, E. and Ford, D.C. (1993). Rillenkarren on gypsum in Nova Scotia. *Géographie Physique et Quaternaire* 47: 239–243.

Stephenson, W.J. and Finlayson, B.L. (2009). Measuring erosion with the micro-erosion meter - contributions to understanding landform evolution. *Earth-Science Reviews* 95 (1-2): 53–62.

Strahler, A.N. (1957). Quantitative analysis of watershed geomorphology. *American Geophysical Union Transactions* 38: 913–920.

Strecker, M.R., Bloom, A.L., Gilpin, L.M. et al. (1986). Karst morphology of uplifted Quaternary coral limestone terraces: Santo Island, Vanuatu. *Zeitschrift für Geomorphologie* 30: 387–405.

Strum, S. (1999). Topographic and hydrogeologic controls on sinkhole formation associated with quarry dewatering. In: *Hydrogeology and Engineering Geology of Sinkholes and Karst. Proceedings of the Seventh Multidisciplinary Conference on Sinkholes and the Engineering and Environmental Impacts of Karst* (ed. B.F. Beck, A.J. Pettit and J.G. Herring), 63–66. Rotterdam: Balkema.

Šušteršič, F. (2006a). A power function model for the basic geometry of solution dolines: considerations from the classical karst of south-central Slovenia. *Earth Surface Processes and Landforms* 31: 293–302.

Swan, C.H. (1978). Middle East. Canals and irrigation problems. *Quarterly Journal of Engineering Geology* 11: 75–78.

Sweeting, M.M. (1972). *Karst Landforms*. London: McMillan Press.

Sweeting, M.M. (1995). *Karst in China. Its Geomorphology and Environment*. Berlin: Springer.

Taheri, K., Gutiérrez, F., Mohseni, H. et al. (2015). Sinkhole susceptibility mapping using the analytical hierarchy process (AHP) and magnitude-frequency relationships: a case study in Hamedan province, Iran. *Geomorphology* 234: 64–79.

Tarhule-Lips, R.F. and Ford, D.C. (1998a). Condensation corrosion in caves on Cayman Brac and Isla de Mona. *Journal of Cave and Karst Studies* 60: 84–95.

Tharp, T.M. (1999). Mechanics of upward propagation of cover-collapse sinkholes. *Engineering Geology* 52: 23–33.

Tharp, T.M. (2001). Cover-collapse sinkhole formation and piezometric surface drawdown. In: *Geotechnical and Environmental Applications of Karst Geology and Hydrology* (ed. B.F. Beck and J.G. Herring), 53–58. Lisse: Balkema.

Thomas, T.M. (1974). The South Wales interstratal karst. *Transactions of the British Cave Research Association* 1: 131–152.

Tihansky, A.B. (1999). Sinkholes, West-Central Florida. In: *Land Subsidence in the United States* (ed. D. Galloway, D.R. Jones and S.E. Ingebritsen), 121–140. U.S. Geological Survey Circular 1182.

Tiner, R.W. (2003). Geographically isolated wetlands of the United States. *Wetlands* 23: 494–516.

Trenhaile, A.S. (1987). *The Geomorphology of Rock Coasts*. Oxford: Clarendon Press.

Trenhaile, A.S. (2002). Rock coasts, with particular emphasis on shore platforms. *Geomorphology* 48: 7–22.

Tricart, J. and Da Silva, T.C. (1960). Un exemple d'évolution karstique en milieu tropical sec: Le morne de Bom Jesus da Lapa (Bahia, Brasil). *Zeitschrift für Geomorphologie* 4: 29–42.

Trudgill, S.T. (1985). *Limestone Geomorphology*. London: Longman.

Trzhtsinsky, Y.B. (2002). Human-induced activation of gypsum karst in the southern Priangaria (east Siberia, Russia). *Carbonates and Evaporites* 17: 154–158.

Turnage, K.M., Lee, S.Y., Foss, J.E. et al. (1997). Comparison of soil erosion and deposition rates using radiocesium, RUSLE, and buried soils in dolines in East Tennessee. *Environmental Geology* 29: 1–10.

Twidale, C.R. and Centeno Carrillo, J.D.D. (1993). Landform development at the Ciudad Encantada, near Cuenca, Spain. *Cuadernos do Laboratorio Xeolóxico de Laxe* 18: 257–269.

Twidale, C.R. and Vidal Romaní, J.R. (2005). *Landforms and Geology of Granite Terrains*. Leiden: Balkema.

Uromeihy, A. (2000). The Lar Dam; an example of infrastructural development in a geologically active karstic region. *Journal of Asian Earth Sciences* 18: 25–31.

Vajedian, S. and Motagh, M. (2019). Extracting sinkhole features from time-series of TerraSAR-X/TanDEM-X data. *ISPRS Journal of Photogrammetry and Remote Sensing* 150: 274–284.

Valenzuela, P., Domínguez-Cuesta, M.J., Meléndez-Asensio, M. et al. (2015). Active sinkholes: a geomorphological impact of the Pajares Tunnels (Cantabrian Range, NW Spain). *Engineering Geology* 196: 158–170.

Van Hengstum, P.J., Maale, G., Donnelly, J.P. et al. (2018). Drought in the northern Bahamas from 3300 to 25000 years ago. *Quaternary Science Reviews* 186: 169–185.

Verbovšek, T. and Gabor, L. (2019). Morphometric properties of dolines in Matarsko podolje, SW Slovenia. *Environmental Earth Sciences* 78: 1–16.

Veress, M. (2009a). Rinnenkarren. In: *Karst Rock Features. Karren Sculpturing* (ed. A. Ginés, M. Knez, T. Slabe, et al.), 211–222. Ljubljana: ZRC Publishing.

Veress, M. (2009b). Wandkarren. In: *Karst Rock Features. Karren Sculpturing* (ed. A. Ginés, M. Knez, T. Slabe, et al.), 237–248. Ljubljana: ZRC Publishing.

Veress, M. (2009c). Meanderkarren. In: *Karst Rock Features. Karren Sculpturing* (ed. A. Ginés, M. Knez, T. Slabe, et al.), 223–236. Ljubljana: ZRC Publishing.

Veress, M. (2009d). Trittkarren. In: *Karst Rock Features. Karren Sculpturing* (ed. A. Ginés, M. Knez, T. Slabe, et al.), 151–159. Ljubljana: ZRC Publishing.

Veress, M. (2010). *Karst Environments. Karren Formation in High Mountains*. Dordrecht: Springer.

Veress, M. (2017). Solution doline development on glaciokarst in alpine and Dinaric areas. *Earth-Science Reviews* 173: 31–48.

Veress, M. (2019). The karren and karren formation of bare slopes. *Earth-Science Reviews* 188: 272–290.

Veress, M. and Tóth, G. (2004). Types of meandering karren. *Zeitschrift für Geomorphologie* 48: 53–77.

Veress, M., Tóth, G., Zentai, Z., and Czöpek, I. (2003). *Vitesse de recul d'un escarpement lapiazé (Ile Diego de Almagro, Patagonia, Chili)*, vol. 41, 23–26. Karstologia: Revue de Karstología et de Spéléologie Physique.

Veress, M., Lóczy, D., Zentai, Z. et al. (2008). The origin of the Bemaraha tsingy (Madagascar). *International Journal of Speleology* 37: 131–142.

Veress, M., Tóth, G., Zentai, Z. et al. (2009). The Ankarana tsingy and its development. *Carpathian Journal of Earth and Environmental Sciences* 4: 95–108.

Veress, M., Zentai, Z., Péntek, K. et al. (2013). Flow dynamics and shape of rinnenkarren. *Geomorphology* 198: 115–127.

Veress, M., Zentai, Z., Péntek, K. et al. (2014). The development of the pinnacles (Lena pillars) along Middle Lena (Sakha Republic, Siberia, Russia). *Proceedings of the Geologists' Association* 125: 452–462.

Veress, M., Samu, S., and Miter, Z. (2015). The effect of slope angle on the development of type a and type b channels of rinnenkarren with field and laboratory measurements. *Geomorphology* 228: 60–70.

Vigna, B., Fiorucci, A., Banzato, C. et al. (2010). Hypogene gypsum karst and sinkhole formation at Moncalvo (Asti, Italy). *Zeitschrift für Geomorphologie* Supplement Band 54 (2): 285–306.

Viles, H.A. (2004a). Biokarst. In: *Encyclopedia of Geomorphology*, vol. 1 (ed. A.S. Goudie), 86–87. London: Routledge.

Viles, H.A. (2004b). Biokarstification. In: *Encyclopedia of Caves and Karst Science* (ed. J. Gunn), 304–306. New York: Fitzroy Dearborn.

Viles, H.A. (2004c). Phytokarst. In: *Encyclopedia of Caves and Karst Science* (ed. J. Gunn), 581–582. New York: Fitzroy Dearborn.

Vincent, P.J. (1983). The morphology and morphometry of some arctic Trittkarren. *Zeitschrift für Geomorphologie* 27: 205–222.

Vincent, P.J. (1995). Limestone pavements in the Bristish Isles. A review. *Geography Journal* 161: 265–274.

Vincent, P.J. (2004). Polygenetic origin of limestone pavements in northern England. *Zeitschrift für Geomorphologie* 48: 481–490.

Vincent, P.J. (2009). Limestone pavements in the Bristish Isles. In: *Karst Rock Features. Karren Sculpturing* (ed. A. Ginés, M. Knez, T. Slabe, et al.), 267–274. Ljubljana: ZRC Publishing.

Walker, S.E. and Matzat, J.W. (1999). Planning the replacement of the Beards Creek Bridge. *Engineering Geology* 52: 35–43.

Wall, J.R.D. and Wilford, G.E. (1966). Two small-scale solution features of limestone outcrops in Sarawak, Malaysia. *Zeitschrift für Geomorphologie* 10: 90–94.

Wall, J., Bohnenstiehl, D.R., Wegmann, K.W. et al. (2017). Morphometric comparisons between automated and manual karst depression inventories in Apalachicola National Forest, Florida, and Mammoth Cave National Park, Kentucky, USA. *Natural Hazards* 85: 729–749.

Waltham, T. (1989). *Ground Subsidence*. New York: Chapman and Hall.

Waltham, T. (2005). Tiankengs of the world, outside China. *Cave and Karst Science* 32: 1–12.

Waltham, T. (2008a). Sinkhole hazard case histories in karst terrains. *Quarterly Journal of Engineering Geology and Hydrogeology* 41: 291–300.

Waltham, T. and Smart, P.L. (1988). Civil engineering difficulties in the karst of China. *Quarterly Journal Engineering Geology* 21: 2–6.

Waltham, T., Bell, F., and Culshaw, M. (2005). *Sinkholes and Subsidence*. Chichester: Springer.

Wang, G., You, G., and Xu, Y. (2008). Investigation on the Nanjing gypsum mine flooding. In: *Geotechnical Engineering for Disaster Mitigation and Rehabilitation* (ed. H. Liu, A. Deng and J. Chu), 920–930. Berlin: Springer-Verlag.

Warren, J.K. (2006). *Evaporites: Sediments, Resources and Hydrocarbons*. Heidelberg: Springer-Verlag.

Warren, J.K. (2016). *Evaporites: A Geological Compendium*. Dordrecht: Springer.

White, W.B. (1988). *Geomorphology and Hydrology of Karst Terrains*. New York: Oxford University Press.

White, W.B. (2019). Springs. In: *Encyclopedia of Caves* (ed. W.B. White, D.C. Culver and T. Pipan), 1031–1040. New York: Academic Press.

White, W.B. and White, E.L. (1995). Correlation of contemporary karst landforms with paleokarst landforms: the problem of scale. *Carbonates and Evaporites* 10: 131–137.

White, E.L., Aron, G., and White, W.B. (1986). The influence of urbanization on sinkholes development in Central Pennsylvania. *Environmental Geology and Water Sciences* 8: 91–97.

Whiteman, C.D., Haiden, T., Pospichal, B. et al. (2004). Minimum temperatures, diurnal temperature ranges, and temperature inversions in limestone sinkholes of different sizes and shapes. *Journal of Applied Meteorology* 43: 1224–1236.

Williams, P.W. (1971). Illustrating morphometric analysis of karst with examples from New Guinea. *Zeitschrift für Geomorphologie* 15: 40–61.

Williams, P.W. (1972a). Morphometric analysis of polygonal karst in New Guinea. *Geological Society of America Bulletin* 83: 761–796.

Williams, P.W. (1972b). The analysis of spatial characteristics of karst terrains. In: *Spatial Analysis in Geomorphology* (ed. R.J. Choerly), 135–163. Methuen: London.

Williams, P.W. (1983). The role of the subcutaneous zone in karst hydrology. *Journal of Hydrology* 61: 45–67.

Williams, P.W. (1985). Subcutaneous hydrology and the development of doline and cockpit karst. *Zeitschrift für Geomorphologie* 29: 462–482.

Williams, P.W. (1993). Climatological and geological factors controlling the development of polygonal karst. *Zeitschrif für Geomorphologie* Supplement Band 93: 159–173.

Williams, P.W. (2004a). Dolines. In: *Encyclopedia of Caves and Karst Science* (ed. J. Gunn), 304–310. New York: Fitzroy Dearborn.

Williams, P.W. (2004b). Polygonal karst and paleokarst of the King Country, North Island, New Zealand. *Zeitschrift für Geomorphologie* Supplement 136: 45–67.

Williams, P.W. (2008). The role of the epikarst in karst and cave hydrogeology: a review. *International Journal of Speleology* 37: 1–10.

Williams, P.W. (2009). Arête and pinnacle karst of Mount Kaijende. In: *Karst Rock Features. Karren Sculpturing* (ed. A. Ginés, M. Knez, T. Slabe, et al.), 433–438. Ljubljana: ZRC Publishing.

Williams, P.W. (2011). Karst in UNESCO World Heritage Sites. In: *Karst Management* (ed. P.E. van Beynen), 459–480. Dordrecht: Springer.

Worthington, S.R.H., Ford, D.C., and Beddows, P.A. (2000). Porosity and permeability enhancement in unconfined aquifers as a result of solution. In: *Speleogenesis. Evolution of Karst Aquifers* (ed. A.B. Klimchouk, D.C. Ford, A.N. Palmer, et al.), 463–472. Huntsville: National Speleological Society.

Xiao, H., Kim, Y.J., Nam, B.H. et al. (2016). Investigation of the impacts of local-scale hydrogeologic conditions on sinkholes occurrence in East-Central Florida, USA. *Environmental Earth Sciences* 75: 1274.

Xiao, H., Li, H., and Tang, Y. (2018). Assessing the effects of rainfall, groundwater downward leakage, and groundwater head differences on the development of cover-collapse and cover-suffosion sinkholes in central Florida (USA). *Science of the Total Environment* 644: 274–286.

Xiao, X., Gutiérrez, F., and Guerrero, J. (2020). The impact of groundwater drawdown and vacuum pressure on sinkhole development. Physical laboratory models. *Engineering Geology* 279: 105894.

Xu, W. and Zhao, G. (1988). Mechanism and prevention of karst collapse near mine areas in China. *Environmental Geology Water Science* 12: 37–42.

Yechieli, Y., Abelson, M., Bein, A. et al. (2006). Sinkhole "swarms" along the Dead Sea coast: reflection of disturbance of lake and adjacent groundwater systems. *Geological Society of America Bulletin* 118: 1075–1087.

Yin, S.X. and Zhang, J.C. (2005). Impacts of karst paleo-sinkholes on mining and environment in northern China. *Environmental Geology* 48: 1077–1083.

Yizhaq, H., Ish-Shalom, C., Raz, E. et al. (2017). Scale-free distribution of Dead Sea sinkholes: observations and modeling. *Geophysical Research Letters* 44: 4944–4952.

Youssef, A.M., Pradhan, B., Sabtan, A.A. et al. (2012). Coupling of remote sensing data aided with field investigations for geological hazards assessment in Jazan area, Kingdom of Saudi Arabia. *Environmental Earth Sciences* 65: 120–130.

Youssef, A.M., Al-Harbi, H.M., Gutiérrez, F. et al. (2016). Natural and human-induced sinkhole hazards in Saudi Arabia: distribution, investigation, causes and impacts. *Hydrogeology Journal* 24: 625–644.

Youssef, A.M., Zabramwi, Y.A., Gutiérrez, F. et al. (2020). Sinkholes induced by uncontrolled groundwater withdrawal for agriculture in arid Saudi Arabia. Integration of remote-sensing and geophysical (ERT) techniques. *Journal of Arid Environments* 177: 104132.

Zaidner, Y., Frumkin, A., Porat, N. et al. (2014). A series of Mousterian occupations in a new type of site: the Nesher Ramla karst depression, Israel. *Journal of Human Evolution* 66: 1–17.

Zambo, L. and Ford, D.C. (1997). Limestone dissolution processes in Beke doline Aggtelek National Park, Hungary. *Earth Surface Processes and Landforms* 22 (6): 531–543.

Zarei, M. and Raeisi, E. (2010). Karst development and hydrogeology of Konarsiah salt diapir, south of Iran. *Carbonates and Evaporites* 25: 217–229.

Zarei, M., Raeisi, E., and Talbot, C. (2012). Karst development on a mobile substrate: Konarsiah salt extrusion. *Iran. Geological Magazine* 149: 412–422.

Zeller, J. (1967). Meandering channels in Switzerland. In: *Symposium on River Morphology*, 174–186. Bern: International Union of Geodesy and Geophysics, International Association of Scientific Hydrology.

Zhang, Z. and Chen, Z. (2019). Numerical simulation of bedrock sagging sinkholes in strain-softening rock induced by embankment construction. *Advances in Civil Engineering* 9426029.

Zhang, G., Wang, Z., Wang, L. et al. (2019). Mechanism of collapse sinkholes induced by solution mining of salt formations and measures for prediction and prevention. *Bulletin of Engineering Geology and the Environment* 78: 1401–1415.

Zhao, H.J., Ma, F.S., and Guo, J. (2010). Regularity and formation mechanism of large-scale abrupt large collapse in southern China in the first half of 2010. *Natural Hazards* 60: 1037–1054.

Zhou, W.F. (1997). The formation of sinkholes in karst mining areas in China and some methods of prevention. *Environmental Geology* 31: 50–58.

Zhu, X. and Chen, W. (2006). Tiankengs in the karst of China. *Speleogenesis and Evolution of Karst Aquifers* 4: 1–18.

Zhu, X. and Waltham, T. (2005). Tiankeng: definition and description. *Cave and Karst Science* 32: 75–79.

Zhu, J., Taylor, T.P., Currens, J.C. et al. (2014). Improved karst sinkhole mapping in Kentucky using LIDAR techniques: a pilot study in Floyds Fork watershed. *Journal of Cave and Karst Studies* 76: 207–216.

Zhu, X., Chen, W., and Zhang, Y. (2019). Tiankeng, Definition of. In: *Encyclopedia of Caves* (ed. W.B. White, D.C. Culver and T. Pipan), 1071–1076. New York: Academic Press.

Zhu, J., Nolte, A.M., Jacobs, N. et al. (2020). Using machine learning to identify karst sinkholes from LiDAR-derived topographic depressions in the Bluegrass Region of Kentucky. *Journal of Hydrology* 588: 125049.

Zseni, A. (2009). Subsoil shaping. In: *Karst Rock Features. Karren Sculpturing* (ed. A. Ginés, M. Knez, T. Slabe, et al.), 103–121. Ljubljana: ZRC Publishing.

Zumpano, V., Pisano, L., and Parise, M. (2019). An integrated framework to identify and analyze karst sinkholes. *Geomorphology* 332: 213–225.

Zwolinski, Z. (1996). Morphological types of kamenitzas. In: *Karren Landforms* (ed. J.J. Fornós and A. Ginés), 239–240. Palma de Mallorca: Universitat de les Illes Balears.

7

Other Karst Landforms

7.1 Poljes

7.1.1 General Aspects and Distribution

The word polje in Slavic languages means a flat field with arable soil. In the karst literature polje refers to large flat-floored depressions with underground drainage through karst rocks and commonly elongated in the direction of the structural grain (Sweeting 1972; Gams 1978; Ford and Williams 2007; Bonacci 2013) (Figures 7.1 and 7.2a). Poljes are mostly topographically enclosed basins, but there are also open poljes. These depressions are of great societal and economic importance since they usually host extensive agricultural areas, concentrate human settlements and infrastructure, and may provide direct access to water resources (karst windows), often subject to remarkable engineering projects that involve profound alterations in the hydrological systems (Milanović 2002, 2018; Bonacci 2013). Poljes can reach tens of kilometers in length and areas of hundreds of square kilometers. Livanjsko and Ličko are the largest poljes in Bosnia and Herzegovina, and Croatia, with surface areas of around 450 and 465 km², respectively. Poljes, despite their large dimensions and significance from the applied perspective, are one of the most poorly understood karst features. This is probably due to the limited recent research on the subject, the fact that poljes can have different origins (polygenesis and morphologic convergence), and that full understanding of their origin and evolution requires multidisciplinary studies integrating geomorphic, hydrologic, and geologic data (Gracia et al. 2003; Doğan et al. 2017). Regarding the origin of poljes, a critical issue in a number of cases is to elucidate the passive or active role played by the controlling tectonic structures, and whether the topographic basins are created by differential solutional denudation or tectonic subsidence.

The Dinaric karst in Italy, Slovenia, Croatia, Bosnia and Herzegovina, and Montenegro is the region where poljes are best developed and where they were initially described (Figure 7.1a). Here, there are more than 100 poljes with a prevailing NW-SE orientation that cover approximately 1500 km², which is about 2.5% of the total karst area (Milanović 2004a; Bonacci 2013). Poljes have been documented in numerous carbonate karst regions and in a few evaporite terrains (Calaforra and Pulido-Bosch 1999a; Doğan and Özel 2005; De Waele et al. 2017b; Gökkaya et al. 2021). They are particularly common in orogens of the broad Alpine-Himalayan collision zone, where thick folded and faulted carbonate successions, and often active faults, create favorable litho-structural conditions for their development. Some of the most notable regions include the Zagros Mountains in Iran, with as many as 175 poljes (Nassery et al. 2009; Mohammadi et al. 2019), well-documented examples in the Taurides of Turkey (Doğan et al. 2017, 2019; Doğan

Karst Hydrogeology, Geomorphology and Caves, First Edition. Jo De Waele and Francisco Gutiérrez.
© 2022 John Wiley & Sons Ltd. Published 2022 by John Wiley & Sons Ltd.

Figure 7.1 Examples of poljes with flat alluviated floors. (a) Popovo Polje in Bosnia and Herzegovina. The depression is drained longitudinally by the Trebišnjica River that terminates in a group of ponors with an overall intake capacity of ca. 300 m³ s⁻¹ (see Figure 7.2d). The river has been channelized to prevent water infiltration. *Source:* Francisco Gutiérrez (Author). (b) Rodenas border polje in the Iberian Chain, Spain, controlled by an inactive reverse fault that juxtaposes Jurassic limestones (foreground) against Triassic sandstones and Paleozoic quartzites (background). This polje is inset into an extensive Neogene planation surface and shows a highly irregular edge on the limestone margin. *Source:* Francisco Gutiérrez (Author). (c) Eynif Polje in the Taurus Mountains of Turkey, a neotectonic border polje associated with an active graben. *Source:* Photo by Ugur Doğan. (d) Suğla Polje and neotectonic graben in the Taurus Mountains, Turkey. Note the extremely flat floor generated by sediment aggradation and its linear fault-controlled edges. *Source:* Photo by Ugur Doğan.

and Koçyiğit 2018; Şimşek et al. 2021) (Figure 7.1c, d), the fold and thrust belt of western Greece (Vott et al. 2009; Deligianni et al. 2013), the Peloponnese (Seguin et al. 2019), the Apennines in Italy (Aiello et al. 2007; Giraudi et al. 2011; Galdenzi 2019; Pisano et al. 2020), the Alps (Nicod 1967), the Carpathians in Hungary (Bella et al. 2016), the Betic Cordillera (Lhenaff 1986 and references therein; Gracia et al. 2000; Sanz de Galdeano 2013) (Figure 7.3), the Iberian Chain (Gutiérrez-Elorza et al. 1982; Peña-Monné et al. 1989; Gutiérrez-Elorza and Valverde 1994; Gracia et al. 1999, 2002, 2003) (Figure 7.1b) and the Pyrenees in Spain (Waltham 1981; Jiménez-Sánchez et al. 2013), the Lusitanian Cordillera in Portugal (Martins 1950; Nicod 1996), or the Atlas in Morocco (Ek and Mathieu 1964). Poljes have also been documented in several Caribbean Islands such as Cuba (Viñales Valley, UNESCO World Heritage Site; Lehmann et al. 1956), Jamaica

Figure 7.2 Images of poljes and ponors. (a) Nieguisi Polje, Montenegro, surrounded and underlain by carbonate rocks, with extremely flat floor and relatively gentle slopes at the margins. *Source:* Francisco Gutiérrez (Author). (b) Zafarraya neotectonic polje in the Betics, southern Spain, during the 1996–1997 flood. *Source:* Photo by Joaquín Rodríguez-Vidal. (c) Kovari Ponor in Duvanjsko Polje, Bosnia and Herzegovina. *Source:* Francisco Gutiérrez (Author). (d) Crnulja ponor and cave, one of the main swallow holes in Popovo Polje, Bosnia, and Herzegovina. *Source:* Francisco Gutiérrez (Author).

(Sweeting 1958; Pfeffer 1986) and Puerto Rico (Monroe 1976), in tropical China (Smart et al. 1986), and there are remarkable examples in high-latitude environments with discontinuous permafrost in northern Canada (Brook and Ford 1980).

7.1.2   Geomorphological, Hydrological, and Geological Features of Poljes

From the morphological perspective, poljes are large depressions in karst terrain characterized by extremely flat floors (Figures 7.1 and 7.2a). Gams (1978) proposed that the floor of poljes must be at least 400 m wide, and Cvijić (1893) indicated a minimum dimension of 1 km, although these morphometric criteria are considered arbitrary by numerous authors (Frelih 2003). The flat topography of the polje floors can be interrupted by isolated residuals of bedrock inliers. These protruding hills surrounded by plains are often designated as "hums," a term derived from a village located next to one of such residual hills in Popovo Polje, Bosnia and Herzegovina. The bottom of some poljes is underlain by a relatively thin veneer of unconsolidated deposits (e.g., alluvium, residual soil) covering a subhorizontal corrosion surface beveled in karst bedrock, which typically cuts across the geological structures. These are essentialy mantled rock-strath surfaces mainly developed by solutional planation within the epiphreatic zone. In other cases the polje floors are underlain by a thick sedimentary fill that can bury an irregular bedrock topography, which may correspond to a paleorelief and/or to fault-blocks affected by differential tectonic displacement. Thus, the floor of these poljes corresponds to an aggradation surface. The thick sedimentary fill can be related to the accumulation of sediments in an enclosed depression that functions as a

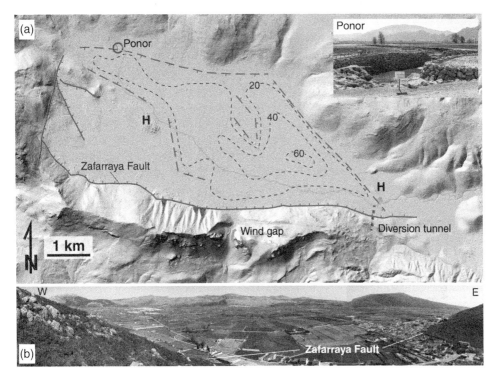

Figure 7.3   The Zafarraya neotectonic polje in the Betic Cordillera, southern Spain. (a) Shaded relief model of the depression showing the main mapped and inferred faults, triangular facets (red triangles), as well as isopachs of the sedimentary fill constructed from borehole and geophysical data. *Source:* Data adapted from Sanz de Galdeano (2013); Francisco Gutiérrez (Author). H stands for hums. Open-source 2 m resolution DEM downloaded from the Spanish Geographical Institute. Inset shows an image on the main ponor, which is the terminus of the allogenic stream that drains the depression. (b) Panoramic view of the Zafarraya Polje from the southern mountain front. *Source:* Francisco Gutiérrez (Author).

passive sediment trap, or as a neotectonic basin where active subsidence creates accommodation space. An example of the latter case is illustrated by the Zafarraya Polje in the Betics of southern Spain, which has been thoroughly investigated from the geomorphological, hydrological, and structural perspective (Lhenaff 1968; López-Chicano and Pulido-Bosch 1993; López-Chicano et al. 2002; Sanz de Galdeano 2013) (Figure 7.3).This is a neotectonic basin in carbonate terrain with underground drainage and bounded by active faults with dominant normal displacement. It has been interpreted as a pull-apart basin developed in a local extensional zone associated with a system of strike-slip faults within a general context of oblique convergence (Sanz de Galdeano 2013). The southern margin of the depression displays a prominent mountain front with some triangular facets controlled by the E-W-trending and N-dipping Zafarraya Fault, a normal fault with some dextral displacement component (Figure 7.3). According to some authors this fault was the source of the Mw ~6.5 Andalucía earthquake of 25 December 1884, which caused more than 800 fatalities (Reicherter et al. 2003). An isopach map of the detrital sedimentary fill constructed from geophysical and borehole data shows an inner rhomboidal structural basin with deposits more than 60 m thick in the depocentral sector. The flat floor of the polje, essentially an aggradation surface, is interrupted on the western sector by inliers of Jurassic limestone (hums) interpreted as small uplifted fault-bounded blocks. The polje is drained longitudinally by an allogenic stream that terminates in a system of ponors situated close to the northwestern edge of the basin (Figure 7.3).

A diversion tunnel was excavated across the southern margin to drain the polje during occasional flooding events related to both surface water inflow and groundwater level rise (López-Chicano and Pulido-Bosch 1993; López-Chicano et al. 2002). The range in the southern margin of the basin displays a wind gap associated with a south-flowing drainage beheaded by vertical displacement along the N-dipping Zafarraya Fault (Figure 7.3).

Poljes are commonly markedly elongated depressions oriented in the direction of the structural grain determined by the dominant geological structures, either compressional, extensional or strike-slip (Figures 7.1 and 7.3). The margins often display steeply rising slopes with a well-defined slope-floor junction. The edges of the bottom can be irregular (Waltham 1981) (Figure 7.1b) or rectilinear and fault-controlled (Figures 7.1c, d and 7.3). In the latter case, the marginal slopes can correspond to fault-line scarps related to differential erosion (e.g., karst bedrock juxtaposed against resistant non-karst bedrock) (Gracia et al. 2002), or to mountain fronts and fault scarps created by dip-slip displacement (Gracia et al. 2003; Doğan et al. 2017). The mountain fronts and scarps associated with active faults in karst terrain tend to display some peculiar features compared to those developed in non-karst rocks. Because of the dominant subsurface drainage and the limited mechanical water erosion, these escarpments tend to degrade slowly and are often scarcely dissected by drainages, lacking the persistent triangular and trapezoidal facets typically found in other less permeable lithologies (Figure 7.3a). The margins of poljes can display a terraced topography with different levels of corrosion surfaces attributable to former polje floors. These stepped sequences record alternating phases of deepening and solution planation (Roglić 1974b; Gospodarič and Habič 1978; Gracia et al. 2002, 2003, 2014). Often the soils and deposits that used to mantle these rock straths have been stripped by erosion and the exposed bedrock displays assemblages of subsoil karren and superimposed free karren. A number of large poljes in the Iberian Chain, NE Spain, inset into extensive Neogene planation surfaces, do not show any clear structural control and display several levels of corrosion surfaces with an overall concentric distribution around the bottom (Gutiérrez-Elorza and Valverde 1994; Gracia et al. 1999). In some cases, the distribution and cartographic relationships of the different corrosion levels and the present-day polje floors allow the recontruction of a long-term evolution that entails the compartmentalization of old large poljes into smaller nested depressions (Gracia et al. 1999, 2002). This contrasts with early ideas whereby poljes are related to the progressive amalgamation of several independent basins (Cvijić 1893).

Poljes are depressions with underground drainage. They can be topographically closed basins or depressions open upstream that receive runoff from a stream that ultimately drains underground (Figure 7.3). Former poljes captured by the external drainage network can display a dominant surface drainage and may be traversed by through-flowing rivers (i.e., poljes open upstream and downstream) (Bella et al. 2016). The flat floor and the subsurface drainage are the main diagnostic criteria used to designate large depressions in karst terrains as poljes, regardless of their origin (Sweeting 1972; Gams 1978; Ford and Williams 2007). They can receive water inflow from springs, runoff related to autogenic and allogenic streams, and direct precipitation. The water mainly leaves the depressions through swallow holes (ponors) (Figure 7.2c, d) and diffuse infiltration. A polje can form part of several hydrological basins so that the divides of the catchments can cut-across the depression. Poljes can perform input, throughput and output hydrological functions, and should be treated as subsystems within larger karst drainage basins with complex underground connections. For instance, in some sectors of the Dinaric karst, poljes display a step-like arrangement and can be hydrologically connected forming part of large and complex karst basins (Figure 7.4). The water drains toward the regional base level mostly via underground routes and eventually at the floor of the poljes. These flat surfaces occupy a small part of the total karst area

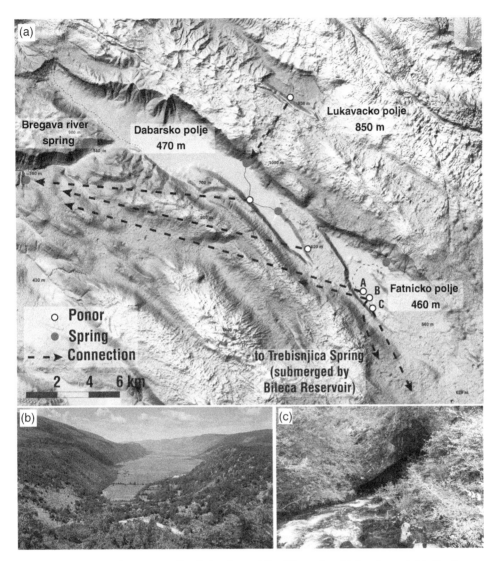

Figure 7.4   Images of the Lukavačko, Dabarsko, and Fatničko poljes. (a) Terrain model indicating underground connections between the main ponors and springs. *Source:* From Bonacci (2013). The approximate elevation at the lowest point of the poljes is indicated. Note the fault escarpments at the margins of the poljes, most probably corresponding to active pull-apart basins in transtensional zones. Terrain model provided by the Karst Research Institute at Postojna, Slovenia. (b) General view looking to the NW of Dabarsko Polje bounded by linear and undissected escarpments. (c) Ponikva Ponor at Dabarsko Polje. This ponor at the SW edge of the polje gives access to a passage offset by faults identifiable in the cave. *Source:* Francisco Gutiérrez (Author).

and are often the only places where surface water can be found, functioning as karst windows. Frequently, springs on one side of the polje feed streams that traverse the floor of the depression and terminate at ponors situated on the outflow side of the depression (Mijatović 1984; Milanović 2004a). This is a very peculiar transverse underground drainage perpendicular to the topographic grain that can be developed largely thanks to the presence of very thick and karstified limestone successions. Figure 7.4 shows three hydrologicaly connected poljes (Lukavačko, 6 km²;

Dabarsko, 31.7 km^2; Fatničko, 5.6 km^2) of Bosnia and Herzegovina. Most probably these enclosed depressions correspond to pull-apart basins bounded by active faults with dominant vertical displacement and expressed as very linear undissected escarpments. The figure indicates elevation data and shows ponor–spring connections demonstrated by groundwater tracing (Milanović 2004a; Bonacci 2013). The water that sinks in the ponor of Lukavačko Polje, the highest depression, drains underground perpendicularly to the structural and topographic grain and emerges at the main spring of Dabarsko Polje. Then it flows across the polje floor along a stream and sinks into Ponikva Ponor (Figure 7.4c) connected with the Bregave River spring to the west. Interestingly, Fatničko Polje has three ponor zones just 2 km apart but with different underground routes (Bonacci 2013). Water swallowed in ponor A reappears at the Bregave River spring, whereas water that sinks in ponor C flows to the Trebišnjica River spring at the Bileća Reservoir. The water evacuated in ponor zone B can reappear at any of those springs depending on groundwater levels. This means that the boundaries of the catchments are not stationary and barely have any relationship with the topography. The construction of the Bileća Reservoir submerged the Trebišnjica River spring altering the flooding regime in the Fatničko Polje and prolonging flood duration (Milanović 1986). Tracing in the Monte Lago polje, Apennines (Italy) proved hydrological connections between the main ponor and multiple springs, suggesting a multidirectional underground flow influenced by the recharge rate (Galdenzi 2019).

Ponors often correspond to cave entrances, cover-collapse sinkholes located in the alluviated floors, or bedrock-collapse sinkholes associated with bedrock exposures. They can have a very high capacity to evacuate both water and sediment load, and their maximum absorption rate during flood events varies depending on the water level (Bonacci 2013). Some poljes have ponors at different elevations, so that the outflow rate can increase sharply when the water level reaches the higher ponors (Drežničko Polje; Bonacci and Plantić 1997). The ponors in Popovo Polje (Figures 7.1a and 7.2d) have a capacity of around 300 m^3 s^{-1} and the largest one, Doljašnica Ponor, has a capacity of ca. 60 m^3 s^{-1}. The inflow in winter used to exceed 1000 m^3 s^{-1} resulting in yearly floods in the polje with depths of 40 m and lasting around 253 days. The excavation of two large tunnels that feed hydroelectric power plants practically eliminated these seasonal floods (Milanović 2004a).

Some poljes are permanently dry, others are affected by temporary flooding (Figure 7.2b), and some host permanent lakes. Polje flooding can occur when the water inflow exceeds water outflow and/or as a consequence of groundwater level rise. The latter situation occurs in poljes which floors are situated within the epiphreatic zone. Flooding can cause significant detrimental effects, but it is a natural process that contributes to regulate the hydrological system by temporarily retaining water in the depressions and prolonging the hydrological response at the springs. When the total water inflow rate (Q_i) exceeds the outflow rate (Q_o), the excess volume ($+\Delta V$) for a time interval (Δt) is stored in the polje and flooding commences. When the outflow rate surpasses the inflow rate, the water budget becomes negative ($-\Delta V$) and flood recession starts, until it ceases. A simple water budget equation describes the inundation process:

$$Q_i - Q_o = \pm \Delta V / \Delta t = \pm Q_r \qquad (6.1)$$

where ΔV represents the positive or negative water storage change at the surface of the depression and Q_r the net inflow or outflow rate responsible for the rise and recession of the flood, respectively (Ristić 1976; Bonacci 1987). In the Zafarraya Polje, López-Chicano et al. (2002) using data on water inflow from an allogenic stream and the retention volume during a flood (Figure 7.2b), estimated the overall infiltration capacity of the polje. This value provides an approximate threshold for the flow rate at which the stream entering the polje can initiate an inundation event. The Planinsko Polje in Slovenia experiences seasonal floods lasting on average 41 days and with water levels commonly reaching 8 m (Blatnik et al. 2017). Stepišnik et al. (2012) identified and dated loamy paleo-flood deposits at the surface (dolines, pocket valleys) and in caves associated with the polje indicating that Holocene floods reached water depths 37 m higher than those recorded in historical times. Flooding can also be related to a rise in groundwater level that results from recharge in the karst aquifer. During these events karst conduits experience reverse flow or backflooding. This can cause perched springs to become active (overflow springs), influent streams to turn into effluent streams, and ponors to function as springs. The ponors that temporarily play the role of springs are known as "estavelles." In the Nahanni karst, northern Canada, Brook and Ford (1980) documented flooding in a sequence of three poljes at different elevations induced by a severe rainfall event and the resulting groundwater level rise. Water depths in the poljes reached as much as 25 m and the lowest polje overflowed. In poljes with a thick sedimentary fill, the deposits can behave as a detrital aquifer with high granular permeability (e.g., alluvium) connected with the karst aquifer (e.g., Zafarraya Polje; López-Chicano and Pulido-Bosch 1993). In contrast, a low-permeability fill can function as an aquitard (Žebre et al. 2016). In these situations, the water that emerges at springs can flow along streams carved in the relatively impervious polje bottom until it reaches a ponor at a site with exposed or shallow karst bedrock. Springs and ponors are often controlled by geological structures (e.g., faults) and contacts (e.g., boundary between karst and non-karst rocks). The springs and ponors associated with active faults, as well as the caves and conduits connected to them, are affected by the displacement of the tectonic structures and experience adjustments in response to such movements (Gams 2005). A good example can be observed in the Ponikva Ponor and Cave on the SW side of Dabarsko Polje (Figure 7.4c), controlled by a active fault with oblique displacement. The cave passage following the ponor is offset by an identifiable fault.

Poljes behave as sediment traps that can receive detrital sediment from multiple sources such as the marginal slopes, autogenic and allogenic streams, piedmont glaciers debouching into the depressions, pyroclastic material from volcanoes (Aiello et al. 2007). Žebre et al. (2016) investigated the glacial and outwash deposits at the Gomance Polje in the northern Dinaric Alps that was reached by outlet glaciers emanating from an icefield. Poljes can also host lacustrine and palustrine environments with deposition of detrital and organic sediments. The sedimentary fill of a number of poljes in the Dinaric karst include thick lacustrine successions and significant coal seams that are locally mined (Mijatović 1984; Gams 2005). Part of the sediment input can be evacuated subterraneously via ponors as detrital particles or as solute load. Sweeting (1972) indicates that as early as 1894, von Groller calculated that 18 480 m^3 of material used to disappear through the ponors of Popovo Polje every year (Figure 7.2d). In actively subsiding neotectonic poljes, the topographic relief depends on the rate at which the margins are lowered by erosion and the balance between subsidence rate and net aggradation rate at the floor. Sediment input and aggradation rates tend to be very low in karst poljes because of the usually restricted topographic catchment area and the limited surface runoff and sediment flux (Figure 7.4). This situation results in starved basins with significant topographic relief; deep hollows in which subsidence largely exceeds aggradation. Some poljes in the classical Dinaric karst have a relatively thin cover underlain by bevelled karst bedrock. For instance, the alluvial cover in Popovo Polje is 1–2 m thick in the upstream sector, increasing to as much as 15–20 m downstream (Milanović 2004a). Nonetheless, numerous poljes in the Dinaric region have thick Neogene fills of terrestrial deposits hundreds to thousands of meters thick, typically decreasing toward the coast: Livanjsko Polje >2300 m, Duvno Polje 2500 m, Duvanjsko Polje 2500 m, Glamočko Polje 700 m, Dabarsko and Fatničko poljes 150–200 m, Imotsko Polje 145 m (Gams 1978; Mijatović 1984; Milanović 2004a).

The bottom of the sedimentary fill is frequently below sea level, the deposits locally show evidence of tectonic deformation (Gospodarič and Habič 1978; Gams 2005), and some polje floors have been submerged by the sea. These data indicate that those poljes are neotectonic basins controlled by faults with a significant vertical displacement component. The main difference between these active fault-bounded basins and those developed in non-karst terrain is that they have underground drainage. Cvijić (1900) argued that "tectonic processes enabled the formation of large poljes through fault activity, although karst erosion gives the final polje shape" (quoted by Mijatović 1984). Grund (1903) stated that "polje is a karstified tectonic field. Only by virtue of its underground drainage does it become a component of the karst phenomenon" (quoted by Lehmann 1959, p. 260). Vrabec (1994) quoted a number of works proposing that some poljes, like those associated with the NW-SE Idrija dextral strike-slip fault (Planinsko, Cerkniško and Loško poljes) correspond to active pull-apart basins. Gams (2005) also supports this concept for some poljes. The 120 km long Idrija Fault (Slovenia), together with other structures of the Dinaric Fault System, accommodate the northward displacement of the Adriatic Plate toward the European continent and has been the source of major surface-rupturing earthquakes (M ~ 6.8, 1511 Idrija earthquake) (Moulin et al. 2014; Grützner et al. 2021). The Dabarsko and Fatničko poljes, shown in Figure 7.4, resemble active pull-apart basins developed in a transtensional zone of a strike-slip fault system. The sedimentary fill of these basins reaches 150–200 m (Milanović 2004a) and the cave associated with the Ponikva Ponor at the SW edge of Dabarsko Polje shows evidence of recent fault displacement. More work is needed to better understand the role played by active faulting on the development of these poljes (e.g., trenching).

Gracia et al. (2003) in the Iberian Chain, Spain interpret the Jiloca Depression as a karst polje associated with a neotectonic half-graben bounded on its eastern margin by three normal faults with an en echelon arrangement (Gutiérrez et al. 2008e). Detailed geomorphic mapping revealed eight levels of corrosion surfaces on the opposite margin that provide evidence of the important role played by alternating phases of solutional lowering and planation in the geomorphic evolution of this neotectonic polje. The San Gregorio Magno polje in the southern Apennines, Italy is a Quaternary graben bounded by active normal faults with a thick sedimentary fill that was affected by the complex surface rupture of the 1980, Ms6.9 Irpina earthquake (Aiello et al. 2007). This depression with internal drainage used to be affected by frequent flooding until the end of the nineteenth century when a tunnel was excavated to connect the depression with a nearby river. Doğan et al. (2017) in the Taurides, Turkey described a system of poljes with underground drainage that correspond to graben depressions bounded by steep and linear mountain fronts, hundreds of meters high, generated by active Quaternary faults. The margins of some depressions display secondary faults that have stepped a perched corrosion surface. The development of the grabens during the Quaternary disrupted a transverse paleodrainage network, recognized as wind gaps and paleovalleys in the intervening horsts. Doğan and Koçyiğit (2018) described the Suğla Polje in the Taurides of Turkey as a border polje within an active graben. One of the faults that bounds the graben beheaded a stream that used to drain the tectonic depression, and now drains underground in a different direction via ponors. The sedimentary fill of the polje-graben reaches 185 m. They identified two levels of corrosion surfaces and deposits of a former pluvial lake. In the Taurus Mountains, Doğan et al. (2019) ascribed the Kestel Polje system covering 521 km^2 to a suite of Quaternary grabens with a sedimentary fill as much as 281 m thick. The development of these neotectonic poljes and the disruption of a transverse paleodrainage created the underground conditions responsible for the formation of the terraced Antalya tufa deposits in the output area, which are considered to be the largest calcareous tufa formation in the world (600 km^2 in area and up to 250 m thick).

7.1.3 Origin and Classification

Karst poljes are a clear example of morphological convergence (equifinality), since they can result from different processes despite showing similar morphological characteristics. They have received several classifications based on multiple criteria. Sweeting (1972), in her comprehensive review on poljes differentiated two main types: (i) poljes completely surrounded by limestones, which in most cases are hydrologically closed basins and (ii) border poljes developed by differential erosion at the contact of limestone with impermeable rocks. Gams (1978, 1994) recognized five types, considering their geological-geomorphological setting and hydrological behavior: (i) *border poljes*, (ii) *piedmont poljes* located at the foot of mountainous terrain, usually glaciated, that supply large amounts of detrital deposits to the depression, (iii) *peripheral poljes* largely underlain by impermeable rocks and with surface runoff that drains through ponors in the karst rocks situated at the margins, (iv) *overflow poljes* with the bottom underlain by impermeable bedrock and water emerging at springs on one side of the depression and sinking at the opposite side, and (v) *poljes in the piezometric level* with the floor completely underlain by karst rocks and situated within the epiphreatic zone, and consequently subject to flooding. Ford and Williams (1989, 2007) considered that Gams' classification can be reduced to three categories: (i) *border poljes* (*Randpolje* in German) dominated by allogenic water and sediment input, (ii) *structural poljes* dominated by geological controls and often associated with fault-bounded basins and relatively impermeable rocks on the floor, and (iii) *base-level poljes* underlain by karst rocks and developed by solutional lowering of the karst surface up to the epiphreatic zone. Here, inspired by the previous works and considering the decisive role played by active faulting in the formation of some poljes (Mijatović 1984; Gracia et al. 2003; Gams 2005; Sanz de Galdeano 2013; Doğan et al. 2017), we propose a simple classification that differentiates two genetic groups and various types for one of the groups (Figure 7.5). Poljes can be split into two groups based on the mechanism responsible for the generation of the topographic depression: (i) *erosional poljes* produced by differential erosion mainly related to solutional denudation of karst rocks; and (ii) *neotectonic poljes* generated by tectonic subsidence related to active faulting. Erosional poljes, depending on the geological and hydrological setting, can be classified into *base-level poljes, border poljes,* and *overflow poljes*. Overflow polje is probably a more adequate term than structural polje, which is rather unspecific and could also include border poljes that are clearly structurally controlled. There is some overlap between these three types. For instance, the bottom of border poljes can be located within the epiphreatic zone and consequently could also be considered as asymmetric base-level poljes.

Neotectonic poljes are associated with tectonic basins bounded by active faults with significant dip-slip displacement (e.g., pull-apart basins, grabens, and half-grabens) (Figure 7.5). Tectonic subsidence is responsible for the creation of the topographic depression and the accomodation space for the deposition of thick sedimentary successions. The presence of karstified bedrock at the margins of these depressions allows the creation of their underground drainage through ponors. Key diagnostic features for the identification of these poljes include thick recent sedimentary fills and the presence of bounding Quaternary faults and associated tectonic landforms (e.g., linear mountain fronts, triangular facets, and disrupted drainages). Tectonic geomorphology and trenching studies can provide valuable information for the characterization of these depressions. Well-documented examples of neotectonic poljes include the Zarraya Polje in Spain (Lhenaff 1968; Sanz de Galdeano 2013) (Figure 7.3), the Jiloca Polje in Spain (Gracia et al. 2003), and the Kembos, Eynif, Suğla, and Kestel poljes in Turkey (Doğan et al. 2017, 2019; Doğan and Koçyiğit 2018) (Figure 7.1c, d).

Base-level poljes are generally surrounded and underlain by permeable karst bedrock, are dominated by autogenic drainage and tend to receive limited detrital input (Figure 7.5). They form where the land surface reaches the epiphreatic zone by differential erosion, and consequently are

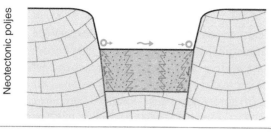

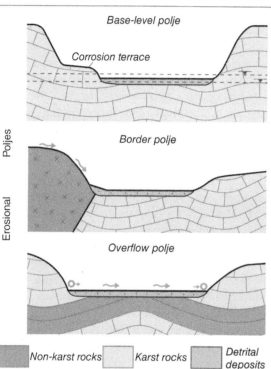

Neotectonic poljes

Base-level polje

Corrosion terrace

Border polje

Overflow polje

Poljes

Erosional

Non-karst rocks   Karst rocks   Detrital deposits

Figure 7.5 Classification of poljes. Poljes can be generated by tectonic subsidence or differential erosion. Three main types of erosional poljes can be differentiated depending on the hydrological and geological setting.

affected by temporary flooding. Dissolution removes irregularities in the bedrock to form a horizontal corrosion surface controlled by the water table. These rock-cut surfaces are typically covered by a relatively thin veneer of residual soils and detrital deposits. Dissolution beneath the soil cover is more efficient because the uptake of biogenic CO_2 from the soil substantially enhances the aggressiveness of the water. The shallow water table that functions as the base level for the corrosion surface prevents the deepening of the depression. Lateral dissolution prevails leading to expansion of the polje floor by the progressive retreat of the marginal slopes (lateral solution planation or corrosion planation). This process is intensified by flooding that causes solutional lateral undercutting. Similar to pediments, these are time-transgressive surfaces that are progressively younger toward the marginal retreating slopes (Ford and Williams 2007). A downward drop in the water table can rejuvenate vertical dissolution deepening the polje to form an inset corrosion surface controlled by the new groundwater level. These poljes may display stepped sequences of corrosion surfaces that record deepening and planation phases related to episodic drops in the epiphreatic zone (Roglić 1972). The overall deepening trend can be interrupted when corrosional lowering reaches non-karst formations. A number of base-level poljes in the Iberian Chain, Spain, displays multiple levels of corrosion surfaces with concentric distribution around the present-day floors (Gutiérrez-Elorza et al. 1982; Gutiérrez-Elorza and Valverde 1994) and corrosional deepening has locally ceased when the polje floors reached impermeable bedrock (Gracia et al. 1996, 2002).

Border poljes are generated by differential solutional lowering acting on karst formations in contact with impermeable rocks (Figure 7.5). The development of these depressions is favored by the intense dissolution that occurs at the contact zone by aggressive allogenic runoff derived from the non-karst bedrock (contact karst). These poljes tend to display an asymmetric geometry (Figure 7.1b). The margins associated with the non-karst rocks typically display higher relief, a linear edge, and alluvial fan aprons consisting of non-soluble detritus. The opposite margin underlain by karst rocks is generally more subdued and frequently shows an irregular edge. Border poljes can also be base-level poljes wherein their bottoms are situated within the epiphreatic zone and can thus display stepped sequences of corrosion surfaces in the margin underlain by karst rocks. The bottom of these poljes can expand by lateral solutional planation and recession of the slopes on the flank underlain by karst rocks. Examples of border poljes include Vistabella Polje, Spain (Gutiérrez-Elorza et al. 1982) Gallocanta Polje, Spain (Gracia et al. 2002), and Baške Ošterije Polje, Croatia (Perica et al. 2002).

The floor of *overflow poljes* is totally or partialy underlain by impermeable bedrock that functions as a hydrogeological barrier (Figure 7.5). This situation can be reached by progressive corrosional lowering of karst formations until the level of erosion reaches an underlying non-karst rock. Water that issues at one of the margins of the polje commonly flows across the depression over impervious bedrock and sinks at the opposite margin. Gams (1978) indicates that there are numerous poljes of this type in the Dinaric karst associated with Paleozoic to Paleogene flysch and detrital formations (e.g., Cerkniško Polje). The Planinsko Polje in Slovenia, underlain by low-permeability Triassic dolomites mantled by a thin alluvial cover is classified by Stepišnik et al. (2012) as an overflow polje.

7.2 Corrosion Plains, Rejuvenation, and Submergence

The term planation surface is used in non-karst terrains to designate large and nearly flat surfaces carved in bedrock. They typically cut across geological structures and can be locally interrupted by residual reliefs. The origin of planation surfaces in non-karstic rocks, despite their simple definition, is one of the most controversial issues in Geomorphology (Huggett 2003; Migoń 2004). Three main modes of formation have been proposed, as well as specific terms with genetic connotations for the suite of formative processes and the resulting plains. *Peneplains*, as formulated by William Morris Davis (1899) in his classical "geographic cycle," are developed following an episode of relief rejuvenation by progressive downwearing and slope decline until the topography is graded to the base level (peneplanation). In this model, local relief and slope gradient decrease through time. *Pediplains* are formed at the foot of slopes that experience backwearing (King 1963). The back slopes undergo progressive parallel retreat and the upland area is replaced by an expanding gently inclined erosional surface. In contrast with peneplains, these are diachronous surfaces that are progressively younger toward the receding slopes. Moreover, the local relief and the slope gradient do not experience significant changes. *Etchplains* form in two stages of etchplanation (Büdel 1957). In the first one, subsurface weathering produces a regolith with a planar boundary on the underlying solid bedrock. In the second stage, a flat etchplain is exposed by the erosional stripping of the weathering mantle. A review on these models of landscape evolution and the views of their followers and detractors can be found in Gutiérrez (2013). Flat erosional surfaces developed in carbonate rocks can be impressively planar and are commonly designated as *corrosion plains* (Figure 7.6).

Figure 7.6 Images of corrosion plains. (a, b) Kistanke corrosion plain cut across deformed carbonate rocks and entrenched by the Krka River, Croatia. This erosional surface used to be mantled by soil as reveal the abundant covered karren, exposed by the stripping of the rock-cut surface. Note strata dipping into the slopes on both sides of Krka Canyon. (c) Neogene planation surface that truncates folding structures affecting Late Cretaceous limestones, Lastra Range, Iberian Chain, NE Spain. *Source:* Photos by Francisco Gutiérrez.

Their formation has significant differences with respect to that of planation surfaces carved in non-karst rocks, due largely to the high permeability of the bedrock and the dominant role played by solutional denudation. Nonetheless, as explained below, the development of corrosion plains shares some features with the different types of planation surfaces described for non-karstic bedrocks.

Corrosion plains were recognized in the early stages of geomorphology when the concepts of the erosion cycle and peneplanation proposed by the American geomorphologist W.M. Davis (1899) were the dominant model of landscape evolution. In 1899, Albrecht Penck, together with his students from Vienna University and Davis, conducted field trips to the Dinaric karst. They documented conspicuous erosional plains in Herzegovina and Dalmatia (Penck 1900; Davis 1901). Jovan Cvijić and Alfred Grund, both disciples of Albrecht Penck, proposed models of karst landscape evolution considering that denudation tends toward a plain controlled by the water table (Grund 1914; Cvijić 1918) (Figure 7.7). Cvijić (1918), most probably influenced by W.M. Davis, sustained that karstification leading to the formation of large plains was preceded by a phase of fluvial erosion. Roglić (1972) and Ford and Williams (2007) present reviews on the evolution of concepts related to karst landscape development. Both, planation surfaces and corrosion plains record protracted periods of denudation under conditions of base level stability and tectonic quiescence. They can occur as stepped sequences of surfaces of different ages that allow the

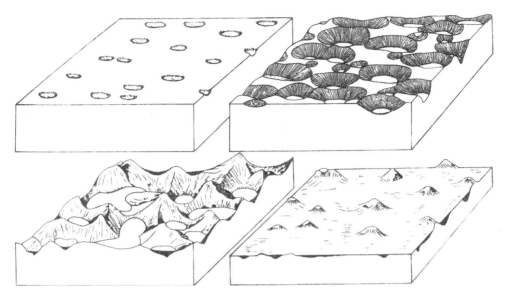

Figure 7.7   Model of karst landscape evolution proposed by Grund (1914).

reconstruction of different cycles of planation and erosional deepening (denudation chronology). Uplifted plains become plateaus that can be dissected and transformed into reliefs with accordant summits. Planation surfaces, corrosion plains, and in some cases their correlative sediments deposited in adjacent basins can be used as markers for the identification and characterization of neotectonic deformations (Gutiérrez-Elorza and Gracia 1997; Benito-Calvo and Pérez-González 2007; Doğan et al. 2017).

Corrosion plains have been reported in a wide range of climatic environments and geomorphic settings, such as tropical tower and cone karst landscapes (Lehmann et al. 1956), the floor of poljes (see Section 7.1), uplifted plateaus in mountain regions (Sweeting 1995), alluvial plains, coastal areas, or the margins of sedimentary basins (Benito-Calvo and Pérez-González 2007). Moreover, they can form in both the input and output zones of karst terrains. The water table, which experiences fluctuations within the epiphreatic zone, is the base level for corrosion plain development. In a karst terrain in which the epiphreatic zone is situated at some depth below the topographic surface and provided the water table remains at a relatively stable position, the development of corrosion plains occurs in two stages with different denudational styles (Figure 7.8):

(1) In the first stage, the ground surface is essentially worn down by water that infiltrates and percolates through the vadose zone. Because of the high permeability and heterogeneity of the karst bedrock, largely determined by discontinuity planes, surface runoff is very limited and the solutional lowering of the surface typically occurs in a differential and focused manner, often resulting in the formation of solution sinkholes. The surface can be also locally lowered by the formation of different types of subsidence sinkholes related to subsurface voids. In carbonate karsts, percolation waters lose a great part of their dissolution capability in the upper few meters of the downward flow path, and hence most of the mass removal occurs at and near the surface within the epikarst (see Section 4.6.1). In this situation, the topographic surface experiences progressive lowering toward the water table, with a vertical denudation trend (downwearing) similar to that proposed for the development of peneplains, although with a limited contribution of mechanical erosion by surface runoff.

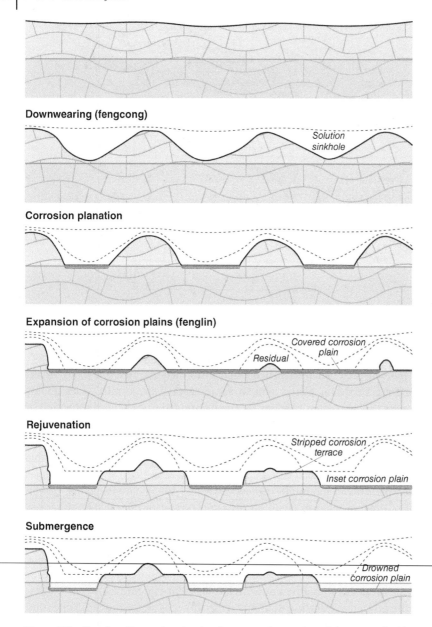

Downwearing (fengcong)

Solution
sinkhole

Corrosion planation

Expansion of corrosion plains (fenglin)

Covered corrosion
plain

Residual

Rejuvenation

Stripped corrosion
terrace

Inset corrosion plain

Submergence

Drowned
corrosion plain

Figure 7.8 Sketches illustrating the development of corrosion plains controlled by a relatively stable water table (epiphreatic zone), with an initial stage dominated by downwearing, followed by a phase of corrosion planation achieved by slope recession (backwearing). A downward shift in the water table rejuvenates downward solutional denudation, leading to the development of a younger corrosion plain inset into perched and stripped corrosion terraces. A water table rise potentially related to a sea-level rise or tectonic subsidence causes the submergence of the corrosion plain.

(2) Once the topographic surface is worn down to the water table, erosion mainly proceeds by slope retreat in the contiguous karst uplands and residual reliefs, resulting in laterally expanding corrosion plains. Downward vadose flow is replaced by slow horizontal phreatic flow under low hydraulic gradient conditions determined by a nearly horizontal water table. The retreat of the

slopes is achieved via two main solutional mechanisms that produce basal undermining and favor the development of slope movements: (a) lateral solutional undercutting, which can be enhanced by periodic flooding and (b) springhead retreat by sapping and the collapse of cave roofs when the springs are fed by underground rivers (Roglić 1957; Gams 1965, 1978; Ford and Williams 2007). This stage of corrosion planation dominated by backwearing has similarities with the concept of pediplanation that produces time-transgressive surfaces.

This morphological evolution of karst landscapes is nicely illustrated by the diagrams produced by Grund (1914) to depict his cycle of karst erosion (Figure 7.7). It is also simulated in a three-dimensional process-response model devised by Ahnert and Williams (1997). This model explains the formation of polygonal karst landscapes that evolve into a corrosion plain with residual hills when the bottoms of the depressions reach the water table, acting as the base level of erosion. Corrosion plains, because of their low elevation and flat topography, are typically mantled by a thin veneer of unconsolidated deposits. This cover may include clayey insoluble residues (e.g., terra rossa) and detrital deposits derived from karst and non-karst rocks (e.g., alluvium). Dissolution of bedrock beneath a soil mantle (i.e., subsoil, subcutaneous, and cryptokarstic dissolution) can be particularly intense because of the aggressiveness enhancement of the water that percolates through CO_2-rich soils and the spatially continuous and prolonged water–rock interaction that occurs beneath damp soils. This potentially rapid dissolution of the rockhead can be reduced by the presence of abundant carbonate particles in the cover. When the corrosion plains shift to a perched position (rejuvenation), the soil cover is generally removed by surface and internal mechanical erosion exposing the rock-cut surfaces. These surfaces typically have an overall planar geometry, but with a rough morphology, the details of which are related to the presence of a wide variety of covered karren features (see Section 6.2.2). The development of planar corrosion surfaces beneath soil cover has similarities to the formation of etchplains, which are generated by weathering beneath a regolith and are subsequently exposed by the stripping of the weathering mantle (Cui et al. 2002).

The karst landscapes that have been transformed into a base-levelled plain can start a *rejuvenation* phase when the water table experiences a downward shift (Figure 7.8). This relative drop in the base level for surface solutional denudation can be related to multiple causes, including local or regional uplift (e.g., tectonic, isostatic), sea-level fall, or fluvial incision that can be induced by the former factors or geomorphic changes such as the capture of endorheic basins and river piracy. The water table decline involves the expansion of the vadose zone and the displacement of the phreatic zone toward deeper less karstified rocks. The bedrock beneath the corrosion plains, previously situated at or near the water table, becomes part of the vadose zone and undergoes renewed corrosional lowering by downward percolation. The ground surface wears down and the corrosion plain can be pockmarked by new solution and subsidence sinkholes that grow vertically until they reach the water table. The lateral growth of the sinkholes and their coalescence progressively result in the development of an inset corrosion plain that expands by slope retreat. The former corrosion plain becomes a perched relict surface and its mantling deposits are commonly stripped by mechanical surface and subsurface erosion, becoming a bare rock strath surface (Figure 7.8). Stepped corrosion surfaces are markers of different paleobase levels and record alternating phases of corrosion planation and corrosional entrenchment that may be accompanied by the development of multilevel cave systems (see Sections 4.6.2 and 11.2.4). An excellent example of an entrenched corrosion plain occurs in the lower course of the Krka River in Croatia, which has incised a deep canyon into an extensive flat plain cut across deformed carbonate formations (Figure 7.6a, b). A dramatic example of base level drop and rejuvenation, which affected karst systems all around the Mediterranean coasts, was the Messinian Salinity Crisis (6–5.3 Myr, late

Miocene), involving the closure of the Mediterranean Sea at the Strait of Gibraltar and a water level drop that reached around 5 km. This was followed by an abrupt submergence phase with the Zanclean flood at ca. 5.3 Myr, when the connection between the Mediterranean and Atlantic was re-stablished. These major paleogeographic changes were accompanied by rapid fluvial incision, development of multi-level cave systems, and probably also corrosion plains that were drowned during the subsequent sea-level rise (Mocochain et al. 2009).

Sea-level variations exert a decisive control on karst evolution in coastal areas and potentially in adjacent regions. These changes can be related to global-scale eustatic variations and to more localized vertical movements caused by tectonic activity or isostatic adjustments. The sea level has experienced large oscillations during the Quaternary related to glacial and interglacial cycles and the accompanying transfer of huge volumes of water from the oceans to the ice sheets, and vice versa. During the Last Glacial Maximum (ca. 22–19 kyr) the sea level was approximately 125 m lower than at the present time. We are currently in an interglacial with a particularly high sea level. This contrasts with the Quaternary period, when for around 70% of the time the sea level was between 30 to 120 m below its current position (Purdy and Winterer 2001) (Figure 7.9). This implies that antecedent karst landscapes, developed in coastal areas under subaerial conditions during sea-level lowstands, are currently drowned by the sea. Note that sea water is saturated or supersaturated with respect to calcite and that inundated karst features tend to be buried by detrital or chemical deposits (e.g., reefs). This is a poorly explored karst topic with a great potential for scientific innovation that requires the application of specific research methods (e.g., bathymetric surveys, submersibles, and SCUBA diving) (e.g., Kan et al. 2015).

A relative sea-level rise in a coastal area involves a positive shift in the water table leading to a number of hydrological effects and geomorphic responses: (i) contraction of the vadose zone, (ii) expansion of the phreatic zone, (iii) upward and landward displacement of the saltwater–freshwater interface, (iv) *submergence* of springs and cave systems (Mocochain et al. 2009; Surić et al. 2010), and (v) drowning of surface landforms, including sinkholes (e.g., blue holes), poljes or plains with residuals (e.g., Ha Long Bay in Vietnam; Waltham and Hamilton-Smith 2004) (Figure 7.10), fluvio-karstic valleys and tufa deposits (Surić 2002); and (vi) a potential change from an erosional to an aggradational trend in both marine and terrestrial environments. Some of the

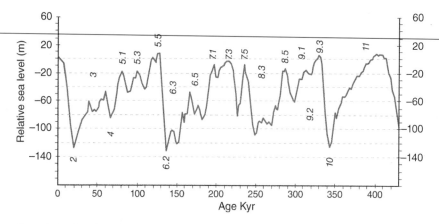

Figure 7.9   Relative sea-level curve reconstructed from oxygen isotope ratios of benthic foraminifera from widely distributed cores using transfer functions generated with records of sea-level data based on corals and other evidence. Numbers denote marine isotope stages. *Source:* Based on Waelbroeck et al. (2002).

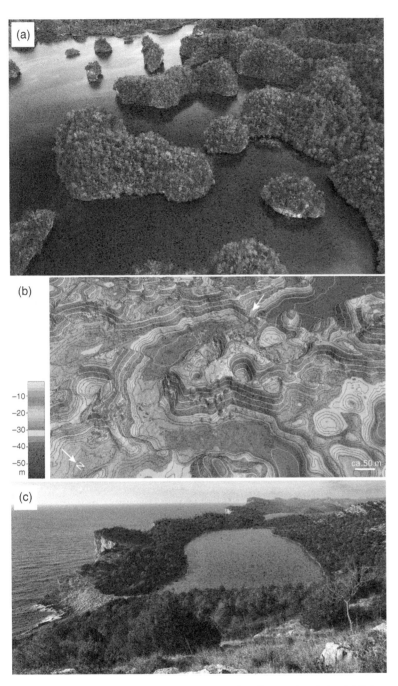

Figure 7.10 Examples of karst submergence related to sea-level rise. (a) Submerged karst topography with residual hills in Kabui Bay, West Papua. *Source:* Photo by Chien C. Lee. (b) Three-dimensional surface model generated by bathymetric data of the drowned karst topography with sinkholes and a fluvio-karst valley in Ishigaki Island, Japan. Arrow points to a threshold within the valley that might have a natural bridge. *Source:* Image from Kan et al. (2015, Figure 4), published with permission of Elsevier. (c) Lake Mir in Dugi Otok Island, Croatia, a coastal lake that used to be a subaerial karst depression during sea-level lowstands. *Source:* Photo by Josipa Grbin.

evidence found in caves for the submergence process include the presence of vadose speleothems (Yucatán, Beddows 2004, Moseley et al. 2015; Mediterranean Sea, Bard et al. 2002) and marine overgrowths (Surić et al. 2005), that in some cases can be used as markers of former water tables (e.g., Mallorca, Tuccimei et al. 2010; Dumitru et al. 2021). Gavazzi (1904) described coastal fresh-water lakes in Croatia (e.g., Vransko Jezero in Cres Island) as poljes deepened when the sea level was at lower position then transformed into lakes when the sea level rose. Lake Mir in Dugi Otok Island, Croatia is a coastal lake that used to be a subaerial karst depression during sea-level low-stands (Surić 2002) (Figure 7.10c). Gracia et al. (2014), based on detailed geomorphic mapping, interpret that the Vendicari Lake system in Sicily, Italy, corresponds to a karst polje with stepped corrosion surfaces formed during the Late Pleistocene sea-level lowstands, subsequently largely drowned during the postglacial sea-level rise. Brunović et al. (2020) reconstruct the morpho-sedimentary evolution of the drowned Lošinj karst depression in the Adriatic Coast (Croatia) using seismic reflection and borehole data. They interpret that this basin functioned as a polje with terrestrial sedimentation during sea-level lowstands in the late Pleistocene. Grigg et al. (2002), using data collected by multibeam high-resolution bathymetric surveys and submersibles, documented drowned karst topography in a limestone bridge that used to connect to islands in Hawaii during sea-level lowstands. They recognized large karst depressions with steep walls indented by notches at different elevations. Kan et al. (2015) used broadband multibeam surveys and SCUBA diving to characterize a submerged karst terrain largely veneered by thin postglacial reef deposits in a bay of Ishigaki Island, Japan. The submerged topography reveals a well-developed tropical karst landscape formed during lowstands, comprising large single and compound solution sinkholes with variable densities (from isolated to polygonal) and an old fluvio-karstic valley with a threshold that might have a natural bridge (Figure 7.10b). Taviani et al. (2012), utilized multibeam bathymetry data to identify submerged sinkholes 50–155 m across and up to 20 m deep offshore the Apulian coast in Italy, formed under subaerial conditions by the collapse of caves in carbonate rocks.

7.3   Residual Hills. Fenglin and Fengcong

The Chinese geomorphologists differentiate two types of landscapes with carbonate residual hills on the basis of morphological and hydrological criteria: fenglin karst and fengcong karst (Ford and Williams 2007; Waltham 2008b; Zhu et al. 2013 and references therein). Fenglin karst (meaning peak forest) refers to a landscape consisting of isolated hills rising from a plain. The plain has a shallow water table and is commonly underlain by a thin veneer of alluvium. Fengcong karst (meaning peak cluster) denotes a terrain comprising residual hills with connected bases and inter-vening internally drained depressions situated well above the water table (Figures 7.11 and 7.12). These terms were first used with a geomorphological connotation by the Chinese geographer Xu Xiake during the Ming Dynasty. Xiake investigated the karst landscapes of southern China from 1636 and 1638 and described them in his book "The Travel Diaries of Xu Xiake" (Zhu et al. 2013). In the fengcong karst, the adjoining hills and the intervening depressions are underlain by a thick vadose zone, drainage is dominated by downward recharge, and corrosional lowering is the pre-vailing denudational style. These terrains often have dendritic cave systems lacking spatial correla-tion with the hills and depressions, and cave rivers may lead to the formation of giant collapse sinkholes (tiankengs) (Figure 6.32a). The terms fengcong-depression and fengcong-valley are often used depending on the degree of interconnection and coalescence between the dolines. In the fenglin karst, the plains are situated within or slightly above the epiphreatic zone and drainage is

Figure 7.11   Images of fenglin and fengcong landscapes. (a) Fenglin and fengcong at Xingyi, Guizhou, South China. *Source:* Photo by Chen Weihai. (b) Fengcong landscape comprising hills with common bases and intervening depressions at the margin of the Li River, Guilin, South China. *Source:* Photo by Chen Weihai. (c) Isolated residual hills, locally known as mogotes, rising from the flat bottom of a border polje in Viñales Valley, Cuba (fenglin landscape). *Source:* Francisco Gutiérrez (Author). (d) Fengcong with linear depressions in Libo, Guizhou, South China. *Source:* Photo by Chen Weihai. (e) Fenglin and fengcong landscapes dominated by conical hills in the Chocolate Hills of Bohol, Philippines. *Source:* Photo by Tony Waltham.

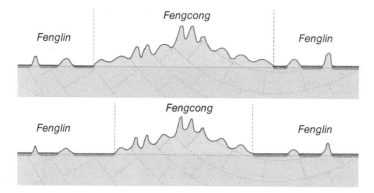

Figure 7.12 Sketches illustrating the relative distribution of fengcong, associated with upland areas with a deep water table, and fenglin in plains with a shallow water table. Under relatively stable water table conditions, fengcong can be transformed into fenglin when the surface is lowered by downward solutional erosion, reaches the water table, and corrosion planation starts to operate. Note that residual hills in fengcong and fenglin can display both tower-like, conical and hemispherical morphologies.

essentially lateral. Corrosion planation and slope retreat are the main erosional mechanisms and the plains are often traversed by allogenic rivers. The terms fenglin-plain, fenglin-valley, and fenglin-polje are used depending on the type and extent of the plain.

Western geomorphologists have described landscapes with carbonate residual hills using the terms tower karst and cone karst (*Turmkarst* and *Kegelkarst* in German). These morphological terms have been used with different meanings creating confusion (Waltham 2008b; Zhu et al. 2013). For instance, some authors use the term tower karst to describe landscapes with tower-like residuals, whereas others consider that it is equivalent to fenglin (plain with isolated hills). However, the isolated hills in fenglin terrains can display a wide range of morphologies (tower-like, cone-like, hemispherical). Similarly, the Western cone karst is considered by some authors as equivalent to the Chinese fengcong, but hills in the latter landscape can display tower-like morphologies. It does not seem coherent to use the word tower for conical hills and cone for tower-like residuals. Cockpit karst and polygonal karst, characterized by an egg-box-like topography comprising densely packed solutional dolines and intervening hills fall within the category of fengcong (e.g., Day and Huang 2009). Moreover, a number of local names that designate the residual hills have also been used by some authors, such as the Spanish words "mogote" and "pepino" meaning hill and cucumber, respectively. We concur with Waltham's (2008b) opinion, who states that "fengcong and fenglin are more useful terms with genetic implications and should take precedence, whereas cones and towers should be used purely as descriptive terms."

Fenglin and fengcong are highly scenic karst landscapes, some of which are designated UNESCO World Heritage Sites (e.g., South China Karst, Viñales Valley in Cuba, Ha Long Bay in Vietnam) (Williams 2011) (Figure 7.11). These landscapes occur in hot humid tropical/subtropical regions, mainly where mean annual rainfall and precipitation exceed 17 °C and 1300 m, respectively (Waltham 2008b). They are the result of long-sustained periods of intense karstification and base level stability in the case of the fenglin karst. The abundant water surplus and the high production of CO_2 in the soils under humid tropical conditions promote rapid rates of solutional denudation (see Section 4.2). The most extensive and diverse fengcong and fenglin landscapes are located in southern China (Zhu 1988; Sweeting 1995). According to Zhu et al. (2013), the fengcong-fenglin system of southern China and northern Vietnam covers around 160 000 km², representing approximately 98% of the fengcong-fenglin of the world. The fengcong is by far the dominant landscape type in China, representing around 92% of the total fencong-fenglin system. Other remarkable fenglin landscapes located in Asia include the drowned residual hills of Ha Long Bay (Waltham 2000), Huong Tich and Bac Son (Khang 1985) in Vietnam, the Chocolate Hills of Bohol in the Philippines (Balázs, 1973a) (Figure 7.11e), the Khammouan karst in Laos (Waltham and Middleton 2000), the Sewu fengcong karst of Java (Lehmann 1936; Haryono and Day 2004; Waltham 2004), some karst areas of northern Borneo (McDonald and Ley 1985). The other main zones are situated in the Caribbean region, including the Viñales Valley in Cuba (Lehmann et al. 1956; Panoš and Štelcl 1968) (Figure 7.11c), the mogotes of Puerto Rico (Monroe 1976; Day 1978), the classical cockpit karst of Jamaica (7500 km^2; Sweeting 1958; Aub 1969), and the karst of Belize (McDonald 1979). Waltham (2008b) and Zhu et al. (2013) provide reviews of fengcong and fenglin karst topography around the world.

The residual hills of both fengcong and fenglin landscapes can show a broad variety of morphologies, from cliff-sided towers to cones and hemispheres (Figure 7.11). Some authors point out that towers are more common in resistant limestones, whereas rounded cones are the dominant shapes in softer limestones, commonly of Cenozoic age (Zhu et al. 2013). Hills often display an elongated form in plan and an asymmetric cross profile, which can be related to structural controls exerted by the orientation of a master joint set and the dip of the strata. The longitudinal profile of the slopes in the fenglin hills can be uniform or irregular, and frequently shows a well-defined slope-plain junction. Some hills can also surmount a basal plinth with gentler sides. Tang and Day (2000), measuring slope profiles in Guilin area, China, found that they show a concave dominated geometry in fengcong hills and a convex–concave profile in fenglin towers. They attributed these differences to the styles of solutional denudation, vertical in fengcong and vertical and lateral in fenglin. Balázs (1973b) classified karst hills on the basis of their diameter/height ratios, but this parameter can show a high variability within each area (Day 1978).

The base of the hills is often undercut by solution notches or foot caves that record preferential basal solutional and/or mechanical undercutting. The rock juxtaposed against cover deposits at the foot of the hill experiences prolonged interaction with subsoil aggressive water. These sites also receive runoff water flowing down the slope of the hills. These subsoil undercuts have similarities with the flared slopes described in granitic rocks. Basal undercutting leads to wall collapse and slope recession (Khang 1985). Jennings (1976) mapped foot caves in 10% of a karst tower perimeter close to Kuala Lumpur in Malaysia, supporting the importance of basal solutional undercutting in tower karst development. He considered that a significant proportion of the existing foot caves, which mainly develop by subsoil dissolution, are likely to be concealed by alluvium and rock fall deposits. In the fenglin karst of Belize, McDonald (1979) documented the important morphogenetic role played by both mechanical and chemical erosion caused by allogenic rivers and ephemeral lakes at the base of towers. Residual hills in tropical areas usually have an indurated weathering crust generated by precipitation of secondary carbonate (case hardening) that closely follows the topography (Sweeting 1972). This outer shell of more resistant and less permeable rock contributes to decelerate the degradation of the hills attenuating both solutional erosion and mass wasting processes. Monroe (1976), thanks to road cuts excavated across limestone hills (mogotes) in Puerto Rico was able to recognize case hardened zones as much as 5–10 m thick, apparently thicker on the upwind side of the mogotes. He attributed the induration of the bedrock to near-surface dissolution and reprecipitation processes governed by wetting and drying cycles.

Two main types of fenglin can be differentiated according to the nature of the plain (Williams 1987; Ford and Williams 2007). The most common type is a horizontal corrosion surface cut across carbonate bedrock and covered by a thin veneer of unconsolidated deposits (Figure 7.12). The other is an aggraded surface underlain by a thick surficial formation that may bury an irregular paleotopography or an old corrosion plain. The latter situation can occur in areas where the base level has experienced an upward shift induced by factors such as sea level rise or tectonic subsidence, inducing sediment aggradation. Some authors also classify plains largely carved in non-karst rocks with residual carbonate hills as fenglin landscapes. However, this is not true fenglin karst since the topography is essentially created by mechanical erosion (Panoš and Štelcl 1968). The morphometry of residual hills can be characterized with the parameters and indexes used for sinkholes, replacing depth by height (Table 6.3). The spatial distribution of the hills can be analyzed using point features given by the summits, or the centroid of the polygons that define their basal edge. An example is provided by Day (1978) with 149 isolated hills up to 52 m high in northern Puerto Rico. Liang and Xu (2014) developed an approach for the automated mapping of fengcong and fenglin in China applying discriminant analysis to a set of morphometric parameters extracted from a DEM. They found that roughness, given by the ratio between the surface area and the planimetric area, was the variable with the greatest discriminant capability, with higher values in fengcong than in fenglin.

The development and distribution of fengcong and fenglin is controlled by the position of the water table, acting as base level for the corrosional lowering of the ground surface. Fengcong terrain essentially occurs in upland areas with a relatively deep water table, whereas the fenglin karst is associated with lowlands lying at the level of the epiphreatic zone (Figure 7.12). Fengcong and fenglin can develop simultaneously, depending on the position of the base level of the karst system in each sector (Zhu et al. 2013). The evolution of the fengcong-fenglin landscapes largely depends on the balance between the rate of corrosional lowering acting in the depressions and the rate and sense at which the water table varies its position, mainly controlled by tectonic uplift or subsidence. The formation of fengcong-fenglin systems integrates the evolutionary models of solution dolines, polygonal karst, and corrosion plains (Figures 6.24, 7.8, and 7.12). Fengcong and fenglin can experience a two-way evolution: (i) fengcong evolves to fenglin as corrosional lowering reaches the base level and (ii) fenglin can transform into fengcong by base level drop and rejuvenation. The direct formation of fenglin is less common and is considered to be strongly influenced by joint-controlled denudation (Williams 1987).

In a tropical karst terrain with a relatively flat topography and a deep water table, differential and focused corrosional lowering generates solution sinkholes, typically at sites of greater fracturing and permeability. The solution sinkholes grow downward and laterally with the interplay of self-accelerating mechanisms (e.g., flow concentration, aggressive enhancement). The expanding and coalescing sinkholes tend to occupy all the space, creating a fengcong landscape of enclosed depressions with intervening hills. The divides of the adjoining depressions can display a mesh pattern forming a polygonal karst, and the hills may show concordant summits as legacies of a former flat topography (Figure 7.11b). If the water table remains stable, or if the corrosional deepening of the depressions proceeds at a rate higher than a water table decline, the topographic lows can reach the epiphreatic zone. Then, the vertical vadose flow is replaced by lateral flow and there is a switch from surface lowering to lateral planation in the bottom of the depressions. Corrosion plains start to form and expand laterally at the expense of the residual hills, and the fengcong evolves progressively into fenglin. The spatial distribution and geometry of the hills are largely derived from a previous fengcong stage (inheritance). Corrosion planation can be significantly aided by throughgoing allogenic rivers that episodically introduce large volumes of aggressive floodwaters. The residual hills are affected by surface dissolution and their erosion is favored by the development of basal notches and foot caves that promote rock falls and cliff retreat. In contrast, case hardening contributes to slow down the degradation of the residual hills.

When the fenglin karst experiences a water table drop (rejuvenation), downward corrosional lowering resumes. If surface lowering can keep pace with the base level drop, the topography maintains its fenglin character and the hills increase their height as the plain is progressively worn down to the epiphreatic zone. However, if the water table drop is more rapid than the rate of corrosional lowering, solution sinkholes and a fengcong topography can be superimposed on the plains of a former fenglin landscape, starting a new cycle of karst erosion. These evolutionary models were satisfactorily simulated by Ahnert and Williams (1997) through three-dimensional numerical and graphic models that consider the topographic control on corrosion rates (flow convergence and divergence) and the shift from lowering to planation when the floor of the depressions reaches the water table. A fenglin landscape could also form directly from a plain if the relative water table drop (rejuvenation) is sufficiently slow for corrosional lowering and corrosion planation to create and maintain a plain with residuals.

The tops of hills in fenglin landscapes progressively wear down by rainfall dissolution. The surrounding plains can experience downward lowering during periods of base level drop. There is destruction and creation of relief at the tops and the bases of the hills, respectively. The height of the hills increases when lowering in the plain exceeds vertical denudation at the top. Thus, the residual hills are time-transgressive landforms, progressively younger toward the base. The height of hills can be also reduced by sediment aggradation in the plain. The hills in fenglin landscapes can host relict caves at different elevations, recording paleobase levels progressively older toward the top. Numerical ages obtained from detrital and chemical deposits accumulated in these caves can be used to reconstruct the evolution of the fenglin landscape and estimate long-term rates of base level lowering. These are in most cases maximum rates based on the minimum ages of cave development, since the deposits mostly postdate the formation of the caves. Examples are provided from towers in the outskirts of Guilin city (Wang 1986; Williams et al. 1986; Williams 1987).

The main factors that favor the development of fenglin karst include (Tang 2002; Waltham 2008b; Zhu et al. 2013): (i) thick successions of pure and resistant limestone, (ii) a hot climate with abundant precipitation that promotes high production of biogenic CO_2 and rapid dissolution rates, (iii) a relatively stable water table that permits the depressions to lower down to the base level, (iv) a cover of unconsolidated deposits (preferably non-calcareous) supporting a soil that contributes to increase the aggressiveness of the water, (v) inflows of aggressive allogenic water, and (vi) fluvial and groundwater flooding that promotes solutional lateral undercutting, rock falls and slope recession.

7.4 Valleys in Karst

The term holokarst, introduced by Cvijić (1893), refers to perfectly developed solutional karst landscapes lacking fluvial valleys. However, most karst terrains display valleys carved by surface waters, indicating that fluvial activity and mechanical water erosion have played a role in the development of the landscape. In fact, Sweeting (1972) indicated that valleys are the most important of the surface landforms in karst which are not produced by true karst processes. The term fluviokarst is often used to refer to landscapes with both karst and fluvial landforms. Gunn (2004a) defines fluviokarst in a more restrictive way as a landscape on soluble rocks dominated by valleys cut by surface runoff. He suggests that fluviokarst can be considered as an end member in a continuous gradation of surface landform assemblages, with holokarst being the opposite end member. A polygonal karst landscape in which the whole area is pockmarked by sinkholes is an extreme case of holokarst.

The formation of valleys requires stream flow with sufficient power and/or aggressiveness to dissect the karst bedrock. The channeled water flow can be derived from surface runoff and groundwater discharge. A number of factors favor the occurrence of stream flow in karst land-scapes, as well as the formation and persistence of valleys: (i) Allogenic input flowing from non-karst terrain. (ii) Low hydraulic gradient and channels lying at the base level, functioning as effluent or gaining streams that receive discharge from the karst aquifer. In contrast, influent streams perched above the groundwater level can rapidly lose flow. These streams with vertical underground flow can be transformed into dry valleys incised by sinks capable of absorbing all the surface flow. (iii) Low hydraulic conductivity of the bedrock and hence limited capacity to drain surface water underground. The low permeability of the bedrock can be related to an immature karst aquifer with poorly developed conduit permeability and the presence of intercalations of non-soluble rocks, impervious formations at shallow depth, or permafrost. (iv) Low-permeability covers that reduce water infiltration. (v) Rapid water input by torrential rains that exceed the infil-tration capacity of the ground, leading to flashy flow events with high geomorphic effectiveness. (vi) Bedrocks with relatively high mechanical and/or chemical erodibility. Some carbonate rocks are rather susceptible to mechanical erosion (e.g., finely bedded and densely jointed limestone, chalk), and chemical denudation by stream flow in evaporites can be very rapid (see Section 4.4). According to Phillips et al. (2004), surface streamflow and underground karst drainage are com-petitive and exclusive processes that can experience bidirectional transitions with geomorphic imprint. In the Inner Bluegrass karst region of Kentucky, Phillips (2017) identified landforms indicative of fluvial-to-karst transition (e.g., losing streams, dry valleys) and of karst-to-fluvial shift (e.g., dolines breached by drainages, pocket valleys).

Cvijić (1893) introduced the term karst valley and differentiated three main types: pocket valley, blind or semi-blind valley, and dry valley. Sweeting (1972) and Gunn (2004b) proposed a fourfold classification adding through (or allogenic) valleys. Here, we differentiate: (i) through (or allo-genic) valleys, (ii) blind (and semi-blind) valleys, (iii) pocket valleys, (iv) cave-collapse valleys, and (v) dry valleys. There can be some overlap between the different types. For instance, a pocket valley fed by a cave river can also be a cave-collapse valley. Moreover, the term dry valley does not have a genetic meaning, but rather refers to the condition of the valley, which can change through space and time.

7.4.1 Through Valleys

Through or allogenic valleys are formed by rivers that originate in impervious lithologies. These drainages can entirely traverse the karst over long distances to the output boundary (Gunn 2004a, 2004b; Ford and Williams 2007). The Cares Gorge in Picos de Europa National Park, northern Spain (Figure 7.13a), and the Tarn Gorge in the Grands Causses, southern France are fine examples. An important factor for the flow regime and evolution of these valleys is the relationship of the allogenic river with the karst aquifer and its effluent (gaining) or influent (losing) behavior (Milanović 1981; Bonacci 1987). Effluent rivers receive water flow from the aquifer, and thus increase their discharge downstream and tend to be perennial. Groundwater input may occur dif-fusely along the valley floor or very often at localized springs that can lead to substantial increases in discharge. Moreover, effluent rivers function as the regional base level for groundwater flow and surface erosion and may control the development of subhorizontal phreatic cave passages. The occurrence of through valleys at the base level is favored by: (i) a reduced elevation difference between the input and output boundaries of the karst (low hydraulic gradient) and (ii) rivers with sufficiently high discharge and stream power to cut the valleys and adjust to base level changes

Figure 7.13 Images of limestone canyons and blind valleys. (a) The Cares Canyon, an allogenic river that traverses the Picos de Europa limestone massif in northern Spain. *Source:* Francisco Gutiérrez (Author). (b) The Vero Canyon carved by an allogenic transverse river that flows perpendicularly across a fold and thrust belt in the southern Pyrenees, Spain. *Source:* Francisco Gutiérrez (Author). (c) The narrow and deeply incised Torrent de Pareis carved by an autogenic drainage in the Tramuntana Range of Mallorca Island, Spain. *Source:* Francisco Gutiérrez (Author). (d) Hillshade of the Ojo Guareña blind and allogenic valley in northern Spain. The ponor connects with the longest cave system in the Iberian Peninsula with more than 100 km of mapped passages. *Source:* Model generated with LiDAR data of the Spanish Geographical Institute. (e) The Hurón River blind valley which flows into the Agua Cave in Burgos Province, northern Spain. Note perched beheaded valley above and beyond the stream-sink. *Source:* Francisco Gutiérrez (Author). (f) Zalmska blind valley and Biogradski ponor in the outflow side of Nevesinje Polje, Bosnia and Herzegovina. *Source:* Francisco Gutiérrez (Author). (g) A blind valley in the Taurus Mountains generated by the underground piracy of a paleodrainage. The arrow points to the location of the main stream-sink. The wind gap in the background indicates the position of the abandoned and beheaded downstream valley section cut in limestone. The blind valley is carved in non-karst formations and lateral planation has resulted in the development of the border Sariot Polje. *Sources:* Photo by Ugur Doğan.

(e.g., uplift). Good examples of allogenic rivers that function as the regional base level for karst systems are the Green River in Kentucky, the Krka River in Croatia, and the Kizilirmak River in the gypsum karst of Sivas, Turkey.

Influent drainages are essentially perched above the groundwater level and consequently lose flow underground along the channel and/or at stream-sinks. The discharge tends to decrease downstream and the valley may become temporarily or permanently dry. Fine examples of influent rivers are the upper Danube River in Germany and the Krčić River in Croatia (Bonacci 1985).

Base level drops related to phenomena such as uplift or sea level lowering induce downcutting and the rejuvenation of the fluvial system. Rivers that keep pace with the base level drop can maintain their effluent condition and may control the development of multilevel cave systems. The different cave storeys can be used to reconstruct the incision history of the related rivers, and estimate downcutting rates when chronological data are available (see Section 4.6.2). If uplift rates exceed the incision capability of the fluvial system, effluent streams can turn into perched influent channels. Bočić et al. (2015) mapped a paleodrainage network in the Una-Korana plateau, Croatia, the largest of the Dinaric karst. They identified around 5800 km of valley sections, 90% of which correspond to relict and dry valleys. The plateau used to be dissected by a well-developed effluent drainage network formed under low hydraulic gradient conditions. Subsequently, neotectonic uplift at a rate higher than the downcutting capability of the fluvial systems led to the development of vertical underground karst drainage, sinkhole development, and disruption of the valley network. They observed a spatial correlation between the highest sinkhole densities and the density of relict drainage sections. This plateau records a former fluviokarst landscape that is currently at an advanced stage of transformation into a holokarst landscape dominated by sinkholes and underground drainage.

Fluvial incision in carbonate rocks commonly produces steep-sided gorges that may reach more than 1 km in depth (Figure 7.13a–c). The development of these impressive canyons with vertical walls is favored by the high mechanical resistance of carbonate rocks and the limited surface runoff that restrict slope decline by water erosion and mass wasting processes (Jennings 1985). Incision in limestone bedrock channels and gorges can be related to both mechanical and chemical denudation. Mechanical erosion can be extremely effective in narrow canyons during flood events, in which water flow experiences dramatic increases in depth and velocity and has a huge erosion and transport capacity (De Waele et al. 2010). The main mechanical erosion processes include quarrying (plucking of rock blocks) and abrasion by particles carried in rapid flows. Corrosion in carbonate bedrock channels occurs under open conditions, in which the CO_2 consumed by carbonate dissolution can be replenished by diffusion from the atmosphere. Thus, the capability of the water to dissolve carbonate minerals (saturation concentration) is not necessarily reduced by the depletion of CO_2 associated with the dissolution process (see Section 3.3.6). However, turbulent water in contact with limestone can reach near-equilibrium conditions over relatively short distances, losing great part of its dissolution capability with the consequent decrease in the dissolution rate (see Section 3.8.1). The flow distance at which allogenic flow can reach near-saturation conditions will largely depend on: (i) the initial degree of saturation, (ii) the ratio between the area of soluble material (bedrock and particles) in contact with water and the volume of water, which can experience large decreases during flood events, and (iii) changes in the saturation index related to the incorporation of underground and surface waters.

The relatively rapid decrease in the aggressiveness of the water in limestone canyons is proved by hydrochemical data and depositional evidence, including precipitation of carbonate coatings on rock surfaces, particles and vegetation, or even the accumulation of significant calcareous tufa deposits. The river water in valleys with calcareous tufa deposition can remain supersaturated with respect to calcite over very long distances and consequently lack the ability to cause solutional denudation (Drysdale et al. 2002; Auqué et al. 2013, 2014; Arenas et al. 2014). Tufa accumulations can form barrier-cascade systems with upstream lakes that differentially rise the valley floor by aggradation, creating a stepped longitudinal profile (e.g., Plitvice Lakes in Croatia; Huangguoshu waterfalls in China) (Figure 7.16a). The relative height of tufa terraces above the thalweg is commonly an inadequate tool for correlating deposits and reconstructing the incision and aggradation history of these systems with nongraded profiles. Factors that can induce or enhance tufa deposition include CO_2 degassing at sites of increased turbulence, CO_2 depletion by photosynthetic activity, common-ion effect by water mixing (e.g., incorporation of Ca-rich water that has interacted with gypsum), evaporation, or temperature increase. Positive feedbacks occur in tufa barriers alike in rimstones, where more turbulent and shallower flows boost CO_2 depletion, carbonate precipitation and the vertical growth of cascades. Covington et al. (2013), in a 250 m long reach of an allogenic cave stream near Postojna (Slovenia), to some extent comparable to a surface drainage, observed a downstream decrease in dissolution rate using limestone tablets. They attribute this trend to a decrease in dissolved CO_2 in the water caused by enhanced degassing at steep reaches with more turbulent flow, and a reduction in the CO_2 production rate of the sediments, with lower content of organic matter. These relationships also indicate some geomorphic feedback loops whereby the stream gradient influences CO_2 dynamics and dissolution rates, which in turn have an impact on the evolution of the longitudinal profile of the stream.

The long-term development of gorges is often related to antecedence and/or superimposition. *Antecedent drainages* refer to rivers that traverse an area subject to uplift, but that have been able to maintain their course by downcutting, generating deep gorges. The incision capacity of these fluvial systems is sufficient to keep pace with bedrock uplift. Lepirica (2015) interprets that several inner gorges in the Rakitnica Canyon in the central Dinarides (Bosnia and Herzegovina) are related to fluvial downcutting at restraining bends of strike-slip faults apparently affected by active contraction and uplift. He also indicates that the trace of the gorges is largely controlled by preexisting subvertical fractures of probable extensional nature. *Superimposed drainages* are those that originally developed on a cover of rocks (often impervious) that has been eroded and the valleys currently dissect lower lying formations with a discordant pattern. This is a case of inheritance in which the course of the rivers is derived from previous geological conditions. Note that the antecedence and superimposition concepts are not mutually exclusive. An important aspect from the evolutionary and hydrological perspective is the geometrical relationship between the drainages and the geological structures. *Strike drainages* follow the structural grain and largely flow over a specific sedimentary package (subsequent streams). *Transverse drainages* cut perpendicularly across geological structures and different formations. These can be split into *dip drainages* and *anti-dip drainages*, depending on whether they flow down- or up-dip, respectively. These terms are more descriptive and intuitive than the classical terminology: consequent, subsequent, and obsequent drainages (Ollier and Pain 2000).

7.4.2 Blind Valleys

An influent river flowing over karst bedrock can lose a substantial proportion of its flow through one or several stream-sinks. These swallow holes located in the stream bed may correspond to collapse sinkholes, solution features such as widened fractures, or even cave entrances. The absorption capacity of the conduit system connected with the stream-sink can increase to an extent that the ponor is able to divert the totality of the discharge underground under most flow conditions. In this case, the stream-sink separates an upstream valley reach with surface flow and a downstream

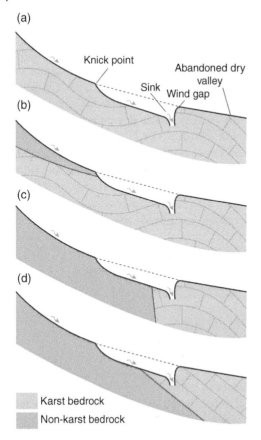

(a)

Knick point

Abandoned dry valley

Sink Wind gap

(b)

(c)

(d)

Karst bedrock
Non-karst bedrock

Figure 7.14 Sketches of blind valleys developed in different geological contexts. (a) Influent valley carved in karst bedrock. (b) Valley initially excavated in non-karst rocks underlain by karst rocks. New sinks can form as the caprock is stripped back by erosion. (c) Allogenic valley in which streamflow reaches karst bedrock juxtaposed against non-karst bedrock. (d) Allogenic valley with a stream-sink associated with the stratigraphic contact between impermeable and soluble formations.

reach that remains dry most of the time (Figure 7.14a). During flood events, the water can overflow the ponor when the stream flow overwhelms the inflow capacity of the stream-sink. This is a semi-blind valley in which there is occasional flow downstream of the sink. The local base level drop associated with the formation and enlargement of the stream-sink may induce the entrenchment of the channel upstream through the headward propagation of an incision wave. Eventually, the valley may become fully blind as the absorption capacity of the sink increases, and the elevation difference of the stream bed at both sides of the ponor grows (Figures 7.13d–g and 7.14). Good examples are Friar's Hole dry valley in West Virginia (Ford and Williams 2007), and the blind valleys located at the outflow side of Planina Polje, Slovenia (Stepišnik et al. 2012). The incision of the river channel upstream of the stream-sink commonly results in an disrupted longitudinal profile including (Figure 7.14): (i) an upper active valley reach with one or several steps (knick points) related to the headward propagation of the incision wave, (ii) the stream-sink that can be expressed as a relatively deep depression nested in the stream bed, especially when karst drainage is dominated by vertical flow, (iii) an upstream-facing scarp at the downstream side of the sink, usually displaying amphitheater-like geometry with the concavity facing upstream, and (iv) an abandoned and dry valley reach beheaded by the sink and expressed as a wind gap (Figure 7.13g).

Blind valleys often occur spatially associated with geological contacts between non-karst and karst bedrocks (contact karst). An allogenic stream flowing over an impervious caprock can lose its flow via sinks once it reaches karst bedrock (Figure 7.14b–d). New stream-sinks can develop as the caprock is stripped back, resulting in the upstream expansion of the lower dry valley (Figure 7.14b).

An alternative situation can be found where the impermeable bedrock is overlain by (e.g., drainages flowing into the Ojo Guareña cave system and the Hurón River that sinks into the Cueva del Agua, both in northern Spain) (Figures 7.13d, e and 7.14d), or juxtaposed against karst bedrock in the downstream reach (Figures 7.13g and 7.14c). The allogenic streams flowing over impervious rocks can be interrupted by sinks once they reach the soluble bedrock. In these contact karst situations, the allogenic water that infiltrates into the sinks is highly aggressive and consequently can enlarge relatively rapidly the conduit system linked to the stream-sink. Doğan (2003) describes a paleod-rainage in the Taurus Mountains of Turkey that used to flow over non-karst and karst bedrock. At some stage, the stream was captured by ponors in the limestone bedrock (underground piracy). The downstream section became a beheaded paleovalley expressed as a wind gap, whereas the upstream stretch started to function as a blind valley with characteristics compatible with a border polje, but essentially developed in an allogenic blind valley underlain by impermeable bedrock (Figure 7.13g). At Yarrangobilly, SE Australia, Jennings et al. (1980) found that the dimensions of blind valleys are influenced by the size of the catchments outside the limestone and the age of the valleys.

7.4.3 Pocket Valleys

Pocket valleys are the reverse of blind valleys, since they form in association with large springs at the foot of a cliff, commonly located at the output margin of the karst. Here, cliff recession is accel-erated by spring sapping, which involves basal undermining and the development of slope move-ments, mainly rock falls. The preferential long-term recession of the rock face results in the development of pocket valleys with amphitheater-like steepheads. The retreat rate of the rock scarp situated above the spring depends on factors such as the height of the cliff, the mechanical resistance of the rocks, and the erosional capability of the spring flow. Groundwater that dis-charges at springs in carbonate aquifers commonly has limited dissolution capability, which in turn can be reduced by CO_2 degassing favored by turbulence and the commonly lower CO_2 partial pressure in the outer atmosphere. Lipar and Ferk (2015) present a review on pocket valleys world-wide, revealing the limited work published so far on these landforms. Some notable examples of pocket valleys include Malham Cove in England, the Urederra Valley in northern Spain, or the steephead of the Causse de Gramat in France. In southern France, Audra et al. (2004) document submerged karst plateaus dissected by deep canyons that terminate abruptly upstream without showing any relationship to any significant drainage. They interpret these relict fluvial landforms as pocket valleys formed during the late Miocene Messinian Salinity Crisis, and subsequently drowned during the Pliocene transgression (Zanclean flood). Lipar and Ferk (2015) analyzed 140 relict pocket valleys in the Nullarbor Plain (South Australia), indented into two escarpments gen-erated by coastal erosion in the Pliocene. These authors interpret that the development of the pocket valleys started in the late Pliocene, when the sea level dropped and coastal erosion ceased. Pocket valleys can also form by the progressive collapse of the roof of caves when the springs are fed by cave rivers. These valleys can be classified as both pocket valleys and cave-collapse valleys.

7.4.4 Cave-Collapse Valleys

The foundering of the roof of underground passages can result in the development of cave-collapse valleys, generally of limited length. Initially, lines of collapse sinkholes form along the trace of the cave. Subsequently, sinkhole expansion and coalescence produce progressively longer linear depressions with breakdown deposits at their bottom, and intervening cave sections (Figure 7.15b). The latter progressively restrict their length by collapse processes to become natural bridges.

Figure 7.15 Images of pocket, cave-collapse and dry valleys in karst. (a) Hillshade of the Urederra spring and pocket valley at the output side of a limestone plateau in the Spanish Pyrenees. P stands for polje floor. *Source:* Model generated from bare ground LiDAR data of the Spanish Geographical Institute. (b) Incipient cave-collapse valley related to the collapse of a cave that connects flooded bedrock collapse sinkholes with the floodplain of the Kizilirmak River in the Sivas gypsum karst, Turkey. *Source:* Photo by Serdar Yeşilyurt. (c) Natural bridge in the cave-collapse Sanqiao gorge in southern China. Circle indicates a group of persons for scale. *Source:* Photo by Alexander Klimchouk. (d) Natural bridge generated by the underground cutoff of an incised meander in the Nela River at Puentedey, northern Spain. *Source:* Francisco Gutiérrez (Author). (e) Dendritic network of flat-bottomed infilled valleys in the gypsum outcrops of the Ebro River valley in NE Spain (average precipitation ca. 300 mm yr^{-1}). These valleys probably record incision during more humid conditions in the past and subsequent alluvial aggradation related to increased sediment supply from the slopes in response to vegetation cover reduction. *Source:* Francisco Gutiérrez (Author).

Finally, the connection of the depressions produces a steep-sided valley inheriting the trajectory from that of the preexisting unroofed cave. Cleland (1910) differentiated between natural bridges, through which a river runs, or used to run, and natural arches, where the span has never bridged a river, but punctures a rock spur as a result of differential weathering and erosion. Natural bridges can also form in tortuous canyons by underground meander cutoff, which can be facilitated by subsurface hydraulic connections and river undermining on the outer side of channel bends (Figure 7.15d). Generally, the formation of cave-collapse valleys is associated with active cave rivers with sufficient erosional capability to remove at least part of the collapse deposits and produce a valley floor with continuous downstream gradient. Important factors for cave-collapse valley development include: (i) the thickness and mechanical resistance of the cave roof, (ii) the discharge of the cave river, and (iii) the competence and aggressiveness of the water flow. Ford and Williams (2007) and Klimchouk (2006) indicate several examples of cave-collapse valleys: the Rak and Škocjan gorges in Slovenia, the Marble Arch Gorge of Ulster, the Da Xiao Cao Kou Gorge of the Yijiehe River in Guizhou (China), the Sanqiao and Tianjing gorges in the Chongqing karst (China) (Figure 7.15c), or the Peruaçú Gorge in Minas Gerais, Brazil.

Klimchouk (2006), based on observations from some karst areas in southern China, suggests that the collapse of large cave river passages probably plays an important role in the development of gorges and valleys in tropical karst areas. He documents fine examples of cave-collapse limestone gorges in the Chongqing karst, with natural bridges as remnants of cave roofs and abrupt terminations in the associated trunk cave passages (Figure 7.15c). Hill et al. (2008) proposed that the sections of the Grand Canyon (Arizona, USA) situated upstream (east) and downstream (west) of the confluence with the Little Colorado River used to correspond to two different drainages, and that they were linked by an underground hydraulic connection that evolved into a cave-collapse gorge. According to this theoretical model, two major rivers used to exist on both sides of the N-trending Kaibab structural arch (open anticline) (see Figure 4 of Hill et al. 2008): (i) on the east, the north-flowing ancestral Little Colorado River, including the present-day Little Colorado River and the eastern Grand Canyon and (ii) on the west, the west-flowing western Grand Canyon. At around 6 Myr ago, the ancestral Little Colorado River became pirated by a large sinkhole initiating an underground connection around 22 km long with the western Grand Canyon to the west (Hill and Polyak 2020). This subsurface flow was established through a limestone aquifer ca. 200 m thick and across the Kaibab arch. The beheaded section of the ancestral Little Colorado River situated downstream of the collapse sinkhole, currently the eastern Grand Canyon, started to reverse its flow by headward erosion. Interestingly, the eastern Grand Canyon displays barved tributaries that support the flow reversal. These are tributaries that join the mainstream in an upstream direction. Hill et al. (2008) hypothesized that groundwater flow created a large cave river that led to the formation of a belt of coalescing collapse sinkholes, which eventually evolved into a canyon connecting the ancestral Little Colorado and the former Colorado River. The development of such cave-collapse canyon could have been facilitated by aggressive upstream knick point migration and spring sapping.

7.4.5 Dry Valleys

Dry valleys lack stream channels on their floor and are commonly found in carbonate and evaporite karst areas. They can be considered as relict features formed by fluvial erosion under past hydrological and/or climatic conditions. Their development and transformation into dry valleys lacking a water course can be attributed to multiple factors: (i) different climatic conditions in the past, with higher or more torrential rainfall (Figure 7.15e), or the presence of permafrost that prevented water infiltration, (ii) superimposition of an inherited drainage developed on a preexisting cover of low-permeability rocks, (iii) a drop in the regional or local groundwater level leading to a shift in the hydrological behavior of the streams from effluent to influent, (iv) progressive increase in the permeability and absorption capacity of the bedrock, including the epikarst, and (v) abandonment of valley reaches by the development of stream-sinks (i.e., underground piracy in blind valleys), tectonic deformation or fluvial capture. The latter two processes can occur in any lithology and fluvial capture involves the connection between the head of an aggressor drainage with a victim stream, with the consequent beheading of the downstream reach of the latter stream.

7.5 Constructional Features. Calcareous Tufas and Travertines

Calcium carbonate can precipitate from supersaturated waters at springs, rivers (autogenic and allogenic), lakes and palustrine environments, generating deposits and constructional features with a wide diversity of facies and geometries known as calcareous tufas and travertines. These continental carbonate deposits can be considered as the sedimentary output of carbonate karst systems, in which the denuded mass is largely transported as solutes by underground and surface water flows, and eventually reprecipitates at the surface under favorable conditions (hydrochemical, hydrological, biological, and climatic) (Viles 2004d). According to the most widely accepted usage (Pedley 1990; Ford and Pedley 1996; Jones and Renault 2010), calcareous tufa refers to carbonate deposits formed in cool water environments (near-ambient temperature) that typically contain remains of microphytes, macrophytes, and invertebrates (Figure 7.16). The term travertine designates carbonate deposits with a dominant crystalline texture accumulated in hydrothermal systems. Travertines are characterized by the predominance of laminated facies and the general lack of macrophytes and invertebrates, which are unable to thrive under hydrothermal conditions (Figure 7.17). Other common differences between calcareous travertines and tufas include (Jones and Renault 2010): (i) higher degree of lithification of travertines, in contrast with tufas that tend to be highly porous and friable, (ii) faster deposition rates of travertines related to rapid CO_2 degassing, (iii) more limited extent of travertine deposits, typically restricted to the proximity of thermal springs, (iv) fewer diversity of macrofacies of travertines, and (v) tufas are dominated by low-Mg calcite, whereas aragonite can be a significant mineral in travertines, which precipitation seems to be favored by temperatures >40 °C, high Mg/Ca ratios, rapid CO_2 exsolution, and high degrees of supersaturation with respect to $CaCO_3$. A major problem related to the differentiation between calcareous tufas and travertines is the lack of a clear temperature boundary and the uncertainties associated with the temperature of the water that produced fossil carbonate deposits (Jones and Renault 2010). Moreover, travertines associated with thermal springs commonly grade distally into tufa deposits as the water gradually cools downstream from the emergence point. Hydrothermal springs are often defined using a relative temperature value (e.g., temperature at the vent >15 °C above the mean ambient temperature). However, Jones and Renault (2010) based on the key role played by the water temperature on the development of biota and travertine facies in spring environments, proposed the use of the absolute temperature value at the exit point to differentiate between cool (<20 °C) and thermal springs (>20 °C), and to subdivide the latter into warm (20–40 °C), mesothermal (40–75 °C), and hyperthermal (>75 °C) springs. Pentecost (2005) and Viles and Pentecost (2007) suggested that calcareous tufas and travertines can be divided into meteogene and thermogene depending on the source of the CO_2 dissolved in the water, either derived from the soil and the atmosphere, or from deep-seated sources within the lithosphere, respectively. This classification has limited applicability in ancient deposits. Moreover, it may lead

Figure 7.16 Images of calcareous tufa deposits. (a) Transverse tufa dams and associated lakes in Plitvice National Park, Croatia. (b) Tufa waterfall mainly consisting of curtains of macrophyte phytoherms in the Krčić River, Croatia. Ellipses indicate persons for scale. (c) Entrenched tufa terrace in the Ebrón River canyon, carved in Mesozoic limestones (Iberian Chain, Spain). (d) Relict tufa dam in a dry valley in "Ar"'Ar area, northwest Saudi Arabia (average rainfall ca. 75 mm yr^{-1}). Paleocurrent is to the right. This carbonate built up records former more humid conditions in the Pleistocene (pluvial period). (e) Tufa waterfall deposits consisting of steeply inclined beds generated by frontal accretion in bryophyte curtains (Salinas Range, Spanish Pyrenees). (f) Tube-shaped accumulation 4 m long formed by tufa deposition around the orifice of a sporadic overflow spring in Pitarque Canyon, Iberian Chain, Spain. (g) Highly irregular and bulbous stalactite-like tufa deposits with plants deposited on an overhanging rock slope near the entrance of Clearwater Cave, Mulu National Park, Malaysia. *Source:* Francisco Gutiérrez (Author).

Figure 7.17 Travertines and sublacustrine tufa. (a) Travertine deposit in Mammoth Hot Springs (Yellowstone National Park, Wyoming, USA) showing inactive waterfalls, undulating cascades and rimstone pools with stalactites and draperies on the overhanging outer face. Note internal lamination concordant with the surface in areas affected by spalling. *Source:* Francisco Gutiérrez (Author). (b) Dry micro-rimstone pools around 10 cm deep in Mammoth Hot Springs. Note fine-grained carbonate sediment on the floor of the pools. *Source:* Francisco Gutiérrez (Author). (c) Active rimstone pools with lobulated downstream edges in Pamukkale Hot Springs, Turkey. *Source:* Francisco Gutiérrez (Author). (d) Frontal edge of the broad travertine apron associated with Thermopolis hot spring, at the bank of the Bighorn River, Wyoming. Note waterfall curtains with overhangs. *Source:* Francisco Gutiérrez (Author). (e) Shelf-constructed ridge-like travertine channel at Taqté Solyman, Zagros Mountains, Iran. *Source:* Photo by Alireza Amrikazemi. (f) Pinnacle-shaped travertine mound 11 m high (Liberty Cap) generated by rising water in Mammoth Hot Springs, Wyoming, USA. *Source:* Francisco Gutiérrez (Author). (g) Travertine fissure ridge at Terme San Giovanni, Rapolano Terma, Siena, Italy. *Source:* Photo by Andrea Brogi. (h) Cluster of chimney-like tufa mounds up to 5 m high formed by sublacustrine springs in Mono Lake, California. Mixing of Ca-rich spring water and bicarbonate-rich lake water results in rapid calcite precipitation around the orifice of the vent. The mounds have been exposed by the human-induced decline of the lake level. *Source:* Francisco Gutiérrez (Author).

to deceptive interpretations because not all travertine deposits originated from thermal springs are thermogene. For instance, hot springs derived from deep-seated artesian aquifers recharged by meteoric waters.

Modern tufas occur in a wide variety of climatic conditions ranging from cool temperate to arid conditions. The climatic range of travertines is even wider since their formation is not restricted by the tolerance conditions of aquatic plants, which play a central role in tufa deposition (Ricketts et al. 2019). Essential requisites for the formation of calcareous tufas and travertines are the presence of calcium- and bicarbonate-rich waters, one or several mechanisms that induce the supersaturation of the water with respect to $CaCO_3$ (e.g., physical or biological CO_2 degassing), adequate hydrological conditions (flow rate and regime) for the accumulation and preservation of the deposits (Drysdale et al. 2002), and the presence of hygrophilous plants and microorganisms in the case of calcareous tufas. Tufa and travertine deposits are scarcely represented in the stratigraphic record and mostly occur in late Cenozoic and especially Pleistocene formations. This can be largely attributed to their limited preservation potential and reduced areal extent and thickness. Tufas and travertines are commonly associated with erosional geomorphic settings where deposits generally have a reduced persistence time, such as slopes (e.g., perched springs), or entrenched fluvial valleys subject to long-term downcutting (Figures 7.16 and 7.17). Regarding the volume of the formations, the calcareous tufa deposits in the coastal piedmont of Antalya (Turkey), located at the output side of a karst catchment largely recharged through a system of neotectonic poljes (Doğan et al. 2019), are considered to be the largest Quaternary tufa accumulation in the world, covering $630\,km^2$ and with a thickness of 245 m (Glover and Robertson 2003; Koşun 2012). These values are minuscule compared with the dimensions that most other continental sedimentary formations of chemical and detrital origin can reach.

Modern tufas and travertines can have a superlative aesthetic value and some constitute major international tourist attractions with significant economic importance (Williams 2011). Some of them have been designated World Heritage Sites by UNESCO, including the tufa-dammed lakes of Plitvice along the Korana River, Croatia (Bonacci and Roje-Bonacci 2004) (Figure 7.16a), the travertines of Pamukkale associated with the thermal springs of the ancient Greek city of Hierapolis, Turkey (Dilsiz et al. 2004; Brogi et al. 2014) (Figure 7.17c), the tufa cascades and barrages of Jiuzhaigou, China (Florsheim et al. 2013), and the travertine terraces with rimstone pools of Huanglong, China (Lu et al. 2000; Yoshimura et al. 2004). Other outstanding examples are the travertines of Mammoth Hot Springs in Yellowstone National Park, Wyoming, USA (Fouke et al. 2000) (Figure 7.17a, b, f), the tufa dams as much as 450 m long of the Band-e-Amir National Park, Afghanistan (Brooks and Gebauer 2004), the sequence of tufa-dammed lakes in the Lagunas de Ruidera Natural Park, Spain (Ordóñez et al. 2005), and the large and intensively quarried travertines near Tivoli (Central Italy), important building stone (*lapis tiburtinus*) for nearby imperial Rome since Emperor Augustus (Faccenna et al. 2008).

A number of physicochemical (abiotic) and biological processes can induce or enhance the precipitation of calcium carbonate in tufas and travertines by increasing the supersaturation degree of the water with respect to calcite and/or aragonite (see Section 3.7 and Figure 3.24): (i) *CO_2 depletion*. The main process involved in the deposition of most calcareous tufas and travertines is CO_2 degassing, which may be accelerated by abiotic processes such as increased turbulence, or induced by biological activity, chiefly photosynthesis that involves the consumption of CO_2 (Figure 3.6). A decrease in the CO_2 partial pressure of the water results in higher pH values, a drop in the solubility of the carbonate minerals, and hence an increase in the saturation state. The groundwater that emerges at springs enters into an open-air environment with lower P_{CO_2}, experiencing CO_2 exsolution to reach equilibrium conditions with the atmosphere. This outgassing process can be very rapid in the proximal sectors of CO_2-rich thermal springs derived from deep high-pressure conditions, leading to rapid abiotic precipitation of calcium carbonate. The degassing of CO_2 is enhanced at sites of higher turbulence and where the specific water surface in contact with the atmosphere is greater (e.g., water flowing over the lip of rimstones or waterfalls, lake shores). The increases in water turbulence are commonly associated with local steepened slopes or obstructions along the flow path (Florsheim et al. 2013; Auqué et al. 2014). These slope breaks may be related to litho-structural controls (e.g., knick points controlled by lithological contacts, active faults), local accumulation of detrital deposits (e.g., log jams), the growth of carbonate build-ups (e.g., tufa dams, phytoherms), or hydraulic changes at tributary junctions (e.g., rapids at sites of lateral detrital input). The positive feedback loop that occurs along the crest of rimstones and tufa dams, whereby deposition enhances turbulence and hence vertical accretion, explains the growth of these constructional landforms. The CO_2 content of the water can also be depleted under daylight conditions by the photosynthetic metabolism of microphytes and macrophytes, inducing carbonate precipitation. These organisms not only contribute to increase the saturation state, but also provide an adequate substrate for the nucleation and growth of carbonate crystals (Golubić et al. 2008). This biomediated deposition is particularly important on biofilms of photosynthetic microorganisms (e.g., cyanobacteria). (ii) *Temperature rise*. The solubility of carbonate minerals decreases with increasing temperature (retrograde solubility). This inverse relationship between the solubility of the calcium carbonate minerals and temperature is magnified by the fact that the solubility of CO_2 gas in water decreases with rising temperatures (Table 3.4) (Figures 3.9 and 3.25). Consequently, tufa and travertine deposition may be favored or inhibited by water temperature rises and drops, respectively, at different spatial and temporal scales (e.g., diurnal, seasonal, and long-term climatic periods) (Arenas et al. 2014; Arenas and Jones 2017). Temperature, together with atmospheric precipitation, plays an important indirect role in tufa systems, controlling the development of vegetation and its metabolic activity (e.g., photosynthesis) and decay, both at the catchment and the carbonate depositional environment scale. Favorable climatic conditions for the development of vegetation and soils in the karst drainage basin can be accompanied by recharge of more aggressive waters (higher uptake of soil CO_2) and the discharge of groundwaters with higher contents in dissolved calcium carbonate. (iii) *Evaporation*. Evaporation contributes to increase the saturation state of the water, potentially inducing or boosting carbonate precipitation. The impact of this abiotic process is influenced by temperature, the area of the air/water interface, the hydrodynamic conditions of the water (e.g., moving or static), and the wind regime. (iv) *Water mixing*. The conjunction of undersaturated waters can result in a supersaturated water mixture. For instance, the incorporation of relatively small proportions of calcium- and sulfate-rich water that has interacted with gypsum can shift carbonate karst waters into a supersaturation state because of the common ion effect (see Section 3.7.2). The formation of the sublacustrine tufa pinnacles and mounds at Mono Lake, California, is related to the mixing of rising calcium-rich groundwater with the denser bicarbonate-rich saline lake water, inducing rapid precipitation around the orifice of the vent (Figure 7.17h).

Calcareous tufas and travertines have received multiple classifications based on a wide variety of criteria, such as the hydro-geomorphic setting (Pedley 1990; Ford and Pedley 1996), the morphology of the accumulation (Chafetz and Folk 1984), the water temperature (Jones and Renault 2010), the associated biota (Pentecost and Lord 1988), or the source of the CO_2 dissolved in the water (Viles and Pentecost 2007). In this brief review, following Pedley's (1990) classification, we differentiate four groups based on the hydro-geomorphic setting: (i) fluvial, (ii) lacustrine, (iii) palustrine, and (iv) springs. Each tufa and travertine environment displays a variety of inter-related morpho-sedimentary subenvironments with characteristic facies and biota. Lateral and vertical facies variations reflect the variable distribution of those subenvironments through space and time.

Arenas-Abad et al. (2010) presented a comprehensive review on calcareous tufa facies found in fluvial environments and associated depositional systems (fluvio-lacustrine, lacustrine, palustrine). Carthew et al. (2006), in a study on fluvial tufas in the Barkly karst of northern Australia, provided an account of tufa facies developed in a monsoonal tropical environment. Jones and Renault (2010), in a review on tufa and travertine spring deposits, described the most common facies found in calcareous cool and thermal spring systems. Below we briefly describe the main tufa and travertine facies, divided into autochthonous and allochthonous (or detrital). The former occurs at the site of precipitation, whereas the latter has experienced some transport. Some facies can occur intermixed and a significant proportion of the facies can be found in multiple depositional environments and subenvironments. The reconstruction of the depositional environment of ancient tufas and travertines should be based on knowledge about the geological context, the associations of facies and their spatial relationships, as well as a number of biological, mineralogical, and geochemical features reviewed by Pentecost (2005).

Autochthonous facies: *Macrophyte phytoherms* collectively designate deposits formed by the encrustation of various types of macrophytes in life position. They can be designated more specifically depending on the dominant type of calcified plant. *Bryophyte phytoherms* commonly occur in waterfalls, dams, and channel banks and display centimeter-thick layers that reproduce the geometry of the accretion surface underlain by mosses, often with rounded morphologies (Figure 7.18b). *Phytoherms of stems* are frameworks of hygrophilous vascular plants (e.g., rushes, reeds), mostly upright stems, anchored to their original substrate and coated by carbonate fringes (Figure 7.18a). In overhangs of waterfalls and cascades, the plants that function as the substrate for carbonate deposition can be suspended forming *curtains of hanging stems* (Figure 7.18c). *Charophyte phytoherms* develop in low-energy and relatively shallow ponded areas. The rapid decay of the encrusted plants in the various types of macrophyte phytoherms produces moulds and significantly contributes to increase the porosity and reduce the density of the tufa deposits. *Stromatolites*, ascribed to carbonate precipitation associated with microbial biofilms (e.g., cyanobacteria), are characterized by alternating subcentimetric laminae with different colors and textures. This facies can form sheets draping an aggradation surface, or heads with various geometries, from rounded domes to columnar bodies (Figures 2.15c and 7.18d). The latter geometries develop in calm and stable water bodies such as lakes or pools. Stromatolites of various morphologies (pinnacles, bulbous), and locally resembling mammillary crusts and shelfstones of caves, can develop on steep submerged walls of lakes (Pedley et al. 1996) (Figure 6.32c). *Speleothem facies* are mostly stalactites and draperies formed by dripping water and water trickles in overhanging outer faces of rimstones and caves situated behind waterfall curtains (Figure 7.17a, d). Highly irregular to bulbous tufaceous stalactites also form on overhanging bedrock slopes, especially in humid tropical environments (Figure 7.16g). *Breakdown breccias* consist of clast-supported, angular, and commonly poorly sorted gravel- to boulder-sized particles produced by the collapse of tufa or travertine deposits in caves and overhangs. *Tubes of arthropods* are tubular burrows, cases or nets created by aquatic arthropods (e.g., *Trichoptera*, *Quironomiae*) that become encrusted by calcium carbonate. These biogenic tubes can be found in great numbers both in situ and transported by water currents (Drysdale 1999). *Crystalline travertine* shows layers of crystalline calcite or aragonite concordant with topographic accretion surfaces subject to rapid precipitation from fast water flows. Elongate crystals commonly show their long axes perpendicular to layering.

Allochthonous or detrital facies: *Phytoclastic* facies consist of detrital particles derived from the erosion of macrophyte phytoherms, mostly fragments of coated stems. This facies typically shows grain-supported texture, fine-grained microdetrital to micritic matrix and in some cases oriented fabrics (Figure 7.18e, g). *Oncolithic* facies are composed of gravel-sized, rounded, oblate to

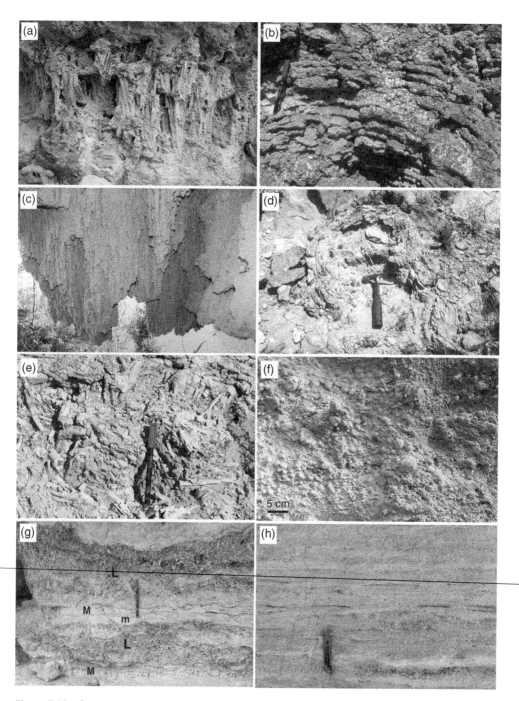

Figure 7.18 Some common autochthonous and allochthonous (detrital) tufa facies. (a) Phytoherm of stems encrusted by calcium carbonate with palisade arrangement. Piedra River Canyon, Iberian Chain, Spain. (b) Bryophyte phytoherm consisting of well-defined and nearly isopachous convex-up beds. Jalón River valley, Iberian Chain, Spain. (c) Curtain of hanging encrusted stems in a tufa waterfall. La Requijada waterfall, Piedra River canyon, Iberian Chain, Spain. (d) Section of dome-shaped stromatolite with internal lamination. (e) Phytoclasts with chaotic fabrics. Piedra River canyon, Iberian Chain, Spain. (f) Massive deposit of oncolitic tufa. Navarrés Neogene Graben, Iberian Chain, Spain (g) Sequences deposited in a tufa-dammed lake consisting of beds of lithoclasts (L), micrite (m) and marls (M). Piedra River Canyon, Iberian Chain, Spain. (h) Thinly bedded microbioclastic and micritic tufa accumulated in a lake upstream of a tufa barrage. Piedra River Canyon, Iberian Chain, Spain. *Source:* Francisco Gutiérrez (Author).

subspherical detrital particles with an internal nucleus (e.g., intraclast, stem fragment) and concentric light and dark, micrite to sparite lamination (Figure 7.18f). They form by the precipitation of successive coatings during transport in non-static water bodies or water currents. *Intraclastic* facies are made up of angular to rounded detrital particles derived from the erosion of tufa/travertine deposits, commonly related to flood events. *Microbioclastic facies* are dominated by sand-sized particles of bioclasts and intraclasts (Figure 7.18h). *Micritic facies* are carbonate muds mostly deposited in lakes and pools that commonly include invertebrates (e.g., gastropods, ostracods). The term micrite generally refers to calcite particles and/or crystals $<4\,\mu m$ (Figure 7.18g, h). *Raft facies* consist of millimeter-thick plates that precipitate at the water–air interface of clam pools or lakes and eventually sink to the bottom. The rafts grow downward and commonly display a smooth upper surface and dentated lower side related to the epitaxial growth of crystals in the water. The formation of these precipitates requires semistatic water bodies and is favored by evaporation and rapid CO_2 outgassing (Taylor et al. 2004). *Lithoclastic* facies are deposits of exotic detrital particles derived from the erosion of the bedrock or non-carbonate surficial formations. Other allied facies commonly found in lacustrine and palustrine environments include *marls* (Figure 7.18g) and *organic-rich muds and peats* deposited in poorly drained areas with reducing conditions that strongly inhibit the decay of organic matter.

Fluvial and fluvio-lacustrine tufas are the most widespread continental carbonate deposits. Pedley (1990) and Ford and Pedley (1996) proposed two fluvial models that represent end members of a continuous spectrum: (i) the braided fluvial model, characterized by shallow and poorly confined braided channels with deposition of calcareous particles in bars and channels, dominated by oncolithic and phytoclastic facies; and (ii) the barrage fluvial model developed in laterally confined entrenched valleys (e.g., canyons), where the flow is strongly influenced by phytoherm barrages dominated by bryophytes. These dams hold lakes that function as sediment traps with deposition of a wide diversity of fluvio-lacustrine and lacustrine facies (e.g., micritic, microdetrital, oncolithic, phytoclastic, and stromatolites) (Figure 7.16a). Arenas-Abad et al. (2010), in their comprehensive sedimentological description of tufa deposits formed in fluvial environments and associated subenvironments (lacustrine and palustrine), differentiate two main sedimentary models that largely coincide with those proposed by Pedley (1990): (i) low-gradient, non-stepped fluvial and fluvio-lacustrine systems; and (ii) high-gradient and stepped fluvial systems. The low-gradient and non-stepped fluvial systems develop on broad and graded alluvial surfaces with mobile channels and associated interchannel and floodplain subenvironments. Oncolithic facies dominate in the channels, forming bars and channel fills. Phytoherms of stems proliferate in palustrine areas within the floodplain, acting as the source of phytoclasts during flood events. Pools in abandoned channels or shallow depressions within the floodplain are prone to deposition of fine-grained detrital facies (microdetrital, micritic) and stromatolites. Sedimentological and geomorphic accounts of tufas ascribable to this model can be found in Arenas et al. (2000) and Gutiérrez et al. (2019c). The high-gradient and stepped systems occur in relatively narrow and confined valleys with breaks in the longitudinal profile (i.e., knick points) mainly related to the construction of tufa barrages. These systems show a succession of subenvironments including: (i) tufa dams with waterfalls, cascades, and caves, (ii) lakes upstream of the dams that function as hydrologic buffers that prevent the destruction of the transverse barrages, and (iii) valley sections with channels and a restricted floodplain. This type of fluvial environment, which essentially fits with the barrage system of Pedley (1990), is widely represented in modern tufas (e.g., Plitvice Lakes in Croatia, Lagunas de Ruidera in Spain), but have a limited long-term preservation potential. Pedley (1990) described separately the paludal and lacustrine environments, although they are frequently associated with fluvial systems. The paludal or palustrine environments are characterized by shallow poorly-drained depressions and valley floors with ephemeral water bodies controlled by variations in the water table and flow rate. Common facies include phytoherms of stems and bryophytes, organic-rich muds and peat. Low-energy conditions limit the occurrence of coarse detrital deposits such as oncolithic facies. Lacustrine environments with perennial water bodies typically show a zonal facies distribution toward the deeper sectors of the lake. Typical facies in the shallow and photic marginal zones include charophyte phytoherms, stromatolite crusts and build-ups, and oncoliths. Micritic and microdetrital facies predominate in the deeper zones, often containing abundant remains of fauna (e.g., gastropods, ostracods) and sand-sized fragments of other tufa deposits (Figure 7.18h).

The overall morphology of tufa and travertine deposits associated with springs is strongly influenced by the geomorphic context. Here, we differentiate the following basic morphological types: (i) waterfalls, (ii) sloping aprons, (iii) terraced aprons, (iv) mounds (subaerial and sublacustrine), and (v) fissure ridges. The first three types develop on sloping terrain commonly associated with perched springs, whereas the latter two are mostly formed at springs located on relatively flat ground and fed by rising groundwater, with the potentiometric surface situated above the topographic surface (artesian springs, pressurized geothermal springs). Perched springs located on subvertical cliffs produce *waterfalls* where the flow is dominated by free fall (Figure 7.16b). In cool springs, precipitation of calcium carbonate from the strongly agitated water produces curtains of encrusted bryophytes and vascular plants that can prograde by frontal accretion. Small blind caves with speleothems often develop behind the overhanging curtains (Gradziński et al. 2018). In some cases, carbonate precipitates at the rim of the orifice to form projecting tube-like contructions (Figure 7.16f). Perched springs situated on non-vertical slopes typically produce fan-shaped aprons with radial water flow. The *sloping aprons*, commonly associated with cool-spring tufas, display a fan-shaped slope with distributary flow constructed by the frontal accretion of bryophyte phytoherms. These slopes can include small terraces with pools and rimstones. The upper part of the aprons may have a shelf with a pool associated with the resurgence point. This morphological type corresponds to the perched springline model of Pedley (1990). Perched thermal springs also produce aprons, but their slopes often display a terraced topography consisting of successions of rimstone pools forming *terraced aprons* (e.g., Pamukkale, Turkey; Mammoth Hot Springs, Wyoming, USA) (Figure 7.17a–c). The rimstones, made up of laminated crystalline calcite, are arcuate dams with downslope-facing convexity and a nearly vertical inner side. The outer side may display botryoidal and pop-corn-like morphologies, micro-rimstone pools, stalactites, and draperies. Confined overland flow can produce shelf-constructed ridge-like channels. Here, vertical accretion results in vertically growing steep-sided ridges with a channel at the top (Figure 7.17e). *Subcircular mounds* mostly develop by rising cool or thermal water at point resurgences located on subhorizontal surfaces. They occur in subaerial and sublacustrine settings and may show a wide range of geometries, ranging from tapered domes or cones to pinnacles (Della Porta 2015) (Figure 7.17f–h). The largest subaerial mound is Takté Soleyman in Iran reaching 70 m above the surrounding ground (Figure 2.15b). Broad tufa mounds may contain an inner depression with a permanent lake (e.g., Basturs Lakes, Spain; Linares et al. 2009; Pellicer et al. 2014, 2016). The mounds in subaqueous environments, where the water enters into a medium denser that the air, tend to be more spatially restricted and with higher height-to-width ratios (e.g., chimney-like pinnacles of Mono Lake, California). Kempe et al. (1991) documented tower-like mounds as much as 40 m high in the bottom of Lake Van, Turkey. *Fissure ridges* develop where thermal water rises along a fracture or a fault, producing linear ridges of crystalline travertine with a triangular cross-section. The crest of the ridges generally displays an open cleft underlain by subvertical veins of banded calcite, flanked by two walls with outward dipping bedded travertine (Figure 7.17g). The crest of the fissures often displays elongated depressions that may host thermal pools (Brogi et al. 2021). These datable deposits can be spatially associated with seismogenic faults and can provide paleoseismological data (e.g., North Anatolian Fault; Karabacak et al. 2019).

Tufa and travertine deposits are potentially datable by various geochronological methods, chiefly U-series and radiocarbon. Carbonaceous layers (peat, sapropels) and charcoal younger than 50 kyr can be dated by radiocarbon. This method can be also applied to some components of tufa deposits, such as shells of arthropods, although ages may be overestimated due to the hard water effect (i.e., incorporation of C atoms derived from bedrock). A common limitation associated with the use of the U-series method on carbonate deposits is the presence of clay and secondary precipitates. Eogenic dissolution and precipitation processes can lead to substantial changes in these soluble and often highly porous sediments, which rarely behave as perfectly closed systems. Preferably, chronological determinations of the tufa deposition intervals should be based on multiple consistent dates and including age data from the lowermost and topmost units.

A number of studies explore the potential impact of climate variability on the temporal distribution of tufa and travertine deposits. Although the relationships between climate and carbonate deposition are complex and depend on a large number of interrelated variables, there seem to be some general climate-controlled temporal patterns (see review by Pentecost 2005). Paleoclimate interpretations indicate that the limiting factors for calcareous tufa formation vary depending on the morpho-climatic environment, largely controlled by latitude (Sancho et al. 2015 and references therein). In humid and relatively cold high-latitude regions, shifts toward warmer conditions (e.g., interglacials) promote tufa deposition. Climate warming can induce effects such as permafrost thawing and higher production of CO_2 in soils, resulting in higher aquifer recharge, greater concentration of dissolved calcium carbonate in the groundwater, and proliferation of photosynthetic vegetation at sites of deposition. In warm drylands, calcareous tufa accumulation occurs or increases during wet periods (e.g., pluvials; Figure 7.16d). A change toward wetter conditions is accompanied by increases in vegetation, soil development, CO_2 production in soils, groundwater recharge, and both water and solute load flux. In the late Holocene, human activy might have some impact on tufa deposition. Goudie et al. (1993) discussed the probable relation between a putative tufa deposition decline in Europe over the last 2.5 kyr and deforestation. Deforestation and the associated removal of soils produce a number of environmental effects that inhibit tufa deposition, notably infiltration of waters with lower dissolution capability, more flashy streams with greater capacity to destroy tufa deposits and thus causing channel incision, or water flows with greater sediment load and turbidity. Subsequently, Dabkowski (2020), using a larger dataset of dated tufas, observed that the late Holocene decline systematically affects fluvial tufas, but not spring and lacustrine tufas, probably due to the higher sensitivity of the former to human activity. Ricketts et al. (2019), using a worldwide compilation of 1649 published travertine ages ranging between 0.21 and 730 kyr, explored relationships between travertine deposition and global climate. The temporal distribution of the ages shows peaks with frequencies of around 100 kyr, mostly associated with glacial terminations and interglacial periods, including a large peak that coincides with the Early Holocene climatic optimum. On a regional scale, there is good correspondence between many peaks and high precipitation periods. This correlation is attributed to higher aquifer recharge during wet intervals.

References

Ahnert, F. and Williams, P.W. (1997). Karst landform development in a three-dimensional theoretical model. *Zeitschrift für Geomorphologie* 108: 63–80.

Aiello, G., Ascione, A., Barra, D. et al. (2007). Evolution of the late Quaternary San Gregorio Magno tectono-karstic basin (southern Italy) inferred from geomorphological, tephrostratigraphical and palaeoecological analyses: tectonic implications. *Journal of Quaternary Science* 22: 233–245.

Arenas, C. and Jones, B. (2017). Temporal and environmental significance of microbial lamination: insights from Recent fluvial stromatolites in the River Piedra, Spain. *Sedimentology* 65: 1597–1629.

Arenas, C., Gutiérrez, F., Osácar, C. et al. (2000). Sedimentology and geochemistry of fluvio-lacustrine tufa deposits controlled by evaporite solution subsidence in the central Ebro depression, NE Spain. *Sedimentology* 47: 883–909.

Arenas, C., Vázquez-Urbez, M., Auqué, L. et al. (2014). Intrinsic and extrinsic controls of spatial and temporal variations in modern fluvial tufa sedimentation: a thirteen-year record from a semi-arid environment. *Sedimentology* 61: 90–132.

Arenas-Abad, C., Vázquez-Urbez, M., Pardo-Tirapú, G. et al. (2010). Fluvial and associated carbonate deposits. In: *Carbonates in Continental Settings* (ed. A.M. Alonso-Zarza and L.H. Tanner), 133–170. Amsterdam: Elsevier.

Aub, C. (1969). The nature of cockpits and other depressions in the karst of Jamaica. *Proceedings of the 5th International Congress of Speleology (Germany)*, 15/1-15/7.

Audra, P., Mocochain, L., Camus, H. et al. (2004). The effect of the Messinian Deep Stage on karst development around the Mediterranean Sea. Examples from Southern France. *Geodinamica Acta* 17: 389–400.

Auqué, L., Arenas, C., Osácar, C. et al. (2013). Tufa sedimentation in changing hydrological conditions: the River Mesa (Spain). *Geologica Acta* 11: 85–102.

Auqué, L., Arenas, C., Osácar, C. et al. (2014). Current tufa sedimentation in a changing-slope valley: the River Añamaza (Iberian Range, NE Spain). *Sedimentary Geology* 303: 26–48.

Balázs, D. (1973a). Karst types in the Philippines. In: *Proceedings of the 6th International Congress of Speleology*, vol. 2 (ed. V. Panoš), 19–38. Prague: Academia.

Balázs, D. (1973b). Relief types of tropical karst areas. In: *Proceedings of the Symposium on karst-morphogenesis*, 16–32. Budapest: International Geographical Union.

Bard, E., Antonioli, F., and Silenzi, S. (2002). Sea-level during the penultimate interglacial period based on a submerged stalagmite from Argentarola Cave (Italy). *Earth and Planetary Science Letters* 196: 135–146.

Beddows, P.A. (2004). Yucatán Phreas, Mexico. In: *Encyclopedia of Caves and Karst Science* (ed. J. Gunn), 786–788. New York: Fitzroy Dearborn.

Bella, P., Veselský, M., Gaál, U. et al. (2016). Jósvafo paleo-polje: morphology and relation to the landform evolution of Aggtelek Karst and Jósva River valley, Hungary. *Zeitschrift für Geomorphologie* 60: 219–235.

Benito-Calvo, A. and Pérez-González, A. (2007). Erosion surfaces and Neogene landscape evolution in the NE Duero Basin (north-central Spain). *Geomorphology* 88: 226–241.

Blatnik, M., Frantar, P., Kosec, D. et al. (2017). Measurements of the outflow along the eastern border of Planinsko Polje, Slovenia. *Acta Carsologica* 46: 83–93.

Bočić, N., Pahernik, M., and Mihevc, A. (2015). Geomorphological significance of the palaeodrainage network on a karst plateau: the Una-Korana plateau, Dinaric karst, Croatia. *Geomorphology* 247: 55–65.

Bonacci, O. (1985). Hydrological investigations of Dinaric karst at the Krčić catchment and river Krka springs (Yugoslavia). *Journal of Hydrology* 82: 317–326.

Bonacci, O. (1987). *Karst Hydrology with Special References to the Dinaric Karst*. Berlin: Springer.

Bonacci, O. (2013). Poljes, ponors and their catchments. In: *Treatise of Geomorphology: Karst Geomorphology*, vol. 6 (ed. A. Frumkin), 112–120. Amsterdam: Elsevier.

Bonacci, O. and Plantić, K. (1997). Hydrology of the Drežničko polje in the karst (Croatia). In: *Karst Waters and Environmental Impacts* (ed. G. Günay and I. Johnson), 303–309. Rotterdam: A.A. Balkema.

Bonacci, O. and Roje-Bonacci, T. (2004). Plitvice Lakes, Croatia. In: *Encyclopedia of Karst and Cave Science* (ed. J. Gunn), 597–598. New York: Fitzroy Dearborn.

Brogi, A., Capezzuoli, E., Alçiçek, M.C. et al. (2014). Evolution of a fault-controlled fissure-ridge type travertine deposit in the western Anatolia extensional province: the Çukurbağ fissure-ridge (Pamukkale, Turkey). *Journal of the Geological Society* 171: 425–441.

Brogi, A., Capezzuoli, E., Karabacak, V. et al. (2021). Fissure Ridges: a Reappraisal of Faulting and Travertine Deposition (Travitonics). *Geosciences* 11: 278.

Brook, G.A. and Ford, D.C. (1980). Hydrology of the Nahanni karst, northern Canada. The importance of extreme summer storms. *Journal of Hydrology* 46: 103–121.

Brooks, S. and Gebauer, D. (2004). Indian subcontinent. In: *Encyclopedia of Karst and Cave Science* (ed. J. Gunn), 442–445. New York: Fitzroy Dearborn.

Brunović, D., Miko, S., Hasan, O. et al. (2020). Late Pleistocene and Holocene paleoenvironmental reconstruction of a drowned karst isolation basin (Lošinj Channel, NE Adriatic Sea). *Palaeogeography, Palaeoclimatology, Palaeoecology* 544: 109587.

Büdel, J. (1957). Die "Dopplelten Einebnungsflächen" in den feuchten Tropen. *Zeitschrift für Geomorphologie* 1: 201–228.

Calaforra, J.M. and Pulido-Bosch, A. (1999a). Gypsum karst features as evidence of diapiric processes in the Betic Cordillera, southern Spain. *Geomorphology* 29: 251–264.

Carthew, K.D., Taylor, M.P., and Drysdale, R.N. (2006). An environmental model of fluvial tufas in the monsoonal tropics, Barkly karst, northern Australia. *Geomorphology* 73: 78–100.

Chafetz, H.S. and Folk, R.L. (1984). Travertines; depositional morphology and the bacterially constructed constituents. *Journal of Sedimentary Research* 54: 289–316.

Cleland, H.F. (1910). North American natural bridges, with a discussion of their origin. *Bulletin of the Geological Society of America* 21: 313–338.

Covington, M.D., Prelovšek, M., and Gabrovšek, F. (2013). Influence of CO_2 dynamics on the longitudinal variation of incision rates in soluble bedrock channels: feedback mechanisms. *Geomorphology* 186: 85–95.

Cui, Z., Li, D., Feng, J. et al. (2002). The covered karst, weathering crust and karst (double-level) planation surface. *Science in China Series D* 45: 366–379.

Cvijić, J. (1893). *Das Karstphänomen. Versuch einer morphologischen Monographie*, Geographische Abhandlungen herausgegeben von A. Penck, Wien, Band V, Heft 3, 1-114.

Cvijić, J. (1900). Karstna polja zapadne Borne I Hercegovine. *Glas Srpske Kraljevske Akademije* 59: 59–182.

Cvijić, J. (1918). Hydrographie souterraine et évolution morphologique du karst. *Revue de Géographie Alpine* 6: 376–420.

Dabkowski, J. (2020). The late-Holocene tufa decline in Europe: Myth or reality? *Quaternary Science Reviews* 230: 106141.

Davis, W.M. (1899). The Geographical Cycle. *Geographical Journal* 14: 481–504.

Davis, W.M. (1901). An excursion in Bosnia, Hercegovina and Dalmatia. *Bulletin of the Geographical Society* 3: 47–50.

Day, M.J. (1978). Morphology and distribution of residual limestone hills (mogotes) in the karst of northern Puerto Rico. *Geological Society of America Bulletin* 89: 426–432.

Day, M.J. and Huang, W. (2009). Reflections on fengcong and fenglin. *Cave and Karst Science* 36: 49–51.

De Waele, J., Martina, M.L., Sanna, L. et al. (2010). Flash flood hydrology in karstic terrain: Flumineddu Canyon, central-east Sardinia. *Geomorphology* 120 (3-4): 162–173.

De Waele, J., D'Angeli, I.M., Tisato, N. et al. (2017b). Coastal uplift rate at Matanzas (Cuba) inferred from MIS 5e phreatic overgrowths on speleothems. *Terra Nova* 29 (2): 98–105.

Deligianni, M.G., Veni, G., and Pavlopoulos, K. (2013). Land use and limitations in the sinkhole and polje karst of the Ksiromero Region, Western Greece. *Carbonates and Evaporites* 28: 167–173.

Della Porta, G. (2015). Carbonate build-ups in lacustrine, hydrothermal and fluvial settings: Comparing depositional geometry, fabric types and geochemical signature. In: *Microbial Carbonates in Space and Time: Implications for Global Exploration and Production, Geological Society of London Special Publication*, vol. 418 (ed. D.W.J. Bosence, K.A. Gibbons, D.P. Le Heron, et al.), 17–68. Geological Society of London.

Dilsiz, C., Marques, J.M., and Carreira, P.M.M. (2004). The impact of hydrological changes on travertine deposits related to thermal springs in the Pamukkale area (SW Turkey). *Environmental Geology* 45: 808–817.

Doğan, U. (2003). Sariot Polje, Central Taurus (Turkey): a border polje developed at the contact of karstic and non-karstic lithologies. *Cave and Karst Science* 30: 117–124.

Doğan, U. and Koçyiğit, A. (2018). Morphotectonic evolution of Maviboğaz canyon and Suğla polje, SW central Anatolia, Turkey. *Geomorphology* 306: 13–27.

Doğan, U. and Özel, S. (2005). Gypsum karst and its evolution east of Hafik (Sivas, Turkey). *Geomorphology* 71: 373–388.

Doğan, U., Koçyiğit, A., and Gökkaya, E. (2017). Development of the Kembos and Eynif structural poljes: Morphotectonic evolution of the Upper Manavgat River basin, central Taurides, Turkey. *Geomorphology* 278: 105–120.

Doğan, U., Koçyiğit, A., and Yeşilyurt, S. (2019). The relationship between Kestel Polje system and the Antalya Tufa Plateau: their morphotectonic evolution in Isparta Angle, Antalya-Turkey. *Geomorphology* 334: 112–125.

Drysdale, R.N. (1999). The sedimentological significance of hydropsychid caddis-fly larvae (order Trochoptera) in a travertine-depositing stream: Louie Creek Northwest Queensland, Australia. *Journal of Sedimentary Research* 69: 145–150.

Drysdale, R.N., Taylor, M.P., and Ihlenfeld, C. (2002). Factors controlling the chemical evolution of travertine-depositing rivers of the Barkly karst, northern Australia. *Hydrological Processes* 16: 2941–2962.

Dumitru, O.A., Austermann, J., Polyak, V.J. et al. (2021). Sea-level stands from the Western Mediterranean over the past 6.5 million years. *Scientific Reports* 11 (1): 1–10.

Ek, C. and Mathieu, L. (1964). La Daya Chiker (Moyen-Atlas, Maroc). Étude Géomorphologique. *Annales de la Société Géologique de Belgique* 87: 65–103.

Faccenna, C., Soligo, M., Billi, A. et al. (2008). Late Pleistocene depositional cycles of the Lapis Tiburtinus travertine (Tivoli, Central Italy): possible influence of climate and fault activity. *Global and Planetary Change* 63: 299–308.

Florsheim, J.L., Ustin, S.L., Tang, Y. et al. (2013). Basin-scale and travertine dam-scale controls on fluvial travertine, Jiuzhaigou, southwestern China. *Geomorphology* 180: 267–280.

Ford, T.D. and Pedley, H.M. (1996). A review of tufa and travertine deposits of the world. *Earth-Science Reviews* 41: 117–175.

Ford, D.C. and Williams, P.W. (1989). *Karst Geomorphology and Hydrology*. London: Unwin Hyman.

Ford, D.C. and Williams, P.W. (2007). *Karst Hydrogeology and Geomorphology*. Chichester, UK: Wiley.

Fouke, B.W., Farmer, J.D., Des Marais, D.J. et al. (2000). Depositional facies and aqueous-solid geochemistry of travertine-depositing hot springs (Angel Terrace, Mammoth Hot Springs, Yellowstone National Park, USA). *Journal of Sedimentary Research* 70: 565–585.

Frelih, M. (2003). Geomorphology of karst depressions: Polje or uvala - A case study of Lučki Dol. *Acta Carsologica* 32: 105–119.

Galdenzi, S. (2019). Monte Lago Polje, a case study regarding the influence of geologic structure and degree of karstification on groundwater drainage in the central Apennines (Italy). *Acta Carsologica* 48: 99–115.

Gams, I. (1965). Types of accelerated karst corrosion. *Proceedings of the International Speleological Conference*, 133–139. Brno.

Gams, I. (1978). The polje: the problem of definition. *Zeitschrift für Geomorphologie* 22: 170–181.

Gams, I. (1994). Types of the poljes in Slovenia, their inundations and land use. *Acta Carsologica* 23: 285–302.

Gams, I. (2005). Tectonic impact on poljes and minor basins (case studies of Dinaric karst). *Acta Carsologica* 34: 25–41.

Gavazzi, A. (1904). Die seen des karstes (karst lakes). *Abhandlungen der KK Geographischen Gesellschaft* 5: 136.

Giraudi, C., Bodrato, G., Lucchi, M.R. et al. (2011). Middle and late Pleistocene glaciations in the Campo Felice Basin (central Apennines, Italy). *Quaternary Research* 75: 219–230.

Glover, C. and Robertson, A.H. (2003). Origin of tufa (cool-water carbonate) and related terraces in the Antalya area, SW Turkey. *Geological Journal* 38: 329–358.

Gökkaya, E., Gutiérrez, F., Ferk, M. et al. (2021). Sinkhole development in the Sivas gypsum karst. Turkey. *Geomorphology* 386: 107746.

Golubić, S., Violante, C., Plenković-Moraj, A. et al. (2008). Travertines and calcareous tufa deposits: an insight into diagenesis. *Geologia Croatica* 61: 363–378.

Gospodarič, R. and Habič, P. (1978). Karst phenomena of Cerknisko polje. *Acta Carsologica* 8: 150–162.

Goudie, A.S., Viles, H.A., and Pentecost, A. (1993). The late-Holocene tufa decline in Europe. *The Holocene* 3: 181–186.

Gracia, F.J., Gutiérrez, F., and Gutiérrez, M. (1996). Los poljes de la región de Layna (Cordillera Ibérica noroccidental). *Cuaternario y Geomorfología* 10: 33–45.

Gracia, F.J., Gutiérrez, F., and Gutiérrez-Elorza, M. (1999). Evolución geomorfológica del polje de Gallocanta (Cordillera Ibérica). *Revista de la Sociedad Geológica de España* 12 (3): 351–368.

Gracia, F.J., Benavente, J., and Anfuso, G. (2000). Implicaciones endokársticas de la evolución geomorfológica de los poljes de Zurraque y Burfo (Sierra de Líbar, Málaga). In: *Actas del I Congreso Andaluz de Espeleología*, 341–351. Federación Andaluza de Espeleología.

Gracia, F.J., Gutiérrez, F., and Gutiérrez, M. (2002). Origin and evolution of the Gallocanta polje. *Zeitschrift für Geomorphologie* 46: 245–262.

Gracia, F.J., Gutiérrez, F., and Gutiérrez, M. (2003). The Jiloca karst polje-tectonic graben (Iberian Range, NE Spain). *Geomorphology* 52: 215–231.

Gracia, F.J., Geremia, F., Privitera, S. et al. (2014). The probable karst origin and evolution of the Vendicari coastal lake system (SE Sicily, Italy). *Acta Carsologica* 43: 215–228.

Gradziński, M., Bella, P., and Holúbek, P. (2018). Constructional caves in freshwater limestone: a review of their origin, classification, significance and global occurrence. *Earth-Science Reviews* 185: 179–201.

Grigg, R., Grossman, E., Earle, S. et al. (2002). Drowned reefs and antecedent karst topography, Au'au Channel, SE Hawaiian Islands. *Coral Reefs* 21: 73–82.

Grund, A. (1903). *Die Karsthydrographie: Studien aus Westbosnien*. Geographische Abhandlungen herausgegeben von A. Penck, Band IX. Vienna.

Grund, A. (1914). Der geographische Zyklus im Karst. *Gesellschaft für Erdkunde* 1914: 621–640.

Grützner, C., Aschenbrenner, S., Jamšek Rupnik, P. et al. (2021). Holocene surface rupturing earthquakes on the Dinaric Fault System, western Slovenia. *Solid Earth Discussions* 1–33.

Gunn, J. (2004a). Fluviokarst. In: *Encyclopedia of Caves and Karst Science* (ed. J. Gunn), 363–364. New York: Fitzroy Dearborn.

Gunn, J. (2004b). Valleys in karst. In: *Encyclopedia of Caves and Karst Science* (ed. J. Gunn), 753–754. New York: Fitzroy Dearborn.

Gutiérrez, M. (2013). *Geomorphology*. Boca Raton: CRC Press.

Gutiérrez, F., Gutiérrez, M., Gracia, F.J. et al. (2008e). Plio-Quaternary extensional seismotectonics and drainage network development in the central sector of the Iberian Chain (NE Spain). *Geomorphology* 102: 21–42.

Gutiérrez, F., Sevil, J., Silva, P.G. et al. (2019c). Geomorphic and stratigraphic evidence of Quaternary diapiric activity enhanced by fluvial incision. Navarrés salt wall and graben system, SE Spain. *Geomorphology* 342: 176–195.

Gutiérrez-Elorza, M. and Gracia, F.J. (1997). Environmental interpretation and evolution of the Tertiary erosion surfaces in the Iberian Range (Spain). In: *Paleosurfaces: Recognition, Reconstruction and Paleoenvironmental Interpretation* (ed. M. Widdowson), 147–158. London: The Geological Society.

Gutiérrez-Elorza, M. and Valverde, M. (1994). El sistema de poljes del Río Guadazaón (Cordillera Ibérica, Prov. de Cuenca). *Cuaternario y Geomorfología* 8: 87–95.

Gutiérrez-Elorza, M., Peña-Monné, J.L., and Simón-Gómez, J.L. (1982). *El polje de Vistabella del Maestrazgo (provincia de Castellón). In: Reunión Monográfica del Karst-Larra*, vol. *82*, 95–104. Pamplona: Servicio Geológico de la Dirección de Obras Públicas de la Diputación de Navarra.

Haryono, E. and Day, M.J. (2004). Landform differentiation within the Gunung Kidul kegelkarst, Java, Indonesia. *Journal of Cave and Karst Studies* 66: 62–69.

Hill, C.A. and Polyak, V.J. (2020). A karst hydrology model for the geomorphic evolution of Grand Canyon, Arizona, USA. *Earth-Science Reviews* 208: 103279.

Hill, C.A., Eberz, N., and Buecher, R.H. (2008). A Karst Connection model for Grand Canyon, Arizona, USA. *Geomorphology* 95: 316–334.

Huggett, R.J. (2003). *Fundamentals of Geomorphology*. London: Routledge.

Jennings, J.N. (1976). A test of the importance of cliff-foot caves in tower karst development. *Zeitschrift für Geomorphologie* Supplement Band 26: 92–97.

Jennings, J.N. (1985). *Karst Geomorphology*. Oxford and New York: Blackwell.

Jennings, J.N., Bao, H., and Spate, A.P. (1980). Equilibrium versus events in river behavior and blind valleys at Yarrangobilly, New South Wales. *Helictite* 18: 39–54.

Jiménez-Sánchez, M., Rodríguez-Rodríguez, L., García-Ruiz, J.M. et al. (2013). A review of glacial geomorphology and chronology in northern Spain: timing and regional variability during the last glacial cycle. *Geomorphology* 196: 50–64.

Jones, B. and Renault, R. (2010). Calcareous spring deposits in continental settings. In: *Carbonate Sediments in Continental Setting. Facies, Environments and Processes* (ed. A.M. Alonso-Zarza and L.H. Tanner), 177–224. Amsterdam: Elsevier.

Kan, H., Urata, K., Nagao, M. et al. (2015). Submerged karst landforms observed by multibeam bathymetric survey in Nagura Bay, Ishigaki Island, southwestern Japan. *Geomorphology* 229: 112–124.

Karabacak, V., Uysal, I.T., Mutlu, H. et al. (2019). Are U-Th dates correlated with historical records of earthquakes? Constraints from coseismic carbonate veins within the North Anatolian Fault zone. *Tectonics* 38: 2431–2448.

Kempe, S., Kazmierczak, J., Landmann, G. et al. (1991). Largest known microbialites discovered in Lake Van, Turkey. *Nature* 349: 605–608.

Khang, P. (1985). The development of karst landscapes in Vietnam. *Acta Geologica Polonica* 35: 305–319.

King, C.A.M. (1963). Some problems concerning marine planation and the formation of erosion surfaces. *Transactions and Papers (Institute of British Geographers)* 33: 29–43.

Klimchouk, A. (2006). Cave un-roofing as a large-scale geomorphic process. *Speleogenesis and Evolution of Karst Aquifers* 4: 1–11.

Koşun, E. (2012). Facies characteristics and depositional environments of Quaternary tufa deposits, Antalya, SW Turkey. *Carbonates and Evaporites* 27: 269–289.

Lehmann, H. (1936). Morphologische Studien auf Java. *Geographische Abhandlungen*, Serie 3 (9): 1–114.

Lehmann, H. (1959). Studien über Poljen in den venezianischen Voralpen und im Hochapennin. *Erdkunde* 13: 249–289.

Lehmann, H., Krömmelbein, K., and Lötschert, W. (1956). Karstmorphologische, geologische und botanische Studien in der Sierra de los Organos auf Cuba. *Erdkunde* 10: 185–204.

Lepirica, A. (2015). Genesis of inner gorges in the Rakitnica Canyon valley (Central Dinarides, Bosnia and Herzegovina). *Zeitschrift für Geomorphologie* 59: 515–545.

Lhenaff, R. (1968). Le poljé de Zafarraya (Province de Granade). *Mélanges de la Casa de Velazquez* 4: 5–25.

Lhenaff, R. (1986). Les grands poljés des Cordillères Bétiques andalouses et leurs rapports avec l'organisation endokarstique. *Karstologia Mémoires* 1: 101–112.

Liang, F. and Xu, B. (2014). Discrimination of tower-, cockpit-, and non-karst landforms in Guilin, Southern China, based on morphometric characteristics. *Geomorphology* 204: 42–48.

Linares, R., Rosell, J., Roqué, C. et al. (2009). Origin and evolution of tufa mounds related to artesian karstic springs in Isona area (Pyrenees, NE Spain). *Geodinamica Acta* 23: 129–150.

Lipar, M. and Ferk, M. (2015). Karst pocket valleys and their implications on Pliocene-Quaternary hydrology and climate: examples from the Nullarbor Plain, southern Australia. *Earth-Science Reviews* 150: 1–13.

López-Chicano, M. and Pulido-Bosch, A. (1993). The Sierra Gorda karstic aquifer (Granada and Málaga). In: *Some Spanish Karstic Aquifers* (ed. A. Pulido-Bosch), 85–93. Granada: Universidad de Granada.

López-Chicano, M., Calvache, M.L., Martín-Rosales, W. et al. (2002). Conditioning factors in flooding of karstic poljes. The case of the Zafarraya Polje (South Spain). *Catena* 49: 331–352.

Lu, G., Zheng, C., Donahoe, R.J. et al. (2000). Controlling processes in a $CaCO_3$ precipitating stream in Huanglong natural scenic district, Sichuan, China. *Journal of Hydrology* 230: 34–54.

Martins, A.F. (1950). Aspectos do relevo calcáiro em Portugal: os poljes de Minde e de Alvados. *Cadernos de Geografia* 1: 25–33.

McDonald, R.C. (1979). Tower karst geomorphology in Belize. *Zeitschrift für Geomorphologie* Supplement-Band 32: 35–45.

McDonald, R.C. and Ley, R.G. (1985). Tower karst geomorphology in northen Borneo. *Zeitschrift für Geomorphologie* 29: 483–495.

Migoń, P. (2004). Planation surface. In: *Encyclopedia of Geomorphology* (ed. A.S. Goudie), 788–792. London: Routledge.

Mijatović, B.F. (1984). Karst poljes in the Dinarides. In: *Hydrogeology of the Dinaric Karst* (ed. B.F. Mijatović), 87–109. Hannover: International Association of Hydrogeologists.

Milanović, P.T. (1981). *Karst Hydrogeology*. Littleton, CO: Water Resources Publications.

Milanović, P.T. (1986). Influence of karst spring submergence on the karst aquifer regime. *Journal of Hydrology* 84: 141–156.

Milanović, P.T. (2002). The environmental impacts of human activities and engineering constructions in karst regions. *Episodes* 25: 13–21.

Milanović, P.T. (2004a). Dinaride poljes. In: *Encyclopedia of Caves and Karst Science* (ed. J. Gunn), 291–293. New York: Fitzroy Dearborn.

Milanović, P.T. (2018). *Engineering Karstology of Dams and Reservoirs*. Boca Raton: CRC Press.

Mocochain, L., Audra, P., Clauzon, G. et al. (2009). The effect of river dynamics induced by the Messinian Salinity Crisis on karst landscape and caves: example of the Lower Ardèche river (mid Rhône valley). *Geomorphology* 106 (1-2): 46–61.

Mohammadi, Z., Mahdavikia, H., Raeisi, E. et al. (2019). Hydrogeological characterization of Dasht-e-Arjan Lake (Zagros Mountains, Iran): clarifying a long-time question. *Environmental Earth Sciences* 78: 1–14.

Monroe, W.H. (1976). *The Karst Landforms of Puerto Rico*. U.S. Geological Survey Professional Paper 899.

Moseley, G.E., Richards, D.A., Smart, P.L. et al. (2015). Early-middle Holocene relative sea-level oscillation events recorded in a submerged speleothem from the Yucatán Peninsula, Mexico. *The Holocene* 25 (9): 1511–1521.

Moulin, A., Benedetti, L., Gosar, A. et al. (2014). Determining the present-day kinematics of the Idrija fault (Slovenia) from airborne LiDAR topography. *Tectonophysics* 628: 188–205.

Nassery, H.R., Alijani, F., and Mirzaei, L. (2009). Environmental characterization of a karst polje: an example from Izeh polje, southwest Iran. *Environmental Earth Sciences* 59: 99–108.

Nicod, J. (1967). Recherches morphologiques en Basse-Provence calcaire. *Études et Travaux de Mediterranée* 5: 1–557.

Nicod, J. (1996). Le poljé de Minde (Portugal central) type de poljé tectonique. *Revue d'Analyse Spatiale Quantitative et Appliquée* 38 (39): 143–151.

Ollier, C. and Pain, C. (2000). *The Origin of Mountains*. London: Routledge.

Ordóñez, S., Martín, J.G., Del Cura, M.G. et al. (2005). Temperate and semi-arid tufas in the Pleistocene to Recent fluvial barrage system in the Mediterranean area: the Ruidera Lakes Natural Park (Central Spain). *Geomorphology* 69: 332–350.

Panoš, V. and Štelcl, O. (1968). Physiographic and geologic control in development of Cuban mogotes. *Zetischrif für Geomorphologie* 12: 117–173.

Pedley, H.M. (1990). Classification and environmental models of cool freshwater tufas. *Sedimentary Geology* 68: 143–154.

Pedley, M., Andrews, J., Ordonez, S. et al. (1996). Does climate control the morphological fabric of freshwater carbonates? A comparative study of Holocene barrage tufas from Spain and Britain. *Palaeogeography, Palaeoclimatology, Palaeoecology* 121: 239–257.

Pellicer, X.M., Linares, R., Gutiérrez, F. et al. (2014). Morpho-stratigraphic characterization of a tufa mound complex in the Spanish Pyrenees using ground penetrating radar and trenching, implications for studies in Mars. *Earth and Planetary Science Letters* 388: 197–210.

Pellicer, X.M., Corella, J.P., Gutiérrez, F. et al. (2016). Sedimentological and palaeohydrological characterization of Late Pleistocene and Holocene tufa mound palaeolakes using trenching methods in the Spanish Pyrenees. *Sedimentology* 63: 1786–1819.

Peña-Monné, J.L., Jiménez-Sánchez, A., and Echeverría-Arnedo, M.T. (1989). Geomorphological cartography and evolutionary aspects of the Sierra de Albarracín poljes (Eastern Iberian Ranges, Teruel, Spain). *Geografia Fisica e Dinamica Quaternaria* 12: 51–57.

Penck, A. (1900). Geomorphologische Studien aus der Hercegovina. *Zeitschrift des Deutschen und Oesterreichischen Alpenvereins* 31: 25–41.

Pentecost, A. (2005). *Travertines*. Berlin: Springer.

Pentecost, A. and Lord, T. (1988). Postglacial tufas and travertines from the Craven district of Yorkshire. *Cave Science* 15: 15–19.

Perica, D., Bognar, A., and Lozić, S. (2002). Geomorphological features of the Baške Oštarije karst polje. *Geoadria* 7: 23–34.

Pfeffer, K.H. (1986). Queen of Spains Valley, Maroon Town, Jamaica: a cross-section of different types of tropical karst. In: *New Directions in Karst* (ed. K. Paterson and M.M. Sweeting), 349–362. Norwich: Geobooks.

Phillips, J.D. (2017). Landform transitions in a fluviokarst landscape. *Zeitschrift für Geomorphologie* 61: 109–122.

Phillips, J.D., Martin, L.L., Nordberg, V.G. et al. (2004). Divergent evolution in fluviokarst landscapes of central Kentucky. *Earth Surface Processes and Landforms* 29: 799–819.

Pisano, L., Zumpano, V., Liso, I.S. et al. (2020). Geomorphological and structural characterization of the Canale di Pirro polje, Apulia (Southern Italy). *Journal of Maps* 16: 479–487.

Purdy, E.G. and Winterer, E.L. (2001). Origin of atoll lagoons. *Geological Society of America Bulletin* 113: 837–854.

Reicherter, K.R., Jabaloy, A., Galindo-Zaldívar, J. et al. (2003). Repeated palaeoseismic activity of the Ventas de Zafarraya fault (S Spain) and its relation with the 1884 Andalusian earthquake. *International Journal of Earth Sciences* 92: 912–922.

Ricketts, J.W., Ma, L., Wagler, A.E. et al. (2019). Global travertine deposition modulated by oscillations in climate. *Journal of Quaternary Science* 34: 558–568.

Ristić, D.M. (1976). Water regime of flooded karst poljes. In: *Karst Hydrology and Water Resources* (ed. V. Yerjevich), 301–318. Fort Collins: Water Resources Publications.

Roglić, J. (1957). Quelques problems fondamentaux du karst. *Information Géographique* 21: 1–12.

Roglić, J. (1972). Historical review of morphologic concepts. In: *Important Karst Regions of the Northern Hemisphere* (ed. M. Herak and V.T. Stringfield), 1–18. Amsterdam: Elsevier.

Roglić, J. (1974b). Les caràcteres specifiques du karst dinarique. In: *Phénomenes Karstiques*, Mémoires et Documents 15, 269–278. Paris: Éditions du Centre national de la recherche scientifique.

Sancho, C., Arenas, C., Vázquez-Urbez, M. et al. (2015). Climatic implications of the Quaternary fluvial tufa record in the NE Iberian Peninsula over the last 500 ka. *Quaternary Research* 84: 398–414.

Sanz de Galdeano, C.S. (2013). The Zafarraya Polje (Betic Cordillera, Granada, Spain), a basin open by lateral displacement and bending. *Journal of Geodynamics* 64: 62–70.

Seguin, J., Bintliff, J.L., Grootes, P.M. et al. (2019). 2500 years of anthropogenic and climatic landscape transformation in the Stymphalia polje, Greece. *Quaternary Science Reviews* 213: 133–154.

Şimşek, M., Öztürk, M.Z., Doğan, U. et al. (2021). Morphometric properties of poljes in the Taurus Mountains, Southern Turkey. *Journal of Geography* 42: 101–119.

Smart, P.L., Waltham, T., Yang, M. et al. (1986). Karst geomorphology of western Guizhou, China. *Cave Science* 13 (3): 89–103.

Stepišnik, U., Ferk, M., Gostinčar, P. et al. (2012). Holocene high floods on the Planina polje, classical dinaric karst, Slovenia. *Acta Carsologica* 41 (1): 5–13.

Surić, M. (2002). Submarine karst of Croatia. Evidence of former lower sea levels. *Acta Carsologica* 31: 89–98.

Surić, M., Juračić, M., Horvatinčić, N. et al. (2005). Late Pleistocene-Holocene sea-level rise and the pattern of coastal karst inundation: records from submerged speleothems along the Eastern Adriatic Coast (Croatia). *Marine Geology* 214: 163–175.

Surić, M., Lončarić, R., and Lončar, N. (2010). Submerged caves of Croatia: distribution, classification and origin. *Environmental Earth Sciences* 61: 1473–1480.

Sweeting, M.M. (1958). The karstlands of Jamaica. *Geographical Journal* 124: 184–199.

Sweeting, M.M. (1972). *Karst Landforms*. London: McMillan Press.

Sweeting, M.M. (1995). *Karst in China. Its Geomorphology and Environment*. Berlin: Springer.

Tang, T. (2002). Surface sediment characteristics and tower karst dissolution, Guilin, southern China. *Geomorphology* 49: 231–254.

Tang, T. and Day, M.J. (2000). Field survey and analysis of hillslopes on tower karst in Guilin, Southern China. *Earth Surface Processes and Landforms* 25: 1221–1235.

Taviani, M., Angeletti, L., Campiani, E. et al. (2012). Drowned karst landscape offshore the Apulian margin (southern Adriatic Sea, Italy). *Journal of Cave and Karst Studies* 74: 197–212.

Taylor, M.P., Drysdale, R.N., and Carthew, K.D. (2004). The formation and environmental significance of calcite rafts in tropical tufa-depositing rivers of northern Australia. *Sedimentology* 51: 1089–1101.

Tuccimei, P., Soligo, M., Ginés, J. et al. (2010). Constraining Holocene sea levels using U-Th ages of phreatic overgrowths on speleothems from coastal caves in Mallorca (Western Mediterranean). *Earth Surface Processes and Landforms* 35: 782–790.

Viles, H.A. (2004d). Tufa and travertine. In: *Encyclopedia of Geomorphology*, vol. 2 (ed. A.S. Goudie), 595–596. Routledge/Taylor & Francis: London-New York.

Viles, H.A. and Pentecost, A. (2007). Tufa and travertine. In: *Geological Sediments and Landscapes* (ed. D. Nash and S. McLaren), 173–199. Oxford: Blackwell.

Vott, A., Bruckner, H., Zander, A.M. et al. (2009). Late Quaternary evolution of Mediterranean poljes-the Vatos case study (Akarnania, NW Greece) based on geo-scientific core analyses and IRSL dating. *Zeitschrift fur Geomorphologie* 53: 145–169.

Vrabec, M. (1994). Some thoughts on the pull-apart origin of karst poljes along the Idrija strike-slip fault zone in Slovenia. *Acta Carsologica* 23: 155–167.

Waelbroeck, C., Labeyrie, L., Michel, E. et al. (2002). Sea-level and deep water temperature changes derived from benthic foraminifera isotopic records. *Quaternary Science Reviews* 21: 295–305.

Waltham, T. (1981). The karstic evolution of the Matienzo depression, Spain. *Zeitschrift für Geomorphologie* 25: 300–312.

Waltham, T. (2000). Karst and Caves of Ha Long Bay. *International Caver* 2000: 24–31.

Waltham, T. (2004). Sewu cone karst, Java. In: *Encyclopedia of Caves and Karst Science* (ed. J. Gunn), 641–642. New York: Fitzroy Dearborn.

Waltham, T. (2008b). Fengcong, fenglin, cone karst and tower karst. *Cave and Karst Science* 35: 77–88.

Waltham, A.C. and Hamilton-Smith, E. (2004). Ha Long Bay, Vietnam. In: *Encyclopedia of Caves and Karst Science*, vol. 413 (ed. J. Gunn). New York: Fitzroy Dearborn.

Waltham, T. and Middleton, J. (2000). The Khammouan karst of Laos. *Cave and Karst Science* 27: 113–120.

Wang, X. (1986). U-series age and oxygen carbon isotopic features of speleothems in Guilin. In: *Proceedings of the 9th International Congress of Speleology*, vol. 1 (ed. Comisión Organizadora del IX Congreso Internacional de Espeleologia), 284–286. Barcelona: IBYNSA.

Williams, P.W. (1987). Geomorphic inheritance and the development of tower karst. *Earth Surface Processes and Landforms* 12: 453–465.

Williams, P.W. (2011). Karst in UNESCO World Heritage Sites. In: *Karst Management* (ed. P.E. van Beynen), 459–480. Dordrecht: Springer.

Williams, P.W., Lyons, R.G., Wang, X. et al. (1986). Interpretation of the paleomagnetism of cave sediments from a karst tower at Guilin. *Carsologica Sinica* 6: 119–125.

Yoshimura, K., Liu, Z., Cao, J. et al. (2004). Deep source CO_2 in natural waters and its role in extensive tufa deposition in the Huanglong Ravines, Sichuan, China. *Chemical Geology* 205: 141–153.

Žebre, M., Stepišnik, U., Colucci, R.R. et al. (2016). Evolution of a karst polje influenced by glaciation: the Gomance piedmont polje (northern Dinaric Alps). *Geomorphology* 257: 143–154.

Zhu, X. (1988). *Guilin Karst*. Shanghai: Shanghai Scientific & Technical Publishers.

Zhu, X., Zhu, D., Zhang, Y. et al. (2013). Tower karst and cone karst. In: *Treatise of Geomorphology. Karst Geomorphology*, vol. 6 (ed. A. Frumkin), 327–340. San Diego: Academic Press.

8

Special Features Associated with Evaporites

Evaporite terrains can display many of the landforms typical of carbonate karst. Bare and covered karren can be widespread, but show less variety. Sinkholes are commonly the most frequent surface karst landform and show higher genetic diversity than in carbonate karsts, including depressions related to sagging subsidence. Poljes, corrosion plains and residual hills are rare in gypsum karst and barely represented in salt bedrock. Nonetheless, other landforms are endemic to evaporite karsts, especially those generated by subsidence related to subjacent dissolution in interstratal and covered karst settings. Subsidence induced by interstratal dissolution of evaporites produces the largest karst landforms on Earth, including fold scarps and belts of depressions associated with dissolution fronts that can reach hundreds of kilometers in length. The special geomorphic features associated with evaporites can be attributed to the distinctive properties of gypsum and salt, mainly halite (Frumkin 2013; Gutiérrez and Cooper 2013): (i) Evaporite minerals are highly soluble and unlike carbonate minerals, the dissolution process does not depend on the acidity of the solution. The equilibrium solubilities of gypsum and halite in pure water are 2.59 and $424 \, g \, L^{-1}$, respectively (mass of solute in $1 \, L$ of solution). These values are around 8000 and 50 times higher than the solubility of calcite in pure water with atmospheric CO_2 partial pressure. This difference can be reduced by several times if the water has percolated through soils with high CO_2 content, increasing its aggressiveness with respect to carbonate minerals (Table 3.1). (ii) Halite and gypsum, under favorable hydrodynamic and hydrochemical conditions, can dissolve very rapidly. Consequently, solutional features in evaporites can develop over much shorter time spans than in carbonate rocks. Halite is characterized by transport-controlled dissolution kinetics; dissolution rate shows a linear relationship with the saturation state of the solution. Gypsum has a mixed behavior, but the change from transport- to surface-controlled kinetics occurs close to equilibrium, when the solution has lost most of its dissolution capability (see Section 3.8.2). Data on the contrasting solutional denudation rates measured in carbonate and evaporite rocks under different conditions are presented in Sections 4.3 and 4.4. (iii) Evaporites can be eroded by mechanical processes more rapidly than the much harder carbonate rocks. Moreover, dissolution and reprecipitation acting on exposures of gypsum and salt can rapidly produce an easily erodible regolith. Additionally, detrital particles derived from the mechanical erosion of evaporites have a very limited persistence in running water. (iv) Some important implications of the geomorphic effectiveness of solutional and mechanical denudation processes acting on evaporites are that: (a) landforms can form and evolve very rapidly, (b) they can experience rapid adjustments in response to changes in the environmental and base level conditions (i.e., short response time), and (c) they often have a limited persistence time, and thus relict geomorphic features are less frequent than in carbonate rocks (i.e., preservation potential). (v) Well-developed karsts can occur in evaporites

Karst Hydrogeology, Geomorphology and Caves, First Edition. Jo De Waele and Francisco Gutiérrez.
© 2022 John Wiley & Sons Ltd. Published 2022 by John Wiley & Sons Ltd.

with lower purity than those formed in carbonate rocks. This results in higher production of clayey residual material that can contribute to reduce the rock surface exposed to dissolution and the overall permeability of the rock mass by clogging fissures and conduits. (vi) The mechanical properties of evaporites are markedly different to those of carbonate rocks. Gypsum has lower mechanical strength, more ductile rheology and spacing between fractures is generally larger than in carbonate rocks. Moreover, dissolution acting along discontinuity planes in gypsum formations can lead to a rapid reduction in the rock mass strength. Salt, due to its mechanical weakness and low yield strength, readily flows by viscous deformation both in the subsurface and at the surface. Salt formations rarely display pervasive joint systems, which are typically the main guiding features for the development of conduit permeability. Moreover, creep deformation of salt, especially under high pressure conditions, anneals discontinuities and voids. (vii) Many evaporitic formations are expressed at and near the surface as successions dominated by gypsum. However, the lithological composition of the non-weathered portions of the formation can be markedly different. Gypsum can grade laterally and vertically into less soluble anhydrite and/or more soluble glauberite in zones with limited groundwater circulation. Moreover, the evaporitic formations may include significant salt units that have been removed by dissolution near the surface. The presence of these salts is commonly unappreciated because of the lack of detailed data on the subsurface stratigraphy. A number of salt karst features may have been erroneously ascribed to gypsum dissolution. (viii) Dissolution in carbonate rocks mostly occurs differentially along groundwater flow paths controlled by discontinuities (i.e., solutional widening of fractures and bedding planes). In contrast, evaporites can experience complete dissolution by the migration of dissolution fronts acting at the edge of an evaporitic body, like a melting block of ice. This is a spatially continuous dissolution style that involves the progressive lateral and/or vertical depletion of the evaporitic formation and the gravitational deformation of the overlying material. It can be considered as a type of contact karst. Because of the high solubility of the evaporites, especially salt, groundwater systems with limited flow rates can remove large volumes of rock in solution if they operate during sufficiently long time spans.

Salt rocks are rarely exposed at the surface because of their extremely high solubility. The limited number of salt outcrops that occur on Earth are essentially associated with actively rising diapiric structures and hot or cold dry climates. Two conditions are met at sites where mobile salt is directly exposed at the surface: (i) the upward salt flow counterbalances or exceeds erosional lowering, mainly related to dissolution; and (ii) erosion removes the non-soluble residual material (capsoils and caprocks) generated by salt dissolution. In many salt structures with positive relief, the salt is overlain by a thick caprock and salt exposures are limited (e.g., Mount Sedom, Israel; Frumkin 1994a) or absent (Calaforra and Pulido-Bosch 1999a; Autin 2002). Some of the most important salt karsts associated with halokinetic structures documented in the literature are: (i) the numerous salt extrusions and namakiers (salt glaciers) of the Hormuz Formation (Bruthans et al. 2010, 2017; Zarei and Raeisi 2010; Zarei et al. 2012; Abirifard et al. 2017), and the Anbal salt pillow of the Gachsaran Formation (Gutiérrez and Lizaga 2016) in the Zagros Mountains, Iran, (ii) Mount Sedom diapir in the Dead Sea Basin, Israel (Frumkin and Ford 1995; Frumkin 1994b, 1998, 2004), (iii) the Cordillera de la Sal salt anticline, Chile (De Waele et al. 2009b, 2020; Pisani and De Waele 2021), (iv) salt diapirs in the Red Sea coast of Saudi Arabia and Yemen (Erol 1989; Davison et al. 1996; Almalki et al. 2015), (v) a number of salt structures developed in Permian and Triasssic salts in Europe (Sirocko et al. 2002; Gutiérrez et al. 2008b; Dahm et al. 2011; Guerrero 2017), (vi) the Cardona salt diapir in NE Spain (Lucha et al. 2008a), and (vii) salt domes in the Gulf of Mexico (Autin 2002), and a few salt domes in Tajikistan (Dzens-Litovsky 1966). There are other areas with significant salt exposures that most probably support interesting karst features, but apparently remain unexplored such as the Great Kavir in Iran (Warren 2008; Abdolmaleki et al. 2014), the Algerian Atlas (Ran el Melah diapir;

Ville 1859), the Kuqua Basin in China, or some Arctic islands in Canada. Regarding interstratal evaporite karst, some of the most remarkable geomorphic manifestations have been documented in Permian evaporites in the southwest of the USA (Hill 1996; Johnson et al. 2021), the Carboniferous Paradox Formation in Utah and Colorado (Cater 1970; Doelling 2000; Gutiérrez 2004), the Carboniferous Eagle Valley Formation in the Rocky Mountains of Colorado (Kirkham et al. 2002; Gutiérrez et al. 2014b), the Devonian Prairie Formation in Canada (De Mille et al. 1964; Broughton 2017), a number of Mesozoic Ca-sulfate formations in the Arabian Platform, especially the late Jurassic anhydrite Hith Formation (Powers et al. 1966; Youssef et al. 2016), the Cretaceous Maha Sarakhan Salt in Thailand and Laos (Warren 2016), or evaporites of different ages in the Teruel and Calatayud Neogene grabens in the Iberian Chain, Spain (Gutiérrez 1996; Gutiérrez et al. 2012a).

Gypsum outcrops are common but are best expressed in areas with low precipitation. In dry areas in which the strata have a subhorizontal attitude, gypsum units often underlie extensive mesas and plains (e.g., the Gypsum Plain in the Delaware Basin; Figure 4.10). Klimchouk et al. (1996b) edited a special issue on gypsum karst, which includes a number of review articles devoted to specific countries or regions. Some of the most important gypsum karsts documented in international publications include: (i) extensive outcrops of sub-horizontally lying gypsum formations in the Permian basins of the southwest of the USA, notably the Delaware Basin (Quinlan et al. 1986; Johnson 1996; Johnson and Neal 2003; Stafford et al. 2008a, 2008c), (ii) several areas of Canada around the Precambrian shield and especially south of the limit of continuous permafrost (Ford 1997), (iii) a belt of Upper Permian gypsum in NE England (Cooper 1996, 2020), (iv) several Cenozoic formations in Spain, remarkably the continental evaporite formations of the Ebro Basin (Gutiérrez et al. 2008b), the epigene and hypogene karsts developed in Eocene marine evaporites in the eastern Spanish Pyrenees (Gutiérrez et al. 2016, 2019b), and the marine Messinian gypsum in the Sorbas Basin (Pulido-Bosch 1986; Calaforra and Pulido-Bosch 2003), (v) the gypsum karst developed in an Eocene formation in the Paris Basin and in scattered outcrops of Triassic evaporites in the northern Alps and Provence, France (Toulemont 1984; Chardon and Nicod 1996; Thierry et al. 2009), (vi) the karst systems developed in Messinian gypsum in Emilia-Romagna and Sicily, Italy (Forti and Sauro 1996; De Waele et al. 2017b), (vii) Upper Permian and Triassic gypsum successions in the Harz Mountains, Germany (Reuter and Stoyan 1993; Kempe 1996), (viii) the Baltic Republics (Paukstys and Narbutas 1996), (ix) the paradigmatic hypogene interstratal karst developed in a Miocene gypsum formation up to 40 m thick in western Ukraine (Klimchouk 1996b), (x) the Oligocene Hafik Gypsum in the Sivas Basin of Turkey (Gökkaya et al. 2021 and references therein), (xi) the platform area around the shield of Saudi Arabia (Youssef et al. 2016 and references therein), (xii) several regions close to Mosul in Iraq (Jassim et al. 1997), (xiii) numerous zones of the Zagros Mountains in Iran with exposures of the widely distributed Miocene Gachsaran Formation (Torabi-Kaveh et al. 2012; Aghdam et al. 2013; Raeisi et al. 2013), and (xiiii) belts of gypsum formations in platform and foredeep structural settings along the western pre-Ural region, Russia (Andrejchuk 1996). Below we describe the distinctive geomorphic features related to evaporite dissolution, grouped according to the setting in which they develop, differentiating: interstratal karst, covered karst, bare karst, and the peculiar situation of salt extrusions, in which dissolution affects a mobile bedrock.

8.1 Interstratal Evaporite Karst

Interstratal karst refers to dissolution acting on a soluble formation overlain by indurated rocks (caprock). The caprocks in interstratal evaporite karsts can include lower-solubility carbonate formations. Dissolution of evaporites, because of their frequent lack of internal permeability, often occurs through the lateral and/or vertical migration of dissolution fronts. The edges in contact with undersaturated groundwater flows retreat inwards and the evaporitic body shrinks like a melting block of ice. The complete or partial dissolution of the evaporites leaves behind: (i) residues consisting of the insoluble (e.g., clays, marls) or less soluble (e.g., anhydrite, gypsum) components of the formation and (ii) solution-collapse breccias related to the fragmentation and settling of overlying and interbedded strata. Interstratal dissolution by migrating solution fronts involves a reduction in the thickness of the stratigraphic succession and the subsidence of the overlying rocks and the ground surface (Figure 8.1a). For instance, complete dissolution of 100 m of the Permian Hartlepool Anhydrite (England), with 0.5–0.8% of insolubles, would produce an interval of solution residues just 50–80 cm thick and a downward displacement of around 100 m in the overlying stratigraphic units (Smith 1972). Given sufficient time and adequate hydrogeological and hydrochemical conditions, evaporite formations, and especially salts, eventually vanish by dissolution. According to Warren (2016), there are probably more intervals of solution residues in the stratigraphic record than remaining salt beds.

Dissolution fronts are commonly associated with the shallow edge of the evaporite formation (e.g., updip edge of dipping units, diapir crest) that is flushed away by meteoric waters in the telogenetic zone. In a dipping stratigraphic succession, the dissolution front can migrate downdip from the feather edge of the evaporite unit. The progressive mass depletion by solution produces a thin dissolution residue overlain by foundered strata that grade at depth into the non-dissolved evaporite unit (Figure 8.1d). Interstratal karst may be also related to deep-seated meteoric or basinal flow systems that can cause hypogene dissolution from below, often controlled by permeability features in the substrata such as fractures or reefs. The spatial and temporal distribution patterns of interstratal dissolution and the associated subsidence phenomena may be controlled by multiple factors, including (Gutiérrez and Cooper 2013): (i) changes in thickness, solubility (e.g., gypsum/anhydrite transition) and depth of the evaporite formation, (ii) permeability of overlying, underlying and interbedded strata (hydrostratigraphy), (iii) configuration of the groundwater flow systems and their hydrochemical features, (iv) geological structure and tectonic activity, (v) geomorphological evolution of the region, base level changes and spatial distribution of recharge and discharge areas, (vi) climate variability, and (vii) hydrogeological changes and deformation induced by ice sheets (Anderson and Hinds 1997).

---

the development of sag basins with centripetal dips in the supra-evaporite strata. The horizontal shortening associated with the ductile bending process tends to be counterbalanced by extensional structures at the margins of the basin. Synsedimentary subsidence is recorded by upward-dip attenuation and cumulative wedge-outs (growth strata) in the basin fill. (c) Basin and dome topography and structure produced by differential evaporite dissolution and subsidence. (d) Monoclinal scarp and associated synclinal trough generated by the downdip migration of a dissolution front within a dipping stratigraphic succession (homocline). Downward flexure of the caprock atop the dissolution front (salt slope) results in local dip reversal associated with the up-dip facing scarp. The dissolution trough and its long-term lateral migration can control the position and shift of morpho-sedimentary systems. (e) Caprock affected by normal faults, horsts and grabens with geomorphic expression resulting from the dissolution of the underlying evaporites. Lateral spreading and subsidence in the caprock blocks can be also induced by lateral flow of salt toward the erosionally unconfined slope (erosional unloading). (f) Crestal graben depression developed along the hinge zone of a salt-cored anticline due to interstratal dissolution in the shallower part of the evaporitic core. (g) Flexural-slip faults generated by evaporite dissolution at the lower part of a steeply dipping stratigraphic succession and valley-ward toppling in the overlying and adjacent caprock. The bedding-parallel faults rupture an unconformable unit generating faults scarps and small half-grabens with geomorphic expression. (h) Ridge and trough topography produced by dissolution acting on evaporite interbeds in a lower steeply dipping stratigraphic succession, and differential sagging in the overlying unconformable unit with an original subhorizontal attitude. (i) Transtratal breccia pipe developed by localized evaporite dissolution and progressive upward collapse (stoping) through the overlying rock formations. Eventually, the hidden breccia pipe ruptures the surface generating a caprock collapse sinkhole. Differential erosion at paleosinkhole sites may produce residual hills consisting of cemented breccia pipes or sinkhole deposits (relief inversion).

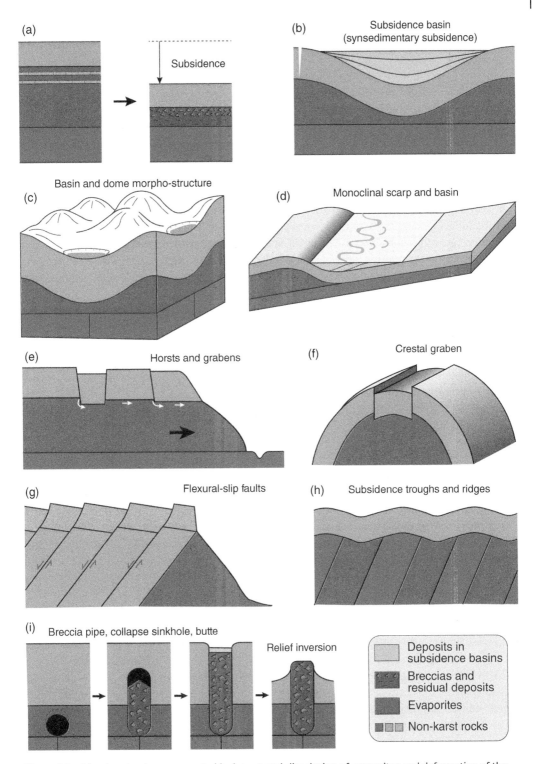

Figure 8.1 Morpho-structures generated by interstratal dissolution of evaporites and deformation of the overlying strata. Note that units underlying the evaporites are not affected by this non-tectonic deformation process (disharmonic structure). (a) Widespread interstratal dissolution produces insoluble residues and solution-collapse breccias, and results in the "condensation" of the stratigraphic succession and the subsidence of the ground surface. (b) Subsidence induced by differential interstratal dissolution can be accommodated by

Interstratal evaporite dissolution is frequently an unappreciated process due to its hidden character, large spatial–temporal scale, and relatively slow rates. Its investigation requires multidisciplinary studies addressing the multiple evidence produced by dissolution and gravitational deformation phenomena (Kirkham et al. 2002; Gutiérrez et al. 2014a): (i) The stratigraphic evidence includes the insoluble residues and solution-collapse breccias produced by the dissolution of the evaporites, the post-depositional thickness changes exhibited by the partially dissolved evaporite units, the sediments accumulated in dissolution-induced basins, and evaporite formations and other precipitates (e.g., cements, nodules) related to the recycling of the dissolved evaporites. For instance, the continental evaporites deposited in the Spanish Cenozoic basins are largely related to the recycling of Mesozoic marine evaporites that experienced substantial subsurface dissolution during the infill of the Cenozoic basins (Utrilla et al. 1992; Gutiérrez et al. 2008b). The age and geometrical features of the sediments filling subsidence basins can provide information on the timing of the dissolution and subsidence episodes. Sedimentary units showing cumulative wedge-outs and progressive dip attenuation (growth strata) record periods of synsedimentary subsidence (Figure 8.1b). Gaining information about the multiple types of stratigraphic evidence may require expensive deep subsurface investigations (e.g., boreholes, seismic profiles). (ii) Structural evidence is provided by a wide variety of ductile and brittle deformation structures developed in the overlying units as a result of subsurface mass depletion. These are non-tectonic gravitational structures restricted to the units situated above the dissolution zone (Figure 8.1). (iii) The geomorphic evidence corresponds to the multiple landforms produced by the subsidence process (e.g., sagging basins and troughs, fold and fault scarps, and grabens) and the effects of the latter on other interconnected geomorphic systems (e.g., disrupted drainages). These landforms may correspond to relict features related to past dissolution or to currently active subsidence morpho-structures. The latter can be investigated by geodetic techniques and may provide indirect data about the ongoing subsurface solution process (e.g., spatial distribution, rate, kinematic style). (iv) Hydrochemical evidence of interstratal dissolution is offered by the brines that have interacted with the evaporites and emerge at springs and output zones (e.g., the salt plains and springs in the Cimarrón and Red rivers of western Oklahoma; Johnson 1981). The flow rate and solute concentration can be used to estimate rates of subsurface dissolution (see Section 4.3.1). Subsidence induced by interstratal evaporite dissolution can generate a variety of morpho-structures largely depending on the spatial patterns of the dissolution process (e.g., widespread, localized), the original structure of the stratigraphic succession (e.g., horizontal, tilted, folded, and fractured), and the rheology of the caprock (e.g., brittle, ductile) (Figure 8.1). The cartographic scale of these structures with geomorphic expression can be of the order of tens or hundreds of kilometers, and strain rates are often significantly higher than those reported for tectonic structures due to the high solubility of the evaporites.

Widespread dissolution of a subhorizontal evaporite unit through the vertical migration of a dissolution front causes a relatively uniform regional subsidence at the surface (e.g., Palo Duro Basin, West Texas; Hovorka 2000) (Figure 8.1a). The insoluble residues and solution-collapse breccias left by the dissolution process are indicators of the former presence of the leached evaporites. Warren (2016) provides a thorough review on these pointers to vanished evaporites. The lack of basal support related to the subjacent dissolution causes the fracturing and foundering of overlying and interlayered beds. The brecciation process affects layers of lithified rocks with brittle rheology that may include less soluble evaporite units such as gypsum or anhydrite. These diagenetic chemical and mechanical processes produce roughly concordant stratiform breccias, also termed blanket breccias, tabular breccias, breccia layers, stratabound breccias, and interstratal breccias. The breccias can be monomict, oligomict, or less commonly polymict, depending on the nature of the rocks associated with the leached evaporites. The particles are mostly angular, but can display some rounding related to post-fragmentation etching. The space between the clasts can be partially or totally filled by fine-grained insoluble residues and cements. Some breccias show multiple phases of dissolution and brecciation, as revealed by geometrical relationships between fractures and interparticle fills. The terminology used for the classification of solution-collapse breccias is presented in Section 2.2.9.2 and Figure 2.19. The stratiform breccias commonly display a well-defined planar or undulating base draped by argillaceous residues, and an irregular and gradational upper boundary with collapse-related failure planes. The fabrics of the angular clasts typically show a gradation from crackle and mosaic packbreccias at the top, to chaotic floatbreccias and packbreccias toward the base (Stanton 1966; Loucks 1999, 2007; Warren 2016) (Figure 2.18e). Collapse breccias may reach hundreds of meters in thickness and can extend laterally across hundreds of kilometers (Ford 1997). Some of these formations are of special economic importance since they can function as prolific karst aquifers (Sánchez et al. 1999) and hydrocarbon reservoirs (Loucks 1999), and may host various types of ores (James and Choquette 1988; Bosák et al. 1989; Sasowsky et al. 2008).

Interstratal dissolution of bedded evaporites acting differentially at specific zones can induce sagging subsidence in the overlying strata and produce structural basins with centripetal/inward dips (Figure 8.1b). These basins can display a subcircular outline or can be elongated (subsidence troughs) reaching tens or hundreds of kilometers in length, respectively (Figure 8.2). Gravitational folding produces a disharmonic structure that solely affects the strata situated above the dissolution zone, including the caprock and evaporite rocks if solution occurs at the base or within the soluble formation. These karstic deformation structures, despite their large scale, should not be misinterpreted as tectonic features related to regional stress fields. The sagging process involves the inward horizontal contraction of the bent strata, which tends to be counterbalanced by extension in the marginal zone of the basin through the development of fissures, normal faults and minor grabens. These extensional structures, in turn, can facilitate downward water percolation and evaporite dissolution at depth (Figure 8.1b). The amount of contraction/extension has a direct relationship with the subsidence magnitude and an inverse relationship with the width of the sagging structure. Subsidence basins, especially those associated with active dissolution, can have geomorphic expression and may function as depositional basins. The sedimentary fill in some cases reaches hundreds of meters in thickness, providing a minimum estimate of the thickness of evaporites removed by subsurface dissolution. The sediments of the basin fill that are coeval with the subsidence phenomenon may be identified by supra-attenuated dips and progressive thickening toward the basin depocenter (cumulative wedge-outs or growth strata) (Figure 8.1b). In the Eastern Mediterranean, Bertoni and Cartwright (2005), using 3D seismic data, produced a detailed characterization of buried circular structures up to 2 km across related to deep-seated dissolution of Messinian (Late Miocene) tabular salt. These subsidence structures are essentially sags with centripetal dips that show small throw concentric collapse faults and growth stratal geometries. These authors, analyzing parameters such as the expansion index (thickness increase of stratigraphic units) and vertical relief (elevation difference of seismic horizons) in the largest structure (2 km across), infer that dissolution of 180 m of salt (0.188 km^3) occurred at depths greater than 250 m in the Pliocene, over a time period of 0.75–1 Myr. The topographic lows created by interstratal dissolution and subsidence can function as discharge zones for the evaporite-dissolving groundwater and may host saline lakes in which evaporite recycling takes place. Subsidence troughs can also control the path and aggradation/incision patterns of fluvial systems (e.g., Pecos River in the Delaware Basin; Lee 1923; Morgan 1941). According to Gustavson (1986), the development of a 200 km long section of the Canadian River valley in the Texas Panhandle is largely related to subsidence due to the downward and lateral migration of dissolution fronts in deep-seated Permian salts, rather than to erosional lowering. This interpretation was based on structure contour maps that reveal thickness changes in the salt formation related to dissolution and the deformation of the overlying strata.

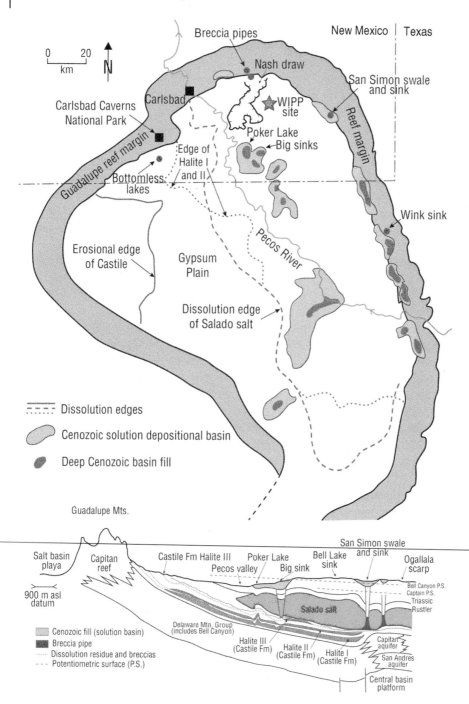

Figure 8.2 Simplified map and cross-section of the east-tilted Delaware Basin in Texas and New Mexico. The sketches show the location of the east-migrating dissolution fronts in the Permian Castile and Salado salt formations, and the distribution of two belts of dissolution-induced depositional basins, mainly located above the eastern section of the Capitan reef and along the Pecos River valley. This fluvial system functions as the regional base level and its position has experienced significant shifts in the Cenozoic. Nash Draw is a dissolution-induced depression with nested collapse structures situated in the vicinity of WIPP site (low-level radioactive waste repository). *Source:* Adapted from Anderson (1981), Hill (2003), and Warren (2016).

In the Delaware Basin, west Texas and southeast New Mexico, interstratal dissolution of the Late Permian Castile, Salado and Rustler formations have generated dissolution-induced depositional basins (subsidence troughs) up to 150 km long filled with late Cenozoic terrestrial sediments as much as 460 m thick (Olive 1957; Anderson et al. 1978; Bachman 1980; Johnson 1989; Hill 1996) (Figure 8.2). The Delaware Basin is an eastward tilted basin, with the Capitan Reef limestone (Guadalupe Mountains) exposed on the western margin. The evaporite units have a regional eastward dip and are affected by down-dip migrating dissolution fronts at their western updip edge. In the center of the basin, the evaporite formations are underlain by relatively permeable carbonates and detrital rocks, including the Capitan Reef limestone. These infra-salt formations include prolific hydrocarbon reservoirs and are the source of upward flows responsible for the hypogene dissolution of the evaporites. The dissolution-subsidence troughs in the Delaware Basin are arranged in two NNW-trending belts (Figure 8.2). The eastern one occurs atop the buried reef margin (e.g., San Simon Swale, the dissolution trough in the Wink Sink area). The western belt at the basin center is associated with the NNW-directed Pecos River valley, which functions as the regional base level and has experienced significant lateral shifts. San Simon Swale is a 260 km^2 topographic and depositional basin with internal drainage that hosts an old playa with nested active collapse sinkholes underlain by breccia pipes (e.g., San Simon Sink, reactivated in 1918) (Figure 8.3a). The inner part of the basin shows concentric fissures with clear geomorphic expression indicative of ongoing dissolution and subsidence (Holt and Powers 2010) (Figure 8.3b). Subsurface data in the dissolution trough located in the Wink Sink area indicate a sharp thinning of about 300 m in the Salado Formation atop the Capitan Reef, and an equivalent thickening in the overlying undifferentiated Cenozoic and Triassic strata (Johnson 1989) (Figure 8.2). Nash Draw, at the northern sector of the Delaware Basin, is a SW-trending depression around 25 km long and up to 15 km wide (Hill 2003) (Figure 8.2). Subsidence in this basin is ascribed to dissolution of halite in the Rustler Formation, around 400 m above the WIPP site. This is a low-level radioactive waste repository located a few kilometers to the east and encased in the deeper Salado Formation (Holt and Powers 2010).

Differential interstratal dissolution of salt in the Devonian Prairie Formation in southern Saskatchewan has generated depositional basins (structural lows) that interrupt the general SW-dipping homoclinal structure of the platform succession (De Mille et al. 1964). These are buried dissolution-induced basins covered by glacial drift and lacking geomorphic expression. The 380 km^2 Rosetown Low is a rectangular structural basin with steep margins, flat bottom, and around 115 m of structural relief. The missing salt has been replaced by solution-collapse breccias and insoluble residues. The 160 km long Regina-Hummingbird trough is a strike-parallel NW-SE-oriented structural basin with a highly irregular outline. Thickness variations in the different supra-evaporite units across the dissolution trough indicate that dissolution of the 100 m thick salt has occurred in multiple phases, starting from the Late Devonian. The location and development of these two dissolution-induced basins seems to be controlled by an underlying reef fringe. These rocks may have supplied aggressive water from below (hypogene dissolution) and may have determined differential compaction and fracturing in the overlying rocks (De Mille et al. 1964). The 45 km long structural basin of the Saskatoon Low is controlled by gravity faults and has a structural relief of 180 m, equivalent to the thickness of salt removed in solution (Christiansen 1967; Christiansen and Sauer 2001). In the Carbondale collapse center, Southern Rocky Mountains of Colorado, Kirkham et al. (2002) documented subsidence basins generated by dissolution of the salt-bearing Carboniferous Paradox Formation and filled with thick alluvium. The 15 km long Sopris Bowl is filled by more than 450 m of late Cenozoic detrital deposits. Depositional basins generated by interstratal evaporite dissolution have been also reported over Permian and Triassic evaporites in Germany (Pfeiffer and Hahn 1972; Reuter and Stoyan 1993). In the Teruel Neogene Graben, Iberian Chain, Spain, the Rio Seco syncline

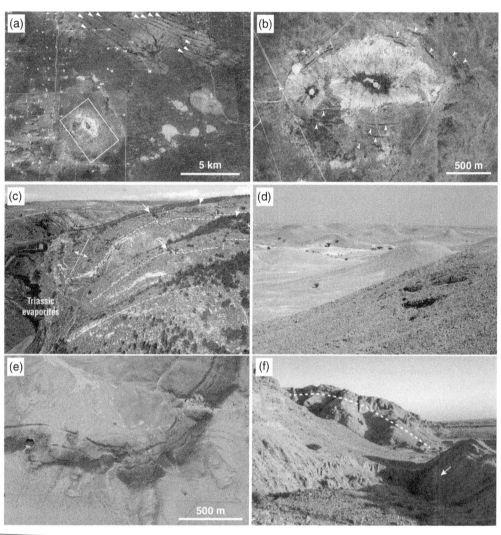

Figure 8.3 Images of subsidence morpho-structures generated by interstratal evaporite dissolution and sagging of the overlying caprock. (a) Partial view of San Simon Swale (Delaware Basin, New Mexico) with a subcircular sagging basin 6 km long (lower left) and a swarm of surface ruptures (arrowheads) expressed as scarps and sinkhole alignments attributable to deep-seated salt dissolution (upper part). Inset polygon indicates area shown in b. *Source:* Image from Google Earth. (b) Close view of the sag basin in San Simon Swale showing a concentric system of extensional fissures (arrowheads) and two sinkholes underlain by breccia pipes. The 70 m sinkhole on the right reactivated in 1918 evacuating underground a large volume of sediment. *Source:* Image from Google Earth. (c) Oblique aerial view of the Rio Seco synform with a crestal graben on the monocline that defines its southern edge. This structure is related to the interstratal dissolution of Triassic evaporites and the subsidence of the overlying Mio-Pliocene sediments of the Turolian stratotype (Teruel Neogene Graben, NE Spain). Arrows point to the edge of the crestal graben. *Source:* Francisco Gutiérrez (Author). (d) Basin and dome morpho-structure generated by differential interstratal dissolution of the Late Jurassic Ca-sulfates and subsidence of the overlying limestones of the Sulaiy Formation, Interior Homocline of Central Saudi Arabia. *Source:* Francisco Gutiérrez (Author). (e) Sinuous monoclinal scarp related to subsidence of caprock strata atop the down-dip migrating dissolution front in the Late Jurassic Hith Formation (Ca-sulfates), Interior Homocline of Central Saudi Arabia. Note caprock collapse sinkhole 70 m across and extensional ruptures (fissures, fault scarps) parallel to the trace of the drape fold that facilitates water percolation. *Source:* Image from Google Earth. (f) Monoclinal scarp associated with the dissolution front of the Hith Formation in the vicinity of Dahal Hith caprock collapse sinkhole, which is the type locality of the anhydrite Hith Formation. Arrow points to caprock collapse sinkhole. *Source:* Francisco Gutiérrez (Author).

is a 1.7 km long gravitational fold produced by interstratal dissolution of Triassic evaporites and the sagging of the overlying Mio-Pliocene units of the Turolian type section, originally with horizontal attitude (Figure 8.3c). The synform has a structural relief of 130 m and the crest of the monocline at one of the margins of the synform displays a graben with geomorphic expression. This secondary extensional structure counterbalances the shortening induced by bending in the inner part of the synform (Calvo et al. 1999; Gutiérrez et al. 2012a).

Differential interstratal dissolution of bedded evaporites and the uneven subsidence of the overlying caprock can produce complex morpho-structures consisting of basins and domes with a variable degree of spatial regularity. This spatially heterogeneous dissolution and subsidence phenomena can be controlled by permeability features such as fracture systems (Figure 8.1c). In the subsidence belt of the Interior Homocline of Central Saudi Arabia, there are extensive tracts of hummocky topography concordant with the underlying basin and dome structure related to the differential interstratal dissolution of calcium-sulfates in the Late Jurassic Hith and Arab formations (Saudi Geological Survey 2007; Memesh et al. 2008) (Figure 8.3d). Differential subsidence can be also related to multiple geometrical scenarios associated with deformed stratigraphic successions including evaporite and non-soluble units. For instance, Figure 8.1h shows a series of antiforms and synforms in an unconformable caprock expressed at the surface as ridges and troughs. These morpho-structures result from interstratal dissolution of evaporite units in an underlying steeply dipping succession with interbedded non-soluble rocks.

In gently dipping successions, the interstratal dissolution of the evaporites commonly occurs by the down-dip migration of dissolution fronts. The consequent gravitational bending of the overlying strata produces strike-parallel monoclinal folds atop the dissolution front (i.e., salt slope) and an adjoining asymmetric synform (dissolution trough) (Gutiérrez and Cooper 2013; Warren 2016) (Figure 8.1d). The supra-evaporite caprock in the monoclinal fold experiences local dip reversal and bedding-parallel extension that increases toward the outer arc of the drape fold. This local stretching is generally accommodated by the development of bedding-normal extensional fissures, normal faults and grabens that rupture the ground surface and may disrupt drainages (Figure 8.3e, f). These fractures compartmentalize the caprock into elongated rock panels that experience subsidence and outward rotation (toppling). Moreover, the over-steepening of both the slope and dip of the strata, together with the intense fracturing, favor the development of slope movements. The monoclinal folds are typically expressed as up-dip facing scarps with fold-parallel surface ruptures, and are often punctured by large caprock collapse sinkholes (Figure 8.3e). The adjacent dissolution trough can control the position and displacement of lake and fluvial systems, and may also function as a discharge zone for the brines derived from the dissolution of the evaporites (Figure 8.1d). These are exceptional mobile karstic folds that migrate laterally over geological time scales, in concert with the down-dip migration of the dissolution front. The dissolution fronts can develop in different evaporite units and dissolution tends to slow down as it propagates toward deeper levels (e.g., gypsum-anhydrite transition). The belts of supra-evaporite gravitational disruption with geomorphic expression attain their largest dimensions along the edge of great evaporite formations. They can reach more than 500 km in length, 250 km in width and hundreds of meters in structural relief (Warren 2016). Some authors indicate that the monoclinal folds associated with dissolution fronts can create good traps for hydrocarbons (Anderson et al. 1988; Anderson and Knapp 1993; Warren 2016). In the hyperarid Interior Homocline of Central Saudi Arabia, the down dip migration of dissolution fronts in the east-dipping anhydrite units of the Late Jurassic Hith and Arab formations has resulted in the subsidence of the overlying sediments. The belt in which the strata have been foundered reaches tens of kilometers in width and extends for more than 550 km (Powers et al. 1966). Here, the geomorphology and the shallow structure are primarily controlled by dissolution-induced subsidence (Memesh et al. 2008). The eastern edge of the subsidence belt is defined by a sinuous west-facing monoclinal escarpment developed by the downward flexure of caprock formations over the dissolution front in the Hith Formation (Figure 8.3e, f). This fold scarp shows abundant evidence of ground instability including caprock collapse sinkholes, gravitational fault scarps that locally disrupt drainages, and landslides. The subsidence belt contains internally drained depressions up to 55 km long that locally host playa-lakes with evaporite deposition (khabras). Ar Riyad, the capital of Saudi Arabia, lies within such a large dissolution-induced subsidence trough (Vaslet et al. 1991). Layla Lakes, 300 km south of Ar Riyad, are located in subsidence depressions that used to be fed by saline groundwater flows. The recent desiccation of the lakes related to groundwater pumping revealed that these are aligned collapse sinkholes with vertical walls covered by unique precipitates of recycled gypsum (Kempe and Dirks 2008). In Teesside, northern England, east-dipping halite and gypsum/anhydrite units of Permian age are bounded up-dip by dissolution fronts and west-facing subsidence monoclines developed on the overlying stratigraphic units (Cooper 2002, 2020). The monoclinal scarp over the gypsum dissolution front, controlled by the gypsum-anhydrite transition, forms a topographic step 20–40 m high equivalent to the thickness of the dissolved gypsum (Cooper 1998).

In western Canada, the west-dipping Middle Devonian Prairie Evaporite, around 200 m thick and dominated by halite, has been affected by multiple phases of interstratal dissolution along its eastern up-dip edge since the Late Devonian (De Mille et al. 1964; Anderson and Knapp 1993). This is probably the largest interstratal evaporite karst feature worldwide, with a dissolution and subsidence belt more than 1800 km long, and as much as 150 km wide (Broughton 2017, 2021). Seismic profiles and borehole data show that the development of subsidence troughs over the dissolution front controls the hydrographic network (e.g., North Saskatchewan River; Anderson and Hinds 1997; Athabasca River; Broughton 2013). Saline springs along the Athabasca River valley (Alberta) with TDS values as high as $100 \, \mathrm{g \, L^{-1}}$ provide hydrochemical evidence for the active salt dissolution by meteoric waters that is currently affecting the Prairie salt, as revealed by the chemical characteristics and isotopic signatures of the brines (Birks et al. 2018). The current hydrogeochemical model is used as an analogue to better understand previous major phases of dissolution-induced subsidence, such as the one that controlled the deposition of the Early Cretaceous oil sands of the McMurray Formation during the Cordilleran tectonism (Broughton 2021). This formation records synsedimentary dissolution-induced subsidence and displays striking sagging and collapse paleosinkholes (Broughton 2013, 2017).

In Kansas, a 160 km long dissolution front occurs along the edge of the SW-dipping Permian Hutchinson Salt, of Permian age and 50–90 m thick. The salt edge has receded around 25 km, locally leaving remnants of salt underlying the subsidence belt due to incomplete dissolution. The migration of the dissolution front has controlled the position of fluvial systems and the associated sedimentation during the Neogene and Quaternary (Anderson et al. 1994). Walters (1978) estimated an average rate of 3.2 km $\mathrm{Myr^{-1}}$ for the horizontal retreat of the dissolution front since the late Pleistocene. In the Holbrook Basin of Arizona, the NW-dipping Permian Supai Salt shows a dissolution edge 70 km long overlain by a monoclinal fold in the overlying strata, known as the Holbrook Anticline. Dissolution of the 150 m thick salt occurs at depths of around 300 m and the anticline has structural and topographic reliefs of 100 and 30 m, respectively. Active or recent deformation along the subsidence belt is expressed by numerous landforms including caprock collapse sinkholes (e.g., McCauley Sinks; Figure 8.7b), subsidence basins with playa-lakes, pressure ridges, fissures, and grabens (Neal 1995; Neal and Colpitts 1997; Neal 1998; Neal and Johnson 2003).

In the Permian Delaware Basin, Texas and New Mexico, the edges of the east-dipping Castile, Salado, and Rustler formations have receded by the down-dip migration of dissolution fronts, causing subsidence in the overlying stratigraphic units (Figure 8.2). According to Anderson

(1981), dissolution in this basin with artesian hydrogeological conditions has reduced the original salt units to 30–40% of their original volume. Bachman and Johnson (1973) calculated a horizontal migration rate for the dissolution front across the basin of 10–12 km Myr^{-1}. The eastern margin of the Pecos River valley, in the section associated with the Bottomless Lakes State Park, New Mexico, is defined by a monoclinal escarpment generated atop a down-dip propagating dissolution front in the gypsum and salt layers of the Permian Seven Rivers Formation. Upward groundwater flow from the underlying San Andrés Limestone of the confined Roswell artesian aquifer generates the breccia pipes and associated sinkhole lakes fed by underwater springs (Land 2003). In South Dakota, a dissolution front has migrated down-dip and radially away from the Black Hills in the Permian Minnelusa Formation (Rahn and Davis 1996; Epstein 2003). The long-term progression of the dissolution edge favored by the erosional lowering of the base level has induced the outward displacement of the sites with active subsidence and brine springs. In the Cretaceous Maha Sarakham Formation of Thailand and Laos, consisting of three evaporite intervals sandwiched between red mudstones, the down-dip migration of dissolution fronts has generated a subsidence trough that controls the path of the Nam Theum River (Supajanya and Friederich 1992; Utha-aroon et al. 1995; Warren 2016).

Subsidence related to interstratal dissolution of evaporites can be accommodated by dip-slip displacement along non-tectonic normal faults, producing scarps, horsts, and grabens, as well as distorting the drainage network. This brittle deformation style with well-defined failure planes typically occurs in thick and competent caprocks with preexisting steeply dipping fractures. In northern lower Michigan, Black (1997) reported karst graben valleys around 500 m wide generated by differential interstratal dissolution of Middle Devonian gypsum controlled by preexisting tectonic faults and joints. Interstratal dissolution of evaporites in the centre of salt-cored anticlines can produce grabens along the crestal zone of the folds (Figure 8.1f). In the Canyonlands section of the Colorado Plateau, Utah and Colorado, the collapse of the crest of a number of salt-cored anticlines caused by dissolution in the evaporitic cores has generated subsidence depressions as much as 50 km long and several hundred meters deep (Cater 1970; Doelling 2000; Gutiérrez 2004; Guerrero et al. 2015). The anticlines are cored by salt walls up to 4 km thick of the Lower Carboniferous Paradox Formation, with a proportion of chloride salt of around 70–80%. These salt structures developed by lateral salt flow between Carboniferous and Late Triassic times, as recorded by thickness variations in the supra-salt formations, and were rejuvenated by compression during the Laramide orogeny in the early Cenozoic. The dissolution-induced subsidence depressions formed along the axial zone of the anticlines are locally traversed by superimposed drainages, producing a paradoxical situation whereby the broad subsidence valleys are perpendicular to the major drainages (e.g., Colorado River in Spanish-Moab Valley, Dolores River in Paradox and Gypsum valleys). The floor of the collapse valleys is underlain by dissolution residues generally more than 150 m thick dominated by gypsum and clay (caprock of salt structure), and locally display blocks of supra-salt formations displaced vertically as much as 1 km from the anticline crest (Cater 1970; Doelling 2000). Most probably, the downward advance of interstratal dissolution in the salt cores was controlled by the entrenchment of the drainage network in response to regional uplift in the Colorado Plateau and the circulation of progressively deeper groundwater flows. The dissolution-induced subsidence structures in the anticline crests display different deformation styles (Gutiérrez 2004): (i) a symmetric axial graben controlled by systems of steeply dipping faults, which may be accompanied by lateral grabens with limited structural relief in the anticline limbs (Figure 8.4a), (ii) a half-graben with a roll-over anticline in the hanging wall, expressed as asymmetric depressions flanked by a prominent escarpment on one side, and a dip slope on the opposite margin (Figure 8.4b), (iii) an axial synform related to sagging subsidence flanked by lateral passive

Figure 8.4 Images of subsidence structures with geomorphic expression generated by interstratal evaporite dissolution and faulting of the overlying caprock. (a) Graben with geomorphic expression on the NE limb of the salt-cored Paradox Anticline, Colorado, related to interstratal dissolution of the Carboniferous Paradox Formation. *Source:* Francisco Gutiérrez (Author). (b) Northwest sector of Moab-Spanish Valley (Arches National Park, Utah) where evaporite dissolution-induced subsidence is accommodated by the development of a half-graben controlled by Moab Fault (arrow), with a throw of 700 m, and downward bending of strata in the footwall. *Source:* Francisco Gutiérrez (Author). (c) Trench excavated across an uphill-facing scarp and the associated trough in the Wasatch Monocline on the eastern margin of Sanpete Valley (Mayfield area, Utah). The trench revealed faulted deposits strongly overprinted by secondary carbonate dated by radiocarbon at 12.6–12.7 kyr. *Source:* Francisco Gutiérrez (Author). (d) Aerial view of Zenzano Fault (Iberian Chain, Spain) generated by interstratal dissolution of Triassic evaporites. The antislope scarp in the dissected dip slope displays triangular facets (arrows point to apexes), defeated streams (DS), and wind gaps (WG) in perched and beheaded stream sections. *Source:* Francisco Gutiérrez (Author). (e) Graben in the Peracalç Range, Spanish Pyrenees, associated with a laterally unconfined escarpment formed by thick halite-bearing Triassic evaporites and a carbonate caprock. These non-tectonic grabens are ascribed to both lateral salt flow toward the debuttressed escarpment and interstratal dissolution of the evaporites. The floor of the depression in around 200 m wide. Note triangular facet (TF) and hanging stream (HS) associated with the fault scarp. Circle indicates a car for scale. *Source:* Francisco Gutiérrez (Author). (f) Heuschkel Park on the eastern margin of the Roaring Fork River (Carbondale collapse center, Rocky Mountains, Colorado) is a synformal-graben depression related to interstratal dissolution of the Eagle Valley Evaporite (EVE) and the collapse of the overlying Paleozoic formations (P) and Miocene volcanics (MV). Lateral flow of the salt-bearing evaporite toward the valley may also have contributed to the extensional deformation in the caprock. *Source:* Image captured from Google Earth.

anticlines, and (iv) complex structures combining both normal faulting and folding, such as those reported in Spanish Valley. In this latter valley, Guerrero et al. (2015) conducted a trenching investigation in a master fault of the collapse depression, obtaining radiocarbon datings and performing a retrodeformation analysis. This work revealed an episodic kinematic style, an average slip rate much higher than those reported for normal faults in the region (3.1 mm yr^{-1}), and a very low average recurrence of around 300 yr. Ge and Jackson (1998) suggested that the depressions developed along the crest of the salt-cored anticlines in Canyonlands are related to extensional tectonics, rather than to evaporite dissolution and collapse. They argued, based on physical laboratory models, that interstratal dissolution should produce a complex structure comprising normal faults in an outer extensional zone, balanced by reverse faults and buckle folds in an inner contractional zone. However, the structure reproduced in the laboratory is not coherent with the deformation styles observed in the multiple regions and contexts where subsidence related to interstratal evaporite dissolution has been documented. Ge and Jackson (1998) simulated the dissolution in the salt-cored anticlines by withdrawing the salt analogue from below (pre-shaped silicone diapir), whereas dissolution occurs at the crest of the salt walls, as revealed by the thick dissolution residues (diapir caprock). Moreover, they used low-cohesion sand, which does not properly reproduce the deformation structures developed in lithified and fractured supra-salt rocks (Paterson 2001). Probably, extensional tectonics has influenced the development of the subsidence depressions, but multiple lines of evidence such as the thickness of the dissolution residues (caprock), the distribution and geometry of the subsidence structures, the fault parameters derived from trenching, or hydrochemical data strongly support the instrumental role played by interstratal salt dissolution (Gutiérrez 2004; Guerrero et al. 2015). In central Utah, between the Colorado Plateaus and the Basin and Range physiographic provinces, Witkind (1994) reported belts of intense localized deformation associated with salt-cored anticlines up to 100 km long generated by the rise of the Middle Jurassic Arapien Shale. This formation, around 3 km thick, has a proportion of salt and Ca-sulfates of around 40% (Picard 1980). Here, the subsidence of the crest of the anticlines induced by the removal of the underlying salt has produced linear valleys with two main types of structures (Witkind 1994): (i) axial grabens bounded by high-angle normal faults; and (ii) paired facing monoclines generated by downwarping of strata into the anticline core. These monoclines are locally ruptured by antithetic normal faults expressed as uphill facing scarps and half-grabens with geomorphic expression, as well as defeated streams and wind gaps (e.g., Wasatch monocline) (Figure 8.4c).

Gravitational faults and graben systems commonly develop associated with erosional escarpments formed by evaporitic formations and an overlying brittle caprock. Non-tectonic faulting in these contexts can be related to interstratal dissolution of the evaporites, lateral flow of salt toward the erosionally unloaded scarp face, or a combination of both. These contexts provide favorable conditions for the development of both types of gravitational deformation phenomena: (i) the formation of stress release or unloading fractures in the debuttressed slopes weaken the rock mass and facilitate water infiltration, (ii) the topographic and hydraulic gradient favor deep groundwater circulation, (iii) differential loading conditions can drive salt flow toward the areas subject to lower load. Interstratal dissolution essentially produces vertical displacement (subsidence) in the caprock, whereas horizontal extension is the dominant displacement component in caprocks affected by lateral spreading (Gutiérrez et al. 2012b). The 1.8 km long Zenzano normal fault in the Iberian Chain, NE Spain, occurs associated with a 475 m high erosional escarpment comprising thick salt-bearing Triassic evaporites capped by brittle rocks. Vertical displacement on this fault induced by interstratal dissolution of the evaporites has produced an antislope fault scarp on a dissected dip slope. The fault scarp is expressed by prominent triangular facets and disrupted transverse drainages that display blocked sections in the downthrown block (defeated streams) and hanging beheaded sections in the footwall (wind gaps) (Carbonel et al. 2013) (Figure 8.4d). The Peracalç

Range in the Spanish Pyrenees displays an active horst and graben morpho-structure associated with a 450 m high erosional escarpment formed by Cretaceous limestones underlain by tectonically thickened halite-bearing Triassic evaporites. These non-tectonic structures are related to both: (i) rock spreading driven by the lateral flow of the evaporites toward the debuttressed mountain front, which shows a bulge at its foot; and (ii) differential subsidence induced by interstratal dissolution. Surface faulting has disrupted a transverse paleodrainage network expressed by wind gaps and defeated streams. Trenching studies revealed episodic displacement on two faults. This unstable area, covering 4.5 km^2 and with a minimum estimated volume of 0.9 km^3, is considered to be the largest mass movement in the Pyrenees (Gutiérrez et al. 2012b) (Figure 8.4e).

In the Ogaden region of eastern Ethiopia, Mège et al. (2013) described gravitational faulting in elongated mesas consisting of a sandstone and limestone cap 100–200 m thick, underlain by a laterally unconfined Cretaceous gypsum formation 300 m thick. Here, the caprock in the mesas displays a horst and graben topography controlled by faults parallel to the axis of the mesas. The underlying gypsum formation shows a dense network of dilated joints filled by satin spar with a dominant direction consistent with the extensional structures in the caprock. Mège et al. (2013) attributed these structures to lateral spreading, interstratal dissolution and diapirism, although the participation of the latter mechanism is debatable given the gypsiferous composition of the evaporite formation. In the Carbondale collapse center of the Rocky Mountains of Colorado, Heuschkel Park and Spring Valley, 3 and 6 km long, respectively, are enclosed graben depressions related to interstratal dissolution of the Eagle Valley evaporite, and possibly also to a rafting effect induced by the flow of the underlying saline formation toward the unloaded Roaring Fork River valley (Kirkham et al. 2002) (Figure 8.4f). The largest documented subaerial graben system related to salt flow, and secondarily to interstratal dissolution, are the Grabens in Canyonlands National Park, Utah, covering around 200 km^2 and with an estimated volume of 60 km^3. This horst and graben system has developed on the eastern margin of the 360–530 m deep Cataract Canyon of the Colorado River, and is affected by active surface deformation as indicated by InSAR, leveling and extensometer data (Furuya et al. 2007; Kravitz et al. 2020). Here, the exposed stratigraphic sequence has a gentle dip toward the canyon (2–4°) and consists of a 460 m thick brittle sandstone sequence lying on the Carboniferous Paradox salt formation. Extension in the upper brittle plate caused by lateral flow of the salt toward the erosionally unloaded canyon has generated an arcuate system of grabens (McGill and Stormquist 1979; Moore and Schultz 1999; Baars 2000) (Figure 8.5a). The built up of flowing salt along the Cataract Canyon has produced the Meander Anticline, an antiform with a sinuous axis that perfectly matches the path of the Colorado River (Huntoon 1982). Interstratal evaporite dissolution beneath The Grabens, largely dominated by subsurface drainage through sinkholes and fissures, contributes to produce the differential subsidence pattern as indicated by the presence of saline springs at the bottom of Cataract Canyon (Reitman et al. 2014), and the large throw documented on some faults (ca. 150 m; Grosfils et al. 2003). The grabens, up to 6 km long and 500 m wide, are bounded by en echelon fault arrays. The floor of the graben depressions shows alignments of cover collapse sinkholes that can function as stream sinks. These sinkholes are related to internal erosion of detrital deposits through dilated fissures and faults in the bedrock (Figure 8.5a). The deeply dissected tributary canyons of the Colorado River expose fissures filled with cemented detrital deposits up to 250 m deep (Ely 1987). Wind gaps at the top of the horsts provide evidence for a distorted and aborted paleodrainage (Trudgill 2002) (Figure 8.5a).

Interstratal evaporite dissolution in steeply dipping successions can induce the rotation of overlying and adjacent strata and the development of flexural-slip faults. These are peculiar bedding-parallel faults with no stratigraphic separation that can produce along-strike fault scarps (Figure 8.1g). This type of faults has been documented in the western margin of the Carbondale

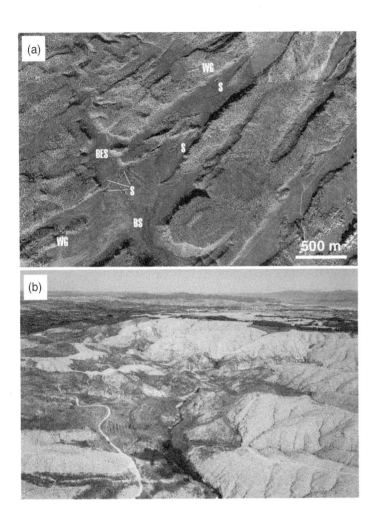

Figure 8.5 Examples of non-tectonic morpho-structures affecting caprocks underlain by salt-bearing evaporites. (a) The Grabens of Canyonlands National Park, Utah, are related to lateral salt flow toward the erosionally unloaded Cataract Canyon of the Colorado River and interstratal evaporite dissolution. Horizontal separation on normal faults generate dilated fissures that allow internal erosion in the overlying deposits and the development of aligned cover collapse sinkholes that can function as stream sinks. S: sinkhole; BS: blind stream; BES: beheaded stream; WG: wind gap. *Source:* Image captured from Google Earth. (b) Partial aerial view of a collapse area in Calatayud Neogene Graben (Iberian Chain, NE Spain), where a caprock composed of carbonate and red detrital sediments has been foundered into the underlying halite- and glauberite- bearing evaporites. Note irregular boundary of the collapse structure and subhorizontal attitude of the evaporites, contrasting with the chaotic deformation in the younger collapsed sediments. *Source:* Francisco Gutiérrez (Author).

collapse center, in the southern Rocky Mountains of Colorado (Gutiérrez et al. 2014a). Here, the steeply dipping stratigraphic succession comprises an upper sequence of Permian to Eocene formations underlain by salt-bearing evaporites of the Carboniferous Eagle Valley Formation, which has been thickened by tectonic deformation and flow (>1.5 km thick). Locally, the steeply dipping supra-evaporite formations are unconformably overlain by Miocene basalts and Quaternary mantled pediments. The debuttressing effect related to active evaporite dissolution induces the rotation (unfolding) of the overlying strata and the development of flexural-slip faults. These faults rupture the unconformable basalt cap and the mantled pediments producing half-graben depressions bounded by antislope scarps parallel to the underlying strata of the monocline (Figure 8.6a). Trenches dug across flexural-slip fault scarps in the basalts and a pediment revealed evidence of multiple surface faulting events with displacement per event values of ≥1 m (Figure 8.6b). The length of the faults (25 km) and the estimated surface rupture area (190 km^2)

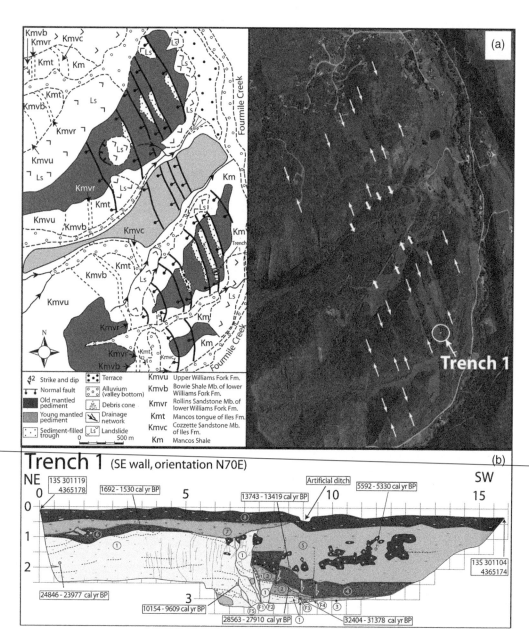

Figure 8.6 Flexural-slip faults related to evaporite dissolution in the Carbondale collapse center, southern Rocky Mountains of Colorado. (a) Geological map and annotated orthoimage showing flexural-slip faults expressed as antislope scarps and troughs in mantled pediments of two different ages unconformably overlying steeply dipping bedrock strata. (b) Log of trench excavated across a flexural-slip fault in the old mantled pediment. Trench location is indicated in subfigure a. The geometrical relationships reveal a minimum of three surface faulting events: the creation of the sediment trap (>32.4 kyr), the formation of colluvial wedge (unit 2) shed for a fault scarp formed in the trough fill (32.4–28 kyr), and the deformation of the colluvial wedge (5.6–1.5 kyr). *Source:* From Gutiérrez et al. (2014a). Reproduced with permission of the Geological Society of America.

suggest that these exceptional karst-related structures might generate damaging earthquakes with moment magnitude (M_w) of around 6 (Gutiérrez et al. 2014a). An important issue from the seismic hazard perspective in evaporitic areas is the differentiation between tectonic faults and non-tectonic faults generated by interstratal evaporite dissolution. The latter faults in most cases are shallow structures not capable of producing damaging earthquakes. Erroneously interpreting a fault related to evaporite dissolution as a seismogenic tectonic structure may lead to significant seismic hazard estimates with adverse economic and societal implications. The US Nuclear Regulatory Commission, in a report on techniques for identifying faults and determining their origin, suggested that evaporite dissolution tends to produce creeping faults characterized by progressive displacement (Huntoon 1999). However, a number of trenching studies carried out in faults related to interstratal evaporite dissolution provide clear evidence of episodic displacement (Gutiérrez et al. 2012a, 2012b, 2014a; Carbonel et al. 2013; Guerrero et al. 2015). Nonetheless, these studies reveal that collapse faults related to evaporite dissolution tend to show different parameters compared to those associated with tectonic faults, and that such parameters can be used to determine their origin: (i) higher maximum displacement to length ratio, (ii) anomalously high average slip rate, (iii) higher displacement per event, and (iv) much lower average recurrence. The stick–slip kinematic regime documented in the collapse faults related to interstratal dissolution contrasts with the creep displacement recorded by extensometers and leveling in a surface fault in the Grabens of Canyonlands, dominated by horizontal extension driven by subjacent lateral salt flow (Kravitz et al. 2020). The criteria that can be used to elucidate the origin of faults in evaporite areas are reviewed by Hanson et al. (1999), Gutiérrez et al. (2012a) and Carbonel et al. (2013).

Interstratal dissolution of evaporites across extensive areas can produce large subsidence morpho-structures with a relatively irregular outline and highly disordered internal structure. In an important paper on interstratal dissolution and subsidence, Kirkham et al. (2002) documented the Carbondale collapse center in the Rocky Mountains of Colorado. This is a 1200 km^2 morpho-structural depression where Miocene volcanic rocks are downdropped as much as 1200 m due to interstratal dissolution of the underlying evaporites. This region was covered by basaltic lava flows at ca. 24–10 Myr, accumulated on a low-relief erosion surface cut across folded Paleozoic-Eocene formations, including the halite-rich Eagle Valley Evaporite of Carboniferous age. After 10 Myr, the entrenchment of the ancestral Colorado River system was accompanied by progressively deeper groundwater circulation, interstratal dissolution of the evaporites, and subsidence of the overlying rocks. The basalt lava flows, dated by Ar/Ar and characterized by paleomagnetic studies for post-depositional rotation, are the markers that have been used for constraining the timing, style, and amount of late Cenozoic gravitational deformation caused by evaporite dissolution. The Carbondale collapse center has a relatively irregular outline controlled by the distribution of evaporites in the subsurface and preexisting tectonic structures and intrusive igneous bodies. The edge is marked by monoclines, swarms of flexural-slip faults (Figure 8.6), and sediment-filled subsidence basins. The interior of the collapse area shows a lower degree of dissection than the margins, with disrupted drainages and large internally drained depressions. Differential subsidence has produced graben and sag depressions, as well as depositional basins with sediment fills more than 400 m thick. Active evaporite dissolution is evidenced by the dramatic increase in dissolved loads experienced by the rivers that traverse the subsidence area. Barret and Pearl (1976) estimated that Yampah Hot Spring in Glenwood Springs, which is a major groundwater output point for the Eagle Valley Evaporite, discharges around 240 metric tons of dissolved halite and gypsum per day. Local salt flow toward the unloaded valleys produces "valley anticlines" with backtilted volcanic rocks and fluvial terraces at the margins (Kirkham et al. 2002). In northwestern Saudi Arabia, Al-Khattabi et al. (2007) mapped the 175 km wide Aba al Qur subsidence depression, resulting from interstratal dissolution of gypsum units of the Late Cretaceous Badanah Formation. The edges of this morpho-structural depression are defined by monoclinal escarpments 30–85 m high. The bottom of the subsidence basin is characterized by chaotic foundered strata, breccias, a poorly developed drainage disrupted by subsidence and ponors, plus numerous subsidence depressions and sinkholes. In the Calatayud Neogene graben, Iberian Chain, Spain, Gutiérrez (1996) documented two collapse areas with highly tortuous boundaries covering 8 and 14 km^2. In these structures, the 100 m thick supra-evaporitic Miocene succession has undergone intense deformation and vertical displacement in excess of 200 m due to the interstratal karstification of the underlying halite- and glauberite-bearing evaporites (Figure 8.5b). Here, the strongly deformed caprock sediments are foundered into older evaporites with subhorizontal attitude, revealing the gravitational nature of the deformation. Initially, evaporite dissolution was ascribed to groundwater derived from the formations overlying the evaporites (epigene dissolution). More recently it has been proposed that dissolution may have been caused by ascending groundwater flows (hypogene dissolution) coming from detrital aquifer units situated beneath the evaporites. Groundwater that recharges at the margins of the basin circulates through coarse-grained detrital units confined by argillaceous facies toward the axis of the basin, ultimately discharging through upward transformational flows across the evaporites (Gutiérrez 2014).

Evaporites can be also affected by localized interstratal dissolution at sites controlled by features that facilitate cross-formational circulation of groundwater (e.g., fractures, faults, and reefs). The solutional removal of the evaporite rock at these sites can induce upward-propagating collapse processes (stoping) generating breccia pipes, as described in Section 6.3.3. Figure 8.1i illustrates a situation in which dissolution generates a large cavity at the base of the evaporite unit. Nonetheless, breccia pipes can also develop without the formation of significant cavities by simultaneous dissolution and collapse. Moreover, dissolution can be initiated at the base or at the upper part of the soluble unit. Breccia pipes are discordant structures with an overall chimney-like geometry that cross-cut the stratigraphic succession (transtratal breccia). They can reach hundreds of meters across and more than 1 km in vertical dimension, and are also designated as breccia chimneys and breccia columns. The ascending foundering process may stop if the cavity roof and the breakdown pile get in contact, due to the bulking effect associated with the rock fragmentation process. Nonetheless, the choked breccia pipe may eventually reactivate and grow upward by dissolution acting in the evaporite unit or in the breccia itself. Eventually, the ascending cavity may intercept the ground surface producing a caprock collapse sinkhole (Figure 6.32d). The resulting caprock collapse sinkholes can reach hundreds of meters across and may function as long-lasting traps for sediment accumulation. Both the breccia pipe and the sinkhole fill can experience cementation, becoming an indurated mass with higher resistance to erosion than the surrounding sediments. Differential erosion in areas with relict breccia pipes and filled caprock collapse sinkholes may result in a relief inversion, with the development of hills consisting of cemented breccias or sinkhole deposits protruding over surfaces carved in the softer host rock (Figure 8.1i). Quinlan (1978) reviewed literature documenting more than 5000 breccia pipes related to interstratal evaporite dissolution in the United States. These breccia pipes have diameters of up to 1000 m, reach depths of 500 m, and some have been sites of uranium deposition. Breccia pipes have been documented throughout the Delaware Basin (southwestern USA) related to interstratal dissolution of the Permian Castile, Salado, and Rustler formations (Anderson and Kirkland 1980; Stafford et al. 2008a). Breccia pipes developed above the Capitan Reef extend vertically across the Castile and Salado formations and are ascribed to hypogene dissolution by groundwater rising from the reef limestones. These breccia pipes can be expressed as subsidence depressions or as mounds related to the topographic inversion of cemented breccias (Bachman 1980; Hill 1996; Stafford et al. 2008a). In western Lake Erie (USA), Carlson (1992) documents breccia pipes more than 460 m deep rooted in the Late Silurian Salina Group, showing foundered blocks indicating as much as 290 m of vertical displacement. This value provides a minimum measure of the thickness of salts removed by dissolution. In the Black Hills of South Dakota, Gott et al. (1974) described Uranium-bearing breccia pipes 430 m high rooted in Ca-sulfate formations. In the coal mines of Shanxi and Hebei, northern China, Yarou and Cooper (1996) reported more than 2800 breccia pipes originated in Ordovician gypsum with interbedded limestones and penetrating through overlying Permo-Carboniferous coal-bearing formations. The breccia pipes reach more than 500 m in vertical dimension and locally show sinuous trajectories. Densities as high as 70 pipes per km^{-2} have been computed in the Shanxi coal basin with more than 1300 pipes. Myers (1962) described doughnut-like residual hills in western Oklahoma attributed to cementation of sandy sinkhole fills in a relict gypsum karst and the subsequent differential erosion of the host rock. The *castiles* in the Gypsum Plain of the Delaware Basin, New Mexico and Texas, are isolated hills of resistant masses of calcite that replaces Ca-sulfates of the Castile Formation and stand up above the surrounding gypsum bedrock (Kirkland and Evans 1976). The secondary calcite is a diagenetic byproduct of the bacterial reduction of sulfate by methane rising from an underlying hydrocarbon reservoir (Hill 1995) (see Section 3.4 and Eq. 3.63).

Transtratal collapse structures can also form above evaporite units by the foundering of coherent blocks bounded by annular failure planes with vertical or outward-dipping attitude. The Crater Lake in Southern Saskatchewan (Canada) is the geomorphic expression of a collapse structure controlled by two concentric cylindrical faults related to the interstratal dissolution of salt in the Devonian Prairie Evaporite Formation. Stratigraphic data gained by boreholes indicate that the inner collapse, 100 m diameter and with a throw of 43 m, experienced subsidence episodes from the Late Cretaceous to the Lower Pleistocene. The younger outer ring fault is 210 m across and has accommodated 15–30 m of vertical displacement during the latest Pleistocene (Christiansen 1971). In the salt-cored anticlines of the Paradox Basin, Colorado and Utah, Weir et al. (1961), Cater (1970) and Sugiura and Kitcho (1981) mapped foundered blocks bounded by cylindrical faults up to 1 km in diameter. These collapse structures with throws greater than 300 m are related to interstratal salt dissolution in the Carboniferous Paradox Formation, situated at depths greater than 600 m (Gutiérrez 2004). In Sant Miquel Valley, NE Spain, Gutiérrez et al. (2019b) mapped and characterized caprock collapse sinkholes hundreds of meters across related to the hypogene interstratal dissolution of an Eocene gypsum formation and the subsidence of an overlying confining marl unit more than 200 m thick. In Gozo Island, Malta, Galve et al. (2015) described coherent cylindrical paleocollapse structures up to 600 m in diameter with internal sagging, potentially induced by deep-seated evaporite dissolution. Growth strata in marine Miocene sediments associated with the paleosinkholes reveal that gradual subsidence operated in the sea floor during the deposition of those formations. The paleocollapses are expressed either as erosional depressions or as prominent buttes, depending on whether the outcropping sediments in the foundered blocks are softer or more resistant than the host rock, respectively (Figure 8.7a). In the Holbrook Basin, Arizona, Neal and Johnson (2003) documented a peculiar collapse morpho-structure generated by interstratal dissolution of Permian salts at a depth of around 200 m. Here, subsidence is expressed as a vaguely-edged sagging depression 3 km in diameter with 50 nested caprock collapse sinkholes up to 100 m across arranged in two concentric rings (McCauley Sinks; Figure 8.7b) and an outer extensional zone with fissures. Probably, the sinkhole rings reflect annular failure zones that contribute to weaken the caprock and provide preferential pathways for downward water percolation and salt dissolution.

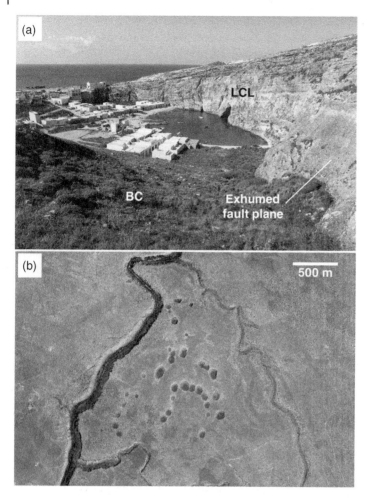

Figure 8.7 Examples of subcircular collapse structures attributed to interstratal dissolution of evaporites. (a) The Qawra paleosinkhole in Gozo, Malta is a collapse structure controlled by a cylindrical fault 370 m across that bounds a downthrown block with limited internal deformation. Differential erosion of the soft collapsed sediments, mainly the Blue Clay (BC), has generated a depression bounded by the more resistant Lower Coralline Limestone (LCL). The scarps that frame the depression largely correspond to the exhumed plane of the master ring fault. The lagoon is connected with the sea through a fault-controlled cave. *Source:* Francisco Gutiérrez (Author). (b) McCauley Sinks in the Holbrook Basin, Arizona form two concentric rings of caprock collapse sinkholes nested in a sag depression related to dissolution of Permian salt at a depth of around 200 m. *Source:* Image captured from Google Earth.

8.2 Covered Evaporite Karst

Alluvial rivers flowing over evaporitic bedrock can be affected by synsedimentary subsidence related to subjacent dissolution. Both upward groundwater discharge in the valley acting as output zone for bedrock aquifers, and water flow in the alluvial aquifer can be involved in the dissolution process. The contact between the alluvial cover and the evaporite bedrock can be highly irregular, and the clayey residual material associated with the rockhead (dissolution residue) is generally interrupted by collapse structures and alluvium-filled solutionally enlarged fissures (cutters) (Figure 4.16).

The alluvium and the karstified evaporite bedrock typically function as interconnected aquifers with contrasting hydrochemistry that can interexchange water in both directions, depending on the relative position of their hydraulic heads. Dissolution-induced subsidence acting on a valley reach disrupts the equilibrium longitudinal profile of the fluvial system, causing a local drop in the base level and leading to a number of morpho-sedimentary adjustments. The river tends to restore the profile by aggrading in the subsidence area (Ouchi 1985; Schumm et al. 2000). Long-sustained subsidence and aggradation result in thickened alluvium that may reach more than 100 m in thickness. If the fluvial system experiences an entrenchment episode followed by the development of a new terrace, the terrace levels may show contrasting morpho-stratigraphic arrangements depending on their position with respect to the alluvium-filled dissolution basin (Figure 8.8). Upstream and downstream of the subsidence area the younger terrace can be inset into the bedrock. However, in the subsidence area, the deposit of the young terrace can be superimposed on and

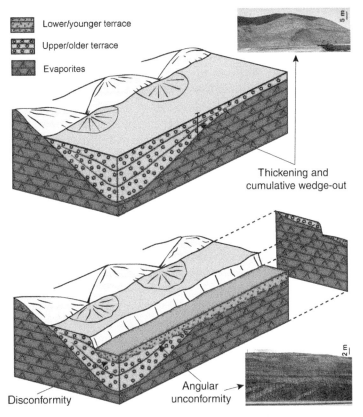

Figure 8.8 Morpho-stratigraphic arrangements of terraces in a valley reach affected by dissolution-induced subsidence counterbalanced by aggradation (subsidence/aggradation ~1). The upper sketch shows an alluvium-filled dissolution-induced basin with growth strata at the margins. The lower sketch displays a subsequent stage with an inset terrace formed after an entrenchment episode. The deposit of the younger terrace is inset into the bedrock upstream of the subsidence bowl, and superimposed on and juxtaposed to the older thickened and deformed deposit in the subsidence basin. The boundary between the two stacked terrace deposits changes from an angular unconformity to a disconformity, depending on the attitude of the bedding in the older unit. Upper inset image shows the thickening of a terrace deposit in the Huerva River valley (Ebro Cenozoic Basin, Spain) coinciding with a lateral facies change from gypsum/anhydrite to halite in the bedrock. The lower inset image illustrates an angular unconformity bounding the deposits of two terraces of the Jalón River in the Neogene Calatayud Graben (Iberian Chain, Spain). *Source:* Francisco Gutiérrez (Author).

juxtaposed to the older thickened and deformed terrace deposit, whereas the terrace treads show an inset relationship. The boundary between the stacked terrace deposits is typically an angular unconformity that grades into a disconformity toward the central part of the subsidence bowl (Gutiérrez 1996; Benito et al. 2000) (Figure 8.8). If the fluvial system does not have sufficient capability to fill the accommodation space created by subsidence, it can also cause incision and degradation upstream, leading to geomorphic responses such as the development of downstream-converging strath terraces that grade into thickened alluvium. Guerrero et al. (2008a) conceived different morpho-stratigraphic configurations of fluvial terraces depending on the subsidence/aggradation ratio (~1, >1, and >>1), which determines the local base level drop created by subsidence in the valley bottom and the local aggradation and degradation patterns.

The thickened alluvium fills complex dissolution-induced basins up to several tens of kilometers long and commonly shows a number of characteristic stratigraphic, sedimentologic, and structural features: (i) growth strata (cumulative wedge-outs) at the margins of the basins recording aggradation coeval to subsidence (synsedimentary subsidence) (Figure 8.8), (ii) common strong carbonate cementation, partially ascribable to the common-ion effect related to calcium ions derived from gypsum dissolution, (iii) high floodplain to channel facies ratio indicating that subsidence bowls, where there is a decrease in the stream power, tend to trap a significant proportion of the sediment load, (iv) common presence of palustrine facies deposited in swamps associated with subsidence basins and paleosinkholes, where the topographic surface used to be intercepted by the water table (Luzón et al. 2008; Gutiérrez et al. 2016), (v) spatial correlation between channels and the deeper parts of sagging basins, indicating that subsidence controls channel paths (i.e., migration, avulsion) (Gutiérrez 1996; Zarroca et al. 2017), (vi) changes in the channel patterns controlled by subsidence (Benito et al. 1998; Guerrero et al. 2008a), and (vii) abundant brittle and ductile non-tectonic deformation, including sagging and/or collapse paleosinkholes, which may affect the underlying bedrock depending on whether dissolution occurs within the evaporite bedrock or at the rockhead. These recent deformation structures can be apparently similar to those related to active tectonics. However, karst-related structures in alluvial karst settings are restricted to the sediments situated above the dissolution zones, typically display annular collapse faults and basins with centripetal dips, and tend to be confined to the valleys. An important aspect from the applied perspective is that the alluvium-filled basins constitute valuable aquifers and a significant source of aggregates.

The great majority of the examples of thickened terrace deposits associated with evaporitic terrains has been documented in Spain, and specially in its Cenozoic basins (Gutiérrez et al. 2001, 2008b). Nonetheless, these karstic phenomena are most probably more widespread than appreciated, and it is likely that subsidence induced by suballuvial evaporite dissolution has exerted a significant control on the evolution of a number of fluvial systems worldwide. The Spanish case studies reveal that the magnitude and extent of the subsidence tends to be significantly larger where the evaporite formations include halite and glauberite units in the subsurface (Gutiérrez 1996; Benito et al. 1998; Guerrero et al. 2008a, 2013; Lucha et al. 2012). The different studies conducted in Spain provide clues on the controls that can influence the dissolution-induced subsidence phenomena and their morpho-stratigraphic record. In the Calatayud Neogene graben (Iberian Chain), the thickened terraces of the Jalón River record a downstream migration of the dissolution-induced subsidence phenomena through time (Gutiérrez 1996). This spatial–temporal pattern suggests the operation of local negative feedbacks, probably related to the depletion of evaporites susceptible to dissolution and the increasing thickness of the alluvial cover. In the central sector of the Ebro Basin, the lower reach of the Gállego River displays a complex dissolution-induced basin around 30 km long and 8 km wide with several depocenters. Here, the thickened terraces record a maximum subsidence of 190 m, and two main phases of dissolution-induced subsidence and aggradation (Benito et al. 1998). Benito et al. (2010) based on OSL ages interpreted that the youngest dissolution-induced subsidence phase, occurred around 140–155 kyr ago (end of MIS 6), was associated with a period of enhanced river discharge derived from extensive glaciers in the Pyrenees. In the lower section of the Huerva River valley (Ebro Cenozoic Basin, Spain), Guerrero et al. (2008a), using deep borehole data drilled for glauberite exploration, documented a spatial correlation between the thickened alluvium and the distribution of a halite unit at least 75 m thick embedded within the subhorizontally-lying evaporite bedrock (upper inset image in Figure 8.8). Periods with a subsidence/aggradation ratio > 1 in this fluvial system characterized by limited sediment flux led to the development of degradation surfaces (strath terraces, pediments) upstream of the subsidence area. A different situation is found in the thickened terraces along the middle reach of the Ebro River (Ebro Cenozoic Basin, Spain), showing parallel longitudinal profiles indicating a subsidence/aggradation ratio ~1. This fluvial system with high sediment flux had the ability to keep pace with subsidence by rapidly filling the accommodation space (Guerrero et al. 2013). This work provides evidence indicating that the lithostratigraphy of the evaporite bedrock and the distribution of salt units can play a decisive role in the timing and spatial distribution of the dissolution and subsidence periods. The progressive entrenchment of the fluvial system may situate the valley floor close to progressively deeper high-solubility units, eventually enabling their dissolution and initiating phases of rapid subsidence. In the evaporite-cored Barbastro-Balaguer Anticline (Ebro Cenozoic Basin, Spain), the thickened alluvium is restricted to the highest terraces of several drainages that traverse this salt anticline. Lucha et al. (2012) proposed that dissolution was more intense during the early stages of development of the fluvial systems following the capture of the Ebro Basin and a change from endorheic to exorheic conditions. Doğan (2005) ascribed local thickenings in terrace deposits of the Tigris River, Turkey, to subsidence caused by dissolution of a Miocene gypsum formation. Some publications briefly mention thickenings and deformations in Quaternary fluvial deposits related to evaporite dissolution in the USA, including the Canadian River in the Texas Panhandle (Gustavson et al. 1982; Gustavson 1986) and the Colorado River in Spanish-Moab Valley, Utah (Doelling 2000; Gutiérrez 2004).

In gypsum karst areas of the western Austrian Alps, Cammeraat et al. (1987) described peculiar debris pillars made up of strongly cemented colluvial breccias as much as 6 m high. The authors proposed the following geomorphological evolution associated with a covered gypsum karst to explain the development of these hoodoo-like landforms: (i) infill of funnel-shaped solution hollows (sinkholes, cutters) by calcareous scree derived from nearby limestone scarps, (ii) strong and rapid induration of the detrital breccia favored by the common-ion effect in waters that have interacted with gypsum, and (iii) differential erosion of the softer gypsum bedrock to form debris pillars (relief inversion). These are most probably special features specific of evaporite karsts that are unlikely to occur in mechanically and chemically more resistant carbonate rocks.

8.3 Bare Evaporite Karst

Gypsum outcrops in semiarid areas often display weathering crusts with characteristic landforms related to dissolution and reprecipitation processes. One of such landforms is the hollow subcircular dome known as gypsum tumulus, generated by local expansion and bulging of the weathering crust (Figure 8.9a, b). They are also designated as gypsum bubbles, gypsum domes, and gypsum blisters, and have been reported in the northern Apennines (Forti 1987; De Waele et al. 2012, 2017b) and Sicily in Italy (Ferrarese et al. 2002), in Sorbas (Calaforra and Pulido-Bosch 1999b) and the Ebro Cenozoic Basin in Spain (Artieda 1996, 2013; Gutiérrez-Elorza and Gutiérrez-Santolalla 1998),

Figure 8.9 Landforms formed in bare evaporite karst areas by dissolution and reprecipitation processes, as well as by hydration of anhydrite. (a) Collapsed gypsum tumuli with macrocrystalline texture on an outcrop of Miocene gypsum in the Ebro Basin, NE Spain. *Source:* Francisco Gutiérrez (Author). (b) Meter-sized gypsum tumuli in the Messinian gypsum of Sorbas, southern Spain. *Source:* Francisco Gutiérrez (Author). (c) Tumuli with concentric structure developed on several layers of bottom-nucleated selenite crystals deposited in an ephemeral salt flat (Hormuz Island, Iran). *Source:* Francisco Gutiérrez (Author). (d) Gypsum ridge developed on a weathering crust in Sorbas, southern Spain. *Source:* Francisco Gutiérrez (Author). (e) Anhydrite-hydration dome and cave at Dingwall, Canada. The dome is 11 m long, 9 m wide and 1.6 m high. *Source:* Photo by Adrian Harzyna. (f) Halite rim at the entrance of an abandoned salt mine in the Ebro Valley, Spain. Chemical analysis indicated a proportion of halite by weight of 95%. Scale is in centimeters. *Source:* Francisco Gutiérrez (Author). (g) Gypsum rim associated with a solution pipe in a gypsum outcrop of the Gachsaran Formation, Zagros Mountains, Iran. Arrows in a, c, and d point to cap lens 8 cm in diameter. *Source:* Francisco Gutiérrez (Author).

and the Delaware Basin, Texas and New Mexico (Stafford et al. 2008a). The domes are commonly several decimeters in diameter and height and the upwarped layer may be up to 50 cm thick. Exceptionally large tumuli up to 10–15 m across and more than 1 m high have been documented in outcrops of macrocrystalline gypsum in Sorbas and Sicily. Although some authors speculated that these domes associated with gypsum outcrops could be related to anhydrite hydration, more in-depth investigations reveal that bulging in these particular areas is related to dissolution and reprecipitation processes acting in a surficial weathering zone (Artieda 1996, 2013; Calaforra and Pulido-Bosch 1999b; Ferrarese et al. 2002). The fresh meteoric water that infiltrates in outcrops of macrocrystalline gypsum fills the cracks in the upper zone of the rock mass causing dissolution. Subsequently, evaporation and the associated capillary rise of the interstitial water lead to migration of the solution toward the surface, its progressive saturation, and gypsum reprecipitation in the outer weathered zone. The volume increase caused by secondary gypsum precipitation together with crystallization pressure cause the local detachment and bulging of the gypsum crust. As the domes grow, the weight of the uplifted crust may exceed its mechanical strength, with the consequent development of concentric and radial cracks. The mechanical strength of the uplifted crust may also be weakened by dissolution. Eventually, the top of the blister collapses producing crater-like depressions (Figure 8.9a, b). The development of gypsum tumuli is favored by the number of wetting and drying cycles, as well as semiarid climates in which evapotranspiration potential exceeds precipitation. Ferrarese et al. (2002) indicate that gypsum tumuli may form in short periods of time (years or decades) and that they have tilted protohistoric graves in western Sicily. Artieda (2013) presents a detailed characterization of gypsum tumuli developed on Miocene gypsum in the central sector of the Ebro Cenozoic Basin, including an enlightening microscopic analysis of the porosity and fabrics of the gypsum crusts. The expansion and differential uplift of the outer weathering zone can produce other landforms, including teepee-like linear and polygonal pressure ridges (Figure 8.9d). These features have been described in Sicily, where they are best developed in macrocrystalline gypsum and in south-facing slopes affected by a larger number of wetting and drying cycles (Macaluso and Sauro 1996b; Ferrarese et al. 2002). Similar domes and pressure ridges also form on exposed bottom-nucleated gypsum and salt layers deposited in the floor of ephemeral saline lakes and sabkhas (Figure 8.9c).

The volume increase associated with the hydration of exposed anhydrite beds and their conversion into gypsum also produces domes and ridges similar to those related to gypsum dissolution and reprecipitation (i.e., morphologic convergence or equifinality) (Figure 8.9e). This near-surface weathering process often occurs on artificial excavations, where anhydrite is suddenly exposed at the surface and subject to interaction with abundant meteoric water. According to the thorough review presented by Jarzyna et al. (2020), anhydrite-hydration domes and ridges have been documented in four areas worldwide: (i) the Harz Mountains of Germany (Kempe 1996), (ii) Alebastrovyye Islands in Russia, (iii) an abandoned gypsum quarry near Dingwall in Canada, and (iv) an active quarry at Pisky, Ukraine. Most probably the list of sites could be expanded considerably by examining recent quarries and other excavations performed in anhydrite-bearing formations (e.g., quarries in Oklahoma). The detachment and upwarping of the hydrated Ca-sulfate layers can create voids more than 10 m across and 1.5 m high that can be considered as a special type of hydration cave (in German Quellungshöhlen). These local deformation features can form over very short time spans of the order of a few years or decades. Jarzyna et al. (2020) indicate that the height of a hydration cave at Dingwall, Canada grew 25 cm in a time lapse of 10 years (average uplift of 2.5 cm yr^{-1}).

Salt rims and blisters are centimeter- to decimeter-scale ephemeral features that form in dry areas along the edge of hollows with air circulation (De Waele and Forti 2010) (Figure 8.9f). The hollows are typically tube-like openings of various origins developed in salt-bearing sediments. When the outside air temperature is lower than that of the opening, moist, and warm air flows upward causing salt dissolution by condensation. The condensation water evaporates along the edge of the hole causing halite precipitation and the formation of a salt rim that grows vertically by capillary rise of water. Under low air flow, the upper edge of the rim can grow inward, eventually producing a salt blister. The outer face of these features typically displays an irregular pop-corn like micromorphology related to condensation solution and reprecipitation. These landforms, with a very limited persistence time, have been documented in the Cordillera de la Sal, Chile (De Waele and Forti 2010), and in Qarhan salt lake in the Quinghai Plateau, China (Yuhua and Lin Hua 1986). Rims composed of gypsum have been reported in Patagonia, Argentina by Forti et al. (1993) (Figure 8.9g). Equivalent speleothems occur in caves associated with the opening of conduits with air circulation (Davis 1995; Hill and Forti 1997). De Waele and Forti (2010) carried out a morphometric study of 35 rims and blisters developed in a salt flat in the Cordillera de la Sal, San Pedro de Atacama, Chile (average annual rainfall <20 mm). They measured a maximum height of 8 cm and lengths in plan of up to 15 cm.

Fluvial valleys excavated in gypsum formations commonly display an asymmetric cross profile with a prominent gypsum escarpment in one of the margins and staircased terraces on the opposite side (Figure 8.10). For instance, most of the valleys excavated in Cenozoic evaporites in Spain feature a linear and unstable gypsum escarpment ≥100 m on one of the flanks (Gutiérrez et al. 2001, 2008b). Other examples include the Salt River in Alberta, Canada (Tsui and Cruden 1984), the Eagle River in the vicinity of Gypsum village in Colorado, the Ure River in Ripon, United Kingdom (James et al. 1981; James 1992), the Sylva River in the fore-Ural region, Russia (Andrejchuk and Klimchouk 2002) or the Kızılırmak River in Sivas, Turkey (Gökkaya et al. 2021) (Figure 8.10e). Gutiérrez (1996) inferred that the asymmetry of the Jalón River valley in Calatayud Neogene Graben (Iberian Chain, Spain) and the development and dynamics of its gypsum escarpment are influenced by dissolution-induced synsedimentary subsidence. The thickened alluvium filling a dissolution trough became more resistant than the gypsum bedrock by cementation, leading to the lateral shift of the valley and preventing its displacement in the opposite direction. Moreover, the lateral migration of the fluvial system is accompanied by renewed dissolution and subsidence at the scarped margin, which enhances landslide activity and tends to perpetuate the lateral displacement of the valley. The gypsum escarpments frequently show numerous landslides, hanging valleys and erosional triangular facets indicative of a rapid retreat rate (Figure 8.10). The cliffs can be very linear and may resemble mountain fronts controlled by active normal faults, but their rapid retreat reveals that they are generated by erosion acting on highly soluble bedrock and frequently affected by penetrative joint sets that control their rectilinear pattern (Gutiérrez et al. 1994; Guerrero and Gutiérrez 2017). The fast scarp recession can be attributed to a number of processes favored by the high solubility and erodibility of gypsum, as well as its relatively low mechanical strength: (i) Long-term entrenchment of the fluvial system, which can be locally influenced by dissolution-induced subsidence or halokinesis in the case of halite-bearing evaporite formations (Gutiérrez and Lizaga 2016). (ii) Rapid undercutting of the cliff by fluvial erosion (Figure 8.10b). (iii) Loss of basal support induced by subsurface dissolution. Groundwater flow lines tend to converge beneath the valley floors resulting in enhanced dissolution at the foot of the scarps. A number of works document a spatial correlation between landslides, springs that cause sapping erosion and active dissolution-induced subsidence at the foot of the slopes (Reuter et al. 1977; Rovera 1993; Gutiérrez 1996; Alberto et al. 2008; Guerrero and Gutiérrez 2017; Denchik et al. 2019). (iv) Frequent occurrence of slope movements favored by the limited mechanical resistance of the rock mass, which can be rapidly reduced by dissolution acting along bedding planes, fractures, and unloading cracks. The type of slope movements is strongly controlled by litho-structural factors (attitude of bedding, jointing, presence of argillaceous units). Karst cavities may favor and guide the development of failure planes (Seijmonsbergen and de Graff 2006). Additionally, the circulation of water

Figure 8.10 Images of gypsum escarpments in asymmetric sections of the Ebro River valley, NE Spain, and in Turkey. (a) Linear gypsum cliff with hanging valleys and triangular facets carved on horizontal Miocene gypsum in Alfajarín village area. Note toppling monolith in the central sector of the scarp. (b) Active landslide induced by fluvial undercutting in an escarpment around 100 m high made up of gypsum with intercalated clay units. At the top of the cliff, remains of El Castellar village, most probably abandoned due to landslide activity. (c) Active landslide in a gypsum scarp with a clay unit at the base of the slope (Villafranca village area). This is a multiple rotational landslide with a lateral spreading component related to plastic deformation of the clay unit, and subsidence induced by subsurface dissolution, mainly acting at the gypsum-clay contact. (d) Cliff on tightly folded Paleogene gypsum at Azagra village, where four rock-fall events have killed a total of 114 people. (e) Gypsum escarpment in the Todürge Canyon of the Kızılırmak River in Sivas region, Turkey. *Source:* Francisco Gutiérrez (Author).

in the karst conduit network, especially when the incoming flow exceeds the discharge capacity, may induce high fluid pressures reducing the normal effective stresses and shear strength of potential failure surfaces. (v) Prompt removal of the landslide deposits by mechanical and chemical erosion, leading to continuous relief rejuvenation. Landslide deposits accumulated at the foot of scarped slopes tend to act as stabilizing and protecting buttresses, but these elements have a limited persistence time when made up of gypsum and associated with river channels. Rapid landslides such as rock fall or topples can pose a significant risk. For instance, recurrent rock fall events occurred in a gypsum cliff at Azagra village (Ebro valley, Spain) in 1856, 1874, 1903, and 1904 killed 11, 100, 1 and 2 people, respectively (Gutiérrez et al. 2008b) (Figure 8.10d).

8.4 Salt Tectonics and Karst in Salt Extrusions

8.4.1 The Special Rheology of Rock Salt. An Exceptional Mobile Bedrock

Rock salt is a very special karst bedrock, not only because of its extremely high solubility, but also because of its mobility (Zarei et al. 2012). Salt can flow laterally and vertically at significant rates under both subsurface and surface conditions producing a wide variety of structures, including emergent salt extrusions. The high mobility of salt is related to its low density, and more importantly to its mechanical weakness and negligible yield strength (Jackson and Hudec 2017).

Salt is a relatively incompressible sediment because of its syndepositional crystalline texture. During burial, it barely changes its density, while the density of other sediments increases significantly by various diagenetic processes (e.g., compaction, dehydration, and cementation). Pure rock salt has an approximate density of $2.0\,g\,cm^{-3}$, which is lower than the density of carbonate and compacted siliciclastic rocks. This leads to density inversion and gravitational instability at depths typically greater than 600–1500 m, whereby light salt is overlain by denser overburden. Buoyant salt is liable to rise unless restrained by the overburden.

Salt, because of its negligible yield strength, can experience viscous strain under very low differential stress and at surface temperature conditions. Time is a critical factor in this creep deformation process because of two reasons: (i) permanent strain occurs continuously as long as the stress operates and (ii) the resisting viscous forces for a given viscosity increase linearly with the strain rate. This means that salt is weaker when flow occurs at slow rates over geological time scales. This is exemplified by the fact that rapidly stirring a viscous fluid like honey takes more effort that doing it slowly. Moreover, the viscosity of rock salt decreases with burial as temperature rises, and it can be reduced significantly if the rock contains small amounts of water and crystal size decreases.

Viscous flow of salt occurs by two main microstructural processes: dislocation creep and solution–precipitation creep (Urai et al. 1986; Schléder and Urai 2007). Dislocation creep mainly involves the deformation of crystals by intracrystalline slip along weak crystallographic planes and grain-boundary migration by recrystallization. Solution–precipitation creep entails dissolution at crystal boundaries subject to higher stress where pressure locally increases solubility (pressure solution). The dissolved ions migrate by diffusion along water films associated with the grain boundaries and precipitation occurs at sites of lower stress. This is a water-mediated process in which the crystals change their shape but do not experience internal strain. Solution–precipitation creep requires the presence of water and its rate can be several orders of magnitude higher in fine-grained salt than in salt with a coarse-grained texture (i.e., the smaller the grain size the lower the viscosity of the damp salt). In contrast, dislocation creep is not affected by the grain size. Solution–precipitation creep is the dominant strain mechanism in extrusive salt sheets (salt glaciers or namakiers), where the salt is damped by meteoric water and the crystal size decreases toward the distal parts of the namakiers by mylonitization. Dislocation creep tends to be the main microstructural deformation process under subsurface conditions (Desbois et al. 2010).

Salt flow can be driven by two mechanisms that may operate in combination; buoyancy related to density inversion of buried salt, and more importantly, differential loading (Hudec and Jackson 2007) (Figure 8.11). Differential loading can be induced by lateral tectonic loading (or

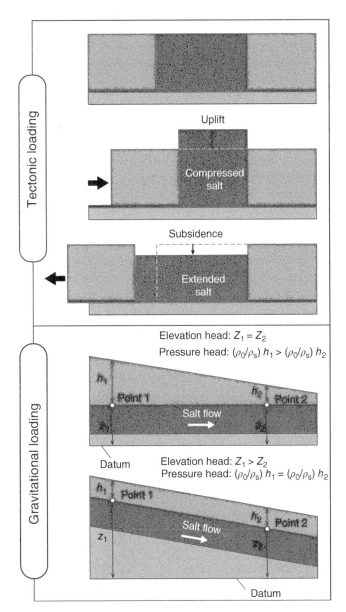

Figure 8.11 Sketches illustrating differential loading conditions related to lateral tectonic loading (or displacement loading) and gravitational loading that can drive salt flow. Tectonic shortening loads horizontally a preexisting diapir. The salt experiences horizontal contraction and uplift when the horizontal tectonic load exceeds the vertical gravitational load. In extension, the salt is unloaded horizontally by divergent horizontal separation of the buttressing side walls. The vertical gravitational load produces lateral expansion and vertical contraction in the salt leading to subsidence. Gravitational differential loading is related to lateral changes in the hydraulic head, which may be due to variations in pressure head or/and elevation head. A laterally varying overburden thickness above a salt layer with horizontal top produces a pressure head gradient. This load variation may be related to sedimentation (e.g., prograding sedimentary wedge), erosion (i.e., erosional unloading), development of ice sheets, or tectonic deformation (e.g., a stack of thrust slices). A salt layer with an inclined top produces an elevation head gradient, inducing down-gradient flow. h: overburden thickness; z: elevation above datum; ρ_s: salt density; ρ_o: average overburden density. *Source:* Adapted from Hudec and Jackson (2007) and Jackson and Hudec (2017).

displacement loading) and vertical gravitational loading. Horizontal tectonic stresses can compress or stretch the mechanically weak salt bodies moving their flanks toward or away from each other, leading to the rise or fall of the salt top. In extensional environments there is overall subsidence (deflation exceeds inflation), whereas in compressional environments net uplift occurs (inflation exceeds deflation) (Figure 8.11). Gravitational loading exerts a continuous vertical force onto the salt layer, which may not be spatially homogeneous. The salt tends to flow toward areas of lower hydraulic head, alike groundwater. Under static conditions, the hydraulic head gradients that drive salt flow may be related to spatial variations in both elevation head and pressure head (Figure 8.11). The flow of salt involves energy expenditure by loss of gravitational potential energy; the loss of potential energy by the sinking overburden is greater than the gain in potential energy by the rising salt. Salt flow induced by gravitational loading can be related to elevation changes in the top-salt topography and/or variations in the weight of the overburden. The changes in the load exerted by the overburden can be due to tectonic and earth surface processes such as the stacking of thrust sheets, prograding sediment wedges (Ge et al. 1997), ice-sheets (Sirocko et al. 2002; Lang et al. 2014), or the excavation of erosional depressions (Huntoon 1982; Schultz-Ela and Walsh 2002). The latter two processes may be of special interest for karst studies since they can induce the rise of the salt resulting in karst rejuvenation. Finite-element modeling conducted by Lang et al. (2014) showed that salt structures respond to ice-sheet loading and unloading by (i) diapir rise when the ice sheet advances close to the diapir (differential loading) and when the ice retreats (unloading) and (ii) diapir fall when the ice sheet covers the diapir. This gravitational loading mechanism has limited duration compared with other processes such as fluvial entrenchment or the build-up and progradation of sedimentary wedges. The development of erosional depressions creates differential loading conditions that may locally drive lateral salt flow toward the unloaded troughs. Gutiérrez et al. (2019c) compiled case studies that illustrate active salt flow related to erosional unloading, providing information on the geomorphic and stratigraphic effects, as well as data on surface displacement rates. The term halokinesis refers to deformation driven by gravitational differential loading without the interplay of significant lateral tectonic forces. An additional potential driver, although not fully demonstrated, is thermal expansion. The thermal expansivity of salt is much higher than that of most other rocks (70 times higher than quartz). Consequently, light hot salt may be overlain by denser cold salt (density inversion), potentially leading to intrasalt convection (Talbot and Pohjola 2009).

8.4.2 Salt Structures and the Geomorphic Impact of Salt Flow

Salt flow can transform an original stratiform body (i.e., the source layer) into salt structures with a wide range of geometries and dimensions (Jackson and Talbot 1986; Hudec and Jackson 2011) (Figure 8.12). When a salt layer thins by flow (salt withdrawal or expulsion), the top of the salt and the overlying overburden subside (salt deflation), potentially creating accommodation space for sediment deposition at the surface. Eventually, the salt can be completely removed and the suprasalt and subsalt rocks get in contact creating a salt weld that prevents salt migration. The salt can flow elsewhere in the subsurface causing salt inflation, and may extrude at the surface. Autochthonous salt lies on its original subsalt rocks, whereas allochthonous salt has migrated from its original source layer through a salt feeder and overlies younger subsalt strata. Salt structures such as salt pillows (e.g., Anbar Ridge, Zagros Mountains), and salt anticlines (e.g., Cordillera de la Sal, Chile) have a concordant contact with the overburden (non-diapiric salt structures). Salt pillows and salt anticlines are buried salt mounds with a length to width ratio in planform (axial ratio) lower and greater than 2, respectively. The rising salt can pierce the overburden (active diapirism) shouldering it aside and generating salt structures with a discordant contact with the enclosing overburden (diapiric salt structures). Eventually, the salt may extrude at the surface (passive diapirism), generally taking advantage of weakness zones or differentially eroded zones (Talbot and Alavi 1996). The most common types of discordant diapiric structures are salt walls (e.g., Mount Sedom, Israel; salt-cored anticlines of the Paradox Basin, Utah and Colorado; Cardona Diapir, Spain), salt stocks (e.g., salt domes in the Zagros Mountains), and salt sheets, either extrusive (e.g., salt glaciers in the Zagros Mountains) or intrusive. Salt walls are elongated and discordant salt ridges with an axial ratio ≥2. Salt stocks are salt plugs with an axial ratio <2 that may display a deep pedestal, a stem and a shallow bulb. Salt sheets are allochthonous salt bodies several times wider than thick that include extrusive salt glaciers (namakiers) and intrusive salt sills injected along a stratification plane. The coalescence of salt sheets results in the formation of salt canopies (Figure 8.12). The term diapir was coined in the Carpathian fold belt of Romania (Mrazec 1907) and is derived from the Greek verb "diaperno", meaning to pierce (Jackson and Hudec 2017). Rocks associated with salt structures typically display high degrees of deformation. This is partially because the mechanically weak salt acts as a strain localizer and enhancer. An additional potential reason is the gravitational deformation of the overburden related to interstratal dissolution of the salt.

The rate at which salt flows depends on a number of extrinsic and intrinsic factors in addition to the amount of differential stress. The strength of the overburden and friction along the edges of the salt bodies tend to impede or retard salt flow (Hudec and Jackson 2007). As indicated above, because of its viscous flow rheology, salt is weaker when affected by slow strain rates. The viscosity of salt slightly decreases with depth as temperature rises. This is the opposite trend to that of most other rocks, that increase their strength as confining pressure increases with burial depth until the brittle-ductile transition zone. The presence of small quantities of water at the crystal boundaries

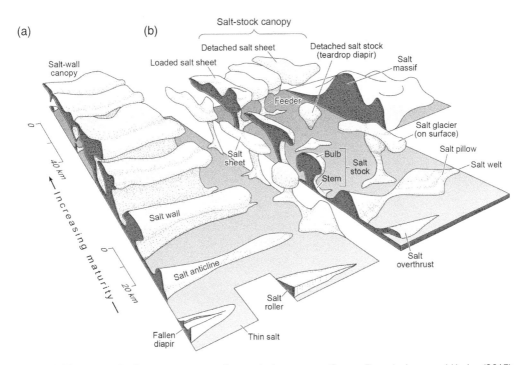

Figure 8.12 Types of salt structures according to their geometry. *Source:* From Jackson and Hudec (2017).

contributes to significantly decrease the viscosity of salt and increase its flow rate. This is because solution–precipitation creep can contribute to the flow of damp salt. The effectiveness of this water-assisted mechanism has an inverse relationship with the size of the halite crystals, and consequently flow rates in fine-grained damp salt can be several orders of magnitude higher than in coarse-grained salt (Urai et al. 2008). The proportion of impurities contributes to increase the viscosity (stiffness) of rock salt, but they seem to have a limited impact if their percentage is below 20–25%.

Jackson and Hudec (2017) compiled displacement rates and strain rates from three types of salt structures: (i) advance rates of subaerial and submarine salt glaciers, (ii) rise rates of salt extrusions in Iran and Israel, and (iii) rise rates of buried diapirs. These rates were obtained by multiple direct (e.g., geodetic techniques) and indirect methods (e.g., long-term deformation of markers) and covering a wide range of time spans. Figure 8.13 shows the range of reported flow rates in each environment and the mean values. The highest rates occur in unconfined salt glaciers despite the low stress and temperature. This is attributed to the incorporation of meteoric water and the down-flow decrease in the size of the halite crystals by mylonitization. Most probably, sliding by basal shear (gravity gliding) also contributes considerably to the advance of namakiers in addition to internal creep deformation by gravitational spreading. The lower flow rates of rising buried diapirs compared to those of emergent diapirs is related to the resisting forces imposed by the sedimentary roof (strength and weight).

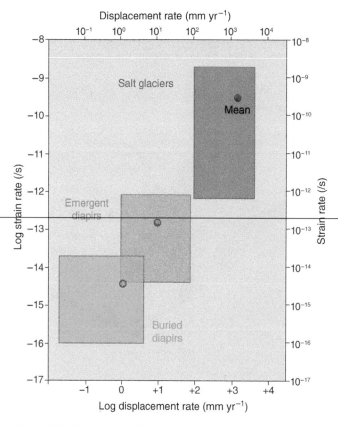

Figure 8.13 Plot showing the range of displacement and strain rates reported for buried diapirs, emergent diapirs and salt glaciers (subaerial and submarine). Mean values are indicated with a circle within each field. Note that rates of salt flow vary over 5–7 orders of magnitude. *Source:* Adapted from Jackson and Hudec (2017).

The movement of salt produces deformation at the ground surface, including subsidence in withdrawal/deflation zones, uplift in accumulation/inflation zones, and horizontal displacement in areas affected by lateral movement such as salt spreading and/or salt gliding. These are shallow deformations that do not produce significant earthquakes, but may have a significant impact on morpho-sedimentary environments and salt karst systems. Like in tectonic geomorphology, surface deformation related to active salt flow can be identified and characterized by using geodetic methods and analyzing Quaternary stratigraphic and geomorphic markers (e.g., terraces, caves) with a known predeformation geometry and position. Numerical dating of the deformed markers allows calculating long-term deformation rates, which may show significant lateral variations. Diapiric geomorphology can be defined as the study of the impact of salt flow on earth-surface processes, landform development, and landscape evolution. The geomorphic responses to the activity of salt structures can be classified into: (i) primary effects related to surface deformation; and (ii) secondary effects including a wide variety of geomorphic processes and features induced by ground deformation produced by salt flow (see review in Gutiérrez et al. 2019a). Common primary geomorphic effects include uplift above inflating buried salt structures, subsidence associated with salt-deflation zones (e.g., withdrawal basins), or the rise of salt extrusions and their lateral displacement across the ground surface in the associated salt glaciers. The main secondary geomorphic effects are changes in the drainage network (deflected, defeated, beheaded, and obliterated streams), the development of multi-level cave systems in salt controlled by diapiric rise, or the formation of subaerial and subaqueous landslides induced by the over-steepening and vertical growth of slopes by active salt flow. Some works conducted in key salt karst sites illustrate the use of deformed and dated geomorphic markers to assess long-term rates of diapiric activity. In Mount Sedom, Israel, Frumkin et al. (2011) and Frumkin (1996) documented multi-level cave passages that record the uplift of the diapir with respect to its adjacent base level, the Dead Sea. These authors estimated long-term uplift rates of ≤6–7 mm yr^{-1} over the past 8 kyr, based on the relative height of the subhorizontal passages and radiocarbon dates from vegetation remains found in cave deposits. Bruthans et al. (2010) documented the geomorphic record of long-term diapiric rise in the extruding Namakdan salt stock (Qeshm Island, Zagros Mountains, Iran), including uplifted marine terraces, fluvial terraces, and cave levels. They calculated uplift rates across the diapir ranging from 4 mm yr^{-1} at 600 m from the diapir edge, to 0.4–0.6 mm yr^{-1} in the surrounding encasing rocks (300 m wide fringe affected by dragging). In extrusions of Miocene salt in the Red Sea coast of Yemen, Davison et al. (1996) calculated uplift rates of 4.6 mm yr^{-1} using coral reefs raised 17 m and radiocarbon dated at 3.7 kyr.

8.4.3 The Morphological Evolution of Salt Extrusions and the Role of Dissolution

The evolution of salt extrusions and the associated glacier-like salt sheets are controlled by two antagonistic processes: (i) the upward supply of salt from the source that tends to rise the salt extrusion above the feeding vent, eventually producing gravity-driven salt sheets; and (ii) the erosion of the exposed salt, which is mainly achieved by dissolution. Interestingly, salt dissolution also plays a critical role in the flow of emergent salt extrusions by reducing the mechanical strength of the salt and participating in the pressure-solution creep mechanism. The Zagros region in Iran, with around 200 salt extrusions of the Neoproterozoic to Cambrian Hormuz salt (Kent 1970; Bosák et al. 1998), is one of the finest natural laboratories for the study of bare salt karst, locally developing on a mobile bedrock. Here, the salt rises from depths of 3–12 km driven by differential overburden pressure and active compression related to the collision between the Arabian and Eurasian plates. The Zagros Mountains mostly display a concordant topography, consisting of anticlinal ridges and intervening synclinal depressions with extensive alluvial fan aprons. A great part of the salt extrusions occurs associated with the anticlinal ridges, which were localized by preexisting diapirs and salt pillows (Hessami et al. 2001). Their position within the anticlines is often controlled by pull-aparts at releasing bends along transverse strike-slip faults. The salt extrusions have pierced the overburden breaching the anticlines and the exposed salt domes are frequently unconfined on one or both sides of the anticlinal ridges. Some diapirs emerge at the nose of anticlines or in alluviated plains. The topography surrounding the salt extrusions plays an important role on the evolution of the domes and salt sheets. The salt extrusions in the Zagros display different morphological stages that allow the reconstruction of their general geomorphic evolution by substituting time by space (i.e., ergodic approach) (Figure 8.14):

1) Growing dome stage. Initially, the salt spelled from the vent forms a dome with steep slopes that grows vertically and laterally. The lateral expansion of the dome can be constrained by the surrounding relief and may attain an elongated or irregular outline depending on its topographic and structural context.

2) Salt fountain stage (Figures 8.15a–c, 8.16a, and 8.17). When the dome reaches a critical height above the vent (ca. 600 m), the weight exceeds the yield strength of the salt and the lower slopes start to spread laterally over the surrounding ground to form salt sheets. These gravity-driven salt flows that resemble glaciers were designated as namakiers by Talbot and Jarvis (1984); the Farsi word namak means salt. At this stage, the salt extrusion develops the shape of a viscous fountain, comprising a summit dome above the vent that grades into namakiers with relatively gentle gradients that flow downslope (Talbot and Pohjola 2009; Aftabi et al. 2010). The height of the summit dome depends on the positive contribution of salt extrusion and the negative role of both lateral flow and salt dissolution. The development of namakiers is restricted to the unconfined sides of the salt fountains. Some salt extrusions display unidirectional or bidirectional salt sheets on the breached sides of the anticlinal ridges (Figure 8.15c). Others like Mesijune (or East Mazyjan) (Figure 8.15a, b) and Syahoo diapirs have multidirectional namakiers

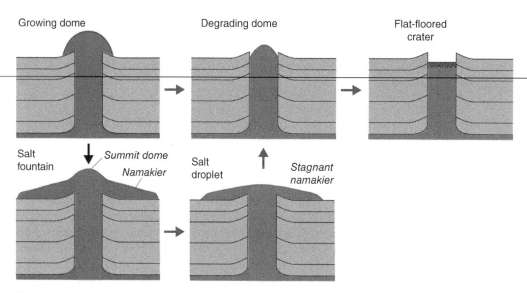

Figure 8.14 Morphological evolution of subaerial salt extrusions observed in the salt stocks of Iran. The red color indicates growing extrusions and the pink color degrading extrusions. The net growth or degradation of the extrusions depends on the rates of upward salt supply, erosion (mainly related to dissolution), and lateral flow in the salt sheets. Drops in the hydraulic pressure of the salt systems can also induce downward salt flow and deflation.

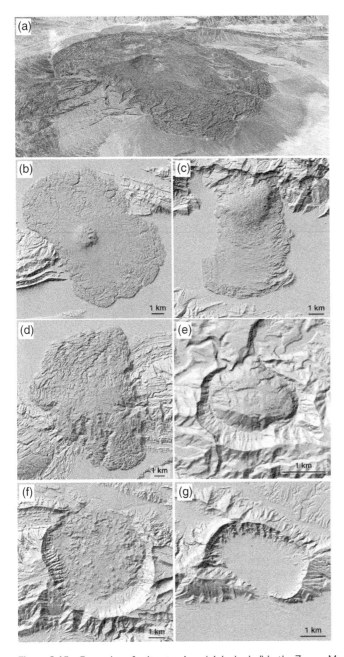

Figure 8.15 Examples of salt extrusions (pink-shaded) in the Zagros Mountains showing different evolutionary stages. (a, b) Mesijune (or East Mazyjan) salt fountain with a well-defined summit dome and multidirectional salt glaciers with gentle slopes. Note flow-transverse ridges and furrows in the distal sector of the namakiers attributable to flow folding. *Source:* Image captured from Google Earth. (c) Gach salt fountain with a summit dome associated with the limb of an anticline, and an unidirectional namakier that flows over unconsolidated alluvial fan deposits in the adjacent piedmont. Note arcuate ridges and furrows in the lower compressional zone. (d) Kalat salt droplet located between the opposing plunging noses of two anticlines. The salt extrusion lacks a summit dome and shows bidirectional namakiers. Note the multiple lobes at the snout of the salt glaciers that control the position of drainages entrenched in the salt. (e) Jalalabad diapir, a protruding degrading dome nested in a crater. Note annular depression developed between the shrinking dome and the walls of the crater, formerly fully occupied by salt. (f) Herang diapir at the stage of a highly degraded dome nested in a crater depression. The salt is completely covered by a mature lithified dissolution residue (caprock) showing a hummocky topography, largely related to the presence of resistant blocks that used to form inclusions in the salt. (g) The fully degraded Kameshk salt structure expressed as a flat-floored crater filled with alluvium. Note collar of overturned strata (flaps) forming an annular ridge. *Source:* Francisco Gutiérrez (Author). Shaded relief models generated with TanDEM-X DEMs from the German Aerospace Center (Project DEM_GEOL288 granted to Francisco Gutiérrez).

Figure 8.16 Images of salt extrusions from the Zagros Mountains of Iran (a-g) and Mount Sedom in the Dead Sea Basin, Israel. (a) Kuh-e-Namak (or Dashti) salt fountain with its summit dome and the southern namakier that descends toward the right of the image. The steep slopes are dominated by densely dissected salt exposures. In the foreground outcrop of dipping Asmari limestone surrounded by salt and two persons for scale. (b) Southern side of Bastak N diapir, a shrinking degrading dome with steep salt exposures located in a breached anticline. (c) Eastern flank of the southern namakier of Jahani salt fountain showing capsoils in the gentler upper part of the salt slopes. The lower sections of the slopes are covered by capsoils and

that flow radially over the surrounding piedmont areas. These boundary conditions have an important impact on the kinematics of the salt glaciers. In laterally confined unidirectional namakiers, the flow rate and the thickness of the salt tend to be greater. In multidirectional salt sheets with diverging streamlines, the flow is distributed in a much wider section and the rate at the distal end tends to decrease as the salt expands over the surrounding ground and the perimeter of the apron increases. The local relief between the summit dome and the distal end of the namakiers can reach as much as 1400 m like in Kuh-e-Namak (or Dashti) salt extrusion (Figure 8.16a). Mesijune Diapir shows the longest namakiers with a maximum length of 6 km (Figure 8.15a, b).

3) Viscous droplet stage. When salt supply ceases due to the exhaustion of the source layer or the pinch off of the vent, the summit dome of the salt fountain falls and the adjoining namakiers become stagnant salt sheets. Then, the salt extrusion attains the morphology of a viscous droplet with a profile similar to that of ice caps (Talbot and Pohjola 2009; Aftabi et al. 2010) (Figure 8.15d). The same situation could be reached if erosional lowering exceeds the rise of the salt, or if the salt flow in the vent stops or reverses its direction (e.g., downward flow to fill space created in the source by compression of the overburden; Dooley et al. 2009). At this stage, the geomorphic evolution of the unfed salt extrusion is dominated by solutional denudation. Salt dissolution is accompanied by the accumulation at the surface of residual deposits made up of the insoluble components of the salt formation (e.g., clay, gypsum, and rock inclusions) (Figure 8.16). Capsoils play a critical role in the geomorphic evolution of salt extrusions since they function as a protective cover that substantially decelerates solutional denudation (Bruthans et al. 2009; Zarei et al. 2012). The salt extrusion experiences progressive erosional lowering and the thin distal edge of the namakiers recedes. Where the salt is completely dissolved, it leaves on the ground surface the insoluble material as dispersed erratic blocks or as a laterally continuous apron of unsorted, polymictic and chaotic deposit similar to glacial melt-out tills (Talbot and Pohjola 2009) (Figure 8.16e). These dissolution-related accumulations are clearly recognizable in the Zagros Mountains because of their dark color, texture and lithological composition, commonly including exotic igneous, metamorphic and sedimentary rocks. They can be found hundreds of meters away from the current salt edge and can be used to reconstruct and date (e.g., cosmogenic nuclides) the former extent of salt extrusions.

4) Degrading dome (Figure 8.16b, e). These are domes dominated by solutional degradation that progressively reduce their height and areal extent. This situation can occur when (i) a waning viscous salt droplet is restricted to a residual salt outcrop associated with the vent, (ii) a growing dome starts to degrade because of the cessation of salt supply or because dissolution rate outpaces the expelling rate, and (iii) the salt extrusion experiences deflation due to loss of

---

colluvium. (d) The Hormuz salt extrusion in an area where the salt is largely covered by a red clay- and iron-rich dissolution residue. The protruding peak in the background corresponds to a large inclusion of marls. Note ephemeral precipitates of halite in the channel bed and the slopes. (e) Eastern flank of the Bastak N degrading diapir, where retreat of the extrusion by dissolution has deposited a mantle of residual deposits resembling a melt-out moraine. Person in the foreground for scale. (f) Frontal zone of the southern namakier of Kuh-e-Namak (or Dashti) salt fountain. Differential solutional erosion controlled by the discontinuous capsoil and isolated blocks has produced a highly irregular landscape of towers, pinnacles and pedestals. (g) Northern namakier of Jahani, covered by relatively thick and indurated capsoil and pockmarked by cover collapse sinkholes. (h) Large caprock collapse sinkhole controlled by a dilated bedding-parallel fault (arrow) in the summit of Mount Sedom, Israel, where the salt is overlain by a thick anhydrite-rich caprock. The crest of the diapir was planated by the pluvial precursor of the Dead Sea lake at ca. 14 kyr. Circle indicates person for scale. *Source:* Francisco Gutiérrez (Author).

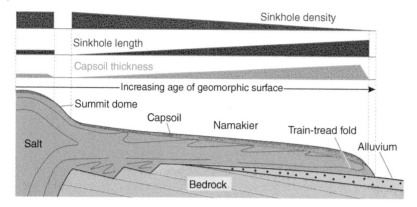

Figure 8.17 Spatial variation in a salt fountain of the capsoil thickness, the size of the sinkholes and their density (number by unit area). These variables are controlled by the age of the geomorphic surface developed on the mobile bedrock and the slope. The highest erosion rates occur in salt exposures associated with steep slopes devoid of protecting residual soils. *Source:* Based on Zarei et al. (2012).

hydraulic pressure in the vent and downward salt flow. These receding domes are often nested in crater-like depressions once fully occupied by the extruded salt (Jahani et al. 2007) (Figure 8.15e). The craters commonly display an annular depression between the dome and the steep rock walls that used to confine the salt extrusion. The progressive degradation of the dome is accompanied by a reduction of its height and a general slope decline, favoring the accumulation of thicker and more widespread residual deposits. The capsoil can become progressively indurated by cementation to form a lithified caprock that typically displays a highly irregular topography related to the presence of sinkholes, gullies and large resistant blocks of non-soluble lithologies (e.g., rock inclusions) (Figure 8.15f). Salt exposures can be restricted to the steep walls of collapse sinkholes. The topographic expression of the dome is progressively degraded by multiple processes including mechanical erosion, aggradation in depressions, dissolution-induced subsidence, and salt deflation.

5) Flat-floored crater. At some stage, the floor of the crater attains a smooth surface underlain by alluvial and colluvial deposits that conceal the residual cap of the salt diapir. The dissolution residues, upturned strata (flaps) at the margins of the crater and resistant rim breccias attached to the walls are some of the evidence of the former presence of a diapir (Figure 8.15g).

The coarse-grained and dry diapiric salt spelled from the vent is too strong to experience viscous flow under subaerial conditions. A 600 m high dome of rock salt with a density of $2 \, g \, cm^{-3}$ exerts a load at its base of just 12 MPa (ca. 120 kg cm^{-2}). This made the mechanics underlying the development of namakiers in an apparently stiff rock difficult to explain in the early studies. However, field observations and laboratory experiments reveal that three main processes make the development of subaerial salt glaciers possible (Talbot and Pohjola 2009):

1) The incorporation of meteoric water in the extruded de-stressed salt, previously subject to high pressure conditions, allows the participation of the pressure-solution creep mechanism, significantly reducing the dynamic viscosity of the rock (Spiers et al. 1990). Laboratory experiments indicate that damp dilated salt flows about 40 times faster than dry confined salt (Cristescu and Hunsche 1998).

2) The expelled salt experiences progressive grain size reduction by mylonitization from the upper part of the domes toward the distal edge of the salt glaciers (Desbois et al. 2010). Mylonitization is a dynamic recrystallization process that involves the transformation of large crystals into

smaller ones during deformation. This grain size reduction promotes pressure-solution creep with the presence of water, and grain-boundary sliding, significantly reducing the dynamic viscosity of the salt. Subaerial salt sheets in Iran display a bimodal grain size distribution consisting of porphyroclasts and mylonitic fine crystals (<2 mm). The porphyroclasts decrease in size and number along the flow path and strain is mainly concentrated in the weaker fine-grained zones.

3) The salt glaciers can slide along discrete mylonitic zones located near the base of the salt. These are smooth shear zones developed over the irregularities of the substratum, underlain by portions of quasi-static salt and overlain by detached salt that slides downslope (Talbot 1981). This field observation indicates that the movement of salt glaciers is related to a combination of lateral spreading and gliding along sliding surfaces, similar to the displacement of wet-based temperate glaciers and some flow-slide-type slope movements.

Gravitational stresses in salt fountains are dominated by radial extension in the crestal zone and flow-parallel compression in the distal part of the namakiers. The lateral gravitational spreading of the salt over a frictional base produces a first order structure consisting of a large recumbent sheath-like fold, with its upper limb subparallel to the top surface of the salt glacier, and its axis perpendicular to the flow direction (Figure 8.17). This major fold develops like a tank tread. In the frontal zone of the advancing salt sheet the upper limb rolls down over the hinge and attaches to the substrate. The upper limb of these tank-tread folds typically display secondary downslope-verging similar-type folds (i.e., layer thickening in the hinge zone). These are asymmetric reclined open folds with the axial plane dipping upslope in the upper part of the glaciers that transform into amplified isoclinal recumbent folds toward the distal sector. Field evidence indicates that in sections with an irregular bed on resistant rocks (e.g., serrated surface on dipping limestone), the salt rather than eroding a smooth surface by plucking and abrasion, tends to fill the irregularities with stagnant salt below intra-salt mylonitic shear zones <1 dm thick (Talbot 1979, 1981) (Figure 8.17). The overlying detached salt moves downslope along these discrete shear zones that function as sliding surfaces. The salt filling the basal irregularities typically displays tight recumbent similar-style folds and thrusts in the upper part of the salt glaciers. Downstream, where the viscosity of the salt decreases by the mylonitization process, the dead space associated with the obstructions is filled by salt showing open drape folds.

Namakiers in the arid Zagros Mountains are characterized by episodic kinematics, with long periods of near-stasis punctuated by short intervals of rapid displacement following rainfall events. Salt sheets display average flow rates of cm to m yr^{-1}, but eventually can undergo surges with flow velocities in the order of dm day^{-1} (Jackson and Talbot 1986) (Figure 8.13). In a namakier in the Zagros, Talbot and Rogers (1980) measured a displacement of around 1 m over two days after a rainfall event. This behavior is attributed to the development of an outer fractured carapace a few meters thick that functions as a restraining corset during dry periods. Rainfall events rapidly weaken this stiff outer shell by dissolution promoting surging events (Talbot and Pohjola 2009). The exposed salt develops a fractured, dilated and high-permeability outer zone typically less than 10 m thick (Figure 4.11d). This near-surface fracturing is related to two mechanisms: (i) de-stressing of the salt coming from a deep source, resulting in the development of unloading joints subparallel to the topographic surface; and (ii) cycles of thermal expansion and contraction, with development of cracks with multiple directions when the temperature drops and the rock shrinks. The thermal expansivity of rock salt is an order of magnitude higher than most other rocks. Talbot and Pohjola (2009) proposed a conceptual model to explain the episodic behavior of salt glaciers. During the long dry periods, salt extrusion causes the inflation of the summit dome, where the salt is stored at high elevation constrained by the strong carapace. Slow gravitational spreading operates through the development the large-scale tank-tread folds along the salt glaciers, with the salt attached to the high frictional base. Rainfall events rapidly weaken the outer carapace by dissolution and the summit dome subsides releasing its potential energy by surging events. The adjoining salt glaciers move by gravity gliding along the near-basal mylonite zone, which can be weakened by the incorporation of meteoric water. A slightly different surface displacement pattern was documented in a salt fountain in the Zagros Mountains of Iran (Syahoo diapir) by DInSAR with data spanning 14 years and including 12 time increments (Aftabi et al. 2010). This subcircular salt extrusion 6.5 km across comprises a central summit dome 2 km in diameter and 150 m high, and a multidirectional salt sheet with active radial flow. The local relief from the summit to the toe of the namakiers reaches 600 m. Soon after the wet intervals the salt fountain experiences a general inflation affecting both the summit dome and the surrounding namakiers. During the dry periods the salt fountain is affected by a general slow subsidence and presumably lateral spreading (i.e., gravitational spreading entails vertical contraction and horizontal extension). The recorded displacement averages $215\,mm\,yr^{-1}$ uplift and $-263\,mm\,yr^{-1}$ subsidence. The deformation maps produced with the multitemporal InSAR data provide spatially continuous displacement values in the line-of-sight (23° from the vertical), but do not allow resolving salt movement in three dimensions. In the Qum Kuh salt fountain of the Great Kavir, Iran, Abdolmaleki et al. (2014) also document seasonal variations in the wet and dry seasons using DInSAR-derived ground displacement data. Talbot and Pohjola (2009) suggest that although rainfall water may reach the mylonitic detachment layers enhancing sliding, the main control is exerted by the strengthening and weakening cycles experienced by the outer weathered zone. This indicates that climate can play an important role on the evolution of salt glaciers, with wetter climate promoting salt flow. There are a number of uncertainties in the proposed kinematic models related to the limited available data, especially concerning the processes that operate in the basal zone of the namakiers. Some relevant questions include: (i) Has the thin and densely fractured outer carapace sufficient strength to constrain gravitational spreading in the salt fountains? (ii) What is the relative contribution of basal sliding in the movement of the salt glaciers, which is mostly attributed to gravitational spreading (viscous flow)? (iii) Do rainfall waters percolate up to the base of the namakiers? The salt below the surficial fractured and weathered zone is generally considered to be impermeable (Zarei et al. 2012). However, water could locally reach the lower boundary through local shafts and fissures, which development may be favored by irregularities in the bedrock (locally exposed within the salt sheets) and lateral variations in the flow rate of the salt. Small quantities of water can rapidly reduce the shear strength of the sliding surfaces by salt dissolution potentially inducing surges in the salt glaciers. Moreover, salt glaciers flow in large sections across easily erodible bedrock (e.g., marls of the Mishan Formation or gypsum of the Gachsaran Formation) and unconsolidated alluvium. The interaction between the active namakiers and the soft substratum remains uncertain. Boreholes with piezometers and inclinometers would provide valuable clues on this issue.

The creation of positive relief by salt extrusion is counteracted by erosion, which is mainly achieved by salt dissolution. In the outer few meters, the salt is transformed into a weak and permeable material by fracturing related to stress-release and thermal expansion-contraction cycles. Water can easily percolate through the fissures causing their solutional enlargement and progressively increasing the permeability of this weathering zone. In contrast with the epikarst developed in fractured carbonate rocks, this thin surficial aquifer is considered to be underlain by impermeable bedrock, especially where the salt is affected by viscous flow, which tends to anneal any discontinuity (Zarei et al. 2012; Abirifard et al. 2017). Dissolution acting on salt exposures generates an in situ residual deposit consisting of the insoluble and less soluble components of the salt bedrock. In the salt extrusions of the Zagros Mountains, this weathering residuum or capsoil typically consists of a clayey, massive, and unsorted deposit that may include angular clasts of a wide range of dimensions and lithologies (Figure 8.16). Studies performed by Bruthans et al. (2009) in eleven salt diapirs of the Zagros Mountains indicate that these capsoils have a high infiltration capacity, despite they become indurated by cementation due to precipitation of gypsum/anhydrite and carbonate minerals. This residual cover prevents the direct interaction between rainfall and overland flow with salt, significantly reducing erosion rates and playing a key role on the geomorphic evolution of the salt extrusions. Bruthans et al. (2008) measured erosion rates in salt exposures and capsoils in coastal diapirs (ca. 170 mm yr^{-1} annual precipitation) and in more humid mountain diapirs located inland (ca. 550 mm yr^{-1} annual precipitation). Overall, the erosion rates were 10 and 35 times lower in the capsoils than in the bare salt surfaces, respectively. The preservation of this protecting cover mainly depends on the slope angle and the presence of vegetation, which development is strongly influenced by the climatic conditions (Figure 8.17). The salt extrusions of the Zagros Mountains located in the more humid coastal areas show thicker and more widespread capsoils than those located the drier coastal zones of the Persian Gulf with limited vegetation cover (Bruthans et al. 2009). On steep slopes (>40–50°), the dissolution residue is rapidly removed by mechanical erosion and the landscape is dominated by dense networks of steep gullies carved in impermeable rock salt (Figure 8.16c). In these highly active badlands, erosion is dominated by dissolution produced by rainfall and overland flow and the slopes are frequently riddled by rillenkarren (Figure 6.4a, c). In areas where the salt is locally protected by discontinuous patches of residual deposits or isolated blocks, differential dissolution can produce an extremely rugged topography of pinnacles and pedestals (Figures 4.8c and 4.18f). An important consequence of the strong control exerted by the slope angle on the persistence of the capsoils and the erosional regime (mechanical versus solutional), is that the role of the different geomorphic processes can be markedly different depending on the evolutionary stage of the salt extrusion and the different sectors within each one. For instance, the flanks of protruding salt domes or the fronts of salt glaciers have steep slopes, whereas degrading salt droplets are dominated by gentle slopes with extensive capsoils (Figures 8.15 and 8.16). Other processes such as fluvial erosion can play a significant role at specific locations. The Firoozabad River cuts through the northern margin of the Jahani salt fountain increasing its average concentration in dissolved halite from 100 to 12000 mg L^{-1}, involving the removal of around 215000 ton yr^{-1} (Abirifard et al. 2017).

The rainfall and runoff water that percolates through the capsoils causes rapid salt dissolution at the rockhead and within the fractured and relatively permeable weathering zone. Subsoil dissolution acting at the rockhead (covered karst) produces a residue that grows at the cover-bedrock interface and leads to progressive ground subsidence. Differential dissolution beneath the soil mantle, which can be controlled by fractures, bedding planes, or more soluble beds, results in the development of cover-collapse and cover-suffosion sinkholes (Figure 8.16g). On the other hand, the collapse of shallow cavities developed within the salt bedrock results in the formation of cover- and bedrock-collapse sinkholes (Figure 8.16h). The surfaces with gentle slopes and mantled by capsoils tend to be pockmarked by densely packed sinkholes, in some cases forming a polygonal karst topography. Zarei and Raeisi (2010) and Zarei et al. (2012), in enlightening papers on the Konarsiah salt extrusion (Zagros Mountains), documented a morphological gradation in the namakiers controlled by the time elapsed since the emergence of the mobile salt and the slope of the topographic surface (Figure 8.16). The thickness of the capsoils shows a progressive increase toward the distal sectors of the salt glaciers, where bedrock has been exposed to weathering for a longer time, provided they have low slopes. The sinkholes show a progressively higher degree of maturity toward the frontal sector of the namakiers. They have larger average lengths and lower densities by number due to the progressive lateral growth and coalescence of the depressions (Figure 8.16). Considering the important role that vegetation plays on the persistence of the capsoils, Bruthans et al. (2009) and Zarei et al. (2012) suggest a possible link between climate variability and the growth/degradation trends of the salt extrusions. A change to wetter conditions can be accompanied by more stable and widespread vegetated capsoils, reducing the overall erosion rates and favoring the growth of the salt extrusions. As explained above, this effect can be also complemented by the weakening of the outer carapace that tends to constrain the subaerial salt flow (Talbot and Pohjola 2009).

Mount Sedom in Israel (ca. 50 mm yr^{-1} annual rainfall), made up of Plio-Pleistocene salt, is the other classical salt diapir (salt wall) intensively investigated from the karst perspective (Frumkin 2004 and references therein). This elongated salt extrusion is associated with a releasing bend on a longitudinal fault of the Dead Sea transform fault system. It covers around 14 km^2 and has a local relief of 250 m above the adjacent southern Dead Sea basin to the east. It shows markedly different geomorphological features with respect to those of the Zagros Mountain due to the following factors: (i) the salt is extensively overlain by a relatively impermeable anhydrite-rich caprock around 50 m thick, (ii) the crest of the salt wall was trimmed by erosion caused by the pluvial precursor of the Dead Sea at around 14 kyr (Lisan Lake), to form a flat erosional surface with patches of laminated lake deposits of the Lisan Formation, (iii) the salt has not experienced lateral flow, (iv) salt exposures are very limited, mainly restricted to an escarpment with hanging valleys and numerous slope movements associated with the Dead Sea shore, (v) the eastern side of the salt diapir has steep topographic gradient and the base level (Dead Sea) has experienced substantial changes in the Late Quaternary, (vi) salt rise is largely accommodated by displacement along subvertical bedding-parallel faults (Zucker et al. 2019). The number of sinkholes is relatively small and they largely correspond to caprock-collapse sinkholes, resulting from the foundering of cavity roofs in the salt beneath the thick caprock. Sinkholes tend to form alignments largely controlled by bedding-parallel faults (Zak and Freund 1980) (Figure 8.16h). According to Frumkin (1994a), 43% of the area corresponds to non-karst catchments, and the rest to drainage basins that drain underground via sinkholes and fissures, mostly connected with integrated cave systems. Most of the salt dissolution occurs beneath the caprock and the dominant geomorphic process at the surface is mechanical erosion. A relatively similar situation has been documented in Anbal Ridge (Zagros Mountains, Iran), an active salt pillow in the Miocene Gachsaran Formation. Here, the salt is largely concealed by a thick gypsum-rich caprock, but the surface is pockmarked by an extremely high density of caprock-collapse sinkholes (Gutiérrez and Lizaga 2016). Salt extrusions seem to occur on other planets. For instance, an elongated dome 23.5 km long and 3.4 km high located in Tithonium Chasma, Mars has been interpreted as a recent diapir of evaporitic sediments. The relatively flat crestal zone of this dome is pockmarked by closed depressions ascribed to solution and collapse sinkholes (Baioni et al. 2009; Baioni and Wezel 2010).

Seawater is around 10% saturated with respect to halite, and consequently can produce karst features in submarine salt diapirs and extrusions. Landforms related to submarine dissolution of outcropping and subcropping salt have been documented in a number of regions, such as the Mediterranean Sea, the Red Sea, the North Sea, and the Gulf of Mexico (Taviani 1984). This is probably an important karst topic that remains poorly explored. Like in subaerial environments, salt dissolution counteracts the creation of relief related to the upward salt flow. The brine derived from salt dissolution generally dissipates as it mixes with seawater. However, because of its higher density, it can form drainages on the sea floor and accumulate in depressions with limited bottom currents forming submarine brine lakes. The brine issuing from vents at the top of the Green Knoll, an isolated rising diapir of the Jurassic Louann Salt in the northern Gulf of Mexico, produces braided drainage patterns on the sloping sea floor (Aharon et al. 1992). The 220 m deep hypersaline submarine lake of the Orca Basin on the continental slope of the Gulf of Mexico results from the dissolution by seawater of about 3.6 billion tonnes of exposed Jurassic Louann Salt, at an estimated rate of nearly 0.5 million t yr^{-1} (Pilcher and Blumstein 2007). In the Florence Rise of the eastern Mediterranean, Stride et al. (1977) ascribe isolated subcircular depressions up to 2 km wide with rough ground, as observed on sonographs, to the summits of salt diapirs affected by dissolution. Using sonar surveys carried out at several sectors of the eastern Mediterranean around Crete and south of Sicily, Belderson et al. (1978) identified subcircular to oval depressions as much as 6 km long with rough-surfaced inner mounds or smooth floors. These features, locally arranged in sinuous belts parallel to the structural grain, were interpreted as extrusions of Messinian salt affected by differential dissolution and subsidence. In the same region of the Eastern Mediterranean Ridge, Kastens and Spiess (1984) documented large depressions 50–100 m deep and several kilometers across above the crest of subcropping diapirs of Messinian evaporites. These basins, with stepped margins presumably controlled by normal faults, were interpreted as collapse grabens related to subjacent dissolution of the diapiric salt beneath a cover of Plio-Quaternary hemipelagic deposits. The basins display a highly irregular floor pockmarked by steep-sided collapse sinkholes 20–50 m across and up to 20 m deep. Kastens and Spiess (1984) suggest that the groundwater flow required for the subjacent dissolution could be explained by a brine-density dissolution model (Anderson and Kirkland 1980), whereby dense brines derived from salt dissolution flow downward causing dissolution at depth and eventually the collapse of the overlying sediments.

References

Abdolmaleki, N., Motagh, M., Bahroudi, A. et al. (2014). Using Envisat InSAR time-series to investigate the surface kinematics of an active salt extrusion near Qum, Iran. *Journal of Geodynamics* 81: 56–66.

Abirifard, M., Raeisi, E., Zarei, M. et al. (2017). Jahani Salt Diapir, Iran: hydrogeology, karst features and effect on surroundings environment. *International Journal of Speleology* 46: 445–457.

Aftabi, P., Roustaie, M., Alsop, G.I. et al. (2010). InSAR mapping and modelling of an active Iranian salt extrusion. *Journal of the Geological Society* 167: 155–170.

Aghdam, J.A., Raeisi, E., Zare, M. et al. (2013). Hydrogeology of non-salt Gachsaran formation in Iran: an example from the Zagros range-tang sorkh valley. *Carbonates and Evaporites* 28: 309–319.

Aharon, P., Roberts, H.H., and Snelling, R. (1992). Submarine venting of brines in the deep Gulf of Mexico: observations and geochemistry. *Geology* 20: 483–486.

Alberto, W., Giardino, M., Martinotti, G. et al. (2008). Geomorphological hazards related to deep dissolution phenomena in the Western Italian Alps: distribution, assessment and interaction with human activities. *Engineering Geology* 99: 147–159.

Al-Khattabi, A.F., Dini, S.M., Wallace, C.A. et al. (2007). *Geologic Map of the Judayyidat 'Ar' Ar Quadrangle (31E) and Faydat Al Adyan Quadrangle (31F)*. Jeddah: Kingdom of Saudi Arabia Geological Survey.

Almalki, K.A., Ailleres, L., Betts, P.G. et al. (2015). Evidence for and relationship between recent distributed extension and halokinesis in the Farasan Islands, southern Red Sea, Saudi Arabia. *Arabian Journal of Geosciences* 8: 8753–8766.

Anderson, R.Y. (1981). Deep-seated salt dissolution in the Delaware Basin, Texas and New Mexico. In: *Environmental Geology and Hydrogeology in New Mexico, New Mexico Geological Society Special Publication*, vol. 10 (ed. S.G. Wells and W. Lamber), 133–145. New Mexico Geological Society.

Anderson, N.L. and Hinds, R.C. (1997). Glacial loading and unloading: a possible cause of rock salt dissolution in the western Canada basin. *Carbonates and Evaporites* 12: 43–52.

Anderson, R.Y. and Kirkland, D.W. (1980). Dissolution of salt deposits by brine density flow. *Geology* 8: 66–69.

Anderson, N.L. and Knapp, R. (1993). An overview of some of the larger scale mechanisms of salt dissolution in Western Canada. *Geophysics* 58: 1375–1387.

Anderson, R.Y., Kietze, K.K., and Rhodes, D.J. (1978). Development of dissolution breccias, northern Delaware Basin, New Mexico and Texas. In: *Geology and Mineral Deposits of Ochoan Rocks in Delaware Basin and Adjacent Areas*, vol. 159, 47–52. New Mexico Bureau of Mines and Mineral Resources, Circular.

Anderson, N.L., Brown, R.J., and Hinds, R.C. (1988). Geophysical aspects of Wabamun salt distribution in southern Alberta. *Canadian Journal of Exploration Geophysics* 24: 166–178.

Anderson, N.L., Hopkins, J., Martínez, A. et al. (1994). Dissolution of bedded rock salt: a seismic profile across the active eastern margin of the Hutchinson salt member, central Kansas. *Computers and Geosciences* 20: 889–903.

Andrejchuk, V.N. (1996). Gypsum karst of the pre-Ural region, Russia. *International Journal of Speleology* 25: 285–295.

Andrejchuk, V.N. and Klimchouk, A.B. (2002). Mechanisms of karst breakdown formation in the gypsum karst of the fore-Ural region, Russia (from observations in the Kungurskaja Cave). *International Journal of Speleology* 31 (1): 89–114.

Artieda, O. (1996). *Génesis y distribución de suelos en un medio semiárido: Quinto (Zaragoza)*. Madrid: Ministerio de Agricultura, Pesca y Alimentación.

Artieda, O. (2013). Morphology and micro-fabrics of weathering features on gyprock exposures in a semiarid environment (Ebro Tertiary Basin, NE Spain). *Geomorphology* 196: 198–210.

Autin, W.J. (2002). Landscape evolution of the Five Islands of south Louisiana: Scientific policy and salt dome utilization and management. *Geomorphology* 47: 227–244.

Baars, D.L. (2000). Geology of Canyonlands National Park, Utah. In: *Geology of Utah's Parks and Monuments*, vol. 28 (ed. D.A. Sprinkel, T.C. Chidsey and P.B. Anderson), 61–83. Utah Geological Association Publication.

Bachman, G.O. (1980). *Regional Geology and Cenozoic History of Pecos Region, Southeastern New Mexico*, U.S. Geological Survey Open-file Report 80-1099, 116. The Survey.

Bachman, G.O. and Johnson, R.B. (1973). *Stability of salt in the Permian salt basin of Kansas, Oklahoma, Texas and New Mexico*, U.S. Geological Survey, Open-file Report 4339-4, 66. U.S. Dept. of the Interior, Geological Survey.

Baioni, D. and Wezel, F.C. (2010). Morphology and origin of an evaporitic dome in the eastern Tithonium Chasma, Mars. *Planetary and Space Science* 58: 847–857.

Baioni, D., Hajna, N.Z., and Wezel, F.C. (2009). Karst landforms in a Martian evaporitic dome. *Acta Carsologica* 38: 9–18.

Barret, J.K. and Pearl, R.H. (1976). Hydrogeologic data of thermal springs and wells in Colorado. *Colorado Geological Survey Information Series* 6: 124.

Belderson, R.H., Kenyon, N.H., and Stride, A.H. (1978). Local submarine salt-karst formation on the Hellenic Outer Ridge, eastern Mediterranean. *Geology* 6: 716–720.

Benito, G., Pérez-González, A., Gutiérrez, F. et al. (1998). River response to Quaternary subsidence due to evaporite solution (Gállego River, Ebro Basin, Spain). *Geomorphology* 22: 243–263.

Benito, G., Gutiérrez, F., Pérez-González, A. et al. (2000). Geomorphological and sedimentological features in Quaternary fluvial systems affected by solution-induced subsidence (Ebro Basin, NE-Spain). *Geomorphology* 33: 209–224.

Benito, G., Sancho, C., Peña, J.L. et al. (2010). Large-scale karst subsidence and accelerated fluvial aggradation during MIS6 in NE Spain: climatic and paleohydrological implications. *Quaternary Science Reviews* 29: 2694–2704.

Bertoni, C. and Cartwright, J.A. (2005). 3D seismic analysis of circular evaporite dissolution structures, Eastern Mediterranean. *Journal of the Geological Society, London* 162: 909–926.

Birks, S., Moncur, M.C., Gibson, J.J. et al. (2018). Origin and hydrogeological setting of saline groundwater discharges to the Athabasca River: Geochemical and isotopic characterization of the hyporheic zone. *Applied Geochemistry* 98: 172–190.

Black, T.J. (1997). Evaporite karst of Northern Lower Michigan. *Carbonates and Evaporites* 12: 81–83.

Bosák, P., Ford, D.C., Glazek, J. et al. (ed.) (1989). *Paleokarst: A Systematic and Regional Review.* Prague: Academia.

Bosák, P., Jaroš, J., Spudil, J. et al. (1998). Salt plugs in the eastern Zagros, Iran. Results of regional geological reconnaissance. *Geolines* 7: 3–180.

Broughton, P.L. (2013). Devonian salt dissolution-collapse breccias flooring the Cretaceous Athabasca oil sands deposit and development of lower McMurray Formation sinkholes, northern Alberta Basin, Western Canada. *Sedimentary Geology* 283: 57–82.

Broughton, P.L. (2017). Hypogene karst collapse of the Devonian Prairie Evaporite basin in western Canada. In: *Hypogene Karst Regions and Caves of the World, Cave and Karst Systems of the World* (ed. A. Klimchouk, A.N. Palmer, J. De Waele, et al.), 617–632. Dordrecht: Springer.

Broughton, P.L. (2021). Alignment of saline springs with evaporite karst structures in northeast Alberta, western Canada: analogue for Cretaceous hypogene brine seeps to the surface. *Acta Carsologica* 50: 119–141.

Bruthans, J., Asadi, N., Filippi, M. et al. (2008). A study of erosion rates on salt diapir surfaces in the Zagros Mountains, SE Iran. *Environmental Geology* 53 (5): 1079–1089.

Bruthans, J., Filippi, M., Asadi, N. et al. (2009). Surficial deposits on salt diapirs (Zagros Mountains and Persian Gulf Platform, Iran): characterization, evolution, erosion and the influence on landscape morphology. *Geomorphology* 107: 195–209.

Bruthans, J., Filippi, M., Zare, M. et al. (2010). Evolution of salt diapir and karst morphology during the last glacial cycle: effects of sea-level oscillation, diapir and regional uplift, and erosion (Persian Gulf, Iran). *Geomorphology* 121 (3-4): 291–304.

Bruthans, J., Kamas, J., Filippi, M. et al. (2017). Hydrogeology of salt karst under different cap soils and climates (Persian Gulf and Zagros Mts., Iran). *International Journal of Speleology* 46: 303–320.

Calaforra, J.M. and Pulido-Bosch, A. (1999a). Gypsum karst features as evidence of diapiric processes in the Betic Cordillera, southern Spain. *Geomorphology* 29: 251–264.

Calaforra, J.M. and Pulido-Bosch, A. (1999b). Genesis and evolution of gypsum tumuli. *Earth Surface Processes and Landforms* 24: 919–930.

Calaforra, J.M. and Pulido-Bosch, A. (2003). Evolution of the gypsum karst of Sorbas (SE Spain). *Geomorphology* 50: 173–180.

Calvo, J.P., Alcalá, L., Alonso-Zarza, A.M. et al. (1999). Estratigrafía y estructura del área de Los Mansuetos (Cuenca de Teruel). Precisiones para la definición del estratotipo del Turoliense. *Geogaceta* 25: 55–58.

Cammeraat, E., de Graaf, L.W.S., Kwadijk, J.K. et al. (1987). On the origin of debris pillars in the Alps of Vorarlberg, Western Austria. *Zeitschrift für Geomorphologie* 31: 85–100.

Carbonel, D., Gutiérrez, F., Linares, R. et al. (2013). Differentiating between gravitational and tectonic faults by means of geomorphological mapping, trenching and geophysical surveys. The case of the Zenzano Fault (Iberian Chain, N Spain). *Geomorphology* 189: 93–108.

Carlson, E.H. (1992). Reactivated interstratal karst-example from the late Silurian rocks of the western Lake Erie (U.S.A.). *Sedimentary Geology* 76: 273–283.

Cater, F. (1970). Geology of the salt anticline region in Southwestern Colorado. *U.S. Geological Survey Professional Paper* 637: 1–80.

Chardon, M. and Nicod, J. (1996). Gypsum karst of France. *International Journal of Speleology* 25: 203–208.

Christiansen, E.A. (1967). Collapse Structures near Saskatoon, Saskatchewan, Canada. *Canadian Journal of Earth Sciences* 4: 757–767.

Christiansen, E.A. (1971). Geology of the Crater Lake Collapse Structure in Southeastern Saskatchewan. *Canadian Journal of Earth Sciences* 8: 1505–1513.

Christiansen, E.A. and Sauer, E.K. (2001). Stratigraphy and structure of a Late Wisconsian salt collapse in the Saskatoon Low, south of Saskatoon, Saskatchewan, Canada: an update. *Canadian Journal of Earth Sciences* 38: 1601–1613.

Cooper, A.H. (1996). Gypsum karst of Great Britain. *International Journal of Speleology* 25: 195–202.

Cooper, A.H. (1998). Subsidence hazards caused by the dissolution of Permian gypsum in England: geology, investigation and remediation. In: *Geohazards in Engineering Geology*, Engineering Geology Special Publications, vol. 15 (ed. J.G. Maund and M. Eddleston), 265–275. London: Geological Society.

Cooper, A.H. (2002). Halite karst geohazards (natural and man-made) in the United Kingdom. *Environmental Geology* 42: 505–512.

Cooper, A.H. (2020). Geohazards caused by gypsum and anhydrite in the UK: including dissolution, subsidence, sinkholes and heave. In: *Geological Hazards in the UK: Their Occurrence, Monitoring and Mitigation*, Engineering Geology Special Publications, vol. 29 (ed. D.P. Giles and J.S. Griffiths), 403–423. London: Geological Society.

Cristescu, N. and Hunsche, U. (1998). *Time Effects in Rock Mechanics, Materials, Modelling and Computation*. Chichester: Wiley.

Dahm, T., Heimann, S., and Bialowons, W. (2011). A seismological study of shallow weak micro-earthquakes in the urban area of Hamburg city, Germany, and its possible relation to salt dissolution. *Natural Hazards* 58: 1111–1134.

Davis, D.G. (1995). Rims, rills and rafts: Shaping of cave features by atmospheric water exchange. *Geo2* 22 (2): 23-29, 32.

Davison, I., Bosence, D., Alsop, G.I. et al. (1996). Deformation and sedimentation around active Miocene salt diapirs on the Tihama Plain, northwest Yemen. In: *Salt Tectonics*, Geological Society Special Publication, vol. 100 (ed. G.I. Alsop, D.J. Blundell and I. Davison), 23–39. London: Geological Society.

De Mille, G., Shouldice, J.R., and Nelson, H.W. (1964). Collapse structures related to evaporites of the Prairie Formation. Saskatchewan. *Geological Society of America Bulletin* 75: 307–316.

De Waele, J. and Forti, P. (2010). Salt rims and blisters: peculiar and ephemeral formations in the Atacama Desert (Chile). *Zeitschrift für Geomorphologie* 54: 51–67.

De Waele, J., Picotti, V., Cucchi, F. et al. (2009b). Karst phenomena in the Cordillera de la Sal (Atacama, Chile). In: *Geological Constraints on the Onset and Evolution of an Extreme Environment: The Atacama Area*, vol. 2 (ed. P.L. Rossi), 113–127. Bologna: Dipartimento di Scienze della Terra e Geologico Ambientali *GeoActa Special Publication*.

De Waele, J., Anfossi, G., Campo, B. et al. (2012). Geomorphology of the Castel de'Britti area (Northern Apennines, Italy): an example of teaching geomorphological mapping in a traditional and practical way. *Journal of Maps* 8: 231–235.

De Waele, J., D'Angeli, I.M., Tisato, N. et al. (2017b). Coastal uplift rate at Matanzas (Cuba) inferred from MIS 5e phreatic overgrowths on speleothems. *Terra Nova* 29 (2): 98–105.

De Waele, J., Picotti, V., Martina, M.L. et al. (2020). Holocene evolution of halite caves in the Cordillera de la Sal (Central Atacama, Chile) in different climate conditions. *Geomorphology* 370: 107398.

Denchik, N., Gautier, S., Dupuy, M. et al. (2019). In-situ geophysical and hydro-geochemical monitoring to infer landslide dynamics (Pégairolles-de-l'Escalette landslide, France). *Engineering Geology* 254: 102–112.

Desbois, G., Závada, P., Schléder, Z. et al. (2010). Deformation and recrystallization mechanisms in actively extruding salt fountain: Microstructural evidence for a switch in deformation mechanisms with increased availability of meteoric water and decreased grain size (Qum Kuh, central Iran). *Journal of Structural Geology* 32: 580–594.

Doelling, H.H. (2000). Geology of Arches National Park, Grand County, Utah. In: *Geology of Utah's Parks and Monuments*, vol. 28 (ed. D.A. Sprinkel, T.C. Chidsey and P.B. Anderson), 11–36. Utah Geological Association Publication.

Doğan, U. (2005). Land subsidence and caprock dolines caused by subsurface gypsum dissolution and the effect of subsidence on the fluvial system in the Upper Tigris Basin (between Bismil-Batman, Turkey). *Geomorphology* 71: 389–401.

Dooley, T.P., Jackson, M.P., and Hudec, M.R. (2009). Inflation and deflation of deeply buried salt stocks during lateral shortening. *Journal of Structural Geology* 31: 582–600.

Dzens-Litovsky, A.I. (1966). *Salt Karst of the USSR*. Leningrad: Nedra (in Russian).

Ely, R.W. (1987). Colluvium-filled fault fissures in the Needles fault zone, Cataract Canyon, Utah. In: *Geology of Cataract Canyon and vicinity* (ed. J.A. Campbell), 69–73. Four Corners Geological Society 10th Field Conference Guidebook.

Epstein, J.B. (2003). Gypsum karst in the Black Hills, South Dakota-Wyoming: Geomorphic Development, Hazards, and Hydrology. In: *Evaporite Karst and Engineering/Environmental Problems in the United States* (ed. K.S. Johnson and J.T. Neal), 241–254. Norman: Oklahoma Geological Survey cicular 109.

Erol, A.O. (1989). Engineering geological considerations in a salt dome region surrounded by sabkha sediments, Saudi Arabia. *Engineering Geology* 26: 215–232.

Ferrarese, F., Macaluso, T., Madonia, G. et al. (2002). Solution and recrystallisation processes and associated landforms in gypsum outcrops of Sicily. *Geomorphology* 49: 25–43.

Ford, D.C. (1997). Principal features of evaporite karst in Canada. *Carbonates and Evaporites* 13: 15–23.

Forti, P. (1987). Le bolle di scollamento: una forma carsica caratteristica dei gessi bolognesi, ancora non sufficiente nota. *Sottoterra* 77: 10–18.

Forti, P. and Sauro, U. (1996). The gypsum karst of Italy. *International Journal of Speleology* 25: 239–250.

Forti, P., Barredo, S., Costa, G. et al. (1993). Two peculiar karst forms of the gypsum outcrop between Zapala and Las Lajas (Neuquen, Argentina). In: *Proceedings of the 14th International Congress of Speleology* (ed. S. Zhang), 54–56. Beijing: China.

Frumkin, A. (1994a). Hydrology and denudation rates of halite karst. *Journal of Hydrology* 162: 171–189.

Frumkin, A. (1994b). Morphology and development of salt caves. *National Speleological Society Bulletin* 56: 82–95.

Frumkin, A. (1996). Uplift rate relative to base-levels of a salt diapir (Dead Sea Basin, Israel) as indicated by cave levels. In: *Salt Tectonics*, vol. 100 (ed. G.I. Alsop, D.J. Blundell and I. Davison), 41–47. London: Geological Society, London Special Publication.

Frumkin, A. (1998). Salt cave cross-sections and their paleoenvironmental implications. *Geomorphology* 23: 183–191.

Frumkin, A. (2004). Sedom salt karst, Israel. In: *Encyclopedia of Caves and Karst Science* (ed. J. Gunn), 637–639. New York: Fitzroy Dearborn.

Frumkin, A. (2013). Salt karst. In: *Treatise on Geomorphology. Karst Geomorphology*, vol. 6 (ed. A. Frumkin and J. Shroder), 407–424. Amsterdam: Elsevier.

Frumkin, A. and Ford, D.C. (1995). Rapid entrenchment of stream profiles in the salt caves of Mount Sedom, Israel. *Earth Surface Processes and Landforms* 20 (2): 139–152.

Frumkin, A., Ezersky, M., Al-Zoubi, A. et al. (2011). The Dead Sea sinkhole hazard: Geophysical assessment of salt dissolution and collapse. *Geomorphology* 134: 102–1117.

Furuya, M., Mueller, K., and Wahr, J. (2007). Active salt tectonics in the Needles District, Canyonlands (Utah) as detected by interferometric synthetic aperture radar and point target analysis: 1992–2002. *Journal of Geophysical Research: Solid Earth* 112: B06418.

Galve, J.P., Tonelli, C., Gutiérrez, F. et al. (2015). New insights into the genesis of the Miocene collapse structures of the island of Gozo (Malta, central Mediterranean Sea). *Journal of the Geological Society* 172: 336–348.

Ge, H.X. and Jackson, M.P.A. (1998). Physical modeling of structures formed by salt withdrawal: implications for deformation caused by salt dissolution. *American Association of Petroleum Geologists Bulletin* 82: 228–250.

Ge, H.X., Jackson, M.P.A., and Vendeville, B.C. (1997). Kinematics and dynamics of salt tectonics driven by progradation. *American Association of Petroleum Geologists Bulletin* 81: 398–423.

Gökkaya, E., Gutiérrez, F., Ferk, M. et al. (2021). Sinkhole development in the Sivas gypsum karst. Turkey. *Geomorphology* 386: 107746.

Gott, G.B., Wolcott, D.F. and Bowles, C.G. (1974). Stratigraphy of the Inyan Kara Group and localisation of Uranium deposits, Southern Black Hills, South Dakota and Wyoming: U.S. Geological Survey Professional Paper 763: 1–57.

Grosfils, E.B., Schultz, R.A., and Kroeger, G. (2003). Geophysical exploration within northern Devils Lane graben, Canyonlands National Park, Utah: implications for sediment thickness and tectonic evolution. *Journal of Structural Geology* 25: 455–467.

Guerrero, J. (2017). Dissolution collapse of a growing diapir from radial, concentric, and salt-withdrawal faults overprinting in the Salinas de Oro salt diapir, northern Spain. *Quaternary Research* 87: 331–346.

Guerrero, J. and Gutiérrez, F. (2017). Gypsum scarps and asymmetric fluvial valleys in evaporitic terrains. The role of river migration, landslides, karstification and lithology (Ebro River, NE Spain). *Geomorphology* 297: 137–152.

Guerrero, J., Gutiérrez, F., and Lucha, P. (2008a). Impact of halite dissolution subsidence on Quaternary fluvial terrace development: Case study of the Huerva River, Ebro Basin, NE Spain. *Geomorphology* 100: 164–179.

Guerrero, J., Gutiérrez, F., and Galve, J.P. (2013). Large depressions, thickened terraces, and gravitational deformation in the Ebro River valley (Zaragoza area, NE Spain): Evidence of glauberite and halite interstratal karstification. *Geomorphology* 196: 162–176.

Guerrero, J., Bruhn, R.L., McCalpin, J.P. et al. (2015). Salt-dissolution faults versus tectonic faults from the case study of salt collapse in Spanish Valley, SE Utah (USA). *Lithosphere* 7: 46–58.

Gustavson, T.C. (1986). Geomorphic development of the Canadian River Valley, Texas Panhandle: an example of regional salt dissolution and subsidence. *Geological Society of America Bulletin* 97: 459–472.

Gustavson, T.C., Smpkins, W.W., Alhades, A. et al. (1982). Evaporite dissolution and development of karst features on the Rolling Plains of the Texas Panhandle. *Earth Surface Processes and Landforms* 7: 545–563.

Gutiérrez, F. (1996). Gypsum karstification induced subsidence: effects on alluvial systems and derived geohazards (Calatayud Graben, Iberian Range, Spain). *Geomorphology* 16: 277–293.

Gutiérrez, F. (2004). Origin of the salt valleys in the Canyonlands section of the Colorado Plateau. Evaporite-dissolution collapse versus tectonic subsidence. *Geomorphology* 57: 423–435.

Gutiérrez, F. (2014). Evaporite karst in Calatayud, Iberian Chain. In: *Landscapes and Landforms in Spain* (ed. F. Gutiérrez and M. Gutiérrez), 111–126. Dordrecht: Springer.

Gutiérrez, F. and Cooper, A.H. (2013). Surface morphology of gypsum karst. In: *Treatise of Geomorphology. Karst Geomorphology*, vol. 6 (ed. A. Frumkin), 425–437. Amsterdam: Elsevier.

Gutiérrez, F. and Lizaga, I. (2016). Sinkholes, collapse structures and large landslides in an active salt dome submerged by a reservoir: the unique case of the Ambal ridge in the Karun River, Zagros Mountains. *Iran. Geomorphology* 254: 88–103.

Gutiérrez, F., Arauzo, T., and Desir, G. (1994). Deslizamientos en el escarpe en yesos de Alfajarín. *Cuaternario y Geomorfología* 8: 57–69.

Gutiérrez, F., Ortí, F., Gutiérrez-Elorza, M. et al. (2001). The stratigraphical record and activity of evaporite dissolution subsidence in Spain. *Carbonates and Evaporites* 16: 46–70.

Gutiérrez, F., Calaforra, J.M., Cardona, F. et al. (2008b). Geological and environmental implications of the evaporite karst in Spain. *Environmental Geology* 53: 951–965.

Gutiérrez, F., Carbonel, D., Guerrero, J. et al. (2012a). Late Holocene episodic displacement on fault scarps related to interstratal dissolution of evaporites (Teruel Neogene Graben, NE Spain). *Journal of Structural Geology* 34: 2–19.

Gutiérrez, F., Linares, R., Roqué, C. et al. (2012b). Investigating gravitational grabens related to lateral spreading and evaporite dissolution subsidence by means of detailed mapping, trenching, and electrical resistivity tomography (Spanish Pyrenees). *Lithosphere* 4: 331–353.

Gutiérrez, F., Carbonel, D., Kirkham, R.M. et al. (2014a). Can flexural-slip faults related to evaporite dissolution generate hazardous earthquakes? The case of the Grand Hogback monocline of west-central Colorado. *Geological Society of America Bulletin* 126: 1481–1494.

Gutiérrez, F., Parise, M., De Waele, J. et al. (2014b). A review on natural and human-induced geohazards and impacts in karst. *Earth-Science Reviews* 138: 61–88.

Gutiérrez, F., Fabregat, I., Roqué, C. et al. (2016). Sinkholes and caves related to evaporite dissolution in a stratigraphically and structurally complex setting, Fluvia Valley, eastern Spanish Pyrenees. Geological, geomorphological and environmental implications. *Geomorphology* 267: 76–97.

Gutiérrez, F., Benito-Calvo, A., Carbonel, D. et al. (2019a). Review on sinkhole monitoring and performance of remediation measures by high-precision leveling and terrestrial laser scanner in the salt karst of the Ebro Valley, Spain. *Engineering Geology* 248: 283–308.

Gutiérrez, F., Fabregat, I., Roqué, C. et al. (2019b). Sinkholes in hypogene versus epigene karst systems, illustrated with the hypogene gypsum karst of the Sant Miquel de Campmajor Valley, NE Spain. *Geomorphology* 328: 57–78.

Gutiérrez, F., Sevil, J., Silva, P.G. et al. (2019c). Geomorphic and stratigraphic evidence of Quaternary diapiric activity enhanced by fluvial incision. Navarrés salt wall and graben system, SE Spain. *Geomorphology* 342: 176–195.

Gutiérrez-Elorza, M. and Gutiérrez-Santolalla, F. (1998). Geomorphology of the Tertiary gypsum formations in the Ebro Depression (Spain). *Geoderma* 87: 1–29.

Hanson, K.L., Kelson, K.I., Angell, M.A. et al. (1999). *Techniques for Identifying Faults and Determining Their Origins*, NUREG/CR-5503. Washington, contract report: U. S. Nuclear Regulatory Commission.

Hessami, K., Koyi, H.A., Talbot, C.J. et al. (2001). Progressive unconformities within an evolving foreland fold-thrust belt, Zagros Mountains. *Journal of the Geological Society* 158: 969–981.

Hill, C.A. (1995). Sulfur redox reactions: hydrocarbons, native sulfur, Mississippi Valley-type deposits, and sulfuric acid karst in the Delaware Basin, New Mexico and Texas. *Environmental Geology* 25: 16–23.

Hill, C.A. (1996). *Geology of the Delaware Basin, Guadalupe, Apache, and Glass Mountains, New Mexico and West Texas.* Permian Basin Section-SEPM. Publication No. 96–39.

Hill, C.A. (2003). Interstratal karst at the waste isolation pilot plant site, southeastern New Mexico. In: *Evaporite Karst and Engineering/Environmental Problems in the United States* (ed. K.S. Johnson and J.T. Neal), 197–210. Norman: Oklahoma Geological Survey cicular 109.

Hill, C.A. and Forti, P. (1997). *Cave Minerals of the World*, 2e. Huntsville, Alabama: National Speleological Society.

Holt, R.M. and Powers, D.W. (2010). Evaluation of halite dissolution at a radioactive waste disposal site, Andrews County, Texas. *Geological Society of America Bulletin* 122: 1989–2004.

Hovorka, S.D. (2000). Understanding the processes of salt dissolution and subsidence. *Solution Mining Research Institute Proceedings*, 11-24. San Antonio, Texas.

Hudec, M.R. and Jackson, M.P.A. (2007). Terra infirma: understanding salt tectonics. *Earth-Science Reviews* 82: 1–28.

Hudec, M.R. and Jackson, M.P.A. (2011). *The Salt Mine: A Digital Atlas of Salt Tectonics*, Bureau of Economic Geology Udden Book Series No. 5, AAPG Memoir 99. The University of Texas at Austin.

Huntoon, P.W. (1982). The Meander anticline, Canyonlands, Utah: An unloading structure resulting from horizontal gliding on salt. *Geological Society of America Bulletin* 93: 941–950.

Huntoon, P.W. (1999). Field-based identification of salt-related structures and their differentiation from tectonic structures. In: *Techniques for Identifying Faults and Determining their Origins*, NUREG/CR-5503. (ed. K.L. Hanson, K.I. Kelson, M.A. Angell, et al.), 1–186. Washington, contract report: U.S. Nuclear Regulatory Commission.

Jackson, M.P.A. and Hudec, M.R. (2017). *Salt Tectonics. Principles and Practice*. Cambridge: Cambridge University Press.

Jackson, M.P.A. and Talbot, C.J. (1986). External shapes, strain rates, and dynamics of salt structures. *Geological Society of America Bulletin* 97: 305–323.

Jahani, S., Callot, J.P., Frizon de Lamotte, D. et al. (2007). The salt diapirs of the eastern Fars Province (Zagros, Iran): a brief outline of their past and present. In: *Thrust Belts and Foreland Basins* (ed. O. Lacombe, F. Roure, J. Lavé, et al.), 289–308. Berlin: Springer.

James, A.N. (1992). *Soluble Material in Civil Engineering*. New York: Ellis Horwood.

James, N.P. and Choquette, P.W. (ed.) (1988). *Paleokarst*. New York: Springer-Verlag.

James, A.N., Cooper, A.H., and Holliday, D.W. (1981). Solution of the gypsum cliff (Permian, Middle Marl) by the River Ure at Ripon Parks, North Yorkshire. *Proceedings of the Yorkshire Geological Society* 43: 433–450.

Jarzyna, A., Bąbel, M., Ługowski, D. et al. (2020). Unique hydration caves and recommended photogrammetric methods for their documentation. *Geoheritage* 12 (1): 27.

Jassim, S.Z., Jibril, A.S., and Numan, N.M.S. (1997). Gypsum karstification in the middle Miocene Fatha Formation, Mosul area, northern Iraq. *Geomorphology* 18: 137–149.

Johnson, K.S. (1981). Dissolution of salt on the east flank of the Permian basin in the southwestern U.S.A. *Journal of Hydrology* 54: 75–93.

Johnson, K.S. (1989). Development of the Wink Sink in Texas, USA, due to salt dissolution and collapse. *Environmental Geology and Water Sciences* 14 (2): 81–92.

Johnson, K.S. (1996). Gypsum karst in the United States. *International Journal of Speleology* 25: 183–193.

Johnson, K.S. and Neal, J.T. (ed.) (2003). *Evaporite Karst and Engineering/Environmental Problems in the United States*, Oklahoma Geological Survey Circular, vol. 109. Norman: Oklahoma Geological Survey.

Johnson, K.S., Collins, E.W., and Seni, S.J. (2003). Sinkholes and land subsidence owing to salt dissolution near Wink, West Texas, and other sites in western Texas and New Mexico. In: *Evaporite Karst and Engineering/Environmental Problems in the United States* (ed. K.S. Johnson and J.T. Neal), 183–195. Norman: Oklahoma Geological Survey.

Johnson, K.S., Land, L., and Decker, D.D. (ed.) (2021). *Evaporite Karst in the Greater Permian Evaporite Basin (GPEB) of Texas, New Mexico, Oklahoma, Kansas, and Colorado*, Oklahoma Geological Survey Circular 113. Norman: Oklahoma Geological Survey.

Kastens, K.A. and Spiess, F.N. (1984). Dissolution and collapse features on the Eastern Mediterranean Ridge. *Marine Geology* 56: 181–193.

Kempe, S. (1996). Gypsum karst of Germany. *International Journal of Speleology* 25: 209–224.

Kempe, S. and Dirks, H. (2008). Layla Lakes, Saudi Arabia: the world-wide largest lacustrine gypsum tufas. *Acta Carsologica* 37: 7–14.

Kent, P.E. (1970). The salt plugs of the Persian Gulf region. *Transactions of the Leicester Literary and Phylosophical Society* 64: 56–88.

Kirkham, R.M., Streufert, R.K., Kunk, M.J. et al. (2002). Evaporite tectonism in the Lower Roaring Fork river valley, westcentral Colorado. In: *Late Cenozoic Evaporite Tectonism and Volcanism in West-Central Colorado*, Geological Society of America Special Paper 366 (ed. R.M. Kirkham, R.B. Scott and T.W. Judkins), 73–99. Denver: Geological Society of America.

Kirkland, D.W. and Evans, R. (1976). Origin of limestone buttes, Gypsum Plain of Texas and New Mexico. *American Association of Petroleum Geologists Bulletin* 60: 2005–2018.

Klimchouk, A.B. (1996b). Gypsum karst in the western Ukraine. *International Journal of Speleology* 25: 263–278.

Klimchouk, A., Lowe, D., Cooper, A. et al. (ed.) (1996b). Gypsum karst of the World. *International Journal of Speleology* 25: 1–307.

Kravitz, K., Mueller, K., Bilham, R.G. et al. (2020). Active steady-state creep on a nontectonic normal fault in Southeast Utah: Implications for strain release in a rapidly deforming salt system. *Geophysical Research Letters* 47: e2020GL087081.

Land, L. (2003). Evaporite karst and regional groundwater circulation in the Lower Pecos Valley of Southeastern New Mexico. In: *Evaporite Karst and Engineering/Environmental Problems in the United States*, vol. 109 (ed. K.S. Johnson and J.T. Neal), 227–232. Oklahoma Geological Survey Circular.

Lang, J., Hampel, A., Brandes, C. et al. (2014). Response of salt structures to ice-sheet loading: implications for ice-marginal and subglacial processes. *Quaternary Science Reviews* 101: 217–233.

Lee, W.T. (1923). Erosion by solution and fill. *U.S. Geological Survey Bulletin* 760-C: 107–121.

Loucks, R.G. (1999). Paleocave carbonate reservoirs: origins, burial depth modifications, spatial complexity and reservoir implications. *American Association of Petroleum Geologists Bulletin* 83: 1795–1834.

Loucks, R.G. (2007). A review of coalesced, collapsed-paleocave systems and associated suprastratal deformation. *Acta Carsologica* 36: 121–132.

Lucha, P., Cardona, F., Gutiérrez, F. et al. (2008a). Natural and human-induced dissolution and subsidence processes in the salt outcrop of the Cardona Diapir (NE Spain). *Environmental Geology* 53: 1023–1035.

Lucha, P., Gutiérrez, F., Galve, J.P. et al. (2012). Geomorphic and stratigraphic evidence of incision-induced halokinetic uplift and dissolution subsidence in transverse drainages crossing the evaporite-cored Barbastro-Balaguer Anticline (Ebro Basin, NE Spain). *Geomorphology* 171: 154–172.

Luzón, A., Pérez, A., Soriano, M.A. et al. (2008). Sedimentary record of Pleistocene paleodoline evolution in the Ebro basin (NE Spain). *Sedimentary Geology* 205: 1–13.

Macaluso, T. and Sauro, U. (1996b). Weathering crust and karren on exposed gypsum surfaces. *International Journal of Speleology* 25: 115–126.

McGill, G.E. and Stormquist, A.W. (1979). The Grabens of Canyonlands National Park, Utah: Geometry, Mechanics, and Kinematics. *Journal of Geophysical Research* 84 (B9): 4547–4563.

Mège, D., Le Deit, L., Rango, T. et al. (2013). Gravity tectonics of topographic ridges: halokinesis and gravitational spreading in the western Ogaden, Ethiopia. *Geomorphology* 193: 1–13.

Memesh, A., Dini, S., Gutiérrez, F. et al. (2008). Evidence of large-scale subsidence caused by interstratal karstification of evaporites in the Interior Homocline of Central Saudi Arabia. European Geosciences Union General Assembly. *Geophysical Research Abstracts* 10: A-02276.

Moore, J.M. and Schultz, R.A. (1999). Processes of faulting in jointed rocks of Canyonlands National Park, Utah. *Geological Society of America Bulletin* 111: 808–822.

Morgan, A.M. (1941). Solution Phenomena in New Mexico. *Symposium on Relations of Geology to the Groundwater Problems of the Southwest*, American Geophysical Union Transactions, 23rd Annual Meeting, 27–35.

Mrazec, L. (1907). Despre cute cu simbure de străpungere (On folds with piercing cores). *Buletinul Societăţii de Ştiinţe din Bucureşci* 16: 6–8.

Myers, A.J. (1962). A fossil sinkhole. *Oklahoma Geology Notes* 22: 13–15.

Neal, J.T. (1995). Supai salt karst features: Holbrook Basin, Arizona. In: *Karst Geohazards. Engineering and Environmental Problems in Karst Terrane* (ed. B.F. Beck), 53–59. Rotterdam: Balkema.

Neal, J.T. (1998). Evaporite karst in the Holbrook Basin, Arizona. In: *Land Subsidence Case Studies and Current Research* (ed. J.W. Borchers), 383–384. Association of Engineering Geologists Special publication 8.

Neal, J.T. and Colpitts, R.M. (1997). Richard Lake, an evaporite karst depression in the Holbrook Basin, Arizona. *Carbonates and Evaporites* 12: 91–98.

Neal, J.T. and Johnson, K.S. (2003). A compound breccia pipe in evaporite karst: McCauley sinks, Arizona. In: *Evaporite Karst and Engineering/Environmental Problems in the United States*, vol. 109 (ed. K.S. Johnson and J.T. Neal), 305–314. Norman: Oklahoma Geological Survey Cicular.

Olive, W.W. (1957). Solution-subsidence troughs, Castile Formation of Gypsum Plain, Texas and New Mexico. *Geological Society of America Bulletin* 68: 351–358.

Ouchi, S. (1985). Response of alluvial rivers to slow active tectonic movement. *Geological Society of America Bulletin* 96: 504–515.

Paterson, M. (2001). Relating experimental and geological rheology. *International Journal of Earth Sciences* 90: 157–167.

Paukstys, B. and Narbutas, V. (1996). Gypsum karst of the Baltic Republics. *International Journal of Speleology* 25: 279–284.

Pfeiffer, D. and Hahn, J. (1972). Karst of Germany. In: *Important Karst Regions of the Northern Hemisphere* (ed. M. Herak and V.T. Stringfield), 189–223. Amsterdam: Elsevier.

Picard, M.D. (1980). Stratigraphy, petrography, and origin of evaporites, Jurassic Arapien Shale, central Utah. *Utah Geological Association Publication* 8: 129–150.

Pilcher, R.S. and Blumstein, R.D. (2007). Brine volume and salt dissolution rates in Orca Basin, northeast Gulf of Mexico. *American Association of Petroleum Geologists Bulletin* 91: 823–833.

Pisani, L. and De Waele, J. (2021). Candidate cave entrances in a planetary analogue evaporite karst (Cordillera de la Sal, Chile): a remote sensing approach and ground-truth reconnaissance. *Geomorphology* 389: 107851.

Powers, R.W., Ramirez, L.F., Redmond, C.D. et al. (1966). *Geology of the Arabian Peninsula; Sedimentary Geology of Saudi Arabia*, vol. 560-D, 1–147. US Geological Survey Professional Paper.

Pulido-Bosch, A. (1986). Le karst dans les gypses de Sorbas (Almeria). Aspects morphologiques et hydrogéologiques. *Karstologia Mémoires* 1: 27–35.

Quinlan, J.F. (1978). Types of karst, with emphasis on cover beds in their classification and development. PhD thesis. University of Texas at Austin.

Quinlan, J.F., Smith, R.A., and Johnson, K.S. (1986). Gypsum karst and salt karst of the United States of America. *Le Grotte d'Italia* 13: 73–92.

Raeisi, E., Zare, M., and Aghdam, J.A. (2013). Hydrogeology of gypsum formations in Iran. *Journal of Cave and Karst Studies* 75: 68–80.

Rahn, P.H. and Davis, A.D. (1996). Gypsum foundation problems in the Black Hills area, South Dakota. *Environmental and Engineering Geoscience* 2: 213–223.

Reitman, N.G., Ge, S., and Mueller, K. (2014). Groundwater flow and its effect on salt dissolution in Gypsum Canyon watershed, Paradox Basin, southeast Utah, USA. *Hydrogeology Journal* 22: 1403–1419.

Reuter, F. and Stoyan, D. (1993). Sinkholes in carbonate, sulphate, and chloride karst regions: principles and problems of engineering geological investigations and predictions, with comments for the construction and mining industries. In: *Applied Karst Geology* (ed. B.F. Beck), 3–25. Rotterdam: Balkema.

Reuter, F., Molek, H., and Bochmann, G. (1977). Slope sliding as secondary process in subsidence areas of chloride-karst. *Bulletin of the International Association of Engineering Geology* 16: 62–64.

Rovera, G. (1993). Instabilité des versants et dissolution des évaporites dans les Alpes internes: l'exemple de la montagne de Friolin (Peisey-Nancroix, Savoie). *Revue de Géographie Alpine* 81: 71–84.

Sánchez, J.A., Coloma, P., and Pérez, A. (1999). Sedimentary processes related to the groundwater flows from the Mesozoic Carbonate Aquifer of the Iberian Chain in the Tertiary Ebro Basin, northeast Spain. *Sedimentary Geology* 129: 201–213.

Sasowsky, I.D., Feazel, C.T., Mylroie, J.E. et al. (ed.) (2008). *Karst from Recent to Reservoirs*, Special Publication 14. Leesburg, VA: Karst Waters Institute.

Saudi Geological Survey (2007). *Geologic Map of the As Sulayyil Quadrangle, shett 20H. Kingdom of Saudi Arabia, scale 1:250,000*. Jeddah: Kingdom of Saudi Arabia Geological Survey.

Schléder, Z. and Urai, J.L. (2007). Deformation and recrystallization mechanisms in mylonitic shear zones in naturally deformed extrusive Eocene-Oligocene rocksalt from Eyvanekey plateau and Garmsar hills (central Iran). *Journal of Structural Geology* 29: 241–255.

Schultz-Ela, D.D. and Walsh, P. (2002). Modeling of grabens extending above evaporites in Canyonlands National Park, Utah. *Journal of Structural Geology* 24: 247–275.

Schumm, S.A., Dumont, J.F., and Holbrook, J.M. (2000). *Active Tectonic and Alluvial Rivers*. Cambridge: Cambridge University Press.

Seijmonsbergen, A.C. and de Graff, L.W.S. (2006). Geomorphological mapping and geophysical profiling for the evaluation of natural hazards in an alpine catchment. *Natural Hazards and Earth System Science* 6: 185–193.

Sirocko, F., Szeder, T., Seelos, C. et al. (2002). Young tectonic and halokinetic movements in the Northern-German-Basin: its effect on formation of modern rivers and surface morphology. *Netherlands Journal of Geoscience* 81: 431–441.

Smith, D.B. (1972). Foundered strata, collapse breccias and subsidence features of the English Zechstein. In: *Geology of Saline Deposits* (ed. G. Richter-Bernberg), 255–269. Paris: UNESCO.

Spiers, C.J., Schutjens, P.M.T.M., Brzesowsky, R.H. et al. (1990). Experimental determination of constitutive parameters governing creep of rocksalt by pressure solution. In: *Deformation Mechanisms, Rheology and Tectonics*, vol. 54 (ed. R.J. Knipe and E.H. Rutter), 509–522. London: Geological Society of London Special Publication.

Stafford, K.W., Land, L., and Klimchouk, A.B. (2008a). Hypogenic speleogenesis within Seven Rivers evaporites: Coffee Cave, Eddy County, New Mexico. *Journal of Cave and Karst Studies* 70 (1): 47–61.

Stafford, K.W., Rosales-Lagarde, L., and Boston, P.J. (2008c). Castile evaporite karst potential map of the Gypsum Plain, Eddy County, New Mexico and Culberson County, Texas: a GIS methodological comparison. *Journal of Cave and Karst Studies* 70: 35–46.

Stanton, R.J. (1966). The solution brecciation process. *Geological Society of America Bulletin* 7: 843–848.

Stride, A.H., Belderson, R.H., and Kenyon, N.H. (1977). Evolving miogeanticlines of the east Mediterranean (Hellenic, Calabrian and Cyprus Outer Ridges). *Royal Society of London Philosophical Transactions* 284: 255–285.

Sugiura, R. and Kitcho, C.A. (1981). Collapse structures in the Paradox Basin. In: *Geology of the Paradox Basin* (ed. D.L. Wiegand), 33–45. Denver: Rocky Mountain Association of Geologists.

Supajanya, T. and Friederich, M.C. (1992). Salt tectonics of the Sakon Nakhon Basin, northeast Thailand. *Journal of the Southeast Asian Earth Sciences* 7: 258–259.

Talbot, C.J. (1979). Fold trains in a glacier of salt in southern Iran. *Journal of Structural Geology* 1 (1): 5–18.

Talbot, C.J. (1981). Sliding and other deformation mechanisms in a salt glacier, Iran. *Geological Society of London*, Special Publication 9: 173–183.

Talbot, C.J. and Alavi, M. (1996). The past of a future syntaxis across the Zagros. *Geological Society London*, Special Publication 100: 89–109.

Talbot, C.J. and Jarvis, R.J. (1984). Age, budget and dynamics of an active salt extrusion in Iran. *Journal of Structural Geology* 6: 521–533.

Talbot, C.J. and Pohjola, V. (2009). Subaerial salt extrusions in Iran as analogues of ice sheets, streams and glaciers. *Earth-Science Reviews* 97: 155–183.

Talbot, C.J. and Rogers, E. (1980). Seasonal movements in an Iranian salt glacier. *Science* 208: 395–397.

Taviani, M. (1984). Submarine "sinkholes": a review. In: *Sinkholes: Their Geology, Engineering and Environmental Impact. Proceeding of the First Multidisciplinary Conference on Sinkholes* (ed. B.F. Beck), 117–121. Rotterdam: Balkema.

Thierry, P., Prunier-Leparmentier, A.M. et al. (2009). 3D geological modelling at urban scale and mapping of ground movement susceptibility from gypsum dissolution: the Paris example (France). *Engineering Geology* 105: 51–64.

Torabi-Kaveh, M., Heidari, M., and Miri, M. (2012). Karstic features in gypsum of Gachsaran Formation (case study; Chamshir Dam reservoir, Iran). *Carbonates and Evaporites* 27: 291–297.

Toulemont, M. (1984). Le karst gypseux du Lutétien supérieur de la region parisienne. Caractéristiques et impact sur le milieu urbain. *Revue de Géologie Dynamique et de Géographye Physique* 25: 213–228.

Trudgill, B. (2002). Structural controls on drainage development in the Canyonlands grabens of southeast Utah. *American Association of Petroleum Geologists Bulletin* 86: 1095–1112.

Tsui, P.C. and Cruden, D.M. (1984). Deformation associated with gypsum karst in the Salt River Escarpment, northeastern Alberta. *Canadian Journal of Earth Sciences* 21: 949–959.

Urai, J.L., Spiers, C.J., Zwart, H.J. et al. (1986). Weakening of rock salt by water during long-term creep. *Nature* 324: 554–557.

Urai, J.L.Z., Schléder, Z., Spiers, C.J. et al. (2008). Flow and transport properties of salt rocks. In: *Dynamics of Complex Intracontinental Basins. The Central European Basin System* (ed. R. Littke, U. Bayer, D. Gajewski and S. Nelskamp), 277–290. Berlin: Springer.

Utha-aroon, C.L., Coshell, L. and Warren, J.K. (1995). Early and late dissolution in the Maha Sarakham Formation: implications for basin stratigraphy. *International Conference on Geology, Geochronology and Mineral Resources of Indochina*, 275–286. Khon Kaen, Thailand.

Utrilla, R., Pierre, C., Ortí, F. et al. (1992). Oxygen and Sulphur isotope compositions as indicators of the origin of Mesozoic and Cenozoic evaporites from Spain. *Chemical Geology* 102: 229–244.

Vaslet, D., Al-Muallem, M.S., Maddah, S. et al. (1991). *Geologic Map of the Ar Riyad quadrangle, sheet 241*. Kingdom of Saudi Arabia: Saudi Deputy Ministry for Mineral Resources Geoscience Map GM-121.

Ville, L. (1859). Notice géologique sur les salines des Zahrez et les gites de sel gemme du Rang-el-Melah et d'Aïn Hadjera (Algérie). *Annales des Mines* 15: 351–410.

Walters, R.F. (1978). Land Subsidence in Central Kansas Related to Salt Dissolution. *Bulletin of Kansas University Geological Survey* 214: 1–82.

Warren, J.K. (2008). Salt as sediment in the Central European Basin system as seen from a deep time perspective. In: *Dynamics of Complex Intracontinental Basins: The Central European Basin System* (ed. R. Littke, V. Bayer, D. Gajewski and S. Nelskamp), 249–276. Dordrecht: Springer.

Warren, J.K. (2016). *Evaporites: A Geological Compendium*. Dordrecht: Springer.

Weir, G.W., Puffett, W.P., and Dodson, C.L. (1961). Collapse structures in southern Spanish Valley, southeastern Utah. *US Geological Survey Professional Papers* 424-B: 173–174.

Witkind, I.J. (1994). The role of salt in the Structural development of Central Utah. U.S. Geological Survey Professional Paper 1528, 145 p.

Yarou, L. and Cooper, A.H. (1996). Gypsum karst in China. *International Journal of Speleology* 25: 297–307.

Youssef, A.M., Al-Harbi, H.M., Gutiérrez, F. et al. (2016). Natural and human-induced sinkhole hazards in Saudi Arabia: distribution, investigation, causes and impacts. *Hydrogeology Journal* 24: 625–644.

Yuhua, G. and Lin Hua, S. (1986). Salt karst in Qinghai Plateau, China. *Le Grotte d'Italia* 12: 337–345.

Zak, I. and Freund, R. (1980). Strain measurements in eastern marginal shear zone of Mount Sedom salt diapir, Israel. *American Association of Petroleum Geologists Bulletin* 64: 568–581.

Zarei, M. and Raeisi, E. (2010). Karst development and hydrogeology of Konarsiah salt diapir, south of Iran. *Carbonates and Evaporites* 25: 217–229.

Zarei, M., Raeisi, E., and Talbot, C. (2012). Karst development on a mobile substrate: Konarsiah salt extrusion. *Iran. Geological Magazine* 149: 412–422.

Zarroca, M., Comas, X., Gutiérrez, F. et al. (2017). The application of GPR and ERI in combination with exposure logging and retrodeformation analysis to characterize sinkholes and reconstruct their impact on fluvial sedimentation. *Earth Surface Processes and Landforms* 42: 1049–1064.

Zucker, E., Frumkin, A., Agnon, A. et al. (2019). Internal deformation and uplift-rate of salt walls detected by a displaced dissolution surface, Dead Sea. *Journal of Structural Geology* 127: 1033870.

9

Cave Geomorphology

9.1 Introduction

In the previous chapters, we learned in which rocks karst and caves form. We explored the chemical processes governing the dissolution of the different karst rocks. We now know that dissolution by liquid water is able to create a distinctive network of underground pathways, connecting recharge areas along the most favorable paths to the outputs of the karst groundwater system. We also realized that this unique groundwater flow is commonly linked to surface landforms, with a reciprocal interaction in which the evolving karst aquifer controls the development of surface features, and vice versa. To get a better idea on how the karst aquifer system and the associated landscape have evolved, we also need to take a look underground. This chapter deals with the morphology of the subterranean world accessible to human beings: caves.

9.1.1 Definition of Cave

Caves have been defined in different ways. General dictionaries indicate definitions such as "a large hole in the ground or in the side of a mountain," or "a natural chamber or series of chambers in the earth or in the side of a hill or cliff," or "a large natural hole in the side of a cliff or hill, or under the ground." Cave entrances can occur in a wide variety of geomorphic settings: mountains, cliffs, high plateaus, glacier fronts or inside glaciers, perfectly flat plains, along coasts, underwater. Lay people typically consider caves as dark underground spaces that may be dangerous, beautiful, mysterious, scary, hostile, just to mention the most common, sometimes rather contrasting, perceptions. The word "cave" is frequently used to designate anthropogenic holes, such as underground quarries, bunkers, or storing rooms, especially when their initial purpose is no longer known. Ancient catacombs and tombs are sometimes called caves by archeologists, as long as they are underground and dark places. Archeologists also use the word "cave" to designate shelters and alcoves in undermined cliffs, regardless of their breadth and depth. So, not all archeological "caves" are dark. In the hydrocarbon and geotechnical industry, large underground voids intercepted by boreholes are often called caves, despite they do not have a natural access. Biologists use "cave" to designate voids that can contain organisms adapted to underground conditions. These caves can be at least one order of magnitude smaller than those accessible to speleologists. Natural voids completely filled with sediments or ore minerals and encountered in underground mines are sometimes also called caves, although they are not explorable. So how should we define a cave?

Karst Hydrogeology, Geomorphology and Caves, First Edition. Jo De Waele and Francisco Gutiérrez.
© 2022 John Wiley & Sons Ltd. Published 2022 by John Wiley & Sons Ltd.

The International Union of Speleology defines a cave as "a natural underground opening in rock that is large enough for human entry" (Ford and Williams 2007, p. 209). This is a rather anthropocentric view lacking any genetic meaning. It is true though, that the human-sized criterion of this definition of cave is the most practical one, since the accessibility of caves constrains the feasibility of carrying out direct observations in their interior, and thus the understanding of their origin. A more scientific definition could be based on the principal process responsible for the formation of the interconnected network of underground passages we call caves: solutional enlargement of underground voids by fluid flow. Following this perspective, cave could be defined as "a natural opening enlarged by dissolution to a width large enough to allow turbulent flow to occur." This definition would thus include a 1-cm wide enlarged fracture or tubular passages that allow groundwater to flow underground at velocities much greater than in granular aquifers. But this definition would not include voids created by other processes than dissolution, such as lava tubes and suffosion caves, for instance.

9.1.2 Cave-Forming Processes

Many geomorphology textbooks adopt a process-landform-based approach to describe the different types of geomorphic features, grouping them according to the most common zonal or azonal geomorphic system in which they develop (e.g., glacial, periglacial, fluvial, and coastal). However, as often happens in geomorphology, landforms are generally the result of a combination of processes. Moreover, different processes may create very similar landforms (morphologic convergence or equifinality), complicating their genetic interpretation. Caves do not escape these rules, and it is often not easy to identify the different processes that have contributed to the formation of a single cave, commonly developed over long time periods.

Caves can form by a variety of processes, which can be categorized into the following groups: dissolution, weathering and erosion, mechanical movement and accumulation, deposition, melting, and solidification (Figure 9.1). Whereas the first process is explained in detail in Chapter 3, the other five require some short explanation here.

Rocks can undergo strain related to tectonic stress fields acting in the Earth's crust or to gravitational forces operating on unstable slopes (slope movements). Competent rock masses are fractured, and the intervening blocks can experience differential movements due to tectonic faulting or near-surface gravitational deformation. The latter is frequently induced or favored by stress release associated with valley entrenchment or glacial unloading. These movements, which are often perpendicular to the fracture planes (dilatation), create narrow crevices which can be explorable. The rock blocks bounded by fractures eventually can fall along steep slopes accumulating in piles of boulders that may have large empty interparticle voids.

Mechanical weathering involves the breakup of rocks into smaller fragments, and generally operates in association with chemical weathering processes. This can be a very effective process in rocks composed of minerals with different behavior in response to physical and/or chemical changes in the environment. For instance, differences in thermal expansion and contraction of minerals cause the formation of tiny cracks between mineral grains. These openings can then be penetrated by fluids and dissolved salts, which can aid in enlarging these voids by chemical reactions (dissolution, hydration) or physical processes (freezing, salt crystallization). The progressive disintegration of the rock produces loose sediments with variable grain sizes (regolith). These soft unconsolidated materials can be eroded and transported by different agents, including runoff, wind, waves, or subsurface water flows.

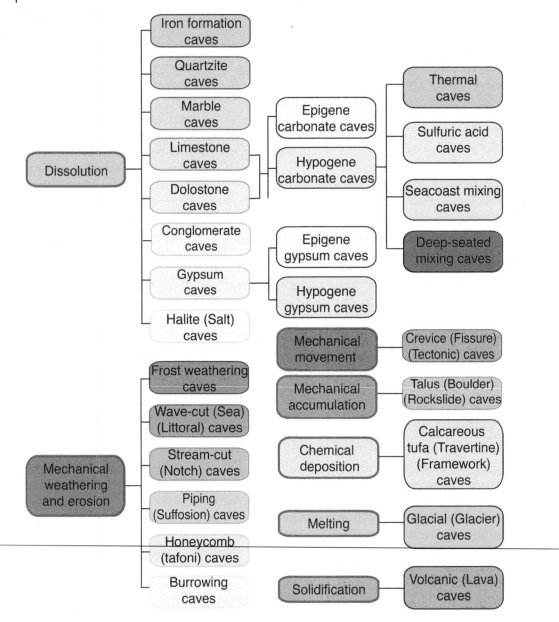

Figure 9.1 Process-based classification of caves. Red-lined boxes represent the six main processes responsible for the formation of caves.

Another process that leads to the formation of caves is chemical deposition. Accumulation of rock occurs for example in growing coral reefs, leaving voids in between the coralline framework. In continental aquatic environments, such as streams and waterfalls, active deposition of carbonate deposits (calcareous tufa, travertine) may create large voids, especially where these sediments accumulate on waterfalls (Gradziński et al. 2018). These caves form coevally to the rock in which they are hosted, and therefore are called "primary" caves, in contrast to "secondary" caves, where the void is carved in a preexisting rock unit.

The other primary caves are the volcanic ones, created during the solidification of lava flows. The last process is ice melting, which produces glacial caves. Water flowing on the surface of glaciers with a temperature above the melting–freezing point can penetrate downward into the glacier. These fluids, because of the high thermal capacity of water, can bring thermal energy relatively deep into the glaciers, continuing to melt ice up to depths of around 150 m (Badino and Piccini 2002). Below this depth, the ice movement is dominated by plastic deformation and continuous processes of melting and freezing causes the openings to close, so that the ice below this depth is almost perfectly impervious. When the ice is less thick than 150 m, the meltwater can reach the bottom of the glacier and form explorable subglacial channels floored by sediments.

9.1.3 Cave Classifications

There are many ways of classifying caves, based on their characteristics and their geological and geographical position. Caves can be long, short, deep, and shallow. They can be characterized by a linear passage, a branching network, a maze, and many other planar forms. Passages can be described as cylindrical, elliptical, canyons, low and wide slots, and a multitude of other shapes. They can be located above the local water table (vadose caves), within the oscillation zone (epiphreatic caves), at the water table (water-table caves), or below it (phreatic caves). They can be filled with sediments (e.g., sand cave), contain ice deposits (ice cave) (Figure 9.2a, b), crystals and mineralizations (crystal cave, barite cave) (Figure 9.2b–e), and be of archeological or paleontological interest.

Caves can be hosted in different rock types (limestone cave, gypsum cave, etc.), and follow geological features (e.g., joint-guided cave, bedding-plane cave). They can be found in high mountain areas (alpine cave), or in plains. They can open up on valley flanks, connect adjacent karst depressions (polje caves, etc.), or be located at the water table level at the foot of residual limestone hills (foot caves, mogote caves) or along the coasts (coastal and submarine caves). Caves occur in different climates (tropical cave, Mediterranean cave, semi-arid cave, etc.). They can be fed by allogenic streams (allogenic stream caves) or be hosted in a pure karst terrain lacking surface streams (holokarst caves). Caves can still host the flowing waters that have formed them (active caves), be activated only during high water level periods (episodic caves, floodwater caves), or be completely offset from the active karst system (relict caves). Old caves can also be exhumed by surface erosion (unroofed caves) (Figure 9.3), or be completely destroyed by multiple agents, including human activity such as quarrying.

Therefore, there is no universal classification, mainly because of the intrinsic complexity of caves regarding their shapes, genetic processes, and their physical context. Often caves owe their origin to a combination or superposition of processes, and they can form in different stages and varying geological timeframes. Here, we mainly adopt a genetic approach to cave classification, which is illustrated in Figure 9.1. In this classification, caves are categorized according to the main process responsible for the formation of the void. This classification is a modified version of the ones proposed by Palmer (2007), Oberender and Plan (2018), and White and Culver (2019) (Figure 9.1).

9.1.3.1 Solution Caves

Caves that owe their origin mainly to dissolution are the most important group, and are called solution (dissolution, karst) caves. They are the main caves of interest in this book and the most abundant on Earth. Solution caves are a significant component of karst systems. Dissolution can take place in many rock types, some of which are very soluble, others are much less. According to the main lithology affected by dissolution, solution caves can be divided into: halite, gypsum, limestone, dolostone, marble, conglomerate, quartzite, and iron formation caves. A further distinction,

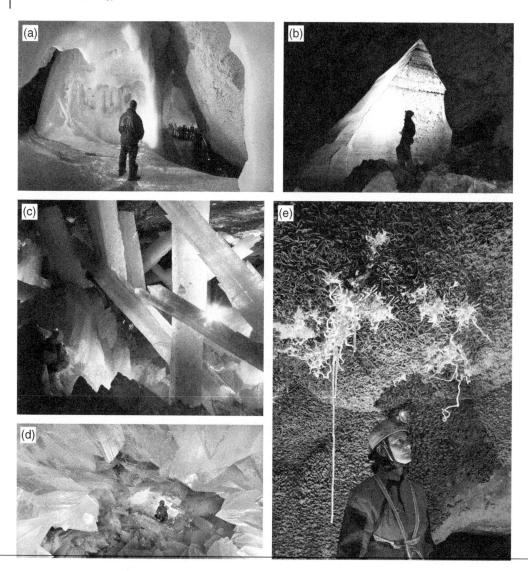

Figure 9.2 Examples of ice and crystal caves: (a) the Eisriesenwelt Cave in Austria, one of the most famous ice caves on Earth. *Source:* Photo by Michael Gruber. (b) A layered ice deposit is melting slowly away in Scarasson Abyss, Piedmont, northern Italy. *Source:* Photo by Bartolomeo Vigna. (c) The Giant Crystal Cave of Naica. *Source:* Photo by Roberta Tedeschi, La Venta Esplorazioni Geografiche. (d) The giant gypsum crystals in the small Geoda de Pulpí, in Andalusia, southern Spain. *Source:* Photo by Victor Ferrer. (e) Honey brown barite crystals with white aragonite bushes, Santa Barbara Cave, Sardinia. *Source:* Photo by Victor Ferrer.

mainly applied to carbonate and gypsum caves is related to the provenance of the fluid and its solutional potential. When they derive from surface sources the caves can be categorized as epigene, and when they have a deep origin they are classified as hypogene caves. A further distinction is made for the hypogene carbonate caves, depending on the source of acidity or the mechanisms responsible for the aggressiveness of the water: caves can be related to thermal CO_2-bearing fluids, to the presence of sulfuric acid (thermal or nonthermal), to the mixing of seawater and freshwater in coastal areas, or to the mixing of different fluids in deep environments.

Figure 9.3 Unroofed caves in Slovenia: (a) Unroofed cave of Loza shown as a valley-like elongated depression in the middle of the forest, 10 km SSW of Postojna, Slovenia. *Source:* Photos by Andrej Mihevc. (b) Relief model of the Loza unroofed cave system generated with LiDAR data with a horizontal resolution of 1 m (Courtesy of Karst Research Institute, Postojna). This unroofed cave, with its 3.5 km length, is the longest in Slovenia. (c) A large protruding stalagmite in the middle of the forest shows that this depression was once a cave. Lipove doline near Divača, Slovenia. *Source:* Photos by Andrej Mihevc.

Halite (or rock salt) is the most soluble mineral that hosts caves. This rock is so soluble that it only survives at the Earth's surface in arid climates. Halite deposits are composed of NaCl, but can contain other minerals of the halide group, such as sylvite (KCl). The caves can form very rapidly, and most have ages of less than 7000 years (Frumkin et al. 1991; Frumkin 1994a, 1994b; De Waele et al. 2020). The most important caves in halite are reported from salt extrusions related to active diapirs, in which the salt formation emerges at the surface as a viscous fluid in response to differential loading conditions. Diapirs with important caves occur in the Zagros Mountains, Iran (Bosák et al. 1999; Bruthans et al. 2008), in Mount Sedom, along the shore of the Dead Sea in Israel (Frumkin and Ford 1995), and in Tajikistan (Dzens-Litovsky 1966). The longest cave is Malham Cave in Mount Sedom, around 10 km long (Kutleša et al. 2019) (Figure 9.4a), whereas 3 N Cave in the Namakdan Diapir, Qeshm Island, Iran is 6580 m long (Bruthans et al. 2010) (Figure 9.4b, c). The salt caves in the Cordillera de la Sal in Chile are formed in a N-S oriented salt anticline. Over 50 caves

Figure 9.4 Halite caves: (a) Steeply dipping rock salt and notch in Malham Cave, Mt. Sedom, Israel. *Source:* Photo by Amos Frumkin. (b) Speleothem formation called the Octopus in 3 N Cave, Qeshm Island, Iran. *Source:* Photo by Michal Filippi, Project NAMAK. (c) One of the largest caverns in the 3 N Cave, named "Megadomes" by the discoverers, with typical flat ceiling and varicolored salt layers. *Source:* Photo by Marek Audy and Richard Bouda, Project NAMAK. (d) A 40-m deep shaft in Cressi Cave system, Atacama, Chile. *Source:* Photo by Alessio Romeo, La Venta Esplorazioni Geografiche. (e) Meandering passage and solution notch in the finely stratified salt deposits of Forat Micò, Cardona Diapir, NE Spain. *Source:* Photo by Victor Ferrer. (f) Salt cave passage with multiple notches in Cressi Cave system. *Source:* Photo by Riccardo De Luca, La Venta Esplorazioni Geografiche.

have been explored so far in this hyperarid climate, with an aggregate length of more than 20 km, including the deepest explored salt cave in the world (Cressi Cave system, 230 m deep) (De Waele et al. 2017a) (Figure 9.4d, f). A peculiar kind of salt caves are found in the Cardona Diapir, NE Spain, where preexisting natural caves and mine galleries experienced a dramatic enlargement due to the interception of phreatic conduits, subsidence activity, and the flooding of the mine by a river (Lucha et al. 2008a) (Figure 9.4e).

The most abundant sulfate found at the surface is gypsum ($CaSO_4 \cdot 2H_2O$), whereas the anhydrous mineral anhydrite ($CaSO_4$), although occurring naturally at the surface in certain conditions, is mainly found underground. When anhydrite is brought back to the surface and interacts with meteoric water, it hydrates and transforms into gypsum before it dissolves (Klimchouk 1996a, 2000a). Caves can form rather rapidly in gypsum under favorable hydrological and hydrochemical conditions (decennia, centuries) due to its high solubility.

In epigene gypsum karst systems, because surface waters rapidly come close to saturation in gypsum, the dissolving power of fluids decreases substantially over relatively short distances. Only fissures with initial apertures wide enough to allow concentrated flow will eventually enlarge to become accessible caves. This results in single-passage caves with a poorly branching pattern, developed close to the water table or at the contact with underlying low-permeability rocks, and conduit sizes adapted to the highest occurring flows (Figure 9.5b, c). These passages are typically controlled by fractures and/or bedding planes, and collapses can create relatively large rooms (Figure 9.5a) (Pisani et al. 2019).

Where the soluble sulfate rock is sandwiched between less soluble and permeable beds, gypsum can be dissolved in a (hypogene) intrastratal setting. In this case, water slowly flows upward from the underlying confined aquifer, through the discontinuities in the gypsum bed, to the overlying rock unit. This is the case of transverse or cross-formational hypogene speleogenesis that has created the remarkable gypsum maze caves of western Ukraine (Klimchouk 2019c), some being among the longest caves on Earth (Table 9.1).

Sulfate rocks are more ductile than other rocks such as limestone, causing the fractures to be more widely spaced. Fracture density also decreases rapidly with depth, explaining why caves in gypsum are often not very deep, barely exceeding 200 m in depth. The deepest known cave is Monte Caldina in Italy, 265 m deep (Franchi and Casadei 1999). Most sulfate caves are in gypsum, although some caves in anhydrite are known in the Harz Mountains (e.g., Barbarossa Cave) (Kempe et al. 2017) (Figure 9.5d) and in the northern Apennines (Upper Secchia Valley, De Waele et al. 2017c) (Figure 9.5e). A special type of cave is that related to the hydration of anhydrite into gypsum, sometimes involving a considerable increase in volume and the bulging of beds to create human-sized voids underneath domes and ridges. These peculiar, small, and rapidly evolving caves have been named hydration caves (Jarzyna et al. 2020). Good overviews of gypsum caves in the world are given in Klimchouk et al. (1996b), Calaforra (1998), and Klimchouk (2019a).

Most caves in the world are developed in carbonate rocks, including the sedimentary rocks limestone and dolostone, and the metamorphic rock marble. Limestone is mainly composed of calcite ($CaCO_3$), dolostone is chiefly made up of dolomite ($CaMg(CO_3)_2$), whereas marble can be constituted of calcite and/or dolomite. Limestone and dolostone display the typical sedimentary textures and structures (i.e., bedding) and to some extent an appreciable porosity (between 5 and 25%). In contrast, marble is composed of interlocked recrystallized minerals with very low matrix porosity (usually of some percent), and is therefore almost impermeable if not fractured. Speleogenesis and karst in limestone is treated in detail in this book, so here we only mention some important characteristics of dolostone and marble. Although rarely reaching volumes large enough for hosting significant solutional caves, some examples of magnesite ($MgCO_3$) caves also exist, and are mentioned below.

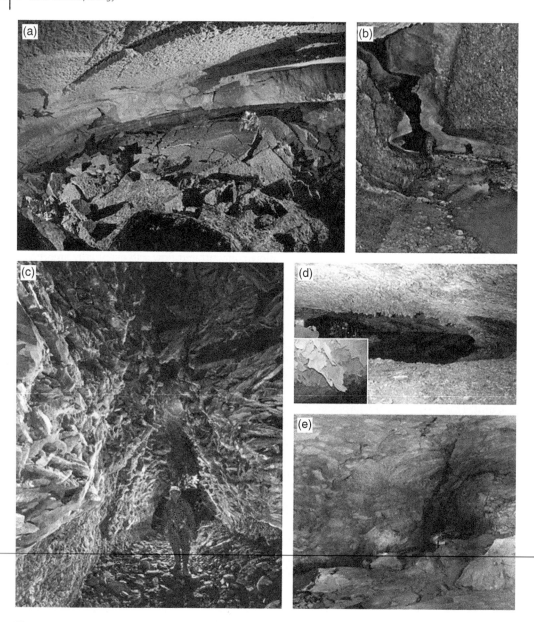

Figure 9.5 Gypsum caves: (a) Sala Giordani, a large collapse hall in one of the longest epigenic gypsum caves of the world, Spipola Cave, Italy. *Source:* Francisco Gutiérrez (Author). (b) The main river passage in Rio Basino Cave, Vena del Gesso Regional Park, Italy. *Source:* Photo by Piero Lucci. Note close-to-horizontal notches and the steep dip of the beds. (c) Large gypsum crystals in a cave in Sorbas, southern Spain. *Source:* Photo by Victor Ferrer. (d) Barbarossa Cave in anhydrites in the Harz Mountains, Germany. Inset shows sheets of hydrated anhydrite (gypsum) flaking off the walls. *Source:* Photos by Stephan Kempe. (e) The large entrance hall in Tanone della Gacciolina Cave, in Triassic gypsum-anhydrites of the Upper Secchia Valley, Northern Italy. *Source:* Photo by Piero Lucci.

Table 9.1 The world's longest caves.

Rank	Cave	Country	Length (m)	Depth (m)	H/E
1	Mammoth Cave System (N.P.)	U.S.A.	675,924	124.1	E
2	Sistema Sac Actun (Nohoch Nah Chich, Aktun Hu) (Underwater+Dry)	Mexico	376,700	119.2	H
3	Jewel Cave (N.M.)	U.S.A.	335,564	248.4	H
4	Sistema Ox Bel Ha (Underwater)	Mexico	318,040	57.3	H
5	Suiyang Shuanghe Dongqun	China	311,500	665.0	E
6	Optymistychna	Ukraine	264,576	15.0	H
7	Wind Cave (N.P.)	U.S.A.	260,231	193.9	H
8	Lechuguilla Cave (C.C.N.P.)	U.S.A.	242,045	484.2	H
9	The Clearwater System (Gua Air Jernih)	Malaysia	238,046	355.1	E
10	Fisher Ridge Cave System	U.S.A.	212,111	4.5	E
11	Hölloch	Switzerland	208,000	1033.0	E
12	Sistema del Alto Tejuelo	Spain	177,574	626.0	E
13	Siebenhengste-hohgant Höhlensystem	Switzerland	164,500	1340.0	E
14	Schoenberg-Höhlensystem (RaucherkarHöhle-Feuertal-höhlensystem)	Austria	153,107	1061.0	E
15	Sistema del Mortillano	Spain	146,500	950.0	E
16	Ozerna	Ukraine	140,490	35.0	H
17	Schwarzmooskogel-höhlensystem, Loser-Augsteck	Austria	135,772	1125.0	E
18	Bullita System (Burke's Back Yard)	Australia	120,400	23.0	E
19	Reseau Felix Trombe / Henne-Morte	France	117,200	1004.0	E
20	Sistema del Gandara	Spain	116,740	814.0	E
21	Toca da Boa Vista	Brazil	114,000	50.0	H
22	Hirlatzhöhle - Schmelzwasserhöhle	Austria	113,550	1560.0	E
23	Marosakabe	Madagascar	113,019	0.0	E
23	Systeme de Ojo Guarena	Spain	110,000	193.0	E
25	Sistema K'oox Baal - Sistema Tux Kupaxa (Underwater+Dry)	Mexico	103,100	34.4	H

The list comprises all caves longer than 100 km. H and E stand for hypogene or epigene, referring to the origin of the cave. Note that almost half of the longest caves in the world are of hypogene origin. *Source:* Adapted from Bob Gulden's list, www.caverbob.com (accessed 21 March 2022).

Dolomite is characterized by a slower dissolution kinetics than calcite, but it is often more densely fractured. In mixed situations, where dolostone and limestone beds alternate, the limestone dissolves more readily than dolostone, so that the largest voids are created in the former, and dolostone beds can project out of the cave walls. When dolostone is massively bedded, caves develop less easily, unless a large amount of water infiltrates into major discontinuities. Dolostone underlying limestone will often form a sort of barrier for karstification, since infiltration waters can be already close to saturation with respect to the carbonate minerals. Only major open fractures can allow water to flow across these barriers, penetrating deeper into the dolostone, but eventually these pathways will tend to taper downward. Caves in dolomites are known in many

Figure 9.6 Caves in dolostone: (a) Worm burrows (bioturbation) in Ordovician dolomite on the wall of Mystery Cave, Minnesota (USA). *Source:* Photo by Arthur Palmer. (b) The intensely weathered and variously colored dolostone bedrock with pure white speleothems in Spider Cave, a sulfuric acid cave in New Mexico (USA). *Source:* Photo by Lukas Plan. (c) One of the large and wide passages in Toca da Barriguda Cave, Bahia (Brasil). *Source:* Photo by Luciana Alta and Vitor Moura. (d) One of the main conduits along a shaly interbed in the Piani Eterni cave system, the longest cave in the Italian Dolomites UNESCO Park. *Source:* Photo by Francesco Sauro.

regions, famous examples being Crevice and Mystery Cave (Figure 9.6a), the former being the longest cave in Missouri (USA), several caves in the Permian dolostone of the Guadalupe Mountains (Figure 9.6b), the Toca da Barriguda and Toca da Boa Vista maze systems in the Bahia State, Brazil (Figure 9.6c) (Klimchouk et al. 2016), the last being the longest cave in the southern hemisphere (Auler et al. 2017), the Piani Eterni Cave System, the longest and most important cave in the Italian Dolomites (Figure 9.6d) (Sauro et al. 2013b; Columbu et al. 2018), just to mention a few examples.

Marbles often form much smaller outcrops than limestones when related to contact metamorphism associated with igneous intrusions, but they can also be extensive in areas subdued to regional metamorphism. When marbles form narrow bands in between non-soluble rock units they can give rise to exceptional examples of stripe karst, such as in Norway (Lauritzen and Skoglund 2013) (Figure 9.7a, b) and Crystal Cave in California (Despain and Stock 2005). The largest cave in California, Lilburn Cave (34 627 m long), is carved in a 4 km^2 patch of marbles surrounded by granites, quartzites, and schists (Figure 9.7c, d) (Abu-Jaber et al. 2001; Tobin and Doctor 2009). Bigfoot Cave in the Marble Mountains of California (Davis and Serefeddin 2004) is the deepest cave of the state reaching −367 m. Good examples of marble caves also occur in France (Figure 9.7f) and in New Zealand (Figure 9.7g). In marble stripe karst, speleogenesis mostly occurs at the boundary between the soluble marble and an adjacent non-soluble rock. In larger marble outcrops, speleogenesis is greatly controlled by open fractures and by the former and current

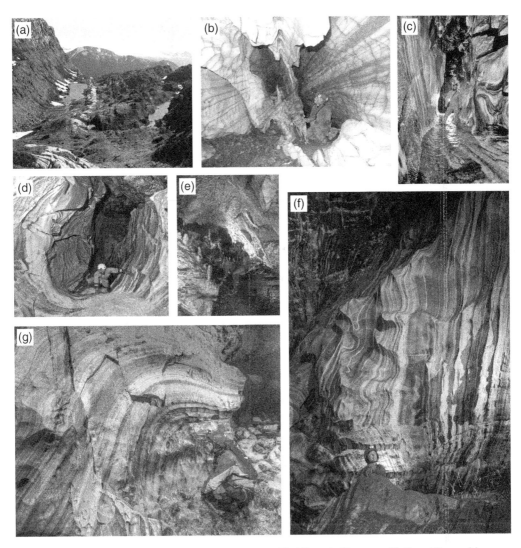

Figure 9.7 Caves in marble: (a) A stripe karst area near Mo I Rana in Norway, with the white marble belt running vertically through the picture and (b) an epiphreatic passage with scallops in marble. *Source:* Photos by Stein-Erik Lauritzen. (c, d) Lilburn, the longest marble cave in California. *Source:* Photo by Dave Bunnell. (e) An epiphreatic passage in the white marbles of Milazzo Cave, Apuan Alps, Italy. *Source:* Photo by Leonardo Piccini. (f) Devonian metalimestones with white and blue beds exposed in the Roquebleue Cave, Hérault, France. *Source:* Photo by Philippe Crochet. (g) Folded marble in Ironstone Cave, Takaka Hill, Northern Island, New Zealand. *Source:* Photo by Neil Silverwood.

position of the regional groundwater level governed by processes such as uplift and fluvial incision. One of the largest cave systems in Italy, Monte Corchia (Apuan Alps), is formed in marble and has three major cave levels that provide evidence indicating that the recharge areas have changed during mountain uplift. The Apuan Alps host a great number of horizontal marble cave systems (Figure 9.7e) and some of the deepest caves of Italy (Piccini 2011).

Magnesite is a much rarer carbonate rock, often occurring in lenses within low-grade metamorphic carbonate complexes together with dolomite. It often occurs along tectonic contact zones where hydrothermal Mg-rich fluids interacted with the carbonate host rock, and metasomatic

Figure 9.8 Jelsava Cave, Slovakia, developed in magnesite. *Source:* Photo by Pavel Bella.

replacement of Mg in calcite and/or dolomite took place. The magnesite bodies are often small, rarely have surface exposures, and are often targeted for magnesite ore exploitation. Magnesite is also much less soluble than calcite and dolomite. Small solutional caves in magnesite have been reported from California, Brazil, and Tasmania. In Slovakia, 16 caves have been mapped in five different magnesite bodies. The length of the mapped caves is generally some tens of meters, the longest reaching 76 m, and they are mostly irregular geodes (Figure 9.8), or elongated fissures in enlarged fractures (Gaál et al. 2017).

Other caves related to the dissolution of carbonates are those hosted in conglomerates. These caves can reach lengths of several tens of kilometers (Bol'shaja Oreshnaja in central Siberia reaches 58 km; Filippov 2004) and are formed by a combination of dissolution, which reduces the inter-granular cohesion, and the erosion of the loosened gravels and matrix. The conglomerates are often composed of a mixture of calcareous and crystalline pebbles, mixed with sands and finer particles, plus carbonate cement, but can also be entirely carbonatic. Good examples of these caves are known in Slovenia (Lipar and Ferk 2011) (Figure 9.9a), Switzerland (Lapaire et al. 2007), the Montello Hill in northern Italy (Ferrarese and Sauro 2005) (Figure 9.9b), in Burgos (Figure 9.9c), Catalonia (Figure 9.9d), and Mallorca (Figure 9.9e) in Spain, and in Honduras (Finch and Pistole 2011).

Dissolution does not only take place in rocks readily soluble in water without acidity (salt and sulfates) or with acidity (generally CO_2), such as carbonates. Very poorly soluble rocks such as quartz sandstones (quartzite) or iron ores can also host caves formed by processes that include dissolution as a fundamental mechanism. Quartzite caves have been known to exist for over 100 years, but the first significant cave systems were explored in the 1970s, and the most important discoveries are less than 20 years old. Although the longest cave hosted in hybrid quartzites (with small amounts of carbonate cements) is located in Meghalaya, India (Krem Puri Cave, over 25 km long) (Figure 9.10b) (Sauro et al. 2020), the most outstanding silicate karst areas are in South America, mainly Venezuela (Figure 9.10c), Brazil (Figure 9.10a), and Colombia. These caves are

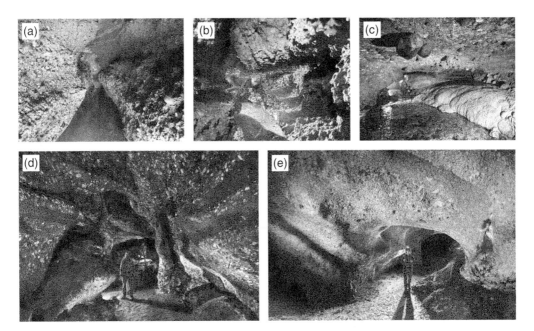

Figure 9.9 Caves in conglomerate: (a) Underground river passage in the Bus del Fun, in the Messinian conglomerates of the Montello Hill, Treviso, Italy. *Source:* Photo by Sandro Sedran, S-Team. (b) A stream passage in the Arneževa Luknja Cave on the west side of Udin Boršt conglomerate terrace, Slovenia. *Source:* Photo by Matej Lipar. (c) Underground river in the Fuentemolinos Cave, Burgos, Spain. *Source:* Photo by Victor Ferrer. (d) Cova de Mura Cave in reddish Eocene conglomerates, Montserrat Massif, Spain. *Source:* Photo by Victor Ferrer. (e) Passage in the conglomerates in the Cova den Canet, Mallorca, Spain. *Source:* Photo by Philippe Crochet.

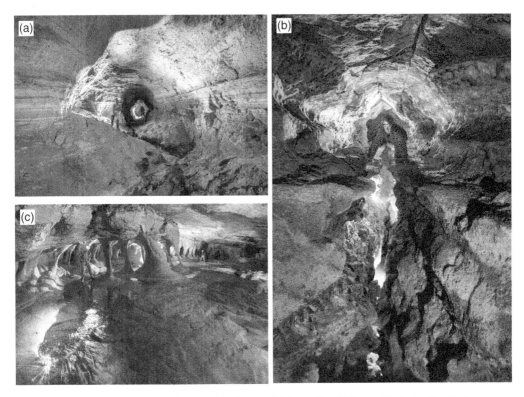

Figure 9.10 Quartzite caves: (a) Gruta do Martignano, Ibitipoca, Brazil. *Source:* Photo by Alessio Romeo. (b) Krem Puri, Meghalaya, India. *Source:* Photo by Mark Burkey, Caving in the Abode of the Clouds Project. (c) Imawari Yeuta, Auyàn Tepui, Venezuela. *Source:* Photo by Vittorio Crobu, La Venta Esplorazioni Geografiche.

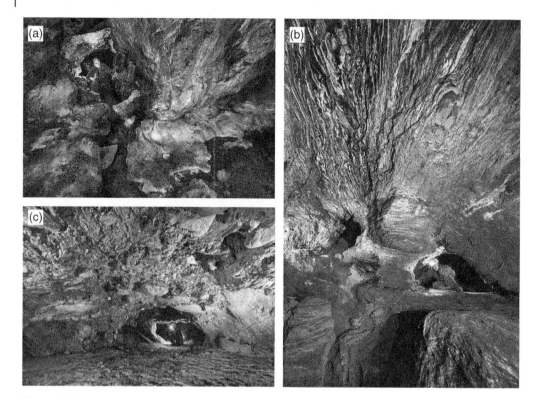

Figure 9.11 Images of a cave developed in an iron formation in the municipality of Conceição do Mato Dentro, state of Minas Gerais, Brazil. *Source:* Photo by Luciana Alt and Vitor Moura.

created by a combination of processes, generally grouped under the name of arenization, in which dissolution of the quartz cement and the consequent reduction of the rock cohesion is a fundamental step (Wray and Sauro 2017).

Brazil is also the country in which caves hosted in banded iron formations (alternating layers of Fe(III)oxides, chert, and silica layers) or their weathering products (known as "canga") are most abundant (Figure 9.11). These iron formation caves are important from a paleontological and archeological point of view. Moreover, due to their association with valuable iron ores, they are often endangered and therefore protected by Brazilian laws. Iron ore caves often have a spongework plan pattern, related to local lithological and petrophysical differences in the host rock. Their origin is generally attributed to deep-seated hypogene dissolution of calcareous or silica cement, leaving a vuggy porosity (Auler et al. 2019). The mobility of iron itself appears to be mediated by bacterial activity, which reduces Fe^{3+} into the more mobile Fe^{2+} species (Parker et al. 2013, 2018; Calapa et al. 2021).

9.1.3.2 Caves Related to Mechanical Movement and Accumulation
The formation and opening of fractures in rock masses due to Earth's crustal movements or gravitational processes can produce cracks, especially on steep unstable slopes. Most of these planar openings are not wide enough for humans to enter, but some can be accessible, and they are normally close-to-vertical, narrow, and rectilinear passages with almost parallel walls. These caves are called crevice or fissure caves (Figure 9.12a), and in the speleological literature, they are sometimes also called, inappropriately, tectonic caves. Crevice caves can occur in any type of compact rock

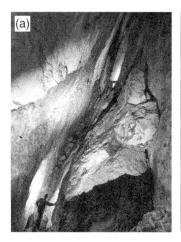

Figure 9.12 Crevice and talus caves: (a) The eastern slope of mount Großer Otter (SW of Gloggnitz, Lower Austria) has a series of slope-parallel crevice caves in Triassic dolomite, such as this cave known as Markiteres Windloch Cave. *Source:* Photo by Lukas Plan. (b) A small talus cave in sandstone boulders in the Elbsandsteingebirge, East Germany. *Source:* Photo by Piotr Migón.

(granites, basalt, and sandstone), but also in soluble rocks, in which case they can be enlarged by dissolution. The floor of these caves is often covered with debris fallen from above (blocks, soil, and vegetation litter), and their roof can be composed of solid rock, or soil.

Rock-fall deposits derived from cliffs and escarpments accumulate at the foot of the slopes forming openwork chaotic breccias. If the rock fragments are large, these deposits can contain large voids in between the boulders, sometimes forming talus caves, also known as boulder caves (Halliday 2004). These are intricate caves, often with multiple entrances and vaguely defined and highly irregular cave perimeters made of blocks and soil. The voids within the boulder deposits are often enlarged by a combination of weathering processes and subsurface water flows (Figure 9.12b) (Duszyński et al. 2018). These caves are particularly important from a biospeleological point of view, especially in areas where soluble rocks (and thus large subsurface voids) are scarce or absent. Touchy Sword of Damocles (TSOD) Cave (Adirondack Mountains, New York State), with its ca. 4 km length in between anorthosite boulders and rock slabs, appears to be the longest talus cave in the world (Cooper and Mylroie 2015).

9.1.3.3 Caves Related to Mechanical Weathering and Erosion

In some cases, the dominant cave-forming process is erosion by moving water, which aids in the removal of previously formed weathered bedrock (regolith) and/or debris. The erosional agent responsible for the removal of the loose material can be a river, the sea, or sheet flow. In frost weathering caves, on the other hand, the dominant formative process is related to freezing and thawing (frost shattering) (Figure 9.13). These are generally called rock shelters or shelter caves. This kind of caves is relatively common in alpine mountain regions, with many examples from Austria (Oberender and Plan 2015). These rock shelters can also be related to breakdown processes induced by salt crystallization (salt weathering) in arid climates, or where sea spray is abundant in hot and wet climates (i.e., Mediterranean region, see tafoni below).

Waves have enough energy to erode coastal cliffs, especially along weaknesses in the rock (fractures, bedding planes, or weaker rock strata). This type of erosion creates the so-called sea caves or littoral caves. Once an opening (inlet) is formed, the energy of the waves can amplify if

Figure 9.13 Frost-weathering caves: the huge Wetzsteinloch Cave (Hochschwab, Styria, Austria) whose formation has been ascribed to frost shattering and erosion of volcanic ash from the core of a fold in Middle Triassic limestone (Oberender and Plan 2015). Arrow points to person for scale. *Source:* Photo by Monika Hölzel.

constructive interference occurs, or simply because of the greater tidal effect on inlets. The presence of boring organisms can help in weakening the rock, especially in carbonates. Many coastal caves are hosted in volcanic rocks (Figure 9.14), which are often well fractured and thus more susceptible to erosion by wave action. The world's most famous sea cave in basalts is without any doubt Fingal's Cave, in the Scottish Island of Staffa (Hebrides), a profusely depicted cave that inspired musician Mendelssohn (Todd 1993) (Figure 9.14a, b). Many examples of sea caves are known in southern California, including the longest one, Painted Cave on Santa Cruz Island (Bunnell 1995). The most voluminous sea cave (over 220 000 m³) appears to be Riko Riko Cave in New Zealand, carved by waves along a series of fractures in Upper Miocene rhyolitic breccia and tuff deposits (Bunnell 2004) (Figure 9.14c).

Fluvial erosion along the banks of bedrock river channels can undermine the flanking steep slopes producing stream-cut caves. These caves of the through-flow type are usually rather small and are often inundated during floods. They are typically formed in hard sandstones, schists, granites, and diorites. Good examples of this type of caves have been documented in the Sierra Nevada, California, such as the 1.4 km long Greenhorn Cave System. These caves are characterized by very smooth walls and typical forms of stream erosion such as wall furrows and concavities, riffles, pools, and potholes (Figure 9.15a, b) (Breisch 1986). These caves often start as simple notches that deepen by a combination of processes, including weathering and erosion.

Notches also form by the combined action of dissolution, salt weathering, and differential erosion on well-layered soluble rocks. These notches often occur in valley margins well above the stream channels (Shtober-Zisu et al. 2015) or along coastal cliffs high above sea level. Notches are

Figure 9.14 Sea or littoral caves: Fingal Cave on the Island of Staffa, Hebrides (Scotland) is probably the most documented sea cave in the world. (a) The cave represented on one of the Liebig trade cards. (b) Fingal Cave painted in the nineteenth century on a small glass window of a magic lantern. *Source:* Both images from Centro Italiano di Documentazione Speleologica (CIDS) Franco Anelli Bologna/Public Domain). (c) Riko Riko Cave in New Zealand, probably the largest erosional littoral cave in the world. *Source:* Photo by Dave Bunnell. (d) Coastal cave carved by wave action in close-to-horizontal and fractured volcanic rocks, Vestman Islands, Iceland. *Source:* Francisco Gutiérrez (Author).

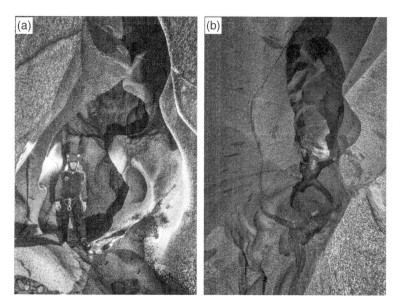

Figure 9.15 Still active river-cut caves carved in granite, California: (a) Millerton Cave System, and (b) "Take it for Granite" Cave. *Source:* Photos by Dave Bunnell.

Figure 9.16 Suffosion caves: (a) Loess cave in Serbia. *Source:* Photo by Piotr Migoń. (b) Gully system carved in marls of the Miocene Mishan Formation, with numerous suffosion caves and collapse sinkholes (see arrows, note persons for scale). This drainage network mainly expands headwards through the development of pipes by subsurface erosion at the steep gully heads and their subsequent collapse. Stars Valley, Qeshm Island UNESCO Geopark, Iran. *Source:* Francisco Gutiérrez (Author).

very well developed at the foot of residual karst towers in tropical climate, where they are known as foot caves (McDonald and Twidale 2011). These, however, are mainly solution caves.

Deep weathering in fractured crystalline rock, such as granites, can transform the rock mass associated with the joints into a friable sandy regolith (grus) that surrounds cores of hard unweathered rock (corestone). The loose sand can be removed selectively by surface and subsurface waters producing the typical granite landscape of bornhardts, tors, kopjes, and boulder fields. In rare cases, the combined differential subsurface weathering and erosion can generate caves in the spaces left between residual boulders (Twidale and Vidal Romaní 2005). A good example of these granite weathering caves is *Grotte du Diable* near Huelgoat, in Brittany, France, that together with the nearby "*roche tremblante*" located in a dense forest has given rise to numerous legends.

Subsurface water erosion (piping, suffosion) is particularly important in fine-grained unconsolidated deposits such as loess and some valley fills. It forms networks of shallow conduits that may reach several meters in diameter that can sometimes be explored and mapped (Figure 9.16). Some small loess caves have been used since prehistoric times by people as dwellings in China, and the collapse of these voids can pose a significant hazard to human-built structures (Peng et al. 2018). In general, caves formed by piping processes, where subsurface water erosion is the principal genetic process, are named piping caves or suffosion caves.

On steep rock slopes and boulder faces, differential weathering can lead to the formation of hollows through a combination of processes collectively designated as cavernous weathering. The term tafoni (singular: tafone) is used to designate decimeter- to meter-scale niches, whereas centimeter-sized hollows are termed alveoli (singular: alveole), or honeycomb weathering when they form a mesh of closely spaced hollows (Figure 9.17a, b). The main mechanisms involved in the formation of these features are granular disintegration and flaking, which may be related to various processes such as salt weathering (crystallization and hydration of salts), chemical alteration, wetting and drying, and frost shattering. The resulting weathering products can be removed by wind action, occasional water flow, and animals. These landforms are frequent in arid and coastal regions and in granular and crystalline rocks containing minerals that are easily altered

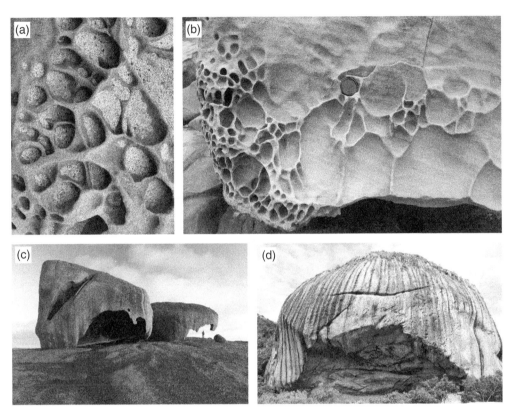

Figure 9.17 Honeycomb and tafoni caves: (a) Honeycomb features in fine-grained granite, Sardinia. *Source:* Photo by Jo De Waele. (b) Cavernous weathering in a coastal cliff made up of sandstone in Kota Kinabalu area, Malaysia. Lens cap 62 mm in diameter. *Source:* Francisco Gutiérrez (Author). (c) Cave-sized tafoni in large boulders of Paleozoic granite on Kangaroo Island, northern Australia. Note person for scale. *Source:* Photo by Philippe Crochet. (d) Giant tafone in massive Precambrian granite, Pedra da Boca locality, Paraíba, NE Brazil. *Source:* Photo by Piotr Migoń. Note the linear rills carved on the granite slope around the tafone, and honeycomb morphologies in the roof of the cave.

into clays (feldspars in granites, arkoses, and volcanic rocks). Some of these features are believed to be, at least partially, of hypogene origin (Klimchouk et al. 2017a). When these weathering landforms reach sizes of several meters, creating true caves, they are named tafoni caves (Huinink et al. 2004). Large tafoni are especially abundant in coastal areas, where the combination of sea spray, wetting and drying cycles, and strong winds create the ideal conditions for their formation (Figure 9.17c). They are also often found in hot and wet climates (Figure 9.17d).

A special kind of erosional caves are those formed, at least partially, by the burrowing action of animals (burrowing cave or salt-ingestion cave). Well-studied examples are the caves located on the flanks of Mount Elgon, a Miocene stratovolcano on the border between Kenya and Uganda. The soft pyroclastic rock contains some sodium rich layers that are "mined" by several animals, including elephants; the caves are called "the elephant caves of Mount Elgon". Making'eny Cave is 250 m long, whereas the more famous Kitum Cave reaches 165 m in length (Figures 9.18a, b). The caves are not excavated by underground water flow, but are created by a combination of weathering, sapping, collapse, and geophagy by animals, including elephants (Lundberg and McFarlane 2006b). The small Rock House Cave, 5 m deep and developed in salt-containing sandstone in Mississippi

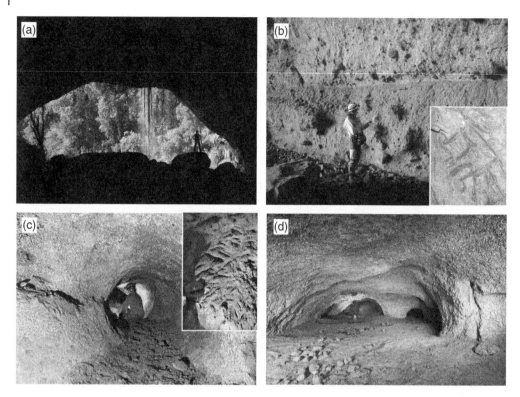

Figure 9.18 Burrowing or salt-ingestion caves: (a) Making'eny Cave entrance looking from inside, with a waterfall in the center (note person to the right for scale). *Source:* Photo by Donald McFarlane and Joyce Lundberg. (b) Main passage of Making'eny Cave, showing two distinct pyroclastic beds. Emergent clasts and the ashy matrix (preferred for salt ingestion) are clearly visible. Inset shows elephant tusk marks in the ash beds in the nearby Kitum Cave. *Source:* Photos by Donald McFarlane and Joyce Lundberg. (c, d) Probable Pleistocene burrowing cave made by an extinct giant Glyptodon, Minas Gerais and Pará States, Brazil. Inset shows claw marks on the walls. *Source:* Photo by Vitor Moura and Luciana Alt.

State, also appears to have been formed due to salt ingestion by deer and maybe bison and cattle (Lundquist and Varnedoe 2006). Other examples of human-sized caves made by animals are the large burrows carved by extinct giant ground sloths and glyptodonts in South America (Mylodont and similar vertebrates) (Figure 9.18c, d). These animals reached up to 3 m height, and dug out underground burrows several tens of meters long (Dondas et al. 2009; Lopes et al. 2017).

9.1.3.4 Depositional Caves

Chemical deposition can be accompanied by the formation of voids, large enough for men to enter (primary megaporosity). The most important examples of such primary or syngenetic caves, formed coevally to the deposition of the host rock, are those associated with calcareous tufa and travertine waterfalls, that can form overhanging shelves and curtains underlain by empty voids. In certain conditions, caves associated with prograding tufa waterfalls can reach lengths of several tens of meters (Figure 9.19a). These calcareous tufa or travertine caves are called progradational, in contrast with the aggradational caves that form within tufa or travertine mounds associated with artesian springs. In this latter case, accretion of carbonate sediment around the spring vent creates an inner crater (Figure 9.19b) (Gradziński et al. 2018). The largest example of aggradational travertine cave is Zendané Soleyman in Iran, not active anymore, with a 109-m deep vertical chimney and long and minor axes at the surface of 35 and 74 m, respectively

Figure 9.19 Tufa and travertine caves: (a) The Labante tufa deposit (Emilia-Romagna, Italy) and its caves. *Source:* Photos by Danilo Demaria. (b) Primary aggradational travertine cave in a mound with central crater near Vysny Sliac, Slovakia. *Source:* Photo by Pavel Bella. (c) Fossil coral reef with small decimeter-sized primary isolated cavities at Cockburn Town, San Salvador Island, Bahamas (20 cm pouch on top of the corals for scale). *Source:* Jo De Waele (Author).

(Figure 2.15b). Progradational tufa or travertine caves can reach several 100 m in length when different primary voids are connected to form intricate caves, comprising isolated chambers, crawlways, and even artificial tunnels. A special kind of tufa or travertine cave is the so-called tree-mold cave, a void created by the decay of a tree trunk that was encrusted by calcareous deposits, leaving the cast of the trunk. Tree-mold caves also form in volcanic ash deposits (Bella and Gaál 2007). Framework caves are small voids formed during the growth of coral structures in reefs (Figure 9.19c). These voids are rarely accessible to man, but dissolution can enlarge them up to a few meters in size.

9.1.3.5 Caves Related to Melting

Glacial caves are created by the melting of ice in glaciers. Their exploration requires caving techniques, which is why they are regarded as true caves, although they are ephemeral and not truly "underground." The largest englacial cave was once the Paradise Ice Cave on Mount Rainier (USA), with 13.25 km of surveyed passages, but the glacier and its cave have slowly disappeared by melting (Anderson et al. 1994a). The scientific study of glacial caves has experienced great progress in the last 40 years, especially regarding the understanding of water flow inside glaciers, the mechanisms of cave formation, and the evolution of voids in ice (Badino 2007; Gulley et al. 2009). Liquid surface

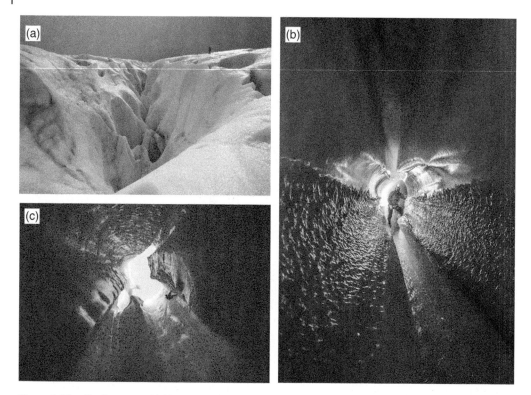

Figure 9.20 Glacier caves: (a) Moulin on Tyndall Glacier in the Torres del Paine National Park, Chile, in 2017. (b) Tunnel in the Gorner Glacier photographed in 2014, Switzerland. (c) A 50-m deep shaft on Grey Glacier in the Torres del Paine National Park, Chile, in 2016. *Source:* Photos by Alessio Romeo, La Venta Esplorazioni Geografiche.

water has temperatures above 0 °C, and is able to melt the ice. The shafts in which these supraglacial rivers disappear are known under the French name *moulins* (Figure 9.20a), and although they are seasonal features, they tend to form every year in approximately the same spots on the glaciers, probably due to the morphology of the glacial surface and irregularities in the bedrock on which the glacier flows (Gulley and Fountain 2019).

Water flow can be guided by permeable layers of rock debris within the ice, or fractures generated by differential ice flow. Englacial caves are rarely deeper than around 200 m (Figure 9.20b, c), because at that depth the water reaches temperatures too low to be able to melt the surrounding ice, and the ice acquires a plastic rheology due to the high pressure that favors the closure of voids (Badino and Piccini 2002). In thin ice bodies, melt waters can reach the bottom of the glacier and form subglacial tunnels and channels, with the floor covered by debris.

9.1.3.6 Caves Related to Solidification

Caves related to solidification refer to volcanic caves. Kazumura Cave on the Kīlauea Volcano, Hawai'i, is currently the Earth's longest and deepest lava cave known to exist (Figure 9.21c, d). A total of 65.5 km of passages have been surveyed with a vertical extent of 1101.5 m (Allred and Allred 1997). Its main passage amounts to 41.86 km, the longest distance that one can travel anywhere else underground. Delissea System, on the Hualālai Volcano, Hawai'i, consisting of a maze of only partly connected caves, is the longest lava cave system with over 80 km of surveyed passages (pers. comm., Peter Bosted).

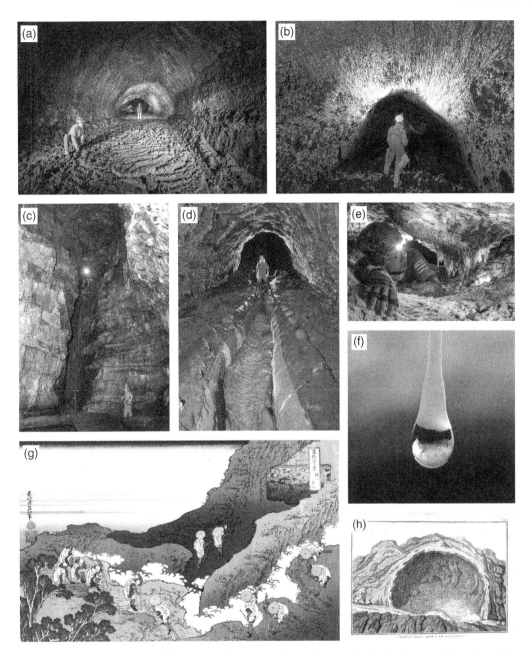

Figure 9.21 Volcanic caves: (a) Typical pyroduct, in the form of a lava tube, in the Cassone Cave, Mount Etna, Sicily. *Source:* Photo by Alessio Romeo. (b) Roof of a pyroduct in the Tre Livelli Cave, Mount Etna, Sicily, with striations. *Source:* Photo by Marco Vattano. (c) Big lava fall in Kazumura Cave, Hawai'i with the large plunge-pool room below. Note the variable resistance of the lava sheets exposed in the slot of the lava fall. This is an extreme example showing headward erosion of lava causing the transformation of a more "tubular" cross-section upstream to a slot-like canyon downstream. *Source:* Photo by Peter Bosted. (d) More recent reddish lava flow in a pyroduct in Hawai'i. *Source:* Photo by Dave Bunnell. (e) Cueva de los Minerales, Irazú Volcano, Costa Rica, is probably the most mineralogically diverse volcanic cave in the world with over 30 types of sulfates. *Source:* Photo by Victor Carvajal. (f) Microbial snottite in Cueva de los Minerales, Irazù volcano, Costa Rica. *Source:* Photo by Scott Trescott. (g) Old Japanese painting of Mount Fuji and its sacred cave by Katsushika Hokusai (1823). *Source:* Katsushika Hokusai/Wikimedia Commons/Public Domain. (h) Old print (Spallanzani 1792) of the Alum Cave on Vulcano Island, Italy. *Source:* mindat.org/Public Domain.

According to Kempe (2012, 2019), unlike the common explanation that attributes these long volcanic caves to the "overcrusting of lava flows," closer inspection shows that most of them are formed by "inflation." According to the most common conception, development of a crust over surface lava channels with levees can occur by two, often concomitant, mechanisms: (i) the welding together of floating clasts; and (ii) the inward growth of lava shelves. Overcrusted lava flows, however, are rather rarely documented. In the alternative process of inflation, several stacked lava sheets cool and are uplifted (bulged) by new lava pulses from below, resulting in a multilayered solid lava roof in which the oldest flow is the one on top, underlain by progressively younger sheets. These lower layers, insulated by the upper ones, can stay hot enough over longer distances, forming lava-filled conduits (tubes), often in parallel arrays and occasionally interconnected. One of these conduits will inevitably have the steepest gradient, concentrating the lava flow and eroding downward, thereby draining the higher-lying parallel conduits. Alongside this entrenchment in the main tunnel (Figure 9.21d), erosion can also occur on the sides of the tube and, when the lava flow is obstructed downstream, the tube can even have rising tracts breaching the roof to reach a downstream higher lying tube. How much of this erosion is mechanical or caused by remelting remains open to debate. The lava flowing in a tube is insulated against heat loss and can sustain flows at low gradients over many kilometers. Within the conduit, the lava flows as a viscous non-Newtonian fluid at the bottom of a gas-filled canyon-like cave passage. This conduit, or pyroduct, may show highly variable cross sections (Figure 9.21a–d). Lava falls, for example, can erode large plunge-pool rooms, such as the ones discovered in Kazumura Cave in Hawai'i (Figure 9.21c).

Besides pyroducts, other volcanic caves can form due to the deformation of lava sheets. These include partings (caves formed by upwarping of solidifying lava sheets), tumuli, and pressure ridge caves (Kempe 2019). Smaller lava caves include imprints of trees or even animal bodies, peripheral lava rise caves, and drained lobes. Volcanic vents, from which lava or gases have emerged, can be very deep, with examples in Hawai'i reaching depths of over 300 m. Mineralogical studies have shown that volcanic caves, and particularly young ones, host a wide variety of rare mineral associations, including a great diversity of sulfates (Forti 2005; Ulloa et al. 2018) (Figure 9.21e).

Some of these minerals have been exploited since Roman times, such as the case of the alum group minerals ($XAl(SO_4)_2 \cdot 12H_2O$, where X stands for K or Na) in the cave of the same name (Alum Cave) on Vulcano Island, Italy (Figure 9.21h). Lava caves are also interesting analogues of potential extraterrestrial habitats for life (e.g., Northup et al. 2011). The microbiology of lava caves is an expanding field of science with the prospect of finding substances of medical applicability (Cheeptham et al. 2013; Riquelme et al. 2017) (Figure 9.21f). Some volcanic caves, such as the one on Mount Fuji, have become part of traditional stories and culture. This cave, halfway to the top, was used by pilgrims for shelter during the night (Figure 9.21g).

9.2 Macromorphology

This section is focused on the gross morphology of dissolution caves, which are the great majority of the caves on Earth. When water starts its underground travel within the phreatic zone, it tends to follow the hydraulic gradient along descending or ascending pathways (i.e., epigenic or hypogenic), enlarging the most suitable openings and connecting recharge areas to outputs (springs). The passages created by the solutional activity of the flowing water can connect with each other forming more or less complex patterns, which can be examined in plan, profile, and three-dimensional view. Caves comprise entrances, passages (in one or more levels), larger voids called rooms, and terminations. Entrances can correspond to water sinks, occasional openings into the cave, or where groundwater flow resurfaces (spring caves).

The geomorphological study of caves starts with the true core of speleology, the passion that drives most cavers to venture underground: exploration. Cavers not only explore for fun, but they are also geographers of the underground. While they explore the many passages in a cave system, they make attempts to understand the cave, to know where it goes and where it comes from, and how the spaces they traverse have formed. The basic information they produce is a cave survey, a map and a profile view of the cave, indicating the characteristic shape of the passages and the features they contain. They make underground geographical maps.

9.2.1 Cave Surveying

A number of caves such as most show caves are of easy access, and scientific work in them is not more complicated than in many other natural environments. However, caves can have a difficult access, and an in-depth scientific knowledge of these sites requires the participation of experienced speleologists. One might say that karst scientists are one-handed without cavers. The ideal cave scientist is also a true cave explorer, well acquainted with cave progression and surveying techniques. The description of how to explore caves in a safe way is beyond the scope of this book, and the reader is referred to specialized textbooks and caving manuals available in many languages, often edited by national speleological organizations.

Cave mapping is an essential part in documenting caves, somewhat similar to the topographic mapping of elements in the landscape. The first known map of a cave in Europe is that of Santa Rosalia Cave, above Palermo (Italy), published in 1651 (Cascini 1651) (Figure 9.22a, b). Both a plan view (map) and a vertical cross-section of the cave give an idea of the size of the void that extended behind the sanctuary. Whereas surface topographic mapping can rely on aerial and satellite images, cave mapping requires the participation of speleologists, often using rudimental and low cost instruments. Cave maps are traditionally drawn starting from measurements made in the cave with a compass, an inclinometer, and a measuring tape or a laser distance meter. Survey lines connect two survey points in the cave, with measurements of direction, slope, and distance. To record additional data on the geometry of the cave, left–right and up–down measurements are made at each survey station. Additional measurements are then linked to the main survey line through the so-called splay shots. Cavers make sketches in the cave itself to record sediments, speleothems, or particular morphological features. All the data are then transferred onto paper or using vector graphic software once back home, producing a map in plan view, a longitudinal profile (projected or extended), and cross-sections at every survey station, in the best cases. The accuracy of cave maps is reported using an international grading system (Häuselmann 2011).

In recent years, these analog cave mapping techniques have been substituted by digital methods and equipment, including laser range finders (e.g., Disto-X), paperless surveying with cellphones or portable devices (small computers), and laser scanners. The latter are still quite expensive, but will be increasingly employed in the near future as their size, weight, and cost decrease. Cave representations also start to be upgraded from the two-dimensional views (plans and profiles) to the 3D visualizations, overlain onto a digital terrain model (DTM) together with neighboring cave systems. For an overview on modern mapping techniques see Trimmis (2018), whereas laser scanning and structure from motion (SfM) photogrammetry techniques are described in Oludare and Pradhan (2016), Fabbri et al. (2017), De Waele et al. (2018b), and Alessandri et al. (2020). An example of a cave survey, including the plan view, longitudinal profile (extended), and cross sections with geomorphologic details is shown in Figure 9.22c.

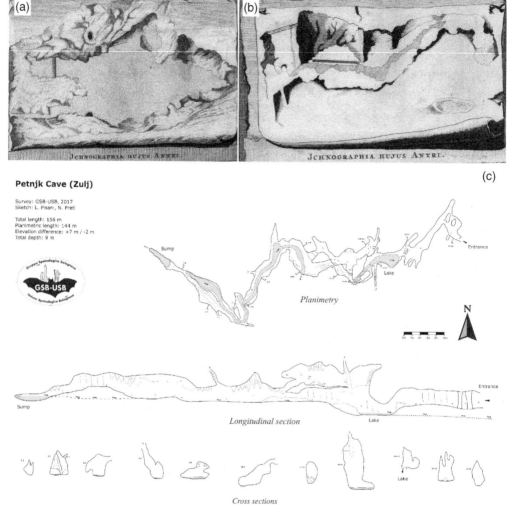

Petnjk Cave (Zulj)

Survey: GSB-USB, 2017
Sketch: L. Pisani, N. Preti

Total length: 156 m
Planimetric length: 144 m
Elevation difference: +7 m / -2 m
Total depth: 9 m

Planimetry

Longitudinal section

Cross sections

Figure 9.22 (a, b) The first known map of a cave in the western world (there are some cave maps drawn in China in medieval times), was produced in 1651, depicting Santa Rosalia Cave in Mount Pellegrino, overlooking Palermo (Sicily, Italy). The upper figures show a plan (map) view (a) and a profile (b) of the cave. *Source:* Cascini 1651. (c) A typical modern cave survey made with the "paperless method" of Heeb (2009), using a laser distance meter and a portable computer or cellphone. *Source:* Survey by Gruppo Speleologico Bolognese/Unione Speleologica Bolognese 2017, courtesy of Luca Pisani and Nevio Preti.

9.2.2 Cave Entrances

Caves are known because they have natural or artificial entrances. Their existence can also be indirectly inferred by geophysical surveying or boreholes. Natural entrances to cave systems can correspond to water recharge and discharge points (swallow holes and springs, respectively) (Figure 9.23a–c). They can also be located on sinkhole bottoms (Figure 9.23d) or can be local interceptions of the land surface with the underlying cave conduits or enlarged fractures, such as shafts in high mountain areas (Figure 9.23e), or openings related to slope retreat (Figure 9.23f), surface

Figure 9.23 Cave entrances: (a) Sinking stream of Totora, La Jalca District, Chachapoyas Province, Department of Amazonas, Peru. *Source:* Photo by Jean-Yves Bigot. (b) A stream sinking into an open fissure named Hunt Pot, Northern Dales, UK. *Source:* Photo by Tony Waltham. (c) The entrance (spring) of the Grotte de Fontaine Plaisir, Doubs, France. *Source:* Photo by Nicola Tisato, La Salle 3D. (d) Entrance of Frio Cave at the bottom of a sinkhole, La Jalca District, Chachapoyas Province, Department of Amazonas, Peru. *Source:* Photo by Jean-Yves Bigot. (e) Dolines and shaft in the Siebenhengste karst, Interlaken, Canton of Bern, Switzerland. *Source:* Photo by Jean-Yves Bigot. (f) Entrance of Cavallone Cave, a sulfuric acid cave. The cave portal is generated by erosion and slope retreat, Abruzzo, Italy. *Source:* Jo De Waele (Author). (g) Artificially enlarged cave entrance in Riu Flumineddu Canyon, Sardinia. *Source:* Photo by Carla Corongiu.

lowering, or artificial excavations. Caves with entrances are the exceptions, and many cave systems have no natural entrance at all. Nowadays, especially in countries where caving is well organized, many new caves are being made accessible after excavation campaigns that can sometimes last for months or years (Figure 9.23g).

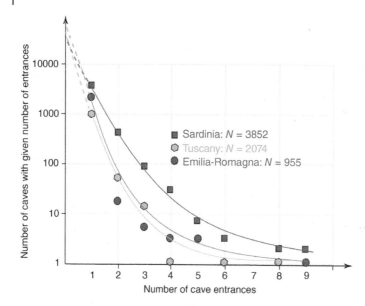

Figure 9.24 Number of caves versus number of entrances in three Italian regions. Data from cave cadasters of Sardinia, Tuscany, and Emilia Romagna. Note that in Emilia Romagna 90% of the caves are carved in gypsum, whereas in Tuscany half of the caves are carved in marble, and in Sardinia almost all caves are in limestone and dolostone.

Cave entrances often have smaller sections than the passages situated beyond them. Downstream entrances (active or former springs) can be large openings, adjusted to the maximum flow rate of the underground river that exits from it. If these rivers are very large, and collapsed blocks are removed by solutional and mechanical erosion by the flowing river, some of these entrances can become huge. The biggest known cave entrance is that of the Gruta Casa de Pedra, in the Parque Estadual Turístico do Alto Ribeira in Iporanga (Brazil), 215 m high.

The fact that there are many caves that still remain hidden underneath the surface, simply because they are not accessible, can be supported probabilistically, considering that most known caves have only one known entrance, and the numbers of systems with two, three, or more entrances decrease according to a Poisson distribution (Curl 1958) (Figure 9.24). Larger cave systems have a higher probability of having at least one natural entrance, or multiple entrances, with respect to small caves. In general, it is estimated that we can access around 10% of the caves some decameters in length (90% of these small caves have no human-sized entrance). Cave explorers probably have discovered and documented less than 10% of these accessible cave systems worldwide, with large geographical regions that remain completely unexplored up to now. This obviously sounds very promising to cave explorers, and the future generations of cavers.

Although it is not possible to give a precise number, it is estimated that today there are around 50 000 km of explored cave passages in the world, roughly representing 10% of the cave passages that can be accessed by cavers. Most cave entrances are still waiting to be discovered, or excavated, whereas in many known caves there are a large number of still unexplored cave passages. We should then also bear in mind all those caves that do not have a human-sized entrance, and those that lie deep below the surface.

The longest cave system is Mammoth Cave, in Kentucky (USA), with almost 676 km of explored and mapped passages, followed by the submarine Sac Actun System in Yucatán peninsula (Mexico), with roughly 376 km of mostly submerged passages, and Jewel Cave in the Black Hills (South Dakota, USA) (Table 9.1). The deepest caves are located in the Arabika Massif, in Abkhazia (Georgia), and exceed 2 km in depth: Veryovkina (−2212 m) and Krubera (Voronja) Cave (−2199 m) (Table 9.2).

Table 9.2 The world's deepest caves.

Rank	Cave	Country	Length (m)	Depth (m)
1	Veryovkina	Georgia	12 700	2212
2	Krubera (Voronja) Cave	Georgia	16 058	2199
3	Sarma	Georgia	6370	1830
4	Illyuzia-Mezhonnogo-Snezhnaya	Georgia	24 080	1760
5	Lämprechtsofen Höhlensystem	Austria	60 000	1727
6	Gouffre Mirolda/Lucien Bouclier	France	13 000	1733
7	Reseau Jean Bernard	France	25 512	1625
8	Torca del Cerro del Cuevón (T.33)–Torca de las Saxifragas	Spain	7060	1589
9	Hirlatzhöhle – Schmelzwasserhöhle	Austria	113 550	1560
10	Sistema Huautla	Mexico	89 000	1560
11	Sistema Cheve (Cuicateco)	Mexico	76 735	1537
12	Shakta Vjacheslav Pantjukhina	Georgia	5530	1508
13	Sima de la Cornisa – Torca Magali	Spain	6445	1507
14	Cehi 2	Slovenia	5536	1505

The list contains all caves deeper than 1.5 km. *Source:* Adapted from Bob Gulden's list, www.caverbob.com (accessed 25 March 2022).

9.2.3 Cave Passages

A cave system is formed by a series of human-sized voids, named cave passages, and a large number of conduits, many of which are not explorable because they are too small for people to enter. Cave passages can be huge if they have been formed by mechanical erosion and dissolution caused by large rivers. The entrance of Brejões Cave, although "only" ~100 m high, is one of the largest in Brazil, preluding to the enormous passages deeper inside the cave (Figure 9.25a). One of the largest cave passages in the world is believed to be the cave called Hang Son Doong in the Phong Nha Ke Bang National Park (Quang Binh province, Vietnam) (Figure 9.25c, d), discovered by local people in 1991 and explored in 2009 by British cavers (Limbert et al. 2016). The main passage of this cave, carved in the last couple of million years by a large tropical river, is 5 km long, and up to 150 m wide and 200 m high (Hoang Kim et al. 2017). The principal part of the huge Deer Cave in Malaysia, Borneo, is roughly 1 km long, never narrower than 90 m, and with a roof almost everywhere higher than 100 m (Figure 9.25b).

Such single large passages are however more the exception than the rule, and often caves comprise an array of different interconnected passage types that may form an intricate network of galleries. The shape of the accessible parts of a cave and the interconnections between passages are a direct consequence of the processes and factors that have contributed to their formation. Analyzing these forms can thus provide insight into the origin of the passages, and consequently the cave system (Bretz 1942). Cave passages can have different shapes, from vertical to descending or ascending, large and wide or narrow and high, rectilinear or winding, rounded or squared, and many other forms.

Figure 9.25 Huge cave entrances and passages: (a) The 100-m high entrance of Gruta Brejões, Bahia, Brazil. *Source:* Photo by Ataliba Coelho. (b) The largest passage on Earth in volume, Deer Cave, Borneo. *Source:* Photo by Robbie Shone. (c, d) The huge passage of Hang Son Doong Cave in Vietnam. *Source:* Photos by Dave Bunnell. Circles indicate cavers for scale.

9.2.3.1 Shafts

Shafts are vertical or steeply inclined passages, most of which form by the downward flow of vadose water. Sometimes the name domepit is used to indicate vadose shafts with thin water films along the walls and no free-falling water, or shafts that intersect a cave passage or vice versa that are visible only from below (domes). Nonetheless, we discourage the use of this term. Shafts can also be formed by upward propagation of underground voids due to progressive roof collapse (i.e., stoping process) and the removal of the fallen blocks by underground water flow (dissolution and mechanical erosion). The widening of joints by dissolution (vadose flows, condensation–corrosion) or by crystallization (e.g., pyrite oxidation with sulfuric acid production and substitution of calcite by gypsum in fissures) can contribute to the weakening of the rock mass and the upward development of these shafts.

Eventually, these large upward migrating voids can intersect the land surface forming giant collapse sinkholes with the rivers flowing below (tiankengs) (Xuewen and Waltham 2005) (Figures 6.32a and 9.26c), exposing the water table, such as in the Blue and Red Lakes (Modro and Crveno Jezero,

Figure 9.26a) in Croatia (Gabrovšek and Stepišnik 2011), or simply generating huge vertical openings with rockfall deposits at their bottom, like the Sótano de las Golondrinas in Mexico (Figure 9.26b) and Kačna Jama in Slovenia (Figure 9.26f). Some of these giant collapse pipes or chimneys also develop in non-carbonate rocks, such as those found in the quartz sandstones of Sarisariñama tepui in Venezuela (Sauro et al. 2019) (Figure 9.26d). Vertical shafts are also typical in sulfuric acid caves, mostly formed by rising of H_2S-laden vapors (Figure 9.26e). Most of the enlargement occurs above the water table, by oxidation of H_2S in condensation films, where H_2S transforms into sulfuric acid and causes the dissolution of calcite, which may be replaced by gypsum (De Waele et al. 2016).

When infiltration waters enter the soluble rock, they tend to follow vertical or steeply dipping fractures (vadose flow). If a single fracture is enlarged, the cross section of the shaft tends to attain a lenticular shape, with its long axis along the direction of the guiding fracture. Less elongated cross-section shapes form (circular, elliptical) when the development of the shaft is guided by the intersection of two fractures or fracture sets. Shafts can be created by a relatively constant low water flow, which is capable of wetting the shaft walls entirely, forming a thin film of flowing water. This is typical of shafts that are diffusely recharged by the epikarst, or by an overlying permeable non-soluble rock (e.g., sandstone). This undersaturated (aggressive) water film dissolves the rock, enlarging the cross section of the shaft and increasing its height both by upward migration of the dissolution front at the roof, and erosion of the floor of the shaft (Baroň 2002). These processes ultimately lead to the typical vertically elongated shaft with close-to-vertical and tapering-upward walls. The walls can be carved by solution flutes, similar to vertical karren at the surface (solution runnels), where more concentrated flow occurs at least occasionally. Flow along the walls occurs in the supercritical laminar regime, and any protruding part slows down the flow, causing enhanced dissolution, thus leveling the rock to a perfectly smoothed surface (White 2000b). If the shaft is recharged at the top (and discharged below) by a low-gradient passage (e.g., widened bedding plane), the water-rock contact surface and overall the dissolution rate will be greater along the vertical shaft walls than in the feeding low-gradient opening, which explains why shafts often have much larger cross-sectional area than the passages that feed and drain them. Shaft wall retreat can occur at rates around one millimeter per year, and they are thus often much younger than the passages they intersect. Many of these vadose shafts have no true genetic relationship with the active cave passages they intersect.

Shafts that are recharged by a large amount of water (a cave stream) contain free-falling water in the form of droplets, showers, or roaring waterfalls. An elliptical plunge pool several meters deep usually forms at the foot of large waterfalls. The falling water can be nebulized, and the microdroplets wet the walls with aggressive spray, contributing to the enlargement of the shaft. The spray can also penetrate into discontinuities of the rock (bedding planes, fractures) contributing to dissolution, rock mass strength reduction and breakdown processes. The water has also a mechanical erosional effect, enlarging the base of the shaft and frequently undercutting the lower part. Mechanical and solutional erosion by the running water cause the headward retreat of the waterfall, elongating the cross-section of the shaft to become a canyon-shaft first, and subsequently a knickpoint-retreat canyon. As this evolutionary pattern continues, a canyon passage interrupted by a series of vertical drops (steps) can eventually be created.

The world's deepest shafts are created by vadose erosion and dissolution of falling water along major fractures in massive or thickly bedded limestones, where the vertical range of the soluble rock is large, and the altitude difference between the recharge areas (high mountains) and discharge points (springs associated with deeply entrenched rivers or at sea level) is considerable (Table 9.3). Many of the deepest caves in the world (see Table 9.2) are successions of vadose shafts connected by short and narrow section of less inclined or almost horizontal passages.

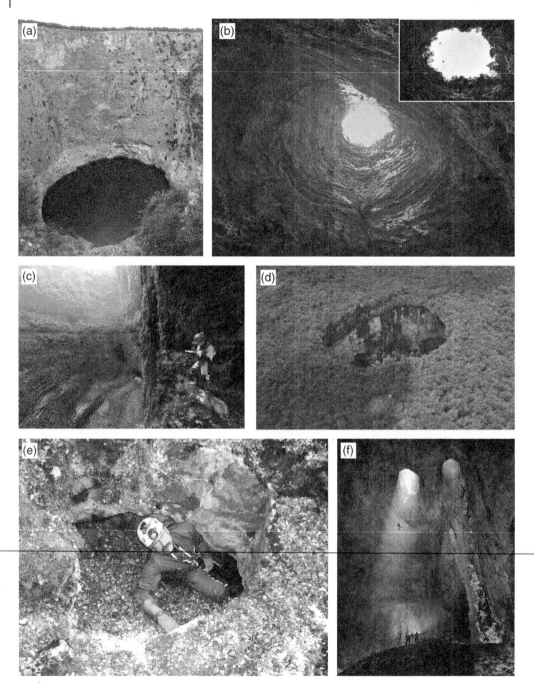

Figure 9.26 Open shafts: (a) Crveno Jezero (Red Lake), Imotski, Croatia. *Source:* Photo by Arthur Palmer.
(b) Sótano de las Golondrinas, San Luis de Potosí, Mexico. *Source:* Photo by Stein-Erik Lauritzen. (c) Tiankeng.
Source: Photo by Robbie Shone. (d) The open shaft of Sima Major, in the Sarisariñama tepui, Venezuela. Note
the small orange tent for scale. *Source:* Photo by Robbie Shone, La Venta Esplorazioni Geografiche. (e) The
small vertical entrance of Pigette Cave, France, caused by the intersection with the lowering ground surface
of a sulfuric acid chimney. *Source:* Photo by Jean-Yves Bigot. (f) The scenic Kačna Jama entrance and shaft in
Slovenia, giving access to an extensive cave system below, eventually connecting, after a rather long series
of passages, to the Reka-Timavo River. Persons for scale. *Source:* Photo by Robbie Shone.

Table 9.3 The world's deepest shafts and single free-fall shafts.

Deepest shafts

1	Vrtoglavica	Slovenia	603
2	Ghar-e-Ghala	Iran	562
3	Patkov gust	Croatia	553
4	Da Keng	China	519
5	Lukina jama	Croatia	516
6	Velebita	Croatia	513
7	Brezno pod velbom	Slovenia	501
8	Miao Keng	China	491
9	Melkboden-Eishöhle	Austria	451
10	Hollenhöhle Hades	Austria	450
11	Baiyu Dong	China	424
12	Meduza (Bojim-bojim shaft)	Croatia	420
13	Abisso di Monte Novegno	Italy	418
14	Minye	Papua New Guinea	417
15	El Sotano (de El Barro)	Mexico	410
16	Abatz	Georgia	410
17	Kocadagini (Airstrike)	Turkey	401

Deepest single free-fall shaft

1	Velebita	Croatia	513
2	Baiyu Dong	China	424
3	Las Golondrinas	Mexico	333
4	Aphanize	France	328
5	El Sotano	Mexico	310
6	Poza Tras la Jayada	Spain	306

The list includes all shafts deeper than 400 and 300 m, respectively. *Source:* Adapted from Bob Gulden's list, www.caverbob.com (accessed March 12, 2022).

9.2.3.2 Canyons

Cave rivers in the vadose zone, driven by gravity, dissolve, and erode at their floors. These passages are enlarged preferentially downward, increasing their height, whereas their width tends to remain relatively stable. In the absence of significant vertical openings along the river path, which would cause the formation of vertical shafts, a vadose canyon develops. Canyons may also form by the headward retreat of waterfalls or knickpoints, creating often a succession of vertical shafts with waterfalls connected by narrow canyons. If water discharge is very large, canyons can reach tens of meters wide and more than 100 m high. A good example is the Hanke Kanal in Škocjanske Jama, Slovenia (Figure 9.27a). Canyons can be rectilinear passages, following major vertical fractures (Figure 9.27b), or can have a sinuous pattern in plan, similar to surface bedrock streams with incised meanders, forming loops that can even intersect at various heights (Figure 9.27e). Cave meanders form by erosion and dissolution, and migrate in the downstream direction as the cave

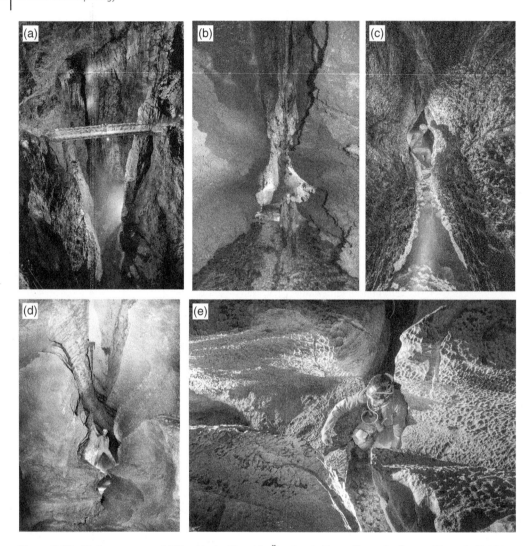

Figure 9.27 Vadose canyons: (a) The Hanke Kanal in Škocjanske Jama, Slovenia. *Source:* Photo by Robbie Shone. (b) White Nile, a rectilinear canyon passage in the Su Palu Cave, Sardinia. *Source:* Photo by Riccardo De Luca. (c) Canyon passage in Buso della Rana, Vicenza, Italy. *Source:* Photo by Sandro Sedran, S-Team. (d) Canyon passage with notches at various heights in the Décamagne Cave, Sainte-Anastasie, Gard, France. *Source:* Photo Jean-Yves Bigot. (e) Right-angle crossing meanders in the Seich Cave, Ariège, France. Note the small scallops. *Source:* Photo Jean-Yves Bigot.

river incises its floor. This can be observed by producing plan-view maps at different elevations above the river bed. Their three-dimensional shape can be extremely complex, with each single meander bend being similar to an inclined tube plunging in the downstream direction. If no vertical or lateral changes in rock type occur, the width of the canyon is largely a function of the discharge responsible for its creation, with greater widths corresponding to higher flow rates. Wavelength (distance between successive meander bends) and canyon width are generally proportional, similar to surface meandering channels (Deike and White 1969). However, in some cases, such as in the Burren, Ireland, and at Waitomo, New Zealand, meander wavelength decreases with increasing channel width, but this is accompanied by a decrease in channel depth as the river gets closer to the local water table (downstream), probably causing this unusual pattern (Smart and Brown 1981). Canyon passage widths are not always regular (Figure 9.27b–d), and enlargements can occur associated with small waterfalls (creating aggressive spray) or lithostructural differences in the rock. Meanders can cut their way far from the main channel path along penetrable discontinuity planes, eventually forming an independent route that can converge to the main passage further downstream, or take a completely independent path in certain cases. Further irregular erosion of the canyon walls can be controlled by the protecting action of sediments, that focus channel widening in certain portions of the walls.

9.2.3.3 Tubes

In the phreatic zone, passages typically take the form of tubes with ideally circular cross sections (Figures 9.28a and 9.29a–c), but often elongated along the bedding planes, specific beds (Figures 9.28d and 9.29d, e) or guiding fractures (Figures 9.28e and 9.29f, g), or with more complicated forms depending on lithological and structural changes in the bedrock (Figures 9.28b, c, f–h and 9.29h). Phreatic tubes form at or below the water table, and their shape indicates that they form by dissolution and erosion acting along the entire perimeter of the passage in water-filled conditions. They do not need to be always completely filled with water, but at least periodically (epiphreatic zone). Their enlargement progresses in all directions around their central axis, and can occur isotropically (homogeneous rocks), or anisotropically, with higher rates along more soluble or erodible beds, fractures or bedding planes (heterogeneous rocks) (Figure 9.28).

The size of phreatic conduits, also called pressure tubes, ranges between a cm to some tens of meters, and mainly depends on the maximum water discharge that flows through them, at least seasonally, and the time elapsed since the onset of their formation. The tubes tend to follow the most favorable discontinuities, which are commonly bedding planes, especially in subhorizontal to gently dipping sedimentary successions, and less often fractures. Phreatic passages often have curving plan geometries, and in long profile they tend to display a looping pattern, with sections going up and down as in a rollercoaster. These ups and downs are designated as phreatic loops, often having some tens of meters of elevation difference, but some can be deeper than 100 m. Phreatic tubes rarely have a perfectly horizontal profile, although their overall slope mimics that of

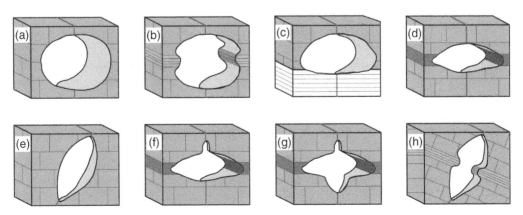

Figure 9.28 Some examples of phreatic tube cross-sections: (a) Perfectly cylindrical tube in homogeneous soluble rock. (b) Influence of a less soluble bed at the center of the tube. (c) Tube with less soluble rock at the bottom. (d) Tube elongated along a more soluble bed. (e) Tube elongated along a steeply dipping fracture. (f–h) Tubes with more complicated forms controlled by fractures and beds with variable solubility.

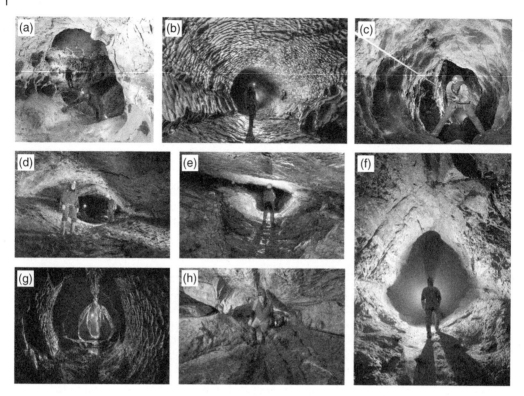

Figure 9.29 Some examples of phreatic tubes: (a) Perfectly cylindrical tube, Su Palu Cave, Sardinia, Italy. *Source:* Photo by Vittorio Crobu. (b) Cylindrical tube in Markov Spodmol Jama, Slovenia. *Source:* Photo by Sandro Sedran, S-Team. (c) Vertical conduit in the Sakany Cave, Quié, Ariège, France. *Source:* Photo by Jean-Yves Bigot. (d) Tube formed along a guiding bedding plane in the Hölloch Cave System, Switzerland. *Source:* Photo by Nicola Tisato, La Salle 3D. (e) Phreatic passage elongated along a dipping bedding plane in the Prérouge Cave, Bauges Massif, France. *Source:* Photo by Jean-Yves Bigot. (f) Tube controlled by a vertical fracture and a bedding plane, Hirlatzhöhle, Austria. *Source:* Photo by Lukas Plan. (g) Multilevel phreatic tube developed along a vertical fracture, Buso dei Pisaroti, Lessini Mountains, Italy. *Source:* Photo by Sandro Sedran, S-Team. (h) Phreatic tube with complex cross-section showing differential dissolution–erosion along various beds and fractures, Hölloch Cave System, Switzerland. *Source:* Photo by Nicola Tisato, La Salle 3D.

the water table. The walls of the tubes are essentially smooth and sculptured with large scallops, generally indicating rather slow flow, although this is not always the case (Figure 9.29b, g).

Not all tubes are located in the phreatic zone. Some passages have all the characteristics of phreatic tubes but are situated in the vadose zone, and are called perched tubes. These are localized features where geological conditions hinder passage enlargement and create local water-filled flow conditions in the vadose zone. These perched phreatic-like tubes can be recognized because the passages before and after have a clear vadose origin (e.g., canyons).

9.2.3.4 Fissures

Cave passages formed along discontinuity planes (bedding planes, fractures) are named fissures or fissure passages. They are straight and narrow passages of lenticular cross section, and cavers often are obliged to squeeze through them (Figure 9.30). Many fissures in cave systems become too narrow to pass through, and sometimes only their central parts are accessible, with their lateral extremities pinching out. Most fissures are created close to or below the water table, and their enlargement has occurred completely underwater (permanently or during flood events). These

Figure 9.30 Some examples of fissures: (a) Vertical fissure and differential corrosion along bedding planes, Guernica Cave, Lessini Mountains, Italy. *Source:* Photo by Sandro Sedran, S-Team. (b) Horizontal fissure developed along a guiding bedding plane, Mandara 'e s'Uru Manna Cave, Sardinia, Italy. Note river gravels on the floor. *Source:* Photo by Carla Corongiu.

fissures are the extreme version of structurally-guided tubes, characterized by more rounded (lenticular) cross sections. There is a continuous spectrum of phreatic forms, grading from fissures with parallel walls to cylindrical tubes. Fissure passages also develop in the vadose zone as structurally-guided narrow passages, only marginally widened by dissolving waters.

9.2.3.5 Mixed Phreatic-Vadose Forms

In many caves, the cross sectional shape of the passages records both phreatic and vadose conditions, generally with a rounded tube on top, entrenched by a vadose canyon in the floor. In most cases, the tube width is larger than that of the canyon width, and the cross-section resembles a keyhole (keyhole passage; Figure 9.31). The simplest explanation for the formation of this type of passage is that the cave passed from phreatic to vadose conditions, thought to occur because of a decrease in water flow. This can be due to climate change (less precipitation), or to the deviation of the water flow to alternative routes upstream of the passage. Alternatively, water flowing more rapidly occupies a smaller cross section, and can excavate downward creating the narrow canyon. Such an increase in flow velocity and/or discharge is often caused by a lowering of the base level (a water table drop), and thus a shift from phreatic conditions to vadose ones. The change from keyhole passages to phreatic tubes in a looping cave system is thus a good indicator of the position of the former water table, and its variations through time because of base level lowering (e.g., fluvial entrenchment, sea-level changes). However, keyhole passages can also be formed by periodic flooding. During floods the entire passage is water-filled, enlarging in all directions, whereas receding floodwater carves the basal canyon. Care should thus be taken in interpreting keyhole passages, since they do not always indicate regional conditions (they can indicate local water table levels), nor the transition between phreatic to vadose flow.

9.2.3.6 Phantom (Ghost-Rock) Passages

Dissolution in soluble rocks not always leads to the formation of accessible caves in a single-stage process (i.e., dissolution and total removal of the rock occur in the same time frame). In pure limestone, the bulk of the rock is removed by dissolution and solute transport even in low

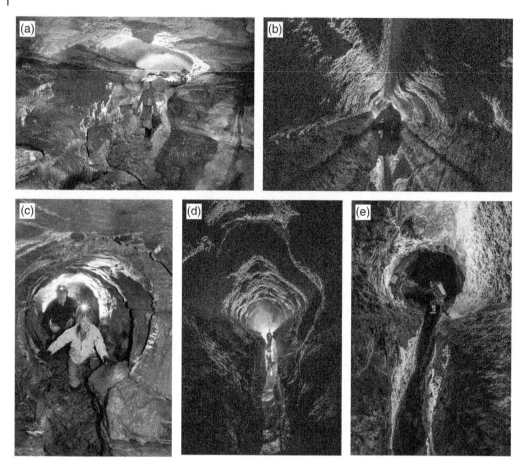

Figure 9.31 Keyhole passages: (a) Keyhole passage in the entrance part of Travers du Four Cave, Le Clapier, Aveyron, France. *Source:* Photo by Jean-Yves Bigot. (b) Phreatic tube with narrow and deep entrenchment in the Buso delle Anguane, Lessini Mts., Vicenza. *Source:* Photo by Sandro Sedran, S-Team. (c) Typical keyhole passage in Knox Cave, New York State. *Source:* Photo by Arthur Palmer. (d) Phreatic tube passage guided by a vertical fracture with a narrow and deep lower canyon and differential corrosion along limestone beds, Buso della Rana, northern Italy. *Source:* Photo by Sandro Sedran, S-Team. (e) Spectacular keyhole passage with a deep and narrow canyon in Bulmer Cave, Mount Owen, New Zealand. *Source:* Photo by Neil Silverwood.

hydrodynamic settings. When the soluble rock contains a significant proportion of impurities, such as in sandy limestones, dissolution removes the carbonate cement in low hydrodynamic conditions, leaving the insoluble sand grains in place (Dubois et al. 2014). These residual sediments are often composed of quartz, clay minerals, and iron and manganese oxides. The result is a rock mass with a network of non-accessible pockets of low-cohesion weathered material, unless mechanical erosion opens major flow routes. The sediments in the voids are not allochthonous infillings, but in-place residual material derived from the host rock. Bedding can often be traced from the residual material to the rocky cave wall, and the weathering residue can preserve portions of undissolved original rock, such as macrofossils composed of larger less soluble carbonate crystals, or chert nodules, similar to those observable in the unweathered bedrock. When the hydrodynamic conditions change (e.g., by water table lowering) and water flow reaches enough power to erode the residual alterite away, the ghost-rock networks are emptied and become accessible cave

Figure 9.32 Some examples of ghost-rock passages: (a) Ghost-rock passage in which the continuity of the strata is visible in the ghost-rock weathering material. *Source:* Photo by Paola Tognini. (b) Set of ghost-rock passages in the chert-rich Moltrasio Limestones, Mt. Bisbino, northern Italy. *Source:* Photo by Paola Tognini. (c) Madrona Cave, Bisbino Mts., one of the most important ghost-rock caves in this area. *Source:* Photo by Mauro Inglese.

systems. Their pattern is generally a network of enlarged fractures with irregular cross-sectional areas where dissolution acted more vigorously. These abrupt changes in cross-sectional areas along the passages are a characteristic that allows ghost-rock caves to be distinguished from normal floodwater mazes and phreatic maze networks (Häuselmann and Tognini 2005). Ghost-rock caves often display remnants of less soluble rock portions, forming pendants, pillars and arches, and many passages pinch out or stop at uneroded remains of in situ weathered rock (Figure 9.32).

9.2.4 Cave Rooms

Speleologists, while exploring caves, often enter voids much larger than the passages that lead into and out of them. These larger spaces are called cave rooms, and can be related to a number of factors, including local widening due to the intersection of several passages, coalescence of adjacent vertical shafts, breakdown processes, and enlargement of a passage by solution at a specific location. The first case is the most common one, but cannot produce large rooms alone, whereas the last two mechanisms are able to create among the largest natural underground voids in the world. Breakdown alone does not lead to the formation of large voids, since the collapsed material piled up on the cave floor occupies more space than the original unbroken rocks (bulking effect). The fallen material needs to be removed by flowing waters (a river, or a fluctuating floodwater table). Removal commonly occurs by a combination of dissolution and mechanical transport. Large rooms can also form when an easily erodible material is present in the rock succession (e.g., marls), and its removal by running waters creates a void and undermines overlying soluble strata favoring collapse (Gilli 1986; Antonellini et al. 2019). Many large cave rooms have a structural control, including for example bounding faults or the core of folded strata (Gilli 1986; Pisani et al. 2019). The largest known rooms in the world owe their formation to these mechanisms. The record is contended between Sarawak Chamber in Lubang Nasib Bagus (Good Luck Cave) in Borneo (Malaysia) (Figure 9.33a), and Miaos Room in the Gebihe Cave System in China (Figure 9.33c), both located in tropical climate areas. The first is the largest in planimetric area ($168\,870\,\text{m}^2$, against $145\,040\,\text{m}^2$ of Miaos), whereas the second has a greater volume ($10\,570\,\text{m}^3$ against $9810\,\text{m}^3$ of Sarawak).

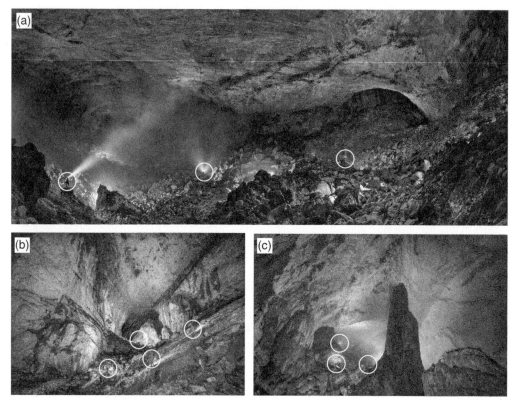

Figure 9.33 Large cave rooms: (a) Sarawak Chamber in Borneo (Malaysia). *Source:* Photo by Robbie Shone. (b) Martelova Dom in Škocjanske Jama, Slovenia. Note Reka River flowing on the bottom left. *Source:* Photo by Mark Burkey. (c) Miaos Room, Gebihe Cave System, China. *Source:* Photo by Mark Burkey. White circles show where cavers are standing.

These tropical caves have beaten, by three to four times in surface and volume, the historical large cave chambers of Martelova Dom in Škocjanske Jama, Slovenia (Figure 9.33b), Grotta Gigante in Italy, and Salle de la Verna in the Pierre Saint Martin Cave, in the French Pyrenees. Large volumes solely created by dissolution are much more difficult to form, and are exclusive to hypogene caves, such as mixing solution caves or sulfuric acid caves. The Big Room in Carlsbad Caverns is a good example of the last category (Figure 9.25b). Most large natural voids tend to display a classical shape with a more or less stable dome or arch-like roof (White and White 1969, 2000). The arched tension zone created above the cavity roofs by the deflection of gravitational stresses controls the development of curved failure planes. These voids are great natural laboratories to study the stability of underground spaces, useful for engineering works (Gilli 1986).

9.2.5 Cave and Passage Terminations

Caves and their passages, from an anthropocentric point of view, end where speleologists are no longer able to pass through. The karst network obviously extends far beyond these human limits, and the water responsible for carving the voids passes through these tiny passages, connecting the recharge areas to the springs. Passage terminations can correspond to different features, including collapsed material, sediment fills, passage constrictions, or simple flooding (the passage becomes submerged). Cavers often try progressing beyond the narrow openings or obstructions by enlarging cave sections or diving the submerged passages (sumps). Cave terminations can therefore be temporary, and cave passages can extend in several directions by subsequent explorations. Cave maps are thus partial representations of the underground karst network, and these carefully drawn maps can also guide cavers in their endeavors to search for new yet undiscovered passages. Understanding the reason for a passage termination is thus important to concentrate exploration efforts in the right directions.

The pinching out of cave passages generally occurs in the upstream directions, especially if the passage divides into different branches (tributaries) that carry progressively less water. Commonly, only the main water-carrying passages convey enough fluid to allow the formation of a human-sized cave from the sinking point to the spring. In the downstream direction, a large stream progressively increases its discharge, due to additional inflow from tributaries, and passage size either remains rather constant or enlarges downstream. These passages often end in a breakdown pile produced by the collapse of the ceiling and/or walls (Figure 9.34a). If the rock fragments are large enough and lack interstitial matrix, these breakdown sections are often explorable with considerable efforts. If the material is composed of small or poorly sorted fragments, the chances to get through are smaller. In some cases, the collapsed material indicates the presence of a higher lying void that can sometimes offer a way to overcome the obstruction. River passages can also terminate in flooded sections, where the passage continues underwater (Figure 9.34b). These submerged passages, named sumps or siphons, can be located high above the water table, so they are hanging and their exploration can lead to the discovery of long dry cave segments on the other side of the sump. Their exploration can be tedious and challenging when they are located close to the water table. Cave diving explorations can be highly demanding both from a physical and technical point of view in cave systems with deep loops (alpine systems). In some cave systems, the downstream main passages can pinch out also because the water has lost its dissolution capacity, or because of the presence of less soluble rock. Restrictions in passage size can also be related to the accumulation of clastic or chemical deposits blocking the way forward (Figure 9.34c, d).

In hypogene cave systems, the acidity of the water can be localized and can change abruptly in certain directions. The largest voids are carved where the dissolving capacity of the fluids is greatest, and the passage size diminishes away from the point where aggressive water enters the system, or aggressiveness is created (e.g., incorporation of CO_2) (Figure 9.34e). This happens farther away from the feeding fractures in sulfuric acid cave systems. The same often occurs in thermal cave systems, where deeply rooted master tectonic structures guide the rising of thermal CO_2-rich waters. In coastal flank margin cave settings water aggressiveness is caused by saltwater/freshwater mixing, so voids tend to be concentrated in the coastal fringe, pinching out landwards. Coastal caves can only penetrate over large distances into the continent when they are recharged by major freshwater flows, and the passages are enlarged by normal (non-mixing) solutional and erosional processes.

9.2.6 Cave Ground Plans

Analyzing cave surveys in their vertical projection on a horizontal plane (cave maps) immediately allows to discern simple caves from more complex ones, and among the complex caves to recognize different ways in which cave passages interconnect and arrange forming the so-called cave patterns. These different plan-view patterns are related to a combination of geological and hydrological factors. The simplest form is a single passage that can be linear (Figure 9.35a), angular (Figure 9.35b), sinuous (Figure 9.35c), or a combination of these geometries. Usually, these caves

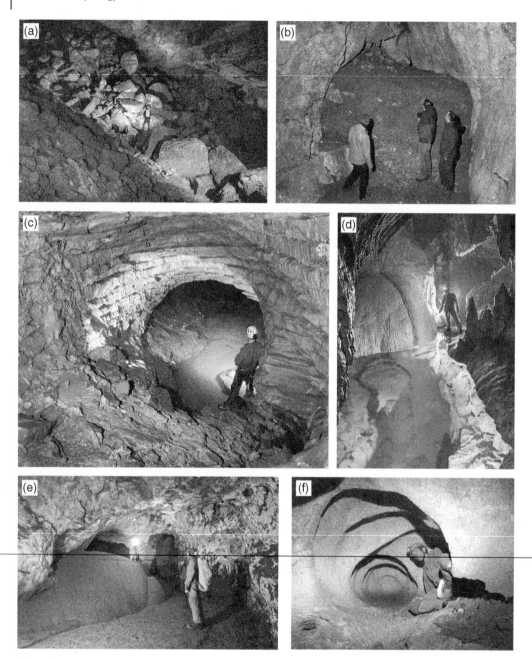

Figure 9.34 Some typical cave terminations: (a) Upstream passage in Su Eni 'e Istettai Cave, Sardinia, blocked by a breakdown pile. *Source:* Photo by Vittorio Crobu. (b) Sediment filling in the phosphate mine of the Mas de Dégot (an ancient filled cave), Bach, Lot, France. *Source:* Photo by Jean-Yves Bigot. (c) Sump in the Combe du Buis Cave, Saint-Guilhem-le-Désert, Hérault, France. *Source:* Photo by Jean-Yves Bigot. (d) Sa Ciedda sump in Su Palu Cave, Sardinia, Italy. *Source:* Photo by Vittorio Crobu. (e) Big flowstone obstructing an abandoned phreatic tube in Su Palu Cave, Sardinia. *Source:* photo Francesco Sauro. (f) Upward tapering void in a sulfuric acid cave, Kraushöhle, Austria. *Source:* Photo by Lukas Plan.

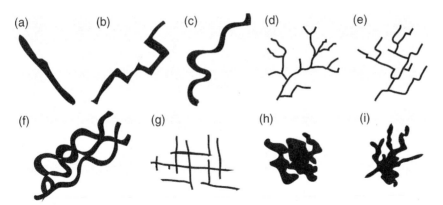

Figure 9.35 Cave ground plans: (a) Linear; (b) Angular; (c) Sinuous; (d) Sinuous branchwork; (e) Angular branchwork; (f) Anastomosing maze; (g) Network maze; (h) Spongework; and (i) Ramiform.

are small, comprising a single sinking point, the throughflow passage, and only one output (spring), or otherwise they are remnants of larger cave systems that have been partially destroyed by surface erosion, or the small accessible part of a much larger cave.

More generally, different types of passages interconnect giving rise to more or less complex cave patterns, the most important of which are branchwork, maze, spongework, and ramiform caves (Palmer 2007). A cave system can show different patterns in various parts, or even a single portion of a cave can display different superposed patterns. The most typical cave pattern is the branchwork type, which is the underground equivalent to the dendritic or angular drainage networks at the surface. Multiple recharge points give rise to cave passages that converge further downstream feeding progressively larger passages until they reach a main drainage passage. In plan view, these cave patterns resemble a branching tree, where the downstream part is the trunk and the upstream parts are the upper branches of the tree. Sinuous branchwork patterns are typical in well-bedded homogeneous rocks (Figure 9.35d), whereas angular ones are indicative of anisotropic fractured rocks (Figure 9.35e).

Caves with maze pattern consist of a network of passages intersecting at various angles. The end members in the maze pattern are anastomotic, network, spongework, and ramiform caves. Anastomotic caves are the most frequent form of maze, composed of curving passages intersecting numerous times and forming a succession of divergences and convergences (distributary and contributory flow) (Figure 9.35f). They generally form along a single bedding plane or a gently inclined fracture by concentrated water flow (floodwater or concentrated infiltration). They also form to overcome passage constrictions or collapse deposits in a master passage, by water flowing along a series of alternative pathways. Network mazes differ from the anastomotic ones in that their passages intersect at sharp and rather constant angles, following the main joint sets in the rock (Figure 9.35g). All the available discontinuity planes have been enlarged at similar rates by slowly flowing waters, coming from the surface (e.g., overlying permeable but non-soluble rock units such as sandstones) or from below (e.g., hypogene intrastratal karstification). Floodwaters can also enlarge all available fractures to bypass constrictions in the main passage, and also mixing of different waters may lead to the homogeneous enlargement of all penetrable fractures. However, dissolution will be more effective in the areas where the recharge, from above or below, is most abundant and able to dissolve more rock. Maze systems of this type are rather regular at a local scale, and solutional fissures related to the widening of fractures pinch out away from the main recharge area of dissolving waters.

A spongework maze is characterized by the interconnection of single dissolution voids of different sizes, resembling the pattern of a sponge (Figure 9.35h). These patterns are mainly formed in massive rocks lacking fractures, and the solutional macroporosity is often the result of local differences in primary porosity. They are frequent in highly porous, rather young limestones, when mixing corrosion is the main process at work (e.g., in coastal areas). This pattern is also typical of sulfuric acid caves, or of caves mainly formed by the mixing of different fluids at depth.

Ramiform caves are characterized by a combination of large rooms connected by more or less small passages with a three-dimensional distribution (Figure 9.35i). Most of these caves are hypogene, often of the sulfuric acid type, and large rooms form where aggressive waters enter the system, whereas the branches are fracture-guided passages that tend to pinch out away from the source of aggressiveness. Variations of the local potentiometric surface, and the three-dimensional architecture of the openings in the host rock, can cause the inlet points in the system to change their position, creating a series of rooms, often merging with adjacent voids.

9.2.7 Cave Profile and Cave Storeys

Although cave profiles are less informative than planimetric views regarding speleogenesis, they offer some clues for the interpretation of how the cave developed through time in an evolving landscape. The vertical arrangement of the cave passages mainly depends on the altitudinal difference between the recharge area and the outputs (springs). In flat lying and scarcely incised landscapes, where the hydraulic gradient is very low, the cave profile is close to horizontal (e.g., Mammoth Cave, Western Kentucky). In high mountain areas with deeply entrenched valleys many cave systems develop in a vertical manner, with a succession of vadose shafts and canyons (alpine caves). Vertical vadose shafts mainly form in thickly bedded and deeply fractured rocks, whereas thinly bedded and less fractured successions are less penetrable in depth and tend to host gently inclining passages. It is therefore not always true that caves tend to be vertical in their upstream parts, despite the fact that groundwater tends to reach the water table as soon as possible. To reach the local water table, water in the vadose zone follows the most efficient pathways, which are not necessarily the shortest (and steepest). There is always a significant geological control on the vertical arrangement of cave passages in the vadose zone, with water following the steepest penetrable openings. These are often bedding planes, and water tends to flow down dip, perpendicularly to the strike of the bedding, unless steep fractures provide a subvertical short cut. In steeply dipping successions, the vadose channels commonly follow the dip direction, rapidly reaching the water table level. In well bedded and gently inclined strata cave passages often maintain a low-gradient profile, following a specific bedding plane over long distances, occasionally interrupted by shafts where vertical fractures disrupt the dipping bed. In mountainous areas where thinly bedded rocks overlie massive and highly fractured formations, the upper part of the cave system is commonly characterized by low-gradient passages, changing into deep vadose shafts where the more massive portion of the stratigraphic succession is reached.

In the phreatic zone, the water has no reason anymore to follow the steepest available discontinuities, but instead tends to flow in the direction dictated by the hydraulic gradient, choosing the most efficient openings and staying more or less close to the water table level. These flowpaths often correspond to bedding planes, with flow roughly parallel to the strike of the bedding, in case of dipping strata and with favorable orientation with respect to the hydraulic gradient. However, the most favorable flow routes are not always located close to the water table, and phreatic passages generally show a sinuous pattern both in vertical projection and in map view. The vertical pattern of phreatic tubes is characterized by ups and downs, called phreatic loops, in which there are descending and rising tubes (called phreatic lifts). Phreatic loops can extend far below the water table, reaching depths of more than 200 m. The depth below the local water table of phreatic conduits appears to depend on three main variables, based on the analysis of data from 20 cave systems in different climate settings (from tropical to cold temperate): (i) the flow length of the system (linear distance between recharge area and spring), (ii) the dip of the beds, and (iii) the maximum elevation difference between the recharge and the discharge areas (Worthington 2004). This can be expressed with the following equations:

$$D = 0.047 L^{0.85} \sin \theta^{0.64} E^{0.11} \tag{9.1}$$

$$D = 0.061 L^{0.91} \sin \theta^{0.72} \tag{9.2}$$

where D is the mean depth of the conduit below the water table, L is the flow length of the system, θ is the mean dip of the bedding (in degrees), and E is the elevation difference between the highest recharge point and the lowest spring. Since elevation E does not seem to improve the correlation with the 20 examples used by Worthington (correlation coefficients in both equations are 0.90) the following simpler equation with a slightly lower correlation coefficient (0.79) can be used as an alternative (Worthington 2004):

$$D = 0.18 \left(L \sin \theta \right)^{0.81} \tag{9.3}$$

The "rollercoaster" pattern of cave passages is not typical only for truly phreatic conduits, but also develops in epiphreatic cave evolution, where floodwater loops are a common feature (Gabrovšek et al. 2014). These caves, typical of alpine and tropical (monsoon) regions, form in karst aquifers characterized by a flashy hydrological behavior, with large and sudden changes in flow rate. During floods, cave passages are rapidly filled by fast flowing aggressive waters. The passages are often developed along sets of intersecting bedding planes and fractures enlarged in temporary phreatic conditions during the flood pulses. During flood recession, only the lower parts of the loops still contain flowing water, and the epiphreatic passages are drained by smaller-sized phreatic conduits (known as "soutirages"). These floodwater loops are not easy to distinguish from true phreatic loops, but they generally have more vadose morphologies in their downstream descending branches (vertical dissolution flutes, potholes, and small basal incisions), contrasting with the pure phreatic morphologies in their rising upstream sections, and are located above the corresponding spring.

Many cave systems have passages preferentially distributed along several more or less horizontal levels that are not guided by the stratigraphy or the structure, but are related long-lasting stable positions of the base level and its variations. These passages distributed at different altitudinal intervals are known under the name of cave levels, although they are not exactly "level." The terms cave tiers or cave storeys should therefore be preferred (Palmer 1987). These tiers are controlled by episodic variations in the position of river valleys in mountainous areas and sea level in coastal areas (Mylroie and Carew 1988; Florea et al. 2007).

Cave storeys form because of base level lowering or rising, the latter occurring less frequently and giving rise to a series of flooded and/or sediment-filled passages stacked at different altitudes, with the upper storey being the youngest one. When a river experiences entrenchment in response to drivers such as tectonic uplift or sea-level drop, the spring shifts to lower altitude, and the cave rivers tend to adjust to the new base level by incising deep continuous canyon passages, or finding a more favorable underground pathway to the spring. The creation of a new, lower lying passage is known as vertical piracy, in which an underground stream diverts and deepens its path, leaving its former upper passage as a dry relict cave section. The passages that connect the different cave storeys are also known under the French name "soutirages" (Häuselmann et al. 2003).

In karst areas where dissolution is very fast (evaporites), rapid bedrock uplift and the associated fast base level lowering can be recorded by cave storeys with datable deposits. Good examples of such multilevel caves are those developed in the actively rising diapir of Mt. Sedom, Israel (Frumkin 1996), and those reported in the gypsum areas in the Northern Apennines, Italy (Columbu et al. 2015, 2017) (Figure 9.36b). These adjustments require longer times in carbonate karst, but cave tiers can also form and have a higher preservation potential. They can be helpful in reconstructing landscape evolution in geodynamically active regions, good examples being the Alps (Audra et al. 2007a; Gabrovšek et al. 2014), the Pyrenees (Calvet et al. 2015), and the Dolomites (Columbu et al. 2018) (Figure 9.36a). In these cases, larger passages indicate longer active dissolution and underground erosion, recording prolonged periods of relatively stable base level position.

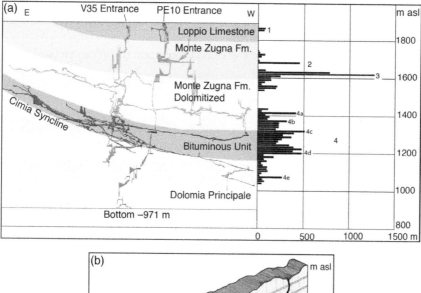

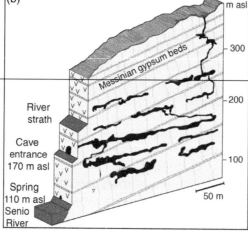

Figure 9.36 Cave storeys or tiers: (a) Cave storey distribution with respect to the stratigraphy in the Piani Eterni Cave system, Italian Dolomites. Four main cave tiers have been identified in this stratigraphic succession dominated by dolomites, all related to interglacial or warm periods. *Source:* Modified from Columbu et al. (2018). (b) Cave tiers in the Re Tiberio cave system developed in Messinian gypsum in the Northern Apennines (Italy), related to mountain uplift and discontinuous fluvial entrenchment recorded by river terraces. *Source:* Adapted from Columbu et al. (2019).

In very slowly uplifting (or almost stable) continental settings, such as the Green River region in Western Kentucky, where bedding is close to horizontal, a minor incision of the external low-gradient river can cause the lateral migration of the spring of several kilometers. Here, four cave storeys developed within a vertical range of 50 m record alternating phases of base level stability and fluvial downcutting over a period of 3.5 Myr in a low-gradient landscape (Granger et al. 2001).

9.3 Speleogens: Small-Scale Passage Morphologies

The roofs, walls, and floors of cave passages can be sculpted by medium- to small-scale morphological features formed by solution and mechanical erosion, collectively designated as speleogens. These morphologies carved in the rock can provide valuable information on the processes that were active in the cave passages, especially in the latest (more recent) stages of their development (Bretz 1942). They are the erosional counterpart of speleothems (secondary mineral deposits), which together with clastic sediments form the depositional morphologies in caves. Both speleogens and cave deposits (which are described in Chapter 10) are fundamental pages of the book that describe the history of the cave and of the landscape in which the cave is located.

White (1988) classifies speleogens into four broad morphological categories: channel, hydraulic, etched, and residual features. Zhu (1988) differentiates five morphological categories of speleogens: flow marks, pits, grooves, tubes, and wall protrusions. One of the most comprehensive works on rock-carved features in caves is the monograph of Slabe (1995). This author classifies the morphologies according to their basic shapes in: (i) channels, (ii) pockets and niches, (iii) bulges, and (iv) broken-off pieces and fragments. From a genetic point of view, Slabe distinguishes: (i) rocky forms due to water flow, (ii) features formed at the interface between sediments and the rock wall, (iii) features related to water trickling down vertical walls, (iv) forms related to the dissolving action of condensation waters, (v) features developed below ice, (vi) biogenic furrows, and (vii) morphologies related to rock weathering and breakdown.

The diverse classifications proposed in different regions of the world (mainly United States, Western Europe, Eastern Europe, and China) have resulted in a proliferation of terms, creating great confusion especially in speleological publications, which are often the basis of the work of many scientists (Bini 1978). It is not rare to find multiple names for identical morphologies or the same name given to different features with distinct origins. It is true that, since many forms are created by a combination of processes, their genetic classification is not straightforward. Furthermore, certain shapes can form in various ways and conditions (morphologic convergence or equifinality), making genetic classifications almost impractical. Some forms still have an uncertain origin, and can be explained by different mechanisms with their pros and cons.

In most cases, each speleogen is the result of a combination of processes caused by erosional agents (water, ice, airflow, animals, and plants) acting on the bedrock. The importance of the dissolution processes in shaping the speleogens depends on the availability of water and its aggressiveness, which are largely a function of climate. The amount of rainfall and its temporal distribution, and the mean daily and annual temperature determine the biome, and thus the flux of biogenic compounds (e.g., soil-derived carbon dioxide) and sediments, and the hydrodynamics of the cave systems. Not all passages in a cave hold water flow under the same hydrodynamic regimes in a given period of time, and cave branches can undergo different hydrological conditions (phreatic, epiphreatic, and vadose) over geological times. In certain climates and topographical conditions, cave passages can experience large temperature changes, and air entering or leaving the caves can cause condensation and/or evaporation. Sediments can accumulate in caves shielding rock surfaces from the action of running waters. Those sediments can be removed, replaced, and transported many times, influencing water and air movement, and the amount of rock exposed to the action of water flow or condensation–evaporation processes. It is obvious that there is a countless number of possible combinations of processes, factors, and conditions leading to the creation of speleogens in caves (Bini 1978). Therefore, we have decided not to adopt a genetic or a morphological classification scheme in this book, but we simply describe the speleogens one by one, starting from the largest forms to the smaller ones, and from those formed at the ceiling to those developed along the walls and the floor, although some features can occur in the whole perimeter of cave passages. Note that when a feature is ascribed to a process or a factor, it means that it was mainly formed in that way, not excluding the minor influence of other processes and factors. For each speleogen type, the range of morphological characteristics, their mechanisms of formation, and their speleogenetic meaning are briefly described.

Cupolas, sometimes called solution domes, are generally meter-sized, rounded, dome-shaped solution hollows occurring in the roofs and sometimes along the walls of caves (Osborne 2004). These morphologies display a wide range of dimensions, with the smaller ones grading into solution pockets, which are less than 1 m wide, and the larger ones reaching several tens of meters in diameter. The Temple of Baal cupola in Jenolan Caves, Eastern Australia, is 50 m wide and 45 m high. Cupolas occur in both carbonate and evaporite caves. The name cupola derives from the dome-shaped morphology of the roof of old churches in Italy. Cupolas were defined as spherical or semispherical enlargements in caves that are larger than or similar to the cave passages that lead into them (Lauritzen and Lundberg 2000). However, meter-sized dome-shaped solution cavities on the roofs of large rooms should also be classified as cupolas. In fact, Osborne (2004) defines cupolas as "solution cavities with a dome-shaped ceiling and a circular to elliptical plan with a diameter or long axis in plan greater than 1.5 m." Collapse domes (in which solution plays a minor role) are not to be included among cupolas, nor are bell holes, which are smaller and normally much deeper than wider (cigar-shaped). Cupolas can be guided by fractures (Figure 9.37a) and display an elongated elliptical plan shape, but some have no controlling fracture at all (Figure 9.37b). Cupolas can generate in different ways and in different conditions. In vadose cave passages, they can form by condensation–corrosion induced by (i) convectional air movements in closed thermal caves above the thermal water pools, or (ii) by advective air flow between two cave entrances at different altitudes or in large cave entrance areas. To be able to form by condensation–corrosion, the air flow and the thermal gradient between the cave atmosphere and the wall has to remain stable over sufficient time to produce enough condensation water. The aggressive water condensed on the rock surface causes its progressive dissolution, to form large solution features such as cupolas. The thermal contrast can be created in caves close enough to the surface and influenced by seasonal temperature changes, or be maintained over long periods by thermal rising waters, or by the presence of guano deposits that decay through an exothermal reaction providing moisture and CO_2 to the cave atmosphere and the wall/roof. Good examples of condensation–corrosion cupolas are those documented in Grotta dell'Orso in northern Italy (Fabbri et al. 2017) (Figure 9.37c), Kraushöhle, Austria (Plan et al. 2012) (Figure 9.37d), and Jewel Cave in the Black Hills, USA (Figure 9.37e). Condensation–corrosion can create a water film that becomes saturated in calcium carbonate on the rim of the cupola, where evaporation and/or degassing is highest, causing the deposition of speleothems (Figure 9.37g). Moreover, adjacent cupolas can enlarge and coalesce generating windows that connect them (Figures 9.37d, h).

In phreatic conditions cupolas form underwater, generally by cellular convection driven by thermal or density gradients, the latter related to differences in solute content. The origin of cupolas solely by mixing corrosion, as proposed by Bögli (1980), has lost credibility. Dilution of fissure

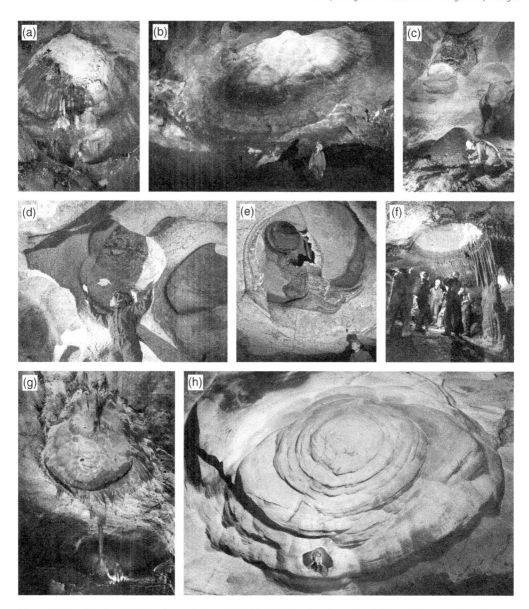

Figure 9.37 Cupolas. (a) Cupola guided by a joint in the Aiguèze Cave, Gard, France. *Source:* Photo by Jean-Yves Bigot. (b) Cupola in a homogeneous limestone bed, with no visible joints, in Saint-Marcel Cave, Ardèche, France. *Source:* Photo by Jean-Yves Bigot. (c) Cupola developed above an active guano pile in Grotta dell'Orso, Ponte di Veja, northern Italy. *Source:* Photo by Sandro Sedran, S-Team. (d) An intricate set of cupolas created by condensation–corrosion above former sulfuric-acid pools in Kraushöhle, Styria, Austria. *Source:* Photo by Lukas Plan. (e) A high condensation–corrosion cupola cutting through old speleothems (brownish layered deposits) and with a white coralloid (popcorn) rind generated by evaporation-induced precipitation from the descending condensation water films, Jewel Cave, South Dakota, USA. *Source:* Photo Dave Bunnell. (f) Cupola of epiphreatic origin, with the brown colored parts stained with organic material from floodwaters, and the white bare limestone roof etched by condensation–corrosion. Claysse Cave System, Ardèche, France. *Source:* Photo by Didier Cailhol. (g) Cupola with rind of speleothems in Saint-Marcel Cave, Ardèche, France. *Source:* Photo by Philippe Crochet. (h) A large cupola with window in Gruta Dente de Cão, Mato Grosso do Sul, Brazil. *Source:* Photo by Philippe Crochet.

water in the much larger mass of conduit waters, and the difficulty of having significantly different geochemical compositions in the feeding-fissure water and that in the main phreatic conduit, make this process highly improbable to allow the formation of such large-scale corrosion forms. This does not exclude that mixing corrosion can cause localized enhanced dissolution along tiny fractures that feed infiltration waters into the main cave passage. Density gradients are particularly important in hypogene gypsum caves, where rising waters become gradually saturated in gypsum, and dense fluids then tend to sink again creating closed cellular convection cells (Klimchouk 2009). Cupolas can also form and enlarge in epiphreatic conditions: air can be trapped in pockets on the cave roof during floods, and the high pressure causes P_{CO_2} to increase, leading to the formation of a mist (condensation). The condensed water film on the walls and the water layer in contact with the pressurized air become more aggressive and dissolve the rock (Lismonde 2000) (Figure 9.37f).

Solution pockets are smaller versions of cupolas, having diameters of less than 1 m, and generally of a few decimeters. As cupolas, they are dead-end solutional cavities, often but not always associated with discontinuities (Figure 9.38a). Their shape is often oriented along the direction of the fractures or bedding planes that have controlled their growth (Figure 9.38b). They can display a wide range of morphologies, including simple hemispherical geometries, hollows that narrow gradually toward the blind end, solution pockets with a stepped shape comprising small hemispherical shapes nested into larger ones, or composite, with many hemispherical pockets intersecting laterally and internally. They are often visible along the roofs and upper walls of spring caves, or in the sections of passages situated upstream and downstream of constrictions (Slabe 1995). This indicates that these pockets are formed at locations of increased turbulence by fast flowing waters during floods. The association of solution pockets on the higher parts of passage walls, and scallops prevailing in the lower parts are probably a reflection of the different flow velocities, the time during which the rock is in contact with the flowing water, and the erosional power of entrained fine- to medium-sized particles. Turbulence is often caused by small irregularities in the cave walls, such as joints and bedding planes or small concavities. Some authors believe that the fissures are a proof that mixing corrosion might be involved, but as explained above, this process probably has only a small effect; however, mixing corrosion can widen the fissure mouths enough to increase turbulence. Another mechanism that causes fissures to be enlarged is related to floods. During flooding of the cave, undersaturated waters are injected into all available fissures. After the flood recedes, the outward flow of this undersaturated water from the fissures leads to a more or less short period in which the pocket walls are covered with a thin film of water able to dissolve substantial amounts of rock (Palmer 2007).

In stream passages which are frequently flooded by highly aggressive waters, solution pockets can intersect each other forming an intricate network of intersecting potholes, and their dividing sharp-edged blades, named **echinoliths** (Aley 1964) (Figures 9.38c, d and 9.47f). Solution pockets are also common in some thermal caves, where they generate underwater by rising, warm and less dense fluids on overhanging walls or roofs or along feeding fractures. These solution pockets are aligned in the direction of discontinuities (stratification, fractures), indicating differential corrosion (Figure 9.38e), ruling out their genesis by condensation–corrosion in aerate conditions, which leads to more homogeneous corrosion in small convection cells less influenced by compositional or structural heterogeneities in the rock.

Bell holes are cylindrical, cigar-shaped, ceiling pockets with round tops that extend vertically upward. They are typically found in the entrance parts of caves in warm and humid climates (Figure 9.39a). Their depth is generally greater than their diameter, penetrating into the cave roof up to several meters and with widths at their mouths ranging between 25 cm and 1.3 m (Tarhule-Lips and Ford 1998b). In certain shallow tropical caves, they pierce the roof reaching the ground

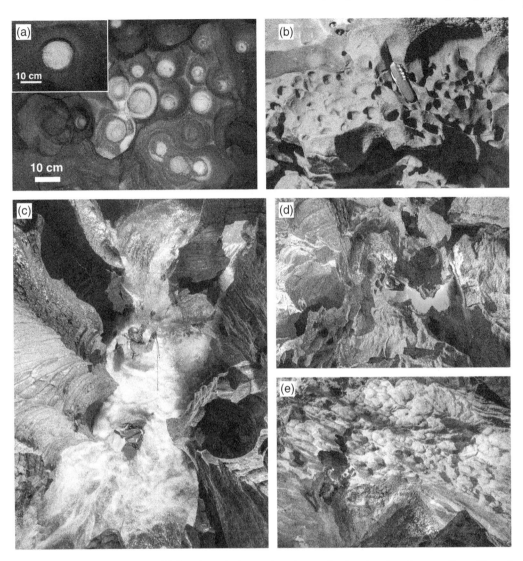

Figure 9.38 Solution pockets. (a) Regular solution pockets in non-fractured rock surfaces on the ceiling of Špilja Vjetrenica, Bosnia and Herzegovina, some of which appear to trap air during floods (white roof). *Source:* Photo by Darko Bakšić. (b) More irregular solution pockets guided by both lithology and structure in the entrance area of Ponikva ponor, in Dabarsko Polje, Bosnia and Herzegovina. They mainly develop in a limestone bed aligned along joints, whereas more scallop-like features are carved in the overlying bed. *Source:* Photo by Lukas Plan. (c) Echinoliths and portions of potholes in Mangawhitikau Cave, Waitomo, New Zealand. *Source:* Photo by Neil Silverwood. (d) Downward view of the lakes and echinoliths in Su Bentu Cave, Sardinia. *Source:* Photo by Antonio Danieli and Michel Renda, La Salle 3D. (e) Phreatic thermal-convection solution pockets in the upper parts of Pál Völgyi Barlang, Budapest. *Source:* Photo by Csaba Egri, La Salle 3D.

surface, which may have experienced downward erosion eventually intersecting the underlying cave roof with bell holes (Figure 9.39b). Bell holes are sometimes occupied by bats, in which case their mouths are sometimes covered with brownish crusts composed of hydroxylapatite or other material related to excreta of the bats (guano and urine) (Figure 9.39a). Their depth/width ratio is typically greater than 1, which distinguishes them from cupolas. Their vertical-upward development in cave roofs differentiates them from normal solution pockets. Bell holes normally do not

Figure 9.39 Bell holes. (a) Large bat cluster inside a bell hole in Runaway Bay Cave, Jamaica. Note the typical brown crust of excreta on the lower edges of the hole. *Source:* Photo by Joyce Lundberg and Donald McFarlane. (b) Bell holes have pierced upward and now intersect the external surface affected by corrosional lowering, Runaway Bay Cave, Jamaica. *Source:* Photo by Joyce Lundberg and Donald McFarlane. (c) Closely spaced bell holes close to the eastern entrance of Fruit Bat Cave, Mulu, Borneo. *Source:* Photo by Joyce Lundberg. (d) Bell holes on the roof carved in Eocene limestone in Australian's Inlet, Puerto Princesa Underground River National Park, Palawan, Philippines. *Source:* Photo by Marco Vattano, La Venta Esplorazioni Geografiche. (e) Bell holes in Al Hoota Cave, Oman. *Source:* Photo by Lukas Plan. (f) Bell hole on the roof and the corresponding underlying bell pit (both pointed by the caver), Al Hoota Cave, Oman. *Source:* Photo by Lukas Plan. (g) Condensation–corrosion cusps on the lower part of a flowstone in Gruta da Marota, Bahia, Brazil. *Source:* Jo De Waele (Author). (h) Two large flowstones carved by condensation–corrosion caused by abundant vampire guano, Paixão Cave, Bahia, Brazil. *Source:* Jo De Waele (Author).

appear to be influenced by local geological structures, cutting across bedding planes, fractures, faults, and speleothems (Lundberg and McFarlane 2009). Bell holes often cover large areas in the roof of entrance zones of tropical caves, developing an evenly spaced polygonal pattern (Figure 9.39c). In roof areas with initial lithostructural heterogeneities, bell holes can develop along the most favorable areas, sometimes coalescing while they increase their diameters (Figure 9.39d). In some cases, bell holes are in association with underlying depressions carved on the floor, named **bell pits** (Lauritzen and Lundberg 2000) (Figure 9.39e, f). The association of bell holes with bats might induce to think that these forms are purely biogenic, but bats might also occupy these favorable sites opportunistically. Although some bell holes might have experienced some kind of phreatic dissolution (Dogwiler 1998), it is now commonly accepted that most have a vadose origin related to condensation–corrosion, both by normal convective air flow, sometimes aided by biogenically produced CO_2, heat and moisture, and the aggressive nature of bat excreta (Lundberg and McFarlane 2009; Dandurand et al. 2019).

Condensation–corrosion appears to be a very efficient process able to dissolve large amounts of bedrock and speleothems (Tarhule-Lips and Ford 1998a). To have this efficiency there has to be a continuous or seasonal inflow of warm and moist air that condenses on the colder cave walls, and a way of keeping the rock colder than the incoming air. These conditions can occur in very shallow caves and the near-entrance parts of caves, where temperature differences between day and night allow cooling down the rock enough at night to act as a condensing surface during the day. Warm and moist air masses are frequent in coastal or lacustrine environments, so condensation–corrosion is often seen in coastal caves close to the entrances. Condensation can also occur at the interface between a cool and dense (cave) air body and the moving warm and moist air above. This causes condensation, and thus corrosion, to occur at a certain height interval (Figure 9.39g), forming **condensation–corrosion cusps**. Prolonged production of warm and moist air is also possible in presence of decaying guano deposits (Figure 9.39h), producing condensation–corrosion above the guano.

Spongework is an intricate network of intersecting hemispherical cavities and tubes forming a three-dimensional Swiss-cheese-like pattern, resembling, as the name suggests, a sponge. These features mainly occur in caves where there is no sign of fast flowing waters (absence of scallops), such as coastal mixing (Figure 9.40a), sulfuric acid (Figure 9.40b), and thermal caves (Figure 9.40d). However, spongework mazes can also locally occur in floodwater caves, where highly aggressive water is forced through all available openings due to the temporary high hydrostatic head (Palmer 1975). In floodwater caves, however, the spongework maze is typically accompanied by scalloped rock walls. In all cases, the three-dimensional spongework pattern forms because of the enlargement of all available pores or closely spaced and variably oriented fissures, or because it is conditioned by small differences in the solubility of the rock (Figure 9.40c). In sulfuric acid caves, spongework forms in the vicinity of zones where sulfuric acids are produced in the cave (close to feeders), or where mixing of different fluids occurs. In coastal flank margin caves, the spongework pattern extends laterally along the mixing zone, with a rather uniform horizontal extension and with the vertical range influenced by the tidal oscillations.

There are some other pocketing morphologies that have different origins than those described above, many of which are related to sulfuric acid speleogenesis. These include small, up to one decimeter wide, bell hole-like features that have been found on subaqueous mammillary calcite deposits (cave clouds) in a Sardinian sulfuric acid cave intercepted by mine galleries (De Waele and Forti 2006). They occur in association with large upward developing grooves around 2 cm deep, corresponding to large bubble trails (Figure 9.41a). These pockets, around one decimeter deep, do not necessarily have vertical orientations, are associated with small fissures, and are always the starting point of the bubble trail channels (Figure 9.41b). The core of the mammillary calcite speleothems is occupied by sulfides (mainly galena; PbS), that oxidizes in the shallow oxygenated

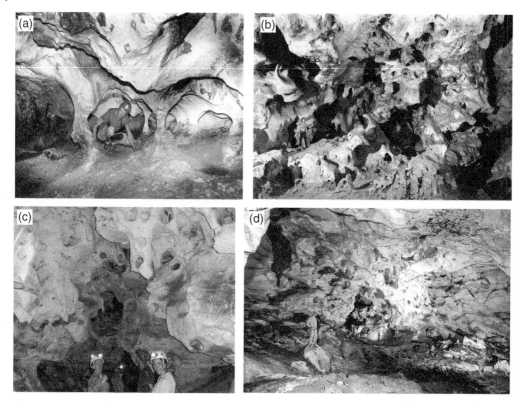

Figure 9.40 Spongework. (a) Spongework dissolution with alcoves and pillars in Ten Bay Cave, Eleuthera, Bahamas. *Source:* Photo by Arthur Palmer. (b) Carlsbad Boneyard, New Mexico, USA, formed in massive reef limestone of Permian age by deeply sourced H_2S-rich water, resulting in intricate spongework patterns. *Source:* Photo by Arthur Palmer. (c) Spongework in a heterogeneous Miocene limestone along the roof of Su Coloru Cave, Sardinia. *Source:* Photo by Laura Sanna. (d) Cupola-like morphologies, megacusps and intertwining solution pockets forming a spongework pattern in Eremita Cave, Sicily. *Source:* Photo by Marco Vattano.

waters, forming sulfuric acid that reacts with the dolomitic host rock and the calcite speleothems, thus releasing CO_2 and replacing the carbonate with sulfates, which are then slowly dissolved. These very rare morphologies have been named **oxidation vents** (De Waele and Forti 2006).

In one of the side passages of the giant Carlsbad Caverns, New Mexico, vertical centimeter-wide cigar-shaped holes have developed at various altitudes on the roof and close-to-horizontal overhanging limestone ledges (Figure 9.41c). These holes, named **zenithal ceiling tube-holes**, resemble lake-shore karren (tube karren or Röhrenkarren; Simms 2002, see Section 6.2.4), but are attributed to fluctuating water levels in a sulfuric acid environment, trapping air in pockets and involving fast condensation–corrosion and replacement of calcite by gypsum and production of CO_2 (Calaforra and De Waele 2011).

More generally, the action of H_2S degassing, its dissolution and oxidation in condensation waters, and the reaction of the produced sulfuric acid with the carbonate host rock, causes the formation of the typical pocketing in the vadose zone of sulfuric acid caves (Figure 9.41d). These morphologies, known as **replacement pockets**, are one of the most diagnostic features of sulfuric acid caves, especially when they are found in association with replacement gypsum, a byproduct of carbonate dissolution by sulfuric acid (Egemeier 1981; Galdenzi and Maruoka 2003; Plan et al. 2012; De Waele et al. 2016). The secondary gypsum precipitate can remain attached to the

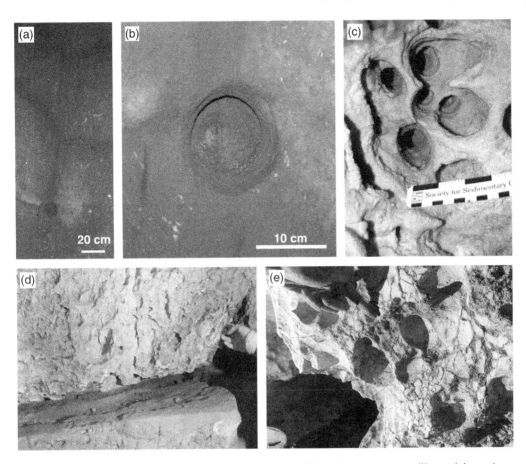

Figure 9.41 Sulfuric acid pocketing. (a) Oxidation vent carved in subaqueous mammillary calcite and associated with a rising bubble trail, Santa Barbara 2 Cave, Sardinia. *Source:* Photo by Paolo Forti. (b) Detail of the oxidation vent with the feeding fissure and the start of the large bubble trail (above). *Source:* Photo by Paolo Forti. (c) Upward view of the zenithal ceiling tube-holes in the Sand Passage in Carlsbad Caverns, USA. *Source:* Jo De Waele (Author). (d) Replacement pockets on the subvertical wall above a remnant of a flat corrosion table, Stephanshöhle, Bad Deutsch Altenburg, Austria. *Source:* Photo by Lukas Plan. (e) Detail of replacement pockets in a limestone breccia in Shpella Kaceverrit, Holtas Gorge, Albania. *Source:* Jo De Waele (Author).

wall in the pockets, and its hygroscopic behavior is able to attract more condensation water, thus boosting the dissolution-replacement reaction and the deepening of the pockets (Figure 9.41e). The original texture and sedimentary structures (i.e., fossils) of the host rock can be preserved in the replacement gypsum (Plan et al. 2012).

In many caves, the ceilings of the passages are carved by sinuous channels, a sort of "upside-down half-tubes," which can occur isolated (Figure 9.42a), but sometimes form intricate networks of anastomosing channels (Figure 9.42b) and intervening residual rock projections, known as rock or **ceiling pendants** (Figure 9.42b, c). The larger forms are known as **ceiling channels**, or ceiling half-tubes, and can have an arcuate (Figure 9.42a, c) or a flat roof (Figure 9.42d, e). The networks of intertwining, generally smaller ceiling channels are called **anastomoses** (Figures 9.42b and 9.43c–e). Their presence in caves was described in the late 1930s by Zdeněk Roth in Domica Cave (Slovak Republic), where they are particularly abundant and well

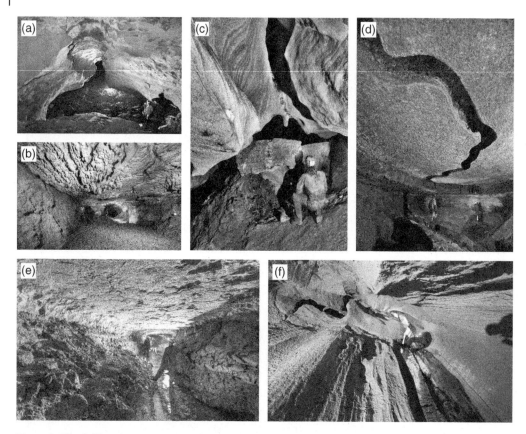

Figure 9.42 Ceiling channels, pendants and anastomoses. (a) Ceiling channel with rounded roof in the Camelié Cave, Lussan, Gard, France. *Source:* Photo by Jean-Yves Bigot. (b) The flat antigravitative ceiling with anastomoses and a few large flat-roofed ceiling channels carved in steeply dipping Messinian gypsum (bedding plane dipping ca. 45° is visible on the left), Ca' Castellina Cave, northern Apennines, Italy. *Source:* Photo by Piero Lucci. (c) Rounded ceiling channels in steeply dipping Messinian gypsum beds, and a central smooth rock pendant in Rio Basino Cave, northern Apennines, Italy. *Source:* Photo by Piero Lucci. (d) Winding ceiling channel in the Covadura Cave formed in Messinian gypsum and marls, Sorbas, southern Spain. A remnant of the initial meandering conduit is now visible on the roof because of mechanical erosion of the underlying argillaceous bed. *Source:* Photo by Victor Ferrer Rico. (e) Flat-roofed wide ceiling channel forming the entire roof carved in limestone, Busa della Pisatela, Veneto, northern Italy. Note detrital sediments almost reaching the roof. *Source:* Photo by Sandro Sedran, S-Team. (f) Meandering ceiling channel produced by headward erosion in a vadose canyon-shaft in Busa della Pisatela, northern Italy. *Source:* Photo by Sandro Sedran, S-Team.

developed (Bella and Bosák 2015). Ceiling channels can have different origins (Pasini 1967): (i) by interception of a preexisting or contemporary phreatic tube by an underlying phreatic passage, (ii) by antigravitative or paragenetic erosion, which involves upward growth of a water-filled passage caused by sediment accumulation on the floor (Cooper and Covington 2020), (iii) by removal of a sedimentary bed underneath a more resistant one, leaving the upper part of the original conduit on the roof (Figure 9.42d), (iv) by regressive erosion of a canyon passage, leaving the initial meandering channel exposed on the roof of the shaft (Figure 9.42f), and (v) by condensation–corrosion caused by rising warm and moist air in thermal caves (Figure 9.44a). Most ceiling channels are related to antigravitative erosion, also known as paragenesis, a term not to be confused with mineral paragenesis, which is why we recommend the use of antigravitative erosion. Some spectacular examples are found in epigenic gypsum caves (Figure 9.42b, c), where climate-driven infilling and entrenchment phases have left many of them visible in different cave storeys. Antigravitative ceiling channels have been described in detail by both Pasini (1967) and Renault (1968). Some spectacular deeply carved ceiling channels and residual pendants can be seen also in alpine limestone caves, such as in Mammuthöhle, Dachstein (Austria) (Figure 9.43a). They are also abundant in Brazil, where alluviation is a widespread phenomenon in many caves (Laureano et al. 2016).

A special type of anastomoses is the one developing along bedding planes or close-to-horizontal fracture planes, known as bedding-plane and joint-plane anastomoses, respectively (Figure 9.43b, c).

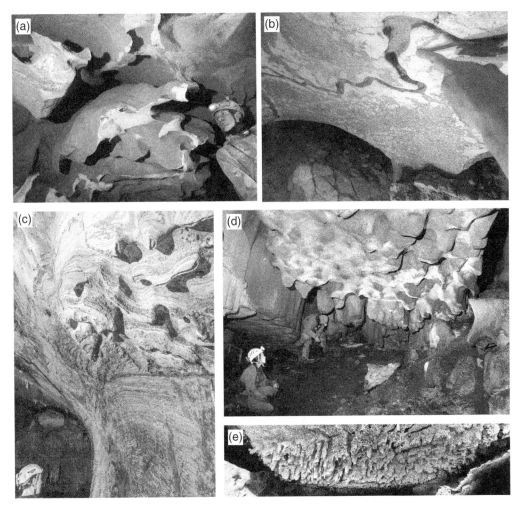

Figure 9.43 Anastomoses and pendants. (a) Deeply carved meandering ceiling channels and remnants of rock pendants in the limestone of Mammuthöhle, Dachstein, Austria. *Source:* Photo by Lukas Plan. (b) Single anastomosis on a fallen limestone block in Saint Marcel Cave, Ardèche, France. *Source:* Jo De Waele (Author). (c) Rounded bedding-plane anastomoses in thinly layered Neoproterozoic limestones in Gruta da Torrinha, Iraquara, Bahia, Brazil. *Source:* Jo De Waele (Author). (d) Smooth anastomotic channels and pendants in the carbonate ceiling of Palma Cave, Palo Grande, Bagua Province, Peru. *Source:* Photo by Jean-Yves Bigot. (e) Anastomoses and pendants in a large block in the Messinian gypsum cave of Rio Basino, northern Apennines, Italy. *Source:* Photo by Piero Lucci.

These are intricate bidimensional networks of hemicylindrical tubes carved upward in the soluble bed, because they are often confined downward by insoluble residue or less soluble rock. These differ from normal paragenetic anastomoses in that they are confined by geological boundaries (bedding planes or joints), whereas the paragenetic ones can be carved through dipping soluble strata cutting across bedding planes (Figures 9.42b and 9.43d, e). Bedding-plane anastomoses are often remnants of early speleogenetic evolutionary phases (Figure 9.43b), or can be formed by floodwaters injected along planar discontinuities.

Channels can also be carved along the walls and ceiling by convectional movements of fluids in hypogenic settings, where warm fluids or condensation waters from moist air masses that dissolve the rock along the uppermost parts of the void (Figure 9.44a, b). They normally have a wavy long

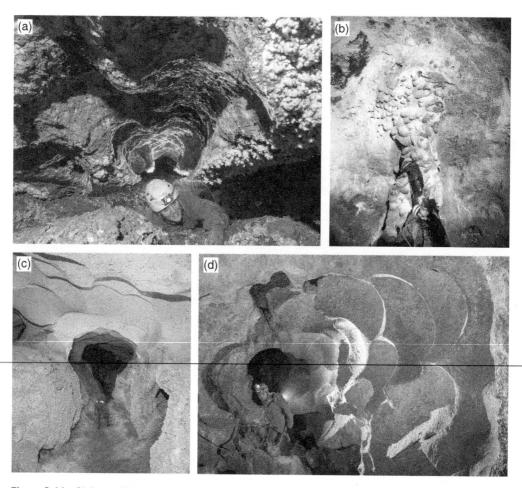

Figure 9.44 Rising wall channels and megacusps. (a) Rising channel generated by condensation–corrosion in the thermal Eisensteinhöhle, Austria. *Source:* Photo by Lukas Plan. Note the roof corroded by condensation waters, and the coating of popcorn (coralloids) in the lower part, where evaporation prevails. (b) Dissolution pockets (small cupola) in a rising wall channel acting as feeder in Ürrütxordokiko Lezea (gouffre OX655), Camou-Cihigue, Atlantic Pyrenees, France. *Source:* Photo by Jean-Yves Bigot. (c) Megacusps forming the wavy roof of Cocci Abyss, Sicily. The roof and upper walls in this thermal cave are smoothed by condensation–corrosion, whereas the lower parts are covered with popcorn due to evaporation-induced precipitation. *Source:* Photo by Marco Vattano. (d) Wide cupolas and megacusps on the upper walls of a passage in Kraushöhle, a sulfuric acid cave formed in the Late Pleistocene, Austria. *Source:* Photo by Lukas Plan.

profile, with chains of large cusps along the ascending path of the fluids that formed them. These channels have been termed **rising wall channels** by Klimchouk (2009), and are part of the so-called "morphological suite of rising flow" that distinguishes ascending hypogene speleogenetic settings from the epigenic ones dominated by downward carving. They start from, or are parts of **feeders**, which continue as rising channels, megacusps, and cupolas, and end as outflow channels, or in large blind cupolas or rising passages. These channels can form underwater, or as condensation–corrosion channels in the aerated part of the cave (Figure 9.44c, d). In all cases, the flow velocities are low, and convectional cells are larger than a couple of decimeters, and generally of a metric scale. These cusps resemble scallops, and that is why they are often called megascallops. It would however be convenient to call these cuspate wall and roof features with their proper name, **megacusps**, avoiding confusion with scallops caused by turbulent flow in streams (D'Angeli et al., 2019a, 2019b).

The ceilings and walls of cave passages are not always corroded and eroded uniformly by running waters or condensing air flows, attaining a smooth and curvy morphology. The bedrock is often heterogeneous, with variations in solubility and/or erodibility that control the differential solutional and mechanical carving of the rock surfaces (Figure 9.45). Whereas mechanical erosion is a less selective process, and morphologies are mainly related to differences in hydrodynamic conditions (e.g., turbulence, flow velocity), dissolution is a much more selective process. Differential dissolution etches deeply the more soluble parts of the bedrock, leaving the less soluble parts sticking out of the walls (Figure 9.45c, e). In a succession of dolomite and limestone beds, the less soluble dolostone beds tend to form protruding ledges (convex parts), and the pure limestone beds the recesses (concave parts). Shaly interbeds, which are less soluble, may protrude in cave passages characterized by slow-moving water, whereas fast running water erodes them away much more easily. Carbonate successions often contain chert nodules, typically arranged in more or less parallel bands that form protrusions on the cave walls, being less soluble and also more resistant to mechanical erosion (Figure 9.45e). Fossils are another common feature sticking out of cave walls, because they are frequently composed of less soluble and erodible coarse crystalline calcite (Figure 9.45d, f), or they are partially or completely replaced with other minerals such as dolomite or silica (quartz or chert). If the host rock contains veins of quartz, dolomite or calcite, these are often also less soluble than the bedrock in which they are hosted. Vein networks are often remnants of hydrothermal deposition in a deep-seated environment, generally long before the caves formed (Figure 9.45d). If the veins are thick and hard enough, slowly moving waters are not able to mechanically wear them, and their differential resistance to dissolution will leave them projecting out of the cave wall. Thinner blades of mineral veins are more easily destroyed by running water, but they can preserve in cave passages where condensation–corrosion is the main process at work. Vein networks often form complex polygonal patterns resembling a set of boxes, and are known under the name **boxwork**. Famous examples are known from the polygenetic caves of the Black Hills (Wind and Jewel Cave, South Dakota) (Figure 9.45a, b).

Along the walls of caves, it is often possible to identify more or less horizontal or undulating notches carved by solution and/or erosion related to underground water bodies or streams. **Solution notches**, also called **waterline notches**, are perfectly horizontal and laterally extensive indents formed at the water level (Figure 9.46a and inset). A vadose pool of standing water in an air-filled cave can absorb CO_2 from the cave atmosphere increasing the aggressiveness of the surface layer of the water body and dissolving more rapidly the walls than the deeper submerged parts. Solution notches are normally a couple of decimeters high, depending on how much the pool level changes, and can get as deep as 1 m. Perfectly horizontal notches form rapidly in thermal

Figure 9.45 Morphologies related to differential dissolution. (a) Extensive boxwork of brown macrocrystalline calcite veins in Wind Cave (Wind Cave National Park). *Source:* Photo by Arthur Palmer. (b) Thick calcite veins forming subvertical protruding fins in Wind Cave, South Dakota, USA. *Source:* Photo by Arthur Palmer. (c) Differential corrosion in metalimestone beds with variable solubility in Roquebleue Cave, Hérault, France. *Source:* Photo by Philippe Crochet. (d) Differential dissolution has left microbially mediated iron-oxide tubes projecting from a wall in a hypogenic cave, Iboussières Cave, France. *Source:* Photo by Jean-Yves Bigot. (e) Chert nodules sticking out of the limestone wall in Guernica Cave, northern Italy. *Source:* Photo by Sandro Sedran, S-Team. (f) A marine urchin fossil in El Jarrito Cave, Matanzas, Cuba. *Source:* Jo De Waele (Author).

Figure 9.46 Solution ramps, bevels (facets), and notches. (a) Horizontal water level notch in Buso del Vento (Lessini Vicentini, Veneto, Italy). *Source:* Photo by Sandro Sedran, S-Team. Inset shows a water level notch in the Seich Cave, Ariège, France. *Source:* Photo by Jean-Yves Bigot. (b) Inclined ramps in Rumbling Falls Cave, Tennessee, USA. *Source:* Photo by Arthur Palmer. (c) Water level notches carved by convection cells above the water level in a thermal sulfuric acid cave, Kraushöhle, Austria. *Source:* Photo by Lukas Plan. (d) Wavy solution ramps in the Santa Ninfa gypsum cave in Sicily. *Source:* Photo by Marco Vattano. (e) Solution bevel and facets in Moncalvo gypsum cave, northern Italy. *Source:* Photo by Bartolomeo Vigna. (f) Nested solution ramps in the Cressi Cave system, a salt cave in Atacama, Chile. *Source:* Photo by Riccardo De Luca, La Venta Esplorazioni Geografiche.

sulfuric acid caves, where small convection cells can originate immediately above the pool levels, causing condensation–corrosion to occur in a highly acidic environment (Figure 9.46c).

These notches also develop at the foot of karst towers in tropical areas, where their bases are in contact with water (e.g., shallow phreatic zone, swamps) that continuously renews its aggressiveness by the input of soil-derived CO_2. These large and deep notches are known as foot caves, and can be as deep as a couple of tens of meters. Similar notch caves can be carved in coastal limestones by a combination of mixing corrosion, biocorrosion, and wave action. These notches are known as tidal notches (De Waele and Furlani 2013).

Solution ramps, also known as vadose and antigravitative (paragenetic) wall notches, differ from solution notches in that they are not perfectly horizontal, but fall and rise along the length of a passage (Figure 9.46b, d, f). Their general slope mimics that of the cave stream that created them. They develop above or below former sediment banks by a prolonged period of dissolution. If the shielding effect of the sediment prevails, and flood waters can readily dissolve the walls above, the notch will be located above the sediments. When flood waters rise slowly, wetting and saturating the sediment but without flowing above the top surface of the deposits, the dissolving action at the sediment-rock contact will prevail, so the higher limit of the notch will indicate the upper limit of the sediments. In meandering cave rivers, where deposits tend to accumulate in the inner side of bends, these notches formed at the sediment-rock contact can shift from one side to the other of the cave passage. Solution ramps generally cut across geological structures, such as bedding planes or fractures, although these structures are often differentially dissolved.

Another relatively reliable water table marker is the flat horizontal ceiling and the underlying passage walls with inward inclination of around 45°, which can be encountered in highly soluble rocks such as halite and gypsum (Figure 9.46e), and sometimes also in limestone. The flat roof is known with the German name "**laugdecke**", whereas the sloping walls are called facets (facetten), or **solution bevels** (Kempe et al. 1975). These passages with sometimes a perfectly triangular cross-section are carved in situations with vertical density gradients in which water flow is extremely slow, with less dense and more undersaturated waters flowing at the top of the passage, and denser and more mineralized waters flowing below. Dissolution will thus be greater at the roof, and when these waters become saturated, their higher density will cause them to descend, thus creating a flow cell.

The corrosive action of waters, and especially of the vapors in sulfuric acid cave environments, creates a series of characteristic morphologies. Sulfidic waters enter these caves from below along feeding fractures. These features are often visible, if not covered by sediments, as narrow fissures (named **slots**) on the floor of the cave (Figure 9.47a). The rising fluids acquire greater corrosive power when H_2S is oxidized, and this process occurs close to the water surface and especially in the cave atmosphere. For this reason, feeding slots tend to get narrow very fast because the dissolving power of rising fluids is very low at depths of only a couple of meters. The slots may not taper downward when the feeding fracture has been active over a long period of time, has been subject to significant water level variations, and carries significant water flow. If the sulfidic fluids rising into the caves are thermal, both water vapor and H_2S escape from the water surface, and condensation occurs along the cave walls and the roof. Thermal convection creates air flow cells in the cave atmosphere immediately above the water table, resulting in the formation of notches along the water level (Figures 9.46c and 9.47b), sometimes in a series of inset levels when the water body descends episodically, leaving upper notches exposed. Water condenses on the upper roofs of the notches and descends along the notch walls, often concentrated in runnels, designated as **sulfuric karren** (Figure 9.47b).

The thermal contrast between the water body and the cave walls is greatest along the roof, situated at the largest distance from the thermal water, and condensation–corrosion will be most active here. Since rising vapor and gas diffuse from the feeding slots, corrosion is more intense

Figure 9.47 Sulfuric acid corrosion tables and feeders (feeding slots). (a) Feeder slot in Acqua Fitusa Cave, Sicily, Italy. *Source:* Photo by Marco Vattano. Note the convectional notch, the dome-shaped passage, and the almost perfect flat corrosion table. (b) Detailed view of the convection notch in Acqua Fitusa Cave, with sulfuric karren. (c) Dome-shaped central chamber of Grotte du Chat, Provence, France. *Source:* Photo by Jean-Yves Bigot. Note fracture-controlled feeding slots on the floor, corrosion tables (on which the central person stands), and the water level notches and cupolas generated by condensation–corrosion.

above the feeders, thus giving rise to a dome-shaped chamber with convection niches and cupolas rising above the feeding fissures (Figure 9.47c).

The acidic condensation waters that flow along the floor back to the feeding slot (or pool) in the lower part of the cave passages are responsible for carving planar surfaces in the rock, known under the name of **corrosion tables** (Egemeier 1981) (Figure 9.47c). Their perfect shape, with the creation of an equilibrium close-to-horizontal slope, is caused by the balanced processes of turbulence and oxygenation (De Waele et al. 2016).

In the lower parts of passages in epigenic caves, and especially those that are permanently active or become active sporadically, the morphology is largely influenced by the erosive and corrosive action of flowing vadose waters. Fast running water causes the formation of bedrock channel morphologies

Figure 9.48 Vadose plunge pools and potholes. (a) Plunge pool at the foot of a waterfall in Scalandrone Cave, Campania, Italy. *Source:* Photo by Francesco Maurano. (b) A large rounded pool with numerous scouring pebbles and cobbles along the stream in Buso della Rana Cave, Veneto, Italy. *Source:* Photo by Sandro Sedran, S-Team. (c) Small waterfall and plunge pool in Eolo Resurgence Cave, Friuli, northern Italy. *Source:* Photo by Sandro Sedran, S-Team. (d) Rapid and swirling water flow in a pool in Chamois Cave, Provence, France. *Source:* Photo by Jean-Yves Bigot. (e) A 4-m deep pothole carved in gneiss underlying marble in Höhle beim Spannagelhaus (Zillertal, Austria). *Source:* Photo by Lukas Plan. (f) Potholes and echinoliths in a stream passage in Sullivan Cave, Indiana, USA. *Source:* Photo by Arthur Palmer.

that can also be found in bedrock streams at the surface. The larger **plunge pools** at the base of waterfalls (Figure 9.48a), smaller **potholes** ("marmites" in French) (Figure 9.48e, f), and their intermediate forms (Figure 9.48b–d) are the most striking of these bedrock channel forms, created by the abrasive action of coarse particles that swirl within vortices in these kettle-like morphologies, acting as a rock mill. In soluble rocks, potholes are much better developed than in non-soluble lithologies such as granites, basalts or sandstones, because of the combined action of abrasion and dissolution. To allow gravels to swirl around in a hole, the water flow should exert sufficient shear stress on the particles, which depends on factors such as flow velocity and the depth of the pothole. As the flow rate and velocity diminish during flood recession, the power of the falling water will no longer be able to activate the rock mill, and normal vadose entrenchment will eventually obliterate the pothole or leave it abandoned and perched on the side of the stream passage (Ford 1965).

Scallops are probably the most characteristic hydraulic features in karst caves. These are asymmetric oyster-shell-, or spoon-shaped solutional concavities occurring in compact rock walls and created by turbulent water flow. They do not develop in impure, heterogeneous and coarse-grained soluble rocks, where beds are too thin or where voids and insoluble or coarse components (e.g., fossils, chert nodules) create irregularities on the rock surface that disrupt the flow, interfering with the scallop geometry. Scallops are not exclusive to caves, and were first described in external limestone bedrock streams (Lugeon 1915). Bretz (1942) was the first to propose their use for determining the direction of water flow in a cave passage, a concept further developed in detail by Coleman (1949). Their longitudinal asymmetric profile shows a steeper side that indicates their upstream end, from where the water flow came. They occur in large groups, forming a scalloped pattern sometimes extending along the floor, walls and roof of sediment-free cave passages (Figure 9.49a–d). They commonly range in size between half a centimeter to several decimeters,

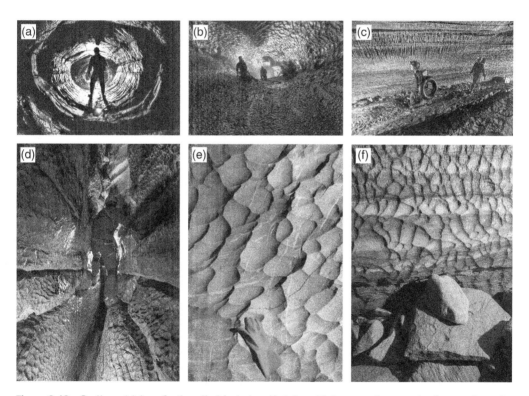

Figure 9.49 Scallops. (a) A perfectly cylindrical phreatic tube with large scallops on the floor, walls, and roof. Aven de Peyrejal, Ardèche, France. *Source:* Photo by Bartolomeo Vigna. (b) Bedding-plane guided phreatic tube with smaller scallops on the floor, and larger ones on the roof. Markov Spodmol Jama, Slovenia. *Source:* Photo by Sandro Sedran, S-Team. (c) Scallops in the Pink Passage in Ellis Basin Cave system, Mount Arthur, Kahurangi National Park, New Zealand. *Source:* Photo by Neil Silverwood. (d) Scallops carved on the walls of a fissure, and differential dissolution along bedding planes in the Serra Carpineto sinking stream, Alburni Mountains, Central-South Italy. *Source:* Photo by Francesco Maurano. (e) Scalloped surface (water flow from right to left) on massive Devonian metalimestone in Lurgrotte, Styria, Austria. *Source:* Photo by Lukas Plan. (f) Detail of the scallops on the red and white wall in the Pink Passage in Ellis Basin Cave system, New Zealand. Steeper sides are on the right side of the scallops and face to the left, indicating flow direction (from right to left). *Source:* Photo by Neil Silverwood.

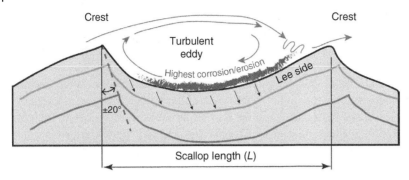

Figure 9.50 Longitudinal profile of a scallop and the associated flow pattern. *Source:* Modified from Curl (1974) and Ford and Williams (2007). Water flow is from left to right. The green profiles indicate the progressive downstream migration of the scallop as the rock surface recedes.

exceptionally reaching lengths up to 2 m. Their length is approximately twice their width, but their dimension is often disturbed by neighboring scallops (Figure 9.49e, f). The size of scallops is a function of flow velocity; the smaller the scallops the more rapid the flow. The theory of scallop formation was formulated in detail by Curl (1966, 1974) and Blumberg and Curl (1974) (Figure 9.50). The main water flow detaches from the boundary layer at the upstream crest of a scallop, and within a short distance, depending on flow velocity, becomes turbulent and distorts the fluid in the lee side of the scallop. Part of the flow thus returns forming a turbulent eddy and rejoins the main flow in the upstream side. Maximum dissolution occurs associated with this turbulent eddy, causing the scallops to wear and migrate downstream with time (Figure 9.50). Scallop formation appears to be more complex than just dissolution by turbulent water flow, and most probably includes mechanical erosion as one of the contributing processes (Covington 2014). Scallops appear to form in waters flowing at velocities between 1 and 3 m s^{-1}. The lower value corresponds to a threshold for turbulent flow initiation and the higher one for the prevalence of mechanical erosion due to high sediment load (leveling the surface).

Measuring the average length of a scallop population (at least 30 randomly selected scallops) allows the estimation of the most recent largest flow velocities. There are several problems with this method, and caution should be taken in thrusting the estimated flow velocities (Springer and Hall 2020). Lauritzen et al. (1985) estimated that flow velocity based on scallop length represents roughly three times the average flow velocity in the cave passage. Flow velocities in a cave passage change through time, are not uniform across the passage cross section and the passage length, and depend on many factors. Ideal conditions are straight passages with perfectly cylindrical shape (tubes), or perfectly rectangular fissures with constant width. Of course, water flow velocity decreases from the center of the passage to cave walls because of shear resistance. Mean flow velocity derived from scallop length thus depends on both passage size and passage shape; higher velocity values are obtained in larger passages, and mean velocity based on scallop size is greater in fissures (with smaller scallops) than in tubes (since water–rock contact is greater in tubes respect to fissures).

Mean velocity estimates based on scallop size are also temperature-dependent, since temperature determines the fluid viscosity. Fluids with higher viscosities (lower temperatures) require higher flow velocities to carve similar-sized scallops. This is why scallops carved in ice by low-viscosity air are much larger than those carved in limestone by higher viscosity water flowing at the same velocity. On average, ice scallops carved by melting at 0 °C caused by warm air flow indicate flow velocities about 7.4 times greater than similar-sized scallops in limestone (Palmer 2007). The graph in Figure 9.51 allows the estimation of average flow velocity based on mean scallop

Figure 9.51 Relationship between mean scallop length and mean flow velocity in cave passages of different sizes (passage diameter) and shapes and at different temperatures. *Source:* Based on the equations of Curl (1974), adapted from Palmer (2007). See text for explanations.

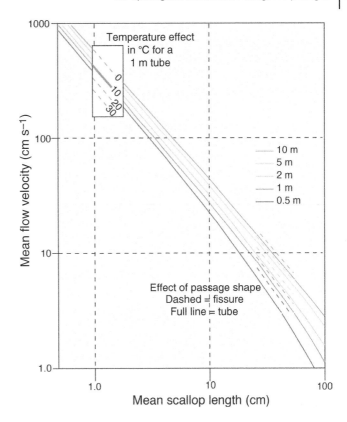

lengths measured in cave passages, adopting different graphical corrections for water temperature, passage diameter, and passage shape.

A practical way of roughly estimating mean flow velocity based on scallop lengths is given by the empirical expression:

$$v \approx X / L \tag{9.4}$$

where v is the mean flow velocity expressed in cm s^{-1}, L is the mean scallop length in centimeters, and X is a number that depends on water temperature (375 at 0 °C, 275 at 10 °C, 210 at 20 °C, and 170 at 30 °C) (Palmer 2007).

Scallops can be very variable in size even in the same passage, depending on local irregularities in the walls, and variations in flow conditions at different water depths. For example, scallops can be larger in the lower parts of the walls and smaller close to the roof, indicating that the latter formed during high-discharge events, when the passage was completely water-filled and flow velocity was much greater. In contrast, the lower scallops indicate more frequent flow conditions with lower velocities. In other cases, scallops on the roof are larger than those carved on the floor, indicating lower velocities during high flow periods. This can be caused by ponding of water upstream of conduit constrictions. During high flow, water has a high velocity, carving small scallops in the lower parts of the upstream passages and the floodwaters do not interact with the roof. However, when the flow rate exceeds the conveying capacity of the constriction, water builds up in the upstream sector of the cave filling completely the passage, and the flow velocity decreases, resulting in larger scallops along the roof. Where cave walls have protrusions into the passage, the water can form local eddies with scallops forming in reverse flow conditions. This is illustrated by scallops developed in the upper walls of deep potholes, where dissolution prevails over abrasion and scallops form indicating the direction of the water rotation in the whirlpool.

Sometimes scallops display a very elongated geometry. These features are known as elongate **ceiling grooves**, and occur in antigravitative (paragenetic) conditions, when the sediments shielding the lower part of the passages almost reach the roof. These rare forms indicate the direction of water flow as do scallops. Palmer (2007) reports an example from Collins Avenue in Mammoth Cave (figure 6.25 in Palmer 2007). In stream passages where turbulent eddies remain stable over long times, a form similar to ripples but carved in the rock can be observed on the walls and roof of cave passages. These rare elongated erosional forms known as **(phreatic) flutes** (Curl 1966), are perpendicular to the flow direction, and like ripples, have their steep side facing upstream, and wavelengths decreasing with higher water velocities.

In vadose conditions water tends to flow vertically, along the steepest available path, and when it is still undersaturated it tends to carve the surface and form **vadose flutes**. These are parallel, close-to-vertical grooves, several centimeters wide and deep, that can cut across bedding planes and other discontinuities (Figure 9.52a). They are often developed in the lower parts of vadose shafts or retreating canyons in massively bedded rocks, where undersaturated water flows in small rivulets down the walls (Figure 9.52b). Similar morphologies also occur on variously sloping surfaces, and are the subterranean equivalent of Rinnenkarren (solution runnels) (Figure 9.52d, e), Wandkarren (wall karren) (Figure 9.52c), and Mäanderkarren (meandering runnels). These forms often occur below bedding planes, indicating that they are formed from highly undersaturated waters flowing out of these fissures, similar to decantation runnels. These aggressive waters were injected into the bedding-plane partings during floods, and can drain out of these fissures for many days, or even weeks, once the flood water has receded.

Drainage grooves (Figure 9.52f) resemble subsoil rounded runnels (Rundkarren), but are found on the overhanging parts of roofs and walls. Their development is not perfectly straight, and their rounded forms differentiate them from the sharp-edged vadose flutes and runnels described above. They often grade upward into ceiling pendants, whereas their downward continuation can connect with ceiling channels in lateral lower-lying passages. These drainage grooves are carved by vadose waters that drain downward along the contact between a sediment fill and the rock walls.

Photokarren, instead, occur in the twilight zones of large cave entrances, where light penetrates at certain angles. They are characterized by deep holes and intervening pinnacles, often sharp, all oriented toward the source of light (the entrance of the cave) (Figure 9.52g, h). These karren features are biologically mediated (phototrophic corrosion, a type of phytokarst), with algae and photosensitive life forms thriving in the moist hollows, decaying and producing CO_2 in these spots and thus deepening the holes, leaving small pinnacles in between.

Bubble trails are channels carved on overhanging walls, but showing an upward development related to rising fluids (Figures 9.41a and 9.53a–c). These channels range in diameter from a millimeter to several centimeters, follow the easiest paths upward, and are often sinuous. These grooves, connecting small hemispherical concavities (small gas chambers), mark the tracks followed by rising gas bubbles in water-filled sections of the cave passage. The gases, generally CO_2 but sometimes also H_2S in sulfuric acid caves, are released by rising fluids due to the lowering of the pressure at shallower water depths, but can also be produced within the water bodies by organic decay or the oxidation of sulfides. This last process forms sulfuric acid that dissolves the carbonate rocks releasing CO_2. These gases concentrate along rising channels, and by dissolving in water again can enhance its aggressiveness and carve the trails.

In thermal caves, but also where air convection continuously causes humid and warm air to move through the cave, condensation waters form an aggressive film and droplets on the cooler

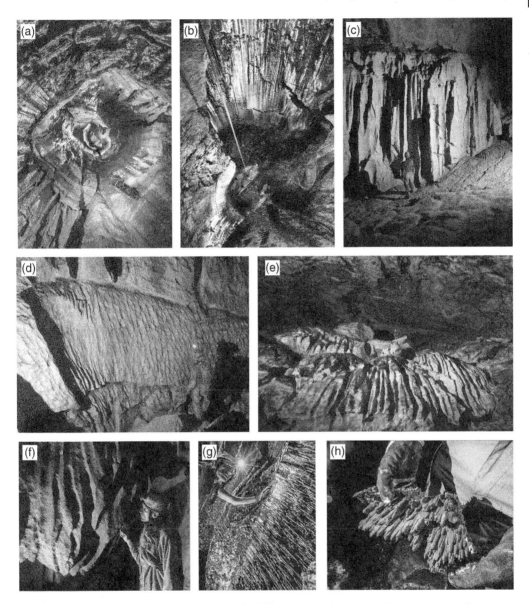

Figure 9.52 Vadose flutes and runnels: (a) Vertical flutes carved along the white and massive limestone walls of Abisso Cabianca, Lessini Mountains, northern Italy. *Source:* Photo by Sandro Sedran, S-Team. (b) Vertical flutes in Ireby Fell Cavern, Lancashire, UK. *Source:* Photo by Mark Burkey. (c) Solution grooves (subterranean Rinnenkarren) on an almost vertical wall in Hirlatzhöhle, Dachstein Gebirge, Austria. *Source:* Photo by Lukas Plan. (d) Pseudo-scallops (center of the picture) and runnels (subterranean Rinnenkarren) on an inclined wall below a bedding plane, which probably drains undersaturated water during flood recession. Hirlatzhöhle, Dachstein Gebirge, Austria. *Source:* Photo by Lukas Plan. (e) Rinnenkarren in the Rock 'n' Rillen Room in Lechuguilla Cave, New Mexico. *Source:* Photo by Max Wisshak. (f) Drainage grooves (similar to rounded runnels or Rundkarren in covered karst settings) on an overhanging wall, carved at the contact with a sediment cover which has been removed. Mammuthöhle, Dachstein Gebirge, Austria. *Source:* Photo by Lukas Plan. (g, h) Photokarren with sharp pinnacles from Tara Cava, Palawan, the Philippines. *Source:* Photo by Alessio Romeo, La Venta Esplorazioni Geografiche.

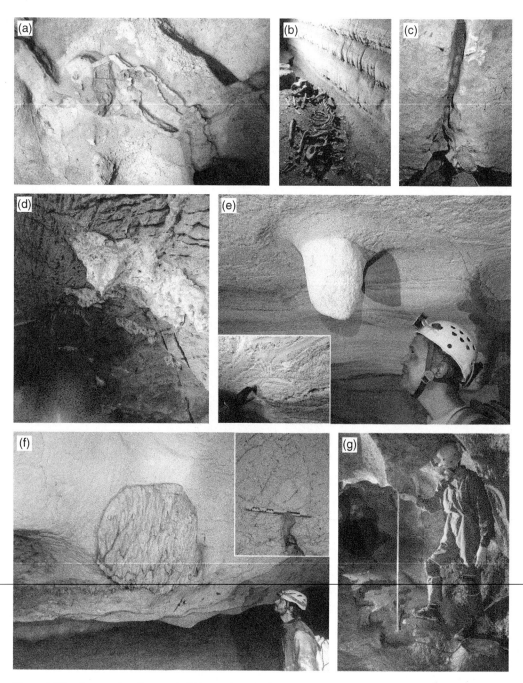

Figure 9.53 Forms related to gas bubbles and to condensation–corrosion: (a) A set of bubble trails on the ceiling of Monello Cave, Eastern Sicily. *Source:* Photo by Jean-Yves Bigot. Note the sea urchin fossil (4 cm in diameter, center) standing out in relief. (b) Bubble trails on the wall of a dry pool with a skeleton of a jaguar, Bahia, Brazil. *Source:* Jo De Waele (Author). (c) Rising bubble trail on the roof of a room in Frasassi Cave, Italy. *Source:* Photo by Arthur Palmer. (d) Etchpits and grooves (1 cm wide) on Triassic dolomites and a white speleothem in the entrance passage of Bàsura Cave, Toirano, northern Italy. *Source:* Jo De Waele (Author). (e) Flat ceiling corroded by condensation waters, and a stump of residual stalactite, Diva de Maura Cave, Bahia, Brazil. *Source:* Jo De Waele (Author). The inset shows the soft dissolution residue on the walls nearby. (f) Limestone wall corroded by condensation waters, and a relict shield speleothem in Saint Marcel Cave, Ardèche, France. The inset shows a portion of a speleothem (center) entirely corroded and level with the rest of the wall. *Source:* Jo De Waele (Author). (g) Ceiling-pendant drip hole (floor) below a pendant in Kraushöhle, Austria. *Source:* Photo by Lukas Plan.

cave walls and overhanging roofs. This occurs in the entrance areas of caves or at the junction between two cave passages with different air temperatures, where mixing clouds form. The same phenomenon occurs at the foot of active shafts, where vaporization occurs, or in passage sections with strong air current before and after constrictions, where air is compressed or expanded, thus causing condensation or evaporation, respectively. The resulting amount of water will never be great enough to drain freely along runnels, instead, it forms droplets and slowly flowing water films. The droplets will eventually create **etchpits**. These are rounded semicircular holes of 1 mm to a couple of centimeters in diameter and depth, separated by rather sharp boundaries (Figure 9.53d). The walls and roofs display a typical pitted appearance, or is covered with a soft dissolution residue (Figure 9.53e inset) and vermiculations (see Section 10.3.9). The condensation–corrosion weathering can dissolve the rock differentially (Zupan Hajna 2003), especially where it is heterogeneous, leaving less soluble components such as fossils and macrocrystalline speleothems as protruding features (Figures 9.45f and 9.53e, f).

If the amount of condensing waters is large enough, at least seasonally, to exceed adhesive forces, the dripping water falls to the floor. Especially in sulfuric acid caves, these acid droplets will create a depression below known as **ceiling-pendant drip hole** (Plan et al. 2012) (Figure 9.53g).

Decaying guano also releases heat, water vapor, and gases including CO_2, leading to condensation on the walls above to form rising **condensation–corrosion channels** (Figure 9.54a). The excess water can concentrate on residual pendants and fall to the ground again, being still corrosive and forming **guano pots** (Figure 9.54a). Similar but smaller guano-related corrosion morphologies are **guano holes**, created in tropical climate by a cyclic accumulation and removal of fresh bat guano, and its corrosive action on underlying close-to-horizontal rock surfaces (Calaforra et al. 2018) (Figure 9.54b).

Figure 9.54 Forms related to condensation–corrosion and guano: (a) Guano pots (lower dark spots), coated with a phosphate crust, and the feeding protrusions and rising condensation channels above. *Source:* Photo by Marco Vattano, La Venta Esplorazioni Geografiche. (b) Guano holes carved in a block on the floor of a large chamber in Saint Paul's Underground River, Palawan, Philippines. Some guano is left in the holes, but the rest has been washed away by abundant dripping during monsoon rains. *Source:* Photo by José Maria Calaforra, La Venta Esplorazioni Geografiche.

References

Abu-Jaber, N., Hess, J.W., and Howcroft, W. (2001). Chemical erosion of the Lilburn Cave system, Kings Canyon National Park, California. *Ground Water* 39 (2): 223–229.

Alessandri, L., Baiocchi, V., Del Pizzo, S. et al. (2020). A flexible and swift approach for 3D image–based survey in a cave. *Applied Geomatics* 1–15. https://doi.org/10.1007/s12518-020-00309-4.

Aley, T. (1964). Echinoliths–an important solution feature in the stream caves of Jamaica. *Cave Notes* 6 (1): 3–5.

Allred, K. and Allred, C. (1997). Development and morphology of Kazumura Cave, Hawaii. *Journal of Cave and Karst Studies* 59 (2): 67–80.

Anderson, C.H. Jr., Vining, M.R., and Nichols, C.M. (1994a). Evolution of the Paradise/Stevens glacier ice caves. *Journal of Cave and Karst Studies* 56 (2): 70–81.

Antonellini, M., Nannoni, A., Vigna, B. et al. (2019). Structural control on karst water circulation and speleogenesis in a lithological contact zone: The Bossea cave system (Western Alps, Italy). *Geomorphology* 345: 106832.

Audra, P., Bini, A., Gabrovšek, F. et al. (2007a). Cave and karst evolution in the Alps and their relation to paleoclimate and paleotopography. *Acta Carsologica* 36 (1): 53–67.

Auler, A.S., Klimchouk, A.B., Bezerra, F.H.R. et al. (2017). Origin and evolution of Toca da Boa Vista and Toca sa Barriguda Cave System in North-eastern Brazil. In: *Hypogene Karst Regions and Caves of the World* (ed. A.B. Klimchouk, A.N. Palmer, J. De Waele, et al.), 827–840. Cham, Switzerland: Springer.

Auler, A.S., Parker, C.W., Barton, H.A. et al. (2019). Iron formation caves: genesis and ecology. In: *Encyclopedia of Caves* (ed. W.B. White, D.C. Culver and T. Pipan), 559–566. New York: Academic Press.

Badino, G. (2007). *Caves of Sky: A Journey into the Heart of Glaciers*. Treviso: La Venta Esplorazioni Geografiche.

Badino, G. and Piccini, L. (2002). Englacial water fluctuation in moulins: an example from Tyndall Glacier (Patagonia, Chile). *Nimbus* 23-24: 125–129.

Baroň, I. (2002). Speleogenesis along sub-vertical joints: a model of plateau karst shaft development: a case study: the Dolný Vrch Plateau (Slovak Republic). *Cave and Karst Science* 29 (1): 5–12.

Bella, P. and Bosák, P. (2015). Ceiling erosion in caves: early studies and Zdeněk Roth as author of the concept. *Acta Carsologica* 44 (1): 139–144.

Bella, P. and Gaál, L. (2007). Tree mould caves within the framework of cave genetic classification. *Nature Conservation* 63: 7–11.

Bini, A. (1978). Appunti di geomorfologia ipogea: le forme parietali. In: *Atti del V Convegno regionale di Speleologia del Trentino-Alto Adige*, 19–46. Lavis (Trentino, Italy): Gruppo Speleologico SAT Lavis.

Blumberg, P.N. and Curl, R.L. (1974). Experimental and theoretical studies of dissolution roughness. *Journal of Fluid Mechanics* 65 (4): 735–751.

Bögli, A. (1980). *Karst Hydrogeology and Physical Speleology*. Berlin: Springer-Verlag.

Bosák, P., Bruthans, J., Filippi, M. et al. (1999). Karst and caves in salt diapirs, SE Zagros Mts. (Iran). *Acta Carsologica* 28 (2): 41–75.

Breisch, R.L. (1986). Greenhorn Caves - The granddaddy of granite caves. *NSS News* 44 (4): 86–88.

Bretz, J.H. (1942). Vadose and phreatic features of limestone caverns. *Journal of Geology* 50: 675–811.

Bruthans, J., Asadi, N., Filippi, M. et al. (2008). A study of erosion rates on salt diapir surfaces in the Zagros Mountains, SE Iran. *Environmental Geology* 53 (5): 1079–1089.

Bruthans, J., Filippi, M., Zare, M. et al. (2010). Evolution of salt diapir and karst morphology during the last glacial cycle: effects of sea-level oscillation, diapir and regional uplift, and erosion (Persian Gulf, Iran). *Geomorphology* 121 (3-4): 291–304.

Bunnell, D. (1995). Preliminary list of the long (>100m) sea caves of the world. *Geo2* 21: 38–39.

Bunnell, D. (2004). Riko Riko Cave, New Zealand - world's largest sea cave? *NSS News* 65 (5): 145–147.

Calaforra, J.M. (1998). *Karstologia de Yesos. Monografìa Ciencias y Tecnologia*. Almeria: University of Almeria.

Calaforra, J.M. and De Waele, J. (2011). New peculiar cave ceiling forms from Carlsbad Caverns (New Mexico, USA): the zenithal ceiling tube-holes. *Geomorphology* 134 (1-2): 43–48.

Calaforra, J.M., De Waele, J., Forti, P. et al. (2018). The guano holes: a new corrosion form from Natuturingam Cave (Palawan, Philippines). *Travaux de l'Institut Speologia Emil Racovitza* 57: 35–47.

Calapa, K.A., Mulford, M.K., Rieman, T.D. et al. (2021). Hydrologic alteration and enhanced microbial reductive dissolution of Fe (III)(hydr)oxides under flow conditions in Fe (III)-rich rocks: contribution to cave-forming processes. *Frontiers in Microbiology* 12: 1932.

Calvet, M., Gunnell, Y., Braucher, R. et al. (2015). Cave levels as proxies for measuring post-orogenic uplift: evidence from cosmogenic dating of alluvium-filled caves in the French Pyrenees. *Geomorphology* 246: 617–633.

Cascini, G. (1651). *Di S. Rosalia Vergine Palermitana, Libri Tre*. Palermo: Appresso Cirilli.

Cheeptham, N., Sadoway, T., Rule, D. et al. (2013). Cure from the cave: volcanic cave actinomycetes and their potential in drug discovery. *International Journal of Speleology* 42 (1): 35–47.

Coleman, J.C. (1949). An indicator of water-flow in caves. *Proceedings of the University of Bristol Spelaeological Society* 6: 57–67.

Columbu, A., De Waele, J., Forti, P. et al. (2015). Gypsum caves as indicators of climate-driven river incision and aggradation in a rapidly uplifting region. *Geology* 43 (6): 539–542.

Columbu, A., Sauro, F., Lundberg, J. et al. (2018). Palaeoenvironmental changes recorded by speleothems of the southern Alps (Piani Eterni, Belluno, Italy) during four interglacial to glacial climate transitions. *Quaternary Science Reviews* 197: 319–335.

Columbu, A., Chiarini, V., De Waele, J. et al. (2017). Late Quaternary speleogenesis and landscape evolution in the northern Apennine evaporite areas. Earth Surface Processes and Landforms 42(10): 1447–1459.

Columbu, A., Drysdale, R., Hellstrom, J. et al. (2019). U-Th and radiocarbon dating of calcite speleothems from gypsum caves (Emilia Romagna, North Italy). *Quaternary Geochronology* 52: 51–62.

Cooper, M.P. and Covington, M.D. (2020). Modeling cave cross-section evolution including sediment transport and paragenesis. *Earth Surface Processes and Landforms* 45 (11): 2588–2602.

Cooper, M.P. and Mylroie, J.E. (2015). Pseudokarst and non-dissolutional caves. In: *Glaciation and Speleogenesis* (ed. M.P. Cooper and J.E. Mylroie), 49–66. Cham: Springer.

Covington, M.D. (2014). Calcite dissolution under turbulent flow conditions: a remaining conundrum. *Acta Carsologica* 43 (1): 195–202.

Curl, R.L. (1958). A statistical theory of cave entrance evolution. *National Speleological Society Bulletin* 20: 9–22.

Curl, R.L. (1966). Scallops and flutes. *Transactions of the Cave Research Group of Great Britain* 7 (2): 121–160.

Curl, R.L. (1974). Deducing flow velocity in cave conduits from scallops. *National Speleological Society Bulletin* 36 (2): 1–5.

Dandurand, G., Duranthon, F., Jarry, M. et al. (2019). Biogenic corrosion caused by bats in Drotsky's Cave (the Gcwihaba Hills, NW Botswana). *Geomorphology* 327: 284–296.

D'Angeli, I.M., Ghezzi, D., Leuko, S. et al. (2019a). Geomicrobiology of a seawater-influenced active sulfuric acid cave. *PLoS One* 14 (8): e0220706.

D'Angeli, I.M., Nagostinis, M., Carbone, C. et al. (2019b). Sulfuric acid speleogenesis in the Majella Massif (Abruzzo, Central Apennines, Italy). *Geomorphology* 333: 167–179.

Davis, J. and Serefeddin, F. (2004). The Marble Mountains of Northern California. *Alpine Karst* 1: 119.

De Waele, J. and Forti, P. (2006). A new hypogean karst form: the oxidation vent. *Zeitschrift fur Geomorphologie* Supplement Band 147: 107–127.

De Waele, J. and Furlani, S. (2013). *Seawater and biokarst effects on coastal limestones*. In: *Treatise on Geomorphology. Karst Geomorphology*, vol. 6 (ed. A. Frumkin), 341–350. Amsterdam: Elsevier.

De Waele, J., Audra, P., Madonia, G. et al. (2016). Sulfuric acid speleogenesis (SAS) close to the water table: examples from southern France, Austria, and Sicily. *Geomorphology* 253: 452–467.

De Waele, J., Carbone, C., Sanna, L. et al. (2017a). Secondary minerals from salt caves in the Atacama Desert (Chile): a hyperarid and hypersaline environment with potential analogies to the Martian subsurface. *International Journal of Speleology* 46 (1): 51–66.

De Waele, J., Piccini, L., Columbu, A. et al. (2017c). Evaporite karst in Italy: a review. *International Journal of Speleology* 46 (2): 137–168.

De Waele, J., Fabbri, S., Santagata, T. et al. (2018b). Geomorphological and speleogenetical observations using terrestrial laser scanning and 3D photogrammetry in a gypsum cave (Emilia Romagna, N. Italy). *Geomorphology* 319: 47–61.

De Waele, J., Picotti, V., Martina, M.L. et al. (2020). Holocene evolution of halite caves in the Cordillera de la Sal (Central Atacama, Chile) in different climate conditions. *Geomorphology* 370: 107398.

Deike, G.H. and White, W.B. (1969). Sinuosity in limestone solution conduits. *American Journal of Science* 267 (2): 230–241.

Despain, J.D. and Stock, G.M. (2005). Geomorphic history of Crystal Cave, Southern Sierra Nevada, California. *Journal of Cave and Karst Studies* 67 (2): 92–102.

Dogwiler, T.J. (1998). Analysis of bell hole morphology and distribution: a tool for evaluating formational processes. MSc thesis. Starkville: Mississippi State University.

Dondas, A., Isla, F.I., and Carballido, J.L. (2009). Paleocaves exhumed from the Miramar formation (Ensenadan stage-age, Pleistocene), Mar del Plata, Argentina. *Quaternary International* 210 (1-2): 44–50.

Dubois, C., Quinif, Y., Baele, J.M. et al. (2014). The process of ghost-rock karstification and its role in the formation of cave systems. *Earth-Science Reviews* 131: 116–148.

Duszyński, F., Jancewicz, K., and Migoń, P. (2018). Evidence for subsurface origin of boulder caves, roofed slots and boulder-filled canyons (Broumov Highland, Czechia). *International Journal of Speleology* 47 (3): 343–359.

Dzens-Litovsky, A.I. (1966). *Salt Karst of the USSR*. Leningrad: Nedra (in Russian).

Egemeier, S.J. (1981). Cavern development by thermal waters. *National Speleological Society Bulletin* 43: 31–51.

Fabbri, S., Sauro, F., Santagata, T. et al. (2017). High-resolution 3-D mapping using terrestrial laser scanning as a tool for geomorphological and speleogenetical studies in caves: an example from the Lessini mountains (North Italy). *Geomorphology* 280: 16–29.

Ferrarese, F. and Sauro, U. (2005). The Montello hill: the "classical karst" of the conglomerate rocks. *Acta Carsologica* 34 (2): 439–448.

Filippov, A.G. (2004). Siberia, Russia. In: *Encyclopedia of Caves and Karst Science* (ed. J. Gunn), 645–647. New York: Fitzroy Dearborn.

Finch, R. and Pistole, N. (2011). Honduras: caving in conglomerate. *NSS News* 5: 4–9.

Florea, L.J., Vacher, H.L., Donahue, B. et al. (2007). Quaternary cave levels in peninsular Florida. *Quaternary Science Reviews* 26 (9-10): 1344–1361.

Ford, D.C. (1965). Stream potholes as indicators of erosion phases in limestone caves. *Bulletin of the National Speleological Society* 27 (1): 27–32.

Ford, D.C. and Williams, P.W. (2007). *Karst Hydrogeology and Geomorphology*. Chichester, UK: Wiley.

Forti, P. (2005). Genetic processes of cave minerals in volcanic environments: an overview. *Journal of cave and Karst Studies* 67 (1): 3–13.

Franchi, M. and Casadei, A. (1999). II sistema carsico di Monte Caldina. Alta Valle del Fiume Secchia, Reggio Emilia. *Speleologia Emiliana* 10: 19–27.

Frumkin, A. (1994a). Hydrology and denudation rates of halite karst. *Journal of Hydrology* 162: 171–189.

Frumkin, A. (1994b). Morphology and development of salt caves. *National Speleological Society Bulletin* 56: 82–95.

Frumkin, A. (1996). Uplift rate relative to base-levels of a salt diapir (Dead Sea Basin, Israel) as indicated by cave levels. In: *Salt Tectonics*, vol. 100 (ed. G.I. Alsop, D.J. Blundell and I. Davison), 41–47. *Geological Society, London Special Publication*.

Frumkin, A. and Ford, D.C. (1995). Rapid entrenchment of stream profiles in the salt caves of Mount Sedom, Israel. *Earth Surface Processes and Landforms* 20 (2): 139–152.

Frumkin, A., Magaritz, M., Carmi, I. et al. (1991). The Holocene climatic record of the salt caves of Mount Sedom Israel. *The Holocene* 1 (3): 191–200.

Gaál, Ľ., Németh, Z., Bella, P. et al. (2017). Caves in magnesite-the case study from Slovakia. *Mineralia Slovaca* 49: 157–168.

Gabrovšek, F. and Stepišnik, U. (2011). On the formation of collapse dolines: a modelling perspective. *Geomorphology* 134 (1-2): 23–31.

Gabrovšek, F., Häuselmann, P., and Audra, P. (2014). 'Looping caves' versus 'water table caves': the role of base-level changes and recharge variations in cave development. *Geomorphology* 204: 683–691.

Galdenzi, S. and Maruoka, T. (2003). Gypsum deposits in the Frasassi Caves, central Italy. *Journal of Cave and Karst Studies* 65 (2): 111–125.

Gilli, É. (1986). Les grandes cavités souterraines, études et applications. *Karstologia* 7 (1): 2–10.

Gradziński, M., Bella, P., and Holúbek, P. (2018). Constructional caves in freshwater limestone: a review of their origin, classification, significance and global occurrence. *Earth-Science Reviews* 185: 179–201.

Granger, D.E., Fabel, D., and Palmer, A.N. (2001). Pliocene-Pleistocene incision of the Green River, Kentucky, determined from radioactive decay of cosmogenic ^{26}Al and ^{10}Be in Mammoth Cave sediments. *Geological Society of America Bulletin* 113 (7): 825–836.

Gulley, J.D. and Fountain, A.G. (2019). Glacier caves. In: *Encyclopedia of Caves* (ed. W.B. White, D.C. Culver and T. Pipan), 468–473. New York: Academic Press.

Gulley, J.D., Benn, D.I., Screaton, E. et al. (2009). Mechanisms of englacial conduit formation and their implications for subglacial recharge. *Quaternary Science Reviews* 28 (19-20): 1984–1999.

Halliday, W. (2004). Talus caves. In: *Encyclopedia of Karst and Cave Science* (ed. J. Gunn), 721–724. New York: Fitzroy Dearborn.

Häuselmann, P. (2011). UIS mapping grades. *International Journal of Speleology* 40 (2): IV–VI.

Haüselmann, P. and Tognini, P. (2005). Kaltbach cave (Siebenhengste, Switzerland): phantom of the sandstone? *Acta Carsologica* 34 (2): 383–396.

Häuselmann, P., Jeannin, P.Y., and Monbaron, M. (2003). Role of epiphreatic flow and soutirages in conduit morphogenesis: the Bärenschacht example (BE, Switzerland). *Zeitschrift für Geomorphologie* 47 (2): 171–190.

Heeb, B. (2009). An all-in-one electronic cave surveying device. *Cave Radio & Electronics Group Journal* 72: 8–10.

Hoang Kim, Q., Nguyen Thi Thanh, B., Nguyen Minh, P. et al. (2017). Terrestrial 3D laser scanning to Rebuild Son Doong cave. In: *FIG Working Week, Surveying the World of Tomorrow - From Digitalisation to Augmented Reality*, 18. Helsinki, Finland.

Huinink, H.P., Pel, L., and Kopinga, K. (2004). Simulating the growth of tafoni. *Earth Surface Processes and Landforms* 29 (10): 1225–1233.

Jarzyna, A., Bąbel, M., Ługowski, D. et al. (2020). Unique hydration caves and recommended photogrammetric methods for their documentation. *Geoheritage* 12 (1): 27.

Kempe, S. (2012). Volcanic rock caves. In: *Encyclopedia of Caves*, 2e (ed. W.B. White and D.C. Culver), 865–873. Amsterdam: Academic Press/Elsevier.

Kempe, S. (2019). Volcanic rock caves. In: *Encyclopedia of Caves* (ed. W.B. White, D.C. Culver and T. Pipan), 1118–1127. New York: Academic Press.

Kempe, S., Brandt, A., Seeger, M. et al. (1975). "Facetten" and "Laugdecken", the typical morphological elements of caves developed in standing water. *Annales des Spéléologie* 30 (4): 705–708.

Kempe, S., Bauer, I., and Glaser, S. (2017). Hypogene Caves in Germany, geological and geochemical background. In: *Hypogene Karst Regions and Caves of the World* (ed. A.B. Klimchouk, A.N. Palmer, J. De Waele, et al.), 329–347. Cham, Switzerland: Springer.

Klimchouk, A.B. (1996a). The dissolution and conversion of gypsum and anhydrite. *International Journal of Speleology* 25: 21–36.

Klimchouk, A.B. (2000a). Dissolution and conversions of gypsum and anhydrite. In: *Speleogenesis: Evolution of Karst Aquifers* (ed. A.B. Klimchouk, D.C. Ford, A.N. Palmer, et al.), 160–168. Huntsville, USA: National Speleological Society.

Klimchouk, A.B. (2009). Morphogenesis of hypogenic caves. *Geomorphology* 106: 100–117.

Klimchouk, A.B. (2019a). Gypsum caves. In: *Encyclopedia of Caves* (ed. W.B. White, D.C. Culver and T. Pipan), 485–495. New York: Academic Press.

Klimchouk, A.B. (2019c). Ukraine giant gypsum caves. In: *Encyclopedia of Caves* (ed. W.B. White, D.C. Culver and T. Pipan), 1082–1088. New York: Academic Press.

Klimchouk, A., Lowe, D., Cooper, A. et al. (ed.) (1996b). Gypsum karst of the World. *International Journal of Speleology* 25: 1–307.

Klimchouk, A.B., Auler, A.S., Bezerra, F.H. et al. (2016). Hypogenic origin, geologic controls and functional organization of a giant cave system in Precambrian carbonates, Brazil. *Geomorphology* 253: 385–405.

Klimchouk, A., Amelichev, G., Tymokhina, E. et al. (2017a). Hypogene Speleogenesis in the Crimean Piedmont, the Crimea Peninsula. In: *Hypogene Karst Regions and Caves of the World* (ed. A.B. Klimchouk, A.N. Palmer, J. De Waele, et al.), 407–430. Cham, Switzerland: Springer.

Kutleša, P., Malenica, M., Čepelak, M. et al. (2019). Međunarodna speleološka ekspedicija "Mount Sedom 2019"-istraživanje najdulje špilje u soli na svijetu-Malham cave, Izrael. *Subterranea Croatica* 17 (26): 64–71.

Lapaire, F., Becker, D., Christe, R. et al. (2007). Karst phenomena with gas emanations in early Oligocene conglomerates: risks within a highway context (Jura, Switzerland). *Bulletin of Engineering Geology and the Environment* 66 (2): 237–250.

Laureano, F.V., Karmann, I., Granger, D.E. et al. (2016). Two million years of river and cave aggradation in NE Brazil: implications for speleogenesis and landscape evolution. *Geomorphology* 273: 63–77.

Lauritzen, S.E. and Lundberg, J. (2000). Meso-and micromorphology of caves. In: *Speleogenesis Evolution of Karst Aquifers* (ed. A.B. Klimchouk, D.C. Ford, A.N. Palmer, et al.), 408–426. Huntsville: National Speleological Society.

Lauritzen, S.E. and Skoglund, R.Ø. (2013). Glacier ice-contact speleogenesis in Marble stripe karst. In: *Treatise on Geomorphology. Karst Geomorphology*, vol. 6 (ed. A. Frumkin), 363–396. Amsterdam: Elsevier.

Lauritzen, S.E., Abbott, J., Arnesen, R. et al. (1985). Morphology and hydraulics of an active phreatic conduit. *Cave Science* 12 (3): 139–146.

Limbert, H., Limbert, D., Hieu, N. et al. (2016). The discovery and exploration of Hang Son Doong. *Boletín Geológico y Minero* 127 (1): 165–176.

Lipar, M. and Ferk, M. (2011). Eogenetic caves in conglomerate: an example from Udin Boršt, Slovenia. *International Journal of Speleology* 40 (1): 53–64.

Lismonde, B. (2000). Corrosion des coupoles de plafond par les fluctuations de pression de l'air emprisonne. *Karstologia* 35 (1): 39–46.

Lopes, R.P., Frank, H.T., Buchmann, F.S.D.C. et al. (2017). Megaichnus igen. nov.: giant paleoburrows attributed to extinct Cenozoic mammals from South America. *Ichnos* 24 (2): 133–145.

Lucha, P., Cardona, F., Gutiérrez, F. et al. (2008a). Natural and human-induced dissolution and subsidence processes in the salt outcrop of the Cardona Diapir (NE Spain). *Environmental Geology* 53: 1023–1035.

Lugeon, M. (1915). Le striage du lit fluvial. *Annales de Geographie* 24: 385–393.

Lundberg, J. and McFarlane, D.A. (2006b). Speleogenesis of the Mount Elgon elephant caves, Kenya. *Geological Society of America Special Papers* 404: 51–63.

Lundberg, J. and McFarlane, D.A. (2009). Bats and bell holes: the microclimatic impact of bat roosting, using a case study from Runaway Bay Caves, Jamaica. *Geomorphology* 106 (1-2): 78–85.

Lundquist, C.A. and Varnedoe, W.W. Jr. (2006). Salt ingestion caves. *International Journal of Speleology* 35 (1): 13–18.

McDonald, R.C. and Twidale, C.R. (2011). On the origin and significance of basal notches or footcaves in karst terrains. *Physical Geography* 32 (3): 195–216.

Mylroie, J.E. and Carew, J.L. (1988). Solution conduits as indicators of late Quaternary sea level position. *Quaternary Science Reviews* 7 (1): 55–64.

Northup, D.E., Melim, L.A., Spilde, M.N. et al. (2011). Lava cave microbial communities within mats and secondary mineral deposits: implications for life detection on other planets. *Astrobiology* 11 (7): 601–618.

Oberender, P. and Plan, L. (2015). Cave development by frost weathering. *Geomorphology* 229: 73–84.

Oberender, P. and Plan, L. (2018). A genetic classification of caves and its application in eastern Austria. *Geological Society of London*, Special Publications 466 (1): 121–136.

Oludare, I.M. and Pradhan, B. (2016). A decade of modern cave surveying with terrestrial laser scanning: a review of sensors, method and application development. *International Journal of Speleology* 45 (1): 71–88.

Osborne, R.A.L. (2004). The troubles with cupolas. *Acta Carsologica* 33 (2): 9–36.

Palmer, A.N. (1975). The origin of maze caves. *National Speleological Society Bulletin* 37: 56–76.

Palmer, A.N. (1987). Cave levels and their interpretation. *National Speleological Society Bulletin* 49 (2): 50–66.

Palmer, A.N. (2007). *Cave Geology*. Dayton, Ohio: Cave Books.

Parker, C.W., Wolf, J.A., Auler, A.S. et al. (2013). Microbial reducibility of Fe (III) phases associated with the genesis of iron ore caves in the Iron Quadrangle, Minas Gerais, Brazil. *Minerals* 3 (4): 395–411.

Parker, C.W., Auler, A.S., Barton, M.D. et al. (2018). Fe (III) reducing microorganisms from iron ore caves demonstrate fermentative Fe (III) reduction and promote cave formation. *Geomicrobiology Journal* 35 (4): 311–322.

Pasini, G. (1967). Osservazioni sui canali di volta delle grotte bolognesi. *Le Grotte d'Italia* 4 (1): 17–74.

Peng, J., Sun, P., and Igwe, O. (2018). Loess caves, a special kind of geo-hazard on loess plateau, northwestern China. *Engineering Geology* 236: 79–88.

Piccini, L. (2011). Speleogenesis in highly geodynamic contexts: the Quaternary evolution of Monte Corchia multi-level karst system (Alpi Apuane, Italy). *Geomorphology* 134 (1-2): 49–61.

Pisani, L., Antonellini, M., and De Waele, J. (2019). Structural control on epigenic gypsum caves: evidences from Messinian evaporites (Northern Apennines, Italy). *Geomorphology* 332: 170–186.

Plan, L., Tschegg, C., De Waele, J. et al. (2012). Corrosion morphology and cave wall alteration in an Alpine sulfuric acid cave (Kraushöhle, Austria). *Geomorphology* 169: 45–54.

Renault, P. (1968). Contribution à l'étude des actions mécaniques et sédimentologiques dans la spéleogenèse. *Annales de Spéléologie* 22: 5–21. & 209-267; 23: 259-307 & 529-596; 24: 313-337.

Riquelme, C., Dapkevicius, M.D.L.E., Miller, A.Z. et al. (2017). Biotechnological potential of Actinobacteria from Canadian and Azorean volcanic caves. *Applied Microbiology and Biotechnology* 101 (2): 843–857.

Sauro, F., Zampieri, D., and Filipponi, M. (2013b). Development of a deep karst system within a transpressional structure of the Dolomites in north-east Italy. *Geomorphology* 184: 51–63.

Sauro, F., Mecchia, M., Tringham, M. et al. (2020). Speleogenesis of the world's longest cave in hybrid arenites (Krem Puri, Meghalaya, India). *Geomorphology* 359: 107160.

Sauro, F., Mecchia, M., Piccini, L. et al. (2019). Genesis of giant sinkholes and caves in the quartz sandstone of Sarisariñama tepui, Venezuela. Geomorphology 342: 223–238.

Shtober-Zisu, N., Amasha, H., and Frumkin, A. (2015). Inland notches: implications for subaerial formation of karstic landforms - an example from the carbonate slopes of Mt. Carmel, Israel. *Geomorphology* 229: 85–99.

Simms, M.J. (2002). The origin of enigmatic, tubular, lake-shore karren: A mechanism for rapid dissolution of limestone in carbonate-saturated waters. *Physical Geography 23* (1): 1–20.

Slabe, T. (1995). *Cave Rocky Relief and its Speleogenetical Significance*, Zbirka ZRC 10. Ljubljana: Znanstvenoraziskovalni Center SAZU.

Smart, C.C. and Brown, M.C. (1981). Some results and limitations in the application of hydraulic geometry to vadose stream passages. In: *Proceedings of the 8th International Congress of Speleology* (ed. B.F. Beck), 724–725. Bowling Green: National Speleological Society.

Spallanzani, L. (1792). *Viaggi alle Due Sicilie ed in alcune parti dell'Appennino*, vol. *2*, 202–206. Pavia: Stamperia Baldassare Comini.

~~Springer, G.S. and Hall, A. (2020). Uncertainties associated with the use of erosional cave scallop~~ lengths to calculate stream discharges. *International Journal of Speleology* 49 (1): 27–34.

Tarhule-Lips, R.F. and Ford, D.C. (1998a). Condensation corrosion in caves on Cayman Brac and Isla de Mona. *Journal of Cave and Karst Studies* 60: 84–95.

Tarhule-Lips, R.F. and Ford, D.C. (1998b). Morphometric studies of bell hole development on Cayman Brac. *Cave and Karst Science* 25 (3): 119–130.

Tobin, B. and Doctor, D. (2009). Estimating karst conduit length using conductivity and discharge measurements in Lilburn Cave, Kings Canyon National Park, California. In: *Proceedings of the 15th International Congress of Speleology*, vol. 3 (ed. W.B. White), 1702–1706. Kerrville USA: National Speleological Society.

Todd, R.L. (1993). *Mendelssohn: The Hebrides and other overtures*. London: Cambridge University Press.

Trimmis, K.P. (2018). Paperless mapping and cave archaeology: A review on the application of DistoX survey method in archaeological cave sites. *Journal of Archaeological Science: Reports* 18: 399–407.

Twidale, C.R. and Vidal Romaní, J.R. (2005). *Landforms and Geology of Granite Terrains*. Leiden: Balkema.

Ulloa, A., Gázquez, F., Sanz-Arranz, A. et al. (2018). Extremely high diversity of sulfate minerals in caves of the Irazú Volcano (Costa Rica) related to crater lake and fumarolic activity. *International Journal of Speleology* 47 (2): 229–246.

White, W.B. (1988). *Geomorphology and Hydrology of Karst Terrains.* New York: Oxford University Press.

White, W.B. (2000b). Speleogenesis of vertical shafts in the Eastern United States. In: *Speleogenesis Evolution of Karst Aquifers* (ed. A.B. Klimchouk, D.C. Ford, A.N. Palmer, et al.), 378–381. Huntsville: National Speleological Society.

White, W.B. and Culver, D.C. (2019). Cave, definition of. In: *Encyclopedia of Caves* (ed. W.B. White, D.C. Culver and T. Pipan), 255–259. New York: Academic Press.

White, E.L. and White, W.B. (1969). Processes of cavern breakdown. *National Speleological Society Bulletin* 31 (4): 83–96.

White, E.L. and White, W.B. (2000). Breakdown morphology. In: *Speleogenesis. Evolution of Karst Aquifers* (ed. A.B. Klimchouk, D.C. Ford, A.N. Palmer, et al.), 427–429. Huntsville: National Speleological Society.

Worthington, S.R.H. (2004). Hydraulic and geological factors influencing conduit flow depth. *Cave and Karst Science* 31 (3): 123–134.

Wray, R.A.L. and Sauro, F. (2017). An updated global review of solutional weathering processes and forms in quartz sandstones and quartzites. *Earth-Science Reviews* 171: 520–557.

Xuewen, Z. and Waltham, T. (2005). Tiankeng: definition and description. *Cave and Karst Science* 32 (2/3): 75–79.

Zhu, X. (1988). *Guilin Karst.* Shanghai: Shanghai Scientific & Technical Publishers.

Zupan Hajna, N. (2003). *Incomplete Solution: Weathering of Cave Walls and the Production, Transport and Deposition of Carbonate Fines.* Ljubljana: Založba ZRC.

10

Cave Deposits

10.1   Introduction

During their entire life, caves are favorable environments for the flux, deposition, and preservation of sediments. Because of their relative isolation and high diversity of sheltered environments, they contain a wide spectrum of physical, organic, and chemical deposits that can be preserved over very long periods of time. Among the continental sedimentary environments, caves are certainly the most diverse and long-living sediment traps. Whereas surficial continental deposits are continuously exposed to weathering and erosional agents (e.g., running water, wind, glacial ice, freezing and thawing, animal, and plant activity) that can cause their removal or disruption, in certain underground voids the conditions can be extremely stable, increasing the preservation potential of primary depositional features. It is not a surprise that some of the most exceptional Middle Pliocene-Lower Pleistocene hominid discoveries have occurred in caves; South Africa (Dirks and Berger 2013; Berger et al. 2015) (Figure 10.1a), and Atapuerca, northern Spain (Pérez-González et al. 2001; Rodríguez et al. 2011) (Figure 10.1b), just to mention a few. The pristine spelean sedimentary archives, which can often be dated, can give valuable clues to reconstruct past environmental and climatic conditions during their deposition. Whereas detrital sediments are more difficult to interpret, because of their intrinsic nature of being more easily eroded, removed and reworked, chemical deposits (i.e., speleothems) (Fairchild and Baker 2012), and cave ice (Luetscher 2005) are among the most important continental paleoenvironmental and paleoclimate archives.

In this chapter, we will only deal with the physical, organic, and chemical deposits found in the dark or semi-dark areas of caves, excluding those found in cave entrances or rock shelters. The latter are often of great archeological interest, but are much more disturbed, similar to other deposits formed in external sedimentary environments, and usually do not share the typical characteristics of most cave interior deposits. It must however be borne in mind that also cave sediments are generally not pristine, and can be extremely complex to interpret. Episodes of erosion, transport, and deposition can occur several times, and multiple processes can modify the original deposits (slumping, burrowing, diagenetic cementation, dissolution). The pathways through which water and sediment enter into the caves (e.g., solutionally enlarged joints), as well as their interior morphology, act as filters that selectively winnow the fines and block larger particles. Detrital sediment can be sealed with flowstone or cohesive clay, increasing its resistance to erosion and enabling its persistence over longer time frames. The sedimentary record in caves is often fragmentary (Sasowsky 2007), with numerous hiatuses and complex geometrical relationships; deposits of different ages often lying side by side, and even older deposits overlying younger ones

Karst Hydrogeology, Geomorphology and Caves, First Edition. Jo De Waele and Francisco Gutiérrez.
© 2022 John Wiley & Sons Ltd. Published 2022 by John Wiley & Sons Ltd.

Figure 10.1 Important cave deposits containing hominid fossils: (a) The Member 2 breccia in which Little Foot (inset) was found in Silberberg Grotte (Sterkfontein, South Africa). The white flowstones, dated with U/Pb at 2.2 Myr, appear to have precipitated later in the breccia member, and the fossil (and breccia) are older than 3 Myr. The fossil appears to belong to an *Australopthecus* species, and with 90% of bone recovery is the most complete early hominid skeleton found to date. *Source:* Photo by Laurent Bruxelles. (b) Image of the Gran Dolina excavations in detrital cave deposits exposed in the walls of an old railway trench in the Atapuerca archeological and paleontological site, Burgos, northern Spain (UNESCO World Heritage Site). The site has yielded remarkable remains of hominids (e.g., *Homo* sp., *Homo antecessor*) as old as 1.2 Myr. Note fining-upward succession with syndepositional dip accumulated in a debris cone within the cave, mainly by sediment-gravity flow. *Source:* Francisco Gutiérrez (Author).

(e.g., undermining and filling). Studying cave sediments is like reading a book with lots of missing pages, but some pages may allow the reader to unravel the big picture.

10.2 Classification of Cave Sediments

Cave interior sediments refer to all types of natural material accumulated in caves, including clastic, organic, and chemical deposits, plus cave ice (Bögli 1980). White (1988) mainly divides cave sediments into clastic and chemical, distinguishing autochthonous (weathering, breakdown, and organic debris) and allochthonous (transported) materials within the clastic group, and travertines, evaporites, phosphates, nitrates, resistates, and ice within the chemical sediments. Gillieson (1996) uses a classification similar to the one of Bögli (1980), but considers bat and bird guano to be autogenic organic deposits. In a certain sense, this is true, since guano is produced within the caves, but chemically these organic materials come from allogenic sources. Ford and Williams (2007) distinguish 23 types of cave interior deposits, using a double classification based on the source (allogenic, autogenic, or precipitated), and the type of deposit (clastic, organic, and mineral). These authors include the organic deposits correctly under the allogenic source types, since vegetal debris or the excreta of animals have an origin external to the cave. Ice is placed within the precipitates, together with speleothems and cave minerals. In this book, we use a similar approach although with some slight changes (Table 10.1).

Physical deposits refer to material that is physically entrained into the cave environment (clastic sediments) or is deposited by physical processes acting within the cave (e.g., breakdown, water currents, and air flow). Chemical deposits, instead, are the result of chemical processes acting in the cave environment, causing the precipitation/crystallization of a suite of different minerals, including ice. They are also related to the physical transportation and sedimentation of chemical sediments from the exterior of the cave. Organic sediments refer to biological material, mainly of vegetal and animal

Table 10.1 Classification of deposits in caves.

	Allogenic (Allochthonous)	Autogenic (Autochthonous)
Physical (Clastic and ice)	Gravitational filtrates Fluvial Lacustrine Marine Aeolian Glacial injecta Fluvioglacial injecta Dejecta-colluvium-debris flow Wind- or waterborne tephra Glacial ice intrusion Snow (glacière) and firn Anthropic waste	Breakdown Fluvial Weathering (speleosols) Aeolian
Organic	Gravitational vegetal debris Gravitational animal remains Wind- or waterborne organics Guano	Microbial mats Cave fauna remains
Chemical	Tufa and travertine fragments Iron nodules Phosphorites	Carbonate speleothems Sulfates Phosphates Nitrates Halides Silica and silicates Oxides and hydroxides Ice speleothems Frozen water bodies Hoarfrost Ground ice Extrusive ice

origin, but also microbial and fungal matter produced in, or transported into the caves. A distinction is made between material that is formed inside the cave (autogenic or autochthonous sediments), and that is formed in other places and transported into the cave (allogenic or allochthonous). The first type is characterized by minerals and chemical signatures related to the cave and karst environments, whereas the latter can contain exotic components and display a much wider geochemical spectrum.

10.3 Clastic Sediments

Clastic sediments in caves, excluding those found at entrances, have been the subject of numerous studies, mainly during the last 60 years. Pioneering studies include Schmid (1958), Davies and Chao (1959), and the fine review by Kukla and Ložek (1958). Excellent early contributions are those of White and White (1968) and the extensive work of Renault (1968). A good overview on clastic sediments in caves is given in Gillieson (1996), and a comprehensive volume on cave sediments was edited by Sasowsky and Mylroie (2004).

10.3.1 The Cave Sedimentary System

The composition, texture and, structure of a clastic sediment in a cave mainly depend on its origin, transportation history, and the process(es) that deposited them (e.g., hydraulics of water flow). The sedimentary environments in caves are highly dynamic, in a certain sense similar to fluvial systems with channels and their floodplains (Miall 1985), although with substantial differences related to the fact that a cave river flows in a closed channel with a potentially unstable roof (conduit): (i) Cave passages usually have greater roughness than bedrock channels at the surface, and they generally lack floodplains. A number of factors can cause dramatic changes in flow velocities in cave passages such as morphological changes (constrictions, bends, roughness, etc.) or deposition. These changes have important effects also on water levels, resulting in large local variations in hydraulic gradient over relatively short distances along a cave passage. This causes sedimentary facies to have greater lateral and longitudinal variations than those deposited in surface channels. (ii) Significant recharge and the resulting flow events can partially or even completely remove earlier sediments. Some parts of the previously deposited sediments are more easily eroded, leading to a selective transport and sorting of the particles in successive recharge pulses. This means that detrital cave sediments are not necessarily deposited by a single event and statically preserved over long periods. Instead, they tend to move through the cave system in multiple cycles of erosion, transportation, and deposition (reworking). Throughout these cycles the sediment tends to decrease in volume and increase its sorting, and part of it can eventually emerge at the spring.

10.3.2 The Origin and Flux of Clastic Sediments in Caves

Cave systems receive clastic material from various sources, brought into the cave by different agents, mainly by running waters, but also by simple gravitational movement, wind, or glacial intrusion. A typical karst system is schematically shown in Figure 10.2, based on an early model of White and White (1968) and its modified version in Bosch and White (2004), where both autogenic and allogenic recharge are contemplated. In such conditions, the sediments can come from the following sources: (i) Allogenic streams. The type and quantity of sediment load that will be carried into the karst system will depend on the area of the catchment and its geological, geomorphological, climatic, and hydrological characteristics. The rock types exposed in the surface drainage

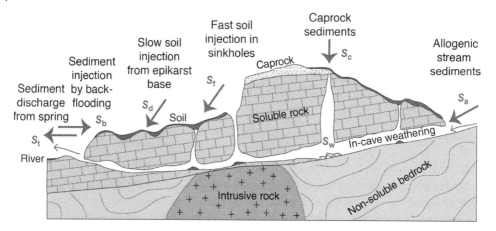

Figure 10.2 A typical fluviokarst system with the different types of allogenic and autogenic sediment inputs.

basin can be highly variable (metamorphic, intrusive, volcanic, or sedimentary). Glacial deposits can be reworked and can be brought into the cave systems, containing exotic clasts derived from areas located at large distances. Allogenic detrital particles can be weathered to a variable degree, depending on the mineralogical composition of the original rocks, climate, and time. Coarse sediments can be entrained more easily into the caves in areas of high topographic gradient with steep allogenic streams, whereas in low relief landscapes the sediments are mostly fine-grained (silts and clays). If the river flow entirely penetrates into the karst system, and no surface overflow is present, the totality of its sediment flux is carried into the underground drainage system. (ii) Autogenic recharge by storm runoff and fast soil injection from a cover (e.g., bottom of dolines) carrying locally derived surficial material into the subsurface. If the karst area is covered by allochthonous deposits such as till, loess or volcanic ash, these materials will also be introduced into the cave systems. Injection into the underlying caves can sometimes occur instantaneously if there is a direct connection between the soil cover and the underlying karst, but often requires multiple storm events to flush the sediments through the epikarst. (iii) Slow soil injection through the epikarst, mainly by percolation water that follows a network of interconnected discontinuities enlarged by dissolution. The material flushed downward is similar to the previous one, and the aperture of the fractures determines the degree of filtering of this sediment flux. (iv) Injection of caprock sediments through vertical fractures and open shafts, often without participation of fluvial processes. Caprock can simply collapse into underground voids. (v) Mobilization of in-cave weathering residue dominated by insoluble materials. These are generally clays and silt, but also quartz sands, silicified fossils, and chert nodules. The quantity of this material depends on the original purity of the soluble rock. (vi) Sediment injection by backflooding. For example, during flood events in large base level rivers, sediment-laden floodwaters may reach a high stage and penetrate into adjacent perched caves depositing large amounts of detrital material. Factors such as the frequency of floods, flood stage and duration, flow velocity, and sediment load control how much material is carried inside the cave system and how far it reaches. Depending on the source of the sediment, this input could be considered as autogenic and/or allogenic.

These different sediment fluxes all contribute to the overall sediment budget of a karst system, which can be expressed by the following expression:

$$S_a + S_f + S_d + S_c + S_w + S_b \pm S_s = S_t \tag{10.1}$$

where, S_a is the allogenic sediment input, S_f is the autogenic sediment recharge by storm runoff, S_d is the slow soil injection through the epikarst, S_c is the caprock-derived input, S_w is the in-cave weathering residue, S_b is the sediment introduced by backflooding, and S_t is the total sediment discharged at the karst spring(s). S_s is the sediment stored in the karst system, with the $\pm$ sign indicating that it can increase or decrease. S_s can experience substantial changes through time, with periods (or events) of net sediment storage and periods of net sediment removal. Over long geological time scales, S_s, which is the net balance between sediment input and output should be close to zero or negative, otherwise the cave system will tend to be entirely filled up with sediments, and water flow will divert to other surface or subsurface routes. It should be borne in mind that the equation does not incorporate removal of rock mass from the cave by dissolution and solute transport, which results in the creation of accommodation space for sediment storage. In carbonate areas, cave formation is commonly much slower than deposition and/or erosion of cave sediments.

10.3.3 Fluvial Erosion and Transport

Transport of clastic sediment through a cave is a complex process, comparable with the movement of particles in a pipe filled with water, or a half-pipe in which water flows in contact with air, depending on whether the flow is phreatic or vadose. This simplified model is complicated by the much higher irregularity of the cave conduit and the variable nature of its perimeter, either rock or unconsolidated deposits. To be able to initiate the movement of a particle, its entrainment threshold (critical shear stress) needs to be exceeded. This threshold mainly depends on the fluid density, which is temperature dependent, the square of flow velocity, and the friction factor that depends on a series of variables such as particle density, shape, orientation, packing, slope, among others. The critical flow velocities needed to entrain particles increases with grain size (large cobbles and boulders are not easy to transport), but relatively large velocities are also needed for smaller particles such as silts and clays, because they tend to have smoother surfaces, creating less turbulence, and have greater cohesive forces that hold the grains attached together. The velocity required to keep a particle moving in a stream channel is lower than that required to entrain it from the more or less cohesive bed. In common cave streams, flow velocities are capable of keeping fine-to-medium sand, silt and clay particles suspended, so, once eroded, they can travel over long distances. Sediments with larger grain sizes are rarely eroded and transported along significant portions of a cave stream. Deposition of suspended particles starts when the stream flow velocity falls below a threshold value. Then, the settling velocity v_t depends on the density difference between the particle ρ_s and the fluid ρ, grain size (the square of its diameter, D^2), and the dynamic viscosity μ (the lower this value is, the faster the particle will settle). This is given by the Stokes' Law:

$$v_t = \frac{1}{18}\frac{(\rho_s - \rho)}{\mu}gD^2 \tag{10.2}$$

The empirical diagram of Hjulström (1935) shows flow velocities needed to erode, transport, and deposit particles of various sizes, and can be used as a rough approximation, since the actual velocities depend on multiple factors in addition to particle size (Figure 10.3). At flow velocities above the red curve the bed material is eroded or entrained, below the blue line particles settle down. The area between the two curves represents flow velocities at which particles are transported, but theoretically cannot be eroded or deposited. This diagram is based on measurements carried out in open channels less than one meter deep. The red curve in Figure 10.3 shows that

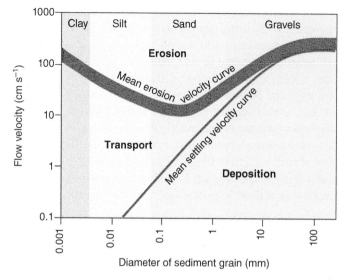

Figure 10.3 Empirical relationship between particle size and stream velocity for the erosion, transport, and deposition of particles. *Source:* After Hjulström (1935).

fine to medium sand is more easily entrained and transported than clay. Fine to medium sands are generally eroded at flow velocities of around $20\,\text{cm s}^{-1}$ and can be transported at velocities 10 times smaller.

Once entrained, the particles can be transported in the water current in different ways, mainly depending on the grain size, flow velocity, and the roughness of the channel bed. These transport modes include rolling, sliding, saltation, and suspension (Figure 10.4). The part of the sediment load moving along the bottom of the channel and most of the time in contact with the floor is known as bedload (rolling, sliding, and saltation). The finer and less dense particles can be transported as suspended load. At low flow velocities, only some small grains may be moved as bedload, with finer particles in suspension, whereas during very high flow most of the stream load can be transported as suspended load, in rare occasions even stripping away all the unconsolidated deposits down to the underlying bedrock (stable bed in Figure 10.4). The boundary between the bedload and suspended load zones is vague. Moreover, cave streams typically experience rapid changes

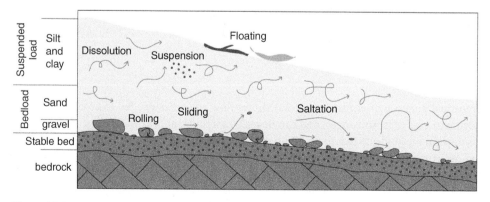

Figure 10.4 Bedload and suspended load in a detrital fluvial system.

in flow velocity and water depth along the channel, related to variations in the geometry of the passages (e.g., narrowings and widenings). The different transport processes, and the corresponding deposition mechanisms lead to distinctive textures and sedimentary structures.

10.3.4 The Cave Sedimentary Facies

The different inputs of clastic sediment (allogenic, autogenic) are to various extents mixed within the cave system. Some sediments (autochthonous) are deposited in situ, without any transport, but most are to some degree displaced from their original position. The particles that are transported through the cave system under various hydrological conditions eventually settle and form a detrital deposit, which can become stabilized, or can be entrained again during a subsequent event, to settle once again in a site further downstream. As mentioned above, there is frequent sediment reworking and recycling in active caves, but eventually some sediments leave the underground environment flowing out at the springs, whereas some of them can remain as stable cave deposits. These clastic cave deposits, which can record valuable information on the past environmental and climatic conditions, can be classified on the basis of their texture (e.g., grain size, sorting), the sedimentary structures, and architecture, which record the depositional processes and their spatio-temporal variations. The widely accepted classification of lithofacies and architectural elements proposed by Miall (1985) for fluvial environments can be used for the analysis of a great proportion of the detrital water-laid deposits found in caves (Figure 10.5). Other clastic deposits such as those generated by gravitational processes (e.g., breakdown, sediment-gravity flow, and slope wash) require specific analyses considering their geomorphic setting within the cave.

Based on the transport mechanisms, five main groups of sedimentary facies can be differentiated (Bosch and White 2004): channel, thalweg, slackwater, regolith, and diamicton. These facies can be recognized because of their different grain size distributions, the degree of sorting, and the sedimentary structures. The diagram in Figure 10.6 schematically illustrates the dominant textural characteristics of the different groups of facies. They are displayed as separated fields, but can overlap and show continuous gradations in nature. For example, if the sediment source area does not contain gravels, the thalweg facies may consist of sands similar to those of the channel facies.

Thalweg facies refer to the generally coarser facies that accumulate in the deepest part of channels (e.g., channel lag). These are coarse particles that may include rip-up clasts transported as bedload (gravels, boulders) in channels where the flow velocity is fast enough to winnow the finer particles (sand and silt) (Figure 10.7a). The channel facies are the most typical stream sediments in caves, commonly composed of stratified and well-sorted sands, and gravels that may display a wide range of fabrics (e.g., imbrication) and sedimentary structures (Figure 10.7b,c). Most of this material is transported as bedload, with only minor amounts as suspended load. The textural characteristics of the deposits, the sedimentary structures (e.g., horizontal bedding, planar, and trough cross bedding) and the architectural elements in which they occur reflect the variable flow conditions and the setting in which they were deposited (Miall 1985) (Figures 10.5 and 10.7d). For instance, lateral accretion bars accumulate in the inner side of migrating meanders. These deposits typically show limited lateral continuity, complex geometrical relationships, and sharp facies changes. This is due to the fact that flow velocity in cave streams varies greatly through time and over short horizontal distances. The slackwater facies are well-sorted fine sediments that accumulate from suspension at sites of reduced flow velocities (Springer and Kite 1997) (Figure 10.7e). These sediments often accumulate during the peak flood, when water stage is highest and velocities are very low at specific sites. Slackwater deposits can be used as markers of paleoflood stage. The well-sorted fine-grained sediments that accumulate during the recession of flow events commonly form a

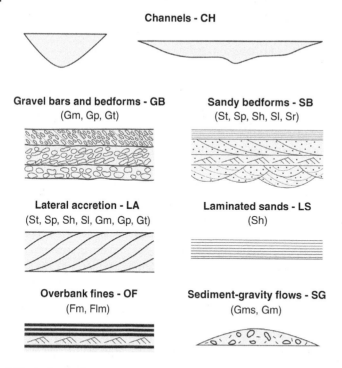

Figure 10.5 Architectural elements and the associated lithofacies proposed by Miall (1985) for the three-dimensional sedimentological analysis of fluvial deposits. This scheme can be applied to detrital sediments accumulated by cave rivers. The lithofacies codes indicate the dominant grain size (upper case) and the internal sedimentary structure (lower case). *Source*: Modified from Miall (1985).

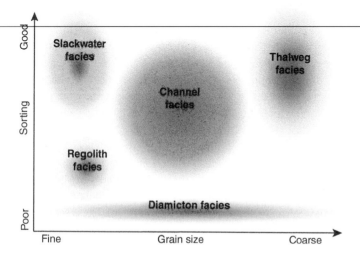

Figure 10.6 The five most common groups of clastic sedimentary facies deposited in caves, based on the degree of sorting and particle size. *Source:* Modified from Bosch and White (2004).

Figure 10.7 Detrital sedimentary facies of cave interior deposits: (a) Well-sorted rounded cobbles of the thalweg facies washed by the river, at the base of Gaping Gill Cave, North Yorkshire, UK. *Source:* Photo by Philippe Crochet. (b) Sands in the entrance passages of Krem Phlang Karun, Meghalaya, India. *Source:* Photo by Mark Burkey. (c) Channel facies exposed in an erosional bank with sandy beds overlain by a clayey layer on top, Mammuthöhle, Austria. *Source:* Photo by Lukas Plan. (d) Channel facies showing, from base to top, cobbles (dark bed), silts (brown layer), and light-colored cross-bedded sands. *Source:* Photo by Jean-Yves Bigot. (e) Perched sloping surfaces underlain by muddy slackwater facies in Stake Pot, Ease Gill Cave system, Cumbria, UK. *Source:* Photo by Mark Burkey. (f) Regolith deposits (red residual clays) draping the floor and boulders in Aven des Neiges, Vaucluse, France. *Source:* Photo by Jean-Yves Bigot. (g) Old indurated and unsorted diamicton in Mammoth Cave, Jenolan Karst Conservation Reserve, western Australia. *Source:* Photo by Csaba Egri.

blanket covering previous high-energy coarser facies. These relatively erosion-resistant clay "coatings" record the final phase of flood events.

The regolith facies is the only autogenic clastic sedimentary facies, composed of the insoluble material that is produced in place by the dissolution of the host rock (Figure 10.7f), and the infiltrates derived from the epikarst and mantling soils, both with little or no lateral transport. They can accumulate only in settings in which water hardly ever flows, or moves at very low velocities (e.g., phreatic maze caves). Although these sediments are generally characterized by fine-grained texture, they differ from the slackwater deposits because of their very poor sorting, and the lack of clear layering. In fact, weathering of the host rock generally releases insoluble clays, but also sand-sized particles, chert fragments, and recrystallized or silicified fossils. In highly impure limestones, this process can produce large quantities of sediment in caves. Diamictons are chaotic, unsorted, and non-stratified deposits generated by sediment-gravity flows, mostly debris flows. These are water-saturated masses of unconsolidated material, including large and fine particles, that flow downslope as a viscous and high-density non-Newtonian fluid (Gillieson 1986) (Figure 10.7g). They typically display matrix-supported texture, reverse grading, and chaotic fabrics.

10.3.5 Other Allogenic Clastic Sediments

Although rivers and running water are the main transporting agents in caves, there are other processes that can introduce allogenic material in caves. These include marine (or lacustrine) sediments, sometimes penetrating into caves with large entrances by wave action. Whereas lacustrine deposits are generally fine-grained, the marine facies range in size from fine sands to boulders, occasionally forming internal beaches (Figure 10.8a). Some typical coastal depositional landforms, such as normal beach berms and storm berms, can be present. The characteristics of these sediments depend on the geology and morphology of the coast and the wave and tidal regimes. They normally contain organic material (e.g., shells, driftwood) and, unfortunately, all kinds of resistant wastes (plastic). It is also common to find aeolian deposits in the entrance parts of caves, and especially in present or former coastal caves, where these sands are often composed of quartz- and feldspar-rich sands. These sediments can be used for optically stimulated luminescence (OSL) dating, which allows the estimation of the age of deposits and archeological sites that are beyond the C^{14} dating techniques (Jacobs et al. 2003). Such type of studies carried out on sands in Bue Marino Cave have allowed the reconstruction of Late Quaternary environments in the central-eastern coast of Sardinia (Andreucci et al. 2017) (Figure 10.8b).

Volcanic ash (tephra) can also be brought deep into caves by air currents, although larger volumes are washed in by rivers, infiltration waters, and waves (Bruins et al. 2019). These allogenic materials, containing minerals rich in potassium, are very suitable for K-Ar dating, and where such layers are found in association with archeological sequences, they give precious and very precise chronological constraints, especially in the Mediterranean region, where ashes from the Campanian super-eruption (ca. 40 kyr) (Figure 10.8c) and Santorini eruption (ca. 3.6 kyr) provide well-established bracketing ages. The introduction of volcanic material into caves also enriches their geochemical environment. This explains why the salt caves in Atacama, where volcanic ashes surround the salt outcrops, contain five times more mineral species than the salt caves in Iran or Israel (De Waele et al. 2017a).

Another way of incorporating allogenic clastic sediments into caves is by active glaciers. Glacial ice can penetrate into large cave openings, bulldozing chaotic and unsorted debris into the cave (Figure 10.8d). Normally, these glacial deposits tend to clog the entrance, and the wedged boulders act as filters for meltwaters, letting only smaller and relatively sorted particles to get through.

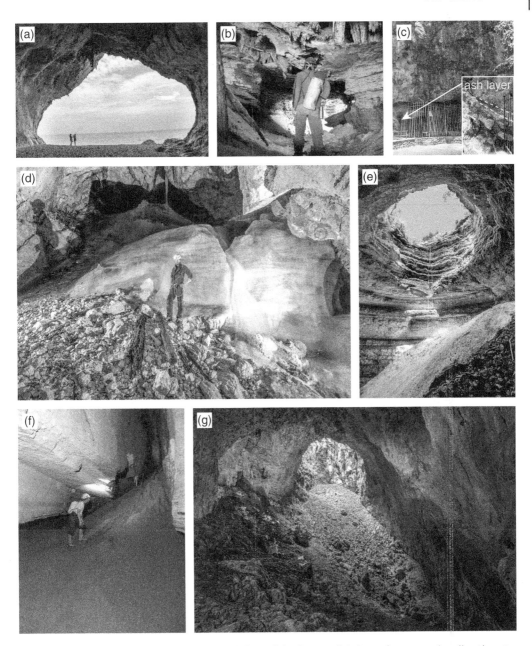

Figure 10.8 Other allogenic cave deposits: (a) One of the famous Cala Luna Caves, opening directly onto a marvelous beach in Sardinia. *Source:* Photo by Csaba Egri. (b) Aeolian sands in the entrance area of Bue Marino Cave, Sardinia. *Source:* Photo by Laura Sanna. (c) Volcanic ash layer of the Campanian eruption (ca. 40 kyr) at the entrance of Castelcivita Cave, central Italy. Inset shows a detail of the ash layer. *Source:* Photos by Antonio Santo. (d) Cave ice and wooden debris in Ledena Jama, Slovenia. *Source:* Photo by Peter Gedei. (e) Perennial snow and ice accumulations at the bottom of the collapse sinkhole and shaft of Buso del Vallon, Lessini Mountains, northern Italy (Note caver descending the rope on top of the shaft for scale). *Source:* Photo by Sandro Sedran, S-Team. (f) Terra rossa infiltrates from the epikarst forms a cone of reddish clayey deposit below the feeding fissure, Gruta do Ioiô, Bahia, Brazil. *Source:* Jo De Waele (Author). (g) Talus cone at the entrance of Golokratna Jama, Slovenia (see person for scale). *Source:* Photo by Borut Lozej and Iztok Cencič.

Glacial ice likely will never penetrate deep into caves, whereas the action of meltwaters can be much more important, but referable to allogenic fluvial input, as described in the preceding chapters. Snow and firn can fall into open shafts, introducing fine particles into the cave environment (snowflakes are often formed by crystallization around a fine airborne particle, such as pollen, volcanic ash or desert and anthropogenic dust) (Figure 10.8e). These in-cave snow deposits are also known under the French name *glacière*.

The gravitational movement of particles through tiny fissures and cracks is another way whereby allogenic surface material can get into caves. The network of fissures and openings operates as a filter, so mostly fine soil-derived material penetrates deep into the caves, with limited or no chemical alteration. *Terra rossa* infiltrates can frequently be found in caves, forming sometimes cone-shaped deposits at the base of the feeding fissures (Figure 10.8f). If the openings are large (open shafts), they can function as pitfalls. Large material can be introduced, including vegetal and animal remains, and fallen rock blocks. Colluvial material and scree can enter caves on sloping terrains, involving mostly large entrance areas, and is often called entrance talus (Figure 10.8g). The description of this mostly external material, which is very different from the typical cave interior deposits, is out of the scope of this book. It must be remembered, though, that these entrance deposits have been widely investigated in archeological and paleontological studies (Figure 10.1).

10.3.6 Provenance of Allogenic Clastic Cave Sediments

The petrographic, mineralogical, and geochemical study of cave sediments can allow us to infer their provenance, transport paths, and the degree of wearing they underwent before reaching their final destination, although some weathering also occurs in the caves, but much less intense than at the surface. In fact, diagenesis of cave interior deposits is generally very weak, due to the rather constant values of humidity, temperature, and the lack of light that inhibits biological activity and bioturbation. Commonly, the sediments found in caves are a mixture of both autochthonous and allochthonous material. In caves where recharge is mainly autogenic, the allochthonous components are scarce, mainly fines brought in by the wind, or infiltrates from the surface (Iacoviello and Martini 2012). In multilevel cave systems, where abandoned cave passages of different ages are accessible, their sedimentary contents can reveal changes in the recharge area. The allogenic input in the Corchia Cave system in Tuscany (Italy) started probably in the Late Pliocene and lasted until the Early Pleistocene, when the original recharge area was cut off by surface uplift, as shown by the different lithological composition of the allogenic stream sediments in the upper levels (Piccini 2011). The content in different clay minerals can also provide insight into the weathering of the parent siliciclastic material, with illite being indicative of colder periods, montmorillonite indicating drier periods, and kaolinite and vermiculite pointing to warmer and wetter climate. The presence of these clay minerals can indicate both the allogenic provenance (supergene weathering of siliciclastic rocks) and the degree of weathering, and thus climatic conditions under which they were produced (Arriolabengoa et al. 2015).

The detailed investigation of a finely laminated clay deposit in a cave pool environment with calcite rafts and cave clouds in Lechuguilla Cave (New Mexico), situated over 250 m below the present land surface, revealed that the clay was brought in suspension from the surface into this deep hypogene sulfuric acid cave (Foos et al. 2000). These studies have thus allowed discerning this surface-derived clay from the typical autochthonous clays (containing alunite and 10 Å-halloysite) of this active sulfuric acid cave environment. This finding shows that even these deep hypogene systems can receive small amounts of clastic material from the surface.

A study on the fine particles released at Barton Springs, Texas, one of the main karst springs of the Edwards karst aquifer, and the main source of drinking water for the city of Austin, showed a large part of these sediments to have an allogenic origin, deriving from soil erosion tens of kilometers away from the recharge zone (Lynch et al. 2004). These findings have implications for unraveling the possible origin of pollutants and bacteria detected in springs.

Another provenance study has used radionuclides, including ^{137}Cs and ^{226}Ra, to verify the origin of fine sediments in Jenolan caves, western Australia (Stanton et al. 1992). The analyses indicate that these sediments mostly entered the caves since the early 1950s, corresponding to the commencement of forest clearance in the recharge areas, in a period during which the amounts of radionuclides derived from atomic bomb experiments in the 1950s and nuclear plant disasters (e.g., Chernobyl disaster in 1986) experienced a significant increase.

10.3.7 Autogenic Clastic Sediments

Part of the clastic sediments found in caves derives from local sources, and in most cases experiences no or almost no lateral displacement. These can mainly be classified into weathering-derived detritus and breakdown or collapse deposits. Considerable lateral movement can occur when these materials are reworked by in-cave fluvial and aeolian processes.

Most soluble rocks are not pure, and contain insoluble components that are left in place as residual material after dissolution of the host rock. Limestones often contain various proportions of chert nodules or silicified fossils, which form the coarse fraction of the insoluble residue. Quartz grains are often present in carbonate rocks, leaving a sandy residue, which may also include less soluble dolomite grains. The finer autogenic particles are generally fine quartz grains, illite, kaolinite, and hematite, corresponding to the most common minerals found in the insoluble residue of limestones. Generally, the amount of autogenic weathering material is small compared to allogenic inputs, and often they mix forming the typical sediments found in caves away from active or past fluvial activity.

Truly autogenic weathering residues have been called speleosols, since they have undergone modifications through leaching, chemical, and microbial weathering, as occurs in the formation of soils (pedogenesis) (Spilde et al. 2009). Speleosols are not simply the insoluble remains of the host rock, but contain neoformation minerals and are also chemically distinct from the original parent rock. In addition to dissolution, leaching (or lixiviation) is also important, decreasing the concentrations of the most mobile elements, and leaving progressively higher concentrations of Fe, Al, and Mn, depending on the degree of weathering and microbial activity. In certain conditions (e.g., sulfuric acid caves) mineralogical by-products are created, such as 10 Å-halloysite (formerly named endellite), trioctahedral smectite, and montmorillonite, which are typical alteration products of sulfuric acid speleogenesis (Polyak and Güven 2000).

Much coarser autogenic clastic sediments are produced by breakdown (incasion in the European literature), related to the gravitational collapse of cave walls and roof. Breakdown modifies or completely obliterates the original phreatic or vadose morphologies of cave passages. The resulting breakdown piles are composed of openwork, angular, unsorted rock fragments (packbreccias), the size and shape of which are largely controlled by the rock structure (i.e., spacing and orientation of discontinuities). Davies (1949) differentiated block, slab, and chip breakdown, when more than one bed, only one, or a portion of a single bed is involved, respectively. Chip breakdown involves generally centimeter-to-decimeter-sized fragments with different shapes, depending on how much of the bed is broken off, and by which process. Stress-release, frost shattering and crystal wedging are common chip-forming processes occurring in caves. Block breakdown frequently occurs in

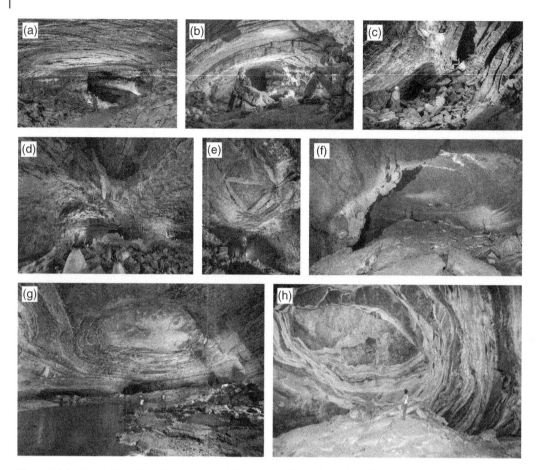

Figure 10.9 Breakdown deposits in caves: (a) Typical slab breakdown from finely layered limestones in Mammoth Cave, Kentucky. *Source:* Photo by Dave Bunnell. (b) Block breakdown in a small passage developed in the hinge zone of an anticline, Bruire Cave, Ain, France. *Source:* Photo by Jean-Yves Bigot. (c) Collapsed blocks from dipping beds, Forgers Cave, Doubs, France. *Source:* Photo by Jean-Yves Bigot. (d) Old collapse in the Hall of the Mountain King, Ogof Craig a Ffynnon, Wales, UK. Note the speleothems covering the angular breccia and the flowstones growing on the edges of the collapsed scar in the roof. *Source:* Photo by Mark Burkey. (e) Triangular collapse scar in the roof of Ellis Basin Cave System, New Zealand. *Source:* Photo by Neil Silverwood. (f) Collapse dome in Grotte du Trabuc, Languedoc, France. *Source:* Photo by Victor Ferrer. (g) Collapse dome in Gruta dos Brejões, Bahia, Brazil. *Source:* Photo by Ataliba Coelho. (h) Block breakdown of thinly bedded salt in the wall of N1 salt cave in Namakdan Diapir, Qeshm Island, Iran. *Source:* Francisco Gutiérrez (Author).

thickly bedded sequences, and often leads to the formation large collapses (Figure 10.9d–h). Slab breakdown is typical of sequences where bed thickness is in the order of some decimeters and bedding plane partings are well developed (Figure 10.9a–c).

Breakdown occurs because of mechanical failure of the rock forming the roof or the walls of a cave passage. The mechanics of breakdown in caves is similar to that in mine tunnels, and was first developed by Davies (1951) and later adapted by White and White (1969, 2000) and White (2019). Differently from mines, where the stress pattern around the freshly opened void is still out of equilibrium, caves are naturally evolving voids that have acquired their equilibrium stability, and their horizontal stresses have mostly been released. Instability thus mainly arises from the gravitational

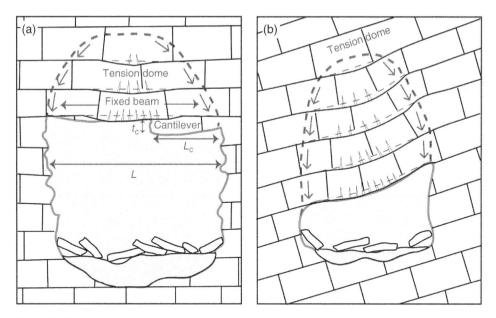

Figure 10.10 Tension dome developed in horizontal and dipping strata in the roof of a cavern. The red arrows indicate that the weight of the rock within the tension zone is mainly transferred to the walls of the cave. Beam length L, cantilever length L_c, and bed thickness t_c are shown in the left panel.

loading of the cave ceiling by the weight of the overlying rocks, and eventually by external forces, such as dynamic loading related to seismic activity. The simplest model is that of a rectangular void in horizontally bedded rocks, in which the roof is composed of beds of a certain thickness that span from one side to the other, forming a beam. The weight of the overburden plus the lack of basal support, cause these beams to sag a little bit, separating the beds from each other. This sagging contributes to slightly shorten the length of the beds, a contraction which may be counterbalanced by the development of dilated joints at the margins of the bent layer. Every bed is supporting the overlying one in some way, and the deflection around the void of the gravitational forces creates a dome-shaped tension zone above the cave void (Figure 10.10). The gravitational load of the overlying rocks is thus directed to the walls of the void, similar to what occurs in a Roman arch. The beam can thus be considered separately, and will fail under its own weight.

According to the simple model proposed by Waltham et al. (2005) and White (2019), given a beam length (span of the cave void from one wall to the other) L for a bed thickness of t, and considering a length-normal width of 1 m and a rock density ρ, the rock unit weight W will be

$$W = Lt\rho \tag{10.3}$$

and the bending moment M_b for a fixed beam, extending from one side to the other side of the cave void without being divided into fragments, will be

$$M_b = \frac{WL}{12} = \frac{L^2 t\rho}{12} \tag{10.4}$$

and the section modulus being

$$Z = \frac{t^2}{6} \tag{10.5}$$

assuming elastic behavior, failure will occur when the tensile strength T is reached in the lower (exposed) surface of the deforming beam

$$T = \frac{M_b}{Z} \qquad (10.6)$$

or, in other terms

$$T = \frac{M_b}{Z} = \frac{L^2 \rho}{2t} \qquad (10.7)$$

so, the stability of the beam is determined by the critical beam thickness t_{crit}

$$t_{crit} = \frac{L^2 \rho}{2T} \qquad (10.8)$$

In the case of a cantilever with length L_c, equivalent to a beam (layer) that is divided into two parts by open fissures, the bending moment M_c becomes

$$M_c = \frac{WL_c}{8} \qquad (10.9)$$

and the critical cantilever thickness t_{crit}

$$t_{crit} = \frac{3L_c^2 \rho}{4T} \qquad (10.10)$$

The main factors determining roof stability in horizontally bedded sequences are thus the length of the fixed beam (i.e. span of the void) or cantilever, the thickness of the bed, rock density, and tensile strength. Rocks with higher tensile strengths will be stable with thinner beds. The actual breakdown mechanisms are much more complicated, mainly because the strength of the rock, which can be highly anisotropic, is largely determined by the discontinuity planes, including bedding planes and joints with multiple orientations (tectonic, stress release). Because of the lower tensile strength and more ductile rheology of gypsum, chamber roofs can reach much larger spans in limestones, given equal bed thickness (Andrejchuk and Klimchouk 2002). Figure 10.11 shows the stability fields in horizontally bedded limestones in relation to bed thickness and cave passage width considering a typical tensile strength for karstified limestones of 5 MPa (Waltham et al. 2005). This graph contains data from 43 caves (stable or collapsed) and shows a reasonable agreement with the modeled curves for intact rock beams. The relationship (blue line) showing the stability threshold of a roof span 17 times the bed thickness ($L = 17t$) is an approximation to ideal conditions. For horizontal beds 2 m thick the roof is expected to be stable up to cavity widths of 34 m.

In natural settings, the attitude of the beds can be steeply dipping (Figures 10.9c), folded (Figure 10.9b) and affected by different sets of joints (Figure 10.9e), complicating the simplified model explained above. Steeply dipping beds, depending on their orientation with respect to the passage, can be more prone to collapse, creating asymmetric breakdown rooms. In these cases, the cohesion between bedding planes and joint faces will greatly influence the gravitational movements. The roof stability is reduced by jointing and cracking, overburden load, and changes in rock characteristics due to dissolution or weathering. Mineral crystallization and hydration, as well as frost shattering, are other processes that tend to dilate joints and create instability. Vadose processes that contribute to weaken the rocks and promote breakdown are described in detail by Osborne (2002). Roof stability can increase by mineral precipitation (calcite, gypsum) that creates

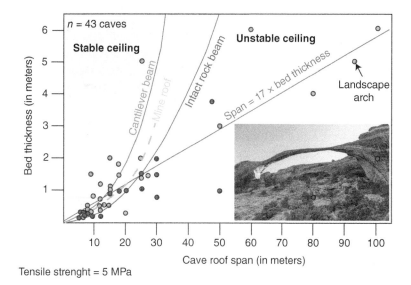

Figure 10.11 Relationship between bed thickness and width of the chamber showing stability fields in horizontally bedded limestones. *Source:* Modified from Waltham et al. (2005). Green dots are stable cave rooms, whereas the red ones indicate collapsed chambers. The extreme case of the 93-m wide Landscape Arch (Utah) (inset, Photo by Francisco Gutiérrez), carved in sandstones, is also included. The curves represent, from left to right, the less stable cantilevers (orange), mine roofs (dashed blue), intact rock beams (pink), and the general relationship $L = 17t$ (dark blue line). *Source:* Francisco Gutiérrez (Author).

binding cements increasing cohesion (Figure 10.9d). Irregular fracturing may allow adjoining blocks to interlock (Figure 10.9e), and the formation of compression arches involving multiple beds (Figure 10.9c). Sarawak chamber, the largest cave room in the world, 600 m long by 435 m wide, has a stable wide arched cave roof in massive reef limestone with a bed thickness of 15–20 m (Gilli 1993; Waltham et al. 2005) (Figure 9.33a).

In subhorizontal stratigraphic successions, repeated collapses cause the roof to propagate upward, ultimately forming a relatively stable arch mimicking the tension dome, with the beds in a cantilever situation. In wide sections of cave passages, or in large rooms, this process can create collapse domes (or breakout domes) with an oval or circular plan form (Figures 10.9f–h). Progressive collapses in these domes lead to the upward migration of the void by a process known as stoping. This localized breakdown phenomenon can proceed over vertical distances of hundreds of meters, as long as part of the collapsed material, which has larger volume and lower density than the parent rock (bulking effect) is removed by dissolution, otherwise the void would become choked (Šušteršič 2006a; Gabrovšek and Stepišnik 2011). The result is the development of a transtratal breccia pipe or collapse chimney filled with a collapse breccia. These highly permeable areas are important for petroleum exploration and mining safety (water inrushes, stability). When these breccia pipes reach the surface steep-walled bedrock or caprock-collapse sinkholes and open shafts can form. A spectacular example of such collapse shafts is Sótano de Las Golondrinas, San Luis Potosí, Mexico (Figure 9.26b).

10.3.8 Anthropic Waste

Cave entrances and dolines have always been considered ideal places to "hide" wastes, or just to get rid of them in a cheap and easy way. With the massive introduction of plastic in the market after World War II, and the growing population and human impact on the environment, the amount of waste that ends up in karst features has increased dramatically. Shafts and vertical entrances have often been used to drop dead animals and garbage of all kinds. Easily accessible descending cave entrances are common targets for such illegal dumping (Figure 10.12a). Waste is often also carried by sinking rivers into caves or sinkholes, with the lightest materials (e.g., plastic, Figure 10.12b) eventually ending up deep into karst systems (e.g., up to 2 km from the entrance in Škocjanske Jama, Slovenia). Old vehicles and household appliances can easily be dropped into vertical shafts (Figure 10.12c). On the border between Italy and the former Yugoslavia, between 1943 and 1945, many civilians and soldiers were killed and thrown into vertical shafts, known as "foibe," and their remains are often still lying on the bottom of the pitches, or buried nearby (Figure 10.12d). Severe illegal dumping also occurs in cities, where artificial voids were carved in limestones in the remote past (aqueducts, tombs, living spaces, and water cisterns) and in recent times (mines) (Figure 10.12e). Moreover, it is not rare to build landfills on carbonate and evaporite terrains, especially where cities are built on karst areas (e.g., Trieste in Italy), or where natural holes (dolines) can be filled easily, with later actions to rehabilitate the area (e.g., Vall d'en Joan near Barcellona, Figure 10.12f). Wastes, in the form of organic, chemical, clinical, plastic or metal materials, or in microscopic forms (microplastics) (Balestra and Bellopede 2022), are increasingly found in karst areas and caves, and actions should be taken to educate people on the importance of karst areas and aquifers and their vulnerability.

10.3.9 Sedimentary Structures

Detrital sediments can show different structures depending on grain size, the hydraulics of the water flow, the bedforms developed during deposition, and modifications occurred after their accumulation. When sedimentary sequences are exposed, they show bedding, changes in grain size and color, depositional and erosional features, and sedimentary structures that reveal flow velocity and direction. Cross beds form on sloping bedforms (see Figures 10.5 and 10.7d), in most cases indicating a paleocurrent in the direction of the dipping beds. Sequences often include convex-downward and channel-shaped sedimentary bodies, recording the excavation of a channel by high-energy flows and subsequent filling with detrital material. These so-called cut-and-fill structures provide evidence of open channel flow and flow events with different stream power and/or sediment load. Fine-grained sediments hosted in caves of high mountain or periglacial areas can show rhythmic sequences of couplets of darker and lighter-toned silts and clays, similar to varved sediments in glacial lakes (Maire 1990). These couplets are ideally seasonal (winter and summer), representing the deposition of one year. Lignier and Desmet (2002) have shown that the varved rhythmites found in a cave in the Jura Mountains (France) are similar to those found in a nearby glacial lake, and may correspond to some centuries of deposition. These deposits also show deformation structures attributable to paleoseismic activity (paleoseismites), with at least four events with an average recurrence of about one century. A similar varved sequence was also studied in Hirlatzhöhle, Dachstein, Austria (Figure 10.13a). These sediments were deposited probably at the end of the last glaciation, and also show features attributable to seismic shaking related to strike-slip faults in the Eastern Alps (Salomon et al. 2018).

Bedforms observable on the surface of cave deposits can be used to reconstruct hydraulic conditions and flow direction at the time of their deposition. A common feature found in caves are ripple marks, small centimeter-scale dune-like features on fine-to-medium sands with an upstream gentle slope and a steep downstream side (Figure 10.13b). Their asymmetry is sometimes nicely

Figure 10.12 Anthropic wastes in caves and karst areas: (a) Waste dump in the entrance of a cave along the roadside near Slinfah, Syria. *Source:* Photo by Jean-Yves Bigot. (b) Plastic and other wastes brought into a gypsum cave in Sicily. *Source:* Photo by Marco Vattano. (c) The door of a car on the bottom of the 200-m deep shaft of Kačna Jama, Slovenia. *Source:* Photo by Marco Vattano. (d) The helmets of soldiers and remains of bombs in the same cave. *Source:* Photo by Borut Lozej. (e) Cavers haul out wastes from an artificial cave carved in limestone in Sicily. *Source:* Photo by Marco Vattano. (f) Vall d'en Joan waste dump in 2008, Garraf Massif, Barcelona area, northeastern Spain. Between 1974 and 2006, the karst valley of Vall d'en Joan was filled with garbage to a height of 80–100 m and a length of about 3 km. In 30 years, 26.7 million tonnes of waste from the city of Barcelona (Spain) has been buried in this karst site. *Source:* Photo by Jean-Yves Bigot.

marked by dark organic material lying on the steep sides, where flow velocity is dampened. In cave passages hosting mud deposits, the highest floodwater level can be recorded by surge marks on these fine-grained sediments on sloping surfaces, whereas the fast, downward evacuation of water at sinking points creates mud runnels (Figure 10.13c). Mud can also be accumulated

Figure 10.13 Sedimentary structures in caves: (a) The varved sediments in Hitlatzhöhle, Austria. The inset shows soft-sediment deformation structures (convolute bedding, flame structures) attributed to seismic shaking. *Source:* Photo by Lukas Plan. (b) Ripples on the sandy floor of Mammoth Cave, Kentucky. *Source:* Photo by Dave Bunnell. (c) Surge marks (right) and mud funnels in Xe Bang Fai Cave, Laos. *Source:* Photo by Chris Howes. (d) Mud pillars carved in organic-rich fine sediments in Škocjan Cave. *Source:* Photo by Borut Lozej. (e) Clay pillars in the Swiss Village in Agen Allwed Caverns, Wales, UK. *Source:* Photo by Mark Burkey. (f) The curious "sandmen" sculptures in fine sands and silts on the floor of a passage in Krem Puri Cave, Meghalaya, India. *Source:* Photo by Marcel Dikstra, Caving in the Abode of the Clouds Project. (g) Mud cracks in the silty floor of Mammuthöhle, Dachstein, Austria. *Source:* Photo by Lukas Plan. (h) Old mud cracks with white calcite coralloids in a dried-out pool in Mitjina Jama, Slovenia. *Source:* Photo by Peter Gedei. (i) Silty-clayey vermiculations on the limestone wall in Pertosa-Auletta Cave, Central Italy. *Source:* Photo by Rosangela Addesso.

during floods on the cave walls forming a thin coating with a horizontal upper limit that marks the maximum flood stage, known as high-water mark.

Intense dripping on clay and silt deposits including coarser particles (pebbles, organic debris) can cause differential erosion and the formation of mud pillars (pyramids) with a capping clast on top, similar to the earth pillars in glacial deposits, also known as pedestals or hoodoos. These mud pillars can create spectacular underground landscapes, covering several square meters of cave floor (Figure 10.13d, e). In homogeneous fine-grained erodible sediments, dripping causes the formation of centimeter-scale deep holes with water in the bottom. The splashing causes part of the eroded mud to accumulate on the borders of these small craters. The resulting rims of mud bordering the holes are often cemented with calcite, making them more resistant to erosion. These hardened fine sediments can finally stand proud above the surrounding, more erodible sediments forming mud stalagmites, characterized by a central hole. Similar forms can be also created by sediment-laden droplets, depositing viscous mud on the floor forming irregular mud stalagmites with crenulated surfaces (Malott and Shrock 1933). Some of these cohesive clay-silt sculptures often occur in caves that are flooded by very slowly flowing waters. Krem Puri Cave (Meghalaya, India) contains some very enigmatic depositional–erosional clay features, named "sandmen," covering wide floors in different passages and which origin is still poorly understood (Sauro et al. 2020) (Figure 10.13f).

When the muds on cave floors dry out, they experience a volume reduction, which is accommodated by the development of cells affected by inward contraction, resulting in the development of a series of polygons, known as mud cracks or desiccation cracks (Figure 10.13g, h). The downward-tapering cracks can be filled with calcite-precipitating waters, forming a carbonate cast of the network of cracks once the mud is eroded away.

Clays can also be found stuck to cave walls and roofs, as a uniform layer, but often also displaying a worm-like pattern in which clay particles concentrate along a network of sinuous lines. These vermiculations, as they are called, mostly form on smooth surfaces where wetting and drying occurs on a regular basis, often not far from entrance areas (Figure 10.13i). These clay patterns have also been found in artificial subterranean environments, such as painted catacombs, and their development can cause the deterioration of decorated walls. For instance, vermiculations and their detrimental effects have been found on Paleolithic rock art in Lascaux Cave, France (Hoerlé et al. 2011). The clay particles can be brought by flooding, or are simply airborne dust, later rearranged to form the worm-like patterns. Vermiculations have been described in great detail by Bini et al. (1978), explaining their genesis by colloidal concentration of the charged clay particles, a theory recently investigated and partly confirmed by Freydier et al. (2019). Their chemistry and mineralogy indicate that they are composed of a mixture of autogenic and allogenic components, including microbial material (Addesso et al. 2019) (Figure 10.13i). Vermiculations in sulfuric acid caves are clearly associated with microbial activity, as shown in Frasassi caves in Italy (Jones et al. 2008) (Figure 10.14h).

10.4 Organic Sediments

Caves are dark and inhospitable places for life and special adaptations are needed to survive in this normally oligotrophic (nutrient-poor) environment. This means that there is almost no local production of biomass, with the exception of cave adapted arthropods and microbiota, but the volumes are extremely low compared to those produced at the surface in temperate and tropical climates. The majority of the organic matter in caves is thus allogenic.

10.4.1 Vegetal and Animal Debris

Vegetal and animal debris can be brought into the underground environment by running waters, wind or animals, or they can simply fall through open cave entrances or penetrate with infiltration waters through fissures. Most large-sized debris (trunks, big animal bodies, and remains) transported by running waters are deposited close to the cave entrances, unless subterranean stream passages are large and the flow has high competence. Large tree trunks are transported to Mrtvo Jezero, or "Dead Lake" in Škocjan Caves, Slovenia, more than 2 km away from the site where the Reka River sinks (Figure 10.14a). Smaller particles (leafs, acorns, seeds, and pollen) are carried farther inside, sometimes giving rise to spontaneous growth of plants that soon are doomed to succumb. Pollen can be brought deep inside caves by airflow. The river-transported organic matter can reach relatively large volumes if the stream is large and vegetation outside is lush, and this organic debris can be a long-lasting supply of food for a series of cave dwelling species.

Vegetal and animal remains can fall into open entrances, sometimes forming large debris cones. Open vertical cave entrances and fissures act as pitfalls, and some of these sites, because of the high preservation potential, are extremely important from a paleontological and paleoanthropological point of view (Simms 1994) (Figure 10.1). Fossil bone remains of extinct bats have been concentrated inside Slaughter Canyon Cave in Carlsbad National Park (New Mexico) by slowly flowing waters (Figure 10.14b), and the geochronological analysis of a flowstone on top has given an age of ca. 210 kyr, which also allowed the estimation of a maximum incision rate for the canyon (Lundberg and McFarlane 2006a). In shallow caves, the roots of plants can penetrate deep into the ground searching for humidity in the cave atmosphere (Figure 10.14c), or the hosted water bodies. Strings of roots up to several tens of meters long hang as curtains from some cave roofs, and are sometimes encapsulated in precipitating calcite sheets and tubes (Figure 10.14d). In rare cases, these roots can also penetrate deep into caves and locally grow upward from the cave floor, forming a sort of stalagmites, searching for the dripping from the roof (Figure 10.14e) (Du Preez et al. 2015).

Some animals use caves as shelters or hibernation places, the most famous of which are certainly cave bears (*Ursus spelaeus)*, and other similar Pleistocene bears (Figure 10.14f). The use of a cave by many generations of bears has sometimes given rise to the formation of important phosphate deposits that were mined especially during World War II. One of the caves with the most spectacular and voluminous bone deposits is Drachenhöhle, near Mixnitz, Austria (Frischauf et al. 2014).

Guano refers to the droppings of bats or birds occurring in caves, often in entrance areas but also very deep into the cave (bats and swiftlets penetrate for 5 km inside the Puerto Princesa Underground River Cave, Palawan, the Philippines (Agnelli et al. 2018)). We categorized these deposits among the allogenic organic sediments, since their composition reflects food sources external to the cave environment, albeit the guano being produced by the animal and ultimately deposited inside caves. These deposits are particularly abundant in tropical caves, where both bats and swiftlets occur in great numbers, and also because of the large availability of food in these environments. In temperate areas, the rate of accumulation of these droppings is lower, but these guano deposits can attain an important thickness if the caves are used over long periods of time. Bat guano decays in the cave releasing water vapor, CO_2, heat, and a series of acids. This has important morphological consequences on the vadose cave environment (Dandurand et al. 2019; Merino et al. 2019) (Figure 10.14g) and causes the formation of a wide variety of phosphate minerals (Audra et al. 2019).

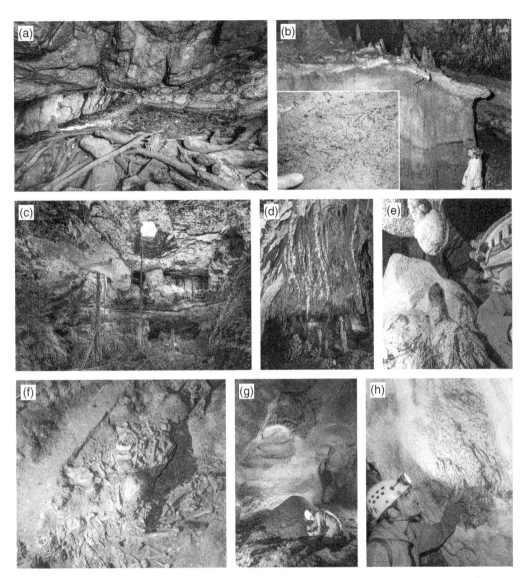

Figure 10.14 Organic sediments in caves: (a) Stranded logs close to Dead Lake (Mrtvo Jezero) in Škocjan Caves, Slovenia. *Source:* Photo by Borut Lozej. (b) Water-laid sediments in Slaughter Canyon Cave, Carlsbad National Park (New Mexico). The inset shows a detail of these sediments with fossil bones of the extinct molossid bat *Tadarida constantinei* (Lawrence 1960). *Source:* Photos by Jean-Yves Bigot. (c) Skylights and long roots in Toca da Boa Vista Cave, Bahia, Brazil. *Source:* Photo by Philippe Crochet. (d) Distorted stalactites formed by carbonate coating around stone oak (*Quercus ilex*) roots, Elighes Artas Cave, Sardinia. *Source:* Photo by Vittorio Crobu. (e) Root stalagmite directed upward toward the source of dripping water in Kaceverrit Cave, Holtas Canyon, Albania. *Source:* Jo De Waele (Author). (f) Bones of *Ursus spelaeus* in Bàsura Cave, Toirano, northern Italy. *Source:* Jo De Waele (Author). (g) Fresh guano heap below a cupola occupied by a bat colony in Grotta dell'Orso, Lessini Mountains, northern Italy. *Source:* Photo by Sandro Sedran, S-Team. (h) Vermiculations in the active sulfuric acid cave of Grotta Bella, Frasassi, Central Italy. *Source:* Photo by Laura Sanna.

With time guano transforms into a granular material, rich in phosphate, nitrate, ammonium, potassium, sulfur, and carbon, and is one of the best natural fertilizers. It has been extensively exploited in tropical areas, especially in the nineteenth century. Bat guano was mined in caves in the USA to produce gunpowder, especially during the Revolutionary Wars of 1775–1783 and 1812 and the Civil War (1861–1865) (Hubbard 2019). These mining operations often had a disastrous impact on the bat colonies and on the cave fauna relying on bat guano as a major food source. Guano is still being exploited in caves in developing countries, and the International Union for Conservation of Nature has issued recommendations for sustainable guano mining, including extraction during periods in which bats are not in the caves, and leaving a portion of fresh guano for the cave dwelling species to feed on (IUCN 2014).

Coprolites are fossilized feces, which can be preserved especially in caves in arid areas, where the lack of humidity prevents the decay of the organic remains, often mediated by fungi and bacteria. These fossils can reach abundant volumes when derived from animals that used caves on a frequent basis. Coprolites of ground sloths, because of their large size, have been found in many caves from the Southern United States to South America, and in Cuba (Hunt and Lucas 2018). These remains allowed constraining the period of extinction of these Pleistocene animals, inferring their diet, and deriving paleoenvironmental and paleoclimatic interpretations (Poinar et al. 1998). The early migration of humans across the Bering Strait into Northern America has also been dated to 14.27–14.00 kyr ago, based on human coprolites found in the Paisley caves along the coasts of Chewaukan Lake, Oregon, USA (Gilbert et al. 2008).

10.4.2 Autogenic Organic Sediments

The organic material that is actually produced in the caves, and can thus be defined as truly autogenic, accounts for a very minor amount of biomass. This includes the remains of troglobitic fauna, including all kinds of arthropods, and some vertebrates. This organic matter becomes part of the food chain in the oligotrophic cave environments, and is recycled almost completely. A portion may escape the cave system under the form of CO_2 derived from its decay. Biofilms are included into the autogenic category, being another important part of the subterranean food chain. These can be rather abundant in areas close to entrances or in shallow cave environments, where infiltration water can bring the necessary food supply to heterotrophic communities. Microbial and fungal growth has important practical implications, especially in rock art caves, where these living colonies can cause deterioration of the rock wall paintings, including famous UNESCO world heritage sites (e.g., Lascaux Cave, France, Bastian et al. 2010; Altamira Cave, Spain, Saiz-Jimenez et al. 2011). Although invisible, microbes appear to be abundant in caves (Barton 2006; Engel 2010) and constitute important elements of the cave ecosystem, often also involved in geological and geomorphological processes (Jones 2010), including speleothem growth (e.g., moonmilk, Sánchez-Moral et al. 2012; Miller et al. 2018), silica precipitation (Ghezzi et al. 2021), and the development of some types of helictites (Tisato et al. 2015) and biovermiculations (Jones et al. 2008; Jurado et al. 2020; Addesso et al. 2021). The microbial biomass increases in sulfuric acid cave environments, where walls are sometimes coated with slime-like substances, named snottites, stream beds are covered with white microbial filaments, and a wide variety of microbial vermiculations decorate the walls and ceiling (Hose and Pisarowicz 1999; Macalady et al. 2006; Jones et al. 2008) (Figure 10.14h). In these cases, microbes are often chemoautotrophic, obtaining energy from inorganic compounds (feeding on the rock), in contrast to the heterotrophic microorganisms occurring where organic surface-derived food sources are available.

10.5   Transported Chemical Deposits

A special kind of material found in caves, halfway between clastic and chemical deposits, are fragments of minerals formed in the caves or in the surface environment of karst terrains and transported inside the cave. These include fragments of speleothems, calcareous tufa and travertine that often form accumulations not far from the site where they precipitated and were broken off. These clasts are soluble and fragile, and are thus best preserved in areas with limited water flow and stream power.

In some Austrian caves (e.g., Mammuthöhle in Dachstein) important accumulations of dark nodules are found, known under the name "bohnerz" (meaning bean ore). These iron nodules, composed of a nucleus of pyrite or marcasite and a weathering rind (oxidized) of goethite and lepidocrocite, are often residual products of limestone dissolution, incorporated into the cave environment as clastic particles (Al-Malabeh and Kempe 2005). However, these types of iron minerals can also form within the cave (autogenic). The sulfides pyrite and marcasite can form by anaerobic bacterial reduction of iron sulfates, and the oxide magnetite by partial bacterial reduction of iron hydroxides in an environment rich in organic matter. This occurs inside fine-grained sediments, and once eroded, magnetite is then readily oxidized to maghemite in oxygen-containing flowing waters (Seemann 1987).

Phosphorite is another chemical deposit that can be deposited in caves. It is generally related to the interaction of guano and the products of its decomposition with the host rock, and can be autogenic (phosphate minerals in guano deposits or at the contact between guano and the cave hostrock), or transported as granular material over short distances (Glenn et al. 1994; Zanin et al. 2005; Audra et al. 2019).

10.6   Chemical Deposits

Caves constitute very peculiar minerogenetic environments in which local chemical and physical conditions (water chemistry, temperature, pH, and Eh) create a wide range of situations for mineral precipitation. It is increasingly more obvious that many of these mineral-forming processes are not only driven by physical and chemical changes, but are also often mediated by the activity of microbiota involved in redox reactions (Barton et al. 2001). Most caves are formed in carbonate rocks, and only some are developed in evaporites (gypsum-anhydrite, halite) or other rocks (mainly quartz sandstones). Therefore, these host rocks are characterized by a limited geochemical diversity, compared to that found in some metamorphic or magmatic rocks. An exception is found in volcanic caves (lava tubes) that show the highest mineralogical variety among all caves (Forti 2005; Ulloa et al. 2018). However, even in the monomineralic rocks, the chemistry of the substrate on which minerals precipitate can be enriched with allogenic sediments, locally produced guano and other organic deposits which give rise to a wide suite of secondary cave minerals. Hypogenic fluids can also introduce various chemical species into the cave systems, mainly sulfates derived from H_2S oxidation, but also different alkali and alkali earth metals, as well as other metals. Finally, aerosols (microparticles and microdroplets suspended in the cave air) can locally be important, bringing nutrients, organic compounds, microorganisms, and pollutants into the cave environment (Dredge et al. 2013). However, most fine dust settles down in the generally quiet atmosphere of caves before reaching far inside. The large majority of chemical deposits found in caves are carbonates (calcite and aragonite), followed by gypsum. All other cave minerals are found in tiny amounts, as crusts, individual crystals, or powdery material. Some form crystal aggregates with a typical morphology, and these often-laminated secondary mineral deposits are known under the name of "speleothems" (Moore 1952).

10.6.1 Minerogenetic Mechanisms in Caves

There are many ways whereby minerals can form in caves, including high-temperature and freezing-induced precipitation, which are restricted to very special environments (e.g., lava and glacial caves). The most common mechanisms are, in order of importance, CO_2 degassing, evaporation, oxidation, alteration by acids, reduction, changes in pH, cooling and warming, and the common-ion effect (Palmer 2007). For more details on all possible minerogenetic processes in caves, the reader is referred to Onac and Forti (2011).

The most common cave minerals are the carbonates calcite and aragonite, polymorphs of calcium carbonate ($CaCO_3$). These minerals are dissolved in slightly acidic waters enriched in CO_2, commonly derived from the soils. The cave atmosphere has a lower concentration of CO_2 than the air in soils, due to the fact that cave air is constantly exchanged with the external low-CO_2 air. Consequently, the CO_2-rich percolation water that enters a cave is not in equilibrium, and the dissolved gas diffuses into the cave atmosphere. The decrease in the P_{CO_2} can result in the supersaturation of the solution with respect to a $CaCO_3$ mineral and the formation of a speleothem (Figure 10.15a). Since CO_2 is more soluble in cold waters, an increase in the water temperature can also induce the release of the dissolved gas and the precipitation of calcium carbonate. Carbonate precipitation is also aided by the retrograde solubility of calcite, aragonite, and dolomite, which solubility decreases as temperature rises (Figure 3.35). For instance, these mechanisms lead to the precipitation of carbonate crusts in pipes with hot water. This process is easy to understand through the typical bidirectional dissolution–precipitation reaction of calcium carbonate:

$$CaCO_{3(solid)} + CO_{2(gas)} + H_2O \Leftrightarrow Ca^{2+}_{(aqueous)} + 2HCO_3^-_{(aqueous)} \tag{10.11}$$

if the concentration of dissolved CO_2 increases, the solution is able to dissolve a larger amount of calcium carbonate and the reaction is displaced to the right (Le Chatelier's Principle), whereas loss of CO_2 can cause the net precipitation of $CaCO_3$ when the solution becomes supersaturated (speleothem formation). It is estimated that approximately 95% of all calcite and aragonite speleothems form by CO_2 exsolution. CO_2 is continuously produced in soils by organic matter decay and root respiration, and the ion Ca^{2+} (or Mg^{2+}) is commonly incorporated in the solution by the dissolution of the host rock, normally limestone or dolostone, but also gypsum. The water that percolates into lava tubes often has low Ca^{2+} concentrations, which explains why they are generally devoid of the classical carbonate speleothems. There are exceptions though, such as volcanic caves where windblown calcareous dust is present in the overlying soils, or where the volcanic rocks are covered by limestone (as occurs in Jeju Island, Korea, Woo et al. 2008).

Evaporation is another process that causes solutions to become supersaturated with respect to one or more minerals. For evaporation to occur in a cave, the air in contact with the water needs to have a relative humidity lower than 100%, so that additional water molecules can transfer to the gas phase as water vapor. These conditions are rarely met in caves located in temperate and tropical areas, where water condensation from the high-humidity atmosphere is more likely to occur instead. Only cold air masses entering the caves are often below the dew point and can locally cause evaporation from the wet cave walls. Evaporation initially induces the precipitation of the less soluble minerals (alkaline earth carbonates, gypsum), ending with highly soluble salts such as

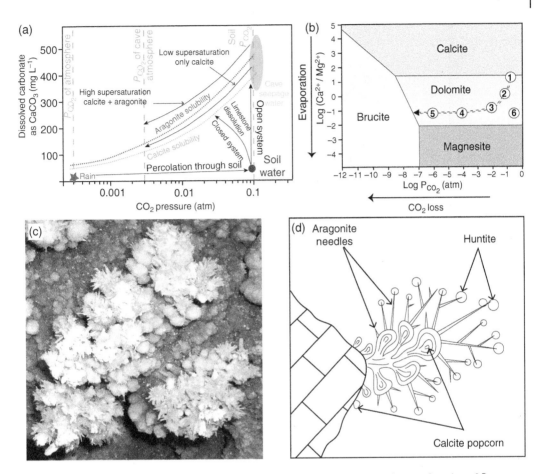

Figure 10.15 Carbonate minerals in caves: (a) Solubility of calcite and aragonite as a function of P_{CO_2} showing the ideal geochemical pathway of rain water percolating through the soil (CO_2 increase), dissolving $CaCO_3$ in open or closed system conditions and depositing calcite and/or aragonite. *Source:* Modified from White (1988). (b) Phase diagram in the system Ca-Mg-CO_2-H_2O with stability fields of calcite, dolomite, magnesite, and brucite as a function of the Ca/Mg ratio and P_{CO_2} in the solution. The hypothetical evolution pathway of cave dripwater with increasing evaporation and CO_2 degassing is shown with the formation, in chronological order, of (1) calcite, (2) Mg-calcite, (3) aragonite, (4) huntite ($CaMg_3(CO_3)_4$), and (5) hydromagnesite ($Mg_5(CO_3)_4(OH)_2 \cdot 4H_2O$). The mineral (6) nesquehonite ($MgCO_3 \cdot 3H_2O$), which forms only very rarely in caves, is also shown. (c) Cave popcorn (coralloid) showing the typical precipitation sequence calcite (botryoidal), aragonite (needles), and huntite (white toothpaste-like material). *Source:* Photo by Jean-Yves Bigot. (d) Schematic representation of the calcite–aragonite–huntite sequence of (c).

halite and sylvite, or the extremely soluble sulfates epsomite and mirabilite. These salts are so soluble that they can be preserved over long times only in extremely dry caves (arid climate areas). The calcium chloride antarcticite has been found to precipitate in a salt cave in the Atacama Desert, Chile, a few days following one of the extremely rare rain events in this desert area, but its hygroscopic properties and extreme solubility caused it to disappear after a couple of months, even in such dry conditions (De Waele et al. 2017a). The most typical cave mineral that forms by evaporation is gypsum. In limestone environments with sulfides (e.g., pyrite, FeS_2), the calcium in solution combines with the sulfate ions produced by oxidation of sulfides when the water evaporates, leading to gypsum precipitation (White and White 2003).

Both CO_2 degassing and evaporation increase in areas where air flow occurs, causing the saturation indices of minerals to rise in the water films. In these relatively wet conditions the first stable minerals that form are calcite and aragonite. So, in typical cave settings, where CO_2 is degassed from the solution and evaporation occurs at low to moderate rates, the less soluble calcite will precipitate first. The remaining solution will have a progressively lower concentration in Ca, increasing its Mg/Ca ratio, especially when evaporation dominates over CO_2 degassing. When the Mg/Ca ratio in the solution exceeds 0.5, at least in caves in temperate areas (cave temperatures of 10–15 °C), calcite precipitation starts to be inhibited, and aragonite deposition begins (Rossi and Lozano 2016). Since aragonite, on the contrary to calcite, does not incorporate the small Mg ions into its crystal lattice, the Mg/Ca ratio increases even faster, causing the precipitation of carbonates with higher Mg contents, in the typical order calcite–aragonite–huntite–hydromagnesite–magnesite (Figure 10.15b–d).

Cave environments are generally aerated and thus have an atmosphere similar to the external one, with the main differences being related to their greater air humidity and higher CO_2 concentrations (often 0.1–1%). The ca. 21% of O_2 in cave air produces a typical oxic environment where oxidation is a common process. Infiltrating waters are also enriched in CO_2 (generally of some %) and are relatively rich in dissolved oxygen, thus minerals such as pyrite and other sulfides oxidize in this mildly acidic environment in air-filled or shallow phreatic caves. This produces sulfuric acid, which reacts almost immediately with the carbonate host rock, acting as a buffer and leading to the production of gypsum, CO_2 and Fe oxides, and hydroxides with a typical orange and reddish color (see Eq. 3.76). Oxidation of H_2S also produces sulfuric acid (see Eqs. 3.68–3.71), and if H_2S is continuously supplied (by hypogene fluids), the environment can keep a very low pH over long time spans, forming alteration products typical of sulfuric acid (SAS) cave environments. Alteration of clays, which are often contained and intercalated in limestone sequences, produces typical SAS minerals such as alunite, 10 Å-halloysite, and jarosite (Polyak and Güven 2000).

Reducing conditions are rarely met in caves, and occur essentially in anoxic (underwater) environments. Since reduction processes are generally endothermic, an energy supply must be present in the cave, often given by the oxidation of organic matter, which consumes oxygen and releases chemical energy. These redox reactions, including those involving Fe, Mn, S, and N, are almost always mediated by microorganisms. Changes in pH can also be responsible for the formation of some cave minerals. In a limestone or dolostone environment, where the buffering effect of the carbonates inhibits large pH changes to occur, this mechanism is rarely the main trigger for the deposition of minerals. It can become influential in sulfuric acid caves, where pH can change drastically over very short distances, as also happens in acid mine drainage tunnels. Different minerals precipitate in these cave streams depending on the distance from the input of the highly acidic waters. pH changes are more relevant in quartz sandstone caves, where there is no buffering effect, and changes of several units of pH can occur. Quartz and opal are more soluble at pH >9 because of dissociation of silicic acid (see Eqs. 3.86–3.89), and if such waters experience a drop in alkaline pH, opal, a hydrated silica mineraloid, can precipitate (Sauro et al. 2018) (Figure 3.23).

Cooling has a significant effect on the solubility of a number of compounds, and this process is particularly important in caves in which thermal fluids rise and cool, causing different minerals to precipitate. The formation of quartz, barite, celestine, and sulfides are often induced by cooling. Naica Cave (Mexico) is an interesting example of how slow cooling and supersaturation can cause the formation of giant gypsum crystals in a cave environment. The solubilities of anhydrite and gypsum are roughly the same at 59 °C, which is the temperature of the mine waters at Naica, but as temperature drops, the solubility of gypsum decreases and this mineral starts to precipitate (García-Ruiz et al. 2007). The limestones at Naica contain lenses of anhydrite that dissolve slowly with cooling temperatures (below 59 °C), delivering Ca^{2+} and SO_4^{2-} to the solution. As a consequence of this constant supply of ions, gypsum crystals grew very slowly in the cave, producing these exceptional (over 10-m-long) blades.

Finally, the common-ion effect can be an important mechanism for the precipitation of minerals, especially in gypsum caves. This occurs when the dissolution of two minerals adds a common ion to the solution. For instance, if water percolates through the soil and comes in contact with limestone, it will dissolve $CaCO_3$ until reaching equilibrium conditions with respect to calcite. If this saturated water interacts with a gypsum layer, the dissolution of the sulfate mineral will introduce additional Ca^{2+} (common ion) into the solution, becoming supersaturated and causing the precipitation of calcite (Figure 3.24). This explains why gypsum caves in temperate areas, where CO_2 levels are high and evaporation processes are less relevant, often contain calcite instead of gypsum speleothems (Calaforra et al. 2008; Columbu et al. 2015, 2017). The same occurs when both limestone, dolomite, and gypsum (or anhydrite) interact with dissolving waters, giving rise to a rather complex process called dedolomitization, explained in Section 3.5 (Schoenherr et al. 2018). Surface water normally first becomes saturated with respect to calcite. If this water gets in contact with gypsum and dissolves it, due to the common ion effect the solution will become supersaturated with respect to calcite, which precipitates. This will lower concentrations of dissolved Ca^{2+} and HCO_3^- causing both gypsum and dolomite to dissolve further (Bischoff et al. 1994). This process, if the fluid remains in contact with both gypsum (or anhydrite) and dolomite, may increase the gypsum solubility by 1.5 and that of dolomite by 7. The remaining fluids, at the end of the process, will be characterized by higher SO_4^{2-} and Mg^{2+} concentrations (Worthington and Ford 1995b). Dedolomitization can be described by the overall reaction

$$MgCa\left(CO_3\right)_2 + CaSO_4 \cdot 2H_2O \rightarrow 2CaCO_3 + Mg^{2+} + SO_4^{2-} + 2H_2O \tag{10.12}$$

Dedolomitization involves thus both dissolution (gypsum and dolomite) and precipitation (calcite), thus causing an increase or a decrease in porosity, depending on where these processes occur (Nader et al. 2008). In near-surface environments (typical of epigenic karst settings), the initial rock volume occupies more space than the dedolomitized rock, involving an increase in secondary porosity that may boost speleogenesis. However, this side-effect of economic importance (petroleum and drinkable water) decreases with burial due to compaction and cementation (porosity and permeability values due to dedolomitization in deep settings tend to decrease) (Schoenherr et al. 2018).

10.6.2 Carbonates

Carbonates are the most common minerals found in caves, largely because most caves are hosted in limestones and dolostones. However, they are often found also in caves hosted in other rocks, such as gypsum or basalts, having important paleoclimatic and paleoenvironmental significance (Woo et al. 2008; Calaforra et al. 2008; Columbu et al. 2015, 2017, 2019). Among these minerals, calcite is by far the most frequent, followed by the other common $CaCO_3$ polymorph aragonite. Both calcite and aragonite probably represent more than 95% by mass of all cave minerals, with a few percent corresponding to gypsum, and less than 1% to the remaining cave minerals, mainly other carbonates, sulfates, and phosphates. The most common mechanisms by which carbonates precipitate from cave waters have been explained in the previous section. The most frequent carbonates found in caves are summarized in Table 10.2.

In typical cave settings, at temperatures of 0–30 °C, normal air pressure, and P_{CO_2} in the range of 0.001–0.01 atm, calcite is the most stable form of $CaCO_3$. Aragonite is more soluble than calcite in normal cave conditions, so one would expect to find only calcite. Instead, it is not uncommon to encounter aragonite speleothems in caves, precipitation of which is favored by high Mg/Ca ratios

Table 10.2 The most important carbonate minerals in caves.

Name	Formula	Crystal system	Characteristics
Aragonite	$CaCO_3$	Orthorhombic	Reacts strongly with 5–10% HCl Sinks in bromoform Acicular habit, various speleothems Twinning common Cleavage parallel to elongation
Calcite	$CaCO_3$	Trigonal	Reacts strongly with 5–10% HCl Floats in bromoform Rhombohedrons, scalenohedrons (dogtooth), prismatic (nailhead), various speleothems Double refraction Rhombohedral cleavage
Dolomite	$CaMg(CO_3)_2$	Trigonal	Reacts in warm 5–10% HCl Moonmilk, crusts, speleothems Rare
Huntite	$CaMg_3(CO_3)_4$	Trigonal	Powdery, fine-grained when dry Milky when wet Moonmilk, crusts, powders Rare
Hydromagnesite	$Mg_5(CO_3)_4(OH)_2 \cdot 4H_2O$	Monoclinic	Powdery, fine-grained when dry Milky when wet Common in moonmilk
Magnesite	$MgCO_3$	Trigonal	Powdery, fine-grained when dry Milky when wet Moonmilk Rare
Monohydrocalcite	$CaCO_3 \cdot H_2O$	Hexagonal	Very rare Moonmilk and crusts
Nesquehonite	$MgCO_3 \cdot 3H_2O$	Monoclinic	Very rare Moonmilk and crusts
Siderite	$FeCO_3$	Trigonal	Rare Yellowish-brown to amber Often in the core of calcite spar or as coatings
Vaterite	$CaCO_3$	Hexagonal	High-temperature $CaCO_3$ polymorph In old carbide wastes

in the solutions, for instance related to prior calcite precipitation (PCP) (Choudens-Sánchez and González 2009). It has in fact been shown that high Mg/Ca ratios inhibit calcite precipitation, thus promoting the formation of aragonite. The nucleation of aragonite, in contrast with that of calcite, is also promoted by very slow precipitation rates (low degree of supersaturation), or on the contrary, when supersaturation is very high, favoring precipitation of vaterite (the third polymorph of

$CaCO_3$) that rapidly transforms into the more stable aragonite. Aragonite is often found in caves hosted in dolostones or dolomitic limestones with Mg-rich waters, in situations where rapid evaporation can lead to a high degree of supersaturation (windy cave areas), or in cave areas with high CO_2 concentrations in the air, determining slight supersaturation conditions.

A special type of calcite forms in cold caves due to freezing of $CaCO_3$-rich waters. The transformation of liquid water into ice causes the CO_2 to be expelled from the solution, and the contemporary increase in the concentration of dissolved ions in a progressively smaller volume of liquid water, both inducing the precipitation of cryogenic calcite (Žák et al. 2008).

The Mg-rich carbonates (dolomite, magnesite, hydromagnesite, and huntite) are minerals commonly found in moonmilk, and form in caves hosted in dolomite rocks by progressive loss of CO_2, evaporation and consequent deposition of calcite, Mg-calcite, and aragonite (Figure 10.15b). Due to the precipitation of the $CaCO_3$ phases, the Mg/Ca ratio in the residual solution increases progressively, allowing the different Mg-rich carbonates to precipitate sequentially in the stability field of dolomite, with huntite followed by hydromagnesite (Figure 10.15b).

Calcite and aragonite usually form speleothems, such as stalagmites, stalactites, and flowstones, which are aggregates of crystals arranged in a characteristic macroscopic shape (morphology). The size and morphology of the crystals, and their arrangement in aggregates forming different types of speleothems depends on the environmental conditions in which nucleation, initiation, growth, alteration, and disintegration of crystals occur (Stepanov 1997; Self and Hill 2003).

Calcite and aragonite crystals can be very small (micrometric), or reach centimeters in length, depending on the growth rate. Rapid precipitation produces tiny crystals, whereas slow deposition leads to the formation of few but larger crystals. Crystallization starts from a stable crystalline nucleus, requiring certain conditions of supersaturation, controlled by factors such as pH, temperature, composition of the solution, etc. At a low degree of supersaturation, crystals tend to nucleate on preexisting crystal surfaces, resulting in the formation of a small number of large crystals. When supersaturation is high, or impurities are introduced into the fluids, more crystals can nucleate, leading to a large number of small crystals with different orientations. Crystals develop stable faces corresponding to the crystalline lattice when growth is slow and no impurities disturb the development of a perfect habit. At faster growth rates, or in the presence of higher amounts of impurities, the symmetry and shape of the forming crystals is increasingly disturbed. The final crystals accumulate large amounts of crystalline defects, resulting in less perfect crystal shapes and uneven crystal faces. Impurities and imperfections can also cause the splitting of crystals into several crystallites. Adjoining crystals that start growing on a specific surface will mutually interact, with ideally oriented crystals (close to perpendicular to the growth surface) impinging on the growth of neighboring less favorably oriented crystals. This is known as competitive (or impingement) growth, which in the end reduces the number of crystals that continues to grow.

The geochemistry, texture, and fabrics of the $CaCO_3$ crystals forming speleothems are an indication of the environmental conditions at the time of deposition, or of postdepositional changes such as aragonite–calcite transformations, aggrading neomorphism, or dissolution and reprecipitation processes (Frisia 2015). Primary speleothem texture and fabrics are influenced by multiple factors such as drip or flow rate, degree of supersaturation, Mg/Ca ratio of the depositing water, and the presence of impurities. Speleothem texture and fabrics are increasingly used to support the interpretation of the geochemical signals (stable isotopes and trace elements) in the paleo-environmental and paleoclimatic reconstructions based on speleothem archives (Railsback 2000; Frisia and Borsato 2010; Fairchild and Baker 2012; Frisia 2015). The most common speleothem textures and fabrics for calcite have been grouped in the following categories: columnar calcite, dendritic calcite, micrite, microsparite, and mosaic calcite (Figure 10.16). Moreover, there are different subtypes of fabrics (e.g., columnar fabric can be subdivided into compact, open, porous, elongated, fascicular

Less common fabric types

Variable drip rates

Dendritic (D)

Branching crystal aggregates forming "scaffold-like" extinction domains characterized by highly irregular boundaries. Very porous.

Microsparite (Ms)

Equant crystals > 2 µm and < 30 µm in diameter associated with micrite, replacing micrite or replacing aragonite rays/needles. Most commonly a result of dissolution-reprecipitation.

Micrite (M)

Crystals with dimensions of < 2 µm. Crystals can not be resolved under the optical microscope. Common stromatolite-like structures. Common association with microsparite.

Mosaic calcite

Mosaic of equant crystals with a diameter > 30 µm. If replacement after aragonite precursor it contains needle-like aragonite relicts. Contact angles between crystals should be 120°. This is most likely a diagenetic fabric resulting from the dissolution and re-precipitation of a precursor, be it calcite micrite or microsparite and/or aragonite.

The Columnar s.s. (sensu stricto as in McDermott et al. 1999) Fabrics

Columnar compact (Cc)

l/w ratio < 6 : 1; competitive growth at interfaces; straight to serrated boundaries; uniform extinction; common "flat" terminations or protruding rhombohedra terminations (~2 µm high).

Columnar open (Co)

l/w ratio < 6 : 1; competitive growth at interfaces; incomplete coalescence of crystals; high intercrystalline porosity, commonly linear; uniform extinction.

Columnar porous (Cp)

...if particulate is present

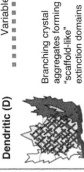

Similar to Cc, but with pervasive microporosity within crystals. Pores can be filled by air, water or particulate (both organic and inorganic). Cc and Cp may alternate in a stalagmite. Co and Cp may also alternate

Drip Rate

Slow and constant ← → Fast and less regular

...if drip water is rich in Mg

Seasonal dripping

Columnar microcrystalline (Cm)

Polycrystalline aggregates with uniform extinction (Frisia et al. 2000) and highly irregular boundaries between extinction domains. Highly porous, typical of speleothems with organic laminae.

Columnar elongated (Ce)

Similar to Cc, Cp, and Co, but with l/w ratio > 6 : 1. When it has lateral overgrowths, because of the presence of impurities, it is classified as Ce$_{lo}$. It can be compact (constant drip rate), open (variable drip rate) and porous (particulate).

Columnar Fascicular Optic (Cfo) and Radiaxial (Crf)

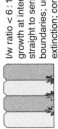

Consists of bundles of fiber-like crystal aggregates, with undulatory extinction diverging away from the growth direction (counterclockwise CCW in fascicular optic and converging clockwise CW in radiaxial. Convex cleavage in Cfo and concave in Crf.

The Columnar type Fabrics (Frisia 2015; Chiarini et al. 2017)

Figure 10.16 Calcite fabrics based on the classifications of McDermott et al. (1999), Frisia (2015), and Chiarini et al. (2017). Scheme prepared by Silvia Frisia and Veronica Chiarini. *Source:* Microscopy photos are courtesy of Kathleen Johnson, Pauline Treble, and Silvia Frisia.

optic, radiaxial, and microcrystalline) that are characteristic of specific depositional environments with different drip rates, Mg/Ca ratios, saturation indices for calcite, pH, and concentrations of impurities in the depositing fluids. Columnar compact fabrics are generally indicative of relatively slow and constant dripping, calcite saturation indices (*SIcc*) below 0.35, Mg/Ca ratio lower than 0.3, pH up to 8.4 and low amounts of impurities. Columnar open fabric forms at higher drip rates, *SIcc* up to 0.35, Mg/Ca well below 0.3, and pH between 7.5 and 8. Columnar porous calcite forms under the same conditions of the previous two fabrics but at pH up to 9 and in presence of higher amounts of particulate. Columnar elongate fabric forms when Mg/Ca is higher than 0.3, and can be compact (constant drip rate), open (variable drips), and porous (when particulates are present). Columnar fascicular optic and radiaxial fabrics form typically when infiltration waters have interacted with Mg-rich calcite or dolomite, with feeding waters having a Mg/Ca ratio greater than 1.5 and a *SIcc* of around 0.5. Columnar microcrystalline calcite fabrics indicate seasonally variable drip rates, *SIcc* up to 3.5, Mg/Ca below 0.3, pH ranging between 7 and 9, and higher concentration of impurities (e.g., colloidal particles). Dendritic calcite forms under variable flow rates and can be explained by the presence of particulates or foreign ions, and possibly biomediation during crystallization. Microsparite can be a product of ripening of micrite, or can be indicative of diagenesis (dissolution–precipitation of an aragonite precursor). Micrite is often biomediated or related to condensation–corrosion processes, and is evidence of high *SIcc* values favoring the formation of many crystal nuclei. The mosaic fabric is most likely diagenetic resulting from the dissolution and reprecipitation of a precursor that can be calcite micrite or microsparite and/or aragonite. Detailed applications of calcite textures and fabrics in paleoclimate studies, with nicely illustrated tables, can be found in Frisia and Borsato (2010), Frisia (2015), and Chiarini et al. (2017).

More recently the speleothem architectural analysis, inspired by the stratigraphical methods used in sedimentology (architectural element analysis and sequence stratigraphy) has been developed to help in unraveling the paleoclimatic significance of speleothems (Martín-Chivelet et al. 2017). Six architectural elements have been distinguished, each formed in a different time ranges (from seasons, over centuries, and to millennia), the higher order elements incorporating the lower orders. These orders are: (i) individual crystallites, (ii) single growth layers, (iii) speleothem fabric, (iv) stacking pattern sets, (v) morphostratigraphic units, and (vi) unconformity-bounded units and major unconformities. This microstratigraphic approach combines all the crystallographic, petrographic and stratigraphic features of speleothems, and can be very useful to compare speleothems from different sites, rationalize geochemical sampling, and better understand the geochemically-based paleo-environmental proxies.

10.6.3 Other Cave Minerals

Cave mineralogy has made huge progress since the first specialized book on the topic was published (Hill 1976). The first edition reported nearly 80 mineral species, mainly from American caves, a number that rapidly rose to 173 in the first "Cave Minerals of the World" edition (Hill and Forti 1986), reaching 255 in the second edition (Hill and Forti 1997). Other 63 new cave minerals were added in the following 14 years (Onac and Forti 2011), and today this number has surpassed 350 (Onac 2019), and continues to rise year after year. Alongside the almost 40 species of carbonate minerals, the most numerous mineral classes are the sulfates (over 100), the phosphates–arsenates–vanadates (around 60), the oxides–hydroxides (over 40), and the halides (a dozen). The other groups (native elements, sulfides, nitrates and borates, molybdates, organic compounds, and silicates) are less numerous and often less cave-specific. For instance, sulfides are often found, but not formed in caves. Table 10.3 reports the most common and interesting non-carbonate secondary

Table 10.3 Common and significant non-carbonate secondary minerals found in caves and their main mechanisms of formation.

Mineral	Dana class.	Formula	Main mechanism
Native elements			
Sulfur	1.3.5.1	S	Bacterial/Oxidation
Oxides–Hydroxides			
Ice	4.1.2.1	H_2O	Freezing
Goethite	6.1.1.2	$FeO(OH)$	Oxidation
Gibbsite	6.3.1.1	$Al(OH)_3$	Oxidation
Birnessite	7.5.3.1	$(Na,Ca,K)_x(Mn^{4+}, Mn^{3+})_2O_4 \cdot 1.5H_2O$	Oxidation
Todorokite	7.8.1.1	$(Na,Ca,K)_2(Mn^{4+}, Mn^{3+})_6O_{12} \cdot 3\text{-}4.5H_2O$	Oxidation
Halides			
Fluorite	9.2.1.1	CaF_2	Hydrothermal
Halite	9.1.1.1	$NaCl$	Evaporation
Nitrates			
Nitratine	18.1.1.1	$NaNO_3$	Guano-Evaporation
Niter	18.1.2.1	KNO_3	Guano-Evaporation
Gwihabaite	18.1.2.2	$(NH_4)NO_3$	Guano
Sveite	19.1.3.1	$KAl_7(NO_3)_4Cl_2(OH)_{16} \cdot 8H_2O$	Quarzite-soil
Mbobomkulite	19.1.4.1	$(Ni,Cu)Al_4(NO_3)_2(SO_4)(OH)_{12} \cdot 3H_2O$	Guano-shale
Hydrombobomkulite	19.1.4.2	$(Ni,Cu)Al_4(NO_3)_2(SO_4)(OH)_{12} \cdot 14H_2O$	Guano-shale
Sulfates			
Misenite	28.1.2.1	$K_8(SO_4)(SO_3OH)_6$	Volcanic
Thenardite	28.2.3.1	Na_2SO_4	Evaporation
Barite	28.3.1.1	$BaSO_4$	Hydrothermal
Celestine	28.3.1.2	$SrSO_4$	Hydrothermal
Anhydrite	28.3.2.1	$CaSO_4$	Evaporation
Sabieite	28.3.5.1	$(NH_4)Fe^{3+}(SO_4)_2$	Guano-pyrite
Millosevichite	28.4.5.1	$(Al,Fe^{3+})_2(SO_4)_3$	Volcanic
Mirabilite	29.2.2.1	$Na_2SO_4.10H_2O$	Evaporation
Lonecreekite	29.5.5.4	$(NH_4)(Fe_3^{3+},Al)(SO_4)_2 \cdot 12H_2O$	Guano-pyrite
Bassanite	29.6.1.1	$2CaSO_4 \cdot H_2O$	Evaporation
Kieserite	29.6.2.1	$MgSO_4 \cdot H_2O$	Evaporation
Gypsum	29.6.3.1	$CaSO_4 \cdot 2H_2O$	Evaporation
Hexahydrite	29.6.8.1	$MgSO_4 \cdot 6H_2O$	Evaporation
Epsomite	29.6.11.1	$MgSO_4 \cdot 7H_2O$	Evaporation
Alunite	30.2.4.1	$KAl_3(SO_4)_2(OH)_6$	SAS
Jarosite	30.2.5.1	$KFe_3^{3+}(SO_4)_2(OH)_6$	SAS
Clairite	31.10.8.1	$(NH_4)_2Fe_3^{3+}(SO_4)_4(OH)_3 \cdot 3H_2O$	Guano-pyrite

Table 10.3 (Continued)

Mineral	Dana class.	Formula	Main mechanism
Phosphates–Vanadates			
Monetite	37.1.1.1	$Ca(PO_3OH)$	Guano
Whitlockite	38.3.4.1	$Ca_9(Mg,Fe^{2+})(PO_4)(PO_3OH)$	Guano
Brushite	39.1.1.1	$Ca(PO_3OH) \cdot 2H_2O$	Guano
Newberyite	39.1.6.1	$Mg(PO_3OH).3H_2O$	Guano
Swaknoite	39.3.4.1	$(NH_4)_2Ca(PO_3OH)_2 \cdot H_2O$	Guano
Hannayite	39.3.5.1	$(NH_4)_2Mg_3(PO_3OH)_4 \cdot 8H_2O$	Guano in lava cave
Francoanellite	39.3.5.2	$(K,Na)_3(Al,Fe^{3+})_5(PO_4)_2(PO_3OH)_6 \cdot 12H_2O$	Guano-terra rossa
Taranakite	39.3.6.1	$K_3Al_5(PO_3OH)_6(PO_4)_2 \cdot 18H_2O$	Guano
Dittmarite	40.1.2.1	$(NH_4)Mg(PO_4) \cdot H_2O$	Guano in lava cave
Variscite	40.4.1.1	$Al(PO_4) \cdot 2H_2O$	Guano
Hydroxylapatite	41.8.1.3	$Ca_5(PO_4)_3OH$	Guano
Ardealite	43.1.1.1	$Ca_2(PO_3OH)(SO_4) \cdot 4H_2O$	Guano
Niahite	40.1.2.2	$(NH_4)(Mn^{2+},Mg)(PO_4) \cdot H_2O$	Guano
Tyuyamunite	40.2a.26.1	$Ca(UO_2)_2(V_2O_8) \cdot 5–8H_2O$	Schists and SAS
Rossiantonite	n.p.	$Al_3(PO_4)(SO_4)_2(OH)_2(H_2O)_{10} \cdot 4H_2O$	Quarzite
Sasaite	43.5.4.1	$(Al,Fe^{3+})_6(PO_4,SO_4)_5(OH)_3 \cdot 35–36H_2O$	Clays-water-guano
Organic compounds			
Mellite	50.2.1.1	$Al_2C_6(COO)_6 \cdot 16H_2O$	Clay-fire
Silicates			
Halloysite-10 Å	71.1.1.2	$Al_2Si_2O_5(OH_4) \cdot 2H_2O$	SAS
Allophane	71.1.4.1	$Al_2O_3(SiO_2)_{1.3–2.0} \cdot 2.5–3.0H_2O$	SAS
Quartz	75.1.3.1	SiO_2	Evaporation/cooling
Opal	75.2.1.1	$SiO_2 \cdot nH_2O$	Evaporation/pH

SAS = sulfuric acid speleogenesis; n.p. = Dana classification number not identified.
Minerals in italics have been first described or are found only in caves, those in bold were named after the first cave where the mineral was discovered. Sveite, Swaknoite, and Sasaite were named after caving organizations.

minerals formed in caves and described in the text. Some of these have been discovered in caves (20), and have taken the name of the cave (7), or even the name of a speleological group (3) (see Table 10.3). Three native elements, or probably four, counting the recent find of aluminum in a quartzite cave in Venezuela, have been found to form in caves (bismuth, gold, and sulfur), sulfur being the most common. It occurs in association with the sulfides pyrite or marcasite (Audra et al. 2015) in sulfuric acid cave environments, where oxidation of H_2S or bacterial reduction of sulfates causes sulfur to precipitate (D'Angeli et al. 2018), or in volcanic caves (Ulloa et al. 2018). Mellite is a very rare organic mineral found in an archeological Italian cave and formed by the reaction of charcoal with Al-rich terra rossa at high temperatures (fire hearths). The following chapters will describe the most important secondary cave oxides-hydroxides, halides, nitrates, sulfates, phosphates–vanadates, and silicates.

10.6.3.1 Halides

Eleven halides are known to form in natural caves, mostly found in salt diapirs (Iran, Israel, Romania, or Atacama), or in lava tubes. The caves of the Cordillera de la Sal (Chile) host, besides the common halite, atacamite (Figure 10.17a, b), and antarcticite (Figure 10.17c) (De Waele et al. 2017a). Halite is by far the most common salt, occurring in a wide variety of forms especially in salt caves (Filippi et al. 2011; De Waele et al. 2017a). It forms massive crusts, irregular sinuous stalactites (Figure 10.17d), or helictites (Figure 10.17e), sometimes showing the typical cubic habit of the crystals (Figure 10.17f).

The other rather common halide is fluorite, occurring in cubic crystals (Figure 10.17g), or crystalline banded masses. Although this mineral is often associated with Mississippi Valley Type ore deposits (galena and sphalerite), and thus formed in the host rock before the development of the cave, it has been found in caves as a true secondary cave mineral, often related to thermal rising fluids, or to highly acidic environments (sulfuric acid caves) (Hill and Forti 1997). Fluorite, if well crystallized, is an ideal mineral to study fluid inclusions, allowing the reconstruction of the temperature and the chemistry of the solution from which it precipitated (Bottrell et al. 2001).

10.6.3.2 Sulfates

More than 100 sulfate minerals have been described as cave minerals, forming the largest group of secondary minerals in the natural subterranean environment. Thirty-seven different sulfate minerals have been identified in two volcanic caves in the Irazú volcano of Costa Rica, making this location the richest in the world for what concerns sulfates (Ulloa et al. 2018) (Figure 10.18a).

Gypsum is the most important sulfate found in caves, and has been reported from carbonate, evaporite, quartzite, and volcanic environments. This mineral occurs as large crystalline forms in evaporite caves (De Waele et al. 2017c) (Figure 10.18b), pasty moonmilk (D'Angeli et al. 2019a), crystalline deposits in sulfuric acid caves (Galdenzi and Maruoka 2003; Plan et al. 2012), gypsum flowers (Palmer and Palmer 2003) (Figure 10.18c), isolated gypsum needles (Figure 10.18d), and extraordinary shapes, including the large chandelier-shaped speleothems of Lechuguilla Cave (Davis 2000) (Figure 10.18e), or the giant 11-m long crystals of Naica, Mexico (García-Ruiz et al. 2007; Sanna et al. 2011) (Figure 9.2c). It is often related to the oxidation of sulfides (e.g., pyrite) or H_2S in hypogene sulfuric acid caves, but gypsum also occurs on guano, and in volcanic environments.

In cave environments close to entrances or where air flow is strong on cave walls, evaporation dominates and the deliquescent "hairy" minerals mirabilite and epsomite are commonly encountered (Figure 10.18f). These minerals have a slightly bitter-salty and very bitter taste, respectively, and are easily distinguished from the tasteless gypsum hair. At low relative humidity levels (below 70%) mirabilite turns into the anhydrous thenardite, whereas epsomite transforms into hexahydrite and then into kieserite. These last three minerals are more typical of lava tube caves. Bassanite and anhydrite can survive only in very dry caves, since they are readily converted into gypsum. Both have been reported from Atacama salt caves; anhydrite as a powder of exceptional purity (De Waele et al. 2017a).

Barite and celestine are among the less soluble sulfates and occur in both thermal and sulfuric acid caves. The most famous cave (geode) filled with bluish celestine crystals is located in Ohio, and is a tourist attraction (Wright 1898). This mineral has also been reported from Lechuguilla Cave, New Mexico (Polyak and Provencio 2001) (Figure 10.18g), and many others, mainly sulfuric acid or thermal caves. Barite is less rare than celestine, and has been reported in many caves worldwide, mainly in thermal and sulfuric acid hypogene settings. The most spectacular barite occurrences in caves are found in Temple of Doom, South Africa (Martini and Marais 1996), and Santa Barbara Cave in Sardinia (Pagliara et al. 2010), where tabular crystals are of centimeter size

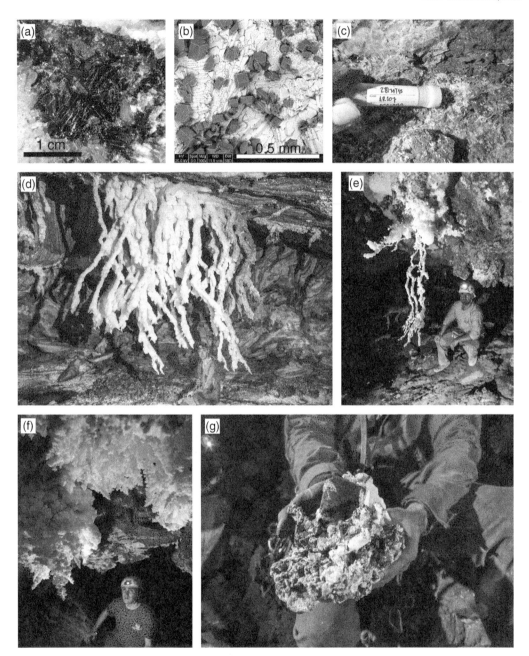

Figure 10.17 Halides in caves: (a, b) Microscopic and SEM image of atacamite from Chulacao Cave, San Pedro de Atacama, Chile. *Source:* Photos by Antonio Rossi. (c) Antarcticite in Cueva dell'Arco de la Paciencia, San Pedro de Atacama, Chile. *Source:* Photo by Marco Vattano, La Venta Esplorazioni Geografiche. (d) Irregular halite stalactites, a formation called "the octopus," in 3N Cave, Namakdan Diapir, Iran. *Source:* Photo by Michal Filippi, Project NAMAK. (e) Halite helictites in Cressi Cave System, Atacama, Chile. *Source:* Photo by Riccardo De Luca, La Venta Esplorazioni Geografiche. (f) Speleothems made up of cubic halite crystals in Cueva Apollo, Atacama, Chile. *Source:* Photo by Marco Vattano, La Venta Esplorazioni Geografiche. (g) Violet fluorite cubes from Is Murvonis Cave, Sardinia. *Source:* Photo by Vittorio Crobu.

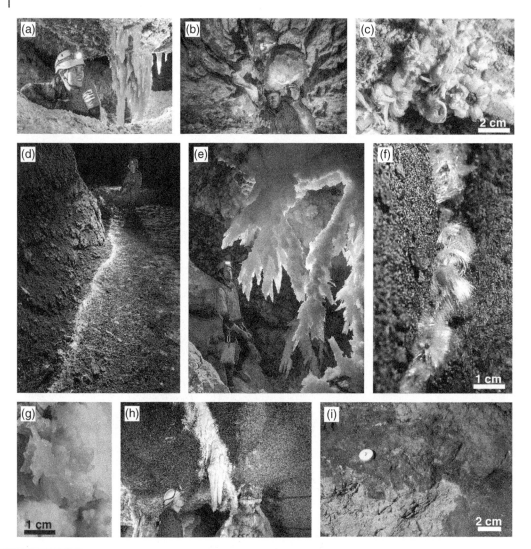

Figure 10.18 Sulfates in caves: (a) Szomonolkite stalactite in Cueva de los Minerales, Irazù volcano, Costa Rica. *Source:* Photo by Victor Carvajal. (b) Yellow gypsum crystal in Mlynky Cave, Ukraine. *Source:* Photo by Csaba Egri. (c) Gypsum flowers related to oxidation of pyrite in Buso del Vento, Veneto, Italy. *Source:* Photo by Sandro Sedran, S-Team. (d) Gypsum needles grown on organic rich clays in Torrinha Cave, Bahia, Brazil. *Source:* Photo by Philippe Crochet. (e) The famous gypsum chandeliers of Lechuguilla Cave, New Mexico. *Source:* Photo by Lukas Plan. (f) Epsomite flowers in Pelagalli Cave, gypsum area of Bologna, Italy. *Source:* Photo by Paolo Forti. (g) Bluish celestine crystals on gypsum in Lechuguilla Cave. *Source:* Photo by Lukas Plan. (h) Brown barite crystals on cave clouds overgrown with a white calcite speleothem in Santa Barbara Cave, Sardinia. *Source:* Photo by Victor Ferrer. (i) Alunite (grey) and jarosite (red) in the sulfuric acid cave of Kraushöhle, Austria. *Source:* Photo by Lukas Plan.

(Figure 10.18h). Barite in Lechuguilla Cave has also been ascribed to nonthermal processes of evaporation or mixing of fluids (Wisshak et al. 2020).

Alunite and jarosite, together with the silicate halloysite-10 Å, are less eye-catching cave minerals, being often powdery or earthy weathering products of sulfuric acid speleogenesis (Polyak and Provencio 2001) (Figure 10.18i). The sulfates containing K and Ar can be used for dating using radiometric techniques (K-Ar or Ar-Ar), being a method through which the exact timing of speleogenesis (when sulfuric acid was weathering the clays) can be estimated (Polyak et al. 1998).

10.6.3.3 Phosphates and Nitrates

Phosphates are the second largest group of cave minerals with around 60 species found to form in the natural subterranean environment. Five of them are rather common in caves, whereas the rest are rare and often related to very special conditions and environments, where ore minerals can be the source of different metals. The formation of most phosphates is associated with guano or bone deposits (Figure 10.19a, b). Cave phosphates are rather dull minerals, often forming crusts, nodules, lenses, and earthy masses (Figure 10.19c), but can sometimes form stalactites or stalagmites

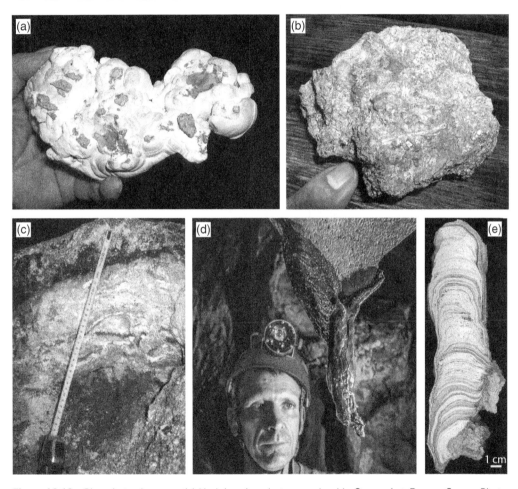

Figure 10.19 Phosphates in caves: (a) Nodular phosphate ore mined in Quercy, Lot, France. *Source:* Photo by Jean-Yves Bigot. (b) Phosphatic bone breccia from the Quercy Mine, France. *Source:* Photo by Jean-Yves Bigot. (c) A typical guano deposit (dark brown material on top) and the weathering products resulting from reaction with the carbonate rock (whitish part) and its leachates with reddish residual clays (terra rossa, below). Abisso dei Cocci Cave, Sicily. *Source:* Jo De Waele (Author). (d) Hydroxylapatite stalactite in the Puerto Princesa Underground River Cave, Palawan, Philippines. *Source:* Photo by Marco Vattano, La Venta Esplorazioni Geografiche. (e) Hydroxylapatite stalagmite (longitudinal cut) over 30-cm-long from Cucchiara Cave, Sicily. *Source:* Jo De Waele (Author).

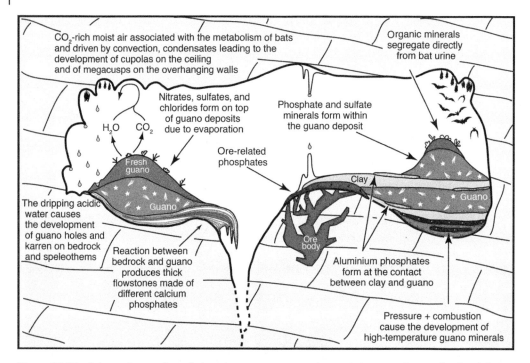

Figure 10.20 Schematic overview of phosphate occurrence and genesis in caves. *Source:* Forti and Onac (2016)/Schweizerbart Science Publishers.

(Figures 10.19d, e). The guano generally reacts with the carbonate host rock or with the clays hosted in the caves, forming phosphates containing C, Al, and Mg, and several phosphates also contain the ammonium ion (NH_4^+) derived from bat urea. Hydroxylapatite is the most stable apatite mineral, with the F- and Cl-rich variety being less commonly found. In acidic, wet and rather fresh guano deposits the typical mineral association is composed of brushite, ardealite, and gypsum. Taranakite and variscite form where guano comes in contact with Al-rich clays in slightly higher acidic conditions (less buffered by carbonates) (Audra et al. 2019) (Figure 10.20). Spontaneous combustion of guano can lead to high temperatures in the guano causing dehydration and, when temperatures reach very high values (exceeding 550 °C), some rare high-temperature phosphates can form (e.g., hydroxylellestadite and berlinite) (Onac et al. 2007).

Nitrates are highly soluble minerals with a limited preservation potential in the normally wet cave environments. They occur in evaporative environments, in warm and arid climates, or in cave areas where air flow causes relative humidity to remain rather low and evaporation to occur constantly. The sodium (nitratine) and potassium (niter) nitrates are the most common minerals. The source of the nitrate is generally guano, but it can also be the soil or basaltic rock (Hill 1981a). These nitrates have been extensively mined during the different wars of the nineteenth century in the United States for the production of gunpowder, and arid conditions have allowed the preservation of most mining structures built over a century ago (De Paepe and Hill 1981).

10.6.3.4 Oxides and Hydroxides
It is quite common to encounter iron and manganese oxides in caves, often forming poorly crystalline reddish, brownish or blackish coatings on walls, floors, and cobbles (Figure 10.21a, b), and sometimes forming speleothems (stalactites, flowstones, and stalagmites) entirely composed of

Figure 10.21 Oxides, hydroxides, and silica in caves: (a) Black manganese oxide coating the walls of a subterranean lake, Sas Venas Cave, Sardinia. *Source:* Photo by Vittorio Crobu. (b) Black Mn coating on boulders in the streamlet in Buso del Vento, Vicenza, northern Italy. *Source:* Photo by Sandro Sedran, S-Team. (c) Goethite flowstone in Guacamaya Cave, Auyàn Tepui, Venezuela. *Source:* Photo by Francesco Sauro, La Venta Esplorazioni Geografiche. (d) Bed of reddish iron hydroxides (below the hand of the person) that controlled the speleogenesis (inception horizon) of Guacamaya Cave, Auyàn Tepui, Venezuela. *Source:* Photo by Francesco Sauro, La Venta Esplorazioni Geografiche. (e) Rippled opal stalactites in Imawari Yeuta Cave, Auyàn Tepui, Venezuela. *Source:* Photo by Riccardo De Luca, La Venta Esplorazioni Geografiche. (f) Opal mushrooms in Imawari Yeuta Cave, Auyàn Tepui, Venezuela. *Source:* Photo by Vittorio Crobu, La Venta Esplorazioni Geografiche.

oxides. The most common iron hydroxide is goethite (Figure 10.21c, d), whereas the most frequent manganese oxides are todorokite and birnessite. These minerals form where reducing and anoxic fluids, often rich in organic material, bring Fe and Mn in solution into the aerated cave environment where oxidation occurs rapidly. Manganese (Mn^{2+}) forms soluble compounds with organic acids (humic, citric, oxalic, and acetic) in anoxic waters, and is oxidized in presence of O_2 into Mn^{4+}, a process which is often mediated by chemoautotrophic bacteria (Northup et al. 2003; Vaccarelli et al. 2021). Fe oxides precipitate at mildly acidic pH values above 6, whereas Mn oxides need more alkaline conditions (pH>8.5) (Bernardini et al. 2021).

Iron is soluble as the ferrous ion (Fe^{2+}) in anaerobic and reducing conditions, often at low pH. Also in this case, in the oxidizing environment of a cave and often with mediation of microorganisms, the ferrous ion oxidizes into the ferric one (Fe^{3+}), precipitating at higher pH levels in the buffering carbonate environment.

In Al-rich environments, the final and most stable mineral product is gibbsite, formed by oxidation of Al-rich soils or terra rossa, often in slightly alkaline conditions (Polyak and Provencio 2001). This mineral often occurs in sulfuric acid caves, but in this case gibbsite is a weathering product of alunite related to alteration occurred once the acid conditions were no longer active (D'Angeli et al. 2019b).

10.6.3.5 Silica and Silicates

Halloysite-10 Å, also named endellite in the past, is another typical by-product of the alteration of montmorillonite clays in a sulfuric acid cave environment (Polyak and Güven 1996). It has also been found in a quartzite cave where the oxidation of sulfides (galena and sphalerite) has created low pH conditions (Sauro et al. 2014). Allophane, another amorphous hydrous aluminum silicate mineral, is also typical of rather acid environments, and mostly occurs where acidic waters are in contact with silicic rocks (granites or volcanic rocks), depositing allophane when they come in contact with the buffering carbonate environment. The most common silicate cave mineral is opal, occurring often in lava caves and being the main mineral forming biomediated speleothems of various peculiar shapes in the South American quartzite caves (Aubrecht et al. 2012; Wray and Sauro 2017; Sauro et al. 2018) (Figure 10.21e, f). Opal also occurs in limestone caves, often interlayered within aragonite and calcite flowstones and passing unrecognized. It frequently forms coralloids, related to evaporation that induces the deposition of low-crystalline silica. Quartz requires more energy to form, and is typical of thermal conditions. It occurs as large crystals in Wind and Jewel Caves (South Dakota) (Palmer et al. 2016) and geodes with meter-sized quartz crystals have been encountered during the excavation of tunnels in the Swiss Alps. The temperature of formation of these crystals, and thus the hydrothermal phase during which the cavities were coated with the crystals, can be inferred studying the fluid inclusions.

10.6.4 Secondary Ice

Ice as a secondary deposit can be classified as a true cave mineral, being composed of the oxide water, (H_2O) that crystallizes in the hexagonal crystal system. It obviously occurs in cave areas where the temperature reaches temperatures below 0 °C, and is mostly an ephemeral deposit that forms during the cold season and disappears in the warm season. In certain conditions (e.g., high elevation), where the cave air remains below or close to the freezing point of water all year round, perennial ice bodies can persist over longer periods. Cooling is generally achieved by a combination of different mechanisms, including trapping of cold winter air, cooling by evaporation of water (during warmer periods this mechanism allows the cave air to keep cool), and cooling by ventilation (the chimney effect).

These, sometimes, large bodies of ice differ from the glacière type described earlier in the clastic sediments section (Figure 10.8d,e and Table 10.1), in that they form in situ. They are not derived from snow transported into the cave, transformed into firn first, and ice later. The ice can form by freezing of dripping and infiltrating waters or by direct freezing of condensing vapor. These perennial ice deposits, as most glacière ones, are generally layered, recording seasonal freezing and thawing. They commonly include fine windborne or water laid particles and dust, or local residues derived from limestone dissolution (Figure 10.22a, b). These internal ice bodies are often accompanied by the

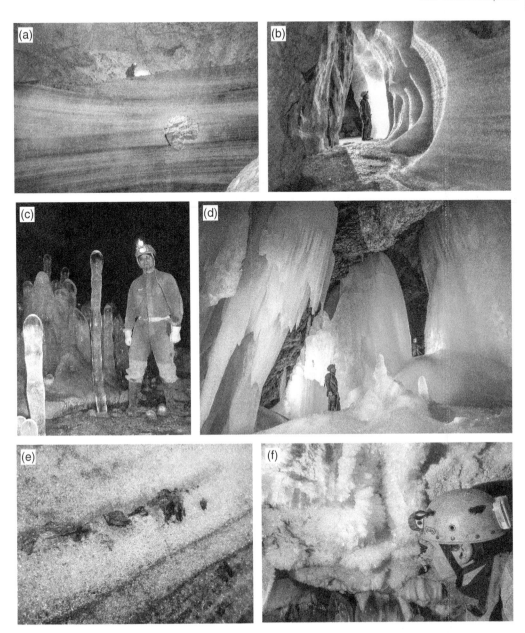

Figure 10.22 Ice in caves: (a) Stratified ice body in Dachstein Mammuthöhle, Austria, as it was in 2006. This internal ice mass has since almost completely melted away. *Source:* Photo by Lukas Plan. (b) Layered ice body in one of the most southern alpine ice caves, Rem del Ghiaccio, Piedmont, northern Italy. *Source:* Photo by Bartolomeo Vigna. (c) Bamboo-like ice stalagmites in Dachstein Mammuthöhle, Austria. *Source:* Jo De Waele (Author). (d) Blue ice falls, stalagmites, and stalactites in Eiskogelhöhle, Austria. *Source:* Photo by Csaba Egri. (e) Organic remains (leafs) within ice suitable for C^{14} dating. Lambda 21 Cave, Mongioie Massif, northern Italy. *Source:* Photo by Bartolomeo Vigna. (f) Hoarfrost composed of hundreds of hexagonal ice crystals, Dachstein Mammuthöhle, Austria. *Source:* Photo by Lukas Plan.

typical speleothems formed by dripping and flowing waters, such as ice stalactites, stalagmites, and flowstones (Figure 10.22c, d). Stalactites, although documented, are less likely to persist because of the presence of warmer air flowing close to the ceiling, whereas the colder air remains trapped in the lower parts of the cave passages. Ice stalagmites often display a bamboo-like shape (Figure 10.22c), consisting of a succession of wider transparent ice knobs formed during warmer (close to 0 °C) periods, when more water fed the stalagmite, and intervening narrow portions of opaque ice formed more rapidly during colder periods, when less dripping reached the stalagmite top.

The freezing of water causes CO_2 to be expelled and solutes to concentrate, boosting the precipitation of dissolved carbonates (calcite). These very special speleothems, named cryogenic carbonates, are easy to recognize because of their very high C^{13} values (Žák et al. 2008). Very special gypsum speleothems can also form by repeated freezing and thawing of ice in gypsum areas (i.e., Pinega, Russia, Korshunov and Shavrina 1998). Moreover, the seasonal freezing and thawing causes the formation of annual ice couplets (similar to varves in glacial lakes), which are however often disrupted by climate-induced irregular melting and freezing (Figure 10.22a, b, e).

Ice can be dated using a variety of techniques, including but not limited to, layer counting, mass turnover rates (ratio between ice thickness and average basal melting rates), and radiometric methods, including C^{14} dating of organic remains, and U/Th or C^{14} dating of cryogenic calcite deposits trapped in ice (Kern 2018). When plant remains are used, it is preferable to use leaves or small twigs, and not large tree logs, because the smaller fragments decay faster at the surface, and thus bracket the age of deposition more tightly. It was thought that these internal ice bodies were remnants of the last ice age, but dating of organic remains trapped in the ice normally delivered ages of some millennia (Figure 10.22e). The glacière (snow fed) ice deposits in Scărişoara Ice Cave in Romania reach basal ages of 10 500 years B.P. at 22.5 m depth in an ice core (Perşoiu et al. 2017), and A294 ice cave in the central Spanish Pyrenees has a basal age of 6100 years B.P. (Sancho et al. 2018). Good reviews on ice caves, the occurrence and formation of ice deposits, and their paleoclimatic importance can be found in Luetscher (2005), Perşoiu and Lauritzen (2018), and Perşoiu and Onac (2019).

Another type of ice deposit corresponds to frozen water bodies such as lakes that are gradually transformed into ice during the cold season. The ice is normally crystal clear and can be transparent right to the bottom of the frozen lake, although some air bubbles are normally trapped. When moist air penetrates in the cave and comes in contact with the cave walls, floor and roof at temperatures below the freezing point it forms hoarfrost (Figure 10.22f). Depending on the temperature difference between air and walls and the moisture supply, the ice crystals can be very fine (like sugar), or form needles, small plates, or up to centimeter-sized hexagonal intertwined whitish crystals. When water trapped in sediments freezes, it forms lens-shaped or wedge-shaped bodies of ground ice, or cracks filled with needle-ice. Since the ice occupies greater volumes than the original liquid water (it expands approximately 9%), it can be extruded forming beautiful flower-like fibrous ice crystals (extrusive ice).

10.7 Speleothems

Speleothems are secondary mineral deposits that form in caves by flowing, dripping, ponded, or seeping water and take on a typical shape. They are mostly composed of minerals such as calcite, aragonite, or gypsum, but other minerals can also form entirely or partially speleothems (Moore 1952). They belong to the chemical sedimentary rocks deposited by water, and similar to most sediments, display more or less evident rhythmic variations in texture, fabrics, or mineralogy that give them a laminated appearance. These laminae represent successive stages of their growth, and can thus be used to reconstruct temporal variations in the physical and geochemical conditions associated with the deposition of each single lamina.

Speleothems can be classified according to their external shape (morphological classification), the type of waterflow that created them (morphogenetic classification), and their internal texture and structure, which depends mainly on the environment in which they formed (ontogenetic classification). Although theoretically this last classification would be the most adequate for inferring the environmental conditions in which the speleothem formed, it is not very practical since it requires disturbing the deposit and carrying out detailed petrographic investigations on thin sections using a polarizing microscope (Self and Hill 2003; White 2012). The classification based on morphological characteristics, although easy to use by speleologists, does not tell much about how the speleothems formed (White 1976). The most widely used and practical classification scheme is the one proposed by Hill and Forti (1997), which describes the morphology in a first classification level, and considers origin, formation mechanisms, and crystallographic features as additional criteria to refine the classification. These authors describe almost 40 types of speleothems, some of which are very special and restricted to extremely rare and unusual conditions. Palmer (2007) proposes a classification scheme considering the main mechanism (and environment) of formation of the speleothems. This is a very practical subdivision because in similar environments cavers can expect to find typical associations of speleothems. We have adopted this practical criterion here, although we will describe only the most common speleothem forms. To know more about the rare and special speleothem types, the reader is referred to Hill and Forti (1997). Additional details and nice examples of different calcite, aragonite, and gypsum speleothems can be found in Cabrol and Mangin (2000), and the most significant speleothems used in paleoclimate research are described in Fairchild and Baker (2012). Figure 10.23 shows an ideal cave with the main types of speleothems, whereas Table 10.4 lists the most frequent forms and their main characteristics.

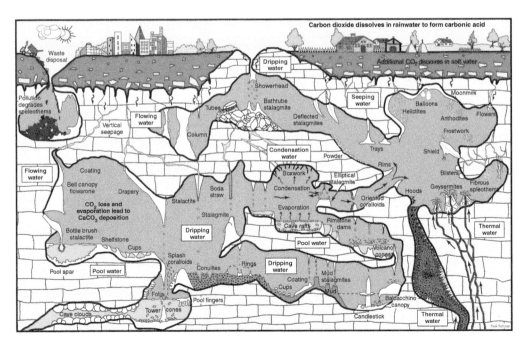

Figure 10.23 Diagram of an ideal cave illustrating the main types of speleothems and depositional settings. *Source:* Modified from Hill and Forti (1997).

Table 10.4 The most common types of carbonate speleothems and their main characteristics.

	Depositional agent	Speleothem	Brief description
Aerate zone	Slow dripping	Soda straw	Ca. 5 mm wide and 0.01–9 m long tubular, hollow, and vertically downward growing speleothem, often fed by slow and constant dripping
	Slow to moderate dripping	Stalactite	Conical, vertically downward growing speleothem tapering to a rather sharp tip. They mostly have an internal feeding tube and are characterized by irregular dripping
	Slowly flowing trails of drops	Drapery (curtain)	Planar and vertically downward-growing speleothem with curvilinear plan form and growth layers parallel to the surface on which it grows
	Fast dripping on the floor	Conical tapering stalagmite	Massive speleothem with no internal feeding tube that grows from the floor, tapering upward and ending with a rounded top
	Slow and constant dripping on the floor	Candlestick stalagmite	Tall cylindrical speleothem with rather constant diameter growing upward, ending with a rounded top and lacking central feeding tube
	Variable dripping from high roofs	Corbelled or terraced stalagmite	Leaved or plated stalagmite with a rather constant central diameter (candlestick) on which various types of protuberances grow outward
	Intense dripping from high roofs	Flat-topped stalagmite	Large stalagmite with a flat top surface
	Unconfined sheet flow	Flowstone	Massive, irregularly layered speleothem covering the floor or walls of caves
	Dripping or flowing water	Bell canopy	Hood-like flowstone with irregular bottom (grown on sediments now eroded) or with perfectly flat underside (grown on a water level)
	Dripping and flowing water	Column	Pillar-like speleothem extending from the bottom to the roof of the cave, resulting from the merging of a stalactite and a stalagmite
	Highly variable dripping water from large holes	Showerhead	Bell-shaped speleothem, hollow inside and with drapery- and flowstone-like walls surrounding a large dripping hole
	Dripping on soft sediments	Conulite/Birdbath	Cone-shaped or a more complex bowl-shaped calcite cast in soft sediments
	Water film	Normal coralloid	Popcorn-like subspherical speleothems with irregular laminae growing on wet walls and roof

Table 10.4 (Continued)

	Depositional agent	Speleothem	Brief description
	Splash water film	Splash coralloid	Popcorn-like often coalescing speleothems with irregular laminae growing on wet walls and roof close to intense dripping points
	Capillary water droplets	Frostwork	Needles of minerals (generally aragonite) with no internal feeding channel
	Capillary water droplets	Helictite (eccentric)	Independently growing branches of mm- to cm-sized diameter with an internal capillary feeding channel (0.01–0.5 mm)
	Capillary water droplets	Anthodite	Clusters of radiating branches a few mm in diameter with an internal capillary feeding channel (0.01–0.5 mm)
	Capillary water film	Welt	Planar excrescence projecting out of a fracture in the wall, often forming collars around columns
	Capillary water film	Shield	Oval or circular planar speleothems composed of two parallel plates separated by the capillary feeding fissure
	Capillary water film	Rim	Tube-like features with a smooth inner side and often a coralloid-covered outer shell. Associated with openings with air circulation
	Capillary water film	Blister	Bubble-like features with a smooth inner side and often a coralloid-covered outer shell. Associated with openings with air circulation
	Water film	Tray	Elephant-feet-like formations covered with coralloids and with a flat underside cut by condensing air masses below
	Water film	Moonmilk	Tooth-paste like, soft microcrystalline material. Powder when dry
Air-water interface	Dripping in shallow pools	Cave pearl	Spherical, ellipsoidal, cylindrical, cubical, or irregular free speleothem forming in shallow cave pools. They have concentric layering, often around a central nucleus that can be foreign (e.g., sand grain). Can reach 20 cm in diameter
	Unconfined sheet flow	Rimstone dam (gours)	Speleothem barrier with perfectly horizontal crest enclosing water pools. Their plan pattern is generally wavy with their convex parts pointing downstream
	Standing water	Shelfstone	Speleothem ledge on borders of pools, with upper flat surface in contact with air, and lower overhanging side underwater

(Continued)

Table 10.4   (Continued)

	Depositional agent	Speleothem	Brief description
	Standing water	Cave raft	Floating thin speleothem flake with an upper flat surface and a lower submerged side coated with crystals projecting into the water
	Standing water under drips	Raft cone/Double raft cones	Underwater accumulation of sunken rafts under dripping points
	Shallow standing water	Tower coral	Crystalline, upward growing accumulations in shallow pools with fluctuating water level
Underwater	Water table pools	Cave cloud (mammillary calcite)	Layered smooth and rounded calcite crusts covering the walls of (often thermal) water table pools
	Oscillating water pools (often thermal)	Folia	Stacked levels of speleothems in the form of bracket fungi, formed by fast CO_2 degassing in pools with a variable water level
	Standing water pools	Poolfinger	Vertically downward-growing smooth finger-like layered speleothem a couple of centimeters in diameter
	Shallow standing water	Pool spar (dogtooth and nailhead)	Calcite crystals growing on the submerged sides and floor of shallow pools with scarce recharge
	Shallow standing water	Subaqueous coralloid	Popcorn-like (globular) speleothem with regular concentric layered texture forming on the walls and floor of shallow pools
	Standing water pools	Chenille spar	Like poolfingers, but characterized by a crystalline palisade-like shape

10.7.1   Speleothems Formed by Dripping and Flowing Water

One of the most common sensations that visitors experience in a cave is the silence interrupted by the rhythmic sound of dripping water. This water has percolated from above, passing through the soil and epikarst, dissolving CO_2 and limestone (or gypsum) along its more or less winding flow path, and becoming saturated or nearly saturated with respect to the dissolved minerals. Upon its arrival on a cave roof, the water tends to attain chemical equilibrium with the cave atmosphere. Since the amount of dissolved CO_2 (in the range 0.01–0.1 atm) is commonly at least one order of magnitude greater than the P_{CO_2} in the cave atmosphere (ranging between 0.01 and 0.001 atm), the water releases carbon dioxide, becomes supersaturated and $CaCO_3$ precipitates. Water droplets falling from the roof and splashing onto the ground can infiltrate, form rivulets, or can flow as a laminar sheet on the floor of the cave driven by gravity. These downward moving droplets of water can thus deposit $CaCO_3$ along their path: (i) where they emerge from the roof, (ii) along their flow path on the roof and walls before they fall vertically, and (iii) at the site below the dripping where they impact and splash. Speleothems formed by dripping water are often called by the general term "**dripstones**". The most characteristic dripstone forms are **stalactites** (hanging from the roof), **stalagmites** (building upward from the floor), and **columns** (when stalactites and stalagmites join forming a speleothem that connects the floor to the roof). The water then can continue flowing in the cave forming **flowstones**.

A droplet surfacing from a tiny crack on the roof hang there attached by surface tension and grows in size. The droplet releases CO_2 and deposits $CaCO_3$. The initial form is a slender cylindrical tube known as "soda straw" (fistuleuse in French, spaghetti or cannula in Italian). Note that soda straws can also form at low points on the ceiling (e.g., overhangs), where droplets tend to hang and grow (not necessarily from a crack). This explains the growth of soda straws at the downflow tip of draperies, fed by water running over the speleothem and not from a fissure.

Soda straws are subvertical tubes with very thin walls (0.1–0.3 mm) made up of only a few crystals that grow at their tips, their c-axis oriented vertically downward. The inner feeding channel of active soda straws is around half a centimeter in diameter, roughly corresponding to the size of the droplet just before it detaches from the speleothem tip. The diameter of soda straws is almost constant, and although thinner and wider straws are known, ranging between 2 and 9 mm, they are mostly around 5 mm in diameter. The minimum diameter of a soda straw mainly depends on the surface tension of water, which is temperature-dependent and keeps the droplet attached to the stalactite tip, and the gravitational force on the droplet, which tends to cause its free fall. This can be expressed by the equation

$$B_o = \frac{d^2 . \rho . g}{\sigma} \tag{10.13}$$

where B_o (the Bond number) is a dimensionless number, d is the soda straw diameter (in cm), ρ is the density of the fluid (in g cm^{-3}), g is the gravitational acceleration (in cm s^{-2}), and σ is the surface tension (in 0.001 N m^{-1}). Experiments on capillary glass tubes of different diameters performed by Curl (1972) have determined the Bond number for which the two forces (surface tension and gravitational force) counterbalance, and thus the minimum diameter of the droplet. The minimum diameter for a temperature of 22 °C and normal tap water is 0.51 cm, with greater values for colder water and smaller diameters for warmer water characteristic of hot tropical areas. Soda straws can grow several meters long (the longest known soda straw reaches over 9 m) (Figure 10.24a), but most are less than one meter long (Figure 10.24b), since they break under their own weight or by small perturbations (e.g., wind, seismic shaking). They form when recharge is steady and rather slow, with the droplet growing slowly and attached to the tip of the speleothems for a certain amount of time. The thickness of the walls of soda straws can vary. It can increase by precipitation along the inner wall or by water extruded through the crystal boundaries to the external part of the speleothem. Soda straws are in fact often laminated recording changing chemical conditions, which are often annual (Fairchild et al. 2001) (Figure 10.40a).

If water flows along the outer side of soda straws it also deposits carbonate (or other minerals), increasing the diameter and transforming the cylindrical speleothem in a true tapered **stalactite** (Figure 10.24c). These speleothems form where the recharge becomes less constant, and drip rate is often greater than that of nearby soda straws. The amount of carbonate that precipitates depends on the water chemistry and the time the droplets hang on the speleothem before falling to the ground, and thus the period during which CO_2 degassing can take place. Fast-flowing droplets produce stalactites that grow at low rates, in favor of the stalagmites growing below. Typical stalactites have a central tube related to the crystallization of carbonate around the edges of hanging droplets, and not necessarily corresponding to the true feeding fissure in the bedrock. As for soda straws, stalactites can start forming on low parts of the ceiling, where water droplets are naturally concentrated. These speleothems can grow through a positive feedback process, whereby more water trickles from their vicinity are collected as they become wider. These stalactites grow both laterally (in diameter) and at their tip by flow along their external surfaces, without being fed from

Figure 10.24 Speleothems formed by dripping and flowing water, stalactites and draperies: (a) Probably the longest soda straw in the world (9.12 m) in Wonder Cave, Mulu National Park, Borneo. *Source:* Photo by Jeff Wade and the Mulu Caves Project. (b) A densely packed group of meter-long soda straws in Shuttleworth Poth, Yorkshire, UK. *Source:* Photo by Mark Burkey. (c) Carrot-like stalactites in Savi Cave, Trieste, Italy. *Source:* Photo by Csaba Egri, La Salle 3D. (d) Gypsum stalactites in a gypsum cave in Sorbas, southern Spain. *Source:* Photo by Victor Ferrer. (e) Stalactite transforming into a drapery in Baradla Barlang, Hungary. *Source:* Photo by Csaba Egri. (f) Long transparent bacon-like drapery in Pál Völgyi Barlang, Budapest, Hungary. *Source:* Photo by Peter Gedei. (g) Profusely decorated room with inactive stalagmites and stalactites covered with crystal coatings of subaerial origin (evaporation), cave in southern France. *Source:* Photo by Andreas Schober.

a central tube. Stalactites often concentrate along water-recharging fractures forming alignments. These slender speleothems can form by precipitation of various minerals, including gypsum, which are found rarely in gypsum caves because the less soluble mineral calcite is more readily deposited (common ion effect), preventing the solution to become supersaturated with respect to gypsum. Gypsum stalactites form more easily in caves in arid areas, where evaporation can induce trickling waters to become supersaturated (Calaforra et al. 2008) (Figure 10.24d).

Ideal stalactite forms have the shape of icicles, with a circular cross-section and a conical geometry tapering downward to a more or less sharp tip. This is due to the fact that both forms, icicles and stalactites, are generated by similar processes (gravitational laminar water flow), although the former are built by a physical process (freezing) and the latter by more complex physical and chemical processes (CO_2 degassing or water evaporation causing precipitation of carbonate) (Meakin and Jamtveit 2010). Stalactites normally have a central tube (like a soda straw) bounded by a thin tubular layer of crystals with the c-axes elongated vertically downward, and a series of concentric layers of radially oriented crystals with their c-axes perpendicular to the central tube. Their outer surface can be very smooth and regular, but is sometimes crenulated and wavy. The shape of stalactites has been studied mathematically by Short et al. (2005). More complex forms (crenulations, bulbous geometries, and diameter variations) can be explained by directional recharge (water flows only on one side), greater evaporation on one side, changes in flow regime, or extrusion of water from cracks, just to mention a few of possible causes. Stalactites cannot reach the sizes of the underlying stalagmites, mainly because they break under their own weight. The longest known free-hanging stalactite has been found in Lapa do Janelão (Minas Gerais, Brazil) and is 28 m long.

Not always droplets grow rapid enough to reach their detachment size, but form trickles of water that run down the overhanging walls or speleothems (Figure 10.24e), clinging to the rock by surface tension. The trails of these rivulets are adapted to the irregularities of the rock surface on which they flow, thus following the steepest, more or less straight pathways with some deviations. Since calcite is preferentially deposited on the outside (convex) parts of the bends (where CO_2 degassing and evaporation are favored), these inflections increase their curvature progressively. Water releases CO_2 along its way depositing a trail of calcite crystals, until the droplet reaches its maximum size, detaching and falling down. A stalactite will often form on the detachment point. This linear trail of calcite builds up layers, often of different colors, which are parallel to the surface on which the speleothem grows. The resulting planar and vertically downward growing speleothems are called **draperies** (or curtains) because of their curving pattern and their thin and translucent appearance (Figure 10.24f). They have a thickness similar to that of soda straws (half a centimeter), making them translucent to strong light (which is why cavers often call them "cave bacon"). The crystal growth direction is perpendicular to the crystallization surface due to impingement or competitive growth, and some crystals grow faster than others. Draperies can display irregularities along their edge, especially on less inclined ones and those exposed to air flow (enhancing CO_2 degassing and evaporation), causing droplets to get trapped and deposition to occur in a saw tooth pattern. Stalactites and draperies can become less active due to local changes in water and/or air circulation, and can be coated by evaporative crystallizations (Figure 10.24g).

The drops that fall to the ground splash and can start flowing as a water film on the cave floor (Figure 10.25a). These dripping waters still contain more or less dissolved carbonate, depending on the initial composition, on whether it has deposited carbonate on the roof or not, and the recharge rate (fast dripping water has less time to deposit on the roof). Not all **stalagmites** have a corresponding stalactite above, and some start forming from a speleothem-free roof where dripwater surfaces without depositing any mineral (Figure 10.25d). The splashing of water causes it to lose CO_2 or, in certain conditions, to evaporate, thus leading to supersaturation and carbonate

Figure 10.25 Speleothems formed by dripping and flowing water, stalagmites: (a) Water splashing on a stalagmite tip in a Brazilian cave. *Source:* Photo by Philippe Crochet. (b) White candlestick stalagmites in LP2 Cave, Sežana, Slovenia. *Source:* Photo by Peter Gedei. (c) Five-meter tall candlestick stalagmite in Aven de la Salamandre, Gard, France. *Source:* Photo by Victor Ferrer. (d) Large conical stalagmite formed by abundant dripping without corresponding stalactite, Grotte des Lombrives, Ariège, France. *Source:* Photo by Philippe Crochet. (e) Stalactites, draperies, and stalagmites grown on fallen blocks in a French cave. *Source:* Photo by Andreas Schober.

deposition (Figure 10.25a). This process forms convex speleothems that grow upward from the floor and have a rounded top. The internal structure of stalagmites is characterized by layering with crystals growing perpendicularly to the surface. Stalagmites, on the contrary to stalactites, have no internal feeding tube and can reach heights of more than 60 m. The highest is located in Cueva San Martín Infierno, Cuba, and is over 67 m high.

Three main morphologies of stalagmites are found in caves (Franke 1965): **candlestick stalagmites**, conical tapering stalagmites, and terraced or corbelled stalagmites. If dripping is slow, most of the mineral that precipitates accumulates close to the impact point, and stalagmites grow tall and slender, like candlesticks (Figure 10.25b, c, e). If dripping is faster, more water reaches the stalagmite and a supersaturated water film can flow along its sides, widening it (Figure 10.25d). These **conical tapering stalagmites** are the most common forms found in caves. In **terraced** or **corbelled stalagmites**, the speleothem has a stem with a rather constant diameter on which different types of protuberances similar to leaves or blades grow (Figure 10.26a–c). These stalagmites resemble giant pine cones or the trunks of petrified palm trees. Stalagmites looking like a pile of plates also belong to this category (Figure 10.26d). These leaved (or plated) stalagmites form under drips falling from high roofs with high variability in flow rate.

Water drops are variable in size, depending mainly on the morphology of the substrate from which they hang, temperature, the chemistry of the water, and flow rate. Typical drop sizes in caves

Figure 10.26   Speleothems formed by dripping and flowing water, stalagmites: (a) A palm tree trunk stalagmite (pigne) in Aven Armand, Lozère, France. *Source:* Photo by Philippe Crochet. (b, c) Leaved stalagmite formed by intense dripping from a high roof in Xe Bang Fai Cave, Laos. *Source:* Photo by Dave Bunnell. (d) "Pile of plates" stalagmite in Aven Orgnac, Ardèche, France. *Source:* Photo by Victor Ferrer. (e) Fast dripping from a high roof generating a flat-topped stalagmite with leaves, Grotte de Vallorbe, Switzerland. *Source:* Photo by Philippe Crochet. (f) Flat stalagmite surface grading into flowstone and containing cave pearls, Xe Bang Fai Cave, Laos. *Source:* Photo by Chris Howes. (g) Strangely shaped stalagmite formed in periods of varying drip rate (being larger now) Grotte de Bournillon, France. *Source:* Photo by Andreas Schober. (h) Bell canopy along the course of the Reka River, in Škocjanske Jama, Slovenia. *Source:* Photo by Borut Lozej.

are ca. 6 mm corresponding to drop volumes ranging between 10 and 15 ml (Genty and Deflandre 1998; Collister and Mattey 2008). During their fall these drops reach the critical velocities (the speed at which the drop starts fragmenting into smaller droplets) at around $9 \, m \, s^{-1}$, within a few meters of their detachment point. At such speeds, roughly 20% of the water drop will be splashed upon impact on the smooth stalagmite surface, forming thousands of microdroplets (Figure 10.25a). During their fall the drops also lose micro-droplets due to turbulence. Other factors (e.g., air flow, irregularities on the roof and at the feeding point) also cause drops to deviate from their vertical path, thus impacting over greater surfaces. Very intense dripping from a roof high above the cave floor causes falling drops to spread over larger surfaces and to create a cloud of micro-droplets around the impact point (Figure 10.26e). Carbonate is thus deposited over a larger surface, causing the stalagmite top to become flattened (Figure 10.26e, f). The micro-droplets that remain suspended in the air, on the other hand, slowly descend to the ground more or less vertically (in very quiet cave environments with no or very slow air currents) or directionally (more or less deviated by slowly moving air). These microdroplets will have greater possibility to deposit on protruding lateral parts of the stalagmite; on all sides in still cave atmospheres and on the windward side where air slowly moves in one direction. This causes the formation of leaves and plates that shield the underlying parts from receiving water, and thus deposition of carbonate. This also explains why leaves do not overlap vertically over short distances, from centimeters to decimeters, depending on how disturbed the environment is, shorter distances indicating greater air movement. Plates might represent varying periods of high drip rates (forming the larger plates) and lower dripping (narrower parts).

In nature, a wide variety of intermediate types of stalagmites are encountered, some of which show the most bizarre shapes and combinations between stalagmites, draperies, and flowstones (Figure 10.26g). **Bell canopies** are hood-like flowstones formed over sediments that have later been eroded away. These canopies have irregular lower edges mimicking the shape of the top of the preexisting sediments (Figure 10.26h). Bell canopies can also form above more or less stable water levels, in which case they have a perfectly horizontal lower edge.

Multiple factors influence the final morphology of stalagmites (Gams 1981b), including but not limited to: (i) supersaturation degree of the drip water, (ii) the flow rate and its variability, (iii) the splashing effect of the water droplets, which causes part of the water to be excluded from the stalagmite deposition system, (iv) the height from which water falls, (v) the changes in the position of the dripping point, and (vi) local cave meteorology (relative humidity, temperature, CO_2 concentration, air flow). This complexity makes modeling stalagmite growth very difficult, and explains the wide variety of morphologies found in these speleothems.

This complexity is also encountered in the internal petrographic fabric and microstratigraphy of stalagmites. Given their paleoclimatic significance, a clear understanding of the genetic meaning of the petrographic and microstratigraphic changes in stalagmites is of great importance to derive sound interpretations from these continental records, coupled with numerical dating and geochemical proxies (Martín-Chivelet et al. 2017) (see Section 10.6.2 for some more detail). In general, stalagmite layers are thicker at the top, and tend to wedge out along their sides.

Stalagmites appear to have a minimum diameter, and it is extremely rare to find diameters smaller than 3–4 cm. From a water film with constant chemistry and in stable conditions of CO_2 degassing, experimental and theoretical work has shown that the minimum diameter can be approximated by the equation (Curl 1973)

$$d = 2\sqrt{\frac{V}{\pi \cdot \delta}} \tag{10.14}$$

where, d is the minimum diameter of the candlestick stalagmite, V is the volume of the falling water droplet, and δ is the thickness of the water film on top of the stalagmite. For typical film thicknesses ranging between 0.02 and 0.07 mm and water drop volumes of 0.14–0.25 ml, the minimum diameter of stalagmites ranges between 1.4 and 4 cm, as predicted by Curl (1973). Modeling stalagmite growth and shape is also an important issue for paleoclimate studies, and has been attempted by several authors (Kaufmann and Dreybrodt 2004; Romanov et al. 2008). Difficulties in modeling are related to the changing chemistry of waters during their movement on the stalagmite apex, mixing between newly dripping water and the water film on the stalagmite, losses by splashing, differences in water film thickness, and many other factors.

Water flowing over inclined cave walls and floor (unconfined sheet flow) deposits carbonate along its path in layers approximately parallel to the surface on which it flows. On close-to-vertical walls, **flowstones** are transitional forms to stalactites and draperies, and sometimes resemble frozen waterfalls (Figure 10.27a, e) or organ pipes (Figure 10.27b, d). They often show a rippled or ribbed surface, due to variable flow conditions producing areas of enhanced (CO_2 degassing) or attenuated deposition (Figure 10.27c, d) (Forti et al. 2017). On less inclined surfaces, they tend to follow the irregularities of the substrate on which they grow, creating zones of faster speleothem growth. On more inclined areas with greater turbulence, CO_2 degassing and speleothem growth increase. Evaporation and/or CO_2 degassing can be described by the escape of molecules from a water film surface. On a flat water surface the angle open to the air is 180°, with molecules escaping at high angles having greater probability to diffuse to the gaseous phase. On the border of a right-angle stone, the escape angle is 270°, and molecules will have greater possibilities to pass to the gaseous phase. Following this scheme, on convex parts of the flowstone CO_2 degassing and/or water evaporation is greater than that on flat surfaces, so the solution becomes supersaturated more easily at the former sites favoring speleothem growth. This differential speed in carbonate precipitation on variously inclined (convex and concave) flowstone surfaces causes layering to be much more irregular in flowstones than in stalagmites, with the occurrence of frequent hiatuses, thickenings, and wedging outs in the laminae over short lateral distances.

Flowstones and stalactites growing from the roof can merge with underlying upward-growing stalagmites or flowstones (Figure 10.27e, f), eventually forming **columns** or pillars (Figure 10.27g), some of which can reach heights in excess of 60 m (a column in Than Sao Hin Cave, Thailand, is 61.5 m high) and diameters of more than 20 m.

Water coming down the roof does not always form a normal stalactite, nor does water hitting the floor always form a conventional stalagmite. Two examples are shown in Figure 10.28a, b. When dripping is very intense and extremely variable, springing out from large openings (some decimeters across), the depositional conditions can be highly variable. During low flow, water trickles can flow out of the feeding channel forming a concentric array of rudimental draperies, normally developing at some outward angle (around 15–45°) from vertical. During high flow, the water arrives undersaturated, dissolving the central part of the speleothem. In the end a kind of bell-shaped, hollow speleothem forms, which is known as **showerhead**, or dribbler when they are smaller. They are typical of shallow, tropical cave systems (Figure 10.28a).

When dripping occurs on a soft erodible material (normally silts and clays) the impact can cause the formation of a hole first, followed by the deposition of a calcite lining. This can create a cone-shaped calcite cast (single drip point) or a more complex bowl-shaped irregular form. The first is known under the name **conulite** (resembling the cone of an ice cream) and the second as **birdbath** (an irregular type of conulite). If the drip water becomes more supersaturated and the sediments are not washed away fast enough a conulite can transform into a stalagmite (Figure 10.28b).

Figure 10.27 Speleothems formed by dripping and flowing water, flowstones (a) Cream-colored flowstone contrasting with the dark limestone in Markov Spodmol Jama, Slovenia. *Source:* Photo by Peter Gedei. (b) Huge flowstone (like a pipe organ) in Ternovizza Cave, Trieste, northern Italy. *Source:* Photo by Sandro Sedran, S-Team. (c) Large ribbed flowstone-drapery in the Avenc dels Pouetons, Montserrat conglomerate cave, Catalonia, northern Spain. *Source:* Photo by Victor Ferrer. (d) Flowstone with ribbed organ pipes in Xe Bang Fai Cave, Laos. *Source:* Photo by Dave Bunnell. (e) Large massive stalactite feeding a small stalagmite, Grotte des Lombrives, Ariège, France. *Source:* Photo by Philippe Crochet. (f) Large flowstone almost touching the underlying stalagmitic flowstone, Buraco do Sopradeiro, western Bahia. *Source:* Photo by Mirjam Widmer. (g) Slender column formed by the coalescence of a stalactite and a candle-stick stalagmite, Umbertinijeva jama, Slovenia. *Source:* Photo by Peter Gedei.

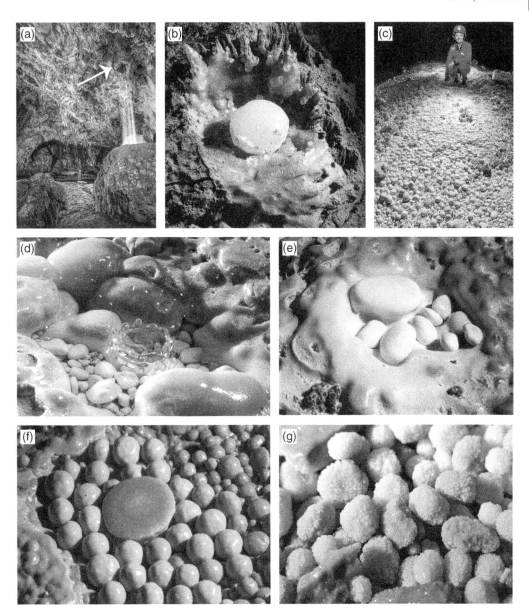

Figure 10.28 Speleothems formed by dripping and flowing water, special forms and cave pearls: (a) A showerhead stalactite (white arrow) in Gruta Temimina, São Paolo, Brazil. *Source:* Photo by Kevin Downey. (b) Conulite in clay deposits and incipient stalagmite, from Mammoth Cave, Jenolan National Park, western Australia. *Source:* Photo by Csaba Egri, La Salle 3D. (c) Blanket of cave pearls in Lapa do Janelão, Minas Gerais, Brazil. *Source:* Photo by Philippe Crochet. (d) Drop splashing in a small pool with cave pearls, Otoska Jama, Slovenia. *Source:* Photo by Philippe Crochet. (e) A nest of partially cemented cave pearls in Baradla Barlang, Hungary. *Source:* Photo by Csaba Egri, La Salle 3D. (f) Regular rounded cave pearls and a larger, flat-topped conical cave pearl in a cave in southern France. *Source:* Photo by Andreas Schober. (g) Cave pearls coated with secondary evaporative calcite crystals in a Slovenian cave. *Source:* Photo by Borut Lozej.

Flowstones and cave floors are typically irregular and can hold temporary or permanent water pools, sometimes covering large surfaces with water depths of only a few centimeters or less. These shallow water bodies often contain oncoid-like calcite or aragonite speleothems resembling pearls, from which their name, **cave pearls**, was derived (Figure 10.28c). These free speleothems are nested inside their own carbonate cup, or occur as groups in a larger calcite (or aragonite) floored nest. The fact that they do not attach to the floor is due to regular dripping or water flow, causing enough oscillation to prevent them from being amalgamated or cemented to the floor (Figure 10.28d). They can be perfectly spherical, ellipsoidal, cylindrical, cubical (Roberge and Caron 1983), or irregular in shape, depending mostly on the shape of the central nucleus from which they started forming, the nature of water agitation, and the available space for each pearl to grow. Most cave pearls start with a central nucleus, often a sand grain or a rock fragment, which is progressively coated by concentric layers of carbonate precipitates. Cave pearls can form very rapidly, especially where CO_2 degassing is favored by airflow, such as in ventilated mine tunnels (Melim and Spilde 2011). Although they are believed to form mainly by abiotic mineral precipitation, there is sometimes evidence of biogenic mediation in their growth (Jones 2009). A misconception is that cave pearls are round because they rotate, so that protuberances are smoothed away by erosion. Cutting cave pearls reveals that they tend to become spherical because they grow at the same speed in all directions as long as they are underwater, smoothing the irregularities of the nucleus and becoming perfectly spherical if no other impediments occur. In case of isotropic impediment, perfectly hexagonal cave pearls should form (the best packing configuration), but often intermediate forms are found (e.g., cubical). Different geometries develop when the top of the pearl comes out of the water and stops growing (Figure 10.28f), or when the adjacent cave pearls or wall do not allow the pearl to expand in that direction. Pearls can be coated by calcite or gypsum crystals when their upper parts stay dry over longer periods, and evaporation and capillary rise of water from below causes the growth of tiny crystals on their wet surfaces sticking out of the water (Figure 10.28g). Cave pearls commonly range from a few millimeters to around 5 cm in diameter, although pearls as big as 20 cm have been found in tropical caves, such as Hang Son Doong in Vietnam. Cave pearl nests are often characterized by pearls with a rather regular size, which is mainly due to the flow rate (and thus energy) occurring in the pools. If dripping becomes very rare, the larger cave pearls will not experience sufficient movement and will cement to the cave floor, leaving only pearls of smaller size as free-evolving speleothems. If occasionally an intense dripping (or water flow) occurs, the energy of the water will remove the smaller pearls from the nest, and only the large pearls will remain.

10.7.2 Speleothems Formed in Vadose Standing Water Bodies

Rimstone dams (in French gours) are barriers, generally of porous calcite, with a perfectly horizontal upper level enclosing shallow pools. Their height ranges between a few millimeters to several meters, the tallest dam being 22 m high in Actun Kotob Cave in Belize. Their length ranges between some centimeters to tens of meters, with the longest dam in the world reaching 61 m in Xe Bang Fai Cave, Laos (Figure 10.29d). They can form upstream or downstream of flowstones, or in any place where water flows as a thin sheet over the cave floor or on inclined side walls. They can also form in stream passages, but only where the flow of water is not strong enough to cause mechanical erosion to prevail over carbonate precipitation on the dam.

When a thin water film enters a cave, it can deposit a flowstone in the upstream part because the high degree of supersaturation allows calcite to deposit anywhere. Downstream the supersaturation decreases and carbonate deposition occurs preferentially in places where CO_2 diffusion is

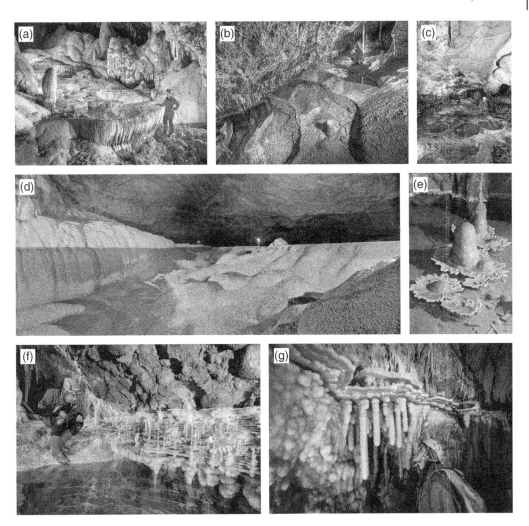

Figure 10.29 Speleothems formed in standing water, rimstones, and shelfstones: (a) Rimstone dam originating from an upstream flowstone and formed in slowly or intermittently flowing water. Gruta Fazenda America, Brazil. *Source:* Photo by Csaba Egri. (b) Rimstone dams in Martinska Jama, Slovenia. *Source:* Photo by Chris Howes. (c) Pool with rimstone dams, shelfstone, and isolated cups in Cova de les Dones, Valencia, Spain. *Source:* Photo by Victor Ferrer. (d) Probably the largest rimstone dam and pool in the world, in Xe Bang Fai Cave, Laos. Cavers (headlamps) for scale. *Source:* Photo by Dave Bunnell. (e) Shelfstones growing around stalagmites in a pool in Grotta delle Margherite, Trieste, Italy. *Source:* Photo by Sandro Sedran, S-Team. (f) Various levels of fragile shelfstone formed in a small pool in Cueva Fuentemolinos, Burgos, Spain. *Source:* Photo by Sergio Laburu, Adrián Vázquez and Roberto F. Gracia. (g) Curious subaqueous, probably overgrown stalactites (called phalagmites by cavers), hanging from shelfstone in a now dry pool in Ellis Basin Cave System, Kahurangi National Park, South Island, New Zealand. *Source:* Photo by Neil Silverwood.

favored. Enhanced CO_2 loss prevails where the water flow is more turbulent or where a convex shape of the substrate creates a larger specific contact surface with the air for degassing to occur. In these more convex and more prominent parts of the substrate, the water film is also thinner than in the more planar or concave areas, also contributing to enhanced CO_2 degassing. Once carbonate has started depositing on the most favorable place, this point with growing relief will become even more propitious for degassing, and carbonate will preferentially precipitate building up the dam through a positive feedback mechanism (Hammer et al. 2010). The horizontal dam crests are also a nice example of a self-repairing process in caves. If a notch is incised, water flow from the upstream basin will be concentrated at this lower spot, causing carbonate precipitation mostly there and not on the other parts of the rim, ultimately repairing the irregularity. On the contrary, if one part becomes higher, less water will flow over it, thus arresting its growth and letting the other parts catch up again. In the long run, growth rate is essentially uniform all along the rim crest. Rimstone dams will thus increase their height constantly, until the hydraulic gradient becomes too low for water to flow downstream, thus causing deposition everywhere within the pool, turning rimstone dams into a normal horizontal flowstone pavement. Rimstone dams thus acquire their greater heights on more inclined slopes and create shallower pools on less inclined substrates. The rimstone pools occupy smaller surfaces but have greater depths on steep substrates, and are much larger but shallower in less inclined areas. Carbonate deposition tends to be greater in places where flow is faster (i.e., enhanced CO_2 degassing), as occurs on the steep downstream parts of the rims. This causes rimstone dams to migrate downstream (frontal accretion), and their convex plan form to point downstream (Figure 10.29a). In low-gradient flow, the planimetric shape of rimstone dams can become much more complicated, with tightly curved and tortuous plan patterns, sometimes even isolating parts of the dams to form disconnected cups (Figure 10.29c). In long profile, the dam can be perfectly vertical, or overhanging upstream or downstream. When the water flow and net precipitation are substantial, the rim may attain an overhanging upstream side, and a downstream convex shape where carbonate precipitation is faster (Figure 10.29b, d). When water flow is occasional and small, the downstream side of the rimstone dam tends to be vertical to overhanging (Figure 10.29a).

The edges of relatively static cave pools, commonly fed by dripping water or small water trickles, can be bordered by horizontal ledges of calcite, known as **shelfstones** (Figure 10.29c). Stalagmites or raft cones emerging from the water surface of a pool can be also bordered by these calcite shelves (Figure 10.29e). The thickness of shelfstones ranges from 1 mm (see cave rafts below) to over 30 cm. Evaporation and/or degassing of CO_2 causes the surficial layers of water in the pools to become supersaturated in carbonate leading to the precipitation of calcite or aragonite. A shelfstone forms if deposition occurs on the margins of the pools or on substrates protruding from the water surface (rocks, stalagmites), whereas free-floating carbonate flakes called rafts form at some distance from the pool edges. Shelfstones grow both underwater and at the water–air interface, with the faster growth occurring at the water surface. This causes shelfstones to grow horizontally away from the wall, always controlled by the water level. With increasing water depth, supersaturation is less easily reached, and less carbonate will deposit, which causes shelfstones to develop overhanging walls below the ledge. They can grow thicker in pools where the water level varies during the year. The top of the shelfstone corresponds to the highest water level reached in the pool, whereas the lower edge indicates the lowest level normally reached by the water. If the pool level has changed many times, a series of tiered shelfstone levels can form (Figure 10.29f, g), the thickness of which is an indicator of both water level oscillations and the time during which the water level stayed within a certain range. Old shelfstone levels can be suspended at high elevation on cave walls in now dried-up pools, and can be decorated with superposed dripstones (stalactites). If these were submerged again for enough time in a supersaturated water pool, they can be coated with mammillary calcite (Figure 10.29g), forming peculiar speleothems called "phalagmites" for obvious reasons (Thomas and Silverwood 2017). More often stalactites in contact with supersaturated pool waters are coated by calcite crystals and pool spar (Figure 10.32b).

In almost static cave pools barely disrupted by water flow, the surface can be partly or entirely covered with thin sheets of floating minerals, mainly aragonite and calcite (Figure 10.30a), named

Figure 10.30 Speleothems formed in standing water, rafts, and raft cones: (a) Large cave pool with floating calcite rafts, Lapa do Convento, Bahia, Brazil. *Source:* Photo by Philippe Crochet. (b) Old calcite rafts thickened by carbonate precipitation while lying on the pool floor, Szemlő-Hegyi Barlang, Budapest, Hungary. *Source:* Photo by Csaba Egri. (c) Undermined calcite raft deposit now forming the ceiling of Pál-Völgyi Barlang, Budapest, Hungary. *Source:* Photo by Csaba Egri. (d) Tall active (underwater) raft cone in Abismo Anhumas, Mato Grosso do Sul, Brazil. *Source:* Photo by Marcelo Krause. (e) Sharp-tipped raft cones in Pál-Völgyi Barlang, Budapest, Hungary. *Source:* Photo by Csaba Egri. (f) Double raft cones in Sima de la Higuera, Murcia, Spain. *Source:* Photo by Victor Ferrer.

cave rafts. Their upper surface is smooth and microcrystalline, whereas their lower part is coated with small crystals projecting into the water. Calcite (or aragonite) precipitates because of CO_2 degassing and/or water evaporation from the surface of the water body, and is enhanced in cave areas where airflow removes the released CO_2 and/or water vapor. Crystal growth at the edges in contact with the air is fast and parallel to the water surface (i.e., rafts grow laterally), whereas on the lower part of the rafts crystals grow more slowly and vertically downward. Rafts can reach thicknesses of 1 mm for diameters of over a decimeter. Although they are denser than water, they can float due to surface tension, until they become too heavy, sinking to the bottom of the pool. They can also sink because of perturbations, like the ones produced by dripping, or because the entire cave pool dries out. When the pool gets filled again with water, some of the rafts can start floating again. Once they sink and stay under water, they can continue to grow in all directions, cementing to the bottom or the walls of the pool (Figure 10.30b). Cave raft accumulations can be rather thick if the process continues over long times, and erosion sometimes removes part of these raft blankets or the underlying sediments, leaving hardened cave rafts on the roof of the cave passage (Figure 10.30c).

If cave rafts sink to the bottom below a dripping point, they can stack on top of each other forming a **raft cone** (volcano cones, tower cones, and comet cones are varieties). These resemble stalagmites, but are not characterized by parallel laminations, instead are accumulations of thin calcite (or aragonite) platelets. Since they accumulate underwater, they are progressively cemented and coated with subaqueous calcite deposits, obscuring their internal flaky structure. In general, they show side angles of around 60° when the rafts are not cemented (Figure 10.30e), but can attain steeper and higher slopes with sides inclined at 70-80° when cemented (tower cones, Figure 10.30d). Raft cones can reach several meters high when pools are deep (Figure 10.30d), although most are a few decimeters tall (Figure 10.30e). Cones developed in slowly flowing waters can become elongated in plan view, with tapered tails pointing downstream (comet cones) (Polyak and Provencio 2005). When they reach the surface of the water pools, they stop growing vertically, and their tops are eroded by the falling water drops, forming a hole (volcano cones). Raft cones can sometimes form in different phases, resulting in two generations of cones superposed on each other, forming double-tower cones, such as in Sima de la Higuera in Spain (Gázquez and Calaforra 2013) (Figure 10.30f).

In isolated standing water pools fed only by dripping or small trickles of water, quantities of precipitating carbonate are usually scarce. Since CO_2 degassing and evaporation occur at the pool surface, the highest supersaturation occurs there (Figure 10.31a–c), but diffusion within the water column also creates supersaturation in calcite (or aragonite) deeper in the water body. Most carbonate thus precipitates in the shallow parts of the pools, with nucleation of numerous crystals but of small sizes. In deeper water levels less carbonate is deposited, but crystals are larger (Figure 10.31c). Competitive (impingement) growth of crystals favors degassing or evaporation reaches the bottom those with the c-axis perpendicular to the substrate on which they grow, resulting in palisade, or columnar textures. Calcite can precipitate in two types of crystalline forms: **dogtooth spar** (scalenohedral crystals) (Figure 10.31c, d) and **nailhead spar** (rhombohedral and mixed forms) (lower crystals in Figure 10.31e). These spars can be used to reconstruct ancient depositional deep phreatic (often thermal) cave conditions and the related landscape evolution (Decker et al. 2018a, 2018b; Gázquez et al. 2018).

Where drips constantly bring enough new dissolved carbonate in shallow pools, supersaturation induced by CO_2 degassing or evaporation reaches the bottom of the water body, forming calcite crystal flowers on the pool floor (Figure 10.31d). If these crystals reach the water surface, and the quantity of carbonate precipitating is small enough, crystals develop only partially, generating hopper forms in which growth occurs at the angular edges of the crystal, leaving most of the crystal faces hollowed (Figure 10.31e). In some rare cases, different branching crystal structures can form, resembling small Christmas trees or leaves (Figure 10.31f). In thermal caves, the walls of phreatic-level pools can be covered with mammillary crusts or coatings of dogtooth calcite spar, which is the typical crystalline form in high-temperature fluids, resembling the surface of a pineapple (Figure 10.31g).

The bottom of shallow isolated (non-phreatic) lakes can sometimes be covered with small cone-like vertically growing calcite structures resembling a collection of toy towers (Figure 10.31h). These are, in reality, a type of **subaqueous coralloids** (a speleothem mostly related to evaporation, see below), but dominated by a vertical growth direction. They appear to form in shallow pools where the water surface is subjected to intense evaporation, and the water level fluctuates

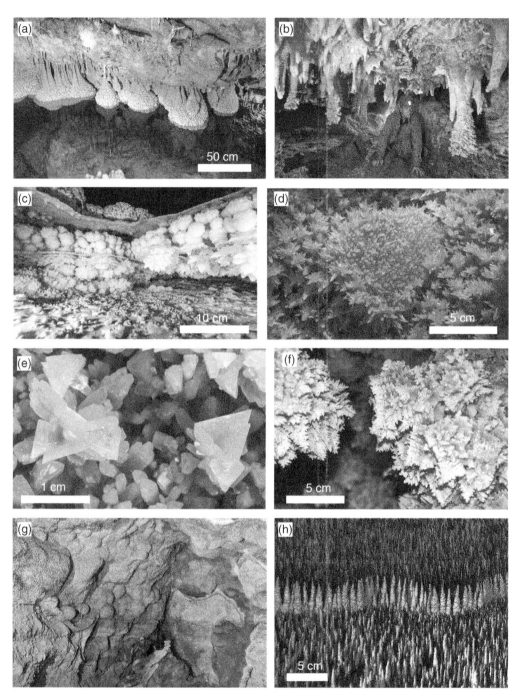

Figure 10.31 Speleothems formed in standing water, spar: (a) Globular pool spar (phreatic overgrowths on speleothems) in an anchialine lake in a coastal cave at Cala Varques, Mallorca, indicating the present oscillating sea level. *Source:* Photo by Jean-Yves Bigot. (b) Stalactites covered with pool spar in Cueva Manolón, Valencia, Spain. *Source:* Photo by Victor Ferrer. (c) Graded pool spar, showing increasing crystal size with water depth, in a gour in Abisso Milazzo, marbles of the Apuan Alps, Tuscany, Italy. *Source:* Photo by Sandro Sedran, S-Team. (d) Scalenohedric calcite pool spar in Grotte des Ecossaises, southern France. *Source:* Photo by Andreas Schober. (e) Calcite hopper crystals and normal calcite spar in a pool in the Puerto Princesa Underground River cave, Palawan, Philippines. *Source:* Photo by Vittorio Crobu, La Venta Esplorazioni Geografiche. (f) Dendritic calcite poolspar in Cueva de Nerja, Andalusia, Spain. *Source:* Photo by Victor Ferrer. (g) The walls in Jewel Cave are entirely covered with thermal calcite spar. *Source:* Photo by Dave Bunnell. (h) Tower coral (cent milles soldats; one hundred thousand soldiers) 5–10 cm tall on a rimstone pool in the Grotte du Trabuc, Cévennes, France. *Source:* Photo by Victor Ferrer.

over a few centimeters, but never dries out completely. In these conditions, evaporation drives capillary movement of the water along the surface of the growing speleothems and within the small cones, and the top of the **tower coral** speleothems corresponds to the maximum water level.

10.7.3 Speleothems Formed in Phreatic (Thermal) Water Bodies

If the pools are associated with the phreatic water level and are thus constantly renewed with new incoming carbonate-rich water, the quantity of calcite that can precipitate increases. These situations are typical (although not exclusive) of hypogene (thermal) cave settings. In such cave pools, the type of subaqueous calcite deposit depends on the chemistry of the water and the ability of the cave atmosphere to exchange air with the atmosphere at the surface, controlling how fast CO_2 degassing and/or evaporation can occur. In thermal caves evaporation can be very important, and if the rising water is rich in dissolved CO_2 both processes give rise to a rapid supersaturation of the water and the deposition of a large volume of calcite, as long as the CO_2 and the water vapor can be removed from above the pools. In these cases of rapid supersaturation of the water bodies, the walls of the pools are covered by subaqueous calcite crusts known under the name **mammillary crusts** or **cave clouds** (Figure 10.32a) (Polyak et al. 2008). Since the overcrusting of the rock surfaces occurs in a more or less homogeneous manner, the irregularities of the substrate (projections) will coalesce and be smoothed causing the cloud-like appearance of these crusts. Cave clouds can be extremely smooth (made of small crystals), but often have a knobby appearance, in which the irregularities are caused by the crystal terminations sticking out of the surface of the cloud (Figures 10.32b and 10.31g).

Folia are another peculiar type of speleothem found in rapidly degassing cave pools, where the water level of the pools appears to change in time. They look like bracket fungi growing in stacked levels on overhanging walls at or below the pool level, but close to it (Figure 10.32c, d). They are commonly found in thermal caves (Budapest, France, Audra et al. 2009c), but occur also in normal epigenic caves such as Mystery Cave in Minnesota (Palmer 2007), or Santa Catalina Cave in Cuba (D'Angeli et al. 2015a) (Figure 10.32d). Although their origin is still partly unresolved, fast CO_2 degassing, rising bubbles, and a fluctuating water level appear to be the necessary ingredients. Extreme examples of probably bio-mediated folia influenced by changes in the halocline are those found in El Zapote cenote in Yucatán, Mexico (Ritter et al. 2019).

10.7.4 Speleothems Formed by Dominant Evaporation

Caves are frequently wet places where the relative air humidity is greater than 90% and often approaches saturation. Cave walls, floor, and speleothems are often wet and covered with a thin water film. In water-saturated air there is a dynamic equilibrium between evaporation and condensation: for each water molecule escaping into the air, another molecule incorporates to the liquid phase. Where unsaturated air flow replaces the saturated air in contact with the wet walls, evaporation can occur. Air flow enhances both evaporation and CO_2 loss from thin water films, so the effects of these processes on carbonate precipitation combine. Where water evaporates, the loss of liquid water tends to be compensated by capillary movement of water from nearby areas. The less soluble mineral, normally calcite, forms first, and because of the increasing Mg/Ca ratio in the remaining solution, the typical precipitation sequence will be calcite–aragonite–huntite–hydromagnesite (Figure 10.15).

The most common evaporative speleothem is the **coralloid**, resembling **popcorn**, cauliflower, grapefruit, or botryoidal corals. This speleothem is probably the most abundant after the

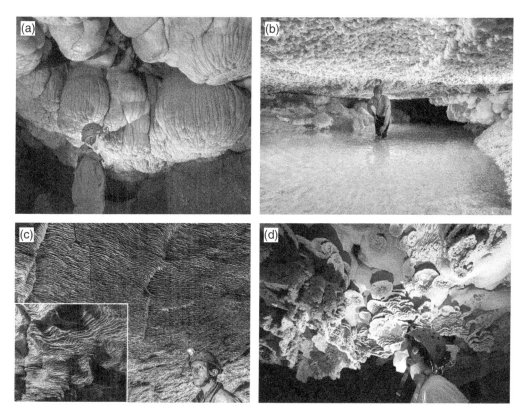

Figure 10.32 Speleothems formed in standing water, mammillary crusts, and folia: (a) Mammillary calcite (cave clouds) in Sima de la Higuera, Murcia, Spain. *Source:* Photo by Michel Renda. (b) Subaqueous calcite coatings in Gouffre Berger, Vercors, France. *Source:* Photo by Csaba Egri. (c) The famous folia of Pál-Völgyi Barlang, Budapest, Hungary. *Source:* Photo by Csaba Egri. (d) Folia in the shallow, nonthermal, pools of Santa Catalina Cave, Matanzas, Cuba. *Source:* Photo by Nicola Tisato, La Salle 3D.

dripstones (stalactite-stalagmite-column). Although coralloids can form in a variety of settings, including under water (Figure 10.33a) and in splash zones (Figure 10.33e) (Vanghi et al. 2017), most are subaerial in origin, with evaporation being the dominant process responsible for mineral precipitation (Caddeo et al. 2015). Subaqueous coralloids tend to have a more regular concentric texture, since they grow at similar speeds in all directions, whereas subaerial coralloids display irregular laminae. Subaqueous coralloids are also associated with the typical speleothems that form in standing water bodies, such as shelfstones, rafts, cave pearls, and spar, and have a clear horizontal upper limit for their growth (the water level). Coralloids formed by splashing water have an origin more similar to that of flowstones, where CO_2 degassing is the dominant process (Vanghi et al. 2017). Since water can reach almost any spot in splash zones, the globular forms tend to coalesce among each other, forming bulbous surfaces.

In dominantly evaporative conditions, the protruding parts of the substratum and/or surfaces subjected to air flow, are the places where the water films experience greater evaporation (and CO_2 degassing) (Figure 10.33e, f). The loss of water by evaporation attracts water from adjacent areas by capillary movement. Coralloids grow preferentially in these favorable places, forming sub-spherical speleothems that often do not experience coalescence because they inhibit the growth of each other (Figure 10.33c). When two spherical speleothems approach, the water molecules

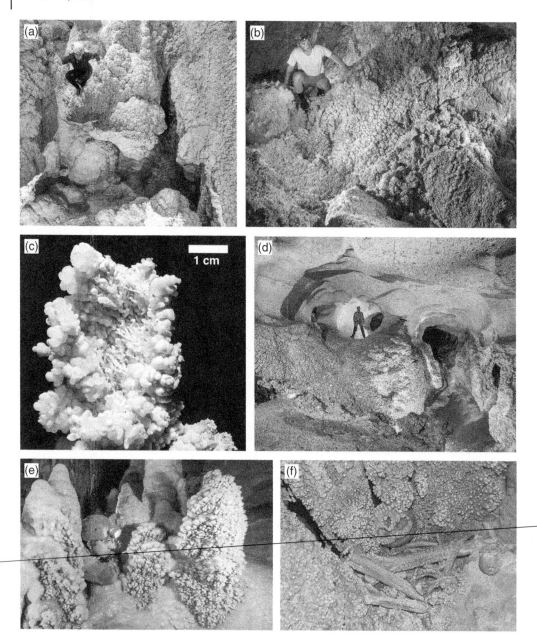

Figure 10.33 Speleothems formed by dominant evaporation and capillary movement, coralloids: (a) A combination of subaqueous (in the lake) and subaerial (above) coralloids in Beremendi-kristály Barlang, Hungary. *Source:* Photo by Csaba Egri. (b) Subaerial coralloids in Gouffre Berger, France. *Source:* Photo by Csaba Egri. (c) Subaerial spherical coralloids in Nagyharsányi-kristály Barlang, Hungary. *Source:* Photo by Csaba Egri. (d) Passage in Abisso dei Cocci (Sicily) with smooth roof carved by condensation waters, and lower part covered with coralloids formed by evaporation. *Source:* Photo by Victor Ferrer. (e) Wind-oriented splash and capillary rising water coralloids in Grotta Savi, Trieste, northern Italy. *Source:* Photo by Sandro Sedran, S-Team. (f) Coralloids growing on the remains of a Neanderthal skeleton, Lamalunga Cave, Apulia, southern Italy. Note large coralloids are growing on the protruding brow ridge. *Source:* Photo by Paolo Forti.

evaporating from one of the spherules tends to incorporate into the water film on the adjacent coralloid, thus slowing down evaporation and speleothem growth.

Movement of air masses at different temperatures, and energy budgets caused by water evaporation (endothermic) and condensation (exothermic) cause subtle differences, that may determine where minerals will deposit and where dissolution will take place. Warm air contains higher amounts of water vapor and is less dense (and thus stays along the roof), whereas colder and drier air stays in the lower parts of the cave passages (density and humidity stratification). The warmer and moister air can condense on the roof releasing heat, and this highly undersaturated water descends as a thin water film, dissolving limestone. In the lower parts of the wall, this water film, that may have reached near-saturation conditions, comes in contact with denser, cooler and drier air, causing evaporation (consuming heat) and deposition of carbonate under the form of coralloids (Figure 10.33d). The movements of cool and dry air also cause the directional (upwind) growth of coralloids on wet surfaces (Figure 10.33e). These speleothems are transitional between true coralloids and tower coral. Evaporation can also cause the precipitation of aragonite needles, forming clusters with radial arrangement (Figure 10.34a, b). This type of needle-shaped speleothem is known under the name of **frostwork**, resembling helictites, but contrary to these they have no internal feeding channel.

Figure 10.34 Speleothems formed by dominant evaporation, frostwork, trays, and rims: (a) Aragonite frostwork on the walls of Abisso dei Cocci, Sicily. *Source:* Photo by Victor Ferrer. (b) Aragonite frostwork in Beremendi-kristály Barlang, Hungary. *Source:* Photo by Csaba Egri. (c) Roof covered with aragonite coralloids and growing hollow gypsum stalagmite on the underlying floor, Lechuguilla Cave, New Mexico. *Source:* Photo by Lukas Plan. (d) Calcite trays in Sima de la Higuera, Murcia, southern Spain. *Source:* Photo by Victor Ferrer. (e) Gypsum stalactite underlain by a growing gypsum Christmas tree in a cave in Sorbas, southern Spain. *Source:* Photo by Victor Ferrer. (f) Rim of gypsum in Lechuguilla Cave, New Mexico. *Source:* Photo by Lukas Plan.

A special association of two different speleothems is that in which the upper part consists of aragonite coralloids, whereas gypsum precipitates on the cave floor. CO_2 degassing dominates over evaporation on the roof, leading to carbonate deposition. On the other hand, evaporation dominates in the lower parts of the cave passage, so gypsum precipitation prevails. In reality, drips are initially undersaturated with respect to gypsum, so no gypsum is deposited immediately below the drip, but evaporation induces gypsum precipitation from the water film around the drip site. Capillary rise of water and evaporation will ultimately result in a hollow gypsum stalagmite forming below the aragonite coralloid formation (Figure 10.34c).

Coralloids do not always occur in the lower parts of cave voids. In certain conditions, they mainly develop in the upper parts of passages and terminate downward in more or less horizontal, often wavy surfaces. This can be related to the presence of water bodies below, causing low-lying air masses to be saturated in water vapor, thus preventing evaporation and growth of the coralloids in the lower parts of the passages. This process can sometimes cause the coralloid-covered speleothems to have a horizontal tray-like surface, as if cut by an invisible knife, controlled by the moist air mass below. These curious elephant-feet-like formations are known under the name of trays (Figure 10.34d).

In gypsum caves in arid areas, evaporation is so strong that both stalactites and stalagmites of gypsum can form (Figure 10.34e). The dripping stalactite will have a large (drop-sized) inner hollow, with gypsum precipitating around the dripping point because of evaporation. The water trickling to the ground evaporates entirely, forming an upward growing stalagmitic mass of gypsum, in some cases resembling a Christmas tree (Figure 10.34e).

The combination of condensation, capillary movement, and evaporation is the cause of the formation of **rims** and **blisters**. These are tube-like and bubble-like features with a smooth inner side and often a coralloid-covered outer shell (Figure 10.34f) (Davis 1995). They grow around small openings with air circulation connecting two large rooms at different levels: the lower passages often contain warm water bodies, whereas the higher passages are colder. Rims are commonly made up of calcite, aragonite, or gypsum. The warm and moist air from below condenses on the inner wall of the connecting small passages, dissolving minerals. Evaporation on the lips of these openings in the upper passages causes the condensed water to migrate by capillary motion from the inner wall to the outer one of the rim where precipitation occurs. There is thus a constant transfer of mineral from inside the rim to the outer shell. Equivalent features formed by salt and gypsum precipitation have been described in surface environments (see Section 8.3 and Figure 8.10f, g) (De Waele and Forti 2010).

Most sulfates precipitate because of evaporation, forming needles or fibers, resembling hairy molds growing on organic residues (e.g., epsomite and mirabilite, Figure 10.18f). Halite can also form such acicular crystals. The most common evaporative mineral in caves is by far gypsum, occurring in a variety of shapes and sizes (Forti 1996, 2017) (Figure 10.35). Gypsum in carbonate caves mainly forms due to the reaction of sulfuric acid with limestone, the acid derived from the oxidation of H_2S or pyrite. Other sources of sulfate can be guano, seawater, or evaporite rocks. Due to the fact that the precipitation of gypsum is mainly related to evaporation, and not to CO_2 diffusion, most gypsum speleothems do not have a feeding channel and are oriented toward the air flow (where evaporation is highest). Instead of forming helictites that have a central feeding tube, well-defined crystals tend to develop.

Gypsum normally forms crusts made up of small sugar-like crystals with a shiny appearance that look like a fresh snow accumulation. Since the mineral is highly soluble, seeping waters can dissolve it and form larger crystals in another place where they reach supersaturation conditions. Depending on the growth speed, crystals of different sizes and shapes can precipitate. The most

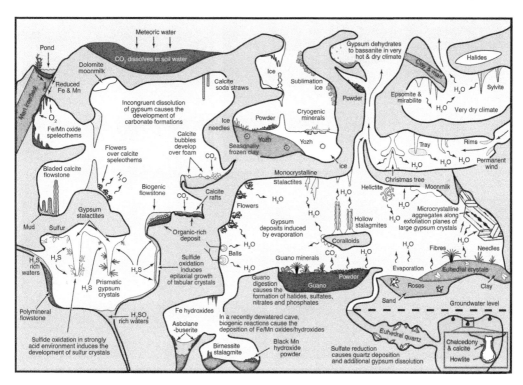

Figure 10.35 Ideal cave with the different types of gypsum deposits and speleothems. *Source:* Forti (2017)/Società Speleologica Italiana.

delicate form is the elongated fiberglass, with needles sometimes several meters long but less than a millimeter thick. The needles look like white curving hair or cotton, and can occur as single crystals, but more often as clusters (Figure 10.36a). If growth is slower, the crystals tend to form winding and curving cords a couple of millimeters thick (Figure 10.36b). **Gypsum flowers** occur where calcium sulfate concentrations are high, such as on weathering pyrite nodules or crystals. These large fiber-like crystals show clear longitudinal streaks, are curved like the petals of a lily flower, and grow outward from the sediment or rock, thus having their youngest parts at their base and the oldest on their tips (Figure 10.18c). With sufficient availability of calcium and sulfate, gypsum forms larger crystals, looking like knife blades hanging from the roof (Figure 10.36c). If these crystals become large and transparent or white, forming complex fingering clusters, they look like **chandeliers**. Small examples of these are known from József-hegyi Barlang (Budapest) (Figure 10.36d), but the most exceptional ones are those of Lechuguilla Cave, in New Mexico (Figures 10.18e and 10.36e). Differently from gypsum flowers, chandeliers grow at their tips from tiny hanging drops subject to slow evaporation. The most recent crystals at the tips are generally transparent, whereas the older ones are coated with microcrystalline gypsum becoming translucent or opaque, as if covered with a snow-like glittering crust. For such large chandeliers of Lechuguilla Cave to form, gypsum is dissolved in an above lying branch of the cave and precipitated from seeping water in the Chandeliers Ballroom below.

Four speleothems are unique to gypsum: **gypsum balls** and **hollow stalagmites** (Figure 10.36f, g), both found in the arid gypsum caves of Sorbas (southern Spain) (Calaforra 1996), **gypsum sails** (Bernabei et al. 2007), and **yozh** (Korshunov and Shavrina 1998). The first are layered spheroidal

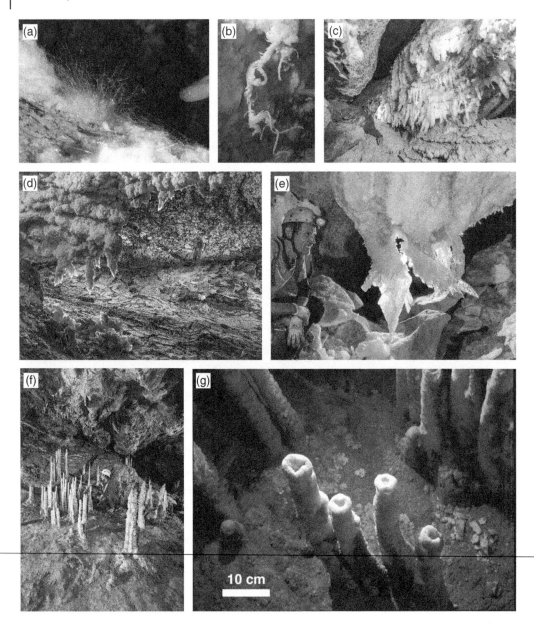

Figure 10.36 Gypsum speleothems formed by dominant evaporation: (a) Gypsum hair in Lechuguilla Cave, New Mexico (finger for scale). *Source:* Photo by Lukas Plan. (b) Gypsum flower in Lechuguilla Cave, New Mexico (finger for scale). *Source:* Photo by Lukas Plan. (c) Gypsum crystals growing on the roof in a windy passage, Grotta Salinella, Sicily. *Source:* Photo by Marco Vattano. (d) Gypsum chandeliers in József-hegyi Barlang, Budapest, Hungary. *Source:* Photo by Csaba Egri. (e) Large gypsum crystals in Lechuguilla Cave, New Mexico. *Source:* Photo by Lukas Plan. (f, g) Hollow gypsum stalagmites in a cave in Sorbas, southern Spain. *Source:* Photo by Victor Ferrer.

speleothems up to 1 dm in diameter forming from some kind of feeding fissure or bedding plane in the wall. Hollow stalagmites (Figure 10.36f, g) are up to 1.5 m tall and 4–6 cm in diameter with a central tube perforating the speleothem up to the floor. Rare infiltration events cause drip water to drill the hole, and water is kept for a certain amount of time on the bottom. Evaporation on the stalagmite top will cause water to move upward from inside the stalagmite by capillary forces to the top depositing gypsum. Gypsum sails are elongated skeleton crystals growing epitaxially on the tip of large selenite crystals and have been found only in Cueva de las Velas, in Naica (Mexico). Yozh are hedge-hog-like gypsum aggregates formed in clay deposits by freezing, and have been described from gypsum caves in Pinega (northern Russia).

10.7.5 Speleothems Formed by Capillary Water

Where water flow is low, it moves as a thin film along walls and ceilings and does not produce dripping. Instead, it evaporates before the drops get big enough to fall to the ground. In such circumstances, the movement of water is driven by the internal cohesion of the water (attraction between water molecules) and adhesive forces between the rock and water film, similar to the capillary flow of water in narrow tubes or fissures (like in vascular plants). The capillary movement can be driven by evaporation, since the water will be drawn toward the driest places, where the molecules pass into the vapor state. Water can also be pushed out of narrow fissures or pores by capillary movement, evaporating fast enough to prevent drops to become large enough to detach from the substrate.

Speleothems formed by this capillary water do not necessarily grow vertically downward or upward controlled by gravity, instead they grow in the most favorable direction for CO_2 degassing and evaporation, or are guided by crystalline structures and defects. Note that many **coralloids** form by movement and evaporation of capillary waters, but they can also be related to wetting and drying on surfaces without capillary movement being involved (i.e., splash, infiltrating waters). Also, **frostwork** is mainly due to evaporation, with capillary movements being of secondary importance, and is therefore included in the earlier category.

If water is pushed out of pores in the substrate, the capillary flow is concentrated in specific points, and **helictites** form (Figure 10.37). Helictites are also named eccentrics, erratics, or anthodites, but we prefer to consider all these subtypes as varieties of a single speleothem type. Upward growing helictites are sometimes called heligmites, but we strongly advice not to use this misleading name; helictites can grow in any direction by the same mechanism. They are mostly composed of calcite or aragonite, with the first being more typical of helictites (more independently growing branches) and the second of **anthodites** (clusters of radiating branches). All share a common characteristic: they have a narrow feeding channel (0.01–0.5 mm) that supplies the tip of the speleothem with water. This differentiates helictites from frostwork that is composed of crystals lacking a central feeding tube. Helictites can be as thin as hair and very short, but some reach diameters of several centimeters and lengths of over a meter. Experiments with other salts (sodium thiosulfate) in controlled laboratory settings have shown them to grow by hydrostatic pressure feeding capillary water to their tips, excluding an influence of air flow (Huff 1940). Of course, to form carbonate helictites, CO_2 degassing, and/or evaporation leading to supersaturation and deposition of carbonate need to be invoked. Helictites grow thicker by water drawn to the surface from the feeding tube through the crystal boundaries by evaporation, or simply by precipitation from a film of water flowing on the outer side. If water flow increases, droplets can grow large enough to fall, and helictites can turn into soda straw stalactites. The contrary can also happen: if flow on a soda straw diminishes, the growth of a helictite can be promoted. Cyclic changes

Figure 10.37 Speleothems formed by dominant capillary water, helictites, and anthodites: (a) Helictites growing on a stalactite in São Mateus Cave, Mato Grosso do Sul, Brazil. *Source:* Photo by Philippe Crochet. (b) Spectacular aragonite helictites in Trou Souffleur de Salindre, Gard, southern France. *Source:* Photo by Sandro Sedran, S-Team. (c) Branching helictites in Cueva El Panal, Cuba. *Source:* Photo by Antonio Danieli, La Salle 3D. (d) The blue aragonite helictites in Grotte de l'Asperge, Montagne Noire, southern France. *Source:* Photo by Nicola Tisato, La Salle 3D. (e) Green aragonite anthodites in Aven de Mont Marcou, France. *Source:* Photo by Andreas Schober. (f) Beaded blue aragonite anthodites in Grotte de l'Asperge. *Source:* Photo by Andreas Schober.

of growth rate at the tip can cause helictites to have a beaded morphology, resembling beads on a string. Other typical helictite forms are hairlike (filiform), wormlike (vermiform), and forking (antler helictites). Although most helictites are subaerial in origin, subaqueous ones have been found in Lechuguilla Cave, New Mexico (Davis 2000). Helictites are described in detail by Slyotov (1985), Hill and Forti (1997), Tisato et al. (2015), and Davis (2019).

If water is pushed out of a linear fissure and evaporates before it can start dripping, planar speleothems form. These are known as **welts** in their initial stage and **shields** when they reach a well

Figure 10.38 Speleothems formed by dominant capillary water, welts, and shields: (a) Welt developed in a fracture on a large flowstone with stalactites on the underside. The stalactites were submerged and coated with subaqueous calcite, transforming them into poolfinger-like speleothems (note thick calcite rafts resting on top of the welt), Básura Cave, northern Italy. *Source:* Jo De Waele (Author). (b) The famous butterfly helictites (in reality monocrystalline welts) in Sonora Caverns, Texas, USA. *Source:* Photo by Lukas Plan. (c) Welt (known by cavers under the name "les cymbales") separated by vertical downward movement of the lower part of the speleothem, TM71 Cave, southern France. *Source:* Photo by Victor Ferrer. (d) Shield with stalactites, Buso della Pisatela, northern Italy. *Source:* Photo by Sandro Sedran, S-Team. (e) Triple shield with dripstones in Aven Orgnac, Ardèche, southern France. *Source:* Photo by Philippe Crochet. (f) Isolated fragile shield in Buso della Pisatela, northern Italy. Note the feeding fracture in the cave wall. *Source:* Photo by Sandro Sedran, S-Team.

developed disc-like morphology. A welt is a planar excrescence projecting out of a fracture in the wall (Figure 10.38a), or more generally in a speleothem (column, flowstone, and stalagmite) (Figure 10.38c). If they grow around cylindrical or conical speleothems they form a sort of collar of calcite, whereas if they develop on walls they form a sort of linear ridge. If evaporation and/or CO_2 degassing is rapid, welts are composed of a large amount of tiny crystals, resembling the speleothems on which they grow. If supersaturation is reached very slowly (because of high humidity and CO_2 levels in the cave atmosphere), the crystals can become large, forming some very nice examples of special monocrystalline planar helictites such as the "butterfly" in Sonora Caverns (Texas, USA) (Figure 10.38b).

Shields are "mature" welts, where the supply of capillary water through the feeding fissure has been long enough to form a semi-circular speleothem sticking out of the wall in the direction of the feeding fracture. They take on an oval or circular shape growing around the point of major outflow from the feeding fissure, where calcite precipitates more rapidly than in the remaining more distal parts, often clogged with calcite. Shields are composed of two parallel plates separated by a narrow fissure through which capillary water moves. These fissures seem to be wider in the interior part of the shield, toward the substratum from which it grows, indicating the water to be slightly undersaturated there and dissolving some carbonate. This water becomes supersaturated where it comes into contact with the cave atmosphere, on the edges of the shield, thus causing its radial growth in a concentric manner (Figure 10.38f). Different adjacent fractures can give rise to adjoining shields, growing together, and forming complex forms (Figure 10.38e). If water flow increases substantially, it can start flowing and dripping turning the shield into the supporting substratum of gravitational speleothems such as flowstones, draperies, and stalactites (Figure 10.38d). Shields are not common in caves, since capillary movement of water is possible only when the fissure remains very narrow. Caves where shields are found normally have large numbers of them, and this might be explained by local structurally favorable conditions that allow fissures to be narrow and sustain capillary flow over long periods.

10.7.6 Speleothems Formed by Microorganisms

The idea that microorganisms constitute the most important volume of biomass in the subsurface is now widely accepted (Summers Engel 2019), and the knowledge on cave microbial diversity and metabolism has increased enormously in the last few decades (Northup and Lavoie 2001; Barton and Northup 2007; Miller et al. 2013), largely thanks to the great technological improvements in the study of the microbial world (i.e., gene sequencing and metagenomics). Nowadays, most cave microbial diversity studies are no longer culture-based, but generally start with the evaluation of 16S rRNA gene sequences specific for bacteria and archaea. The combination of these methods with (meta)transcriptomics, geochemical analyses (stable isotopes of C, N, S), and advanced microscopic techniques (light microscopy, TEM and SEM, SIMS, and FISH) have allowed to achieve deeper insights into microbial activity and metabolism.

The notion that microorganisms were implicated in speleothem growth is over 50 years old (e.g., Went 1969 for fungi, and Danielli and Edington 1983 for bacteria), but it is not until the early twentyfirst century that compelling evidence for microbially mediated speleothems and cave mineral deposits were substantiated (Northup et al. 2000; Barton et al. 2001; Melim et al. 2001; Northup et al. 2003). It is not easy to demonstrate that microorganisms are actively involved in the precipitation of speleothems or cave minerals, and not just entombed by the precipitating material. Contamination is one of the issues that can lead to erroneous conclusions about the biological origin of a speleothem, and not all micromorphologies that look like microorganisms are really what we think they are and can be inorganic structures. Barton et al. (2001) presented an overview on the criteria to determine whether a biological trace is a proof of the biological origin of a speleothem or not. The combined presence of mineralized microbial cells, fabrics commonly attributed to microbial activity (stromatolite-like structures), and geochemical proxies such as stable isotopes or lipid biomarkers, are strong indications of a biomediated mineral deposit (Jones 2010). It is now widely accepted that some speleothems are formed with the essential mediation of microorganisms (microscopic bacteria, fungi, and algae) (Tisato et al. 2015; Sauro et al. 2018, Jones and Northup 2021).

Biomineralization, the formation and transformation of minerals by microorganisms, has been documented in a wide variety of environments, such as coral reefs and shallow water environments (stromatolites), but also in extreme habitats such as hot springs, deep-ocean hydrothermal vents, and caves (Lowenstam and Weiner 1989; Jones 2010). The involvement of microorganisms in the precipitation of a mineral can be indirect or direct. The first is also known as biologically induced mineralization (BIM), in which minerals are formed because of changes in the environment related to metabolic activity (pH and/or Eh). The direct involvement of microorganisms in mineral precipitation is known as biologically controlled mineralization (BCM), in which the organisms control the nucleation site, growth, and final morphology of the crystals. Microorganisms can also passively influence mineral precipitation, since their extracellular substances composing the cell walls are electrically charged and provide binding sites for metal ions.

Microbial processes often involve redox reactions (oxidation–reduction), and can occur in aerobic or anaerobic conditions. In the presence of oxygen, inorganic compounds can be oxidized by chemolithoautotrophs (e.g., oxidation of S^0 to S^{6+}, from Fe^{2+} to Fe^{3+}, or from Mn^{2+} to Mn^{4+}), which then use the energy released by these oxidation reactions (electron loss) to convert inorganic carbon (CO_2 or HCO_3^-,) into organic compounds for their metabolic needs. Microorganisms gaining their energy and carbon requirements from organic compounds are known as heterotrophs. In the presence of oxygen (O_2) they use it to oxidize the organic compounds, whereas in anoxic environments they can get their energy from organic electron acceptors, or from the reduction (electron gain) of nitrate, sulfate, Mn^{4+}, or Fe^{3+}. In the often nutrient-poor or oligotrophic cave environment, chemolithoautotrophs can be the dominant microorganisms, but often some organic carbon is present in caves, brought in by bats, streams, infiltration, and aerosols, so heterotrophy can also be at play (Northup et al. 1997).

Different speleothems are now believed to be true biothems (biomediated cave mineral precipitates) (Queen and Melim 2006), including: poolfingers and U-loops, chenille spar, moonmilk, some cave pearls, helictites, and silica speleothems. **Poolfingers** are finger-like, often vertically hanging speleothems that form in cave pools (Figure 10.39a, b). They have diameters of up to 2 cm and can extend for half a meter, although most poolfingers are much smaller. In Hidden Cave (New Mexico) their internal structure is laminated, with layers of darker micrite alternating with layers of dogtooth calcite crystals (Melim et al. 2001). The micritic layers contain filaments roughly 1 micron in diameter and with lengths between 5 and 50 μm, some of which are reticulated, all belonging to enigmatic microbial species. These micrite layers have a depleted $\delta^{13}C$, which can be an indication of microbially mediated carbonate deposition. Similar speleothems have also been found in Cottonwood Cave in the same area (Melim et al. 2009). Poolfingers in these locations are often accompanied by web-like filamentous calcite structures, some forming **U-loops** connecting two poolfingers (Figure 10.39c). Although rather rare, poolfingers are known from several caves in different geographic and climatic conditions, and their origin has always been thought to be by microbial activity (Adolphe et al. 1991).

Chenille spar is a more crystalline palisade-like version of poolfinger. It consists of vertically oriented calcite spar grown around filaments hanging from shelfstones or the overhanging edges of standing pools. This special kind of microbially mediated pool spar is very rare, with the best examples found in Lechuguilla Cave, New Mexico (Melim et al. 2001). **Cave pearls** have sometimes been considered as biothems, based on the fact that they contained mineralized spores, mucus, and filamentous microbes (Jones 2009). Studies on pearls in a cave in Gran Cayman have shown them to have undergone alternating abiotic and biotic growth periods. Another example of typical biothem is **moonmilk** (of calcite, aragonite, huntite, hydromagnesite, gypsum, or other minerals) (Cacchio et al. 2004; Sánchez-Moral et al. 2012). This pasty speleothem (Figure 10.39d) is often composed of microscopic fibers of calcite crystals, and generally contains filamentous microbes and/or bacteria. Convincing evidence for the microbial origin of these microcrystalline speleothems, such as presence of lipids, stable isotope ratios typical of biological fractionation, or RNA/DNA ratios have been reported (Blyth and Frisia 2008; Sánchez-Moral et al. 2012). In some situations it has been shown that the metabolic activity of microorganisms causes changes in environmental conditions (rise in pH, often related to CO_2 depletion), which induces calcite moonmilk to precipitate (Portillo and González 2011). In most cases, once calcite starts precipitating, the metabolic activity of the microorganisms decreases, demonstrating that conditions become less favorable for these organisms (Portillo and González 2011;

Figure 10.39 Biothems: (a) Poolfingers in Hidden Cave, New Mexico. *Source:* Photo by Lukas Plan.
(b) Poolfingers in a cave in the French Jura Mountains. *Source:* Photo by Andreas Schober. (c) Poolfingers and U
loops in a cave in southern France. *Source:* Photo by Andreas Schober. (d) Irregular calcite moonmilk stalactites
in Grotta Nera, Majella National Park, central Italy. *Source:* Jo De Waele (Author). The inset shows gypsum
moonmilk (photo is 3 cm wide) in Fetida Cave, Santa Cesarea Terme, southern Italy. *Source:* Photo by Marco
Vattano. (e) Anomalous "jumping" helictites in Asperge Cave, southern France. Inset shows the white
mycobacterium colonies found close to these helictites. *Source:* Photos by Mirjam Widmer. (f) Enigmatic opal
helictites and (g) opaline crusts forming "mushrooms", both of probable microbial origin in Imawari Yeuta Cave,
Venezuela. *Source:* Photos by Francesco Sauro (f) and Riccardo De Luca (g).

Sánchez-Moral et al. 2012). Gypsum moonmilk in a sulfuric acid cave in southern Italy has also
shown a great microbial diversity, supporting the biogenic hypothesis (D'Angeli et al. 2019a)
(Figure 10.39d inset). Although moonmilk, no matter its mineralogy, is often considered to form
because of microbial mediation, this toothpaste-like speleothem can also form by completely
abiotic processes.

In a cave in southern France, special **helictites** have often been found associated with mycobacterial colonies on the clay walls, and their study has provided evidence supporting a microbial mediation in the formation of these eccentric speleothems (Tisato et al. 2015) (Figure 10.39e). The bacterial goo occurs inside the central feeding channel of the helictites, and especially on their tips.

Microbes are also most probably involved in the formation of other mineral deposits in caves, including native sulfur, nitrates, iron and manganese oxides, and silica speleothems. Most of these deposits are powdery or microcrystalline, and do not form true speleothems. Exceptions are the rusticles of Lechuguilla Cave (New Mexico), composed of iron oxides coating microbial filaments (Davis et al. 1990; Provencio and Polyak 2001). The opal speleothems of the quartzite caves of Venezuela are also most probably formed with the mediation of microorganisms (Figure 10.39f, g) (Aubrecht et al. 2008; Sauro et al. 2018; Ghezzi et al. 2021), similar to the silica coralloids found in lava tubes (Miller et al. 2014).

10.7.7 The Color of Speleothems

The most common cave minerals (calcite, aragonite, and gypsum) normally are colorless, transparent (e.g., gypsum needles), translucent, or white because of light scattering. In most caves though, speleothems have colors varying between white, yellowish, orange, brownish, and even red (Figures 10.24, 10.25, and 10.40a–c). In rare occasions, also green calcite (Figure 10.37e) or blue aragonite speleothems (Figures 10.37d, f, and 10.40d) can be found. Orange-reddish colors are often (erroneously) attributed to the presence of iron, while black and grey tones are wrongly

Figure 10.40 Color of speleothems: (a) Yellowish and whitish banded soda straws in Malaval Cave, southern France. (b) Brown-white colored shield and dripstones in a cave in southern France. (c) White flowstone with more recent humic- or fulvic-rich depositing waters. (d) Blue aragonite in Asperge Cave, southern France. *Source:* Photos by Andreas Schober.

associated with manganese. Although it is true that small amounts of iron- and/or manganese oxides can cause speleothems to be colored, these substances are generally too insoluble in an oxidized cave environment. If they stain speleothems, they are incorporated as fine particles, or form coatings on them.

The creamy, yellowish and orange color typical of calcite speleothems is mostly due to the presence of humic and fulvic acids, which are formed in the soil and transported in solution in the slightly acidic cave drip water (Gascoyne 1977; White 1997a). Only very small amounts of these organic substances are needed to provide color to a speleothem. Humic acids, which give darker colors to calcite, have higher molecular weights than fulvic acids, which give rise to the creamy-yellowish tones. In warm tropical climates, organic acids are degraded to lighter molecules more easily, giving lighter tones to the speleothems compared with those found in high latitude climates. Aragonite, contrary to calcite, is less suitable for incorporating organic molecules in its crystal lattice, and is therefore generally white.

The exotic green and blue colors of calcite and aragonite often derive from the incorporation of metal ions into the crystal structures, mainly Cu^{2+} or less often Co^{2+}, Ni^{2+}, Fe^{2+}, Cr^{2+}, or Mn^{2+}. The metal Zn^{2+} does not give color to the speleothems. In some cases, these foreign metal ions are not detected in colored speleothems, and colors are probably due to crystalline defects.

10.7.8 Speleothem Growth Rate

Growth rates of speleothems are highly variable, and greatly depend on the type of mineral that is depositing. In the case of evaporative speleothems, if dissolved concentrations are high and evaporation is fast, crystals can grow several centimeters in a matter of hours. In halite caves, salt speleothems can grow as much as half a meter in the first year after an important rain event (Filippi et al. 2011). Also, gypsum fiber crystals (or flowers) can grow rather fast, at rates of around one centimeter per year, especially if the sulfate ions are readily made available (e.g.,oxidation of sulfides).

For carbonate speleothems, the precipitation is given by the reverse reaction of equation 10.11

$$Ca^{2+}_{(aqueous)} + 2HCO_3^-_{(aqueous)} \Leftrightarrow CaCO_{3(solid)} + CO_{2(gas)} + H_2O \qquad (10.15)$$

in which degassing of CO_2 or in some cases evaporation of H_2O cause the solution to become supersaturated in $CaCO_3$. Although the chemical–physical processes are complex (see Dreybrodt 2019 for a summary and detailed references on the subject), the precipitation rate depends on a sequence of processes and their kinetics, occurring in the water film and in the more or less static diffusion boundary layer (DBL), which is the liquid–solid interface where molecular diffusion dominates. In general, $CaCO_3$ deposition mainly depends on the degree of calcite supersaturation, the thickness of the water film, the concentration of CO_2 in the cave atmosphere, the temperature, and the flow regime (laminar or turbulent). At high supersaturation, many crystallites form, whereas at low degree of supersaturation the existing crystallites act as nucleation sites causing the growth of larger crystals. The thickness of the water film also plays an important role, with thicker films depositing more calcite than thinner ones, all boundary conditions being the same. However, CO_2 diffusion into the cave atmosphere is more effective in thin water films, thus increasing supersaturation and accelerating deposition of calcite. The higher the P_{CO_2} of the cave atmosphere, the smaller the difference with the P_{CO_2} in the water, and consequently the slower the CO_2 exsolution process becomes. High CO_2 caves (with poor air exchange) slow down calcite precipitation, and often are places where monocrystalline speleothems (e.g., helictites) are more easily found. On the contrary, CO_2 loss from the water film to the cave atmosphere is more rapid at higher temperatures. Finally, under laminar flow conditions, the DBL is less disturbed and thinned, and precipitation is mainly controlled by the diffusion-limited reactions. When turbulent flow conditions prevail, diffusion of the molecules across the DBL to the bulk solution increases, as do the calcite precipitation rates (Liu and Dreybrodt 1997). In the case of stalagmites, another important controlling factor is the drip rate of the feeding water, which will control the supersaturation of the water that will reside on the speleothem. For low drip rates, there will be periods of no deposition, for medium drip rates a maximum growth speed will be reached, if dripping is too fast, less time will be available for the water to dissolve $CaCO_3$ before entering the cave and to deposit $CaCO_3$ while flowing over the speleothem surface. However, the growth of stalagmites is controlled by a complex suite of processes and factors, and there cannot be a global rule for determining stalagmite growth rates (Railsback 2018).

Theoretical deposition rates on stalagmites were confronted with true growth rates obtained by in situ measurements on limestone tablets of known weight (Buhmann and Dreybrodt 1985), or using speleothems grown over known time spans or with annual lamination, showing a relatively good agreement ($R^2 = 0.63$) (Baker and Smart 1995; Baker et al. 1998; Genty et al. 2001a). Growth rates measured on stalagmites from five European countries also show a good correlation with mean annual surface temperature ranging between 7 and 14 °C, given by the regression

$$R = 0.193T - 0.67 \, (R^2 = 0.63) \tag{10.16}$$

in which R is the growth rate expressed in mm yr^{-1} and T the mean cave air temperature (Genty et al. 2001a). This temperature dependence is related to the higher CO_2 production in soils and the higher dissolution capability of waters that have percolated through soils with higher CO_2 concentration. This equation is applicable in regions where vegetation is present, generally characterized by precipitation throughout the year, resulting in soil-covered cave sites, absence of summer drought, and low amounts of prior calcite precipitation (PCP). Although highly variable, stalagmite growth rates normally range between 0.01 and 1 mm yr^{-1}. So, 1 cm of flowstone or stalagmite top can roughly take from 10 to 1000 years to form. However, the study of stalagmites for paleoclimate reconstructions from all over the world has shown speleothem growth to be very variable, with periods of fast deposition alternating with no deposition periods, and even episodes of dissolution (Atsawawaranunt et al. 2018). Speleothems such as stalagmites and flowstones generally have grown over thousands of years, and some can be hundreds of thousands of years old (Woodhead et al. 2019).

In certain conditions speleothem growth can be extremely fast. For example, in artificial tunnels (with strong airflow), where recent objects (shoes, bottles) are covered with several centimeters-thick flowstones in less than 50 years (Figure 10.41a), or cover an empty wine bottle left on an active stalagmite top in a matter of a few years (Figure 10.41b). Soda straw and stalagmite growth can be very rapid also on concrete walls and roofs, since the calcium hydroxide created by the weathering of the cement reacts with the CO_2 causing calcite to precipitate (Figure 10.41c, d).

10.7.9 Decay of Speleothems

Speleothems are not always actively growing, but can pass from active phases to completely inactive ones, or become completely abandoned by the feeding solution (dripping, capillary, running, or standing water). With the natural evolution of caves, including stream erosion, collapse,

Figure 10.41 Fast speleothem growth: (a) Young speleothems covering a bottle, a shoe (left), and pipes in an aqueduct in Sassari, Sardinia. *Source:* Photo by Alessio Romeo. (b) A wine bottle covered with calcite in less than 30 years in Tiscali Cave, Sardinia. *Source:* Jo De Waele (Author). (c) Stalagmites are taking over a 40-year-old military underground bunker in western Slovenia. *Source:* Jo De Waele (Author). (d) Two small stalagmites growing on the concrete path of Pertosa-Auletta Cave, southern Italy. *Source:* Photo by Orlando Lacarbonara.

infilling and progressive abandonment of active cave passages, many of the previously deposited speleothems become dry, are eroded or corroded, or are buried beneath cave sediments. As most sediments, speleothems tend to disappear from the geological record as time passes by. It has been shown that sampling random speleothems from caves in a karst area would yield many more young speleothems than old ones (Scroxton et al. 2016).

If dripwater changes from supersaturation to undersaturation state, the drops corrode the previously deposited speleothem, sometimes drilling a hole (Figure 10.42a). Stalactites can become too large and heavy and eventually fall to the cave floor, where they can be covered by other speleothems or sediments (Figure 10.42b). Stalagmites and columns can be undermined by erosion, break or collapse on the ground (Figure 10.42c), or can also break during earthquakes (speleoseismology) (Becker et al. 2006) (Figure 10.42d). Speleothem decay can occur for a variety of reasons, including physical processes such as drying, frost shattering, salt wedging, but also chemical ones

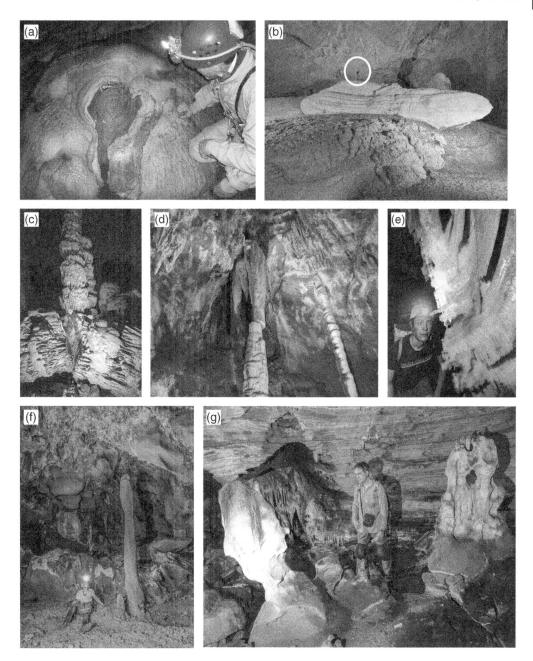

Figure 10.42 Speleothem decay: (a) Corrosive dripping on a massive flowstone in Barega Cave, Sardinia, Italy. *Source:* Jo De Waele (Author). (b) Large stalactite (person standing above it for scale) fallen on the ground in Xe Bang Fai Cave, Laos. *Source:* Photo by Dave Bunnell. (c) Broken column collapsed due to undermining in Grotta Impossibile, northern Italy. *Source:* Photo by Sandro Sedran, S-Team. (d) Broken column in Básura Cave, Toirano, northern Italy. Also note corroded column on the right (white: corroded side, red: airborne terra rossa). *Source:* Photo by Bartolomeo Vigna. (e) Corroded drapery in Gruta da Torrinha, Bahia, Brazil. *Source:* Jo De Waele (Author). (f) Large corroded stalagmite in Puerto Princesa Underground River Cave, Palawan, Philippines. *Source:* Photo by Marco Vattano, La Venta Esplorazioni Geografiche. (g) Two stalagmites corroded by guano in Gruta do Paixão, Bahia, Brazil. *Source:* Jo De Waele (Author).

such as dissolution (changing drip water chemistry, condensation corrosion) (Figure 10.42e, f), biomediated (bacterial) processes, and corrosion by different acids (nitric, sulfuric, phosphatic, and organic) (Figure 10.42g).

10.7.10 Dating Clastic and Chemical Cave Sediments

Caves are an integral part of the landscape in which they develop, forming over time spans that range from a few years (halite caves) to many millions of years (quartzites). They host chemical and detrital sediments formed inside these voids that record changes occurred at the surface in response to variations in the regional and local geological and environmental conditions. Dating these sediments is of primary interest to geologists and paleoclimatologists, in order to put these proxies of environmental variability in a chronological framework, and decipher the timing and rate of the changes that occurred in and around the cave. The age of cave deposits can also indicate a minimum age for the voids they fill, in the absence of a direct method for measuring the age of a cave, which is essentialy a denudational feature (Sasowsky 1998). The oldest known cave deposits appear to be those found in Transvaal, South Africa, which are 2.2 billion years old (Martini 1981).

The dating method depends on the type and origin of material to be dated and on its presumable age. The older a sample is, the lesser methods suitable to date it. Some materials cannot be numerically dated, but any deposit can be used for relative dating on the basis of geometrical relationships with other sediments (e.g., interdigitation, superposition, cross-cutting, and inset relationships). We distinguish numerical and relative dating methods. The first gives an age estimate for the dated material with a variable error margin, whereas the second provides a chronological order for a succession of events, without delivering precise ages.

It has to be underlined that karst records can be extremely difficult to interpret. Cave sediments can be deposited, eroded and redeposited many times, with reworking and mixing of sedimentary components of different ages. In caves, the general concept whereby the underlying sediments are older than the ones resting above them (law of superposition) is not always valid, and detailed morpho-stratigraphic observations are essential for establishing a reliable relative chronology, before designing a sampling strategy for geochronological (numerical) dating. The cross-cutting relationships used in different fields of geology are instead usually valid in caves, with the features cutting deposits being younger than those that are cut.

For a general overview on the various dating techniques applied to Quaternary materials and landforms, the reader is referred to the general textbooks of Noller et al. (2000) and Walker (2005). Specific chapters on dating techniques in caves can be found in Bosák (2002), White (2004), Palmer (2007), and Ford and Williams (2007), and specifically concerning speleothem dating in Dorale et al. (2007) and Fairchild and Baker (2012). This section briefly explains the most currently used methods for dating cave sediments and speleothems (Table 10.5 and Figure 10.43).

10.7.10.1 Radiocarbon

Carbon is among the most common elements in karst regions, being part of the carbonate rocks themselves (the hostrock and speleothems), and obviously participating in the organic carbon cycle (wood, shells, teeth, and bones). Carbon occurs under the form of three main isotopes: the stable isotopes ^{12}C and ^{13}C, corresponding respectively to 98.9 and 1.1% of the total amount of carbon, and the radioactive and unstable ^{14}C isotope (radiocarbon). The latter isotope is present in very minor amounts, but is constantly produced in the upper atmosphere by interaction of cosmic rays, and thus slow moving neutrons, with atmospheric nitrogen (^{14}N). The ^{14}C produced is widely

Table 10.5 Dating methods most often used in caves.

Dating method	Material dated	Age range (years)
^{14}C	Charcoal, bones, wood, shells	300–55 000
^{14}C (bomb-spike)	Speleothems	Post 1960 samples
Amino Acid Racemization (AAR)	Teeth, shells	1000–5 000 000
Paleomagnetism (Pal)	Magnetic minerals in sediments or speleothems	Samples >780 000
^{40}K/^{40}Ar	K-bearing minerals	1000–5 000 000
^{40}Ar/^{39}Ar	K-bearing minerals	1000–4 500 000 000
^{26}Al/^{10}Be	Quartz	10 000–8 000 000
OSL-TL (Lum)	Quartz-Feldspar	1000–700 000
ESR	Teeth, bones, shells, quartz	500->5 000 000
U-Th	Speleothems	100–600 000
U-Pb	Speleothems	100 000->100 000 000
Lamina counting (LC)	Speleothems	Variable
Fluorescence (Fluo)	Speleothems	Variable
Trace elements (Trac)	Speleothems	Variable

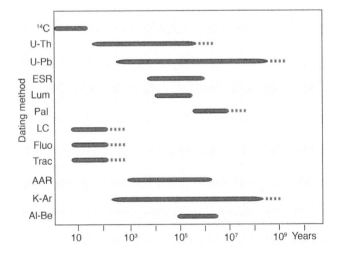

Figure 10.43 Dating methods applied to cave sediments and their approximate age ranges. (See Table 10.5 for the meaning of the acronyms).

dispersed throughout the atmosphere, and under the form of CO_2 is dissolved in the rain and surface waters and taken up by photosynthetic organisms (Figure 3.6).

The half-life of ^{14}C, which is the time needed for half of the mass of ^{14}C to decay to ^{14}N, is 5730 ± 30 years, making it suitable to date rather accurately organic material up to 55 000 years old (around nine half-lives). This makes ^{14}C dating one of the most powerful tools for dating archeological material and sediments with organic material. However, there are several sources of errors associated with ^{14}C dating. The cosmic ray flux, and consequently the amount of radiocarbon produced in the upper atmosphere, has changed over time, introducing a first source of error. The ^{14}C concentrations in the atmosphere also show regional and climatic variability, being more rapidly or slowly brought to the troposphere depending on the precipitation regime and atmospheric circulation patterns. The boost in burning fossil fuels since the nineteenth century has introduced larger quantities of ^{14}C-poor carbon into the atmosphere, thus lowering the natural $^{14}C/^{12}C$ ratio, and resulting in potentially overestimated ages for organic remains from this period. This will make dating of "recent" archeological material challenging in the future. On the contrary, the emission or production of artificial ^{14}C from nuclear plants and nuclear bomb tests, respectively, caused the $^{14}C/^{12}C$ ratio to increase drastically in the 50s. At present, values are still decreasing and turning to pre-1950 equilibrium values because of the nuclear test-ban treaty signed by the United States, the former Soviet Union, and Great Britain in the summer of 1963, CO_2 uptake by the biosphere, the exchange with older CO_2 in the oceans, and burning of fossil fuels.

Organic material in caves is suitable for ^{14}C dating, as long as calibration curves are used to correct for atmospheric $^{14}C/^{12}C$ ratio variations based on independent dating methods (U/Th dates, tree ring counting, varves, corals, etc.; e.g., IntCal20, Reimer 2020). However, caution should always be paid in the interpretation of ^{14}C dates even on organic material, since shells, bones, wood, or charcoal are not completely closed systems, and new ^{14}C can be taken up from infiltrating groundwaters, resulting in younger radiometric ages.

On the other hand, ^{14}C dating is severely limited in carbonate speleothems because of the unknown dead carbon proportion (dcp) derived from the rock and the soil (Macario et al. 2019). In fact, sources of carbon in speleothems include, besides the atmospheric carbon component (which is in equilibrium with the atmospheric $^{14}C/^{12}C$ ratio), the bedrock, the soil, and sometimes the cave atmosphere itself. The ^{14}C in rocks has generally long since decayed, supplying an amount of dead (completely ^{14}C-free) carbon. Also, the carbon present in soils can be rather old, deriving from respired CO_2 (plant roots and soil microbes) and decaying organic matter, with variable percentages of ^{14}C ("live" carbon). Following Eq. (3.52) (the calcite dissolution reaction), one half of the carbon in a speleothem appears to come from the host rock, and the other half from the soil air CO_2, which would lead to a 1 : 1 proportion of dead and live carbon. However, the high reactivity of soil CO_2 and HCO_3^- in water increases the live carbon in both carbonate containing organisms and speleothems. Comparison with U-Th analyses on Holocene stalagmites has shown that the dcp ranges between 5 and 40%, with typical values of 12–20% (Genty et al. 1999, 2001b; Columbu et al. 2019), but values approach or even surpass 50% in exceptional cases, such as in Corchia Cave in Italy, where a combination of closed-system conditions and sulfuric acid dissolution might explain such high values (Bajo et al. 2017). Dcp values are also susceptible to significant variations (around 5–10%) over time periods in the order of centuries, often severely limiting the use of this dating method to speleothems, and largely favoring U/Th dating instead.

The nuclear bomb ^{14}C peak, on the other hand, can be of some help in determining the age of a sample. High ^{14}C levels in modern stalagmites are a clear sign indicating that they have formed in the last few decades, and variations in these levels in modern stalagmites can indicate changes in hydrology (more or less rapid infiltration, water–rock interaction, and mixing effects in the soil-epikarst storage system).

10.7.10.2 Uranium–Thorium

Uranium has three naturally occurring isotopes: ^{238}U (99.28%), ^{235}U (0.72%), and ^{234}U (0.005%). The first two are the starting points of two radioactive decay chains comprising a series of unstable isotopes including Proactinium, Thorium, Actinium, Radium, Radon, Polonium, Bismuth, and

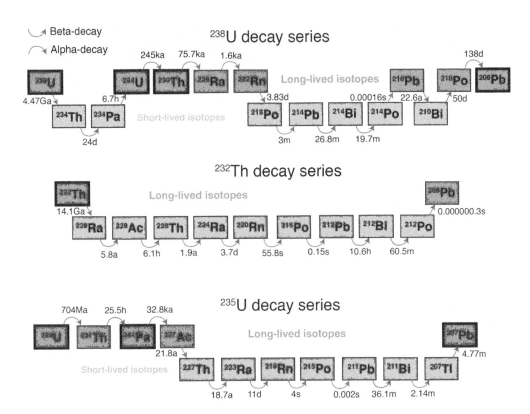

Figure 10.44 The main decay series for ^{238}U into ^{206}Pb, ^{232}Th into ^{208}Pb, and ^{235}U into ^{207}Pb, used in the numerous dating methods, including Uranium–Thorium and Uranium-Lead dating. Most commonly measured long-lived isotopes are reported in a bold box. *Source:* Modified from Fairchild and Baker (2012).

Thallium, ending with the stable isotopes of Lead (Figure 10.44). The unstable isotopes along the radioactive decay chains are characterized by different half-lives, some in the range of millions of years, others decaying very rapidly in a matter of seconds. Different isotope ratios can thus be used for dating purposes, applying those with long half-lives for older samples, and those with shorter half-lives to determine the age of younger samples. Since the isotopic abundances depend on the decay laws, the decay constants of the different isotopes need to be very well known, and technological advancements are continuously refining these values.

The most commonly applied method uses the decay of ^{234}U to ^{230}Th, with both elements having half-lives of 245 000 and 75 400 years, respectively. Since ^{238}U has a much longer half-life (4.47 billion years), the ratios ^{234}U/^{238}U and ^{230}Th/^{238}U can be used to obtain an absolute age. Depending on the initial ^{234}U/^{238}U ratio (the higher this ratio the later secular equilibrium will be reached) and the laboratory equipment used, this method allows to date samples up to 600–700 kyr old, after which the isotopes are in a state of secular equilibrium. Uranium is derived from the weathering of the rocks, and during these processes a higher proportion of ^{234}U is generally mobilized with respect to that of the other two isotopes (this is called "daughter excess"). In other circumstances, if percolating waters are interacting with an older, already depleted source rock, the depositing speleothem will show a deficiency in ^{234}U, making dating more challenging. All Uranium isotopes are rapidly oxidized and easily form complex ions with bicarbonate ($UO_2(CO_3)_2^{2-}$ and $UO_2(CO_3)_3^{4-}$), and as such are transported in the groundwater. In this manner, these ions can then precipitate in the calcite and aragonite lattice, with aragonite being able to contain ten times more Uranium than calcite because of its larger crystalline structure. ^{230}Th (and ^{232}Th) is essentially insoluble, so this radioisotope cannot travel as a solute and coprecipitate with Uranium into the calcite or aragonite speleothems. The only way Thorium can get trapped into the speleothem is by adsorption onto clays or other fine particles, which are then incorporated into the speleothem layers. This is called detrital contamination. In general, the ^{230}Th/^{232}Th ratio is used to determine whether a speleothem is contaminated and whether the age yielded by a sample can be considered reliable or not: for ratios higher than 20 the radiogenic ^{230}Th completely predominates, and detrital contamination can be assumed insignificant. In case this ratio is lower, corrections are applied (Hellstrom 2006).

In ideal conditions, speleothems should contain enough Uranium (at least 10 μg g^{-1}), they should remain closed, so that no Uranium has been added or leached after the speleothem has precipitated, and initial Thorium content should be very low. In these cases, the Uranium incorporated into the crystal lattice will start decaying, producing radiogenic ^{230}Th deriving only from the decay of ^{234}U, increasing its concentration with time.

Initially (1960–1980), the concentration of the radiogenic isotopes was measured using alpha-spectrometry, which needed rather large amounts of samples to be used and a lot of time for the analyses. This method was replaced first by thermal ionization mass spectrometry (TIMS) in the late 1980s, and in the late 1990s by the inductively coupled plasma mass spectrometers (ICP-MS) with a single-collector, and then with a multi-collector. Required sample sizes are now in the order of some mg, with greater accuracies and much shorter time for analyses. Uranium and Thorium are isolated from each other and from the carbonates and enriched using ion chromatography with specific organic resins, ultrapure acids and several other chemicals in dedicated super clean laboratories. More details on the U-Th dating techniques and other methods using Uranium and Thorium radiogenic isotopes can be found in Dorale et al. (2007), Scholz et al. (2012), Cheng et al. (2013), and Spötl and Boch (2019).

10.7.10.3 Uranium-Lead and Other Methods

The age limits of the U/Th method of <1 Myr can be overcome using the U-Pb decay chain, which theoretically enables samples as old as the Earth itself to be dated, given the half-life of ^{238}U (4.47 billion years) (Richards et al. 1998, Woodhead et al. 2006, 2012; Woodhead and Petrus 2019). The great improvements in Multi Collector and Laser Ablation Inductively Coupled Plasma Mass Spectrometry (respectively MC-ICP-MS and LA-ICP-MS) allow for the accurate determination of very small quantities of Uranium (ca. 1 ppm) and Lead (100 ppb) isotopes, overcoming one of the initial problems of this dating technique. Other problems such as the difficulty in obtaining a suitable range of parent/daughter isotope ratios for isochron reconstruction, and the exact determination of the initial ^{234}U/^{238}U ratio are still challenging to overcome. ^{234}U excess will lead to erroneously older U-Pb ages, whereas ^{234}U deficiency will cause the contrary to occur. The approximate range of the ^{234}U/^{238}U ratio can be obtained from U/Th dating of young samples where secular equilibrium has not yet been reached, or from ^{238}U/^{204}Pb isochrons from samples where the ^{204}Pb can be precisely measured. As for U/Th dating, samples need to be "clean", having low detrital Th, and consequently Pb, and thus showing rather high ^{230}Th/^{232}Th ratios. More details on the possible solutions for the dating problems are reported in Engel et al. (2019).

U/Pb dating has been successful in an increasing number of case studies. Lundberg et al. (2000) obtained an age of 91.3 ± 7.8 Myr for calcite filling voids (geodes) developed in the earliest karst phases in the Guadalupe Mountains in New Mexico. Polyak et al. (2008) obtained ages between 17 and 0.8 Myr from mammillary calcite coatings of different caves vertically distributed over an altitudinal range of 1200 m above the Colorado River, inferring information on the Colorado Plateau uplift and the correlative incision history of the Grand Canyon. U/Pb dating has also been successfully used in dating up to 3 Myr flowstones in South African early hominid sites (Pickering et al. 2019) and the over 4 Myr old Nullarbor cave systems (Woodhead et al. 2019). Speleogenetic (sulfuric-acid derived) dolomite has been dated with the U/Pb method in Big Room, Carlsbad Carverns, USA, delivering a similar age of ca. 4 Myr previously obtained by ^{40}Ar/^{39}Ar alunite dating (Polyak et al. 2016). More recently, U/Pb dating of stalagmites from Corchia cave, Italy, has allowed constraining the glacial terminations up to almost 1 Myr ago (Termination XII) (Bajo et al. 2020).

The decay of ^{235}U over short-lived ^{231}Th to ^{231}Pa, with a half-life of 708 kyr, can also be used to date Pleistocene carbonates. The shorter half-life of ^{231}Pa (32.8 kyr) with respect to ^{230}Th (75.7 kyr) allows dating material up to approximately 250 kyr old only. The lower content of ^{235}U in speleothems is another limiting factor, although it can be easily overcome by the much-improved mass spectrometers. This dating method can be used as an independent means of checking U/Th dates in carbonate samples (Edwards et al. 1997).

The decay of ^{230}Th to ^{226}Ra, with half-lives of 75.7 kyr and 1.6 kyr, respectively, can be used to date samples up to 10 kyr old, and can potentially be used as an independent method alongside ^{14}C dating. Latham et al. (1986) used it on a 2000-year-old stalagmite in Mexico, and although this method is increasingly used by archeologists, it is still not popular in speleothem studies.

The even more rapid decay of ^{226}Ra to ^{210}Pb (half-lives of 1.6 kyr and 22.6 years, respectively) can be used for very modern samples (a few centuries old), where both ^{14}C and U/Th dating are not suitable. This method has been successfully applied on modern and actively growing speleothems (e.g., Baskaran and Iliffe 1993; Condomines and Rihs 2006) and, combined with lamina counting, to carefully reconstruct the climate variability over the last 150 years in central China, recorded in a stalagmite (Paulsen et al. 2003).

10.7.10.4 Aluminum–Beryllium

Cosmic rays strike exposed rocks producing cosmogenic radioisotopes such as ^{3}He, ^{10}Be, ^{14}C, ^{21}Ne, ^{26}Al, and ^{36}Cl. Improvement in mass spectrometric techniques, and a better understanding of the physical processes behind cosmogenic nuclide formation and decay, has allowed these methods to become increasingly important in geomorphological studies (Gosse and Phillips 2001, Granger and Muzikar 2001). Cosmogenic nuclides allow for both exposure and burial dating, the last being of more interest in karst studies.

The isotopes used in burial dating are ^{10}Be (half-life of 1.39 Myr) and ^{26}Al (half-life of 0.72 Myr) present in small quantities in quartz gravel and sand. When these sediments are exposed to cosmic rays at the Earth's surface they will acquire a fixed ^{26}Al/^{10}Be ratio of 6 : 1, and since these isotopes will decay at different rates once they are shielded from further radiation, this burial time can be dated. For instance, burial dating can be used when the sediments are flushed underground in a cave and shielded by at least 10 m of rock above. The low amounts of these isotopes are measured with expensive equipment (e.g., accelerator mass spectrometers), causing the cost/sample ratio to become rather high. Actual age range for Al-Be burial dating is between 100 000 and 5 million years, making it a powerful method for obtaining minimum ages of the cave levels in which these sediments are found, extending well beyond the age limits of U/Th dating.

Caution must be taken while interpreting burial ages of cave sediments. The dated sediment is not always the last being deposited in a cave passage before it was abandoned. The possibilities of previously shielded sediments coming from older upper cave passages or from shielded portions of thick external sedimentary deposits, and of younger sediments entering older cave passages can occur, leading to over- and underestimated ages, respectively.

Despite these possible biases, burial dating has helped a lot in understanding landscape evolution and river entrenchment in many areas of northern America, such as in New River in Virginia (Granger et al. 1997) and the Green River (Mammoth Cave area) in Kentucky (Granger et al. 2001), but also in more rapidly evolving landscapes such as in the Swiss Alps (Haüselmann et al. 2007) and in the Pyrenees (Calvet et al. 2015). Al-Be dating has also been successfully applied on quartz-containing windborne volcanic ashes in caves in the Bighorn Basin, Wyoming (Stock et al. 2006). Another application in which the cosmogenic dating of sediments older than 600 kyr is of great interest is in the field of paleoanthropology and archeology, with good examples from Atapuerca Cave in Spain (Carbonell et al. 2008) and Sterkfontein in South Africa (Granger et al. 2015). General information on the use of cosmogenic nuclides for dating cave sediments can be found in Granger and Fabel (2019).

10.7.10.5 Potassium–Argon

Alunite ($KAl_3(SO_4)_2(OH)_6$) and jarosite ($KFe_3(SO_4)_2(OH)_6$) are weathering by-products formed by the reaction between sulfuric acid and clay minerals such as illite, montmorillonite or kaolinite (Polyak and Provencio 2001). Although these minerals only occur in small quantities, filling replacement pockets or forming thin layers on cave floors, the fact they contain potassium makes them suitable targets for $^{40}K/^{40}Ar$ and the derivative $^{40}Ar/^{39}Ar$ dating. Alunite and jarosite are poorly soluble in carbonate waters and form compact and very small pseudo-rhombohedral crystals that function as closed systems in which the decaying ^{40}K produces ^{40}Ar without possibilities of being depleted or enriched by infiltrating waters. ^{40}K has a half-life of 1.248 billion years so that the method can also be applied to very old caves. This dating method is well established for dating K-rich volcanic and plutonic rocks, but was first applied in a karst setting in 1998 in Carlsbad Cavern, New Mexico. Here, alunite dating revealed caves to have ages ranging between 12 and 4 Myr, with the youngest caves being at the lower levels (Polyak et al. 1998). The great advantage of this dating method is that it is, together with the U/Pb dating of speleogenetic dolomite (Polyak et al. 2016), the only one that exactly pinpoints the active speleogenetic phase, whereas all other dating methods on chemical and physical sediments only give minimum ages for the voids they occupy. Alunite dating has been carried out in a few sulfuric acid caves, including Kraushöhle in Austria (Plan et al. 2012), Provalata Cave in Macedonia (Temovski et al. 2013), and Cavallone Cave in central Italy (D'Angeli et al. 2019b).

10.7.10.6 Electron Spin Resonance, Optically Stimulated Luminescence, and Thermoluminescence

Electron spin resonance (ESR), Optically Stimulated Luminescence (OSL), and thermoluminescence (TL) are based on the detection of trapped charges (unpaired electrons and positive "holes") induced by radiation deriving from radioactive decay or solar/cosmic rays producing defects in crystals. The amount of these charge defects accumulates until all crystalline traps are filled. The speed at which this occurs depends on the type of material, and on the dose rate (DR) of radiation. The latter depends on variable solar and cosmic ray irradiation and the radioactivity of the natural environment surrounding the sample (external dose rate), plus the radioactivity produced inside the sample (internal dose rate). The accumulated dose (AD) in a sample is measured by an additive technique (irradiating the sample), successively extrapolating back to zero. The external and internal dose rates are based on theoretical modeling based on in situ (or sample) measurements, and are the most critical ones to evaluate. The ESR age is derived from the ratio between AD and DR. Further details on the ESR method can be found in Blackwell (2006).

In TL, samples are heated to 450 °C and the luminescence curve is measured. In OSL, samples are irradiated with an argon laser at a wavelength of 514 nm releasing only the accumulated dose in light sensitive traps. This method has a greater precision, but can be applied only on quartz and feldspar grains. More details on luminescence dating can be found in Bateman (2019).

10.7.10.7   Amino Acid Racemization

The amino acid racemization (AAR) method is based on the predictable post-mortem breakdown of biological amino acids (L-enantiomers) into non-biological counterparts (D-enantiomers). This method has been applied for over 50 years on different kinds of biological materials (wood, textiles, teeth, bones, and different kinds of carbonate biominerals). It can potentially also be used to date organic acids (humic and fulvic) present in speleothems, but only few studies have attempted this approach (Lauritzen et al. 1994). Racemization mainly depends on the amino acid considered and its position in the protein analyzed, as well as environmental factors such as temperature, pH, salinity, microbial activity. In closed systems (e.g., the inside of a tooth), the ratio between D and L forms of a specific amino acid (D/L value) is an indication of the time elapsed since the animal (or plant) died. In living organisms, D-enantiomers are completely lacking (D/L value = 0), and with increasing time since the death of the organism the D/L value will increase up to racemic equilibrium, when L-and R-configurations are equal (D/L value = 1). Because the racemization rate depends on a series of factors, AAR is mostly compared with other dating methods (^{14}C, U/Th, or other methods), but it can potentially be used to date material beyond the limits of U/Th dating. More details on the method can be found in Demarchi and Collins (2014).

10.7.10.8   Paleomagnetism

When a sediment is deposited, or a rock crystallizes, any magnetic material will orient with the magnetic field of the Earth. The Earth's magnetic field has not always been as it is today, and besides the drifts (changes of the position of the magnetic poles over some decimals of degrees every year), the magnetic field has reversed (the N and S poles have flipped) several times during geologic history. These N-S pole (field) reversals have occurred rapidly, and have a recurrence interval of 0.1 to 1 million years. Periods in which the North (magnetic) Pole was in the Northern hemisphere (as it is today) are called "normal," whilst the inverse situation is called "reverse." The long-lasting periods of normal or reverse polarity are named "Chrons," and are the principal paleomagnetic time units. Chrons are also punctuated by short-lived (decennial to centennial) periods of polarity changes, named "subchrons" (Figure 10.45). These changes in polarity combined with

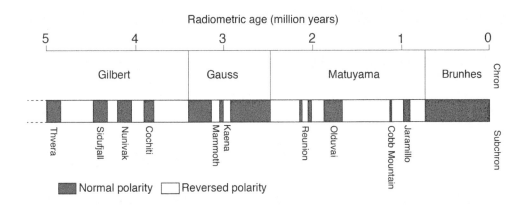

Figure 10.45   Magnetostratigraphic scale for the last 5 million years.

absolute datings of rocks (mainly K-Ar and U-Pb dating of volcanic rocks) have allowed researchers to create the magnetostratigraphic timescale. More general details on paleomagnetism can be found in Tauxe (2010).

Paleomagnetism can be used in both cave clastic sediments and speleothems. Sediments carried into caves by streams generally contain magnetic mineral grains which are oriented according to the Earth's magnetic field. This orientation is fixed when the sediment is covered by successive layers. The collection of oriented sediment samples in many locations in a cave system, especially along vertical sediment profiles, allows measuring the detrital remanent magnetism (DRM) of the samples in a magnetometer, obtaining both the magnetic direction and intensity. In speleothems magnetite can be included in the chemical deposit, as well as small amounts of detrital magnetic sediments. Although the magnetic signal in speleothems is much weaker than in detrital sediments, in this case both the chemical remanent magnetism (CRM) and the DRM can be measured. The advantage of this method is that the paleomagnetic record can be chronologically constrained by U/Th analyses (Latham et al. 1979; Latham and Ford 1993), and can even bring to light short-lived geomagnetic excursions (Pozzi et al. 2019).

The first thorough paleomagnetic study on cave sediments was carried out in Mammoth Cave, Kentucky, finding several reversals over the different cave levels, thus showing the upper passages of the cave to be older than 1 million years (Schmidt 1982). Similar studies have been carried out worldwide, with very detailed work in eastern Europe especially (Zupan Hajna et al. 2010). The most suitable sediments investigated are clays and silts, since they are deposited in still waters and the magnetic particles are thus best-oriented, but sometimes also coarser sediments can be used with good results if no obvious deformation structures are observed. Cave sedimentation is rarely continuous over long time scales, in contrast with lake and sea floor sediments, making the paleomagnetic record fragmentary. Postdepositional weathering, infiltration of fines, and bioturbation can also be major problems. A reversed polarity sample in a cave sediment tells us the cave passage cannot be younger than 780 kyr (the sediment was deposited before the Brunhes Chron started), likewise a normal polarity sediment indicates either a younger age (deposited during the Brunhes Chron), or older than 900 kyr (Jaramillo or older). If multiple stacks of sediments are present in different cave levels (or caves) and at different locations of the same level, a large number of samples can allow to reconstruct a more complete magnetostratigraphy (see for example Zupan Hajna et al. 2010). More details on paleomagnetic studies in caves can be found in Bosák and Pruner (2011) and Sasowsky (2019).

10.7.10.9 Lamina Counting-Fluorescence-Trace Elements

Speleothems grow by the chemical precipitation of calcium carbonate from infiltrating waters (drips), which chemistry is modulated by hydrological factors and the cave environment, both depending on climate variability. If these changes in discharge and cave meteorology follow a seasonal pattern, these fingerprints can be encountered in speleothems under the form of annual fluorescence (Proctor et al. 2000), petrological changes (Railsback et al. 1994) or trace element variations (Fairchild and Treble 2009, Smith et al. 2009). This has been the basis for annual-layered-counting chronologies especially for Late Holocene speleothems, although not all layering is necessarily annual, leading to small age variations which can only be corrected by high-precision absolute dating (U/Th) (Shen et al. 2013). Layer counting is often done manually, but in recent years more automatic methods have been devised (Meyer et al. 2006; Smith et al. 2009).

If speleothem banding is truly annual, lamina-counting chronology is obviously the most precise dating method, with a resolution of 1 year. In some regions, there are two very distinct groundwater recharge periods each year, and banding can be semi-annual. Some years can also have scarce precipitation, and speleothem growth can be interrupted for a year. These circumstances need to be well understood, and annual lamina counting should be confirmed by radiometric dating. On the contrary, in difficult-to-date speleothems, annual lamina counting can help to validate the obtained U/Th or U/Pb ages. More details are found in Baker et al. (2008).

10.8 Cave Sediments as Paleoclimate and Paleo-Environmental Archives

The sedimentary record in caves can provide valuable information regarding the processes that occurred at the surface immediately before and during their deposition. They can record former short- and/or long-term environmental conditions (climate, hydrology, and vegetation cover) and the corresponding erosional and/or depositional processes occurring in the surrounding landscape (Figure 10.46). These sedimentary processes are greatly influenced by climatic factors. Caves are

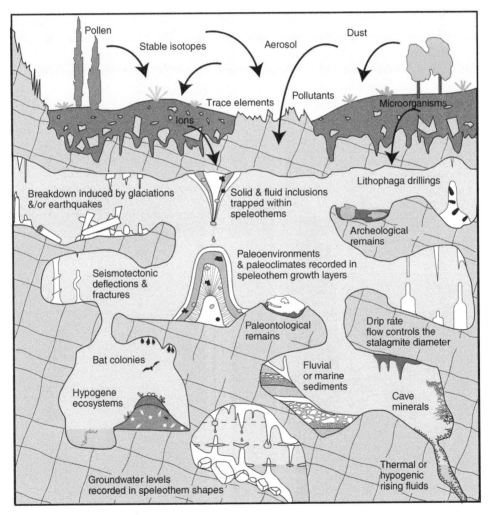

Figure 10.46 Caves are sedimentary traps that record environmental changes at the surface. These can be studied in both chemical and detrital sediments. *Source:* Drawing modified from Paolo Forti.

also very conservative sedimentary repositories, with many subterranean environments sheltered from high energy erosional events. Air temperature tends to be almost constant (annually varying less than 1 °C), and the relative humidity of the air is generally close to saturation, with evaporation occurring only locally or sporadically in cave sectors distant from flowing streams or in ventilated passages. Although clastic cave sediments can record climate-related events, the best paleoclimate records are obtained from speleothems.

The climate signal of the surface is transferred into speleothems by infiltrating waters, which carry in solution or suspension chemical and organic substances that are embedded in the precipitating minerals. Three types of speleothems are best suited for paleoclimate research, mainly because of their simple internal structure, rather continuous accretion, and the stable conditions under which they form. These are stalagmites, flowstones, and subaqueous more or less massive speleothems. The layering of these chemical deposits coupled with absolute dating methods allows the reconstruction of their growth over time, and the precipitated minerals preserve the geochemical signal at the moment of their crystallization, as long as no dissolution or recrystallization has taken place afterward. Sampling in stalagmites is best performed close to the central axis of growth (Figure 10.47), whereas flowstones or subaqueous mammillary calcite deposits are investigated through drilled cores, studying variations perpendicular to the laminae. The proxies used to determine past variations in temperature, recharge, and vegetation cover are speleothem growth rate and petrography, stable isotopes (C and O), trace elements (Mg, Sr, P, and others), and content in organic material (fluorescence, luminescence, pollen, and organic acids).

Oxygen isotopes are useful to infer temperature changes in the water that deposited the speleothems. $\delta^{18}O$ (see Eq. 5.24) values decrease as temperature rises, and vice versa. Although these isotopes do not allow determining the exact temperature at the time of formation (fluid inclusion

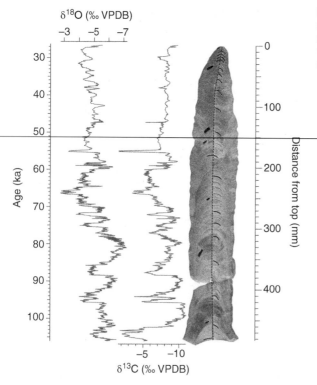

Figure 10.47 An example of a stalagmite used for paleoclimatic reconstruction in the Central Mediterranean. *Source*: Courtesy of Andrea Columbu.

studies are better for this), they can give approximations and show cooling and warming trends. In areas where the temperature effect is negligible, $\delta^{18}O$ reflects changes in rainfall amount (Columbu et al. 2017, 2018, 2019, 2021).

Variations in $\delta^{13}C$ help to determine the type of vegetation that dominated the landscape above the cave. C3 plants give lighter (lower) $\delta^{13}C$ values and are common in cool and wet areas, whereas C4 plants give heavier (higher) values that are instead more typical of warm and dry areas. This, in turn is related to changes in hydrology (rainfall). The interpretation of these proxies is not always straightforward, since many factors can have influence on the complex processes involving these stable isotopes (see Lachniet 2009 for details).

Other methods are generally applied to help interpreting the stable isotope variations. Fluid inclusion analysis in carbonates allows determining the exact composition of the carbonate precipitating water, and estimating a paleotemperature value (see for example Affolter et al. 2019). Trace element variations can tell more about the hydrological conditions during which the speleothem grew. Calcite precipitation increases the concentration of dissolved Mg and Sr in the residual solution, and this is reflected in the Mg/Ca and Sr/Ca ratios measured in speleothems. So, generally, high amounts of Mg and Sr in speleothems indicate drier periods, whereas lower values are indicative of wetter periods (Fairchild and Treble 2009). These interpretations can be improved analyzing petrographic changes in the carbonate material, leading to hydrological and geochemical inferences (Frisia 2015). These environmental and climatic changes revealed by the different proxies can then be placed in a chronological timeframe using the dating methods described above.

Speleothems, as paleoclimate archives, have several advantages with respect to other Quaternary archives: (i) they can be precisely dated by U-series methods, (ii) they contain multiple very high-resolution proxies (see above), (iii) they occur in many regions in the world and in different climates, (iv) they may often be correlated to present and past human presence, and can be used to reconstruct local climate variations, and (v) they occur in very conservative environments, so are better and more easily preserved. Their weaknesses lie in their complexity, because many factors are involved in the transfer of the climate signal into speleothems, and conservation issues, since speleothems are often unique and irreplaceable.

Speleothem science is one of the most rapidly advancing branches in cave research and it is hard to keep pace with the large number of scientific articles published every year. Good comprehensive overviews on the topic are found in Fairchild and Baker (2012) and Polyak and Denniston (2019).

References

Addesso, R., Bellino, A., D'Angeli, I.M. et al. (2019). Vermiculations from karst caves: the case of Pertosa-Auletta system (Italy). *Catena* 182: 104178.

Addesso, R., Gonzalez-Pimentel, J.L., D'Angeli, I.M. et al. (2021). Microbial community characterizing vermiculations from karst caves and its role in their formation. *Microbial ecology* 81 (4): 884–896.

Adolphe, J.P., Choppy, J., Choppy, B. et al. (1991). Biologie et concretionnement: un exemple, les baguettes de gours. *Karstologia* 18 (1): 49–55.

Affolter, S., Häuselmann, A., Fleitmann, D. et al. (2019). Central Europe temperature constrained by speleothem fluid inclusion water isotopes over the past 14,000 years. *Science Advances* 5 (6): eaav3809.

Agnelli, P., De Vivo, A., De Waele, J. et al. (2018). Preserving an astonishing ecosystem while improving tourism: The case of Natuturingam Cave (Palawan, Philippines). *NSS News* (June): 4–10.

Al-Malabeh, A. and Kempe, S. (2005). Origin of iron ore nuggets ("Bohnerze") through weathering of basalt as documented by pebbles from the Herbstlabyrinth, Breitscheid-Germany. *Acta Carsologica* 34 (2): 459–470.

Andrejchuk, V.N. and Klimchouk, A.B. (2002). Mechanisms of karst breakdown formation in the gypsum karst of the fore-Ural region, Russia (from observations in the Kungurskaja Cave). *International Journal of Speleology* 31 (1): 89–114.

Andreucci, S., Sechi, D., Buylaert, J.P. et al. (2017). Post-IR IRSL290 dating of K-rich feldspar sand grains in a wind-dominated system on Sardinia. *Marine and Petroleum Geology* 87: 91–98.

Arriolabengoa, M., Iriarte, E., Aranburu, A. et al. (2015). Provenance study of endokarst fine sediments through mineralogical and geochemical data (Lezetxiki II cave, northern Iberia). *Quaternary International* 364: 231–243.

Atsawawaranunt, K., Comas-Bru, L., Amirnezhad Mozhdehi, S. et al. (2018). The SISAL database: a global resource to document oxygen and carbon isotope records from speleothems. *Earth System Science Data* 10 (3): 1687–1713.

Aubrecht, R., Brewer-Carías, C., Šmída, B. et al. (2008). Anatomy of biologically mediated opal speleothems in the World's largest sandstone cave: Cueva Charles Brewer, Chimantá Plateau. *Venezuela. Sedimentary Geology* 203 (3-4): 181–195.

Aubrecht, R., Barrio-Amorós, C.L., Breure, A.S.H. et al. (2012). *Venezuelan Tepuis: Their Caves and Biota, Acta Geologica Slovaca Monograph*. Bratislava: Comenius University.

Audra, P., Mocochain, L., Bigot, J.-Y. et al. (2009b). Morphological indicators of speleogenesis: hypogenic speleogenesis. In: *Hypogene Speleogenesis and Karst Hydrogeology of Artesian Basins* (ed. A.B.Klimchouk and D.C.Ford), 23–32. Kiev: Ukrainian Institute of Speleology and Karstology.

Audra, P., Gázquez, F., Rull, F. et al. (2015). Hypogene Sulfuric Acid Speleogenesis and rare sulfate minerals in Baume Galinière Cave (Alpes-de-Haute-Provence, France). Record of uplift, correlative cover retreat and valley dissection. *Geomorphology* 247: 25–34.

Audra, P., De Waele, J., Bentaleb, I. et al. (2019). Guano-related phosphate-rich minerals in European caves. *International Journal of Speleology* 48 (1): 75–105.

Bajo, P., Borsato, A., Drysdale, R. et al. (2017). Stalagmite carbon isotopes and dead carbon proportion (DCP) in a near-closed-system situation: An interplay between sulphuric and carbonic acid dissolution. *Geochimica et Cosmochimica Acta* 210: 208–227.

Bajo, P., Drysdale, R.N., Woodhead, J.D. et al. (2020). Persistent influence of obliquity on ice age terminations since the Middle Pleistocene Transition. *Science* 367 (6483): 1235–1239.

Baker, A. and Smart, P.L. (1995). Recent flowstone growth rates: field measurements in comparison to theoretical predictions. *Chemical Geology* 122 (1–4): 121–128.

Baker, A., Genty, D., Dreybrodt, W. et al. (1998). Testing theoretically predicted stalagmite growth rate with recent annually laminated samples: implications for past stalagmite deposition. *Geochimica et Cosmochimica Acta* 62 (3): 393–404.

Baker, A., Smith, C.L., Jex, C. et al. (2008). Annually laminated speleothems: a review. *International Journal of Speleology* 37 (3): 193–206.

Balestra, V. and Bellopede, R. (2022). Microplastic pollution in show cave sediments: first evidence and detection technique. *Environmental Pollution* 292: 118261.

Barton, H.A. (2006). Introduction to cave microbiology: a review for the non-specialist. *Journal of Cave and Karst Studies* 68 (2): 43–54.

Barton, H.A. and Northup, D.E. (2007). Geomicrobiology in cave environments: past, current and future perspectives. *Journal of Cave and Karst Studies* 69 (1): 163–178.

Barton, H.A., Spear, J.R., and Pace, N.R. (2001). Microbial life in the underworld: biogenicity in secondary mineral formations. *Geomicrobiology Journal* 18: 359–368.

Baskaran, M. and Iliffe, T.M. (1993). Age determination of recent cave deposits using excess 210Pb - A new technique. *Geophysical Research Letters* 20 (7): 603–606.

Bastian, F., Jurado, V., Nováková, A. et al. (2010). The microbiology of Lascaux cave. *Microbiology* 156 (3): 644–652.

Bateman, M.D. (ed.) (2019). *Handbook of Luminescence Dating*. Caithness: Whittles Publishing.

Becker, A., Davenport, C.A., Eichenberger, U. et al. (2006). Speleoseismology: a critical perspective. *Journal of Seismology* 10: 371–388.

Berger, L.R., Hawks, J., de Ruiter, D.J. et al. (2015). *Homo naledi*, a new species of the genus *Homo* from the Dinaledi Chamber, South Africa. *eLife* 4: e09560.

Bernabei, T., Forti, P., and Villasuso, R. (2007). Sails: a new gypsum speleothem from Naica, Chihuahua, Mexico. *International Journal of Speleology* 36 (1): 23–30.

Bernardini, S., Bellatreccia, F., Columbu, A. et al. (2021). Morpho-Mineralogical and Bio-Geochemical Description of Cave Manganese Stromatolite-Like Patinas (Grotta del Cervo, Central Italy) and Hints on Their Paleohydrological-Driven Genesis. *Frontiers in Earth Sciences* 9: 642667.

Bini, A., Cavalli Gori, M., and Gori, S. (1978). A critical review of hypotheses on the origin of vermiculations. *International Journal of Speleology* 10: 11–33.

Bischoff, J.L., Juliá, R., Shanks, W.C.III et al. (1994). Karstification without carbonic acid: bedrock dissolution by gypsum-driven dedolomitization. *Geology* 22 (11): 995–998.

Blackwell, B.A. (2006). Electron spin resonance (ESR) dating in karst environments. *Acta Carsologica* 35 (2-3): 123–153.

Blyth, A.J. and Frisia, S. (2008). Molecular evidence for bacterial mediation of calcite formation in cold high-altitude caves. *Geomicrobiology Journal* 25 (2): 101–111.

Bögli, A. (1980). *Karst Hydrogeology and Physical Speleology*. Berlin: Springer-Verlag.

Bosák, P. (2002). Karst processes from the beginning to the end: how can they be dated. In: *Evolution of Karst: From Prekarst to Cessation* (ed. F.Gabrovšek), 191–223. Ljubljana: Zalozba ZRC.

Bosák, P. and Pruner, P. (2011). Magnetic Record in Cave Sediments: A Review. In: *The Earth's Magnetic Interior*, IAGA Special Sopron Book Series, vol. 1 (ed. E.Petrovský, D.Ivers, T.Harinarayana and E.Herrero-Bervera), 343–360. Dordrecht: Springer.

Bosch, R.F. and White, W.B. (2004). Lithofacies and transport of clastic sediments in karstic aquifers. In: *Studies of Cave Sediments* (ed. I.D.Sasowsky and J.R.Mylroie), 1–22. New York: Kluwer Academic/Plenum Publishers.

Bottrell, S.H., Crowley, S., and Self, C. (2001). Invasion of a karst aquifer by hydrothermal fluids: evidence from stable isotopic compositions of cave mineralization. *Geofluids* 1 (2): 103–121.

Bruins, H.J., Keller, J., Klügel, A. et al. (2019). Tephra in caves: distal deposits of the Minoan Santorini eruption and the Campanian super-eruption. *Quaternary International* 499: 135–147.

Buhmann, D. and Dreybrodt, W. (1985). The kinetics of calcite dissolution and precipitation in geologically relevant situations of karst areas: 1. Open system. *Chemical Geology* 48 (1–4): 189–211.

Cabrol, P. and Mangin, A. (2000). *Fleurs de Pierre: Les Plus Belles Concrétions des Grottes de France*. Lausanne: Delachaux et Niestlé.

Cacchio, P., Contento, R., Ercole, C. et al. (2004). Involvement of microorganisms in the formation of carbonate speleothems in the Cervo Cave (L'Aquila-Italy). *Geomicrobiology Journal* 21 (8): 497–509.

Caddeo, G.A., Railsback, L.B., De Waele, J. et al. (2015). Stable isotope data as constraints on models for the origin of coralloid and massive speleothems: the interplay of substrate, water supply, degassing, and evaporation. *Sedimentary Geology* 318: 130–141.

Calaforra, J.M. (1996). Some examples of gypsum karren. In: *Karren Landforms* (ed. J.J.Fornós and A.Ginés), 253–260. Palma de Mallorca: Universitat de les Illes Balears.

Calaforra, J.M., Forti, P., and Fernández-Cortès, A. (2008). Speleothems in gypsum caves and their paleoclimatological significance. *Environmental Geology* 53 (5): 1099–1105.

Calvet, M., Gunnell, Y., Braucher, R. et al. (2015). Cave levels as proxies for measuring post-orogenic uplift: evidence from cosmogenic dating of alluvium-filled caves in the French Pyrenees. *Geomorphology* 246: 617–633.

Carbonell, E., Burmúdez de Castro, J.M., Pares, J.M. et al. (2008). The first hominin species of Europe. *Nature* 452: 465–470.

Cheng, H., Edwards, R.L., Shen, C.-C. et al. (2013). Improvements in ^{230}Th dating, ^{230}Th and ^{234}U half-life values, and U-Th isotopic measurements by multi-collector inductively coupled plasma mass spectrometry. *Earth and Planetary Science Letters* 371-372: 82–91.

Chiarini, V., Couchoud, I., Drysdale, R. et al. (2017). Petrographical and geochemical changes in Bosnian stalagmites and their palaeo-environmental significance. *International Journal of Speleology* 46 (1): 33–49.

Collister, C. and Mattey, D. (2008). Controls on water drop volume at speleothem drip sites: an experimental study. *Journal of Hydrology* 358 (3-4): 259–267.

Columbu, A., De Waele, J., Forti, P. et al. (2015). Gypsum caves as indicators of climate-driven river incision and aggradation in a rapidly uplifting region. *Geology* 43 (6): 539–542.

Columbu, A., Chiarini, V., De Waele, J. et al. (2017). Late Quaternary speleogenesis and landscape evolution in the northern Apennine evaporite areas. *Earth Surface Processes and Landforms* 42 (10): 1447–1459.

Columbu, A., Sauro, F., Lundberg, J. et al. (2018). Palaeoenvironmental changes recorded by speleothems of the southern Alps (Piani Eterni, Belluno, Italy) during four interglacial to glacial climate transitions. *Quaternary Science Reviews* 197: 319–335.

Columbu, A., Drysdale, R., Hellstrom, J. et al. (2019). U-Th and radiocarbon dating of calcite speleothems from gypsum caves (Emilia Romagna, North Italy). *Quaternary Geochronology* 52: 51–62.

Columbu, A., Audra, P., Gázquez, F. et al. (2021). Hypogenic speleogenesis, late stage epigenic overprinting and condensation-corrosion in a complex cave system in relation to landscape evolution (Toirano, Liguria, Italy). *Geomorphology* 376: 107561.

Condomines, M. and Rihs, S. (2006). First ^{226}Ra-^{210}Pb dating of a young speleothem. *Earth and Planetary Science Letters* 250 (1–2): 4–10.

Curl, R.L. (1972). Minimum diameter stalactites. *National Speleological Society Bulletin* 34 (4): 129–136.

Curl, R.L. (1973). Minimum diameter stalagmites. *National Speleological Society Bulletin* 35 (1): 1–9.

D'Angeli, I.M., Ghezzi, D., Leuko, S. et al. (2019a). Geomicrobiology of a seawater-influenced active sulfuric acid cave. *PLoS One* 14 (8): e0220706.

Dandurand, G., Duranthon, F., Jarry, M. et al. (2019). Biogenic corrosion caused by bats in Drotsky's Cave (the Gcwihaba Hills, NW Botswana). *Geomorphology* 327: 284–296.

D'Angeli, I.M., De Waele, J., Melendres, O.C. et al. (2015a). Genesis of folia in a non-thermal epigenic cave (Matanzas, Cuba). *Geomorphology* 228: 526–535.

D'Angeli, I.M., Carbone, C., Nagostinis, M. et al. (2018). New insights on secondary minerals from Italian sulfuric acid caves. *International Journal of Speleology* 47 (3): 271–291.

D'Angeli, I.M., Nagostinis, M., Carbone, C. et al. (2019b). Sulfuric acid speleogenesis in the Majella Massif (Abruzzo, Central Apennines, Italy). *Geomorphology* 333: 167–179.

Danielli, H.M.C. and Edington, M.A. (1983). Bacterial calcification in limestone caves. *Geomicrobiology Journal* 3 (1): 1–16.

Davies, W.E. (1949). Features of cavern breakdown. *National Speleological Society Bulletin* 11: 34–35.

Davies, W.E. (1951). Mechanics of cavern breakdown. *National Speleological Society Bulletin* 13: 36–43.

Davies, W.E. and Chao, E.C.T. (1959). *Report on sediments in Mammoth Cave, Kentucky*. U.S. Geological Survey Administrative Report to the National Park Service.

Davis, D.G. (1995). Rims, rills and rafts: Shaping of cave features by atmospheric water exchange. *Geo2* 22 (2): 23–29. 32.

Davis, D.G. (2000). Extraordinary features of Lechuguilla Cave, Guadalupe Mountains, New Mexico. *Journal of Cave and Karst Studies* 62 (2): 147–157.

Davis, D.G. (2019). Helictites and related speleothems. In: *Encyclopedia of Caves* (ed. W.B.White, D.C.Culver and T.Pipan), 514–520. New York: Academic Press.

Davis, D.G., Palmer, M.V., and Palmer, A.N. (1990). Extraordinary subaqueous speleothems in Lechuguilla cave, New Mexico. *National Speleological Society Bulletin* 52 (2): 70–86.

De Choudens-Sánchez, V. and González, L.A. (2009). Calcite and aragonite precipitation under controlled instantaneous supersaturation: elucidating the role of $CaCO_3$ saturation state and Mg/Ca ratio on calcium carbonate polymorphism. *Journal of Sedimentary Research* 79(6): 363–376.

Decker, D.D., Polyak, V.J., Asmerom, Y., and Lachniet, M.S. (2018a). U–Pb dating of cave spar: A new shallow crust landscape evolution tool. *Tectonics* 37(1): 208–223.

Decker, D.D., Polyak, V.J., and Asmerom, Y. (2018b). Spar caves as fossil hydrothermal systems: Timing and origin of ore deposits in the Delaware Basin and Guadalupe Mountains, New Mexico and Texas, USA. *International Journal of Speleology* 47 (3): 263–270.

De Paepe, D. and Hill, C.A. (1981). Historical Geography of United States Saltpeter caves. *National Speleological Society Bulletin* 43 (4): 88–93.

De Waele, J. and Forti, P. (2010). Salt rims and blisters: peculiar and ephemeral formations in the Atacama Desert (Chile). *Zeitschrift für Geomorphologie* 54: 51–67.

De Waele, J., Carbone, C., Sanna, L. et al. (2017a). Secondary minerals from salt caves in the Atacama Desert (Chile): a hyperarid and hypersaline environment with potential analogies to the Martian subsurface. *International Journal of Speleology* 46 (1): 51–66.

Demarchi, B. and Collins, M. (2014). Amino acid racemization dating. In: *Encyclopedia of Scientific Dating Methods* (ed. W.Rink and J.Thompson), 1–22. Dordrecht: Springer.

Dirks, P.H.G.M. and Berger, L.R. (2013). Hominin-bearing caves and landscape dynamics in the Cradle of Humankind, South Africa. *Journal of African Earth Sciences* 78: 109–131.

Dorale, J.A., Edwards, R.L., Alexander, E.C. et al. (2007). Uranium-series dating of speleothems: current techniques, limits, and applications. In: *Studies of Cave Sediments* (ed. I.D.Sasowsky and J.R.Mylroie), 177–197. Dordrecht: Springer.

Dredge, J., Fairchild, I.J., Harrison, R.M. et al. (2013). Cave aerosols: distribution and contribution to speleothem geochemistry. *Quaternary Science Reviews* 63: 23–41.

Dreybrodt, W. (2019). Speleothem deposition. In: *Encyclopedia of Caves* (ed. W.B.White, D.C.Culver and T.Pipan), 996–1005. New York: Academic Press.

Du Preez, G.C., Forti, P., Jacobs G. et al. (2015). Hairy Stalagmites, a new biogenic root speleothem from Botswana. *International Journal of Speleology* 44 (1):7–47.

Edwards, R.L., Cheng, H., Murrell, M.T. et al. (1997). Protactinium-231 dating of carbonates by thermal ionization mass spectrometry: implications for Quaternary climate change. *Science* 276 (5313): 782–786.

Engel, A.S. (2010). Microbial diversity of cave ecosystems. In: *Geomicrobiology: Molecular and Environmental Perspective* (ed. L.Barton, M.Mandl and A.Loy), 219–238. Dordrecht: Springer.

Engel, A.S. (2019). Microbes. In: *Encyclopedia of Caves* (ed. W.B.White, D.C.Culver and T.Pipan), 691–698. New York: Academic Press.

Engel, J., Woodhead, J., Hellstrom, J. et al. (2019). Corrections for initial isotopic disequilibrium in the speleothem U-Pb dating method. *Quaternary Geochronology* 54: 101009.

Fairchild, I.J. and Baker, A. (2012). *Speleothem Science: from Process to Past Environments.* Chichester: Wiley.

Fairchild, I.J. and Treble, P.C. (2009). Trace elements in speleothems as recorders of environmental change. *Quaternary Science Reviews* 28 (5-6): 449–468.

Fairchild, I.J., Baker, A., Borsato, A. et al. (2001). Annual to sub-annual resolution of multiple trace-element trends in speleothems. *Journal of the Geological Society* 158: 831–841.

Filippi, M., Bruthans, J., Palatinus, L. et al. (2011). Secondary halite deposits in the Iranian salt karst: general description and origin. *International Journal of Speleology* 40 (2): 141–162.

Foos, A.M., Sasowsky, I.D., LaRock, E.J. et al. (2000). Detrital origin of a sedimentary fill, Lechuguilla cave, Guadalupe Mountains, New Mexico. *Clays and Clay Minerals* 48 (6): 693–698.

Ford, D.C. and Williams, P.W. (2007). *Karst Hydrogeology and Geomorphology.* Chichester, UK: Wiley.

Forti, P. (1996). Speleothems and cave minerals in gypsum caves. *International Journal of Speleology* 25 (3): 91–104.

Forti, P. (2005). Genetic processes of cave minerals in volcanic environments: an overview. *Journal of cave and Karst Studies* 67 (1): 3–13.

Forti, P. (2017). Chemical deposits in evaporite caves: an overview. *International Journal of Speleology* 46 (2): 109–135.

Forti, P. and Onac, B.P. (2016). Caves and mineral deposits. *Zeitschrift für Geomorphologie, SupplementaryBand* 60 (2): 57–102.

Forti, P., Badino, G., Calaforra, J.M. et al. (2017). The ribbed drapery of the Puerto Princesa Underground River (Palawan, Philippines): morphology and genesis. *International Journal of Speleology* 46 (1): 93–97.

Franke, H.W. (1965). The theory behind stalagmite shapes. *Studies in Speleology* 1 (2-3): 89–95.

Freydier, P., Martin, J., Guerrier, B. et al. (2019). Rheology of cave sediments: application to vermiculation. *Rheologica Acta* 58 (10): 675–685.

Frischauf, C., Liedl, P., and Rabeder, G. (2014). Revision der fossilen Fauna der Drachenhöhle (Mixnitz, Steiermark). *Die Höhle* 65: 47–55.

Frisia, S. (2015). Microstratigraphic logging of calcite fabrics in speleothems as tool for palaeoclimate studies. *International Journal of Speleology* 44 (1): 1–16.

Frisia, S. and Borsato, A. (2010). Karst. In: *Carbonates in Continental Settings - Facies, Environments and Processes, Developments in Sedimentology*, vol. 61 (ed. A.M.Alonso-Zarza and L.H.Tanner), 269–318. Amsterdam: Elsevier.

Gabrovšek, F. and Stepišnik, U. (2011). On the formation of collapse dolines: a modelling perspective. *Geomorphology* 134 (1-2): 23–31.

Galdenzi, S. and Maruoka, T. (2003). Gypsum deposits in the Frasassi Caves, central Italy. *Journal of Cave and Karst Studies* 65 (2): 111–125.

Gams, I. (1981b). Contribution to morphometrics of stalagmite. In: *Proceedings of the Eight International Congress of Speleology*, vol. I (ed. B.F.Beck), 276–278. Bowling Green: National Speleological Society.

García-Ruiz, J.M., Villasuso, R., Ayora, C. et al. (2007). Formation of natural gypsum megacrystals in Naica. *Geology* 35 (4): 327–330.

Gascoyne, M. (1977). Trace element geochemistry of speleothems. In: *Proceedings of the 7th International Speleological Congress*, 205–207. Sheffield: UK.

Gázquez, F. and Calaforra, J.M. (2013). Origin of double-tower raft cones in hypogenic caves. *Earth Surface Processes and Landforms* 38 (14): 1655–1661.

Gázquez, F., Columbu, A., De Waele, J. et al. (2018). Quantification of paleo-aquifer changes using clumped isotopes in subaqueous carbonate speleothems. *Chemical Geology* 493: 246–257.

Genty, D. and Deflandre, G. (1998). Drip flow variations under a stalactite of the Pere Noel cave (Belgium). Evidence of seasonal variations and air pressure constraints. *Journal of Hydrology* 211 (1-4): 208–232.

Genty, D., Massault, M., Gilmour, M. et al. (1999). Calculation of past dead carbon proportion and variability by the comparison of AMS ^{14}C and TIMS U/Th ages on two Holocene stalagmites. *Radiocarbon* 41 (3): 251–270.

Genty, D., Baker, A., and Vokal, B. (2001a). Intra- and inter-annual growth rate of modern stalagmites. *Chemical Geology* 176 (1-4): 191–212.

Genty, D., Baker, A., Massault, M. et al. (2001b). Dead carbon in stalagmites: carbonate bedrock paleodissolution vs. ageing of soil organic matter. Implications for ^{13}C variations in speleothems. *Geochimica et Cosmochimica Acta* 65 (20): 3443–3457.

Ghezzi, D., Sauro, F., Columbu, A. et al. (2021). Transition from unclassified Ktedonobacterales to Actinobacteria during amorphous silica precipitation in a quartzite cave environment. *Scientific reports* 11 (1): 1–18.

Gilbert, M.T.P., Jenkins, D.L., Götherstrom, A. et al. (2008). DNA from pre-Clovis human coprolites in Oregon, North America. *Science* 320 (5877): 786–789.

Gilli, É. (1993). Les grands volumes souterrains de Mulu (Bornéo, Sarawak, Malaisie). *Karstologia* 22 (1): 1–14.

Gillieson, D. (1986). Cave sedimentation in the New Guinea highlands. *Earth Surface Processes and Landforms* 11 (5): 533–543.

Gillieson, D. (1996). *Caves: Processes, Development and Management*. Oxford: Blackwell Publishers.

Glenn, C.R., Föllmi, K.B., Riggs, S.R. et al. (1994). Phosphorus and phosphorites: sedimentology and environments of formation. *Eclogae Geologicae Helvetiae*87: 747–788.

Gosse, J.C. and Phillips, F.M. (2001). Terrestrial in situ cosmogenic nuclides: theory and application. *Quaternary Science Reviews* 20 (14): 1475–1560.

Granger, D.E. and Fabel, D. (2019). Dating cave sediments with cosmogenic nuclides. In: *Encyclopedia of Caves* (ed. W.B.White, D.C.Culver and T.Pipan), 348–352. New York: Academic Press.

Granger, D.E. and Muzikar, P.F. (2001). Dating sediment burial with in situ-produced cosmogenic nuclides: theory, techniques, and limitations. *Earth and Planetary Science Letters* 188(1-2): 269–281.

Granger, D.E., Kirchner, J.W., and Finkel, R.C. (1997). Quaternary downcutting rate of the New River, Virginia, measured from differential decay of cosmogenic 26Al and 10Be in cave-deposited alluvium. *Geology* 25 (2): 107–110.

Granger, D.E., Fabel, D., and Palmer, A.N. (2001). Pliocene-Pleistocene incision of the Green River, Kentucky, determined from radioactive decay of cosmogenic ^{26}Al and ^{10}Be in Mammoth Cave sediments. *Geological Society of America Bulletin* 113 (7): 825–836.

Granger, D.E., Gibbon, R.J., Kuman, K. et al. (2015). New cosmogenic burial ages for Sterkfontein member 2 Australopithecus and member 5 Oldowan. *Nature* 522: 85–88.

Hammer, Ø., Dysthe, D.K., and Jamtveit, B. (2010). Travertine terracing: patterns and mechanisms. *Geological Society of London*, Special Publications 336 (1): 345–355.

Haüselmann, P., Granger, D.E., Jeannin, P.Y. et al. (2007). Abrupt glacial valley incision at 0.8 Ma dated from cave deposits in Switzerland. *Geology* 35 (2): 143–146.

Hellstrom, J. (2006). U-Th dating of speleothems with high initial ^{230}Th using stratigraphical constraint. *Quaternary Geochronology* 1 (4): 289–295.

Hill, C.A. (1976). *Cave Minerals*. Huntsville (AL): National Speleological Society.

Hill, C.A. (1981a). Origin of cave saltpeter. *The Journal of Geology* 89 (2): 252–259.

Hill, C.A. and Forti, P. (1986). *Cave Minerals of the World*. Huntsville, Alabama: National Speleological Society.

Hill, C.A. and Forti, P. (1997). *Cave Minerals of the World*, 2e. Huntsville, Alabama: National Speleological Society.

Hjulström, F. (1935). Studies of the morphological activity of rivers as illustrated by the River Fyris. *Bulletin of the Geological Institute of the University of Uppsala* 25: 221–527.

Hoerlé, S., Konik, S., and Chalmin, E. (2011). Les vermiculations de la grotte de Lascaux: identification de sources de matériaux mobilisables par microanalyses physico-chimiques. *Karstologia* 58 (1): 29–40.

Hose, L.D. and Pisarowicz, J.A. (1999). Cueva de Villa Luz, Tabasco, Mexico: reconnaissance study of an active sulfur spring cave and ecosystem. *Journal of Cave and Karst Studies* 61 (1): 13–21.

Hubbard, D.A.Jr. (2019). Saltpeter mining. In: *Encyclopedia of Caves* (ed. W.B.White, D.C.Culver and T.Pipan), 885–888. New York: Academic Press.

Huff, L.C. (1940). Artificial helictites and gypsum flowers. *Journal of Geology* 48 (6): 648–659.

Hunt, A.P. and Lucas, S.G. (2018). The record of sloth coprolites in North and South America: implications for terminal Pleistocene extinctions. *New Mexico Museum of Natural History and Science Bulletin* 79: 277–298.

Iacoviello, F. and Martini, I. (2012). Provenance and geological significance of red mud and other clastic sediments of the Mugnano cave (Montagnola Senese, Italy). *International Journal of Speleology* 41 (2): 317–328.

IUCN (2014). IUCN SSC guidelines for minimizing the negative impact to bats and other cave organisms from guano harvesting. Version 1.0. Gland: IUCN.

Jacobs, Z., Wintle, A.G., and Duller, G.A. (2003). Optical dating of dune sand from Blombos Cave, South Africa: I - multiple grain data. *Journal of Human Evolution* 44 (5): 599–612.

Jones, B. (2009). Cave pearls - the integrated product of abiogenic and biogenic processes. *Journal of Sedimentary Research* 79 (9): 689–710.

Jones, B. (2010). Microbes in caves: agents of calcite corrosion and precipitation. *Geological Society of London, Special Publications* 336 (1): 7–30.

Jones, D.S., Lyon, E.H., and Macalady, J.L. (2008). Geomicrobiology of biovermiculations from the Frasassi cave system, Italy. *Journal of Cave and Karst Studies* 70 (2): 78–93.

Jones, D. S. and Northup, D. E. (2021). Cave Decorating with Microbes: Geomicrobiology of Caves. *Elements: An International Magazine of Mineralogy, Geochemistry, and Petrology* 17 (2): 107–112.

Jurado, V., Gonzalez-Pimentel, J.L., Miller, A.Z. et al. (2020). Microbial communities in vermiculation deposits from an Alpine cave. *Frontiers in Earth Science* 8: 635.

Kaufmann, G. and Dreybrodt, W. (2004). Stalagmite growth and palaeo-climate: an inverse approach. *Earth and Planetary Science Letters* 224 (3-4): 529–545.

Kern, Z. (2018). Dating cave ice deposits. In: *Ice Caves* (ed. A.Persoiu and S.E.Lauritzen), 109–122. Amsterdam: Elsevier.

Korshunov, V.V. and Shavrina, E.V. (1998). Gypsum speleothems of freezing origin. *Journal of Caves and Karst Studies* 60: 146–150.

Kukla, J. and Ložek, V. (1958). K problematice výzkumu jeskynních výplní [To the problems of investigation of the cave deposits]. *Československý Kras* II: 19–83.

Lachniet, M.S. (2009). Climatic and environmental controls on speleothem oxygen-isotope values. *Quaternary Science Reviews* 28: 412–432.

Latham, A.G. and Ford, D.C. (1993). The paleomagnetism and rock magnetism of cave and karst deposits. In: *Applications of Paleomagnetism to Sedimentary Geology*, SEPM Special Publication 49(ed. D.M.Aïssaoui, D.F.McNeill and N.F.Hurley), 149–155.

Latham, A.G., Schwarcz, H.P., Ford, D.C. et al. (1979). Palaeomagnetism of stalagmite deposits. *Nature* 280 (5721): 383–385.

Latham, A.G., Schwarcz, H.P., and Ford, D.C. (1986). The paleomagnetism and U-Th dating of Mexican stalagmite, DAS2. *Earth and Planetary Science Letters* 79 (1-2): 195–207.

Lauritzen, S.E., Haugen, J.E., Løvlie, R. et al. (1994). Geochronological potential of isoleucine epimerization in calcite speleothems. *Quaternary Research* 41 (1): 52–58.

Lawrence, B. (1960). Fossil *Tadarida* from New Mexico. *Journal of Mammalogy* 41 (3): 320–322.

Lignier, V. and Desmet, M. (2002). Les archives sédimentaires quaternaires de la grotte sous les Sangles (Bas-Bugey, Jura méridional, France). Indices paléo-climatiques et sismo-tectoniques. *Karstologia* 39(1): 27–46.

Liu, Z. and Dreybrodt, W. (1997). Dissolution kinetics of calcium carbonate minerals in H_2O-CO_2 solutions in turbulent flow: the role of the diffusion boundary layer and the slow reaction $H_2O + CO_2 \rightarrow H^+ + HCO_3^-$. *Geochimica et Cosmochimica Acta* 61 (14): 2879–2889.

Lowenstam, H.A. and Weiner, S. (1989). *On Biomineralization*. New York: Oxford University Press.

Luetscher, M. (2005). *Processes in Ice Caves: And Their Significance for Paleoenvironmental Reconstructions*. Zurich: University of Zurich.

Lundberg, J. and McFarlane, D.A. (2006a). A minimum age for canyon incision and for the extinct molossid bat, *Tadarida constantinei*, from Carlsbad Caverns National Park, New Mexico. *Journal of Cave and Karst Studies* 68 (3): 115–117.

Lundberg, J., Ford, D.C., and Hill, C.A. (2000). A preliminary U-Pb date on cave spar, Big Canyon, Guadalupe Mountains, New Mexico, USA. *Journal of Cave and Karst Studies* 62: 144–148.

Lynch, F.L., Mahler, B.J., and Hauwert, N.N. (2004). Provenance of suspended sediment discharged from a karst aquifer determined by clay mineralogy. In: *Studies of Cave Sediments*(ed. I.D.Sasowsky and J.R.Mylroie), 83–93. New York: Kluwer Academic/Plenum Publishers.

Macalady, J.L., Lyon, E.H., Koffman, B. et al. (2006). Dominant microbial populations in limestone-corroding stream biofilms, Frasassi cave system, Italy. *Applied and Environmental Microbiology* 72(8): 5596–5609.

Macario, K.D., Stríkis, N.M., Cruz, F.W. et al. (2019). Assessing the dead carbon proportion of a modern speleothem from central Brazil. *Quaternary Geochronology* 52: 29–36.

Maire, R. (1990). La haute montagne calcaire (karsts. Cavités. Remplissage. Quaternaire. Paléoclimats). *Karstologia Mémoire* 3: 1–731.

Malott, C.A. and Shrock, R.R. (1933). Mud stalagmites. *American Journal of Science* 5 (25): 55–60.

Martín-Chivelet, J., Muñoz-García, M.B., Cruz, J.A. et al. (2017). Speleothem Architectural Analysis: Integrated approach for stalagmite-based paleoclimate research. *Sedimentary Geology* 353: 28–45.

Martini, J.E.J. (1981). Early Proterozoic paleokarst of the Transvaal, South Africa. In: *Proceedings of the Eighth International Congress of Speleology*, vol. 8 (ed. B.F.Beck), 6–8. Bowling Green: National Speleological Society.

Martini, J.E.J. and Marais, J. (1996). Grottes hydrothermales dans le Nord-Ouest de la Namibie. *Karstologia* 28: 13–18.

McDermott, F., Frisia, S., Huang, Y. et al. (1999). Holocene climate variability in Europe: evidence from δ18O, textural and extension-rate variations in three speleothems. Quaternary Science Reviews 18 (8–9): 1021–1038.

Meakin, P. and Jamtveit, B. (2010). Geological pattern formation by growth and dissolution in aqueous systems. *Proceedings of the Royal Society A: Mathematical, Physical and Engineering Sciences* 466 (2115): 659–694.

Melim, L.A. and Spilde, M.N. (2011). Rapid growth and recrystallization of cave pearls in an underground limestone mine. *Journal of Sedimentary Research* 81 (11): 775–786.

Melim, L.A., Shinglman, K.M., Boston, P.J. et al. (2001). Evidence for microbial involvement in pool finger precipitation, Hidden Cave, New Mexico. *Geomicrobiology Journal* 18 (3): 311–329.

Melim, L.A., Liescheidt, R., Northup, D.E. et al. (2009). A biosignature suite from cave pool precipitates, Cottonwood Cave, New Mexico. *Astrobiology* 9 (9): 907–917.

Merino, A., Fornós, J.J., Mulet, A. et al. (2019). Morphological and mineralogical evidence for ancient bat presence in Cova des Pas de Vallgornera (Llucmajor, Mallorca, Western Mediterranean). *International Journal of Speleology* 48 (2): 115–131.

Meyer, M.C., Faber, R., and Spötl, C. (2006). The WinGeol Lamination Tool: new software for rapid, semi-automated analysis of laminated climate archives. *The Holocene* 16 (5): 753–761.

Miall, A.D. (1985). Architectural-element analysis: a new method of facies analysis applied to fluvial deposits. *Earth-Science Reviews* 22 (4): 261–308.

Miller, A.Z., Dionísio, A., Jurado, V. et al. (2013). Biomineralization by cave dwelling microorganisms. In: *Advances in Geochemistry Research* (ed. J.Sanjurjo Sanchéz), 77–105. Hauppauge: Nova Science Publishers.

Miller, A.Z., Pereira, M.F., Calaforra, J.M. et al. (2014). Siliceous speleothems and associated microbe-mineral interactions from Ana Heva Lava Tube in Easter Island (Chile). *Geomicrobiology Journal* 31(3): 236–245.

Miller, A.Z., Garcia-Sanchez, A.M., Martin-Sanchez, P.M. et al. (2018). Origin of abundant moonmilk deposits in a subsurface granitic environment. *Sedimentology* 65 (5): 1482–1503.

Moore, G.W. (1952). Speleothem - a new cave term. *National Speleological Society News* 10 (6): 2.

Nader, F.H., Swennen, R., and Keppens, E. (2008). Calcitization/dedolomitization of Jurassic dolostones (Lebanon): results from petrographic and sequential geochemical analyses. *Sedimentology* 55 (5): 1467–1485.

Noller, J.S., Sowers, J.M., and Lettis, W.R. (2000). *Quaternary Geochronology: Methods and Applications*, vol. *4*. Washington: American Geophysical Union.

Northup, D.E. and Lavoie, K.H. (2001). Geomicrobiology of caves: a review. *Geomicrobiology Journal* 18 (3): 199–222.

Northup, D.E., Reysenbach, A.L., and Pace, N.R. (1997). Microorganisms and speleothems. In: *Cave Minerals of the World* (ed. C.A.Hill and P.Forti), 261–266. Huntsville: National Speleological Society.

Northup, D.E., Dahm, C.N., Melim, L.A. et al. (2000). Evidence for geomicrobiological interactions in Guadalupe caves. *Journal of Cave and Karst Studies* 62 (2): 80–90.

Northup, D.E., Barns, S.M., Yu, L.E. et al. (2003). Diverse microbial communities inhabiting ferromanganese deposits in Lechuguilla and Spider Caves. *Environmental Microbiology* 5: 1071–1086.

Onac, B.P. (2019). Minerals in caves. In: *Encyclopedia of Caves* (ed. W.B.White, D.C.Culver and T.Pipan), 699–709. New York: Academic Press.

Onac, B.P. and Forti, P. (2011). Minerogenetic mechanisms occurring in the cave environment: an overview. *International Journal of Speleology* 40 (2): 79–98.

Onac, B.P., Effenberger, H.S., and Breban, R.C. (2007). High-temperature and "exotic" minerals from the Cioclovina Cave, Romania: a review. *Studia UBB Geologia* 52 (2): 3–10.

Osborne, R.A.L. (2002). Cave breakdown by vadose weathering. *International Journal of Speleology* 31(1): 37–53.

Pagliara, A., De Waele, J., Forti, P. et al. (2010). Speleothems and speleogenesis of the hypogenic Santa Barbara cave system (South-West Sardinia, Italy). *Acta Carsologica* 39 (3): 551–564.

Palmer, A.N. (2007). *Cave Geology*. Dayton, Ohio: Cave Books.

Palmer, A. N. and Palmer, M. V. (2003). Geochemistry of capillary seepage in Mammoth Cave. Speleogenesis and Evolution of Karst Aquifers 1 (4): 1–8.

Palmer, A.N., Palmer, M.V., and Paces, J.B. (2016). Geologic history of the Black Hills caves, South Dakota. *Geological Society of America*, Special Papers 516: 87–101.

Paulsen, D.E., Li, H.C., and Ku, T.L. (2003). Climate variability in central China over the last 1270 years revealed by high-resolution stalagmite records. *Quaternary Science Reviews* 22 (5-7): 691–701.

Pérez-González, A., Parés, J.M., Carbonell, E. et al. (2001). Géologie de la Sierra de Atapuerca et stratigraphie des remplissages karstiques de Galería et Dolina (Burgos, Espagne). *L'Anthropologie* 105 (1): 27–43.

Perşoiu, A. and Onac, B.P. (2019). Ice in caves. In: *Encyclopedia of Caves* (ed. W.B.White, D.C.Culver and T.Pipan), 553–558. New York: Academic Press.

Perşoiu, A., Onac, B.P., Wynn, J.G. et al. (2017). Holocene winter climate variability in Central and Eastern Europe. *Scientific Reports* 7 (1): 1196.

Piccini, L. (2011). Speleogenesis in highly geodynamic contexts: the Quaternary evolution of Monte Corchia multi-level karst system (Alpi Apuane, Italy). *Geomorphology* 134 (1-2): 49–61.

Pickering, R., Herries, A.I., Woodhead, J.D. et al. (2019). U-Pb-dated flowstones restrict South African early hominin record to dry climate phases. *Nature* 565 (7738): 226–229.

Plan, L., Tschegg, C., De Waele, J. et al. (2012). Corrosion morphology and cave wall alteration in an Alpine sulfuric acid cave (Kraushöhle, Austria). *Geomorphology* 169: 45–54.

Poinar, H.N., Hofreiter, M., Spaulding, W.G. et al. (1998). Molecular coproscopy: dung and diet of the extinct ground sloth Nothrotheriops shastensis. *Science* 281 (5375): 402–406.

Polyak, V.J. and Denniston, R.F. (2019). Paleoclimate records from speleothems. In: *Encyclopedia of Caves* (ed. W.B.White, D.C.Culver and T.Pipan), 784–793. New York: Academic Press.

Polyak, V.J. and Güven, N. (2000). Clays in caves of the Guadalupe mountains, New Mexico. *Journal of Cave and Karst Studies* 62 (2): 120–126.

Polyak, V.J. and Güven, N. (1996). Alunite, natroalunite and hydrated halloysite in Carlsbad Cavern and Lechuguilla Cave, New Mexico. *Clays and Clay Minerals* 44: 843–850.

Polyak, V.J. and Provencio, P. (2001). By-product materials related to H_2S-H_2SO_4 influenced speleogenesis of Carlsbad, Lechuguilla, and other caves of the Guadalupe mountains, New Mexico. *Journal of Cave and Karst Studies* 63: 23–32.

Polyak, V.J. and Provencio, P.P. (2005). Comet cones: a variety of cave cone from Fort Stanton Cave, New Mexico. *Journal of Cave and Karst Studies* 67 (2): 125–126.

Polyak, V.J., McIntosh, W.C., Güven, N. et al. (1998). Age and origin of Carlsbad Cavern and related caves from $^{40}Ar/^{39}Ar$ of alunite. *Science* 279 (5358): 1919–1922.

Polyak, V.J., Hill, C., and Asmerom, Y. (2008). Age and evolution of the Grand Canyon revealed by U-Pb dating of water table-type speleothems. *Science* 319 (5868): 1377–1380.

Polyak, V.J., Provencio, P.P., and Asmerom, Y. (2016). U-Pb dating of speleogenetic dolomite: a new sulfuric acid speleogenesis chronometer. *International Journal of Speleology* 45 (2): 103–109.

Portillo, M.C. and González, J.M. (2011). Moonmilk deposits originate from specific bacterial communities in Altamira Cave (Spain). *Microbial Ecology* 61 (1): 182–189.

Pozzi, J.P., Rousseau, L., Falguères, C. et al. (2019). U-Th dated speleothem recorded geomagnetic excursions in the lower Brunhes. *Scientific Reports* 9 (1): 1114.

Proctor, C.J., Baker, A., Barnes, W.L. et al. (2000). A thousand-year speleothem proxy record of North Atlantic climate from Scotland. *Climate Dynamics* 16 (10-11): 815–820.

Provencio, P.P. and Polyak, V.J. (2001). Iron oxide-rich filaments: possible fossil bacteria in Lechuguilla cave. *New Mexico. Geomicrobiology Journal* 18 (3): 297–309.

Queen, J.M. and Melim, L.A. (2006). Biothems: biologically influenced speleothems in caves of the Guadalupe Mountains, New Mexico, USA. *New Mexico Geological Society Guidebook, 57th Field Conference*, 167-174. Socorro: New Mexico Geological Society.

Railsback, L.B. (2000). *An Atlas of Speleothem Microfabrics*. Athens: Department of Geology, University of Georgia.

Railsback, L.B. (2018). A comparison of growth rate of late Holocene stalagmites with atmospheric precipitation and temperature, and its implications for paleoclimatology. *Quaternary Science Reviews* 187: 94–111.

Railsback, L.B., Brook, G.A., Chen, J. et al. (1994). Environmental controls on the petrology of a late Holocene speleothem from Botswana with annual layers of aragonite and calcite. *Journal of Sedimentary Research* 64 (1a): 147–155.

Reimer, P.J. (2020). Composition and consequences of the IntCal20 radiocarbon calibration curve. *Quaternary Research* 96: 22–27.

Renault, P. (1968). Contribution à l'étude des actions mécaniques et sédimentologiques dans la spéleogenèse. *Annales de Spéléologie* 22: 5–21. & 209-267; 23: 259-307 & 529-596; 24: 313-337.

Richards, D.A., Bottrell, S.H., Cliff, R.A. et al. (1998). U-Pb dating of a speleothem of Quaternary age. *Geochimica et Cosmochimica Acta* 62 (23-24): 3683–3688.

Ritter, S.M., Isenbeck-Schröter, M., Scholz, C. et al. (2019). Subaqueous speleothems (Hells Bells) formed by the interplay of pelagic redoxcline biogeochemistry and specific hydraulic conditions in the El Zapote sinkhole, Yucatán Peninsula, Mexico. *Biogeosciences* 16 (11): 2285–2305.

Roberge, J. and Caron, D. (1983). The occurrence of an unusual type of pisolite: the cubic cave pearls of Castleguard Cave, Columbia Icefields, Alberta, Canada. *Arctic and Alpine Research* 15 (4): 517–522.

Rodríguez, J., Burjachs, F., Cuenca-Bescós, G. et al. (2011). One million years of cultural evolution in a stable environment at Atapuerca (Burgos, Spain). *Quaternary Science Reviews* 30 (11-12): 1396–1412.

Romanov, D., Kaufmann, G., and Dreybrodt, W. (2008). Modeling stalagmite growth by first principles of chemistry and physics of calcite precipitation. *Geochimica et Cosmochimica Acta* 72 (2): 423–437.

Rossi, C. and Lozano, R. (2016). Hydrochemical controls on aragonite versus calcite precipitation in cave dripwaters. *Geochimica et Cosmochimica Acta* 192: 70–96.

Saiz-Jimenez, C., Cuezva, S., Jurado, V. et al. (2011). Paleolithic art in peril: policy and science collide at Altamira Cave. *Science* 334 (6052): 42–43.

Salomon, M.L., Grasemann, B., Plan, L. et al. (2018). Seismically-triggered soft-sediment deformation structures close to a major strike-slip fault system in the Eastern Alps (Hirlatz cave, Austria). *Journal of Structural Geology* 110: 102–115.

Sánchez-Moral, S., Portillo, M.C., Janices, I. et al. (2012). The role of microorganisms in the formation of calcitic moonmilk deposits and speleothems in Altamira cave. *Geomorphology* 139: 285–292.

Sancho, C., Belmonte, Á., Bartolomé, M. et al. (2018). Middle-to-late Holocene palaeoenvironmental reconstruction from the A294 ice-cave record (Central Pyrenees, northern Spain). *Earth and Planetary Science Letters* 484: 135–144.

Sanna, L., Forti, P., and Lauritzen, S.E. (2011). Preliminary U/Th dating and the evolution of gypsum crystals in Naica caves (Mexico). *Acta Carsologica* 40 (1): 17–28.

Sasowsky, I.D. (1998). Determining the age of what is not there. *Science* 279 (5358): 1874–1874.

Sasowsky, I.D. (2007). Clastic sediments in caves - imperfect recorders of processes in karst. *Acta Carsologica* 36 (1): 143–149.

Sasowsky, I.D. (2019). Magnetism of cave sediments. In: *Encyclopedia of Caves* (ed. W.B.White, D.C.Culver and T.Pipan), 658–664. New York: Academic Press.

Sasowsky, I.D. and Mylroie, J.E. (ed.) (2004). *Studies of Cave Sediments: Physical and Chemical Records of Paleoclimate*. New York: Kluwer Academic/Plenum Publishers.

Sauro, F., De Waele, J., Onac, B.P. et al. (2014). Hypogenic speleogenesis in quartzite: the case of Corona'e Sa Craba Cave (SW Sardinia, Italy). *Geomorphology* 211: 77–88.

Sauro, F., Cappelletti, M., Ghezzi, D. et al. (2018). Microbial diversity and biosignatures of amorphous silica deposits in orthoquartzite caves. *Scientific Reports* 8 (1): 17569.

Sauro, F., Mecchia, M., Tringham, M. et al. (2020). Speleogenesis of the world's longest cave in hybrid arenites (Krem Puri, Meghalaya, India). *Geomorphology* 359: 107160.

Schmid, E. (1958). Höhlenforschung und Sedimentanalyse. Ein Beitrag zur Datierung des alpinen Paläolithikums. *Schriften des Instituts für Ur- und Frühgeschichte der Schweiz* 13: 1–185.

Schmidt, V.A. (1982). Magnetostratigraphy of sediments in Mammoth Cave, Kentucky. *Science* 217(4562): 827–829.

Schoenherr, J., Reuning, L., Hallenberger, M. et al. (2018). Dedolomitization: review and case study of uncommon mesogenetic formation conditions. *Earth-Science Reviews* 185: 780–805.

Scholz, D., Hoffmann, D.L., Hellstrom, J. et al. (2012). A comparison of different methods for speleothem age modelling. *Quaternary Geochronology* 14: 94–104.

Scroxton, N., Gagan, M.K., Dunbar, G.B. et al. (2016). Natural attrition and growth frequency variations of stalagmites in southwest Sulawesi over the past 530,000 years. *Palaeogeography, Palaeoclimatology, Palaeoecology* 441: 823–833.

Seemann, R. (1987). Mineralparagenesen in österreichischen Karsthöhlen. *Mitteilungen der Österreichischen Mineralogischen Gesellschaft* 132: 117–134.

Self, C.A. and Hill, C.A. (2003). How speleothems grow: an introduction to the ontogeny of cave minerals. *Journal of Cave and Karst Studies* 65 (2): 130–151.

Shen, C.C., Lin, K., Duan, W. et al. (2013). Testing the annual nature of speleothem banding. *Scientific Reports* 3: article 2633.

Short, M.B., Baygents, J.C., and Goldstein, R.E. (2005). Stalactite growth as a free-boundary problem. *Physics of Fluids* 17 (8): 083101.

Simms, M.J. (1994). Emplacement and preservation of vertebrates in caves and fissures. *Zoological Journal of the Linnean Society* 112 (1-2): 261–283.

Slyotov, V.A. (1985). Concerning the ontogeny of crystallictite and helictite aggregates of calcite and aragonite from the karst caves of southern Fergana. *Cave Geology* 2 (4): 196–208.

Smith, C.L., Fairchild, I.J., Spötl, C. et al. (2009). Chronology building using objective identification of annual signals in trace element profiles of stalagmites. *Quaternary Geochronology* 4(1): 11–21.

Spilde, M.N., Kooser, A., Boston, P.J. et al. (2009). Speleosol: a subterranean soil. In: *Proceedings of the 15th International Congress of Speleology* (ed. W.B.White), 338–344. Kerrville, Texas: National Speleological Society.

Spötl, C. and Boch, R. (2019). Uranium series dating of speleothems. In: *Encyclopedia of Caves* (ed. W.B.White, D.C.Culver and T.Pipan), 1096–1102. New York: Academic Press.

Springer, G.S. and Kite, J.S. (1997). River-derived slackwater sediments in caves along Cheat River. *West Virginia. Geomorphology* 18 (2): 91–100.

Stanton, R.K., Murray, A.S., and Olley, J.M. (1992). Tracing the source of recent sediment using environmental magnetism and radionuclides in the karst of Jenolan caves, Australia. In: *Erosion and Sediment Transport Monitoring Programmes in River Basins* (ed. J.Bogen, D.E.Walling and T.Day), 125–133. Wallingford UK: IAHS.

Stepanov, V.I. (1997). Notes on mineral growth from the archive of VI Stepanov (1924–1988). *Proceedings of the University of Bristol Spelaeological Society* 21 (1): 25–42.

Stock, G.M., Riihimaki, C.A., and Anderson, R.S. (2006). Age constraints on cave development and landscape evolution in the Bighorn Basin of Wyoming, USA. *Journal of Cave and Karst Studies* 68 (2): 74–81.

Šušteršič, F. (2006a). A power function model for the basic geometry of solution dolines: considerations from the classical karst of south-central Slovenia. *Earth Surface Processes and Landforms* 31: 293–302.

Tauxe, L. (2010). *Essentials of Paleomagnetism*. San Diego: University of California Press.

Temovski, M., Audra, P., Mihevc, A. et al. (2013). Hypogenic origin of Provalata Cave, Republic of Macedonia: a distinct case of successive thermal carbonic and sulfuric acid speleogenesis. *International Journal of Speleology* 42 (3): 235–246.

Thomas, M. and Silverwood, N. (2017). *Caves: Exploring New Zealand's Subterranean Wilderness*. Auckland: Whio Publishing.

Tisato, N., Torriani, S.F., Monteux, S. et al. (2015). Microbial mediation of complex subterranean mineral structures. *Scientific Reports* 5: 15525.

Ulloa, A., Gázquez, F., Sanz-Arranz, A. et al. (2018). Extremely high diversity of sulfate minerals in caves of the Irazú Volcano (Costa Rica) related to crater lake and fumarolic activity. *International Journal of Speleology* 47 (2): 229–246.

Vaccarelli, I., Matteucci, F., Pellegrini, M. et al. (2021). Exploring Microbial Biosignatures in Mn-Deposits of Deep Biosphere: A Preliminary Cross-Disciplinary Approach to Investigate Geomicrobiological Interactions in a Cave in Central Italy. *Frontiers in Earth Sciences* 9: 590257.

Vanghi, V., Frisia, S., and Borsato, A. (2017). Genesis and microstratigraphy of calcite coralloids analysed by high resolution imaging and petrography. *Sedimentary Geology* 359: 16–28.

Walker, M. (2005). *Quaternary Dating Methods*. Chichester: Wiley.

Waltham, T., Bell, F., and Culshaw, M. (2005). *Sinkholes and Subsidence*. Chichester: Springer.

Went, F.W. (1969). Fungi associated with stalactite growth. *Science* 166 (3903): 385–386.

White, W.B. (1988). *Geomorphology and Hydrology of Karst Terrains*. New York: Oxford University Press.

White, W.B. (1997a). Color of speleothems. In: *Cave Minerals of the World* (ed. C.A.Hill and P.Forti), 239–244. Huntsville: National Speleological Society.

White, W.B. (2004). Paleoclimate records from speleothems in limestone caves. In: *Studies of cave sediments* (ed. I.D.Sasowsky and J.R.Mylroie), 135–175. New York: Kluwer Academic/Plenum Publishers.

White, W.B. (2012). Speleothem microstructure/speleothem ontogeny: a review of Western contributions. *International Journal of Speleology* 41 (2): 329–359.

White, W.B. (2019). Springs. In: *Encyclopedia of Caves* (ed. W.B.White, D.C.Culver and T.Pipan), 1031–1040. New York: Academic Press.

White, E.L. and White, W.B. (1968). Dynamics of sediment transport in limestone caves. *National Speleological Society Bulletin* 30 (4): 115–129.

White, E.L. and White, W.B. (1969). Processes of cavern breakdown. *National Speleological Society Bulletin* 31 (4): 83–96.

White, E.L. and White, W.B. (2000). Breakdown morphology. In: *Speleogenesis. Evolution of Karst Aquifers* (ed. A.B.Klimchouk, D.C.Ford, A.N.Palmer, et al.), 427–429. Huntsville: National Speleological Society.

White, W.B. and White, E.L. (2003). Gypsum wedging and cavern breakdown: studies in the Mammoth Cave System, Kentucky. *Journal of Cave and Karst Studies* 65 (1): 43–52.

Wisshak, M., Barton, H.A., Bender, K.E. et al. (2020). Active growth of non-hydrothermal subaqueous and subaerial barite ($BaSO_4$) speleothems in Lechuguilla cave (New Mexico, USA). *International Journal of Speleology* 49 (1): 11–26.

Woo, K.S., Kim, J.C., Choi, D.W. et al. (2008). The origin of erratic calcite speleothems in the Dangcheomul cave (lava tube cave), Jeju Island, Korea. *Quaternary International* 176: 70–81.

Woodhead, J. and Petrus, J. (2019). Exploring the advantages and limitations of in situ U–Pb carbonate geochronology using speleothems. *Geochronology* 1 (1): 69–84.

Woodhead, J., Hellstrom, J., Maas, R. et al. (2006). U-Pb geochronology of speleothems by MC-ICPMS. *Quaternary Geochronology* 1 (3): 208–221.

Woodhead, J., Hellstrom, J., Pickering, R. et al. (2012). U and Pb variability in older speleothems and strategies for their chronology. *Quaternary Geochronology* 14: 105–113.

Woodhead, J.D., Sniderman, J.K., Hellstrom, J. et al. (2019). The antiquity of Nullarbor speleothems and implications for karst palaeoclimate archives. *Scientific Reports* 9 (1): 1–8.

Worthington, S.R.H. and Ford, D.C. (1995b). High sulfate concentrations in limestone springs: an important factor in conduit initiation?*Environmental Geology* 25 (1): 9–15.

Wray, R.A.L. and Sauro, F. (2017). An updated global review of solutional weathering processes and forms in quartz sandstones and quartzites. *Earth-Science Reviews* 171: 520–557.

Wright, G.F. (1898). A recently discovered cave of Celestine crystals at Put-in-Bay, Ohio. *Science* 8: 502–503.

Žák, K., Onac, B.P., and Perşoiu, A. (2008). Cryogenic carbonates in cave environments: a review. *Quaternary International* 187 (1): 84–96.

Zanin, Y.N., Tsykin, R.A., and Dar'in, A.V. (2005). Phosphorites of the Arkheologicheskaya cave (Khakassia, east Siberia). *Lithology and Mineral Resources* 40 (1): 48–55.

Zupan Hajna, N., Mihevc, A., Pruner, P. et al. (2010). Palaeomagnetic research on karst sediments in Slovenia. *International Journal of Speleology* 39 (2): 47–60.

11

Speleogenesis: How Solutional Caves Form

Chapter 9 explains how a "cave" can be defined, and the different processes that can lead to the formation of a "cave." Caves can occur in a wide variety of rocks and can be generated by different processes and their combinations (Figure 9.1). The most important ones are dissolution caves, mainly developed in carbonate rocks (limestones, dolostones, and marbles) and sulfate rocks (gypsum and anhydrite), but also in the extremely soluble rock salt (halite), in the poorly soluble quartzites or iron formations, and in intermediate rocks (conglomerates). This chapter, based on the concepts and processes illustrated and explained in the previous ones, mainly deals with the formation of caves in carbonate rocks, with final sections dedicated to halite, gypsum, quartzite, and iron formation caves. There is no single model for the formation of caves in soluble rocks; all theories are based on the complex interactions between geological factors (lithology, age of the rocks, geological structure, and stratigraphic setting), their change through time (landscape and base level evolution, surface erosion and sedimentation, and tectonics) and the time-dependent local conditions at the surface (temperature, precipitation, vegetation, and chemical composition of the water), which are also the factors that control the development of karst landscapes (Figure 1.3). One might well state that unraveling speleogenesis (the formation of caves) involves a detailed understanding of local geological conditions, the chemical and physical processes at work, and the hydrological context.

Much has been done in the field of speleogenesis over the last decades. More detailed information can be found in the excellent book titled "*Speleogenesis: evolution of karst aquifers*", produced by the International Union of Speleology and the National Speleological Society (USA) (Klimchouk et al. 2000). Further work has been gathered in the volume "*Evolution of karst: from prekarst to cessation*", edited by the Karst Research Institute of Postojna (Slovenia) (Gabrovšek 2002). Good overviews are also given by Palmer (2007) and Ford and Williams (2007), whereas details on hypogene cave genesis can be found in Klimchouk et al. (2017b) and Klimchouk (2019b). Entries in the Encyclopedia of Caves on gypsum cave genesis (Klimchouk 2019a), telogenetic speleogenesis (Gabrovšek 2019), sulfuric acid caves (Palmer and Hill 2019), quartzite caves (Auler and Sauro 2019), and iron ore caves (Auler et al. 2019) are also valuable references.

11.1 The Growth of Ideas About Cave Genesis

Caves are parts of the geological landscape and the explanation of their genesis has been accompanied by a long series of conflicting ideas. The central role played by water is indicated in the writings of several Greek philosophers, although the origin of the cave-forming water and its movement

Karst Hydrogeology, Geomorphology and Caves, First Edition. Jo De Waele and Francisco Gutiérrez.
© 2022 John Wiley & Sons Ltd. Published 2022 by John Wiley & Sons Ltd.

were not so clear. The origin of caves was often linked to karst springs (rivers coming out of the mountains), which have been known and used for water supply since immemorial time in the Mediterranean area (Greeks, Romans, and Arabs), but also in China. Stalactites were believed to form from dripping waters in China since the eleventh century; Shen Kuo (1031–1095) first, and Fan Chengda (1126–1193) a century later (Barbary and Zhang 2004). These early ideas on how karst systems work were based on extensive descriptive studies on caves and karst areas. Xu Hongzu, who published under the name of Xu Xiake (1586–1641), travelled through most of southern China describing many of the over 300 caves he visited. He is considered the "father of karst science" in China (Figure 11.1a, b). In Europe, the earliest detailed descriptions of karst date back to the same period, such as those from the Balkan area of Nikola Gučetić (1585), from Dubrovnik, and Janez Vajkard Valvasor (1687). Although the idea of water being responsible for cave formation was already mentioned by Faventies (Italy) in 1561, the importance of the action of

Figure 11.1 Some famous pioneers in speleology and speleogenesis: (a) A series of Chinese stamps issued in 1987 in honor of the 400th anniversary since the birth of Xu Xiake. Note the second stamp showing a person in a cave. (b) The statue of Xu Xiake in front of the Institute of Karst Geology in Guilin, China. *Source:* Photo by Shengrong Du; beibaoke/Shutterstock. (c) Portrait of Jovan Cvijić by the Serbian artist Uroš Predić (1857–1953). *Source:* Uroš Predić/Wikimedia Commons/Public Domain. (d) A picture of Alfred-Edouard Martel (National Library of France). *Source:* Bibliothèque nationale de France/Wikimedia Commons/Public Domain. (e) A picture of the American geologist J Harlen Bretz, one of the pioneers in cave and karst studies. *Source:* US Hanna Holborn Gray Special Collections Research Center, University of Chicago Library.

flowing groundwater in the formation of caves was first clearly laid out in Catcott's writings (1761), based on observations of caves in the Mendip Hills (UK). However, he, as many scientists in those days, connected this genetic concept to the "biblical flood," which supposedly involved enough water to carve the large voids in less than 6000 years. This "biblical flood" belief severely limited scientific progress until the end of the eighteenth century. Despite this, current ideas on cave formation derive mostly from observations and thoughts that find their roots in the eighteenth century. The erosional power of streams carrying sediments were believed to be capable of carving large caves by most scientists. The idea that dissolution of limestone could also be responsible for at least part of the void formation encountered more resistance up to the end of the nineteenth century, with scientists dividing between "erosional" and "dissolutional" supporters. Nevertheless, the famous Scottish geologist James Hutton (1795), by many considered one of the fathers of modern Geology, clearly attributed the origin of limestone caves to the dissolution of rock by water, and the deposition of carbonate dripstones to the loss of CO_2. The true importance of CO_2 in the dissolution of limestone became clear 35 years later, with the contemporaneous, but apparently independent publications of Charles Lyell (Scotland) and Charles Edouard Thirria (France) in 1830. This opened the dispute between "vadose" and "phreatic" dissolution, which was brought into the first part of the twentieth century. At the turning point between the nineteenth and twentieth century, a time during which two of the most famous karstologists were most active and produced very detailed descriptions of both surface karst phenomena and caves – Cvijić (1893) and Martel (1894) (Figure 11.1c, d) – theories on cave formation generally combined mechanical erosion, solutional denudation of rock under vadose conditions, and phreatic dissolution.

The first half of the twentieth century was characterized by the so-called "water table debate", which started in Europe and was later carried forward in the United States. Alfred Grund (1903) divided karst aquifers into two simple zones: one above the water table and the other below. Water was thought to flow through a network of caves getting narrower at depth and degenerating into a large number of narrow fissures and pores, with flow eventually concentrated at or slightly below the water table following Darcy's principles. This contrasted with Katzer's (1909) and Martel's (1921) views of underground rivers, whereby water was believed to flow through interconnected channels (cave rivers) and the general water table did not exist at all. Cvijić (1918), on the other hand, distinguished three hydrological zones in karst aquifers: dry (vadose), transitional, and saturated (phreatic) zones. Most groundwater flow would have occurred at or immediately below the water table, in his transitional zone.

There were two schools of thought in the United States. William Morris Davis (1930) in a work published during the final period of his career (at the age of 80 years), later supported by Harlen Bretz (1942) (Figure 11.1e), believed that caves form in the deep phreatic environment and are later enlarged by vadose flow and erosion. On the contrary, Swinnerton (1932), although recognizing that dissolution in the phreatic zone is possible, argued that cave formation is likely to be favored where water flow is strongest, close to the water table, and used a hydrological division of the karst aquifer in three zones, similar to the ones proposed by Cvijić. Rhoades and Sinacori (1941) were the first to combine both theories in a model that presumed slow enlarging pathways in both deep phreatic and shallow or vadose zones, but with faster evolution close to the discharge points. A master conduit would cut backward from the spring along the water table, controlling its position and eventually concentrating most flow along this zone.

The increased knowledge on caves (morphology, structural control, and infilling), a better understanding of the physical and chemical processes at work in the underground, and the evolution of hydrology as a science brought about a better understanding of speleogenesis in the last 60 years. Nowadays, there is no single speleogenetic model able to explain the formation of all caves, but most processes and the associated conditions are rather well known. Besides the "normal" carbonate cave systems, the formation of which can be explained by relatively general speleogenetic processes, there are several alternative mechanisms of cave formation, including seacoast mixing, the rising of thermal waters, and the presence of sulfuric acid. These theories of cave genesis in carbonate rocks will be dealt with in the following sections. On the other hand, the formation of caves in other rock types (halite, gypsum, quartzite, and iron formation) will be treated separately at the end of this chapter.

11.2 Geological Controls on Cave Genesis in Carbonate Rocks

Regardless of the processes involved in cave genesis, lithology and geological structure always exert some control on the final outcome in terms of cave pattern and morphology at macro- and mesoscale. This section is focused on the geological properties of carbonate rocks (limestones and dolostones) that host most of the cave systems in the world.

Both lithological and structural features evolve through time, from the sedimentation to early diagenesis (compaction, cementation, and mineral replacement), shallow burial (eogenesis), deep burial (mesogenesis), and uplift/exposure (telogenesis) (Klimchouk and Ford 2000a). Through this typical evolution from soft and poorly lithified sediment, to deeply buried hard fractured rock and to its final exposure at the surface, the heterogeneity of porosity and fissure permeability increases drastically, with an increasing importance of the fissure network over primary (or matrix) porosity. Most cave systems occur in exposed and relatively mature limestones and their telogenetic stage, where the fissure network plays a significant control on the final cave pattern. However, in young eogenetic limestones, the primary porosity and permeability of the rock play a fundamental role on groundwater flow and dissolution. In deep-seated settings, where soluble rocks have not yet been exposed at the surface, cave patterns tend to be less heterogeneous and more predictable. This has important applications in the petroleum industry.

11.2.1 Influence of Lithology

Carbonates are the most variable group of karst rocks because they show a wide range of compositional and stratigraphic characteristics, reflecting different depositional and diagenetic settings. They can form in deep sea environments (basin and slope), open and rimmed platforms that may include lagoons, and terrestrial environments (e.g., lakes, rivers, dunes, springs, and caves) (see Chapter 2). They can be formed by in place biogenic structures such as reefs, a wide range of allochem particles (e.g., bioclasts, pellets, and lithoclasts), and micritic or sparitic matrix, all variously cemented and recrystallized, and more or less dolomitized. See Section 2.2 for more details.

11.2.1.1 Influence of Rock Purity

Carbonate rocks are composed of a mixture of minerals including calcite (with variable amounts of Mg), aragonite, dolomite, and other mostly non-carbonate minerals. The most soluble calcium carbonate mineral is high-Mg calcite, followed by aragonite, pure calcite, and low-Mg calcite (Figure 3.11). Ordered pure dolomite is slightly less soluble than calcite and aragonite. However, nonstoichiometric dolomite with an excess in Ca (calcian dolomites) and disordered dolomites with crystallographic dislocations and substitutions may reach solubilities higher than that of aragonite (Figure 3.11). With all other variables (discussed below) remaining the same, pure limestones composed of high-Mg calcite are more soluble than pure dolostones, and more importantly, dissolve much faster due to the much slower dissolution kinetics of dolomite (see Section 3.8.1). Where caves are developed in alternations of dolostones and limestones, the voids in the former are typically much smaller than those carved in pure limestone.

The most common insoluble minerals occurring in carbonate rocks are clay minerals and silica (chert nodules, quartz grains). Karst and speleogenesis are greatly inhibited in carbonate rocks with more than 20–30% of clays, since these insoluble small particles tend to clog the initial protoconduits. Sand-sized quartz grains have a lower limiting effect on karstification, and some caves can be formed in sandy limestones and calcareous sandstones. In general, however, carbonate rocks with more than 30% of impurities rarely allow the formation of important cave and karst systems (see Section 3.9.1).

11.2.1.2 Influence of Grain Size and Texture

Grain size (or crystal size) greatly influences dissolution rate; smaller particles have greater specific surface exposed to dissolution, so they dissolve more readily. Although this is generally a valid statement, this is not always true, since the potential surface that can interact with water also depends on the shape of the particles; spherical grains have lower specific surface than angular ones. The grain size distribution has also some influence, especially on the porosity (discussed below), and therefore the ability of the water to come into contact with the grain boundaries. Packing, which influences the surface exposed to dissolution, is another factor to be taken into account. Porcellaneous limestones, composed of very fine and well-sorted micritic carbonate particles, dissolve less faster than coarser or fossiliferous limestones, as illustrated in the Gaping Gill cave system in Yorkshire where cave passages are perched upon these fine-grained limestones and preferentially follow the coarser fossiliferous limestones above (Glover 1974).

11.2.1.3 Influence of Matrix Porosity

Matrix (or primary) porosity (as used in Chapter 4) is also called fabric porosity, and is related to factors such as grain size, particle shape and sorting, packing, and sedimentary fabrics. The primary porosity of most limestones is not homogeneously distributed, with portions (beds) of rock showing higher initial porosity. This has an influence on fluid flow since the early diagenesis and potential changes in the initial porosity (e.g., dissolution and dolomitization increase porosity, whereas cementation decreases it). This eventually can lead to focused fluid flow along particular layers or sedimentary packages, which can later be used for cave inception.

Matrix primary porosity in young carbonates not obliterated yet by diagenesis ranges between 25 and 80%. This is why cave passages hardly develop in such porous limestones (e.g., emerged reefs, non-fractured chalk), unless the mechanisms of dissolution are related to other processes than turbulent water flow, such as water mixing phenomena (e.g., caves in seacoast mixing zones). Diagenesis (compaction, cementation, recrystallization, and other processes) transforms these young and highly porous limestones into hard and compact rocks with much lower matrix porosity. Micrites often have porosities of less than 2%, whereas sparites reach values of 5–10%. Dolomitization in general increases the matrix porosity by about 5%, whereas when recrystallization is almost complete (by metamorphism, such as in marbles) porosity decreases to a few percent.

11.2.2 Influence of Stratigraphic Position of Soluble Rocks

The position of the soluble rocks (e.g., exposed, overlain by caprocks) and their relationships with the adjacent less soluble or insoluble units have an important influence on where caves will form and which kind of karst network will develop. Several stratigraphic settings can be distinguished (Figure 11.2).

The most favorable condition occurs where the soluble rocks are entirely exposed at the surface, largely covered by soils and locally forming bare rock outcrops (Figure 11.2a). Water will flow from higher areas to lower lying ones as soon as a hydraulic gradient is created. Enough elevation difference is needed for water to develop preferential underground routes (e.g., caves). Karst springs will be located in the topographic lows that function as base levels (coastlines, river valleys) and cave systems will develop from the sinking points located in the recharge area toward the springs. Extensive bare karst terrains can be dissected by rivers and karst systems which may enter in competition forming independent karst drainage basins (Figure 11.2b). It is interesting to note that karst groundwater divides do not necessarily coincide with the topographic divides.

Soil covers play the role of increasing the CO_2 contents in the infiltrating waters boosting their dissolution capacity. Sinking points can form preferentially where CO_2-rich waters create concentrated runoff, often leading to the development of caves. There is often a positive correlation between the size of the sinking points (and associated caves) and the thickness and areal extent of the soils, since these parameters rule the amount of aggressive water that enters the soluble bedrock at discrete points. Thick, low-permeability soils, however, often cause a diminution of sinking points, since more water is retained in the soil and lower amounts of it infiltrate causing a more subdued dissolution of the underlying soluble rock.

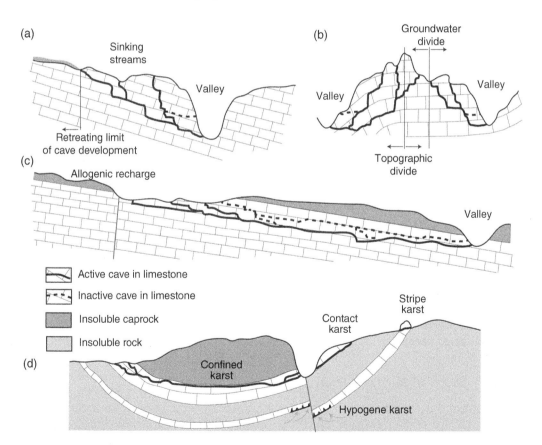

Figure 11.2 Influence of stratigraphic position of the soluble rock units on speleogenesis. (a) Bare and soil-covered karst. (b) Competing karst systems. Note the difference between groundwater and topographic divides. (c) Unconfined karst systems developed adjacent to (allogenic recharge) or underneath an insoluble but permeable caprock. (d) Karst system developing in confined, contact, and stripe karst settings.

In general, insoluble rocks adjacent to soluble ones have an important impact on cave development (contact karst). Large areas of insoluble rocks lying at higher elevation (upstream) and next to the soluble units can collect large amounts of allogenic water that enters the karst bedrock through a few discrete points (sinking streams) (Figure 11.2c). When insoluble rocks overlie the soluble ones, the thickness and permeability of the caprock will determine whether and how efficiently caves will form in the underlying soluble units. If the caprock is too thick and impervious, the quantity and aggressiveness of the waters reaching the buried karst rocks decreases, and so does cave development. On the other hand, insoluble caprock helps to protect the underlying caves from surface erosion, increasing their preservation potential.

When the base of the impervious caprock is situated below the local base level, cave development is possible only in confined conditions (Figure 11.2d). Epigenic cave development occurs only over short distances where infiltration points are located at elevations high enough to allow phreatic flow from the recharge area to the springs. Confined settings are typical of artesian karst development, such as in gypsum beds in sequences consisting of carbonates, sulfates, and fine clastic sediments. In these deep settings, limestones and dolostones are often more permeable than the evaporites (e.g., poorly fractured and low porosity anhydrite or gypsum often behave more as aquicludes), so these carbonate rocks initially act as aquifers. On the other hand, solutional aggressiveness of waters toward carbonates in these deep settings is generally very limited, whereas sulfates are readily dissolved creating confined maze networks of caves, which then start to behave as karst aquifers. The formation of large voids by hypogene dissolution in these evaporites can cause the collapse of the overlying rocks, creating breccia pipes by stoping. If the thickness of the dissolved evaporites is large, these transtratal subvertical breccia pipes can reach hundreds of meters in height. The dissolution of thinner evaporite beds, on the other hand, can create more permeable horizons (e.g., solution-collapse breccias) that can have a great influence on later epigenic speleogenetic phases.

Insoluble rocks can also underlie the soluble formations, thus forming a less permeable substratum above which karst groundwaters flow (Figure 11.2d). These insoluble rocks limit downward flow and karstification, creating perched groundwater flow and associated cave systems (contact karst). Karst springs at these lithological contacts can remain hanging high above valley floors.

A final stratigraphical position is that of a narrow dipping soluble unit hosted in a dominantly insoluble sequence, forming elongated surface exposures (stripe karst) (Figure 11.2d). If recharge is high enough, caves can form between higher lying sinks and downstream springs often following the strike of the soluble bed. These springs are generally perched high above the nearby valleys, at the contact between the soluble band of rock and the underlying less permeable formation.

The development of important cave systems requires the thickness of the soluble beds to be of at least 5 m for epigenic cave systems, and less for artesian mazes. In soluble beds less than 2 m thick, no sizeable cave systems can develop.

11.2.3 Influence of Geological Structures

Cave patterns are greatly influenced by geological structures, since even minor changes in hydraulic conductivity can determine the paths taken by groundwater flow. This structural guidance on caves is most clearly visible in plan view. Cave passage directions often show numerous abrupt changes at acute angles, with a tendency to follow the easiest flow paths. These more penetrable openings are often only minor fractures and bedding planes and are not necessarily the most important structures. Only in rare occasions major cave passages can be aligned along important geological structures, developing along extensive straight lines. Surface streams, on the other hand, often align along major disturbances and are less influenced by minor structural elements.

In deep hypogene settings, where fissure widths are smaller and less variable and flow is diffuse and slow, speleogenesis is rather homogeneous, generally resulting in maze-like patterns. In these settings, groundwater tends to flow along all the available discontinuity planes, widening them by dissolution. On the contrary, in shallow environments, where fissure openings are much more heterogeneous and flow is often fast, subsurface flow selects the larger pathways and with more favorable orientations. This often results in branchwork or ramiform cave patterns.

The main structural elements that guide cave development are bedding planes, joints, and faults. These planar breaks in the soluble formations guide around 95% of all the solutional cave voids, the remaining 5% being controlled by changes in primary (matrix) porosity. This is because groundwaters preferably penetrate these planar openings and water–rock interaction (dissolution and precipitation) mainly occurs within these voids and in their immediate surroundings. Larger structural elements such as folds (synclines and anticlines) and their associated fracture patterns also influence underground water movement and karst development. This can lead to very complex situations, especially in highly deformed mountain belts.

11.2.3.1 Influence of Bedding Planes

Bedding-plane partings are related to changes in the sedimentation regime, including interruptions and erosional events. These variations can be very small (e.g., minor grain size change), or major (e.g., from silt-sized carbonate grains to coarse oolithic carbonate sands), or even drastic (carbonate beds alternating with paper-thin clay intervals). Interruptions may occur when a shallow marine environment experiences a short emergence episode, recorded by evidence of erosion, dissolution, beach rock deposition or hardground formation, later turning back to marine conditions.

Prominent bedding planes are generally visible in the field and define what geologists call a bed (the smallest lithostratigraphic unit). Single beds range in size from less than a centimeter (in which case they are called laminae) to over 1 m (very thick beds) (Table 3.5). Bedding planes in thinly bedded rock formations (bed thickness of 3–10 cm) tend to have limited areal extents, whereas very thick beds (1–10 m) can have bedding planes that are continuous for kilometers.

Not all bedding planes are easily penetrable by groundwaters, and flow will choose the most permeable ones. The bedding-plane partings more prone to let water pass initially are those that have significant changes in grain size, lithology, or mineralogy, such as shale beds in a limestone succession, or the beds containing impurities such as pyrite or chert nodules (see inception horizons in Section 11.2.3.5). In deep settings, bedding planes that are initially less permeable will tend to be preferential sites for diagenetic processes (cementation, mineralization, and stylolite formation), diminishing their permeability. On the other hand, more permeable bedding planes will tend to increasingly concentrate water flow, maintaining their higher permeability. Original bedding planes perfectly match each other and are poorly permeable unless associated with important textural or lithological changes. When rocks are subject to tectonic deformation, differential slippage can disrupt the matching surfaces, causing brecciation and the development of slickensides (e.g., striations, steps) that increases permeability even when displacement is only of a few centimeters. In most deformed strata (tilted or folded beds), some differential slippage occurs. During telogenesis (uplift/exposure) stress release opens up discontinuity planes, including some of the less permeable ones.

Water flow in caves follows the most penetrable bedding planes, jumping from major to minor ones depending on how easily water can flow through. In thinly bedded to laminated rock units water is able to flow along many closely spaced bedding planes, causing solutional widening of voids to be dispersed. In such rocks, dispersed dissolution together with the low mechanical strength of the thin beds does not allow the formation of penetrable caves. The most extensive cave systems controlled by bedding planes develop in medium to massively bedded sequences, where solutional and erosional processes are focused along a few planes, and the mechanical strength of the beds allows large voids to remain stable over longer geological periods (Figures 10.9a and 11.3a). However, where thinly-bedded portions are interbedded between thick and massive beds of soluble rock, water flow is mostly concentrated in the closely spaced bedding planes, resulting in cave voids guided by these thinly bedded sections.

Figure 11.3 Influence of geological structures on cave development. (a) A bedding-plane guided passage in "New Discovery" in Mammoth Cave, Kentucky, constrained by cosmogenic burial dating at 1.3 Myr. *Source:* Photo by Arthur Palmer. (b) Joint-guided passage in Buso della Rana, Veneto (N-Italy). *Source:* Photo by Sandro Sedran, S-Team. (c) The giant fault plane with the vertical orange stains of infiltrating water, characterizing the "Grandissima Frana" in Su Bentu Cave, Sardinia. Note cavers and their flashlights for scale. *Source:* Photo by Riccardo De Luca. (d) A fault plane with slickensides in Buso della Rana. *Source:* Photo by Sandro Sedran, S-Team.

11.2.3.2 Influence of Joints

Lithified sediments are crossed by planar ruptures that show no visible shear displacement (slippage), but only a slight opening between the two opposing blocks. These ruptures are called joints, and veins when they are filled with secondary minerals (normally calcite or quartz in carbonate rocks). The term shear fracture is often used to designate joints in which there has been some displacement between the blocks, but this separation is too small to be visible to the naked eye (here treated as joints).

The rupture of the rocks is often accompanied by the formation of irregularities along the fracture plane (e.g., plumose marking), but later dissolution–cementation often obliterates these initial irregularities, especially along the most important joints. With increasing lithostatic pressure (burial depth) most joints tend to be more closed, and thus less prone to fluid flow, whereas larger (master) joints tend to extend both laterally and vertically, focusing water flow and dissolution and becoming very important during early speleogenesis (Klimchouk and Ford 2000a).

Dolostones are generally more susceptible to fracturing than limestones (Ding et al. 2012). Stiff chert inclusions tend to concentrate joints, causing less fracturing to occur in the hosting relatively more ductile carbonate rocks (Spence and Finch 2014; Antonellini et al. 2020).

In contrast to bedding planes, which are primary sedimentary features, joints can be created in multiple stages once sediments have experienced some degree of lithification (induration) and acquire a brittle rheology. Joints can locally form at near-surface conditions by erosion-induced unloading (e.g., extensional joints associated with cliffs) (Sasowsky and White 1994), or by enhanced gravitational stress (e.g., cave passage walls). Joints generally cut perpendicularly across bedding planes, but they can also traverse layering at high angles. Some joints extend across several beds and are often called master joints. These larger fractures can be important paths for underground water flow and speleogenesis.

Densely jointed soluble rocks tend to have a more dispersed fluid flow and lower potential for the development of large cave passages, unless the entire fissured rock is removed by dissolution and mechanical erosion. In thickly bedded soluble rocks joint spacing is larger and fluid flow occurs in a more concentrated manner. Large cave systems are thus better developed in thickly bedded soluble rocks with widely spaced joints. Master joints crossing multiple beds will be the most favorable paths for underground water flow (Figure 11.3b). In gently dipping strata fractures are often oriented along the dip direction of the bedding, with secondary strike-parallel joints. In steeply dipping successions the joint systems are commonly more complicated. Where sedimentary beds are folded, higher densities of strike-parallel joints tend to occur in the hinge zone, affected by local extension during folding (bending-moment fracturing).

11.2.3.3 Influence of Faults

Faults are fractures with visible shear displacement. Faults can be vertical, variously inclined, or even horizontal, and although they often have a uniform angle of inclination and direction over short distances, they do show variations in dip and strike. The length of faults ranges between a few centimeters to many hundreds of kilometers, whereas their movement can be vertical (up and down), horizontal, or a mixture of these (Figure 3.46).

In general, faults created in a compressional setting tend to be sealed by recrystallized rock, and are therefore unfavorable locations for groundwater flow. Their permeability can be increased by slickensides (Figure 11.3c, d), fault breccias, and by later erosional unloading which opens all types of faults by dilation. Normal faults, formed by extension, are intrinsically more open than compressional faults, although this is not always true. Most large faults of all types with significant displacements are characterized by the presence of less permeable material. Large faults therefore often act as barriers more than as preferred underground pathways for water flow.

Large faults never come alone and often many secondary faults feather off at acute angles from the main rupture surface. Some of these minor fractures can be created by local extension and can be more penetrable for fluids. These zones of contrasting permeability can guide cave development, causing underground passages to follow these feathering fractures, with the less permeable core of the fault zone acting as a lateral barrier.

11.2.3.4 Influence of Folds

Soluble beds and formations are often folded in a wavy pattern. There are many types of folds, with the fold plane and axis variously inclined (vergence, plunge). We mainly differentiate anticlines (convex upward structures with the oldest strata in the core) and synclines (concave upward with the youngest layers in the core). These structures can be very small (centimeter-sized) or can be hundreds of kilometers across. Rock folding is essentially a slow process occurring in rocks with ductile rheology. Plastic behavior in soluble rocks such as limestones and dolostones is favored by high depths, where both temperature and lithostatic pressure are high.

The crests of anticlines and troughs of synclines are zones where strike-parallel joint sets show a high density. Dissolution by groundwater flow thus tends to be focused in these more permeable areas, and especially in the core of synclines in vadose conditions, and in the core of anticlines in deep (artesian) settings (Figure 10.9b). Differential slippage along bedding planes on the limbs of folds can also create high permeability pathways (Figure 10.9c).

In complex layered sequences, where soluble beds are intercalated with layers of insoluble rocks of varying permeability, the controlling factors can become extremely more complex. These folded successions are often also crossed by a variety of faults and joints, increasing the heterogeneity. Favorable speleogenetic settings, such as syncline troughs bounded by lateral faults, can cause the compartmentalization of the karst aquifer in several sub-basins feeding different springs. Conditions of artesian confinement can be created where the soluble strata are sandwiched between impermeable beds, allowing groundwater to flow over very large distances with descending and ascending directions toward areas of lower hydraulic head.

11.2.3.5 Inception Horizons

Most caves hosted in carbonate successions appear to have developed along a restricted number of beds or stratigraphic contacts (Rauch and White 1970) (Figure 11.4). For example, in Mammoth Cave (Kentucky, USA) only five out of the 24 recognized horizons in the 100-m-thick limestone sequence have focused cave development (Palmer 1989). In the more complex structural environment of the Dinaric Karst, in Velika Dolina, which is the largest of the collapse dolines of Škocjanske Jama, the relict cave passages visible along the walls are developed along only three main bedding planes, out of a total of 62 (Knez 1998). Investigations using published cave maps and profiles and a 3D statistical analysis of 15 alpine cave systems (Switzerland, Austria, and Germany), Ogof Draenen (South Wales, UK), Shuanghedong (Guizhou, China), and Mammoth Cave (Kentucky, USA) have recognized the concentration of the main cave passages along just a few bedding planes or horizons (Filipponi et al. 2009) (Figure 3.42).

The question that arises is why only some bedding planes or horizons are the favored locations for cave development, and others are less susceptible to be enlarged by groundwater flow. It is assumed that in the earliest phases of speleogenesis, or even during the diagenesis of the carbonate sequence, dissolution mainly takes place at specific horizons for a variety of reasons, known as "inception horizons". These horizons are favorable to water flow and dissolution because of physical, lithological, or mineralogical deviations from the surrounding rocks in the carbonate sequence (Lowe and Gunn 1997). Lowe (2000) mentioned the following possible reasons for these changes

Figure 11.4 Inception horizons. (a) A bedding-plane with terra rossa (probably a paleokarst surface) that guided speleogenesis in Cocci Abyss, Sicily. Note the other bedding planes with terra rossa in the cupola on the roof, which did not function as preferential paths during early speleogenesis. (b) Detail of the guiding bedding plain with pockets of terra rossa. *Source:* Photos by Marco Vattano.

in rock properties: (i) higher primary porosity and initial permeability of the bedding plane or horizon, (ii) presence of more soluble minerals such as gypsum in a particular bed, (iii) bedding plane or horizon with presence of sulfide minerals (e.g., pyrite) that release sulfuric acid when oxidized, and (iv) presence of insoluble material such as shales or volcanic ashes that focuses water flow and dissolution above (epigene) or below (hypogene). Low-temperature metamorphism of dolomite in contact with clays releases CO_2 that may also enhance the development of solutional porosity at certain horizons (Pezdič et al. 1998).

Detailed studies in the Siebenhengste Cave System (165 km of passages, 1340 m deep, Switzerland) revealed the nature of the seven main inception horizons in the over 200 m thick Cretaceous Schrattenkalk Formation and one in the underlying Drusberg marls. These inception horizons are (from bottom to top): (i) limestone layer 2 m thick in the marly Drusberg sequence, (ii) marl layer some decimeters thick with large amounts of secondary gypsum intercalated in limestones, (iii) a centimeter-thick marl horizon with siliceous nodules in an otherwise pure limestone succession, (iv) a marl horizon some centimeters thick in limestone, (v) a blocky limestone layer, (vi) the contact between oolithic and massive limestones, and (vii) fossil-poor bedding plane with a network of anastomosing channels (Filipponi et al. 2009).

Inception horizons (Lowe 1992b, 2000) are important in the late phases of diagenesis and the early phases of speleogenesis, guiding the development of successive cave passages. These horizons create preferential pathways for the first dissolving fluids (speleoinception), even long before true cave genesis starts, and pave the way for the later true stages of speleogenesis, which are guided by these stratigraphical elements and the later structural imprint of tectonic and lithogenetic fissures.

11.2.4 Topography, Base-Level and Climate

The formation of solutional caves requires undersaturated groundwater to flow toward areas of lower hydraulic head. Groundwater flow can be downward or upward, depending on the geological and hydrological conditions. A great majority of caves is formed by meteoric waters circulating freely in soluble rocks, moving from topographically higher-lying recharge areas toward lower-lying discharge points. These are the so-called epigenic caves. The main requirements for the development of these caves are: (i) the presence of undersaturated meteoric water in liquid state, (ii) adequate exposures of soluble rocks, (iii) a hydraulic gradient related to the elevation difference between the recharge and discharge area, and (iv) sufficient time. Significant portions of the soluble rock need to be above the local base level, which is normally determined by the altitude of nearby rivers, the ocean, or a lake. Topographic relief does not necessarily have to be high; very extensive cave systems can develop in low gradient plateaus. A good example is the longest cave system in the World, Mammoth Cave in Kentucky (USA), with a vertical extent of only 120 m and a horizontal distance between the most distant points of ca. 11 km (Palmer 1989; Audra and Palmer 2011).

Most cave systems develop in geological time scales, ranging between 10 years for halite caves to several hundred thousand of years or one million for limestone and quartz sandstone caves, respectively. Conditions such as climate and landscape can vary substantially during the life cycle of a cave influencing speleogenetic processes. Many continental areas are subject to uplift of tectonic or non-tectonic origin (e.g., glacial unloading, magmatic intrusion). Uplift and erosion are intimately linked: when an area rises, meteoric processes tend to erode it and sediments accumulate in subsiding areas. Erosion is mainly operated by streams, which deepen their valleys in response to uplift. This causes karst springs emerging in the valley to adjust to this base level lowering, shifting to lower elevations, sometimes accompanied also by a lateral displacement of several kilometers. The cave systems feeding these karst springs also adjust to the new base level, expanding downward (and sometimes laterally) and carving new cave levels (Audra and Palmer 2011) (Figure 11.5a–d). In this common situation, the higher cave levels are the oldest. The opposite is also possible, where there is fluvial aggradation induced by sea-level rise or tectonic subsidence. When this aggradation is important and causes the base level to rise over substantial altitudinal ranges and over long periods of time, such as during the Pliocene aggradation phase after the Messinian Salinity Crisis in the Mediterranean area, younger cave levels can be carved at higher altitudes (Mocochain et al. 2009) (Figure 11.5e, f). Where base-level rise is less important, such as during glacial advances, surface drainage networks can be partially or entirely disrupted, causing discharge areas to migrate or change. Coeval alluviation can cause antigravitative erosion (paragenesis) or eventually the complete filling of previously carved cave passages. Deglaciation and renewed river erosion can then reactivate parts of the preglacial karst drainage system, or carve new underground routes. The shifts of base-level have an effect on the hydraulic gradient of the karst system, and thus the potential energy of the water that flows through it, favoring sedimentation during base-level rise and erosion during base-level lowering.

Climate change, especially during the Quaternary, is cyclic and in phase with cave development, over timescales of around 100 kyr. Relatively long cold phases (glacials or ice ages) are interrupted by shorter warm phases (interglacials) (Figure 4.3). During colder periods mean global annual temperature drops by several centigrades, the ice sheets and polar fronts advance toward lower latitudes, global precipitation decreases together with the vegetative cover (and CO_2 in the soils), and relative sea level drops (up to around 120 m). These periods are generally less favorable for cave formation in temperate areas, but can have positive effects on karstification in arid areas (pluvial phases), because the shift of the polar fronts causes more cyclones to reach these otherwise dry areas. Interglacials have the opposite effect. At a regional scale, climate changes can have enormous effects on the potential of groundwater to carve cave systems, and these changes in water availability, CO_2 concentration, and temperature are clearly recorded in the karst record. The periodic alternation of interglacial and glacial periods causes shifts in base level (not only in coastal

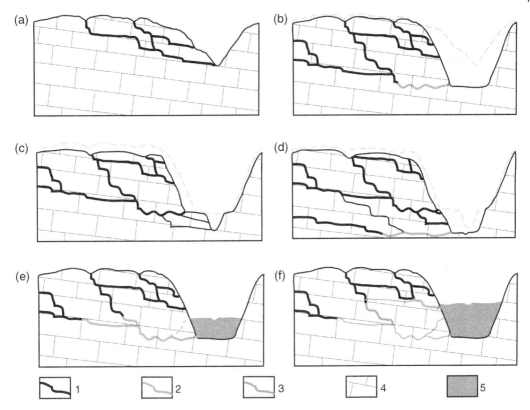

Figure 11.5 Response of cave systems to changing base level. (a) Cave system adjusted to local base level (the nearby valley floor). Most passages will be young and vadose. (b) When the valley experiences an entrenchment episode, the cave system adjusts to the new base level, carving lower levels. If base level remains stable for enough time, the valley widens, forming a floodplain underlain by a surface carved in bedrock, and the lowest active cave branch adjusts to the water table. (c) Renewed entrenchment leaves a strath terrace and causes "soutirages" to form, until (d) stability is reached again, with the formation of a new active lower cave level. *Source:* Based on Audra and Palmer (2011). (e) If base level rises (e.g., sea level rise) the river aggrades, and the sediments force the active karst system to find its way through the alluvial infilling, forming alluviated springs. In rare occasions, when alluviation is slow enough, new alternative pathways at higher elevation may be formed. (f) If aggradation continues, the lower cave levels are filled with alluvial sediments and higher cave levels are reactivated again. *Source:* Based on Mocochain et al. (2009).

areas), and together with the vertical tectonic and non-tectonic movements of the Earth's crust can cause the formation of cave storeys (Palmer 1987).

Caves formed by rising groundwater flow (confined caves) are concentrated in areas where the water can escape, such as along important fault zones, where the upper confining unit is thinner and breached, or where there is a relative low in the potentiometric surface, often corresponding to river valleys.

11.3 Simple Models of Initial Cave Development

Before computers started dominating our lives, the initial development of underground karst networks was simulated using physical models. These included analog models based on the propagation of electric currents (e.g., through a copper sulfate solution in a shallow container,

Bedinger 1966), or dyes (colored water flowing through a sandy matrix packed between two parallel and transparent glass plates, Ewers 1982), and models based on dissolution on artificial gypsum (plaster of Paris) or salt (Ewers 1982).

The growing access to personal computers in the early 1980s allowed karst scientists to create the first simplified digital models of cave development. These early models were based on the rather detailed knowledge of the chemical processes at work in the system $CaCO_3$-H_2O-CO_2 (see Chapter 3) and the fluid mechanics in porous and fractured rock aquifers (see Chapter 5 and Dreybrodt and Gabrovšek 2002). The simplest case considered in these pioneering modeling attempts was that of a single fissure in limestone (James and Kirkpatrick 1980; Palmer 1984; Dreybrodt 1990). Modeling was extended also to two-dimensional networks of fissures in soluble rocks some years later (Groves and Howard 1994; Howard and Groves 1995). A review on these rapidly progressing modeling attempts in speleogenesis and karst aquifer understanding is given in Romanov et al. (2004) and especially in Dreybrodt et al. (2005b). Kaufmann (2009) was among the first to attempt a 3D modeling approach, in which realistic boundary conditions could be introduced (Borghi et al. 2012). Nowadays, the increased capabilities of the hardware and software easily allow to run 3D models on underground karst network evolution, and thus speleogenesis (Li et al. 2020b), but it still remains a great challenge to satisfactorily validate these predictive models. Although process modeling of karst systems is beyond the scope of this book (the reader can refer to Dreybrodt et al. (2005b) for details), explaining some simple models allows to gain insight into the processes and mechanisms involved in the creation of complex karst networks.

11.3.1 Hardware Models of a Single Input

The plaster and salt hardware models of Ewers (1982) were among the first to show the influence of conduit growth on flow patterns (Figure 11.6a). These models are not perfectly comparable to karst aquifers, since they are different in scale, hydraulic gradient and, especially, dissolution kinetics, but they have been very useful in the early stages of modern speleogenetic ideas. The simplest model is the single input one, where a head of water is applied to a point on one side of the bedding plane analog (soluble rock plate), and the discharge occurs on the opposite side. Flow through this analog soluble bed is Darcian. The aggressive fluid starts dissolving a radially distributed conduit network with the branches developing along initially more permeable paths, more or less in the direction of the highest hydraulic gradient (Figure 11.6b). Initially, the fluid becomes saturated at the extremities of these pathways, but newly still aggressive water progressively arrives more downstream into the dissolving block. All the more permeable routes connected to these main conduits (e.g., intersections with joints) are enlarged, but at a much lower rate than the main ones. One of the main tubes, which is often (but not always) the one most favorably oriented with respect to the hydraulic gradient, will be ahead of the others. The equipotential field will deform in response, reducing the hydraulic gradients at the ends of the competing tubes, in practice slowing down their solutional advancement. The master tube eventually reaches the discharge boundary, completing breakthrough. From this moment on, most of the flow is directed through this victor tube, boosting its enlargement. This will lead, at a certain time, to the onset of turbulent flow, further increasing the growth rate of this conduit. The simple fact that this tube is the most efficient flow route causes the equipotential field to be governed by this tube, creating the great heterogeneity typical of karst aquifers. The fast enlargement of the main tube swallows up the minor parallel ones, or short lateral routes, but some of the main lateral secondary tubes may continue evolving slowly, and may be important in later stages of cave growth. Often less favorable sideways will get gradually abandoned by water flow and will end up clogged with sediments. Some of the main

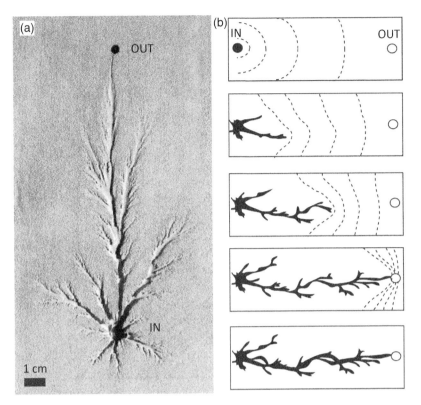

Figure 11.6 Hardware model of a single input. *Source:* Based on Ewers (1982). (a) The plaster model of the single input, showing the radial pattern, and the "victor" tube connecting the input point (IN) to the spring (OUT). (b) Schematic representation of the evolution of the conduit system from a single input. The dashed lines are equipotentials. Note the evolution from a branchwork (distributary) to a more anastomotic (contributory) pattern in the late stage (lower panels).

parallel side tubes can continue evolving, becoming alternative routes for water flow, and transforming the initial dendritic network of tubes in an anastomotic one (from distributary to contributory flow). The master tube generally does not follow a straight line, but its winding pattern is a reflection of the heterogeneity in initial permeability of the dissolving rock. However, directions with the steepest hydraulic head are preferred, and slight deviations from this ideal path are mostly due to initially high permeability routes. In nature, these proto-networks of tubes are very small, a few centimeters in diameter at their upstream ends. They are sometimes preserved and visible on cave roofs with inception bedding planes when the underlying beds have fallen off. A good example of this is Kentucky Avenue in Mammoth Cave, USA.

11.3.2 Computer Modeling of a Single Fissure

The simplest element in a karst aquifer is the fissure (Figure 11.7). Such an idealized fissure can be described by its length L, its width w, and long dimension of its cross-section b. The shape of the fissure has an influence on the resistance to flow R, with rectangular fissures allowing water to flow through easier than ellipsoidal ones. For water to flow through the fissure, a hydraulic gradient i has to be present, given by the altitudinal difference h divided by the fissure length L. The fluid is characterized by a viscosity η and a density ρ, which are temperature-dependent and

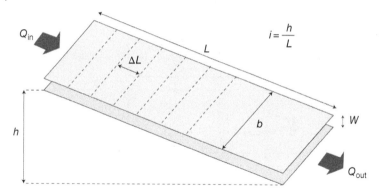

Figure 11.7 The typical fissure considered in models and its characteristics (see text for details). Flow is from left to right, with $Q_{in} = Q_{out}$.

depend on the chemical composition of the water. In general, for modeling purposes, the fissure length, the long-dimension of its cross-section, and the hydraulic gradient are considered time-independent. The initial concentration in dissolved carbonate is generally also assumed to be zero. Over long time periods and in nature these conditions are never met. Water entering a karst system always has some dissolved calcium carbonate (and other chemical species), and the hydraulic gradient changes due to valley entrenchment, surface denudation, and other factors.

To start modeling, the fissure is divided into short equal-size segments ΔL (discretization) and the initial conditions are defined, including length and altitude difference (hydraulic gradient), width (aperture) of the hypothetical fissure, its morphology (rectangular or ellipsoidal) and its long-dimension width (b), carbon dioxide content in the water, and temperature. The flow rate Q through a fissure is then calculated, together with the linear and nonlinear kinetics constants (k_l and k_n), and the order of the nonlinear kinetics n. See Dreybrodt et al. (2005b) for more details.

The water passing through the fissure interacts with the walls, dissolving the rock and widening the fissure. For each fissure segment the dissolution rate, dissolved load, saturation ratio, carbon dioxide loss, and fissure enlargement is computed. The water exiting the fissure segment then passes through the second segment, but with different initial chemical and physical characteristics. All parameters are calculated again for this second segment, and these computations are repeated for all remaining fissure segments. These operations are applied for discretized time periods (e.g., 100 years) resulting in a final fissure width. At every time step, the initial aperture of the fissure is different (larger), and new discharge rates are calculated, repeating the entire process for additional time steps.

Before the 1970s, experiments on dissolution by water flowing through fractures or tubes in limestone brought scientists to believe that the dissolution process was governed by linear kinetics (Weyl 1958). If this were true, water flowing through limestone fissures would get saturated in calcite after a few meters, preventing dissolution in downstream fissures. In the 1970s, scientists started realizing that nonlinear rate laws apply for solutions with high saturation ratios (White 1977; Palmer 1984). This sharp decrease in dissolution rate of solutions close to saturation allows waters to penetrate deeply into narrow fractures, enlarging fissures very slowly but rather uniformly up to the discharge boundary (see Section 3.8.1 and Figure 3.36). It is this slow enlargement rate by water near to saturation that enables breakthrough to occur in geologically reasonable times, and once this happens, larger amounts of less saturated fluids can enlarge these pathways rapidly, eventually forming caves (Palmer 1991).

Different modeling results have shown that the flow rate, and thus fissure aperture, increases slowly over relatively long times (typically some 10 kyr in limestones), and then suddenly accelerates drastically due to the positive feedback between increasing flow and dissolution. The time at which this happens is known as breakthrough time T_B. The modeling results have also shown that breakthrough time mainly depends on initial fissure aperture, the hydraulic gradient and the fissure length. The most important factor appears to be the initial fissure aperture. All other variables remaining equal, a fissure that is twice as wide will reach the breakthrough time around eight times faster. If hydraulic gradients are doubled, breakthrough times will approximately be reduced by 2.5 times, whereas for fissures twice as long T_B will roughly be doubled (Palmer 2007).

Temperature and CO_2 concentrations play more complicated roles in breakthrough times. Higher temperature increases the reaction rates, and because of the lower viscosity of water at higher temperatures, also the flow velocity will be greater. On the other hand, this higher reactivity causes dissolution to proceed faster in upstream sectors, decreasing the solutional capacity of the fluids downstream. Furthermore, CO_2 has lower solubility in warmer water, so less carbonate rock is dissolved (Figure 3.17). This temperature effect can be partly counterbalanced by the higher CO_2 production rates in warmer environments, decreasing breakthrough times. In general, however, breakthrough times increase with temperature.

Another factor that should be taken into account is that in natural conditions the solutional growth of prototubes normally occurs in closed conditions (in the phreatic zone or in water-filled fissures in the vadose zone), so CO_2 consumed by dissolution is not replenished (Figures 3.12 and 3.13). In such closed system conditions, breakthrough times are five times longer than in open systems. The dissolutional capacity of flowing waters can also be increased along the flow path if additional acidity is introduced, such as through the oxidation of organic matter. Another way of rendering saturated solutions aggressive again is by mixing (Figure 3.27), but it has been shown that large differences in CO_2 contents are necessary to decrease the breakthrough times significantly (Gabrovšek and Dreybrodt 2000).

This idealized modeling gives a rough approximation of what would happen in an irregular natural fissure. Modeling the flow of a dissolving fluid along a rough fissure causes it to widen along preferential flow paths, where the initial fissure aperture is greater, forming "wormholes." The competition between these wormholes will lead to the selection of a dominant wormhole, where solutional widening gets focused. This reduces the breakthrough time considerably, even by up to one order of magnitude (Upadhyay et al. 2015).

11.3.3 Hardware Modeling of Multiple Inputs

Although the single input hardware models on soluble rocks in unconfined settings gave very useful insights into the problem of early cave genesis, more complex models were needed to better approximate the real-world with multiple inputs in different hydrological and topographic situations. The first models were developed by Ewers (1982) using multiple aligned inputs, a common situation in many karst settings (Figure 11.8a). This model is somehow representative of a plain with allogenic recharge in which multiple rivers come into contact with soluble rocks along a sharp geological boundary. The different inputs will cause solutional enlargement in the heterogeneous bedding plane or fissure, and because of a variety of possible reasons (e.g., different hydraulic heads, more open initial fissures in one of the inputs, and differences in the time of initiation) some input branches will go ahead of the others. As happened in the single input models, this situation will modify the equipotential field, decelerating the advancement of the competing inputs. Competition between adjacent inputs diminishes with increasing spacing and with lower hydraulic gradients. When one of the tubes reaches breakthrough conditions, most flow will be directed toward it, forcing adjacent tubes to connect to the trunk conduit forming tributaries (Figure 11.8a). A good real-world example of such system is the Domica-Baradla Cave System in the UNESCO World Heritage Site of "Caves of Aggtelek Karst and Slovak Karst" (Figure 11.9a).

Another experiment was carried out using multiple inlets in a row, feeding an inclined bedding plane fissure but with drainage occurring along the strike (90° on one side of the model) (Figure 11.8b–g). In the case of a rather tight fissure, where resistance to flow is high (Figure 11.8b–d), the connection between parallel passages once the nearest-to-the-spring branch reaches breakthrough, takes place rather randomly, induced by a combination of favorable hydraulic gradients and nearest distance between adjacent branches. The resulting pattern is one of more or less strike-oriented main cave

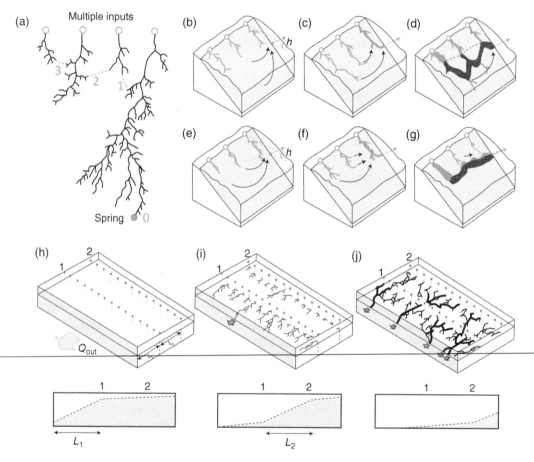

Figure 11.8 Hardware modeling with multiple inputs. *Source:* Modified from Ewers (1982). (a) Competitive development of branching cave systems with multiple inputs in a row. Zero (0) indicates breakthrough of the "victor" tube, whereas 1–3 indicate the sequence of connections between adjacent cave systems following breakthrough. (b–g) Multiple inputs in a single rank with lateral flow along strike controlled by the position of the water table (wt). Note the enlarging inclined fissure. (b–d) Shows the situation in which the fissure has high resistance to flow, whereas (e–g) shows the case where the fissure is more conductive to lateral flow. (h–j) The situation of multiple inputs on two ranks (rows of circles) with a constant hydraulic head (outflow is to the left), seen in perspective view (above) and in profile (below). The connections between cave systems occur through a headward sequence, with the area of steep hydraulic head receding upstream.

passages, where water has chosen the less resistant flow routes. If the fissure is larger, flow will be easier, and shorter flow routes in the direction of the hydraulic gradient will have more possibility to form. This leads to a straighter strike-oriented main cave branch (Figure 11.8e–g). The over 200 km long Hölloch Cave System in Muotatal (Switzerland) is a good example of such systems (Figure 11.9b).

Another common situation in nature is the introduction of water in multiple ranks. This situation is illustrated in Figure 11.8h–j. In this model two ranks of inputs are placed parallel to the output boundary at distances L_1 and $L_2 = 2L_1$. Water will descend vertically to the underlying horizontal plane which acts as the permeable fissure. The initial hydraulic gradient of the inputs nearest to the output boundary is high and will allow proto-caves to develop here, whereas the hydraulic gradient of the rear inputs is very low inhibiting the initiation of proto-cave development. The breakthrough of proto-caves in the first rank will cause the headward retreat of the steep hydraulic gradient area, thus enhancing permeability development upstream between the second and first rank inputs. From this point onward, both lateral connections between adjacent first-rank inputs and longitudinal connections between second- and first-rank inputs will proceed more or less simultaneously forming branchwork networks of passages.

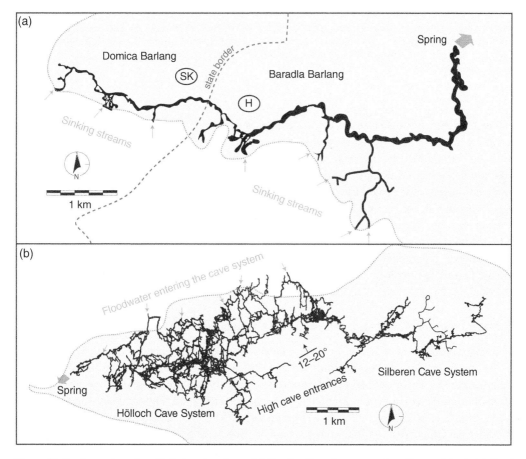

Figure 11.9 Examples of multiple input systems. (a) Domica-Baradla cave system, Caves of Aggtelek Karst and Slovak Karst UNESCO World Heritage Site, Slovak Republic-Hungary. *Source:* Based on Bella et al. (2019). (b) Hölloch and Silberen Cave systems, Muotatal, Switzerland. *Source:* Based on Jeannin (2016). The grey-shaded area represents soluble rocks.

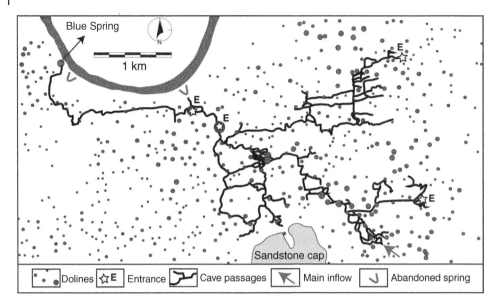

Figure 11.10 Blue Spring Cave, Indiana (USA). The map shows the distribution of sinkholes on the plain and the underlying cave passages in black. Size of circles is proportional to sinkhole depth, ranging between 3 and 25 m. The cascading joint-guided dendritic cave pattern with its multiple inputs is clearly visible. *Source:* Modified from Palmer (2007).

Cave development thus occurs in a cascading pattern, with areas nearest to springs developing caves first, and progressively inducing cave development from inputs farther away from the spring area. Good examples of such systems are those in the sinkhole plains of the Midwest USA, such as Blue Spring Cave in Indiana (Figure 11.10). This cave shows a joint-guided branching pattern formed in the densely jointed Mississippian Salem Limestone and fed by multiple doline inputs in the well-bedded St. Louis Limestone. The famous Mammoth Cave area in Central Kentucky is also a cascading cave system, although complicated by the lowering and lateral shifting of the spring area controlled by factors such as the entrenchment of the Green River, aggradation periods during glacial periods, progressive dismantling of the sandstone cover, and lithological changes. This has resulted in a three-dimensional network of cave passages of different ages (stages), complicating the visualization of the overall cascading pattern of each single speleogenetic stage.

11.3.4 Computer Modeling of Two- and Three-Dimensional Fissure Networks

Modeling the evolution of two-dimensional karst networks commenced in the early 1990s (e.g., Groves and Howard 1994; Howard and Groves 1995) and has been developed by several research teams over the following decades (Siemers and Dreybrodt 1998; Kaufmann and Braun 1999, 2000). Three-dimensional karst evolution models have also experienced substantial development in the last 15 years (e.g., Kaufmann et al. 2010; Starchenko et al. 2016). The details of these modeling efforts are well illustrated in the specialized book by Dreybrodt et al. (2005b), and a recent chapter in the Encyclopedia of Caves by Kaufmann et al. (2019), to which the readers are referred. Figure 11.11 shows an example taken from Dreybrodt and Gabrovšek (2002).

In general, these models have allowed us to get a rough idea on epigene cave development, especially in the initial phases, and confirm the main results of the single fissure modeling. Modeling

(a) (b) (c)

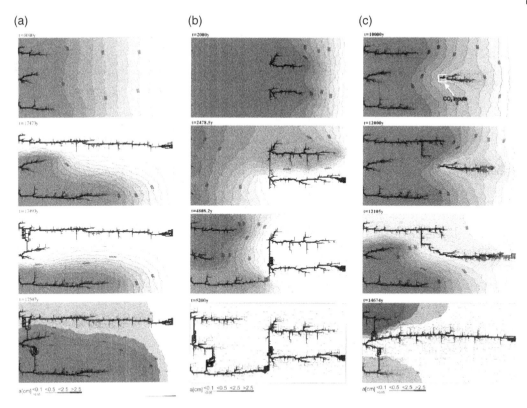

Figure 11.11 Computer model outputs of some of the most common situations found in nature. *Source:* Modified from Dreybrodt and Gabrovšek (2002). Hydraulic head is set at 50 m, input water is at equilibrium with P_{CO_2} = 0.05 atm, initial fissure width is 0.02 mm, and length and width of the flow field are 2000 and 500 m, respectively. (a) Multiple inputs in a single rank, showing breakthrough of the upper cave branch, and following connection of neighboring caves. (b) Multiple inputs in two ranks, with hydraulic head of 50 m between the middle inputs and the right, and negligible initial head between the first rank of inputs and the middle ones. Breakthrough in caves to the right triggers the development and connection of caves to their left. (c) Multiple inputs in two ranks, with a CO_2 source (additional P_{CO_2} set at 0.05 atm) in the middle of the flow field, propagating cave evolution of the inputs to its left.

has clearly demonstrated the relationships between initial fissure width, breakthrough time, hydraulic gradient, and length of the flow path. Breakthrough times are mostly influenced by initial fissure apertures, but also by altitudinal difference between inflow and outflow areas and by distance from input to output (flow path length), both determining hydraulic gradient. The wider initial fissures with the highest hydraulic gradient and the shortest flow paths will likely be able to transmit greater quantities of water, and will thus enlarge more rapidly than others. Initial fissure width is the main controlling factor, but the main flow path will follow these larger fissures only if they are favorably oriented with respect to the hydraulic gradient. Shorter flow paths will be chosen only if they are wide enough to be favorable for flow, otherwise lengthier paths might be chosen.

Of course, the chemistry of the waters (saturation degree) has also an important influence on breakthrough times and cave development. Since most of the solutional capacity of the water is consumed in the initial part of the flow path, fissure widening is very slow in the downstream sectors. This is even more pronounced if the waters that enter the main flow path are already nearly saturated. On the contrary, if undersaturated waters are somehow incorporated in these downstream parts, fissure widening accelerates and breakthrough times shorten. Mixing corrosion can decrease breakthrough times only if differences in P_{CO_2} are significant, and especially if mixing occurs at least half-way along the fissure length. Open-system conditions, where CO_2 consumed in the dissolution process can in some way be replenished, are more favorable than closed-system ones. This is why most solutional widening occurs close to the water table, not only because this is the shortest path to the output. Deep phreatic loops are preferred flowpaths only if large initial fissures have favored these longer routes before the fissures associated with the water table had the possibility to widen enough. It has been shown that with long breakthrough times, simple, mainly linear cave patterns tend to form, whereas cave patterns become branched and more complicated with decreasing breakthrough time (changing other initial modeling parameters) (Dreybrodt and Siemers 2000).

11.4 Hydrogeological Controls on Cave Genesis

Palmer (1991), based on an extensive literature review covering a representative set of limestone caves in different geomorphological settings, recognized that cave patterns are mainly controlled, in addition to the local and regional geology (lithology, position of the soluble rocks, and structural elements), by the mode and nature of groundwater recharge. This refers to where and how much water is introduced into the soluble rocks, the ascending or descending direction of water flow, the location of the discharge areas, or the origin of the solutional power. These hydrogeological factors can be influenced by the topographic characteristics of the area, climate, and the long-term landscape evolution occurring over the period between initial enlargement of fissures and the final configuration of the cave.

Groundwater can be introduced into the soluble rocks in multiple ways, it can flow under different hydrodynamic conditions, and it can acquire solutional aggressiveness by various chemical and physical processes. Ford and Williams (2007) distinguished caves formed by normal meteoric waters, deep waters of different chemistries, brackish waters, and their combinations. They further divided the first category into caves formed in unconfined and confined settings, and the second was subdivided into caves formed by CO_2-rich fluids, or those enriched in H_2S.

There has been a general distinction between epigenic and hypogenic caves in recent years, the first being formed by descending waters and the second by rising fluids (Klimchouk 2007). There were actually two schools of thought regarding this dichotomy, one based on geochemical criteria and the other on hydrogeological factors. According to Palmer (2000) hypogenic caves are those formed by water that acquired its aggressiveness beneath the surface, independent of surface or soil CO_2, or other near-surface acid sources. The other mainstream of karst researchers defines hypogenic speleogenesis as the formation of solutionally enlarged permeability structures (void-conduit systems) by upwelling fluids that recharge the cavernous zone from hydrostratigraphically lower units, whereas fluids originate from distant or deep sources, independent of recharge from the overlying or immediately adjacent surface (Klimchouk 2017, 2019b). According to the first definition, caves formed in the mixing zone of fresh-saline waters, or by mixing of different waters at the phreatic–vadose interface would belong to the hypogene karst type, whereas the artesian cave development in evaporite sequences by transformational flows would be considered epigenic. The hydrogeological approach, on the contrary, would place the artesian gypsum caves in the hypogenic category, and the coastal or vadose-phreatic mixing caves in the epigenic one. We believe that this broad distinction in two main categories (i.e., epigenic and hypogenic), although useful, does not describe comprehensively the multiple ways in which caves can form.

In this book, we use a practical subdivision of cave genesis based on both their geochemical and hydrogeological characteristics, following the concepts proposed by Palmer (2007). Although many caves have a much more complicated origin, the cave-forming processes are essentially related to the different conditions described below.

The main hydrogeological–geochemical conditions are: (i) recharge via sinkholes, (ii) sinking streams, (iii) diffuse recharge through a permeable caprock, (iv) diffuse recharge from an underlying insoluble formation, (v) mixing conditions in coastal areas, (vi) rising (thermal) fluids rich in CO_2 or H_2S, and (vii) deep-seated acidic fluids. Conditions (vi) and (vii) are commonly related to hydrocarbon reservoirs and are often encountered in deep drilling operations (Figure 3.16). Acidity at high depths is often obtained by mixing between fluids of different origin and chemistry, by cooling of thermal fluids, or formation of sulfuric acid. Other acid-forming processes are related to the maturation of hydrocarbons into petroleum, or the oxidation of hydrocarbons in presence of iron oxides, both of which release a variety of organic acids. Porosity enhancement related to such fluids has been proposed by Mazzullo and Harris (1992), but compelling evidence is still lacking, and geochemical principles make the development of such solutional porosity rather difficult to sustain (Ehrenberg et al. 2012).

11.4.1 Multiple-Point Recharge

Infiltration of water through sinkholes and small sinking streams is the most common type of recharge occurring in karst areas. This type of groundwater input gives origin to the most abundant type of cave pattern: branchwork caves. These consist of merging tributary passages forming progressively higher order passages, similar to a dendritic pattern in surface streams. Not all tributaries are explorable cave branches, but can be narrow fissures feeding water to the main cave passages. This means that branchwork caves can sometimes be composed of single stream passages, in which tributaries are not seen on cave maps because they are mostly not penetrable.

These caves commonly occur where soluble rock forms extensive outcrops of bare bedrock or covered by a rather thin soil cover. Surface waters infiltrate underground following all the available openings, enlarging the fissures and forming the typical epikarst. This "external skin" of more karstified bedrock generally less than 10 m thick is characterized by a dense network of enlarged fissures commonly filled with soil. Since the solutional aggressiveness of infiltrating waters dies out rather rapidly, widened fissures tend to taper out downward, becoming less susceptible to water flow. The underlying transmission zone is therefore less permeable with respect to the epikarst, which forms a perched aquifer where water is stored and slowly transmitted downward (Williams 2008). Some fissures in the transmission zone are wider than others, and function as preferential drains of the epikarst, especially during recharge periods when the inflow exceeds the infiltration capacity of the transmission zone (Figure 5.25). These larger pathways enlarge faster than others, since they convey greater quantities of water and at higher speed, allowing the water to maintain its solutional aggressiveness over longer distances. These preferential drainage paths are commonly expressed at the surface as solution dolines (or sinkholes) in bare karst areas (Figures 6.22 and 6.23). In covered karst zones, these epikarst drains may eventually allow the internal erosion and/or deformation of the mantling deposits generating cover-subsidence sinkholes by various mechanisms (suffosion, collapse, and sagging) (Figure 4.16). Each of these larger openings will eventually give rise to the formation of an underlying cave passage. These are often very narrow and not explorable in their initial parts, but as they receive inflow from adjacent infiltration pathways both the flow rate and the passage size increase.

Upstream passages are generally characterized by vadose morphologies (widened vertical fissures), with canyons and shafts, sometimes interrupted by short perched phreatic tubes. They eventually reach the water table, taking on the typical rounded phreatic morphology. These passages can be permanently filled with water (below the water table), or be located in the oscillation zone (epiphreatic zone), thus becoming submerged only during floods.

Groundwater flows toward the spring, which is often located in the most favorable position with respect to the geological and topographical setting. In many cases, these springs occur in the topographically lowest points, such as the floor of a valley entrenched into the soluble rock formation, or at the contact between the soluble rocks and an underlying less permeable and insoluble formation. Springs can also be guided by large structural elements such as syncline axes, or major fault zones. It is very rare to have multiple springs, and if they occur they are generally not very distant from each other. These nearby springs can be related to collapse and sediment filling in the conduit close to the outlet, or simply by the fact that the rocks at the valley margin are affected by dilation and stress release jointing, allowing water to choose different pathways and exits.

Cave passages tend to converge by various natural causes. For instance, the guiding fissures intersect each other, or widened bedding planes in folded strata converge toward the syncline axis. The most permeable flow route also has the lowest hydraulic head, thus causing adjacent flow paths to converge toward this most efficient drainage route. As illustrate the modeling experiments performed by Ewers (1982), the "victor" tube that reaches the spring becomes the most efficient flow path with the lowest hydraulic head, causing lateral flow to converge toward it. This simplified pattern can be complicated by base-level changes accompanied by variations in the position of springs and correlative flow diversions, and by localized floodwater maze patterns (such as those related to internal collapses).

Probably, one of the best examples of multiple-point recharge branchwork caves is the almost 50-km-long Crevice Cave in Perry County, Missouri (USA), carved in Ordovician dolostones (Moss 2013) (Figure 11.12a). Multiple inlets through sinkholes feed vadose stream passages descending northeastward roughly along the dip of the strata, joining the main stream passage with phreatic sections, which flows along the strike in a southeast direction. The cave comprises two distinct but closely spaced cave levels in response to the entrenchment of the Cinque Hommes Creek, where the spring is located, slightly complicating the dendritic pattern. Blue Spring Cave (Indiana) is another good example in fractured limestone (Figure 11.10).

The Ease Gill Cave, which is part of the over 86-km long Three Counties Cave System in the Yorkshire Dales (UK), is another very fine example of branchwork cave. A series of sinking points along the Ease Gill stream gives rise to different cave branches developed on two levels, all leading into the main passage roughly oriented along the strike of the dipping Lower Carboniferous limestones (Figure 11.12b). The Codula Ilune Cave System in Sardinia, with its over 70 km of development and one of the longest in Italy, has multiple inlets along the influent stream, feeding the main underground river that follows the strike of the Jurassic limestone beds dipping ca. 30° eastward (Figure 11.12c). This is a good example of a branchwork cave with an important structural guidance.

11.4.2 Concentrated Recharge

When recharge to a cave system is concentrated, the overall cave pattern deviates from the branchwork configuration. A typical situation is that of an allogenic sinking stream that collects water from a poorly permeable and insoluble terrain before entering the area where the soluble bedrock is exposed. A similar situation can occur when a surface river experiences a significant water level

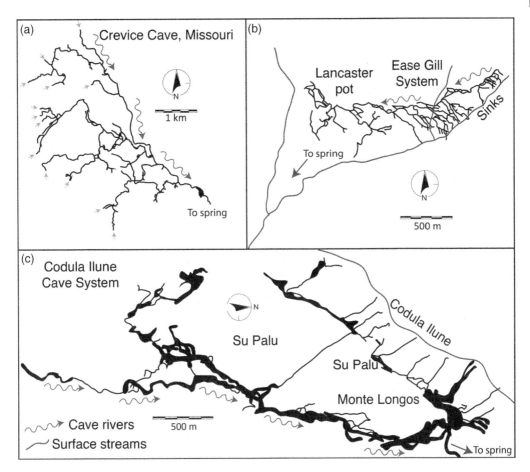

Figure 11.12 (a) The dendritic pattern with sinuous passages of Crevice Cave, Missouri, with downdip vadose canyons and the main strike-parallel river passage in the well-bedded Ordovician Joachim Dolomite. *Source:* Based on Moss (2013). (b) The Ease Gill part of the Three Counties Cave System in the Yorkshire Dales, the longest system of the United Kingdom, showing the typical branchwork pattern with multiple inlets along the Ease Gill stream. *Source:* Modified from Red Rose Cave and Pothole Club (1992). (c) The upstream part of the Codula Ilune Cave System, one of the longest in Italy (>70 km). Multiple inlets (sinks) follow major NW-SE fractures joining in N-S strike-oriented main drains. *Source:* Survey courtesy of Sardinian Speleological Federation.

rise causing floodwaters to penetrate in a concentrated manner into the adjacent soluble rocks. In Alpine karsts, where extensive bare areas are exposed to rainfall or snowmelt, the underlying cave systems can experience sudden and extreme water level variations producing internal and dangerous floods.

Floodwaters differ from normal groundwaters in four ways: (i) they have greater discharges (flow rate can increase rapidly by three orders of magnitude), (ii) they flow under higher hydraulic gradient, (iii) floodwaters flow much faster, and (iv) they can maintain their solutional power over longer distances. This produces a set of distinct cave wall and ceiling morphologies, and the typical cave pattern produced by these sudden floodwaters is a maze (Palmer 2001).

The water level rises rapidly during a flood, ascending as much as hundreds of meters in alpine caves. The vertical zone in which these water level oscillations take place is known as the epiphreatic zone (Audra and Palmer 2011). Despite the fact that floods may occur only a couple of times a year, their effectiveness in shaping the cave passages is enormous. Floodwaters are much more aggressive than normal groundwaters, their greater discharge allows transporting more dissolved load, and their high flow velocities and competence make them capable of mobilizing sediment load with a wide range of sizes, increasing enormously their mechanical erosion capability. In caves subject to periodic flooding most of the passage enlargement occurs during these extreme events.

The water is more readily transferred through the major flow routes during a flood pulse, rapidly transforming the partly air-filled passages into high-pressure tubes filled with undersaturated floodwater. This aggressive water is forced into the adjacent fissure network, enlarging these planar discontinuities rapidly (Figure 5.20).

In rapidly uplifting mountain areas, where fluvial downcutting rates are high, entrenchment in the vadose caves may not keep pace with the external valley incision. This is especially true for caves where regular flow is subdued and most of the cave enlargement occurs during the seasonal floods. In this case, incision of the external river outpaces that of the small cave stream. This causes the hydraulic gradient to build up in the cave, accelerating the formation of a looping cave profile (Gabrovšek et al. 2014). Undersaturated water fills the entire active cave passage during floods, but when the flood declines, aggressive floodwater will mainly remain trapped in the lower sectors of the cave. These will slowly drain through the fissure network creating "soutirages," narrow passages often barely explorable by cavers that connect the main cave passage to the active phreatic conduits below. These subvertical passages slowly drain the floodwaters after floods, but also function as inputs of aggressive floodwater into the main conduit when the water table rises.

A good example of such caves is the Saint-Benoît Cave, Alpes de Haute-Provence (southern France) (Audra and Bigot 2009) (Figure 11.13a). This 2-km-long cave system is developed along strike in a steeply dipping nummulithic Eocene limestone unit some tens of meters in thickness. The upper epiphreatic level (Grotte de la Lare) is characterized by a large tunnel around 8 m wide with a typical rounded phreatic cross-section, showing a long profile with a series of loops (ups and downs) with altitudinal difference of up to 20 m. In dry periods, these passages are completely dry, but during floods the conduits of the Perles overflow spring and the perennial Fontani spring are not capable of draining the abundant inflow (snowmelt and concentrated infiltration upstream from allogenic terrains). Floodwater rises from a narrow passage and starts filling the conduits. The upstream parts of the loops only show phreatic morphologies (Figure 11.13b), whereas the downstream sectors are first eroded by free-running waters (Figure 11.13c), and only in a later stage of the flood become completely filled with water. During flood recession, the lower parts of the loops are drained by narrow enlarged fissures ("soutirages") which can feed the lower lying springs, but sometimes are only lower bypasses between the loops.

Where the water table rises in the caves due to fast recharge through extensive karren fields, such as in bare alpine karst plateaus, the aggressive inflow tends to enlarge all available openings, forming anastomotic mazes. The most extensive alpine cave systems, Hölloch in Muotatal and Siebenhengste-Hohgant höhlensystem (Switzerland), and Schönberg höhlensystem in Totesgebirge, and Hirlatzhöhle-Schmelzwasserhöhle in Dachstein (Austria), have mostly formed in this way (Figure 11.9b).

Close to rivers, where flooding causes water table rises of a few meters and backflooding in the underground karst network at the valley margins, all the available fissures larger than 0.1 mm can be enlarged by these undersaturated waters. This can cause the formation of floodwater maze systems, such as Lapa do Bode Cave along the flanks of a large meander of Rio Una (Bahia, Brazil) (Figure 11.14a) (Pereira 1998). Lapa Vermelha (Figure 11.14b) and Escrivânia Cave System (Lagoa Santa karst region, Minas Gerais, Brazil) (Figure 11.14c) are other examples of floodwater maze

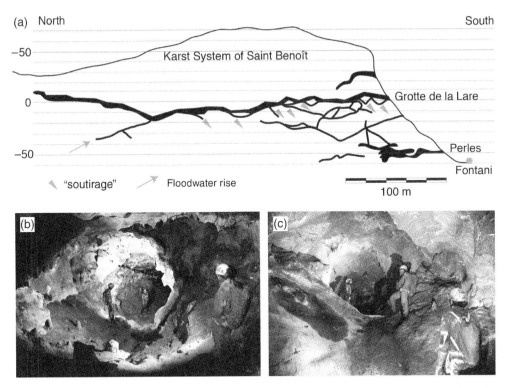

Figure 11.13 (a) Long profile of the Saint-Benoît Cave showing the looping pattern of the main drain (Grotte de la Lare, 640 m asl), and the many "soutirages" connecting the upper level with the Perles overflow spring (590 m asl) and the perennial Fontani spring on the valley floor (567 m asl). *Source:* Audra and Bigot (2009). (b) An upstream view of the rising part of a phreatic loop, with the typical rounded cross-section. (c) On the downstream, descending part of the loop the phreatic morphology is still visible on the roof, but the floor displays evidence of vadose flow and shallow entrenchment. *Source:* Photos by Jean-Yves Bigot.

caves formed along an underground river meander and a lake, respectively (Auler 2020). Diversion of surface water from a river through underground pathways can result in the cut off of water flow in meanders and the formation of spectacular floodwater mazes, often superposed on previously existing cave patterns. One of the best examples of this type of cave is Sof Omar in Ethiopia, carved in Jurassic limestones and modified into a floodwater maze by the Weib River, which enlarged most fissures and bedding planes during flood events (Asrat 2015) (Figure 11.14d). Also, Mystery Cave in Minnesota is a fissure-guided floodwater maze, possibly overprinting a previous hypogenic maze system (Klimchouk 2007; Palmer 2011) (Figure 11.14e).

When the rocks bordering the river valleys are composed of evaporites, undersaturated floodwater dissolves the rock along all the available and penetrable fissures very rapidly, forming complex maze-like systems characterized by important instability phenomena. These floodwaters can also invade previously formed artesian gypsum caves (see Section 11.4.4). Such systems occur in the Arkhangelsk area (Perm, Northern Russia), where a large number of caves are carved in a succession of gypsum and dolostone layers. An example of these is the 17.5 km long Kulogorskaya-Troya system on the right bank of the Pinega River, which is the longest epigenic gypsum cave in the world (Franz et al. 2013) (Figure 11.15a), and Ordinskaya Cave, on the flanks of the Kungur River, 4800 m long, of which 4500 m are completely underwater (Figure 11.15b, d). The flooded passages of this cave show a large number of feeders through which the river-derived floodwaters enter the

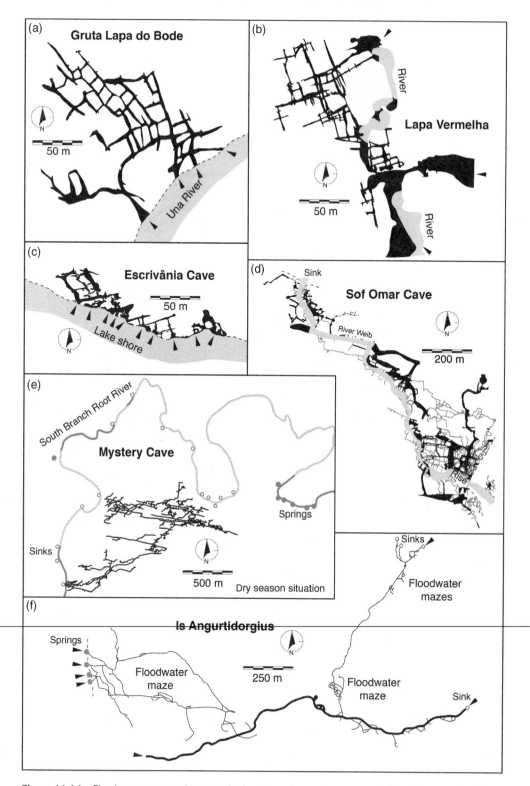

Figure 11.14 Floodwater caves and patterns in limestones (cave entrances are indicated by arrowheads): (a) Gruta Lapa do Bode (Bahia, Brazil), a maze system formed by floodwaters along the banks of the Una River. *Source:* Modified from Pereira (1998). (b) Lapa Vermelha (Minas Gerais, Brazil), a famous hominid site characterized by an extensive floodwater maze at different heights above a river. *Source:* Modified from Auler (2020). (c) Escrivânia Cave System, consisting of a network of enlarged fissures on the border of a lake. *Source:* Modified from Auler (2020). (d) The Sof Omar Cave, cutting a long bend of the Weib River and showing a floodwater maze morphology developed along several levels (the blue trace shows the underground river, and the grey lines represent higher lying cave passages). *Source:* Modified from Asrat (2015). (e) Mystery Cave in Minnesota (USA), possibly a hypogene joint-guided maze now functioning as a floodwater maze cutting off several meanders in the Root River. *Source:* Modified from Alexander (1996). (f) The Is Angurtidorgius cave system (Sardinia) with several floodwater mazes bypassing constrictions and collapse debris. *Source:* Modified from Bartolo et al. (1980).

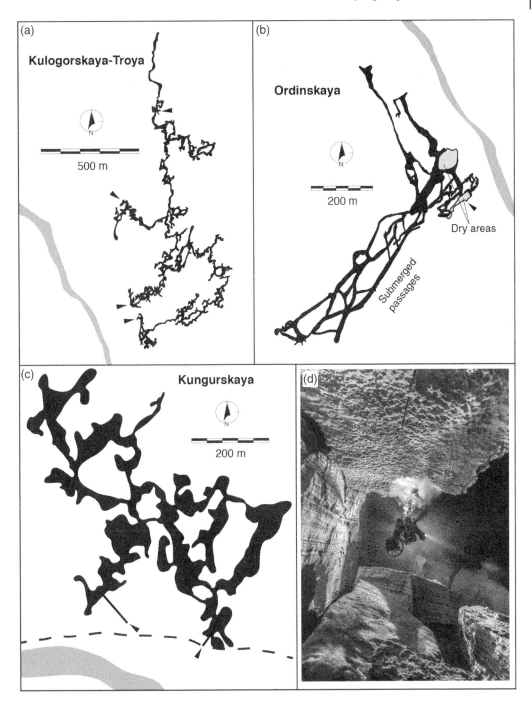

Figure 11.15 Floodwater caves in gypsum in the Perm region, Russia (arrowheads indicate cave entrances) (a) Kulogorskaya-Troya cave system, the longest epigenic gypsum cave in the world. (b) Ordinskaya Cave containing the longest sumps in Russia. Grey areas indicate the air-filled portions and all black areas are submerged. (c) Kungurskaya, or Kungur Ice Cave, one of the most important show caves in Russia. (d) One of the huge submerged passages in Ordinskaya Cave, with the layered gypsum rock clearly visible. *Source:* Photo by Andrey Gorbunov.

system, carving large and rather unstable voids (Sivinskih 2009). The famous Kungurskaya Cave formed in a similar manner, explaining the anomalously large dimensions of its underground rooms in gypsum, and the frequent occurrence of instability phenomena (Andrejchuk and Klimchouk 2002) (Figure 11.15c).

Floodwater action is also possible in restricted areas of normal branchwork caves, where cave passages are obstructed or constricted by collapse debris, sediment infilling, or the presence of less soluble beds. Small amounts of water may pass through these cave sections during base flow, but discharge during floods can be several orders of magnitude greater than the flow capacity of the constrictions. Consequently, during high flow conditions, the hydraulic gradient builds up in the upstream part. Undersaturated floodwater penetrates all available openings in the rock around the restrictions, carving many alternative flow pathways. Differences in the sectional area of these multiple passages causes the hydraulic gradients in them to be very variable, leading to water flows in many directions (down gradient), eventually interconnecting many passages and forming a floodwater maze. Is Angurtidorgius Cave in Sardinia (Italy), formed in well bedded and jointed Eocene nummulithic limestones, is a nice example of this type of floodwater mazes developed around constrictions (Bartolo et al. 1980) (Figure 11.14f).

Floodwaters not only influence the general or local pattern of caves, but they also generate a series of typical meso- and small-scale solutional features (speleogens) which are diagnostic of these highly undersaturated waters. These morphologies are often initial stages of floodwater maze formation. Typical floodwater morphologies include widened and tapering fissures, blind fracture-controlled solution pockets (Figure 9.38a, b), bedding-plane anastomoses, solution grooves (subterranean karren) (Figure 9.52c, d). Cupolas, albeit not diagnostic for floodwater origin, can also form due to air trapping by rising water levels (Figure 9.37b, f).

11.4.3 Diffuse Recharge from Above

Caves can also form by diffuse and slow recharge through a permeable karstic or non-karstic rock formation. Two main situations can be distinguished: (i) slow infiltration from the epikarst to a soluble formation underlain by insoluble and poorly permeable beds and (ii) slow diffuse recharge from an insoluble caprock.

Where the epikarst is well developed, the widespread vertical infiltration pathways pinch out downward and connections between adjacent enlarged fissures are poor. No significant cave systems form in alpine epikarst areas because of the lack of integration of the vertical pathways into extensive horizontally developing cave systems. However, where the epikarst is underlain by insoluble and less permeable rock formations, lateral water flow is promoted at this lithological contact (i.e., contact karst), interconnecting the voids into a true epikarstic cave system. The finest example of such epikarst caves is the 120-km long Bullita Cave system in the Northern Territories, which is the longest cave of Australia (Grimes and Martini 2016) (Figure 11.16a, b). Here, the 30-m thick Proterozoic Supplejack dolostone member is underlain by a 3-m thick shale bed which guided horizontal cave development and limited downward percolation, forming an orthogonal network of shallow cave passages.

A much more common type of network caves is that formed in a soluble rock formation recharged from above through an overlying permeable but insoluble caprock, such as a sandstone. Water descending through the permeable caprock reaches the top of the soluble formation with a rather uniform chemistry and flow rate. These waters are generally highly aggressive (undersaturated) and enlarge all available penetrable fissures. Flow rate through fissures of different initial width is rather uniform because of the diffuse recharge mechanism. Many Appalachian network

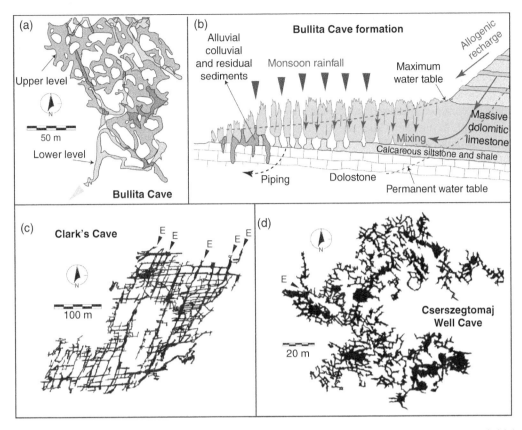

Figure 11.16 Caves formed by diffuse flow from above (arrowheads with E indicate cave entrances): (a) A portion of the Bullita Cave network in plan view and, (b) schematic cross-section showing the influence of the epikarst recharge and a siltstone and shale bed on cave genesis. The cave voids shown in white at the contact between dolomitic limestone (blue) and calcareous siltstone and shale (yellow). *Source:* Grimes and Martini (2016)/Instituto Geológico y Minero de España. (c) The plan view of Clark's Cave in Virginia (USA), a good example of a maze cave formed in Devonian limestone recharged by diffuse infiltration through the overlying Oriskany sandstones. *Source:* Palmer (2009a)/National Speleological Society. (d) Cserszegtomaj Well Cave, probably formed by descending waters from the Pannonian sandstone covering the Triassic dolostone. The cave maze probably is a paleokarst surface composed of a network of grikes and intervening clints (buried karst pavement). *Source:* Modified from Bolner-Takács (1999).

caves are formed by diffuse infiltration into limestones through a relatively thin caprock of permeable sandstones (Palmer 2009b) (Figure 11.16c).

   Dissolution beneath a more or less permeable sandstone cover has also been reported in Upper Triassic dolostones in Hungary, where the over 2-km long Cserszegtomaj Well Cave was accidentally intercepted by a borehole at around 50 m depth (Bolner-Takács 1999) (Figure 11.16d). Here, an intricate maze network of sizeable caves has developed at the contact between dolostones and the overlying cemented sandstones of Pannonian (Late Miocene–Pliocene) age on the roof, leaving a dolomite sand on the floor of the cave. The cave exposes a buried paleokarst surface with clint-and-grike features that was filled and covered by low permeability sand and clay layers, probably in early Miocene times. Although attributed to rising, maybe thermal, waters, the fact that voids are concentrated at the contact between dolostones and an overlying sandstone unit above support an origin by descending waters.

11.4.4 Diffuse Recharge from Below

Diffuse upward recharge from a non-karstic aquifer into overlying soluble rocks can also give rise to the formation of important cave systems. Two main situations can be envisaged involving basal injection of meteoric waters (i) into limestones and (ii) into evaporites. The first situation is rather uncommon, since the fluids are often close to saturation after the long underground pathways typical of confined aquifers. The second, although not very common, includes some of the longest cave systems in the world, such as Optymistychna Cave, Ukraine, developed in a gypsum unit just around 30 m thick, being the sixth longest cave in the world.

Water circulation occurs under confined conditions, in which the aquifer unit is capped by low-permeability formations and conditions are phreatic, with the potentiometric surface lying above the contact between the two hydrostratigraphic units. Artesian aquifers are a special type of confined aquifers in which the potentiometric surface of the confined aquifer is above the topographic surface, allowing water to spontaneously flow out of wells that penetrate into the water-bearing formation.

Brod (1964) probably was the first to report basal injection caves in carbonate rocks. In Eastern Missouri (USA), some dozens of small fissure caves occur in 60-m thick Ordovician limestones and dolostones overlying a 40-m thick sandstone formation which functions as the recharging basal aquifer. The local geological situation, with a partially denuded and fractured anticlinal ridge 60 km long, has caused groundwaters flowing through the Ordovician sandstone beds to discharge upward through the soluble carbonate formations, forming these localized fissure caves (Figure 11.17a). The fluids flowing through the sandstones retained their undersaturated condition over long flowpaths, still being able to dissolve the carbonate rocks when flowing upward across them.

Another well-known example of basal injection cave in limestones is the Botovskaya Cave in Eastern Siberia (Filippov 2000). This maze system is the longest cave in the Russian Federation, with almost 70 km of mapped passages (Figure 11.17b). It has formed in a 6 to 12 m thick limestone bed of Lower Ordovician age sandwiched between sandstone beds and dipping gently (6–8°) to the northeast. Infiltration from above is excluded because of the presence of low permeability units, so the dissolving fluids must have risen from below. Solutional capacity was probably enhanced by mixing between rising deep-sourced artesian waters and shallower meteoric artesian waters.

The Krem Puri Cave (Meghalaya), with over 25 km of mapped passages showing an orthogonal maze pattern, is the second longest cave in India. It is carved in a thin calcareous sandstone bed within a prevalently quartz sandstone succession of Late Cretaceous age (Figure 11.17c). It is not clear whether recharge occurred from below or from above, but this cave appears to have formed in a deep phreatic setting, and basal injection is the most likely mechanism. The acidity of the waters was probably obtained, at least partly, and in a late (oxygenated) stage by the oxidation of sulfides in the sandstone succession (Sauro et al. 2020).

Kalahroud Cave (Isfahan, Iran) is a 4.5-km long basal injection cave in a 40-m thick bed of Cretaceous massive limestones, underlain by Early Cretaceous conglomerates and sandstones and overlain by less permeable marly limestones (Bahadorinia et al. 2016) (Figure 11.17d). Morphologies clearly indicate rising flow, but intersection with the surface and river entrenchment has caused the deactivation of the cave, probably rather recently.

In the arid Judean Desert (Israel-Palestine), several isolated chambers (Frumkin and Fischhendler 2005) and maze networks are attributed to artesian groundwater flow rising into Upper Cretaceous limestone, confined by an overlying thick marl and chalk succession (Figure 11.17e). These caves, which are now deactivated and dry, appear to have formed during

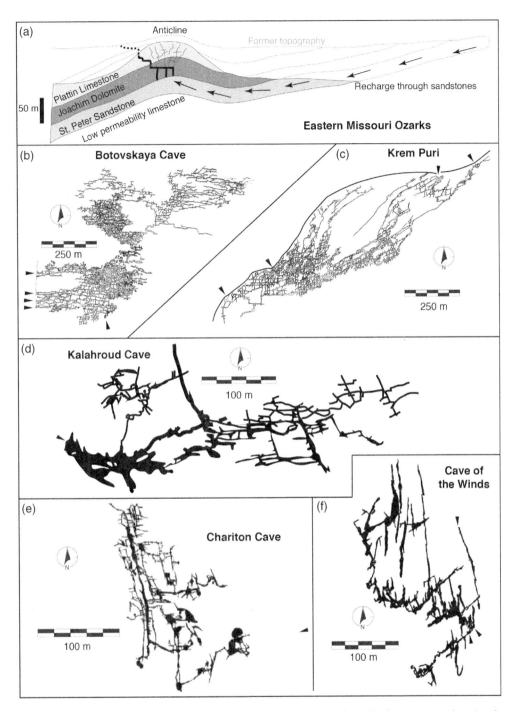

Figure 11.17 Caves formed by diffuse flow from below: (a) Schematic geological cross-section showing the formation of basal injection fissure caves in Eastern Missouri, USA. *Source:* Modified from Brod (1964). (b) Botovskaya Cave in Siberia, the longest limestone cave in Russia, formed by basal injection of deep artesian fluids and their mixing with meteoric shallower artesian waters. *Source:* Modified from Vaks et al. (2013). (c) Krem Puri Cave, Meghalaya (India), an over 25 km-long maze system carved in a carbonate-cemented sandstone bed within a thick Late Cretaceous quartz sandstone succession. *Source:* From Sauro et al. (2020). (d) Kalahroud Cave near Isfahan (Iran), a relict basal injection cave in Cretaceous limestones. *Source:* Modified from Bahadorinia et al. (2016). (e) The 3.5 km long Chariton Cave, the largest maze cave in the Judean Desert, Israel. *Source:* Modified from Frumkin et al. (2017). (f) Cave of the Winds (Colorado) formed by mixing between slightly thermal rising waters and meteoric fluids. *Source:* Modified from Luiszer (2009). The arrowheads indicate cave entrances.

Oligocene-Early Miocene times, with recharge taking place in the distant Nubian sandstones. Acidity might have been produced, at least locally, by sulfuric acid generated by bacterial sulfur reduction, H_2S production and its oxidation in the aerate cave environments (Frumkin et al. 2017).

Another example of basal injection cave formed by slightly thermal waters is Cave of the Winds (Colorado, USA) (Figure 11.17f) (Luiszer 2009). This 3.2-km long cave has formed in Ordovician and Early Carboniferous limestones, overlain by low permeability arkosic sandstones of Late Carboniferous age that inhibit water from flowing upward. Upward recharge occurred from the underlying fractured Pikes Peak granite and Sawatch quartzite into the crest of an anticline, and solutional aggressiveness of the water was obtained by mixing these rising fluids with surface-derived waters.

The most famous and profusely studied basal injection (artesian) caves are those occurring in western Ukraine, and representing the longest gypsum caves of the world: Optymistychna, 264 km; Ozerna, 140 km (Figure 11.18a); and Zoloushka, 92 km, between Ukraine and Moldova (Andrejchuk and Klimchouk 2017; Klimchouk and Andrejchuk 2017). These multistorey cave systems are developed in a 10–40 m thick Middle Miocene (Badenian) gypsum unit sandwiched between an underlying 10–30 m thick calcareous and biohermal permeable formation, and the overlying 0.5–25 m thick Ratynsky limestone also functioning as an aquifer. This succession is in turn covered by low permeability argillaceous sediments up to 100 m thick that confine the underlying aquifer units. Rising artesian flow is focused in areas where the hydraulic gradient is directed upward, corresponding to the flanks of the entrenched valleys of the Dniester and Prut Rivers and their tributaries. Initial flow was slow and mainly driven by convection, in which less mineralized fluids rose because of their lower density, then becoming denser due to gypsum dissolution, and descending again closing density current loops. In order to carve the two-dimensional multistorey voids, flow must also have been lateral, connecting multiple inputs (below) to a few outputs (above). Once the flowpaths traversing the gypsum bed connected the lower and upper aquifers (breakthrough), flow rate could not increase significantly because of the limited recharge and discharge, so competition between the available flowpaths was less important than in epigenic speleogenesis. All possible flowpaths continued to be enlarged at some rate, producing the observed mazes arranged in different storeys, controlled by packages with differing texture and joint systems. This flow pattern is confirmed by morphological observations, the occurrence of lateral flow morphologies (the morphologic suite of rising flow, Klimchouk 2007) and by modeling exercises (Birk et al. 2005; Rehrl et al. 2008). The breaching of the confining aquitard by surface denudation or river entrenchment caused upward flow to become more concentrated, but rarely to the point to produce linear caves.

Zoloushka Cave is the most spectacular example of a recently drained artesian gypsum cave, due to its intersection in 1946 by a gypsum quarry and dewatering for mining. Because of its recent drainage and its particular setting, confined and active until a few decades ago, this cave was one of the most important natural laboratories for the study of speleogenetic and geomicrobiological underground processes, showing its transformation into an unconfined gypsum cave over the last 75 years (Andrejchuk and Klimchouk 2001, 2017; Kotula et al. 2019).

Artesian multistorey caves are also known from other regions, with probably the most interesting example being Coffee Cave in New Mexico (USA) (Stafford et al. 2008a) (Figure 11.18b). This rectilinear maze comprises four levels, controlled by the alternation of meter-thick gypsum beds and less soluble dolostones of the Seven Rivers Formation of Permian age. Recharge is probably from the underlying San Andres limestones (Permian) toward the Pecos River potentiometric low.

Some artesian gypsum caves have also been discovered in the Messinian evaporite formations in Piemonte (Northern Italy), where underground gypsum mines have intercepted large water-filled

(a)   (b)

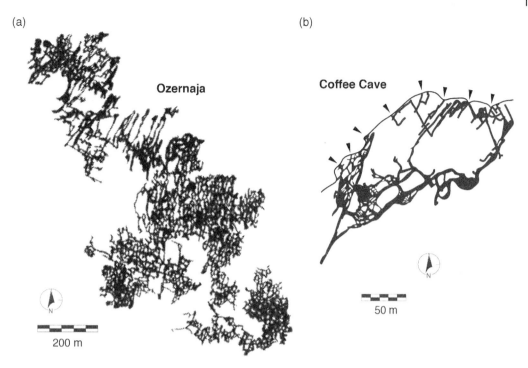

Figure 11.18 Multistorey gypsum network caves (arrowheads indicate cave entrances): (a) Ozernaja Cave, western Ukraine. *Source:* Modified from Klimchouk and Andreychuk (2017). (b) Coffee Cave, New Mexico (USA), formed by groundwaters rising toward the Pecos River. *Source:* Modified from Stafford et al. (2008a).

passages. Morphologies and the geochemistry of waters point to upward recharge from an underlying limestone bed (Vigna et al. 2010). A small 650-m long maze cave has also been found thanks to an underground Messinian gypsum mine near Monticello d'Alba, with clear evidences of basal injection origin (Banzato et al. 2017b).

11.4.5 Rising Thermal Fluids

Temperature rises at a rate of around 25–30 °C km^{-1} in the upper Earth's crust, mostly due to heat flow from the much hotter underlying mantle mainly by natural decay of radioactive elements. This temperature gradient is higher (ca. 50 °C km^{-1}) in areas where mantle upwelling occurs (mid-ocean ridges or mantle plumes) and can be as much as 100 °C km^{-1} in areas of active volcanism. This rate of temperature change with increasing crustal depths is known as the geothermal gradient. Groundwater that flows along deep paths is heated up by the surrounding rocks as a consequence of this gradient.

There are different definitions for thermal waters. One definition classifies all waters with temperatures higher than 20 °C, often used for therapeutic purposes (such as the Roman "thermae"), as thermal waters. A more scientific definition requires thermal waters to be around 5 °C warmer than the normal average temperature of groundwaters in the surrounding region (White 1957). This first definition is not applicable in warm climates, where all spring waters may have temperatures above 20 °C, whereas the second definition works well in temperate and tropical climates, but not in cold (permafrost) regions where the ground temperature is well below 0 °C and most spring waters fall in the "thermal" category. Here, following the second definition, we consider that cave-forming thermal waters are those at least 5 °C warmer than the host rock, creating the thermal differences that lead to the typical conditions in which evaporation and condensation–corrosion processes and thermal convection can take place and have relevant imprint on the resulting cave morphologies and deposits.

The most important thermal groundwater resources outside volcanic areas are related to deep carbonate aquifers. Many of the ancient Roman baths are located at the border of karst areas, and thermal springs are linked to deep regional groundwater flow systems. These deeply rooted carbonate aquifers are important geothermal resources that can be used for CO_2 sequestration. The injection of CO_2 in these reservoirs causes carbonate to dissolve (acting as a CO_2 sink) often improving permeability and transmissivity and increasing the efficiency of the geothermal energy supply. Depleted oil and gas carbonate reservoirs can also be used as CO_2 sequestration sites. Despite the fact that carbonate rocks have a high tensile strength, these operations, which cause carbonate dissolution to increase deep underground, can potentially lead to subsidence phenomena at the surface (Goldscheider et al. 2010). With the increasing public awareness that CO_2 is an important greenhouse gas, the use of carbonate reservoirs as CO_2 sinks will surely increase in the future (Raza et al. 2019).

In a certain sense, all groundwater present at a certain depth falls in the category of thermal waters. In epigenic karst regions, the heat of the host rock is carried away rather efficiently by fast flowing waters of meteoric origin (Badino 2005). This explains why waters at the bottom of Krubera Cave (Arabika Massif, Abkhazia), over 2 km deep, are still cold (around 1 °C), and do not reach temperatures of 50–60 °C as occurs in 2 km deep South African gold mines. On the contrary, if deep flowpaths occur and enough time is given to the water to heat up (slowly moving waters), waters can reach the surface at temperatures well above those of the host rock in equilibrium with surface-derived groundwaters. If these hot (or warm) waters play an important role in cave genesis, the resulting accessible voids are called hydrothermal caves (Ford 1995). Since these thermal fluids rise from depth, and acidity derives from mixing between different fluids or from deep-seated sources, all caves formed in this way can be classified as hypogenic (sensu Palmer 2007 and Klimchouk 2007).

Three different settings can be recognized (Dublyansky 2019): (i) the deep endokarst, at depths of more than 4 km, where fluid pressures are high, producing the solutional porosity encountered in deep drilling of oil and gas reservoirs, (ii) the deep-seated (genuine) hydrothermal karst (up to 4 km deep), where temperature gradients are low and fluids of different chemistry and origin slowly dissolve carbonate rocks creating characteristic collapse breccias, and (iii) the shallow hydrothermal karst, where rising thermal fluids of various origin and chemistry come in contact with shallow meteoric waters, temperature gradients are high, and pressures are close to those of the atmosphere.

Active deep-seated caves have been found during drilling operations in the Rhodope Mountains (Madan tin-zinc ore deposit, Bulgaria). A giant cavern estimated to be 800 m long, 620 m wide, and almost 500 m in vertical dimension filled with thermal water (between 86 and 130 °C) was discovered in the late 60s of the last century (Dublyansky 2000a). This is actually the largest natural cave chamber in the world, at least 22 times bigger than Miaos Room in China!

The shallow hydrothermal karst setting is the best known and most important, creating most of the large and accessible thermal caves we know today. Different fluids can participate in the speleogenetic processes: (i) juvenile waters related to igneous activity, (ii) fluids produced by metamorphic processes, (iii) connate waters initially trapped in sediments and expelled during burial, and (iv) deeply circulating meteoric waters. The chemistry of these fluids can vary, with CO_2- and H_2S-rich waters being the most important (Dublyansky 2000b). In this section, we describe the role of cooling CO_2-rich fluids in cave genesis, leaving the H_2S-rich fluids for the following section. Solutional aggressiveness can also be produced by the mixing between waters of different chemistry, mainly different amounts of dissolved CO_2 (solubility of which is temperature dependent), variable amounts of H_2S, or presence of other salts (e.g., brines). These mixing effects are also dealt with later, in Sections 3.7.5 and 11.4.7.

Note that "pure" thermal caves are probably very rare, and the cave formation mostly depends on a combination of processes, including mixing corrosion, sulfuric acid speleogenesis (SAS), and cooling of rising thermal waters. To be purely thermal in origin, the thermal gradient needs to remain stable over long-enough periods of time for the speleogenetic processes to be able to create man-sized voids. The heat released by the rising cooling fluids is transferred to the rock, and then to the surface above. This slow process diminishes the thermal difference between the host rock and the rising fluids gradually, until thermal equilibrium is attained. The solutional aggressiveness attained by the rising fluids due to cooling diminishes with time, and eventually expires completely once thermal equilibrium is reached.

The rising of thermal water causes these fluids to cool down gradually, while simultaneously the pressure drops. This causes two antagonistic effects: (i) the decrease in pressure causes CO_2, and other gases such as CH_4 and H_2S, to escape into the gaseous phase, like the degassing process that occurs when opening a bottle of sparkling water, and (ii) CO_2 solubility increases with decreasing temperature, so part of the CO_2 enters into the water again and contributes to increase the dissolution capability of the water. Consequently, the solubility of $CaCO_3$ increases gradually along the ascending path of the thermal CO_2-bearing fluids, as long as the effect of the CO_2 degassing process does not predominate. This means that dissolution of carbonate rocks prevails in the deeper thermal zone. In contrast, close to the water table, where CO_2 degassing occurs more rapidly, the fluids become supersaturated in dissolved $CaCO_3$, resulting in net carbonate deposition instead. These two zones of net dissolution and net deposition are analyzed later on.

The rather low flow rates of thermal fluids and the gradual creation of acidity, normally produces network mazes in fractured limestones, as long as the flow rate is high enough. This is confirmed through computer modeling by Andre and Rajaram (2005); a fluid at 60 °C with P_{CO_2} of 0.03 atm flowing through a 500-m long fracture of initial width of 0.05 mm will cause its almost uniform widening with a maturation time (similar to the breakthrough time in epigenic speleogenesis) of around 6600 years. Once a fracture is sufficiently enlarged, the increased flow of thermal water causes the temperature gradient to become smaller and water cools down less efficiently, thus causing a negative feedback mechanism which retards its enlargement, in favor of nearby narrower fissures (Figure 3.26; see more detailed explanation about this modeling experiment in Section 3.7.1). This explains why networks are the predominant patterns in hydrothermal (CO_2-based) caves.

As mentioned above, thermal fluids can dissolve carbonate rock deep below the water table, but become supersaturated when they approach the aerated zone due to rapid CO_2 exsolution. Since the water table is not stable over geological times, its changes can cause net dissolution zones to become net deposition zones and vice versa, and these variations can occur multiple times. Most known hydrothermal caves are now accessible, and thermal fluids responsible for their formation have long disappeared. Their drainage often has occurred long time ago, when the thermal cave was brought closer to the surface, and surface denudation has caused them to be breached (unroofing) and accessible. When thermal caves come close to the surface, epigenic waters enter the voids overprinting the original thermal morphologies. On the other hand, the presence of thermal water in the caves is not necessarily a proof of their hydrothermal origin. Rising fluids often choose the easiest paths, potentially taking advantage and invading preexisting normal meteoric caves. The processes involved in hydrothermal cave formation are also variable, with various degrees of mixing between different fluids, and several possible sources of acidity. Recognizing the thermal origin of a cave is therefore far from being a simple task!

In general, a combination of morphological, mineralogical, and geochemical evidence can support the hypothesis of the hydrothermal origin of a cave. First of all, the presence of hot springs in the area is an indication of the general fluid flow system typical of hydrothermal caves (Mádl-Szőnyi et al. 2017). Certain caves still host thermal water bodies and are still actively enlarging in the deep submerged thermal aquifer, and depositing typical warm-water mineral suites close to and above the thermal water bodies. Typical hydrothermal cave deposits are large euhedral calcite crystals up to 1 m long formed in the deep-seated hydrothermal setting, where low and steady supersaturation can be maintained over long time scales. Typical crystal morphologies are scaleno-hedral crystals known as dogtooth spar (Figure 11.19a), and crystals ending with an obtuse

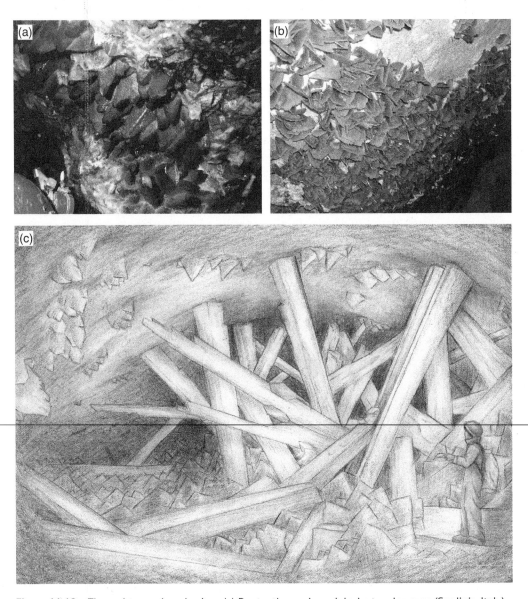

Figure 11.19 Thermal cave mineral suites: (a) Dogtooth spar in an Iglesiente mine cave (Sardinia, Italy). *Source:* Photo by Paolo Forti. (b) Barite crystals in Santa Barbara Cave, Sardinia. *Source:* Jo De Waele (Author). (c) Sketch of the Cueva de los Cristales in Naica Mine, Mexico, with gypsum crystals several meters long. *Source:* Drawing by Laura Sanna, La Venta Esplorazioni Geografiche.

rhombohedron known as nailhead spar. These spars often have a typical short (0.3 seconds) but strong orange-reddish luminescence glow after being illuminated with a powerful light (for crystals grown in water with T > 60 °C), and a longer (1 to 20 seconds) blue-white glow for crystals grown in thermal waters with T < 60 °C (Dublyansky 1995). Fluid inclusions in these minerals can offer further evidence of the thermal and chemical characteristics of the water from which they precipitated. Thermal calcites also often have higher amounts in trace elements, including Mn, Fe, Pb, and Sr, and more negative values of $\delta^{18}O$ (often between −20 and −6‰) than nearby cold-water precipitates. $\delta^{13}C$ in thermal calcites shows little or no variation along the growth axis, since geochemical conditions can remain very stable through time and fluids are in equilibrium with the carbonate host rock, in contrast to the changing $\delta^{13}C$ values of meteoric infiltration waters. When calcites precipitate in a thermal cave where mixing with meteoric water is important, $\delta^{13}C$ will show variations along the growth direction of the crystals, indicating variable importance of isotopically light soil-derived organic carbon sources.

Besides calcite, other minerals can be indicative of a thermal water origin, including quartz, barite (Figure 11.19b), fluorite, and various types of sulfides. The uranium–vanadium mineral tyuyamunite, together with calcite and barite, precipitated from thermal waters with temperatures between 30 and 60 °C in Fersmana Cave, a karst void intercepted by a mine in the Tyuya-Muyun Massif (Kyrgyzstan) (Dublyansky et al. 2017). In Cueva de los Cristales (Naica, Mexico), the giant gypsum crystals, one of which reaches the length of 11 m, have grown at a constant temperature of ca. 54 °C by a solution-mediated anhydrite-gypsum phase transition, since anhydrite becomes the unstable calcium sulfate phase below around 56 °C (García-Ruiz et al. 2007) (Figure 11.19c).

Of course, these minerals, to be indicative of a thermal origin of the caves, need to be genetically linked to the void-creating stage, and not simply to the thermal water invasion of a normal meteoric cave. In shallow thermal settings, CO_2 degassing becomes important and causes the formation of calcite rafts, raft cones, and folia. These speleothems, however, also form in cold-water caves, where degassing is important (e.g., Cueva Grande de Santa Catalina in Cuba, D'Angeli et al. 2015a), so their association with other indicators of thermal origin (such as stable isotopes) is needed to conclusively prove their origin from thermal waters. Stable isotopes can also be used on the cave rock walls to determine whether hydrothermal alteration has occurred or not (Dublyansky et al. 2014; Spötl et al. 2021).

All these indicators of hydrothermal origin must be accompanied by a series of morphological features typical of hypogene settings, in which slowly flowing waters, convective flow mechanisms, and subaerial condensation–corrosion are dominant, and where typical fast-flow features such as scallops and coarse sediments are lacking. Hydrothermal caves by definition lack a genetic relationship with surface features (dolines, sinking streams, and vadose shafts). In one of the most important hydrothermal cave areas of the world, the Rózsadomb in Budapest (Hungary), virtually all caves have been discovered by quarrying or excavations for houses (Bolner-Takács and Kraus 1989) (Figure 11.20).

The main cave patterns encountered in hydrothermal caves are individual isolated chambers, and 2D and 3D mazes. The Grotte des Champignons in Provence (southern France) is a good example of the first category (Audra et al. 2002). Bush-like caves with upward penetrating cupolas are another typical pattern found in caves of hydrothermal origin. One of the best examples of such type of cave is the Sátorkö-Puszta Cave near Dorog, in the Pilis Mountains of Hungary (Jakucs 1977; Bolner-Takács and Kraus 1989). A 350-m long vertical chain of interpenetrating cupolas resembles a bunch of grapes (Figure 11.21a). Gypsum deposits found in the lower chambers suggest that sulfuric acid was present during the cave formation stages. The Rancho Guadalupe

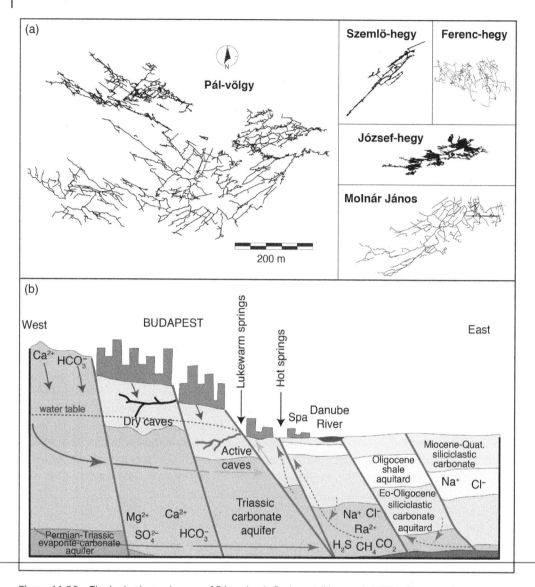

Figure 11.20 The hydrothermal caves of Rózsadomb, Budapest (Hungary). (a) The five most important caves (scale and North for all caves). (b) Schematic cross-section through the Rózsadomb area showing the position of active and dry caves and the general groundwater flow (blue = cold, orange = lukewarm, and red = hot). Dotted arrows indicate basinal contribution to flow. *Source:* Modified from Mádl-Szőnyi et al. (2017).

Cave in Cuatro Ciénegas (Mexico) shows similar shapes, but smaller in size (Piccini et al. 2007) (Figure 11.21b). Most caves in the Rózsadomb, Budapest, are of the two- and three-dimensional network and ramiform maze type (Figure 11.20). These are characterized by irregular rooms, dead-end fissures, solution pockets, and spongework, with rounded rooms hosting cupolas (Figure 11.22a), and abrupt changes in cross-sectional dimensions (Leél-Őssy 2017). Most of these

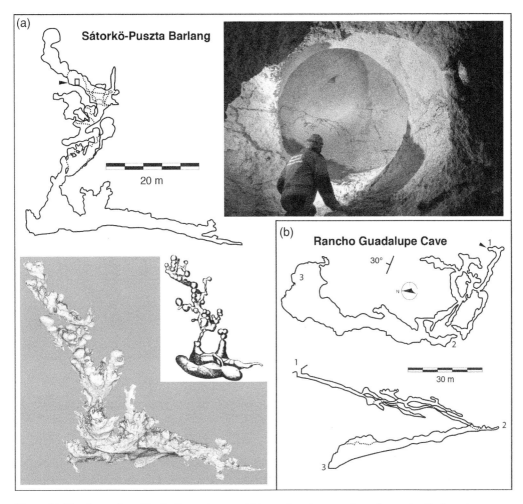

Figure 11.21 Examples of hydrothermal caves. (a) Large basal room where thermal sulfuric waters probably entered the cave, and upward developing interpenetrated cupolas in Sátorkö-Puszta Barlang, Hungary. Speleological cross-section (upper left), a cupola (upper right). *Source:* Photo by Csaba Egri, a 3D laser scan survey (lower right, courtesy of Dr. Surányi Gergely, laser scan survey by Burken Ltd., Hungary) and schematic sketch (inset). *Source:* Modified from Jakucs (1977). (b) The plan and cross-section of Rancho Guadalupe Cave in Cuatro Ciénegas, Mexico (note dip of strata). *Source:* Modified from Piccini et al. (2007).

caves have many different types of speleothems typically related to thermal waters that experience CO_2 degassing, such as thick deposits of calcite rafts, and raft cones (Figure 11.22b, d).

Several thermal caves have been found in Murcia region (SE Spain), such as Sima de la Higuera and Sima Destapada, east of Murcia, and Cueva del Agua near Cartagena. The first two are typical three-dimensional maze networks with many of the distinctive thermal cave morphologies (cupolas, bubble trails, blind-end widened fissures, spongework, and boxwork) and speleothems (calcite spars, calcite rafts, tower cones, and folia). Sima de la Higuera (Figure 11.22c) has been abandoned by the thermal waters, whereas Sima Destapada still contains a thermal lake (31 °C) with over 400 m of explored submerged passages. Cueva del Agua near Cartagena also still contains a brackish

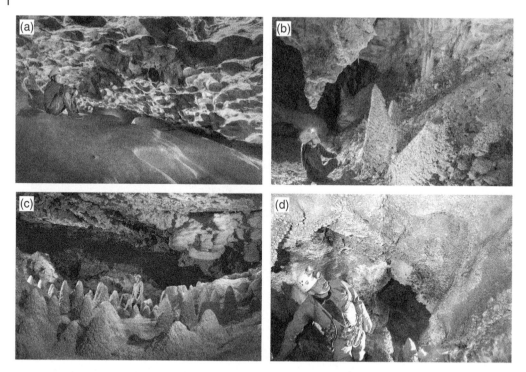

Figure 11.22 Hydrothermal caves. (a) Ceiling pockets and cupola in the Mátyás-hegyi Barlang, Budapest. *Source:* Photo by Csaba Egri. (b) Raft cones in Pál-völgyi Barlang, Budapest. *Source:* Photo by Csaba Egri. (c) Cave clouds (roof and walls) associated with raft cones in Sima de la Higuera, Murcia (southeastern Spain). *Source:* Photo by Victor Ferrer. (d) Folia and raft cones in the Adaouste Cave (S-France). *Source:* Photo by Jean-Yves Bigot.

thermal water body (ca. 30 °C, 42 mS cm^{-1}) underlying a freshwater layer of variable thickness (0–6 m). Its connection to the sea makes this cave a good example of a thermal cave in which speleogenesis is also conditioned by freshwater-saltwater mixing corrosion (Gázquez et al. 2017).

At Monsummano Terme (Tuscany), another well-known thermal cave (Grotta Giusti) is used for therapeutic purposes. This cave, partially elongated along an enlarged fracture, is largely occupied by a thermal lake (30–34 °C), dived to a depth of over 50 m. This cave is especially known for its speleothems, with cave clouds, calcite spars, raft cones, and folia (Piccini 2000).

Relict caves of almost certain thermal origin are those of Kef el Kaous (Traras Massif along the coast of Algeria) (Collignon 1984) and the caves of Nooitgedag in the marble belt of Damara, northwest of Windhoek (Namibia), and Temple of Doom Cave in Kaokoland, northern Namibia. The first two are characterized by maze networks and contain typical cave morphologies (cupolas, blind chimneys) and speleothems (raft cones and calcite rafts) of probable hydrothermal origin (Martini 2017). The Temple of Doom, on the contrary, is a fissure connected to a large room from where a passage of rounded phreatic shape descends to a final pool at −84 m from the entrance. Cupolas, blind chimneys, and spongework occur in the higher parts of this passage, whereas cm-sized barite crystals once covered the walls and floor (Martini and Marais 1996).

Provalata Cave (Republic of Macedonia) is instead a nice example of a multiphase cave in which thermal CO_2-rich water first carved the main cave voids (probably during the Pliocene), which were filled by pyroclastic-derived clays (ca. 1.8 Myr), later emptied by vadose water and then modified by SAS (around 1.6–1.46 Myr ago, based on alunite $^{40}Ar/^{39}Ar$ dating) (Temovski et al. 2013).

11.4.6 Sulfuric Acid Fluids

The role of sulfuric acid in the genesis of some caves has been known for quite a long time in Europe. Rising warm (ca. 40 °C) H_2S-rich waters were put in relation with the presence of actively developing karst voids in the thermal spring area of Aix-les-Bains (Savoie, SE France) at the very beginning of the nineteenth century (Socquet 1801), and Martel mentioned the importance of sulfuric acid in speleogenesis before World War II (Martel 1935). In Central Italy, an underground tunnel excavated for a spa at Triponzo (Southern Umbria) intercepted a small cave with sulfuric waters actively dissolving the limestone (Principi 1931).

The role of sulfuric acid in cave formation was first described as a rather exceptional process in Lower Crevice Cave (Dubuque, Iowa, USA). Here the acid is produced by the oxidation of pyrite, marcasite, and galena in the Middle Ordovician Galena Dolomite (Morehouse 1968). The breakthrough in the understanding of the SAS arrives some years later, with the PhD Thesis of Egemeier on Lower and Upper Kane caves in Wyoming (USA), in the lower of which a H_2S-rich stream flows with abundant white microbial filaments (Egemeier 1981). In this case, the H_2S appears to be produced by the reduction of sulfate beds in petroleum-bearing units of the stratigraphic sequence. The same mechanisms were the basis of the still valid sulfuric acid theory for the formation of the cave systems in the Guadalupe Mountains (Davis 1980), subsequently validated by stable isotope analyses on replacement gypsum, which gave negative values compatible with an origin from the reduction of dissolved sulfate in the presence of hydrocarbons, and not related to dissolution and subsequent redeposition of the Castile Formation evaporites (Hill 1987, 1990). From the early 1980s, Carlsbad Caverns, the surrounding caves and, later on, Lechuguilla Cave (discovered in 1986) became the paradigm of the SAS model (Jagnow et al. 2000).

The first African SAS cave, Rhar es Skhoun (Azerou Massif, Algeria) was described in the early 1980s, with an initial thermal speleogenetic phase followed by a subaerial sulfuric acid phase (Collignon 1983). Unfortunately, this cave, together with several other SAS caves in the area, are seriously endangered due to intense quarrying activities (Audra 2017). The discovery of Movile Cave in southern Romania in the early 1990s shed new light onto very special geo-ecosystems, in which sulfuric acid is still actively enlarging the cave (Sarbu et al. 1996). The genesis of the Frasassi Cave System in Italy was explained with the sulfuric acid mechanisms in the same period (Galdenzi and Menichetti 1990), and in the following years several other caves along the Apennines in Central Italy were ascribed to this process (Galdenzi and Menichetti 1995). Today, Italy is the richest country in the world regarding both active and inactive documented SAS caves (D'Angeli et al. 2019c).

The fluids interacting in SAS caves are generally derived, at least in part, from a deep (hypogenic) source (Figure 11.23). If the origin of these rising waters is deep enough, these fluids can also be thermal, as documented in many active SAS caves (e.g., Acquasanta Terme, Italy, Galdenzi et al. 2010; Montecchio Cave, Tuscany, Italy, Piccini et al. 2015; Chevalley Aven and Serpents Cave, France, Audra et al. 2007b; Cerna Valley caves, Romania, Wynn et al. 2010), but this is not necessarily the case (e.g., the sulfuric acid waters flowing in Frasassi caves, Italy, show normal temperatures around 13 °C, Galdenzi 2012). In many cases, this thermal imprint is rather weak, such as in Movile Cave (Romania), where waters are only 8 °C warmer than the mean annual temperature (Sarbu et al. 1996), or Cueva de Villa Luz (Tabasco, Mexico) where they are only 5 °C warmer (Hose et al. 2000).

Sulfuric acid is produced by the oxidation of sulfides in the oxygenated part of the karst aquifer. Sulfides can be metallic (e.g., pyrite, FeS_2) or hydrogen sulfide (H_2S). However, most SAS caves known in the world are formed by the oxidation of hydrogen sulfide. H_2S involved in SAS can

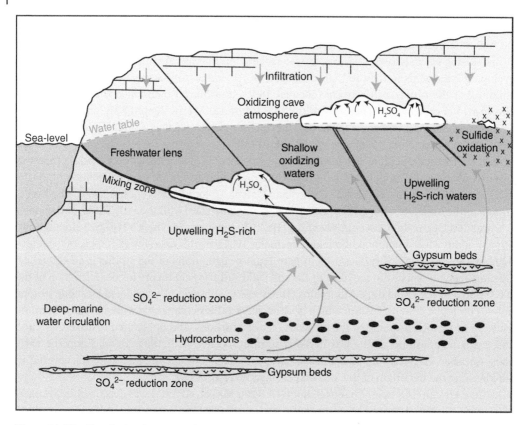

Figure 11.23 Sketch showing the typical continental and coastal settings in which sulfuric acid caves form. *Source*: D'Angeli et al. (2019c) Reproduced with permission of Elsevier.

sometimes have a volcanic origin (e.g., Movile Cave, Sarbu et al. 1996), but is generally formed by bacterial or thermochemical sulfate reduction (BSR and TSR, respectively). The first occurs in low-temperature diagenetic environments (0–80 °C, and < 2.7 km depth in areas with normal geothermal gradient), whilst the second takes place at greater depths (3.3–4.7 km) with temperatures between 100 and 140 °C (Machel 2001) (Figure 3.6). Sulfates can be found deep in the crust as dissolved anions in connate water, deeply circulating marine waters, or hydrocarbon reservoirs, and in solid form as evaporites (gypsum and anhydrite). Reduction of these sulfates takes place in presence of organic compounds (typically hydrocarbons) hosted in the sedimentary sequences (see Section 3.4). In comparable geochemical situations, sulfate reduction mediated by bacteria (BSR) is generally faster than that induced by heating (TSR). In general, the H_2S produced with the interplay of hydrocarbons migrates as a dissolved gas elsewhere and calcite precipitates (see Eq. 3.63). Some authors report the replacement of the original gypsum or anhydrite rocks into bio-epigenetic calcite (Hill 1995). Elemental sulfur may form by bacterially mediated oxidation when dissolved H_2S comes into contact with oxygenated meteoric waters. If descending meteoric waters are very poor in dissolved oxygen, only minor parts of the H_2S is consumed by this process, and the remaining H_2S can migrate upward and react with metal-rich chloride waters present in the reducing zone forming the typical Pb-Zn Mississippi Valley Type ore deposits (Hill 1995). If dissolved hydrogen sulfide comes into the aerate environment, it interacts with oxygen and oxidizes producing sulfuric acid, which is the main speleogenetic agent in SAS caves.

H_2S oxidation typically occurs in two cave environments: (i) where ascending H_2S-rich fluids mix with shallow oxygen-rich waters of meteoric or marine origin or (ii) above the water table in the cave atmosphere. The oxidation of H_2S is more efficient if mediation by sulfur-oxidizing bacteria occurs (Engel et al. 2004; Palmer 2013; Jones and Northup 2021). Since most SAS caves are developed in buffering carbonate environments (Palmer 2013), these reactions generally occur at pH close to 7.

Once sulfuric acid (H_2SO_4) comes into contact with the carbonate rock, it immediately reacts dissolving the carbonate minerals and producing sulfates and CO_2. Replacement-solution is often used to describe this reaction, in which calcite is replaced by gypsum (see Eq. 3.72), whereas dolomite is converted into gypsum and epsomite (Egemeier 1981). This replacement is often volumetric, and original textures and fossils in the bedrock are sometimes perfectly preserved (Plan et al. 2012). Gypsum, being more soluble than calcite, can then be dissolved by running waters, allowing the voids to enlarge. The CO_2 produced by the acid reaction is released and can contribute to further promote the dissolution of limestone and other carbonate rocks.

Sulfuric acid can also be produced by oxidation of other sulfides, mainly pyrite (FeS_2), but also sphalerite (ZnS), galena (PbS) and other sulfides common in Mississippi Valley Type ore deposits. This process is often very localized, creating a scattered porosity, but there are examples in which sulfuric acid produced by oxidation of dispersed but abundant pyrite has created large cave systems. The large maze cave systems in Campo Formoso (NE Brazil) are believed to have formed mainly by mixing corrosion and oxidation of metal sulfides hosted in the Precambrian dolomite hostrock (Auler and Smart 2003; Klimchouk et al. 2016) (Figure 11.24).

In Transvaal (South Africa), the Mbobo Mkulu Cave is located between a thick layer of Archean dolostone and an overlying chert breccia covered with a black shale rich in sulfides. It is the oxidation of these sulfides by percolating waters that is responsible for most of the dissolution in the underlying dolostone (Martini et al. 1997).

However, normally these acid dissolution phenomena are very localized, and can occur in epigenic cave systems, such as the Pisatela-Rana cave system in northern Italy, where gypsum formation due to the oxidation of pyrite causes haloclastic phenomena (Tisato et al. 2012). Gypsum also occurs widely in Mammoth Cave, the largest (epigenic) cave system in the world, and is derived from pyrite oxidation on the external surface of the hostrock. Besides breakdown caused by gypsum crystal wedging in the rock fissures, sulfuric acid corrosion is only a very minor component in speleogenesis in this cave (Garrecht Metzger et al. 2015). In Baume Galinière, in the Vaucluse area (southern France), intense pyrite oxidation has created an almost 200-m long maze cave with an exceptional occurrence of SAS-derived secondary minerals (Audra et al. 2015).

As explained above, the reaction of sulfuric acid with limestone gives rise to the formation of gypsum (Galdenzi and Maruoka 2003). Because of the exothermic reaction between calcite and sulfuric acid and the low activity coefficient of water, anhydrite crystals also form along the reaction front, but are rapidly transformed into gypsum in the moist cave environment. Gypsum often occurs as pure saccharoid white masses, or crusts, in contrast to SAS-related gypsum formed by oxidation of pyrite (or other sulfides) that is often stained with yellow-brownish iron oxides.

Gypsum deposits in SAS caves normally occur as up to 30 cm thick wall rinds (white crystalline crusts) or, when these fall to the floor, a few meters thick floor deposits (Polyak and Provencio 2001).

Alongside speleogenetic gypsum, a series of other SAS byproducts has been reported from the many sulfuric acid caves in the world, and especially from those in the Guadalupe Mountains in New Mexico and the Italian SAS caves. The most important of these minerals are hydrated halloysite (also known as endellite), and the alunite-jarosite group minerals (Polyak and Provencio 2001; Audra

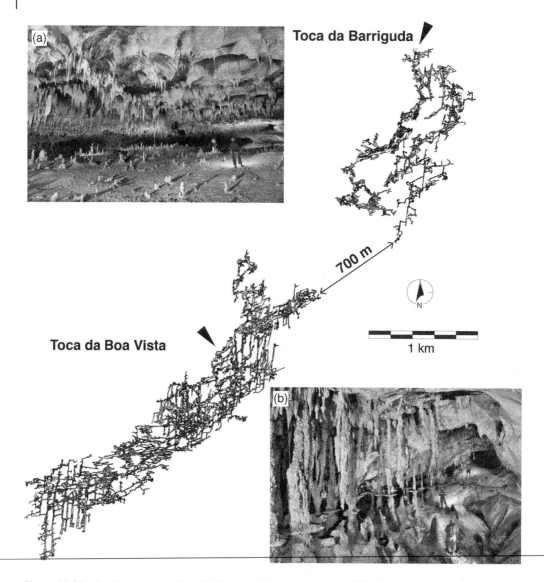

Figure 11.24 The Toca da Boa Vista (114 km) and Toca da Barriguda (35 km) cave systems in northern Bahia, Brazil. Plan view of the two caves, which are only 700 m apart (Map courtesy of Grupo Bambuí de Pesquisas Espeleológicas). (a) Large passages in Toca da Barriguda Cave. Note the cupola and elongated rock projections on the roof. *Source:* Photos by Luciana Alt. (b) Dry pool with shelfstones, cave rafts and raft cones in Toca da Boa Vista Cave. *Source:* Photos by Victor Moura.

et al. 2015; D'Angeli et al. 2018). Some of these minerals, and especially alunite and jarosite, can be used to date the timing of speleogenesis using the ^{40}K—Ar^{39} or Ar^{40}-Ar^{39} methods (Polyak et al. 1998).

In many SAS caves the incomplete oxidation of H_2S also produces elemental sulfur. These yellow coatings, crusts, and crystalline materials have been reported from many sulfuric acid caves. Elemental sulfur is mainly stable at low pH, so it often occurs on gypsum, and rarely on the buffering carbonate substrate.

Another common byproduct of SAS processes in carbonate rocks is dolomite, derived from the preferential removal of Ca in limestone weathering. The enriched Mg-solution can then precipitate dolomite as soon as pH levels are high enough to allow this carbonate to be preserved. This speleogenetic dolomite has been found in Lechuguilla Cave and Carlsbad Caverns, embedded in replacement gypsum, and derives from the partial leaching of the original limestone. This dolomite can thus also be used as an independent geochronometer alongside alunite and jarosite, using the U—Th and U—Pb methods (Polyak et al. 2016).

The action of sulfuric acid on the hostrock leaves a typical geomorphological imprint on the cave walls, both at a macroscopic and microscopic level. In a certain sense, these morphologies are shared between most hypogenic cave systems, since they derive from the fact that aggressive waters come from below, and move slowly upward (Klimchouk 2007, 2009). In hypogenic SAS caves, the acidity is introduced through discrete points (discharge feeders) and is consumed rapidly around these aggressive inputs (Figure 11.25a), and cave enlargement can continue upward prevalently through condensation–corrosion processes (Hill 1987). However, the strength of sulfuric acid, compared to that of carbonic acid, creates a series of sculpturing features very typical of sulfuric acid caves (Audra et al. 2009a, 2009b).

Sulfuric acid caves have very typical patterns that mostly depend on the type of recharge; in this case, fractures and joints bringing rising sulfidic waters (Palmer 2011). This gives most SAS caves a typical elongated plan form, if the feeding fracture (in short, feeder) was an isolated one (i.e., Lower Kane Cave, Egemeier 1981), and an anastomotic or maze pattern, in case of a denser network of recharging fractures (Carlsbad Caverns and Lechuguilla Cave, Hill 1987) (Figure 11.26). The size of the passages is an indication of both the quantity of acid that reacted with the host rock, and the time during which fluids continued rising into the cave (Figure 11.25b). Cave voids are thus larger in correspondence of the points in which H_2S rose into the cave over longer time periods, or at intersections of fractures, and tend to diminish in size farther away from the injection points, eventually pinching out and ending in blind passages. If more spatially separated injection points give rise to adjacent caves, their enlargement over sufficient periods of time can

Figure 11.25   Common sulfuric acid cave morphologies: (a) A feeder and the typical upward taperingpassage in Sette Nani Cave (Calabria, South Italy). Note white gypsum crusts on walls and roof. *Source:* Photo by Orlando Lacarbonara. (b) The Big Room in Carlsbad Caverns, along the main tourist trail, is thelargest void of the cave, carved close to the former water table in the massif reef limestone of Permian age(tourist in white circle for scale). *Source:* Photo by Arthur Palmer.

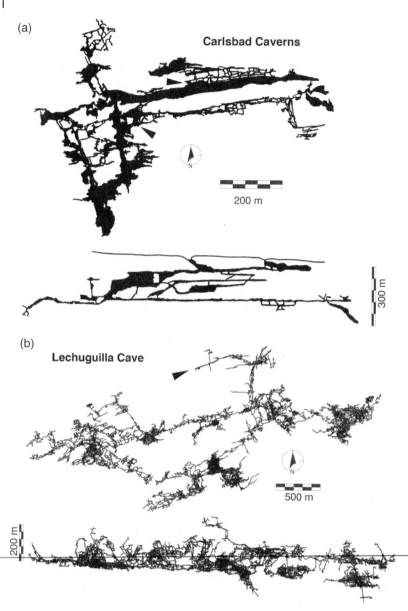

Figure 11.26 Map and profile of (a) Carlsbad Caverns, and (b) Lechuguilla Cave, in Carlsbad CavernsNational Park, New Mexico. *Source:* Courtesy of Cave Research Foundation and National Park Service.

cause these initially isolated cave systems to intercept and merge into complex cave systems. This merging between adjacent dissolutional voids causes the formation of sometimes very fragile and thin rock partings, or forms such as pillars, pendants, blades, projecting corners, arches, and half-tubes.

Sulfuric acid caves are thus mainly influenced by the spatial distribution of fractures that allow the rising fluids to reach the aerate zone. Bedding planes and other primary structures only condition the shape of the passages at a very local scale, and often cave rooms and voids cut straight through these structures, assuming a close-to-horizontal development around the injection points (Figure 11.25a). If the above lying beds are less permeable, though, they act as barriers that guide the H_2S flow along their lower boundaries. In this way, the cave passages tend to follow these hydraulic boundaries, since H_2S mainly oxidizes in these stratigraphic positions. A good example is offered by Acquasanta Terme caves (Rio Garrafo, Galdenzi 2017).

It has to be borne in mind, however, in contrast to most epigenic caves where phreatic dissolution is overwhelmingly important, that most of the rock dissolution occurs above the water level, by condensation–corrosion processes boosted by the oxidation of H_2S in the oxygen-rich cave atmosphere. This is especially true if sulfuric waters are also thermal, and temperature gradients between the water bodies and overlying walls and roof of the cave are significant. Many roof and wall morphologies, although often believed to be of phreatic origin, are indicative of enlargement in the aerated environment. Condensation–corrosion normally attacks the host rock uniformly, regardless of most primary and secondary structures in the rock (bedding planes and fractures). However, microcrystalline calcite dissolves more rapidly than that of macrocrystalline texture, leaving mineral veins (boxwork) and fossils in positive relief. The slow and relentless dissolution of the rock also leaves a fine-grained, powdery residue composed of the less soluble components of the rock (e.g., chert, quartz grains) and weathering products (e.g., sulfates and clays).

Most void volumes in Kraushöhle (Austria) and Grotte du Chat (Provence, France) are characterized by the widespread occurrence of impressive ceiling cupolas and large wall convection niches, the coalescence of which sometimes gives rise to rounded dome-like chambers (De Waele et al. 2016). Rising air flow also carves megacusps greater than half a meter. These upward developing ceiling cupolas and spheres represent subaerial convection cells in which condensation is likely to occur prevalently at the cooler ceiling, where dissolution is thus most active. These rising domes can breach the surface, thus creating occasional entrances to otherwise inaccessible caves (i.e., Kraushöhle in Austria, Pigette Cave, France, Figure 9.26e). These stacked spherical roofs are well known also from thermal caves such as those described in Hungary (Szunyogh 1990), where they have been interpreted, probably erroneously, as a result of phreatic convection (e.g., Figure 9.21). Instead, the warmer rising air follows the overhanging walls forming a condensation path that eventually evolves into a condensation–corrosion channel. These winding smooth channels differ from most paragenetic ceiling half-tubes because of their inclined surfaces, smooth and wavy roof profile, and the lack of sediments in the channels. Condensation–corrosion channels are also characterized by the presence of a succession of megacusps that give them their wavy long profile (D'Angeli et al. 2019b). It is true, however, that condensation–corrosion morphologies, with their rounded shapes, are often confused with passages generated by phreatic water flow.

Condensation–corrosion morphologies are especially well developed above the fissures through which H_2S-rich water (or air) enters the aerated cave environment. These often-narrow fissures are known as discharge feeders, a name used in all hypogenic cave systems (Figure 11.25a). Where visible, the discharge feeders are typically elongated features with a near-vertical downward development and often too small to be explored (e.g., Acqua Fitusa Cave in Sicily, De Waele et al. 2016; D'Angeli et al. 2019c). This is an indication that corrosion by sulfuric acid mainly occurred above the feeder in the cave atmosphere, through condensation–corrosion processes, and dissolution is much less intense below the water level. Many of the active sulfuric acid caves have explorable underground streams, but the fissures from which H_2S rises are generally not accessible, and often even masked by sediments (i.e., Cueva de Villa Luz, Hose et al. 2000; Santa Cesarea Terme caves, D'Angeli et al. 2017b).

In some caves (e.g., Grotte du Chat in France, Bad Deutsch Altenburg in Austria, and Acqua Fitusa in Sicily) the floor close to the feeding fissures is almost perfectly horizontal (Figure 11.25a).

The sulfuric acid-rich waters are able to corrode carbonate rocks of different characteristics in the same manner and at the same rate. These horizontal corrosional forms, which are known under the name of corrosion tables, were already described from Lower Kane Cave by Egemeier (1981), and are exclusive to sulfuric acid caves (Audra et al. 2009a; De Waele et al. 2016). They show very gentle slopes (generally <1%) toward the feeding fissures. Corrosion tables develop entirely above the water level in the cave, and the position of their lower margins indicates the upper limit of the water level at the time of their formation. Water level changes (i.e., lowering), if occurring gradually, cause different corrosion tables to develop, forming a stepped series of levels, as in Grotte du Chat (southern France), where up to twelve levels occur within a vertical range of 6 m (Audra 2008).

Upon draining, when the sulfuric water level falls because of regional uplift or local fluvial entrenchment, feeders become dry and can still act as thermal vents through which vapor-laden air rich in H_2S rises, as long as a thermal water source stays close enough. The external edges of these discharge feeders are often lined with calcite popcorn and gypsum crusts, caused by evaporation processes occurring in the expanding and cooling air (Audra et al. 2007b).

Generally, sulfuric waters can be seen in active caves as gently running rivers (i.e., Lower Kane Cave, USA, Egemeier (1981); Cueva de Villa Luz, Mexico, Hose et al. (2000); Acquasanta Terme, central Italy, Galdenzi et al. (2010)) or, more frequently, as very slowly running waters or more or less still standing pools (e.g., Triponzo, central Italy, Principi (1931); Grotta delle Ninfe, Calabria, Italy, Galdenzi and Maruoka (2019); Frasassi, central Italy, Galdenzi and Jones (2017); Movile Cave, Romania, Sarbu et al. (1996)). In these, often, large water bodies the uppermost part of the water column normally has higher aggressiveness than the deeper one because of greater levels of oxygen, required to oxidize the rising H_2S into sulfuric acid. Poorly mineralized condensation waters, especially in thermal SAS caves, also feed the water pools further increasing acidity of the surface layers. Most of the time, these water levels enlarge the cave passages laterally at discrete levels, sometimes generating solutional notches at various heights (e.g. Kraushöhle, Austria, Plan et al. 2012). These notches are formed both immediately below and above the water level by the uppermost, more aggressive water layers and by the rising air above the standing water body.

In general, the width of the passages is largest close to the water level (floor), showing typical triangular cross-section profiles (Figure 11.25a). The lateral extension of these water levels also causes many SAS caves to have a markedly close-to-horizontal profile (i.e. Carlsbad Caverns, USA, Hill (1987); Grotte du Chat in France and Kraushöhle in Austria, De Waele et al. (2016)).

In areas where uplift is important, SAS cave systems are arranged in different levels adjusted to the present and former base levels, which are normally the local level of the river or the nearby sea or lake (Figure 11.27). The Frasassi cave system is composed of an active lower cave level (Grotta del Fiume), where sulfidic waters rise from fracture-guided feeders into an anticlinal ridge and flow to the Sentino River, and several (probably seven) higher lying levels which are now abandoned by the sulfidic waters. These include the Grotta Grande del Vento, open to tourists, and the much higher lying Buco Cattivo and Grotta di Frasassi-Traversata del Mezzogiorno caves.

The most diagnostic morphological features of sulfuric acid caves are replacement pockets, which are hemispherical corrosion forms some centimeters wide and deep carved into the walls in a more or less regular pattern. The distribution of these cup-like depressions is denser on vertical walls and the upper part of inclined rock surfaces, and becomes sparser on overhanging walls and roofs. Their formation is related to focused acid corrosion and simultaneous replacement of calcite by microcrystalline gypsum (Galdenzi and Maruoka 2003), which is often still visible inside the pockets. The presence of this hygroscopic sulfate causes the acid fluids to persist and corrosion to proceed, functioning like an acid sponge and leading to the deepening of the concavities. This can ultimately result in the formation of a Swiss-cheese like morphology in the corroded bedrock,

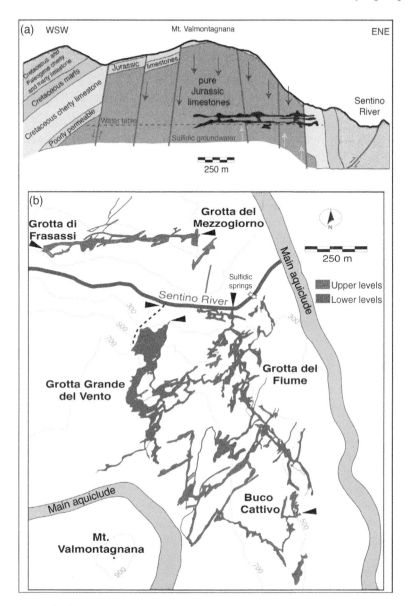

Figure 11.27 The Frasassi cave system, Central Italy: (a) Schematic geological cross-section across the Frasassi Anticline. *Source:* Modified from Galdenzi and Jones (2017). (b) Plan view of the Frasassi cave system on both sides of the Sentino River gorge. *Source:* Maps from the Federazione Speleologica Marchigiana caving groups.

forming a spongework macro-porosity. They are less developed on overhanging walls and roofs due to the fact that the neo-formed gypsum falls off more easily than from vertical or steeply inclined walls. As for all condensation–corrosion morphologies, the dissolved rock surface is generally very smooth and regular.

Acidity of drip waters can be very high, reaching pH values as low as 0 in some cases (Cueva de Villa Luz in Mexico, Hose et al. 2000; Jones and Northup 2021), especially in places where direct contact between the sulfuric acid fluid and the carbonate host rock is prevented (drips hanging from gypsum or from acidic biofilms). These extremely acid droplets are at the origin of sulfuric karren and associated solution pans (Cueva de Villa Luz, Grotte du Chat, Acqua Fitusa) (De Waele et al. 2016), and of the unique decimetric bowl-shaped basins in the floor of Kraushöhle in Austria ("ceiling pendant drip holes" in Plan et al. 2012).

11.4.7 Coastal Mixing

Cave development can be extremely intense in coastal areas, especially in young and porous (eogenetic) limestones. This is due to the mechanical action of waves and tides, which produces littoral caves (formed within the range of tides), and the special type of solutional cave-forming environment characterized by coastal mixing corrosion. This section deals with solutional caves associated with the freshwater–seawater mixing zone, which are morphologically very different from both the littoral (mechanically enlarged) caves, and the inland karst caves formed by underground freshwater. As these coastal mixing caves are formed by aggressiveness created at some depth and often show poor relationship with surface landforms, they can be regarded as hypogenic caves in the sense of Palmer (2007). If not breached by coastal erosion or surface denudation, they often have poor direct connections with the sea and with the overlying surface, and their inner morphologies often lack signs of turbulent flow (i.e., scallops and stream sediments).

These caves are not related to downward infiltrating waters that commonly lose most of their dissolution capability within the first 10 m of their flow path (see Section 4.6.1). Neither seawater is capable of dissolving carbonate rocks, since it is saturated in $CaCO_3$. It is the mixing of two different waters (Wigley and Plummer 1976) that creates an aggressive solution capable of creating the large solution spongework caves typical of coastal karst areas (Smart et al. 1988) (see Section 3.7.5 and Figure 3.18). There are two mixing zones in coastal aquifers: (i) an upper zone where percolating (vadose) waters mix with phreatic freshwater and (ii) the interface within the phreatic zone between the freshwater and the underlying and denser seawater. Seawater has an average density of $1.025\,\mathrm{g\,cm^{-3}}$, which is 1/40 times denser than freshwater (i.e., $1.0\,\mathrm{g\,cm^{-3}}$ at 4 °C). The boundary between these two types of groundwater can be sharp in calm conditions, called the halocline, or can be disturbed by tides, intense recharge, pumping, etc. Cave divers can clearly see this halocline as a hazy interface due to the different refraction of the light in the water layers with sharp density contrast (Figure 11.32b). Often, instead of the well-defined halocline, water gradually passes from the underlying seawater to less dense and brackish water, up to the freshwater on top. This so-called mixing zone can be several meters thick, and mixing is more important closer to the coast, where the tidal influence is greater.

Mixing corrosion is generally greater in the lower seawater–freshwater boundary than in the upper vadose-phreatic boundary, mainly because of CO_2 degassing (which increases pH) at the upper boundary into the aerate zone. In addition to the mixing phenomena, which cause waters to become undersaturated in carbonates, additional acid inputs may also occur. The boundaries between the vadose and phreatic zones, and the freshwater and seawater are also density interfaces, where incoming organic particulate material can concentrate. The oxidation of these organics produces CO_2 which can enhance the aggressiveness of the water and its capacity to dissolve limestone. In some cases (e.g., the Yucatán peninsula), this source of acidity can be the most important driver of dissolution, largely overriding the simple mixing corrosion (Gulley et al. 2014, 2015, 2016). If the decaying organic matter consumes all available oxygen in the water, anoxic conditions will prevail. If low-oxygen levels persist in time, especially in the underlying saltier sulfate-rich water, sulfate reducing anaerobic bacteria can produce H_2S. When this gas diffuses upward and comes into contact with oxygen again, sulfuric acid is produced causing localized replacement-dissolution. These processes can occur in higher oxygen-rich water layers, in the aerate environment, or during storm pulses that supply oxygen-rich floodwaters. This sulfuric acid dissolution is especially active in cenotes and blue holes, where large amounts of organic debris can be locally introduced into the underlying mixing-corrosion cave environment (Bottrell et al. 1991; Ritter et al. 2019).

Water infiltrating inland will pile up and create the necessary gradient to flow gently toward the coast (base level). Because of the density difference (freshwater–seawater density ratio of 40/41), the freshwater body will float upon the denser brackish and salt water. For every centimeter of freshwater piling up onto the seawater below, the landward dipping freshwater–seawater boundary will be 40 cm lower. The freshwater body wedges toward the coast and the freshwater–seawater interface becomes shallower. At the discharge boundary associated with the coastline, the vadose–phreatic and freshwater–seawater interfaces converge, and the discharging freshwater flows faster as the area of the section decreases. These increased flow velocities bring in and out reactants faster than in any other area of the lens. Moreover, both vadose–phreatic and freshwater–seawater mixing zones overlap, causing this area to be the most undersaturated volume of water. This seaward edge of the freshwater lens is also the most influenced by tides, where mixing between different waters is greatest. Especially on small islands and in young porous limestones, such as in the Caribbean and western Pacific, where limestones are often less than a few million years old and inland recharge is not too high, cave formation especially occurs in this coastal fringe, at the tapering edge of the freshwater lens (Mylroie and Mylroie 2013a; Mylroie 2019). This explains why these coastal mixing caves are called "flank margin caves" (Mylroie and Carew 1990) (Figure 11.28).

Typical flank margin caves on such small tropical islands made up of eogenetic limestones are formed by slow diffuse flow, so they lack the characteristic morphologies of turbulent waters (i.e., scallops, allogenic "river" sediments). They are instead characterized by isolated chambers or irregular spongework or ramiform patterns, in which the chambers and passages connecting them have width to height ratios of $\geq 10:1$ (Figure 11.29a–c). Typical meso-morphologies include curvilinear and smooth walls (Figure 11.29d), isolated rock pillars (spongework) (Figure 11.29e), globular and interpenetrated rooms with maze-like areas, and passages tapering out away from to the coast. Cave passages can penetrate deeper into the landmass only where important freshwater flows are present. If freshwater flow is subdued also these fracture-guided passages tend to taper out rapidly. Enlargement occurs mainly along the mixing zones, which causes the voids to extend

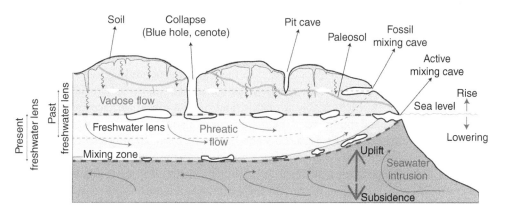

Figure 11.28 Schematic representation of a coastal mixing zone area. *Source:* Based on Mylroie and Carew (1990).

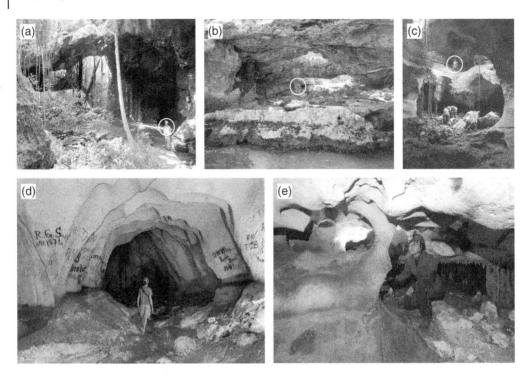

Figure 11.29 Pictures of flank margin cave morphologies: (a) The large partially unroofed chambers of Indian Cave, (b) Boatyard Cave, and (c) Sonny's Cave, Bahamas. *Source:* Photos by Michael Lace. (d) The large cave passage with undulating walls and roof in Hatchet Bay Cave, Eleuthera, Bahamas. *Source:* Photo by Arthur Palmer. (e) Spongework morphologies in Miocene limestones in Pellegrino Cave, Siracusa, Sicily. *Source:* Photo by Marco Vattano.

laterally, and not vertically, with a decreasing trend inland, away from the coastline. The position of the mixing zone is the main factor controlling cave genesis, and the solutional voids cut across bedding planes and fractures. Clearly bedding- or fracture-controlled passages do develop when these are feeders of freshwater into the mixing zone. This especially occurs in more lithified and less porous limestones, where cave passages are often elongated along these permeable pathways (Figure 11.34c). As permeability increases, the mixing and dissolution front extends further inland, expanding mainly where larger (mainly horizontal) freshwater recharge occurs. This gives these caves the typical spongework pattern, in which passage size and density decreases away from the mixing front (i.e., coastline at the time of formation).

The longer a mixing zone stays at a specific position, the higher the permeability and the thinner the freshwater lens become, and the more precisely flank margin caves become reliable sea-level markers (Mylroie and Carew 1988). Stable sea levels (and mixing zones) also account for larger volumes of dissolved rock, and thus larger flank margin caves. The cave with the longest development (almost 20 km) is Sistema del Faro (Isla de Mona, Puerto Rico), apparently resulting from the connection between three previously separated caves; from north to south, Cueva Losetas, Cueva al lado del Faro, and Cueva del Lirio (Lace et al. 2016) (Figure 11.30a). This cave, now located 40 m above sea level and opened through over 40 perched entrances on a coastal cliff, is believed to be almost 2 million years old. This age is in agreement with the large size of the cave and its interconnected chambers, formed over long periods of stable mixing zone position, before the onset of Quaternary glaciations and related sea-level oscillations. The cave was preserved from complete

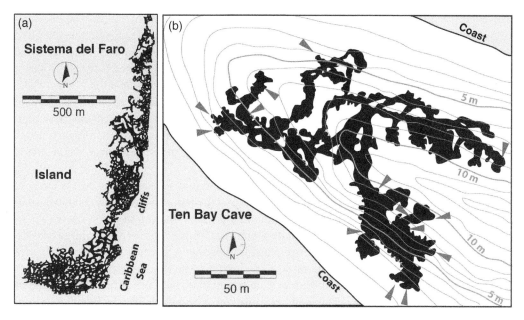

Figure 11.30 Surveys of flank margin caves: (a) Sistema del Faro (Isla de Mona, Puerto Rico), the longest flank margin cave in the world formed during early Pleistocene times and now opening high on a coastal cliff. *Source:* Survey from Marc Ohms, Isla de Mona Project, National Speleological Society. (b) Ten Bay Cave (Eleuthera, Bahamas) shown on the topographic map, illustrating its development around 125 kyr ago on both flanks of a fossil coastal dune, along the margin of the freshwater lens. *Source:* From Mylroie and Mylroie (2013a).

erosion because of tectonic uplift that placed it well beyond the influence of waves and tides. What remains of this cave is probably only part of a much larger system, which now develops parallel to the coastline and penetrates only around 300 m inland.

Some of these old flank margin caves can be found far from the present-day coastline, such as in the 11 km long and 1 km wide strip of Pleistocene dune ridges of Naracoorte (South Australia). This area hosts over 150 flank margin caves at relative heights above the present sea level between 60 and 70 m, formed in the freshwater lens margin between 1.1 and 0.8 Myr ago (White and Webb 2015). Regional tectonic uplift has brought these caves at higher altitudes, causing a seaward displacement of the coastline of over 80 km.

Active flank margin caves often have no direct connection to the coast, but are intercepted by later surface erosion or cliff retreat. A good example of this is Ten Bay Cave, on Eleuthera Island, Bahamas (Figure 11.30b). This cave developed during the MIS5e highstand on both sides of a fossil dune, with cave branches converging in the nose of the dune. It is an extremely shallow cave, with several entrances breaching the cave roof, but apparently, no lateral connection to the coast at the time of its formation (Mylroie and Mylroie 2013a).

From the two examples above, it is clear that flank margin cave development is greatly influenced by sea level changes, which determine the position and stability of the freshwater lens. In recent geological times (i.e., Quaternary), both global glacioeustatic changes and local tectonic activity have caused sea level to change position, vertically shifting the main locus of dissolution in coastal areas (Figure 7.9). The typically slow variations of sea level related to tectonics cause the mixing solution voids to increase their height and thus decrease their width to height ratios. On the other hand, glacioeustatic changes are too rapid to allow flank margin caves to form. Only some relatively stable periods during highstands and lowstands allowed sufficient time for sizeable flank margin caves to form. In the Bahamas, which are generally believed to be passively subsiding at a rate of 0.01–0.02 mm yr^{-1}, most flank margin caves have formed during the last interglacial (ca. 125 kyr) in a single cave-forming phase, and some submerged caves are located between depths of 105 and 125 m below mean present sea level, ascribed to the Last Glacial Maximum (LGM). There are, however, several mixing caves with wavy roofs at heights above the present mean sea level of up to 24 m (Mylroie et al. 2020), which might indicate that some areas of the Bahamas are uplifting, or, as some ages from speleothems seem to indicate (ca. 260 kyr in Conch Bar Cave, Caicos), that flank margin cave development has occurred also during highstands earlier than MIS5e (Smart et al. 2008).

Blue holes in the Bahamas provide access to some long and well-developed linear caves at depths of around 20 m below present sea level. Whereas, the Bahamas today are a collection of rather small islands distributed along the Bahama Banks, a 20-m drop of the sea level would cause the emergence of most of this carbonate platform, forming a series of very large islands. The area of emerged land would experience a quadratic growth, whereas the island perimeter would increase only linearly, so that the surface-to-coastline ratio would rise rapidly, and higher inland recharge might have caused turbulent flow to dominate over the diffuse flow typical of flank margin cave development (Vacher and Mylroie 2002). This explains why long linear integrated networks of conduits have formed in the Bahamas during the long periods in which most of the Bahamas Banks were exposed to subaerial conditions. The Santa Catalina Cave near Matanzas (Cuba), and other nearby cave systems (Bellamar, El Jarrito) are all flank margin caves with considerable extent, suggesting long periods of sea-level stability and important mixing corrosion associated with considerable amounts of freshwater coming from their large recharge areas (De Waele et al. 2017b). Speleothems in Santa Catalina Cave demonstrate coastal mixing to have occurred since MIS11, with alternating periods of flank margin cave development (when sea levels were high) and speleothem formation (during sea-level lowstands) (De Waele et al. 2018a).

These same conditions are encountered in the large coastal carbonate areas of Yucatán and Florida, with the difference that these are not large islands, but are extensive coastal eogenetic carbonate aquifers. These areas are characterized by vast underwater cave systems, and especially in Yucatán by the presence of a large number of cenotes, which are often karst windows allowing access to the underlying integrated underground drainage network. An interesting feature of this area is the presence of the Chicxulub impact crater, which has influenced the local hydrogeology and is at the origin of the so-called "ring of cenotes" (Perry et al. 1995). Speleogenesis in these areas is largely determined by the combination of important inland recharge in a tropical climate, the highly porous nature of the carbonate rocks, structural elements (regional faults and fracture trends) and lithological variations, and coastal mixing corrosion (Back et al. 1986; Smart et al. 2006). The aquifer is characterized by a density and salinity stratification of fresh, brackish, and salty waters, large inflow of continental recharge, localized inputs of organic matter (through cenotes), and more intense mixing corrosion effects in the coastal areas (Kambesis and Coke 2013; Coke 2019).

In Yucatán State, to the north of the peninsula, many of the large bedrock-collapse sinkholes, knowns as cenotes (from the Maya word "dzonot"), are deep, flooded and often bell-shaped shafts that rarely allow access to long horizontal passages. Most cenotes become very wide and reach great depths, and are thus technically very challenging to explore. The freshwater–seawater boundary is often found at −70 m depth from the cenote surface, and below this depth, salt water is often anoxic and characterized by a hydrogen sulfide cloud. The deepest of these shafts is Cenote Sabak

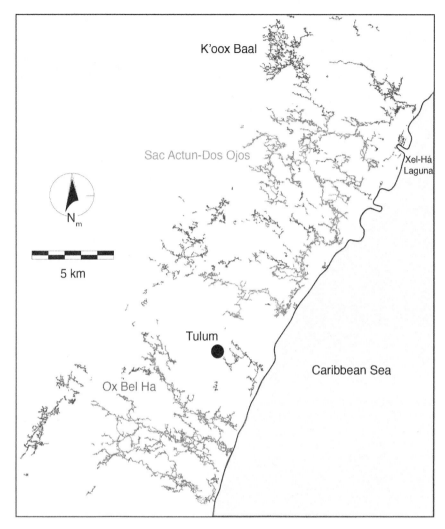

Figure 11.31 The coastal karst area around Tulum (Quintana Roo, Mexico) as for the situation in 2018. *Source:* Coke (2019)/With permission of Elsevier.

Ha (−165 m), meaning "smoke water" in Yucatec Maya language, the bottom of which has not yet been reached (Coke 2019).

The best-studied examples of extensive coastal karst systems are located in the regions of Tulum and Xel-Ha, Quintana Roo (Mexico) (Figure 11.31). This area contains the longest caves in Mexico, Sac Actun-Dos Ojos (the first meaning "White Cave" in Yucatán Maya), around 377 km long, mostly underwater, and still under intense exploration, Ox Bel Ha ("Three Paths of Water" in Yucatán Maya), "only" 318 km long but entirely underwater, and K'oox Baal, over 103 km long and located north and in a more inland position with respect to the other systems (Figures 11.31 and 11.32). Sac Actun-Dos Ojos System, north of Tulum, extends along a 13-km long strip of limestone associated with the coast, forming an intricate maze network that penetrates inland for over 7 km. It has a distinct flank margin cave imprint in the coastal fringe (<1 km from the Caribbean) developed in Pleistocene porous limestones, grading into a more fracture-controlled pattern in the older and more lithified limestones situated between 1 and 7 km from the coastline. The Ox Bel Ha

Figure 11.32 Pictures of the third longest cave system in Quintana Roo, K'oox Baal, over 103 km long: (a) Most passages are characterized by very wide but low galleries, often with submerged speleothems. (b) In still areas, the halocline is clearly visible. (c) Stalactites and flowstones reveal that these caves formed when the sea level was at a lower relative position. (d) Cave diving requires severe safety protocols and intensive training. Note the fretted rock surfaces due to mixing corrosion. *Source:* Photos by Radoslav Husák.

System has a distinctively more inland character, lacking the extensive mixing cave maze network that characterizes Sac Actun-Dos Ojos. All caves tend to penetrate inland for no more than 12 km, where they terminate on the Holbox Fracture Zone, which appears to be the main recharge area of these exceptionally long underwater caves (Kambesis and Coke 2013).

At least five main cave levels have been identified in this region: around 6 m a.p.s.l., at present sea level, and at depths of 10, 20, and 90–110 m b.p.s.l. The last level is related to glacial sea-level lowstands and has only rarely been explored because of technical difficulties. Most caves develop at rather shallow depth below the surface. Two main passage shapes can be identified: (i) wide and low elliptical tubes with large lateral extent (Figure 11.32a) and (ii) narrow but high vertical fissure passages which extend over shorter distances. Whereas speleothems are rather scarce in the coastal maze networks where mixing corrosion is very active (Figure 11.32d), the more inland submerged passages can be profusely decorated (Figure 11.32a, c), showing these passages to have been above sea level (during colder periods) over rather long time spans, to allow such large speleothems to form.

The substantial dimensions of the integrated cave systems in Quintana Roo can be attributed to: (i) large freshwater recharge from inland, (ii) the presence of stable coastal discharge areas, (iii) an important mixing zone, (iv) influx of organic matter from the multiple cenotes as an additional source of CO_2, and (v) the Pleistocene sea-level fluctuations which caused the mixing dissolution zone to migrate vertically and laterally.

Flank margin caves also occur in non-tropical areas, such as in the rather porous Late Miocene calcarenites of Mallorca (Spain). Karst on the southern and eastern coast (Migjorn) of this Mediterranean area is hybrid, representing a good example of the influence of mixing corrosion, important inland recharge, and the lithological and structural characteristics of the hostrock. Caves in Mallorca show variable mixtures of patterns with large collapse chambers connected by short passages (e.g., Cova Genovesa, Figure 11.33b), ramiform and spongework patterns, and fracture-guided elongated mazes. Large cave systems such as the Cova des Pas de Vallgornera straddle different limestone facies, with ramiform and spongework pattern in the reef front, where the porous limestone is dominated by large fragments of corals (Figure 11.34a, b), and elongated maze passages in the more compact back reef lagoonal facies (Figures 11.33a and 11.34c) (Ginés et al. 2013, 2014).

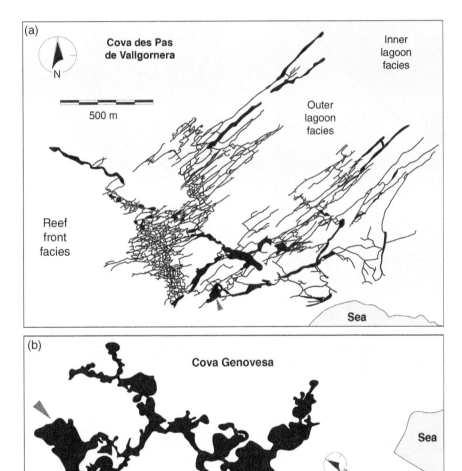

Figure 11.33 Plan view of two typical mixing caves in Mallorca Island, Spain: (a) Cova des Pas de Vallgornera, with different cave patterns in the reef front and back reef lagoonal facies. *Source:* Reproduced from Ginés et al. (2014). (b) Cova Genovesa formed entirely in reef front facies and characterized by large breakdown chambers connected by short passages.

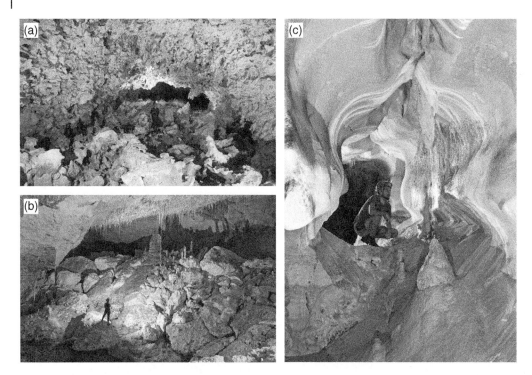

Figure 11.34 Cova des Pas de Vallgornera, Mallorca (Spain): (a) Passage in the highly porous reef front facies, without any structural control. Solutional spongework features, produced by coastal mixing corrosion, cover the whole cross-section. (b) Main breakdown chamber developed in the reef front facies. (c) Wavy mixing corrosion notches in a fracture-controlled passage carved in the less porous and compact back reef facies. *Source:* All photos by Tony Merino.

Coastal mixing corrosion also occurs in older and less permeable telogenetic limestones, but its effects become predominant over those related to normal turbulent flow especially where the rocks are characterized by high secondary permeability. On the west coast of the Island of Cres (Croatia) some small flank margin caves have formed in an Upper Cretaceous limestone breccia, whereas the adjacent non-brecciated limestone with much lower porosity lacks evidence of flank margin cave development (Otoničar et al. 2010). In New Zealand, small flank margin caves tens of meters long have locally formed in densely fractured telogenetic Oligocene and Pliocene limestones in both the North and South Island (Mylroie et al. 2008). A detailed study on Fico Cave in Sardinia (Italy), hosted in the massive Jurassic limestones of the Gulf of Orosei, a location known for its extensive karst systems, revealed passages developed along major extensional fractures parallel to the coast, with clear morphologies of slow diffuse flow in a coastal mixing environment. The cave shows a series of clearly distinct levels related to past sea level highstands, recording the uplift experienced by the coast during the Quaternary (D'Angeli et al. 2015b). An even more striking example of old flank margin caves in telogenetic limestones that have allowed the reconstruction of Quaternary coastline changes and uplift has been reported from San Vito Lo Capo-Custonaci (west of Palermo, Sicily, Italy). Rumena Cave, now located at around 100 m a.p.s.l., and the nearby Fantasma Cave lying at 70 m a.p.s.l., show clear evidence of coastal mixing processes in a diffuse flow setting. Dating of a stalagmite in the highest cave indicated that these systems have formed over the last 1.2 Myr, displaced to their present position by gradual uplift (Ruggieri and De Waele 2014; Stocchi et al. 2017).

11.5 Caves in Non-Carbonate Rocks

11.5.1 Halite Caves

Halite is at least three orders of magnitude more soluble than calcite (ca. $424 \, \text{g L}^{-1}$ against $0.3 \, \text{g L}^{-1}$ for calcite in water with P_{CO_2} of 0.05 atm at 25 °C (mass of solute in 1L of solution); Table 3.1). Therefore, rock salt rarely crops out at the surface because it is readily dissolved leaving an insoluble residue, mainly consisting of clays and marls. Salt layers do occur in numerous sedimentary successions around the world and their subsurface dissolution leads to a number of subsidence features and hydrochemical effects described in Chapter 8.

Although salt karst is widely documented in many countries, caves have rarely been explored and reported. Where extensive outcrops are lacking, salt mining can allow the discovery of natural dissolution voids at depth, such as in the famous Wieliczka mine in Poland (Przybyto 2000), or cause surface waters to flow in contact with the salt bedrock, as in Cardona diapir and mine, Spain (Lucha et al. 2008a) (Figure 11.35a). Rock salt can also be exposed in temperate areas by landslides in mountainous regions or along the flanks of rapidly evolving rivers, such as in Romania. In such conditions, salt caves and associated surface landforms develop very rapidly and change continuously (Povara et al. 1982).

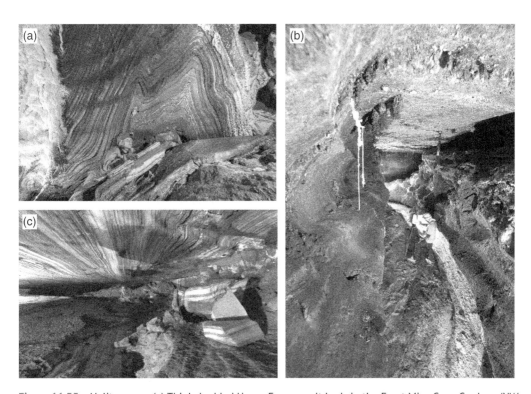

Figure 11.35 Halite caves: (a) Thinly bedded Upper Eocene salt beds in the Forat Mico Cave, Cardona (NW Spain). *Source:* Photo by Victor Ferrer. (b) Canyon passage with several lateral notches in Lechuza del Campanario Cave, Atacama (Chile). *Source:* Photo by Vittorio Crobu, La Venta Esplorazioni Geografiche. (c) The colored and well layered Precambrian Hormuz salt exposed on roof and floor of the Namakdan (3 N) Cave in Qeshm Island, Iran. *Source:* Photo by Marek Audy and Richard Bouda, Project NAMAK.

In normal conditions, rock salt can only survive at the surface where climate is extremely arid. Often outcrops are rare and of limited extent, mainly related to rising salt diapirs that have pierced the overburden rocks (see Section 8.4). Caves reaching several kilometers of underground development can form in rock salt areas, especially in rising extrusions. Important salt karst areas hosting significant cave systems are known for Mount Sedom (Israel) (Frumkin 1994a, 1994b), the Zagros Mountains (Iran) (Bosák et al. 1999), in the Cordillera de la Sal close to San Pedro de Atacama in Chile (De Waele et al. 2009b, 2020), and in the salt domes of Khodja-Mumyn and Khodja-Sartis near Kuljab in Tajikistan (Dzens-Litovsky 1966; Klimchouk 2004a).

Mount Sedom is the exposed upper part of an elongated (11 km long and 1 km wide) salt wall located along the southwestern edge of the Dead Sea (Israel), and rising 280 m above the lake level (as of 2020). The salt is of Mio-Pliocene age piercing through younger lake evaporites and clastic sediments. Dissolution of the salt when it was submerged by the pluvial Lisan Lake at ca. 14 kyr produced a horizontal dissolution surface (salt table) overlain by an anhydrite-rich caprock that reaches 50 m in thickness and forms the substrate for small allogenic streams. The annual rainfall of around $50 \, mm \, yr^{-1}$ allows cave development, especially through flood events, some of which are capable of dissolving as much as 2 cm of halite in the active cave floors in a single event (Frumkin and Ford 1995). Around 120 caves are known from this area with a total development of around 24 km, including the longest salt cave in the world, Malham Cave, with 10.2 km of accessible underground passages (Kutleša et al. 2019) (Figure 11.36a). Most caves show branchwork patterns, with multiple steep vadose inlets piercing through the thick caprock and rapidly reaching a more or less horizontal cave floor active during rain events. Slowly infiltrating waters reach saturation, but fast flowing floodwaters often remain undersaturated over long distances. It is usual to see waters entering stream sinks to increase their dissolved content from $10 \, g \, L^{-1}$ to around $300 \, g \, L^{-1}$ at the springs. Many salt caves have a meandering pattern (Figure 11.36) and show deep erosional/dissolutional notches at different heights above the present cave floor (Figure 11.35b).

Extensive caves with similar internal morphology have been documented in both Iranian and Chilean salt outcrops (Figure 11.35b). The longest cave in Iran is 3 N (Namakdan) Cave, with its 6.5 km length (Figures 11.35c and 11.36b). In the Cordillera de la Sal (Chile), the longest caves belong to the Cressi cave system, which comprises three meandering cave segments separated by short skylight canyon passages, all together around 4.5 km long (Figure 11.36c). Other long caves in this area are Cueva del Aire (2.2 km) and Cueva de l'Arco de la Paciencia (1.95 km, Figure 11.36d). All the world's known halite caves are Holocene in age. The oldest vegetation remains found in caves in Mount Sedom, Atacama, and Namakdan are 7000, 4000, and 2500 years old respectively (Frumkin et al. 1991; Bruthans et al. 2010; De Waele et al. 2020). Older cave levels are rapidly destroyed by surface denudation, collapse due to the poor tensile strength of halite (half that of limestones), and their infilling by detrital sediments. Salt caves can be extraordinarily beautiful, with ephemeral salt stalactites and contorted speleothems (Filippi et al. 2011; De Waele et al. 2017a) (Figure 9.15).

11.5.2 Gypsum–Anhydrite Caves

Gypsum and anhydrite are much more soluble in pure water than any of the main carbonate minerals (calcite, aragonite, and dolomite), and much less soluble than halite; at 25 °C gypsum and halite have solubilities of $2.6 \, g \, L^{-1}$ and $424 \, g \, L^{-1}$, respectively, against a solubility for calcite of $0.3 \, g \, L^{-1}$ in water with atmospheric P_{CO_2} of $0.005 \, g \, L^{-1}$ (mass of solute in 1 L of solution); Table 3.1). The solubility of calcite and other carbonate minerals has a direct relationship with the CO_2 content in the water, whereas that of gypsum, anhydrite, and halite remain unaltered. In general,

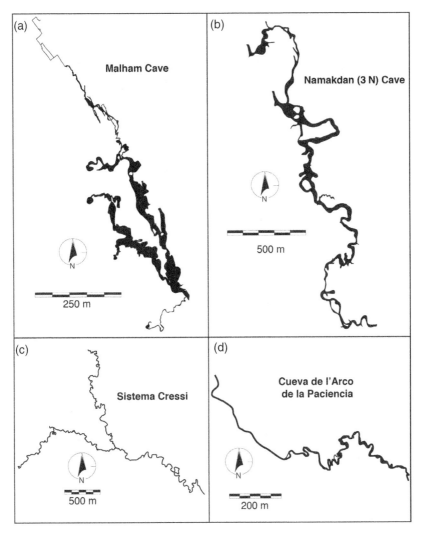

Figure 11.36 Maps of halite caves: (a) Malham Cave, Mount Sedom, the longest salt cave in the world. *Source:* Modified from Kutleša et al. (2019). (b) Namakdan (3 N) Cave, Iran. *Source:* Modified from Bruthans et al. (2010). (c) Cressi Cave System, and (d) Cueva de l'Arco de la Paciencia, both in the Cordillera de la Sal in Atacama, Chile. *Source:* Courtesy of Commissione Grotte Eugenio Boegan Trieste).

under common meteoric conditions, gypsum is around 8–50 times more soluble than calcite (i.e., water with the P_{CO_2} of the atmosphere and CO_2-rich soils, respectively). The solubility of anhydrite at shallow meteoric conditions is not well constrained because this mineral is unstable at surface conditions and in the presence of liquid water dissolution is preceded by conversion into gypsum (Klimchouk 1996a). Anhydrite is the more stable calcium sulfate phase at temperatures above 60 °C and at high burial depths (Figure 2.22) (see Section 2.3.1). Most of the gypsum formations, which have been buried at high depths and were later brought close to the surface during telogenesis, are secondary, derived from the hydration of anhydrite because of the interaction with meteoric water in shallow geological environments. It is generally believed that the anhydrite–gypsum conversion involves an increase in volume (anhydrite is denser than gypsum), especially in shallow (open-system) environments. If this were true, this volume increase would seal fissures in the sulfate rocks, thus inhibiting water circulation and karstification (Chiesi et al. 2010), but this concept is still controversial and may be true only in certain conditions (Klimchouk 2000a). On the contrary, the gypsum-to-anhydrite conversion in deep settings releases water, increasing the hydraulic pressure in the surrounding rocks and potentially causing deformation (see Section 2.3.1).

The solubility of gypsum (or anhydrite), especially in deep settings, can increase substantially by (i) the presence of other salts, which decreases the activity of calcium and sulfate ions (ionic strength effect, see Section 3.7.3), (ii) anaerobic reduction of sulfates in presence of organic matter, which consumes sulfates, (iii) dedolomitization, which decreases Ca ions and increases Mg ions in the solution, and (iv) lithostatic pressure increase.

Of special interest from the speleogenetic perspective is the solubility and dissolution kinetics of gypsum in shallow (epigenic) conditions, since this is the most common situation for the formation of caves in evaporites. The kinetics of gypsum dissolution is slower from that occurring in carbonates. Carbonate dissolution at normal pH conditions is mainly controlled by the relative slow reaction at the rock–water interface, which is the rate-controlling factor rather than the more rapid transfer of ions to the bulk solution through the diffusion boundary layer (surface-controlled kinetics). In other words, the reactions occurring between water, CO_2, and the carbonate minerals are the rate-limiting factors. In contrast, gypsum is characterized by a mixed kinetics, in which the rate-limiting step varies depending on the saturation state of the solution (see Section 3.8.2). Gypsum rapidly dissociates into the water and the rate-limiting factor is diffusional transport through the boundary layer (transport-controlled kinetics). However, close to equilibrium (saturation ratio ~ 0.9), the surface-reaction becomes slower than diffusional transport and the regime changes from transport- to surface-controlled. In the transport-controlled stage, which is responsible for most of the solutional work, the thickness of the diffusion boundary layer is an important factor; the thinner it becomes the faster gypsum is dissolved. The thickness of the boundary layer mainly depends on flow velocity and regime, with a linear dependence between dissolution rate and flow velocity in laminar flow conditions, and a disruption of the boundary layer in turbulent flow conditions, accompanied by a significant increase in dissolution rate. As a consequence, caves in gypsum preferably form where conditions allow water to flow fast, as occurs in presence of steep hydraulic gradients, relatively large initial fissure widths, and limited production of insoluble material that tends to clog solutional openings. A distinction needs to be made between epigenic (unconfined) conditions, where water flow is fast and over long distances, and the confined conditions typical of artesian (basal injection) speleogenesis, where flow is much slower and over shorter lengths.

In epigenic settings, the direct relationship between flow velocity and dissolution rate causes selection of more favorable flow paths to be much more pronounced in gypsum than in limestones. Initially, larger flow paths enlarge much faster than the tinier ones, reaching breakthrough very rapidly, and once it is achieved this positive feedback becomes even more pronounced. Therefore, in epigenic gypsum karst settings, where flow velocities can be high, gypsum caves tend to be characterized by simple (linear) cave patterns connecting recharge points (sinking streams, large dolines) to discharge points (springs) (Klimchouk 2019a). Significant dissolution occurs where the water initially comes in contact with the gypsum. In the vadose (percolation) zone, water entering the few open vertical fissures rapidly enlarges them, forming vertical shafts. The transmission of water flow to the base level (e.g., the spring in the nearest valley) mostly occurs close to the water table, since this is where flow reaches the greatest flow velocities. This explains why phreatic loops are very rare, and if they occur they are typically very shallow and related to local structural conditions. At the water table, most conduit enlargement occurs during high flow events (i.e., floods) through a combination of dissolution by rapidly flowing floodwaters that can maintain undersaturated conditions over longer distances, and mechanical erosion boosted by entrained sediment particles and favored by the high erodibility of the gypsum. The dimensions of the few "victor" conduits generally reflect the maximum discharge of the flood events that created them, but other processes such as collapses do not allow these conduits to survive for long times (Pisani et al. 2019).

Since evaporite successions often contain alternations of gypsum and insoluble beds (i.e., marls), or are underlain by fine-grained clastic sediments (i.e., claystones), cave streams often follow these lithological contacts. A good example of this type of contact karst is Covadura Cave (Sorbas, Spain) composed of six cave levels clearly following the gently dipping lithological contacts between Messinian gypsum beds and intervening marly interbeds. At first, speleogenesis appears to have occurred under phreatic conditions with water dissolving small sinuous conduits at the base of the gypsum beds in contact with marls. These protoconduits are still visible on the ceiling of the main cave passages. At a later stage, when the base level and the water table gradually dropped in response to fluvial downcutting, subsurface mechanical erosion in the marly interbeds created the large accessible passages with flat-topped triangular cross-section (Calaforra 1998; Calaforra and Pulido-Bosch 2003; Calaforra and Gázquez 2017) (Figures 3.48 and 11.38c). Although initial studies attributed the development of the phreatic protoconduits to a single phase, the different levels might well be related to successive stages controlled by climate within a general context of tectonic uplift, as support the stepped sequences of terraces developed by the fluvial systems in the area (Harvey et al. 2014). Probably, episodic fluvial entrenchment was accompanied by the alternation of short epiphreatic episodes followed by major subsurface erosional phases. The creation of protoconduits can better be explained by epiphreatic flow close to the water table during short flood pulses, probably with intervening phases of antigravitative erosion. An increase of hydraulic gradient and flow rate in these proto-conduits then causes mechanical erosion to prevail over dissolution, thus removing the easily erodible marls more efficiently. These successive phases of epiphreatic flow in protoconduits and mechanical erosion of the marls would thus have occurred in successive periods during the Pleistocene. The different cave levels are now connected by vadose shafts carved along major vertical discontinuities and generally fed by large dolines at the surface.

A similar mechanism also explains the multilevel epigenic gypsum cave systems in Emilia Romagna region (northern Apennines, Italy), and Sicily (De Waele et al. 2017c). The best example of such cave systems is Re Tiberio in the Vena del Gesso Regional Park (northern Apennines), carved during the last 130 kyr, with cave level formation mainly occurring at the end of cold periods, and very limited speleogenesis and some deposition of calcite speleothems during warmer episodes (Columbu et al. 2015). This climate-controlled cave evolution, coupled with episodic fluvial entrenchment, river terrace formation, and paragenetic episodes in the previously carved cave levels has contributed to the understanding of landscape evolution in this region of the Apennines (Columbu et al. 2017).

Karstification can occur in a single phase in many epigenic gypsum cave systems, with water entering the soluble evaporite rock, reaching base level quickly, and exiting at the spring. These through-flow systems are sometimes characterized by an underground river flowing on an insoluble (or less soluble) and less permeable underlying unit (e.g., carbonates, marls, and shales). These single-level caves, or the tiers of multilevel caves, generally show a rather simple plan view, with linear and poorly branching patterns often strongly influenced by stratigraphic and/or structural elements (i.e., bedding planes or major fractures). D.C. Jester Cave in Greer County, Oklahoma (USA) is the longest gypsum cave in America, reaching 10 065 m in length (Figure 11.37a). This cave is developed in a 5-m thick gypsum bed in the Permian Blaine Formation and consists of angular passages, straight sections, and sinuous portions controlled by several sets of joints or

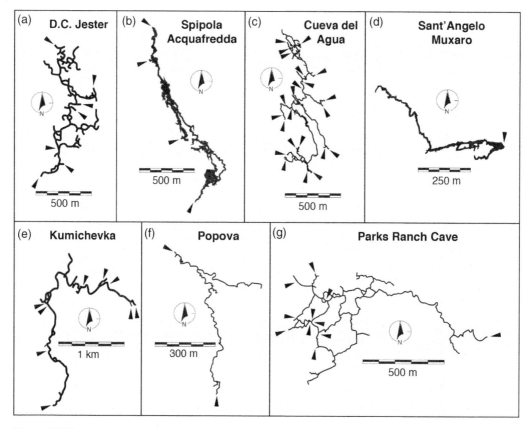

Figure 11.37 Maps of epigenic gypsum caves: (a) D.C. Jester Cave, Oklahoma (USA), the longest epigenic gypsum cave system of the United States. (b) Spipola-Acquafredda cave system, Bologna (Italy); (c) Cueva del Agua, Sorbas (Spain). (d) Sant'Angelo Muxaro Cave, Sicily (Italy). (e) Kumichevka-Vizborovskaya Cave, Perm (Russia). (f) Popova Cave, northern Caucasus (Russia). (g) Parks Ranch Cave, New Mexico (USA).

fractures. Most passages have a flattened elliptical cross-section generally three times wider than high (Bozeman and Bozeman 2002; Johnson 2018).

Many large epigenic cave systems are known from the Messinian gypsum areas of Italy, especially those of the northern Apennines, Calabria, and Sicily (De Waele et al. 2017c). The longest of these is the Spipola-Acquafredda cave system (ca. 11 km, Bologna, northern Italy) (Figure 11.37b), carved in steeply dipping gypsum beds and following the strike of the strata, with some sections following down-dip or along NW-SE-oriented joints (Pisani et al. 2019). This cave can be explored from the upstream sink (Acquafredda), almost to the spring, and has a river flowing through it. Up to five horizontal storeys can be recognized, corresponding to ancient base levels that controlled the position of the underground river. Grave Grubbo-Vallone Cufalo (Calabria, southern Italy) is 2830 m long and connects an active sinking stream to the spring (Parise and Trocino 2005). It consists of a rather simple linear passage following the bedding and some joint systems. In Sicily, the longest gypsum caves are Santa Ninfa (1500 m), Monte Conca (2400 m), Sant'Angelo Muxaro (1760 m) (Figures 11.37d and 11.38a), and Rocca Entella (700 m). The first three are active caves in which an underground river can be followed over long distances, whereas Rocca Entella, comprising three distinct levels following the strike of the gypsum beds, was carved by a river that is now completely dry. All these caves are characterized by rather simple linear passages guided by the bedding planes and some local joint systems. Monte Conca cave system presents the most complex pattern, comprising two main cave levels and some intermediate ones, showing how the cave system rapidly adapted to the changing boundary conditions at the surface (Madonia and Vattano 2011) (Figure 11.38b).

The Messinian plateau of Sorbas is the karst area with the largest density of caves in Spain. Some caves, like Covadura, are developed along six levels and have a complex geometrical pattern, but others, such as the 8.9 km long Cueva del Agua (longest gypsum cave in Spain) are shallow linear branchworks with a large number of entrances (dolines), and the main branches connecting large dolines to the underground river (Figures 11.37c and 11.38c, d).

Figure 11.38 Examples of epigenic gypsum caves: (a) River passage in the microcrystalline Messinian gypsum of Sant'Angelo Muxaro Cave, Sicily. *Source:* Photo by Marco Vattano. (b) Cupolas along the main branch in Monte Conca Cave System, Sicily. *Source:* Photo by Marco Vattano. (c) Flat-topped triangular passage carved in the marly interbed of GEP Cave, Sorbas (southern Spain). Note the protoconduit in the gypsum roof. *Source:* Photo by Victor Ferrer. (d) Intersection between an enlarged vertical fracture and a marly interbed in Cueva del Tesoro, Sorbas. *Source:* Photo by Victor Ferrer. (e) An active passage in the Permian microcrystalline gypsum of Parks Ranch Cave, New Mexico. *Source:* Photo by Lukas Plan.

In Russia, vast gypsum outcrops occur in older (Permian and Jurassic) evaporite successions, often interstratified with dolostones. The most important of these gypsum regions are located in Perm (e.g., Pinega Karst). Kulogorskaya-Troya (17.5 km long), Ordinskaya, and Kungurskaya caves (Figure 11.15) have much more complex cave patterns, displaying mazes probably related to an early artesian speleogenetic phase, later overprinted by floodwaters from the nearby rivers (Kungur and Pinega). On the other hand, the Kumichevka-Vizborovskaya cave system, 4500 m long, is a simple linear cave system carved by an underground stream (Figure 11.37e). In the northern Caucasus, several simple linear gypsum cave systems with underground rivers, some of which longer than 2 km, have been explored in Jurassic evaporite beds. Pshashe-Setenay (2700 m) and Popova (2000 m) are the most important ones (Figure 11.37f).

Other important gypsum caves are described from the Gypsum Plains in New Mexico (USA). Here, the 6600 m long Parks Ranch Cave is the most important one, showing a relatively complex anastomotic pattern (Figures 11.37g and 11.38e). This cave is traversed by several streams and is clearly adapted to the present-day geomorphology and climate, but its anastomotic character and several lines of evidence found in the system suggest that it was initially a hypogenic (artesian) gypsum cave, subsequently exploited by infiltrating waters from several dolines and the nearby ephemeral Black River (Stafford et al. 2008b).

11.5.3 Quartzite Caves

Caves can form in a variety of quartz-rich lithologies, including quartzites, quartz sandstones, orthoquartzites, and sandstones with a wide range of metamorphic grades (Wray and Sauro 2017). In this chapter, we do not distinguish between these rock types, since speleogenetic processes are essentially the same. All these rocks are generally composed of quartz grains with an interlocked texture of particles overgrown or held together by an amorphous silica cement intermixed with variable amounts of phyllosilicates (pyrophyllite, kaolinite, biotite, etc.) and iron–aluminum hydroxides. Speleogenetic processes in sandstones with carbonate cement are more similar to those occurring in any carbonate rock. Dissolution of the carbonate cements loosens the quartz grains, which are washed away. The longest cave in quartz-rich rocks, Krem Puri in India (25 km), has formed in this way (Sauro et al. 2020).

Quartz is a very poorly soluble mineral (6–10 mg L^{-1} at standard temperatures), whereas the amorphous hydrated silica form ($SiO_2.nH_2O$) is over 10 times more soluble (over 100 g L^{-1} in comparable conditions) (Table 3.1). Both increase their solubility at higher temperatures and pH (see Section 3.6 and Figure 3.23). At the typical temperatures of the high quartzite mountains of Venezuela (16–20 °C) or in the lower lying more arid climate of Central Brazil (20–26 °C), surface waters are often rich in a variety of organic acids, and consequently are slightly acidic (pH 4–5). In such natural conditions, because of the slow kinetics of quartz dissolution, the development of large cave systems requires protracted geological-scale time spans. The slow dissolution of quartz by meteoric waters in near surface conditions explains why these rocks generally form resistant mountains (known as "tepuis" in the Guyana shield between Venezuela and Brazil) that stand out over a planated landscape, with surface denudation rates as low as 1 mm kyr^{-1}. On the contrary, the long residence times of infiltration waters flowing along fractures and bedding planes in these rocks allows them to reach saturation in dissolved silica, as shown by chemical analysis of drip waters in quartzite caves in Venezuela (Mecchia et al. 2014). This slow but continuous process of underground dissolution along discontinuity planes contributes to reduce the intergranular cohesion, a process known as arenization (Martini 2004). The loosened quartz grains are then removed by subsurface mechanical erosion (Mecchia et al. 2019). In quartz-rich rocks with other minerals such as phyllosilicates (e.g., kaolinite, biotite), their weathering can also create the necessary permeability and weakening effect, loosening the quartz grains involved in the arenization process (Auler and Sauro 2019).

Because of the very slow rate of these processes, most caves in quartzite or quartz sandstone are hosted in very old (Proterozoic) or old (Paleozoic) rocks that have been at or close to the surface over long periods of time. Rainfall in these areas is often greater than $3500\,mm\,yr^{-1}$, although it might be less than $1000\,mm\,yr^{-1}$ in the arid areas of Central Brazil. Computer modeling based on the chemical constraints of the processes involved have shown that caves of the size of those found in the Guyana shield of South America would require some tens of millions of years to form (Mecchia et al. 2019), but direct numerical dating of these voids (and the chemical and physical sediments they contain) is still lacking.

Caves in quartz-rich lithologies have been reported from numerous regions. The most important are located in stable cratonic areas within the tropical climate zone, where old and scarcely deformed quartz-rich rocks form large outcrops, generally standing out from the surrounding landscape (Wray and Sauro 2017). The world's most important and longest quartzite caves have been found in South America, especially in Venezuela, Brazil, and Colombia (Aubrecht et al. 2012; Auler and Sauro 2019). Despite the fact that explorations in these remote areas started in the 1970s (Zawidzki et al. 1976), many of these large cave systems have been explored only in the last 20 years, and today more than 20 caves exceed 2 km of development, most of them located in the tepuis of Venezuela and Brazil. However, most of the tepuis are very difficult to reach, and future explorations will surely bring this number of large quartzite caves to increase steadily. The longest known quartzite cave is the 18.7 km long Imawarì Yeuta Cave, on Auyán Tepui (Venezuela) (Figure 11.39a), followed by the 10.8 km long Roraima Sur-Ojos de Cristal Cave in Roraima Tepui, also in Venezuela (Figure 11.39b). Roof collapses penetrating up to the surface often have separated longer underground cave systems in a sequence of smaller caves, such as the overall 20-km long Charles Brewer-Muchimuk-Colibrí System on Chimantha Tepui, Venezuela (Figure 11.39c).

Quartzite caves are of two main types: (i) those controlled by stratigraphic features (i.e., bedding planes, specific beds) and (ii) those exploiting vertical fractures. The first are generally the largest systems, following extensive close-to-horizontal stratification planes and or layers, which can often be regarded as favorable inception horizons (Sauro 2014). Depending on the local geological and hydrological conditions, these systems can be a single stream passage (e.g., Caverna Aroe-Jari, Brazil, and Cueva Guacamaya, Venezuela, Figure 11.39d, e), a branchwork of different underground streams that do not always converge (e.g., Cueva Imawarì Yeuta, Figure 11.39a, Roraima Sur, Figure 11.39b), or an anastomotic network (e.g., Muchimuk-Colibrì, Figure 11.39c). Generally, the voids are carved by mechanical erosion along a favorable horizon, such as a layer of iron-hydroxides, or kaolinite- and mica-rich stratigraphic intervals. Underground passages are much wider than high, with erosional remnants, such as sandstone pillars, often forming groups (Sauro 2014) (Figure 11.40a–d).

Close to the vertical cliffs of the quartzite massifs dilated stress-release joints are often enlarged by infiltrating waters, sometimes forming very deep (>300 m) crevices. The widening of these tapering-downward fissures is related to a combination of chemical weathering, flaking and gravitational spalling processes, involving the falling of quartzite slabs detached from the walls. Often, different sets of fractures can enlarge forming angular cave patterns (Figure 11.39f–i). In some cases, a horizontal cave storey can be connected at depth, usually following a favorable stratigraphic level (e.g., Sima Aonda and Cueva Akopan-Dal Cin, Figure 11.39f, g).

Interestingly, most of the quartzite caves in which streams are actively flowing today are suspended high above the regional base level, which corresponds to the rivers flowing in the

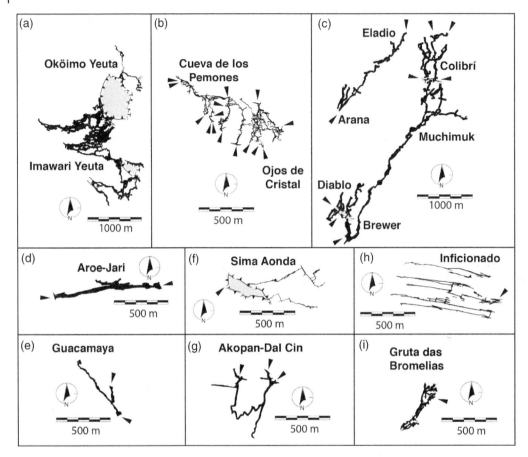

Figure 11.39 Examples of patterns in quartzite caves: (a) Imawari Yeuta Cave, Auyán Tepui, Venezuela; (b) Roraima Sur Cave system (Cueva de los Pemones and Ojos de Cristal), Roraima Tepui, Venezuela; (c) Charles Brewer Cave System (Colibrí, Muchimuk, Brewer and Diablo caves), Chimantha Tepui, Venezuela; (d) Caverna Aroe-Jari, Mato Grosso, Brazil; (e) Guacamaya Cave, Auyán Tepui, Venezuela; (f) Sima Aonda, Auyán Tepui, Venezuela; (g) Akopan-Dal Cin Cave, Chimantha Tepui, Venezuela; (h) Inficionado system, Minas Gerais, Brazil; (i) Bromelias Cave, Minas Gerais, Brazil. *Source:* Modified from Wray and Sauro (2017). Note the larger scale bar for the longest caves (Imawari Yeuta and Charles Brewer cave system, a and c). The grey pattern indicates large collapse dolines.

surrounding lowlands. These active caves are closely related to the present-day geomorphic setting, with entrances located on the sidewalls of large collapse sinkholes, and stream profiles adapted to the bottom of these depressions, often controlled by stratigraphical elements. They are related to suspended aquifers in equilibrium with the local geological and geomorphological conditions, but not adjusted to the overall regional drainage systems.

11.5.4 Iron Formation Caves

Solutional caves have been reported from iron-rich geological materials such as ferricrete in tropical regions (e.g., India, Australia, and several countries in tropical Africa), and especially in the large iron-rich mining regions of Brazil. Despite the fact that the small caves in iron formations have been reported by travelers and mining companies from the Iron Quadrangle (Minas Gerais,

Figure 11.40 Examples of quartzite caves: (a) River passage in Imawarì Yeuta Cave. *Source:* Photo by Vittorio Crobu, La Venta Esplorazioni Geografiche. (b) The typical structurally-controlled quartzite columns in Imawarì Yeuta Cave, Venezuela. *Source:* Photo by Riccardo De Luca, La Venta Esplorazioni Geografiche. (c) The giant "Paolino" room in Imawarì Yeuta Cave. *Source:* Photo by Vittorio Crobu, La Venta Esplorazioni Geografiche. (d) Large and wide stratigraphically-controlled river passage in Gruta de Torras, Chapada Diamantina, Brazil. *Source:* Photo by Alessio Romeo, La Venta Esplorazioni Geografiche. Note the iron oxides on the roof.

Brazil) since the early nineteenth century, it was not until the early 1960s that cave geologists first dedicated attention to these very special caves. Early descriptions attributed the genesis of these rather small caves, at least for the larger ones, to the dissolution of carbonate constituents present in minor quantities in these iron ore deposits (Simmons 1963). Because of their small size, these caves were neglected over the following 40 years both by scientists and cavers, and it was not until the Brazilian cave protection law was passed in 2008, that mining companies were obliged to carry out systematic surveys in these caves (Auler et al. 2019). Today, over 3000 iron formation caves (IFCs) have been mapped, mainly in five regions of Brazil: Carajás area in the Amazon basin (northern Brazil), the Quadrilátero Ferrífero (Iron Quadrangle) in Minas Gerais (south-central Brazil), the southern Espinhaço mountain range (Minas Gerais), the Salinas region in the northern Espinhaço mountain range, and the Caetité area, both in Bahia (Central Brazil). The first two areas contain the highest-grade iron ores, and host 80% of the IFCs known today. Both areas are characterized by Archean rocks (2.7–2.5 Gyr), including the original Banded Iron Formations (BIFs) and their weathering products. One of the byproducts of the weathering of the BIFs under tropical conditions is a sort of conglomerate containing chunks of the original iron-rich rock held together by a ferruginous cement, and known under the local name of "canga." This very resistant alteration product forms a blanket of some meters thick that protects the underlying and softer rock from further weathering. Although ages are heterogeneous, some canga appears to be Eocene in age (ca. 55 Myr) (Monteiro et al. 2014), which is compatible with the very low weathering rates (0.13–0.46 mm kyr^{-1}) obtained by cosmogenic ^{3}He analyses (Shuster et al. 2012). These weathering surfaces might well be among the oldest on Earth, and caves entirely carved in canga can thus be the world's oldest still accessible caves.

IFCs can form entirely in canga, in original BIFs, in other iron weathering products, or at the contact between canga and BIF or between BIF and other rock types. Most caves are rather small (the longest known today is less than 400 m), and very shallow (often less than 5 m below the topographic surface), roughly following the direction of the gentle topographic gradient. Since these ferruginous rocks are very resistant to erosion, they generally occur in the high sectors of ridges or plateaus, where the caves are generally found. The access to IFCs is often accidental, at the foot of small scarps of the ferruginous rocks. Most IFCs are dry and situated well above the local water table, and they are only occasionally affected by vadose infiltration and local water flow. The plan pattern of the caves is sponge-like, with irregular, more or less circular chambers, sometimes isolated, and often interconnected by small passages (Figure 9.11). These caves form by the slow subsurface weathering of the iron-rich rocks, and the removal of the weathering products by occasional water flows. Dissolution of the iron oxides appears to be mediated by Fe(III)-reducing microorganisms, which metabolize the host rock allowing the more soluble Fe(II) forms to be taken away in solution by the occasional groundwater flows (Parker et al. 2018). IFCs typically do not contain large speleothems, coralloids being the most common forms. These secondary precipitates are often composed of iron oxides and phosphates (Albuquerque et al. 2018).

11.6 Condensation–Corrosion

Condensation of water vapor from moist air masses onto colder cave walls, or at the contact between warm humid and cold air masses in the cave atmosphere are responsible for the formation of water droplets and films, which are initially very low in dissolved load. This water quickly absorbs gases from the cave air, generally CO_2 but sometimes also H_2S (in sulfuric acid caves), becoming aggressive toward the carbonate minerals. Condensation waters are highly aggressive even without absorbing CO_2 in halite and gypsum caves (Gázquez et al. 2015). This condensation–corrosion progressively dissolves the rocks producing a dissolution residue. Part of the residual material can be washed away if sufficient water condenses to form a flowing water film or it can be left in place as loose mineral grains that eventually can fall to the floor. This soft weathering rind can occasionally be over a centimeter thick (Figure 11.41a) and is often used by cave visitors to engrave their names or leave signs (Figure 11.41b). Weathering patterns of cave walls by these subtle condensation–corrosion processes can be extremely complex, depending on the original rock texture and mineralogy, intensity and duration of the condensation processes, their alternation with evaporation periods, the air flow in the cave, and many other physical and chemical factors. The book by Zupan Hajna (2003) offers a wealth of information on the "incomplete dissolution" related to condensation–corrosion, weathering of cave walls and the production, transport and deposition of fine residual material, including less soluble carbonate particles.

Condensation of water in caves has been experimentally investigated by de Freitas and Schmekal (2003, 2006) in Glowworm Cave, New Zealand, showing evaporation and condensation to have both diurnal and seasonal changes, with higher condensation rates occurring in the warmer periods. Five main processes can lead to the condensation of water vapor in the underground environment (Figure 11.42): (i) adiabatic cooling of rising air masses, (ii) mixing of saturated air masses

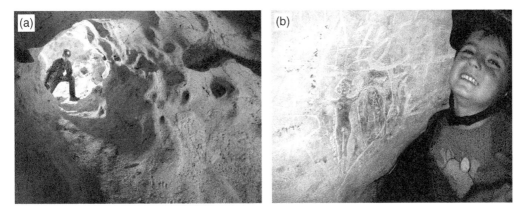

Figure 11.41 The weathering of cave walls by condensation–corrosion: (a) A two million-year-old phreatic passage high above the active cave levels of Codula Ilune system, Sardinia (southern Italy). Note the signs of fingers and hands on the cave wall on the lower right. The soft powdery material covering the walls is 1 cm thick. *Source:* Photo by Vittorio Crobu. (b) Grotte de Saint-Sébastien (southern France) with drawings and names carved in the weathered wall. "Winnie the Pooh" for scale. *Source:* Jo De Waele (Author).

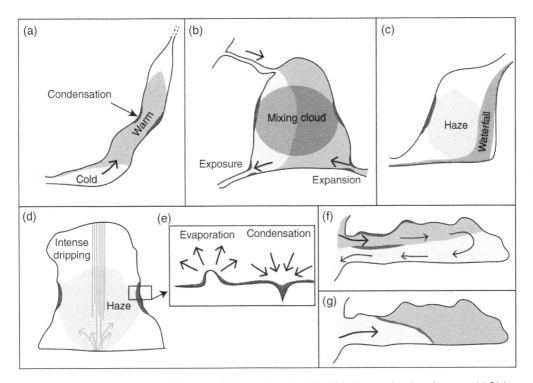

Figure 11.42 The various conditions in which condensation (dark blue) can take place in caves: (a) Rising and cooling of moist air masses in the cave, (b) Mixing clouds with condensation occurring on the nearby walls, and condensation by adiabatic air expansion or by exposure in the lower passages (interaction with cooler rock), (c) Condensation from a haze caused by a waterfall, and (d) by intense dripping, (e) detail of (d) showing preferential sites for evaporation (protruding and convex parts) and condensation (grooves and concave parts). (f) Daily (and seasonal) movements of warm and moist air masses into a cave, and (g) of cold air entering the cave.

(100% of relative humidity) but with different temperatures and concentrations in water vapor, (iii) adiabatic expansion of air masses, (iv) condensation from hazes, and (v) Raoult condensation (Badino 2010).

The first process (adiabatic cooling of rising air masses) is related to the fact that in caves the temperature of rising air cools at a rate of 3.5 °C km^{-1}, which is around half of the average adiabatic lapse rate in the Earth's atmosphere (ca. 6.5 °C km^{-1}) (Badino 2010). A mass of air saturated in water vapor cools down when it rises, leading to condensation. Since air flow in a cave depends on the differences between external and internal temperatures, and the configuration of the cave and the landscape, condensation thus occurs when moist air masses rise in altitude and cool down. This situation generally occurs when the external temperature drops below that of the cave, and cold air enters the cave system through lower entrances, forcing moist and warmer resident air masses to rise in the cave. Even slight cooling of the rising air masses leads to condensation along the flow path (Figure 11.42a).

The second process (mixing of saturated air masses) is related to the mixing of two air masses saturated in water vapor at different temperatures that produces a supersaturated mixing cloud and condensation; this is a mechanism similar to the mixing corrosion effect in $CaCO_3$-saturated solutions, the so-called Bögli effect. Clouds in caves can often be observed at the intersection of two cave passages bringing together air coming from two different entrances at different altitudes (and thus temperatures) (Figure 11.42b). These intersections where mixing clouds form are often enlarged by corrosion caused by the continuous or seasonal creation of condensation waters on their walls.

Local adiabatic air expansion causes air masses to cool down in restricted areas, inducing condensation. In meteorology, this phenomenon causes the formation of banner clouds on the leeward (downwind) side of mountains. In cave environments, this happens at the downstream end of passage constrictions, where the air exiting the narrow section expands rapidly into the larger room beyond. Narrow passages in caves induce condensation on the downwind side, where the air expands and cools down, and on the windward (upwind) exposed parts of the narrowing passages, where warm air interacts with cooler rock (Figure 11.42b).

Condensation from haze occurs where small water droplets stay suspended in the cave atmosphere (as an aerosol). Evaporation from convex water surfaces (the droplets) occurs more easily than from flat water films, so the atmosphere around these suspended aerosol water droplets is always supersaturated in water vapor. Hazes can be created by mechanical fragmentation of water drops at the foot of waterfalls (Figure 11.42c), or around intense dripping sites (Figure 11.42d). These hazes cause the surrounding walls to be constantly wetted, not by direct splash, but by condensation from the supersaturated air masses around fragmentation clouds. Condensation from these supersaturated air masses preferentially occurs on concave or flat surfaces in the immediate surroundings of these hazy clouds, and especially in small fissures in the walls, where evaporation is inhibited and condensation is promoted (Figure 11.42e).

Raoult condensation is a less important process, occurring because salty water evaporates less than pure water, so water vapor condenses more easily onto salty solutions or surfaces. This is one of the reasons why salt, which is also hygroscopic (attracts water molecules), is always wetted by condensation waters when exposed to moist air masses.

Condensation of water is an exothermic process that releases a significant amount of heat into the environment (around 586 cal g^{-1} at 20 °C), and this heat causes the local air, rock wall and moisture to warm up. In closed environments, such as a thermal aerated void isolated from the external atmosphere (i.e., with no entrances), this heat is slowly dissipated through the rock, which is a bad conductor. The system thus tends to reach an equilibrium, in which the thermal energy flux released by condensing waters is counterbalanced by the energy dissipated through the rock. Condensation thus tends to attenuate through time, because of the decreasing thermal gradients between rock and air masses (Dublyansky and Dublyansky 2000; Dreybrodt et al. 2005a). Final recession rates in limestone walls for a thermal difference of 3 °C maintained between the cooler rock wall and the warmer moist air, a situation only possible in thermal caves, have been estimated at 0.05–0.2 mm yr^{-1} by Szunyogh (1990) and 0.09 mm yr^{-1} by Lismonde (2003). This would mean that a spheroidal pocket of 2 m diameter, typical of such thermal cave environments (e.g., Sátorkö-puszta Cave, Figure 11.21a) would form in a time range of 10–40 kyr.

Condensation–corrosion is particularly intense in shallow thermal caves (Audra et al. 2007b), where substantial temperature gradients exist between the thermal water bodies and the walls and roof above, and heat can be dissipated through the rock to the above lying surface, cooled down by colder infiltrating waters, or taken away by significant air currents in the caves. The warm and moist air above a thermal pool, being less dense, rises, comes in contact with the cool cave roof and produces condensation. The higher the roof is, the farther away from the thermal source, and the cooler it will be. Therefore, the highest cave roofs are the best place for condensation waters to form, but condensation also occurs along the walls, thus increasingly making the rising air mass less supersaturated in water vapor.

In non-stationary conditions, where air temperature differences between the external atmosphere and that of the cave show large daily and seasonal fluctuations, condensation and the associated corrosion can occur, but with limited intensity. Modeling has shown that daily temperature differences of 10 °C between cave air and the entering external air masses can cause condensation–corrosion-induced rock wall retreat at rates of 0.003 mm yr^{-1}, whereas recession rates related to seasonal differences of the same range would be one order of magnitude lower (0.00035 mm yr^{-1}) (Dreybrodt et al. 2005a). These values appear to be slightly higher in Movile Cave, Romania (0.001 mm yr^{-1}) (Sarbu and Lascu 1997), probably due to the presence of H_2S in the cave atmosphere. In Central Brazil, where many caves have formed under hypogenic conditions in a semi-arid climate, the overprinting caused by condensation–corrosion can dominate over the action of gently flowing or standing waters, forming characteristic condensation–corrosion pockets and even airflow scallops (Figure 11.43a, b). These caves occur in a stable geodynamic setting, which allows their preservation over long periods of time due to low denudation rates. Condensation–corrosion processes probably have acted over different glacial–interglacial cycles, especially during periods that were wetter than today.

Condensation–corrosion is particularly widespread in areas where warm and moist external air masses can penetrate into the cooler cave environment. This is common in tropical areas, and especially close to large external water bodies like oceans or lakes (Figure 11.42f, g). Moreover, in these caves the air can be enriched in CO_2 related to the high productivity of soils, or simply to the presence of large bat colonies. In Cayman Brac (Cayman Islands), warm and moist air entering the caves along the roof causes condensation to occur especially in the entrance parts. This corrosion is dominant and precludes speleothem development and preservation in the entrance parts, slowly balancing out deeper into the cave, where speleothems can develop and be pristine. The formation of bell holes, occurring mostly close to the entrance areas of the caves, is mainly attributed to these diurnal cycles of condensation–corrosion (Tarhule-Lips and Ford 1998b). In Runaway Bay Cave in Jamaica, located 500 m from the coast at 20 m a.s.l., the formation of the numerous bell holes deep inside the cave has been attributed to a condensation–corrosion process aided by the metabolic activity of bats (Lundberg and McFarlane 2009) (Figure 9.39). It is roughly estimated that a 1-m deep bell hole can form in a time span of ca. 50 kyr by this process alone.

Figure 11.43 Signs of condensation–corrosion in old caves: (a) Condensation–corrosion pockets on the lower and middle part of a calcite flowstone in Lapinha de Iramaia Cave, Bahia (Central Brazil). *Source:* Jo De Waele (Author). These corrosion areas mostly occur at 1–4 m height, depending on how high the passages are, and appear to be correlated to the interface between entering warm and moist air that stays close to the cave roof, and the less mobile colder air that resides in the lower parts of the cave. These signs of corrosion tend to diminish the farther one goes from the large entrance parts. (b) Airflow scallops in Lapinha, Andaraí, Bahia (Central Brazil). Note their symmetrical form due to alternating air flow direction. (c) Limestone pillar in Colombo Cave, Toirano (northern Italy), isolated by condensation–corrosion of incoming coastal air masses in the large entrance parts, aided by biocorrosion (bat guano on the floor). *Source:* Photo by Jean-Yves Bigot.

Condensation and evaporation are also important in non-tropical coastal areas. These processes mainly influence the morphology of the initial parts of caves with their entrance exposed to the sea, as documented in Mediterranean areas (De Waele et al. 2009a), but also older caves located some kilometers away from the coast and at higher elevations, where fog can rise along valleys and enter large cave entrances. The Colombo Cave in Toirano (northern Italy) shows very clear evidence of long-lasting condensation–corrosion processes (Columbu et al. 2021) (Figure 11.43c). Condensation–corrosion is often boosted by the presence of large bat colonies and their guano deposits, the exothermic decay of which increases temperature gradients, as well as water vapor and CO_2 contents. Examples in southern Africa (Dandurand et al. 2019) and Cyprus (Cailhol et al. 2019) show that these biocorrosion processes can have a great impact on the final morphology of the caves, and obliterate important archeological evidence from our past. This process might have caused the loss of rock art in many caves opening on the valley flanks of the Ardèche River (southern France), for example. In contrast, the Chauvet Cave in this valley has been preserved from this natural condensation–corrosion destruction by a rockfall that has completely closed its natural entrance shortly after the cave's use by prehistoric artists (Jones and Elliott 2019), thus preventing warm external air to enter the cave.

11.7 A Summary: Life Cycle of Solution Caves

This integration section describes the long-term evolution of solution caves from the early opening of fissures and conduits to the erosional removal of an old cave passage. No matter which type of rock is dissolved, or what kind of fluid is responsible for the creation of the voids, all caves undergo a sequence of evolutionary stages, from the initial very slow widening of narrow fissures (initiation), passing through the rapid enlargement phase, the mature cave stage (transition and stagnation), and the final abandonment and decay of the cave passages. The time needed to pass from one phase to the other depends on a large variety of factors, mainly rock solubility, the fluids involved, climate, tectonics, and the surrounding landscape. These stages are shown in the conceptual graph of Figure 11.44, showing the evolution of a single epigenic cave passage. An indication of the time needed to pass from one phase to the other is shown below, based on the mean dissolution rates of the different rock types. The Y-axis is logarithmic and roughly indicates conduit dimensions, which mainly depend on factors such as the solubility of the rock, flow rate, and the tensile strength of the rocks in which the voids are carved.

11.7.1 Initiation of a Cave

The initiation stage commences when undersaturated water starts to flow through a soluble rock unit. This slow water movement tends to occur through all penetrable fissures in the rock, which spatial distribution and density depend on the geological history of the rock itself and the lithology. Groundwater flows toward areas of lower hydraulic head governed by the hydraulic gradient.

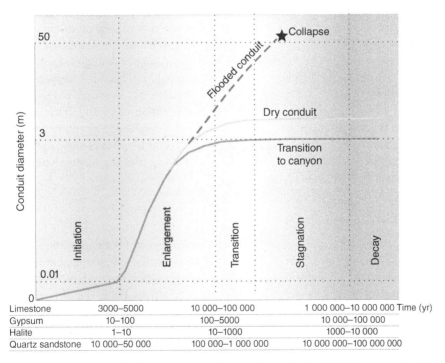

Figure 11.44 Stages of epigenic cave development from the tiny little fissures on the left to the large cave passages on the right. The indicative time intervals are based on mean dissolution rates of the different soluble rocks. *Source:* Modified from White (1988).

In epigenic karst settings, these differences in hydraulic head are most commonly related to uplift and surface erosion. Precipitation falling on higher lying areas partly infiltrates and flows through the available permeability pathways toward the lower lying discharge areas, generally local river valleys or the coast. Most initial fissures (joints, faults, and bedding-plane partings) are very narrow, and allow only small amounts of water to flow through. The long water–rock interaction causes these slow-moving fluids to reach near-saturation conditions over short distances, and most of the solutional work occurs in the initial portion of the underground flowpaths. However, some minor solutional aggressiveness can be maintained over longer distances. All of the solutional potential is expressed over the entire flow length: this means that in shorter flow paths more rock is dissolved per unit length than in longer ones (i.e., shorter fissures enlarge faster than longer ones). Also, higher hydraulic gradients cause water to flow a little bit faster, which causes rock removal by dissolution to be faster too in steeper fissures. Thus, shorter, steeper and more direct routes are favored over longer and gentler sloping ones. More importantly, since fissures have variable initial widths, some are able to carry higher flow rates from the early beginning and enlarge more rapidly. In a certain sense, the selection of the most favorable flow routes occurs right from the start, largely depending on the initial aperture of the fissures and how favorable their orientation and distribution with respect to the local hydraulic gradient is.

In general, one can say that the rate of early fissure enlargement mainly depends on the flow rate (more undersaturated water moves through, more rock is dissolved) and the length of the path along which dissolution is distributed (shorter paths enlarging faster than longer ones). Or, in other words, fissure enlargement in the initial phases of growth is a function of the ratio of discharge Q to flow length L (Q/L). In conditions in which spatial differences in water discharge are limited, such as where recharge occurs through an overlying or underlying low permeability formation, flow rate will be similar in all fissures, no matter their initial aperture, and almost all fissures will tend to widen at a similar rate. This is the reason why network maze patterns prevail in such conditions, and water flow through the caves is always rather homogeneous and slow.

The initiation stage ends when the most favorable flowpaths have widened enough to focus most of the water flow through the system (the so-called "breakthrough"). When this happens, the waters that flow through the system can retain solutional aggressiveness over much longer distances, often emerging at the springs with some remaining dissolution capability. Initiation of epigenic cave formation is very fast in halite (1–10 years), between a decade and a century for gypsum caves, and some thousands or some tens of thousands of years for limestones and quartzites, respectively. These breakthrough times will be up to one order of magnitude greater where slow and homogeneous descending or rising recharge occurs.

11.7.2 Rapid Cave Enlargement

Once breakthrough is achieved, the chosen flow paths will rapidly enlarge because of the associated substantial increase in flow rate and solutional work. The much higher enlargement rate of these few flow paths determines the locations where cave passages will form, with the narrower fissures barely enlarging anymore. The increased hydraulic conductivity along the few selected flow paths also causes the hydraulic head to decrease (i.e., local water table lowering), thus causing water from neighboring fissures to converge toward these main drainage paths. The initial size of these selected preferential flow paths is between some millimeters to 1 cm. In such conditions, flow changes from laminar, typical of flow in narrow fissures, to turbulent. This causes three different processes to concur in passage enlargement: turbulent flow, enhanced dissolution, and sediment transport causing abrasion.

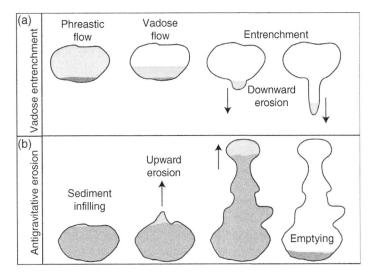

Figure 11.45 Vadose entrenchment and antigravitative erosion. When a cave passage passes from phreatic conditions to aerated ones, two main situations can occur: (a) the water flow is always able to transport the entrained sediments and carves a canyon or, (b) when sediments are continuously deposited on the floor, an upward growing antigravitative canyon develops by dissolution along the upper walls and roof.

In highly soluble rocks (i.e., halite and gypsum), the dissolution rate is fast enough to allow cave passages to adjust rapidly to the highest available flow rate. The size of phreatic passages in salt and gypsum caves indicates the highest amount of water that was flowing through them. In carbonate rocks and in other less soluble rocks the dissolution rate may not be fast enough to keep pace with the amount of water flowing, becoming the limiting factor for enlargement rate (i.e., conduits enlarge too slowly to be able to adjust to the maximum possible flow rate). Conduits in limestone will grow at a maximum enlargement rate, which can be estimated between 1 and 100 cm kyr^{-1}, and their size will generally not be adapted to the largest possible water flow. Whereas significant salt and gypsum passages can be carved under favorable conditions in a matter of 10 and 100 years, respectively, the development of cave passages in limestones normally requires some thousands of years.

Given enough time, limestone cave passages can eventually reach their maximum size, adjusted to the highest available flow. Subsequently, the passages start to be partially air-filled, at first periodically, and later on at all times, and dissolution and erosion mainly occurs along the floor and the lower walls of the cave. This leads to the formation of canyons and keyhole passages, where downward carving is aided by sediments entrained by the waters, and by periodic injection of floodwaters, which are more aggressive (Figure 11.45a). If sediments accumulate on the floor, dissolution is focused on the upper walls and roof of the passage, leading to the formation of upward growing antigravitative (or paragenetic) canyons (Figure 11.45b).

11.7.3 Cave Maturation

Once a mature cave stream has formed, connecting a recharge area (e.g., a sinking stream) to a spring, water rapidly reaches the local water table and flows downstream along a very gentle hydraulic gradient. The efficient transmission of flow tends to lower the hydraulic head along this main channel, a phenomenon confirmed by both physical and computer modeling. The water in nearby fissures thus tends to converge toward this more efficient underground drainage channel. This situation is rarely stationary in nature. The lowering of the hydraulic head in the main channel causes the hydraulic gradient to steepen upstream and on the sides of the channel, favoring the growth of incipient channels in these upstream or adjacent areas. Where the soluble rock extends over wide areas and is not covered by insoluble formations, this leads to the opening of new sinking points in the surface channels upstream, or of new underground drainage routes connecting upstream doline clusters to the main channel. The more a growing underground passage below a doline becomes efficient in draining water and evacuating the subsiding soil, the more rapidly the sinkhole increases its size and its catchment area. These processes entail a positive feedback mechanism, whereby the larger a doline becomes, the higher the recharge and flow rate, and the more rapidly the underlying drainage channel enlarges. These new underground tributaries will generally converge to the main channel, since this is the direction of highest hydraulic gradient. This is how the typical branchwork caves develop, expanding headwards and draining water from progressively more distant recharge areas. Upstream parts are generally younger and smaller, with less water flowing through them. There is a general increase in the number of contributing tributaries, the flow rate and the size of the main passages in the downstream direction. Generally, the chemistry of the waters flowing through a mature cave system is rather uniform, but the downstream increase in flow rate and sediment load may be accompanied by greater mechanical erosion. In a mature karst system, all water eventually finds its way underground, whereas the surface is characterized by dense doline fields, short and small sinking rivers on the edges of the karst area, and in some cases, through flowing canyons with major effluent allogenic drainages.

Landscapes, however, are dynamic systems, and uplift causes rivers to entrench and hydraulic gradients to increase. Fluvial downcutting causes the lowering of base level and a displacement of the discharge areas, leading springs to migrate vertically and often also laterally downstream over long distances. New diversion routes are carved by the underground waters at greater depths to adjust to this new situation, and the old cave levels become abandoned.

Once an underground flow path adjusts to the base level, the hydraulic head settles almost to the level of the spring. The passage continues enlarging slowly adjusting to the highest available flow, but at a certain stage it becomes oversized (i.e., flow capacity exceeds maximum discharge), and vadose conditions start occurring. The passage first experiences alternating phases of phreatic and vadose flow, and finally it carries a river permanently flowing in contact with rock and air. Whereas the water flowing in phreatic conditions was unable to lose CO_2, once the river flows in an air-filled channel the water tends to release CO_2 into the cave atmosphere, losing part of its solutional aggressiveness. These open-system conditions can lead these cave river waters to deposit thin sheets of calcite (flowstones), or a series of rimstones.

11.7.4 Cave Abandonment and Decay

Once cave passages are entirely abandoned by the streams that carved them (often gradually), they can remain in a stationary phase without experiencing any significant change over long time periods (tens of thousands of years in limestones). Whereas dissolution can occur locally by undersaturated infiltration or condensation waters (see Section 11.7), these trickles can form secondary chemical cave deposits (speleothems), which tend to cover the fluvial sediments accumulated in the previous stage.

Airflow and small inflowing waters bring sediments into the cave covering the floors and walls, and weathering of the limestone causes the accumulation of residual clays and favors collapse processes by reducing the rock mass strength. This sedimentation phase can last over very long periods of time, and the sediments, trapped and protected in the conservative underground environment, offer often continuous records of environmental and climate changes occurring at the surface. In the meantime, archives of environmental variability at the surface may be lost by weathering and erosion processes. As surface denudation proceeds, the cave levels are intersected gradually by the land surface, resulting in the development of collapse sinkholes, natural bridges, unroofed caves, and cave-collapse valleys. Surface lowering gradually removes preferentially the more soluble parts, often leaving the insoluble cave sediments, and the large macrocrystalline, and compact speleothems as patches or loose rubble in the landscape. Large unroofed caves have frequently been found in the Slovenian karst (Mihevc 2007), and the paleomagnetic dating of their sediments indicates that they were active as much as 5 million years ago (Zupan Hajna et al. 2020). In the Bologna gypsum karst, calcite speleothems are found scattered on the surface, and represent relics of now completely eroded gypsum caves, some of which must have been older than the oldest age of the speleothems (i.e., 600 kyr) (Columbu et al. 2017).

References

Albuquerque, A.R.L., Angélica, R.S., Gonçalves, D.F. et al. (2018). Phosphate speleothems in caves developed in iron ores and laterites of the Carajás Mineral Province (Brazil) and a new occurrence of spheniscidite. *International Journal of Speleology* 47: 53–67.

Alexander, E.C. Jr. (1996). *The Underground Rivers of Mystery Cave*. University of Minnesota Digital Conservancy https://hdl.handle.net/11299/185397.

Andre, B.J. and Rajaram, H. (2005). Dissolution of limestone fractures by cooling waters: early development of hypogene karst systems. *Water Resources Research* 41 (1): W01015.

Andrejchuk, V.N. and Klimchouk, A.B. (2001). Geomicrobiology and redox geochemistry of the karstified Miocene gypsum aquifer, western Ukraine: the study from Zoloushka Cave. *Geomicrobiology Journal* 18 (3): 275–295.

Andrejchuk, V.N. and Klimchouk, A.B. (2002). Mechanisms of karst breakdown formation in the gypsum karst of the fore-Ural region, Russia (from observations in the Kungurskaja Cave). *International Journal of Speleology* 31 (1): 89–114.

Andrejchuk, V.N. and Klimchouk, A.B. (2017). Zoloushka Cave (Ukraine-Moldova) - A Prime Example of Hypogene Artesian Speleogenesis in Gypsum. In: *Hypogene Karst Regions and Caves of the World* (ed. A.B. Klimchouk, A.N. Palmer, J. De Waele, et al.), 387–406. Cham: Springer.

Antonellini, M., Del Sole, L., and Mollema, P.N. (2020). Chert nodules in pelagic limestones as paleo-stress indicators: A 3D geomechanical analysis. *Journal of Structural Geology* 132: 103979.

Asrat, A. (2015). Geology, geomorphology, geodiversity and geoconservation of the Sof Omar Cave System, southeastern Ethiopia. *Journal of African Earth Sciences* 108: 47–63.

Aubrecht, R., Barrio-Amorós, C.L., Breure, A.S.H. et al. (2012). *Venezuelan tepuis: their caves and biota, Acta Geologica Slovaca Monograph*. Bratislava: Comenius University.

Audra, P. (2008). Hypogenic sulphidic speleogenesis. *Berliner Höhlenkundliche Berichte* 26: 5–30.

Audra, P. (2017). Hypogene Caves in North Africa (Morocco, Algeria, Tunisia, Libya, and Egypt). In: *Hypogene Karst Regions and Caves of the World* (ed. A.B. Klimchouk, A.N. Palmer, J. De Waele, et al.), 853–864. Cham: Springer.

Audra, P. and Bigot, J.-Y. (2009). Les grottes de Saint Benoît. *Spelunca* 114: 17–27.

Audra, P. and Palmer, A.N. (2011). The pattern of caves: controls of epigenic speleogenesis. *Géomorphologie: Relief, Processus, Environnement* 17 (4): 359–378.

Audra, P., Bigot, J.-Y., and Mocochain, L. (2002). Hypogenic caves in Provence (France). Specific features and sediments. *Acta Carsologica* 31 (3): 33–50.

Audra, P., Hobléa, F., Bigot, J.-Y. et al. (2007b). The role of condensation-corrosion in thermal speleogenesis: study of a hypogenic sulfidic cave in Aix-les-Bains, France. *Acta Carsologica* 36 (2): 185–194.

Audra, P., Mocochain, L., Bigot, J.-Y. et al. (2009a). Hypogene cave patterns. In: *Hypogene Speleogenesis and Karst Hydrogeology of Artesian Basins* (ed. A.B. Klimchouk and D.C. Ford), 17–22. Kiev: Ukrainian Institute of Speleology and Karstology.

Audra, P., Mocochain, L., Bigot, J.-Y. et al. (2009b). Morphological indicators of speleogenesis: hypogenic speleogenesis. In: *Hypogene Speleogenesis and Karst Hydrogeology of Artesian Basins* (ed. A.B. Klimchouk and D.C. Ford), 23–32. Kiev: Ukrainian Institute of Speleology and Karstology.

Audra, P., Gázquez, F., Rull, F. et al. (2015). Hypogene Sulfuric Acid Speleogenesis and rare sulfate minerals in Baume Galinière Cave (Alpes-de-Haute-Provence, France). Record of uplift, correlative cover retreat and valley dissection. *Geomorphology* 247: 25–34.

Auler, A.S. (2020). Caves and speleogenesis in the Lagoa Santa Karst. In: *Lagoa Santa Karst: Brazil's Iconic Karst Region* (ed. A.S. Auler and P. Pessoa), 167–186. Cham: Springer.

Auler, A.S. and Sauro, F. (2019). Quartzite and quartz sandstone caves of South America. In: *Encyclopedia of Caves* (ed. W.B. White, D.C. Culver and T. Pipan), 850–860. New York: Academic Press.

Auler, A.S. and Smart, P.L. (2003). The influence of bedrock-derived acidity in the development of surface and underground karst: evidence from the Precambrian carbonates of semi-arid northeastern Brazil. *Earth Surface Processes and Landforms* 28: 157–168.

Auler, A.S., Parker, C.W., Barton, H.A. et al. (2019). Iron formation caves: Genesis and ecology. In: *Encyclopedia of Caves* (ed. W.B. White, D.C. Culver and T. Pipan), 559–566. New York: Academic Press.

Back, W., Hanshaw, B.B., Herman, J.S. et al. (1986). Differential dissolution of a Pleistocene reef in the ground-water mixing zone of coastal Yucatan, Mexico. *Geology* 14: 137–140.

Badino, G. (2005). Underground drainage systems and geothermal flux. *Acta Carsologica* 34 (2): 277–316.

Badino, G. (2010). Underground meteorology- "What's the weather underground?". *Acta Carsologica* 39 (3): 427–448.

Bahadorinia, S., Hejazi, S.H., Nadimi, A. et al. (2016). The morphology and development of Kalahroud Cave, Iran. *International Journal of Speleology* 45 (3): 243–257.

Banzato, C., Vigna, B., Fiorucci, A. et al. (2017b). Hypogene gypsum caves in Piedmont (N-Italy). In: *Hypogene Karst Regions and Caves of the World* (ed. A.B. Klimchouk, A.N. Palmer, J. De Waele, et al.), 211–224. Cham: Springer.

Barbary, J.P. and Zhang, S. (2004). Historique et point sur les recherches spéléologiques en Chine. In: *Spéléo-karstologie et environnement en Chine (Guizhou, Yunnan, Liaoning), Karstologia Mémoires*, vol. 9 (ed. R. Maire, J.P. Barbary, S. Zhang, et al.), 459–474.

Bartolo, G., Dore, M., and Lecis, A. (1980). *Is Angurtidorgius*. Cagliari (Italy): Gia Editrice.

Bedinger, M.S. (1966). Electric analog study of cave formation. *National Speleological Society Bulletin* 28 (3): 127–132.

Bella, P., Bosák, P., Braucher, R. et al. (2019). Multi-level Domica-Baradla cave system (Slovakia, Hungary): middle Pliocene-Pleistocene evolution and implications for the denudation chronology of the Western Carpathians. *Geomorphology* 327: 62–79.

Birk, S., Liedl, R., Sauter, M. et al. (2005). Simulation of the development of gypsum maze caves. *Environmental Geology* 48 (3): 296–306.

Bolner-Takács, K. (1999). Paleokarst features and other climatic relics in Hungarian caves. *Acta Carsologica* 28 (1): 27–37.

Bolner-Takács, K. and Kraus, S. (1989). The results of research into caves of thermal water origin. *Karszt és Barlang* (Special Issue): 31–38.

Borghi, A., Renard, P., and Jenni, S. (2012). A pseudo-genetic stochastic model to generate karstic networks. *Journal of Hydrology* 414: 516–529.

Bosák, P., Bruthans, J., Filippi, M. et al. (1999). Karst and caves in salt diapirs, SE Zagros Mts. (Iran). *Acta Carsologica* 28 (2): 41–75.

Bottrell, S.H., Smart, P.L., Whitaker, F. et al. (1991). Geochemistry and isotope systematics of sulphur in the mixing zone of Bahamian blue holes. *Applied Geochemistry* 6 (1): 97–103.

Bozeman, J. and Bozeman, S. (2002). Speleology of gypsum caves in Oklahoma. *Carbonates and Evaporites* 17 (2): 107–113.

Bretz, J.H. (1942). Vadose and phreatic features of limestone caverns. *Journal of Geology* 50: 675–811.

Brod, L.G. (1964). Artesian origin of fissure caves in Missouri. *National Speleological Society Bulletin* 26 (3): 83–114.

Bruthans, J., Filippi, M., Zare, M. et al. (2010). Evolution of salt diapir and karst morphology during the last glacial cycle: effects of sea-level oscillation, diapir and regional uplift, and erosion (Persian Gulf, Iran). *Geomorphology* 121 (3-4): 291–304.

Cailhol, D., Audra, P., Nehme, C. et al. (2019). The contribution of condensation-corrosion in the morphological evolution of caves in semi-arid regions: preliminary investigations in the Kyrenia Range, Cyprus. *Acta Carsologica* 48 (1): 5–27.

Calaforra, J.M. (1998). *Karstologìa de Yesos*. Monografìa Ciencias y Tecnologìa. Almeria: University of Almeria.

Calaforra, J.M. and Gázquez, F. (2017). Gypsum speleogenesis: a hydrogeological classification of gypsum caves. *International Journal of Speleology* 46 (2): 251–265.

Calaforra, J.M. and Pulido-Bosch, A. (2003). Evolution of the gypsum karst of Sorbas (SE Spain). *Geomorphology* 50: 173–180.

Catcott, A. (1761). *A treatise on the Deluge*. London: Withers.

Chiesi, M., De Waele, J., and Forti, P. (2010). Origin and evolution of a salty gypsum/anhydrite karst spring: the case of Poiano (Northern Apennines, Italy). *Hydrogeology Journal* 18 (5): 1111–1124.

Coke, J.G. (2019). Underwater caves of the Yucatan peninsula. In: *Encyclopedia of Caves* (ed. W.B. White, D.C. Culver and T. Pipan), 1089–1095. New York: Academic Press.

Collignon, B. (1983). Spéléogenèse hydrothermale dans les Bibans (Atlas Tellien, Algérie). *Karstologia* 2: 45–54.

Collignon, B. (1984). Étude morphologique et structurale de la grotte de Kef el Kaous (Algérie). *Spelunca Mémoires* 13: 88–92.

Columbu, A., De Waele, J., Forti, P. et al. (2015). Gypsum caves as indicators of climate-driven river incision and aggradation in a rapidly uplifting region. *Geology* 43 (6): 539–542.

Columbu, A., Chiarini, V., De Waele, J. et al. (2017). Late Quaternary speleogenesis and landscape evolution in the northern Apennine evaporite areas. *Earth Surface Processes and Landforms* 42 (10): 1447–1459.

Columbu, A., Audra, P., Gázquez, F. et al. (2021). Hypogenic speleogenesis, late stage epigenic overprinting and condensation-corrosion in a complex cave system in relation to landscape evolution (Toirano, Liguria, Italy). *Geomorphology* 376: 107561.

Cvijić, J. (1893). *Das Karstphänomen. Versuch einer morphologischen Monographie*, Geographische Abhandlungen herausgegeben von A. Penck, Wien, Band V, Heft 3, 1-114.

Cvijić, J. (1918). Hydrographie souterraine et évolution morphologique du karst. *Revue de Géographie Alpine* 6: 376–420.

Dandurand, G., Duranthon, F., Jarry, M. et al. (2019). Biogenic corrosion caused by bats in Drotsky's Cave (the Gcwihaba Hills, NW Botswana). *Geomorphology* 327: 284–296.

D'Angeli, I.M., De Waele, J., Melendres, O.C. et al. (2015a). Genesis of folia in a non-thermal epigenic cave (Matanzas, Cuba). *Geomorphology* 228: 526–535.

D'Angeli, I.M., Sanna, L., Calzoni, C. et al. (2015b). Uplifted flank margin caves in telogenetic limestones in the Gulf of Orosei (Central-East Sardinia - Italy) and their palaeogeographic significance. *Geomorphology* 231: 202–211.

D'Angeli, I.M., Vattano, M., Parise, M. et al. (2017b). The coastal sulfuric acid cave system of Santa Cesarea Terme (southern Italy). In: *Hypogene Karst Regions and Caves of the World* (ed. A.B. Klimchouk, A.N. Palmer, J. De Waele, et al.), 161–168. Cham: Springer.

D'Angeli, I.M., Carbone, C., Nagostinis, M. et al. (2018). New insights on secondary minerals from Italian sulfuric acid caves. *International Journal of Speleology* 47 (3): 271–291.

D'Angeli, I.M., Nagostinis, M., Carbone, C. et al. (2019b). Sulfuric acid speleogenesis in the Majella Massif (Abruzzo, Central Apennines, Italy). *Geomorphology* 333: 167–179.

D'Angeli, I.M., Parise, M., Vattano, M. et al. (2019c). Sulfuric acid caves of Italy: a review. *Geomorphology* 333: 105–122.

Davis, W.M. (1930). Origin of limestone caverns. *Geological Society of America Bulletin* 41: 475–628.

Davis, D.G. (1980). Cave development in the Guadalupe Mountains: a critical review of recent hypotheses. *National Speleological Society Bulletin* 42: 42–48.

De Freitas, C.R. and Schmekal, A. (2003). Condensation as a microclimate process: measurement, numerical simulation and prediction in the Glowworm Cave, New Zealand. *International Journal of Climatology* 23 (5): 557–575.

De Freitas, C.R. and Schmekal, A. (2006). Studies of condensation/evaporation processes in the Glowworm Cave, New Zealand. *International Journal of Speleology* 35 (2): 75–81.

De Waele, J., Mucedda, M., and Montanaro, L. (2009a). Morphology and origin of coastal karst landforms in Miocene and Quaternary carbonate rocks along the central-western coast of Sardinia (Italy). *Geomorphology* 106: 26–34.

De Waele, J., Picotti, V., Cucchi, F. et al. (2009b). Karst phenomena in the Cordillera de la Sal (Atacama, Chile). In: Geological Constraints on the Onset and Evolution of an Extreme Environment: The Atacama Area (ed. P.L. Rossi). *GeoActa Special Publication* 2: 113–127.

De Waele, J., Audra, P., Madonia, G. et al. (2016). Sulfuric acid speleogenesis (SAS) close to the water table: examples from southern France, Austria, and Sicily. *Geomorphology* 253: 452–467.

De Waele, J., Carbone, C., Sanna, L. et al. (2017a). Secondary minerals from salt caves in the Atacama Desert (Chile): a hyperarid and hypersaline environment with potential analogies to the Martian subsurface. *International Journal of Speleology* 46 (1): 51–66.

De Waele, J., D'Angeli, I.M., Tisato, N. et al. (2017b). Coastal uplift rate at Matanzas (Cuba) inferred from MIS 5e phreatic overgrowths on speleothems. *Terra Nova* 29 (2): 98–105.

De Waele, J., Piccini, L., Columbu, A. et al. (2017c). Evaporite karst in Italy: a review. *International Journal of Speleology* 46 (2): 137–168.

De Waele, J., D'Angeli, I.M., Bontognali, T. et al. (2018a). Speleothems in a north Cuban cave register sea-level changes and Pleistocene uplift rates. *Earth Surface Processes and Landforms* 43 (11): 2313–2326.

De Waele, J., Picotti, V., Martina, M.L. et al. (2020). Holocene evolution of halite caves in the Cordillera de la Sal (Central Atacama, Chile) in different climate conditions. *Geomorphology* 370: 107398.

Ding, W., Fan, T., Yu, B. et al. (2012). Ordovician carbonate reservoir fracture characteristics and fracture distribution forecasting in the Tazhong area of Tarim Basin, Northwest China. *Journal of Petroleum Science and Engineering* 86: 62–70.

Dreybrodt, W. (1990). The role of dissolution kinetics in the development of karst aquifers in limestone: a model simulation of karst evolution. *The Journal of Geology* 98 (5): 639–655.

Dreybrodt, W. and Gabrovšek, F. (2002). Basic processes and mechanisms. In: *Evolution of Karst: From Prekarst to Cessation* (ed. F. Gabrovšek), 115–154. Ljubljana: Založba ZRC.

Dreybrodt, W. and Siemers, J. (2000). Cave evolution on two-dimensional networks of primary fractures in limestone. In: *Speleogenesis: Evolution of Karst Aquifers* (ed. A.B. Klimchouk, D.C. Ford, A.N. Palmer, et al.), 201–211. Huntsville, Alabama: National Speleological Society.

Dreybrodt, W., Gabrovšek, F., and Perne, M. (2005a). Condensation corrosion: a theoretical approach. *Acta Carsologica* 34 (2): 317–348.

Dreybrodt, W., Gabrovšek, F., and Romanov, D. (2005b). *Processes of Speleogenessis: A Modeling Approach*. Ljubljana: Založba ZRC.

Dublyansky, Y.V. (1995). Speleogenetic history of the Hungarian hydrothermal karst. *Environmental Geology* 25 (1): 24–35.

Dublyansky, V.N. (2000a). A giant hydrothermal cavity in the Rhodope Mountains. In: *Speleogenesis: Evolution of Karst Aquifers* (ed. A.B. Klimchouk, D.C. Ford, A.N. Palmer, et al.), 317–318. Huntsville, Alabama: National Speleological Society.

Dublyansky, Y.V. (2000b). Dissolution of carbonates by geothermal waters. In: *Speleogenesis. Evolution of Karst Aquifers* (ed. A.B. Klimchouk, D.C. Ford, A.N. Palmer, et al.), 158–159. Huntsville, Alabama: National Speleological Society.

Dublyansky, Y.V. (2019). Hydrothermal caves. In: *Encyclopedia of Caves* (ed. W.B. White, D.C. Culver and T. Pipan), 546–552. New York: Academic Press.

Dublyansky, V.N. and Dublyansky, Y.V. (2000). Role of condensation in Karst hydrogeology and speleogenesis. In: *Speleogenesis: Evolution of Karst Aquifers* (ed. A.B. Klimchouk, D.C. Ford, A.N. Palmer, et al.), 100–112. Huntsville, Alabama: National Speleological Society.

Dublyansky, Y.V., Klimchouk, A.B., Spötl, C. et al. (2014). Isotope wallrock alteration associated with hypogene karst of the Crimean Piedmont, Ukraine. *Chemical Geology* 377: 31–44.

Dublyansky, Y.V., Michajljow, W., Bolner-Takács, K. et al. (2017). Hypogene Karst in the Tyuya-Muyun and the Kara-Tash Massifs (Kyrgyzstan). In: *Hypogene Karst Regions and Caves of the World* (ed. A.B. Klimchouk, A.N. Palmer, J. De Waele, et al.), 495–507. Cham: Springer.

Dzens-Litovsky, A.I. (1966). *Salt Karst of the USSR*. Leningrad: Nedra (in Russian).

Egemeier, S.J. (1981). Cavern development by thermal waters. *National Speleological Society Bulletin* 43: 31–51.

Ehrenberg, S.N., Walderhaug, O., and Bjørlykke, K. (2012). Carbonate porosity creation by mesogenetic dissolution: reality or illusion? *American Association of Petroleum Geologists Bulletin* 96 (2): 217–233.

Engel, A.S., Stern, L.A., and Bennett, P.C. (2004). Microbial contributions to cave formation: new insight into sulfuric acid speleogenesis. *Geology* 32: 269–273.

Ewers, O.R. (1982). Cavern development in the dimensions of length and breadth. Hamilton PhD thesis. McMaster University.

Filippi, M., Bruthans, J., Palatinus, L. et al. (2011). Secondary halite deposits in the Iranian salt karst: general description and origin. *International Journal of Speleology* 40 (2): 141–162.

Filipponi, M., Jeannin, P.Y., and Tacher, L. (2009). Evidence of inception horizons in karst conduit networks. *Geomorphology* 106 (1-2): 86–99.

Filippov, A.G. (2000). Speleogenesis of the Botovskaya Cave, Eastern Siberia, Russia. In: *Speleogenesis: Evolution of karst aquifers* (ed. A.B. Klimchouk, D.C. Ford, A.N. Palmer, et al.), 282–286. Huntsville, Alabama: National Speleological Society.

Ford T.D. (1995). Some thoughts on hydrothermal caves. *Cave and Karst Science* 22(3): 107–118.

Ford, D.C. and Williams, P.W. (2007). *Karst Hydrogeology and Geomorphology*. Chichester, UK: Wiley.

Franz, N., Sorokin, S., Inshina, I. et al. (2013). Field Measurements of Gypsum Denudation rate in Kulogorskaya cave system. In: *Proceedings of the 16th International Congress of Speleology*, vol. 3 (ed P. Bosák and M. Filippi), 185–189. Brno: Czech Speleological Society.

Frumkin, A. (1994a). Hydrology and denudation rates of halite karst. *Journal of Hydrology* 162: 171–189.

Frumkin, A., Magaritz, M., Carmi, I. et al. (1991). The Holocene climatic record of the salt caves of Mount Sedom Israel. *The Holocene* 1(3): 191–200.

Frumkin, A. (1994b). Morphology and development of salt caves. *National Speleological Society Bulletin* 56: 82–95.

Frumkin, A. and Fischhendler, I. (2005). Morphometry and distribution of isolated caves as a guide for phreatic and confined paleohydrological conditions. *Geomorphology* 67 (3-4): 457–471.

Frumkin, A. and Ford, D.C. (1995). Rapid entrenchment of stream profiles in the salt caves of Mount Sedom, Israel. *Earth Surface Processes and Landforms* 20 (2): 139–152.

Frumkin, A., Langford, B., Lisker, S. et al. (2017). Hypogenic karst at the Arabian platform margins: implications for far-field groundwater systems. *Geological Society of America Bulletin* 129 (11-12): 1636–1659.

Gabrovšek, F. (ed.) (2002). *Evolution of Karst: From Prekarst to Cessation*. Ljubljana: Založba ZRC.

Gabrovšek, F. (2019). Speleogenesis: Telogenetic. In: *Encyclopedia of Caves* (ed. W.B. White, D.C. Culver and T. Pipan), 989–995. New York: Academic Press.

Gabrovšek, F. and Dreybrodt, W. (2000). Role of mixing corrosion in calcite-aggressive H_2O-CO_2-$CaCO_3$ solutions in the early evolution of karst aquifers in limestone. *Water Resources Research* 36 (5): 1179–1188.

Gabrovšek, F., Häuselmann, P., and Audra, P. (2014). 'Looping caves' versus 'water table caves': the role of base-level changes and recharge variations in cave development. *Geomorphology* 204: 683–691.

Galdenzi, S. (2012). Corrosion of limestone tablets in sulfidic ground-water: measurements and speleogenetic implications. *International Journal of Speleology* 41: 25–35.

Galdenzi, S. (2017). The Thermal Hypogenic Caves of Acquasanta Terme (Central Italy). In: *Hypogene Karst Regions and Caves of the World* (ed. A.B. Klimchouk, A.N. Palmer, J. De Waele, et al.), 169–182. Cham: Springer.

Galdenzi, S. and Jones, D.S. (2017). The Frasassi Caves: A "Classical" Active Hypogenic Cave. In: *Hypogene Karst Regions and Caves of the World* (ed. A.B. Klimchouk, A.N. Palmer, J. De Waele, et al.), 143–159. Cham: Springer.

Galdenzi, S. and Maruoka, T. (2003). Gypsum deposits in the Frasassi Caves, central Italy. *Journal of Cave and Karst Studies* 65 (2): 111–125.

Galdenzi, S. and Maruoka, T. (2019). Sulfuric acid caves in Calabria (South Italy): Cave morphology and sulfate deposits. *Geomorphology* 328: 211–221.

Galdenzi, S. and Menichetti, M. (1990). *Il Carsismo della Gola di Frasassi*. Bologna: Istituto Italiano di Speleologia.

Galdenzi, S. and Menichetti, M. (1995). Occurrence of hypogenic caves in a karst region: examples from central Italy. *Environmental Geology* 26: 39–47.

Galdenzi, S., Cocchioni, F., Filipponi, G. et al. (2010). The sulfidic thermal caves of Acquasanta Terme (central Italy). *Journal of Cave and Karst Studies* 72: 43–58.

García-Ruiz, J.M., Villasuso, R., Ayora, C. et al. (2007). Formation of natural gypsum megacrystals in Naica, Mexico. *Geology* 35 (4): 327–330.

Garrecht Metzger, J., Fike, D.A., Osburn, G.R. et al. (2015). The source of gypsum in Mammoth Cave, Kentucky. *Geology* 43: 187–190.

Gázquez, F., Calaforra, J.M., Forti, P. et al. (2015). The role of condensation in the evolution of dissolutional forms in gypsum caves: study case in the karst of Sorbas (SE Spain). *Geomorphology* 229: 100–111.

Gázquez, F., Calaforra, J.M., Rodríguez-Estrella, T. et al. (2017). Evidence for regional hypogene speleogenesis in Murcia (SE Spain). In: *Hypogene Karst Regions and Caves of the World* (ed. A.B. Klimchouk, A.N. Palmer, J. De Waele, et al.), 85–97. Cham: Springer.

Ginés, A., Ginés, J., and Gràcia, F. (2013). Cave development and patterns of caves and cave systems in the eogenetic coastal karst of southern Mallorca (Balearic Islands, Spain). In: *Coastal Karst Landforms* (ed. M.J. Lace and J.E. Mylroie), 245–260. Dordrecht: Springer.

Ginés, J., Fornós, J.J., Ginés, A. et al. (2014). Geologic constraints and speleogenesis of Cova des Pas de Vallgornera, a complex coastal cave from Mallorca Island (Western Mediterranean). *International Journal of Speleology* 43 (2): 105–124.

Glover, R.R. (1974). Cave development in the Gaping Ghyll System. In: *Limestones and Caves of Northwest England* (ed. T. Waltham), 343–384. Newton Abbot: David and Charles.

Goldscheider, N., Mádl-Szőnyi, J., Erőss, A. et al. (2010). Thermal water resources in carbonate rock aquifers. *Hydrogeology Journal* 18 (6): 1303–1318.

Grimes, K.G. and Martini, J.E.J. (2016). Bullita cave system, Judbarra/Gregory Karst, tropical Australia. *Boletín Geológico y Minero* 127 (1): 21–44.

Groves, C.G. and Howard, A.D. (1994). Early development of karst systems: 1. Preferential flow path enlargement under laminar flow. *Water Resources Research* 30 (10): 2837–2846.

Grund, A. (1903). *Die Karsthydrographie: Studien aus Westbosnien*. Geographische Abhandlungen herausgegeben von A. Penck, Band IX. Vienna.

Gučetić, N. (Gozze di Vito Nicolò)(1585). *Quatro Giornate sopra le Metheore di Aristotele*. Venice: Stamperia Francesco Ziletti.

Gulley, J., Martin, J.B., and Moore, P.J. (2014). Vadose CO_2 gas drives dissolution at water tables in eogenetic karst aquifers more than mixing dissolution. *Earth Surface Processes and Landforms* 39 (13): 1833–1846.

Gulley, J.D., Martin, J.B., Moore, P.J. et al. (2015). Heterogeneous distributions of CO_2 may be more important for dissolution and karstification in coastal eogenetic limestone than mixing dissolution. *Earth Surface Processes and Landforms* 40 (8): 1057–1071.

Gulley, J.D., Martin, J.B., and Brown, A. (2016). Organic carbon inputs, common ions and degassing: rethinking mixing dissolution in coastal eogenetic carbonate aquifers. *Earth Surface Processes and Landforms* 41 (14): 2098–2110.

Harvey, A.M., Whitfield, E., Stokes, M. et al. (2014). The late Neogene to Quaternary drainage evolution of the uplifted sedimentary basins of Almeria. In: *Landscapes and Landforms of Spain* (ed. F. Gutiérrez and M. Gutiérrez), 37–61. Heidelberg: Springer.

Hill, C.A. (1987). Geology of Carlsbad Cavern and other caves in the Guadalupe Mountains, New Mexico and Texas. *Bulletin of the New Mexico Bureau of Mining and Mineral Resources* 117: 1–150.

Hill, C.A. (1990). Sulfuric acid speleogenesis of Carlsbad Cavern and its relationship to hydrocarbons, Delaware Basin, New Mexico and Texas. *American Association of Petroleum Geologists Bulletin* 74: 1685–1694.

Hill, C.A. (1995). Sulfur redox reactions: hydrocarbons, native sulfur, Mississippi Valley-type deposits, and sulfuric acid karst in the Delaware Basin, New Mexico and Texas. *Environmental Geology* 25: 16–23.

Hose, L.D., Palmer, A.N., Palmer, M.V. et al. (2000). Microbiology and geochemistry in a hydrogen-sulphide-rich karst environment. *Chemical Geology* 169: 399–423.

Howard, A.D. and Groves, C.G. (1995). Early development of karst systems: 2. Turbulent flow. *Water Resources Research* 31 (1): 19–26.

Hutton, J. (1795). *Theory of the Earth, with Proofs and Illustrations*, vol. 2. Edinburgh: Cadell.

Jagnow, D.H., Hill, C.A., Davis, D.G. et al. (2000). History of the sulfuric acid theory of speleogenesis in the Guadalupe Mountains, New Mexico. *Journal of Cave and Karst Studies* 62: 54–59.

Jakucs, L. (1977). Genetic types of the Hungarian karst. *Karszt es Barlang* (Special Issue): 3–18.

James, A.N. and Kirkpatrick, I.M. (1980). Design of foundations of dams containing soluble rocks and soils. *Quarterly Journal of Engineering Geology and Hydrogeology* 13 (3): 189–198.

Jeannin, P.Y. (2016). Main karst and caves of Switzerland. *Boletino Geológico y Minero* 127 (1): 45–46.

Johnson, K.S. (2018). Gypsum caves of North Texas and Western Oklahoma. In: *Hypogene Karst of Texas* (ed. K.W. Stafford and G. Veni), 111–122. Austin (USA): Texas Speleological Survey.

Jones, W.K. and Elliott, L.F. (2019). Art in European caves. In: *Encyclopedia of Caves* (ed. W.B. White, D.C. Culver and T. Pipan), 71–75. New York: Academic Press.

Jones, D.S. and Northup, D.E. (2021). Cave Decorating with Microbes: Geomicrobiology of Caves. *Elements: An International Magazine of Mineralogy, Geochemistry, and Petrology* 17 (2): 107–112.

Kambesis, P.N. and Coke, J.G. (2013). Overview of the controls on eogenetic cave and karst development in Quintana Roo, Mexico. In: *Coastal Karst Landforms* (ed. M.J. Lace and J.E. Mylroie), 347–373. Dordrecht: Springer.

Katzer, F. (1909). Karst und Karsthydrographie. *Zur Kunde des Balkanhalbinsel, Reisen und Beobachtenungen* 8 (3): 1–94.

Kaufmann, G. (2009). Modelling karst geomorphology on different time scales. *Geomorphology* 106 (1-2): 62–77.

Kaufmann, G. and Braun, J. (1999). Karst aquifer evolution in fractured rocks. *Water Resources Research* 35: 3223–3238.

Kaufmann, G. and Braun, J. (2000). Karst aquifer evolution in fractured, porous rocks. *Water Resources Research* 36: 1381–1391.

Kaufmann, G., Romanov, D., and Hiller, T. (2010). Modeling three-dimensional karst aquifer evolution using different matrix-flow contributions. *Journal of Hydrology* 388 (3-4): 241–250.

Kaufmann, G., Romanov, D., and Dreybrodt, W. (2019). Modeling the evolution of karst aquifers. In: *Encyclopedia of Caves* (ed. W.B. White, D.C. Culver and T. Pipan), 717–724. New York: Academic Press.

Klimchouk, A.B. (1996a). The dissolution and conversion of gypsum and anhydrite. *International Journal of Speleology* 25: 21–36.

Klimchouk, A.B. (2000a). Dissolution and conversions of gypsum and anhydrite. In: *Speleogenesis: Evolution of karst aquifers* (ed. A.B. Klimchouk, D.C. Ford, A.N. Palmer, et al.), 160–168. Huntsville, USA: National Speleological Society.

Klimchouk, A.B. (2004a). Asia, Central. In: *Encyclopedia of Caves and Karst Science* (ed. J. Gunn), 196–201. London Fitzroy: Dearborn.

Klimchouk, A.B. (2007). Hypogene speleogenesis: hydrogeological and morphogenetic perspective. *National Cave and Karst Research Institute*. Special paper 1: 106.

Klimchouk, A.B. (2009). Morphogenesis of hypogenic caves. *Geomorphology* 106: 100–117.

Klimchouk, A.B. (2017). Types and settings of hypogene karst. In: *Hypogene Karst Regions and Caves of the World* (ed. A.B. Klimchouk, A.N. Palmer, J. De Waele, et al.), 1–39. Cham: Springer.

Klimchouk, A.B. (2019a). Gypsum caves. In: *Encyclopedia of Caves* (ed. W.B. White, D.C. Culver and T. Pipan), 485–495. New York: Academic Press.

Klimchouk, A.B. (2019b). Speleogenesis-Hypogene. In: *Encyclopedia of Caves* (ed. W.B. White, D.C. Culver and T. Pipan), 974–988. New York: Academic Press.

Klimchouk, A.B. and Andreychuk, V. (2017). Gypsum karst in the southwest outskirts of the Eastern European Platform (Western Ukraine): a type region of artesian transverse speleogenesis. In: *Hypogene Karst Regions and Caves of the World* (ed. A.B. Klimchouk, A.N. Palmer, J. De Waele, et al.), 363–385. Cham: Springer.

Klimchouk, A.B. and Ford, D.C. (2000a). Lithologic and structural controls of dissolutional cave development. In: *Speleogenesis: Evolution of Karst Aquifers* (ed. A.B. Klimchouk, D.C. Ford, A.N. Palmer, et al.), 54–64. Huntsville: National Speleological Society.

Klimchouk, A.B., Ford, D.C., Palmer, A.N. et al. (2000). *Speleogenesis: Evolution of Karst Aquifers*. Huntsville: National Speleological Society.

Klimchouk, A.B., Auler, A.S., Bezerra, F.H. et al. (2016). Hypogenic origin, geologic controls and functional organization of a giant cave system in Precambrian carbonates, Brazil. *Geomorphology* 253: 385–405.

Klimchouk, A.B., Palmer, A.N., De Waele, J. et al. (ed.) (2017b). *Hypogene Karst Regions and Caves of the World*. Cham: Springer.

Knez, M. (1998). The influence of bedding-planes on the development of karst caves (a study of Velika Dolina at Škocjanske jame Caves, Slovenia). *Carbonates and Evaporites* 13: 121–131.

Kotula, P., Andreychuk, V., Pawlyta, J. et al. (2019). Genesis of iron and manganese sediments in Zoloushka Cave (Ukraine/Moldova) as revealed by $\delta^{13}C$ organic carbon. *International Journal of Speleology* 48 (3): 221–235.

Kutleša, P., Malenica, M., Čepelak, M. et al. (2019). Međunarodna speleološka ekspedicija" Mount Sedom 2019"-istraživanje najdulje špilje u soli na svijetu-Malham cave, Izrael. *Subterranea Croatica* 17 (26): 64–71.

Lace, M.J., Kambesis, P.N., and Mylroie, J.E. (2016). Sistema Faro, Isla de Mona, Puerto Rico: speleogenesis of the world's largest flank margin cave. *Boletín Geológico y Minero* 127 (1): 205–217.

Leél-Őssy, S. (2017). Caves of the Buda thermal karst. In: *Hypogene Karst Regions and Caves of the World* (ed. A.B. Klimchouk, A.N. Palmer, J. De Waele, et al.), 279–297. Cham: Springer.

Li, S., Kang, Z., Feng, X.T. et al. (2020b). Three-Dimensional hydrochemical model for dissolutional growth of fractures in Karst Aquifers. *Water Resources Research* 56 (3): e2019WR025631.

Lismonde, B. (2003). Limestone wall retreat in a ceiling cupola controlled by hydrothermal degassing with wall condensation. *Speleogenesis and Evolution of Karst Aquifers* 1 (4): 1–3.

Lowe, D.J. (1992b). The origin of limestone caverns: an inception horizon hypothesis. PhD thesis. Manchester Metropolitan University.

Lowe, D.J. (2000). Role of stratigraphic elements in speleogenesis: the speleoinception concept. In: *Speleogenesis, Evolution of Karst Aquifers* (ed. A.B. Klimchouk, D.C. Ford, A.N. Palmer, et al.), 65–76. Huntsville: National Speleological Society.

Lowe, D.J. and Gunn, J. (1997). Carbonate speleogenesis: an inception horizon hypothesis. *Acta Carsologica* 26 (2): 457–488.

Lucha, P., Cardona, F., Gutiérrez, F. et al. (2008a). Natural and human-induced dissolution and subsidence processes in the salt outcrop of the Cardona Diapir (NE Spain). *Environmental Geology* 53: 1023–1035.

Luiszer, F.G. (2009). Speleogenesis of cave of the winds, Manitou Springs, Colorado. In: *Select Field guides to cave and karst lands of the United States*, Karst Waters Institute Special Publication 15 (ed. A. Summers Engel and S.K. Engel), 119–132. Leesburg, Va.: Karst Waters Institute.

Lundberg, J. and McFarlane, D.A. (2009). Bats and bell holes: the microclimatic impact of bat roosting, using a case study from Runaway Bay Caves, Jamaica. *Geomorphology* 106 (1-2): 78–85.

Machel, H.G. (2001). Bacterial and thermochemical sulfate reduction in diagenetic settings - old and new insights. *Sedimentary Geology* 140: 143–175.

Mádl-Szőnyi, J., Erőss, A., and Tóth, Á. (2017). Fluid flow systems and hypogene karst of the Transdanubian Range, Hungary-with special emphasis on Buda Thermal Karst. In: *Hypogene Karst Regions and Caves of the World* (ed. A.B. Klimchouk, A.N. Palmer, J. De Waele, et al.), 267–278. Cham: Springer.

Madonia, G. and Vattano, M. (2011). New knowledge on the Monte Conca gypsum karst system (central-western Sicily, Italy). *Acta Carsologica* 40 (1): 53–64.

Martel, E.A. (1894). *Les Abîmes*. Paris: Delagrave.

Martel, E.A. (1921). *Nouveau Traité des Eaux Souterraines*. Paris: Librairie Octave Dion.

Martel, E.A. (1935). Contamination, protection et amélioration des sources thermominérales. *Congrès international des mines, de la métallurgie et de la géologie appliquée*, 7e session, 2: 791–798.

Martini, J.E.J. (2004). Silicate karst. In: *Encyclopedia of Caves and Karst Science* (ed. J. Gunn), 1385–1393. London: Fitzroy Dearborn.

Martini, J.E.J. (2017). Hypogene Karst in Southern Africa. In: *Hypogene Karst Regions and Caves of the World* (ed. A.B. Klimchouk, A.N. Palmer, J. De Waele, et al.), 865–878. Cham: Springer.

Martini, J.E.J. and Marais, J. (1996). Grottes hydrothermales dans le Nord-Ouest de la Namibie. *Karstologia* 28: 13–18.

Martini, J.E.J., Wipplinger, P.E., and Moen, F.G. (1997). Mbobo Mkulu Cave, South Africa. In: *Cave Minerals of the World* (ed. C.A. Hill and P. Forti), 336–339. Huntsville: National Speleological Society.

Mazzullo, S.J. and Harris, P.M. (1992). Mesogenetic dissolution: its role in porosity development in carbonate reservoirs. *American Association of Petroleum Geologists Bulletin* 76 (5): 607–620.

Mecchia, M., Sauro, F., Piccini, L. et al. (2014). Geochemistry of surface and subsurface waters in quartz-sandstones: significance for the geomorphic evolution of tepui table mountains (Gran Sabana, Venezuela). *Journal of Hydrology* 511: 117–138.

Mecchia, M., Sauro, F., Piccini, L. et al. (2019). A hybrid model to evaluate subsurface chemical weathering and fracture karstification in quartz sandstone. *Journal of Hydrology* 572: 745–760.

Mihevc, A. (2007). The age of karst relief in West Slovenia. *Acta Carsologica* 36 (1): 35–44.

Mocochain, L., Audra, P., Clauzon, G. et al. (2009). The effect of river dynamics induced by the Messinian Salinity Crisis on karst landscape and caves: example of the Lower Ardèche river (mid Rhône valley). *Geomorphology* 106 (1-2): 46–61.

Monteiro, H.S., Vasconcelos, P.M., Farley, K.A. et al. (2014). (U-Th)/He geochronology of goethite and the origin and evolution of cangas. *Geochimica et Cosmochimica Acta* 131: 267–289.

Morehouse, D.F. (1968). Cave development via the sulfuric acid reaction. *National Speleological Society Bulletin* 30: 1–10.

Moss, P. (2013). Recharge area delineations and hazard and vulnerability mapping in Perry County, Missouri. *Carbonates and Evaporites* 28 (1-2): 175–182.

Mylroie, J.E. (2019). Coastal caves. In: *Encyclopedia of Caves* (ed. W.B. White, D.C. Culver and T. Pipan), 301–307. New York: Academic Press.

Mylroie, J.E. and Carew, J.L. (1988). Solution conduits as indicators of late Quaternary sea level position. *Quaternary Science Reviews* 7 (1): 55–64.

Mylroie, J.E. and Carew, J.L. (1990). The flank margin model for dissolution cave development in carbonate platforms. *Earth Surface Processes and Landforms* 15 (5): 413–424.

Mylroie, J.E. and Mylroie, J.R. (2013a). Caves and karst of the Bahama Islands. In: *Coastal Karst Landforms* (ed. M.J. Lace and J.E. Mylroie), 147–176. Dordrecht: Springer.

Mylroie, J.E., Mylroie, J.R., and Nelson, C.S. (2008). Flank margin cave development in telogenetic limestones of New Zealand. *Acta Carsologica* 37 (1): 15–40.

Mylroie, J., Lace, M., Albury, N. et al. (2020). Flank margin caves and the position of Mid-to Late Pleistocene sea level in the Bahamas. *Journal of Coastal Research* 36 (2): 249–260.

Otoničar, B., Buzjak, N., Mylroie, J.E. et al. (2010). Flank margin cave development in carbonate talus breccia facies: an example from Cres Island, Croatia. *Acta Carsologica* 39 (1): 79–91.

Palmer, A.N. (1984). Recent trends in karst geomorphology. *Journal of Geological Education* 32 (4): 247–253.

Palmer, A.N. (1987). Cave levels and their interpretation. *National Speleological Society Bulletin* 49 (2): 50–66.

Palmer, A.N. (1989). Stratigraphic and structural control of cave development and groundwater flow in the Mammoth Cave region. In: *Karst Hydrology, Concepts from the Mammoth Cave Area* (ed. W.B. White and E.L. White), 293–316. Von Nostrand Reinhold: New York.

Palmer, A.N. (1991). Origin and morphology of limestone caves. *Geological Society of America Bulletin* 103 (1): 1–21.

Palmer, A.N. (2000). Hydrogeologic control of cave patterns. In: *Speleogenesis: Evolution of Karst Aquifers* (ed. A.B. Klimchouk, D.C. Ford, A.N. Palmer, et al.), 77–90. Huntsville: National Speleological Society.

Palmer, A.N. (2001). Dynamics of cave development by allogenic water. *Acta Carsologica* 30 (2): 13–32.

Palmer, A.N. (2007). *Cave Geology*. Dayton, Ohio: Cave Books.

Palmer, A.N. (2009a). Cave exploration as a guide to geologic research in the Appalachians. *Journal of Cave and Karst Studies* 71 (3): 180–192.

Palmer, A.N. (2011). Distinction between epigenic and hypogenic maze caves. *Geomorphology* 134 (1-2): 9–22.

Palmer, A.N. (2013). Sulfuric acid caves. In: *Treatise of Geomorphology. Karst Geomorphology*, vol. 6 (ed. A. Frumkin), 241–257. Amsterdam: Elsevier.

Palmer, A.N. and Hill, C.A. (2019). Sulfuric acid caves. In: *Encyclopedia of Caves* (ed. W.B. White, D.C. Culver and T. Pipan), 1053–1062. New York: Academic Press.

Parise, M. and Trocino, A. (2005). Gypsum Karst in the Crotone Province (Calabria, Southern Italy). *Acta Carsologica* 34 (2): 369–382.

Parker, C.W., Auler, A.S., Barton, M.D. et al. (2018). Fe (III) reducing microorganisms from iron ore caves demonstrate fermentative Fe (III) reduction and promote cave formation. *Geomicrobiology Journal* 35 (4): 311–322.

Pereira, R.G.F.A. (1998). Caracterização Geomorfológica e Geoespeleológica do Carste da Bacia do Rio Una, Borda Leste da Chapada Diamantina (Município de Itaetê, Estado da Bahia). MSc thesis. Universidade de São Paulo.

Perry, E., Marin, L., McClain, J. et al. (1995). Ring of cenotes (sinkholes), northwest Yucatan, Mexico: its hydrogeologic characteristics and possible association with the Chicxulub impact crater. *Geology* 23 (1): 17–20.

Pezdič, J., Šušteršič, F., and Mišič, M. (1998). On the role of clay-carbonate reactions inspeleo-inception: a contribution on the understanding of the earliest stage of karst channel formation. *Acta Carsologica* 27 (1): 187–200.

Piccini, L. (2000). Il carsismo di origine idrotermale del Colle di Monsummano (Pistoia-Toscana). *Le Grotte d'Italia* 5 (1): 33–43.

Piccini, L., Forti, P., Giulivo, I. et al. (2007). The polygenetic caves of Cuatro Ciénegas (Coahuila, Mexico): morphology and speleogenesis. *International Journal of Speleology* 36 (2): 83–92.

Piccini, L., De Waele, J., Galli, E. et al. (2015). Sulphuric acid speleogenesis and landscape evolution: Montecchio cave, Albegna river valley (Southern Tuscany, Italy). *Geomorphology* 229: 134–143.

Pisani, L., Antonellini, M., and De Waele, J. (2019). Structural control on epigenic gypsum caves: evidences from Messinian evaporites (Northern Apennines, Italy). *Geomorphology* 332: 170–186.

Plan, L., Tschegg, C., De Waele, J. et al. (2012). Corrosion morphology and cave wall alteration in an Alpine sulfuric acid cave (Kraushöhle, Austria). *Geomorphology* 169: 45–54.

Polyak, V.J. and Provencio, P. (2001). By-product materials related to H_2S-H_2SO_4 influenced speleogenesis of Carlsbad, Lechuguilla, and other caves of the Guadalupe mountains, New Mexico. *Journal of Cave and Karst Studies* 63: 23–32.

Polyak, V.J., McIntosh, W.C., Güven, N. et al. (1998). Age and origin of Carlsbad Cavern and related caves from $^{40}Ar/^{39}Ar$ of alunite. *Science* 279 (5358): 1919–1922.

Polyak, V.J., Provencio, P.P., and Asmerom, Y. (2016). U-Pb dating of speleogenetic dolomite: a new sulfuric acid speleogenesis chronometer. *International Journal of Speleology* 45 (2): 103–109.

Povara, I., Cosma, R., Lascu, C. et al. (1982). Un cas particulier de karst dans les dépôts de sel (Slanic Prahova, Roumanie). *Travaux de l'Institut Spéologique Emile Racovitza* 21: 8793.

Principi, P. (1931). Fenomeni di idrologia sotteranea nei dintorni a Triponzo. *Le Grotte d'Italia* 1: 45–47.

Przybyto, J. (2000). Kopalnia Wieliczka - nowe Groty Krystalowe? *Jaskinie* 3 (20): 25–27.

Rauch, H.W. and White, W.B. (1970). Lithologic controls on the development of solution porosity in carbonate aquifers. *Water Resources Research* 6: 1175–1192.

Raza, A., Gholami, R., Rezaee, R. et al. (2019). Significant aspects of carbon capture and storage – a review. *Petroleum* 5 (4): 335–340.

Rehrl, C., Birk, S., and Klimchouk, A.B. (2008). Conduit evolution in deep-seated settings: Conceptual and numerical models based on field observations. *Water Resources Research* 44 (11): w11425.

Rhoades, R. and Sinacori, N.M. (1941). Patterns of groundwater flow and solution. *Journal of Geology* 49: 785–794.

Ritter, S.M., Isenbeck-Schröter, M., Scholz, C. et al. (2019). Subaqueous speleothems (Hells Bells) formed by the interplay of pelagic redoxcline biogeochemistry and specific hydraulic conditions in the El Zapote sinkhole, Yucatán Peninsula, Mexico. *Biogeosciences* 16 (11): 2285–2305.

Romanov, D., Gabrovšek, F., and Dreybrodt, W. (2004). Modeling the evolution of karst aquifers and speleogenesis. The step from 1-dimensional to 2-dimensional modeling domains. *Speleogenesis and Evolution of Karst Aquifers* 2 (1): 2–28.

Ruggieri, R. and De Waele, J. (2014). Lower-to Middle Pleistocene flank margin caves at Custonaci (Trapani, NW Sicily) and their relation with past sea levels. *Acta Carsologica* 43 (1): 11–22.

Sarbu, S.M. and Lascu, C. (1997). Condensation corrosion in Movile cave, Romania. *Journal of Caves and Karst Studies* 59: 99–102.

Sarbu, S.M., Kane, T.C., and Kinkle, B.K. (1996). A chemoautotrophically based cave ecosystem. *Science* 272: 1953–1955.

Sasowsky, I.D. and White, W.B. (1994). The role of stress release fracturing in the development of cavernous porosity in carbonate aquifers. *Water Resources Research* 30 (12): 3523–3530.

Sauro, F. (2014). Structural and lithological guidance on speleogenesis in quartz-sandstone: evidence of the arenisation process. *Geomorphology* 226: 106–123.

Sauro, F., Mecchia, M., Tringham, M. et al. (2020). Speleogenesis of the world's longest cave in hybrid arenites (Krem Puri, Meghalaya, India). *Geomorphology* 359: 107160.

Shuster, D.L., Farley, K.A., Vasconcelos, P.M. et al. (2012). Cosmogenic ^3He in hematite and goethite from Brazilian "canga" duricrust demonstrates the extreme stability of these surfaces. *Earth and Planetary Science Letters* 329–330: 41–50.

Siemers, J. and Dreybrodt, W. (1998). Early development of karst aquifers on percolation networks of fractures in limestone. *Water Resources Research* 34 (3): 409–419.

Simmons, G.C. (1963). Canga caves in the Quadrilátero Ferrífero, Minas Gerais, Brazil. *The National Speleological Society Bulletin* 25: 66–72.

Sivinskih, P. (2009). Features of geological conditions of the Ordinskaya underwater cave, fore-Urals, Russia. In: *Hypogene Speleogenesis and Karst Hydrogeology of Artesian Basins* (ed. A.B. Klimchouk and D.C. Ford), 267–269. Kiev: Ukrainian Institute of Speleology and Karstology.

Smart, P.L., Dawans, J.M., and Whitaker, F.F. (1988). Carbonate dissolution in a modern mixing zone. *Nature* 335 (6193): 811–813.

Smart, P.L., Beddows, P.A., Coke, J. et al. (2006). Cave Development on the Caribbean Coast of the Yucatan Peninsula, Quintana Roo, Mexico. In: *Perspectives in Karst Geomorphology, Hydrology, and Geochemistry* (ed. R. Harmon and C. Wicks), 105–128. Geological Society of America, Special Paper 404.

Smart, P.L., Moseley, G.M., Richards, D.A. et al. (2008). Past high sea-stands and platform stability: evidence from Conch Bar Cave, Middle Caicos. In: *Developing Models and Analogs for Isolated Carbonate Platforms-Holocene and Pleistocene Carbonates of Caicos Platform, British West Indies*, vol. 22 (ed. W. Morgan and R.M. Harris) SEPM Core Workshop, 203–210.

Socquet, J.M. (1801). *Analyse des Eaux Thermales d'Aix (en Savoie), Département du Mont-Blanc (Analysis of Thermal Waters at Aix, in Savoy, Mont-Blanc Department)*. Chambéry: Cleaz.

Spence, G.H. and Finch, E. (2014). Influences of nodular chert rhythmites on natural fracture networks in carbonates: an outcrop and two-dimensional discrete element modelling study. *Geological Society of London*, Special Publications 374 (1): 211–249.

Spötl, C., Dublyansky, Y., Koltai, G. et al. (2021). Hypogene speleogenesis and paragenesis in the Dolomites. *Geomorphology* 382: 107667.

Stafford, K.W., Land, L., and Klimchouk, A.B. (2008a). Hypogenic speleogenesis within Seven Rivers evaporites: Coffee Cave, Eddy County, New Mexico. *Journal of Cave and Karst Studies* 70 (1): 47–61.

Stafford, K.W., Nance, R., Rosales-Lagarde, L. et al. (2008b). Epigene and hypogene gypsum karst manifestations of the Castile Formation: Eddy County, New Mexico and Culberson County, Texas, USA. *International Journal of Speleology* 37 (2): 83–98.

Starchenko, V., Marra, C.J., and Ladd, A.J. (2016). Three-dimensional simulations of fracture dissolution. *Journal of Geophysical Research: Solid Earth* 121 (9): 6421–6444.

Stocchi, P., Antonioli, F., Montagna, P. et al. (2017). A stalactite record of four relative sea-level highstands during the Middle Pleistocene Transition. *Quaternary Science Reviews* 173: 92–100.

Swinnerton, A.C. (1932). Origin of limestone caverns. *Geological Society of America Bulletin* 43: 662–693.

Szunyogh, G. (1990). Theoretical investigation of the development of spheroidal niches of thermal water origin. Second approximation. In: *Proceedings of the 10th International Congress of Speleology* (ed. A. Kósa), 766–768. Budapest: Hungarian Speleological Society.

Tarhule-Lips, R.F. and Ford, D.C. (1998b). Morphometric studies of bell hole development on Cayman Brac. *Cave and Karst Science* 25 (3): 119–130.

Temovski, M., Audra, P., Mihevc, A. et al. (2013). Hypogenic origin of Provalata Cave, Republic of Macedonia: a distinct case of successive thermal carbonic and sulfuric acid speleogenesis. *International Journal of Speleology* 42 (3): 235–246.

Tisato, N., Sauro, F., Bernasconi, S.M. et al. (2012). Hypogenic contribution to speleogenesis in a predominant epigenic karst system: a case study from the Venetian Alps, Italy. *Geomorphology* 151: 156–163.

Upadhyay, V.K., Szymczak, P., and Ladd, A.J. (2015). Initial conditions or emergence: What determines dissolution patterns in rough fractures? *Journal of Geophysical Research: Solid Earth* 120 (9): 6102–6121.

Vacher, H.L. and Mylroie, J.E. (2002). Eogenetic karst from the perspective of an equivalent porous medium. *Carbonates and Evaporites* 17 (2): 182–196.

Vaks, A., Gutareva, O.S., Breitenbach, S.F. et al. (2013). Speleothems reveal 500,000-year history of Siberian permafrost. *Science* 340 (6129): 183–186.

Valvasor, J.W. (1687). *Die Ehre des Hertzogthums Crain*, 4 Vols. Ljubljana: Endter.

Vigna, B., Fiorucci, A., Banzato, C. et al. (2010). Hypogene gypsum karst and sinkhole formation at Moncalvo (Asti, Italy). *Zeitschrift für Geomorphologie* Supplement Band 54 (2): 285–306.

Weyl, P.K. (1958). The solution kinetics of calcite. *The Journal of Geology* 66 (2): 163–176.

White, D.E. (1957). Thermal waters of volcanic origin. *Geological Society of America Bulletin* 68: 1637–1658.

White, W.B. (1977). Role of solution kinetics in the development of karst aquifers. In: *Karst Hydrogeology*, vol. 12 (ed. J.S. Tolson and F.L. Doyle), 503–517. *Memories of the International Association of Hydrogeologists.*

White, W.B. (1988). *Geomorphology and Hydrology of Karst Terrains*. New York: Oxford University Press.

White, S. and Webb, J.A. (2015). The influence of tectonics on flank margin cave formation on a passive continental margin: Naracoorte, Southeastern Australia. *Geomorphology* 229: 58–72.

Wigley, T.M.L. and Plummer, L.N. (1976). Mixing of carbonate waters. *Geochimica et Cosmochimica Acta* 40 (9): 989–995.

Williams, P.W. (2008). The role of the epikarst in karst and cave hydrogeology: a review. *International Journal of Speleology* 37: 1–10.

Wray, R.A.L. and Sauro, F. (2017). An updated global review of solutional weathering processes and forms in quartz sandstones and quartzites. *Earth-Science Reviews* 171: 520–557.

Wynn, J.G., Sumrall, J.B., and Onac, B.P. (2010). Sulfur isotopic composition and the source of dissolved sulfur species in thermo-mineral springs of the Cerna Valley, Romania. *Chemical Geology* 271: 31–43.

Zawidzki, P., Urbani, F., and Koisar, B. (1976). Preliminary notes on the geology of the Sarisariñama plateau, Venezuela, and the origin of its caves. *Boletín de la Sociedad Venezolana de Espeleología* 7: 29–37.

Zupan Hajna, N. (2003). *Incomplete Solution: Weathering of Cave Walls and the Production, Transport and Deposition of Carbonate Fines*. Ljubljana: Založba ZRC.

Zupan Hajna, N., Bosák, P., Pruner, P. et al. (2020). Karst sediments in Slovenia: Plio-Quaternary multi-proxy records. *Quaternary International* 546: 4–19.

Index

Italics refer to figures, bold to tables

Karst Hydrogeology, Geomorphology and Caves, First Edition. Jo De Waele and Francisco Gutiérrez.
© 2022 John Wiley & Sons Ltd. Published 2022 by John Wiley & Sons Ltd.